D0579990

GEOLOGY

Thomas Moran (1837–1926), who was already well known for an impressive painting of the cliffs along the Yellowstone River, was a guest artist accompanying John Wesley Powell's 1873 expedition to the Grand Canyon. After viewing the Canyon from a mountain overlook, Moran wrote:

The whole gorge for miles lay beneath us and it was by far the most awfully grand and impressive scene that I have ever yet seen. . . . Above and around us rose a wall of 2000 feet and below us a vast chasm 2500 feet in perpendicular depth and 1/2 mile wide. . . .

The color of the Grand Canyon itself is red, a light Indian Red, and the material sandstone and red marble and is in terraces all the way down. All above the canyon is variously colored sandstone mainly a light flesh or cream color and worn into very fine forms. . . .

Moran returned to and painted views of the Grand Canyon numerous times. His paintings were widely exhibited in cities back east. He also made prints of Grand Canyon scenes, which were published in various magazines and journals, thus making the grandeur of the continent's western scenery accessible everywhere.

GEOLOGY

An Introduction to Physical Geology

Second Edition

Stanley Chernicoff
University of Washington, Seattle

Houghton Mifflin Company Boston New York

Editor-in-Chief: Kathi Prancan
Senior Associate Sponsor: Susan Warne
Senior Project Editor: Chere Bemelmans
Editorial Assistant: Joy Park
Senior Production/Design Coordinator: Jill Haber
Associate Production/Design Coordinator: Jodi O'Rourke
Manufacturing Manager: Florence Cadran
Marketing Manager: Penelope Hoblyn

Cover Design: Walter Kopec.

Cover Image: *A Miracle of Nature* by Thomas Moran, 1913, oil on canvas, 20⅛″ × 30⅛″.
Private Collection/SuperStock.

Photo credits begin on page A-31.

Printed in the U.S.A.

Library of Congress Catalog Card Number: 98-72010

ISBN: 0-395-92351-4

23456789-VH-02 01 00 99

About the Author

Born in Brooklyn, New York, Stan Chernicoff began his academic career as a political science major at Brooklyn College of the City University of New York. On graduation, he intended to enter law school and pursue a career in constitutional law. He had, however, the good fortune to take geology as his last requirement for graduation in the spring of his senior year, and he was so thoroughly captivated by it that his plans were forever changed.

After an intensive post-baccalaureate program of physics, calculus, chemistry, and geology, Stan entered the University of Minnesota–Twin Cities, where he received his doctorate in Glacial and Quaternary Geology under the guidance of one of North America's preeminent glacial geologists, Dr. H. E. Wright. Stan launched his career as a purveyor of geological knowledge as a senior graduate student teaching physical geology to hundreds of bright Minnesotans.

Stan has been a member of the faculty of the Department of Geological Sciences at the University of Washington in Seattle since 1981, where he has won several teaching awards. At Washington, he has taught Physical Geology, the Great Ice Ages, and the Geology of the Pacific Northwest to more than 20,000 students, and he has trained hundreds of graduate teaching assistants in the art of bringing geology alive for non-science majors. Stan studies the glacial history of the Puget Sound region and pursues his true passion, coaching his sons and their buddies in soccer, baseball, and basketball. He lives in Seattle with his wife, Dr. Julie Stein, a professor of archaeology, and their two sons, Matthew (the midfielder, second baseman, two-guard) and David (the striker, second baseman, point guard).

Contents in Brief

Contents

Preface

The introductory course in physical geology, taken predominantly by non-science majors, may be the only science course some students will take during their college years. What a wonderful opportunity this provides us to introduce students to the field we love and show them how fascinating and useful it is. Indeed, much of what students will learn in their Physical Geology course will be recalled throughout their lives, as they travel across this and other continents, dig in backyards, walk along a beach, or sit by a mountain stream. For this reason, our book team—author, illustrators, photo researchers, and editors—have expended the best of our abilities to craft an exciting, stimulating, and enduring introduction to the field.

The Book's Goal

The book's goal is basic—to teach what everyone should know about geology in a way that will engage and stimulate. The book embodies the view that this is perhaps the most useful college-level science class a non-science major can take—one that we believe all students should take. Physical geology can show students the essence of how science and scientists work at the same time as it nurtures their interest in understanding, appreciating, and protecting their surroundings. In this course they can learn to prepare for any number of geologic and environmental threats, and see how our Earth can continue providing all of our needs for food, shelter, and material well-being as long as we don't squander these resources.

Content and Organization

The unifying themes of plate tectonics, environmental geology and natural resources, and planetary geology are introduced in Chapter 1 and discussed in their proper context within nearly every chapter. Chapter 1 also presents the important groups of rocks, the rock cycle, and geological time—building a foundation for the succeeding chapters. Chapters 1 through 8 introduce the basics—minerals, rocks, and time—to prepare the reader for the in-depth discussions of structural geology, geophysics, and tectonics that follow in Chapters 9 through 12. After the Earth's first-order features—ocean basins, continents, and mountain systems—have been discussed, the processes that sculpt these large-scale features are addressed. Chapters 13 through 19 present the principal geomorphic processes of mass movement, streams, groundwater, glacial flow, arid region and eolian processes, and coastal evolution. Chapter 16, Caves and Karst, is a brief chapter covering material that is usually embedded in or appended to groundwater chapters. Because caves are among the natural settings that many students visit, and because karst environments are particularly sensitive to environmental damage, we have expanded the discussion as a separate chapter. The final chapter, Human Use of the Earth's Resources, ties together earlier discussions from throughout the book. It reinforces principles that relate to the origins of resources, especially energy-producing ones, and stresses our responsibility to manage them wisely.

New to the Second Edition of *Geology*

A second edition is a wonderful opportunity to build on the first:

- To weave in the latest discoveries in the geosciences.
- To offer up-to-the-minute examples of exciting geological processes, such as the most recent volcanic eruptions and earthquakes.
- To rethink how concepts have been presented in the first edition—to clarify and illustrate them more effectively.

This—the second edition of *Geology*—attempts to accomplish all of these goals, all to ensure that our students have the very best introductory experience with our science. Toward these ends, the following is just a sampling of the new and exciting areas that this edition of *Geology* explores:

- An expanded discussion of the proposed origin of the Moon from a collision between the Earth and a Mars-sized impactor.
- A discussion of Hawai'i's next island, Loihi.
- Iceland's spectacular subglacial eruption of 1996 and its resulting outburst floods.
- Updates on recent and ongoing eruptions in the Caribbean (Montserrat), Mexico City, and New Zealand.
- Dealing with "VOG"—Hawai'i's volcanic smog.
- An expanded discussion of weathering on Mars—with evidence from Pathfinder.
- New insights into the origin of life on Earth.
- A new section on using cosmogenic isotopes to date the Earth's surface.
- The moment magnitude scale—an alternative to the Richter Scale.
- The tsunami of 1700—evidence of catastrophic earthquakes in the Pacific Northwest.
- Monitoring nuclear testing with seismology.
- An expanded discussion of seismic tomography and its wide-ranging uses in deep-Earth studies.
- Using the global positioning system to track plate motion.
- Reconstructing Rodinia and other pre-Pangaea supercontinents.
- Evidence of massive submarine landslides from the flanks of K'ilauea.
- The artificial floods of Glen Canyon, 1997.
- Studies of the groundwater of Yucca Mountain, our proposed nuclear-waste repository.
- Heinrich events and North Atlantic Deep Water circulation and their effects on global glaciation.
- The glaciation of Europa.
- Global warming, sea level, and coastal destruction.

These topics and many more constitute a substantial effort to ensure that a new edition of *Geology* brings new ideas to its readers.

This edition of *Geology* also benefits significantly from a change in text design. The new two-column format has enabled the book's designers to offer much-enlarged photos and illustrations—a concern from the first edition. More than two hundred new photos have been selected (under the outstanding direction of Photo Researcher Townsend P. Dickinson) that illustrate most vividly the processes described in the text.

The Artwork

The drawings in this book are unique. Ramesh Venkatakrishnan is an experienced and respected geology professor and consultant. He is also a highly gifted artist. His drawings evolved along with the earliest drafts of the manuscript, sometimes leading the way for the text discussions. We have worked together since we were graduate teaching assistants at the University of Minnesota–Twin Cities. The desire to illuminate what we want introductory students to know about the Earth is shared by both of us.

As you will see when you leaf through this book, the art explains, describes, stimulates, and teaches. It is not schematic; it shows how the Earth and its geological features actually look. It is also not static; it shows geological processes in action, allowing students to see how geological features evolve through time. Every effort has been made to illustrate accurately a wide range of geological and geomorphic settings, including vegetation and wildlife, weathering patterns, even the shadows cast by the Sun at various latitudes. The artistic style is consistent throughout, so that students may become familiar with the appearance of some features even before reading about them in subsequent chapters. For example, the stream drainage patterns appearing on volcanoes in Chapter 4, Volcanoes and Volcanism, set the stage for the discussion of drainage patterns in Chapter 14, Streams and Floods. The colors used and the map symbols keyed to various rock types follow international conventions and are consistent throughout.

The second edition builds on the strengths of the art program of the first. The images in this edition have been enhanced digitally by renowned geology illustrators George Kelvin and John Woolsey to sharpen their focus, deepen their colors, and lend additional clarity and simplicity to their subjects.

Pedagogy

Nearly every chapter contains one or more Highlights—in-depth discussions of topics of popular interest that provide a broader view of the relevance of geology. In many cases, the Highlights comprise a late-breaking story that also shows the reader that the Earth's geology and its effects on us are changing daily.

To help readers learn and retain the important principles, every chapter ends with a Summary, a narrative discussion that recalls all of the important chapter concepts. Key terms, which are in boldface type in the chapter, are listed at the chapter's end and also appear in boldface in the Summary. Also at the end of every chapter are two question sets: *Questions for Review* helps students retain the facts presented, and *For Further Thought* challenges readers to think more deeply about the implications of the material studied.

The author and illustrators have tried to introduce readers to world geology. This book emphasizes, however, the geology of North America (including the offshore state, Hawai'i), while acknowledging that geological processes do not stop at national boundaries or at the continent's coasts. Wherever data are available—from the distribution of coal to the survey of seismic hazards—we have tried to show our readers as much of this continent, and beyond, as feasible. Photos and examples have been selected from throughout the United States and Canada and from many other regions of the world.

The metric system is used for all numerical units, with their English equivalents in parentheses, so that U.S. students can become more familiar with the units of measurement used by virtually every other country in the world.

The Supplements Package

Geology is accompanied by an array of materials to enhance teaching and learning.

Students who wish additional help mastering the text can use the Study Guide by W. Carl Shellenberger (Montana State University—Northern). For each chapter, the Guided Study section helps students focus on and review in writing the key ideas of each section of the chapter as they read. The Chapter Review, arranged by section and composed of fill-in statements, enables them to see if they have retained the ideas and terminology introduced in the chapter. The Practice Tests and the Challenge Test, which consist of multiple-choice, true/false, and brief essay questions, test their mastery of the material. All answers are accompanied by page references for easy review.

The Instructor's Resource Manual by Chip Fox (Texas A&M University—Commerce) features an outline lecture guide with teaching suggestions embedded in it and student activities and classroom demonstrations. Answers to the end-of-chapter questions in the textbook are also provided. Also included is a comprehensive Test Bank, compiled by Chip Fox, that contains more than one thousand questions. There are at least 40 multiple-choice questions per chapter, classified as either factual or conceptual/analytical. There are also ten short essay questions, complete with answers, for each chapter. A computerized version of the Test Bank is available in both IBM and Macintosh formats.

Also available with this edition is the *Geology Laboratory Manual* by James D. Myers, James E. McClurg, and Charles L. Angevine of the University of Wyoming. This inexpensive manual is closely tied to the text and offers twenty physical geology labs on topics such as maps, plate tectonics, sedimentary and metamorphic rocks, streams, and groundwater. Each lab contains multiple activities to develop and hone students' geological skills. Worksheets are designed to be torn from the manual and submitted for grading.

More than 130 of the text's diagrams and photographs are available for classroom use as full-color slides or transparencies.

The book is supported further by its award-winning website, GEOLOGYLINK (found at www.geologylink.com), maintained and updated regularly by its webmaster, Rob Viens of the University of Washington. This site will tell you what of geological import has happened overnight while you slept. It also contains expanded discussions of "hot topics" in the field of geology and an exhaustive encyclopedia of links to all things geological. For the second edition of *Geology* GEOLOGYLINK contains chapter quizzes and tutorials as well as an on-line version of the *Peterson Field Guide to Rocks and Minerals* by Frederick Pough. These outstanding teaching and learning aids help the student learn physical geology through multimedia technology, study physical geology in a stimulating, yet thoughtful way, and master the principles of physical geology.

Acknowledgments

Some remarkably talented, dedicated people have helped me accomplish far more than I could have done alone. A "committee" of top-flight geologists has been assembled who have dramatically clarified definitions and explanations, eliminated ambiguities, corrected factual errors and fuzzy logic, and, in general, helped the author hone the manuscript in countless ways and helped the illustrator select what to show and how best to do it. Special thanks must go to Kurt Hollocher of Union College and L. B. Gillett of SUNY-Plattsburgh for their extremely insightful critiques of the first edition. In addition, for their constructive criticism at various stages along the way, we wish to thank these excellent reviewers:

From the first edition of *Geology:*

Gail M. Ashley, *Rutgers University, Piscataway*

David M. Best, *Northern Arizona University*

David P. Bucke, Jr., *University of Vermont*

Michael E. Campana, *University of New Mexico*

Joseph V. Chernosky, Jr., *University of Maine, Orono*

G. Michael Clark, *University of Tennessee, Knoxville*

W. R. Danner, *University of British Columbia*

Paul Frederick Edinger, *Coker College (South Carolina)*

Robert L. Eves, *Southern Utah University*

Stanley C. Finney, *California State University, Long Beach*

Roberto Garza, *San Antonio College*

Charles W. Hickox, *Emory University*

Kenneth M. Hinkel, *University of Cincinnati*

Darrel Hoff, *Luther College (Iowa)*

David T. King, Jr., *Auburn University*

Peter T. Kolesar, *Utah State University*

Albert M. Kudo, *University of New Mexico*

Martin B. Lagoe, *University of Texas, Austin*

Lauretta A. Miller, *Fairleigh Dickinson University*

Robert E. Nelson, *Colby College (Maine)*

David M. Patrick, *University of Southern Mississippi*

Terry L. Pavlis, *University of New Orleans*

John J. Renton, *West Virginia University*

Vernon P. Scott, *Oklahoma State University*

Dorothy Stout, *Cypress College (California)*

Daniel A. Sundeen, *University of Southern Mississippi*

Allan M. Thompson, *University of Delaware*

Charles P. Thornton, *Pennsylvania State University*

From the Second Edition of *Geology:*

William W. Atkinson, *University of Colorado, Boulder*

Joseph Chernosky, *University of Maine, Orono*

Cassandra Coombs, *College of Charleston*

Peter Copeland, *University of Houston*

Katherine Giles, *New Mexico State University*

L. B. Gillett, *SUNY-Plattsburgh*

Kurt Hollocher, *Union College*

Kathleen Johnson, *University of New Orleans*

Judith Kusnick, *Cal-State Sacramento*

Bart Martin, *Ohio Wesleyan University*

Ronald Nusbaum, *College of Charleston*

Meg Riesenberg

Roger Stewart, *University of Idaho*

Donna L. Whitney, *University of Minnesota*

I would also like to thank William A. Smith (Charleston Southern State University) for his sharp eye in reviewing the art for accuracy.

At Houghton Mifflin, developmental editors Virginia Joyner and Marjorie Anderson, working under oppressive time constraints, performed the arduous task of reining in the author's long-windedness with extraordinary grace and intelligence and brought organization wherever they found disorder. Senior Associate Sponsor Sue Warne, Senior Project Editor Chere Bemelmans, Art Editor Charlotte Miller, Copyeditor Jill Hobbs, and Editorial Assistant Joy Park polished each chapter of prose and every rough sketch, working with all the elements of the book until they formed a coherent whole. Photo research was handled masterfully by Photo Researchers Townsend P. Dickinson and Mardi Welch Dickenson. The book's pleasing appearance was created under the supervision of Senior Production/Design Coordinator Jill Haber, Associate Production/Design Coordinator Jodi O'Rourke, and Layout Designer Penny Peters. I very much appreciate Editor-in-Chief Kathi Prancan's support and behind-the-scenes hard work and Marketing Manager Penny Hoblyn's energetic marketing support. Thanks are due also to Associate Editor Marianne Stepanian, who coordinated and edited the supplements.

Finally, I also wish to acknowledge with deep appreciation the role of Ron Pullins (formerly of Little, Brown and now of Focus Publishing) and Worth Publisher's Kerry Baruth who championed the cause of *Geology* with their respective companies during its early gestation.

After they leave our classrooms, students may well forget specific facts and terminology of geology, but they will still retain the general impressions and attitudes they formed during our course. We hope that our words and illustrations will help advance the goals of those teaching this course and contribute to their classes. We have used our teaching experiences to craft a textbook that we think our own students will learn from and enjoy. We hope your students will, too. We invite your comments; please send them to the author, whose e-mail address is sechern@u.washington.edu.

Stan Chernicoff

To the Student

Geology is the scientific study of the structure and origin of the Earth, and the processes that have formed it over time. This book was created to bring you some of the excitement of that study through words and illustrations. Over the years, I have derived much pleasure from introducing geology to thousands of my own students at the Universities of Washington and Minnesota. I have also been a student and understand that some topics will be more interesting to you than others. I have worked to make every aspect of the book as fascinating and useful to you as possible.

Unlike some subjects you might study, your Physical Geology course is not over after the final exam. Wherever you live or travel, geology is far more than just the scenery—although your appreciation for the landscape will be much enhanced by a basic knowledge of geology. When you drink from a kitchen tap, dig in garden soil, see a forest—or see it being cut down for a construction project—geology has a role. If you gaze at a waterfall, swim in coastal currents, endure an earthquake—or read about those who did—you will understand more about the experience after taking this course. As you will learn, geology is everywhere—in the products you buy, the food you eat, the quality of your environment. The materials and processes encountered in the study of geology supply all of our needs for shelter, food, and warmth. This course will help you understand why earthquakes occur in some places and not in others, where we should build our homes and businesses to avoid floods and landslides, where we can find safe drinking water, and more. Knowledge of geology can also help make us better citizens as we learn how to prevent further damage to our environment and clean up some of our past mistakes.

You will also learn about more distant matters: the age of the Earth, how the planet has changed over its long lifetime, and how some of its creatures have changed along with it. You will explore some geological ideas about our Moon and the planets with which we share our solar system. Highlight boxes in each chapter provide additional information about particularly interesting topics. These include some of the geo-news events you've heard or read about in recent months.

The language geologists use helps us describe natural phenomena with precision and accuracy. We have minimized the new terminology you will need to learn, but to do well in the course you will still need to master some technical terms. We have also included some of their etymology (mostly Latin) so you can learn from where these unusual terms derive. The key terms are in bold type when introduced. They are also listed at the end of each chapter and defined in the glossary at the end of the book.

The drawings in this book are unique, showing you how the Earth really looks both on and below the surface. As you read and examine each illustration, you will find that the words, the photographs, and the drawings are all important—they are expressly designed to give you a full picture of geology. The text and illustrations, the key terms, and even the chapter summaries work together to help you learn the concepts and terms. When you read the summary, ask yourself if you know these key points; can you cite examples beyond those given in the brief summary?

Each chapter also includes two sets of questions, one testing your retention of the facts and one challenging you to think more deeply about some of their implications. Test yourself, and then go back and reread any material you are not sure you understand.

Finally, with the advancements in communications and multimedia technology, we have crafted two additional means by which you can learn an amazing amount about the Earth and, in doing so, strongly enhance your prospects for "doing well" in your geology course. The first—GEOLOGYLINK (available at www.geologylink.com), the text's web site—is a remarkable device for keeping you up-to-date regarding all things geological. Visiting the site for a few minutes each morning will prepare you well for geology class; in fact, you'll probably be more current than your professor regarding the major geo-events of the day. Use this resource and dazzle your professors and friends with your geo-knowledge. In addition to the latest information on geologic events, GEOLOGYLINK contains quizzes and tutorials on the major concepts in the course as well as an on-line version of the *Field Guide to Rocks and Minerals* by Frederick Pough.

The second teaching aid is the *Geology* CD-ROM that accompanies this book. This tool is designed to help you master the principles of physical geology and test your mastery of them. I strongly encourage you to spend time with both of these wonderful educational tools; they are ideal learning companions to the textbook.

As the author of this book, I hope you will enjoy it and gain an appreciation for geology that will enrich your life. This book has been written, illustrated, and designed to compel you to keep it on your bookshelf (although a few of you may actually choose to sell it to "Used Books"). It is our hope that 5, 10, or 20 years from now, when you see something that sparks your geological curiosity, you'll use your old geology book as a reference. If you have any comments, complaints, or compliments, please send them to me or to Houghton Mifflin. My e-mail address is sechern@u.washington.edu.

Features of

GEOLOGY,

Second Edition

The rich detail and technical accuracy of the illustrations help convey complex concepts to introductory students.

Photos are often paired with art to emphasize a point.

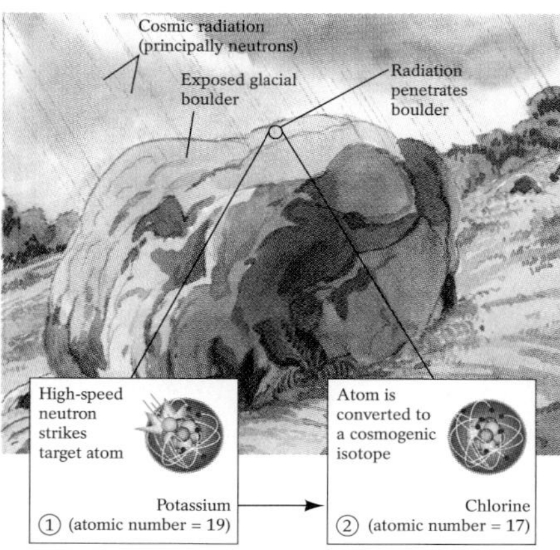

Figure 8-30 As soon as it is exposed, any surface rock begins to be bombarded with cosmic radiation. This converts target atoms into radioactive "cosmogenic" isotopes whose decay rate can then be measured to determine the time of initial exposure.

Figure 8-31 This massive basaltic boulder was plucked from the Columbia River basalt flows of eastern Washington by a moving glacier about 15,000 years ago. Since the moment it melted from the ice, it has been bombarded by cosmic radiation; its time of initial exposure can be determined by surface-exposure dating.

This method measures *cosmogenic* isotopes, those produc in extremely small quantities at the surfaces of newly posed rocks and other surface materials from cosmic- bombardment.

Intergalactic cosmic radiation, predominantly in form of neutrons, constantly bombards the Earth at p nomenal speeds. (Countless numbers of these particles passing through your body as you read this passage.) Th neutrons strike the rocks and soils at the Earth's surface, p etrating to a maximum depth of 2 to 3 meters (7 to 10 fe before being absorbed. As these high-speed neutrons hit t get elements in rocks and soils—such as oxygen, silicon, a minum, iron, magnesium, calcium, and potassium—t knock other atomic particles from the nuclei of these e ments, converting these elements into new "cosmogenic" dioactive isotopes (Fig. 8-30). Silicon, magnesium, iron, a aluminum, for example, are typically converted to *berylliu 10;* potassium, calcium, and chlorine become *chlorine-*

The atoms of these radioactive substances, like ot radioactive atoms, immediately begin to decay. Unlike in case of non-cosmogenic isotopes, whose number of par atoms is fixed (such as uranium in the Earth's crust), n atoms of these cosmogenic isotopes are constantly being p duced, even as others decay. Thus a simple parent–daugh decay relationship does not apply to these isotopes. Inste we must know and then subtract their production rates a accurately determine their decay rate, or half-life. Much r orous testing and study have revealed that chlorine-36 ha half-life of roughly 300,000 years; beryllium-10's half-lif

approximately 1.5 million years. Knowing these half-lives enables geologists to date newly exposed surface boulders (Fig. 8-31) whose ages fall between the ranges of carbon-14 dating and other isotope-dating schemes.

Of course, any new dating method faces technical challenges. The first involves the extremely small amounts of these cosmogenic isotopes present in the rocks. During the

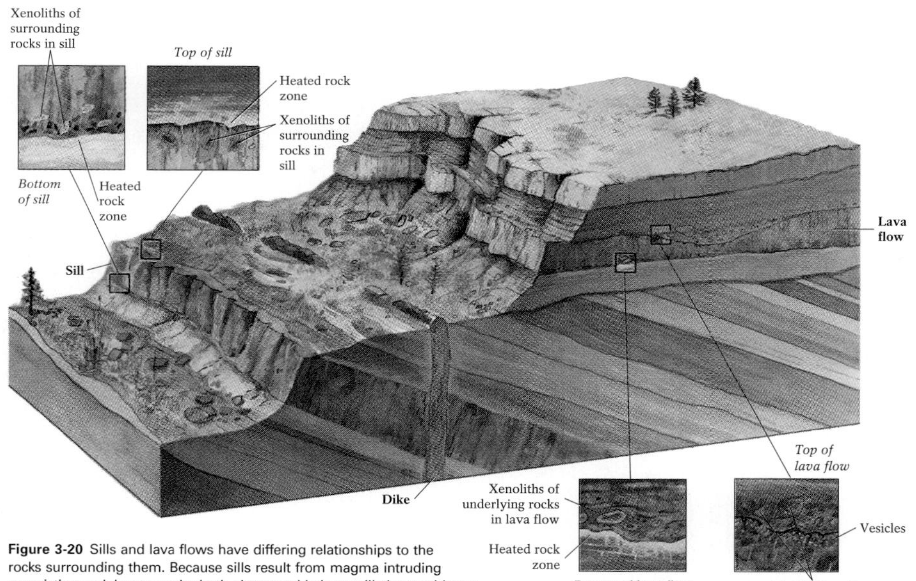

Figure 3-20 Sills and lava flows have differing relationships to the rocks surrounding them. Because sills result from magma intruding preexisting rock layers, rocks *both above and below* a sill show evidence of heating by the magma. Only the rock *below* a lava flow is affected, as overlying layers are not deposited until after the flow has solidified. In addition, because they occur at the surface, lava flows show evidence of having been exposed to air, whereas sills do not.

1. Look at the adjacent surfaces of the surrounding rocks. When lava is extruded, there are no overlying rocks and only the rocks *beneath* it will show evidence of heating. When hot magma intrudes between two layers of rock to form a sill, it heats the adjacent surfaces of *both* layers before cooling.
2. Look at the top and bottom surfaces of the igneous layer. Both surfaces of a sill will contain fragments of the surrounding rock that were pried loose as magma intruded, whereas only the bottom of a lava flow incorporates preexisting rock.
3. Look at the top surface of the igneous layer. Because the top of a lava flow is exposed to the air for some time before being overlain by other lava or sediment, its gases are free to escape. Consequently, the surface of a lava flow is often pockmarked by cavities called *vesicles* that were formerly occupied by escaping gas bubbles. The tops of most sills, which were never exposed to the air, display few, if any, vesicles.
4. Look for signs of weathering (discussed in Chapter 5). The upper surface of a lava flow would appear somewhat weathered from its exposure to the atmosphere before being overlain, whereas a sill would not show signs of weathering because it was never exposed.

Batholiths and Other Large Plutons

Large concordant plutons are commonly several kilometers thick and tens or even hundreds of kilometers across. They may be mushroom-shaped or saucer-shaped, close to the surface or deep beneath it.

When thick, viscous magma intrudes between two parallel layers of rock and lifts the overlying one, it cools to form a mushroom-shaped or domed concordant pluton, or **laccolith** (from the Greek *lakkos*, or "reservoir") (see Fig. 3-17). Laccoliths tend to form at relatively shallow depths, where there is little pressure to keep the overlying rock in place. Many are granitic, formed from felsic magma that flows so slowly that it tends to bulge upward instead of spreading

Artwork shows geologic features in a naturalistic context, to give students a sense of how these features actually look.

Highlight 4-3 Fire and Ice: Iceland's Subglacial Eruption of 1996

On September 29, 1996, an earthquake shook the ground beneath the 600-meter (2000-foot)-thick Vatnajökull icecap of southeastern Iceland. Vatnajökull, covering 8300 square kilometers (3200 square miles), is the largest glacier in Europe. The quake signaled the onset of a fissure eruption emanating from Iceland's Bardarbunga and Grimvötn volcanic centers. On the morning of October 1, scientists discovered a deep subsidence basin in the glacier's surface, close to the location of a major subglacial eruption in 1938. Throughout the day, the basin grew and three additional basins opened. The development of these aligned basins indicated that intensive melting was occurring at the base of the glacier along a 5- to 6-kilometer (3- to 4-mile)-long fissure. By October 2, steam rising from the glacier's surface indicated that the eruption had broken through the ice. Over the next several days, a steam column rose to more than 10 kilometers (6 miles), the surface fissure extended roughly the same distance, and black ash was being thrown intermittently into the air to heights of 300 to 500 meters (1000–1600 feet) (Fig. 4-26).

The greatest danger from a subglacial eruption is a *jokulhlaup,* or outburst flood. These floods occur when a great volume of subglacial meltwater, draining through subglacial tunnels, finds an outlet at the glacier's margin and discharges catastrophically (Fig. 4-27). Vatnajökull, overlying a segment of the mid-Atlantic ridge in southeastern Iceland, has been the source of numerous jokulhlaups in the past. By October 12, Vatnajökull's surface had risen 15 to 20 meters (50 to 65 feet) directly above the Grimsvötn caldera. This change suggested that meltwater from the eruption was flowing into and filling the 10-kilometer (6-mile)-diameter caldera. Engineers, anticipating a flood of monumental proportions, worked around the clock to reinforce dikes and cut diversionary channels to steer the expected torrents away from Iceland's main, island-ringing road, about 50 kilometers (30 miles) south of the icecap.

The long, unnerving wait for the jokulhlaup ended on November 5, when Iceland's largest flood in 60 years burst from the edge of the icecap. For two days, the flow formed the second-largest river in the world, destroying three bridges, severing vital telephone lines, and, despite valiant efforts, washing away Iceland's south-coast road. Such are the rigors of life in Iceland—the self-proclaimed Land of Fire and Ice.

Figure 4-26 Steam and ash rising from the subsidence basin on the surface of the Vatnajökull icecap, directly above the subglacial Grimvötn volcano.

Highlight boxes introduce high-interest topics, often recent geologic events.

Subglacial Eruptions When polar regions coincide with zones of volcanic activity, two of geology's most pronounced extremes—fire *and* ice—collaborate in spectacular geological

ıts. For example, in areas where large ice sheets overlie anically active fissures, hot mafic lava periodically erupts ɛath the ice, melting vast amounts of it and producing strophic outbursts of meltwater. One such drama captid geologists for six weeks in the autumn of 1996 when olcano erupted beneath a massive icecap in Iceland hlight 4-3).

oclastic Eruptions

oclastic eruptions, usually involving viscous, gas-rich magmas, vary from moderately to spectacularly explosive and tend to produce a great deal of solid volcanic fragments. Moderately explosive eruptions on the Italian island of Stromboli in the eastern Mediterranean occur almost continuously. A

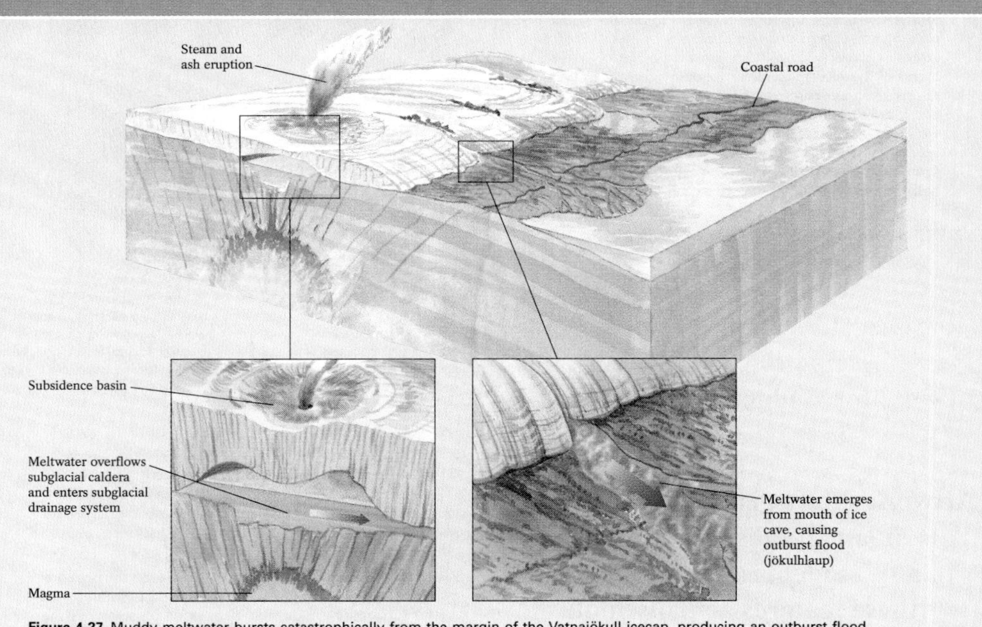

Figure 4-27 Muddy meltwater bursts catastrophically from the margin of the Vatnajökull icecap, producing an outburst flood, or jokulhlaup.

Every chapter has been updated to include fascinating developments and significant discoveries of the last 3 years.

32 Part I Forming the Earth

Chapter Summary

Geology is the scientific study of the Earth. Geologists, like other scientists, systematically collect data derived from experiments and observations. They analyze and interpret their findings and develop **hypotheses** to explain how the forces of nature work. The hypotheses that are consistently supported by further study and investigation may be elevated to the status of a widely accepted explanation, or **theory.** A theory that withstands rigorous testing over a long period of time may be declared a **scientific law.** In order to be accepted, all scientific investigations must conform to these **scientific methods.** The hypotheses that have undergone such scrutiny include two that attempted to explain the evolution of the Earth's geologic features. **Catastrophism,** which was popular until the mid-eighteenth century, held that the Earth had evolved through a series of immense worldwide upheavals; **uniformitarianism,** proposed that the Earth has evolved slowly and gradually by small-scale processes that can still be seen operating. Today scientists recognize the effects of both slow processes and catastrophic events in the evolution of the Earth.

The present universe began with the "Big Bang" roughly 12 billion years ago. The Sun, which is a star, formed about 5 billion years ago from the collapse of a gas cloud, the center of which heated up as particles drawn inward by gravity collided and produced nuclear reactions. As the out region of the gas cloud cooled, the Earth and other plane developed (the Earth about 4.6 billion years ago) by accr tion of colliding masses of matter, some perhaps as large the planet Mars. Earth's collision with one such mass m have spawned the Moon.

During the Earth's first few tens of millions of years existence, the impact of these accreted masses along with t heat produced by radioactive decay warmed its interior u til the accumulated heat was sufficient to melt much of t planet's constituents. This period of internal heating caus the Earth to become layered, or differentiated. We base mu of our knowledge about the origin of the Earth's interior la ers on the study of **chondrules,** small nuggets of rocky m terial found in many meteorites that are believed to droplets of matter that condensed directly from the origin solar nebula. By comparing the composition of the Earth crust today to that of these chondrules, we note that t Earth's crust is quite deficient in iron. The early period heating must have caused the Earth's densest elements, p marily iron, to sink toward its interior while its lightest el ments rose and became concentrated closer to its surfac

Today, the Earth has three principal concentric laye of different densities: the thin, least dense outer layer, call the **crust;** a thick, denser underlying layer, called the **mantl** and a much smaller **core,** which is the most dense of Earth layers. Over the last 4 billion years, the Earth has cool slowly from its initial higher temperatures. Enough heat remains in its interior to generate currents of flowing mantle rock that have kept the outer portion of the Earth mobile. The Earth's **lithosphere,** a composite layer made up of the crust and the outermost segment of the mantle, is solid and brittle and forms large rocky plates; these plates move along at the Earth's surface atop the warm, flowing **asthenosphere** beneath them.

Time plays an important role in the evolution of the Earth's geologic features and materials. The Earth is believed to be about 4.6 billion years. Over such a vast amount of time, many gradual geological changes can take place that occur too slowly to be perceived on a human time scale. The Earth's three principal types of rocks undergo such changes, actually turning from one type into another, depending on the environmental forces acting on them. **Rocks,** which are defined as naturally occurring aggregates of inorganic materials **(minerals),** are categorized according to the way in which they form. The three basic rock groups are **igneous rocks,** which solidify from molten material; **sedimentary rocks,** which are compacted and cemented aggregates of fragments of preexisting rocks of any type; and **metamorphic rocks,** which form from any type of rock when its chemical composition is altered by heat, pressure, or chemical reactions in the Earth's interior. The continual transformation of the Earth's rocks from one type into another over time is called the **rock cycle.**

End-of-chapter summaries present an overview of the content in narrative form to help students review.

238 Part I Forming the Earth

which evidence of life began to be abundantly preserved as fossils in rocks. The Phanerozoic Eon is divided into the **Paleozoic** ("ancient life") **Era, Mesozoic** ("middle life") **Era,** and **Cenozoic** ("recent life") **Era;** the Paleozoic was dominated by marine invertebrates (such as primitive clams, snails, and corals), and later fish and amphibians, and the Mesozoic and Cenozoic were dominated by reptiles (such as dinosaurs) and mammals, respectively.

Key Terms

geochronology (p. 207)	parent isotope (p. 220)
historical geology (p. 208)	daughter isotope (p. 220)
relative dating (p. 208)	radiometric dating (p. 221)
absolute dating (p. 209)	half-life (p. 222)
principle of uniformitarianism (p. 209)	uranium–thorium–lead dating (p. 225)
principle of original horizontality (p. 210)	potassium–argon dating (p. 225)
principle of superposition (p. 210)	rubidium–strontium dating (p. 225)
principle of cross-cutting relationships (p. 210)	carbon-14 dating (p. 225)
principle of inclusions (p. 210)	fission tracks (p. 227)
fossils (p. 212)	dendrochronology (p. 227)
principle of faunal succession (p. 212)	varves (p. 227)
index fossils (p. 213)	lichenometry (p. 229)
unconformities (p. 213)	geologic time scale (p. 232)
correlation (p. 217)	Paleozoic Era (p. 233)
	Mesozoic Era (p. 233)
	Cenozoic Era (p. 234)

Questions for Review

1. Briefly explain the difference between relative and absolute dating.

2. Discuss three of the basic principles that serve as the foundation of relative dating.

3. What qualifies a species to become an index fossil? How are index fossils used in the correlation of sedimentary rock strata?

4. Sketch and label two different types of unconformities.

5. Name three parent–daughter radiometric dating systems, and give the half-lives of each parent isotope, as well as the rocks or sediments that are most likely to be dated by each.

6. Briefly discuss two potential problems that may diminish the reliability of an isotopically derived date.

7. Briefly explain how carbon-14 enters the cells of living organisms.

8. Select three absolute-dating methods. Describe their basic principles, and the materials that can be dated by each technique.

9. Using the geologic time scale, state when each of the following great events in Earth history occurred: the origin of the world's iron ores; the first appearance of a protective atmosphere; the origin of flowering plants; the origin of birds; the age of reptiles; the current ice age.

10. If the oldest rocks found on Earth are less than 4.0 billion years old, what evidence suggests that the Earth is actually 4.6 billion years old?

For Further Thought

1. Why are obsidian and basalt more susceptible to the development of hydration and weathering rinds than granite and andesites?

2. Look at Figure 8-13 on page 216, the geologic profile of the Grand Canyon. The Cambrian Tapeats Sandstone lies unconformably above several different bodies of rock. Identify two different types of unconformities that separate the Tapeats Sandstone from the underlying rocks.

3. Using the various dating methods discussed in Chapter 8, derive the sequence of events that produced the hypothetical landscape below. (Go slowly, and don't jump to premature conclusions. Consider all the principles that we've discussed.) Which of the layers might be dated absolutely?

4. Although geologists claim that "the present is the key to the past" (the principle of uniformitarianism), the Earth has certainly changed throughout its 4.6-billion-year history. Think of two geological processes that operate differently today than they did in the past, and discuss how they vary.

5. Suppose you decided not to accept the 4.6-billion-year age of the Earth that geologists have determined (primarily from the ages of Moon rocks and meteorites and the evolution of lead isotopes on Earth). Devise an alternative strategy for determining the age of the Earth, assuming that you have unlimited funds.

The Key Term list is a tool for quick review and gives the page number for the full discussion, for students who need to reread the material. (The terms also appear in the glossary.)

Questions for Review help students review the factual content of the chapter, and For Further Thought questions encourage them to think critically about the implications of the information they have learned.

GEOLOGY

Part 1 Forming the Earth

1

A First Look at Planet Earth

At 5:03 P.M. Pacific daylight time on October 17, 1989, baseball fans across North America were settling down in front of their television sets to watch game 3 of the World Series from San Francisco. At that moment, violent movement along a small segment of California's San Andreas fault (Fig. 1-1) caused the screens to go blank. When service was restored, instead of baseball, millions of people viewed grim scenes of collapsed buildings and freeways (Fig. 1-2), broadcast live. Widespread destruction took the lives of scores of Bay Area residents and injured hundreds more. People in other earthquake-prone regions wondered whether their own towns and cities might be the next to suffer a life-threatening quake.

Four years later, at 4:31 A.M. Pacific standard time on January 17, 1994, millions of Southern Californians were jolted awake by a powerful earthquake that took 57 lives, buckled numerous freeways, and, with estimated cleanup and repair costs of more than $15 billion, proved to be one of the most expensive natural disasters ever to occur in the United States.

Figure 1-2 The collapse of the Nimitz Freeway in Oakland, California, during a major earthquake along the San Andreas fault on October 17, 1989. The San Andreas fault, a fracture in the Earth's crust that cuts northwest–southeast across much of California, is responsible for some of North America's most powerful earthquakes.

Figure 1-1 The San Andreas fault, as seen from the air over Carrizo Plain in California.

Figure 1-3 The Kobe, Japan, earthquake of January 1995 was a moderate-to-strong quake, not nearly as powerful as many that have struck California during the twentieth century.

Exactly one year later, another earthquake—this one over 9000 kilometers (5700 miles) away near Kobe, Japan—put the effects of the 1989 Loma Prieta and 1994 Northridge earthquakes into a new perspective. The Kobe quake killed more than 5000, injured 30,000, and left 300,000 homeless (Fig. 1-3). The destruction wrought by the Kobe earthquake was a frightening reminder that the Loma Prieta and Northridge earthquakes may have been only preludes to the long-awaited "Big One" that looms in the minds of 36 million Californians.

Do the citizens of Boston, Massachusetts, Memphis, Tennessee, and Charleston, South Carolina, need to worry quite as much about earthquakes as do the citizens of San Francisco and Los Angeles? Considering only recent events, perhaps not. But destructive tremors struck the Boston area in 1755, Memphis in 1811, and Charleston in 1886—and could well do so again. North Americans would be wise to understand the geological forces that cause phenomena such as earthquakes so that they can prepare for them. With this practical aspect of geology in mind, let us look at what geology is and what geologists do.

Geology is the scientific study of the Earth's processes and materials. Geologists speculate about the origin of the Earth, and they examine the wide variety of materials it contains and the processes, such as earthquakes, that act at or near its surface. The subjects of their investigations range from the smallest atoms to entire continents and ocean basins. Geologists study erupting volcanoes and windswept sand dunes, raging floods and creeping glaciers, the slow, silent enlargement of underground caverns and the remains left behind by titanic meteorite impacts. Some collect and interpret evidence of past life on Earth—from the vestiges of the earliest known single-celled algae found in the Australian outback in 3.3-billion-year-old rocks to the bones of our earliest human ancestors excavated from 3-million-year-old volcanic ash layers in East African valleys. Geologists today are not even Earthbound: They study lunar rocks collected by astronauts and analyze data, gathered by space probes, from the farthest planets in our solar system (Fig. 1-4).

Everything we use comes from the Earth, and so geology has an enormous practical impact on our daily lives. The natural resources that shape modern life have come to us, in part, through geological knowledge. Geologists help locate the ingredients for cement, concrete, and asphalt, with which we build our cities and highways, as well as the oil, gas, and coal that fuel our cars and light and heat our homes. In recent years, many geologists have searched for a rapidly dwindling resource—fresh, uncontaminated underground water for industrial, agricultural, and domestic use. Geological study also helps us to predict and avoid some of nature's life-threatening hazards. Geologists identify slopes that may produce landslides and sites that are too unstable to contain dams or nuclear power plants safely. They detect potential earthquake locations and recommend ways to avoid damage from flooding rivers. They warn us away from eroding shorelines and devise ways to clean up the mess we've made through decades of environmental ignorance or negligence.

Other geologists probe the most fundamental mysteries of our planet: How old is the Earth? How did it form? When did life first appear? Why do some areas suffer from devastating earthquakes, while others are spared? Why are some regions endowed with breathtaking mountains and others with fertile plains? Some of the same questions have engaged human curiosity for millennia. (You may have asked some of them yourself as you walked into the lecture hall for your first geology class.) We will begin to answer these questions by describing how the science of geology operates, introducing some of the basic concepts and standards that underpin this discipline. We will discuss how our planet may have formed and speculate about some changes it has gone through since. Finally, we will examine some of the Earth's large-scale geological processes and determine how scientists deduced the nature of these processes.

Geology and the Methods of Science

Scientists make one basic assumption: The world works in an orderly fashion in which natural phenomena will recur given the same set of conditions. As scientists understand it, every effect has a cause. The principal objective of science is to discover the fundamental patterns of the natural world.

Geologists, like other scientists, begin their investigations with a question or set of questions about how some

(a)

(b)

(c)

Figure 1-4 Geologists at work. **(a)** Sampling 1200°C lava from Hawaiian volcanoes; **(b)** unearthing dinosaur remains in Dinosaur National Monument, Utah; **(c)** studying rocks on the Martian surface, with a little mechanical assistance (the Pathfinder mission of 1997).

part of the natural world functions. Why do earthquakes occur in some places and not in others? Why are some places blessed with great mineral wealth—diamonds, gold, platinum—and others not? Using **scientific methods,** scientists gather all available data bearing on their subject—taking measurements, describing the phenomena they observe, compiling the results of laboratory experiments. They then propose a **hypothesis,** a logical but tentative explanation that fits all the data collected and is expected to account for future observations as well. Good scientific research requires that investigators be receptive to *whatever* they discover, even if their data refute their hypothesis. Often a number of hypotheses are proposed to explain the same set of data. For example, within the past 20 years, more than 50 hypotheses have been proposed to explain why the Earth's climate periodically plunges into ice ages.

Once a hypothesis is introduced to the scientific community, interested parties begin to test it—often vigorously. Scientists conduct further experiments or make further observations to see if the hypothesis can stand up to intense scrutiny. Some hypotheses undergo numerous modifications and evolve over many years until they fully account for all relevant findings. Others simply fail the test and are cast aside. New information and technologies capable of more accurate observations often reveal fatal flaws in once-reasonable hypotheses. For example, high-powered telescopes shattered the hypothesis that the Earth is the center of our solar system. The history of science is littered with discarded hypotheses that were once quite popular but after closer analysis proved to be false. A flawed hypothesis usually gives way to one that encompasses more of the available data.

A hypothesis that is repeatedly confirmed by observation and experimentation becomes more widely accepted within the scientific community. Ultimately, a comprehensive hypothesis that consistently explains accumulating data may become a **theory**—a generally accepted explanation for a given set of data or observations. An example of how multiple hypotheses are developed and tested, possibly to become theories, is given in Highlight 1-1.

Even when a hypothesis survives testing and becomes an established theory, new technology and new data may require it to be updated—perhaps even replaced. For example, when we are able to drill more deeply into the Earth we may discover evidence that will cause geologists to revise current widely accepted theories about the composition of the Earth's interior. A theory that meets rigorous testing over a long period of time may ultimately be declared a **scientific law.** For example, any free-falling object within the Earth's gravitational field *always* falls toward the Earth's surface, demonstrating the *law* of gravitational attraction.

Geologists apply scientific methods in ways that are unique to their science. Unlike some chemistry or physics experiments, which may involve reactions that last but a millionth of a second and that occur on an atomic scale, geologists must often deal with the effects of processes acting over vast periods of time, across enormous areas, and sometimes out of sight. Furthermore, geologists can't always test hypotheses through direct observation or experimentation. To supplement their field and laboratory work, they sometimes use scaled-down models to study large-scale geological phenomena, or they rely on computers to calculate the results of mathematical models.

The Development of Geological Concepts

Almost two centuries of observation and hypothesis formation and testing have contributed to our current understanding of how our planet developed. For a long time

Highlight 1-1 What Caused the Extinction of the Dinosaurs?

For years, paleontologists (geologists who study ancient life forms) have wondered what might have caused more than 75% of all the forms of life then on Earth to vanish about 65 million years ago. The most dramatic loss was the extinction of the dinosaurs, a group of animals that had roamed the planet for 150 million years, but numerous other life forms vanished as well—large and small, water- and land-dwelling, plant and animal.

Some early hypotheses focused on only one kind of organism to explain these extinctions. Some proposed that epidemic diseases eliminated dinosaur populations or that egg-stealing mammals, then on the rise, ravaged dinosaur nests. But neither of these hypotheses accounted for the loss of two-thirds of all marine animal species, which led some scientists to propose that the oceans became lethally salty. But this did not explain why some marine creatures survived. To explain the extinction of gigantic terrestrial reptiles, tiny marine organisms, and many life forms in between, a number of hypotheses invoked global environmental change. Did the Earth suffer from a period of drastic cooling 65 million years ago? Did a shift in the planet's protective magnetic field allow harmful solar radiation to reach land and sea, eliminating a wide variety of life forms? Did a nearby star explode, bathing the Earth in cosmic radiation? Surely, each of these events would have affected all life on Earth simultaneously. Why, then, were 25% of the planet's species spared?

Several hypotheses suggest that wholesale extinction followed some catastrophic disruption of the global food chain. The food chain refers to nature's succession of predator–prey relationships, specifying "who eats whom." In this chain, a meat-eating dinosaur such as *Tyrannosaurus rex* would perish if its prey, typically a plant-eating dinosaur, became scarce, which would result if that prey's food supply somehow diminished. Some scientists looking for indications of a global event that could have caused such a disruption in the food chain have noted evidence of widespread fires followed by very rapid cooling at this period, but they disagree on what might have caused this. One group has cited evidence that massive volcanic eruptions of India's Deccan plateau sent a cloud of volcanic ash and gas around the Earth, blocking out sunlight, cooling the planet, and leading to a worldwide decline in vegetation, including microscopic marine plants. These scientists reason that without the plants their diets were based on, many plant-eating animals would have died out, and their extinction would in turn have wiped out the meat-eaters who were their predators.

Another group of scientists, led by geologist Walter Alvarez and his father, Nobel prize–winning physicist Luis Alvarez, have proposed this scenario: A meteorite at least 10 kilometers (6 miles) in diameter crashed into the Earth, releasing a shower of pulverized rock into the atmosphere. The resulting dust veil and the accompanying smoke from widespread fires would have blocked sunlight (in much the same way as volcanic ash would have), cooled the planet, and led to an "impact winter" that may have lasted for decades—long enough to devastate the global food chain.

(a)

Strong evidence supports this impact hypothesis. An unusual 2.5-centimeter (1 inch)-thick layer of clay has been documented at many localities and on all continents in rocks that date from about 65 million years ago (Fig. 1-5a). The clay contains iridium, an element that is extremely rare in rocks of terrestrial origin but is much more abundant in meteorites. The Alvarezes and their associates contend that the iridium-rich layer resulted from the global fallout of pulverized rock. Fossils of numerous species, including many now extinct, have been found in the rocks that formed just before the iridium-rich layer was deposited, whereas only about a fourth as many species are represented in the rocks that formed just after this layer was deposited, suggesting that many extinctions occurred during the time of iridium deposition.

Further evidence of a meteorite impact includes the presence of small glassy spheres called *tektites* in sediment layers around the world dating from this period (Fig. 1-5b). These may have been formed when superheated rocks at an impact site were hurled into the air in a molten state, dispersed in the atmosphere, and cooled rapidly as they fell back to Earth. Mineral grains shattered by very high pressures—as would occur if they had been struck by a meteorite—have also been found at proposed impact sites and throughout the world in rock layers dated at about 65 million years (Fig. 1-5c). Furthermore, the high concentration of carbon soot (a product of burned vegetation) found within the iridium layer could be evidence of global wildfires, which may have been touched off as countless glowing bits of falling debris ignited the Earth's vegetation.

Although this evidence of a 65-million-year-old meteorite impact is certainly compelling, most of it is scattered in various lo-

(b)

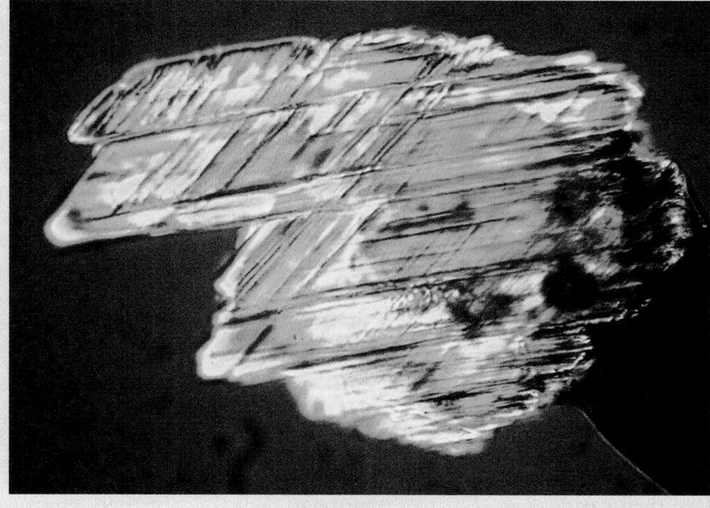

(c)

cations around the world. What may be the final piece to the puzzle, a virtually uninterrupted record of the events of that time, was extracted during the coring of the Atlantic Ocean floor, 320 kilometers (200 miles) east of Jacksonville, Florida, in 1996. The 40-centimeter (16-inch) core of mud contains a whitish fossil-rich layer at the bottom of the core that is overlain by a thin gray-green layer of impact debris topped by an iron-rich band that may constitute remains from the meteorite itself. This layer, in turn, is overlain by a fossil-poor layer that may have been deposited over a period of about 5000 years.

As yet, no extinction hypothesis has achieved theory status. Analysis of the Earth's 65-million-year-old deposits continues today, to document further the percentage of organisms that became extinct at that time and to search for additional evidence of a meteorite strike or a catastrophic volcanic eruption. For more than a decade, the proponents of the impact hypothesis awaited confirmation of a key piece of evidence—definite location of an impact site. A meteorite capable of such vast environmental disruption would have left behind a crater at least 160 kilometers (100 miles) across. The search for the impact site may now be over. A huge partially submerged crater has been mapped and studied along the coast of Mexico's Yucatán Peninsula. Because of the age of its rocks and the presence of tektites in nearby sediments, the Yucatán's Chicxulub crater, 300 kilometers (180 miles) in diameter, is widely considered to be the proverbial "smoking gun" (Fig. 1-5d). Still, the debate over the cause of this mass extinction is expected to continue for years to come. It is important to remember that all geologists do not agree with the meteorite hypothesis and, more importantly, it is acceptable . . . no, make that essential . . . that they do *not* agree. Such debate is the driving force of scientific progress.

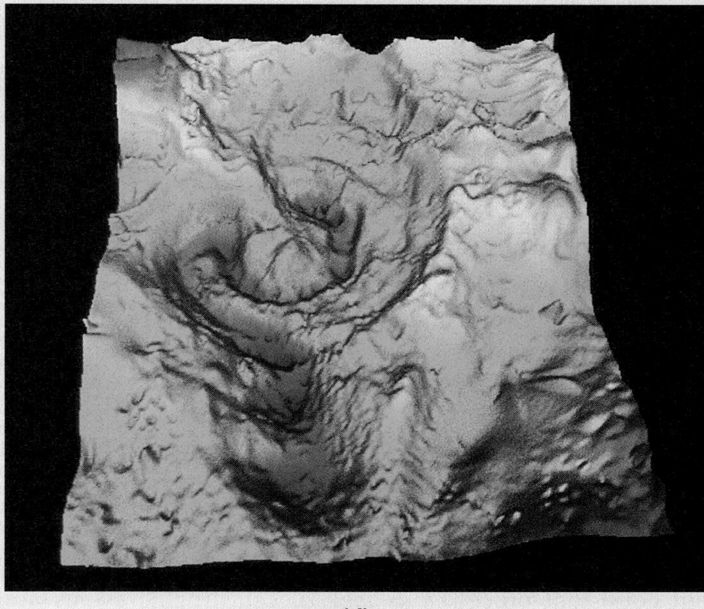

(d)

Figure 1-5 **(a)** An iridium-containing layer of clay (marked by coin) found in Gubbio, Italy. **(b)** Tektites from Thailand. Such aerodynamic spheres, found around the world in sediments of about 65 million years of age, are believed to have been formed when rock melted by meteor impact was hurled into the air and cooled and solidified before falling back to Earth. **(c)** A magnified photograph of a mineral grain that has been shattered by the shock of a meteorite impact. **(d)** This radar image shows the circular outline of an enormous impact crater in Mexico's Yucatán Peninsula. The Chicxulub crater is large enough to account for the effects of a massive meteorite strike.

people believed that the Earth had evolved through episodes of dramatic and rapid change interspersed with long periods of relative stability and little change. Since natural scientists understood best those processes that they could directly observe, such as volcanic eruptions, monumental earthquakes, and raging floods, they believed that these violent events alone explained the origin of such apparently inactive Earth features as mountains, valleys, and fossils. **Catastrophism,** the hypothesis that the Earth evolved through a series of immense worldwide upheavals, was the prevalent view until the mid-eighteenth century.

During the latter part of the eighteenth century, the Scottish naturalist James Hutton (1726–1797) recognized that slow processes, such as rivers cutting through valley floors and loose soil creeping down gentle slopes, acting over a vast amount of time, may affect the Earth more cumulatively than do occasional catastrophic events. This idea met with great resistance because it implied that the Earth was a lot older than most Christians at the time believed it to be. Reckoning based on the Bible put the Earth's age at only a few thousand years. Hutton proposed that the physical, chemical, and biological processes that anyone could see changing the Earth in small ways during one's own lifetime must have worked in a similar manner throughout a very long history. His hypothesis, called **uniformitarianism,** proposed that our observations of current geological processes could be used to interpret the rock record of long-past geological events. By the 1830s, after much debate, uniformitarianism prevailed over catastrophism. Its acceptance has been hailed as the birth of modern geology.

Hutton maintained that "the present is the key to the past." Geologists today do acknowledge that the physical processes we see shaping the Earth's appearance today (such as erosion, glaciation, and mountain building) have probably acted throughout the Earth's history, although probably at significantly variable rates. They also know, however, that some geological events are indeed catastrophic and that much geological change does occur during these brief, spectacular events. A great earthquake may shift a land area more than 6 meters (20 feet) in a single moment. In 1989, Hurricane Hugo eroded more of the Carolina coast in one day than had the preceding century of slow, steady wave action. Thus, slow but consistent processes *as well as* catastrophic events are continuously reshaping our planet.

The Earth in Space

In some respects the Earth seems like nothing special. The planet is a slightly flattened sphere with an average radius of 6371 kilometers (3957 miles), orbiting approximately 150 million kilometers (93 million miles) from the medium-sized star we call the Sun. Our Sun is only one of about 100 billion stars in the Milky Way galaxy, a pancake-shaped cluster of stars that itself is only one of about 100 billion such galaxies in the observable universe. Despite Earth's relative smallness in the universe, it is perfectly positioned to receive just the right amount of the Sun's radiant energy to support life. Because of Earth's composition and geologic past, it has developed a watery envelope and protective atmosphere on which countless living species have relied for billions of years. But how did the Earth come to be what *may* be the only life-sustaining planet in the solar system? (Recent discoveries of what may be evidence of past life on Mars suggest that the jury's still out on this one. More on this in Chapter 8.)

The Probable Origin of the Sun and Its Planets

Cosmologists (scientists who study the origin of the universe) have proposed that the present universe began as a very small, very hot volume of space containing an enormous amount of energy. Many scientists believe the birth of all the matter in the universe occurred when this space expanded rapidly with a "Big Bang" roughly 12 billion years ago. The timing of the hypothetical Big Bang has been estimated recently with the aid of the Hubble Space Telescope, which observes the current positions and speeds of the visible galaxies as they move away from one another. By tracing their paths backward, cosmologists estimate their point and time of origin.

Immediately after the Big Bang, the universe began to expand and cool, as it continues to do today. By a few minutes after the bang, the universe had cooled to a temperature of about 1 billion degrees Celsius. At this time, the universe consisted only of three kinds of subatomic particles (protons, neutrons, and electrons—discussed further in Chapter 2)—remnants of its original composition. These particles eventually combined to form atoms, the building blocks of all the matter in the universe today. At this early stage, however, any atomic particles formed immediately broke apart in violent collisions with other particles.

Only after about a million years had passed, when the universe had cooled to approximately 3000° Celsius (about 6000° Fahrenheit), could atoms of hydrogen and helium—the simplest, lightest elements—begin to exist without being torn apart. At that time the universe consisted of about 75% hydrogen gas and 25% helium gas by weight, a composition that has not changed much to this day. As the universe continued to expand, its matter became less uniformly dispersed. Pockets of relatively high gas concentrations began to attract more gas by the force of gravity. Where enough gas gathered, the resulting gas clouds collapsed inward because of gravity. These accumulations became galaxies, large disk-shaped structures, and clusters of galaxies. Although cosmologists have amassed a great deal of evidence to support these details of the Big Bang hypothesis, galaxy formation is still not well understood.

Within each galaxy, such as our own Milky Way, some gas clouds collapsed further to form stars. Stars continue to

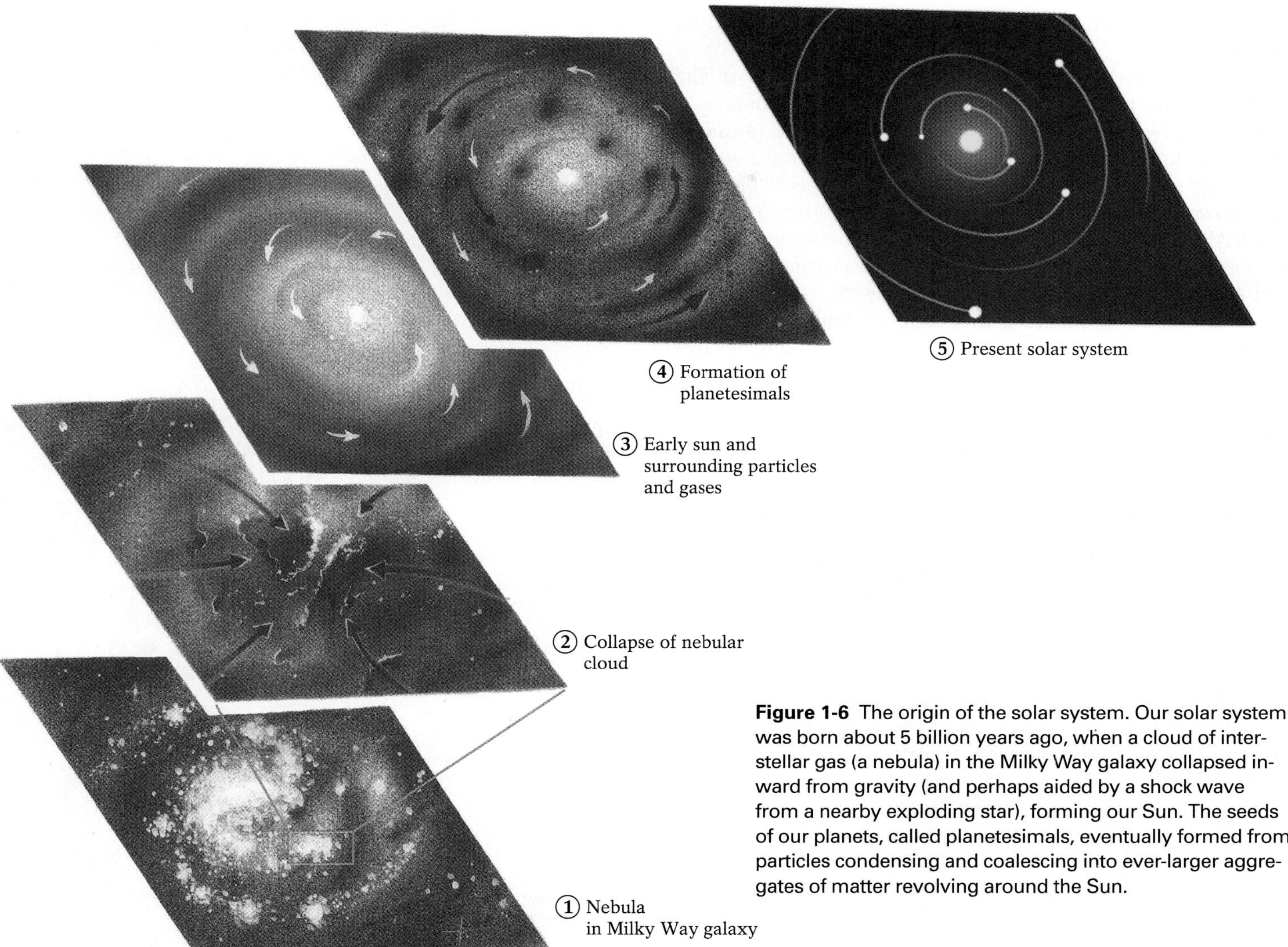

Figure 1-6 The origin of the solar system. Our solar system was born about 5 billion years ago, when a cloud of interstellar gas (a nebula) in the Milky Way galaxy collapsed inward from gravity (and perhaps aided by a shock wave from a nearby exploding star), forming our Sun. The seeds of our planets, called planetesimals, eventually formed from particles condensing and coalescing into ever-larger aggregates of matter revolving around the Sun.

be born in this way in all galaxies, including our own. (Through a telescope, you can see a "star nursery" in the belt of the constellation Orion.) The heat released by the compression resulting from the collision of infalling gas particles causes the particles in the core of each star to move faster and faster. Eventually they move fast enough to collide and engage in nuclear reactions, which release the energy that keeps the star hot and glowing brightly. In these nuclear reactions, the particles fuse, forming larger particles that will become the nuclei of helium and other, heavier elements.

Stars are not only born; they die as well—some slowly and some rapidly. A star that is dying very rapidly is called a *nova* (from the Latin for "new") because it appears as a very bright new star in the heavens. Dying stars are important because they heat up so much that new nuclear reactions occur, producing the nuclei of heavier elements. In this way dying stars act as manufacturing plants for heavy elements such as iron. They are also distribution centers, dispersing these nuclei over the space surrounding each dying star. It is believed that most of the matter in and on the Earth—including that in our bodies—comes from such dying stars.

Our own Sun is a star that came into being well after our galaxy was created, even after earlier stars had already died. Their remnants contributed to the gas cloud, or *nebula*, that eventually became our entire solar system (Fig. 1-6). The interstellar ("between stars") nebula that would become our Sun and planets was originally dispersed across a vast area of space, extending well beyond what would become the outermost planet, Pluto. About 5 billion years ago, however, this nebula began to collapse inward, perhaps due to a shock wave from a nearby exploding star. As its component materials were compressed and heated by the shock wave, they were drawn by gravity toward a hot center. There they collided in nuclear reactions and generated enormous heat, forming the infant Sun.

As heat became more and more concentrated in the center of this new star, some material in the nebula surrounding it began to cool and condense into small grains of matter. Lighter uncondensed substances were swept outward by strong solar winds—the streams of high-speed particles that flowed from the infant Sun. In this way, the first solid materials to form in our solar system became separated into a hot inner zone of denser substances, such as iron and nickel, and a cold outer zone of low-density gases, such as hydrogen and helium. Ultimately, this compositional partitioning would evolve into the four rocky inner planets and the five gaseous outer planets.

As the first bits of matter condensed, they continued to collide and coalesce, forming aggregates that grew to a few kilometers or larger in diameter. As they grew, the increasing gravity of these planetary seeds, or *planetesimals,* attracted other bodies of solid matter. Cosmologists originally believed that this process of planetary growth, or *accretion,* was slow and gradual, much like the way one might create a large aluminum-foil ball by the steady addition of small lumps. Recently, however, our view of planetary accretion has changed drastically. We now believe that as they grew, huge planetesimals, easily the size of Mercury or even Mars, collided violently. Such violent collisions ejected great masses of molten material into space, perhaps forming some of the moons that today orbit the planets of our solar system. (More on this shortly.)

As intense solar radiation warmed the four protoplanets closest to the Sun, their surfaces were heated to high temperatures. Nearly all of their light gases—hydrogen, helium, ammonia, and methane—vaporized and were carried away by solar winds. The matter remaining in these four small, dense inner planets consisted primarily of iron, nickel, and silicate minerals, those that contain a large amount of silicon and oxygen. Together these heavier substances make up most of Mercury, Venus, Earth, and Mars.

The frigid outer planets developed so far from the warmth of the Sun that much less of their matter vaporized. These planets—Jupiter, Saturn, Uranus, Neptune, and Pluto—probably consist predominantly of low-density masses of icy condensed gases, principally hydrogen, helium, methane, and ammonia.

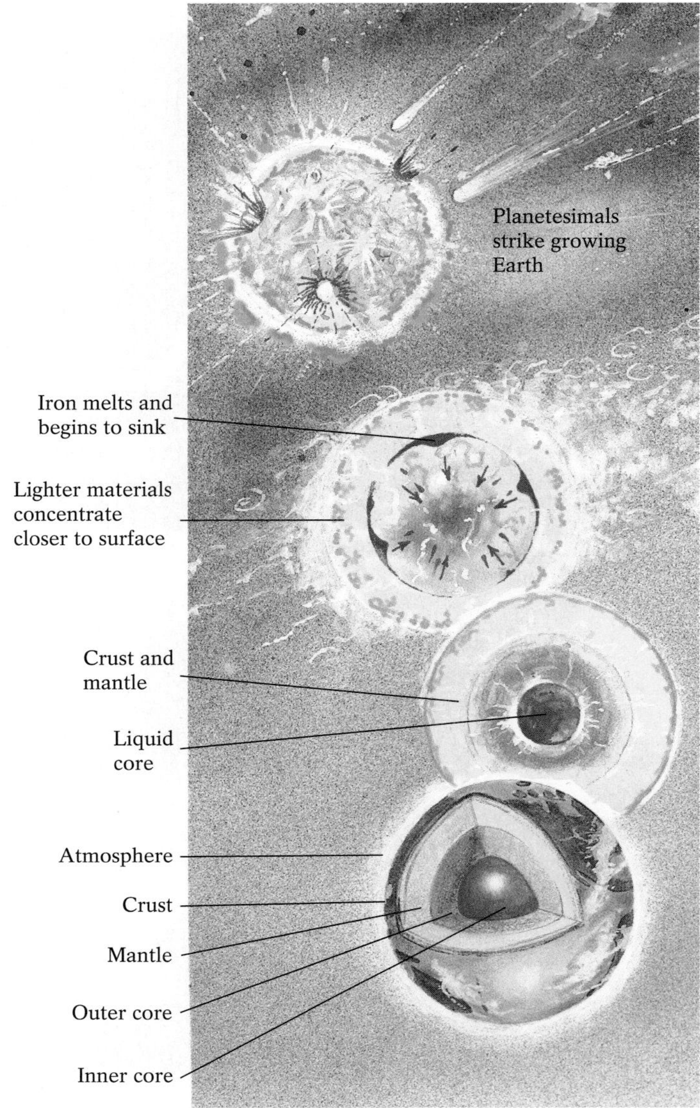

Figure 1-7 Heating and differentiation of the early Earth. Increased temperatures in the planet's interior, caused largely by planetesimal impacts and decay of radioactive substances, are believed to be responsible for the layering of the Earth's interior.

The Earth's Earliest History

No disaster movie could begin to depict the extreme violence and chaos of the first 20 million years or so of Earth history. During this time our planet became heated to the melting point of iron, developed its layered internal structure, collided with one or more other massive planetary bodies, and spewed out a mass of its own substance to form our nearest neighbor, the Moon.

During its first few million years, the gravitational pull of the proto-Earth attracted other planetesimals. The resulting collisions converted the enormous energy of motion to thermal energy upon impact. Some of this energy radiated back into space, but much was retained by the new planet as succeeding planetesimals struck its surface and "buried" the heat from earlier strikes. As the proto-Earth grew, its deep rocks became compressed under the weight of the growing mass of overlying rock. Such "compressional heating" is akin to the heating produced in a bicycle pump as the air in the pump is compressed. Think about how warm the barrel of a bicycle pump becomes as you inflate your tires.

Another source of heat was the atoms of radioactive substances, such as uranium, that released heat as their nuclei split apart, a process known as *fission.* Such heating is described as *radiogenic.* As heat from these two sources accumulated and became trapped within the Earth's interior, the planet's internal temperature began to rise. This heat set in motion the process that has created a layered Earth—one whose constituent substances are sorted into distinct internal layers (Fig. 1-7).

Differentiation What were the initial substances that made up the early Earth? For a clue, geologists look to some of the meteorites that have struck the Earth and survived. Many meteorites contain **chondrules** (KON-drools), small nuggets of rocky material believed to be droplets of matter that condensed directly from the original solar nebula. (The meteorites that contain chondrules are called *chondrites.*) It seems that the composition of the Earth itself, coalescing from the same nebula, should be similar to that of chondritic meteorites. Yet when we analyze such meteorites, we find a puzzling disparity: They are roughly 35% iron in composition, whereas the rocks at the Earth's surface average only about 6% iron. If our assumptions regarding chondrules are correct, where is all the Earth's missing iron?

We now believe that during the Earth's first 10 to 20 million years of existence, as the planet was still accreting, its internal temperature rose to the melting point of iron. As a result, much of the iron liquefied. Because the iron was denser than the surrounding materials, it sank to the proto-Earth's center by the pull of gravity. As the iron sank, lighter materials rose and became concentrated closer to the surface. Thus the matter that had originally made up a *homogeneous* Earth, perhaps similar in composition to meteoritic chondrules, became separated, like oil and vinegar, into concentric zones of differing densities. This separation process is called *differentiation.* The densest materials, probably iron and nickel, formed the Earth's core at the planet's center. Lighter materials, composed largely of silicon and oxygen and other relatively light elements, formed the Earth's outer layers (the mantle and crust). Even lighter materials—gases that had been trapped in the interior—escaped, combining to form the Earth's first atmosphere and oceans. Let's look more closely for a moment at the differentiated Earth before we consider the Earth's next great event—the collision that spawned the Moon.

A Glimpse of the Earth's Interior Contrary to those memorable but misleading scenes in old Hollywood movies and Jules Verne novels, the Earth is not a hollow ball filled with jungles and dinosaurs. Scientists have determined that the Earth's interior consists of three principal concentric layers, each with a different basic composition and related density (Fig. 1-8). (*Density* expresses the quantity of matter in a given volume of a substance. Because its chemical structure is more compact, 1 cubic centimeter of iron is far denser than 1 cubic centimeter of glass.)

The outermost layer of the Earth is a thin **crust** of relatively low-density silicon and oxygen-based rocks. Underlying it is the **mantle**, a thicker layer of denser rocks

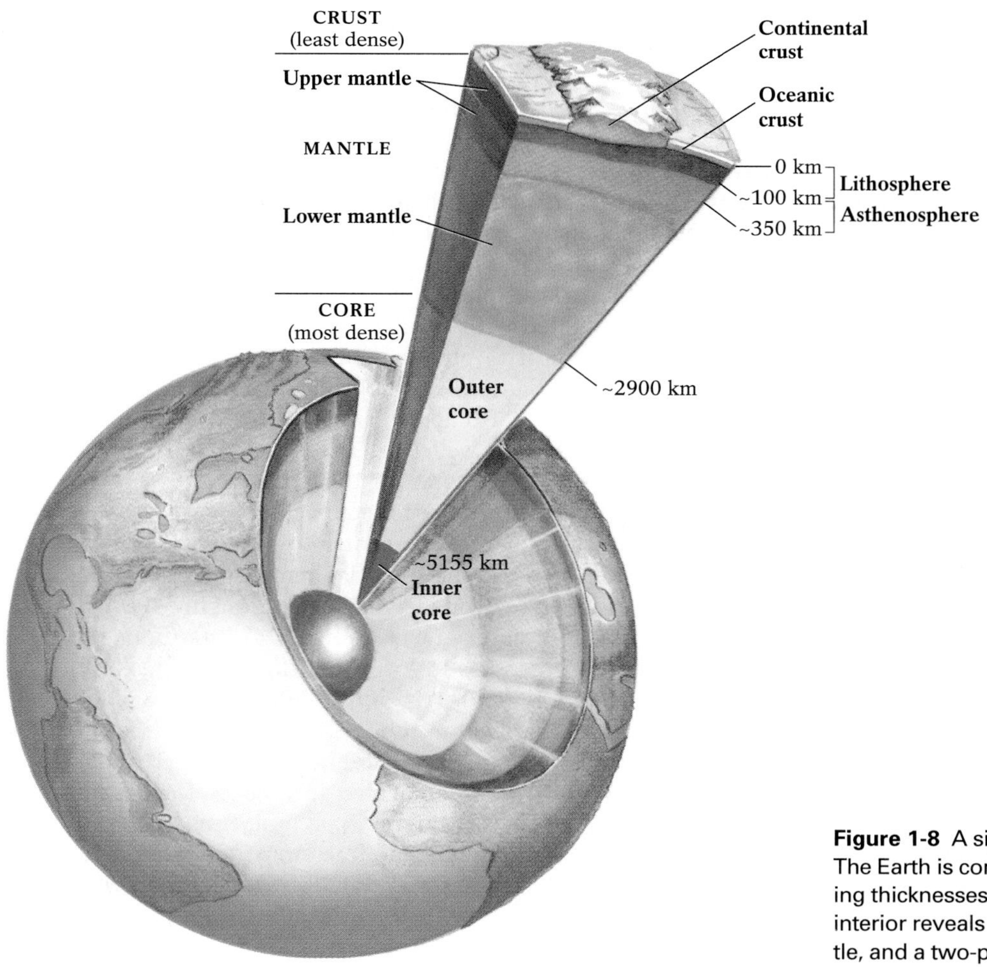

Figure 1-8 A simplified model of the Earth's interior. The Earth is composed of concentric layers of differing thicknesses and densities. A slice of the Earth's interior reveals a thin crust, a massive two-part mantle, and a two-part core.

(still silicon and oxygen-based but containing some heavier elements, such as iron and magnesium). At the Earth's center is its **core**, the densest layer of all, consisting primarily of metals such as iron and nickel. It is important to note that these three layers are made up of materials of three very different compositions. The arrangement of the three layers is somewhat like that of a hard-boiled egg—with its thin shell, extensive white, and small central yolk—but the egg model does not show a number of important *sublayers* that are fundamental to our understanding of our dynamic Earth.

The near-surface zone of the Earth's interior is subdivided into two distinct layers—*based not on their composition but rather on how they behave mechanically.* In a sense, they are differentiated by how strong they are. The outer 100 kilometers (60 miles) of the Earth, encompassing both the Earth's crust and the uppermost portion of the mantle, is a solid, relatively strong, rocky layer known as the **lithosphere** ("rock layer," from the Greek *lithos*, "rock"). Underlying the lithosphere is the **asthenosphere** ("weak layer," from the Greek *aesthenos,* "weak"), a layer of heat-softened, relatively weak, slow-flowing yet still-solid rock located in the upper mantle from about 100 to 350 kilometers (60–220 miles) beneath the Earth's surface. It is within the lithosphere and asthenosphere that such large-scale geological processes as mountain building, volcanism, earthquake activity, and the creation of ocean basins originate. Below the mantle, the core is divided into a liquid outer core and a solid inner core. We will study these layers in detail in Chapter 11.

Overlying the lithosphere is the Earth's atmosphere, composed in part of the gases that have escaped over time from the Earth's interior, primarily during volcanic eruptions. (Recent observations of icy meteorites suggest that some of the Earth's atmospheric water and gas may have come from space. More on this in Chapter 8.) The ongoing degassing of the Earth's interior (yes, it's still happening today) has contributed significantly to the oceans that cover 71% of the Earth's surface and the thick atmospheric envelope of gases that support our planet's living organisms. Fortunately, the Earth's interior probably did not melt completely during the period of planetary differentiation. If it had, the Earth might have lost *all* its gases during its first 20 million years, and its life-supporting atmosphere might never have accumulated. The Earth did, however, lose its earliest atmosphere, which is believed to have been composed mainly of hydrogen, helium, ammonia, and methane. This was no great loss to us since these gases are lethal to most life forms. With this knowledge of the composition of the differentiated Earth, we can now consider the dramatic manner in which we may have lost our first atmosphere—and gained our Moon.

The Origin of the Moon The birth of our Moon has sparked lively debate for centuries. Did it form as a companion planet coalescing independently from the solar nebula at the same time as Earth? Did it form elsewhere, only to be drawn into Earth's orbit and held hostage there by our planet's relatively strong gravity? Or did it form, as some scientists now believe, in a great cataclysmic collision between the Earth and another nearby planetesimal?

The answer may lie in the Moon's composition. It is 36% less dense than the Earth, suggesting that it contains much less iron. This would apparently rule out independent accretion from the solar nebula, for if the Moon did form in the same way as the Earth, its composition would be similar. The Moon's composition, confirmed in part by the rock-collecting efforts of U.S. Apollo astronauts, is actually quite similar to that of the Earth's mantle . . . perhaps for the following reasons.

By roughly 4.55 billion years ago, the Earth had probably attained much of its size and had become layered, with most of its iron dispatched to its core. Its mantle and crust were composed largely of a variety of silicate minerals. The solar system at this time, however, was still more crowded and chaotic than it is today. Numerous speeding objects, many larger than Mercury, some as large as Mars, still hurtled through Earth's neighborhood. Thus the stage was set for a monumental collision between Earth and a Mars-sized impactor that may have jolted and tilted the Earth's axis, certainly reheated the planet's outer layers to their melting points, and most probably propelled massive molten chunks of the Earth into space, ultimately to coalesce as our familiar nighttime companion, the Moon.

Picture this scene. The gravitational pull of the growing Earth, the largest body in this section of the solar system, attracts a Mars-sized planetesimal. The speeding impactor is traveling at supersonic speeds, perhaps as fast as 14 kilometers per second (31,500 miles per hour). At the moment of impact, the Earth's incipient atmosphere is blown away, replaced by a sky filled with a rain of molten iron blobs, remnants of the impactor's iron core. (It, too, must have undergone some differentiation.) The collision vaporizes much of the crust of both the Earth and the impactor and a good portion of their mantles as well (Fig. 1-9), but does not penetrate to the Earth's iron core. As the remnants of the impactor's dense iron core are pulled back to Earth, jets of vaporized crust and mantle are shot into space. The Earth's gravity, however, captures this material before it flees the area altogether, holding it in orbit at a distance of roughly 400,000 kilometers (240,000 miles). Over a few tens of thousands of years, this debris coalesces to form the Moon.

This hypothesis (and it is *only* a hypothesis) explains why the Moon is relatively iron-poor. The impactor failed to reach the iron core of the already-differentiated Earth, and the impactor's dense iron core failed to reach sufficient velocity to escape the Earth's gravitational pull. Thus the Moon may be composed largely of some fraction of the Earth's vaporized mantle and crust that was cast into space during what may have been the most spectacular moment of Earth's early history.

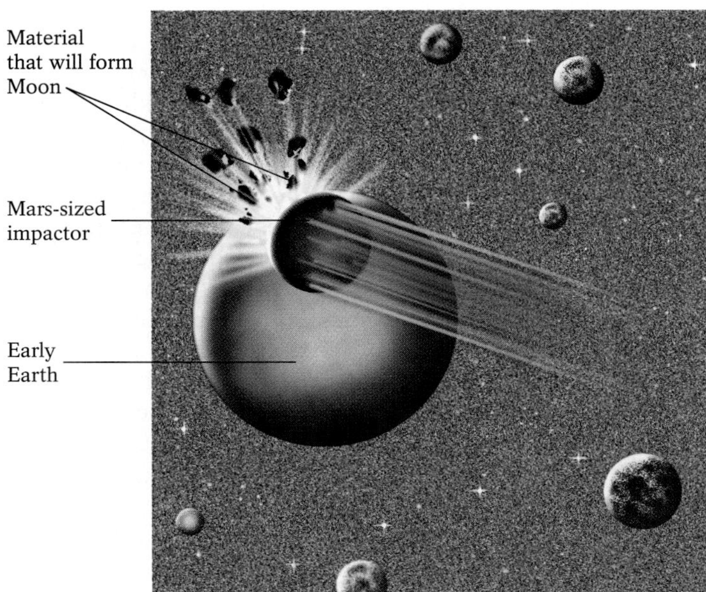

Figure 1-9 A catastrophic impact between the proto-Earth and a Mars-sized impactor is now believed by many scientists to have spawned the Earth's Moon.

Where are the "wounds" of this great collision? Unfortunately, the Earth's surface is perhaps the worst place to look for evidence of the planet's violent past. The Earth's dynamic internal processes—related to volcanism, earthquakes, and mountain building—would have eradicated much of the evidence. And the Earth's atmosphere, which causes the planet's surface rocks to weather away and erode, would have removed the rest. Evidence of much more recent meteorite impacts still remains, not yet erased by the Earth's active geological processes (see Fig. 2-30a). But a search for the evidence of the greatest collision in the Earth's history is unlikely to yield a clue. We are forced to look to the composition of Moon rocks and chondritic meteorites to reveal this fascinating story.

Thermal Energy in the Earth

We have seen how the buildup of thermal energy in the Earth's interior caused some of its component substances to melt and sink to the Earth's center during the planet's earliest days, before the collossal collision that produced the Moon. The phenomenon of heat, which is a form of energy transfer, is closely tied to numerous processes that continue to shape our planet today. Thermal energy is transferred from place to place within the Earth, always moving from warmer to cooler areas. There are three primary methods by which thermal energy is transmitted through the Earth: conduction, convection, and radiation (Fig. 1-10).

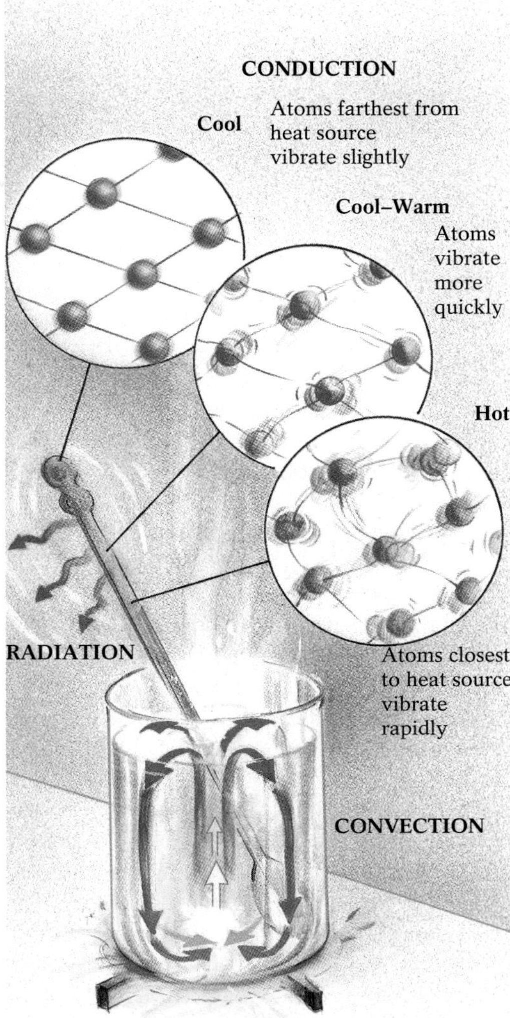

Figure 1-10 Three ways in which heat is transmitted. These phenomena, occurring on a much larger scale within the Earth, are responsible for many geological processes. *Conduction* of heat involves the passage of thermal energy from atom to neighboring atom. In the diagram, the atoms in the part of the spoon that is immersed in the hot soup have begun to vibrate, passing thermal energy to adjoining atoms. The end of the spoon farthest from the heat source is still cool, but eventually the heat will be conducted throughout the length of the spoon. *Convection* involves the movement of heat from place to place by a flowing medium. Because the soup at the bottom of the pot is closer to the flame, it is the first part of the soup to become hot. As it heats up, it expands (becoming less dense and rises), and the cooler (more dense) soup above it sinks to the bottom, displacing it and forcing it upward. When the warm soup arrives at the top, it encounters the relative coolness of the air and contracts in volume (becomes more dense) as it begins to cool. The cooled soup then sinks back toward the bottom, to be reheated and then to rise again. In the diagram heat is also being *radiated* from the warm spoon and from the hot pot to the cooler air surrounding it.

In *conduction,* minute particles such as atoms (discussed in Chapter 2) become "excited" by thermal energy from an outside source until they vibrate rapidly, colliding with neighboring particles and setting them in motion. This generates a chain reaction of vibration that transfers thermal energy. Rocks, however, are generally poor conductors of thermal energy. In fact, if the thermal energy produced in the Earth's deep interior during the planet's creation had flowed by conduction alone, it would not yet have come to the surface today, more than 4.5 billion years later.

When thermal energy is transferred by *convection,* material actually moves from one place to another, carrying the thermal energy with it. When the temperature in the Earth's interior became high enough to melt some of its components, or soften them sufficiently so that they were able to flow, thermal energy began to be transported by moving fluids. Eventually this energy heated surrounding substances as well, causing hot low-density materials to rise toward the surface, carrying thermal energy with them. Convection is a much faster as well as a more efficient way of transferring energy than conduction. The rising hot material carried by the process of convection probably caused the planet's first volcanic eruptions.

All heated objects also *radiate* energy in one form or another. Energy transmitted by radiation moves in the form of one or more different types of electromagnetic waves, such as radio waves, microwaves, infrared waves, visible light waves, ultraviolet light waves, and X-rays. These forms of energy are then converted into thermal energy when they strike and are absorbed by an object. In this way, a microwave oven's radiant energy is converted to thermal energy to thaw a frozen pizza, and light radiated by the Sun is transformed into the thermal energy that heats the Earth's atmosphere when it is absorbed by the Earth's surface. In a similar way, radioactive substances within the Earth emit different forms of electromagnetic radiation. These forms of electromagnetic radiation are converted to thermal energy when they warm surrounding rocks.

As the Earth's internal heat caused it to differentiate into its major concentric layers, some upwelling molten material reached the surface, where it cooled and solidified, forming the Earth's earliest crust. Among the low-density substances that rose toward the surface were oxygen and silicon. These substances combined to form the silicate minerals that abound in the Earth's crust and upper mantle. Some heat-producing radioactive substances, such as uranium and thorium, also moved toward the surface, incorporated within the silicates and other light minerals. The heat radiating from these elements repeatedly melts the rocks of the Earth's crust and reforms them into a wide variety of rocks. This "cycling" of crustal materials is summarized briefly in the next section, Rock Types and the Rock Cycle, and discussed more thoroughly in upcoming chapters.

Rock Types and the Rock Cycle

A **rock** is a naturally formed aggregate of one or more **minerals,** which are naturally occurring inorganic solids that originated within the Earth. Three types of rocks exist in the Earth's crust and at its surface, each reflecting a different fundamental origin. **Igneous rocks** have cooled and solidified from molten material either at or beneath the Earth's surface. **Sedimentary rocks** form when preexisting rocks are weathered and broken down into fragments that accumulate and become compacted or cemented together. They may also form from the accumulated and compressed remains of certain plants and animals, or from chemical precipitates of materials previously dissolved in water. **Metamorphic rocks** form when heat, pressure, or chemical reactions in the Earth's interior change the mineralogy, chemical composition, and structure of *any* type of preexisting rock.

Over the great extent of geologic time and through the dynamism of Earth's processes, rocks of any one of these basic types may eventually evolve into either of the other types or into a different form of the same type. Rocks of any type exposed at the Earth's surface can be worn away by rain, wind, crashing waves, flowing glaciers, or other means, with the resulting fragments carried elsewhere to be deposited as new *sediment.* This sediment might eventually become new sedimentary rock. Sedimentary rocks, in turn, may become buried so deeply in the Earth's hot interior that they may be changed into metamorphic rocks, or they may melt and eventually form new igneous rocks. Under heat and pressure, igneous rocks can also become metamorphic rocks. The processes by which the various rock types evolve and change over time are illustrated in the **rock cycle** (Fig. 1-11).

Time and Geology

To change sedimentary rocks to metamorphic rocks may require a vast stretch of time. Clearly, the Earth, which is believed to be about 4.6 billion years old, has had plenty of time with which to work. Over so vast a period of time, even processes that operate at imperceptibly slow rates (Fig. 1-12) have changed the Earth's appearance dramatically. Fossil evidence of ancient marine creatures in some rocks of the Grand Canyon, for example, show that these rocks, today near the top of a hot, dry plateau 2300 meters (7500 feet) above sea level, once lay at the bottom of an ocean. Over millions of years, the mud at the bottom of this ocean—still containing the remains of sea creatures that died there—gradually solidified into rock, was uplifted to its present elevation, and was cut through and exposed by the Colorado River. As James Hutton knew, rocks, if we study them closely, can bear witness to an unimaginably long history.

Geologists think of time in both *relative* and *absolute* terms. *Relative dating* of the various rocks exposed within

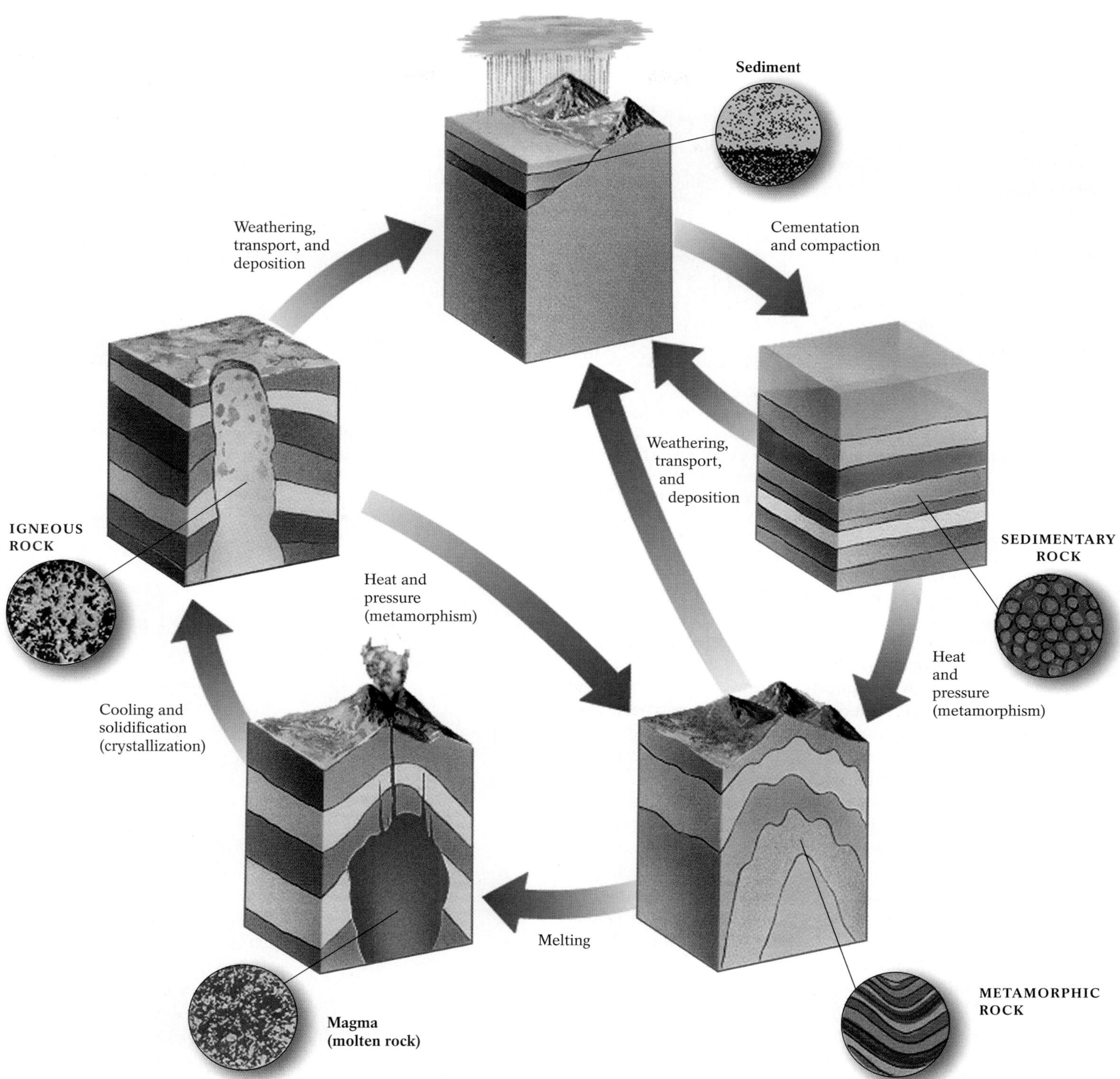

Figure 1-11 The rock cycle—a simplified scheme illustrating the variety of ways that the Earth's rocks may evolve into other types of rocks. For example, an igneous rock may weather away and its particles eventually consolidate to become a sedimentary rock. The same igneous rock may remain buried deep beneath the Earth's surface, where heat and pressure might convert it into a metamorphic rock. The same igneous rock, if it is buried even deeper, may actually melt and form a new igneous rock once it has recooled and solidified. There is no prescribed sequence to the rock cycle. A given rock's evolution may be altered at any time by a change in the geological conditions around it.

(a)

(b)

Figure 1-12 Two photos of the Grand Canyon, taken from the same perspective (looking down the Colorado River, one-half kilometer below Lees Ferry) about 100 years apart. Other than signs of encroaching vegetation, there is no perceptible geological difference between the earlier scene **(a)**, photographed in 1873, and the later scene **(b)**, photographed in 1972. Although geological processes in the arid Southwest are very slow on a human time scale, the Grand Canyon was formed by these very processes over the course of millions of years. Before gradually solidifying, becoming uplifted, and being incised by the Colorado River, the rocks of the canyon originated as sediment on the floors of a succession of ancient inland seas.

a rock outcrop asks and answers the questions: "Which rocks are older?" and "Which rocks are younger?" The answers are typically found by observing the spatial relationships of rock bodies to one another. For example, the *principle of superposition* states that where layers of flat-lying sedimentary rocks or a series of solidified lava flows have not been disturbed since their formation, *younger rocks overlie older rocks.* Quite logically, the older rocks had to be there first for the younger rocks to be laid on top of them (Fig. 1-13).

Absolute dating of rocks asks and answers the question: "How old?" We are just not satisfied by knowing which is older or younger; we want a specific age—*in years*. Rock-dating techniques developed during the twentieth century, based on the constant decay of radioactive elements, now let geologists specify the absolute ages of some rocks in years. The oldest rocks found on Earth, near Yellowknife Lake in Canada's Northwest Territories, have been dated by the known decay rate of uranium at 3.96 billion years. We discuss methods by which geologists date rocks in Chapter 8.

If Earth's history were viewed as a great textbook, its rocks would be the pages on which most of the events are written. Almost everything we know about our past we know because it was preserved, in some way, in rock. Geologists initially divided Earth's history into four major *eras* of varying length and a dozen smaller *periods,* based largely on fossil evidence showing the existence of various key organisms in sedimentary rocks and on spatial relationships such as superposition. With the advent of modern rock-dating technology, however, the geologic time scale now includes absolute ages for its eras and periods (Fig. 1-14).

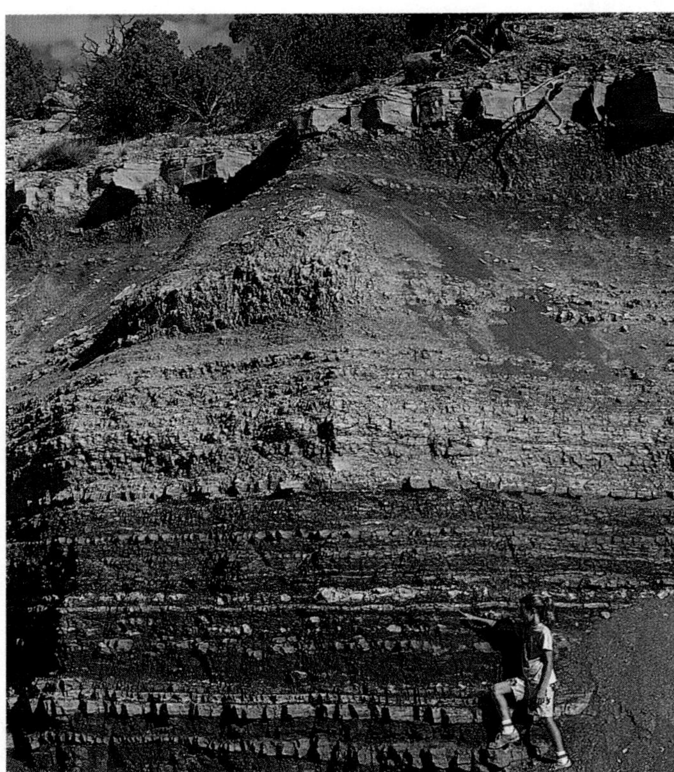

Figure 1-13 These sedimentary rock layers from Grand Junction, Colorado, illustrate the principle of superposition. The child's feet rest on the outcrop's oldest rocks; the thick layer at the top are the outcrop's youngest.

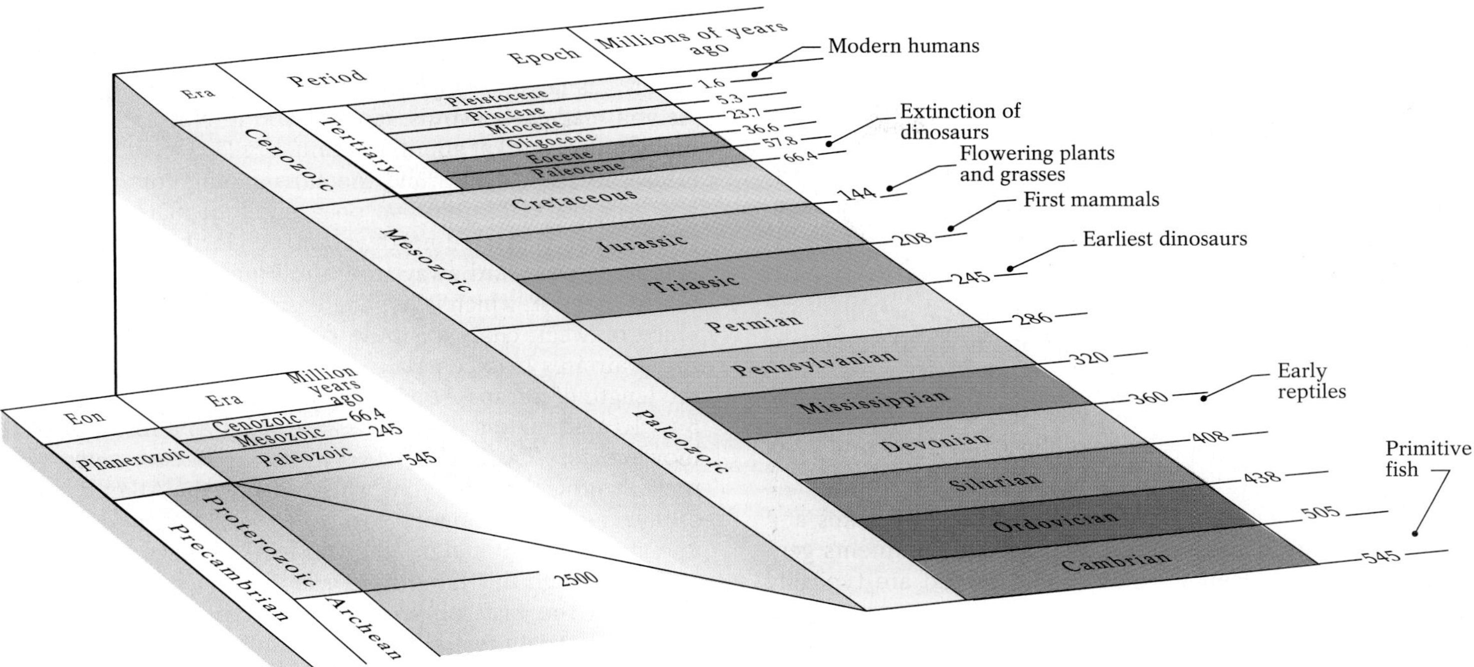

Figure 1-14 A simplified version of the geologic time scale. (See also Figure 8-33.) The different eras of the Phanerozoic Eon derive their names from the nature of the life forms associated with those eras: Paleozoic, "ancient life"; Mesozoic, "middle life"; Cenozoic, "recent life." The names of the various periods of the Paleozoic and Mesozoic Eras are taken primarily from places where their rocks were originally identified. For example, the Devonian Period is named for the rocks in Devonshire, England. Note that the Precambrian Eon comprises about 87% of all of the Earth's history. The most recent era, the Cenozoic, is more finely divided because more rocks of this age are exposed and geologists have more precise ways to date them.

Plate Tectonics

When we tell the life story of the Earth, we discover that certain regions of the world have been periodically devastated by earthquakes, while others remain unscathed. We puzzle over the reason a chain of volcanoes stretches along the west coast of North America from northern California to Alaska but no such chain of volcanic peaks can be found in such mid-continental locations as Minnesota, Iowa, or Missouri. And travel agents surely know that the "slopes" of Ohio and Indiana stir little excitement among skiers, who flock instead to Sun Valley, Idaho, and Vail, Colorado. Each of these examples highlights a definite pattern in the Earth's geological makeup—*it is not random.*

For centuries, geologists sought to explain such large-scale patterns with a multitude of independent hypotheses, each tailored to a specific location. That's why geologists in the Swiss Alps proposed that mountain building occurs when rocks are pushed together, whereas those studying the Grand Tetons of Wyoming proposed that mountains form when rocks are pulled apart. Such hypotheses might have given reasonable pictures of regional processes, but they failed to explain some common underlying cause of *all* mountain ranges. And no unifying hypothesis adequately explained the observed "combination" of geological phenomena—why, for example, earthquakes typically occur where mountains rise and volcanoes erupt.

In the 1960s, an exciting new hypothesis called **plate tectonics** (from the Greek adjective *tektonikos,* or "built," as in archi*tect*ure) revolutionized our understanding of how the Earth functions. This hypothesis changed the way geologists viewed the world as dramatically as, a century earlier, the theory of evolution changed how biologists thought about living things. After only a few decades of observation and testing, the hypothesis of plate tectonics has become a widely accepted theory because it provides answers to questions that earlier hypotheses could not resolve. It has given us a way to understand processes such as mountain building, predict such potential catastrophes as earthquakes and volcanic eruptions, and even find underground reservoirs of oil, natural gas, and precious metals. Finally, the plate tectonic theory enables us to fit our observations about the ancient past into the same conceptual framework as our understanding of the geological phenomena occurring today.

According to the theory of plate tectonics, the Earth's lithosphere consists of large rigid plates that move. Over millions of years, plate movements have created the Earth's ocean basins, changed the shape of our continents, and crafted the planet's great mountain ranges.

Basic Plate Tectonic Concepts

The wide-ranging theory of plate tectonics can be summed up with four basic concepts.

1. The outer portion of the Earth—its crust and uppermost segment of mantle (i.e., its lithosphere)—is composed of rigid units called plates.
2. The plates move . . . slowly.
3. Most of the world's large-scale geological activity, such as earthquakes and volcanic eruptions, occurs at or near plate boundaries.
4. The interiors of plates are relatively quiet geologically, with far fewer and usually milder earthquakes than occur at plate boundaries and little volcanic activity.

Figure 1-15 shows the Earth's seven major plates and a number of its smaller ones. Note that the continents generally are not independent plates, but instead are typically parts of composite plates that contain both continental and oceanic lithosphere. For example, the North American plate includes the North American continent and the adjacent western half of the Atlantic Ocean. Continental portions of plates are composed of thicker, lower-density lithosphere, whereas oceanic portions of plates are thinner and of higher density. Plates are more than 120 kilometers (75 miles) thick in continental regions but only about 80 kilometers (50 miles) thick at ocean basins.

As the Earth's plates move, everything on them, even features as large as continents and oceans, moves with them. As you read these words, most of you in North America are moving westward at about 4 centimeters (1.5 inches) per year. (Those in Los Angeles and the surrounding communities of westernmost California live on a different plate, the Pacific plate, which is headed northward.) The North American plate is moving westward away from the Eurasian plate, its neighbor to the east, which is moving eastward. The increasing distance between cities on these two plates, however, will not cause airfares between Baltimore and London to increase—the length of a trans-Atlantic flight is extended by only about 5 to 10 centimeters (2–4 inches) per year as a result of plate movements. (This is about two times the rate at which your toenails grow.) If Columbus were crossing the Atlantic today, 500 years after his famous voyage, he would have to sail only an extra 25 to 50 meters (80–160 feet) to reach shore.

Although this rate of plate motion may seem insignificant, over the vast course of geologic time it can have large consequences. In just a few tens of millions of years, the Atlantic Ocean has grown a thousand kilometers (about 600 miles) wider. But does this mean that the Earth's overall size is increasing? No, our planet's size has remained essentially constant since its early formative years. Thus, if plates move apart and grow at certain boundaries, they *must* converge and be consumed at others. The Atlantic Ocean has, in fact, opened and closed several times over the course of time.

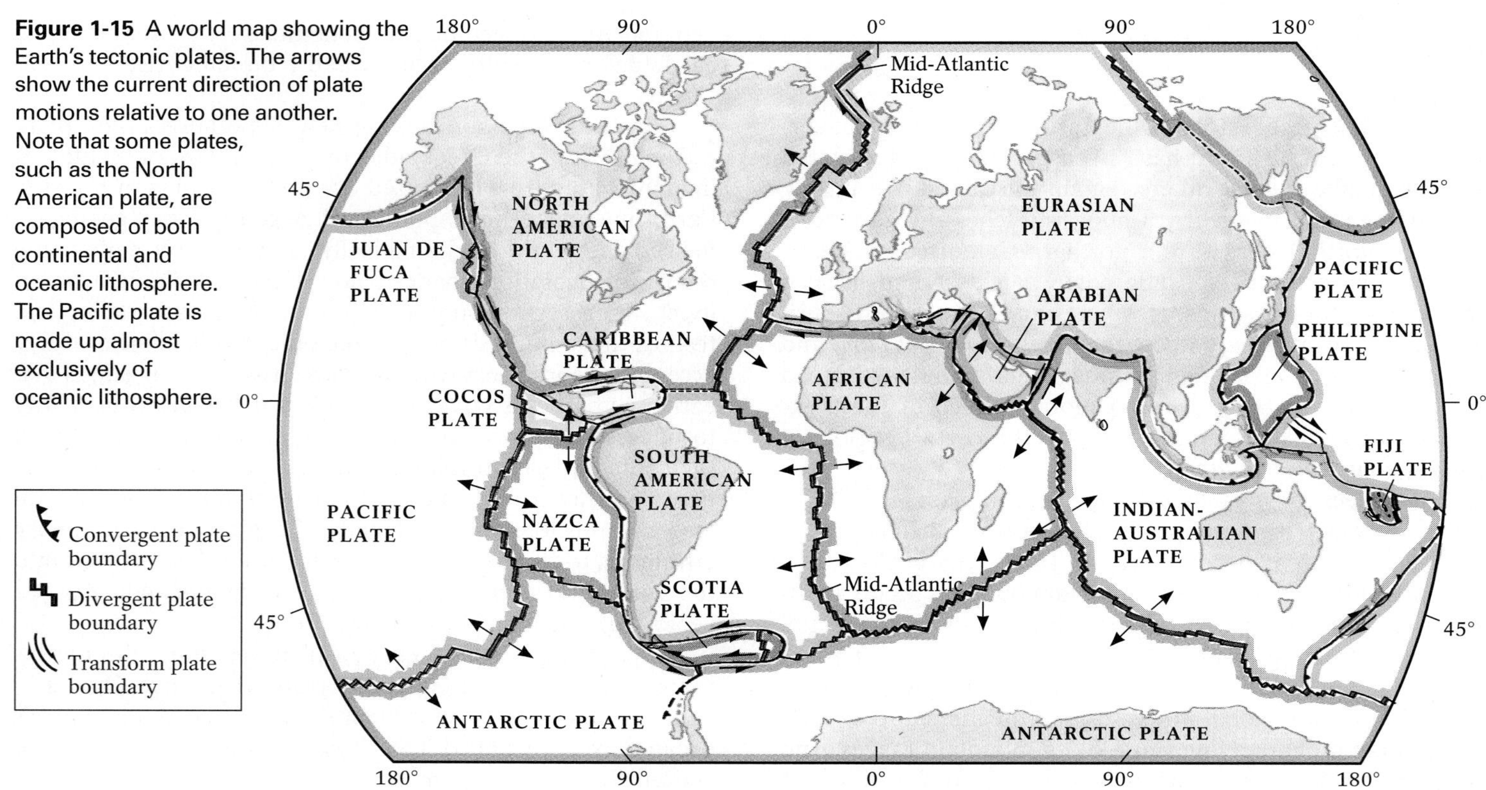

Figure 1-15 A world map showing the Earth's tectonic plates. The arrows show the current direction of plate motions relative to one another. Note that some plates, such as the North American plate, are composed of both continental and oceanic lithosphere. The Pacific plate is made up almost exclusively of oceanic lithosphere.

Figure 1-16 A map of the Pacific Ocean and surrounding continents, indicating some of the sites of volcanic eruptions and zones of earthquake activity recorded over the last 100 years or so. Note the proportion of volcanoes and earthquakes that occur at plate boundaries.

Figure 1-17 This satellite view of the East African rift zone shows how northeastern Africa is tearing away from the Arabian Peninsula. Looking southeast, the Nile River delta is at the bottom center of the photograph, and the Red Sea stretches toward the horizon.

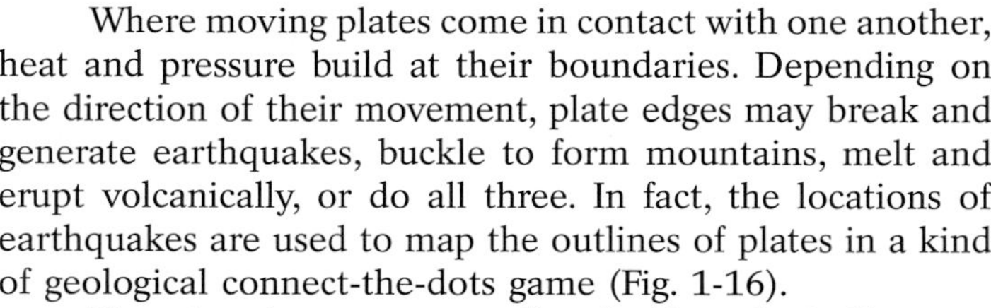

Where moving plates come in contact with one another, heat and pressure build at their boundaries. Depending on the direction of their movement, plate edges may break and generate earthquakes, buckle to form mountains, melt and erupt volcanically, or do all three. In fact, the locations of earthquakes are used to map the outlines of plates in a kind of geological connect-the-dots game (Fig. 1-16).

Plate interiors are generally placid geologically, typically unaffected by distant plate-edge activity. San Francisco has frequent earthquakes because it is located at a moving plate boundary; Chicago has very few because it is located within a relatively inactive plate interior.

Plate Movements and Boundaries

The Earth's plates move in several ways with respect to one another, and plate boundaries are categorized according to which type of movement they demonstrate. There are three major types of plate movements and corresponding boundaries: divergent plate boundaries, where plates move apart; convergent plate boundaries, where plates move together; and transform plate boundaries, where plates move past one another in opposite directions.

Rifting and Divergent Plate Boundaries Plate interiors may become geologically active if slow-flowing currents in the Earth's asthenosphere pull apart and eventually tear a preexisting plate into two or more smaller plates. This process, discussed in detail in Chapter 12, is known as **rifting.** The Great Rift Valley of East Africa, where the African plate has been coming apart, is a prime example of an early stage in rifting (Fig. 1-17). If this rifting continues, there may be two Africas on the world map in the not-too-distant geological future. Farther north, such rifting completely separated the Arabian plate from the African plate over the last 20 million years, forming the crustal depression occupied by the Red Sea and the Gulf of Aden. As their complementary coastlines illustrate (Fig. 1-18), these two plates were almost certainly once attached.

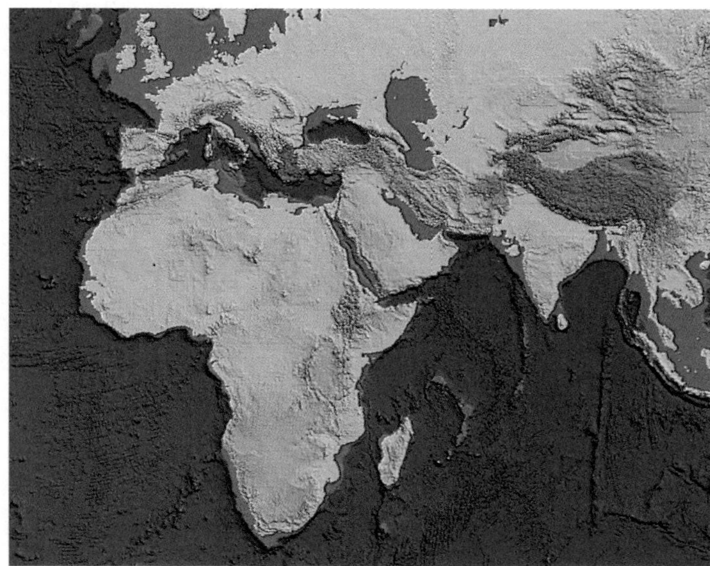

Figure 1-18 The V-shaped tear in northeastern Africa—formed by the Red Sea and the Gulf of Aden—is the Earth's best current example of plate rifting.

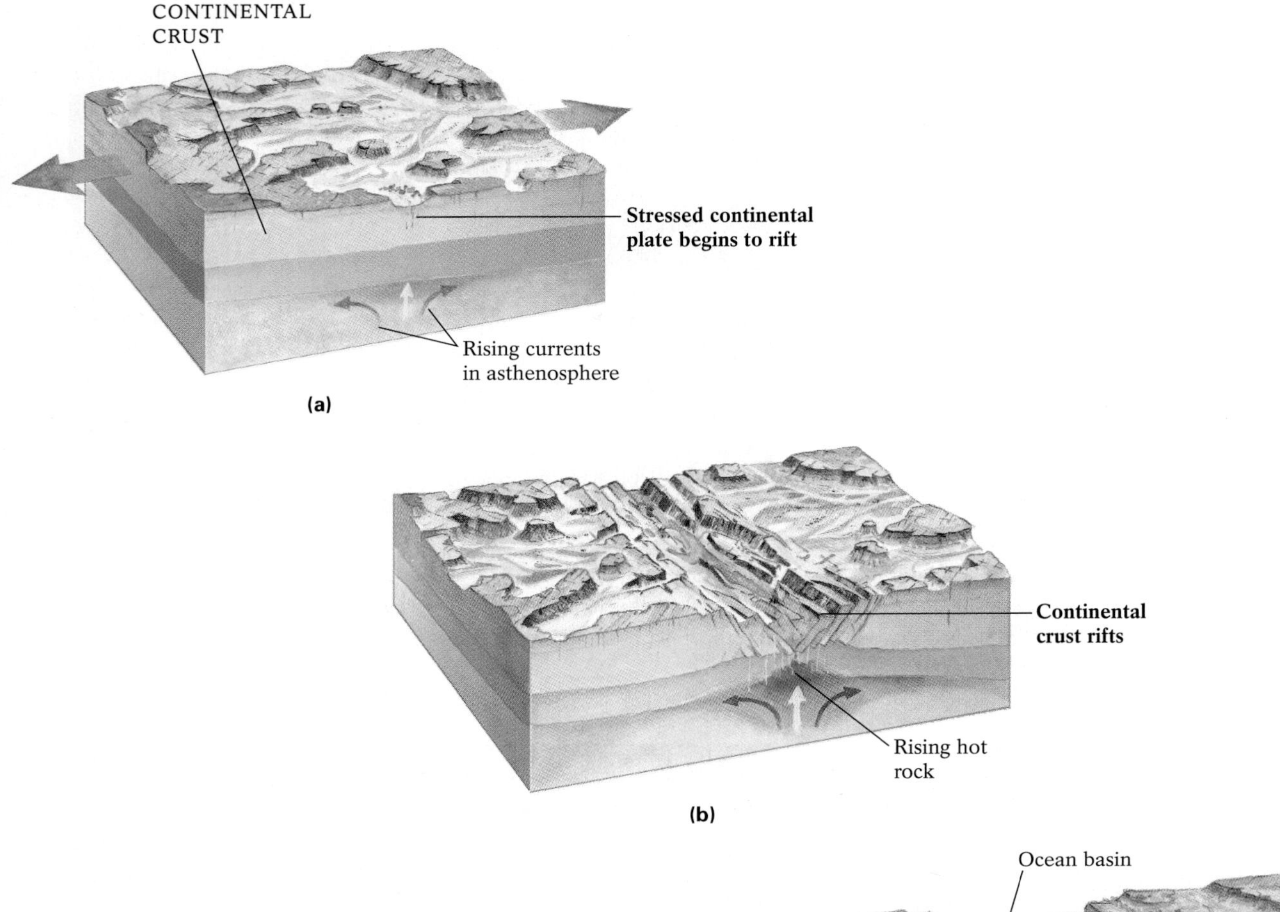

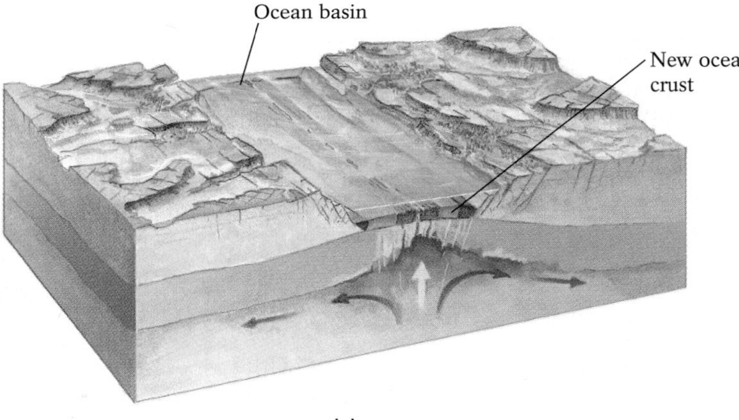

Figure 1-19 Plate rifting and divergence. When currents in the underlying asthenosphere pull one of the Earth's plates in opposite directions **(a)**, the plate is stressed and eventually rifts. **(b)** As the plate fragments continue to move (diverge) farther from one another, molten rock from the mantle rises into the gap and solidifies along the edges of the plates **(c)**, forming new oceanic crust that is eventually covered with water to form a new ocean basin.

Once a plate has been rifted, the resulting smaller plates may continue separating from one another by a type of plate motion described as **divergence** (Fig. 1-19). Divergence proceeds typically at a rate of about 1 to 10 centimeters (0.5–4 inches) per year. As molten rock rises into the fractures between the rifted plates, it cools and solidifies, becoming attached to the edges of the rifted plates. As divergence continues, the older rifted segments separate further and eventually, an ocean basin forms. The ocean basin fills with seawater as rifting opens new connections to other oceans.

Throughout the period of divergence, erupting molten rock expands the ocean basins by creating new oceanic crust. The center of this volcanic activity is the *mid-ocean ridge*, a continuous chain of submarine mountains that meanders around the globe like the stitches on a baseball (Fig. 1-20). The process of plate growth at mid-ocean ridges is known as **sea-floor spreading.** If, for example, the young Red Sea between the African and Arabian plates continues to grow at its present rate, it may someday become a full-blown ocean like the Atlantic or Pacific.

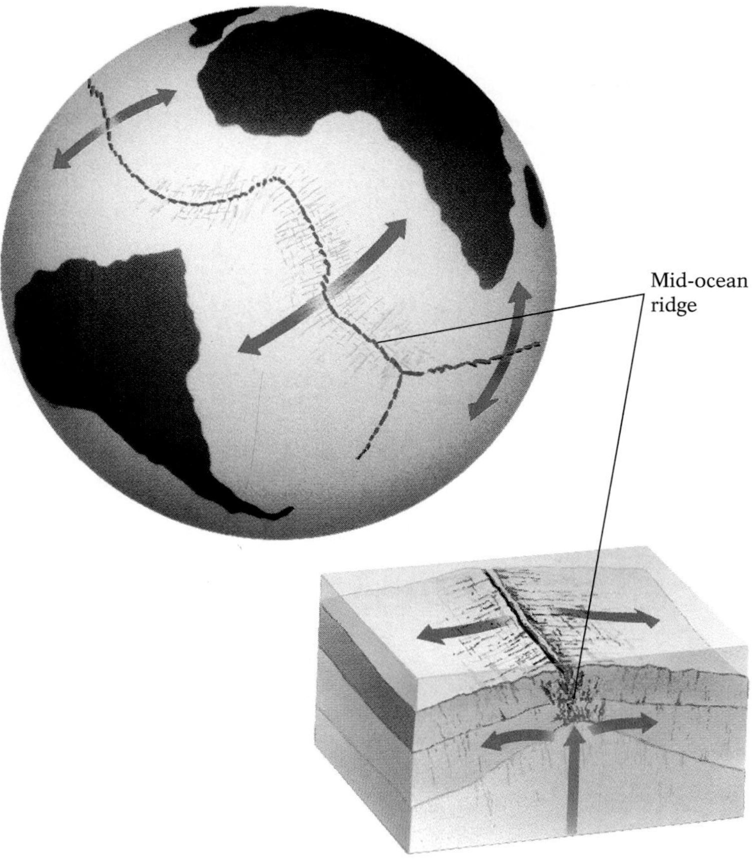

Figure 1-20 Divergent zones—places where plates grow by addition of new volcanic rock—stretch for 65,000 connected kilometers (40,000 miles) principally along the Earth's sea floors. Here, divergence between South America and Africa has created the southern Atlantic Ocean basin.

Plate Convergence and Subduction Boundaries Some plates move *toward* each other in a phenomenon known as **convergence.** Plate convergence may involve two continental plates, two oceanic plates, or one of each (or, in the case of composite plates, the oceanic or continental portions of the two plates). Most often, when one oceanic plate and one continental plate or two oceanic plates converge, the denser of the two dives beneath the other and sinks into the Earth's interior, where it is consumed. This process is known as **subduction.** Because plates of oceanic lithosphere are always denser than those of continental lithosphere, the oceanic plate always subducts when it converges with a continental plate (Fig. 1-21). Some oceanic plates, however, are more dense than others; thus, when two oceanic plates converge, the denser one subducts beneath the less dense one.

Continental plates are generally too buoyant to subduct into the denser underlying mantle. Thus, when two continental plates collide, neither plate can subduct completely, although their edges may be *temporarily* dragged down to depths perhaps as deep as 200 kilometers (120 miles) before being thrown back toward the Earth's surface. Instead of subducting, colliding continental plates become welded together, forming a much larger *single* plate. Such convergence of continental plates, called **continental collisions** (Fig. 1-22), has created many of the Earth's largest mountain ranges. Northern Africa's Atlas Mountains and southern Europe's Alps were formed by past collisions of the African and Eurasian plates; the Himalayas owe their great height to the ongoing collision of the Indian and Eurasian plates. On our continent, North America's Appalachians formed from a series of collisions of the African, Eurasian, and North American plates and several other smaller plates that took place between about 400 and 250 million years ago.

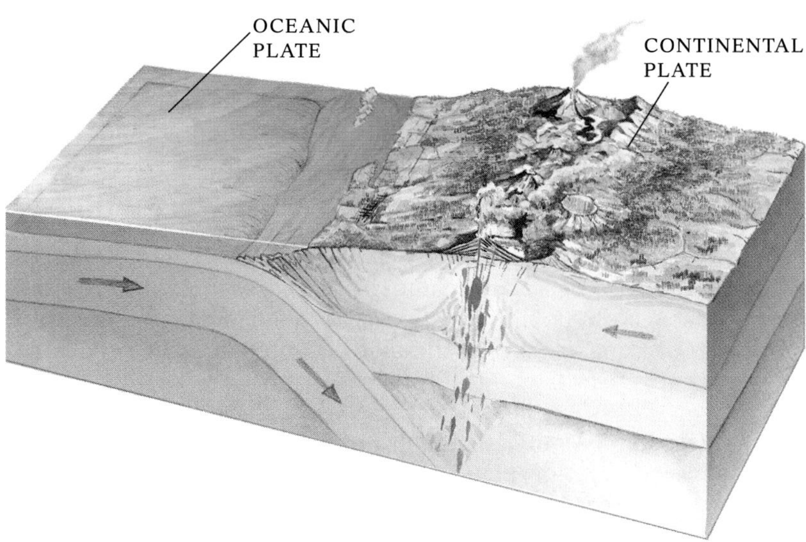

Figure 1-21 Oceanic plate subduction. Converging plates push against one another and crumble, with one plate often sinking, or subducting, below the other.

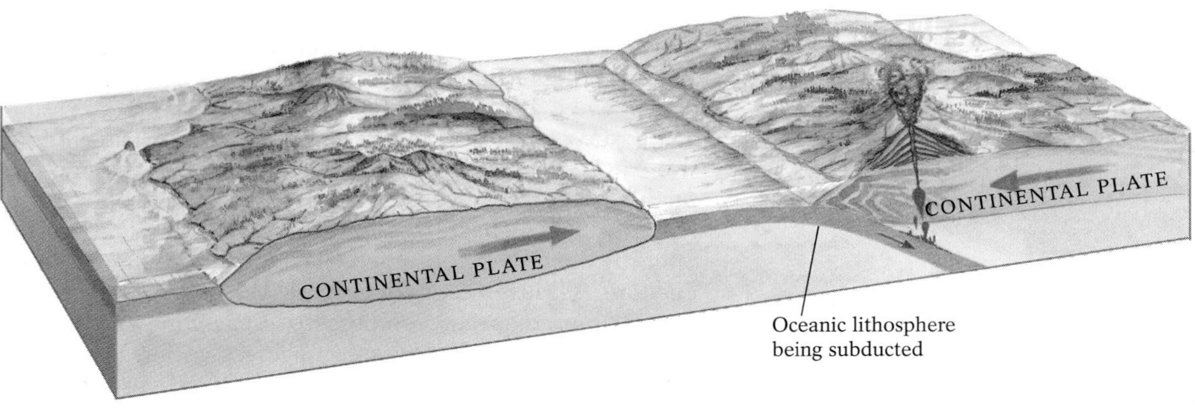

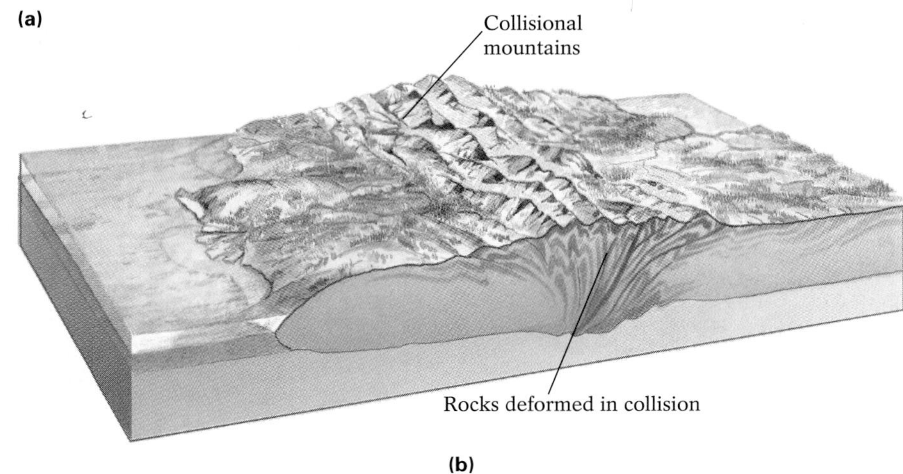

Figure 1-22 Continental collision. **(a)** Two continental plates converge as the oceanic lithosphere between them becomes subducted. **(b)** With the intervening oceanic lithosphere completely subducted, the two continental plates collide and are uplifted, because neither is dense enough to subduct. The result of the collision is a mountain range composed of highly deformed rocks. Note the thickened plate at the point of collision.

The friction of colliding oceanic plates produces earthquakes, and the molten material produced as the subducting plate sinks deep into the Earth's hot mantle causes the growth of chains of volcanoes. Initially these chains form under water. Eventually, however, after millions of years, the volcanoes grow high enough above the sea floor to emerge as a chain of volcanic islands, known as an *island arc*. This has happened in the northern Pacific, where the Pacific plate descends beneath the northwestern oceanic edge of the North American plate to form the earthquake-wracked, explosively volcanic Aleutian Islands of Alaska. The 1990 eruption of Mount Redoubt and the 1996 eruption of Mount Pavlov, both in Alaska, gave evidence of ongoing subduction in this region (Fig. 1-23a).

When an oceanic plate subducts beneath a continental plate, powerful earthquakes also occur and a chain of explosive volcanoes also grows—but on land. For example, at the western shores of North America, the edges of the oceanic Juan de Fuca and Gorda plates are subducting beneath the advancing continental edge of the North American plate (Fig. 1-23b). As these subducting plates descend to warmer depths, melting occurs (by a process discussed in Chapter 3) that fuels the chain of volcanoes of the Cascade Range, which stretches from northern California to southern British Columbia.

Subduction rates can be as high as 15 to 25 centimeters (6–10 inches) per year. This is considerably faster than the average rate of divergence, which might lead us to expect that, over time, oceanic plates would eventually disappear and the Earth itself would shrink. This does not happen, however, because less of the Earth's surface is devoted to subduction than to divergence. As maps of the sea floor show, only about 40,000 kilometers (25,000 miles) of subduction zones exist, whereas about 65,000 kilometers (40,000 miles) of the world's plate boundaries are diverging. It is this balance between the creation of oceanic plates at divergent zones and the destruction of oceanic plates at subduction zones that maintains the Earth's size.

Transform Motion and Transform Plate Boundaries The third major type of plate boundary occurs where two plates, either oceanic or continental, move *past* one another in opposite directions, a process known as **transform motion**

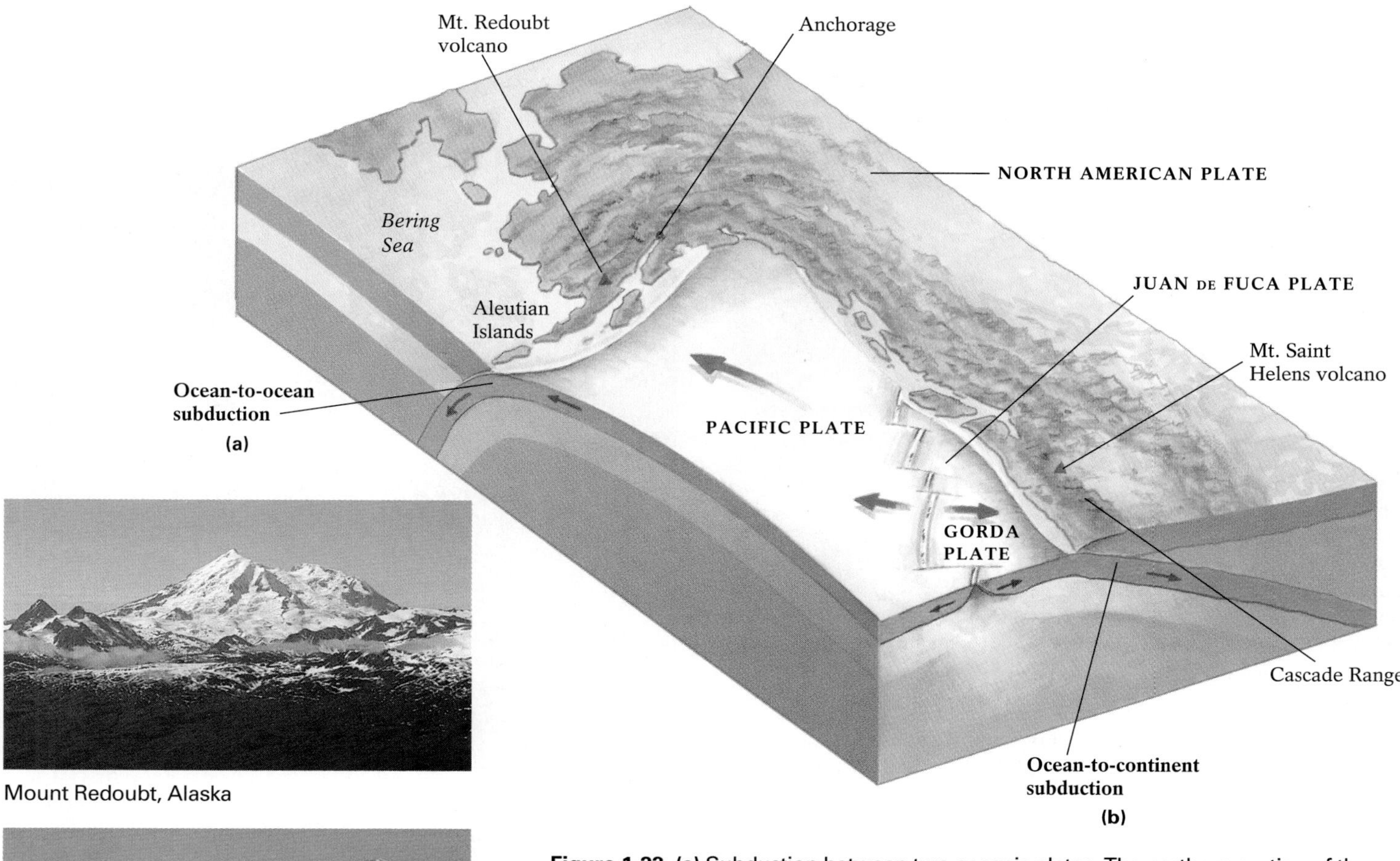

Mount Redoubt, Alaska

Mount St. Helens, Washington

Figure 1-23 **(a)** Subduction between two oceanic plates. The northern portion of the oceanic Pacific plate is currently subducting beneath the oceanic portion of the North American plate that underlies the Bering Sea. Alaska's Mount Redoubt, a product of this subduction, has been disrupting the lives of Anchorage residents intermittently since 1990. **(b)** Subduction between an oceanic plate and a continental plate. Along the coasts of southern British Columbia, Washington, Oregon, and northern California, several small plates of the Pacific Ocean basin are subducting beneath the continental edge of the North American plate. Mount St. Helens, in the subduction-produced Cascade Range mountains of Washington, has erupted numerous times since March 1980.

(Fig. 1-24). At transform boundaries, plates neither grow (as at divergent boundaries) nor shrink (as at convergent boundaries). Because no mantle material wells upward and no plates are subducting into the Earth's interior, transform boundaries generally produce little or no volcanism. However, great friction results as the moving plates grind past each other, so these boundaries do produce earthquakes. Any community caught between two plates at a transform boundary may be periodically devastated. Such is the case with San Francisco and Los Angeles, which are located within the San Andreas transform zone between the North American and Pacific plates. Transform plate motion breaks and displaces any feature, natural or human-made, that traverses a transform boundary (Fig. 1-25).

So those are the basic facts about plate tectonics. The next important question to ask: Why should we believe such a fantastic story? What evidence supports the hypothesis that the Earth's outer layer moves and as it does, creates and consumes oceans, and shuffles the position of the Earth's continents?

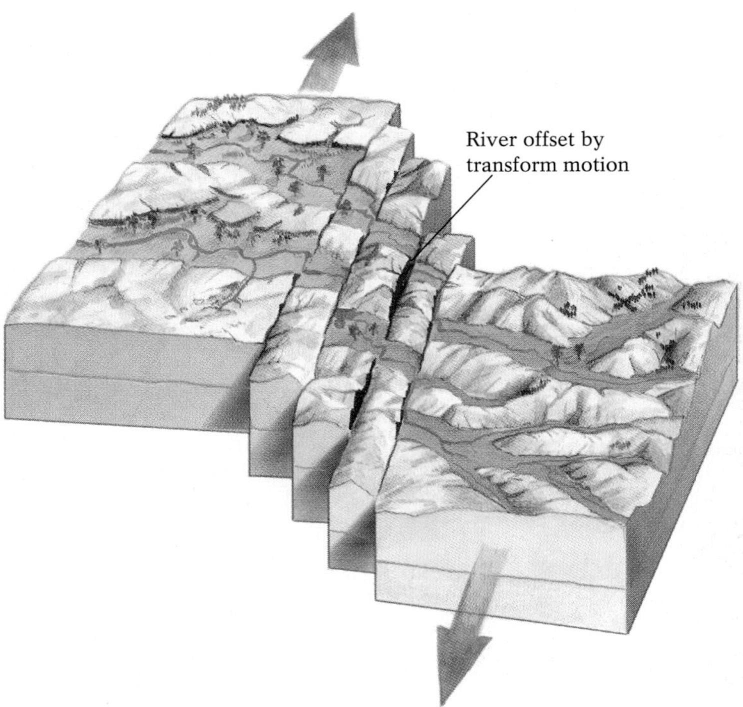

Figure 1-24 Transform motion. Where or when the Earth's plates move past one another in opposite directions, friction builds up at their edges but the plates are neither uplifted nor subducted.

Figure 1-25 An orange grove along the San Andreas transform plate boundary, with its tree rows offset by an earthquake there. Since any feature that cuts across an active transform plate boundary will be broken and displaced by plate movement (as are the rivers in Figure 1-24), plate motion confounds precise planting in orchards.

In Support of Plate Tectonics: A Theory Develops

In 1782, the American scientist–philosopher–statesman Benjamin Franklin hypothesized: "The crust of the Earth must be a shell floating on a fluid interior. Thus the surface of the globe would be capable of being broken and disordered by the violent movements of the fluids on which it rested." Almost 200 years later, that remarkable insight by one of history's finest scientific thinkers embodies some of the key concepts of the theory of plate tectonics. But how was modern plate tectonic theory developed? And what causes all this to happen?

Alfred Wegener and Continental Drift

In the early part of the twentieth century the German geophysicist–meteorologist Alfred Wegener (1880–1930) developed a controversial hypothesis that provided crucial underpinnings for the revolutionary theory of plate tectonics later in the century. Despite blistering ridicule from the leading geologists of his day, Wegener proposed that the continents float on the denser underlying interior of the Earth, and periodically break up and drift apart. He asserted that all of the Earth's continents had been joined together about 200 million years ago as a supercontinent he called *Pangaea* ("all lands") (Fig. 1-26). Wegener hypothesized that Pangaea had covered about 40% of the Earth's surface, most of it in the Southern Hemisphere. While Pangaea existed, what is now the area of New York City sweltered in the lush tropics near the equator, and most of eastern Africa and India shivered under a dome of glacial ice near the South Pole. Pangaea was surrounded by a single ocean, which Wegener called *Panthalassa* (named for the Greek goddess of the sea). Over a vast period of time beginning about 180 million years ago, Pangaea broke up, forming a number of continents that migrated to all regions of the globe. Wegener called this dispersal **continental drift.**

Wegener supported his hypothesis with the observations he made on the shapes of continental margins, patterns of present-day animal life, similarities among far-distant fos-

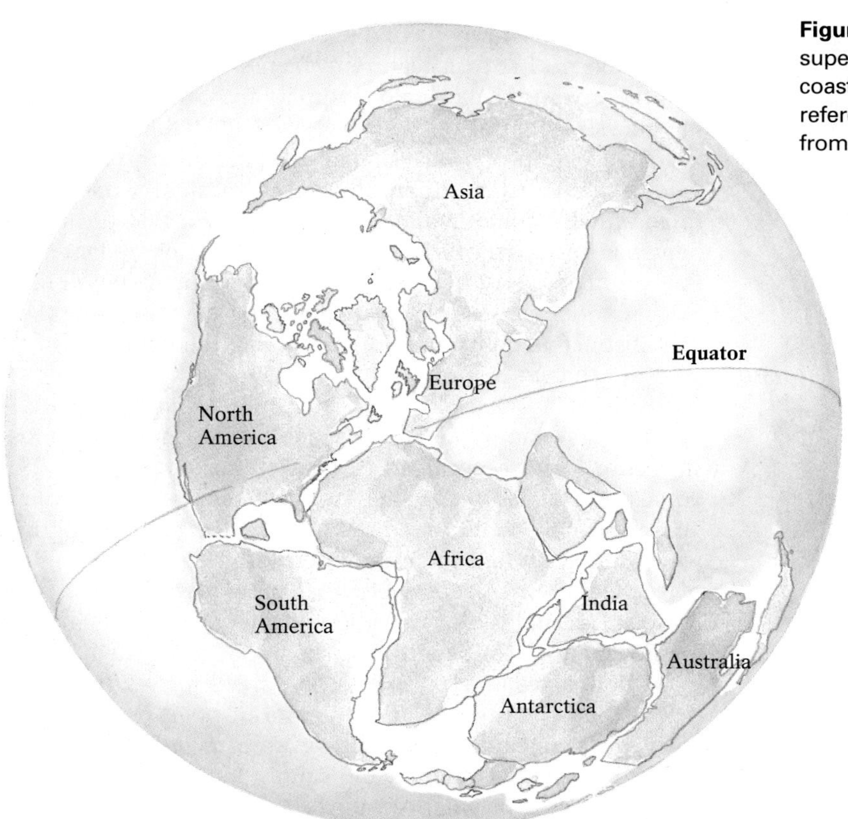

Figure 1-26 A reconstruction of the proposed supercontinent Pangaea, with present-day coastlines and continent names shown for reference. This configuration of plates dates from about 225 million years ago.

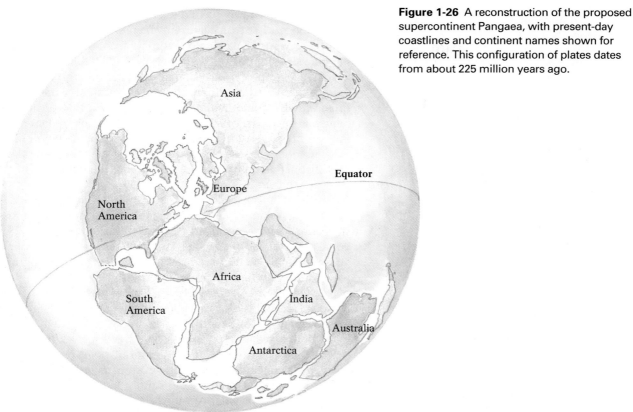

sils and rocks, and evidence of past climates at odds with the climates at present locations.

Continental Fit The English philosopher Sir Francis Bacon was among the first to note that the outlines of the continents of the world could be pieced together jigsaw-puzzle style. Bacon wrote of this in 1620, soon after seeing the new maps that came out of the global explorations of the sixteenth century. This concept reappeared periodically for the next three centuries (Fig. 1-27). Almost 300 years later, Wegener tinkered with the puzzle of continental shapes until he fit them together, forming a model of the landmass he called Pangaea. Today, precise computer fitting has confirmed Bacon's and Wegener's hypotheses, and we can see that the reunited continents would indeed fit together remarkably well. For example, when the borders of South America and Africa

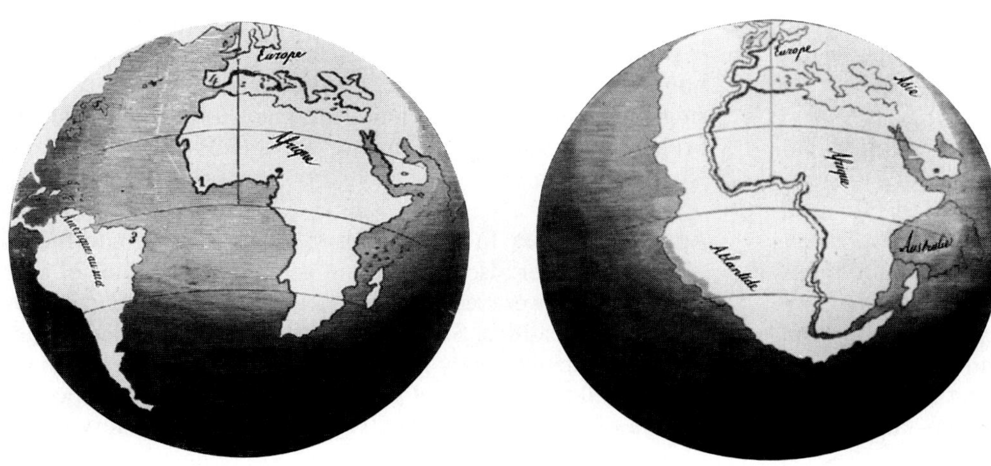

Figure 1-27 An early observation of the continental fit of South America and Africa, which was first proposed by Sir Francis Bacon in 1620. French naturalist Antonio Snider-Pelligrini sketched this diagram in 1858 for his work *La Création et ses Mystères Dévoiles,* which suggested that Noah's Deluge was responsible for the movement of the continents.

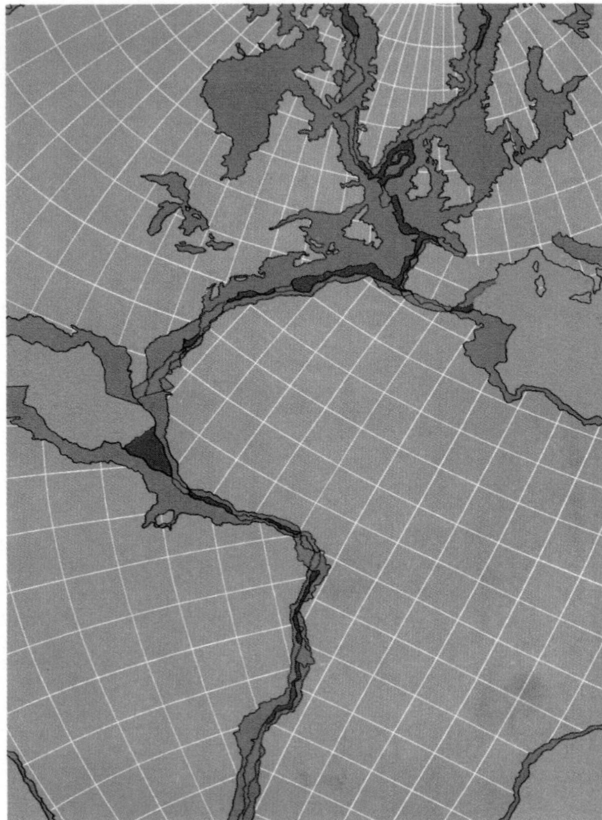

Figure 1-28 Precise matching of the continental shelves of circum-Atlantic continents by computer analysis. (From "The Origin of the Oceans," by Sir Edward Bullard, published 1969. Copyright © 1969 by *Scientific American, Inc.* All rights reserved.) Because the seaward (below sea level) edges of the continental shelves, shaded dark brown, represent the true edges of the continents, their fitting is more precise than simple shoreline fitting. The only places at which the plates appear to overlap, shaded red, are marked by geologic materials that have been deposited *since* the breakup of Pangaea. The precision of this fitting is one of the compelling lines of evidence that led to the wide acceptance of the theory of plate tectonics.

are juxtaposed, South America tucks almost perfectly into the niche in the western coast of Africa (Fig. 1-28).

Habitats of Living Animals Studies of the distribution patterns of certain modern animals helped to convince Wegener that now-separated landmasses were once united into a supercontinent. He noted, for example, that the hippopotamus is today found only in Africa and 500 kilometers (300 miles) to the east on the island of Madagascar. How did a creature that is clearly not built for long-distance, open-ocean swimming get from the mainland to the island? Could a land-bridge have once existed between Africa and Madagascar in the Mozambique Channel between these two landmasses? Oceanographic surveys conducted during the 1940s found no evidence of such a feature. Wegener considered it likely that there was once a single landmass that rifted apart, leaving ancestral hippos on both sides of the channel.

Wegener proposed a similar explanation for the unusual wildlife native to Australia, the only continent with kangaroos, wallabies, and koalas. He hypothesized that this continent was once part of a large Southern Hemisphere landmass that rifted about 40 million years ago. Drifting alone on their island continent, Australia's fauna began to evolve along their own distinctive path, gradually becoming the unique animals they are today.

Habitats of Ancient Animals Wegener also examined the fossil record for rare occurrences of past life forms (Fig. 1-29). Fossils of *Mesosaurus,* a small reptile that lived 240 million years ago, have been found only in Brazil and South Africa, which are separated today by 5000 kilometers (3000 miles) of the southern Atlantic Ocean. The skeletal structure of *Mesosaurus* and the composition of the deposits in which its fossil remains have been found show that it paddled around in shallow lakes and estuaries (the part of a river that empties into an ocean) but did not swim in open oceans. Paleontologists conclude from this that *Mesosaurus* lived on what was then a single landmass and simply traveled overland between the sites of what are now Africa and South America before the landmass rifted. A similar explanation accounts for the occurrence in Antarctica, Africa, Madagascar, and India of the fossil remains of another non–ocean-swimming prehistoric creature, *Lystrosaurus.*

Related Rocks In the Northern Hemisphere, the 390-million-year-old rocks of the mountains of eastern North America are remarkably similar in mineral composition, structure, and fossil content to rocks of the same age in eastern Greenland, western Europe, and western Africa (Fig. 1-30a). Wegener recognized that if North America, Africa, and Europe had been joined in the past, there would have been a continuous chain of mountains from Alabama all the way to Scandinavia (Fig. 1-30b). Iceland would have been a puzzle for Wegener, for it shows no evidence of this mountain chain. We now know, however, that Iceland formed atop the diverging mid-Atlantic ridge only about 60 million years ago, long after Pangaea broke up to form the separate continents of Europe and North America.

Ancient Climates In his attempt to prove that the continents drifted, Wegener also pointed out geological evidence of past climates that were very different from the current climates of certain locations. Such is the case, for example, with glaciers. As glaciers slowly flow across a landscape, they pick up rocks, boulders, and sand grains and, as they melt, deposit a jumble of debris of all sizes (Fig. 1-31a). The bouldery mate-

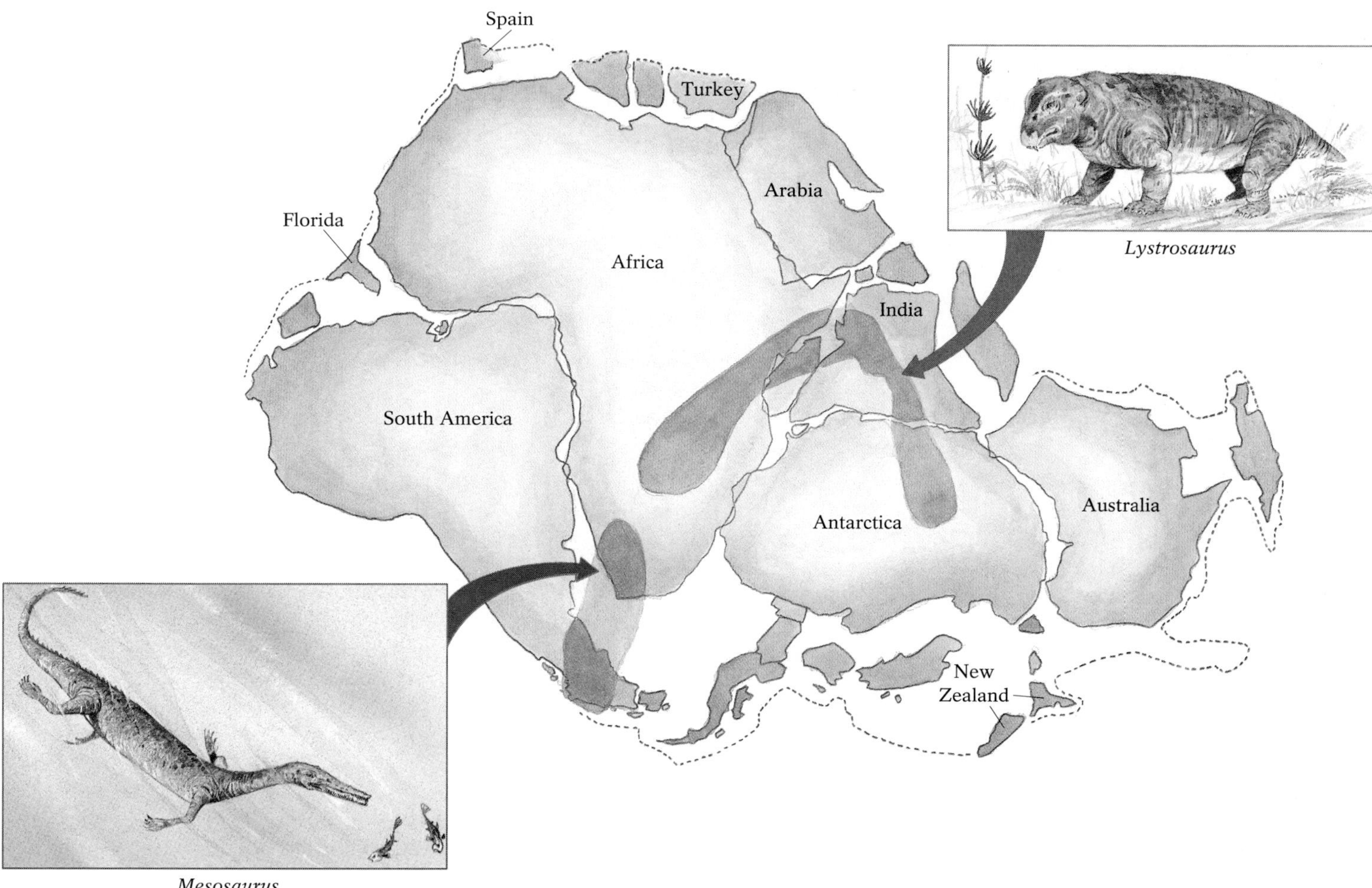

Figure 1-29 The distribution of *Mesosaurus* and *Lystrosaurus* when the Southern Hemisphere continents were joined together as part of Pangaea. Fossils of these reptiles date from the Triassic Period (208–245 million years ago). The less-than-sleek *Lystrosaurus*—a sheep-sized reptile that lived about 225 million years ago in what are now Antarctica, Africa, Madagascar, and India—and *Mesosaurus* (discussed in the text), neither of which were long-distance swimmers, simply strolled to the spots where we find their remains today. That was, of course, long before these once-contiguous lands rifted and diverged, spreading the sea floors that were to become the Indian and the southern Atlantic Oceans, respectively.

rial shown in Figure 1-31b is glacial debris found on the west coast of South Africa, which tells us that South Africa was glaciated about 250 million years ago (the estimated age of that deposit). An icy past for this region would be plausible only if South Africa once occupied the space close to the South Pole now claimed by the continent of Antarctica.

As further evidence of continental drift, the ancient bedrock of such warm places as India, Australia, Africa, and South America is scored by a distinctive pattern of aligned scratches (Fig. 1-32). Such striations form as glaciers drag debris along their beds, scratching the underlying bedrock. These warm regions, then, must have once been located in colder climes.

Similar reasoning explains the discovery of coal in presently cold climates. Coal forms after a great accumulation of swamp vegetation has been buried, compressed, and heated in a warm, moist environment. If we know that a cliff in arctic Spitsbergen, Norway, contains a seam of coal surrounded by kilometers of ice, we can infer that the coal must have formed in a very different climate (in this case, some 300 million years ago) and then journeyed northward a great distance.

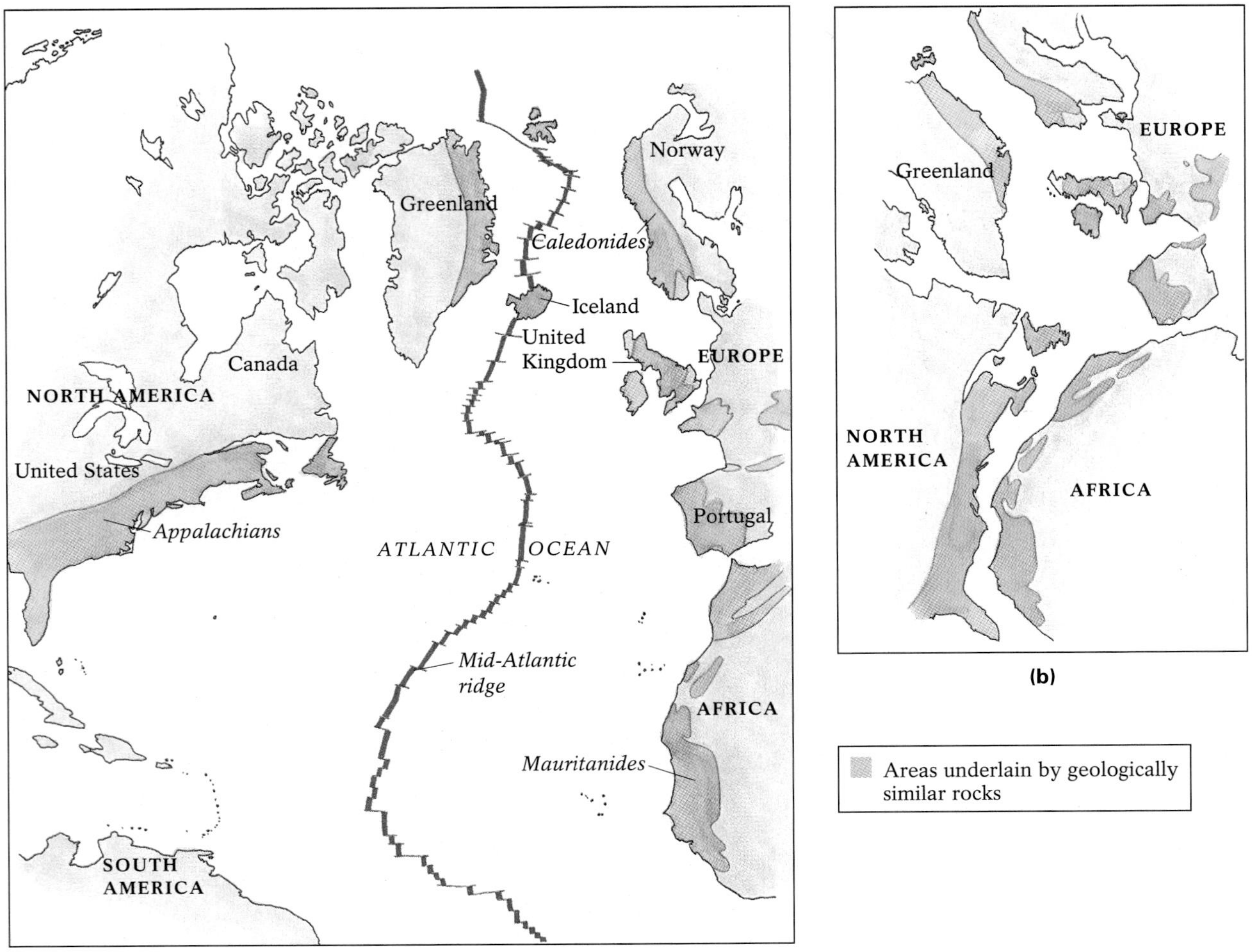

Greenland
Norway
Caledonides
Iceland
United Kingdom
EUROPE
Canada
NORTH AMERICA
United States
Appalachians
ATLANTIC OCEAN
Portugal
Mid-Atlantic ridge
AFRICA
Mauritanides
SOUTH AMERICA

(a)

EUROPE
Greenland
NORTH AMERICA
AFRICA

(b)

Areas underlain by geologically similar rocks

Figure 1-30 The current positions of the Northern Hemisphere continents surrounding the Atlantic Ocean **(a)**, and their pre-rifting positions as part of Pangaea **(b)**. Note the absence of Iceland in the Pangaea reconstruction—Iceland originated from volcanic eruptions at the mid-Atlantic ridge *after* the breakup of Pangaea.

(a) (b)

Figure 1-31 **(a)** Glacial debris at the margin of the Columbia Icefield in Jasper National Park, Alberta, Canada. The debris is in the foreground and on the slope left of the ice flow. **(b)** Glacial debris from the Dwyka area, Cape Province, South Africa.

(a)

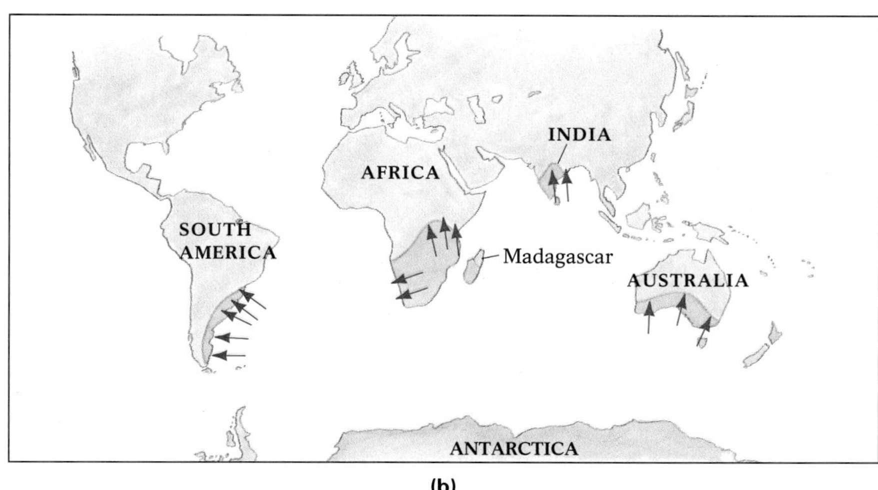

(b)

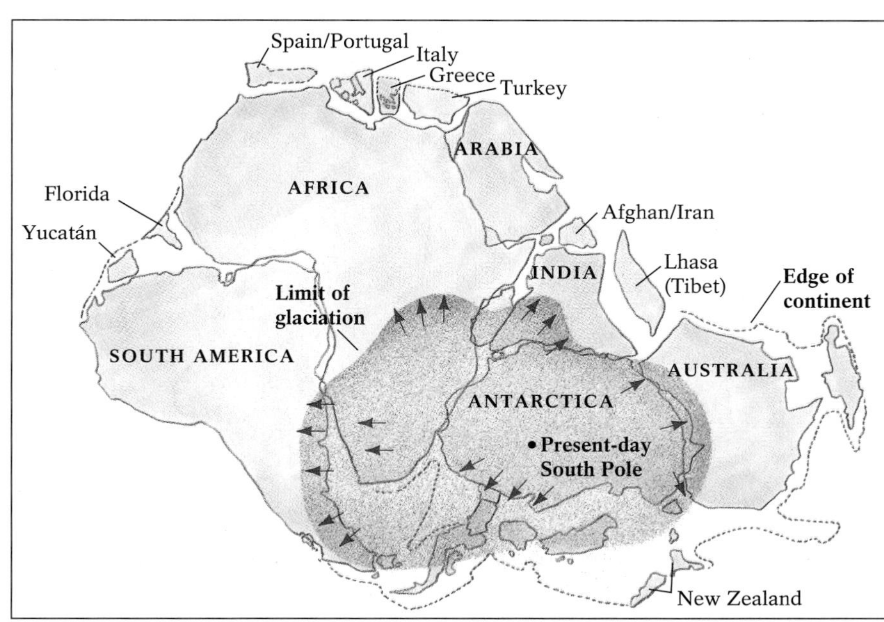

(c)

Figure 1-32 **(a)** Glacial striations in bedrock in Victoria, British Columbia. **(b)** Bedrock striation patterns as they appear today on separated Southern Hemisphere continents: The striations are oriented in what appears to be a random pattern. **(c)** Striation patterns as they would appear on Pangaea, with " reunited" Southern Hemisphere continents occupying the South Pole: The striations form a systematic pattern resembling the spokes of a bicycle wheel, a pattern that typifies the outward flow of a large glacier.

The Driving Force Behind Plate Motion

With such strong evidence to support it, why was Wegener's hypothesis of continental drift initially rejected? The best answer is that Wegener's proposal that the continents, driven by the Earth's rotation, plowed through denser oceanic rocks did not seem physically plausible. At the time, no one knew of the existence of lithospheric plates and the heat-softened underlying asthenosphere, and consequently Wegener was unable to propose a scientifically acceptable mechanism for continental drift. That mechanism would become clear only later, and the scientific community of Wegener's day could not accept so radical an idea that contested the long-held notion of immovable continents and ancient, featureless sea floors without better evidence. In November 1928, only two years before he died, Wegener endured his final rejection when a meeting of esteemed American geologists concluded, "if we are to believe Wegener's hypothesis, we must forget everything which has been learned in the last 70 years and start all over again."

That's where Wegener's hypothesis of continental drift stood for the next three decades. With the advent of deep-sea drilling and the development of other ways to study the Earth's interior (discussed in Chapter 11), scientists discovered the existence of the lithospheric plates and identified

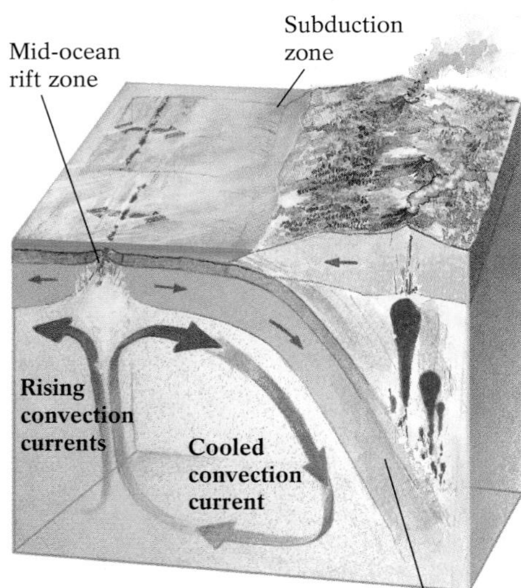

Figure 1-33 Convection cells and plate motion. Heat within the Earth's mantle results in rising ("convecting") currents of warm mantle material, which drag the lighter lithospheric plates along with them as they flow beneath the Earth's surface. As rising mantle material spreads beneath the lithospheric plates, it cools, becomes more dense, and begins to sink back to the deeper interior, where it is reheated to rise again. (A similar dynamic is what causes a pot of soup to heat uniformly as hot low-density soup near the heat source, rises and cool higher-density soup sinks.) Such a cycle, known as a convection cell, is believed by many scientists to be a principal driving mechanism of plate tectonics.

several ways in which the plates might move. During the 1960s and 1970s, a great deal of evidence appeared in support of Wegener's lines of reasoning, enhancing the acceptability of the Pangaea hypothesis and replacing the notion of continental drift with the theory of plate tectonics. Since the 1980s new technologies have been increasing our knowledge of the physics of the Earth's interior, enabling geologists to propose plausible hypotheses to explain what drives plate tectonics.

Many geologists now believe that heat-driven currents in the Earth's mantle are principally responsible for plate movements. These currents, known as **convection cells,** develop when portions of a heated substance become less dense and rise toward the surface, displacing cooler (more dense) portions, which are in turn pulled down by gravity (Fig. 1-33). As convection cells move heated mantle rocks toward the surface in this way, the convecting mantle material encounters the overlying solid lithospheric plates. Unable to reach the surface, the rising material moves laterally beneath the plates, dragging them along as it flows.

Sometimes the drag produced by neighboring convection cells pulls the lithosphere in opposite directions. If the pull is great enough to cause the lithosphere to rift, the rifted plates are then carried along in opposite directions by the slowly convecting currents. A relatively small amount of mantle-derived material does escape to the surface as lava at such divergent plate boundaries, gradually cooling to form the new outer boundaries of the plates. The plates continue to diverge from the site of the rising hot mantle over millions of years, becoming cooler and more dense. Eventually, now far removed from the spreading mid-ocean ridge, the oceanic lithosphere becomes dense enough to sink back into the Earth's interior at a subduction zone, where it will be reheated, eventually to reenter the cycle.

Some geologists believe that convection alone drives plate tectonic processes. Others believe that gravity assists convection, literally pushing the plates away from uplifted mid-ocean ridges and pulling the plates, or *slabs,* down into the Earth's interior at subduction zones. These processes are known respectively as *ridge push* and *slab pull* (discussed in Chapter 12). Convection and gravity, then, may be working together to cycle the Earth's plates.

The Earth's Plate Tectonic Future

Our knowledge of current rates and directions of plate motion enables us to speculate about the Earth's continental configuration millions of years into the future, assuming that those rates and directions remain relatively constant. In Figure 1-34, we see the Earth as it *may* look 100 million years from now. This scenario, of course, assumes that the rates and directions of all plate movement will remain constant. Plate motions, however, have changed in the past and may do so again. Nevertheless, the reunion of the continents to form another supercontinent in the future is a strong possibility.

A Preview of Things to Come

This brief introduction to the principles of plate tectonics should prepare you for the next 11 chapters, which constitute the first two major parts of this text. As Part 1 contin-

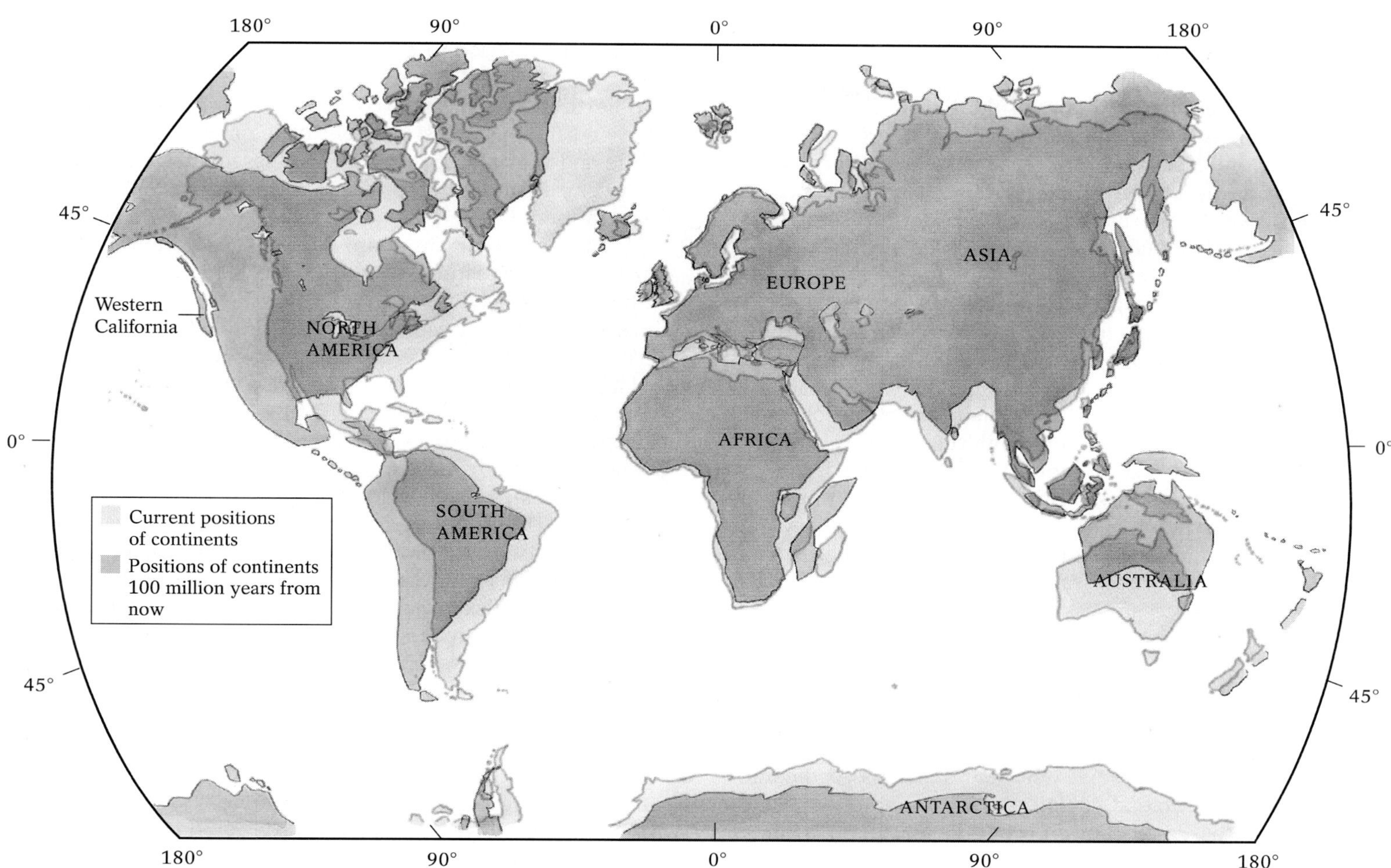

Figure 1-34 The projected position of the Earth's continents 100 million years from now. These positions are based on the assumption that plate velocities and directions will remain as they are today. Note the possible collisions of Africa and Europe (with the loss of the Mediterranean Sea) and Australia and Indonesia, and the movement of western California toward Alaska along a continent-long transform plate boundary.

ues, Chapters 2 through 7 focus on the origins of various Earth materials—from minerals to igneous, sedimentary, and metamorphic rocks—and explain some important processes that produce those rocks, such as volcanism and weathering. Chapter 8 explains how geologists interpret the rock record to unravel the Earth's long history. In Part 2, Chapters 9 through 11 are devoted to the large- and small-scale geologic structures at the Earth's surface and the dynamics of the Earth's interior. Chapter 12 provides a more detailed look at the global effects of plate tectonics.

Finally, Part 3 describes how surface processes sculpt the Earth's large-scale features created by tectonic forces. In Chapters 13 through 19, we will explore how solid rock is turned to dust by the Earth's climate; why some slopes are more stable than others (and precautions to take when building a house in a landslide-prone region); how streams modify the landscape (and why they occasionally flood our homes); where to find clean, drinkable, underground water; how caves form; why glaciers once covered much of North America and when they will be back; where sand dunes form and why the world's deserts grow larger every day; and why crashing waves are eroding some of our coasts at an alarming rate. The concluding chapter focuses on the ways humans use and manage the Earth's natural resources—from the oil that heats our homes and fuels our cars to the fertilizers that enrich our crops and guarantee our next meal.

The relevance of geology will be apparent in every chapter of this text. Pay close attention—what you learn will affect you in one way or another every day of the rest of your life.

Chapter Summary

Geology is the scientific study of the Earth. Geologists, like other scientists, systematically collect data derived from experiments and observations. They analyze and interpret their findings and develop **hypotheses** to explain how the forces of nature work. The hypotheses that are consistently supported by further study and investigation may be elevated to the status of a widely accepted explanation, or **theory.** A theory that withstands rigorous testing over a long period of time may be declared a **scientific law.** In order to be accepted, all scientific investigations must conform to these **scientific methods**. The hypotheses that have undergone such scrutiny include two that attempted to explain the evolution of the Earth's geologic features. **Catastrophism,** which was popular until the mid-eighteenth century, held that the Earth had evolved through a series of immense worldwide upheavals; **uniformitarianism,** proposed that the Earth has evolved slowly and gradually by small-scale processes that can still be seen operating. Today scientists recognize the effects of both slow processes and catastrophic events in the evolution of the Earth.

The present universe began with the "Big Bang" roughly 12 billion years ago. The Sun, which is a star, formed about 5 billion years ago from the collapse of a gas cloud, the center of which heated up as particles drawn inward by gravity collided and produced nuclear reactions. As the outer region of the gas cloud cooled, the Earth and other planets developed (the Earth about 4.6 billion years ago) by accretion of colliding masses of matter, some perhaps as large as the planet Mars. Earth's collision with one such mass may have spawned the Moon.

During the Earth's first few tens of millions of years of existence, the impact of these accreted masses along with the heat produced by radioactive decay warmed its interior until the accumulated heat was sufficient to melt much of the planet's constituents. This period of internal heating caused the Earth to become layered, or differentiated. We base much of our knowledge about the origin of the Earth's interior layers on the study of **chondrules,** small nuggets of rocky material found in many meteorites that are believed to be droplets of matter that condensed directly from the original solar nebula. By comparing the composition of the Earth's crust today to that of these chondrules, we note that the Earth's crust is quite deficient in iron. The early period of heating must have caused the Earth's densest elements, primarily iron, to sink toward its interior while its lightest elements rose and became concentrated closer to its surface.

Today, the Earth has three principal concentric layers of different densities: the thin, least dense outer layer, called the **crust;** a thick, denser underlying layer, called the **mantle;** and a much smaller **core,** which is the most dense of Earth's layers. Over the last 4 billion years, the Earth has cooled slowly from its initial higher temperatures. Enough heat remains in its interior to generate currents of flowing mantle rock that have kept the outer portion of the Earth mobile. The Earth's **lithosphere,** a composite layer made up of the crust and the outermost segment of the mantle, is solid and brittle and forms large rocky plates; these plates move along at the Earth's surface atop the warm, flowing **asthenosphere** beneath them.

Time plays an important role in the evolution of the Earth's geologic features and materials. The Earth is believed to be about 4.6 billion years. Over such a vast amount of time, many gradual geological changes can take place that occur too slowly to be perceived on a human time scale. The Earth's three principal types of rocks undergo such changes, actually turning from one type into another, depending on the environmental forces acting on them. **Rocks,** which are defined as naturally occurring aggregates of inorganic materials **(minerals),** are categorized according to the way in which they form. The three basic rock groups are **igneous rocks,** which solidify from molten material; **sedimentary rocks,** which are compacted and cemented aggregates of fragments of preexisting rocks of any type; and **metamorphic rocks,** which form from any type of rock when its chemical composition is altered by heat, pressure, or chemical reactions in the Earth's interior. The continual transformation of the Earth's rocks from one type into another over time is called the **rock cycle.**

For centuries, scientists have tried to decipher the origin of the Earth's largest geologic features, such as oceans and mountain ranges, and to learn the reasons for geologic catastrophes such as volcanoes and earthquakes. In the late twentieth century, the theory of **plate tectonics** has provided an explanation for all of these. According to this theory, the outermost portion of the Earth (the lithosphere) is composed of seven major and a dozen or more minor plates. The plates consist of relatively dense oceanic lithosphere, relatively light continental lithosphere, or a combination of both types. The Earth's continents and oceans drift from place to place atop the moving plates. The movement of the continents, called **continental drift,** was first recognized by the German geophysicist–meteorologist Alfred Wegener (1880–1930) in what is considered the first step toward the development of the comprehensive plate tectonic theory. The plates move in three ways: away from one another, by **divergence;** toward one another, by **convergence;** or past one another in opposite directions, by **transform motion.**

Divergence is preceded by **rifting,** the process in which a preexisting plate is torn into a number of smaller plates. New oceanic lithosphere forms between rifted plates as molten rock from the Earth's mantle wells upward, cools, and solidifies. In this manner, new oceanic rock is continuously added to the rifted edges of the diverging plates, a process known as **sea-floor spreading.** Convergence involving either two oceanic plates or one oceanic plate and one continental plate results in **subduction,** in which the denser oceanic plate sinks below the other into the Earth's mantle; a **continental collision** occurs when both converging plates are continental, in which case neither is dense enough to subduct completely and both plates' edges are instead eventually uplifted by the pressure of the collision. Because of the hot mantle material that rises at divergent plate boundaries, the pressure that develops at the edges of colliding plates, and the friction that occurs between plates moving past one another at transform boundaries, rocks at plate boundaries may melt (fueling volcanic eruptions), buckle (forming mountains), or break (generating earthquakes). By comparison, plate interiors are relatively quiet geologically.

One of the principal mechanisms that drives the plates appears to be the **convection cells** that circulate the Earth's internal heat. As the deeper rocks are warmed by the Earth's various internal heat sources, they become lighter and rise. Near the surface, the rising currents of flowing material spread laterally beneath the Earth's lithosphere, pulling the plates along atop the flowing asthenosphere. The effect of gravity on dense oceanic plates may also contribute to plate motion, by pushing the plates away from uplifted mid-ocean ridges and pulling the plates down into the Earth's interior at subduction zones.

Key Terms

geology (p. 4)
scientific methods (p. 5)
hypothesis (p. 5)
theory (p. 5)
scientific law (p. 5)
catastrophism (p. 8)
uniformitarianism (p. 8)
chondrules (p. 11)
crust (p. 11)
mantle (p. 11)
core (p. 12)
lithosphere (p. 12)
asthenosphere (p. 12)
rock (p. 14)
minerals (p. 14)
igneous rocks (p. 14)
sedimentary rocks (p. 14)
metamorphic rocks (p. 14)
rock cycle (p. 14)
plate tectonics (p. 17)
rifting (p. 19)
divergence (p. 20)
sea-floor spreading (p. 20)
convergence (p. 21)
subduction (p. 21)
continental collision (p. 21)
transform motion (p. 22)
continental drift (p. 24)
convection cells (p. 30)

Questions for Review

1. Briefly explain the differences between a scientific hypothesis, a scientific theory, and scientific law.

2. Contrast the principles of catastrophism and uniformitarianism.

3. Draw a simple sketch of the major layers that make up the Earth's interior. What parts of the Earth's internal structure form the Earth's plates?

4. Describe the three major types of rocks in the Earth's rock cycle.

5. Draw simple sketches of divergent plate boundaries, two kinds of convergent plate boundaries, and transform plate boundaries.

6. At which type of plate boundary do oceanic plates grow? At which type of plate boundary are oceanic plates consumed?

7. Why are there so few earthquakes in Minneapolis and Indianapolis?

8. How could one use the distribution of modern and ancient animals to support the theory of continental drift?

9. Briefly describe how oceanic plates vary in age as they proceed outward from a diverging mid-ocean ridge.

10. Draw a simple sketch of a convection cell and explain how it works. Explain the difference between convection and conduction of heat.

For Further Thought

1. When geologists find ancient glacial deposits in equatorial Africa, they usually interpret them as polar deposits that have drifted as part of lithospheric plate from a cold place to a warm place. Formulate another hypothesis to explain this phenomenon.

2. Why is there no current volcanic activity along the east coast of North America?

3. Describe how plate tectonic activity might affect the rock cycle.

4. Find the Ural Mountains on a map of eastern Europe. Briefly explain how they might have formed.

5. As recently as 5 million years ago, South America and North America were separated, unattached by Central America. Using Figure 1-15, speculate about how the Central American connection that binds the Western Hemisphere might have formed.

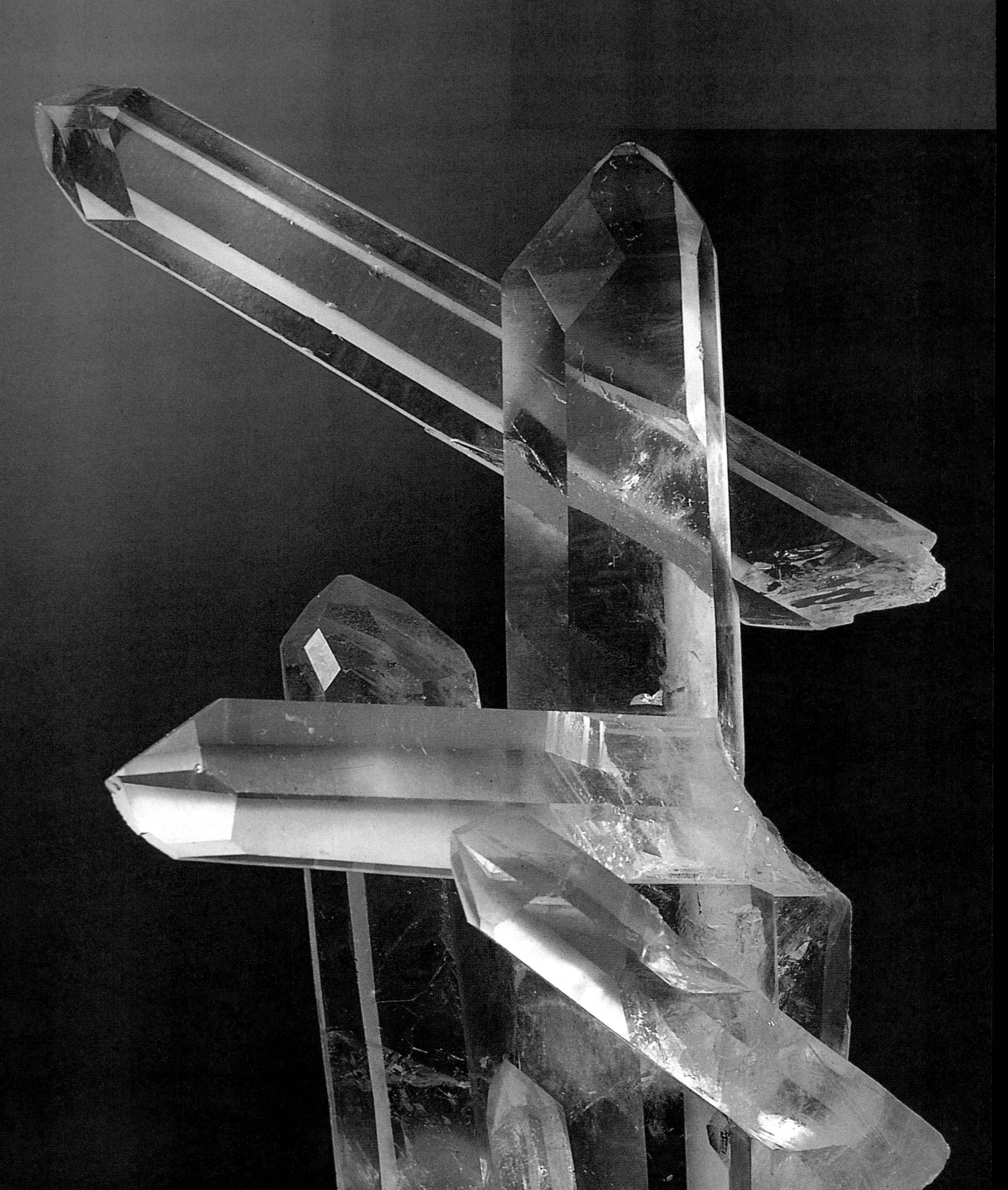

2 Minerals

Figure 2-1 Crystals of quartz, showing the beauty and symmetry for which many minerals are valued.

Many cultures have long valued minerals for their sheer visual appeal—for their stunning colors or luster, or their often-perfect symmetry (Fig. 2-1). But the importance of minerals is hardly just aesthetic, and is not restricted to the "pretty" specimens found in museums or jewelers' showcases. From the simple flint tools made by our ancestors hundreds of thousands of years ago to the quartz crystal in a modern timepiece, minerals have always helped us improve our lot in life (Fig. 2-2).

Even our physical well-being depends on certain types of minerals. Many of the nutrients our bodies need come from minerals in the soil that are incorporated into the fruits and vegetables we eat. In this way, we acquire calcium, phosphorus, and fluorine, which harden our bones and teeth; sodium and potassium, which help regulate blood pressure; and iron, a major component of the hemoglobin in blood, which carries life-sustaining oxygen to our cells.

(a)

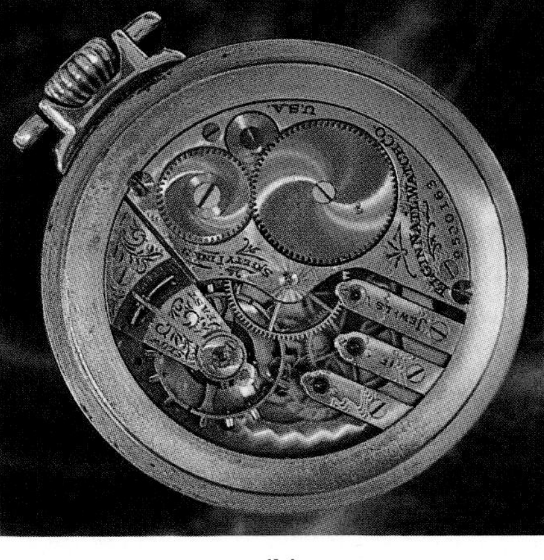

(b)

Figure 2-2 Minerals have been providing us with essential tools for hundreds of thousands of years. **(a)** This flint hand scraper was used by Native Americans more than a thousand years ago, probably to scrape flesh and hair from animal skins. **(b)** This personal timepiece runs with great accuracy on the constant vibration of a quartz crystal.

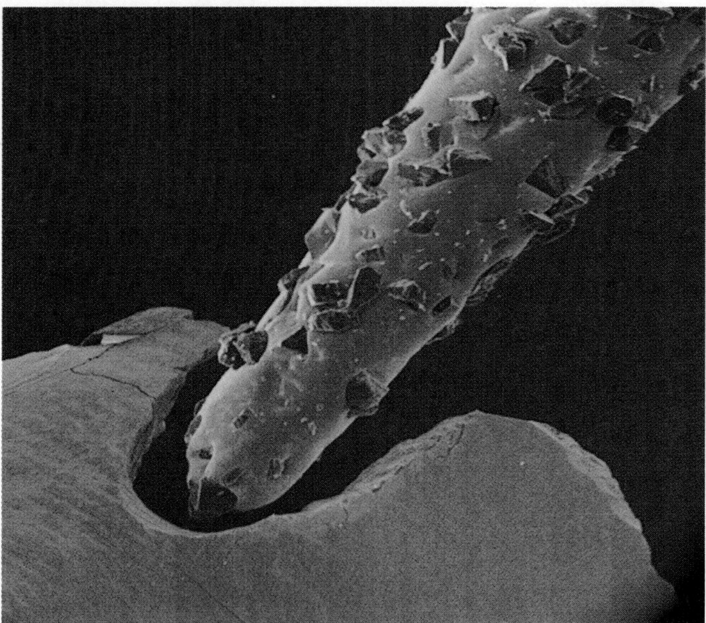

Figure 2-3 Today, because of technological advances and the existence of diamonds, a visit to the dentist is not nearly as long or as painful as it once was. Here we see a drill bit studded with diamonds (the hardest known natural substance) cutting swiftly through the relatively soft enamel of a human molar composed largely of the calcium-phosphorus mineral, apatite. (Magnified 25×)

Every day we use a vast array of minerals in a remarkable number of ways. A common absorbent mineral called talc (used in talcum powder) dries and soothes our skin. Minerals provide the sulfur used to manufacture fertilizers, paints, dyes, detergents, explosives, rubber, synthetic fibers, or a simple book of matches, as well as the fluorine that helps to refrigerate food and cool homes and offices. Certain minerals yield the valuable metal aluminum, an ideal component for airplanes, garden furniture, and beer and soft-drink cans because of its lightness, strength, and resistance to corrosion. When you're laid low by a common intestinal problem, you may run for a spoonful of Kaopectate, a remedy whose active ingredient is the mineral kaolinite. (For another everyday example, see Figure 2-3.)

As you will see, minerals are vastly important to geologists because they make up the rocks we study to interpret the Earth's past. **Rocks** are simply naturally occurring aggregates, or combinations, of one or more minerals, with each mineral retaining its own discrete characteristics. For example, the rock granite shown in Figure 2-4 contains minerals such as quartz and orthoclase feldspar.

In this chapter, we will examine what minerals are and how they are formed, and discuss methods of identifying the different kinds of minerals. We will also look at the distinctive structures of some important minerals and see how their characteristics determine our uses for them.

Figure 2-4 One can clearly see the individual mineral grains in this piece of granite, an igneous rock. How many different minerals can you see in this specimen?

What Is a Mineral?

Minerals are naturally occurring solids consisting of one or more chemical elements in specific proportions, whose atoms are arranged in a systematic internal pattern. For example, rubies, emeralds, and quartz are minerals. Because minerals are *naturally* occurring solids, the thousands of synthetic compounds produced in laboratories do not qualify as minerals. Because minerals have a *systematic internal organization*, substances such as the gemstone opal, which lacks systematic internal organization of its atoms, are not considered minerals.

Minerals are composed of one or more **elements** in specific proportions. An element is a form of matter that cannot be broken down into a simpler form by heat, cold, or reaction with other chemical elements. Aluminum and oxygen are two common elements. **Atoms** (from the Greek *atomos*, or "indivisible") are the smallest particles of an element that retain all of its chemical **properties**. A property is a characteristic of a substance that enables us to distinguish it from other substances. All atoms of a given element are essentially identical, and the atoms of one element differ in fundamental ways from the atoms of every other element.

There are 112 known elements, of which 92 occur naturally and 20 are laboratory creations. A chemical symbol, a one- or two-letter abbreviation, identifies every element. These symbols are usually the first letter or letters of the English or Latin name of the element, such as O for oxygen, Al for aluminum, and Na (from the Latin *natrium*) for sodium. Chemists have arranged the 112 elements into the Periodic Table of Elements (Fig. 2-5).

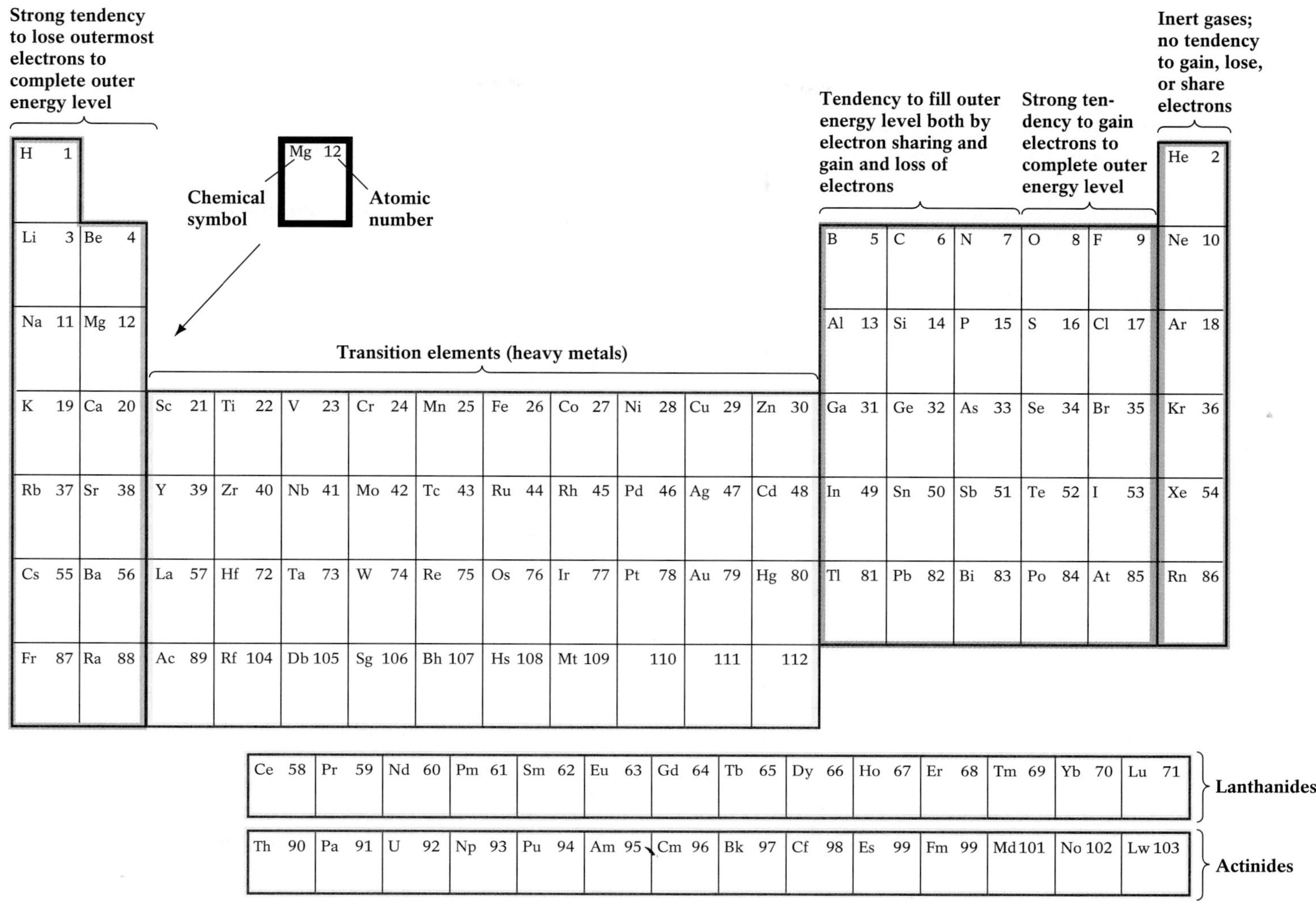

Figure 2-5 The Periodic Table of Elements. The periodic table groups all the elements by similarities in their atomic structures, which result, in turn, in similarities in their chemical properties.

Atoms of one or more elements may combine in specific proportions to form chemical **compounds**. For example, the mineral quartz is a chemical compound that always contains one silicon atom for every two oxygen atoms. We use *chemical formulas* to express the fixed proportions of atoms that make up a compound; for example, the silicon and oxygen combination that makes up quartz has the chemical formula SiO_2. Subscript numerals denote the number of atoms of each element in the chemical formula; thus one can tell from its formula that quartz contains two atoms of oxygen for every atom of silicon.

Geology students sometimes wonder why so much discussion about minerals concerns chemistry. The principal reason is that virtually all minerals are chemical compounds, and it is their chemical compositions and the internal arrangement of their atoms that determine their distinctive characteristics. These characteristics in turn determine their uses in and value to society. To know why diamonds are hard, why gold can be pounded into wafer-thin leaves, and why we build skyscrapers with skeletons of titanium steel, we must understand the chemical makeup of minerals.

The Structure of Atoms

An atom is incredibly small, approximately 0.00000001 centimeter (one hundred-millionth of a centimeter) in diameter. This line of type would contain about 800,000,000 atoms laid side by side. As small as an atom is, it consists of even smaller particles: **protons** and **neutrons**, which are located in the **nucleus**, or center, of the atom; and **electrons**, which move about outside the nucleus. Most of an atom's volume is, in fact, just empty space. If, for example, the volume of an atom's nucleus were expanded to the size of a basketball, the orbiting electrons would be located up to 3 kilometers (2

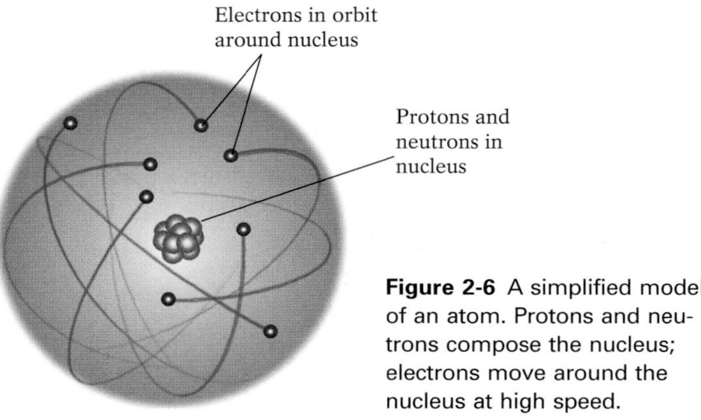

Figure 2-6 A simplified model of an atom. Protons and neutrons compose the nucleus; electrons move around the nucleus at high speed.

miles) away from the basketball! A simplified model of an atom is shown in Figure 2-6.

Each proton in the nucleus carries a single positive charge, expressed as +1, and has a mass of 1.67×10^{-24} gram, which for convenience is referred to as an *atomic mass unit* (AMU) of 1. Neutrons are nearly identical to protons in size and mass, but, as their name suggests, they have no charge—they are neutral. They do, however, contribute to the **atomic mass** of the atom, which is the sum of protons and neutrons within an atom's nucleus. Thus an atom containing one proton and one neutron has an atomic mass of 2 AMU. The third type of atomic particle, the electron, travels around the nucleus at a speed so great that it could orbit the Earth in less than a second. Each electron carries a single negative charge, expressed as –1, and has a mass about 1/1836 that of a proton or neutron. Thus an electron's contribution to an atom's mass is negligible.

The number of protons in an atom's nucleus is its **atomic number**. Every atom of an element has the same number of protons in its nucleus; this number differs from the number of protons in the nuclei of all other elements. For example, every magnesium atom has 12 protons in its nucleus, or an atomic number of 12. An atom with 13 protons in its nucleus is an entirely different element, aluminum. Thus the number of protons determines an atom's identity.

The number of neutrons in an atom's nucleus can vary without causing the atom's identity to change; only the atomic mass changes. Atoms of the same element that have the same number of protons but different numbers of neutrons in their nuclei and that consequently differ in atomic mass are known as **isotopes**. For example, the element oxygen (atomic number = 8) always contains eight protons in its nucleus, but it has three isotopes: $^{16}_{8}O$, with eight neutrons in its nucleus; $^{17}_{8}O$, with nine neutrons; and $^{18}_{8}O$, with ten neutrons. The subscript numeral in these notations is the atomic number and the superscript numeral is the atomic mass. Some isotopes of certain elements contain nuclei that break down spontaneously and emit some of their particles, making them *radioactive*. When the nuclei of radioactive isotopes break down, they also give off a large amount of heat. In nuclear power plants, it is this heat that is used to produce steam and electricity. Two common radioactive isotopes are $^{235}_{92}U$ (uranium-235) and $^{14}_{6}C$ (carbon-14).

The number of electrons in an atom is always the same as the number of protons. For example, hydrogen (atomic number = 1) has one proton and one electron, whereas iron (atomic number = 26) has 26 protons and 26 electrons. Because the number of an atom's protons equals the number of its electrons, its positive charge exactly balances its negative charge and it has no net charge. However, all of an atom's positive charge is concentrated in the protons in its nucleus, whereas its negative charge is distributed among the electrons flying about its periphery.

The negatively charged electrons, attracted by the positively charged protons, tend to congregate about the nucleus in ever-changing orbits, forming an *electron cloud*. Why don't they crash into the nucleus or into one another? The momentum associated with their high speeds keeps them in orbit around the nucleus. At the same time, the electrons in the cloud repel one another because of their like charges. Most of the time, each electron moves within a specific region of space around the nucleus, called an **energy level**. This position maximizes the force of attraction between the electron and the nucleus while minimizing the force of repulsion between the electron and all the other electrons.

Electrons fill an atom's lowest, or first, energy level before any enter the higher energy levels more distant from the nucleus. The lowest energy level in any atom always has a maximum capacity of two electrons. The second energy level can hold a maximum of eight electrons, and succeeding energy levels can each hold eight or more electrons. The number and energy-level positions of some atoms' electrons are shown schematically in Figure 2-7.

How Atoms Bond

To form chemical compounds, atoms combine, or **bond**, by losing, gaining, or sharing one or more electrons during a chemical reaction. The transfer or sharing of electrons between bonded atoms changes the electron configuration of each. Two key factors determine which atoms will unite with which others to form compounds: Each atom should achieve chemical stability, and the resulting compound should be electrically neutral.

An atom has chemical stability when its outermost energy level is filled with electrons. Thus atoms bond by transferring or sharing electrons to fill their outermost energy levels. (For hydrogen and helium atoms, which have only the lowest energy level, this requires two electrons in the outermost energy level; for all other atoms, it requires eight.) Because chemical stability almost always requires eight electrons in the outermost energy level, this is known as the **octet rule**.

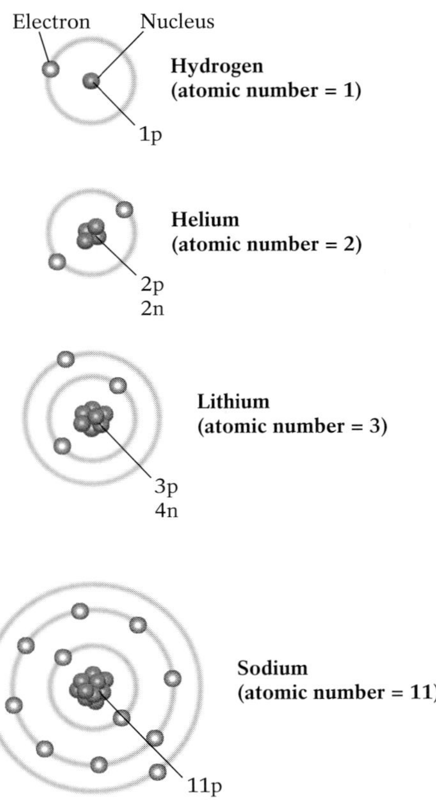

Figure 2-7 Energy-level diagrams of various elements. The nucleus contains the protons (p) and neutrons (n); electrons are shown as balls orbiting the nucleus along concentric circular tracks representing energy levels. (Electron size is exaggerated for clarity; electrons are actually much smaller than protons and neutrons.) Hydrogen and helium, because they have 2 or fewer electrons, have only one energy level; lithium, with 3 electrons has 2 energy levels; sodium, because it has 11 electrons, has three energy levels (2 electrons in the first level, 8 electrons in the second level, and 1 electron in the third level). Note that the number of electrons in an atom always equals the number of protons.

Whether an atom of an element loses outer electrons, gains them, or shares them depends largely on the number of electrons in its outermost energy level. Atoms with only one or two electrons in their outermost energy levels have a strong tendency to give up those electrons. Atoms with six or seven electrons in their outermost energy levels tend to acquire electrons. Atoms with three, four, or five electrons in their outermost energy levels tend to share electrons with other atoms instead of transferring or receiving them. Atoms whose outer energy levels are already filled with eight electrons (or, in the case of helium, which has only the first energy level, with two electrons) are very stable chemically, or *inert*; because they generally do not need to lose, gain, or share electrons, they are far less likely to bond with other atoms.

Elements vary a great deal in the electron configurations of their atoms, so various types of bonding are possible. The atoms that make up the vast majority of the Earth's minerals are most often linked by ionic bonding, covalent bonding, or metallic bonding.

Ionic Bonding An atom does not change its identity when it loses or gains an electron, but it does lose its electrical neutrality and become positively or negatively charged. For example, when a potassium atom loses one electron, it becomes a positively charged particle because it has one more proton than electrons. When a chlorine atom gains a single electron, it becomes a negatively charged particle because it has one less proton than electrons. An atom that has lost or gained one or more electrons to become a charged particle is called an **ion**. A positively charged ion is indicated by the symbol for the element followed by a superscript "+" (for example, K^+). A negatively charged ion is indicated by the symbol for the element followed by a superscript "–" (for example, Cl^-). The number of electrons lost or gained appears as a superscript before the charge (for example, Ca^{2+}).

When an atom with a strong tendency to lose electrons comes into contact with an atom with a strong tendency to gain electrons, they generally transfer electrons so that each atom achieves the chemical stability of a full outer energy level. The donor atom loses one or more electrons and becomes a positively charged ion, and the receiving atom gains one or more electrons and becomes a negatively charged ion. These oppositely charged ions then attract *each other* to form an **ionic bond**. The result is an electrically neutral, chemically stable compound.

Figure 2-8 shows the ionic bonding of sodium and chlorine atoms to form a neutral compound. Sodium has a strong tendency to lose the lone electron in its third, or outer, energy level, thereby completely eliminating that energy level. Bonding turns the sodium atom into a positively charged ion with only two energy levels; the second energy level, now the outer one, is full, with eight electrons. Chlorine has a strong tendency to gain a single electron to add to the seven in its outer energy level, which then becomes filled with eight electrons. Bonding turns the chlorine atom into a negatively charged ion. The neutral compound that results from ionic bonding of sodium (Na) and chlorine (Cl) is sodium chloride (NaCl).

The physical and chemical properties of ionic compounds differ from those of their component elements. Pure sodium, for instance, is a soft silvery metal that reacts vigorously when mixed with water and may even burst into flame. Chlorine usually occurs as Cl_2, a green poisonous gas that was used as a chemical weapon during World War I. Sodium chloride neither ignites nor poisons: It is the common white crystalline mineral halite (the table salt that we sprinkle on french fries)—a substance that regulates some of the biochemical processes essential to all animal life.

Covalent Bonding Atoms whose outer energy levels are about half full, containing three, four, or five electrons, tend to achieve chemical stability by *sharing* electrons with other similarly equipped atoms. In such a case, both atoms fill their outer energy levels with the shared electrons rather than

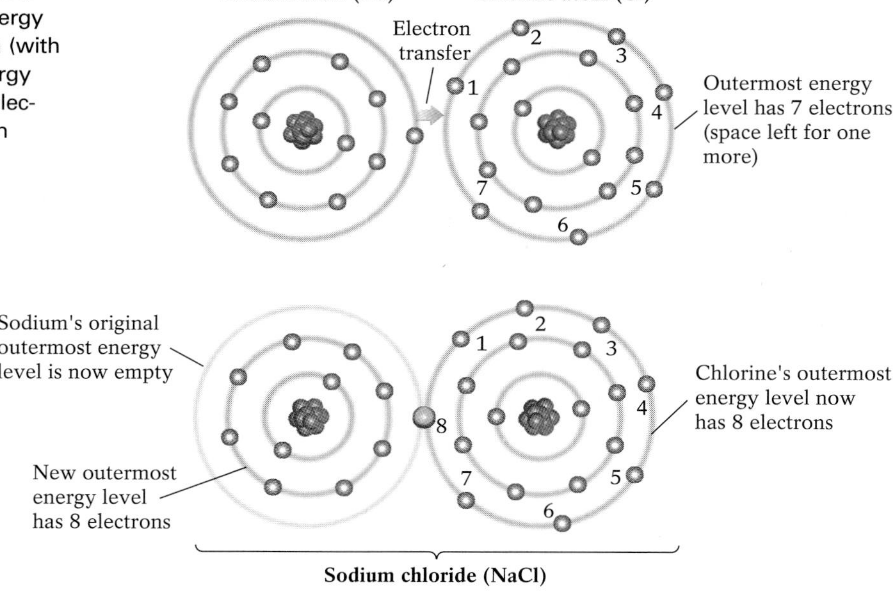

Figure 2-8 Ionic bonding of sodium (Na) and chlorine (Cl). When a sodium atom (with one electron in its outer energy level) donates its outermost electron to a chlorine atom (with seven electrons in its outer energy level), the outer energy level in the new configuration of each atom has eight electrons. The two resulting ions (Na^+ and Cl^-) unite to form sodium chloride (NaCl), a neutral ionic compound.

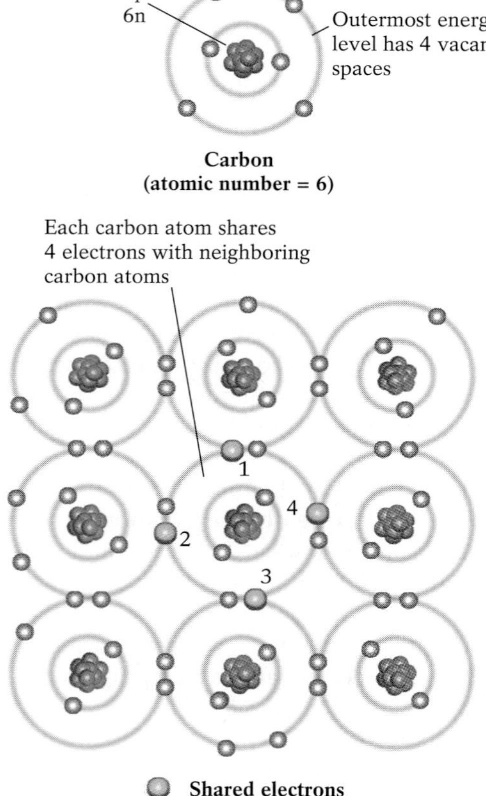

Figure 2-9 Covalent bonding in diamond, a mineral that consists entirely of the element carbon. Each carbon atom in diamond is bonded covalently to four neighboring carbon atoms; the great strength of these bonds accounts for the fact that diamond is the hardest known substance on Earth.

transferring electrons from one to the other. Sharing electrons produces a **covalent bond**, in which the outer energy levels of the atoms overlap. Covalent bonds are generally stronger than any other type of bond. In some cases, two or more atoms of a single element may bond covalently with each other. For example, overlapping carbon atoms bond covalently in the all-carbon mineral, diamond (Fig. 2-9).

Metallic Bonding The atoms of some electron-donating elements tend to pack closely together, with each typically surrounded by either eight or twelve others. This produces a cloud of electrons that roam independently among the positively charged nuclei, unattached to any specific nucleus. The attraction of a negatively charged electron cloud to a cluster of positively charged nuclei is called **metallic bonding**, and is responsible for the properties that define metals. For example, metals are efficient conductors of electricity, which requires freely moving electrons. The mobile electrons of metallically bonded substances explain (among other phenomena) why copper wiring can be used to transmit an electrical current to appliances.

Intermolecular Bonding Another type of bonding affects minerals principally because of the way it bonds compounds that *react* with minerals. Certain compounds (but not minerals) exist as *molecules,* stable groups of bonded atoms that are the smallest particles identifiable as compounds. A prime example is water. One molecule of water, H_2O, consisting of exactly two atoms of hydrogen stably bonded to one atom of oxygen, has all the properties of water, whereas its component elements alone do not.

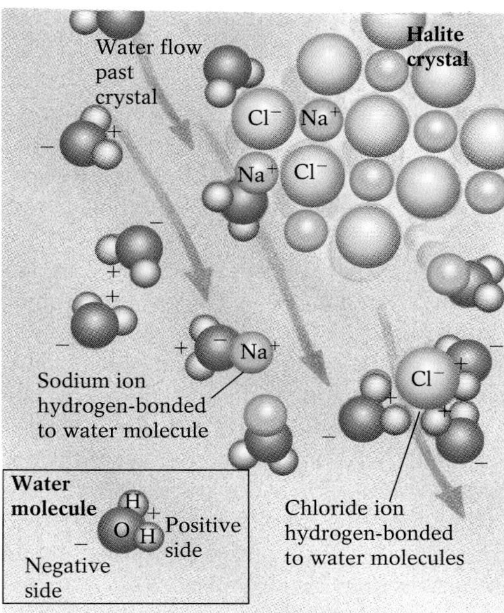

Figure 2-10 Water molecules each have a positively charged side (the hydrogen side) and a negatively charged side (the oxygen side). These regions attract oppositely charged ions from other compounds. The attraction of sodium ions to the negative side of water molecules combined with the attraction of chlorine ions to the positive side causes halite (table salt) to dissolve in water.

The water molecules, which are held together internally by strong covalent bonds, often weakly attach themselves to other molecules by **intermolecular bonds**. This bonding is a result of an uneven distribution of electrons. The positive charge of the oxygen atom's nucleus is greater than that of the hydrogen atoms' nuclei, which means the shared electrons are more attracted to, and spend more time near, the oxygen nucleus. The oxygen side of a water molecule subsequently develops a weak negative charge from the presence of these negatively charged electrons. Because the electrons spend less time near the hydrogen nuclei, a weak positive charge develops on that side of the molecule. These weakly charged regions attract oppositely charged regions of nearby molecules, forming weak **hydrogen bonds** with these molecules.

The formation of hydrogen bonds explains why so many substances, even those that are strongly bonded, dissolve in water. Consider what happens to salt. When halite (NaCl) and water are combined, the positively charged sides of the water molecules attract the chlorine ions and the negatively charged sides attract the sodium ions. Although hydrogen bonds are weak compared to the ionic bonds that hold the Na and Cl together in halite, the combined effect of many water molecules is strong enough to overcome the ionic bonds and split halite into its component Na^+ and Cl^- ions (Fig. 2-10).

Some minerals have many of the same qualities as molecules, especially in an uneven distribution of their charges. These minerals may be subject to a type of intermolecular bond in which a number of electrons momentarily group on the same side of an atom's nucleus, giving that side of the atom a slight negative charge and the electron-poor side a slight positive charge. The positive side may briefly attract electrons of neighboring atoms, and the negatively charged side may attract the nuclei of neighboring atoms. This type of weak intermolecular attraction is called a **van der Waals bond**, after Johannes van der Waals, the Dutch physicist who discovered it in the late nineteenth century. Although weak, van der Waals bonds can be sufficient to bond atoms or layers of atoms together in certain minerals, such as the carbon atoms in graphite.

Hydrogen and van der Waals bonds are much weaker than ionic, covalent, and metallic bonds, and can be broken by the addition of small amounts of heat. For example, the intermolecular bonds between water molecules in ice break when the temperature is raised to 0°C (32°F)—the melting point of ice—changing the water from a solid to a liquid. The strong covalent bonds that form the individual water molecules, however, are unaffected.

Mineral Structure

When a mineral grows unrestricted in an open space, it develops into a regular geometric shape known as a **crystal** (from the Greek *kryos*, or "ice"). This crystal form may be a combination of simple geometric shapes such as cubes, pyramids, and prisms. The geometric shape is the external expression of the mineral's microscopic internal **crystal structure**, the orderly arrangement of its ions or atoms into a latticework of repeated three-dimensional units. Every crystal of a given mineral has the same crystal structure, as can be seen by the consistent orientation of the planar surfaces, or *faces*, in well-formed crystals.

More often than not in nature, minerals grow in restricted regions, often competing with other growing minerals for the available space. In this circumstance, they form interlocking masses of small, irregular grains that bear little resemblance to well-formed mineral crystals. Although these grains do not develop the precise geometric shape of a crystal, their internal crystal structure does remain characteristically regular.

Sometimes molten rock cools too rapidly for its atoms to form even a semblance of an orderly arrangement. The resulting solid, called *glass*, does not qualify as a true mineral: It may have a relatively constant composition, but it lacks a specific crystal structure. Geologists call glass and other similar mineral-like substances **mineraloids**. Obsidian, a natural glass, is one type of mineraloid. Like manufactured glass, obsidian is formed by the rapid cooling of a mixture of molten silica (SiO_2) and small quantities of one or more other

Table 2-1 The Most Abundant Elements in the Earth's Continental Crust

Element	Proportion of Crust's Weight (%)
Oxygen (O)	45.20
Silicon (Si)	27.20
Aluminum (Al)	8.00
Iron (Fe)	5.80
Calcium (Ca)	5.06
Magnesium (Mg)	2.77
Sodium (Na)	2.32
Potassium (K)	1.68
	98.03
Other elements	1.97
Total	100.00

elements. The windowpanes of very old buildings show that glass is essentially a cooled fluid. Over time the window glass gradually flows until the lower portion of each pane is noticeably thicker than the top.

Determinants of Mineral Formation

The kinds of minerals that form in a particular time and place depend on the relative abundances of the available elements, the relative sizes and other characteristics of those elements' atoms and ions, and the temperature and pressure at the time of formation.

Only eight of the 90 naturally occurring elements in the Earth's continental crust are relatively abundant, with oxygen and silicon dominating (Table 2-1). In the entire Earth, iron and oxygen dominate, silicon is fairly abundant, and magnesium is more significant than it is in the crust (Table 2-2). Most minerals in the crust are oxygen-and-silicon–based compounds, and most of those in the upper mantle are oxygen-silicon-iron-magnesium–based compounds.

Table 2-2 The Most Abundant Elements in the Entire Earth

Element	Proportion of Earth's Weight (%)
Iron (Fe)	34.8
Oxygen (O)	29.3
Silicon (Si)	14.7
Magnesium (Mg)	11.3
Sulfur (S)	3.3
Nickel (Ni)	2.4
Calcium (Ca)	1.4
Aluminum (Al)	1.2
	98.4
Other elements	1.6
Total	100.00

Given two elements of equal abundance, the element that will contribute more readily to mineral formation is the one that "fits" better with the other elements present. Atoms and ions in minerals tend to become packed together as closely as their sizes permit. In an ionically bonded mineral, each positive ion attracts as many negative ions as can fit around it and each negative ion does the same with positive ions. Thus their relative sizes determine how many negative ions will surround a positive ion and vice versa. In the mineral halite (NaCl), for example, the relative sizes of the positive sodium ions (Na^+) and the negative chlorine ions (Cl^-), dictate that one sodium ion is always surrounded by six chlorine ions (Fig. 2-11a). Positive ions are generally smaller than negative ions (Fig. 2-11b) because they lose electrons from their outermost energy level, and so lose that energy level (see Fig. 2-8). *In a crystal structure, therefore, smaller positive ions usually occupy the spaces between larger negative ions.*

Ionic Substitution Imagine building a wall of red bricks but running short of them partway through. If you substitute yellow bricks of the same size, the appearance of the wall will be altered but its stability will not be affected. The same is true of **ionic substitution**, when certain ions of similar size and charge replace one another within a crystal structure, depending on which is most available during the mineral's formation. As a result of ionic substitution, some minerals that have the same internal arrangement of ions may vary in composition.

Iron (Fe^{2+}) and magnesium (Mg^{2+}), which are nearly identical in size and charge, substitute freely for one another in the mineral olivine, $(Fe,Mg)_2SiO_4$. (Note: In a mineral's chemical formula, the elements that can substitute for one another in a mineral's crystal structure appear in parentheses and are separated by commas.) The color, melting point, and other physical characteristics of olivine differ depending on whether Fe^{2+} or Mg^{2+} predominates, but its chemical stability and crystal structure are unaffected.

Polymorphism Two minerals may have the same chemical composition but different crystal structures because they formed under different temperature and pressure conditions. Such minerals are known as **polymorphs** (Greek for "many forms"). Graphite (the "lead" in your pencil) and diamond, for example, are polymorphs that consist entirely of carbon. Graphite's structure forms under the low pressure prevalent at shallow depths, whereas diamond's structure results from intense pressure at depths greater than 150 kilometers (about 90 miles). Theoretically, from the moment it reaches the surface, diamond should begin to change into graphite, the carbon polymorph that is more stable in the Earth's low-

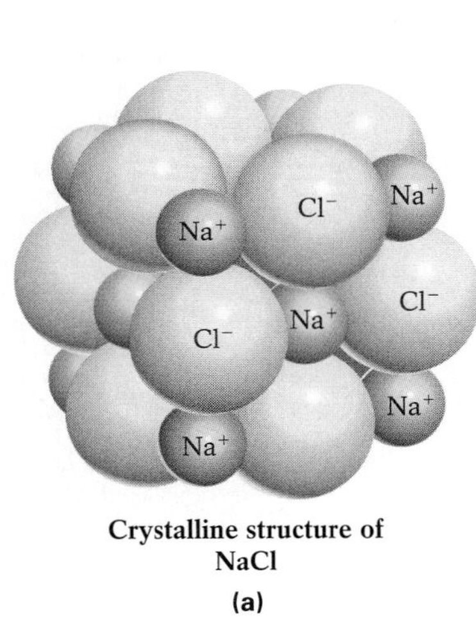

Ionic Charge
+4
+3
+2
+1
−1
−2
C
Si
Mn
Fe^{3+}
Al
Fe^{2+}
Ca
Pb
Mg
K
Li
Na
Cl
F
O
S
Increasing ionic diameter
(b)

Figure 2-11 **(a)** The crystal structure of halite (NaCl), in which small sodium ions (Na^+) are tucked in between larger chlorine ions (Cl^-). **(b)** The relative diameters of ions of the Earth's most common elements. Note that negatively charged ions are generally larger than positively charged ions, which have lost an energy level.

temperature/low-pressure surface environment. Fortunately for owners of diamond jewelry, a great number of strong carbon–carbon bonds must be broken before a diamond becomes graphite. At room temperature, this conversion is too slow to be measured and diamonds are effectively stable. Thus, in terms of human longevity, as advertised, diamonds *are* "forever."

How Minerals Are Identified

A mineral's chemical composition and crystal structure give it a unique combination of chemical and physical properties that distinguish it from all other minerals. Geologists can observe many of these properties readily in the field, whereas other properties require some bulky or sophisticated equipment and can be studied only in laboratories. We can seldom identify a mineral accurately on the basis of only one property; usually several must be established before an identification is considered conclusive.

In the Field

Most geologists and dedicated rockhounds can identify a great many minerals in the field by examining them with the naked eye and performing some very simple tests.

Color Color may be the first thing you notice about a mineral, but it is perhaps the least reliable identifying characteristic. Different-colored minerals may contain the same elements (as is the case with the polymorphs diamond and graphite), whereas minerals that are similar in color may be completely different in composition. As a result, we cannot determine the identity of an unknown mineral on the basis of its color alone.

A mineral's color depends on how much light its chemical makeup and its crystal structure cause it to absorb. When white light, which contains all the colors of the spectrum, strikes a mineral, part of the spectrum is absorbed and the remainder is reflected back from the mineral and perceived by the eye. For example, when we see an emerald, only the green part of the spectrum is transmitted; the mineral absorbs the rest of the colors of the spectrum. Certain elements, such as iron, manganese, chromium, and nickel, absorb a great deal of light and transmit little; minerals containing these elements are commonly dark in color or black. Elements such as sodium, potassium, calcium, and silicon absorb little light; minerals containing them are characteristically light in color.

A few alien atoms in a mineral's structure may completely change its color. For example, pure corundum (Al_2O_3) is generally white or light gray, but even a trace of the element chromium produces a brilliant red variety of corundum called *ruby*. The addition of a little titanium and iron produces another variety of corundum, a deep blue *sapphire*.

(a)

(b)

Figure 2-12 Variations in the color of minerals. **(a)** These two mineral samples have one thing in common, although you'd never guess it from their colors—they are both quartz, composed of silicon dioxide (SiO_2). Their colors differ because they each contain minute traces of different impurities. **(b)** Variations in color within a single crystal of a variety of tourmaline called elbaite (or "watermelon" tourmaline). The color of the crystals of this mineral, typically including pink and green, derives from impurities incorporated into their structures during formation.

Depending on the types of impurities they contain, different samples of a given mineral may exhibit any number of different colors (Fig. 2-12).

Heating or exposure to radiation displaces some electrons, atoms, or ions from their designated sites in a crystal structure and can cause significant color variations. Unprincipled jewel merchants have been known to "improve" the colors of poor-quality gems with heat or X-rays. Those colors may fade, however, as the displaced ions and atoms gradually return to their original sites within the crystal structure.

Because a mineral's color is rarely unique to that mineral and because a mineral's true color can be "disguised" in so many ways, color is not by itself a reliable diagnostic tool for mineral identification.

Luster **Luster** describes how a mineral's surface reflects light. Minerals can exhibit metallic or nonmetallic luster. When any light shines on a metal, its energy stimulates the metal's loosely held electrons and causes them to vibrate. The vibrating electrons emit a diffuse light, giving metallic surfaces their characteristic shiny luster (Fig. 2-13). Thus metals, with their great number of "free" electrons, often display a shiny, metallic luster. Nonmetallic lusters are more varied; they can be pearly, earthy, silky, vitreous (glassy), or adamantine ("like a diamond"), (Fig. 2-14).

Streak **Streak** is the color of a mineral in its powdered form. We can observe streak when we pulverize the mineral's sur-

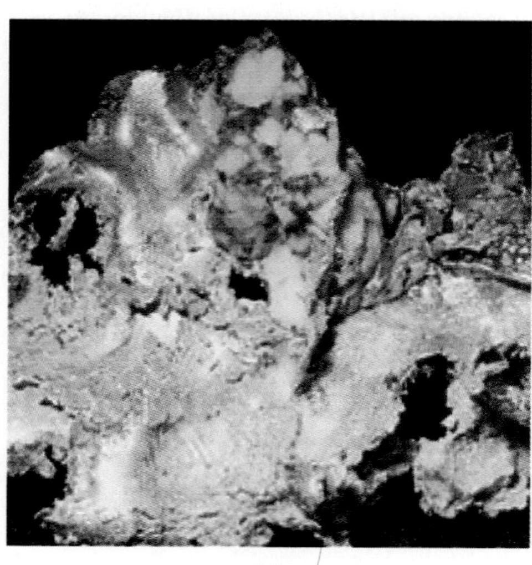

(a)

(b)

Figure 2-13 Gold **(a)** and silver **(b)** exhibit the shiny luster characteristic of metals.

(a) (b) (c) (d) (e)

Figure 2-14 Nonmetallic lusters can be: **(a)** pearly, as in gypsum; **(b)** earthy, as in realgar; **(c)** silky, as in chrysotile; **(d)** vitreous (glassy), as in rose quartz; or **(e)** adamantine, as in anglesite.

face by rubbing it across an unglazed porcelain slab known as a streak plate. The color of a mineral's streak often differs from that of the intact mineral sample. For instance, the mineral hematite (Fe_2O_3), typically steel-gray in color, has a distinctive reddish-brown streak (Fig. 2-15). Streak is often a more accurate indicator of mineral identity, because trace impurities in the sample do not affect it.

Streak can distinguish between similar-appearing minerals that may have considerably different economic values. Pyrite (FeS_2), often called "fool's gold" because of its brassy yellow color, has a black streak, in contrast with the golden-yellow streak of true gold. Several similar-appearing minerals have similar streaks as well, however; additional tests are necessary to identify these confidently.

Figure 2-15 Although many samples of hematite (Fe_2O_3) are steel-gray, the streaks of all samples are reddish-brown.

Hardness Geologists define **hardness** as a mineral's resistance to scratching or abrasion. They test a mineral's hardness by scratching it with a series of minerals or other substances with known hardnesses. (Geological hardness is *not* a function of how easily a mineral breaks—a solid rap with a hammer will easily shatter a diamond, the world's hardest natural substance.) Because every scratch removes atoms from the surface of the mineral, and thus breaks the bonds holding these atoms, a mineral's hardness indicates the relative strength of its bonds.

The Mohs Hardness Scale, named for its developer, German mineralogist Friedrich Mohs (1773–1839), assigns relative hardnesses to several common and a few rare and precious minerals. An unknown mineral that can be scratched by topaz but not quartz has a hardness between 7 and 8 on the Mohs scale. Table 2-3, which lists the minerals and common testing standards used in the Mohs scale, explains why geologists are often found with a few copper pennies, a pocketknife, and well-worn fingernails.

The hardness of a mineral depends on the strength of its weakest bonds. Thus minerals with many covalent bonds are generally harder than those with ionic or other bonds. Because every atom of carbon in a diamond forms a strong covalent bond with four neighboring carbon atoms (see Fig. 2-9),

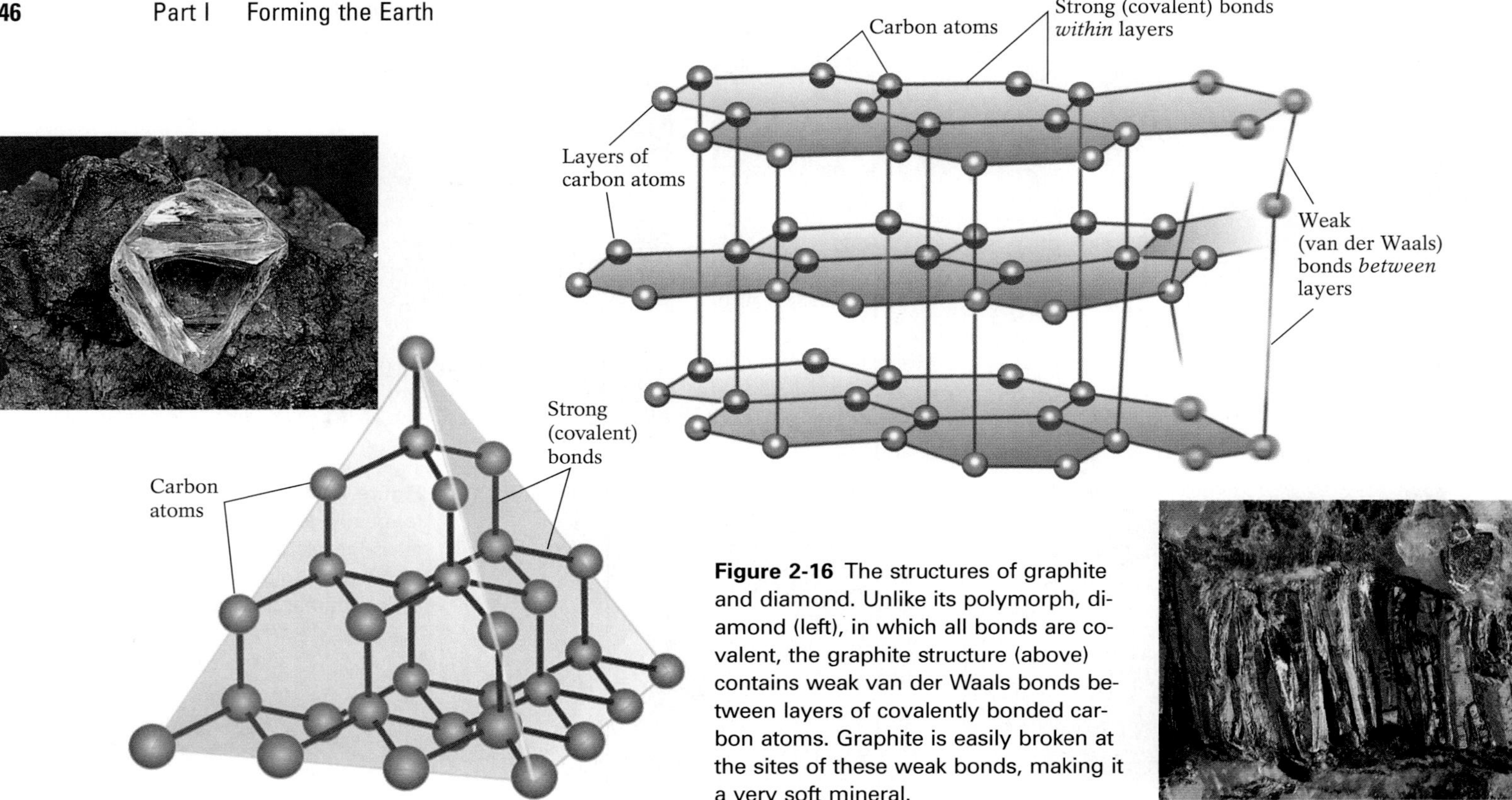

Figure 2-16 The structures of graphite and diamond. Unlike its polymorph, diamond (left), in which all bonds are covalent, the graphite structure (above) contains weak van der Waals bonds between layers of covalently bonded carbon atoms. Graphite is easily broken at the sites of these weak bonds, making it a very soft mineral.

diamond is exceptionally hard and durable. Graphite, diamond's polymorph, is one of the softest minerals. Its sturdy layers of covalently bonded carbon atoms are only weakly bonded to one another by van der Waals forces (Fig. 2-16). When you write with a pencil, pressure on the point breaks these weak bonds, leaving a trail of carbon layers on the paper.

Table 2-3 The Mohs Hardness Scale

Mineral	Hardness	Hardness of Some Common Objects
Talc	1	
Gypsum	2	
		Human fingernail (2.5)
Calcite	3	
		Copper penny (3.5)
Fluorite	4	
Apatite	5	
		Glass (5–6), Pocketknife blade (5–6)
Orthoclase (potassium feldspar)	6	
		Steel file (6.5)
Quartz	7	
Topaz	8	
Corundum	9	
Diamond	10	

Cleavage **Cleavage** is the tendency of some minerals, when hammered or struck, to break consistently along distinct planes in their crystal structures where the bonds are weakest, or fewer in number. Such breaks form smooth, flat surfaces, called *cleavage planes,* on the mineral. When a mineral that tends to cleave is struck along a plane of cleavage, every fragment that breaks off will have the same general shape.

In characterizing cleavage, geologists consider the number of cleavage surfaces produced and the angles between adjacent surfaces. Halite, for example, cleaves in three mutually perpendicular directions, forming cubes, rectangles, and stepped shapes with 90° angles between the fragments' sides (Fig. 2-17a), whereas mica cleaves in only one direction, forming sheets (Fig. 2-17b). Calcite's ($CaCO_3$) three cleavage planes are not perpendicular to one another (Fig. 2-17c).

Minerals that appear similar by other diagnostic criteria can have different cleavage-plane angles. For example, augite and hornblende are two common black minerals that are similar in external form, hardness, and other characteristics. When a crystal of each is broken, however, the two prominent cleavage planes of augite intersect at an angle of about 90°, whereas the two cleavage planes of hornblende intersect at about 120° and 60° angles. The different cleavage angles reflect the different arrangements of atoms and strengths of bonds within the crystal structures of these minerals (Figs. 2-17d and 2-17e).

The cleavage planes in diamonds and certain other minerals enable gem cutters to fashion beautiful jewels from raw,

(a) (b) (c) (d) (e)

Figure 2-17 Distinguishing minerals by their cleavage. **(a)** Halite has three mutually perpendicular cleavage planes, forming cubes, rectangles, and stepped shapes. **(b)** Mica has one perfect cleavage plane, forming sheets. **(c)** Calcite's three cleavage planes are not mutually perpendicular; calcite cleavage produces a geometrical shape called a rhombohedron. **(d)** A photomicrograph of cleavage in augite shows its planes intersecting at nearly 90° angles. **(e)** A photomicrograph of cleavage in hornblende shows its planes intersecting at angles of about 60° and 120°.

uncut stones. Highlight 2-1 explains how valuable specimens of the Earth's hardest substance can be cut without destroying them.

Fracture Minerals that do not cleave—because their bonds are equally strong in all directions and are distributed uniformly throughout the crystal—will break at random, or **fracture**. Unlike a straight, smooth-faced cleavage plane, a fracture appears as a jagged irregular surface or as a curved, shell-shaped (*conchoidal*) surface. In the mineral quartz, composed exclusively of silicon and oxygen, all of the atoms bond covalently in a three-dimensional framework with equal bond strengths in all directions. Thus, when you strike a crystal of quartz, it fractures (Fig. 2-18). Because quartz fractures instead of cleaving, we can distinguish it from similar-looking minerals that cleave. Some minerals, such as pyroxene, may cleave in one or more directions and fracture in others.

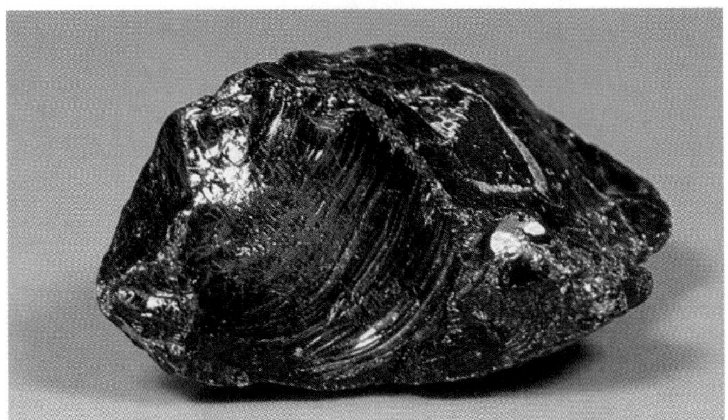

Figure 2-18 A conchoidal fracture surface on a quartz crystal. Quartz, with equally strong covalent bonds in all directions, has no planes of weakness. It therefore fractures irregularly instead of cleaving.

Highlight 2-1 Cutting Diamonds

Not long ago, a television commercial for a luxury automobile featured a diamond cutter seated in the rear of a large sedan, anxiously preparing to cut a huge diamond. One false move, or a jarring bump from inferior shock absorbers, and a precious stone would be shattered. The diamond cutter's anxiety would certainly be warranted in such a situation—diamond cutting is a difficult and exacting skill, requiring steady hands and vision and finely honed tools, as well as a precise knowledge of diamonds' crystal structure.

Actually, cutters have traditionally cleaved (not cut) diamonds along planes within the crystal that have fewer bonds across them per unit area. Knowing the sites of these planes, an experienced diamond cutter can cleave a raw, irregularly shaped diamond into a perfectly symmetrical jewel. A master cutter may analyze a large stone for days or even weeks to determine which cleavage planes would maintain the largest gem while minimizing any visible imperfections. X-rays can locate cleavage planes with precision, and they are sometimes used, but the trained eye of an experienced diamond cutter is usually sufficient.

Diamond can be cleaved in four directions. Thus a diamond cutter can choose among many complex combinations of cleavage faces, or *facets* (Fig. 2-19). The cutter scratches the locations of the crucial planes on the raw stone with a diamond-tipped stylus (what else could scratch a diamond?), and then a sharp blade is positioned at each scratch and struck forcefully with a mallet. The stone cleaves smoothly, exposing the chosen planes. If a plane has been marked inaccurately, or if the blade is off the mark, the stone will shatter instead of cleaving. Because of the great value of diamonds and the unpleasant financial consequences of a cleavage error, diamonds today are also sawed and then ground smooth, a slower but safer way to cut a stone. By using ultrafine saw blades with cutting edges impregnated with diamond dust and rotating at high speeds, a cutter can gradually saw or grind a diamond's surface into the desired shape.

After a diamond has been cut, its flat cleavage surfaces are polished with the only substance hard enough to abrade a diamond, a paste of pulverized industrial-quality diamonds mixed with olive oil.

During cutting, a diamond may lose as much as half its size and weight. It will, however, gain in value: A well-cut diamond will have scores of polished surfaces that admit light, disperse the light into the various colors of the spectrum, and then redirect the light to the viewer's eye as brilliant flashes of color.

(a)

(b)

Figure 2-19 Diamond before and after cutting. **(a)** A raw, uncut diamond. **(b)** A cut and faceted diamond.

Smell and Taste Experienced geologists occasionally sniff and lick rocks to identify minerals with a distinctive smell or taste—not always a pleasant task. Some sulfur-containing minerals, when treated with dilute hydrochloric acid (HCl), emit the familiar rotten-egg stench associated with hydrogen sulfide gas (H_2S). Halite's salty taste distinguishes it from similar-looking minerals such as calcite; sylvite (KCl) is distinctively bitter. Kaolinite absorbs liquid rapidly—when licked, it absorbs saliva and sticks to the tongue. But taste unknown minerals with caution: Realgar and orpiment, for example, which smell like garlic (especially when heated), have as their major element the poisonous metal arsenic.

Effervescence Certain minerals, particularly those that contain carbonate ions (CO_3^{2-}), *effervesce,* or fizz, when mixed with an acid. A few drops of dilute hydrochloric acid (HCl) on calcite ($CaCO_3$) produce a rapid chemical reaction that releases carbon dioxide gas, in the form of bubbles, and water. This helps to distinguish calcite from similar-looking minerals such as halite, which does not effervesce.

Crystal Form Because the three-dimensional geometric form of a crystal is an external expression of a mineral's internal structure, the shape of a well-formed crystal may be distinctive enough to identify the mineral. For example, the angles between adjacent faces of a given mineral crystal are the same in every well-formed, unbroken sample of that mineral. Thus the adjacent faces on a perfect quartz crystal from Herkimer, New York (a good place to find them), are at angles of 120°, the same as on perfect quartz crystals from Hot Springs, Arkansas (another good source).

The crystals of some minerals are not always geometrically shaped. Instead, some crystals may grow together into forms resembling distinctive nonmineral objects, making the minerals easy to identify. A rosette shape characterizes barite ($BaSO_4$); botryoidal malachite ($Cu_2(OH)_2CO_3$) looks like "a bunch of grapes," from which it gets its name (derived from the Greek *botruoeides*); stellate pyrite (FeS_2) resembles a collection of stars (Fig. 2-20).

In the Laboratory

So many minerals have similar physical characteristics that even experienced geologists can only make an educated guess as to their identity. They will then bring samples back from the field to test them in the laboratory. There, they can use special equipment to analyze a variety of physical and chemical properties with greater precision than is ever possible in the field. Figure 2-21 shows a sampling of the methods geologists use to identify minerals in the field and in the laboratory.

Specific Gravity **Specific gravity** is the ratio of a substance's weight to the weight of an equal volume of pure water. For example, a mineral that weighs four times as much as an equal volume of water has a specific gravity of 4. A cubic centimeter of quartz (SiO_2), with a specific gravity of 2.65, is markedly lighter in weight than an equal volume of galena

(a)

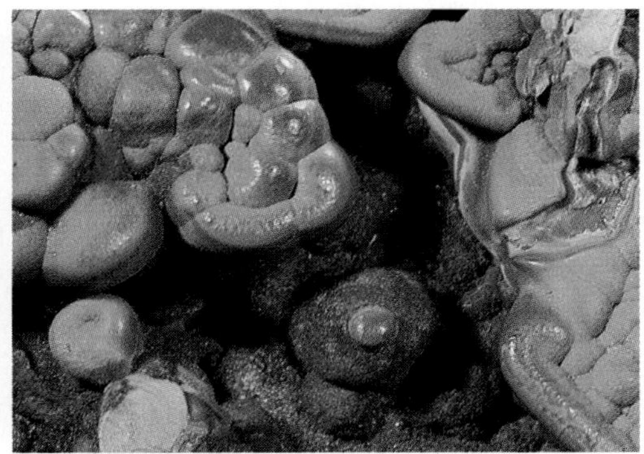

(b)

(c)

(d)

Figure 2-20 Unusual crystal aggregates. **(a)** Barite rosettes; **(b)** botryoidal malachite (green); **(c)** stellate pyrite; **(d)** stibnite needles.

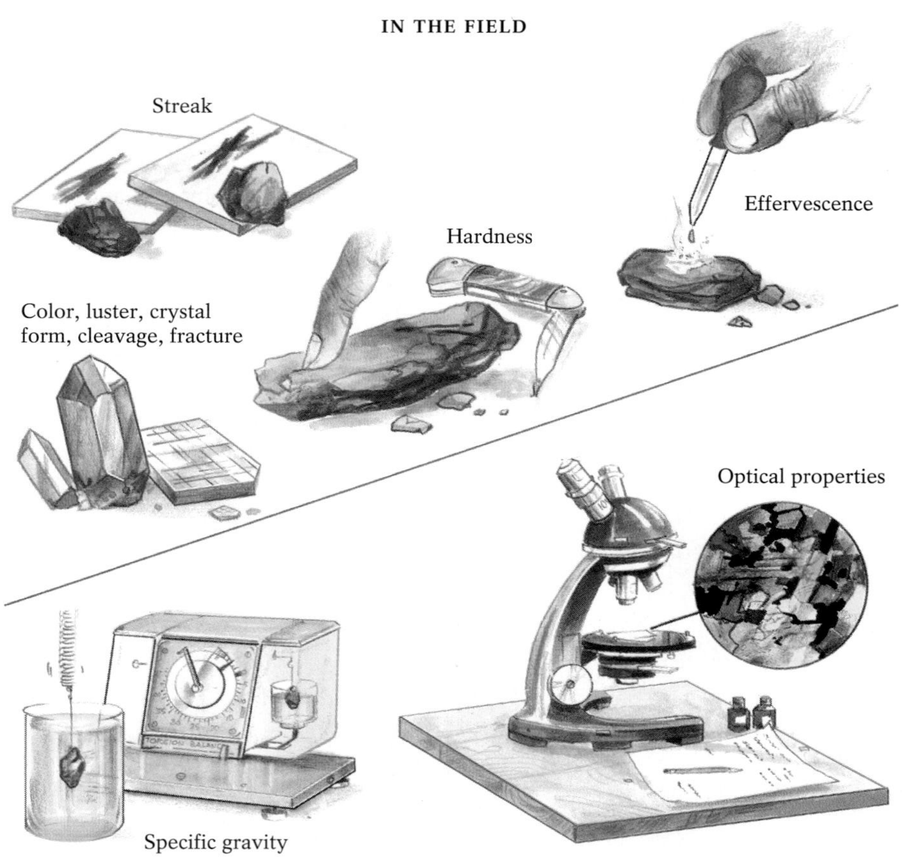

Figure 2-21 Several analyses—some simple and some complicated—may help determine a mineral's identity.

(PbS), with a specific gravity of 7.5. A mineral's specific gravity is essentially the same as its *density*.

Specific gravity can help distinguish between two apparently similar minerals. The specific gravity of 24-karat (pure) gold is 19.3, whereas the specific gravity of pyrite ("fool's gold") is 5. Gold's high specific gravity is the reason prospectors can "pan" for it. When they slosh river sand about in a pan, the heavier gold particles sink to the bottom, while lighter particles, usually of such common minerals as quartz and feldspar, readily wash away.

Mineral polymorphs, with their different atomic arrangements, usually have different specific gravities. Even though both consist exclusively of carbon atoms, graphite has a specific gravity of 2.3, whereas diamond's is 3.5, because of its highly compressed crystal structure.

Other Laboratory Tests A *petrographic microscope* transmits different kinds of light through a thin (0.03-millimeter) slice of a mineral sample. Geologists can identify most minerals by their colors under specific light conditions (such as under polarized or unpolarized light, as in Figure 2-22).

When exposed to ultraviolet light, certain minerals glow in distinctive colors. This property, called **fluorescence**, characterizes fluorite, calcite, scheelite ($CaWO_4$), and willemite (Zn_2SiO_4), among other minerals (Fig. 2-23). A mineral that continues to glow after the ultraviolet light has been removed is said to exhibit **phosphorescence**.

In **X-ray diffraction**, X-rays passed through a mineral sample become scattered, or *diffracted*, into distinctive patterns. The arrangement of the atoms and ions in a mineral's crystal structure determines these patterns, which are unique to each mineral (Fig. 2-24).

Some Common Rock-Forming Minerals

Rock-forming minerals are those that compose the common rocks of the Earth's crust and mantle. Five groups of minerals predominate. Most are *silicates*, which contain silicon, oxygen, and usually one or a few other common elements. Important nonsilicates include *carbonates*, containing car-

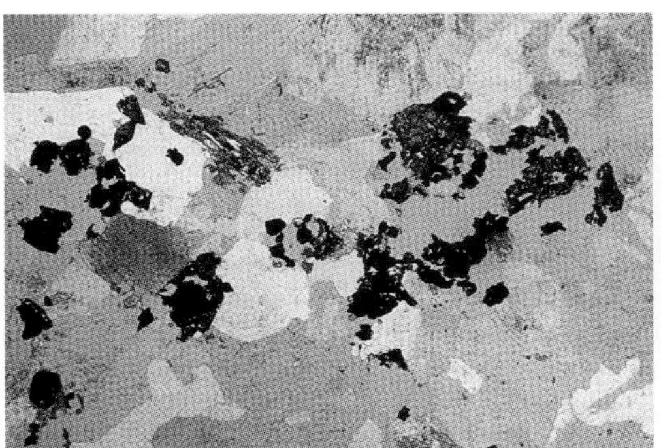

Figure 2-22 The colors that appear in a mineral sample while it is magnified under polarized light can be used to identify the mineral. (left) A thin section of granite as seen in regular light (magnified 70×). (right) The same section as seen under polarized light.

(a)

(b)

Figure 2-23 Minerals that glow in distinctive colors when exposed to ultraviolet light are said to be *fluorescent.* The minerals willemite and calcite in this rock specimen **(a)** glow bright green and red, respectively, when exposed to UV light **(b)**.

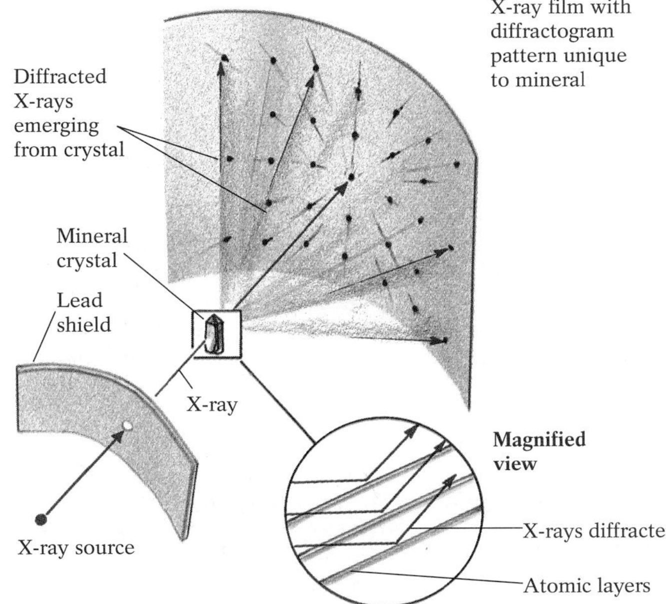

Figure 2-24 Mineral identification by X-ray diffraction. X-rays passed through a mineral scatter in a manner that is unique to each mineral. A *diffractogram* is the photographic record of the pattern of X-rays produced by a mineral.

bon, oxygen, and other atoms; *oxides*, containing oxygen and other atoms; *sulfates*, containing sulfur, oxygen, and various other atoms; and *sulfides*, containing sulfur and other atoms but no oxygen.

The Silicates and Their Structures

Because silicon and oxygen are so abundant and unite so readily, **silicates** make up more than 90% of the weight of the Earth's crust. (See Table 2-1.) Silicates are the dominant component of most crustal rocks, whether igneous, sedimentary, or metamorphic.

Why oxygen and silicon? Oxygen, the primary element in the Earth's crust, is the only common crustal element whose atoms readily accept electrons to form negative ions. Silicon, the second most abundant element in the Earth's crust, is very compatible with oxygen: Its small, positively charged ion fits snugly in the niches among large, closely packed oxygen ions. The crystal structure of all silicates contains repeating groupings of four negative oxygen ions congregated around a single positive silicon ion to form a four-faced structure called a *tetrahedron* (Fig. 2-25).

The four oxygen ions, each with a –2 charge, have a combined negative charge of –8; the one silicon ion has a +4 charge. The result is a **silicon-oxygen tetrahedron** (SiO_4^{4-}) with a –4 charge. To form an electrically neutral compound, a silicon-oxygen tetrahedron must either acquire four positive charges or disperse its negative charge. Positive ions of another element may bond with the tetrahedron to balance its charge, or adjacent tetrahedra may share oxygen ions, claiming half of their negative charge.

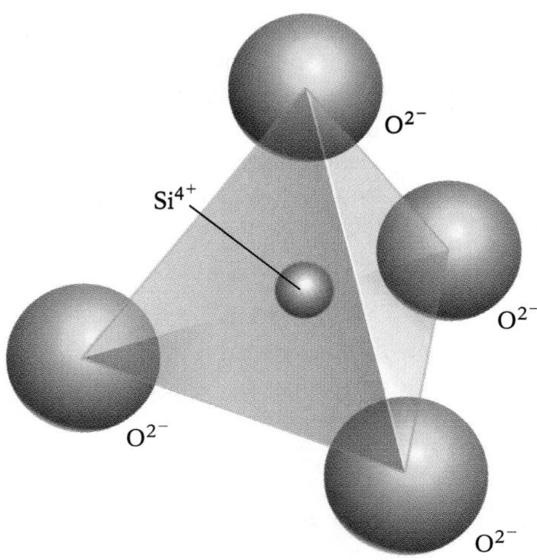

Figure 2-25 The silicon-oxygen tetrahedron. Four oxygen ions occupy the corners of this structure, with a lone silicon ion embedded in the open space at the center. (If you have access to four basketballs and a racquetball, you can easily re-create the SiO_4 tetrahedron.) The silicon-oxygen tetrahedron has an overall charge of –4.

The number of oxygen ions shared by a tetrahedron, expressed as a silicon-to-oxygen ratio, determines the type of silicate and its crystal structure. A tetrahedron that does not share its oxygen ions has a silicon-to-oxygen ratio of 1:4, whereas a tetrahedron that shares all four of its oxygen ions has a silicon-to-oxygen ratio of 1:2 (see Fig. 2-26).

The Earth's crust contains more than a thousand different silicate minerals, testimony to the ease with which SiO_4 tetrahedra combine in nature with various positive ions and with each other. There are five principal silicate crystal structures: *independent tetrahedra, single chains, double chains, sheets,* and *three-dimensional frameworks* (Fig. 2-26a–e). Each structure represents a different means of sharing oxygen ions, and each has its own silicon-to-oxygen ratio. Consequently, each displays distinctive physical characteristics and properties (Table 2-4).

Independent Tetrahedra Independent tetrahedra, bonded with positive ions of other elements and sharing no oxygen ions (Fig. 2-26a), form several prominent silicates. Olivine, for example, contains positive iron and/or magnesium ions whose +2 charges balance the –4 charge of the tetrahedron. Silicates

Table 2-4 Common Rock-Forming Silicates

Silicate	Formula	Silicon-to-Oxygen Ratio* (silicate structure)	Properties
Quartz	SiO_2	1:2 (framework)	Hardness of 7; breaks by fracture; six-sided prismatic crystals; specific gravity 2.65
Alkali feldspars	$KAlSi_3O_8$	1:2 (framework)	Hardness of 6.0–6.5; strong cleavage in two directions at right angles; pink or white in color; specific gravity 2.5–2.6
Plagioclase feldspars	$(Ca,Na)AlSi_3O_8$	1:2 (framework)	Hardness of 6.0–6.5; strong cleavage in two directions at right angles; white to bluish-gray in color; specific gravity 2.6–2.7
Muscovite mica	$K_2Al_4(Si_6Al_2O_{20})(OH,F)_2$	1:2.5 (sheet)	Hardness of 2–3; perfect cleavage in one direction; colorless and transparent to light green-gray; specific gravity 2.8–3.0
Biotite mica	$K_2(Mg,Fe)_6Si_3O_{10}(OH)_2$	1:2.5 (sheet)	Hardness of 2.5–3.0; perfect cleavage in one direction; black to dark brown in color; specific gravity 2.7–3.2
Amphiboles	$(Na,Ca)_2(Mg,Al,Fe)_5(Si,Al)_8O_{22}(OH)_2$	1:2.75 (double chain)	Hardness of 5–6; cleaves in two directions at 56° and 124°; black to dark green in color; specific gravity 3.0–3.3
Pyroxenes	$(Mg,Fe,Ca,Na)(Mg,Fe,Al)Si_2O_6$	1:3 (single chain)	Hardness of 5–6; cleaves in two directions at about 90°; black to dark green in color; specific gravity 3.1–3.5
Olivine	$(Mg,Fe)_2SiO_4$	1:4 (independent tetrahedra)	Hardness of 6.5–7.0; green in color; breaks by fracture; specific gravity 3.2–3.6

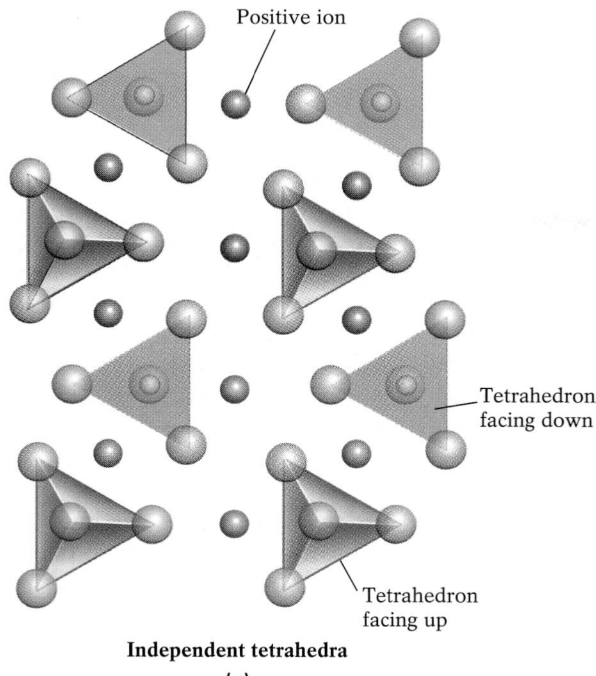

Figure 2-26 Types of silicate structures. **(a)** Independent tetrahedra. Positive ions are positioned between tetrahedra such that each tetrahedron, with a –4 charge bonds to two positive ions, each with a +2 charge, thereby neutralizing their combined charge. No oxygen ions are shared between the tetrahedra; therefore the silicon-to-oxygen ratio is 1:4. Photo: Olivine. As is typical of silicates with independent tetrahedral structures, this mineral fractures rather than cleaving.

with independent tetrahedra are noted for their hardness (6.5–8 on the Mohs scale), a consequence of the strong ionic bonds between the tetrahedra and the interspersed positive ions. Independent tetrahedra always maintain a silicon-to-oxygen ratio of 1:4.

Single Chains A tetrahedron that shares two corner oxygen ions and still has an excess –2 charge forms a linear chain of tetrahedra (Fig. 2-26b). The accumulated negative charges cause the chain itself to act as a negative ion complex, attracting positive ions that neutralize its negative charge and bind adjacent chains. Because the bonds *within* chains are strong and those *between* chains are relatively weak, single-

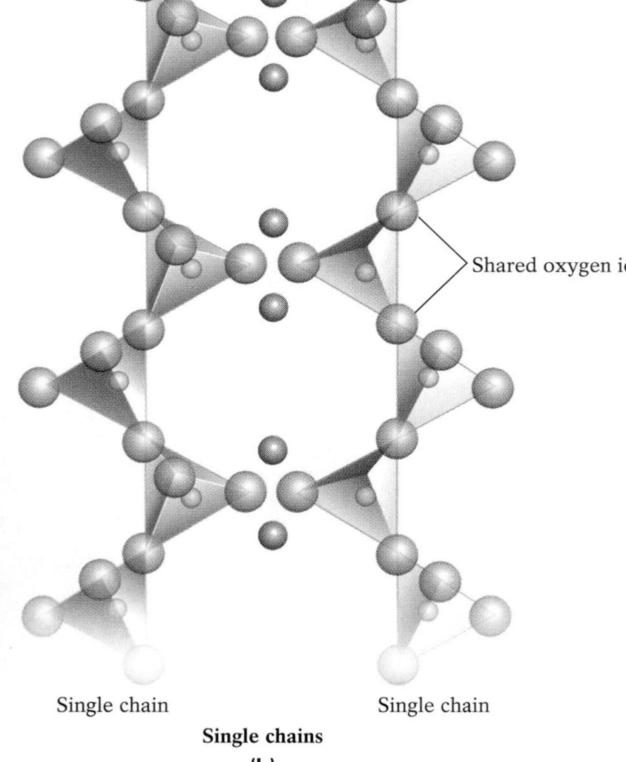

Figure 2-26 Types of silicate structures, continued. **(b)** Single chains. Each tetrahedron shares two of its corner oxygen ions with adjacent tetrahedra, forming a linear chain of tetrahedra with a silicon-to-oxygen ratio of 1:3. Because each tetrahedron still has a –2 charge after sharing two of its oxygen ions, the cumulative negative charge on such chains attracts a variety of positive ions that bind between them and neutralize the negative charge of the chains. Photo: Pyroxene crystals.

chain silicates tend to cleave parallel to the chains. A prominent group of single-chain silicates are the *pyroxenes,* which typically contain positive calcium, iron, and magnesium ions that bind the negatively charged chains together. This pattern of oxygen-sharing produces a silicon-to-oxygen ratio of 1:3.

Double Chains A double chain forms when adjacent tetrahedra share two corner oxygens in a linear chain, and, in addition, some share a third oxygen with a tetrahedron in a neighboring chain (Fig. 2-26c). This shared oxygen binds the two chains together. Because they share more oxygen than do single chains, double chains have a silicon-to-oxygen ratio of 1:2.75. Any of several positive ions (commonly Na, Ca, Mg, Fe, and Al) may link double chains to adjacent double chains, producing the complex silicates known as the *amphiboles.* The most common amphibole is *hornblende,* in which aluminum substitutes for some of the silicon within the tetrahedron because their ions are similar in size (see Fig. 2-11b) and charge.

Hornblende's relatively high iron and/or magnesium content gives it a dark green-to-black color, similar to that of the pyroxenes. These two similar-looking types of silicates have different crystal structures, the distinctive internal planes of weakness, and readily distinguishable cleavage angles.

Sheet Silicates When all three oxygens at the base of a tetrahedron are shared with other tetrahedra, a sheet is formed that can extend indefinitely in the two dimensions of a plane. The fourth oxygen, at the peak of each tetrahedron, projects outward from the sheet, free to bond with available positive ions and thus bind two adjacent sheets together (Fig. 2-26d). The two-sheet layers are in turn bound to adjacent two-sheet layers by other positive ions. The bonds *between* the two-sheet pairs are far weaker than the bonds *within* each sheet or those that hold each two-sheet pair together, and they are easily broken, producing the sheet silicates' characteristic planar cleavage.

The *micas* are sheet silicates with a silicon (and substituted aluminum)-to-oxygen ratio of 1:2.5. In the common mica *muscovite,* aluminum ions securely bond adjacent sheets of tetrahedra to form a two-sheet pair, but the two-sheet pairs are weakly bonded by intervening potassium ions. These weak bonds readily break when we attempt to peel the layers of muscovite, causing sheets of this mica to separate easily.

To visualize the structure of a sheet silicate, consider a stack of peanut-butter sandwiches separated by pieces of wax paper. Each sandwich is analogous to a two-sheet layer of mica; the peanut butter represents the strong aluminum bonds between the sheets, and the wax paper represents the weak potassium bonds between the two-sheet layers. If you pick up the top sandwich by grasping its upper slice of bread, the whole sandwich, bound strongly by the peanut butter, remains intact. However, each sandwich separates readily from the one below. The same is true for each two-sheet pair in a sheet-silicate mineral such as muscovite mica.

Positive ion

Double chains

(c)

Figure 2-26 Types of silicate structures, continued. **(c)** Double chains. Each tetrahedron shares two of its corner oxygens with adjacent tetrahedra, forming a linear chain, and in addition some of the tetrahedra share a third oxygen with tetrahedra in an adjacent chain, thus joining the chains together. Some positive ions are also interspersed between the single chains, as well as adjoining double chains. The silicon-to-oxygen ratio of double chains is 1:2.75. Photo: Hornblende (an amphibole), showing the 56° and 124° cleavage angles characteristic of double-chain silicates.

Framework Silicates When a tetrahedron shares all four of its oxygen ions with adjacent tetrahedra, a structure with a three-dimensional framework results. This highest degree of oxygen-sharing results in the lowest silicon-to-oxygen ratio, 1:2. This is the structure of the two most abundant minerals in the Earth's crust, quartz and feldspar.

Quartz (SiO_2) and its various polymorphs are the second most abundant group of minerals in the continental crust. They are the only silicates composed entirely of silicon and oxygen. In quartz, every oxygen ion in a tetrahedron is shared with an adjacent tetrahedron; thus quartz achieves electrical neutrality without the need for other ions (Fig. 2-26e). Its strong bonds (which include both covalent and ionic bonds) make quartz the hardest of the common rock-forming minerals (7 on the Mohs scale). Because all of its bonds are equally strong, quartz has no weak internal planes and therefore does not cleave but breaks by fracturing.

Pure quartz is transparent and colorless, due to the absence of other ions that would absorb light. Because of its open structure, however, quartz often traps a few stray ions when it crystallizes. These impurities give quartz a wonderful range of colors, as is seen in the gemstones amethyst, rose quartz, and tigereye. When quartz has room to grow undisturbed, it assumes its characteristic crystal shape—prisms with pyramid-like faces on top and bottom (see Fig. 2-26e).

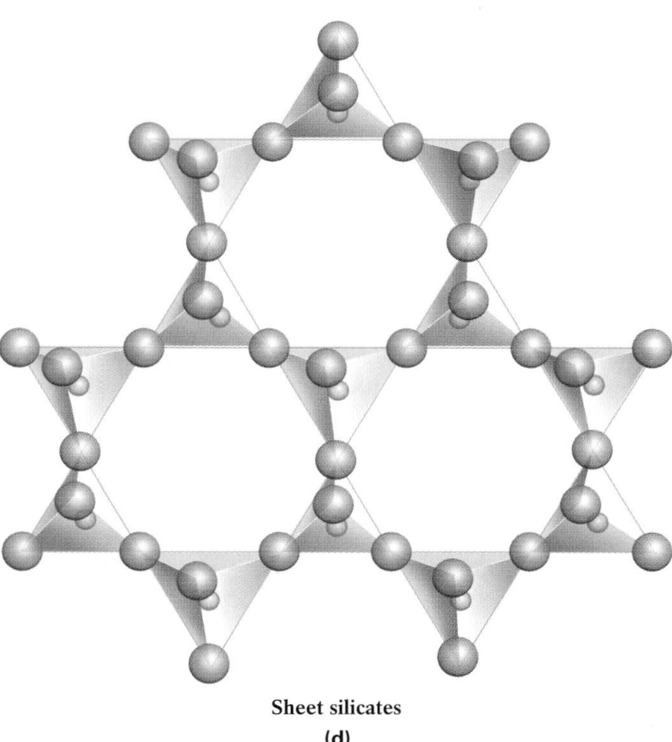

Sheet silicates
(d)

Figure 2-26 Types of silicate structures, continued. (d) Sheet silicates. Each tetrahedron shares all three of its corner oxygens, forming a sheet of adjoined tetrahedra. The fourth oxygen in each tetrahedron extends upward to bond with positive ions and, subsequently, another sheet. (Other positive ions bind two-sheet pairs to adjacent two-sheet pairs.) The silicon-to-oxygen ratio of sheet silicates is 1:2.5. Photo: Muscovite mica, showing the planar cleavage of the sheet silicates.

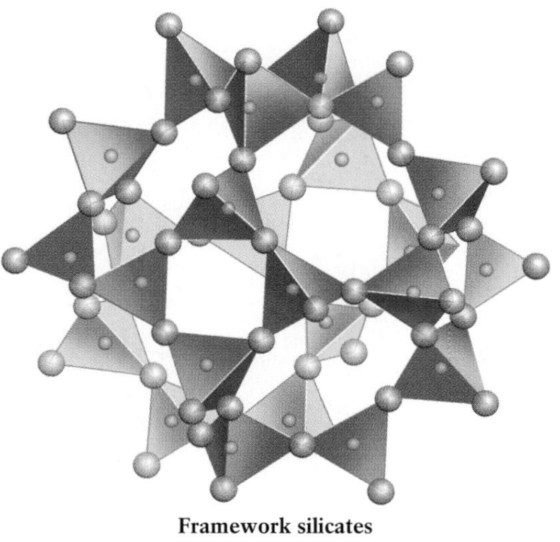

Framework silicates
(e)

Figure 2-26 Types of silicate structures, continued. (e) Framework silicates. Each tetrahedron shares all four of its oxygens, forming a three-dimensional framework structure. Because the charge on each tetrahedron is neutralized by the sharing of all its negative charges, no positive ions bond with the structure. The silicon-to-oxygen ratio of framework silicates is 1:2. Photo: Quartz crystals. Due to its sturdy framework structure, quartz does not cleave.

The *feldspars*, which account for about 60% by volume of continental crust, crystallize from molten rock across a wide range of temperature and pressure. In feldspars, as in quartz, every oxygen ion in a tetrahedron is shared with an adjacent tetrahedron. Unlike in quartz, however, atoms of aluminum regularly replace some of the silicon atoms in the feldspars, and potassium, sodium, and calcium ions occupy open spaces between the tetrahedra.

There are two types of feldspars, based on their chemical composition. The *plagioclase* feldspars contain varying proportions of sodium and calcium ions, which substitute for one another. The all-sodium plagioclase feldspar *albite* is at one end of the range, and the all-calcium plagioclase feldspar *anorthite* is at the other. The *alkali* feldspars are rich in potassium; the large size of the potassium ion produces a slightly different internal structure. All feldspars have two prominent cleavage planes that intersect at nearly 90° angles. They are relatively hard, about 6 on the Mohs scale.

Nonsilicates

Because the common rock-forming silicates are so abundant, no one gets very excited about the discovery of a fresh supply of plagioclase feldspar—we've got plenty already. But nonsilicate minerals constitute only about 5% of the Earth's continental crust. They include native elements—those that are not combined in nature with other elements—such as the precious metals gold, silver, and platinum, as well as such useful metals as iron, aluminum, copper, nickel, zinc, lead, and tin. Nonsilicates also include a few noteworthy gems, such as diamonds, rubies, and sapphires.

Nonsilicates are constructed from a variety of negative-ion groups, most containing oxygen. The most common nonsilicates are the carbonates, oxides, and sulfates, as well as a few groups without oxygen, such as the sulfides (Table 2-5).

Carbonates The **carbonate** ion complex (CO_3^{2-}) contains one central carbon atom with strong covalent bonds to three neighboring oxygen atoms. Because it has an overall negative charge, it typically combines with one or more positive ions. When it combines with calcium, it forms *calcite* ($CaCO_3$), commonly found in the abundant sedimentary rock, *limestone*. When the carbonate complex combines with calcium and magnesium, it forms *dolomite* ($CaMg(CO_3)_2$), found in another common sedimentary rock, *dolostone*. The ionic bonds between the negative carbonate ion group and the positive ions are relatively weak, so calcite and dolomite are relatively soft (3–4 on the Mohs scale). They also dissolve readily in acidic water, with important geological consequences (such as the creation of caves; see Chapter 16).

Table 2-5 Common Nonsilicate Minerals

Mineral Type	Composition	Examples	Uses
Carbonates	Metallic ion(s) plus carbonate ion complex (CO_3^{2-})	Calcite ($CaCO_3$)	Cement
		Dolomite ($CaMg(CO_3)_2$)	Cement
Oxides	Metallic ion(s) plus oxygen ion (O^{2-})	Hematite (Fe_2O_3)	Iron ore
		Magnetite (Fe_3O_4)	Iron ore
		Corundum (Al_2O_3)	Gems, abrasives
		Cassiterite (SnO_2)	Tin ore
		Rutile (TiO_2)	Titanium ore
		Ilmenite ($FeTiO_3$)	Titanium ore
		Uraninite (UO_2)	Uranium ore
Sulfides	Metallic ion(s) plus sulfur (S^{2-})	Galena (PbS)	Lead ore
		Pyrite (FeS_2)	Sulfur ore
		Cinnabar (HgS)	Mercury ore
		Sphalerite (ZnS)	Zinc ore
		Molybdenite (MoS_2)	Molybdenum ore
		Chalcopyrite ($CuFeS_2$)	Copper ore
Sulfates	Metallic ion(s) plus sulfate ion (SO_4^{2-})	Gypsum ($CaSO_4 \cdot 2H_2O$)	Plaster
		Anhydrite ($CaSO_4$)	Plaster
		Barite ($BaSO_4$)	Drilling mud
Native elements	Minerals consisting of a single element	Gold (Au)	Jewelry, coins, electronics
		Silver (Ag)	Jewelry, coins, photography
		Platinum (Pt)	Jewelry, catalyst for gasoline production
		Diamond (C)	Jewelry, drill bits, cutting tools

Oxides Mineral **oxides** are produced when negative oxygen ions combine with one or more positive metallic ions. The resulting minerals include some of our major sources of iron—hematite (Fe_2O_3) and magnetite (Fe_3O_4)—as well as aluminum, tin, titanium, and uranium.

Sulfides and Sulfates With six electrons in its outer energy level, sulfur can either accept or donate electrons. When sulfur accepts electrons, it becomes a negative ion that bonds with various positive ions to form the **sulfides.** This valuable group of minerals, including copper sulfides such as *chalcocite* (Cu_2S) and lead sulfides such as *galena* (PbS), is the source of a number of important metals.

When sulfur donates electrons, it becomes a positive ion that bonds with oxygen to form the negative **sulfate** ion complex (SO_4^{2-}). *Gypsum* ($CaSO_4 \cdot 2H_2O$), a common and useful sulfate mineral, is used to manufacture sheetrock and plaster of Paris, staples in the building construction trade.

Finding and Naming Minerals

Each year, about 40 new minerals are discovered. Most are so small they can hardly be seen without the aid of a microscope. One new find, a fist-sized chunk of one of the most striking blue minerals ever seen, has caused quite a stir among mineralogists. Purchased 15 years ago by geologist Anna Grayson at a roadside market in Morocco, this mysterious specimen remained in Grayson's home until she showed it to several mineralogists in 1996 (Fig. 2-27). Preliminary study of this rare find indicates that it is made of fibrous, asbestos-like crystals composed predominantly of silicon, oxygen, aluminum, calcium, magnesium, and iron. The new mineral's most unusual property is its ability to change color in the polarized light of a mineralogist's microscope—from purple to blue to a deep cream color. Mineralogists are still investigating this as-yet-unnamed mineral, and an expedition is being planned to search for the source area.

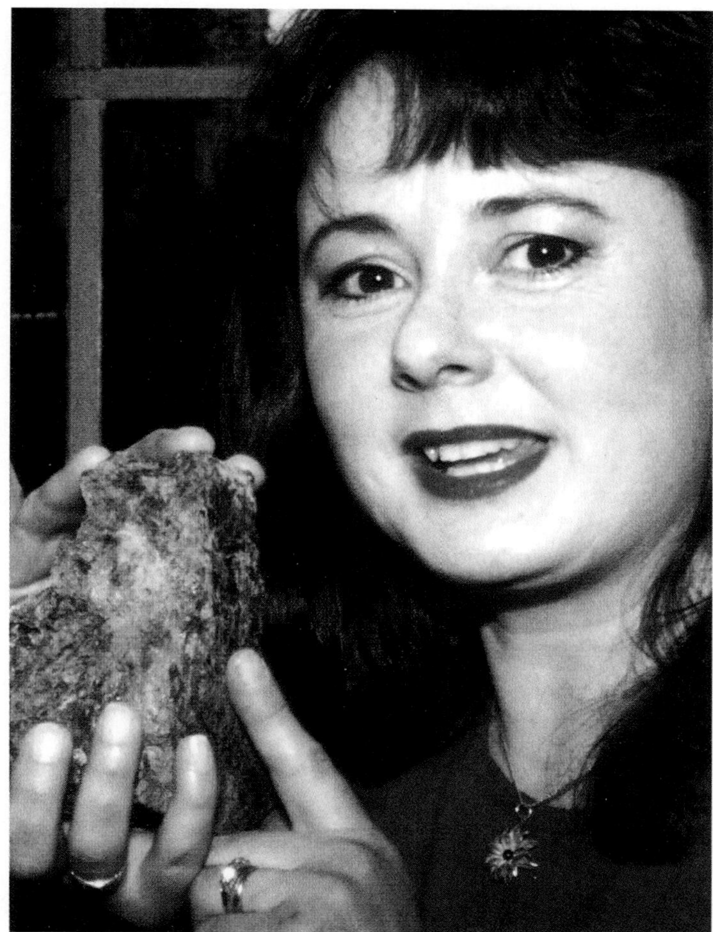

Figure 2-27 British geologist Anna Grayson showing the remarkable specimen of a heretofore unknown mineral she purchased at a roadside market in Morocco.

Once characterized mineralogically and chemically, this new mineral will join the encyclopedia of more than 3000 known minerals. The names of some challenge even the most eloquent among us—try pronouncing yugawaralite, ammoniojarosite, or anabohitsite. Why do some minerals get such tongue-twisting names? Why not simply refer to them in a systematic way based on the elements that form them?

One reason is that atoms of the same chemical compound can be arranged to form more than one mineral. For example, the atoms in calcium carbonate ($CaCO_3$) form several different crystal structures, including those of calcite and its polymorph, aragonite. Another reason is that some minerals' chemical compositions are so complex that a name based on their formulas would not be possible. Consider pumpellyite, a mineral found in rocks that have been slightly heated and compressed during certain types of mountain-building events—its chemical formula is $Ca_4(Mg,Fe^{2+})(Al,Fe^{3+})_5O(OH)_3(Si_2O_7)_2(SiO_4)_2 \cdot 2H_2O$. Finally, many minerals were known by common names for centuries before their chemical compositions were identified, and few of us would choose to say, "please pass the sodium chloride" at the dinner table or wish to receive a 2-carat "high-pressure carbon" engagement ring.

An international commission approves the names proposed for new minerals as they are identified. Some names acknowledge the geographical location where a mineral was discovered, such as labradorite (found in the eastern Canadian province of Labrador); some refer to a distinctive physical characteristic, such as citrine (yellow quartz), named for its color's resemblance to that of a citron, a lemon-like fruit; others honor heroes, such as the Moon mineral armalcolite, named for the first lunar explorers Neil *Arm*strong, Edwin *Al*drin, and Michael *Col*lins.

Gemstones

Several minerals lead a glamorous life as gemstones—precious or semiprecious minerals that display particularly appealing color, luster, or crystal form and can be cut or polished for ornamental purposes. Some gems, such as diamonds and emeralds, are quite rare. Others may be unusually perfect crystals of common rock-forming silicates. *Amethyst* is a variety of quartz whose appealing purple tones are produced by small numbers of iron atoms scattered throughout the crystal structure. The valued gemstone *amazonite* is an abundant alkali feldspar that owes its vivid blue-green color to the substitution of a few lead ions for potassium ions in its crystal structure. Common nonsilicates also can be gems. The aluminum oxide *corundum*, with its hardness of 9, is a popular abrasive used in emery cloth and sandpaper—but when its crystals are perfectly formed, and derive color from a few other ions trapped in their structures, ordinary corundum becomes not-so-ordinary *sapphires* and *rubies* (Fig. 2-28).

How Gemstones Form

Minerals of gemstone quality form under conditions that promote the development of perfect, large crystals. This happens most often in two ways: when molten rock cools and crystallizes underground, or when preexisting rock is subjected to extraordinary pressure and heat.

Molten rock often migrates into fractures in surrounding cooler rocks, where its ions and atoms crystallize in reasonably large spaces, producing perfect crystals and, if the space is big enough, enormous crystals. A single pyroxene crystal excavated in South Dakota was more than 12 meters (40 feet) long and 2 meters (6.5 feet) wide; it weighed more than 8000 kilograms (8 tons). This process also produces such complex silicate gemstones as topaz, tourmaline, and beryl.

Gemstones may also form when the heat and pressure applied along the edges of colliding plates cause the ions and atoms in their rocks to migrate and recombine, creating new minerals that are more stable under the new conditions. For example, heating and compression of carbonate rocks that contain aluminum ions can cause the aluminum to combine with oxygen liberated from calcite. We know some of the resulting aluminum oxides as rubies and sapphires.

And then there is that rare stone, the diamond, which is transformed from unspectacular carbon into a brilliant crystal by extremely high pressures at great depths (greater than 150 kilometers, or 90 miles). Diamonds most often occur where hot gas has propelled fluidized rock rapidly from great depth to the surface, carrying the diamonds along with it. The resulting diamond-rich structures, known as *kimberlite pipes* (named for Kimberley, South Africa), are typically a few hundred meters to a kilometer across and many kilometers deep. Kimberlite pipes are found in Siberia, India, Australia, Brazil, the Northwest Territories of Canada, southern and central Africa, and scattered throughout the Rockies of Colorado and Wyoming.

Many kimberlite pipes apparently formed between 70 and 150 million years ago, when the supercontinent of Pangaea was rifting into the continents that we know today. (See Chapter 1 for a discussion of Pangaea.) As the rifts opened, they apparently provided pathways through which diamond-bearing fluidized rock could rise from great depths to the surface. Other pipes are located in the most ancient rocks of each continent's central nucleus, created when the hotter early Earth was far more dynamic volcanically and tectonically than it is today. Thus, the value of diamonds is in part related to the unique historical circumstances under which they were propelled to the surface.

(a)

(b)

(c)

Figure 2-28 The minerals corundum, sapphire, and ruby each have the chemical formula Al_2O_3. Their different colors are produced by trace impurities in their crystal structure. **(a)** Ruby (red) embedded in plagioclase (colorless). **(b)** Sapphire. **(c)** Ruby.

Highlight 2-2 The Curse of the Hope Diamond

The exquisite 44.5-carat Hope Diamond is the world's largest blue diamond and one of the most popular attractions in the National Museum of Natural History in Washington, D.C. (Fig. 2-29). Yet, although it is virtually flawless and truly priceless, ownership of this remarkable gem has at times proven to be somewhat less than a stroke of great fortune.

The Hope Diamond is believed to be a remnant of a 112-carat Indian diamond stolen from a statue of the Hindu goddess Sita by a Brahman priest. Legend has it that the angry goddess cast an eternal spell of misfortune on any and all who acquired the gem. The spell apparently took effect shortly after the stone was smuggled to France, where it was sold to the French "Sun King," Louis XIV, who named it the "French Blue." Louis had the gem cut into a 67-carat teardrop-shaped jewel; after wearing it but once, he was stricken by a deadly strain of smallpox. The stone passed down to Louis XVI and his wife, Marie Antoinette, both of whom, while possessing the gem, literally lost their heads in the French Revolution.

The gem was stolen again during the chaos at the end of the French Revolution. It reappeared in London 38 years later, recut and bought by a British banker and gem collector, Henry Thomas Hope. We know little about any notable misfortune Hope may have suffered, but in 1890 the Hope Diamond was inherited by Lord Francis Hope, whose wife, an American actress, soon ran off with another man. Lord Hope was later forced to sell the gem to avoid bankruptcy. His unfaithful wife, who had frequently worn the stone, died in poverty.

The ongoing woeful saga of the diamond's subsequent owners continued for decades. One owner, an Eastern European prince, gave it to an exotic dancer—but later began to doubt her affections and shot her to death. Another, a Greek gem merchant, drove off a cliff and perished with his wife and children. Finally, in 1911, the Hope Diamond was purchased for $154,000 by Evalyn Walsh McLean, a wealthy and eccentric American socialite who sometimes had her Great Dane wear the diamond to greet her party guests. Although she scoffed at the diamond's curse, she too suffered personal tragedy during her years of ownership: A son died in a car accident; a daughter died from an overdose of sleeping pills; her husband went insane and was institutionalized.

The diamond's final private owner, diamond merchant Harry Winston, acquired the stone from the late Dame McLean's estate in 1947. For unknown reasons, he donated it to the Smithsonian Institution 11 years later, inexplicably sending it in a plain brown wrapper by registered mail. (The actual package is a popular item in the National Postal Museum.) The Hope Diamond now holds a place of honor in the museum's remodeled Hall of Gems, *perhaps* bringing an end to its colorful yet sometimes tragic history.

Figure 2-29 The exquisite blue Hope Diamond. The gem owes its unusual blue color to a few stray atoms of chromium within its crystal lattice.

Only a few kimberlite bodies produce gem-quality diamonds. In southern Africa, where most of the world's diamonds are mined, only 1 in 200 kimberlite pipes yields enough diamonds to make it worth the high cost of mining them. The story of one of South Africa's most famous diamonds, in Highlight 2-2, illustrates the mystique this most coveted of gems holds for many of us.

Synthetic Gems: Can We Imitate Nature?

The most valuable gemstones are rare, so people have tried for centuries to duplicate nature's feat by producing synthetic gems. In the twentieth century they have had some success, even managing to surpass nature in some cases. Artificial emerald crystals, first created in the 1930s, are more transparent, richer in color, and more exquisitely shaped than natural ones, which are often marred by gas bubbles and other impurities. The high quality of synthetic emeralds contributes to their great market value, which, at several hundred dollars per carat, is still far less than the price of the extremely rare natural ones.

Diamonds can now be made in the laboratory, by subjecting carbon to extreme heat and pressure. Almost any carbon-rich substance will do as a starting point, including sugar or peanuts. On December 12, 1954, scientists in the General Electric research lab in Schenectady, New York, created the first tiny synthetic diamonds. Today, more than 20 tons of industrial-grade synthetic diamonds are produced each year, destined for such practical uses as drill bits in oil-well drilling and modern dentistry.

In 1970, the first gem-quality diamonds were synthesized. Now, some synthetic diamonds even outshine the original. Strontium titanate, a synthetic mineral sold under the trade names of Fabulite and Wellington Diamond, glitters four times more vividly than a real diamond. Being only moderately hard, between 5 and 6 on the Mohs scale, it is not very durable. And because its creators can manufacture tons of it, this synthetic gem is only as rare as they choose. Another synthetic, cubic zirconia, can be manufactured in batches of 50 kilograms (110 pounds) and sold wholesale for a few cents per carat. Its optical qualities are virtually indistinguishable from those of nature's diamonds, and cubic zirconia is quite durable. How, then, do you know if your diamond is real or fake? One of the principal ways to tell is to check the specific gravity, which is higher in zirconia than in a natural diamond.

Minerals as Clues to the Past

After they identify mineral samples, geologists use their knowledge of how different minerals form to determine the geological events and environmental conditions that may have produced them. For example, large deposits of halite, such as those in Michigan, Kansas, and Louisiana, suggest the evaporation of an ancient saltwater sea. Glaucophane, a blue variety of amphibole, is known to form only in high-pressure, low-temperature conditions. These conditions occur together only in subduction zones where oceanic plates descend beneath adjacent plates at convergent plate boundaries. (See the discussion of subduction in Chapter 1.) Thus a rock containing glaucophane most probably formed in an ancient subduction zone.

Polymorphs, with identical chemical compositions but different crystal structures, form under different heat and pressure conditions. Ordinary quartz, with a specific gravity of 2.65, crystallizes at relatively low temperature and pressure. One of its polymorphs, *stishovite*, with a specific gravity of 4.28 and a dense compressed structure, crystallizes at temperatures higher than 1200°C (2200°F) and pressures more than 130,000 times that at sea level. Geologists believe that these extreme conditions were probably produced by geological events of enormous magnitude, such as meteorite impacts. This hypothesis gains support from the presence of stishovite in the fractured rocks of Meteor Crater near Winslow, Arizona, and at Manicouagan Lake, Quebec, which have been proposed as ancient meteorite-impact sites (Fig. 2-30).

Now that we have introduced the basic structure of minerals and explained how they form and are identified, we can begin to examine more closely the common types of rocks that make up the Earth's crust. In the next six chapters, we will discuss how igneous, sedimentary, and metamorphic rocks form (introduced in Chapter 1 with the rock cycle), and how we often use the minerals in these rocks to interpret past geological events.

One final thought about minerals: As a geology student, you may now be tempted to browse from time to time through the bins of mineral specimens at rock shops and museums. Not long ago, a man in Tucson, Arizona, was doing just that. From a barrel of unspectacular stones, he pulled a potato-sized lavender specimen, paid the proprietor $10, and went home with what may have been the largest sapphire (1905 carats) ever found. Although the value and quality of the stone have been hotly debated, a quote from the keen-eyed rockhound still rings true: "The only reason more gems aren't found in this country is that no one is looking for them." Now that you know something about minerals, you too can start looking.

Chapter Summary

Minerals are naturally occurring solids with specific chemical compositions and definite internal structures. **Rocks** are naturally occurring aggregates of minerals. Minerals are composed of one or more chemical **elements**, the form of matter that cannot be broken down to a simpler form by heat, cold, or reaction with other elements. Each element, in turn, is made up of **atoms**, infinitesimally small particles that retain all of an element's chemical **properties**, the characteristics of a substance that enable us to distinguish it from other substances. When atoms of two or more elements combine in specific proportions, they form chemical **compounds**, which have properties different from any of their constituent elements individually. Nearly all minerals are chemical compounds (except for the native elements).

The center of an atom is occupied by its **nucleus**, which contains both positively charged particles called **protons** and uncharged particles of equal mass called **neutrons**. An element's **atomic mass** is the sum of its protons and neutrons. An element's **atomic number** is determined by the number

(a)

(b)

Figure 2-30 (a) Meteor Crater, near Winslow, Arizona, is believed to have been produced by the impact of a relatively small meteorite about 25,000 years ago. **(b)** The circular basin that forms Manicouagan Lake (75 kilometers, or 40 miles, in diameter), in Quebec, Canada, is also thought to be the product of an ancient meteorite strike, dating from about 214 million years ago. Studies of rocks near both sites have revealed the presence of the extremely high-pressure variety of SiO_2, stishovite. Our knowledge of the conditions that form stishovite reinforces hypotheses that propose impact origins for Meteor Crater and Manicouagan Lake.

of protons in its nucleus; every atom of a given element always has the same number of protons and thus the same atomic number. However, atoms of a given element may differ in atomic mass because the number of neutrons in their nuclei may vary. Atoms of a given element that contain different numbers of neutrons are **isotopes** of that element.

A cloud of negatively charged particles called **electrons** surrounds the nucleus and moves about it at high speed. Each electron occupies a specific region of space called an **energy level**; an atom may have one or more energy levels. The electrons occupying the outermost energy level can usually be donated, acquired, or shared with other atoms to achieve the most chemically stable electron configuration, one in which eight electrons typically fill an atom's outer energy level. The tendency of most atoms to fill their outer energy level in this manner is known as the **octet rule**.

Atoms combine to form chemical compounds in a variety of ways known as **bonding**. When an atom donates or acquires electrons, it becomes an electrically charged particle called an **ion**. An atom that donates electrons becomes positively charged; an atom that acquires electrons becomes negatively charged. Positive and negative ions are attracted to one another and may form one of several types of chemical bonds to form electrically neutral chemical compounds.

When electrons are acquired or donated between atoms, an **ionic bond** is formed. When the electrons in the outer energy levels of atoms are shared between atoms, a **covalent bond** is formed. In **metallic bonding**, electrons move continually among numerous closely packed nuclei. **Intermolecular bonds**, including **hydrogen bonds** and **van der Waals bonds**, form from weak attractions between groups of atoms, whose orbiting electrons move in such a manner that they create only temporary charges.

As a mineral forms through chemical bonding, all of its ions or atoms occupy specific positions to form its **crystal structure**, a three-dimensional pattern that repeats throughout the mineral. When mineral growth is not limited by space, a **crystal** may form with a regular geometric shape that reflects the mineral's internal crystal structure. Naturally occurring inorganic solids that lack systematic crystal structures are **mineraloids**. Volcanic glass, which cools too rapidly from molten rock to develop a crystal structure, is an example of a mineraloid.

The types of minerals that will form at a given time and place are determined by which elements are available to bond, the charges and sizes of their ions, and the temperature and pressure under which the minerals form. Ions and atoms of similar size and charge are able to replace one another within a crystal structure; this **ionic substitution** produces minerals whose crystal structures are the same but which vary in composition. Because they form under different environmental conditions, two minerals of the same chemical composition, such as the carbon-based minerals diamond and graphite, may have different crystal structures; such minerals are called **polymorphs**.

Geologists identify minerals out in the field by noting their external characteristics and measuring certain physical properties, including color, **luster, streak, hardness, cleavage, fracture**, and crystal form. Methods used to identify minerals in the laboratory include determining their **specific gravity**, examining them under a petrographic microscope, assessing whether they **fluoresce** or **phosphoresce** under ultraviolet light, and analyzing them by **X-ray diffraction**.

Although scores of substances may have orderly internal patterns, the number of naturally occurring minerals is limited by the availability of the Earth's elements, their electrical charges, and the relative sizes of their ions. The Earth's crust is composed primarily of only eight common elements, which combine to produce the **rock-forming minerals**. The two most prominent elements, oxygen and silicon, readily combine to form the **silicon-oxygen tetrahedron**. This is the basic building block of the most abundant group of minerals in the Earth's crust, the **silicates**, which include more than a thousand different minerals. Silicon-oxygen tetrahedra may be linked in a variety of crystal structures: independent tetrahedra (olivine is an example); single chains of tetrahedra (typified by the pyroxenes); double chains (the amphiboles); sheet structures (the micas); and framework structures (feldspars and quartz).

A number of nonsilicates are also common rock-forming minerals. Among them are the **carbonates** (such as calcite and dolomite); the **oxides** (such as iron-rich magnetite and hematite); and the **sulfides** (such as lead-based galena and iron-based pyrite) and **sulfates** (such as calcium-based gypsum).

Gemstones are minerals that are valued for their particularly appealing color, luster, or crystal form. Some gemstones, such as diamonds and emeralds, are quite rare; others are well-formed specimens of relatively commonplace minerals, usually containing trace impurities in their structures that impart distinctive colors to the crystals. Although naturally formed gemstones are considered the most valuable, less valuable gemstones can be synthesized in the laboratory.

Minerals that form under unique geological circumstances, when identified in ancient rocks, can provide clues to the past. The quartz polymorph stishovite, for instance, is known to form under high-temperature/high-pressure conditions; thus it is considered to be evidence of a possible past meteoric impact.

Key Terms

rocks (p. 36)
minerals (p. 36)
element (p. 36)
atom (p. 36)
property (p. 36)
compound (p. 37)
proton (p. 37)
neutron (p. 37)
nucleus (p. 37)
electron (p. 37)
atomic mass (p. 38)
atomic number (p. 38)
isotope (p. 38)
energy level (p. 38)
bond (p. 38)
octet rule (p. 38)
ion (p. 39)
ionic bonding (p. 39)
covalent bonding (p. 40)
metallic bonding (p. 40)
intermolecular bonding (p. 41)
hydrogen bonding (p. 41)
van der Waals bonding (p. 41)
crystal (p. 41)
crystal structure (p. 41)
mineraloid (p. 41)
ionic substitution (p. 42)
polymorph (p. 42)
luster (p. 44)
streak (p. 44)
hardness (p. 45)
cleavage (p. 46)
fracture (p. 47)
specific gravity (p. 49)
fluorescence (p. 50)
phosphorescence (p. 50)
X-ray diffraction (p. 50)
rock-forming minerals (p. 50)
silicates (p. 51)
silicon-oxygen tetrahedron (p. 51)
carbonates (p. 56)
oxides (p. 57)
sulfides (p. 57)
sulfates (p. 57)

Questions for Review

1. What is a mineral? How does a mineral differ from a rock?

2. Briefly describe the structure of an atom. What is an isotope? What is an ion?

3. How does an atom achieve chemical stability? How does a chemical compound achieve electrical neutrality?

4. Describe three types of chemical bonding.

5. What is a mineral crystal? Describe two circumstances under which a mineral probably will *not* form into a crystal.

6. Define and give an example of a mineral polymorph.

7. Describe four properties of minerals that you could use to identify an unknown mineral.

8. Briefly discuss why silicon and oxygen are so compatible in nature.

9. List four different silicate structures and give a specific mineral example of each.

10. List two types of common nonsilicates and give a specific mineral example of each.

11. Describe the connections between at least two different geological environments and the formation of gemstones.

For Further Thought

1. If some sodium (Na^+) substitutes for calcium (Ca^{2+}) in the plagioclase feldspars, why must some aluminum (Al^{3+}) replace some silicon (Si^{4+}) in the mineral's structure?

2. Sulfur forms a small ion with a high positive charge. Why, then, doesn't sulfur unite universally with oxygen to form the basic building blocks of most crustal minerals?

3. If you were trying to prove that a meteorite struck a particular place on the Earth at some time in the past, what mineralogical evidence would you look for at the proposed impact site?

4. Why doesn't the mineral quartz exhibit the diagnostic property of cleavage? Considering its physical beauty, why isn't a quartz crystal a more valuable gemstone?

5. In the photos at right you can see crystals of real gold and "fool's gold" (pyrite). How would you distinguish these two similar-looking minerals?

3

Igneous Processes and Igneous Rocks

Figure 3-1 Lava fountains from Hawai'i's Halemaumau Crater. This red-hot, liquid rock will cool and harden into solid rock within hours.

On the island of Hawai'i, you can watch a volcano erupt and later touch the warm rock that passed through the volcano in a molten state just hours or days before (Fig. 3-1). In the Sierra Nevada of California, you can walk on rocks that cooled 80 million years ago from molten rock 20 kilometers (12 miles) below the Earth's surface. The remains of ancient volcanic eruptions and vast uplifted regions of formerly sub-surface rocks can be found in almost every state and province of North America. As we saw in Chapter 1, rocks such as these, which cooled and crystallized directly from molten rock, either at the surface or deep underground, are called **igneous rocks** (from the Latin *ignis*, or "fire").

Geologists actually use two terms to classify molten rock. **Magma** is molten rock *within* the Earth. It may be completely liquid or, as is more common, it may be a fluid mixture of liquid, solid crystals, and dissolved gases. When magma reaches the Earth's surface, we call it **lava**, molten rock that flows *above* ground. (Lava and the volcanic structures we associate with it are discussed in the next chapter.)

Most igneous processes occur underground, generally shielded from our view by the loose sediment or sedimentary rocks that cover most of the Earth's surface. Although geologists can observe the lava spewing from a volcano, they cannot see magma moving underground. (Consider how little you know about the maze of cables and pipelines concealed under the soil and concrete of your community.) Thus, to investigate igneous processes, geologists often rely on geophysical studies of the Earth's interior (discussed in Chapter 11), computer models that simulate subterranean temperature and pressure conditions, and laboratory experiments that attempt to reproduce pressure, temperature, composition, and other factors that affect the melting and crystallization of rocks and minerals. Geologists also look for regions where surface rock layers have been eroded (that is, physically worn away by such agents as rivers, glaciers, winds, ocean waves, and gravity), opening up windows to the subsurface rocks.

We can see igneous rocks that solidified underground in some of North America's most scenic places, from Mount Katahdin in northern Maine to the Yosemite Valley of eastern California. Some of our continent's most ancient rocks

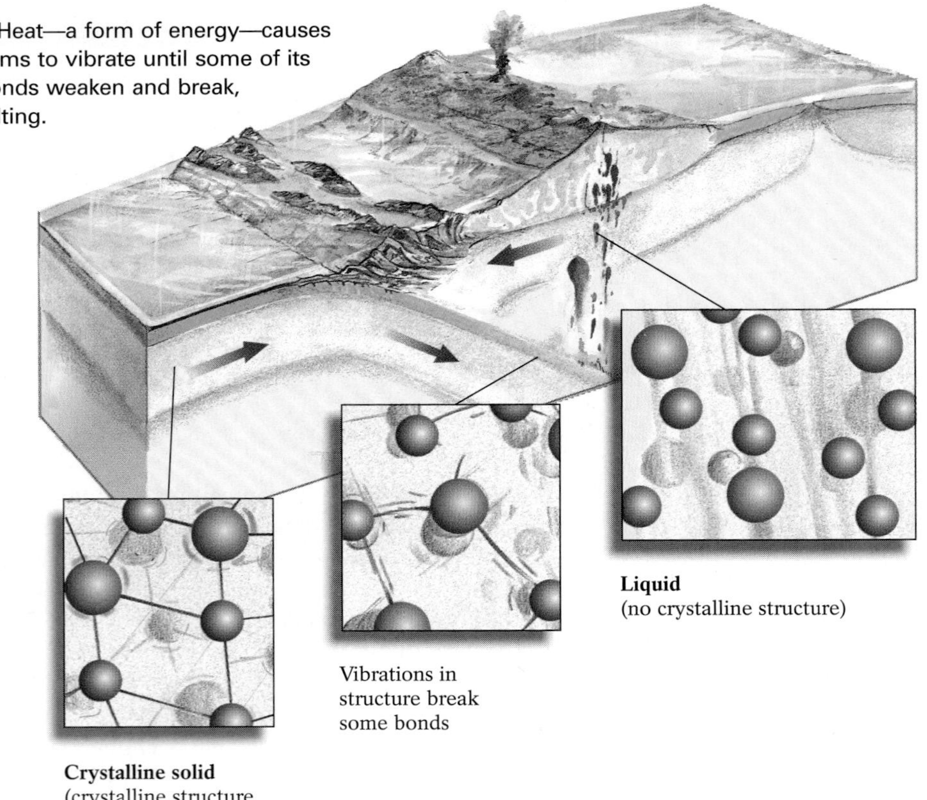

Figure 3-2 Heat—a form of energy—causes a solid's atoms to vibrate until some of its chemical bonds weaken and break, causing melting.

are the igneous roots of one-time mountains that have long since been eroded by downcutting rivers and rasping glaciers to expose the underlying rocks. Such rocks can be seen in northern Minnesota, Ontario, and Quebec.

Our discussion of the Earth's igneous processes and rocks describes first some characteristics of molten rock, and then the types of rocks that form when molten material cools and solidifies. We will examine how and why rocks melt and crystallize, paying special attention to how plate tectonics affects the origin and distribution of igneous rocks on Earth (and how Moon rocks differ in this respect). Finally, we will see that economically valuable materials—such as gold, silver, and copper—often originate from igneous processes.

Melting Rocks and Crystallizing Magma

A crystalline solid melts when the bonds between its ions break, allowing the charged particles to move freely (Fig. 3-2). When underground temperatures become high enough, bonds in minerals are broken; eventually the heated rock, no longer a crystalline solid, becomes magma—a hot liquid, perhaps containing some still-solid fragments. Since the various types of minerals melt at different temperatures, different minerals melt out of the rock as the heat gradually increases. At the same time, the composition of the magma changes as each newly molten mineral enters and enriches it.

When the heat dissipates, the movements and vibrations of the particles in the molten mass begin to slow down; their bonds cease to break, more bonds start to form, and tiny crystals begin to appear. Additional ions and atoms bond at sites on the crystal structure, as described in Chapter 2. In this way, crystals grow until they touch the edges of adjacent crystals. As cooling progresses, different minerals crystallize out of the magma at different temperatures. The magma's composition again changes as each crystallized mineral is removed from the fluid mix. Ultimately, if cooling continues, the entire body of magma becomes solidified.

Geologists classify igneous rocks by their two most obvious properties: their *texture*, which is determined by the size and shape of their mineral crystals and the manner in which these grew together during cooling, and their *mineral content*, which is determined by their chemical composition and their cooling history.

Igneous Textures

A rock's *texture* refers to the appearance of its surface, specifically the size, shape, and arrangement of the minerals within the rock (Fig. 3-3a–e). The most important factor controlling these features in igneous rocks is the rate at which magma or

Figure 3-3 Igneous textures. **(a)** Rocks that cool slowly underground, as did this granite in Yosemite National Park, California, have *phaneritic* (coarse-grained) textures.

lava cools. When a magma's minerals crystallize slowly over hundreds or thousands of years, there is ample time for crystals to grow large enough to be seen by the unaided eye. Rock textures in which we can see the crystals clearly are called *phaneritic* (from the Greek *phaneros*, or "visible") (Fig. 3-3a). Slow cooling typically occurs when magmas enter, or *intrude*, preexisting solid rocks; thus, rocks with phaneritic textures are called **intrusive rocks**. They are also known as **plutonic rocks** (for Pluto, the Greek god of the underworld).

Some igneous rocks develop at relatively low temperatures from magmas with a high proportion of water. Under these watery conditions, ions are free to move quite readily to bond to growing crystals, enabling them to become unusually large (sometimes several meters long). Rocks with such exceptionally large crystals are called **pegmatites** (from the Greek *pegma*, or "fastened together") (Fig. 3-3b). In western Maine, near the towns of Bethel and Rumford, rocks with pegmatitic textures contain 5-meter (17-foot)-long crystals of the mineral beryl. Most pegmatites consist primarily of such common minerals as quartz, feldspar, and mica. Some rare elements, such as beryllium, also occur in pegmatites; pegmatite outcrops that contain beryllium in the form of the gemstones emerald and aquamarine are popular destinations for amateur and professional mineral hunters.

Figure 3-3 **(b)** Extremely coarse-grained *pegmatites*, such as the one shown here, form from ion-rich magmas having a high water content.

Some igneous rocks solidify so quickly that their crystals have little time to grow. In these *aphanitic* (from the

Figure 3-3 **(c)** Volcanic rocks such as this basalt, because they cool rapidly above ground, typically have *aphanitic* (very small-grained) textures.

Greek *a phaneros*, or "not visible") textures, crystals are so small they can barely be seen by the naked eye (Fig. 3-3c). Rocks with aphanitic textures are typically called **extrusive rocks** because they generally form from lava that has flowed out, or been *extruded*, onto the Earth's surface. They are also known as **volcanic rocks**, because lava is a product of volcanoes (named for Vulcan, the Roman god of fire). Aphanitic textures may also form when a thin body of magma intrudes cool near-surface rocks and thus cools rapidly.

In some igneous rocks, large, often perfect, crystals are surrounded by regions with much smaller or even indistinguishable grains. These *porphyritic* textures form as a result of slow cooling followed abruptly by rapid cooling. First, gradual underground cooling produces large crystals (Fig. 3-3d) that grow slowly within a magma. Then the mixture of remaining liquid magma and the early-formed crystals it contains rises nearer to the surface, where it encounters cooler rocks or erupts into the cooler environment at the surface. In either case, the molten material cools rapidly to produce the enveloping body of smaller grains.

Figure 3-3 **(d)** Some rocks have a *porphyritic* texture, marked by large crystals surrounded by an aphanitic matrix.

A given magma can produce igneous rocks having any of the full range of igneous textures—the amount of time available for cooling and crystal growth determines whether the rock textures will be aphanitic, phaneritic, or porphyritic. Aphanitic igneous rocks that cooled rapidly at the Earth's surface, for example, are commonly underlain by phaneritic rock that derived from the same magma but cooled slowly underground.

Volcanic Glass When lava from a volcano erupts into the air or flows into a body of water, much of it cools so quickly that its ions don't have enough time to form any crystals at all. The ions are essentially frozen in place randomly, bonded to any available ions nearby. The texture of the resulting rock is described as *glassy* (Fig. 3-3e).

There are two common types of volcanic glass. *Pumice* (from the Latin *spuma*, or "foam") forms when bubbling, highly gaseous, silica-rich lava cools very quickly. Some pumice has so many tiny cavities that it can float. Large rafts of pumice blown from coastal and island volcanoes have been known to float out to sea for hundreds of kilometers before they finally became waterlogged and sank.

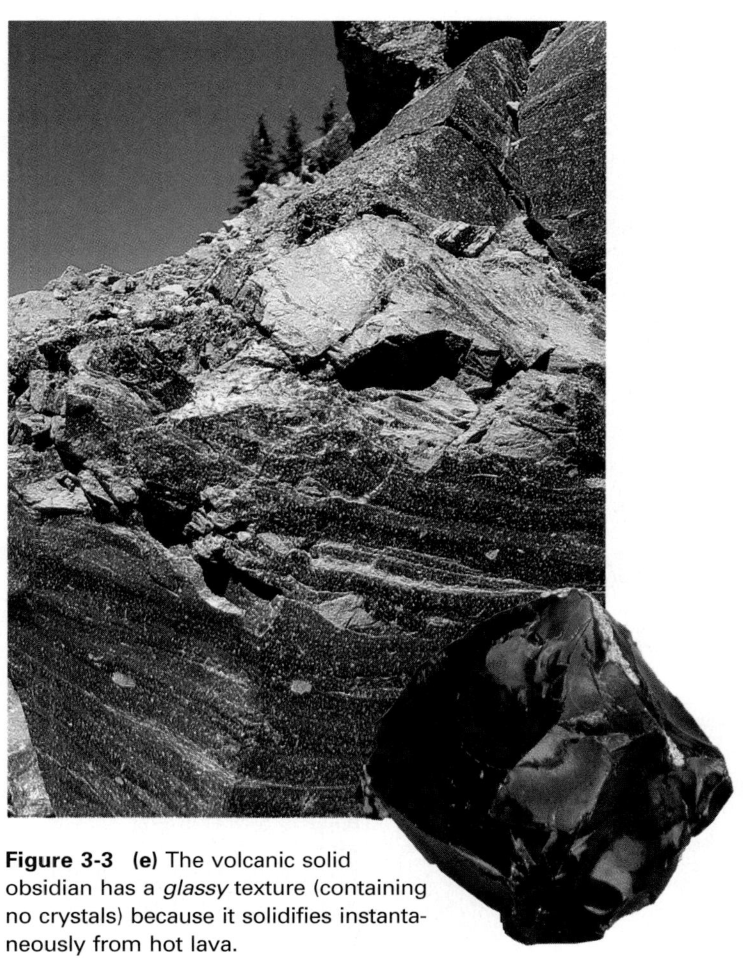

Figure 3-3 (e) The volcanic solid obsidian has a *glassy* texture (containing no crystals) because it solidifies instantaneously from hot lava.

The second type of volcanic glass is *obsidian*. Obsidian forms when very silica-rich lavas, containing less gas than those that produce pumice, cool very quickly. The disordered arrangement of its ions means that obsidian lacks an organized crystal structure as well as the systematic internal planes of weakness that characterize most crystals. For this reason, obsidian breaks by fracture rather than by cleavage. Early human beings worked obsidian to fashion projectile points and sharp-edged cutting tools. To demonstrate the quality of such ancient tools, in 1986 David Pokotylo, the curator of archaeology at the University of British Columbia's Museum of Anthropology, underwent hand surgery with an obsidian microblade scalpel prepared in the ancient style of making obsidian tools. To the surprise of the surgical team, Dr. Pokotylo's incisions healed more rapidly and more cleanly than those made with conventional steel blades.

Igneous Compositions

The Earth's magmas consist largely of the most common elements: oxygen, silicon, aluminum, iron, calcium, magnesium, sodium, potassium, and sulfur. The relative proportions of these components within a body of magma, added to the prevailing temperature, pressure, and its water content, distinctly characterize a magma. Ultimately these factors determine the mineral content of the rocks it will form. Water vapor (H_2O), carbon dioxide (CO_2), and sulfur dioxide (SO_2) are the major dissolved gases in molten rock, accounting for a small percentage of a magma's total volume.

We saw in Chapter 2 that the Earth's crust consists primarily of silicon-and-oxygen–based minerals. These silicates are the major constituents of igneous rocks. Based on their silica content (the amount of silicon and oxygen), igneous rocks and magmas are classified as ultramafic, mafic, intermediate, or felsic (Table 3-1), with ultramafic material having the smallest proportion of silica to other ions and felsic material having the largest. Figure 3-4 illustrates how the mineralogical composition of the common igneous rocks varies in each of these categories. (See also Table 2-5 for chemical compositions of minerals.)

Table 3-1 Common Igneous Compositions

Composition Type	Percentage of Silica	Other Major Elements	Relative Viscosity of Magma	Temperature at Which First Crystals Solidify	Igneous Rocks Produced
Felsic	>65%	Al, K, Na	High	~600–800°C (1100–1475°F)	Granite (plutonic) Rhyolite (volcanic)
Intermediate	55–65%	Al, Ca, Na, Fe, Mg	Medium	~800–1000°C (1475–1830°F)	Diorite (plutonic) Andesite (volcanic)
Mafic	45–55%	Al, Ca, Fe, Mg	Low	~1000–1200°C (1830–2200°F)	Gabbro (plutonic) Basalt (volcanic)
Ultramafic	<40%	Mg, Fe, Al, Ca	Very low	>1200°C (2200°F)	Peridotite (plutonic) Komatiite (volcanic)

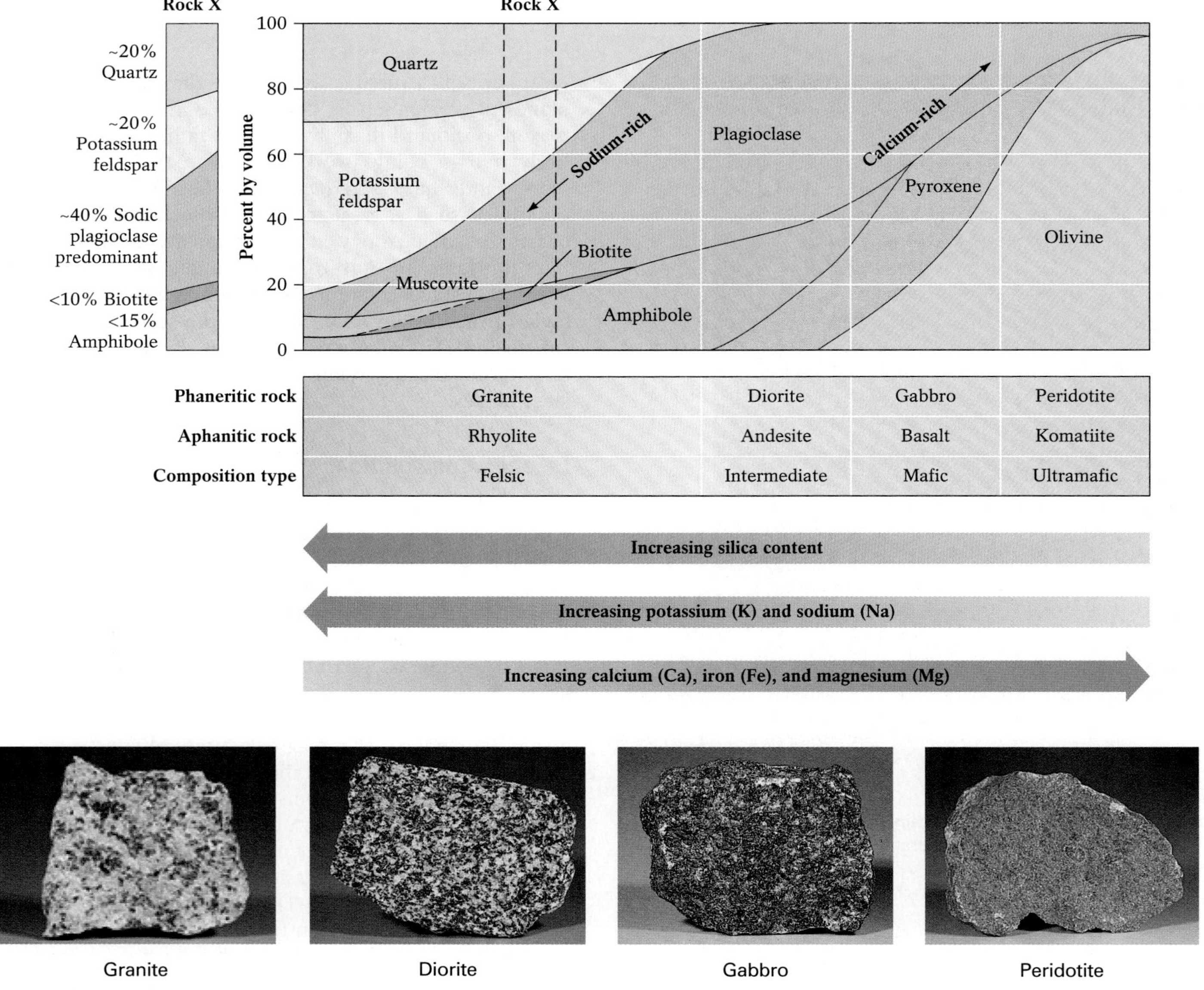

Figure 3-4 An igneous rock classification chart, showing the range of compositional types among the igneous rocks, from felsic to ultramafic. The colored areas in the body of the chart indicate the mineral components of the rocks. The sample segment shows how to interpret the chart, using as an example a rock with a composition of granite.

Ultramafic Igneous Rocks The term "mafic" is derived from *ma*gnesium and *f*errum (Latin for "iron"). Ultramafic igneous rocks are dominated by the iron-magnesium silicate minerals olivine and pyroxene and contain relatively little silica (less than 40%) and virtually no feldspars or free quartz. The most common ultramafic rock, **peridotite** (pe-RID-o-tite), contains between 40% and 100% olivine. Ultramafic rocks generally crystallize slowly deep in the Earth's interior, developing their typically coarse-grained phaneritic texture. (The relatively rare extrusive equivalent of peridotite is called komatiite.) These rocks are dark in color and very dense. Peridotite appears at the Earth's surface only where extensive erosion has removed overlying crustal rocks. It is most likely to be found where converging continental tectonic plates have collided and been uplifted, bringing deep rocks closer to the surface.

Mafic Igneous Rocks Mafic igneous rocks have a silica content between 45% and 55%. These are the most abundant rocks of the Earth's crust, and the mafic rock **basalt** (ba-SALT) is the single most abundant of them. Basalt, whose principal minerals include pyroxene, calcium feldspar, and a minor amount of olivine, is dark in color and relatively dense. Because it forms from molten rock that has cooled fairly rapidly at or near the surface, its texture is typically aphanitic. It is the dominant rock of the world's oceanic plates, making up most of the ocean floor and many islands, including the entire Hawai'ian chain, Samoa and Tahiti in the Pacific, and Iceland in the Atlantic.

Basalt also covers several large areas of our continents, most noticeably in Brazil, India, South Africa, Siberia, and the Pacific Northwest of North America (Fig. 3-5).

When a magma containing the same mix of minerals as basalt cools more slowly underground, its plutonic equivalent, the coarse-grained phaneritic **gabbro**, is produced. Gabbro outcrops are generally seen only where extensive erosion has removed surface rocks. Geologists, drilling deep along mid-ocean ridges, have found that gabbro lies beneath the basalts of the ocean floor.

Intermediate Igneous Rocks Intermediate igneous rocks contain more silica than mafic rocks—between about 55% and 65%. They typically consist of iron and magnesium silicates, such as pyroxene and amphibole, along with sodium- and aluminum-rich minerals, such as sodium plagioclase and mica, and a small amount of quartz. They are generally lighter in color than mafic rocks. The aphanitic intermediate igneous rock **andesite** named for the Andes mountains of South America, where it often dominates the local geology, is the world's second most abundant volcanic rock. (See Figure 3-5.) Andesites contain pyroxene, the hornblende variety of amphibole, a minor amount of quartz, and an abundance of andesine, a plagioclase mineral with about 60% sodium and 40% calcium ions. Andesite's plutonic equivalent is **diorite**, which can be recognized by its coarse-grained, salt-and-pepper appearance (because both dark- and light-colored grains are present).

Felsic Igneous Rocks The term "felsic" is derived from *fel*dspar and *si*lica. Felsic igneous rocks contain more silica—65% or more—than mafic or intermediate igneous rocks. They are generally poor in iron, magnesium, and calcium silicates, but rich in potassium feldspar, aluminum-rich micas, and quartz. The most common felsic igneous rock is the plutonic rock **granite**. Potassium feldspars typically impart granite's usual pinkish cast. In many specimens, sodium feldspars contribute a porcelain-like look to these rocks. Quartz grains and flakes of biotite mica are also scattered throughout granite. The magma that produces granite flows and crystallizes slowly underground, seldom reaching the surface. Thus outcrops of **rhyolite**, granite's volcanic equivalent, are far less common than those of granite. Granitic rocks can generally be seen at the surface only after erosion removes overlying rocks. As you can see in Figure 3-5, granitic igneous rocks are exposed sporadically on the Earth's continents. Rocks of felsic composition have a greater variety of textures than any other igneous rock. They range from aphanitic rhyolites and several glassy rocks to ultra-coarse pegmatites.

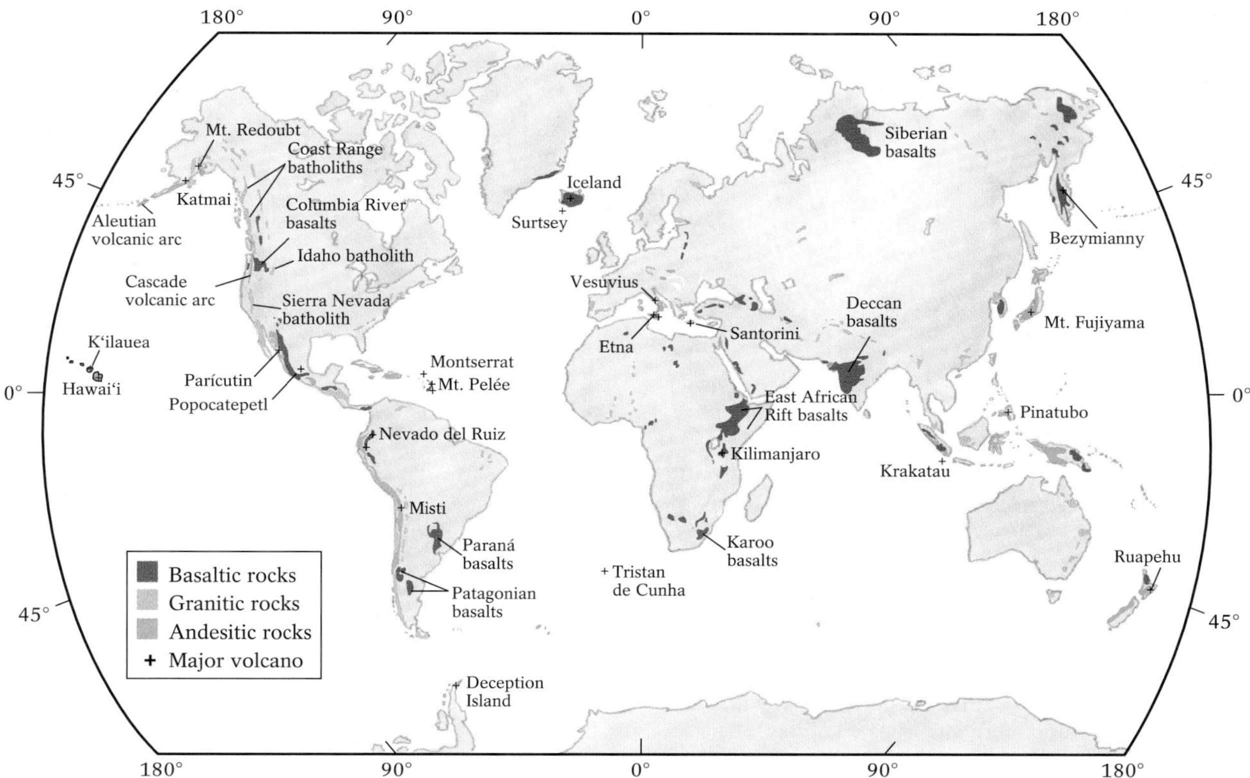

Figure 3-5 Distribution of the Earth's major continental igneous rock provinces. As well as being the most abundant igneous rock of the ocean floors, the mafic igneous rock basalt is widespread on the continents.

The Creation of Magma

When rocks melt to produce new magma, the process is not quite as simple as that of the warm midday sun melting last night's snowfall. Snow consists of a single solid mineral, ice (H_2O), and therefore it all melts at one temperature, 0°C (32°F). Most rocks are composed of several minerals, each with a different melting point reflecting the strength of its own chemical bonds. For example, the bonds between sodium and oxygen in the plagioclase albite are weaker than those between calcium and oxygen in the plagioclase anorthite. Consequently, at Earth-surface pressures, albite's bonds break (and albite melts) at 1118°C (2050°F), whereas anorthite's bonds can withstand a temperature of 1553°C (2800°F) before they break (and anorthite melts).

As rocks are heated within the Earth, not all the minerals within the rocks melt simultaneously. A body of rock typically undergoes **partial melting** when it is heated to the melting points of some but not all of its component minerals. When rocks melt *partially*, some remains solid (the minerals with higher melting points) while the melted portion (minerals with lower melting points) may flow away.

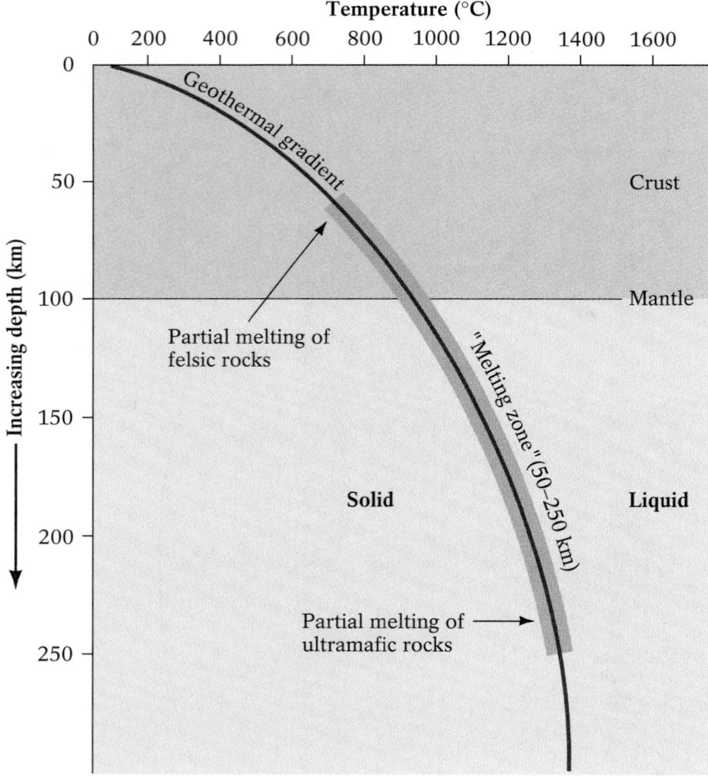

Figure 3-6 A graph showing the geothermal gradient—the rate at which the Earth's internal temperature increases with depth. The temperatures between 50 and 250 kilometers (30–150 miles) in depth exceed the 700°C (1300°F) melting point of felsic rock and the 1300°C (2400°F) melting point of ultramafic rock. Thus rocks tend to melt at the temperatures found in the Earth's lower crust and upper mantle.

Numerous factors control the melting of rocks and the creation of magmas. Heat, pressure, and the amount of water in the rocks all combine to influence the point at which particular rocks melt. As you will see, all these factors affect what type of rocks eventually form from magmas and where on Earth they occur.

Heat As we saw in Chapter 1, the heat in the Earth's interior comes from three primary sources: (1) the heat liberated continuously by decay of radioactive isotopes; (2) the heat produced during the formation of the planet, still rising from the Earth's core and released in part as the liquid outer segment of the core crystallizes; and (3) to a far lesser degree, the frictional heat produced as the Earth's plates move against one another and over the underlying asthenosphere.

Temperatures in the Earth increase with depth, at a rate referred to as the *geothermal gradient*. As you can see in Figure 3-6, interior temperatures rise quite rapidly (from about 50°C [125°F] to more than 1200°C [2200°F]) within the Earth's outer 250 kilometers (150 miles). A high concentration of radioactive isotopes in the rocks of the lower crust seems to be primarily responsible for this great heat, melting a portion of the rocks at these depths.

Pressure When high pressure holds the ions and atoms in a crystalline solid closer together, higher temperature is required to vibrate, weaken, and break their bonds. The

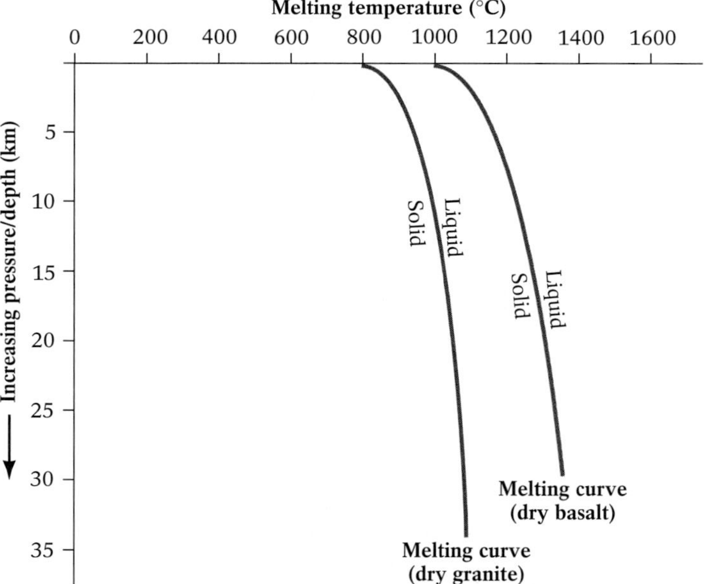

Figure 3-7 Melting-temperature curves for dry basalt and dry granite. (As you will see in Figure 3-9, adding water to rock changes its melting curve.) For both, melting temperatures increase with increasing depth, because the pressure at greater depths stabilizes rock's crystal structure, raising its melting point.

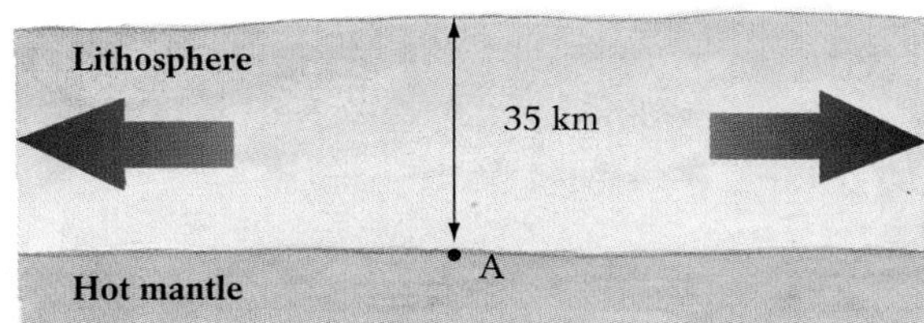

① Divergence begins.

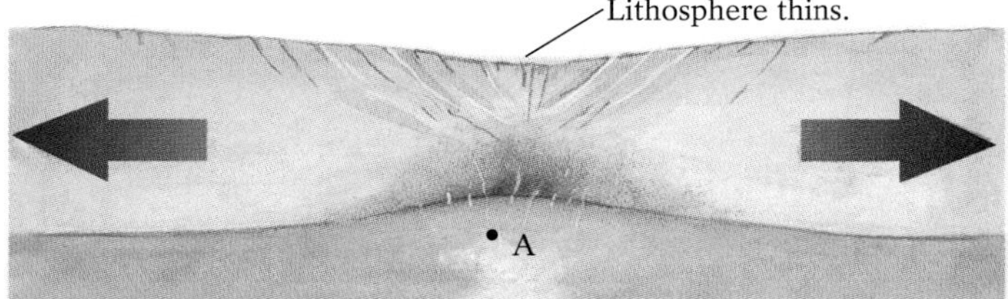

② Pressure at point A relaxes as crust thins. Hot mantle rock at point A begins to melt.

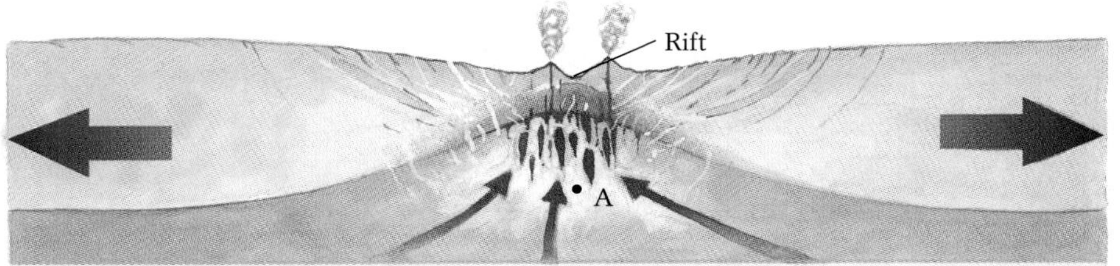

③ As divergence continues, rock at point A melts and begins to rise into the rift as magma.

Figure 3-8 The effect of lithospheric thinning on rock in the Earth's upper mantle. Hot mantle rocks remain solid primarily because of the pressure applied by the weight of overlying rock. This pressure may be removed suddenly through the plate tectonic process of rifting. Directly below the rift zone, where pressure from overlying rocks is reduced, the hot mantle rocks begin to melt and produce new mantle-derived magmas.

deeper a rock lies beneath the Earth's surface, the greater the pressure from the weight of overlying rocks and the higher the melting point. In general, then, as pressure increases, the temperature at which rocks melt also increases (Fig. 3-7). At the Earth's surface, for example, a crystal of the sodium feldspar albite melts at 1118°C (2050°F). At a depth of 100 kilometers (60 miles), however, where the pressure is 35,000 times that on the surface, a temperature of 1440°C (2650°F) is required to melt albite.

If the pressure on rock is somehow reduced or removed—as happens when tectonic plates rift and diverge (Fig. 3-8)—its melting point drops below its current temperature and it begins to melt.

Water Water, even a small amount, lowers the melting point of rocks. Recall from Chapter 2 that every water molecule has a positive and a negative side, which attract oppositely charged regions or ions in other compounds. As water molecules tug on charged ions at the surface of a mineral's crystal structure, the mineral's bonds weaken, allowing lower temperatures to vibrate and then completely break them. Under high pressure, water has an even greater effect on the melting point of a mineral: Whereas dry rocks become more resistant to melting with greater depth, wet rocks become less resistant (Fig. 3-9). This is because high pressure increases

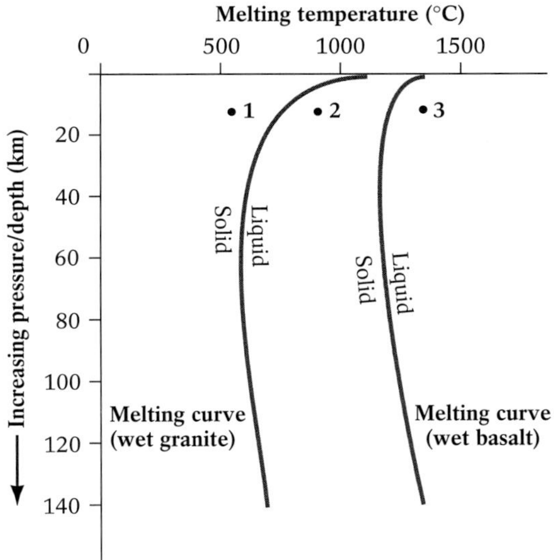

Figure 3-9 Melting-temperature curves for wet basalt and wet granite. For both, melting temperatures decrease with increasing depth, because highly pressured water destabilizes the crystal structures of the minerals in these rocks, lowering their melting points. At point 1, both granite and basalt would be solid; at point 3, both would be liquid; at point 2, granite would be liquid but basalt would still be solid.

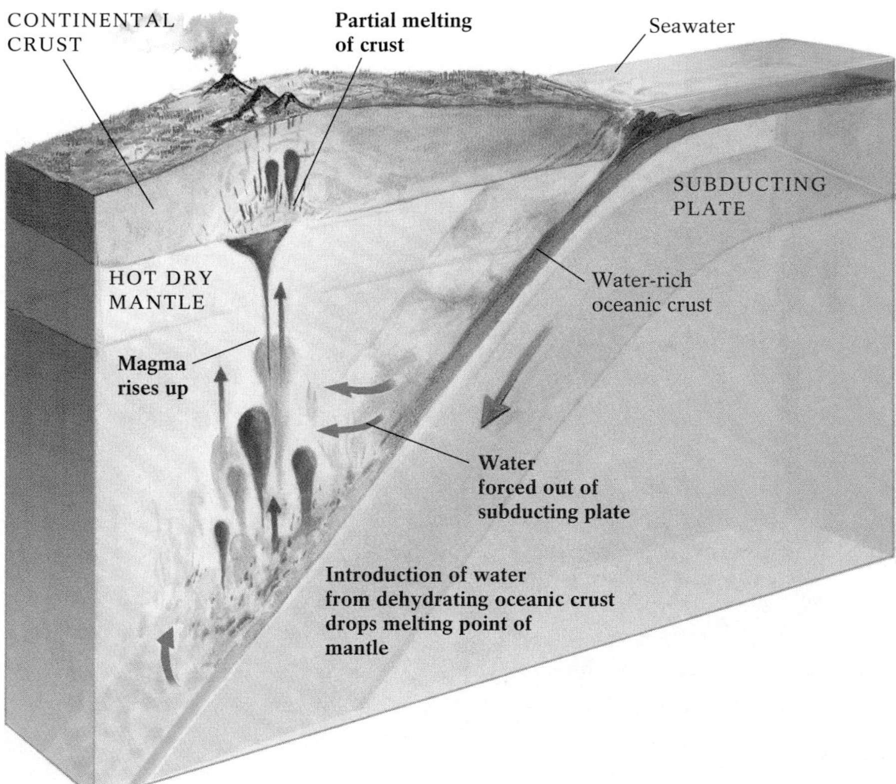

Figure 3-10 As a water-rich oceanic plate subducts, it descends into the Earth's warmer, higher-pressure interior. Increasing pressure and temperature during subduction drive water from the plate's sediments and basalts into the dry, hot mantle rocks above it. Before this water enters the mantle rocks, high pressure prevents them from melting; when water is driven into these rocks, their melting point drops and they begin to melt, producing the new magmas that fuel subduction-zone volcanism.

the bond-breaking ability of water. As we will see, this combination of high pressure and high water content is an important factor in the production of magmas at plate boundaries where oceanic plates subduct (Fig. 3-10).

The Fluidity and Viscosity of Magma

Magmas tend, for several reasons, to rise. When two materials of different densities occupy a space together, the denser is pulled down by gravity and the lighter is forced to rise. Magma rises because it is less dense than the solid rock that surrounds it. In addition, the gases in magma expand outward as it rises, helping to drive the magma upward. Finally, magma can rise when surrounding rocks press on it and squeeze it upward (see Chapter 7), much as toothpaste oozes out when the tube is squeezed.

A magma's ability to rise is largely controlled by its *fluidity*—its ability to flow—which is governed by its temperature and composition. Increased heat invariably increases the fluidity of any substance—be it maple syrup, molasses, or magma—because when the temperature of a substance increases, its ions and atoms move about more rapidly, thus breaking the temporary bonds that inhibit flow. Think of what happens when you pour cold maple syrup—thick and gluey—direct from the refrigerator into a saucepan, warm it up on the stovetop, and then pour hot, runny syrup on your pancakes. **Viscosity**—the term generally applied to describe magma flow—is a fluid's *resistance* to flow (the opposite of fluidity): Viscous fluids flow quite sluggishly; less viscous fluids flow more easily. Viscosity *increases* with *decreasing* temperature. Thus relatively cool magmas flow sluggishly; relatively hot magmas flow more easily.

The viscosity of magma generally increases with silica content (see Table 3-1), because the oxygen ions at the corners of the unbonded silicon-oxygen tetrahedra (discussed in Chapter 2) form temporary bonds with other ions in the magma. Felsic magma is very viscous because it tends to be relatively cool (it crystallizes at low temperatures) and has a high silica content. Conversely, because mafic magma is relatively hot and has a low silica content, it is much less viscous and flows easily. Therefore mafic magmas are more likely to rise to the surface and erupt than are felsic magmas, which tend to cool underground into plutonic rocks. This explains why geologists find more basalt at the Earth's surface than rhyolite. The low-viscosity basalt flows to the surface with relative ease and cools as volcanic rock; viscous rhyolitic magma is less likely to reach the surface, instead cooling underground to form granite.

The Crystallization of Magma

Eventually, every magma cools and solidifies. As it does so, it generally undergoes a number of significant changes in composition.

The temperature at which a mineral melts is the same as the temperature at which it crystallizes. Minerals that melt first (at the lowest temperatures) during heating (such as quartz and orthoclase) are thus the last to crystallize during cooling; minerals that melt last (at the highest temperatures) during heating (such as olivine and pyroxene) are the first to crystallize during cooling. A partially cooled body of magma contains solid crystals of minerals that crystallize at higher temperatures, along with a liquid containing the atoms and ions of minerals that will not crystallize until the temperature is lowered further. As the magma continues to cool, additional ions and atoms crystallize out of the melt, leaving progressively less liquid. At each stage of cooling, the proportion of crystal to liquid changes, as does the chemical interaction between them.

Bowen's Reaction Series In 1922, Canadian geochemist Norman Levi Bowen and his colleagues at the Geophysical Laboratory of the Carnegie Institution in Washington, D.C., determined the sequence in which silicate minerals crystallize as magma cools. Their work made it possible to summarize a complex set of geochemical relationships, **Bowen's reaction series**, in a single diagram (Fig. 3-11). This series demonstrates that a full range of igneous rocks, from mafic to felsic, could be produced from the same, originally mafic, magma. Bowen proposed that if early-forming crystals remained in contact with the still-liquid parent magma, they would continue to react with it (thus the term *reaction series*) and so evolve into new minerals.

Bowen's reaction series shows that the silicate minerals crystallize from mafic magmas in two ways—in a discontinuous series and in a continuous series. Ferromagnesian minerals (the iron- and magnesium-rich silicates) crystallize one after another in a specific sequence. Because each successive type of ferromagnesian mineral crystallized differs in both composition and internal structure from the one before, Bowen called this the *discontinuous series*. As mafic magma cools, the first ferromagnesian mineral to crystallize is olivine, which has a low silica content and a relatively simple structure of independent tetrahedra. The crystallization of olivine removes some iron and magnesium from the parent magma, changing its composition by increasing the proportion of the other major ions in the magma. As long as the scattered olivine crystals remain in contact with this changing liquid, they continue to *react* with it, forming new minerals that are stable within the magma's changing composition. The tetrahedra in these new minerals begin to become linked into the single-chain structure of the pyroxenes.

The discontinuous evolution of the ferromagnesian minerals continues as pyroxene crystals are transformed into the double-chained amphiboles. The series culminates in the formation of the complex sheet silicate, biotite mica, the last ferromagnesian mineral to form. By then, most of the magma's original iron and magnesium ions and atoms are in crystals. Any minerals that crystallize after biotite will contain little or no iron or magnesium.

Meanwhile, calcium plagioclase is also crystallizing at the same high temperatures as olivine and the pyroxenes. As in the case of those ferromagnesian minerals, the early-forming calcium plagioclase crystals continue to interact with the remaining liquid. Gradually sodium ions from the liquid magma replace the calcium ions in the crystals. Because one type of ion replaces a very similar ion (recall the discussion of ionic substitution in Chapter 2), there is little change in the internal structure of the plagioclases formed; therefore Bowen called this the *continuous series*. The resulting sequence of plagioclase feldspars ranges from calcium-rich anorthite, through a variety of calcium–sodium mixtures, to sodium-rich albite.

When magma cools slowly, sodium ions invade anorthite crystals gradually, starting at the surface and then dispersing through the entire crystal. When magma cools rapidly, sodium ions penetrate the

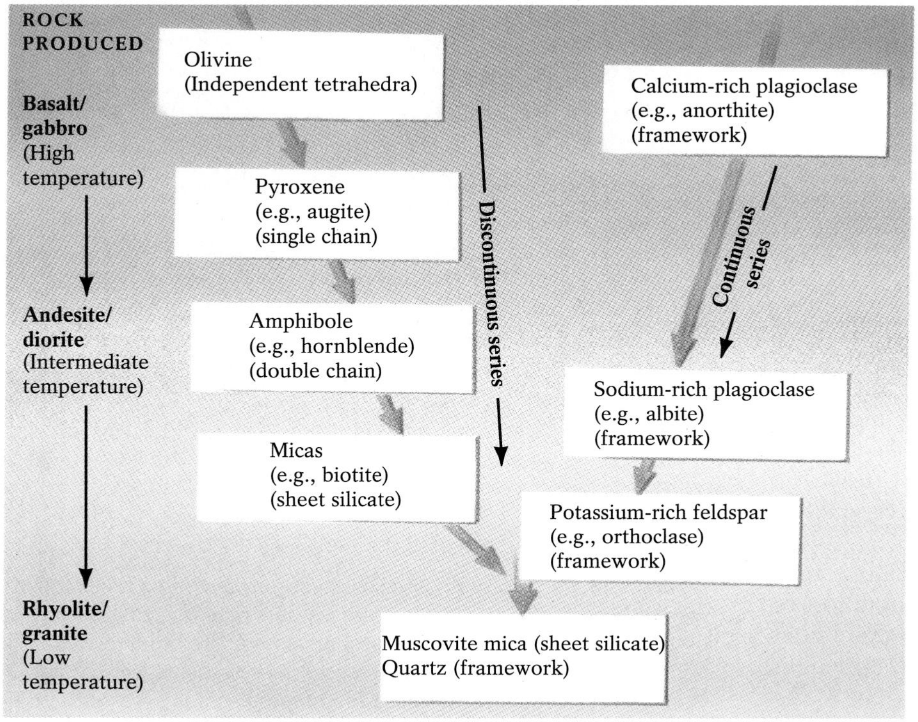

Figure 3-11 Bowen's reaction series, showing the sequence of minerals that crystallize as an initially mafic magma cools under ideal conditions.

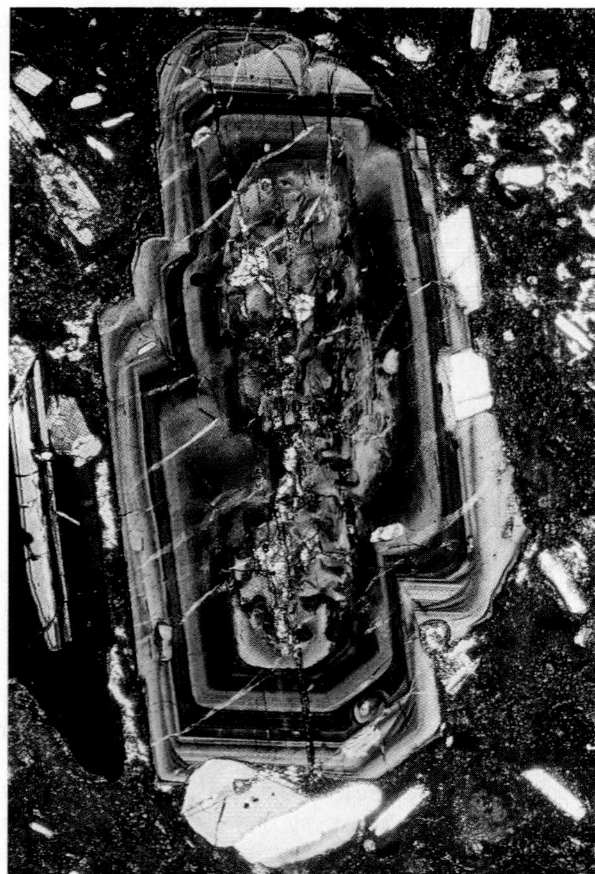

Figure 3-12 A zoned plagioclase crystal formed under rapid cooling conditions. The crystal's outer layer is dominated by sodium ions, which replaced earlier-bonding calcium ions as the crystal reacted with ions in the cooling magma. The crystal cooled too rapidly, however, to allow for complete replacement of calcium ions by sodium ions in the interior.

surface but do not have time to invade the crystal interior, which retains its calcium ions. This rapid cooling results in the formation of *zoned* plagioclase crystals, with sodium-rich rinds and calcium-rich cores (Fig. 3-12).

After the ferromagnesian minerals and the plagioclase feldspars have crystallized from an initially mafic magma, less than 10% of the original liquid remains. Depending on its initial composition, this liquid may now contain high concentrations of silica, potassium, and aluminum. In such a case potassium feldspar, potassium-aluminum mica (typically muscovite), and quartz are the last minerals to crystallize.

How Magma Changes as It Cools Bowen's reaction series, developed in the laboratory, assumes an ideal condition in which all the early-forming crystals remain in the liquid magma and react with it until crystallization is complete. In nature, however, this condition rarely applies (Fig. 3-13). As a magma cools, several things may happen to the crystals that

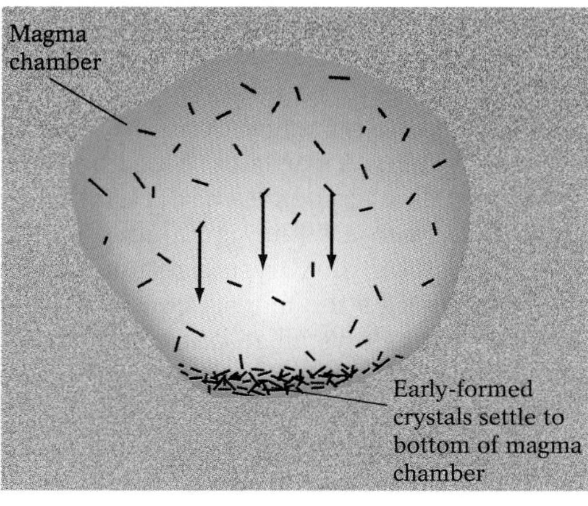

(a)

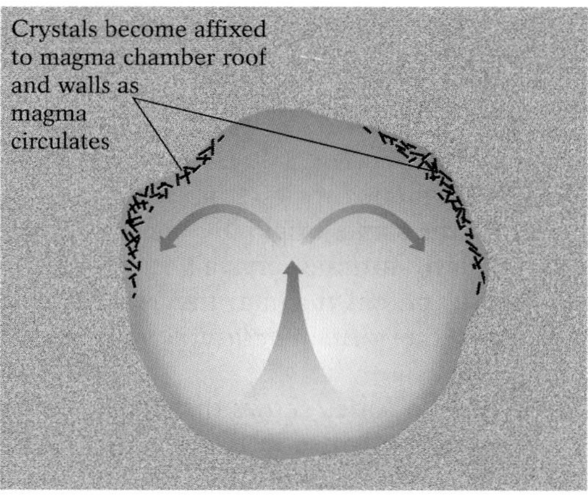

(b)

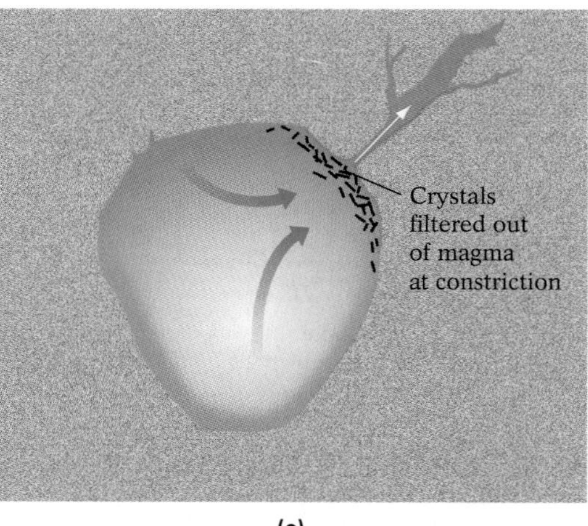

(c)

Figure 3-13 Early-forming crystals do not always remain in contact with the liquid magma. Instead, the crystals may **(a)** settle to the bottom of the magma chamber; **(b)** become affixed to the walls and roof of the magma chamber as magma circulates within the chamber; or **(c)** be filtered out of the magma as it is pressed through small fractures in the surrounding rock.

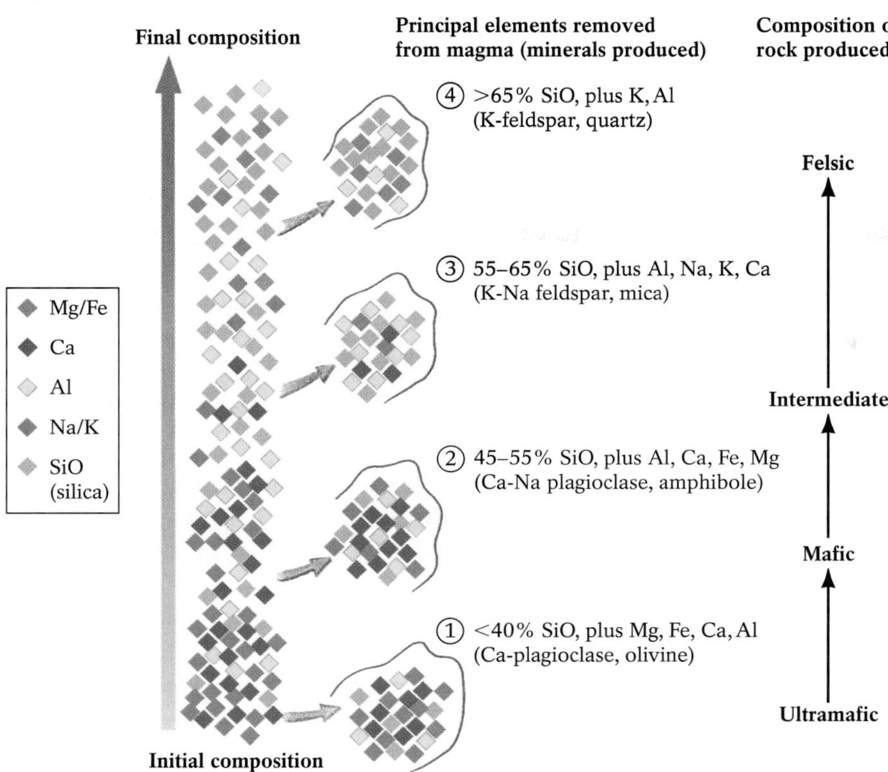

Figure 3-14 When early-forming minerals (such as olivine and calcium plagioclase) are removed from a magma, they deplete the remaining magma of a significant portion of early-crystallizing elements such as magnesium, iron, and calcium; in turn, the magma contains a higher proportion of later-crystallizing elements, such as sodium, aluminum, and potassium. Thus igneous rocks that crystallize late from such evolving magma have different compositions (progressively more felsic) than rocks that crystallize earlier.

form. Some do remain suspended, continuing to exchange ions and atoms with the liquid. But early-forming crystals might also drop out of the magma and no longer react with it. Crystals that are denser than the surrounding liquid *may sink to the bottom of the magma chamber* and become buried by later-settling crystals. The hot rising liquid *may plaster the crystals against the walls or ceiling of the magma chamber*. The largest crystals *may even be filtered out of the melt* entirely as the magma flows into fractures too narrow for them to pass. Removal of crystals in any of these ways substantially affects the composition of the remaining magma and thereby the composition of any rocks that may later form from it (Fig. 3-14).

A magma that loses crystals at various stages of its cooling has in effect become separated into a number of independently crystallizing bodies. The rocks that form from such a magma differ in composition both from one another and from the original magma, with each successive body crystallized being more silica-rich than the last (see Fig. 3-14). By this process, called **fractional crystallization**, a single parent magma can produce a variety of igneous rocks of different compositions. The New Jersey Palisades, a line of cliffs on the west bank of the Hudson River, are a classic example of this phenomenon (Fig. 3-15).

Other Magma Crystallization Processes Bowen believed that all igneous rocks form by fractional crystallization of mafic magmas, evolving according to his reaction series. Geologists building on Bowen's work, however, realized that these processes alone could not account for all igneous rocks. For example, although felsic igneous rocks can crystallize from mafic magmas, too little magma remains after the mafic minerals crystallize to produce large bodies of felsic and intermediate rocks. How, then, does one explain the 1000-kilometer (600-mile) stretches of felsic igneous rock on our continents? They must have been produced by processes other than fractional crystallization of mafic magma.

One process that may account for such vast areas of intermediate and felsic rock is *assimilation*. As magma moves, blocks of rock from the walls of the magma chamber may break free and be wholly or partially melted by the surrounding hot magma. Assimilation of such rock bodies can substantially alter a magma's composition. In addition, two or more different bodies of magma may flow together and mix to form a magma of hybrid composition (a process known as *magma mixing*). The 1912 volcanic eruption in Alaska's Aleutian Islands produced rocks containing both felsic and mafic minerals, suggesting that two distinct bodies of magma had combined to fuel the eruption. A similar style of eruption occurred at Mt. Pinatubo in the Philippines in 1991.

Intrusive Rock Formations

No one has ever actually seen magma move underground, so we can only speculate about how it does from the igneous formations that we can see. For example, in some areas erosion has exposed thousands of square kilometers of solidified intrusive magma. Did these vast regions result from magma flowing into preexisting subterranean cavities? That is unlikely: the weight of overlying rocks at depths below about 10 kilometers (6 miles) would have collapsed and destroyed such large underground spaces. The most likely scenario is that magma moved forcefully into cracks in preexisting rock, actually pushing the rock aside to create its own space. In a similar way, rising magma may force overlying rocks to bulge upward. The resulting igneous rock appears as a domed intrusion within other rocks, a distinctive igneous structure known as a *diapir* (DI-a-pir).

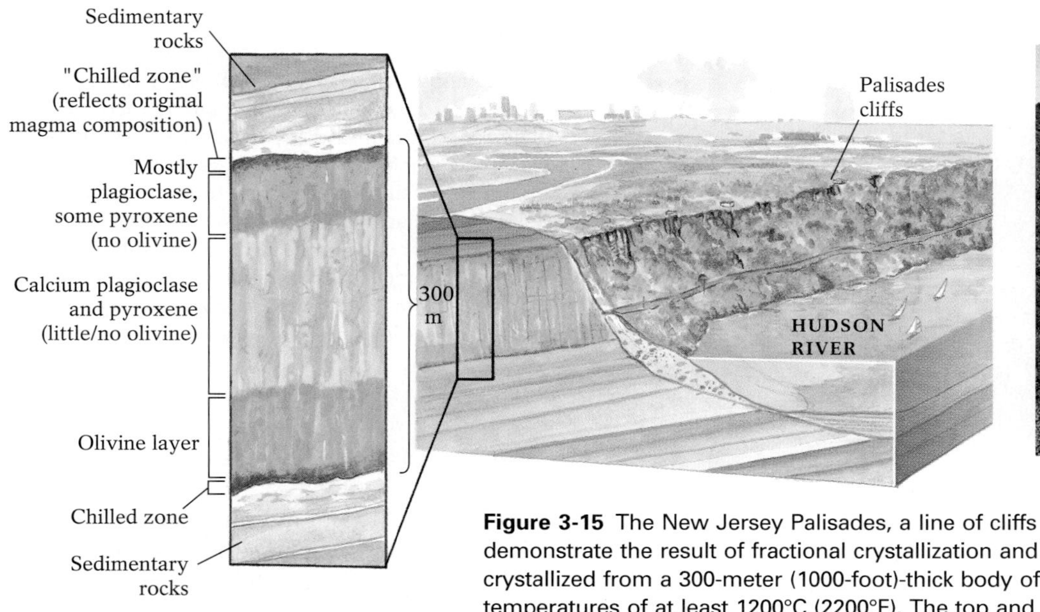

Figure 3-15 The New Jersey Palisades, a line of cliffs in the northeastern part of the state, demonstrate the result of fractional crystallization and crystal settling. The rocks of the Palisades crystallized from a 300-meter (1000-foot)-thick body of magma that intruded preexisting rocks at temperatures of at least 1200°C (2200°F). The top and bottom margins ("chilled zone") of the Palisades solidified very rapidly without undergoing fractional crystallization, because the magma came into contact with cold surrounding rocks; these rocks provide us with a glimpse of the magma's original composition. The bottom third of the Palisades has a high concentration of olivine crystals, the central third is a mixture of calcium plagioclase and pyroxene with no appreciable olivine, and the upper third consists largely of plagioclase with no olivine and little pyroxene. It appears that early-forming olivine crystallized and then settled to the bottom of the magma body; pyroxene and plagioclase crystallized next, with the denser pyroxenes settling and concentrating in the center, and the lighter plagioclases occupying the uppermost section. Since all the magma of the Palisades cooled and solidified fairly quickly, there was no residual magma from which later-forming minerals could crystallize.

When moving magma incorporates preexisting rock as it rises, some of the incorporated rocks melt and become assimilated into the magma; unmelted rocks are carried within the magma. When this magma eventually solidifies, we can see such "foreign" rocks as distinctly different rock masses called **xenoliths** (ZEEN-o-liths)(from the Greek *xenos*, or "stranger," and *lithos*, or "stone") (Fig. 3-16).

The distinctive subsurface intrusive igneous forms created by flowing underground magma are known as **plutons**. Plutons may be classified by their position relative to the preexisting rock, called *country rock*, surrounding it: *Concordant* plutons lie parallel to layers of country rock; *discordant* plutons cut across layers of country rock. Plutons of both varieties come in a range of shapes and sizes, as shown in Figure 3-17. Erosion of overlying rocks may eventually expose these formations.

Figure 3-16 A dioritic xenolith within granite. The granitic magma encompassed the preexisting dioritic rock (seen here as a dark gray mass within the lighter gray granite) but was not hot enough to melt it. As a result, the diorite was preserved as a discrete mass when the magma eventually solidified.

Tabular Plutons

Tabular plutons are slab-like intrusions of igneous rock that are broader than they are thick, like a table top. If magma flows into a relatively thin fracture in country rock, or pushes between sedimentary rock layers, a tabular pluton will result when the magma cools. Tabular plutons may be as small as a few centimeters thick or as large as several hundred meters thick.

A **dike** is a discordant tabular pluton that cuts across preexisting rocks. Dikes are generally steeply inclined or nearly vertical, suggesting that they formed from rising magma, which tends to follow the most direct route upward. Dikes often occur in clusters where magma apparently infiltrated and solidified in a network of fractures.

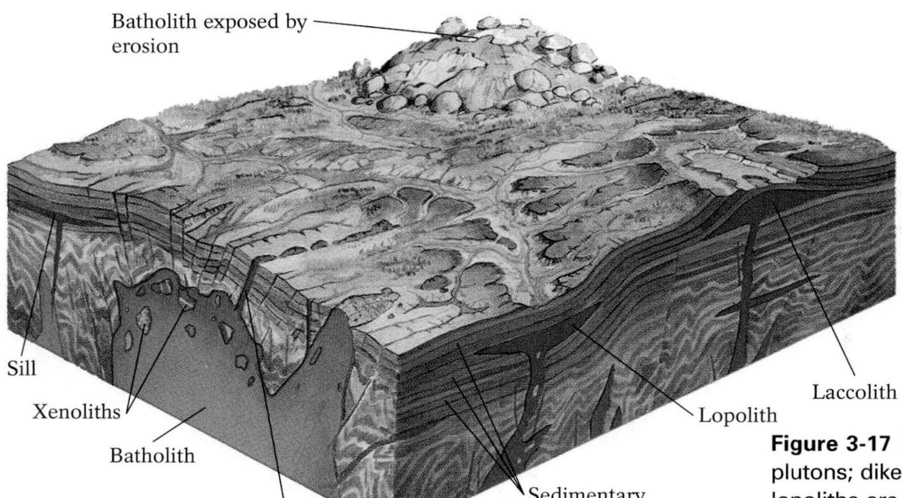

Figure 3-17 Plutonic igneous features. Sills are concordant tabular plutons; dikes are discordant tabular plutons. Laccoliths and lopoliths are larger concordant plutons, and batholiths are even larger discordant plutons.

Many dikes result from magma that rises into volcanoes and then solidifies; we can see these erosion-resistant rocks when the less resistant volcanic material at the surface wears away. Some dikes diverge, like the spokes of a bicycle wheel, from a *volcanic neck*, a vertical pluton remaining in what was once a volcano's central magma pathway. If you travel along Route 64, running through the Navajo and Hopi lands of the Four Corners area (the intersection of Colorado, New Mexico, Arizona, and Utah), you can see more than a hundred such volcanic necks, remnants of ancient volcanic plumbing (Fig. 3-18).

A **sill** is a concordant tabular pluton lying parallel to layers of preexisting rocks. Sills are produced when intruding magma enters a space between layers of rock, melting and incorporating adjacent sedimentary material. Sills generally form within a few kilometers of the Earth's surface; at greater depths, overlying rocks tend to compress and close off any spaces into which magma might flow.

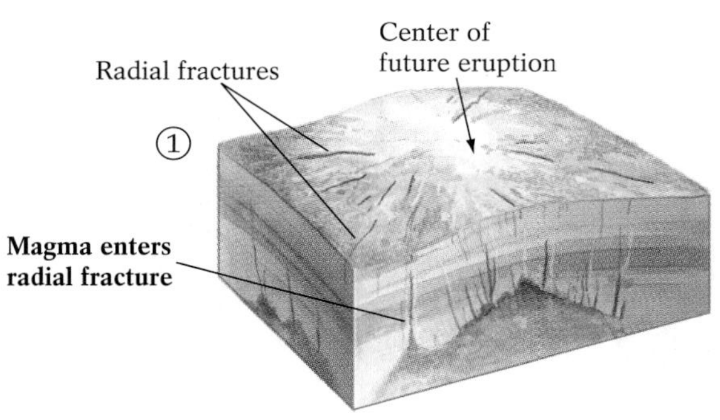

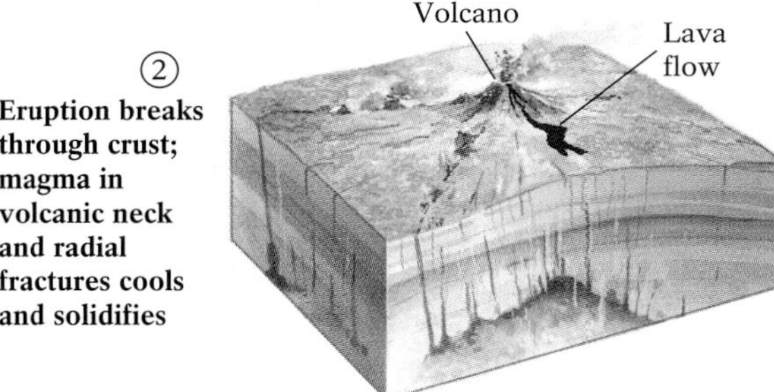

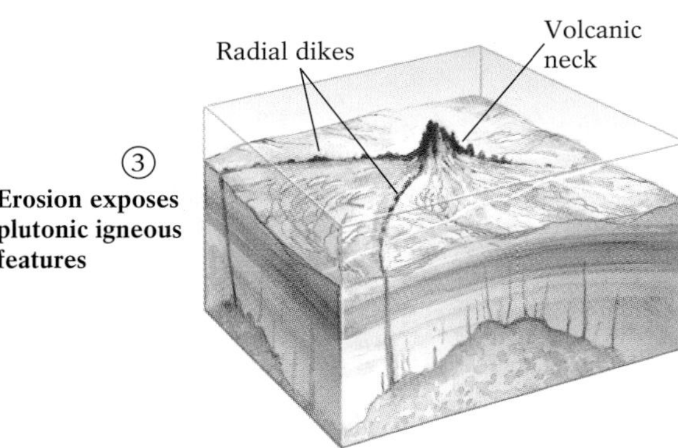

Figure 3-18 Shiprock Peak, in New Mexico, is believed to be a volcanic neck, the congealed lava from the interior of a former volcanic cone. Erosion of the surrounding sedimentary rock and the cone itself has exposed this volcanic neck and radial dikes.

Highlight 3-1 Tabular Plutons Save the Union

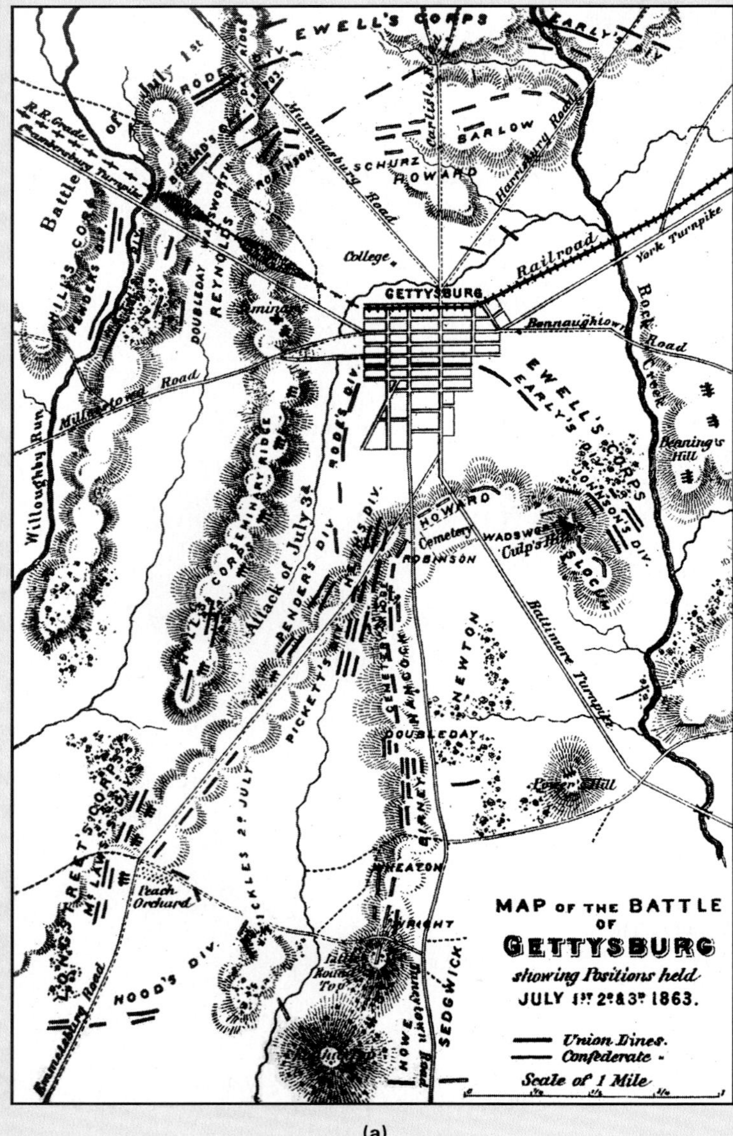

(a)

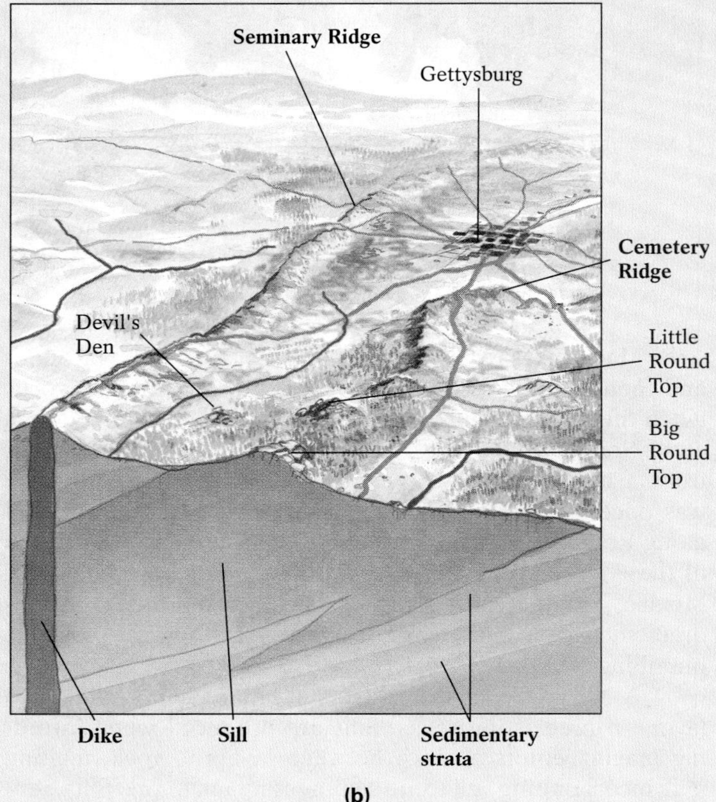

(b)

Figure 3-19 **(a)** A Civil War–era map of the site of the Battle of Gettysburg. Seminary Ridge appears in the upper left; Cemetery Ridge is to its lower right. **(b)** A contemporary artist's rendering of the relevant topographic features.

The Battle of Gettysburg, which lasted three days and took the lives of tens of thousands of Civil War soldiers, was effectively won by the Union on a hot July 3rd in 1863. On this day, Confederate troops ventured forth from their outpost on a narrow dike of resistant basalt called Seminary Ridge to charge against the Union stronghold on the equally resistant but thicker basaltic sill called Cemetery Ridge (Fig. 3-19). (This offensive would become known as "Pickett's charge.") The steep forward slope of the Cemetery Ridge sill impeded the Confederate charge, and a protective wall constructed from basaltic boulders by Union troops concealed them and repelled Confederate shots. Thus, with an assist from a well-placed basaltic sill, Union forces defeated the Confederate offensive at Gettysburg, a turning point in the American Civil War.

A sill and a dike in the south-central Pennsylvania town of Gettysburg provided the setting for an event that affected the course of United States' history. This event is recounted in Highlight 3-1, "Tabular Plutons save the Union."

Sills pose an interesting challenge to geological detectives. How can we distinguish a sill intruded between layers of preexisting rock from a lava flow that is buried under subsequent flows or sedimentary rocks? There are several critical clues (Fig. 3-20):

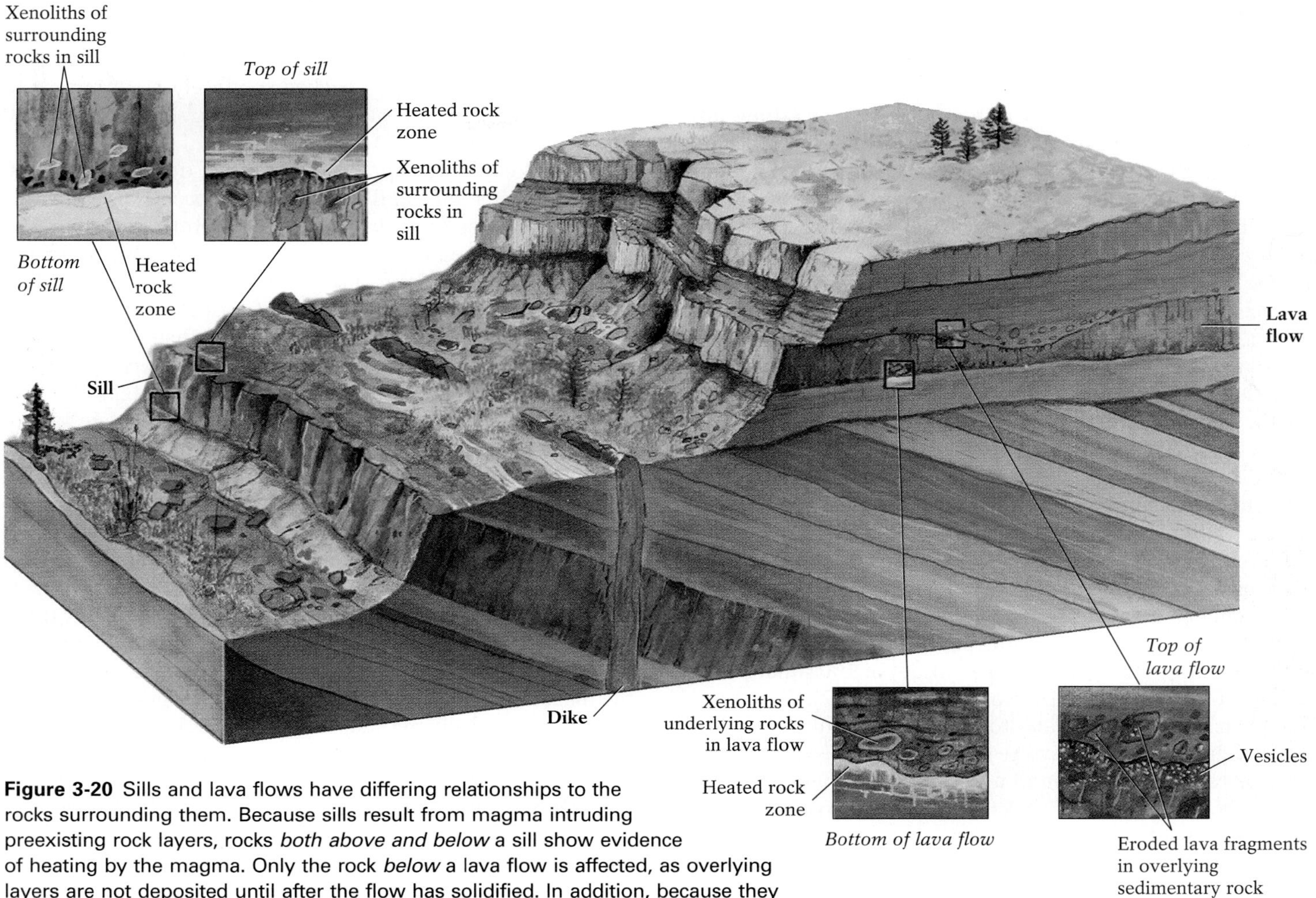

Figure 3-20 Sills and lava flows have differing relationships to the rocks surrounding them. Because sills result from magma intruding preexisting rock layers, rocks *both above and below* a sill show evidence of heating by the magma. Only the rock *below* a lava flow is affected, as overlying layers are not deposited until after the flow has solidified. In addition, because they occur at the surface, lava flows show evidence of having been exposed to air, whereas sills do not.

1. Look at the adjacent surfaces of the surrounding rocks. When lava is extruded, there are no overlying rocks and only the rocks *beneath* it will show evidence of heating. When hot magma intrudes between two layers of rock to form a sill, it heats the adjacent surfaces of *both* layers before cooling.
2. Look at the top and bottom surfaces of the igneous layer. Both surfaces of a sill will contain fragments of the surrounding rock that were pried loose as magma intruded, whereas only the bottom of a lava flow incorporates preexisting rock.
3. Look at the top surface of the igneous layer. Because the top of a lava flow is exposed to the air for some time before being overlain by other lava or sediment, its gases are free to escape. Consequently, the surface of a lava flow is often pockmarked by cavities called *vesicles* that were formerly occupied by escaping gas bubbles. The tops of most sills, which were never exposed to the air, display few, if any, vesicles.
4. Look for signs of weathering (discussed in Chapter 5). The upper surface of a lava flow would appear somewhat weathered from its exposure to the atmosphere before being overlain, whereas a sill would not show signs of weathering because it was never exposed.

Batholiths and Other Large Plutons

Large concordant plutons are commonly several kilometers thick and tens or even hundreds of kilometers across. They may be mushroom-shaped or saucer-shaped, close to the surface or deep beneath it.

When thick, viscous magma intrudes between two parallel layers of rock and lifts the overlying one, it cools to form a mushroom-shaped or domed concordant pluton, or **laccolith** (from the Greek *lakkos*, or "reservoir") (see Fig. 3-17). Laccoliths tend to form at relatively shallow depths, where there is little pressure to keep the overlying rock in place. Many are granitic, formed from felsic magma that flows so slowly that it tends to bulge upward instead of spreading outward. The overlying rock rises to form a dome; when this overlying dome erodes away, the igneous rock below is exposed (Fig. 3-21). Sills form in a similar way, but they are typically basaltic and relatively flat because they form from easier-flowing mafic magmas that can enter small spaces readily and spread outward.

Unlike upward-bulging laccoliths, saucer-shaped concordant plutons called **lopoliths** (from the Greek *lopas*, or "saucer") sag downward (see Fig. 3-17). These are probably produced when mafic magma rises from a deep magma source; the weight of the dense magma then depresses the underlying rock into the space from which the magma has risen. One such structure is exposed along the western shore of Lake Superior, where the overlying country rock has been eroded away (Fig. 3-22). Lopoliths often contain mineralogically distinct layers. The layers on the bottom near the floor of the magma chamber hold the densest early-forming crystals. Several layers near the base of the Bushveld lopolith of South Africa contain the Earth's richest concentration of the dense metal platinum.

Some igneous intrusions are even vaster. **Batholiths** (from the Greek *bathos*, or "deep") are massive discordant plutons with surface areas (when exposed) of 100 square kilometers (40 square miles) or more (see Fig. 3-17). They typically form from intermediate- to high-viscosity magmas (diorites and granites) that, because they flow so slowly, are more likely to remain underground and crystallize than to rise to the surface through small cracks and fissures. Most batholiths are a complex of many large plutons that have intruded one another in stages, often over tens of millions of years. The Sierra Nevada batholith in east-central California, which consists of numerous plutons packed so closely as to appear to be a continuous mass, probably formed over a span of 130 million years (Fig. 3-23).

Most batholiths appear to form at a depth of about 30 kilometers (20 miles). They generally occur in elongated mountain ranges and can be seen where erosion of overlying rocks exposes deep cores of plutonic rocks, as in the Yosemite Valley of California's Sierra Nevada and the White Mountains of New Hampshire. The Coast Range batholith of western British Columbia stretches more than 2000 kilometers (1200 miles) and is as much as 290 kilometers (180 miles) wide. Batholiths also reside in the ancient centers of continents, the deep, erosion-resistant rocks of long-gone mountains; 2 to 3 billion years ago, such mountains may have risen sharply at convergent plate boundaries. In North America, they have since been relegated to their mid-continental positions in what are now Minnesota, Wisconsin, Michigan, and Ontario by vast collisions of ancient lithospheric plates. (Figure 1-22 illustrates this process.)

The texture of batholith rock is relatively aphanitic (fine-grained) at the exterior and coarsens gradually to a phaneritic interior. The fine texture develops where the margins of the magma come in contact with the cooler country rock that it intrudes, crystallizing relatively quickly. The coarse-grained interior texture develops where the magma was able to cool more slowly. Recent studies using earthquake waves passing through batholiths underground suggest that their shape may resemble that of a human tooth, the incisor—increasing with depth to its widest point and then tapering.

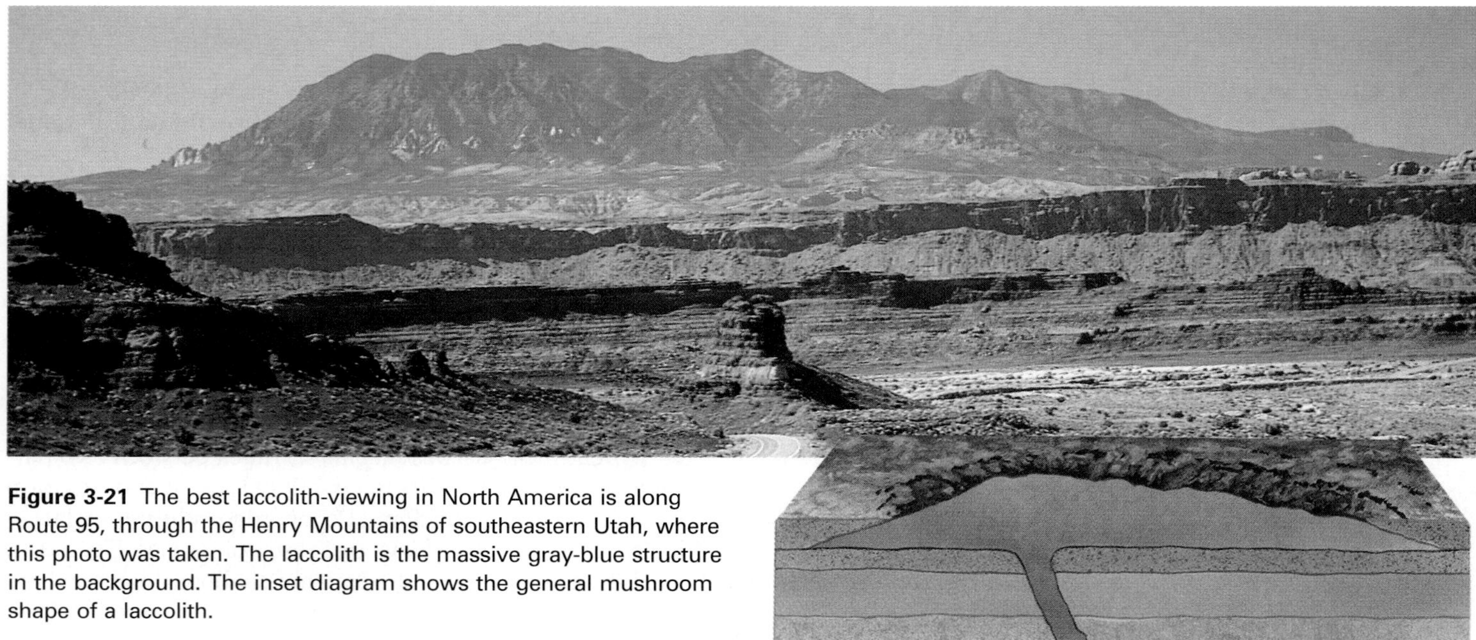

Figure 3-21 The best laccolith-viewing in North America is along Route 95, through the Henry Mountains of southeastern Utah, where this photo was taken. The laccolith is the massive gray-blue structure in the background. The inset diagram shows the general mushroom shape of a laccolith.

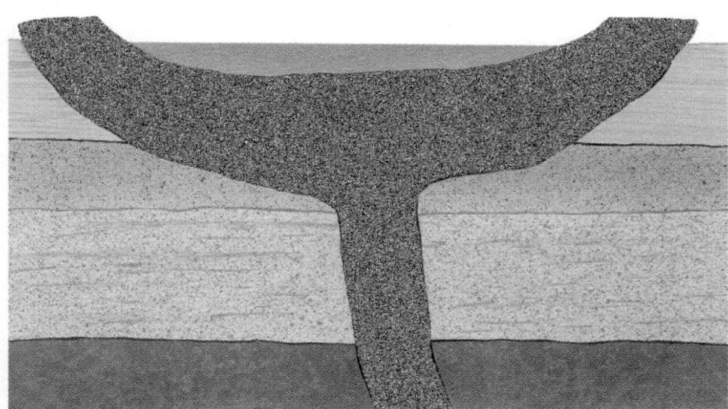

Figure 3-22 The gabbroic Duluth lopolith, a classic North American lopolith on the shores of the western end of Lake Superior in Minnesota, is more than 250 kilometers (150 miles) in diameter and about 15 kilometers (10 miles) thick. This photo shows only a portion of the exposed rock. The inset diagram shows the general mushroom shape of a lopolith.

Figure 3-23 Virtually all the rock exposed in Kings Canyon National Park is composed of only one rock type—diorite. The panorama of dioritic rock here is part of the Sierra Nevada batholith.

Figure 3-24 Plate settings and the various types of basalt. Basalts form at divergent plate boundaries, atop intraplate hot spots, above subduction zones, and in zones of mid-plate rifting. The different compositions of these basalts are mostly a result of the fact that they form from magmas originating at varying depths in the Earth's mantle.

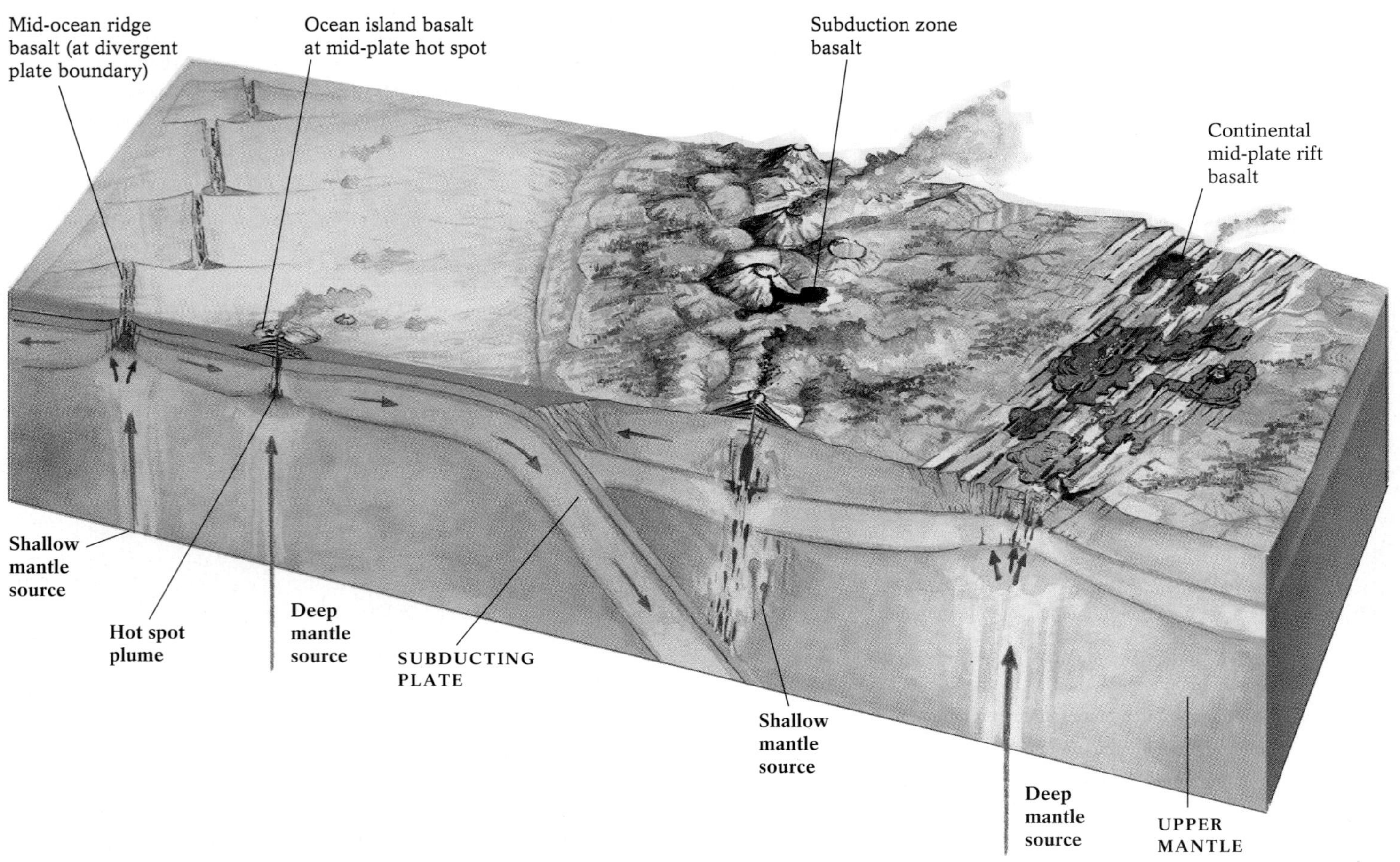

Plate Tectonics and Igneous Rock

The worldwide distribution of igneous structures and rocks *is not random*. Certain structures and rocks consistently appear in some geological settings but not in others. Plutonic igneous structures, for example, tend to form at or near the boundaries of diverging or converging plates. These plate movements provide openings and opportunities for magma to intrude older rocks. Smaller plutonic features, such as dikes and sills, generally occur in divergent or rifting zones, where mafic magmas move as the Earth's brittle outer layers are stretched and pulled apart. Where oceanic plates have subducted, intermediate and granitic batholiths exist, marking many modern and ancient plate boundaries. The chain of western batholiths in North America, stretching from British Columbia through the California Sierras to Baja California, developed through more than 200 million years of oceanic-plate subduction.

The Origin of Basalts and Gabbros

Basalt and gabbro (the intrusive equivalent of basalt) are virtually the only igneous rocks in oceanic crust. Because the thin oceanic crust lies directly over the mantle, the source for the magma must be the ultramafic mantle. But mantle material is not homogeneous in composition, and therefore neither are the world's basalts and gabbros. The varied compositions of these rocks seem to depend on whether they derived from deep- or shallow-mantle magma sources.

During the Earth's earliest millennia, the uppermost segment of the mantle apparently melted partially and its lighter components rose to become the Earth's earliest crust. This process largely depleted the remaining upper mantle of such light elements as sodium, potassium, and aluminum. Small amounts of these elements may still exist below the upper mantle, however. A basalt or gabbro that contains these elements probably derived from the mantle's deeper, undepleted zone, whereas rocks lacking these elements probably crystallized from magma that came from the depleted uppermost mantle.

The low viscosity of mafic magma allows much of it to flow to the Earth's surface and erupt as basalt. Consequently, gabbros are relatively rare, and most of what we know about the origin of mafic rock is from the Earth's basalts. These are grouped into two main categories—oceanic and continental—based on the general environmental setting in which they formed (Fig. 3-24). Within each of the categories, basalts can be further distinguished by the related factors of composition, magma source, and plate-tectonic setting.

Oceanic Basalts *M*id-*o*cean *r*idge *b*asalts, or MORBs, the most abundant volcanic rock, account for about 65% of the Earth's surface area. Eruptions at oceanic divergent boundaries produce these rocks. Because they have low concentrations of sodium, potassium, and aluminum, MORBs probably formed from partial melting of the upper mantle, which is depleted of these elements.

Ocean *i*sland *b*asalts, or OIBs, are found not at divergent plate boundaries but atop *hot spots*, volcanic zones (generally intraplate) that lie over deep-mantle heat sources. Because they contain small but significant amounts of sodium, potassium, and aluminum, OIBs probably originated from a deep part of the mantle that was not depleted of those elements. MORBs generally erupt unwitnessed beneath thousands of meters of seawater, but we can readily see eruptions of OIBs in such places as the Hawai'ian Islands.

Continental Basalts Basalts in continental settings vary more than their oceanic counterparts. They form both where new rifts tear at old continental plates and where oceanic plates subduct. The compositions of basalts associated with continental rifting show that they most likely derived from deep-mantle sources, whereas those at subduction zones tap shallower sources. Both, however, form as hot basaltic magma rises through tens of kilometers of continental crust, incorporating many of the materials in its path. Thus the varied composition of continental basalts may result from melting and assimilation of continental rocks as well as from partial melting of deep and shallow mantle rocks.

The Origin of Andesites and Diorites

Whereas basalts and gabbros are the most common igneous rocks of the ocean basins, the less mafic andesites and diorites abound along the geologically active subductive margins of continents and on oceanic islands that rose from the sea floor when oceanic plates subducted. Nearly continuous regions of andesitic rock are found on virtually all the lands that border the Pacific Ocean, following the nearly continuous pattern of subduction zones surrounding the Pacific Ocean basin (Fig. 3-25).

Geologists believe that a number of processes combine to produce rocks of intermediate composition from a subducting oceanic plate. One likely factor is water, which may be bonded to the hydrated minerals of the rocks of the subducting oceanic crusts. Water is also trapped in spaces within the sediments blanketing the descending plate, within the crystal structures of minerals in the sedimentary muds, and within fractures in the oceanic basalt. As an oceanic plate subducts, its water is pressed out into the overlying mantle, which, although it is quite warm, has been kept solid by the prevailing pressure. Recall that water lowers the melting point of rocks under pressure. Thus water driven from the subducting plate enters the warm mantle and lowers its melting point, promoting partial melting. Because partial melting of the ultramafic rocks in the upper mantle generally produces mafic magma, a new batch of basaltic magma forms above a subducting oceanic slab.

Why, then, does more andesite than basalt erupt from subduction-zone volcanoes? Some other process (or processes) must add felsic material to these basaltic magmas to produce the intermediate composition of andesite. For example, as subduction-zone magmas rise through overlying felsic and

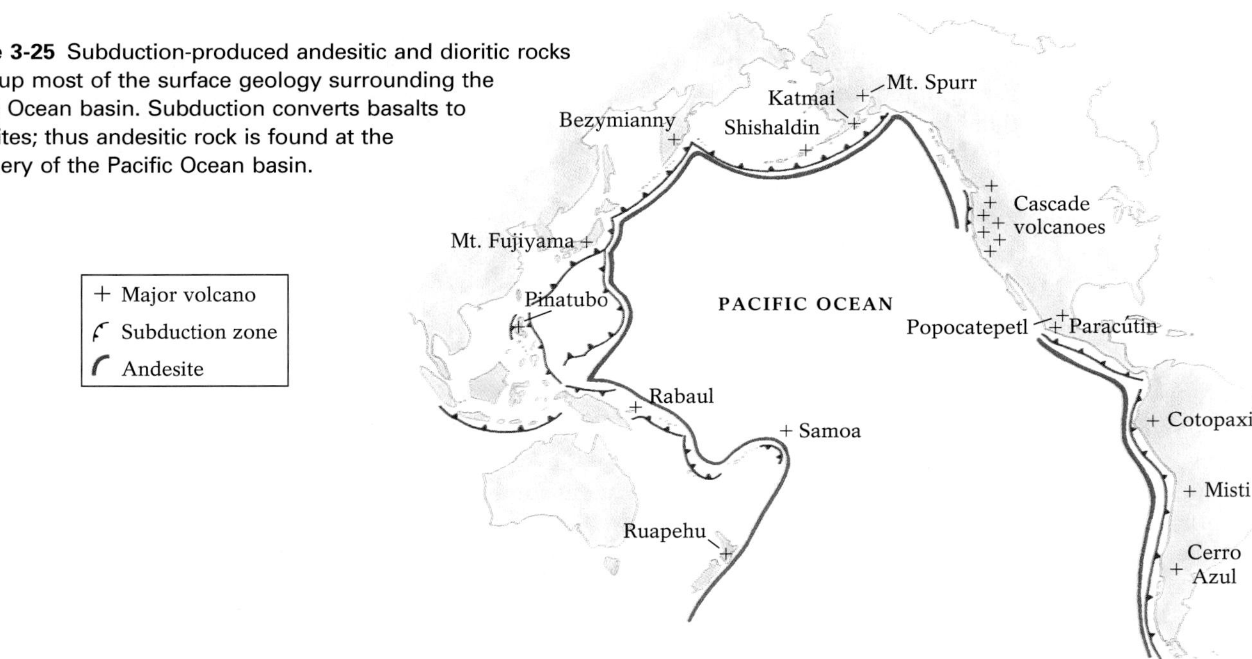

Figure 3-25 Subduction-produced andesitic and dioritic rocks make up most of the surface geology surrounding the Pacific Ocean basin. Subduction converts basalts to andesites; thus andesitic rock is found at the periphery of the Pacific Ocean basin.

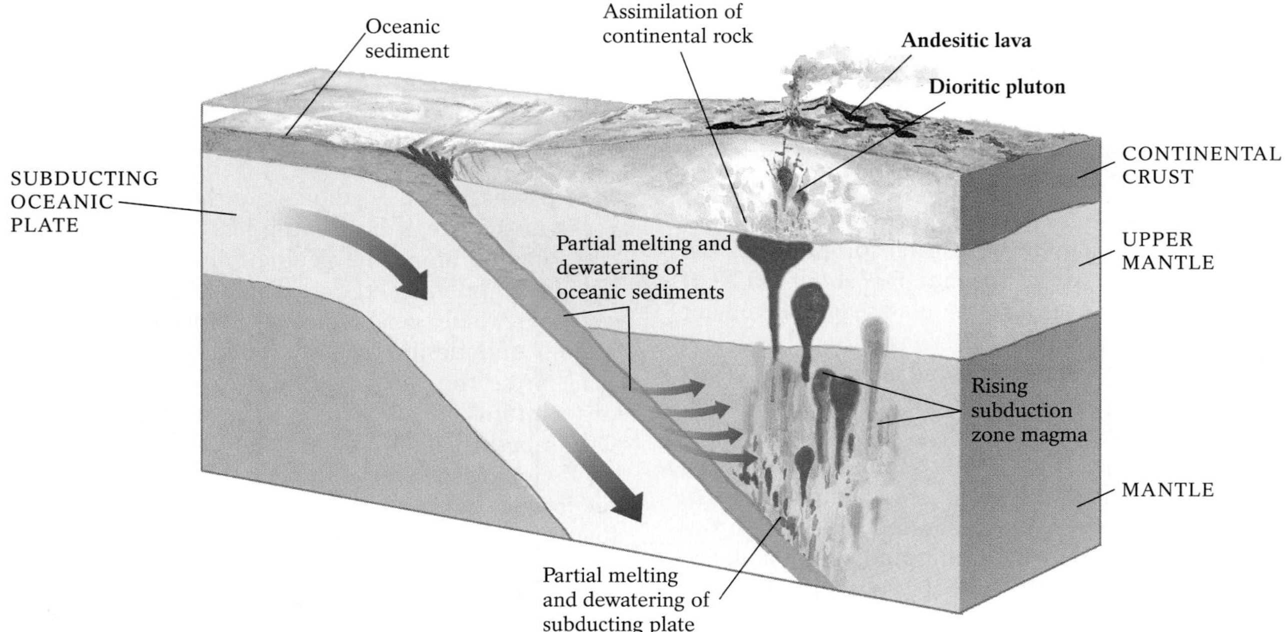

Figure 3-26 The factors involved in the origin of andesite and diorite. Subduction presses water out of the descending plate and its associated oceanic sediment. The water then enters the mantle rock and lowers its melting point, causing it to melt and rise as basaltic magma. The composition of this initially mafic magma is made more intermediate by its mixing with partially melted felsic oceanic sediment and oceanic crust from the subducting plate, as well as with felsic country rock assimilated by the magma as it rises up through the continental crust.

intermediate rocks, they may assimilate these materials, thus changing the composition of the rising magma. Additionally, some small fraction of the 200 meters (650 feet) of oceanic sediment that, on average, covers oceanic plates may melt and enter the magma mix.

Oceanic sediment comes principally from the airborne debris of continental volcanism (usually intermediate to felsic in composition), the felsic minerals transported to the oceans by continental rivers, and the silica-based shells and skeletons of microscopic marine organisms. Carried into the hot mantle on subducting plates, some of these felsic materials may melt and then mix with mantle-derived basalt to produce an intermediate magma that cools to form andesite or diorite. Although this topic is still the subject of intense debate and research, some of the various factors that may contribute to the production of andesite and diorite are summarized in Figure 3-26.

The Origin of Rhyolites and Granites

Nearly all rhyolitic and granitic rocks occur on continents. Many geologists believe that these rocks originate principally from the partial melting of rocks in the lower continental crust, and there is convincing laboratory evidence for this. When pressures and temperatures comparable to those at the lower crustal depth of 35 to 40 kilometers (21–24 miles) are applied to wet rocks of typical continental compositions (similar in composition to andesite/diorite), the rocks partially melt to yield the felsic rhyolitic and granitic magmas.

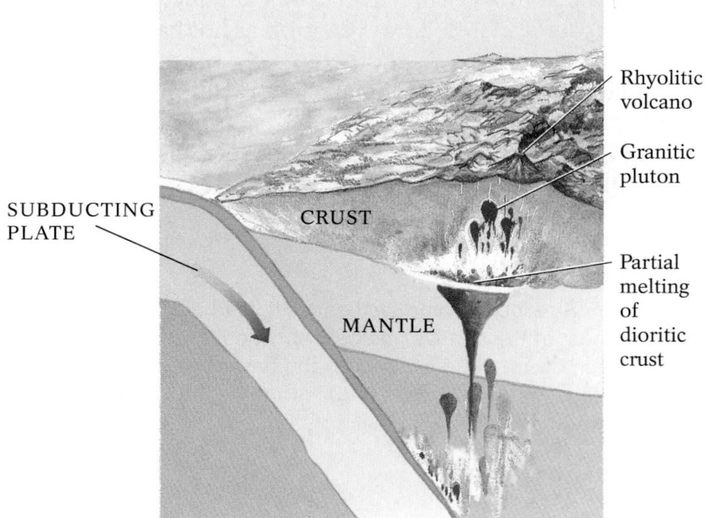

Figure 3-27 The origin of felsic rocks at convergent plate boundaries. Hot rising mafic and intermediate magmas partially melt dioritic rocks in the lower continental crust, producing granitic plutons and granite's volcanic equivalent, rhyolite.

Most granitic intrusions appear at or near modern or ancient subduction plate margins. Apparently, rising hot mafic and intermediate magmas and the frictional heat that accompanies subduction cause partial melting of gabbroic and dioritic rocks in the crust at the base of plate-edge mountain belts. Partial melting of mafic and intermediate rocks pro-

Figure 3-28 A telescopic view of the near side of the Moon, showing its highlands and maria. (Dark areas are the maria; light areas are the highlands.)

duces a new magma that is predominantly felsic. Because these magmas are typically very viscous, they generally rise slowly and tend to cool at depth, producing the felsic plutonic rock granite. When such magmas reach the surface, as they sometimes do when their water content is high, they erupt as rhyolite (Fig. 3-27).

Igneous Rocks of the Moon

Since 1969, geologists who study igneous rocks have extended the reach of their rock hammers some 400,000 kilometers (240,000 miles) to the Moon. Moon rocks collected by Apollo astronauts in the 1970s indicate that the Moon's surface contains at least two distinct types of geological/geographical provinces—the highlands and the *maria* (MAR-ee-a) (plural of Latin *mare*, or "sea") (Fig. 3-28).

The rocks of the lunar highlands date from about 4.0 to 4.5 billion years ago, when the early Moon's interior was apparently hot enough to develop a multilayered internal structure. (As we saw in Chapter 1, a planetary body that is entirely or partially molten separates into distinct layers as its lighter materials rise and denser materials sink toward the interior.) The rocks of the lunar highlands, which probably crystallized as the Moon's earliest crust, consist principally of anorthosite, a coarse-grained plutonic igneous rock composed almost exclusively of the calcium plagioclase anorthite.

Differences between Moon rocks and Earth rocks demonstrate some fundamental geological distinctions between the Earth and the Moon. First, the Moon rocks we've studied have been totally waterless. If there ever was any water on the Moon, it became heated, vaporized, and escaped the Moon's weak gravitational field very early in the Moon's evolution. The lack of water on the Moon explains the absence of such prominent water-containing Earth minerals as the amphiboles and micas. (Recent studies of the Moon's surface suggest that water may actually exist in isolated, protected spots near the Moon's poles.)

Another difference between the Moon and the Earth is that the origin of the Moon's igneous rocks seems unrelated to plate tectonics. Most lunar geologists believe that the Moon has never had moving plates. (Unlike most of the Earth's mountains, lunar mountain ranges apparently did not form from plate convergence.) The Moon's igneous activity also seems unrelated to the Moon's internal heat, much of which was probably lost long ago. The Moon's small size enabled its heat to be readily conducted from its interior to the surface and then into space.

Some lunar igneous rocks appear to derive from a process that is relatively rare on Earth (at least nowadays)—the generation of new magmas by meteorite impact. Most of the Earth's recent incoming meteorites have been incinerated as they passed through the atmosphere, although some of the very large ones back in the Earth's first billion years or so probably survived the early atmosphere and generated impact magmas. (Recall from Chapter 1 that the evidence of their resulting craters has been erased by plate tectonic reshaping of the lithosphere and surficial weathering and erosion.) The crushing force of a large impact on the Moon first produces an enormous quantity of shattered rock, which collects as angular fragments interspersed with bits of glass fused by the heat of the impact. The heat may raise the temperature of the stricken rocks to their melting points, thus generating new magma. The impact also fractures the lunar crust, providing subsurface magmas with an easy path to the surface.

Unlike the Earth's watery oceans, the lunar maria, or "seas," are actually vast solidified basalt flows. They were named when Galileo and other early astronomers, using the

Incoming meteor
Lunar surface
Meteor impact
Lunar crust broken and melted by impact. Upper mantle melts and wells up into fractured crust.
Lunar maria
Basaltic lava fills and overflows impact craters.

Figure 3-29 Lunar maria—the vast solidified basalt flows that are prominent features of our Moon—form from meteor impacts. Above: A photo of lunar craters, mountains, and mare taken during the Apollo 10 mission of May 1969.

crude telescopes of the time, believed that they were looking at true seas. From about 4 to 3.85 billion years ago, intense meteorite activity gouged numerous craters in the Moon's predominantly anorthositic surface. The impacts removed great volumes of overlying rock, instantaneously reducing pressure on the Moon's ultramafic upper mantle and accelerating partial melting there. This produced basaltic magma that rose to the surface through impact-induced fractures in the crust. Basaltic lava flowed into and filled the craters, forming the lunar maria (Fig. 3-29).

Although minor cratering continues today, the eruptions of the mare basalts apparently marked the final episodes of major igneous activity on the Moon. By the close of this period, much of the debris remaining after the origin of the solar system had already been swept up. The lack of younger igneous rocks on the Moon's surface today suggests that its interior is no longer hot enough to produce new magmas.

The Economic Value of Igneous Rocks

The practical uses of igneous materials range from the glittering (gemstones and precious metals) to the utilitarian (crushed basalt for road construction). Any urban center displays one of the principal uses we have found for plutonic igneous rocks—the decorative building stone that adorns the exteriors and lobbies of many banks and office buildings. We see the same appealing polished granites and diorites in cemeteries, where they serve as durable tombstones. On a smaller scale, glassy pumice is the abrasive in grease-removing cleansers and is also used to remove callouses from hands and feet. Until recently, pumice was an ingredient in toothpaste because of its ability to remove dental stains and plaque; since it also claimed its share of tooth enamel, however, milder abrasives have now taken its place. A few other familiar minerals, such as the diamonds found in ultramafic rocks and the emeralds and topazes in felsic pegmatitic rocks, are also of igneous origin.

Gold and silver often appear in or around granitic rocks, as do less shiny but equally valuable ores of copper, lead, and zinc. These late-crystallizing metallic ions typically become concentrated in the hot, watery fluids that circulate after most other minerals have crystallized. When these

Figure 3-30 The Bingham Canyon copper mine near Salt Lake City, Utah, where plutonic igneous rocks are exposed over a 5-by-8-kilometer (3-by-5-mile) area. Subterranean magmas and groundwater deposited the abundant copper found throughout these rocks.

fluids enter fractures in adjoining rocks, the elements crystallize as metal-rich veins. Recently, geologists studying the eruptive potential of Colombia's Galeras volcano discovered what prospectors have always counted on: Volcanoes manufacture gold deposits. Highlight 3-2 illustrates this valuable geologic axiom.

Groundwater that percolates down through an igneous region may redeposit valuable minerals. When the water comes in contact with a magma chamber or a body of still-warm plutonic rock, it is heated and some or all of it may be converted to steam, which then invades surrounding rocks and dissolves any soluble minerals in them. When the steam eventually cools and condenses, its dissolved load recrystallizes to form mineral-rich deposits. The gold of the Homestake Mine in the Black Hills of South Dakota, the lead, silver, and zinc of northern Idaho, the copper of northern Michigan and Bingham Canyon in northern Utah (Fig. 3-30), and the silver of the Comstock Lode of Nevada all accumulated from the action of heat-driven subterranean fluids, known as *hydrothermal fluids*.

We can now build on your general knowledge of the Earth's igneous processes and rocks and expand our discussion of igneous activity to that which occurs above ground. The following chapter focuses on igneous phenomena we can observe—the volcanic eruptions and rocks produced when magma reaches the Earth's surface and escapes as lava.

Chapter Summary

Igneous rocks are the most abundant type of rock in the Earth's crust and mantle. They form when molten rock cools and crystallizes. Molten rock contained beneath the Earth's surface is called **magma**; when it erupts onto the surface, it is called **lava**. The texture of an igneous rock reflects the rate at which its parent magma or lava cooled, and its mineral content reflects the composition and evolution of the molten rock from which it formed.

Intrusive, or **plutonic**, igneous rocks form from magma that cools slowly underground. These rocks are generally coarse-grained, or phaneritic, because ample cooling time allows crystals to grow to visible sizes. Igneous rocks with exceptionally large crystals are called **pegmatites**. **Extrusive**, or **volcanic**, igneous rocks form when lava cools quickly at the Earth's surface. These rocks are generally fine-grained, or aphanitic, because rapid cooling limits crystal growth.

The most common igneous rocks include ultramafic **peridotite**, an iron-and-magnesium–rich plutonic rock containing less than 40% silica; mafic **basalt**, an iron-magnesium-and-calcium–rich volcanic rock (40–50% silica), and its intrusive equivalent, **gabbro**; intermediate **andesite**, an iron-aluminum-and-sodium–rich volcanic rock (about 60% silica), and its intrusive equivalent, **diorite**; and felsic **rhyolite**, a potassium-and-aluminum–rich volcanic rock (70% or more silica), and its intrusive equivalent, **granite**.

Magmas are produced when rocks in the Earth's interior melt. The factors that affect a rock's melting point include local heat and pressure conditions and the water content and composition of the rock. Most magmas are created when preexisting rocks **partially melt**—that is, the minerals with lower melting points liquefy first and start to flow as a molten mass, leaving behind an unmelted portion of rock. Magmas **viscosity**—its resistance to flow—increases with decreasing temperature and decreases with increasing temperature.

As magma cools, different minerals crystallize from it at different temperatures. The silicate minerals as a group crystallize in two specific sequences, known together as **Bowen's reaction series**. In the discontinuous series, mafic

Highlight 3-2 There's Gold in Them Thar Hills

Galeras volcano looms ominously over the peaceful city of Pasto (population 300,000) in the Andes of western Colombia (Fig. 3-31). One of the world's most dangerous volcanoes, Galeras erupted unexpectedly in January 1993, killing six geologists and three tourists and inspiring an urgent vigil that continues today. Teams of geologists regularly probe the volcano's crater, collecting gases, water from hot springs, and samples of lava from past eruptions, searching for signs of an impending blast to avert a volcanic disaster. A team from New Mexico's Los Alamos National Laboratory recently visited Galeras for just that purpose but came away with a valuable discovery—the rocks of Galeras volcano are accumulating gold at a surprisingly rapid rate.

Laboratory tests of rock samples collected around gas vents in the valley revealed that veins in some old volcanic deposits were studded with tiny pencil-point-sized flecks of gold—as much as 7.8 ounces of gold per ton of rock. This concentration is quite impressive, especially when compared with concentrations of less than 1 ounce per ton within numerous active gold mines of the western United States. Geologists estimate that if the volcano remains active for roughly 10,000 years, as much as 200 tons of gold could accumulate within the volcano—a future gold mine in the making.

Figure 3-31 Colombia's Galeras Volcano—one of the world's most active and potentially explosive—is a veritable factory of gold production. Geologists' enthusiasm for searching for and mining this treasure, however, is tempered by Galeras' recent history of tragic eruptions.

The volcano's gold, like the gold in Alaska, the Colorado Rockies, and California's Sierra Nevada range, is probably melted from crustal rocks by the heat of rising magma beneath the volcano. Water vapor and other gases, charged with dissolved gold, rise from the volcano's subterranean magma chamber and seep upward through the rocks of the volcano, disseminating the precious golden flakes. Compared with other volcanoes around the world, Galeras appears to be one of the richest. But with the lure of these riches comes the sobering threat of a volcanic disaster without warning.

silicate minerals evolve in distinct steps, with both their compositions and their internal crystal structures changing at each step. In the continuous series, the plagioclase feldspars evolve as sodium ions gradually replace calcium ions in the developing crystals, resulting in very minor changes in the minerals' internal crystal structures.

As early-forming minerals crystallize, they may remain suspended in the magma, continue to exchange ions with it, and ultimately evolve into later-forming minerals as predicted by Bowen's reaction series. However, some early-forming crystals may separate from the liquid magma by settling to the bottom of the magma chamber (if the crystals are heavier than the magma), becoming plastered to the walls of the magma chamber, or being filtered out as the magma moves into fractures too narrow for the crystals to pass. When early-forming crystals are removed from a magma, the loss of the ions in those crystals changes the magma's composition. As a result, the magma produces rocks that differ in composition from those that would have been produced by the original, unseparated magma. This process is called **fractional crystallization**.

Recent studies of igneous rocks suggest that the composition of a magma may also be modified by assimilation of preexisting rocks or by mixing with another body of magma having a different composition. When masses of preexisting rock are only partly assimilated in a magma, they appear in the solidified rock as distinct bodies known as **xenoliths**.

Bodies of magma that cool underground form **plutons**, igneous features that are distinct from surrounding rocks. Plutons are classified by their shapes, sizes, and orientation relative to the rocks they intrude. Concordant plutons are parallel to the preexisting rock layers; discordant plutons cut across the preexisting rock layers. Tabular plutons are igneous formations that are relatively thin, like a table top. Discordant tabular plutons are called **dikes**; concordant tabular plutons are called **sills**. Large concordant plutons include mushroom-shaped **laccoliths** and saucer-shaped **lopoliths**; large discordant plutons are called **batholiths**.

The principal igneous rock types are typically associated with specific plate tectonic settings. Basalts and gabbros most often appear at divergent plate boundaries (the mid-ocean ridge basalts, or MORBs), atop intraplate hot spots (the ocean island basalts, or OIBs), and where a continental plate is rifting. Andesites and diorites are found where oceanic plates have subducted to form volcanic mountains. Partial melting of the lower portions of the continental crust forms rhyolites and granites. They are often associated with subduction-produced mountains, and they also occur where continental rifting has taken place.

Igneous rocks also exist on the Moon. Lunar rocks collected by the Apollo astronauts in the 1970s indicate that the Moon's highlands consist largely of anorthosite, a coarse-grained plutonic igneous rock composed almost exclusively of the calcium plagioclase anorthite. On the maria are vast areas of basalt. Igneous rocks on the Moon differ fundamentally from those on Earth in that they contain no water and their formation apparently does not involve plate tectonics.

Igneous rocks are valued for the gemstones and precious metals they contain as well as for a variety of practical purposes (including the use of crushed basalt in road construction).

Key Terms

igneous rocks (p. 65)
magma (p. 65)
lava (p. 65)
intrusive rocks (p. 67)
plutonic rocks (p. 67)
pegmatites (p. 67)
extrusive rocks (p. 68)
volcanic rocks (p. 68)
peridotite (p. 70)
basalt (p. 70)
gabbro (p. 71)
andesite (p. 71)
diorite (p. 71)
granite (p. 71)
rhyolite (p. 71)
partial melting (p. 72)
viscosity (p. 74)
Bowen's reaction series (p. 75)
fractional crystallization (p. 77)
xenolith (p. 78)
plutons (p. 78)
dike (p. 78)
sill (p. 79)
laccolith (p. 81)
lopolith (p. 82)
batholith (p. 82)

Questions for Review

1. Briefly describe the textural difference between phaneritic and aphanitic rocks. Why do these rocks have different textures?

2. Some igneous rocks contain large visible crystals surrounded by microscopically small crystals. What is the term for these rocks? How does such a texture form?

3. What elements would you expect to predominate in a mafic igneous rock? In a felsic igneous rock?

4. Name the common *extrusive* igneous rocks in which you would expect to find each of the following mineral types: calcium feldspar; potassium feldspar; muscovite mica; olivine; amphiboles; sodium feldspars. Which *plutonic* igneous rock contains abundant quartz and muscovite mica, but virtually no olivine or pyroxene?

5. What factors, in addition to heat, control the melting of rocks to generate magma?

6. What is the basic difference between the continuous and discontinuous series of Bowen's reaction series?

7. Briefly describe three things that might happen to an early-crystallized mineral surrounded by liquid magma.

8. What is the difference between a sill and a dike? Between a batholith and a lopolith?

9. Briefly discuss two specific types of plate tectonic boundaries and the igneous rocks that are associated with them.

10. What is the basic difference between a MORB and an OIB?

For Further Thought

1. What type of igneous feature is shown in the photo below?

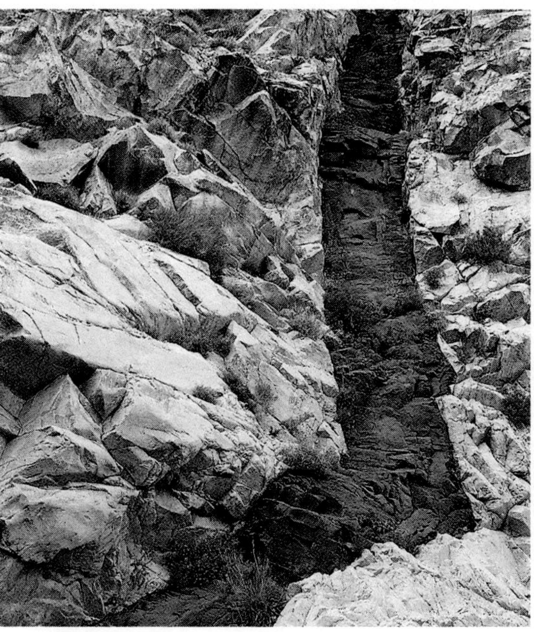

2. Felsic rocks such as rhyolite may occur together with basaltic rocks at locations where continents are undergoing rifting. Give one possible explanation for this.

3. Why do we rarely find batholiths made of gabbro?

4. Speculate about how the distribution of the Earth's igneous rocks will change when the Earth's internal heat is exhausted and plate tectonic movement stops.

5. Why are there virtually no granites or diorites on the Moon? How might small volumes of such felsic rock form under the geological conditions believed to be responsible for the Moon's igneous rocks?

4

Volcanoes and Volcanism

Figure 4-1 Popocatepetl ("El Popo")—Mexico City's neighboring volcano—is one of several volcanoes currently erupting around the world and threatening surrounding communities.

One summer day in 1883, the Indonesian island of Krakatau all but vanished in a spectacular volcanic eruption. Krakatau, an uninhabited volcanic island in the Sunda Straits of the southwest Pacific Ocean, had for many years been a landmark for clipper ships carrying tea from China to England. The volcano, standing 792.5 meters (2601 feet) high, had been inactive for more than 200 years. On the morning of August 27, it suddenly exploded in one of modern history's most violent volcanic eruptions, leaving a crater 300 meters (1000 feet) *below* sea level. Krakatau's eruption equaled in force the detonation of 100 million tons of TNT. It jostled the atmosphere around the globe, causing sharp barometric changes as far away as San Francisco and London. And the sound of the explosion was heard as far as central Australia, 4802 kilometers (2983 miles) away, which is akin to the residents of San Diego, California, hearing an explosion in Boston, Massachusetts.

No one, as far as we know, perished directly from the destruction of Krakatau. However, between 36,000 and 100,000 lives were lost when the resulting waves, up to 37 meters (121 feet) high, pounded coastal villages on the nearby islands of Java and Sumatra. The eruption produced a black cloud of volcanic debris that rose to an altitude of 80 kilometers (50 miles), blocked out all sunlight, and plunged the region into darkness for three days. The cloud's finest particles, swept aloft by wind currents, reduced incoming solar radiation by as much as 10% worldwide, causing a drop of more than 1°C (1.8°F) in global temperatures. The suspended particles also produced years of spectacular crimson sunsets. A few months after the eruption, on October 30, 1883, residents of Poughkeepsie, New York, and New Haven, Connecticut, summoned fire brigades to douse blazes that were, in fact, only the fiery glow of the brilliant evening sky.

As our Krakatau example attests, powerful volcanic eruptions and their aftereffects are among the Earth's most destructive natural events. **Volcanoes** are the landforms created when magma escapes from the Earth's interior through openings, or **vents,** in the Earth's surface and becomes *lava.*

The lava then cools and solidifies around the vents, forming volcanic rock. A recent survey by the Smithsonian Institution found that about 600 volcanoes have erupted in the past 2000 years, some of them many times over. In a single year, approximately 50 volcanoes erupt around the world (Fig. 4-1). Yet volcanoes, despite their dangers, also provide some of the world's most breathtaking scenery. Each year, millions are drawn to the potentially dangerous slopes of Mount Rainier in Washington state, Fujiyama in Japan, and Mount Vesuvius in Italy.

The geological processes that result in the expulsion of molten rock as lava at the Earth's surface are collectively known as **volcanism.** Volcanoes are like windows into the Earth, providing us with information that would otherwise be inaccessible. Ascending magma carries subterranean rock fragments to the surface, giving us a glimpse of actual rocks from the Earth's interior. Volcanoes can also show us a glimpse of the past, wherever volcanic deposits buried and preserved evidence of past organisms. Footprints of the earliest upright-walking hominids, for example, were preserved in fresh volcanic ash in East Africa 3.6 million years ago (Fig. 4-2).

Volcanism also produces other notable benefits. We owe much of the air we breathe and the water we drink to volcanic eruptions, which throughout the Earth's history have released useful gases from the planet's interior. The water liberated by past volcanic eruptions makes up a significant portion of the Earth's *hydrosphere*—its oceans, lakes, rivers, underground waters, glaciers, and clouds. Nitrogen and oxygen from past eruptions have combined with other components to produce the Earth's gaseous *atmosphere*.

Figure 4-2 A footprint in volcanic ash from East Africa. Paleoanthropologists have learned a great deal about human evolution from imprints such as this one, recording the passage of some of our early ancestors more than 3.6 million years ago.

Figure 4-3 Anak Krakatau ("Child of Krakatau"), the small island that emerged from the remains of the volcanic island Krakatau during eruptions in the 1920s. The original Krakatau volcano was demolished in a monumental eruption in 1883.

Volcanic activity constantly adds to the Earth's inventory of habitable real estate. Iceland, Hawai'i, Tahiti, many islands of the Pacific and Caribbean, and nearly all of Japan and Central America are products of volcanism. A new volcanic island is even growing where Krakatau used to be (Fig. 4-3). Volcanic terrains often become prime agricultural lands as fresh volcanic ash replenishes the nutrients in nearby soils. The rich coffees grown in South and Central America sprout from fertile volcanic soils. On the Indonesian island of Java, fine volcanic ash retains water and the abundant nutrients (such as potassium, calcium, and sodium) that nourish plants. This allows Java's population density to be about 200 times that of neighboring Borneo, whose far less fertile soils derive from solid extrusive igneous rock.

Volcanic landscapes may also contain a source of inexpensive, clean energy. During the dark polar winter in Reykjavic, Iceland, the world's northernmost capital, people stroll in shirt-sleeves through warm shopping malls. Iceland has no oil, no coal, no natural gas, and very few trees for fuel, but it does have an abundant underground hot-water supply, thanks to the molten rock that fuels Iceland's nearly constant volcanic activity. By tapping the scalding water just meters beneath their feet, Icelanders can heat more than 80% of their homes and businesses. Similar *geothermal* ("Earth heat") resources are being tapped in other recently active volcanic settings, among them the Geysers area, 150 kilometers (100 miles) north of San Francisco, California.

Thus volcanoes and volcanism are at once both a great hazard and a great boon to humankind. In this chapter, we will explore the causes and characteristics of different types of volcanoes. We will also describe the threats they pose to us and how we have learned to cope with them. Finally, we will examine volcanism on some of our neighboring planets in the solar system.

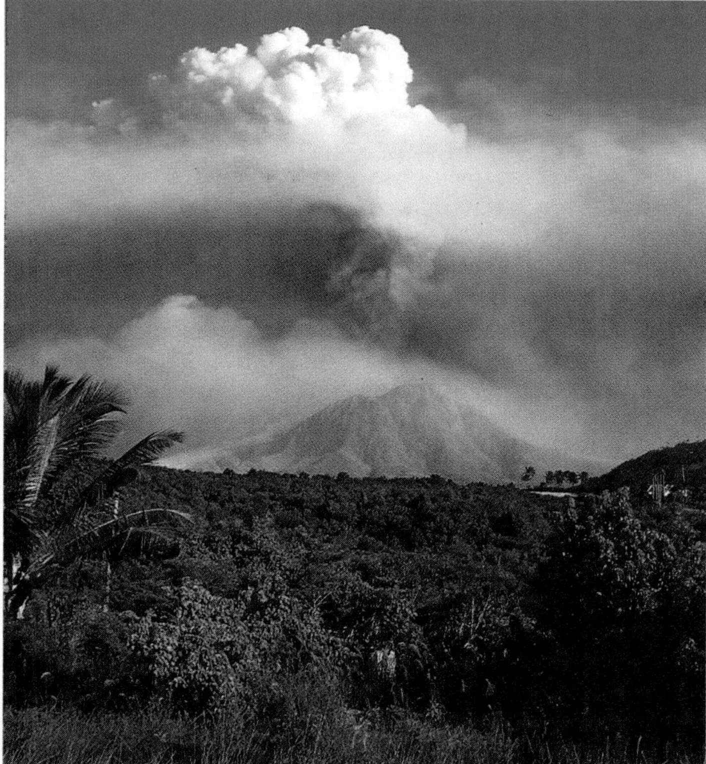

Figure 4-4 The Soufriere Hills on the Caribbean island of Montserrat has been erupting intermittently and explosively for the past three years. Flows of hot volcanic ash cascading down the volcano's slopes have forced the evacuation of most of the region's inhabitants.

The Nature and Origin of Volcanoes

The current status of a volcano is the key to its threat to human life and property. Is it active, dormant, or extinct? An *active* volcano is either currently erupting or has erupted recently (in geological terms). Certain active volcanoes, such as K'ilauea in Hawai'i or Stromboli in the Mediterranean, erupt almost continuously. Others erupt periodically, such as Lassen Peak in northern California, which last erupted in 1917. Active volcanoes can be found on all the continents except Australia and on the floors of all the major ocean basins. Indonesia with 76 active volcanoes, Japan with 60, and the United States with 53 are the world's most volcanically active nations. A quick tour of the planet in the late 1990s finds major eruptions occurring or brewing at Mounts Pavlov, Okmok, and Akutan in Alaska, Ruapehu in New Zealand, Rabaul in New Guinea, Soufriere Hills on the Caribbean island of Montserrat (Fig. 4-4), and K'ilauea in Hawai'i. Such eruptions range from the quiet oozing of molten basalt from K'ilauea to the cataclysmic explosions of Mexico City's fearsome "El Popo" (Popocatepetl).

A *dormant* volcano is one that has not erupted recently (within the past few thousand years) but is considered likely to do so in the future. A number of signs may suggest that a dormant volcano is stirring to wakefulness. A shallow heat source, such as rising magma or heated rocks, might convert underground water into hot-water springs and steam. The discovery in 1975 of new hot springs on the slopes of Mount Baker near the Washington–British Columbia border, for example, has caused anxiety in nearby communities about the prospect of a future eruption there. Rising magma may also create a measurable bulge at the Earth's surface. A bulge beneath the Mammoth Lakes area of eastern California has caused considerable concern among volcanologists, local residents, and thousands of vacationing skiers. The presence of relatively fresh (within about 1000 years old) volcanic rocks in a volcano's vicinity also suggests that it is capable of a repeat performance (Fig. 4-5).

(a)

(b)

Figure 4-5 Which of these two volcanoes is more likely to erupt? **(a)** The slopes of Washington state's Mount Rainier remain deeply scored by repeated episodes of glacial erosion that occurred over hundreds of thousands of years and appear not to have received a fresh covering of lava in thousands of years. **(b)** The slopes of Mount St. Helens (shown here before its 1980 eruption), believed to be the youngest of Washington's Cascade Range volcanoes, are relatively uneroded. Although intermittently dormant for hundreds or thousands of years, its volcanic cone has continued to grow.

A volcano is considered *extinct* if it has not erupted for a very long time (perhaps tens of thousands of years) and is considered unlikely to do so in the future. One indication that a volcano is probably extinct is extensive erosion since its last eruption (see Figure 3-18—Shiprock Peak, the erosional remains of an extinct volcano). A truly extinct volcano is no longer fueled by a magma source. It is wise, however, to view virtually *all* volcanoes with caution. Residents of the Icelandic island of Heimaey, for example, believed their Mount Helgafjell to be extinct until it came to life in a spectacular eruption in 1973, its first in 5000 years.

The Causes of Volcanism

Volcanism begins when magma created by the melting of preexisting rock (discussed in Chapter 3) reaches the surface through fractures in the Earth's lithosphere. (The distribution of the Earth's lithospheric cracks, usually associated with tectonic plate boundaries and with intraplate hot spots, determines where most volcanoes will form.) Magma will erupt if it flows upward rapidly enough to reach the surface before it can cool and solidify. Two characteristics of magma determine its potential to erupt: its gas content and its viscosity.

Gas in Volcanic Magma Magmatic gases make up from less than 1% to about 9% of most magmas. The principal gases are water vapor and carbon dioxide, with smaller quantities of nitrogen, sulfur dioxide, and chlorine. When the magma is tens of kilometers underground, the gases remain dissolved, held in the magma by the pressure of surrounding rocks. But this pressure decreases as magma rises toward the surface, allowing the gases to separate, or *unmix,* from the magma. Because the gases are less dense than their surrounding magma, they migrate upward and expand outward, pushing any overlying magma before them.

The rising, expanding, unmixed gases collect near the top of a magma body and press against the overlying rock. Where something blocks their passage to the surface, such as a plug of old congealed lava, these gases accumulate, exerting great pressure against the overlying rock until it ultimately shatters. As soon as the overlying rock is removed, the pent-up gases expand rapidly, much as the gases in a shaken soft-drink bottle fizz and bubble out when the bottle is opened. A volcano's initial blast typically removes any overlying obstructions, hurling masses of older rock skyward. Shreds of the liberated, gas-driven lava are then sprayed violently into the air as gases expand. The eruption may continue violently for hours or even days as the gases escape. It may then settle down to a relatively placid outpouring of degassed magma, or it may cease altogether.

Magma Viscosity A magma's *viscosity* is its *resistance to flow.* Very viscous magma is not as likely to make it to the surface to erupt as is more fluid magma. On the other hand, the viscous flows that do erupt tend to do so far more explosively. The key factors in determining viscosity are the magma's temperature and composition.

Increased heat typically *reduces* the viscosity of any fluid—be it motor oil or maple syrup. It flows more easily because the ions and atoms in the heated fluid move about more rapidly, breaking the temporary bonds that inhibit flow. Conversely, decreasing temperature typically *increases* a fluid's viscosity, causing it to move more sluggishly.

The viscosity of a magma also generally increases with silica content (see Table 3-1). In silica-rich magma, the oxygen ions at the corners of the unbonded silicon-oxygen tetrahedra (discussed in Chapter 2) form relatively strong temporary bonds that link clusters of chains of the tetrahedra. Felsic magma, for example, tends to be very viscous, and thus slow-flowing, because it has a high silica content *and* it is relatively cool (it crystallizes at a low temperature; Fig. 4-6a). Conversely, mafic magma, with its low silica content and high temperature, is much less viscous, flowing easily. That is why mafic magmas are typically more likely to rise to the surface and erupt than are felsic magmas, which tend to cool underground and form plutonic rocks.

How does viscosity affect explosiveness? Gases do not escape from all magmas with equal ease. In the more fluid, low-viscosity mafic magmas, migrating gases meet with relatively little resistance and therefore escape easily when the magma reaches the surface. They don't accumulate to build up the high pressure that causes explosive eruptions. Mafic magmas, then, tend to erupt quietly, with a relatively gentle outpouring of lava (Fig. 4-6b). On the other hand, because the movement of gases in highly viscous felsic magma is impeded, gas pressure builds up within the molten material and it tends to erupt explosively when it reaches the surface. The viscosity of intermediate magmas, such as those that produce andesites, falls somewhere between mafic and felsic. But these magmas are cool enough and sufficiently rich in silica to erupt explosively.

The Products of Volcanism

A volcanic eruption can produce a flowing stream of red-hot lava, a shower of ash particles as fine as talcum powder, a hail of volcanic blocks the size of automobiles, or any number of intermediate-sized products. The quantity of lava produced by volcanoes ranges from small spurts to vast floods. An immense submarine lava flow, believed to have been extruded within the last 25 years, was recently discovered in the vicinity of the East Pacific rise, off the coast of South America. The flow contains approximately 15 cubic kilometers (9 cubic miles) of basalt, enough to pave over the entire U.S. interstate-highway system to a depth of 10 meters (33 feet).

(a)

(b)

Figure 4-6 The viscosity of a volcano's lava typically controls its eruptive style (quiet or explosive) and the shape of the volcano. Here (a) at the Valley of the Ten Thousand Smokes in Katmai National Park in Alaska, highly viscous, virtually non-flowing, felsic lava produces this steaming dome. In (b), at Hawai'i's K'ilauea Volcano, low-viscosity, basaltic lava flows rapidly as a red-hot lava stream.

Both the type and the amount of material produced by a volcano depend largely on the composition of its lava. In considering the products of volcanism, we will first examine the different types of lava and their properties and then describe the various forms in which volcanic materials accumulate at the surface.

Types of Lava

The composition of a lava resembles that of its parent magma. The lava, however, contains less dissolved gas, as most gases escape into the atmosphere during an eruption. The most common type of lava is basalt—a mafic lava. As we've learned, mafic magmas are the most likely to erupt because they tend to be hot and highly fluid, moving readily to the surface before solidifying. This in part explains why there is probably much more basalt than gabbro—its plutonic equivalent—in the Earth's crust. Magmas of felsic composition tend to be cooler and much more viscous, only rarely reaching the surface—as rhyolite lava—before solidifying. For this reason, there is far less rhyolite than granite (its plutonic equivalent) in crustal rocks. Andesitic lavas are intermediate between basaltic and rhyolitic lavas in both composition and fluidity; they erupt much more frequently than rhyolitic lavas but are less common than basalt.

Basaltic Lava For nearly a century, the Hawai'ian Volcano Observatory on the Big Island of Hawai'i has been observing eruptions, producing much of what we know about *subaerial* ("under air," as opposed to under water) basaltic lava flows and the resulting rocks. The observatory has found the temperature of Hawai'ian flows to be as high as 1175°C (2150°F). Such hot, low-viscosity lava cools to produce two

Figure 4-7 These ropy pahoehoe-type lavas from Hawai'i's K'ilauea volcano cooled only days or even hours before these tourists began trekking around on them.

principal types of basalt, *pahoehoe* (pa-HOY-hoy) and *'a'a* (AH-ah). Pahoehoe, which means "ropy" in a Polynesian dialect, is aptly named. Highly fluid basaltic lava moves swiftly down a steep slope at speeds that may exceed 30 kilometers per hour (20 miles per hour), spreading out rapidly into sheets about 1 meter (3 feet) thick (Fig. 4-7). The surface of such a flow cools to form an elastic skin that is then dragged into rope-like folds by the continuing movement of the still-fluid lava beneath it. The ropy surface of pahoehoe basalt is

Figure 4-8 This relatively slow-moving basaltic lava cools to form a blocky, jagged, ʻaʻa-type surface texture.

generally quite smooth. Native islanders refer to it as "ground you can walk on barefoot," and most old Hawaiʻian foot trails follow ancient pahoehoe flows.

As basaltic lava flows farther from the vent, it cools, loses much of its dissolved gas, and becomes more viscous. A thick, brittle crust develops at its surface and continues to move forward slowly, carried along by the warmer, more fluid lava below it. The cooler outer region breaks up to produce a rough surface having numerous jagged projections sharp enough to cut animals' hoofs or a geologist's field shoes. Flows having these features are called ʻaʻa flows (Fig. 4-8). (*ʻAʻa* is a local term of unknown origin that may recall the cries of a barefoot islander who strayed onto its surface. Ancient foot trails meticulously avoid ʻaʻa fields.) ʻAʻa is often found downstream from pahoehoe, the product of the same flow.

Basalt flows may also contain *lava tubes,* subway-like tunnels that form when lava solidifies into a crust at its surface but continues to flow underneath (Fig. 4-9). Eventually, as the eruption wanes, the still-molten lava drains from the cooled tubes, leaving them hollow. These tubes enable lava to travel great distances from a volcano's vent. Lava Beds National Monument in northeastern California is rife with 300 or more tubes within the area's pahoehoe flows. This natural labyrinth sheltered the Modoc Indians in 1872 as they battled several hundred troops from the United States Army in an unsuccessful attempt to reclaim their native lands. Fewer than 60 Modoc warriors, who knew every crevice in the lava field, inflicted serious harm on their pursuers from their sanctuary within the lava tubes—but eventually were overcome and forced to leave their territory.

Subaerial basalt flows may produce several other distinctive features as they cool. Escaping gas often leaves small, pea-sized vesicles (discussed in Chapter 3), which are preserved at the top of the basalt when it cools. Vesicle-rich basalt is known as *scoria* (Fig. 4-10). As basaltic lava cools, it shrinks in volume, often producing a pattern of cracks known as *columnar jointing.* As cooling proceeds, the cracks extend inward from the top and bottom surfaces of the flow into its interior, creating six-sided polygonal columns of basaltic rock. A side view of these columns suggests a large bunch of pencils; an aerial view suggests oversized ceramic bathroom floor tiles (Fig. 4-11). In North America, the Devil's Postpile in California's Sierra Nevada and Devil's Tower in northeastern Wyoming (site of the climax of the film *Close Encounters of the Third Kind*) are spectacular examples of basaltic columns.

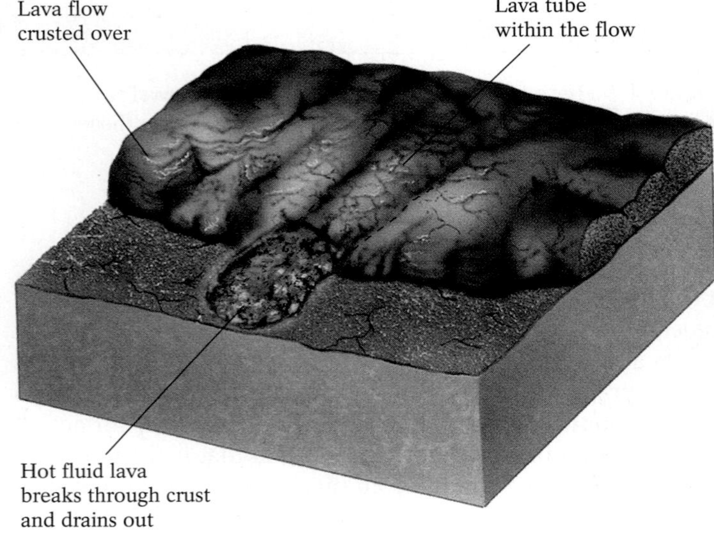

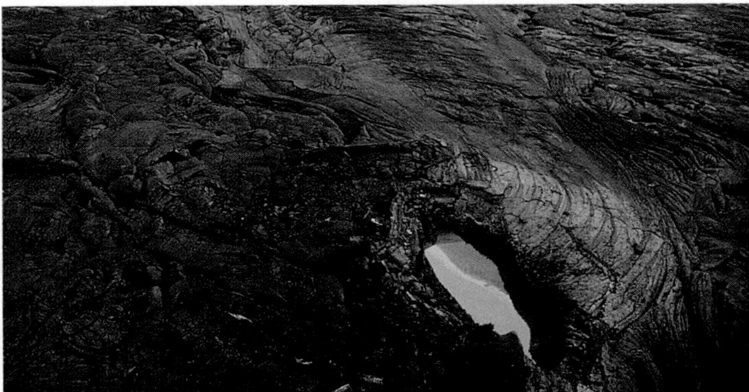

Figure 4-9 (a) Lava tubes form when a lava flow's surface cools and solidifies but the lava continues to flow under the surface in tunnels. When the lava drains from the tunnel, it leaves behind the empty tube. (b) A lava tube forming at K'ilauea.

Figure 4-10 Vesicular basalt, or scoria. These pores remain in the cooled rock after gases burst from bubbles on the surface of highly fluid lava.

Lava vent

Hot flowing lava

Cooler

Hotter

(a)

Cooling lava

Subaqueous ("underwater") basaltic eruptions are much more common than subaerial eruptions but less well understood, because they usually occur beneath thousands of meters of seawater. We do know that when basaltic lava erupts beneath the sea, it develops a distinctive *pillow structure* that, as its name suggests, resembles a stack of bedroom pillows. Cold water instantly chills the extruded lava, forming a thin deformable skin that stretches to resemble an elongated pillow as additional hot lava enters under it. As the pillow shape expands, its surface cracks, allowing some lava to flow out from it and form another pillow, from which yet another pillow might grow, and so on (Fig. 4-12).

Andesitic Lava Andesitic lava, intermediate in composition between mafic basaltic lava and felsic rhyolitic lava, flows more slowly than basaltic lava and typically solidifies before traveling as far from its vent. Like basaltic lavas, andesitic lavas may develop vesicles and ‘a‘a-type surface textures. We rarely see pahoehoe-type andesitic flows, however, because

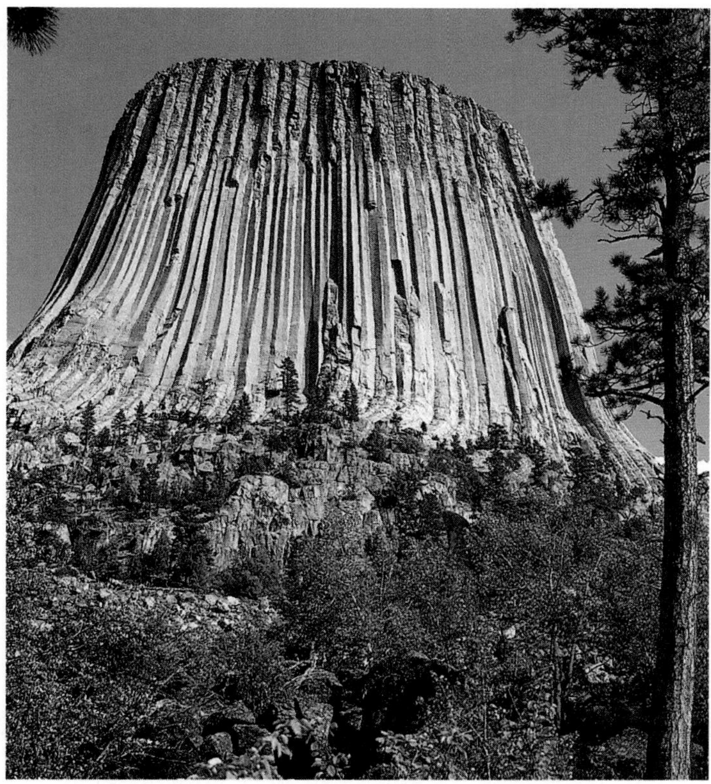

(b)

Figure 4-11 Contraction of basaltic lava flows as they cool **(a)** sometimes produces geometrically patterned joints. **(b)** Such structures can be found in North America in eastern California, eastern Oregon and Washington, southern Idaho, and eastern Wyoming, such as those shown here at Devil's Tower National Monument.

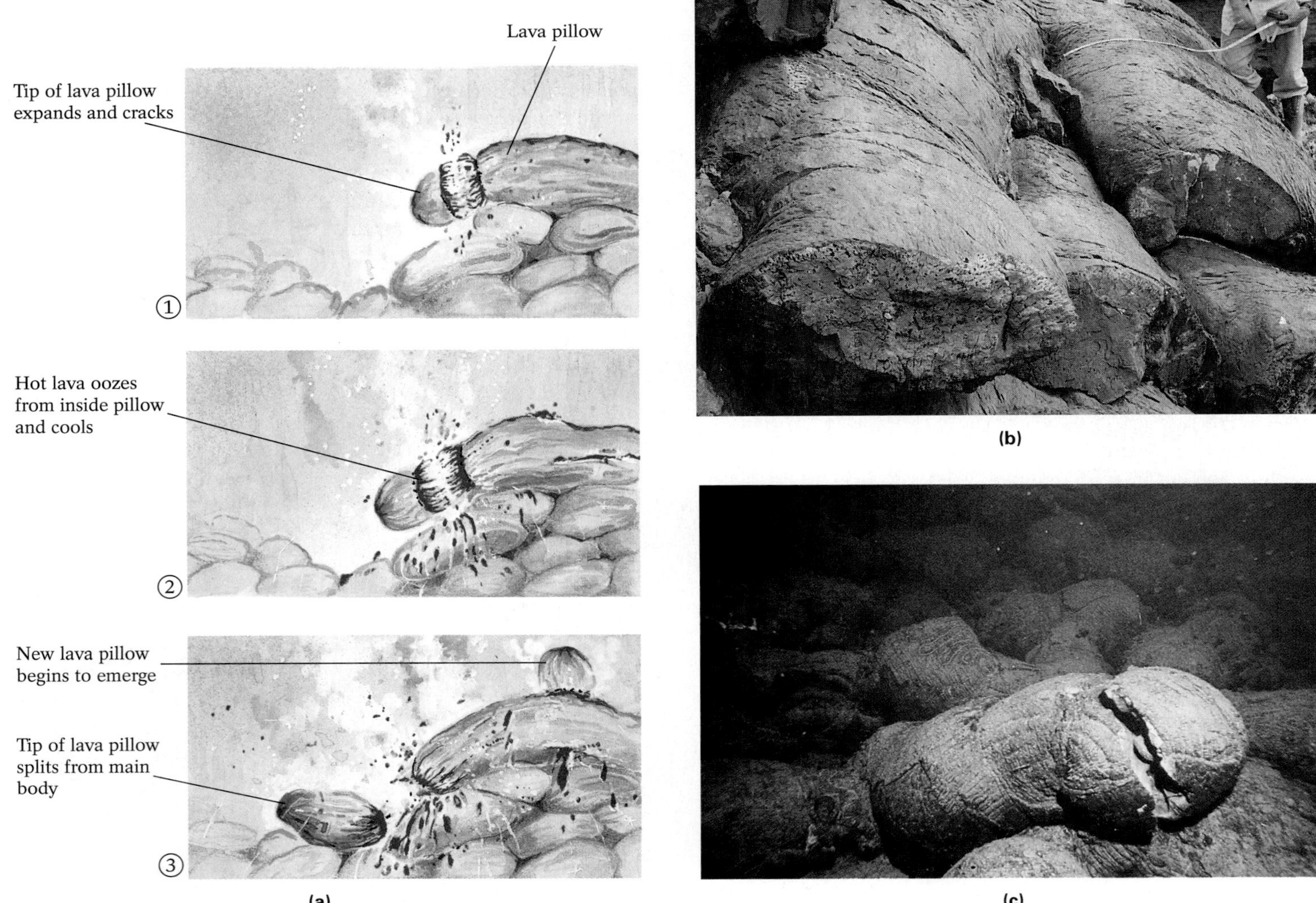

Figure 4-12 **(a)** The formation of basaltic pillow lavas. Our knowledge of these structures comes from studying preserved ancient pillows, **(b)**, such as these in Newfoundland, Canada and from observing modern pillows, such as these in the Galapagos **(c)**, from deep-sea submersibles.

these lavas are simply too viscous to stretch into a ropy structure. Andesitic lavas, particularly those closer to basalt in composition, can also produce columnar jointing and pillow structures. More felsic andesitic magmas can be viscous enough to impede the passage of rising gases and thus erupt in major volcanic explosions.

Rhyolitic Lava Rhyolitic magma, being the most felsic and highly viscous, typically moves so slowly that it tends to cool and solidify underground as plutonic granite. It is less likely to erupt as lava at the Earth's surface. When rhyolite does erupt, however, it usually explodes violently, producing an enormous volume of solid airborne fragments instead of just a lava flow. Viscous rhyolitic lava rarely flows far from the vent and does not produce the structures that typify less viscous lavas. Rhyolitic lava flows are typically short and thick, such as those that flowed as recently as 1000 years ago from the Newberry caldera in central Oregon (Fig. 4-13) or from the Medicine Lake volcano in eastern California.

Felsic magmas with high water and gas content may bubble out of a vent as a froth of lava that quickly solidifies into an extremely porous volcanic rock, *pumice.* Pumice may contain so many vesicles from which gas escaped that it can be lighter than water. When Krakatau erupted in 1883, sailing ships were trapped for three days in the waters of the Sunda Straits by huge rafts of floating pumice. When the pumice finally became waterlogged and sank, the ships could escape the incessant shower of hot ash.

Pyroclastic Materials

An explosive eruption expels lava forcefully into the atmosphere, where it cools rapidly, solidifying into countless fragments of various sizes and shapes. Such an eruption might

Figure 4-13 This young rhyolite flow is located at Paulina Lake in eastern Oregon near the town of Bend.

also shatter some of a volcano's preexisting rock. *Volcanic blocks,* for instance, are chunks of igneous rock ripped from the throat of a volcano during an eruption; they tend to be angular and range from the size of a baseball to the size of a house. Blocks weighing as much as 100 tons have been found as far as 10 kilometers (6 miles) from their source volcano. All such fragmental volcanic products are known as **pyroclastics** (from the Greek *pyro,* meaning "fire," and *klastos,* meaning "fragments"). Pyroclastic materials may travel through the air as dispersed particles, or they may hug the ground as dense flows.

Tephra Those pyroclastic materials that cool and solidify from lava as they are propelled through the air are called **tephra.** Tephra particles are classified by size, ranging from fine dust to massive chunks (Fig. 4-14). *Volcanic dust* particles are only about one-thousandth of a millimeter in diameter with the consistency of cake flour. The dust can travel great distances downwind from an erupting volcano and remain in the upper atmosphere for as long as a year or two. Somewhat grittier is *volcanic ash,* particles less than 2 millimeters in diameter, ranging from the size of a grain of fine sand to that of rice. Ash generally stays in the air for only a few hours or days.

Tephra and modern technology clearly don't mix. During the 1980 eruption of Mount St. Helens in Washington state (see Highlight 4-4, p. 120), ash clogged automobile carburetors and fouled the bearings of farm machinery. An even

(a)

(b)

Figure 4-14 Tephra particles range in size from fine dust to large boulders. **(a)** This range can sometimes be found within a single deposit, as here at Mono Craters, California. **(b)** A volcanic bomb, the largest type of tephra.

greater threat occurred when aircraft that strayed too close to the volcano's eastward-drifting cloud sucked St. Helens' abrasive, gritty tephra into their engines. In the past 12 years alone, the jet engines of more than 60 commercial aircraft have been seriously damaged by inadvertently flying through tephra clouds. Some of the craft even stalled in mid-flight. Steering clear of erupting volcanoes is no guarantee of safety—some near-disasters occurred when aircraft flew through tephra that had drifted more than 1000 kilometers (600 miles) downwind of its volcanic source. Atmospheric scientists are searching for ways to monitor the paths of drifting tephra clouds. One such device tracks the lightning that typically flashes within these clouds.

Coarser tephras fall sooner and closer to the volcanic vent, pulled to Earth by gravity. *Cinders,* or *lapilli* (Italian for "little stones"), range from the size of peas to that of walnuts (2 to 64 millimeters in diameter). *Volcanic bombs* (64 millimeters or more) form when sizable blobs of lava erupt and solidify in mid-air.

Tephras typically accumulate in distinct layers that record the frequency and magnitude of volcanic activity. The first step in "reading" a tephra layer is to link it to a specific volcano, usually by comparing the tephra's chemical composition to samples of known volcanic origin. We can also determine the source of a given tephra layer by mapping its thickness and grain size. In general, the coarser-grained and thicker a tephra layer, the closer it is to its source (Fig. 4-15). Once we can link a region's tephra layers to specific volcanoes, it is possible to date the layers and estimate the individual eruptive histories of each volcano.

Pyroclastic Flows When the amount of pyroclastic material expelled by a volcano is great and the particles are too large to be carried aloft, gravity almost immediately pulls the material back down onto the volcano, where it rushes downslope as a **pyroclastic flow,** or **nuée ardente** (noo-AY AR-dant) (French for "glowing cloud"). Trapped air and magmatic gases keep the flow quite buoyant as the gases sharply reduce the friction barrier between the flow and the ground. With little frictional resistance, pyroclastic flows may reach speeds exceeding 150 kilometers per hour (100 miles per hour), even on gentle slopes (Fig. 4-16).

Particles in a pyroclastic flow travel through the air so briefly that they do not cool significantly. Thus they may still be red-hot when they settle to the ground, capable of melting glass or burning an apple into a smear of carbon. As the flow travels downslope, its gases escape. When the warm particles finally come to rest, they may still be soft enough to fuse with one another, forming a **welded tuff,** a solid volcanic rock composed of solidified tephra. Tuffs also form when a thick pile of accumulated tephra solidifies under the pressure of overlying material.

Volcanic Mudflows Pyroclastic material that accumulates on the slope of a volcano may mix with water, forming a volcanic mudflow, or **lahar** (a term coined on the Indonesian island of Java, where explosive eruptions and abundant loose, moist soil cause frequent disastrous mudflows). A lahar may contain a range of particle sizes from the finest ash to enormous 100-ton boulders. Lahars typically form when an explosive eruption occurs on a snow-capped volcano, and hot

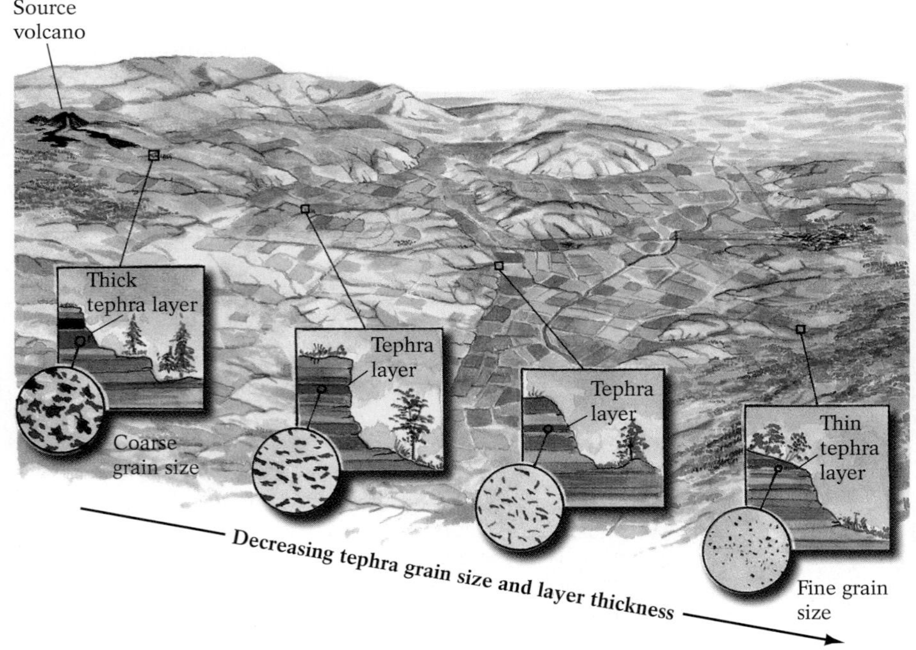

Figure 4-15 **(a)** The thickness and texture (grain size) of tephra layers decrease with distance from their volcanic source. **(b)** Note the variations in color, texture, and thickness of these Cascade tephras.

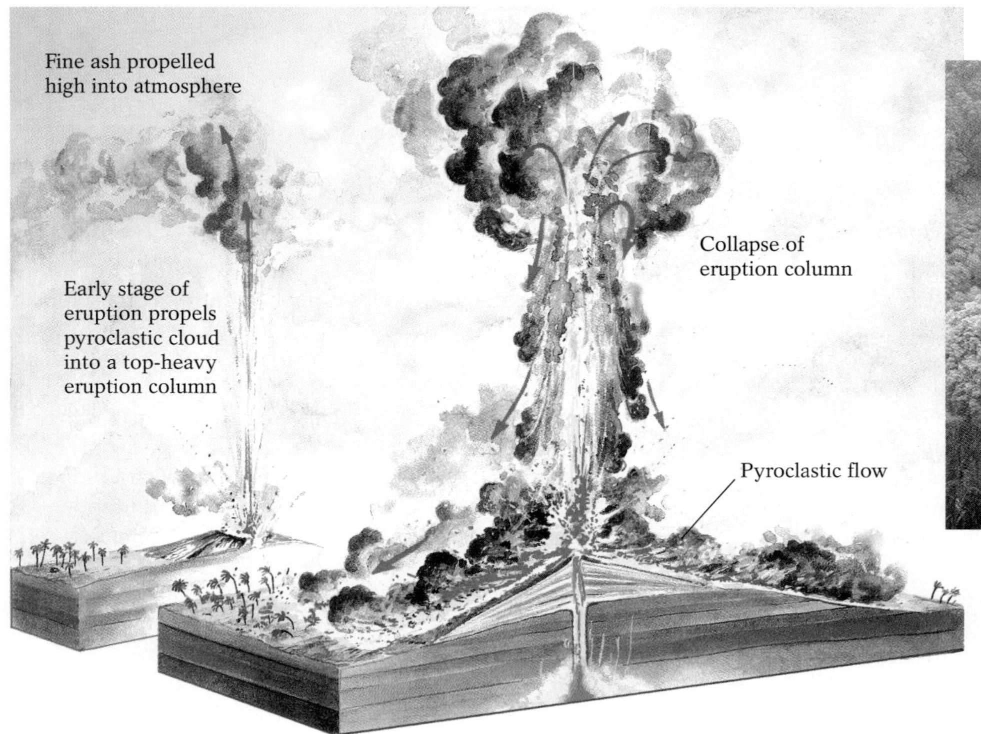
Fine ash propelled high into atmosphere
Early stage of eruption propels pyroclastic cloud into a top-heavy eruption column
Collapse of eruption column
Pyroclastic flow

Figure 4-16 Pyroclastic flows, or nuées ardentes, are produced when a massive amount of airborne pyroclastic material is pulled to Earth by gravity and rushes downslope. Photo: A pyroclastic flow from the May 1991 eruption of Mount Pinatubo, in the Philippines.

pyroclastic material melts a large volume of snow or glacial ice. In November 1985, a small eruption of Nevado del Ruiz in the Andes of Colombia caused the devastating lahar that buried the highland town of Armero and took the lives of 23,000 people (Fig. 4-17).

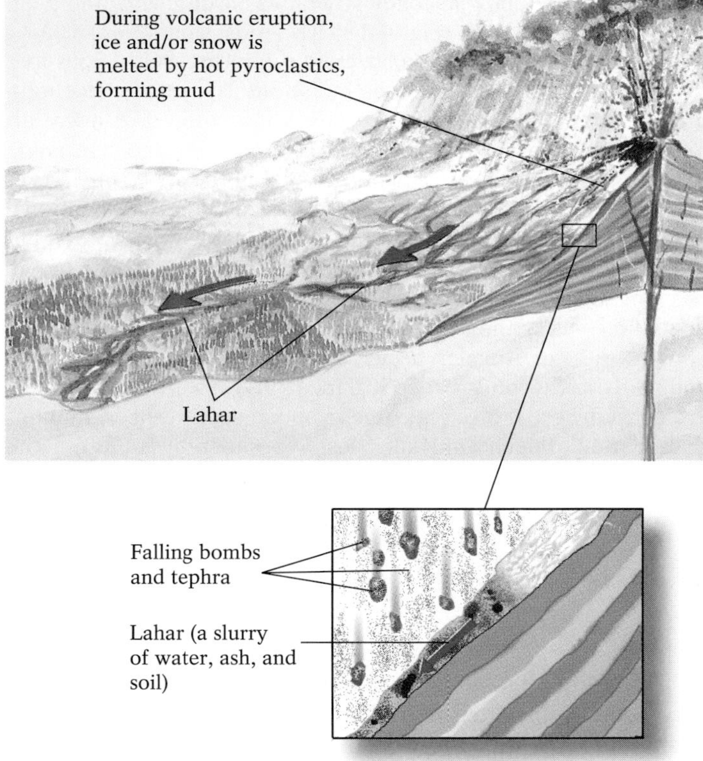
During volcanic eruption, ice and/or snow is melted by hot pyroclastics, forming mud
Lahar
Falling bombs and tephra
Lahar (a slurry of water, ash, and soil)

Figure 4-17 Some lahars occur when pyroclastic eruptions melt snow and ice on volcanic slopes, producing torrents of mud. Photo: This lahar resulted when Colombia's Nevado del Ruiz volcano erupted in 1985, melting about 10% of the snow on the slopes of the volcano and producing a 40-meter (137-foot)-high wall of mud that buried the town of Armero and approximately 23,000 of its residents. Armero was located about 50 kilometers (30 miles) from the summit of Nevado del Ruiz.

Highlight 4-1 Hawai'i's Dreaded VOG

Sulfur dioxide concentrations in the air over the Big Island of Hawai'i often rise to 1000 parts per billion, almost as high as the killer smog that enveloped London in December 1952, taking 4000 lives. This "pollution" has been blamed for a host of physical complaints among the local residents and is a potential threat to local agriculture and the island's economy. Yet the Big Island sits thousands of kilometers from the nearest smokestack and has no coal-burning power plants or copper smelters. There are only a few small communities, several daily busloads of tourists . . . and a hyperactive volcano named K'ilauea. Here is the culprit responsible for the clouds of sulfur dioxide gas that form Hawai'i's "VOG"—the locals' nickname for volcanic smog.

Finding the source of sulfur dioxide is not too hard—simply strap on a gas mask and stroll across the still-warm, freshly hardened black lavas at K'ilauea's summit crater. There, mounds of yellow sulfur crystals encrust the lava's fractures wherever sulfurous gas plumes have been expelled. K'ilauea, the world's most active volcano, has been erupting almost continuously since the mid-1980s, producing a steady stream of basaltic lava and a variety of noxious gases (Fig. 4-18). The most abundant—CO_2—quickly dissipates, but the next most common—SO_2—does not.

Typically, K'ilauea's gaseous emissions last only a few minutes or hours. The gases oxidize quickly as they drift away, affecting only the immediate vicinity of the volcano. On calm, windless days these brief episodes may foul the air for the scores of scientists and other employees of Hawai'i Volcanoes National Park and the U. S. Geological Survey's volcanic observatory, but not at lethal levels. As the sulfur dioxide reacts with water in the atmosphere, tiny molecules of sulfuric acid form, scattering light and creating the haziness reminiscent of unpleasant smoggy days in the Los Angeles basin. During much of the year, prevailing westerly winds sweep K'ilauea's VOG to the island's western coast, where it often becomes trapped against the mountains that overlook the city of Kailua-Kona (population 10,000). From time to time, the VOG even reaches Honolulu, about 260 kilometers (160 miles) to the west.

Hawai'ian VOG has yet to be blamed directly for any serious health hazards, although episodes of high sulfur emissions appear to correlate with increased incidence of bronchitis and other respiratory problems. Short-term exposure need not worry tourists, but longer-term exposure, especially to senior citizens, may be of greater concern. As in air-polluted industrial centers, local residents are advised to refrain from outdoor exercise and avoid smoking on particularly bad VOG days. And unlike at industrial sources, where emissions can be cleansed by scrubbers and filtration systems, little can be done about Hawai'i's pollution source.

Figure 4-18 Gases emitted regularly from Hawai'i's K'ilauea volcano, especially sulfur dioxide, are responsible for the big island's air "pollution."

The fine dust and ash sprayed from explosive eruptions may actually produce rain clouds, as atmospheric moisture, augmented by water vapor from the eruption, condenses around the cooling tephra particles. These clouds may then produce torrential rains, triggering lahars that carry loose ash and soil down a volcano's slope. A mudflow whose water content is particularly high may travel tens of kilometers per hour. Even the swiftest among us could not outrun such flows. Lahars may continue to develop long after a volcanic eruption has ended if normal heavy rains mobilize loose, fresh volcanic ash. For example, lahars continue to flow from the deposits laid down by the 1991 eruption of Mount Pinatubo in the Philippines.

Secondary Volcanic Effects

As if suffocating ash falls, searing nuées ardentes, and enveloping hot lahars were not enough, volcanic eruptions produce secondary effects that alter the environment and affect human, animal, and plant life as well as change the composition of the atmosphere and even, in some cases, the global climate. For a sense of the environmental challenges posed

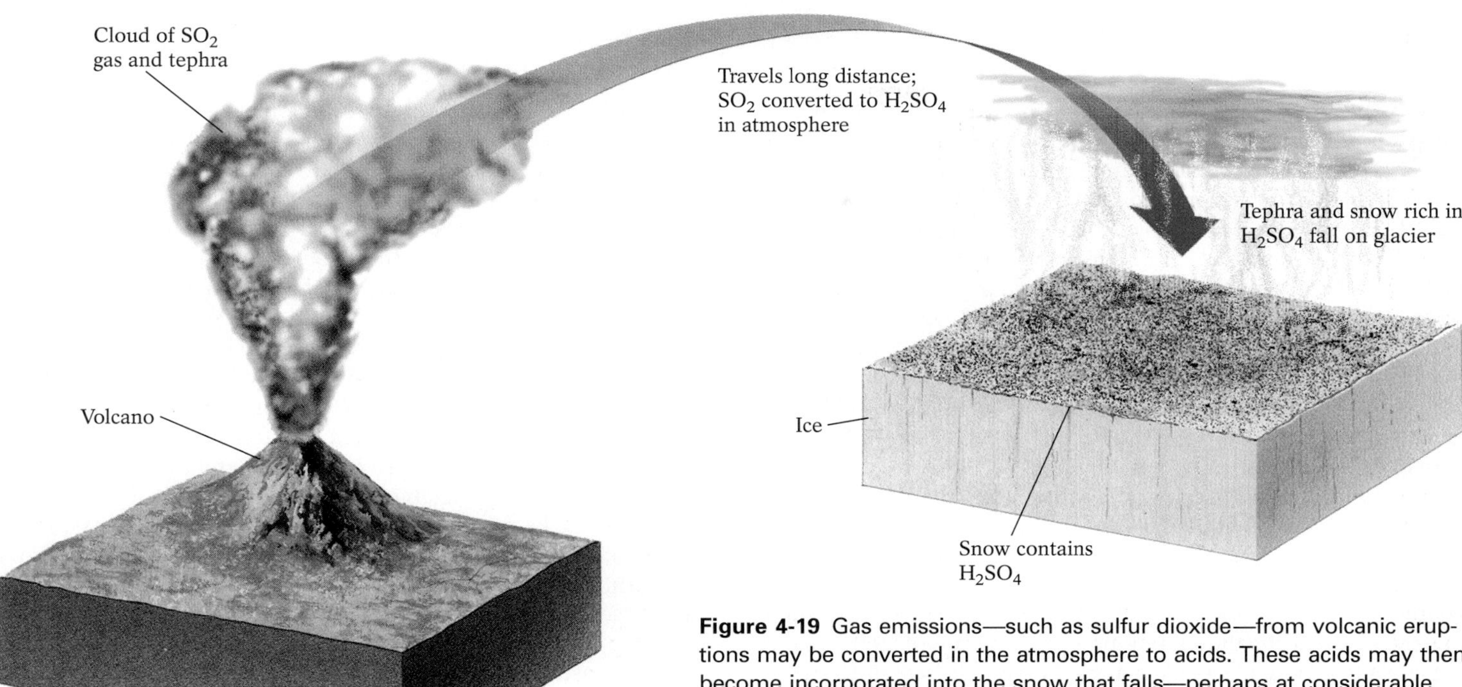

Figure 4-19 Gas emissions—such as sulfur dioxide—from volcanic eruptions may be converted in the atmosphere to acids. These acids may then become incorporated into the snow that falls—perhaps at considerable distances downwind—on glaciers. This acid-rich snow eventually becomes part of the glacier, forming an indirect record of world-wide volcanism within the ice.

by life in the shadow of an active volcano, see Highlight 4-1.

When magmatic gases such as sulfur dioxide (SO_2) escape during and after an eruption, they may combine with atmospheric water vapor and oxygen, forming molecules of airborne acids, such as sulfuric acid (H_2SO_4). Sulfuric acid droplets may remain in the atmosphere for years, producing acidic rain and snowfall, increasing the acidity of local, regional, and global waters, and absorbing incoming solar radiation. Volcanologists who have analyzed ice-core samples from Greenland's thick ice sheets have identified periods of sharply increased acidity. The estimated age of the acidic ice often correlates well with a cataclysmic volcanic eruption. Apparently, when the Earth's atmosphere contains as unusually high amount of SO_2—possibly due to volcanic activity—the snow that falls is markedly more acidic (Fig. 4-19).

Volcanic gas and ash emissions also affect worldwide climate. The dust and ash from a large tephra column can rise into the stratosphere and remain suspended there for as long as five years. These particles reflect incoming sunlight back into space, thus lowering the amount of radiation that can reach and warm the Earth's surface. Droplets of emitted SO_2 in the atmosphere also reflect and absorb radiation. The combination of dust and gas from a single large eruption can drop the Earth's temperature by as much as 2° to 3° Celsius (4–6°F), an effect that may last for more than a decade. Some monumental eruptions can have even greater effects (see the discussion of the Toba eruption later in this chapter).

The first recorded scientific demonstration of the effect of volcanism on climate took place in May 1784, when a blue haze and dry fog hovered over Europe and the weather was unusually cool. Scientist–philosopher–statesman Benjamin Franklin, living in Paris as America's ambassador to France, hypothesized that Europe's unusual climate was caused by an enormous eruption of gassy basalt during the previous year in Iceland. He tested his hypothesis by using a magnifying glass to focus the sun's rays on a sheet of paper. Normally this activity would burn a hole in the paper, but in the spring of 1784 it did not. Franklin concluded correctly that less sunlight was reaching Europe as a result of the Icelandic eruption.

The greater the eruption, the longer its effects will linger. The spectacular eruption of Indonesia's Mount Tambora in 1815, which took an estimated 50,000 lives and decapitated the peak, was followed by what became known as the "Year Without a Summer." In all, 150 to 200 cubic kilometers (60–80 cubic miles) of tephra were ejected into the atmosphere (about 200 times that expelled from Mount St. Helens in 1980). Ash blanketed 1 million square kilometers of southeastern Asia. Airborne tephra blocked the sun's rays and brought two days of total darkness to a 600-kilometer (400-mile) area surrounding the volcano. Globally, summer air temperatures dropped 1.0° to 2.5°C (2–5°F). Snow fell in upstate New York the following June; people in Connecticut celebrated the Fourth of July in overcoats, and August frosts destroyed crops from the Midwest to Maine. Tambora's dust and gas reduced sunlight by more than 10% and produced

Highlight 4-2 Loihi: Growing Pains of the Next Hawai'ian Island

About 30 kilometers (20 miles) southeast of the Big Island of Hawai'i, geologists are watching expectantly and nervously as an oceanic island comes to life. There, a submerged shield volcano named Loihi (low-EE-hee, which means "the long one" in a Polynesian dialect) has grown to a height of roughly 4500 meters (15,000 feet) by erupting a vast volume of basaltic lava over the last 100,000 years. Its summit still resides 1000 meters (3300 feet) below the Pacific Ocean's surface, and local geologists estimate that it won't emerge from the sea for another 50,000 to 100,000 years. Still, Loihi is more than capable of making its presence felt both to the local islanders and to distant points around the Pacific Ocean basin, as the summer of 1996 dramatically demonstrated.

The most intense swarm of earthquake's ever recorded in Hawai'i—more than 5000 quakes—occurred during July and August of 1996 beneath Loihi's summit. Following this activity, oceanographers and geologists from the University of Hawai'i visited the peak to assess any changes in Loihi. They began their underwater investigation along the southern rim of Loihi's summit at Pele's Dome, a peak named for the Hawai'ian fire goddess. To their amazement, the dome, formerly 300 meters (1000 feet) high, had vanished, only to be replaced with a crater more than a kilometer (about six-tenths of a mile) in diameter and more than 300 meters (1000 feet) deep. Further inspection of the area showed that bus-sized boulders littered Loihi's slopes and the surrounding sea floor over a 10-kilometer (6-mile) area.

Apparently, Loihi's summit had collapsed catastrophically, spawning huge rock avalanches and forming a deep crater—Pele's Pit (Fig. 4-21). Geologists hypothesize that the summit's collapse and the associated earthquake swarm occurred when lava erupting from deep fractures near Loihi's base partially drained the magma reservoir beneath the summit, allowing it to collapse inward. The mass of rock that fell into the abyss—some 300 million tons—would have filled 50 million dump trucks.

The violent rearrangement of Loihi's summit left countless steaming cracks, some as large as 6 meters (20 feet) across, spewing a mixture of boiling water and dissolved minerals. Of greater concern, however, was the instability of the entire volcano. The immense cave-in of the summit should have triggered a large seawave, or *tsunami,* with the potential to destroy nearby populated coastal areas, such as Honolulu's Waikiki Beach. Fortunately, no tsunami developed, perhaps because the collapse

a long succession of dreary days. The depressing weather that summer at Lake Geneva in Switzerland has been credited with creating the morbid mood for Mary Shelley's gruesome classic, *Frankenstein.*

Like Krakatau, major volcanic eruptions of oceanic islands and volcanic seamounts (submerged volcanoes) may produce another potentially catastrophic secondary effect—giant seawaves. No place better illustrates this potential than the seamount Loihi, the next Hawai'ian island, discussed in Highlight 4-2.

Eruptive Styles and Associated Landforms

When we think of volcanoes, highly photogenic, snow-capped peaks may come to mind (Fig. 4-20a). We do not usually picture scruffy little hills (Fig. 4-20b), which are also a common type of volcano. Despite their different appearances, nearly all volcanoes have the same two major components: a mountain or hill, or **volcanic cone,** constructed from the products of numerous eruptions over time, and a steep-walled, bowl-shaped

(a)

(b)

Figure 4-20 Volcanoes come in various shapes and sizes. **(a)** Lofty, symmetrical Mount Augustine, in Alaska's Cook Inlet; **(b)** small, stubby cinder cones from the Flagstaff area of northern Arizona.

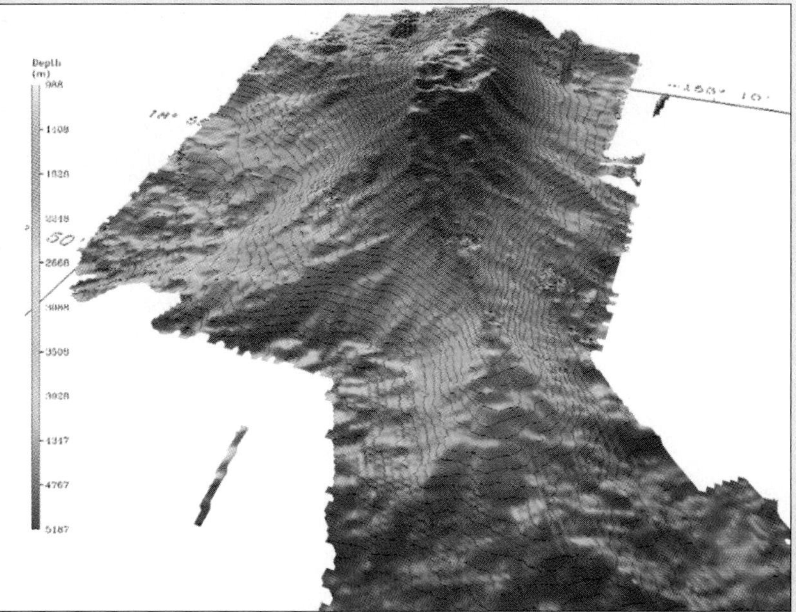

occurred over several days rather than instantaneously. In the future, Hawai'ians may not be so lucky. The sea floor around Loihi and along the southeastern coast of the Big Island of Hawai'i is littered with the rocky remains of countless submarine slides, some of which caused giant waves that may have reached as high as 300 meters (1000 feet). (This type of event will be discussed in greater detail in Chapter 13.) And unlike other tsunami that have struck Hawai'i after long journeys from distant Pacific Rim earthquakes, a giant wave spawned by another, perhaps swifter, collapse of a portion of Loihi could arrive with a warning of only minutes or seconds.

Thus Loihi grows higher in fits and starts—building itself up volcanically and then collapsing and shedding its substance in great rockslides, broadening its base as a foundation for future upward growth. This pattern may be the blueprint for the creation of many, if not all, of the thousands of volcanic islands—from Hawai'i to Tahiti—that adorn the Pacific Ocean floor.

Figure 4-21 Pele's Pit, a depression at the summit of the Loihi seamount, formed when an eruption emptied much of the volcano's underlying magma chamber, collapsing the summit.

depression, or **volcanic crater,** surrounding the vent from which those volcanic products emanate. A volcano's crater forms when lava and pyroclastics that have accumulated in the area around the vent subside to form a depression as the magma chamber empties. Volcanic cones and craters have a variety of shapes and dimensions, depending on the eruptive style and the composition of the volcanic products.

Effusive Eruptions

Effusive eruptions are relatively quiet, nonexplosive events that generally involve basaltic lava. They are seldom accompanied by powerful volcanic blasts and rarely produce large volumes of pyroclastic materials. Basaltic lava, which is highly fluid, flows freely from central volcanic vents, as well as from elongated cracks on land and from fractures at submarine plate boundaries and intraplate hot spots.

Central-Vent Eruptions In central-vent eruptions, basaltic lava flows out—sometimes as a spectacular red-hot fountain—in all directions from one main vent, solidifying in more or less the same volume all around. Over time, successive flows accumulate to form a discernible low, broad, cone-shaped structure, known as a **shield volcano** because it resembles a warrior's shield lying on the ground (Fig. 4-22). Shield volcanoes owe their gentle slopes and broad summits to the low viscosity of basaltic lava, which flows swiftly, sometimes within a network of lava tubes, and far enough from the vent to produce gentle slopes. A vertical slice through a shield volcano would reveal hundreds of layers of basalt that record eruptions throughout the volcano's life.

The classic location for the study of central-vent eruptions and shield-volcano development is on the Big Island of Hawai'i, near the southeastern end of the 2590-kilometer (1600-mile)-long chain of islands and undersea mountains in the north-central Pacific Ocean. This chain developed as the Pacific plate moved over a relatively stationary hot spot in the Earth's mantle that has acted like a blowtorch burning vast welts into the overlying plate. (We will look more closely at the formation of volcanic island chains in Chapter 12.)

Hawai'i's shield volcanoes are the largest single objects on Earth. The Big Island of Hawai'i consists of five major shields, the two largest being Mauna Kea and Mauna Loa, whose summits are 4205 meters (13,792 feet) and 4170 meters (13,678 feet) above sea level, respectively. Because the bases of these volcanoes are on the floor of the Pacific Ocean, thousands of meters below sea level, their total heights—more than 9000 meters (30,000 feet)—are greater than that of Mount Everest, the Earth's highest continental peak. Mauna Loa, with a circumference of 600 kilometers (400 miles), is composed of thousands of layers of lava flows that have erupted over the past 750,000 years.

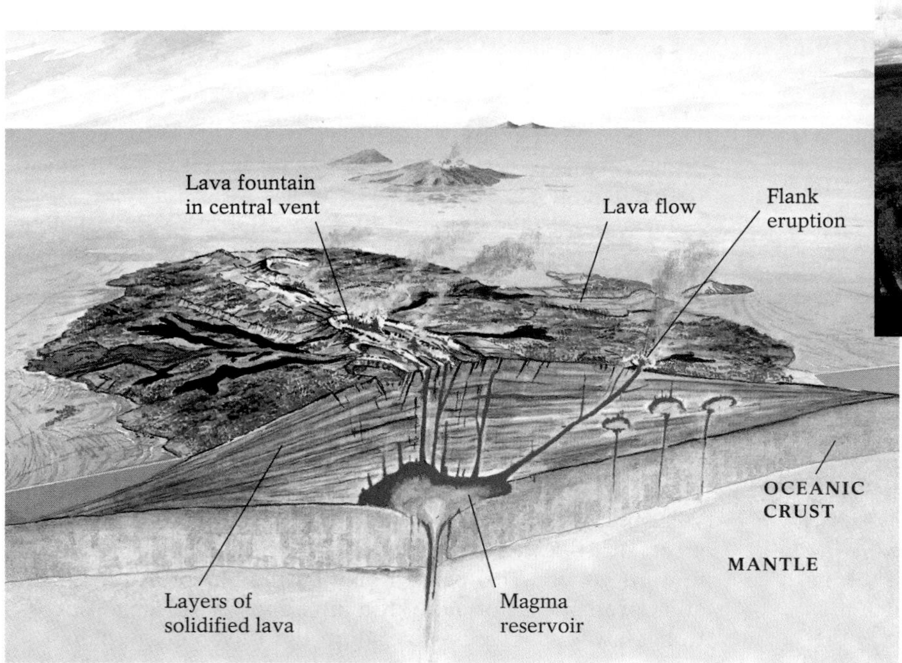

Figure 4-22 Low, broad shield volcanoes form their gentle slopes by the gradual accumulation of numerous basaltic lava flows. Inset: Photo of summit of Hawai'i's K'ilauea volcano, a shield volcano that actually developed on the flank of an even larger shield volcano, Mauna Loa.

At the start of an effusive eruption, lava begins to accumulate in the volcanic crater, sometimes forming a lava lake that may eventually overflow the rim of the crater. If enough lava has erupted to empty or partially empty the volcano's subterranean reservoir of magma, the unsupported summit of the volcanic cone may collapse inward, forming a larger, more-or-less circular summit depression known as a **caldera** (cal-DER-a). Whereas an initial summit crater may be 300 meters (1000 feet) in diameter and several hundred meters deep, a caldera may be a few kilometers to more than 15 kilometers (10 miles) in diameter and more than 1000 meters (3300 feet) deep.

The weight of a collapsed summit may close off a volcano's central vent and divert any remaining underlying magma laterally, producing a *flank eruption* from the side of the volcano. Flank eruptions also occur when congealed lava plugs up the central vent or when the volcanic cone has grown so high that rising magma seeks a lower, more direct route to the surface. Pressurized lava spewing from a secondary vent on a volcano's flank often forms spectacular lava foun-

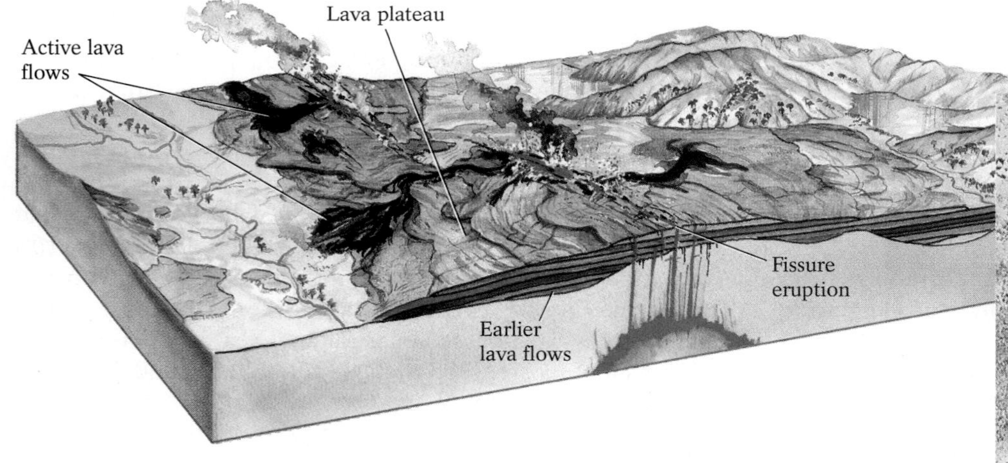

Figure 4-23 Lava-plateau formation. Inset: The 15-million-year-old basaltic Columbia River Plateau of eastern Washington. Geologists have identified more than 60 individual lava flows here, in some places totaling more than 2 kilometers (1 mile) in thickness. The lava from just one of these flows could pave Interstate 90 from Boston to Seattle to a depth of 175 meters (575 feet), about the height of the Washington Monument in Washington, D.C.

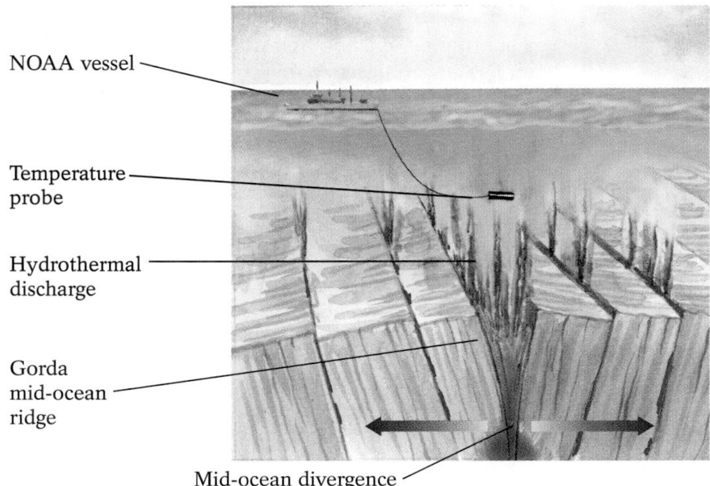

Figure 4-24 In late February and early March 1996, a segment of the Gorda Ridge, located about 200 km (120 miles) due west of Gold Beach, Oregon, erupted a flood of basaltic lava. (top) the NOAA (National Oceanic and Atmospheric Administration) vessel MacArthur, towing a temperature probe, detected sharply elevated water temperatures as steam erupted from the ridge. (bottom) The Gorda and Juan de Fuca Ridges—oceanic divergent zones—have had several periods of flood-basalt eruption during the past five years. The Co-Axial segment of the Juan de Fuca Ridge has been particularly active.

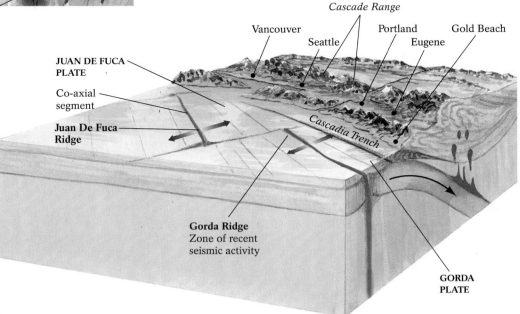

tains 100 to 200 meters (330–660 feet) high. Hawai'i's K'ilauea is a volcanic cone built up by flank eruptions on Mauna Loa's southeastern slope. One of the world's most active volcanoes, K'ilauea's supply of fresh basaltic lava is continuously replenished because it is still located above the Hawai'ian hot spot.

Fissure Eruptions on Land Rising basaltic magma will exploit any available route to the surface. That route may be through a series of linear fractures, or *fissures*, in the Earth's crust, which develop most often where plates are diverging. A large volume of highly fluid basaltic lava may gush out at speeds of 40 kilometers per hour (25 miles per hour) or more from kilometer-long cracks in the crust. These are called *flood basalts.* As lava flows away from a fissure, it may spread out over thousands of square kilometers; successive flows build up immensely thick *lava plateaus,* (Fig. 4-23). A disastrous basaltic flood began in Iceland on June 8, 1783, as a series of lava flows spread out from one 32-kilometer (20-mile)-long fissure. The lava blocked and diverted rivers, melted snow and ice into raging torrents, and destroyed much of Iceland's scarce agricultural land. Many livestock were poisoned after grazing on grass contaminated by hydrogen fluoride emitted during the eruption. As a result, 10,000 people—20% of Iceland's population—perished from starvation.

Shield volcanoes, such as Hawai'i's K'ilauea, can also experience fissure-type eruptions when a linear section of the volcano's flank rifts open. Along K'ilauea's southeastern flank, low-viscosity basaltic lava issues from the volcano's eastern rift zone.

Subaqueous (Underwater) Eruptions Most subaqueous eruptions—especially oceanic, or *submarine,* eruptions—are quiet and effusive, because below about 300 meters (1000 feet) of water, water pressure prevents the water vapor and other gases in lava from expanding or escaping. Effusive submarine eruptions of basalt may form submarine shield volcanoes, when lava flows from a central vent, or flood basalts, when it issues from a linear fissure or network of fissures. In both cases, the basalt exhibits the pillow structures that typify submarine eruptions of basalt (see Fig. 4-12). In late February and early March of 1996, the Gorda ridge, an oceanic divergent zone off Oregon, entered an eruptive phase that will probably culminate in another substantial submarine flood basalt, the third in the past three years along the Pacific Northwest coast (Fig. 4-24).

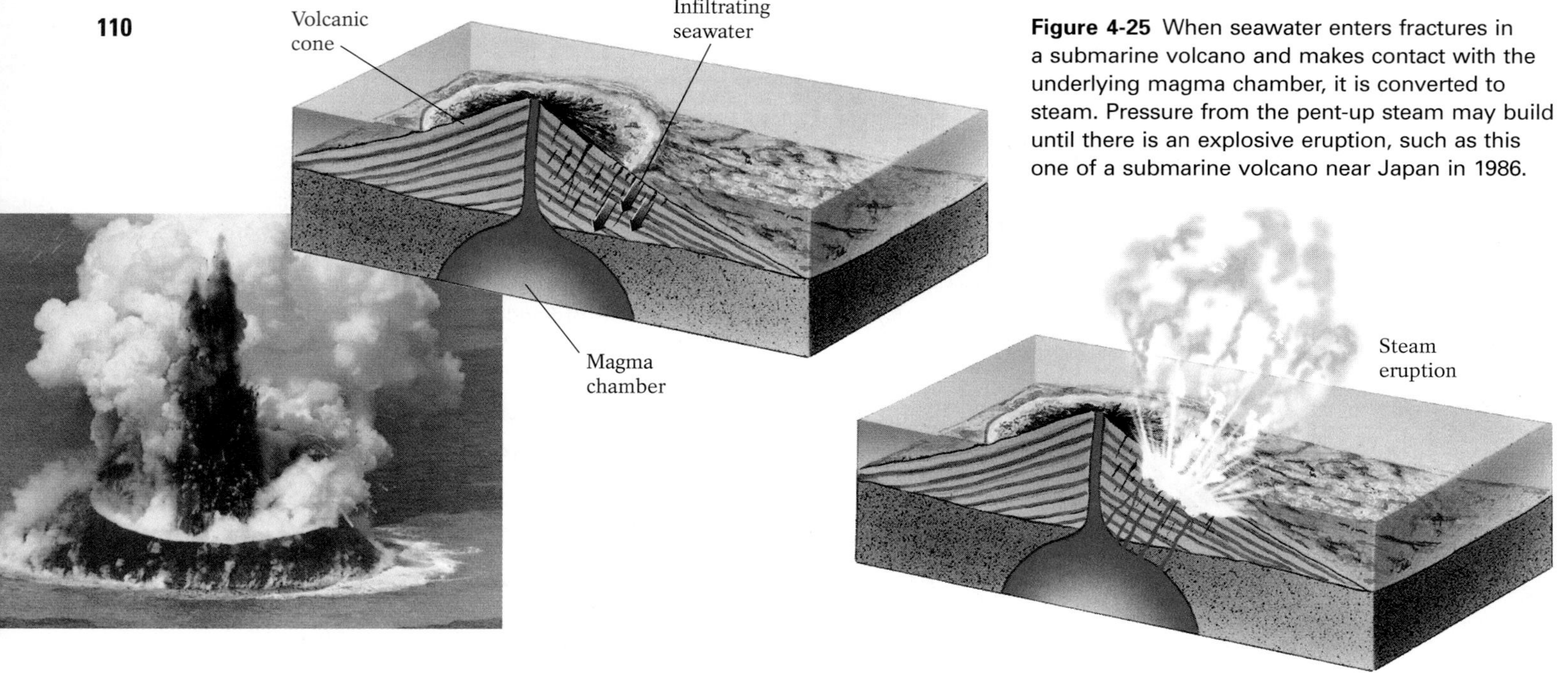

Figure 4-25 When seawater enters fractures in a submarine volcano and makes contact with the underlying magma chamber, it is converted to steam. Pressure from the pent-up steam may build until there is an explosive eruption, such as this one of a submarine volcano near Japan in 1986.

Highlight 4-3 Fire and Ice: Iceland's Subglacial Eruption of 1996

On September 29, 1996, an earthquake shook the ground beneath the 600-meter (2000-foot)-thick Vatnajökull icecap of southeastern Iceland. Vatnajökull, covering 8300 square kilometers (3200 square miles), is the largest glacier in Europe. The quake signaled the onset of a fissure eruption emanating from Iceland's Bardarbunga and Grimvötn volcanic centers. On the morning of October 1, scientists discovered a deep subsidence basin in the glacier's surface, close to the location of a major subglacial eruption in 1938. Throughout the day, the basin grew and three additional basins opened. The development of these aligned basins indicated that intensive melting was occurring at the base of the glacier along a 5- to 6-kilometer (3- to 4-mile)-long fissure. By October 2, steam rising from the glacier's surface indicated that the eruption had broken through the ice. Over the next several days, a steam column rose to more than 10 kilometers (6 miles), the surface fissure extended roughly the same distance, and black ash was being thrown intermittently into the air to heights of 300 to 500 meters (1000–1600 feet) (Fig. 4-26).

The greatest danger from a subglacial eruption is a *jokulhlaup,* or outburst flood. These floods occur when a great volume of subglacial meltwater, draining through subglacial tunnels, finds an outlet at the glacier's margin and discharges catastrophically (Fig. 4-27). Vatnajökull, overlying a segment of the mid-Atlantic ridge in southeastern Iceland, has been the source of numerous jokulhlaups in the past. By October 12, Vatnajökull's surface had risen 15 to 20 meters (50 to 65 feet) directly above the Grimsvötn caldera. This change suggested that meltwater from the eruption was flowing into and filling the 10-kilometer (6-mile)-diameter caldera. Engineers, anticipating a flood of monumental proportions, worked around the clock to reinforce dikes and cut diversionary channels to steer the expected torrents away from Iceland's main, island-ringing road, about 50 kilometers (30 miles) south of the icecap.

The long, unnerving wait for the jokulhlaup ended on November 5, when Iceland's largest flood in 60 years burst from the edge of the icecap. For two days, the flow formed the second-largest river in the world, destroying three bridges, severing vital telephone lines, and, despite valiant efforts, washing away Iceland's south-coast road. Such are the rigors of life in Iceland—the self-proclaimed Land of Fire and Ice.

Figure 4-26 Steam and ash rising from the subsidence basin on the surface of the Vatnajökull icecap, directly above the subglacial Grimvötn volcano.

Some subaqueous eruptions, however, are far from quiet and effusive. Occasionally, shallow submarine eruptions are driven by jets of steam that propel lava above the ocean's surface, where it shatters and cools rapidly to form a substantial tephra cloud. Even more explosive submarine eruptions occur when seawater enters a magma chamber through the ruptured walls of an island volcano. The cold water in contact with the red-hot magma produces a great deal of superheated steam that may then expand violently, shatter the volcanic cone, and propel magma and cone fragments skyward (Fig. 4-25).

Subglacial Eruptions When polar regions coincide with zones of volcanic activity, two of geology's most pronounced extremes—fire *and* ice—collaborate in spectacular geological events. For example, in areas where large ice sheets overlie volcanically active fissures, hot mafic lava periodically erupts beneath the ice, melting vast amounts of it and producing catastrophic outbursts of meltwater. One such drama captivated geologists for six weeks in the autumn of 1996 when a volcano erupted beneath a massive icecap in Iceland (Highlight 4-3).

Pyroclastic Eruptions

Pyroclastic eruptions, usually involving viscous, gas-rich magmas, vary from moderately to spectacularly explosive and tend to produce a great deal of solid volcanic fragments. Moderately explosive eruptions on the Italian island of Stromboli in the eastern Mediterranean occur almost continuously. A

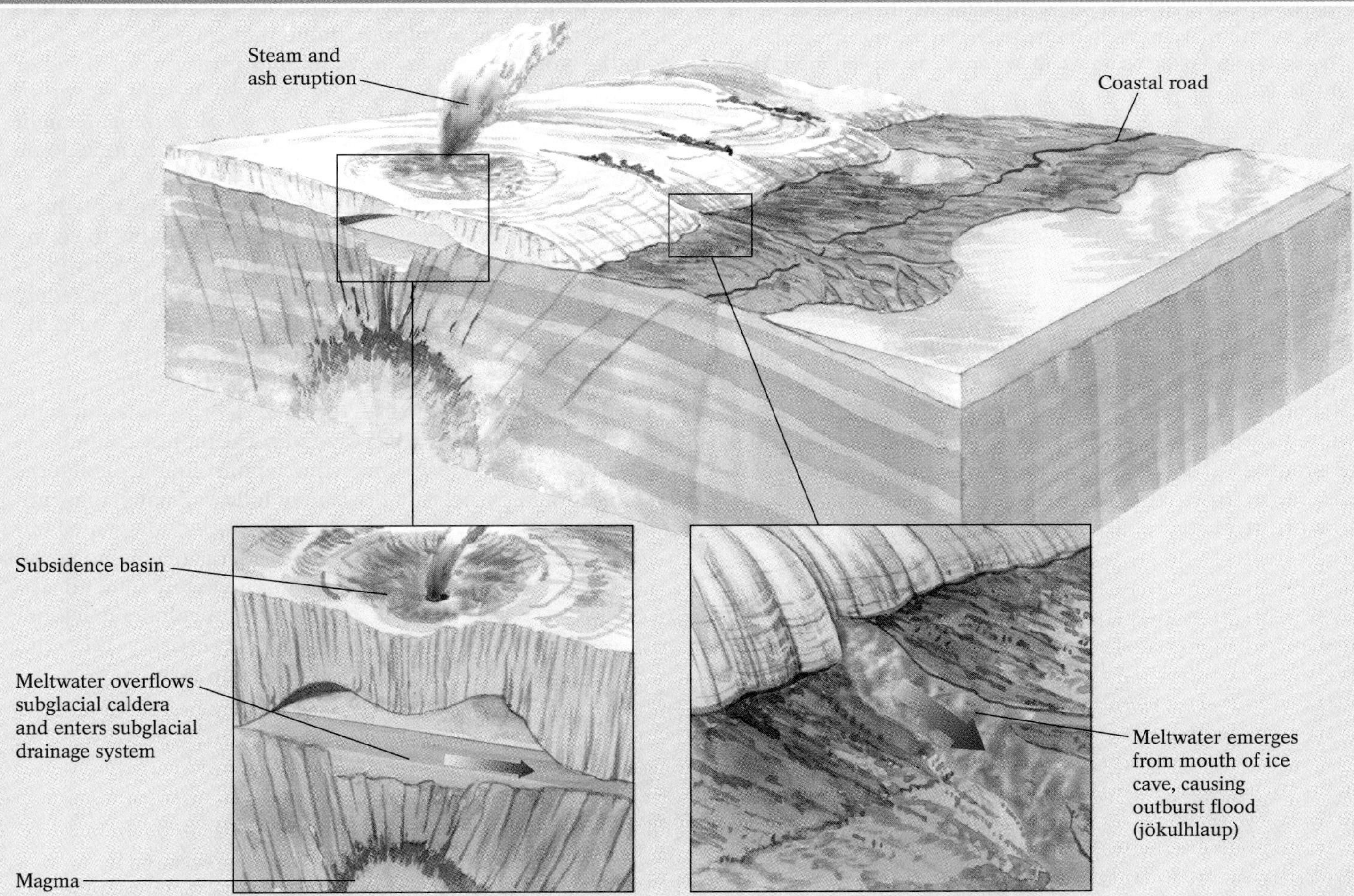

Figure 4-27 Muddy meltwater bursts catastrophically from the margin of the Vatnajökull icecap, producing an outburst flood, or jokulhlaup.

Figure 4-28 When Mount Vesuvius erupted in A.D. 79, the people in nearby Pompeii were trapped and suffocated beneath a layer of volcanic ash as much as 6 meters (20 feet) thick. Archaeologists who excavated Pompeii found cavities lined with an exact imprint of the decomposed bodies; by pouring plaster into the cavities, they were able to make casts that displayed the victims' musculature, agonized facial expressions, and in some cases even the folds in their clothing.

constant muffled thumping mixes with the daily sounds of nature in the region. The blasts cover the region with a cloud of ash and hurl sizzling bombs, some as heavy as 2000 kilograms (4400 pounds), as far as 3 kilometers (2 miles) from the crater.

Fed by large quantities of extremely gas-rich magma, pyroclastic eruptions may periodically produce spectacular vertical columns of tephra rising tens of kilometers into the atmosphere. The surrounding countryside is simultaneously smothered by suffocating showers of hot ash and swirling nuées ardentes and hot lahars. Residents of the prosperous Roman resort town of Pompeii, for example, lived in the shadow of the craggy peak of Mount Vesuvius, which looms 1220 meters (4002 feet) above the Bay of Naples. Unfortunately, they did not realize that this vine-clad mountain was a volcano until it exploded on August 24, A.D. 79, sending a tephra column into the atmosphere, obscuring the midday sun, and shrouding the region in darkness. Most Pompeiians, inhaling the still-warm particles, perished by asphyxiation; others were crushed by falling masonry. Towering waves trapped those who tried to escape by sea. The entire city was entombed beneath a layer of more than 6 meters (20 feet) of volcanic ash. This layer preserved the most minute details of Roman life until it was stripped away by archaeologists in the mid-eighteenth century (Fig. 4-28).

If it is fed by magma containing less gas, a pyroclastic eruption may produce dense clouds of hot ash and pumice instead of towering tephra columns. These pyroclastic materials tend to fall back to the surface almost immediately, with little chance for air-cooling, and then race downslope as nuées ardentes.

A felsic magma that contains fairly little gas or water may become a lava so viscous that it does not even flow out of the volcano's crater. Instead, it cools and hardens within the crater, forming a **volcanic dome** that caps the vent, trapping the volcano's gases and building pressure for another eruption. This pressure might be relieved by the escape of gases laterally through the flank of the volcano, or it might build to such a point that it finally shatters the volcanic dome in a particularly explosive eruption (Fig. 4-29).

The life of an explosive volcano may consist of thousands of years of repeated eruptions marked by towering tephra columns and fiery pyroclastic flows. The eruptive histories of such volcanoes sometimes culminate in an extremely energetic eruption, followed by collapse of the cone and the creation of a huge caldera. This final explosion typically begins with a titanic blast that produces a rapidly moving, ground-hugging blanket of pyroclastic debris—called a *base surge*—combined with a massive vertical tephra column. As the escaping gases lose force, the tephra column collapses and additional pyroclastic flows may follow. Finally, the summit of the cone may subside into the emptied magma chamber, forming a large caldera. A caldera in an oceanic volcano may immediately fill with seawater, as it did at Krakatau. A caldera in a land-based volcano might eventually fill with freshwater to form a sizable lake.

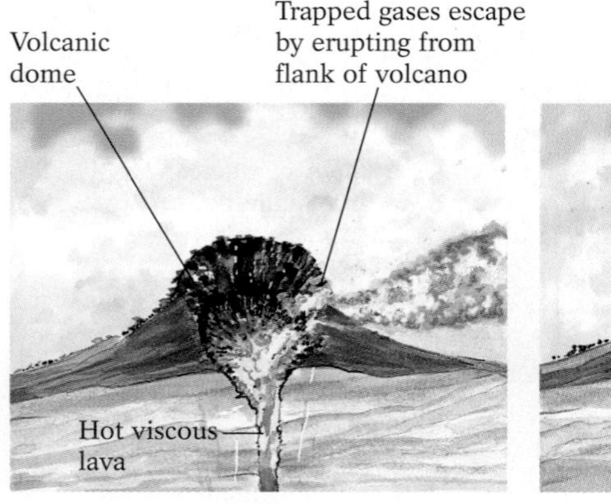
Volcanic dome
Trapped gases escape by erupting from flank of volcano
Hot viscous lava
(a)

Trapped gases explode out, destroying volcanic dome
(b)

Figure 4-29 Volcanic domes are caused by extremely viscous lavas that solidify before they leave their crater, plugging the volcanic vent. Buildup of pressure beneath a dome may result in gases escaping from the flank of the volcano **(a)**, or eventual explosive destruction of the dome **(b)**.

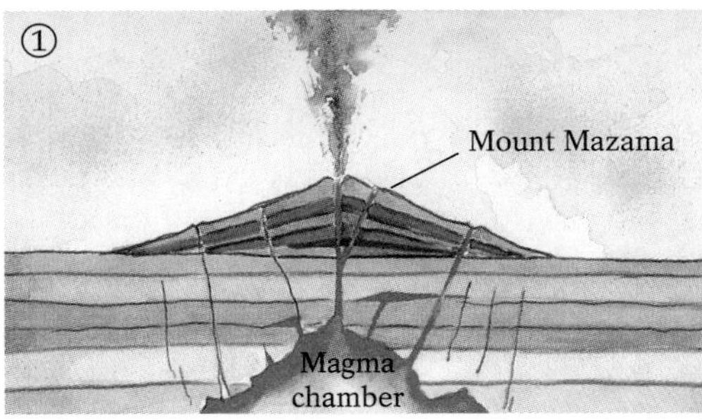

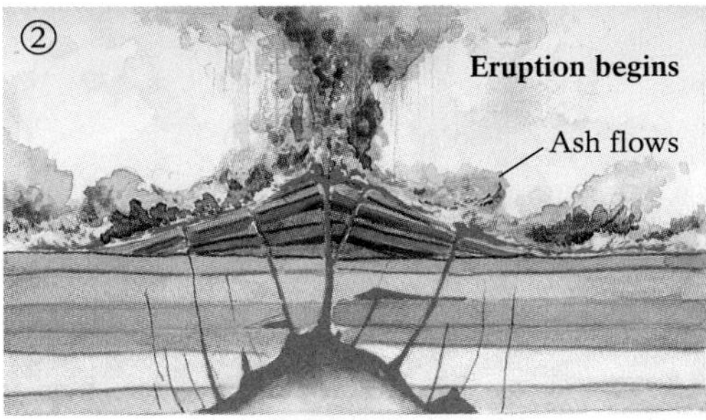

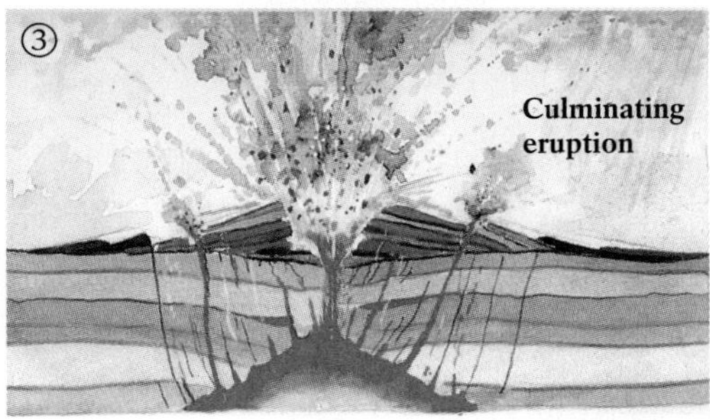

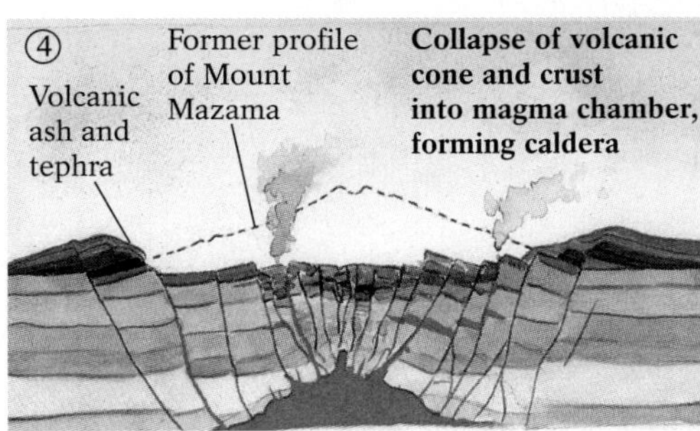

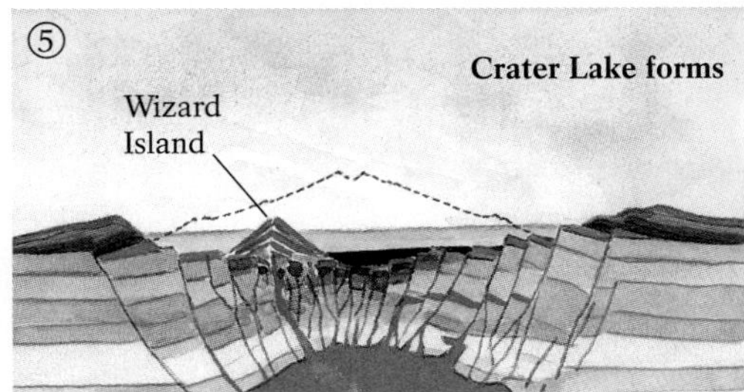

A culminating caldera-forming eruption took place in the southern Oregon Cascades about 6700 years ago, when Mount Mazama buried thousands of square kilometers with tephra and pumice before its summit collapsed to form the caldera we know now as Crater Lake. Before this eruption, the mountain is believed to have been about 3700 meters (12,000 feet) high and glacier-covered, comparable in majesty to Mount Rainier or Mount Shasta. Today, it is a sawed-off, goblet-shaped mountain only 1836 meters (6058 feet) above sea level, but it contains North America's deepest, bluest lake (Fig. 4-30).

This event may not have ended Mount Mazama's excitement, however. Eruptions over the past 1000 years have produced three smaller volcanic cones within the lake, two of which are still below water level. The third, Wizard Island, rises above the surface at the western shore of the lake. Recent dives to the lake bottom reveal that hot water and steam are being vented continuously. Mount Mazama may be entering a new eruptive sequence as new magma enters the plumbing of the volcano. Such rejuvenated volcanoes are called *resurgent calderas*.

Ash-Flow Eruptions Some spectacular and extremely dangerous pyroclastic eruptions even occur where there is no recognizable volcanic cone. An *ash-flow eruption* results when

Figure 4-30 The last eruption of Mount Mazama, a volcano in the southern Oregon Cascades, occurred about 6700 years ago and lowered its height by more than a mile. Because very little rock from the shattered cone of the volcano has been found in the areas surrounding Mount Mazama—including Oregon, Washington, and British Columbia—geologists have concluded that the summit of the volcano must have collapsed inward during its last eruption, rather than exploding outward, and to this day remains buried beneath the caldera produced by its collapse. This caldera (above) now contains Crater Lake, North America's deepest lake (more than 600 meters [1900 feet] deep). Wizard Island, a new cone growing from the lake bottom, appears in this photo in the left-hand side of the lake.

a large reservoir of extremely viscous, gas-rich magma rises to just below the Earth's surface, stretching the crust and forcing it up into a bulge marked by a series of ring-like fractures (Fig. 4-31). The thinned bedrock over such a magma reservoir may collapse, forcing magma into the fractures to produce a circular pattern of tremendous tephra columns. These soon collapse to form numerous flows of hot swirling ash and pumice. As the magma reservoir empties, the surface crust becomes undermined until it collapses to form a caldera.

The Toba ash-flow eruption, which took place approximately 75,000 years ago in the central highlands of the Indonesian island of Sumatra, may have been the greatest single volcanic event of the past million years. It produced a caldera 50 kilometers by 100 kilometers (30 miles by 60 miles) and buried an area of 25,000 square kilometers (about 10,000 square miles) beneath a blanket of pyroclastic material averaging 300 meters (1000 feet) in thickness. The amount of tephra and gas propelled skyward cooled the Earth so profoundly that some geologists believe that the ensuing "volcanic winter" may have hastened the arrival of the last period of world glacial expansion. Global temperatures plummeted by as much as 6°C (12°F).

In North America, a similarly monumental eruption occurred roughly 28 million years ago in what is now southwestern Colorado. About 50 kilometers (30 miles) northeast of Durango sits a crumpled eroded caldera named La Garita, a 75-kilometer (45-mile)-diameter depression that resulted from an eruption of mythic proportions. The La Garita ash flow was at least 20,000 times larger than the 1980 eruption of Mount St. Helens (described later in Highlight 4-4).

Ash-flow eruptions may recur when a collapsed magma chamber refills magma, forcing the caldera floor to arch upward again. At least three times in the last 2 million years, the Yellowstone plateau in Wyoming has erupted in devastating ash flows after being pushed up like a blister by an enormous mass of felsic magma. One of these events created the huge caldera that now contains beautiful Yellowstone Lake. Today, only a few kilometers below Yellowstone's caldera, a new, huge mass of felsic magma may be pooling. Yellowstone National Park's thermal features—its geysers (periodic steam blasts), hot springs (quieter flows of hot water), and gurgling mudpots (pools of boiling mud)—are all heated by this shallow subterranean magma reservoir. Another ash-flow eruption may not be imminent here, but geologists monitor the area continually for signs of increasing volcanic activity.

The most immediate threat of an ash-flow eruption in North America is near Mammoth Lakes, the popular ski resort in eastern California. The Long Valley caldera there formed about 700,000 years ago when an enormous pyroclastic eruption showered much of western North America with hot tephra. A dusting of this tephra can even be found as far east as central Nebraska. Over the past 20 years, the United States Geological Survey has watched the floor of the caldera rise more than 25 centimeters (9 inches). Other

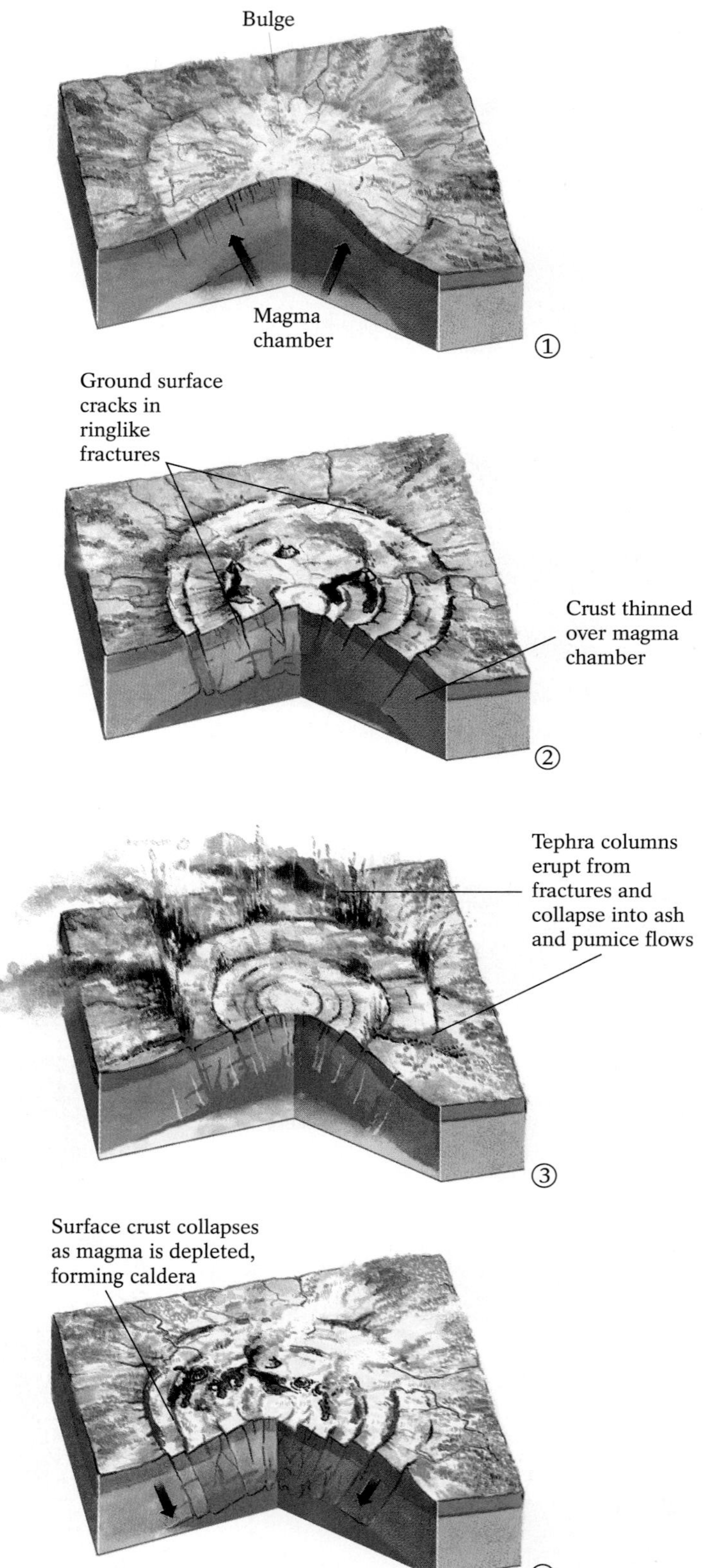

Figure 4-31 Ash-flow eruptions begin when rising viscous magma causes the surface crust to bulge. Cracks develop in the crust, allowing gas-rich tephra columns to erupt and then collapse into hot swirling ash and pumice flows. The partial emptying of the magma chamber weakens its roof and causes the surface to collapse, expelling more pumice in spectacular ash flows and forming a caldera. Later, new magma may refill the chamber and force the caldera upward again until another eruption takes place.

Figure 4-32 The origin of composite cones, steep-sided volcanoes consisting of layers of predominantly andesitic lava flows alternating with deposits of pyroclastic materials (and often associated with one or more flank eruptions). The composite cones in the Cascade Mountains of western North America, such as Mount Shasta shown here, have been built up over tens or even hundreds of thousands of years.

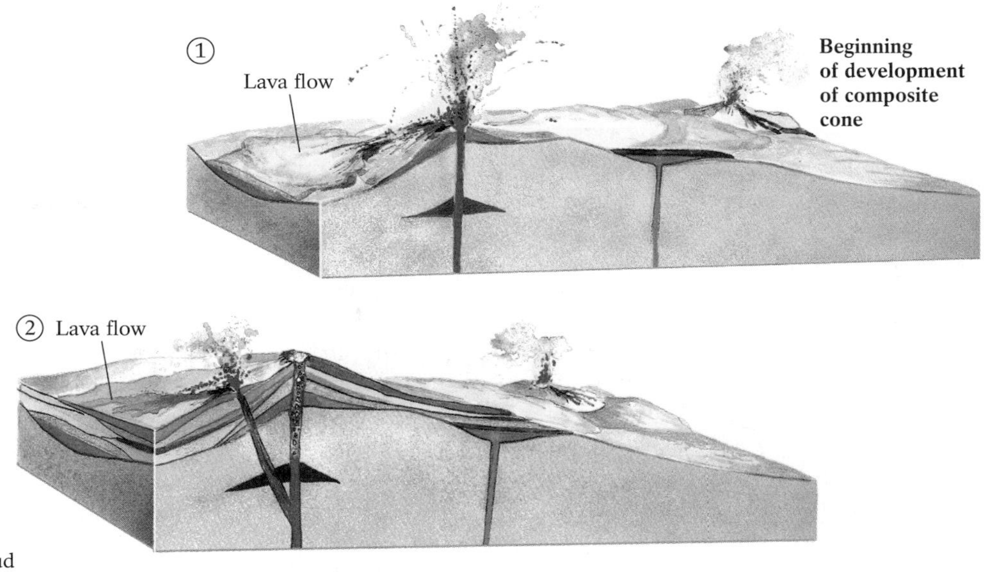

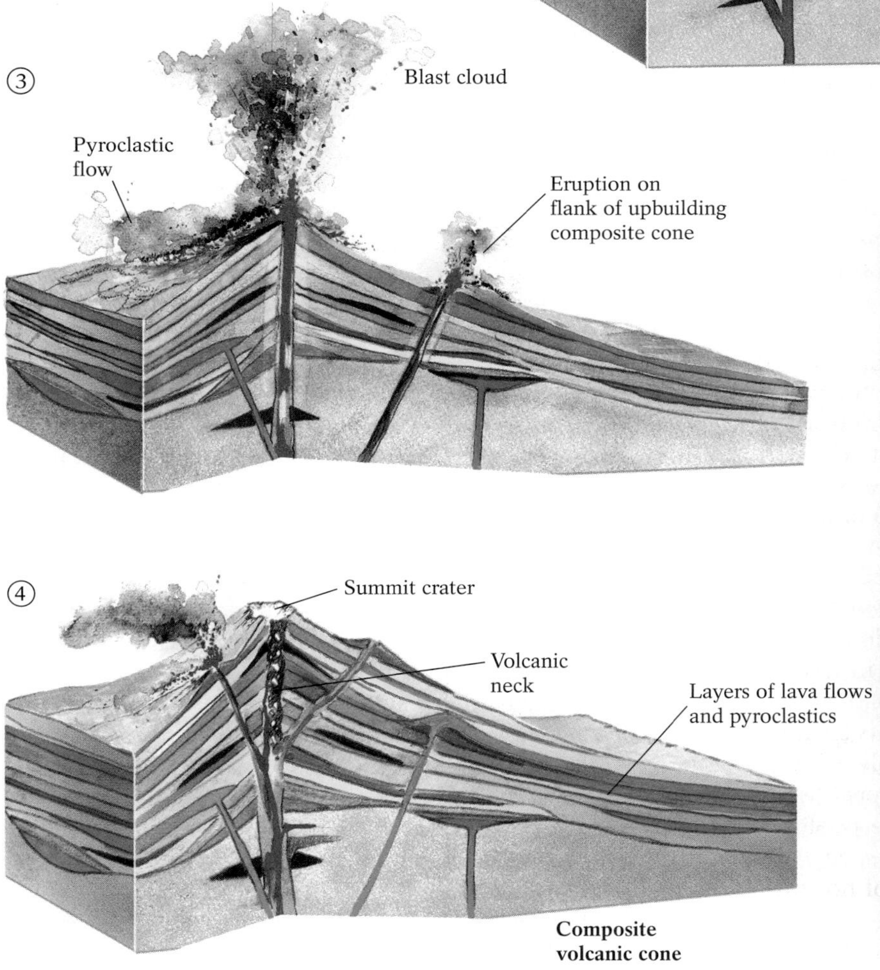

measurements indicate that magma has recently risen from a depth of about 8 kilometers (5 miles) to about 3.2 kilometers (2 miles) beneath the surface. In 1982, the area was designated a potential volcanic hazard, requiring heightened scientific vigilance and a regional plan for coping with an eruption.

Types of Pyroclastic Volcanic Cones Because lavas of intermediate-to-felsic composition are highly viscous, they solidify relatively close to the volcano's vent. The composition of such a magma, and subsequently the volcano's eruptive style, may change periodically over time so that the cone intermittently ejects a large quantity of tephra instead of lava. The larger tephra particles from these eruptions fall near the summit to form steep cinder piles, which are then covered by the next lava flow. This produces the characteristic landform of pyroclastic eruptions—the **composite cone,** or **stratovolcano** (*strato* means "layered"), built up from alternating layers of lava and pyroclastics (Fig. 4-32). Because each pyroclastic de-

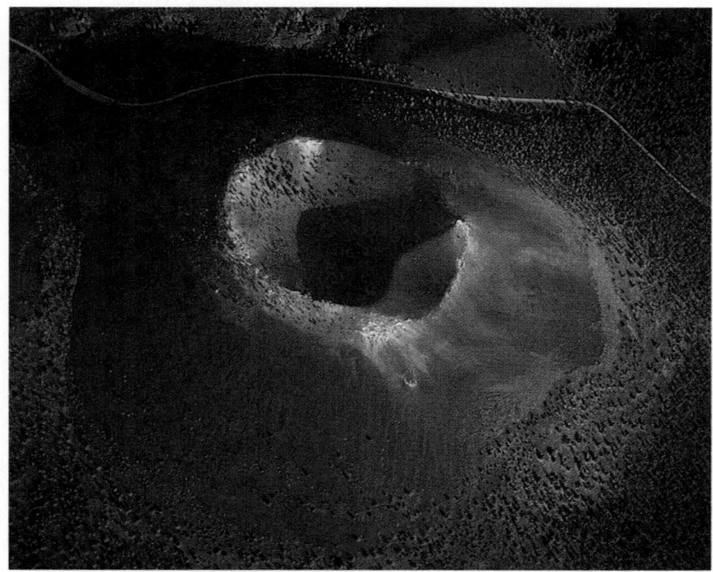

Figure 4-33 Sunset Crater, north of Flagstaff, Arizona. This cinder cone grew to a height of about 300 meters (1000 feet) sometime around A.D. 1065, toward the end of the region's volcanic activity.

posit produces a steep slope that is then protected from erosion by a successive layer of lava, composite cones grow to be very large and have 10° to 25° slopes. They include some of the Earth's most picturesque volcanoes, among them Mounts Rainier, Hood, and Shasta in the western United States.

Unlike composite cones, **pyroclastic cones** consist almost entirely of loose pyroclastic material. When the dominant pyroclastics are cinders, they form **cinder cones** (Fig. 4-33). Pyroclastic cones are typically the smallest volcanoes (less than 450 meters [1500 feet] high) and are generally steep-sided, because pyroclastic materials can pile up to form slopes between 30° and 40°. But no intervening lava flows bind their loose pyroclastics, so pyroclastic cones are readily eroded. Pyroclastic cones may form from a magma of any composition as long as it contains enough gas to shower the lava into the air and create a cone of pyroclastic material around a volcanic vent.

Plate Tectonics and Volcanism

Plate boundaries and hot spots often coincide with volcanic activity. About 80% of the Earth's volcanoes surround the Pacific Ocean, where several oceanic plates are subducting beneath adjacent landmasses. Another 15% are similarly situated above subduction zones in the Mediterranean and Caribbean seas. The remainder are scattered along the ridges of divergent plate boundaries (as in the case of Iceland's 22 active volcanoes) and atop continental and oceanic intraplate hot spots (as in the case of Yellowstone National Park and the Hawai'ian volcanoes). Each type of plate tectonic setting determines the eruptive style and physical appearance of its volcanoes (Fig. 4-34).

Subduction zones foster explosive pyroclastic volcanism because of the intermediate and felsic magmas produced there. As we saw in Chapter 3, water expelled from the descending plate enters the warm, overlying mantle, lowering its melting point and causing it to melt partially. When these materials mix with some of the descending plate's melted silica-rich sediments, an intermediate-to-felsic magma forms. Thus subduction-zone volcanism tends to produce steep-sided composite cones composed primarily of andesite.

Explosive volcanism also occurs at intracontinental hot spots and where felsic continental plates are stretched thin and begin to rift. In both of these cases, felsic and intermediate rocks at the base of the continental crust are melted by contact with hot mafic magma rising from the mantle. Eruptions of viscous felsic magmas create the volcanic domes, calderas, and ash-flow deposits characteristic of intracontinental rift zones and hot spots. Rift zones and hot spots may also produce mafic lava, although not necessarily simultaneously with the felsic eruptions. This type of *bimodal* (two-type) volcanism can be found in the stretched-and-thinned Basin and Range region of southwestern North America. There, powerful blasts of felsic ash and pumice have alternated with quiet effusions of basaltic lava throughout the last 20 million years.

At divergent zones (such as in Iceland), continental rifts (such as the East African rift valley), and oceanic intraplate hot spots (such as Hawai'i), effusive eruptions of low-viscosity basaltic lava generally produce near-horizontal lava plateaus and gently sloping shield volcanoes.

One plate tectonic environment rarely, if ever, produces volcanic activity: a transform boundary such as the San Andreas fault. There, no plates are subducting and generating andesitic lavas, nor are they diverging and allowing mantle-derived mafic lavas to rise. Thus, as a knowledgeable geology student, you would watch with considerable skepticism as lava flows down Wilshire Boulevard in Los Angeles in the entertaining but geologically questionable disaster extravaganza *Volcano*. Table 4-1 on page 118 shows the relationships among lava types, eruptive styles, types of volcanic cones and products, and plate tectonic settings.

Coping with Volcanic Hazards

The Earth is long overdue for a monumental volcanic eruption—one that is at least a thousand times more violent than anything we've seen in the twentieth century. The geological record indicates that such events occur at least once, perhaps twice, every 100,000 years. It has been 75,000 years since the last one—the Toba eruption, discussed earlier in this chapter. That eruption reduced global surface temperatures as much as 6°C (12°F). Such cooling would devastate crops and produce widespread starvation. A worldwide expansion of glacial ice might even ensue.

Because we can't slow subduction, stop divergence, or chill intraplate hot spots, we will probably never be able to prevent such catastrophic volcanic eruptions. In certain cultures, people rely on faith and prayer to protect them from volcanoes. Hawai'ian legend relates that the fire goddess Pele flies into fits of rage when she is defeated in competition with the islands' young athletic chiefs, stamping her feet in anger to cause earthquakes and summoning rivers of destructive lava in revenge. Some Hawai'ians attempt to appease Pele by offering her bushels of sacred ohelo berries, flowered leis, and bottles of gin. For their part, geologists are developing ways to predict volcanic eruptions more accurately, so as to prevent casualties and minimize damage. The first step in such a process is to determine where general volcanic danger zones exist. For purposes of this text, we will focus on North America.

Regions at Risk

Broad regions of North America are at risk of major damage and destruction from volcanism, especially those western U.S. states and Canadian provinces located where the North

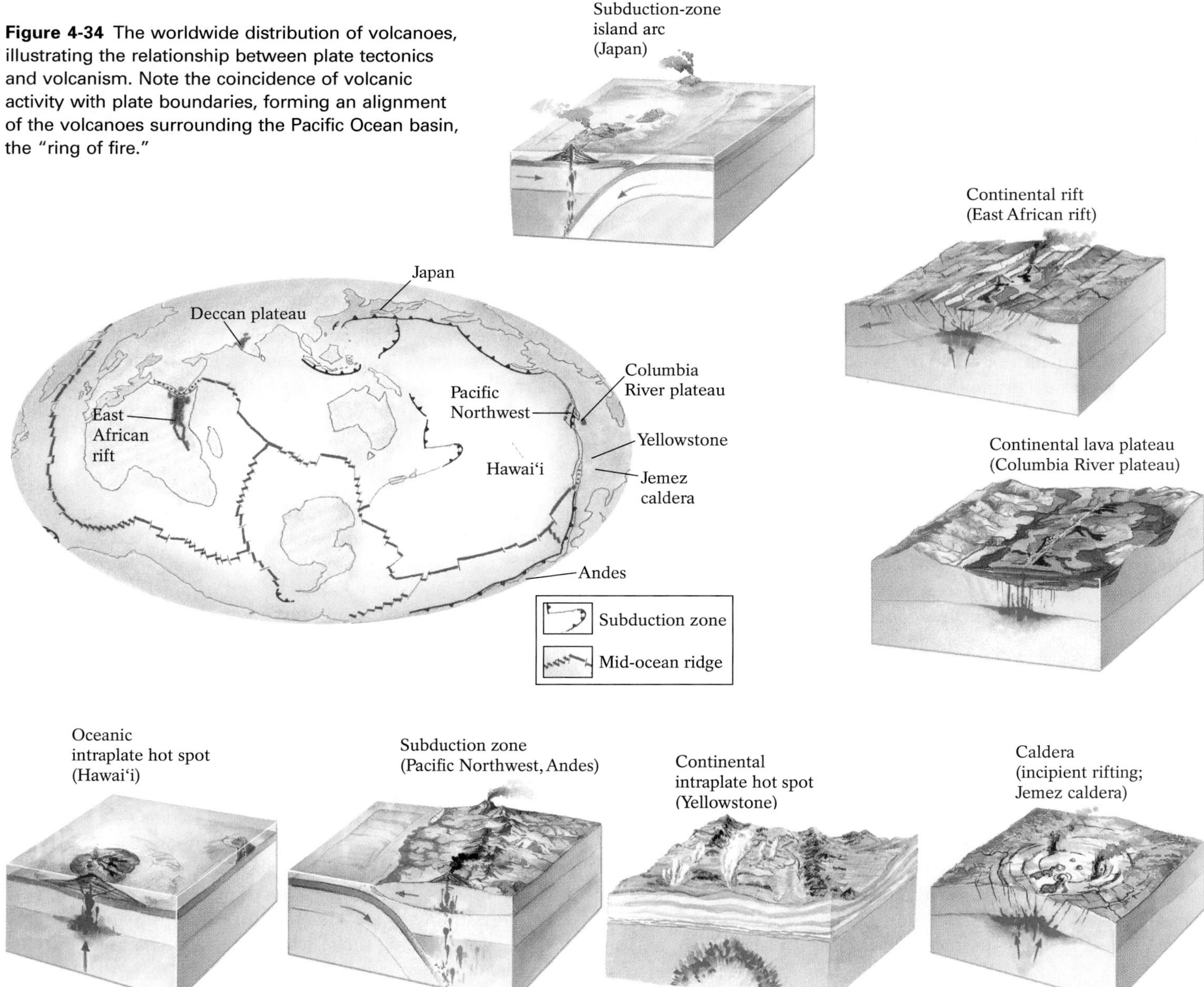

Figure 4-34 The worldwide distribution of volcanoes, illustrating the relationship between plate tectonics and volcanism. Note the coincidence of volcanic activity with plate boundaries, forming an alignment of the volcanoes surrounding the Pacific Ocean basin, the "ring of fire."

Table 4-1 Lava Types, Associated Volcanic Features, and Plate Tectonic Settings

Lava Type	Eruptive Style	Typical Volcanic Landforms	Common Volcanic Products and Effects	Common Plate Tectonic Setting	North American Example(s)
Basaltic (mafic composition)	Quiet, effusive	Lava plateaus, shield volcanoes, occasional cinder cones	‘a‘a lava, pahoehoe lava, vesicular basalts, pillow lavas, columnar basalts	Divergent plate boundaries (such as the mid-Atlantic ridge), oceanic intraplate hot spots (such as underlies Hawai‘i), intraplate rifts (such as the East African rift)	Columbia River lava plateau (Washington and Oregon), Belknap Crater (eastern Oregon), Craters of the Moon (Idaho)
Andesitic (intermediate composition)	Fairly explosive, pyroclastic	Composite cones, cinder cones	Relatively viscous lava, lahars, tuffs (from airborne ash), welded tuffs (from pyroclastic flows)	Subduction zones	Cascades (British Columbia, Washington, Oregon, northern California), Aleutians (Alaska)
Rhyolitic (felsic composition)	Very explosive, pyroclastic	Volcanic domes, calderas	Extremely viscous lava, ash-flow deposits, welded tuffs (from pyroclastic flows	Subduction zones, especially at continental margins, intracontinental rifts, intracontinental hot spots	Yellowstone plateau (Wyoming, Montana), Jemez Mountains (Rio Grande rift, New Mexico), Long Valley caldera (eastern Sierra Nevada, California)

American plate overrides the subducting Juan de Fuca and Gorda plates and where the plate sits atop the continent's scattered intraplate hot spots (Fig. 4-35). Crater Lake in Oregon, Yellowstone National Park in Wyoming, and the Long Valley caldera in California, previously discussed, are examples of recent catastrophic volcanism.

The Valles caldera, one of North America's largest volcanoes, looms over the Los Alamos National Laboratories (the birthplace of the atomic bomb) near Santa Fe, New Mexico. The massive crater—more than 25 kilometers (15 miles) in diameter—developed during two major eruptions 1.6 and 1.2 million years ago. Since that time, it has been the site of numerous explosive ash-flow eruptions, the last occurring about 60,000 years ago. This relatively recent event suggests that the caldera may be entering a new phase of activity. Thus geologists are monitoring the caldera very closely, looking for signs that the crust beneath the caldera is heating up and rising. A large reservoir of new magma appears to be rising to about 20 kilometers (12 miles) below New Mexico's Rio Grande region, believed by some geologists to be an incipient rift zone. An eruption there, with catastrophic effects throughout northern New Mexico, could occur next week or 30,000 years from now.

Many sites in the Cascade Range are more immediately threatening. Although most of the range's volcanoes have been relatively quiet in recent decades, Mount Baker and Mount St. Helens in Washington, Mount Hood in Oregon, and Mount Shasta and Lassen Peak in northern California were all actively erupting between 1832 and 1880. Today, Cascade volcanoes threaten major metropolitan areas in southern British Columbia, Washington, Oregon, and northern California. Mount McLoughlin, 50 kilometers (30 miles) from Medford and Klamath Falls in Oregon, and Mount Hood near Portland, which last erupted in 1865, are closely watched for signs that they are awakening. Mount Baker, near the Washington–British Columbia border, resumed intermittent rumbling and steam emissions so actively that in 1975 the U.S. Geological Survey predicted, albeit incorrectly, that it would be the next Cascade volcano to erupt. Mount Rainier, the Cascade's grandest peak, last erupted in 1882 with a small whiff of brown ash, but its last major eruption occurred some 2000 years ago. In a future eruption, Rainier's greatest threat to nearby towns and cities—such as Tacoma and Seattle—would come from large lahars spawned by steam and tephra emissions onto its glacier-clad slopes. Mount Garibaldi, just north of Vancouver, British Columbia, at the northern end of the Cascade volcanic chain, has eruptive potential as well. Mount St. Helens' explosive 1980 eruption was relatively unspectacular compared with recent volcanic events elsewhere, but it was the greatest volcanic disaster in North America's recent history. Highlight 4-4 documents Mount St. Helens' fury and illustrates some of the hazards facing those who live in the shadow of an active volcano.

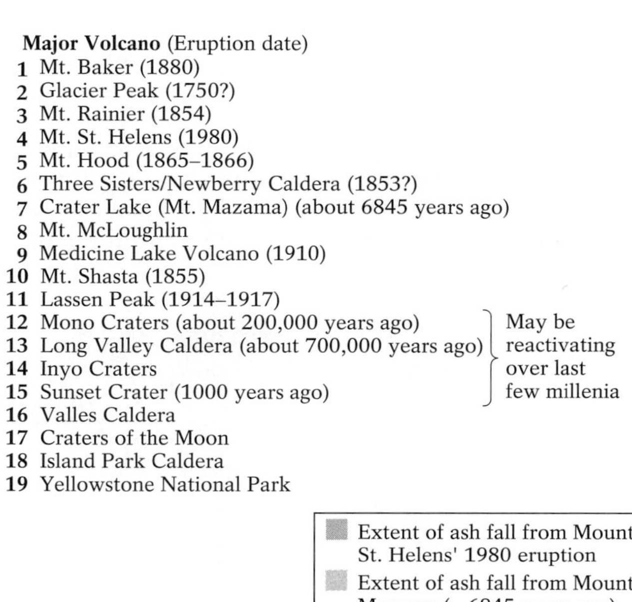

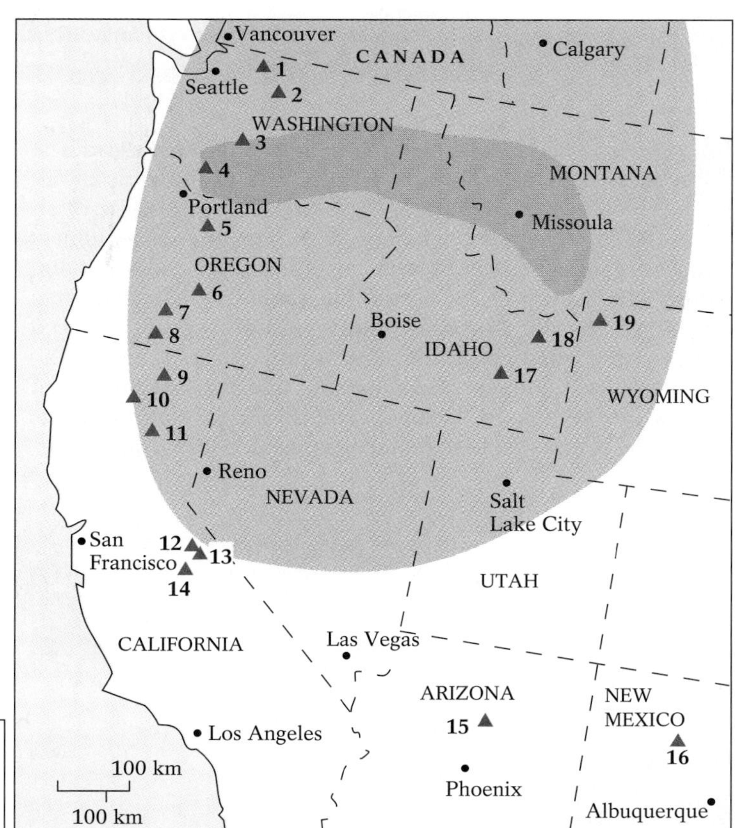

Figure 4-35 A map of western North America showing areas that have experienced recent volcanic activity. These sites may be considered some of the most likely for future volcanic events.

Defense Plans

Today, as much as 10% of the Earth's population lives in the shadow of one of the planet's 600 active volcanoes. Any day now you may hear of a catastrophic eruption of such potentially dangerous volcanoes as Popocatepetl outside Mexico City, Soufriere Hills on the Caribbean island of Montserrat, or Mount Unzen in Japan. We should need no other reminder of the destructive power of volcanoes than the memory of the 23,000 who perished in 1985 in Armero, Colombia, victims of a huge lahar flowing from the slopes of Nevado del Ruiz.

Here in North America (and within clear view of my campus on one of Seattle's rare sunny days), tens of thousands live uneasily within the reach of a similar lahar from Mt. Rainier. Much to the chagrin of its growing number of neighbors, this magnificent Cascade peak has recently been cited as the most volcanically hazardous terrain in the continental United States. With all this volcanic danger lurking, an effective plan to avert volcanic disasters is imperative. Such a plan must start with keeping people out of the most hazardous regions. We must also build structures that protect life and property, and learn how to predict eruptions accurately so as to provide ample warning and allow for timely evacuations.

Volcanic Zoning Many volcanic regions practice *volcanic zoning.* Under such a policy, areas with great potential for danger are set aside as national parks, monuments, and recreation areas, and are closed to residential and commercial development. As a result, there are no large metropolitan areas near Hawai'i Volcanoes National Park, nor are there any on the slopes of Mount Rainier, Mount St. Helens, Crater Lake, Yellowstone National Park, or other hazardous areas throughout the volcanic western states. Occasionally, however, our efforts to zone volcanic areas are less successful. Much of Hilo, Hawai'i, for example, is built atop 30-year-old lava flows. Despite governmental efforts, developmental pressures have placed some residents "in harm's way." Local insurance companies have helped to limit dangerous development in Hilo by simply refusing to insure homeowners in high-risk zones.

Battling Lava and Lahars Where preventive zoning is impractical, other means can avert or reduce potential damage, especially from lava flows and lahars. The first recorded attempt to fight a lava flow took place in Sicily during an eruption of Mount Etna on March 25, 1669. About 50 residents of the town of Catania, wearing wet cowskins for protection, used iron bars to poke holes through the hardening crust at

Highlight 4-4 Mount St. Helens Clears Her Throat

Mount St. Helens' cone, the youngest in the Cascades at less than 10,000 years of age, came to life with an audible boom on March 30, 1980, after 123 years of silence. Weeks of public anxiety and scientific watchfulness followed, ending on May 18 when an eruption blasted about 400 meters (1300 feet) of rock from the summit. What many had considered North America's most beautiful peak changed abruptly to a squat gray crater (Fig. 4-36).

Pre-eruption underground rumblings had signaled the rise of magma into Mount St. Helens' cone. Probably blocked on its way to the central vent by a plug of congealed lava from an earlier eruption, the ascending magma took a sharp turn northward instead. A bulge appeared on the volcano's northern slope; growing at the rate of 1.5 meters (5 feet) per day, it protruded an ominous 122 meters (400 feet) by May 17.

At 8:31 on the morning of May 18, a powerful earthquake beneath the mountain released internal stress that had been building for weeks. It dislodged the bulge and sent it hurtling down the mountain as an avalanche of debris traveling at 400 kilometers per hour (250 miles per hour). A northward-directed jet of hot (500°C [900°F]) ash and gas immediately erupted as a pyroclastic flow, racing downslope with hurricane force

(a)

(b)

(c)

Figure 4-36 **(a)** Mount St. Helens' pre-eruption symmetry made it one of the world's most beautiful and photographed volcanoes. **(b)** Its eruption on May 18, 1980, opened a crater on the volcano's north side, dramatically changing its appearance **(c)**.

(greater than 300 kilometers per hour [200 miles per hour]), cutting a swath of complete destruction 30 kilometers (20 miles) wide. The blast and subsequent nuée ardente buried the nearest 12 kilometers (7 miles) of forest land beneath meters of pyroclastics, blew down entire stands of mature trees like matchsticks to a distance of 20 kilometers (12 miles) (Fig. 4-37), and singed the forest beyond for an additional 6 kilometers (4 miles). More than 26 kilometers (16 miles) away, the heat scalded fishermen, who plunged into lakes and streams. On Mount Adams, 50 kilometers (30 miles) away, climbers felt a gust of intense heat just before being bombarded with hot, ash-blasted pine cones.

Meanwhile, a tephra column rose from the summit vent to an altitude of 25 kilometers (15 miles). Swept eastward by the prevailing winds, the dense cloud of gray ash began to fall on the cities and towns of eastern Washington. Yakima, 150 kilometers (100 miles) to the east, received 10 to 15 centimeters (4–6 inches) of ash. In Spokane, farther east, visibility fell to less than 3 meters (10 feet) and automatic street lights switched on at noon. Proceeding across the continent, the ash cloud dusted every state in its path. Hundreds of downwind communities would later spend millions of dollars cleaning up.

Several hours after the start of the eruption, snow and large chunks of glacial ice trapped within the dislodged debris from the mountain's northern slope began to melt. This enormous volume of meltwater mixed with loose material and the eruption's fresh pyroclastics to produce a lahar that rushed 28 kilometers (17 miles) westward down the Toutle River valley at 80 kilometers per hour (50 mph). The lahar picked up logging trucks and hundreds of thousands of logs along the way, and buried 123 homes beneath 60 meters (200 feet) of mud.

In all, 60 human lives were lost, along with the lives of 500 blacktail deer, 200 brown bear, 1500 elk, and countless birds and small mammals. The only survivors were burrowers such as frogs and salamanders, which fled into the soft sands of lake shores and stream banks. The human toll would have been much higher if state officials had not heeded the warnings issued in March by the U.S. Geological Survey, after the volcano's initial reawakening, and evacuated the area of most of its year-round residents and closed it to seasonal residents and spring hikers. If the eruption had occurred one day later, on Monday, hundreds of loggers at work would have been buried beneath the debris avalanche.

Since 1980, a volcanic dome of highly viscous lava has been rising slowly in Mount St. Helens' crater; it is now 40 stories high. The volcano will probably erupt intermittently, perhaps for decades, as new lavas erupt from the deep scar on its northern slope and eventually restore the mountain's symmetrical cone.

Figure 4-37 The forest north of Mount St. Helens was blown down for miles by the force of the volcano's pyroclastic flow and covered with volcanic ash. The area shown in this photo is 12 kilometers (7 miles) from the crater.

Figure 4-38 A lava wall should be a minimum of 3 meters (10 feet) high and have a wide, sturdy base. Because the wall is meant to divert lava gradually away from inhabited areas, it should be built at an angle to the expected direction of the lava flow. The wall in this photo, hand-built from lava rocks, has successfully diverted a basaltic flow on the island of Hawai'i.

the sides of the advancing flow. They did this to divert the lava, allowing it to drain along a different path away from their homes. The idea was sound, but the lava's new route headed directly toward the watchful town of Paterno. Staunch defenders of Paterno rushed into a somewhat violent confrontation with the Catanians, "convincing" them to discontinue their innovative lava-control efforts.

The twentieth century has provided more effective methods for diverting lava flows. For example, a carefully positioned explosive device can disperse a flow over a wider area so that it thins out, cools, and solidifies more rapidly. The hardening lava blocks the path of still-flowing lava behind it, forcing it to accumulate upstream and flow along a less damaging route. In 1935, a strategically placed bomb coaxed a flow to detour around Hilo.

Other solutions have similar effects. In 1973, Icelanders on the coastal island of Heimaey saved the local port city by cooling a lava flow with seawater pumped by 47 large barge-mounted pumps anchored in a nearby harbor. Japanese engineers have designed steel-and-concrete dams that can trap large boulders and slow mudflows to minimize damage and gain time for evacuation. Anywhere, a simple lava wall—a barrier of boulders or rocks (Fig. 4-38)—can sometimes effectively protect an individual homestead.

In some parts of Indonesia, a series of ropes are strung across valleys that have a history of catastrophic lahars. The first movement of mud triggers a siren, warning downstream communities to evacuate. In Japan, closed-circuit video cameras and other sensing devices vigilantly monitor the slopes of some of that country's most potentially dangerous volcanoes. In the wake of the 1985 Armero tragedy, such a system is now also in place on the slopes of Colombia's Nevado del Ruiz. The Japanese have also engineered debris channels to divert lahars away from populated areas.

Studying Volcanoes and Predicting Their Eruptions In any given year, roughly 50 of the Earth's active volcanoes erupt—usually with some warning. Before they blow, they typically shake, swell, warm up, and belch a variety of gases. Because developing countries rarely have the necessary equipment to monitor an awakening volcano, they call the Volcano Disaster Assistance Program (VDAP)—a scientific SWAT team that rushes to volcanoes to assess their potential for violence and to predict when they might ignite. The VDAP, created a decade ago after the Armero disaster, is based at the U.S. Geological Survey's Cascade Volcanic Observatory in Vancouver, Washington. There team members wait for a call that sends them jetting around the world armed with lasers, seismometers, and other devices used to monitor volcanoes. The VDAP was dramatically depicted in Hollywood's recent *Dante's Peak*.

Geologists have had fair success predicting individual eruptive episodes when they concentrate on a specific volcano *after* an eruptive phase has begun. They monitor changes in a volcano's surface temperature, watch for the slightest expansion in its slope, and keep track of regional earthquake activity (Fig. 4-39). A laboratory at the University of Washington in Seattle is staffed 24 hours a day to monitor the rumblings of Mount St. Helens. If this sounds like we've mastered the art of volcano prediction, remember that the U.S. Geological Survey missed the call on Mount St. Helens' 1980 blast, despite the fact that a large team of scientists armed with the latest in prediction technology were closely watching the mountain. On a more encouraging note, however, volcanologists did successfully predict the eruption of Mount Pinatubo in the Philippines, evacuating virtually everyone within 25 kilometers (15 miles) before the volcano's powerful blast of May 17, 1991, and sharply limiting loss of life.

Before a volcano erupts, hot magma rises toward the surface, giving signs of increased heat. Ongoing surveys can identify new surface hot springs and take the temperature of the water and steam in existing ones. If the escaping steam isn't much hotter than the boiling point of water, surface water is probably seeping into the mountain and being heated by contact with hot subsurface rocks, and all may be well, at least for the present. If the steam is superheated, with temperatures as high as 500°C (900°F), however, it probably derives from shallow water-rich magma, a sign that an eruption may be brewing. Impending eruptions may also be preceded by increased gas emissions from rising magmas. For this reason, volcanologists continuously monitor sulfur dioxide and carbon dioxide emissions on potentially active volcanoes.

As magma rises, the volcanic cone itself begins to heat up. Its overall temperature can be monitored from an orbit-

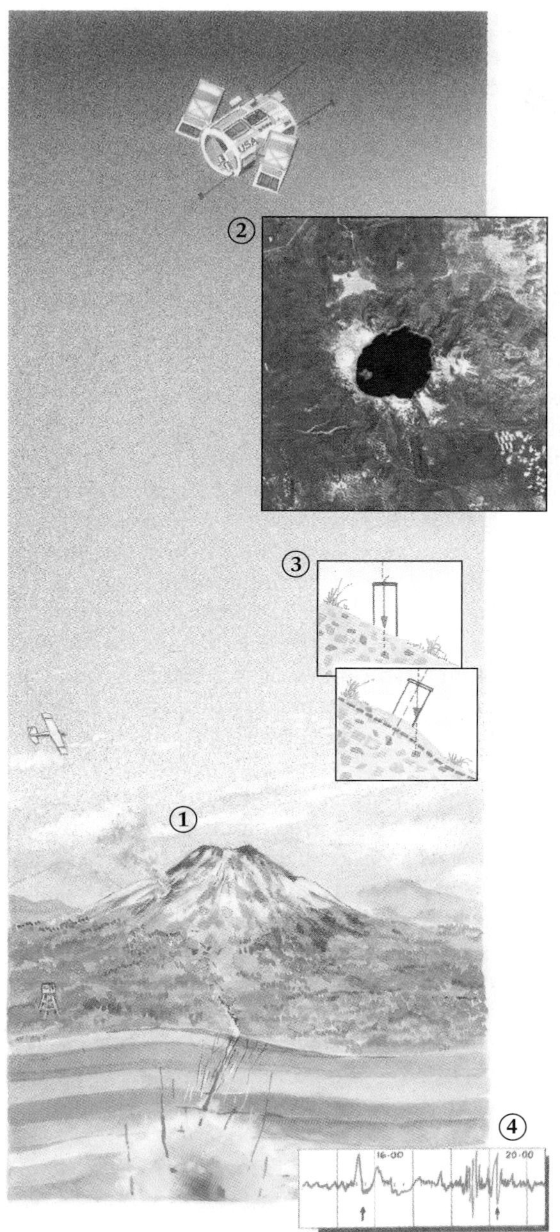

Figure 4-39 Techniques commonly used to predict volcanic eruptions include watching for escape of superheated steam (1), satellite monitoring of volcanic cone temperature (here, Crater Lake in Southern Oregon) (2), detecting volcanic-cone bulges via tiltmeters (3), and locating increased tremor activity via seismographs (4).

ing satellite equipped with infrared heat sensors to detect the slightest change in surface temperature. This high-altitude technology serves as a simultaneous early-warning system for many of the Earth's 600 or so active volcanoes.

Active volcanoes also expand in volume as they swell with a new supply of magma from below. An increase in the steepness or the bulging of a volcano's slope may signal an

Figure 4-40 The remains of the town of San Juan Parangaricutiro, which was engulfed during June and July of 1944 with lava from the eruption of Mexico's Parícutin volcano (visible in the background). The eruption, which lasted nine years, began in February 1943 with the sudden appearance of a small cinder cone in a cornfield about 300 kilometers (200 miles) west of Mexico City. Because the flow that buried the town moved only a few feet per hour, the Mexican Army was able to evacuate the 4000 or so residents before any lives were lost.

impending eruption. A *tiltmeter,* a highly sensitive device somewhat like a carpenter's level, detects the inflation of a volcanic cone and is sensitive enough to measure a change in slope angle equal to the thickness of a dime. Changes in the geographic position of the volcano itself, perhaps in response to movement along faults, are monitored via satellite using the Global Positioning System (the same system that tracks rental cars and cellular phones).

As magma rises, it pushes aside fractured rock, enlarging the fractures as it moves. Because this fracturing causes earthquakes, a distinctive pattern of earthquake activity called *harmonic tremors,* a continuous rhythmic rumbling, often precedes an eruption. Sensitive equipment locates these tremors and documents the changing position of the rising magma. Knowing the rate at which the magma rises enables volcanologists to estimate when an eruption will occur. This is the principal means by which scientists accurately predicted recent eruptions of Mount St. Helens.

If we don't know of a region's volcanic potential, however, efforts to predict eruptions are thwarted. Occasionally a new volcano appears suddenly and rather unexpectedly. In 1943 the volcano Parícutin developed literally overnight in the Mexican state of Michoacan, 334 kilometers (207 miles) west of Mexico City (Fig. 4-40). This area, though, was known to be volcanic, lying just northeast of a subducting segment of the Pacific Ocean.

Make no mistake about it—studying volcanoes up close is a dangerous business. Within the past decade or so, several prominent volcanologists have lost their lives as they worked in the craters of active volcanoes, two of them in a pyroclastic flow spawned by the 1991 eruption of Japan's Mount Unzen, six others during the 1993 blast from the crater of Galeras volcano in Colombia. Because the principal way to study a volcano is on the ground in the active zone, the U.S. Geological Survey and NASA have been working to create "robotic" volcanologists to reduce the human death toll. One such robot is the aptly named Dante (a reference to "The Inferno" a portion of Dante's *Divine Comedy*). Designed to navigate the rubbly slopes of craters, map surfaces with lasers, collect samples, and sniff and analyze gases, these 850-kilogram (1700-pound), spider-legged volcanologist-surrogates have thus far produced decidedly mixed results. The original Dante broke down after "hiking" just a few tens of meters into the crater of Antarctica's Mount Erebus in 1993. In 1994, the new and improved model—Dante II—completed most of its mission into the crater of Alaska's Mount Spurr, even scrambling over a field littered with boulders and a 3-meter (10-feet)-deep snowpack. Unfortunately, Dante II lost its footing and keeled over just 150 meters shy of the crater rim. The plucky little android, lying like a turtle on its back with four of its eight legs broken, had to be retrieved by two human geologists.

Extraterrestrial Volcanism

Volcanism in our solar system occurs (or has occurred in the past) not only on Earth but also on the Moon, Mars, Venus, and the moons of Jupiter and Neptune. There are now no active volcanoes on the Moon, but at least one-fourth of its surface is covered by ancient flood basalts known as lunar *maria* ("seas"). As we saw in Chapter 3, early in its existence, the Moon was struck repeatedly by large meteorites that left deep craters and fractures in its crust. Basaltic lava flowed through those cracks, filling the craters and forming the maria. The smoother-appearing parts of the maria often contain winding trenches, called *rilles,* stretching for hundreds of kilometers. These are probably collapsed lava tubes in the basalt. The Moon's surface also includes a few distinct shield volcanoes, such as those in the Marius Hills.

As much as 60% of the surface of Mars is covered by volcanic rock derived from approximately 20 volcanic centers. Virtually all Martian volcanoes are shields of incredible size. Olympus Mons, for example, is approximately 23 kilometers (14 miles) high, more than twice the height of Mount Everest; its diameter is about equal to the length of the state of California (Fig. 4-41). The size of this and other Martian shields suggests that they have remained stationary over their underlying hot spots for a very long time, allowing enormous mountains of volcanic rock to accumulate. Thus we can hypothesize that either the lithosphere of Mars does not move or it moves far more slowly than the Earth's plates. The volume of volcanic rock in Olympus Mons exceeds the total volume of the Hawai'ian Islands, a *chain* of volcanoes produced because the Earth's Pacific plate moves over its underlying hotspot.

Venus also contains large shield volcanoes, some stretching in long chains along great faults in the surface. Thanks to the remarkably sharp radar images from NASA's Magellan satellite, which began orbiting Venus in August 1990, we know that some volcanoes in the Maxwell Montes region are more than 11 kilometers (6.5 miles) tall, higher than Mount Everest. The shield volcano of Rhea Mons would cover all of New Mexico and much of adjacent parts of Colorado, Texas, and Arizona. Radar also indicates that molten lava lakes may still exist on the Venutian surface. A cluster of seven lava domes discovered in 1991, each more than 15

Figure 4-41 An artist's rendition, based on images from NASA's Viking mission to Mars, of the Olympus Mons volcano. This enormous shield volcano is large enough to cover most of the northeastern section of the United States.

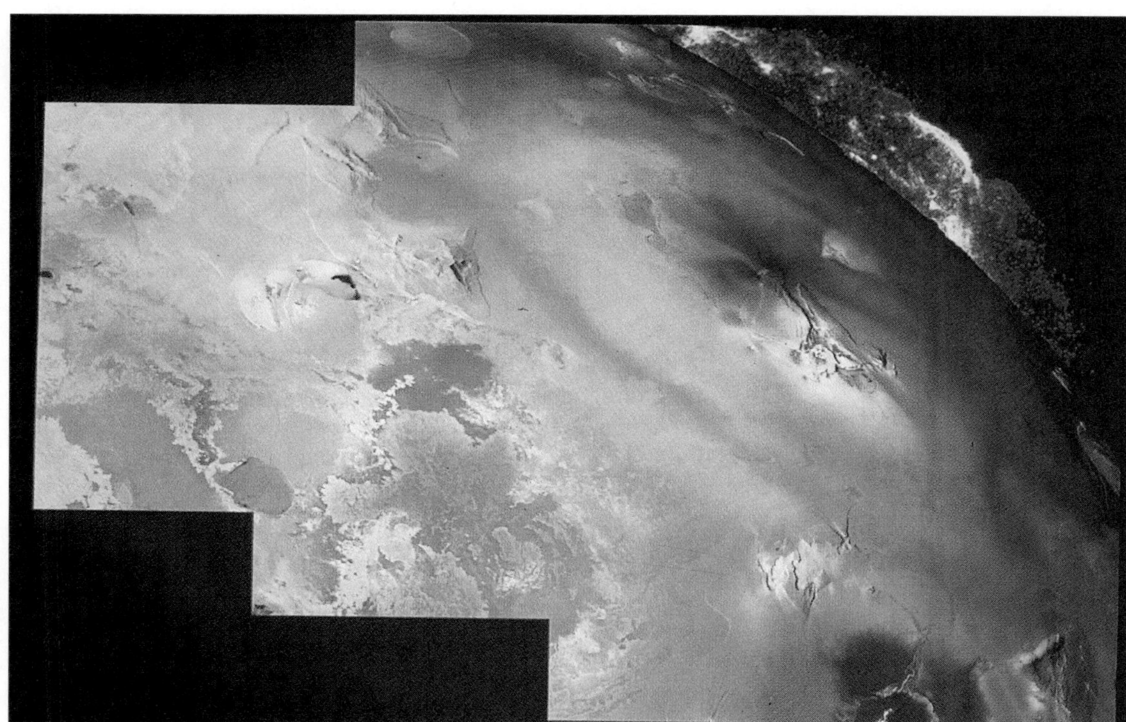

Figure 4-42 A computer-enhanced photo of a volcanic eruption on Jupiter's moon, Io, from data gathered by Voyager 1.

kilometers (10 miles) in diameter, may be similar in structure and composition to the one growing today in the crater of Mount St. Helens. A series of lava flows, in what appear to be rift valleys similar to those of East Africa, suggest a long sequence of multiple volcanic events on Venus. We have not yet determined the ages of these flows or whether they record plate activity on Venus.

One of Jupiter's moons, Io, and Neptune's largest moon, Triton, are believed to be the only other bodies in our solar system showing direct evidence of volcanic activity. NASA's Voyager and Galileo probes have detected eight volcanoes so far on Io, seven of which erupted in one recent four-month period (Fig. 4-42). Io erupts with molten sulfur and enormous clouds of sulfurous gas, which are propelled at speeds approaching 3200 kilometers per hour (2000 miles per hour) to heights of 500 kilometers (300 miles) above the surface. Consequently, Io is noted for its yellow-red snowfalls of sulfur, lakes of molten sulfur, and huge multicolored lava flows of black, yellow, orange, red, and brown. Geologists believe that volcanism on Io may result from the frictional heat generated as this moon's surface rises and falls in response to the enormous gravitational pull of Jupiter.

We have now examined igneous activity at relatively shallow depths and the impact it has on the Earth's surface. In later chapters we will explore the more deep-seated processes that generate such near-surface activity. In the next few chapters, however, we will continue to look at the Earth's outermost layer and the variety of rocks that exist there. In Chapter 5, we will see how these rocks are affected by conditions at the planet's surface.

Chapter Summary

Volcanism is the set of geological processes that mark the ascent of magma to the Earth's surface and its expulsion as lava. A **volcano** is the landform that results from the accumulation of lava and rock particles around an opening, or **vent,** in the Earth's surface from which lava is extruded. Volcanoes may be active, dormant, or extinct.

Magmas flow and erupt in distinctive ways, depending on their gas content and their viscosity, or resistance to flow. Because of its high temperature and relatively low silica content, mafic magma has low viscosity (is highly fluid); it generally erupts (as basaltic lava) relatively quietly, or effusively, because its gases can readily escape from the magma and don't build up high pressure. Magmas of intermediate temperature and composition, which erupt as andesitic lavas, are more viscous; they trap their gases, causing pressures to build up until an explosive eruption occurs. Felsic magma, with its high silica content and relatively low temperature, is highly viscous and generally erupts (as rhyolitic lava) most explosively.

The nonexplosive volcanic eruptions characteristic of basaltic lavas produce lava flows. When they solidify, they create such distinctive features as pahoehoe- and ʻaʻa-type surface textures, columnar joints, lava tubes, and pillow structures. The explosive volcanic eruptions characteristic of rhyolitic lavas typically eject **pyroclastic** material—fragments of solidified lava and shattered preexisting rock ejected forcefully into the atmosphere. Lava that cools and solidifies as it falls back to the surface assumes a wide range of shapes and sizes; all such particles are collectively called **tephra.** Accumulations of soft, still-warm tephra particles sometimes fuse and solidify to form the volcanic rock called **welded tuff.**

Explosive ejection of pyroclastic material is usually accompanied by a number of life-threatening primary effects, such as **pyroclastic flows,** or **nuées ardentes** (high-speed, ground-hugging avalanches of hot pyroclastic material), and **lahars** (hot volcanic mudflows). The secondary effects that often follow explosive eruptions include short-term water and air pollution and longer-term, sometimes global, climatic changes from wind-borne tephra and gas emissions.

Despite their different appearances, nearly all volcanoes have the same two major components: a mountain, or **volcanic cone,** composed of the volcano's solidified lava and pyroclastics, and a bowl-shaped depression, or **volcanic crater,** containing the vent from which the lava and pyroclastics emanate. If enough lava erupts and empties a volcano's subterranean reservoir of magma, the cone's summit may collapse, forming a much larger depression, or **caldera.**

Lavas of different composition form distinctly different landforms. Effusive eruptions, usually involving basaltic lava, form gently sloping landforms, largely because the highly fluid lava flows a great distance from the vent. In Hawaiʻi, basaltic magma generally reaches the surface through distinct conduits, or vents; these central-vent eruptions usually produce broad-based cones called **shield volcanoes.** At divergent plate boundaries, such as in Iceland, basaltic magma generally reaches the surface through long linear cracks, or fissures, in the Earth's crust, and spreads to produce nearly horizontal lava plateaus.

Explosive **pyroclastic eruptions** involve viscous, gas-rich magmas and so tend to produce great amounts of solid volcanic fragments rather than fluid lavas. Rhyolitic lavas tend to be so viscous that they cannot flow out of a volcano's crater; they cool and harden within their craters to form **volcanic domes.** Ash-flow eruptions occur in the absence of any volcanic cone at all; they are produced when extremely viscous, gas-rich magma rises to just below the surface bedrock, first bulging and then collapsing it.

The characteristic landform of pyroclastic eruptions is the **composite cone,** or **stratovolcano,** which is composed of alternating layers of pyroclastic deposits and solidified lava. Pyroclastic eruptions may also produce **pyroclastic cones** or **cinder cones,** composed almost entirely from the accumulation of loose pyroclastic material around a vent. All pyroclastic-type volcanoes produce steep-sided cones, because the materials they eject—solid fragments and highly viscous lavas—do not flow far from the vent.

The various types of volcanic eruptions are associated with different plate tectonic settings. Explosive pyroclastic eruptions of felsic (rhyolitic) lava generally occur within continental areas where plate rifting is taking place, or atop intracontinental hot spots such as the one beneath the Yellowstone plateau of northwestern Wyoming. Most intermediate (andesitic) eruptions occur where oceanic plates are subducting, such as in the Cascades of Washington and Oregon and along the rim of the Pacific Ocean. Effusive eruptions of (mafic) basalt occur generally at divergent plate margins, such as the mid-Atlantic ridge in Iceland, and above oceanic intraplate hot spots, such as the one beneath the Hawaiʻian Islands.

The best way to avoid volcanic hazards is to minimize human use of potentially eruptive locations. Structural and strategic defenses include the construction of lava walls, warning systems to facilitate evacuations, and measurements to predict impending eruptions. Prediction techniques include measuring changes in a volcano's slopes, watching for and recording related earthquake activity, and tracking changes in the volcano's external heat flow, both on land (for example, by the appearance of new hot springs) and from space (with the help of infrared satellite imagery).

Volcanism is not restricted to the Earth; it has occurred elsewhere in the solar system in the past and continues to do so today. Ancient (3-billion-year-old) volcanism is responsible for much of the rock and landform development on the surface of the Earth's Moon. Relatively recent volcanic activity has been detected on Mars and Venus; Io, one of the moons of Jupiter, and Triton, one of Neptune's moons, are currently active.

Key Terms

volcano (p. 93)
vent (p. 93)
volcanism (p. 94)
pyroclastics (p. 101)
tephra (p. 101)
pyroclastic flow (p. 102)
nuée ardente (p. 102)
welded tuff (p. 102)
lahar (p. 102)
volcanic cone (p. 106)
volcanic crater (p. 107)
shield volcano (p. 107)
caldera (p. 108)
pyroclastic eruption (p. 110)

volcanic dome (p. 112)
composite cone (p. 115)
stratovolcano (p. 115)
pyroclastic cone (p. 116)
cinder cone (p. 116)

Questions for Review

1. What criteria do geologists use to designate a volcano as active, dormant, or extinct?

2. Briefly compare basaltic, andesitic, and rhyolitic lava in terms of their composition, viscosity, temperature, and eruptive behavior.

3. Within a single basaltic lava flow issuing from a Hawai'ian volcano, why is pahoehoe lava found closer to the vent than 'a'a lava?

4. Contrast the nature and origin of nuée ardentes and lahars.

5. Describe the basic types of effusive eruptions and the volcanic landforms associated with each.

6. How does a composite cone form? What type of lava is associated with it? How could you distinguish a composite cone from a pyroclastic cone?

7. What types of volcanoes and volcanic landforms are associated with subduction zones? With divergent plate boundaries?

8. Identify three sites in North America that pose a volcanic threat to nearby residents.

9. Describe three techniques that geologists use to predict volcanic eruptions.

10. Compare the volcanism on the Moon with the volcanism on Io, Jupiter's moon.

For Further Thought

1. Look at the photograph below and speculate about the plate tectonic setting where this volcano is found, the composition of the rocks that make up the volcano, and whether eruptions of this volcano tend to be explosive or effusive.

2. The 1980 eruption of Mount St. Helens made a lot of headlines but had no discernible effect on global climate. Conversely, the eruption of the Philippines' Mount Pinatubo in 1991 caused a 1°C drop in global temperature. What differences between these eruptions might explain why one had a sharp effect on climate whereas the other did not?

3. Why do we find andesite throughout the islands of Japan but not throughout the islands of Hawai'i?

4. How would you explain the origin of a volcanic structure composed of 10,000 meters of pillow lava covered by 3000 meters of basalt containing vesicles, columnar jointing, and 'a'a and pahoehoe structure?

5. Under what circumstances might active volcanism resume along the east coast of North America? Within the Great Lakes region of North America?

Weathering Processes

Highlight 5-1
Lake Powell's Puzzling Pits

Soils and Soil Formation

Highlight 5-2
Conserving Our Fragile Soils

Weathering in Extraterrestrial Environments

5

Weathering: The Breakdown of Rocks

Figure 5-1 Monument Valley in northeastern Arizona shows the results of millions of years of exposure to the Earth's environment. The rocks that were removed to produce these landforms were less resistant to weathering and erosion than are the hardy rocks that remain.

More than half a century after they were sculpted, the seemingly eternal faces of Presidents Washington, Lincoln, Jefferson, and Theodore Roosevelt hewn out of the side of South Dakota's Mount Rushmore are in real danger of losing their noses, lips, and mustaches. Quite ordinary environmental factors cause individual mineral grains in rock to be removed or altered, producing end products that look much different from the original rocks. The process by which rocks and minerals break down at or near the Earth's surface is called **weathering,** a slow but potent process that attacks even the hardest rocks.

Weathering plays a vital role in our daily lives, with both positive and negative outcomes. It frees life-sustaining minerals and elements from solid rock so they can become part of our soils, later to pass into the foods we eat and the water we drink. Indeed, without weathering we would have very little food, because weathering produces the very soil in which much of our food is grown. Weathering is also responsible for much of the Earth's most spectacular scenery (Fig. 5-1). But weathering can also wreak havoc on the structures we build. Countless monuments, from the pyramids of Egypt to ordinary tombstones, undergo drastic deterioration from freezing water, hot sunshine, and other climatic forces. Engineers and stonemasons must intervene regularly to secure loosened rocks and sharpen the fading features of the faces on Mount Rushmore (Fig. 5-2).

Rocks weakened by weathering are more vulnerable to **erosion,** the process by which moving water, wind, or ice (or, sometimes, the sheer force of gravity alone) carries pieces of rock and deposits them elsewhere. The fragments removed from rock by erosion have usually been loosened by weathering, although a high-energy erosive event such as a flood may dislodge even unweathered rock. Like many geological processes, weathering and erosion are interrelated, often working in tandem. Together they produce sediment, the

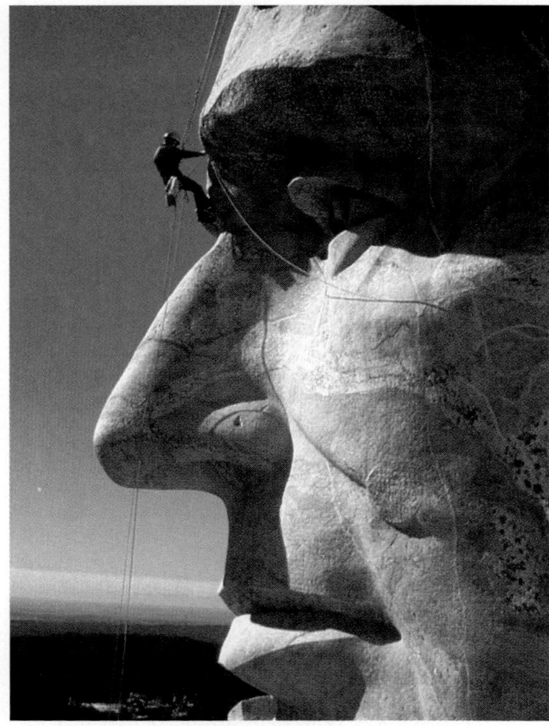

Figure 5-2 A worker repairing weathering damage to the granitic sculptures on Mount Rushmore National Monument, South Dakota.

Surface fractures in rock

loose, fragmented surface material that is the raw material for sedimentary rock (discussed in Chapter 6).

In the following pages, we will explore the different ways in which solid rock weathers, the factors that control weathering, and the various practical products that result from it. In later chapters, we will examine erosion and the most common erosional agents: gravity in Chapter 13, streams in Chapter 14, groundwater in Chapters 15 and 16, glaciers in Chapter 17, wind in Chapter 18, and coastal waves in Chapter 19.

Weathering Processes

Rocks weather in two ways. **Mechanical weathering** breaks a mineral or rock into smaller pieces (*disintegrates* it) but does not change its chemical makeup. The only changes are to physical characteristics, such as the size and shape of the weathered structure. **Chemical weathering** actually changes the chemical composition of minerals and rocks that are unstable at the Earth's surface (*decomposes* them), converting them into more stable substances. Minerals and rocks that are chemically stable at the Earth's surface resist chemical weathering.

Mechanical and chemical weathering go hand in hand in most environments. Mechanical weathering makes rocks more susceptible to chemical weathering by creating more surface area for chemical attack (Fig. 5-3). Think about how crushing a sugar cube with a spoon causes it to dissolve more rapidly in hot water: By mechanically disintegrating the sugar cube and powdering the granules, you vastly increase the surface area exposed to the hot water. Similarly, the area of a

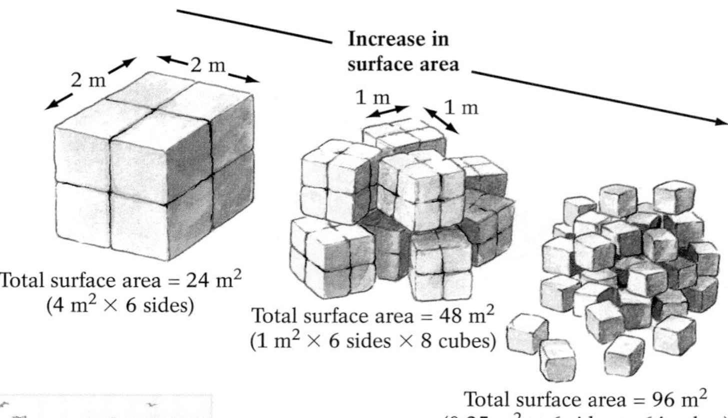

Figure 5-3 Mechanical weathering breaks rocks and minerals into fragments, increasing the total surface area exposed to the processes of chemical weathering.

① Water enters cracks in rock

② Water freezes as temperature drops; expands against walls of rock

③ Cracks are enlarged; intervening rock is dislodged

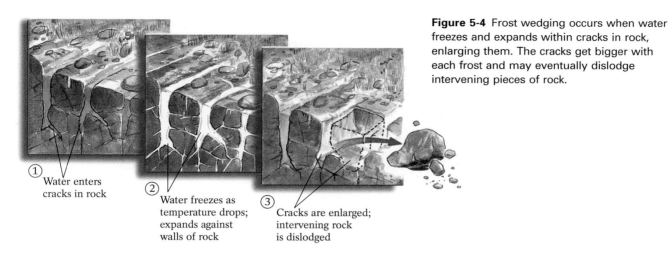

Figure 5-4 Frost wedging occurs when water freezes and expands within cracks in rock, enlarging them. The cracks get bigger with each frost and may eventually dislodge intervening pieces of rock.

boulder that is vulnerable to weathering agents consists only of its outer surface until mechanical weathering breaks it down and increases its surface area.

Mechanical Weathering

Several natural processes reduce rocks to smaller sizes without causing any change in their chemical makeup. In any given location, one or more of these processes may be working at the same time.

Frost Wedging If you have ever put a full water bottle in the freezer, only to find later it has popped its top, you have learned that water expands in volume (by about 9%) when it freezes. This means that when water trapped in the pores or cracks in a rock turns to ice in freezing temperatures, it has to expand. The upper surface of the water freezes first because it is in direct contact with the cold atmosphere. As the cold penetrates downward, the rest of the water freezes as well and, because the surface has already frozen, cannot expand upward. This ice has to expand outward, exerting a force greater than that needed to fracture even solid granite, enlarging the cracks and often loosening or dislodging fragments of rock (Fig. 5-4). As this wedge of ice forms, it draws more water to itself, which, in turn, freezes and enlarges the wedge, putting even more pressure on the surrounding rocks. This process, called **frost wedging,** is one of the most effective types of mechanical weathering.

Frost wedging is most common in environments where surface water is abundant and temperatures often fluctuate around the freezing point of water (0°C, 32°F). When water frozen in a rock's cracks thaws each day in the warm afternoon sun and then refreezes each night, the cracks expand rapidly, causing blocks or slivers of rock to fall and collect at its base as **talus** (Fig. 5-5). Frost wedging also occurs when water seeps into roadway cracks and then freezes, creating the notorious pothole "obstacle courses," found in some northern cities.

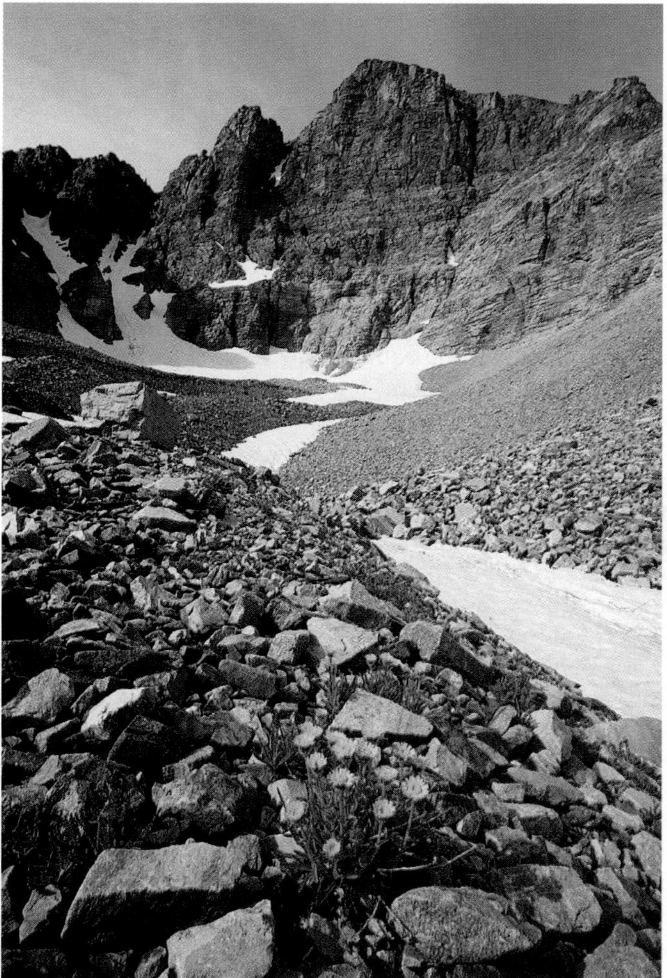

Figure 5-5 Frost wedging is particularly evident on mountains in moist temperate regions, where, even in summer, nighttime temperatures often fall below freezing. Here at Wheeler Peak in Great Basin National Park, Nevada, a cycle of freezing and thawing occurs daily, creating the fresh rock rubble, or talus, that carpets the slopes of many mid- to high-latitude mountains.

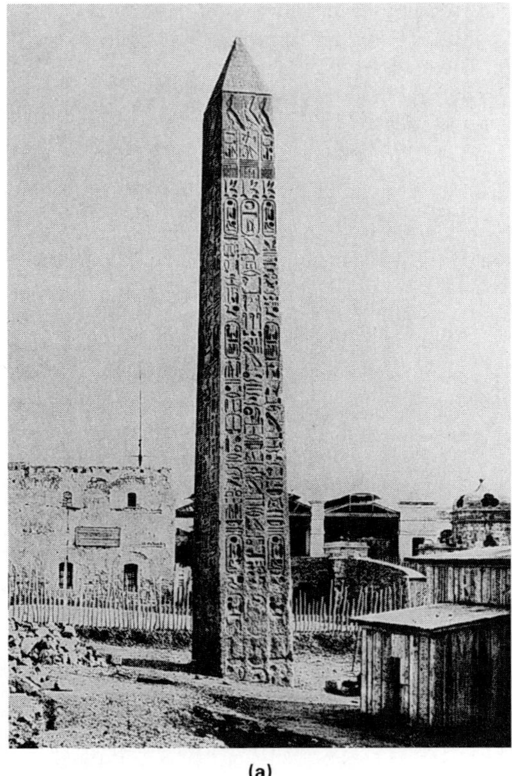

(a)

(b)

Figure 5-6 Cleopatra's Needle as it looked for about 3500 years in Egypt **(a)** and as it looks today, after a century of weathering in New York City's Central Park **(b)**.

Crystal Growth When saltwater enters a crack in a rock and evaporates, the growing salt crystals apply great pressure to the walls of the crack, prying them farther apart. Rocky shores, where salty sea spray soaks into rocks, evaporates, and leaves behind growing salt crystals, are especially susceptible to weathering by crystal growth. This process has also damaged some well-known stone structures of human design. Among them is Cleopatra's Needle, a granite obelisk that stood for more than 3000 years in Egypt without much loss of the finely engraved hieroglyphics on its sides. Before being relocated to Central Park in New York City in 1879, however, the obelisk was stored at a site where salty groundwater readily penetrated the column. Salt crystallized within small water-filled cracks in the granite as the water evaporated in the hot Middle Eastern air, and this crystal growth caused the initial disintegration of the needle's surface. On arrival in New York's humid environment, the same salt crystals absorbed water and expanded, further disrupting the obelisk's surface. Today, little remains of the ancient message inscribed on the obelisk's sides (Fig. 5-6).

Thermal Expansion and Contraction If you have ever sat around a campfire, you may have noticed that thin layers of rock fall off the cracked surfaces of rocks close to the fire. This occurs because of **thermal expansion,** the enlargement of a mineral's crystal structure in response to heat. Minerals vary in the rate and extent to which they expand when heated, largely because of the different strengths of their chemical bonds. A grain of quartz, for example, expands about three times as much as a grain of plagioclase feldspar when subjected to the same increase in heat. If a rock that contains both minerals is heated, the expanding quartz grains push against neighboring feldspar grains, loosening and eventually dislodging them. Because rocks are generally poor conductors of heat, the heat does not penetrate very quickly into the rock; instead, the heated outer portion of a rock tends to break away from the cool inner portion.

Rock surfaces that are exposed to high daytime and low nighttime temperatures probably undergo a similar process at a much slower pace. Their component minerals expand with repeated heating and contract upon cooling, which should eventually cause the rock's outer layer to break apart.

Desert rocks, which are exposed to daily temperature fluctuations of up to 56°C (100°F), would seem to be most susceptible to this process. To test this hypothesis, the effects of thermal expansion and contraction were gauged by heating and then cooling a block of highly polished granite through a range of 38°C (68°F) 89,500 times, the equivalent of 244 years of daily temperature variations. No detectable change occurred in the granite's brightly polished surface as a result. Did this experiment simulate the true environmental conditions of a desert? Is 244 years sufficient to show the effects of thermal expansion and contraction? When the experiment was repeated with a fine mist of water introduced

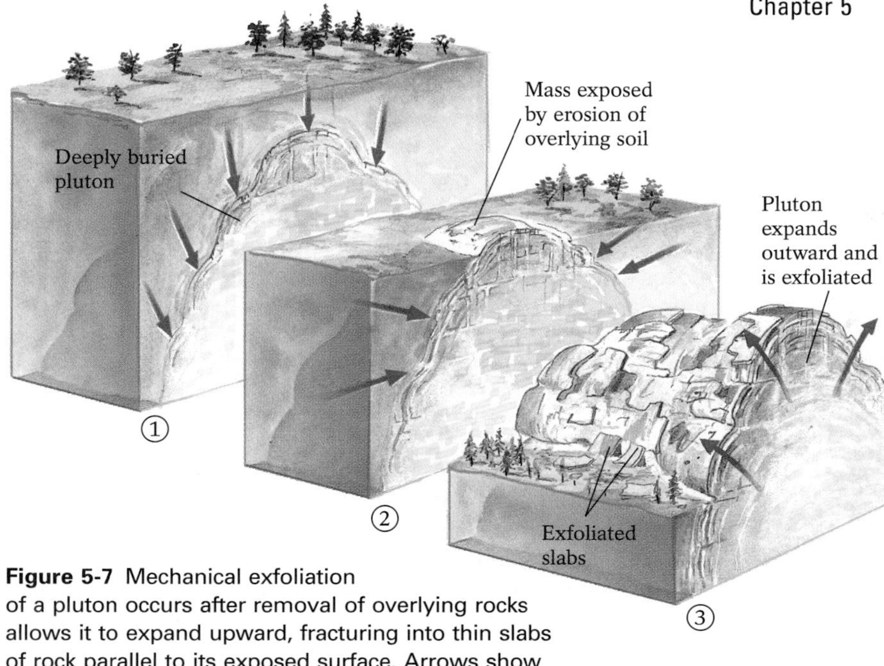

Figure 5-7 Mechanical exfoliation of a pluton occurs after removal of overlying rocks allows it to expand upward, fracturing into thin slabs of rock parallel to its exposed surface. Arrows show the direction of pressure.

Figure 5-8 The "steps" of this mountain in California's Yosemite National Park were produced by mechanical exfoliation of the mountain's granitic rock.

during the cooling cycles to simulate the effects of morning dew, the granite lost its polish, the surfaces of some grains within it became irregular, and cracks appeared near the rock surface. These results suggest that some desert rocks may weather mechanically through the combined agencies of extreme temperature variations and moisture.

Mechanical Exfoliation When erosion of overlying rock and soil exposes a wide area of a large plutonic mass, pressure on the mass is reduced and it expands. Since surrounding rock continues to exert pressure on the structure that remains underground, the expansion takes place outward from the exposed surface. As the rock in the structure expands, it fractures into sheets parallel to its exposed surface. These sheets may then break loose and fall from the sloping surface of the structure, a weathering process known as **mechanical exfoliation** (Fig. 5-7). Many mountains have a dramatic "stepped" appearance produced by large thin slabs of rock, several meters in thickness, exfoliating from underlying rocks (Fig. 5-8).

Other Mechanical Weathering Processes Almost all rocks contain some cracks and crevices. In the case of surface rocks, plants and trees often take root in these cracks. Although rock would seem strong enough to withstand it, the force applied to a crack by a growing tree root is surprisingly powerful and quite capable of enlarging the crack (Fig. 5-9). The buckled and broken sidewalks of some tree-lined boulevards convincingly demonstrate the weathering ability of such root growth.

Large animals may also contribute to mechanical weathering by stepping on stones or pebbles and crushing them into smaller particles. Some birds ingest stones, which are

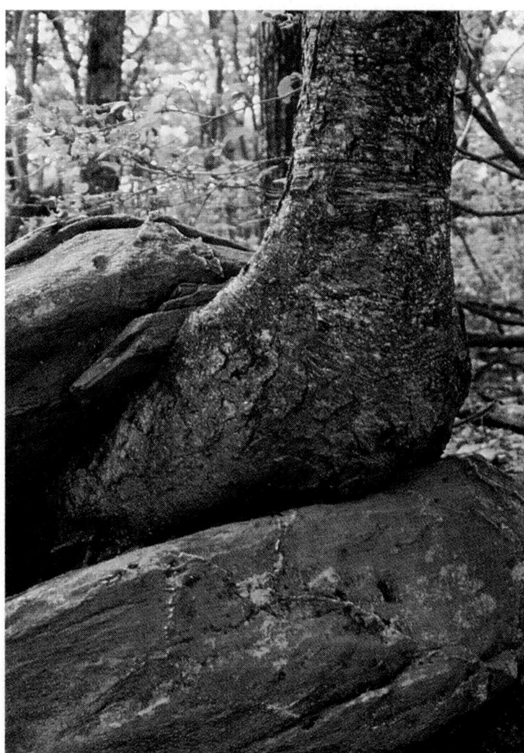

Figure 5-9 Tree roots growing in rock fractures exert an outward force that expands the fractures, mechanically weathering the rocks. Pressure from the growing birch root has split this rock.

used as grinding tools in their digestive processes and eventually released worn down in size. The cumulative activities of even very small creatures such as earthworms and insects can also help break rocks into smaller fragments.

Mechanical weathering by **abrasion** occurs when rocks and minerals collide during transport or when loose transported material scrapes across the exposed surfaces of stationary rock. For example, sediment fragments carried in swirling streams or by gusty winds collide and grind against one another and other rock surfaces, breaking loose fragments into smaller particles and shearing off new fragments as well. Similarly, rocks carried along at the base of a glacier scrape against underlying rocks and are ground to even smaller sizes.

Chemical Weathering

Chemical weathering alters the composition of minerals and rocks, principally through chemical reactions involving water. The water may come from oceans, lakes, rivers, glaciers, or the underground water system; it may also come directly from the atmosphere as rain, snow, or dew.

Water is the single most important factor controlling the rate of chemical weathering, because it carries ions to the reaction sites, participates in the reactions, and then carries away the products of the reactions. This role is made possible by the oppositely charged nature of the two sides of a water molecule (discussed in Chapter 2) and its tendency to break down into its component H^+ and OH^- ions. The three major processes by which rocks weather chemically are dissolution, oxidation, and hydrolysis.

Dissolution In **dissolution,** water removes ions or ion groups from a mineral or rock and carries them away, often without leaving a visible trace. Such ions are said to be *in solution.* When the water carrying them eventually evaporates, these substances remain and are said to have *precipitated* from the solution.

The slightly charged sides of the water molecule are what attract and remove oppositely charged ions from the surfaces of minerals. Some mineral deposits containing gypsum and halite, for example, are readily dissolved by water molecules (see Fig. 2-10). Dissolution of halite (NaCl) occurs when the positive (H^+) side of a water molecule attracts and dislodges chloride ions (Cl^-) and the negative (OH^-) side of a water molecule attracts and dislodges sodium ions (Na^+).

Often water itself does not remove ions from a mineral or rock; instead it reacts with another compound in the environment to form a substance—usually an acid—that does. This happens when the common sedimentary rock limestone weathers chemically by dissolution. The process by which limestone is dissolved begins in the atmosphere and soil, where water and carbon dioxide (CO_2) combine to form carbonic acid (H_2CO_3):

$$\underset{\text{Water}}{H_2O} + \underset{\text{Carbon dioxide}}{CO_2} \rightarrow \underset{\text{Carbonic acid}}{H_2CO_3}$$

Although relatively weak, this acid effectively decomposes calcite, the principal mineral in limestone, to form calcium and bicarbonate ions:

$$\underset{\text{Calcite}}{CaCO_3} + \underset{\text{Carbonic acid}}{H_2CO_3} \rightarrow \underset{\text{Calcium ion}}{Ca^{+2}} + \underset{\text{Bicarbonate ions}}{2\ HCO_3^-}$$

The calcium and bicarbonate ions are then carried off in solution by circulating groundwater, often leaving distinct voids in the parent rock (Fig. 5-10).

These two simple reactions have created most of the world's caves (discussed in Chapter 16). Minute fractures in soluble limestone grow into large underground passages through the gradual dissolution of surrounding rock. Also as a result of these reactions, dissolved calcium and bicarbonate move from the continents to the oceans through the

Figure 5-10 Weathered and unweathered limestone boulders. In humid environments, such as those in most southeastern states, limestone dissolves extensively along cracks and crevices, leaving behind thousands of cavities that range from minor surface depressions to large underground cave systems. In this photo, the limestone boulder above the hammer is weathered; the one below is not.

Earth's groundwater and surface-water systems. In the ocean these ions provide many types of marine creatures with the nutrients that are the raw materials for their shells. The sea creatures discard their carbonate shells onto the world's sea floors, where they may ultimately become consolidated into carbonate rock (see Chapter 6). These rocks may then rise above sea level, perhaps uplifted by plate tectonic forces or exposed by receding seas. In their turn, the newly exposed layers of rock will weather and supply future sea creatures with their calcium and bicarbonate ions, continuing the carbonate cycle between land and sea.

Limestone buildings and statues, like natural limestone structures, are subject to dissolution. The formation of carbonic acid in the atmosphere makes rainwater naturally slightly acidic (natural, unpolluted rainfall has an average pH of 5.4). The more acidic the rain in a given locality, the more rapidly its limestone features, natural or human-made, will dissolve. Downwind of some industrial centers, the rainfall is particularly acidic: Industrial processes such as the burning of sulfurous coal and the smelting of sulfur-rich copper, iron, and nickel ore release sulfur dioxide gas (SO_2). This gas combines chemically with oxygen and water in the atmosphere to form sulfuric acid (H_2SO_4), which is much stronger than naturally occurring carbonic acid. When this acid rain falls, it dissolves carbonate rock and building stone quite rapidly.

Oxidation In **oxidation,** a mineral's ions combine with oxygen to form an *oxide.* For example, when iron ions in mafic rocks bond with oxygen in the atmosphere, the reaction forms the iron oxide Fe_2O_3 (hematite):

$$4\ Fe^{+3} + 3\ O_2 \rightarrow 2\ Fe_2O_3$$

In the presence of water, many other iron oxides are formed: Two of the most common are yellow-brown goethite, FeO(OH), and yellowish limonite, $FeO(OH)\cdot nH_2O$ (where n represents any integer). These iron oxides, all commonly called *rust,* form wherever iron-rich minerals and rocks come into contact with water. Rust often stains the surfaces of minerals with high iron content, such as olivine, pyroxene, and amphibole, and rocks made up of these minerals, such as basalt, gabbro, and peridotite. Like other products of chemical weathering, rust is very stable at the Earth's surface. Thus (much to the chagrin of the owners of aging cars) after rust forms, it does not dissolve.

Other substances, particularly other metals, are also subject to oxidation. Copper, a reddish element with a metallic luster, is more resistant than iron to oxidation; after long exposure to the atmosphere, however, it develops a green surface, or patina, characteristic of copper carbonates and copper sulfates. Recently, tens of millions of dollars were spent to remove some of the accumulated patina from the copper Statue of Liberty in New York's harbor.

Hydrolysis In **hydrolysis,** H^+ or OH^- ions from water molecules displace other ions from a mineral's structure, forming a different mineral. Aluminum-rich silicates such as the feldspars, the most abundant minerals in the Earth's crust, are weathered primarily by hydrolysis.

In a typical hydrolysis reaction, potassium feldspar reacts with water's hydrogen ions. The result is a stable clay mineral (a type of silicate similar to the micas), dissolved silica in the form of silicic acid, and potassium ions liberated by the following reaction:

$$\underset{\text{Potassium feldspar}}{2\ KAlSi_3O_8} + \underset{\text{Hydrogen ions (from water)}}{2\ H^+} + 9\ H_2O$$

$$\rightarrow \underset{\text{Kaolinite clay}}{Al_2Si_2O_5(OH)_4} + \underset{\text{Silicic acid (in solution)}}{4\ H_4SiO_4} + \underset{\text{Potassium ions (in solution)}}{2\ K^{+2}}$$

The clays formed during hydrolysis accumulate within the upper few meters of the Earth's surface, being either incorporated into soils or washed out to sea to become oceanic mud. The silicic acid and the potassium ions are generally transported in solution by underground and surface water. Much of the dissolved silica either cements loose grains of sediment to form sedimentary rocks or serves as the raw material for the shells and skeletons of many marine organisms. Growing plants may extract the dissolved potassium ions from soil water, or the ions may be transported to the sea, eventually to be deposited as potassium salt.

Factors That Influence Chemical Weathering

We have seen that water is a key player in all chemical-weathering reactions. But water's usefulness as a weathering agent may be enhanced by other factors: local temperature and moisture patterns, the activity of living organisms, the amount of time that the mineral or rock has been exposed to weathering, and the chemical stability of the minerals.

Climate As water is instrumental in all dissolution, oxidation, and hydrolytic reactions, a climate with ample moisture already provides the essential ingredient for chemical weathering. Another key climatic factor that accelerates virtually all chemical reactions is heat. Thermal energy from heat excites bonded atoms and ions, causing the bonds to vibrate rapidly until they break; the greater the heat applied, the more bonds that are broken, and, in turn, the more atoms and ions that are freed to participate in other chemical reactions. In general, the rate of chemical reaction doubles with every 10°C (18°F) increase in temperature.

Because of the combined effects of water and heat, chemical weathering occurs more readily in warm, moist climates than in cold, arid ones. For example, the hot, steamy climate of low-latitude tropical areas such as Puerto Rico is

Annual precipitation
Temperature
Equator
Equatorial and tropical rain forests
Savannahs
Low-latitude deserts and semi-deserts
Grasslands (steppes)
Temperate regions and mixed boreal forests
Arctic and tundra regions
40°C
1800 mm
30°C
20°C
600 mm
10°C
Precipitation
Temperature
Evaporation
Bedrock at or very near surface
Soil
Deeply weathered bedrock (~40–50 meters deep)
Soil
Bedrock
Increasing depth of weathering

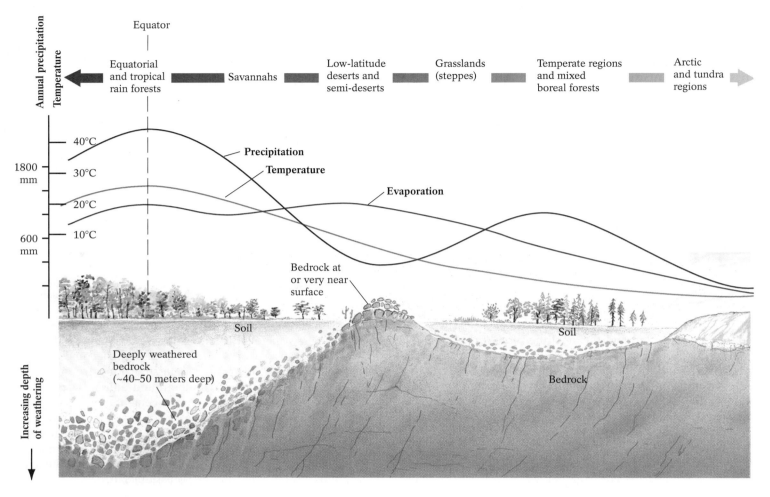

Figure 5-11 The relationship between climate and weathering, shown as a function of the depth to which a region's bedrock (the solid rock immediately underlying the surface soil) is typically weathered. The deepest weathering occurs in the warm, moist tropical zone near the equator, where temperature and precipitation are greatest. Weathering is minimal in the arctic and in the deserts, where water is in short supply as an agent of chemical weathering.

optimal for chemical weathering, whereas cold, dry places such as Antarctica experience little chemical weathering (Fig. 5-11).

Warm, moist climates also promote the growth of lush vegetation; after they die, the decayed plants produce organic acids that enhance chemical weathering. In addition, photosynthesis by plants produces O_2, promoting oxidation, and animal respiration produces CO_2, which combines with water in soil to form chemically reactive carbonic acid.

Living Organisms Organisms can affect the chemical weathering rate of rocks and minerals by exposing them to weathering agents. Burrowing animals, for instance, such as groundhogs, prairie dogs, and even ant colonies, commonly transport unweathered materials from below ground to the surface, where they can be weathered. The common earthworm, which churns up underground minerals in the course of consuming the organic matter in soil, is particularly effective (Fig. 5-12): In humid, temperate climates, an average

Figure 5-12 Earthworms churn up soil as they move through it, bringing underground minerals to the surface to be exposed to weathering.

Figure 5-13 Hills of glacial deposits of substantially different ages in California's Sierra Nevada, each produced by different periods of glaciation. Weathered for more than 100,000 years, the older hills in the foreground are more rounded and subdued and are marked by thicker soils than the much-sharper younger hills behind them, which have weathered for less than 20,000 years. Some pebbles within the older hills have weathered to soft clay, whereas similar ones in the younger hills are still fresh and solid.

earthworm population brings 7 to 18 tons of soil per acre to the surface each year; in the tropics, where worms can grow to be 1.5 meters (5 feet) long, they can stir up 100 tons of soil per acre each year. The burrows of such animals also enhance weathering by allowing the circulation of weathering agents in air and water.

Time A direct relationship exists between time and weathering: The longer a rock or sediment is exposed to a weathering environment, the more it will decompose. For example, within some of the hills of sediment left behind by the glaciers that once covered much of North America's Great Lakes region, pebbles containing minerals suceptible to weathering have been so thoroughly weathered that you can rub them between your fingers into balls of soft clay. Pebbles of the same composition in other glacially deposited hills in the region, however, are still fresh and solid. Assuming the other factors that control weathering affected these hills equally, the hills with the rotted pebbles must have been exposed to the weathering environment for a much longer time (i.e., are much older) than those with the fresh, unweathered pebbles. The glaciated mountains of the North American West (Fig. 5-13) also provide evidence of the link between time of exposure and degree of weathering.

Mineral Composition The effect of chemical weathering on rock is determined largely by the *stability* (resistance to chemical change) of the rock's component minerals (Fig. 5-14). A mineral does not tend to change chemically as long as it remains in an environment similar in temperature and pressure to the one in which it originally crystallized. At the Earth's surface, however, a mineral's environment almost

Figure 5-14 Two gravestone inscriptions from the same cemetery in Williamstown, Massachusetts, showing differential weathering of rock types. The marble gravestone (left), though exposed to the same climate as the granite gravestone (right) and erected 50 years *later,* has suffered noticeably more weathering damage. Marble, composed predominantly of the chemically reactive mineral calcite, is much more susceptible to chemical weathering than is granite, a rock composed of very stable minerals such as quartz and orthoclase.

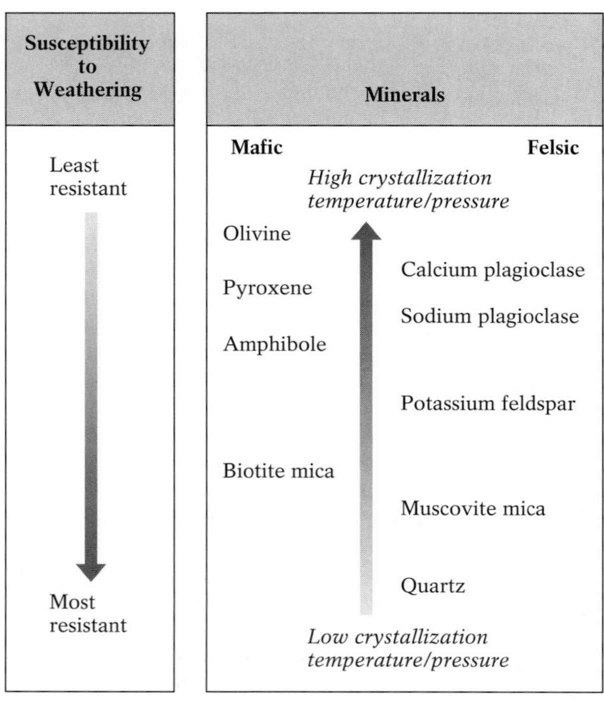

Figure 5-15 Mafic igneous rocks, which crystallize beneath the surface at high temperatures, tend to weather chemically fairly rapidly in the Earth's relatively cooler surface environment. Felsic rocks, which crystallize at lower temperatures—closer to those at the Earth's surface—tend to weather more slowly. Note how this weathering sequence relates to Bowen's reaction series, which was introduced in Chapter 3.

always differs from that at which it crystallized. As a result, the mineral becomes relatively unstable and more susceptible to chemical weathering. As a general rule, the greater the difference between the conditions under which a mineral crystallized and conditions at the Earth's surface, the greater will be the mineral's tendency to change. Minerals that crystallized at high temperatures and pressures are the most unstable at the surface; those that crystallized at lower temperatures and pressures are relatively stable (Fig. 5-15).

When high-temperature minerals such as olivine and pyroxene are exposed at the surface, they quickly begin to oxidize and hydrolyze. The end products of these chemical reactions, such as clay minerals and metallic oxides, are stable at the Earth's surface. When low-temperature minerals such as quartz and mica are exposed, they weather chemically at a much slower rate, because they are more stable in the Earth's surface environment.

Some Products of Chemical Weathering

After dissolution, oxidation, and hydrolysis remove the soluble ("dissolvable") components in rocks, a combination of relatively insoluble, unreactive, and stable new products of chemical weathering remains. These materials include the clay minerals, several economically valuable metal ores, and markedly rounded boulders that shed their outer layers.

Clay Minerals Clay minerals result primarily from the hydrolysis of feldspars and mafic minerals. There are several varieties of clay minerals, each with a distinctive chemical composition and set of physical properties that determine its various practical uses; the variety produced from a given rock depends on the climatic conditions at the time of hydrolysis. For example, kaolinite, $Al_2Si_2O_5(OH)_4$, is the common product of the hydrolysis of feldspars in a warm, humid (tropical) climate. To produce kaolinite, H^+ ions from water must completely displace the large positive ions (K^+, Na^+, Ca^{2+}) in the parent feldspars during hydrolysis. The name kaolinite derives from the Kaoling region of northern China, where the mineral is mined extensively and used in the manufacture of fine porcelain.

In drier, cooler climates, some large positive ions remain after hydrolysis, producing other types of clay minerals. For example, smectites, a highly absorbent group of clay minerals, form when micas and amphiboles are only partially hydrolyzed, leaving some of their calcium, sodium, and magnesium ions unreplaced. Smectites are often used to filter impurities from beer and wine. The negatively charged surfaces of smectite particles attract and bond with the positively charged impurities in those beverages and then settle from the liquids, leaving them clear and pure.

Clay minerals are also used widely in industry as agricultural fertilizers, as lubricants in the boreholes of oil-drilling rigs, in the manufacture of bricks and cement, and in the production of paper. (The slick finish of this book's paper is a clay product.) Several common household staples owe their effectiveness to some properties of clays. Because of its absorbency and deodorizing properties, for instance, the clay mineral smectite is a key ingredient in cat litter. Kaolinite is the active ingredient in Kaopectate®, absorbing the toxins and bacteria that cause intestinal irritation.

Metal Ores *Ores* are aggregates of minerals that have economic value and can be profitably extracted from their surrounding rocks. Most economically valuable minerals—usually metals—rarely occur in pure form. Aluminum, for example, although quite abundant in the Earth's crust, is generally widely dispersed in clays, feldspars, and micas. Pure aluminum metal was once so highly prized that Napoleon had banquet cutlery fashioned from it, instead of from silver or gold.

Intense chemical weathering of feldspar-rich rocks in hot, moist climates produces the aluminum ore bauxite ($Al_2O_3 \cdot nH_2O$). Bauxite forms because aluminum, being insoluble in water, remains concentrated in soils after most common elements, such as calcium, sodium, and silicon, have dissolved out of them. This important ore is found most often in tropical areas, such as the Caribbean islands and parts

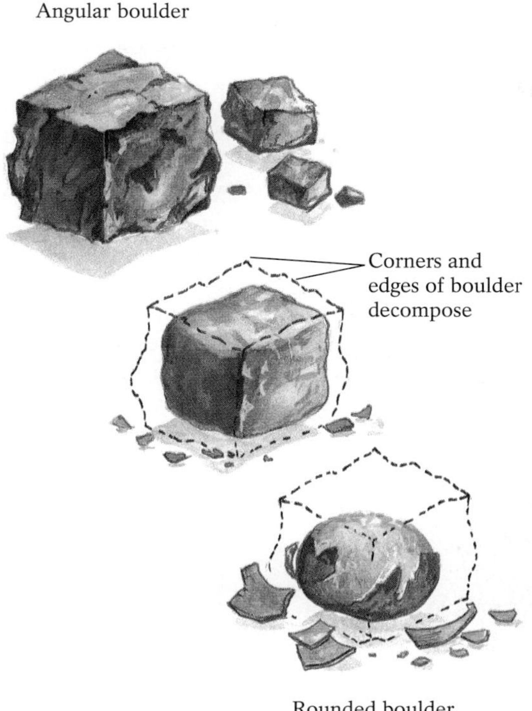

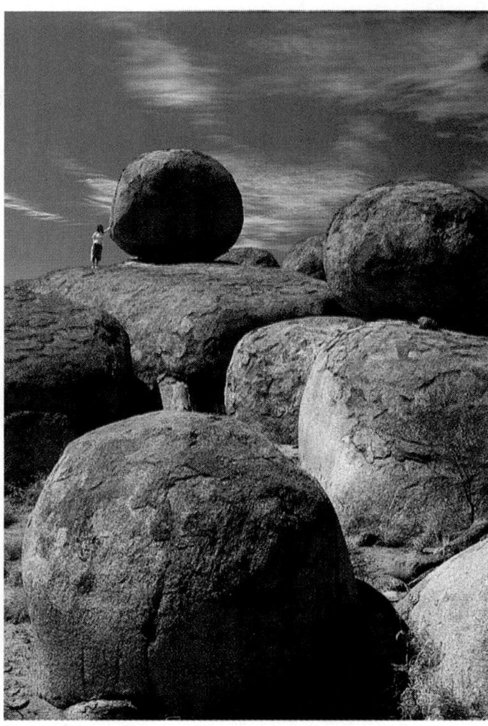

Figure 5-16 Spheroidal weathering. Rounded boulders, such as these in Australia, evolve from angular ones as chemical and mechanical weathering gradually erode the rocks' protruding corners and edges.

of Australia, Africa, and South America. Substantial ancient bauxite deposits can also be found in Georgia, Alabama, and Arkansas, suggesting that the climate in those places was once warmer and moister than it is today.

Weathering processes also produce or enrich much of the world's mineable iron, copper, manganese, nickel, and even silver deposits.

Rounded Boulders Mechanically weathered rocks (and those that are also fractured and jointed by other geologic stresses) are initially quite angular, but in an active chemical-weathering environment, they don't remain so for long. Because the corners and edges of angular rock masses offer more surface area, they are more likely to weather chemically and to do so at a faster rate than the rocks' planar faces. As their corners and edges decompose, angular rocks eventually evolve into smoothly rounded boulders. This process is known as **spheroidal weathering** (Fig. 5-16).

Once a spheroidal shape is attained it remains essentially unchanged, because weathering agents (particularly water) act uniformly across the boulder's entire surface. Weathering of the feldspars on the surface of the boulder produces clay minerals. As these minerals absorb water and increase in volume, they expand outward and separate from the boulder's unweathered interior, peeling off in concentric layers much like the layers of an onion (Fig. 5-17). This process is sometimes called *chemical exfoliation.*

As we saw in this chapter's opening photo, the combination of mechanical and chemical weathering often produces

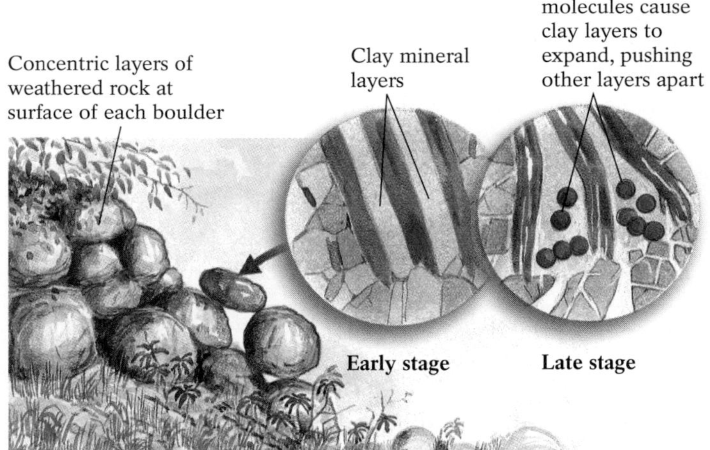

Figure 5-17 A rounded boulder maintains its shape because chemical weathering occurs uniformly over its surface. The surface of the boulder falls off, in concentric layers, because the clay minerals produced by chemical weathering of feldspars at the boulder's surface expand with water and occupy a greater volume than the unweathered feldspars in underlying layers.

unique, exotic landscapes. Although geologists can generally identify weathering as the principal agent acting on a landscape, they are sometimes baffled by which specific weathering process is at work. As we discussed in Chapter 1 (Highlight 1-1, "What Caused the Extinction of the Dinosaurs?"), we are

Highlight 5-1 Lake Powell's Puzzling Pits

Carved into southeastern Utah's 170-million-year-old Entrada Sandstone, Swiss-cheese-style, are dozens of house-sized pits. Located along the shores of Lake Powell, some of these cavernous holes (Fig. 5-18), perhaps the largest such holes in the world, are large enough to swallow a five-story building. The largest measures 21 meters (70 feet) wide and 17 meters (54 feet) deep. Their origin has thus far confounded investigators. Although no one knows for sure, the pits have apparently been excavated from the region's fine-grained, reddish sandstone by a process or set of processes related to weathering and erosion. Confronted with such a mystery, Professor Dennis Netoff of Sam Houston State University in Huntsville, Texas, has led groups of students to these puzzling pits for several field seasons. Dr. Netoff and his students have proposed several working hypotheses that they have been testing for the past five years.

Because the sandstone here has been exposed for an estimated 800,000 to 1.6 million years and the local climate has varied dramatically during this period, many of the mechanical- and chemical-weathering processes discussed in the text may have played a part in the pits' development. Similar but smaller surface pits, known as potholes, commonly develop where the flowing waters of streams, carrying durable grains and pebbles of rock, swirl in a whirlpool action and scour the stream's underlying bedrock by abrasion. Utah's mega-pits, however, may be much too large to have formed by this process alone. Such a process may have been enhanced if the chemical-weathering process of dissolution locally removed the carbonate cement that binds the sandstone's grains, softening the rock and thus accelerating its removal. Dissolution may have been localized in the areas now marked by these huge pits if springs of acidic groundwater rose through them. Alternative hypotheses cite the possibility of wind as an agent that may have aided the pits' growth to their unusual size.

Dr. Netoff and his students will probably continue to probe these baffling cavities for years to come. Like most geological mysteries, these pits will be studied until their investigators propose a logical explanation that is repeatedly confirmed by observation and experimentation. Therein lies the true joy of geological research.

Figure 5-18 These deep weathering pits in the sandstones at Lake Powell, Utah, have thus far defied geologists' attempts to explain their origin.

sometimes temporarily stumped by a geological problem: In other words, we don't have all the answers. Highlight 5-1 illustrates how geologists are applying the scientific method of multiple hypotheses to some puzzling landforms in southeastern Utah.

Soils and Soil Formation

Mechanical and chemical weathering of sediment and the preexisting solid rock, or *bedrock,* underlying it produce **regolith** ("rock blanket"), the fragmented material covering much of the Earth's land surface (Fig. 5-19). Geologists refer to the upper few meters of regolith, which contains both mineral and organic material, as **soil.** Soils, the most common product of weathering on land (oceanic mud is more plentiful), may well be the most valuable of all natural resources (besides clean water). People everywhere depend on fertile soils to support plant growth and provide life-sustaining nutrients in foods.

Influences on Soil Formation

Five factors are particularly important in determining the nature of an area's soil: parent material, climate, topography, vegetation, and time.

Parent Material A soil's **parent material** is the bedrock or sediment from which the soil develops. The parent material's

Figure 5-19 A roadcut in Sarawak, Borneo, showing layers of regolith (weathered surface sediment and bedrock). Extreme chemical weathering can produce regolith that is more than 60 meters (200 feet) deep.

mineral content determines both the nutrient richness of the resulting soil and the amount of soil produced (Fig. 5-20). For example, soils forming from basaltic bedrock differ markedly in composition and rate of development from those forming from sandstone bedrock.

We can clearly see the relationship between parent material and soil development in Java and Borneo, two neighboring Indonesian islands with the same climate. The parent materials for Java's soils are largely fresh, nutrient-rich volcanic ash deposits, whereas the parent materials for Borneo's soils consist of numerous granitic batholiths, gabbroic intrusions, and andesitic flows. Ongoing volcanism on Java replenishes the positive-ion nutrients—such as potassium, magnesium, and calcium—in its soils, whereas Borneo, lacking fresh ash deposits, has soils depleted of nutrients. The islands' population densities reflect their dramatically different soil fertility and agricultural productivity: Java has approximately 460 inhabitants per square kilometer and Borneo about 2.

Climate An area's climate—the amount of precipitation it receives and its prevailing temperature—controls the rate of chemical weathering and consequently the rate of soil formation. Climate also regulates the growth of vegetation and

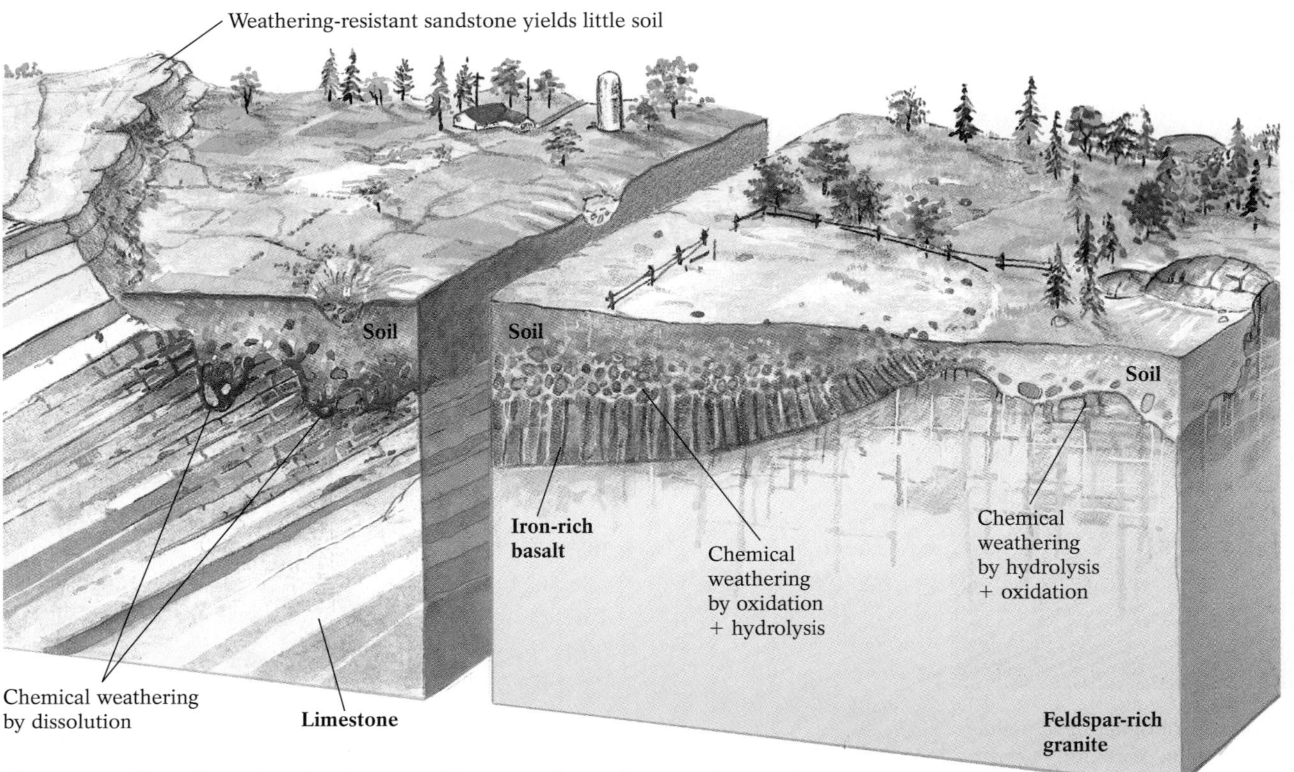

Figure 5-20 The effect of bedrock composition on soil development in a moist, temperate environment. Advanced chemical weathering of iron-rich basalts (oxidation), fractured limestone (dissolution), and feldspar-rich granite (hydrolysis) produces thick soils on these types of rock. Sandstone, however, tends to remain relatively unweathered because it consists primarily of weathering-resistant quartz grains; as a result, any soil produced on sandstone will be sparse.

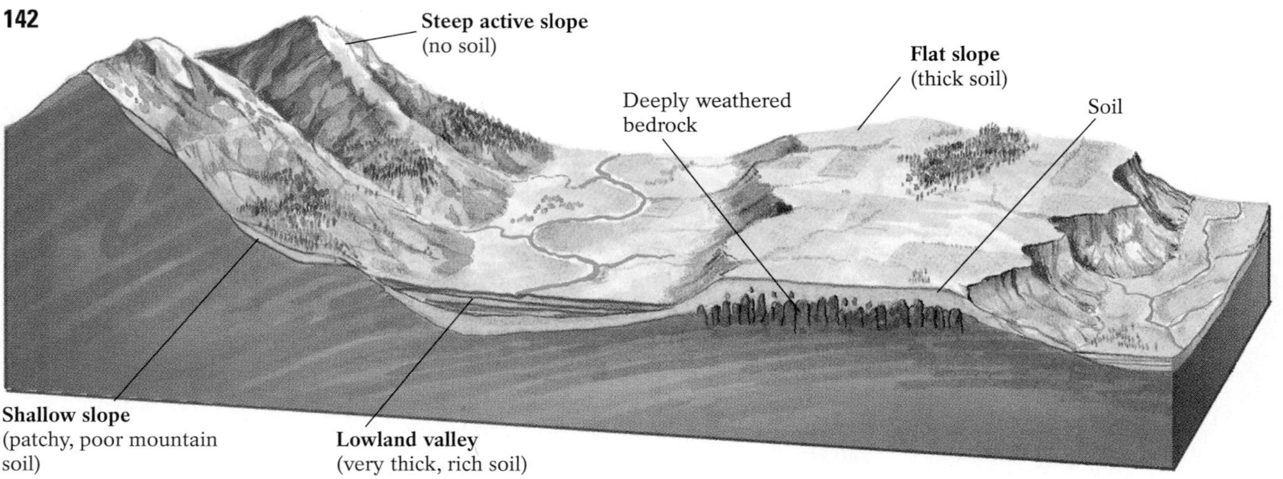

Figure 5-21 The effect of landscape on soil development. Soils are generally thin or nonexistent on steep slopes, because the water required for chemical weathering runs off such slopes (and because any soil that does accumulate would wash away downhill). Soils tend to be thickest in lowland valleys, where water and loose material transported from upland come to rest.

the abundance of microorganisms that contribute CO_2 and O_2 to the processes of dissolution, hydrolysis, and oxidation. Chemical weathering and soil formation are most rapid in warm, moist climates, and slowest in cold or dry climates.

Topography **Topography** refers to the physical features of a landscape, such as mountains and valleys, the steepness of slopes, and the shapes of landforms. Topography influences the availability of water and other weathering factors as well as the rate of soil accumulation. For example, steep slopes allow rainfall and snowmelt to flow away swiftly, leaving little water to penetrate the surface. Consequently, little or no soil develops as a result of chemical weathering. Any soil that does form on steep slopes is usually transported downslope before it can accumulate to a significant depth. Conversely, in level, low-lying areas, water accumulates and readily infiltrates the ground, enhancing the prospects for chemical weathering and soil development (Fig. 5-21).

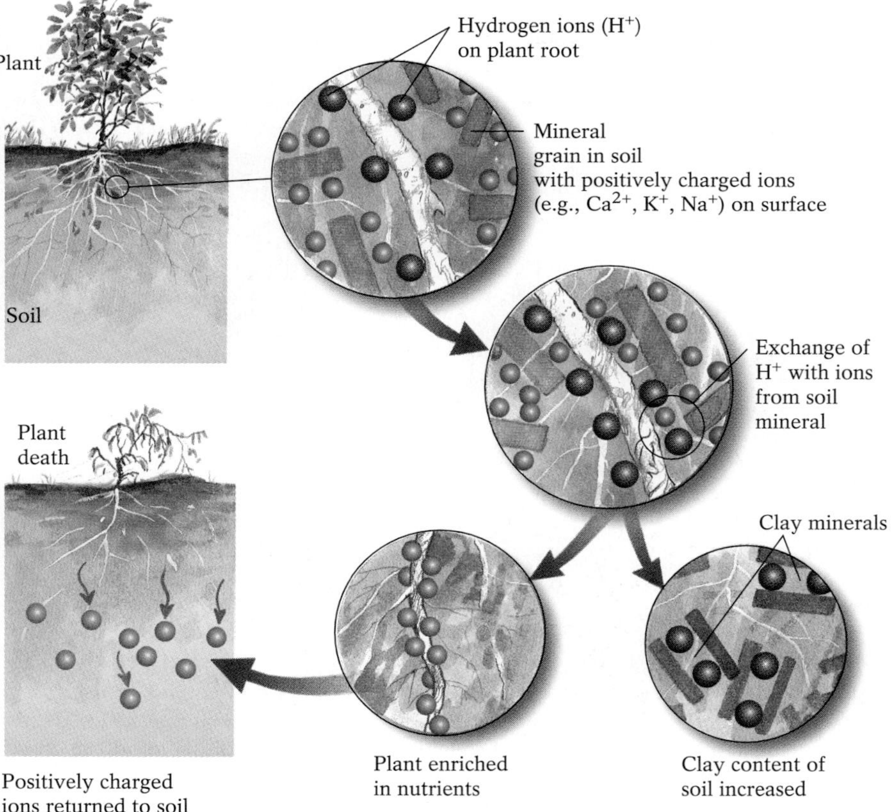

Figure 5-22 Vegetation contributes to soil development through the exchange of H^+ ions (from the surfaces of plant roots) with positive ions (from soil minerals such as feldspars). This process increases the soil's clay content while providing the plant with nutrients, which return to the soil when the plant dies.

Highlight 5-2 Conserving Our Fragile Soils

The United States is not a post-agricultural society. All of us are completely dependent on agriculture to provide the food we need to sustain our lives. Agriculture in turn depends on soil, and especially the uppermost, most fertile layer, or *topsoil*. Yet few of us view the tremendous loss of agricultural topsoil through erosion—about 6 billion tons annually in the United States and many times that amount worldwide—as a major environmental problem (Fig. 5-23a).

Although soil erosion is a natural process that occurs to all soils, it is the rate of erosion that determines the future of agricultural productivity. If topsoil is lost to erosion faster than its nutrients can be replaced by natural weathering processes, a soil's fertility—its capacity to sustain crops—drops dramatically. When land is covered by vegetation, erosion is a relatively slow process. Many of the activities of civilization, however—urbanization, logging, ranching—remove vegetative cover, exposing the land to much more rapid rates of erosion. Likewise, most methods of agriculture operate by clearing the land of all vegetation, plowing the stubble into the ground, and leaving the bare land susceptible to increased rates of erosion by water and wind.

Fortunately, many methods do exist that protect valuable topsoil—some of them traditional farming methods that have been in use for hundreds of years. No-tillage farming and minimum-tillage farming use methods that disturb the soil as little as possible and leave previous crop residue on top of the ground to retain soil (Fig. 5-23b). Specially designed tillers prepare the subsoil without turning over the topsoil. Other machines simultaneously inject fertilizers, herbicides, and seeds into the soil, disturbing neither the groundcover nor the soil. Although economics, both short-term and long-term, is generally the governing factor, these and other soil conservation methods, such as terracing (converting a slope into a series of step-like platforms), contour farming (planting crops in rows that run perpendicular to the slope of the land), and strip cropping (alternating multirow strips of two or more crops within an agricultural field) are becoming more widespread throughout the United States and other parts of the world.

(a)

(b)

Figure 5-23 The fight against soil erosion. **(a)** Agricultural soil being eroded by surface water near Moscow, Idaho. **(b)** One of the weapons against soil erosion, no-tillage farming, minimizes topsoil disruption when preparing the soil for planting.

Vegetation Vegetation contributes organic matter to soils and produces much of the O_2 and CO_2 involved in chemical-weathering reactions. Soils developing on prairie grasslands, for example, receive large quantities of organic matter from surface plant remains and from the decay of extensive subsurface root systems.

Plants also contribute H^+ ions that help weather soils. The H^+ ions, weakly attached to plant roots, replace the large positive ions—such as calcium, potassium, and sodium—in feldspars and other minerals, hydrolyzing them into clay minerals. This exchange helps develop the soil while providing the plants with ions that are nutritionally beneficial to them (and to us) (Fig. 5-22). These ions return to the soil when the plants die and decompose, becoming available for the next generation of plants. If the plants are harvested before they die, however, these ions are permanently removed from the soil. As a result, continuous farming eventually depletes a soil's supply of calcium, sodium, potassium, and other ions; it then becomes necessary to supplement these with natural or synthetic fertilizers and allow a fallow period (a period during which the soil is not farmed) to let the ions reaccumulate. As you can see in Highlight 5-2, the struggle to protect soils requires agricultural skill, advanced technology, and eternal vigilance.

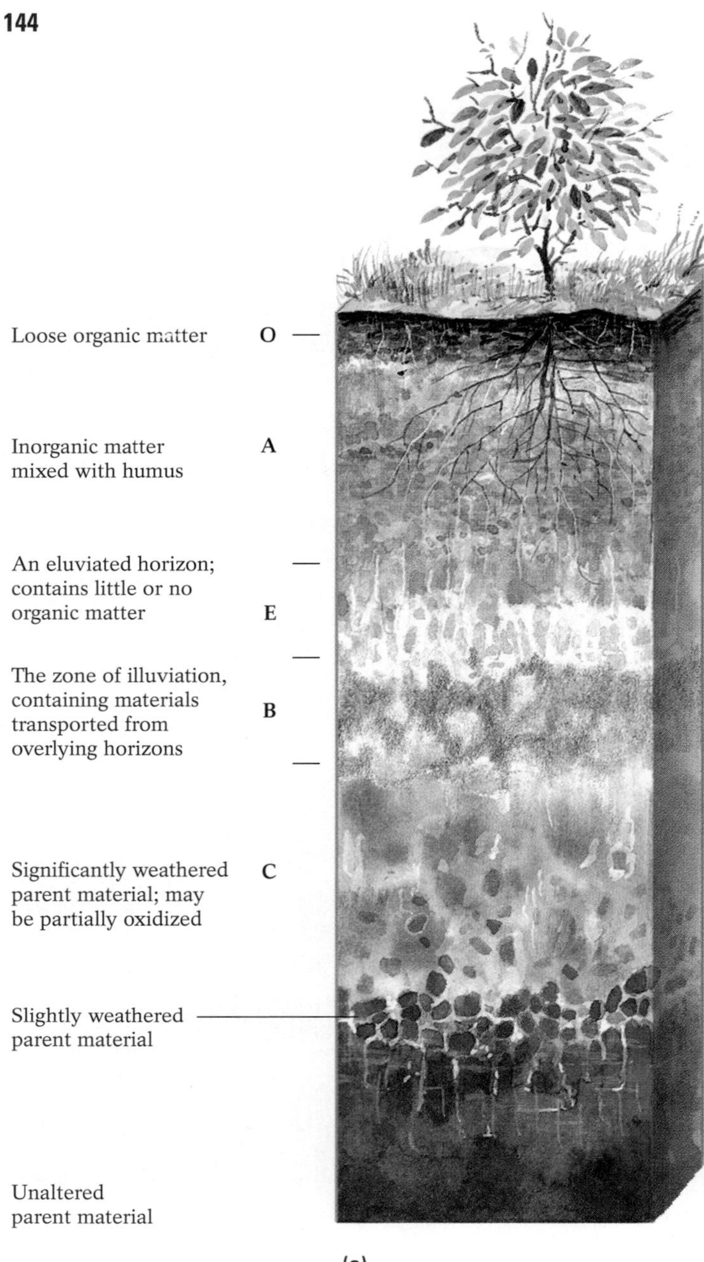

(a)

(b)

Figure 5-24 Profile of a typical mature soil. **(a)** The features of a typical temperate-zone soil profile, including the O, A, E, B, and C horizons. **(b)** A vertical succession of soil horizons is clearly visible even in this small section of soil in Australia.

Time If other factors in soil formation were equal, a landscape exposed to weathering influences over a long period of time would contain a thicker, more well-developed soil than a younger landscape. Other factors, however, almost always vary considerably. Soils may begin to develop within a few hundred years in warm, wet environments but may require thousands or hundreds of thousands of years in arid or polar regions. A thicker soil, then, is not necessarily an older one.

Typical Soil Structure

The effects of weathering are greatest at a soil's surface, which is directly exposed to the weathering environment. Below the surface, distinct weathering zones develop as infiltrating water dissolves substances in the upper layers of the soil and transports them to lower levels in solution or as suspended fragments. These transported substances are then commonly precipitated or deposited in the lower soil layers. Thus the upper part of a developing soil loses some of its original materials, while the lower part gains new components. Each distinct weathering zone is called a **soil horizon.**

Soil scientists have identified dozens of horizon types. Among the most common are the O, A, E, B, and C horizons, which are found virtually everywhere in temperate zones; other horizons develop in more specialized environments. The vertical succession of soil horizons in a given location, the location's **soil profile** (Fig. 5-24), is a product of the local soil-forming conditions. Because different localities vary in the types of horizons and their depth and degree of development, they have different soil profiles.

The upper portion of a soil profile typically consists of the O, A, and E horizons. The O ("organic") horizon in temperate regions consists mainly of organic matter, such as recognizable fibers of plant matter. It teems with life—2 trillion bacteria, 400 million fungi, 50 million algae, and thousands of insects in a single kilogram (2.2 pounds). These organisms contribute CO_2 and organic acids to the developing soil.

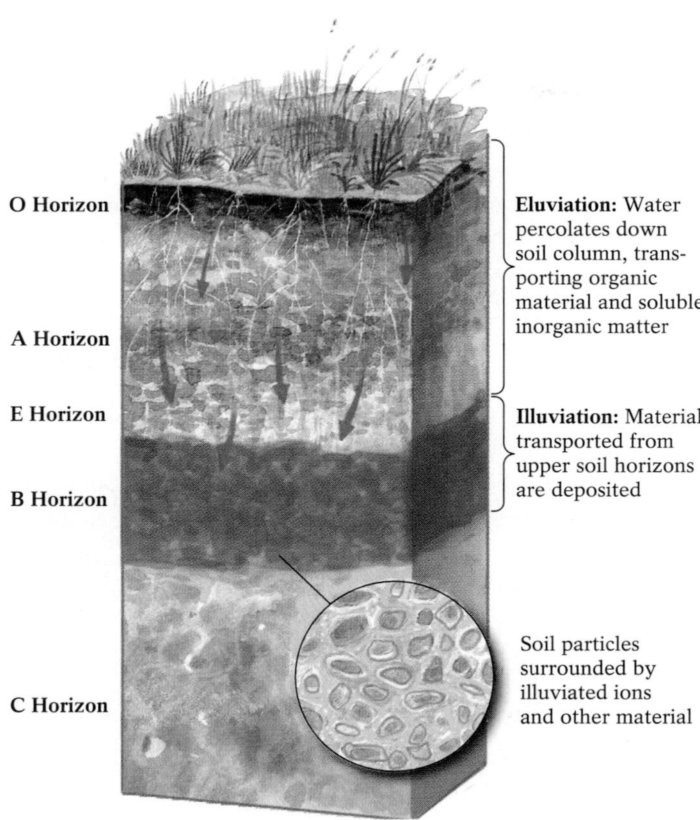
O Horizon
A Horizon
E Horizon
B Horizon
C Horizon
Eluviation: Water percolates down soil column, transporting organic material and soluble inorganic matter
Illuviation: Materials transported from upper soil horizons are deposited
Soil particles surrounded by illuviated ions and other material

Figure 5-25 Eluviation and illuviation in a mature soil profile. Material is removed, or eluviated, as water passes through the O, A, and E horizons and is deposited, or illuviated, as water infiltrates the B horizon.

The A horizon consists mainly of inorganic mineral matter mixed with *humus,* a dark-colored, carbon-rich substance derived from decomposed organic material from the O horizon. The thickness of the A horizon depends on the quantity of decomposed vegetation in the soil. In a tropical environment with lush vegetation, an obvious A horizon develops far more rapidly than in an area with sparse vegetation.

The E horizon is a light-colored zone below the A horizon with little or no organic material. Its light color results from the dissolution and removal of iron and aluminum compounds from the upper few meters of regolith. The E stands for *eluviation,* the process by which water removes material from a soil horizon: Freshwater containing organically produced CO_2 percolates downward from the surface, dissolving soluble inorganic soil components and transporting them along with fine soil particles to lower horizons.

The B horizon, under the O, A, and E horizons, is a zone of *illuviation,* where the materials dissolved or transported mechanically from the upper horizons end up (Fig. 5-25). There are a variety of B-horizon types, classified according to their predominant components: A Bh horizon, for example, has a high concentration of added humus; a Bo horizon has a high concentration of oxides.

In arid and semi-arid areas, where surface water evaporates quickly, a distinct carbonate-rich horizon develops within or below the B horizon at the depth to which annual rainfall penetrates. This **caliche** (ka-LEE-chee) layer forms when brief, heavy rains dissolve calcium carbonate in the upper layers of soil and transport it downward; as the water rapidly evaporates, the carbonate precipitates, creating a markedly white layer in the soil (Fig. 5-26). Particularly arid

Figure 5-26 The white layer in this soil in eastern Washington is caliche, carbonate material dissolved from upper soil horizons and precipitated in lower ones.

regions may require hundreds of thousands of years to form a well-developed layer of caliche.

The O, A, E, and B horizons bear little resemblance to the original parent material. In contrast, the C horizon—the lowest zone of significant weathering—consists of parent material that has been partially weathered but still retains most of its original appearance. It may show signs of oxidation from penetrating oxygen-rich groundwater, or it may be completely unoxidized. The C horizon is very thin where little chemical weathering has taken place—for example, in a desert—but can be as much as 100 meters (330 feet) thick where chemical weathering is extensive—for example, in the warm, wet tropics. Below the C horizon lies unaltered parent material.

Classifying Soils

Until recently, North America was classified by soil scientists as a three-soil continent: The eastern half was covered by *pedalfers* (pe-DAL-fer) (from the Latin root *ped* for "soil," "al" for aluminum, and "fe" for iron), the western half by *pedocals* (PED-o-kal) ("ped," plus "cal" for the caliche layers typical of the arid West), and the southern tropical zone by *laterites* (from the Latin *latere* or "brick," a name that de-

Table 5-1 Classification of Soils

Soil Order	General Properties	Typical Geologic or Geographic Setting
Andisols	Soils develop principally on fresh volcanic ash; high fertility resulting from high weathering rates of glass shards	Young surfaces, in volcanic terranes
Entisols	Minimal development of soil horizons; first appearance of O and A horizons; some dissolved salt in subsurface	Young, newly exposed surfaces, such as new flood or landslide deposits, fresh volcanic ash, or recent lava flow; also found in very cold and very dry climates, or wherever bedrock strongly resists weathering
Inceptisols	Well-developed A horizon; weak development of B horizon, which still lacks clay enrichment; some evidence of oxidation in B horizon; little evidence of eluviation or illuviation	Relatively young surfaces; cold climates where chemical weathering is minimal, or on very young volcanic ash in tropics, in resistant bedrock, and on very steep slopes
Mollisols	Thick, dark, highly organic A horizon; B horizon may be enriched with clays; first appearance of E horizon	Semi-arid regions; generally grass-covered areas having adequate moisture to support grasses but not to cause significant dissolution of soluble materials in upper horizons
Alfisols	Relatively thin A horizon overlying clay-rich B horizon; strongly developed E horizon	Many climates, although most common in forested, moist environments
Spodosols	Much eluviation of A and E horizons, leaving a light-colored, grayish topsoil; aluminum/iron-enriched B horizon stained by dissolved organic material	Moist climates, usually on sandy parent materials (which allow water to infiltrate readily); grasses or trees may provide the organic matter
Aridisols	Thin A horizon with little organic matter overlying thin B horizon with some clay enrichment; caliche layer present in B or C horizons	Arid lands with sparse plant growth
Histosols	Wet, organic-rich soil dominated by thick O and A horizons	Found where production of organic matter exceeds addition of mineral matter, generally where surface is continuously water-saturated; often found in coastal environments
Vertisols	Very high clay content; soil shrinks (upon drying) and swells (upon wetting) with moisture variations	Equatorial and tropical areas with pronounced wet and dry seasons
Oxisols	Shows extensive weathering; highly oxidized B horizon is deep red from layer of oxidized iron	Generally older landscapes in moist climates having tropical rainforests
Ultisols	Shows extensive weathering; highly weathered clay-rich B horizons with high concentrations of aluminum	Very moist, lushly forested climates, often subtropical and tropical

rives from the hard building stones produced when these soils dry out). Pedalfers are relatively fertile, highly organic, iron- and aluminum-rich soils formed in humid temperate environments. Pedocals are relatively infertile, thin, organic-poor soils with high concentrations of calcium carbonate. Laterites are typically thick, extremely infertile soils composed principally of insoluble iron and aluminum.

As we have seen, however, the complex interaction of parent material, climate, topography, vegetation, and time causes local soil development to vary in significant ways. Modern soil classification attempts greater precision than the old "three-soil" system, for very practical reasons: Distinct soil types have specific physical and chemical characteristics that affect our uses for them. Where we locate landfills, how we design buildings, and the ways we cultivate soils for food all depend on accurate soil classification. Our current classification scheme (Table 5-1) names soils according to obvious physical characteristics, describes a soil's clay content, and indicates degree of nutrient depletion. Classification terminology also provides information about moisture content, mean annual air temperature, horizon development, soil chemistry, organic matter content, and even the origin and relative age of the soil. For example, an *entisol* (the root "ent" is derived from *recent*) is a soil that has not yet experienced significant horizon development; it may be a recent flood deposit or fresh volcanic ash. A *vertisol* (which tends to expand *vert*ically) contains clay minerals that swell when moistened and shrink when dried. Vertisols can disrupt most structures built on them.

Extreme weathering in tropical areas produces *oxisols* (named for their high concentration of insoluble iron *ox*ides) or *ultisols* (for their *ult*imate, or most advanced, degree of soil development). In ultisols, even the ordinarily insoluble quartz has been dissolved away, leaving only the most insoluble elements, such as iron and aluminum. (Oxidized iron makes these soils dark red.) When these soils dry out, they are strong enough to serve as building materials in tropical regions (Fig. 5-27); adobe is a fine-grained soil mixture traditionally used to construct dwellings in the American Southwest and Mexico. Oxisols and ultisols are depleted of potassium, calcium, sodium, magnesium, and other nutrients and so are poor prospects for agriculture. Raising crops with significant nutritional value in these soils requires advanced agricultural technology and intensive use of fertilizers. The crops of the tropics therefore tend to be the so-called cash crops with poor nutritional value, such as coffee, tobacco, sugar cane, palm oil, and cacao (the prime ingredient in chocolate). While some of us may subsist on that diet, we are not going to get the nutrients that less weathered and more agriculturally useful soils provide.

Paleosols Sometimes previously buried soils are uncovered that differ from other soils in their regions, suggesting that they formed under different—in particular, ancient—conditions. Buried soils that predate modern soil formation are called **paleosols** ("old soils"). Examples include aluminum-rich bauxite deposits in Georgia, caliche horizons in Connecticut, and deep, highly weathered regolith in southwestern Minnesota. All these presumably formed under climates with very different temperatures and moisture conditions from those that prevail today. Similarly, oxisols and ultisols indicative of humid semi-tropical or tropical climates have been found beneath the soils forming today in the warm, dry climate of Australia. Thus paleosols enable us to identify climate changes that have occurred in the geologic past.

Figure 5-27 The Angkor Wat temple in Cambodia, showing the varying degrees of durability exhibited by different soil types. The bricks that form the temple's foundation have been fashioned from ultisols, soils that have undergone advanced chemical weathering; these bricks remain remarkably fresh because they are composed of the stable products of this weathering, which resist additional chemical weathering. The intensity of chemical weathering in this warm, moist climate is obvious from the eroded condition of the temple's columns and statuary, carved from normally resistant sandstone.

Weathering in Extraterrestrial Environments

Weathering as we know it on Earth does not take place on our celestial neighbors. Long suspected, this fact has been confirmed by recent discoveries about the surface conditions of the Moon, Venus, and Mars. The reasons for the absence of weathering, however, differ in each case.

The Moon has no atmosphere; without atmospheric water, oxygen, or biological activity, no chemical weathering is possible. A mechanical weathering process—the impact of meteorites and micrometeorites—produced the Moon's regolith, which consists primarily of shattered bedrock and glassy fragments expelled from impact craters. The sharp edges of lunar craters, however, even the oldest ones, suggest the absence of Earth-like chemical-weathering and water-related mechanical-weathering processes as well. Thus the footprints left in the lunar dust by the Apollo 11 astronauts—the first to walk on the Moon—are likely to retain their freshness for millions of years (Fig. 5-28).

Figure 5-28 A footprint left on the lunar surface by one of the Apollo 11 astronauts in July 1969. Because no chemical weathering occurs on the Moon, this footprint will remain for millennia. Micrometeorite bombardment and the debris it ejects, however, may obscure the print over millions of years.

Venus has a surface temperature of about 475°C (900°F) and an atmosphere composed almost exclusively of CO_2, which traps heat radiating from the planet's surface in a greenhouse effect. High temperatures normally promote an increased rate of chemical reaction, but the temperature of Venus is so high that it instantly evaporates every trace of water from the planet's surface and atmosphere. Hydrolysis, carbonation, and oxidation cannot take place without water; thus the Venutian landscape appears remarkably unweathered chemically (Fig. 5-29). Mechanical weathering on Venus, as seen on recent radar images, is probably due to thermal

Figure 5-29 Angular rocks on Venus, photographed by the USSR's Venera-13 probe. The sharp, unweathered edges are testimony to the planet's apparent lack of chemical weathering.

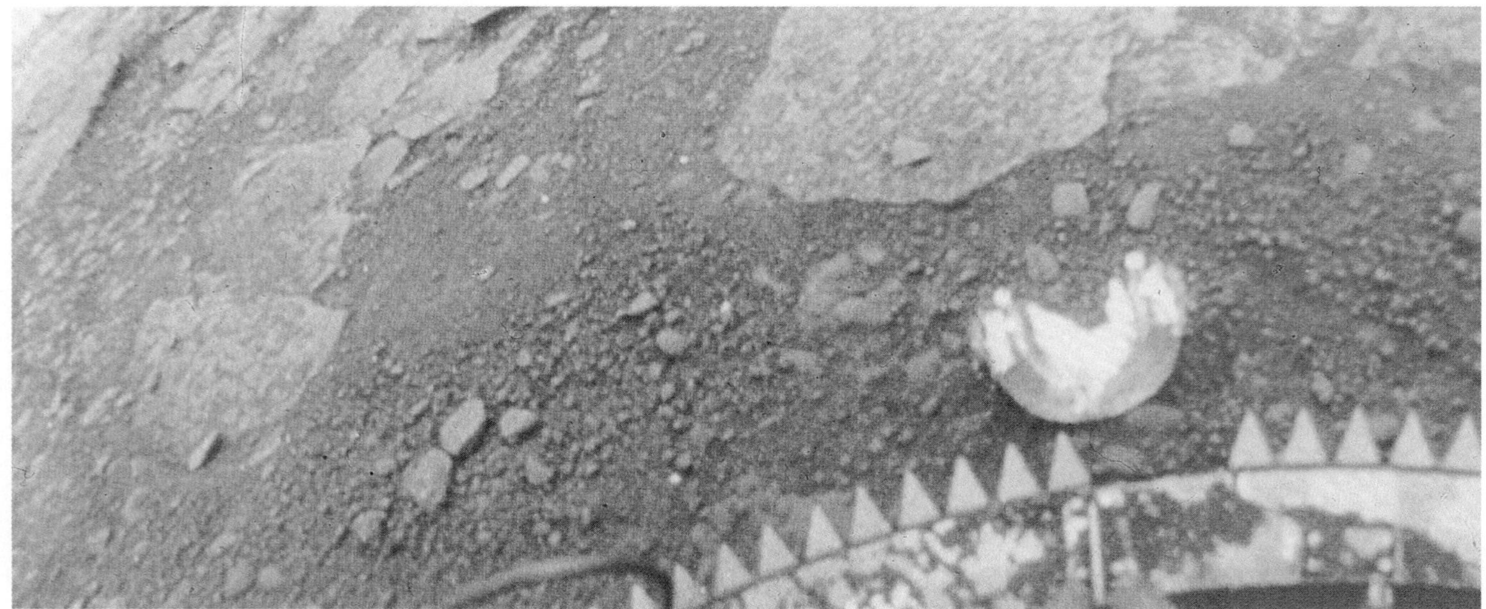

Figure 5-30 An unenhanced color photo of the surface of Mars, taken by the Sojourner lander of the United States' Pathfinder probe in July 1997. The redness of the Martian regolith is most likely due to oxidation of iron-rich rocks and sediments.

expansion and contraction and exfoliation. High winds at the planet's surface may also be a factor.

Of all planets in our solar system, Mars has surface conditions that most closely resemble those on Earth, yet its surface temperatures range from −108°C (−225°F) to 18°C (64°F). The thin atmosphere of Mars consists largely of CO_2, with small amounts of nitrogen and water vapor. Because of the cool temperatures, most surface water is trapped in ice, where it is generally unavailable for chemical reactions (although it may promote mechanical weathering by frost wedging). The lack of heat also reduces the rate of chemical reactions. There is, however, clear evidence of chemical weathering in the planet's past. The oxidation of iron-rich bedrock apparently produced the reddish-brown color of the Martian landscape, which we can see easily with a high-quality telescope from a distance of 78,000,000 kilometers (47,000,000 miles). Soil analyses performed by the versatile Viking lander in 1976 confirmed the high concentration of these reddish iron oxides at the planet's surface (Fig. 5-30).

As we have seen, various mechanical- and chemical-weathering processes can convert solid bedrock to loose, transportable fragments and dissolved ions. Once liberated from their parent bedrock, these fragments and ions are free to move under the influence of such surface-shaping forces as gravity, streams and glaciers, wind, and coastal waves. The next chapter discusses processes that transport weathered material, and their subsequent deposition and conversion to sedimentary rock.

Chapter Summary

Solid rocks can be broken down by weathering, erosion, or, often, a combination of the two. **Weathering** is the slow but constant process whereby environmental factors gradually break down stationary rocks at the Earth's surface. **Erosion** occurs when rock fragments are transported and deposited elsewhere by moving water, wind, or ice. Working together, weathering and erosion produce the Earth's sediment, the loose, fragmented geologic material that is the raw material for sedimentary rock.

Two types of weathering exist: **mechanical weathering,** which results in the physical disintegration of rock into smaller pieces without changing its chemical composition, and **chemical weathering,** which changes the chemical composition of the weathered rock. Rocks and minerals with structures that are chemically unstable at the Earth's surface are most susceptible to chemical weathering, which changes them into more stable substances.

Mechanical weathering may be accomplished by a variety of processes: **frost wedging,** the expansion of cracks in rock as water in the cracks freezes and expands; salt-crystal growth within cracks, which forces the crack's walls farther apart; **thermal expansion and contraction,** the alternate enlargement and shrinking of rock as it is repeatedly heated and cooled; **mechanical exfoliation,** the fracturing and removal of successive rock layers as deep rocks expand upward after overlying rocks have eroded away; penetration of growing plant roots, which expands existing cracks in rock, and the activities of burrowing animals; and **abrasion** of transported particles as they collide with one another or with stationary rock surfaces. Rock fragments that fall from a weathered structure and collect on its slopes and base are called **talus.**

Chemical weathering is largely controlled by climatic factors, such as temperature and, especially, the availability of water. The process of **dissolution,** most effective on soluble rocks such as limestone, occurs when the activity of water or reaction with naturally occurring acid decomposes minerals or rocks and water carries off the products. **Oxidation,** the reaction of certain chemical compounds with oxygen, works most effectively on iron-rich rocks, such as basalt and ultramafic peridotites. **Hydrolysis,** the replacement of major positive ions in minerals (particularly the feldspars) with H^+ ions from water, produces clay minerals, the most common products of chemical weathering.

The rate at which a given rock or mineral weathers chemically depends on a number of factors: climate (hot, moist regions exhibit more weathering than cold or dry regions); the activity of living organisms; the length of time it has been exposed to weathering; and the chemical stability of its components at the Earth's surface. (Minerals that crystallize at very high temperatures, such as olivine and pyroxene, are less stable and more readily weathered than cooler-crystallizing minerals, such as quartz and mica.) Chemical weathering produces the clay minerals and several economically valuable metal ores, and it has the physical effect of rounding previously angular boulders, a phenomenon known as **spheroidal weathering.**

Mechanical and chemical weathering together produce the Earth's **regolith,** the loose, fragmented material that covers much of the Earth's surface, and **soils,** the uppermost, organic-rich portion of the regolith. Soil development is governed by five factors: **parent material** (the bedrock or sediment from which a soil develops), climate, **topography** (the physical features of a landscape), vegetation cover, and time. A developing soil consists of distinct layers having different compositions, called **soil horizons;** the vertical succession of soil horizons in a given location is the location's **soil profile.**

Most temperate-zone soil profiles typically consist of layers designated (from the surface down) as the O, A, E, B, and C soil horizons. **Caliche** is a white carbonate layer, characteristic of desert environments, produced when water carrying dissolved calcium carbonate percolates down from the surface to a lower soil layer and then evaporates, leaving a precipitate of calcium carbonate.

Soil classification helps guide land-use decisions. Soil scientists now distinguish 11 different orders of soil within North America. **Paleosols,** or "old soils," are buried layers of ancient soil that may contain evidence of a past climate different from that of today. For example, a paleosol underlying the modern soil in what is currently a moist, temperate environment may contain a caliche layer, indicating that a warm, arid environment prevailed in that location at some time in the past.

Weathering takes place on other planets in our solar system and on the Moon, albeit differently than it does on Earth. No chemical weathering occurs on the Moon, because it has neither atmosphere (thus no O_2 or CO_2) nor surface water; instead, frequent meteorite impacts weather lunar surface rocks mechanically. Because the high temperatures on Venus instantly vaporize its surface water, little chemical weathering takes place there, although images of its surface regolith suggest that some mechanical weathering occurs on that planet. Although most of the surface water of Mars is now trapped in the ground as ice, the characteristic redness of the planet's surface suggests that in the past climatic conditions did enable water to oxidize its iron-rich bedrock.

Key Terms

weathering (p. 129)
erosion (p. 129)
mechanical weathering (p. 130)
chemical weathering (p. 130)
frost wedging (p. 131)
talus (p. 131)
thermal expansion and contraction (p. 132)
mechanical exfoliation (p. 133)
abrasion (p. 134)
dissolution (p. 134)
oxidation (p. 135)
hydrolysis (p. 135)
spheroidal weathering (p. 139)
regolith (p. 140)
soil (p. 140)
parent material (p. 140)
topography (p. 142)
soil horizon (p. 144)
soil profile (p. 144)
caliche (p. 145)
paleosols (p. 147)

Questions for Review

1. Describe the fundamental difference between mechanical and chemical weathering.

2. Discuss three ways in which rocks can be weathered mechanically.

3. Explain how the process of mechanical exfoliation works.

4. Of granite, limestone, and basalt, which would be most susceptible to the chemical-weathering process of oxidation? To the process of dissolution? Which would weather to produce the most clay?

5. What role does climate play in chemical weathering?

6. Why is quartz a more common mineral in sandstones than plagioclase feldspars? (Hint: See Figure 5-15.)

7. Discuss how soils vary in a region of irregular topography.

8. Describe the principal characteristics of the major soil horizons, and explain how those characteristics develop.

9. In what ways does the weathering environment of the Moon differ from that on Earth?

For Further Thought

1. Since the Industrial Revolution, we have been burning coal, heating oil, and gasoline at an ever-increasing rate. Combustion of these fuels produces carbon dioxide. What would you expect to be the effect of burning these fuels on weathering rates? Explain.

2. Describe what would happen to the physical condition of Cleopatra's Needle (p. 132) if the obelisk were returned to its original home in Egypt.

3. Imagine that the Earth some day becomes devoid of water. How would the nature of chemical weathering change in polar regions? In the arid subtropical deserts? In the equatorial tropics?

4. To the right is a photo of a soil developed on a lava flow in eastern Washington. Judging from the appearance of the soil, what rock type constitutes the lava flow? What weathering processes and products are responsible for the color of the soil? Describe the climate that was most likely responsible for this type of weathering.

5. Look around your community, identify the major building stones, and compare their relative states of weathering. Did the local builders make wise choices when selecting their building materials, considering your local climate? Which rock(s) would work best for construction in your area? Which would be poor choices?

6

Sedimentation and Sedimentary Rocks

Figure 6-1 Layers of sedimentary rock (Navajo Sandstone) in Zion National Park, Utah. These rocks at Checkerboard Mesa are composed of the cemented grains of ancient sand dunes. The fascinating "cross-bedded" patterns they exhibit are typical of wind-blown sands.

Most of us don't live where we can readily view a volcanic eruption or the movement of an active fault, but we probably can walk on a sandy beach or sit by a muddy stream. Even in the heart of New York City, we can watch rivulets of rain collect pebbles and dirt and deposit them into a corner drain. Much of that rainwater will eventually find its way to the Atlantic Ocean, and some of those pebbles will wend their way to a resting place on the Atlantic sea floor. No matter where you live, you don't have to travel far to find **sediment** (from the Latin *sedimentum,* meaning "settling")—fragments of solid material that settle down and accumulate, typically in layers, after being transported some distance by water, wind, or ice, or precipitating out of solution in water.

Sediment, which consists mostly of the products of erosion and weathering (discussed in Chapter 5), can accumulate virtually anywhere on the Earth's surface—from the glaciated summits of the Himalayas, 10 kilometers above sea level, to the deep trenches on the floor of the Pacific Ocean, 10 kilometers below sea level, and everywhere in between. They are continually deposited in lakes, streams, deserts, swamps, beaches, lagoons, and caves, on continental shelves, and at the bases of glaciers throughout the world.

Most sediment is ultimately converted to solid **sedimentary rock** (Fig. 6-1). Sedimentary rocks make up only the thin top layer of the Earth's crust, accounting for barely 5% by volume of the Earth's outer 15 kilometers (10 miles). Nevertheless, they constitute *75% of all the rocks exposed at the Earth's land surface.* Sedimentary rocks are our principal source of coal, oil, and natural gas; much of our iron and aluminum ores; and our cement and other natural building materials. They also store nearly all of our underground freshwater.

In addition, sedimentary rocks contain clues about the condition of the Earth's surface as it existed in the far-distant past. They record the presence of great mountains in areas now monotonously flat, and tell tales of vast seas that once covered the now-dry interior of North America. Some sedimentary rocks hold the fossil remains of past life,

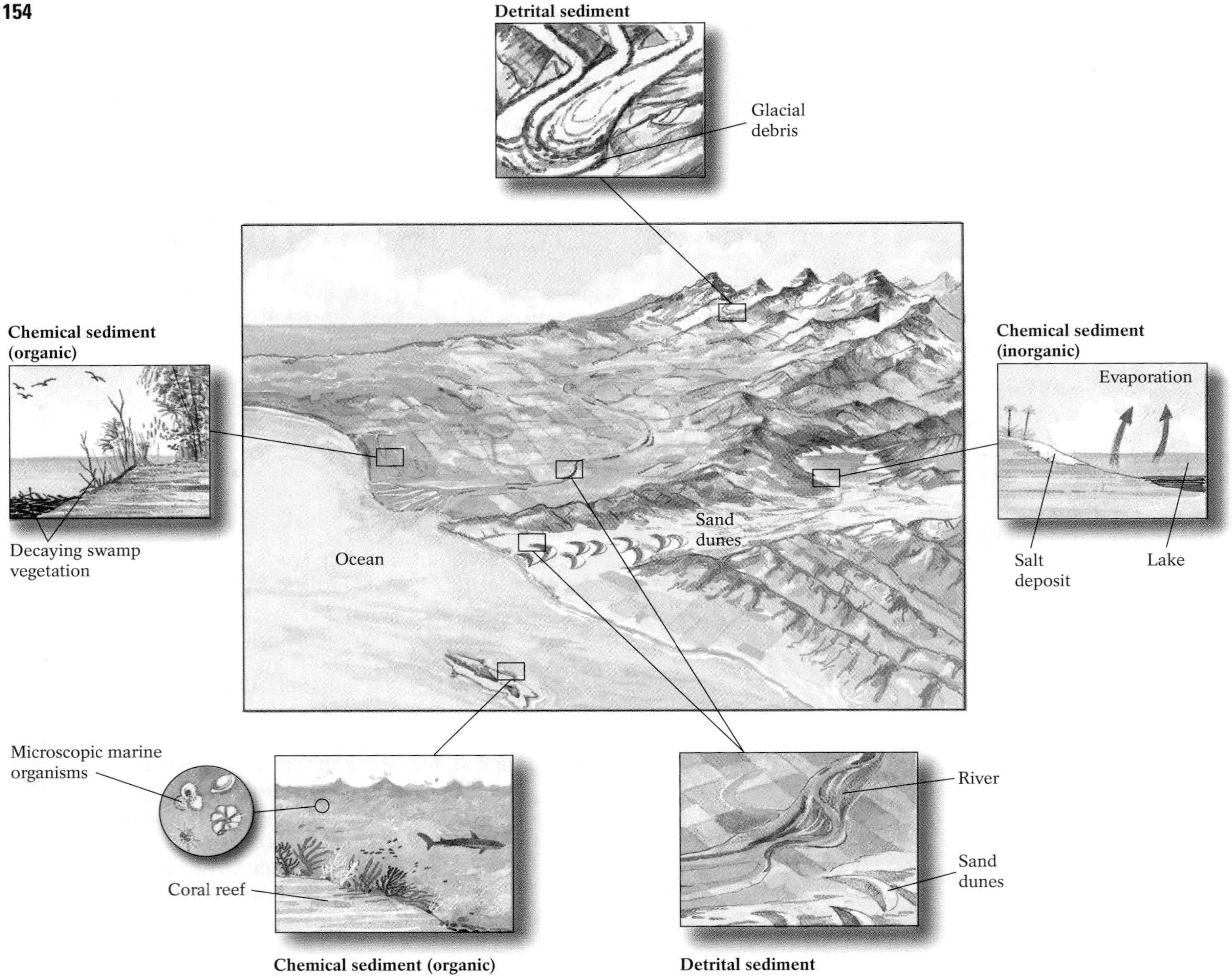

Figure 6-2 Various types of sediments and their origins. Detrital sediments consist of preexisting rock fragments, such as glacial debris or river-channel sand. Chemical sediments often consist of minerals precipitated directly from water, such as salt deposits produced by the evaporation of small, temporary lakes; they may also be composed of organic debris, such as partially decayed swamp vegetation or the shells of small marine organisms.

which tell us much about the planet's evolution through its history. (Chapter 8, "Telling Time Geologically," discusses fossils in detail.)

This chapter examines the origins of sedimentary rocks, their classification, and the ways in which geologists use them to reconstruct past surface environments. The chapter concludes by showing the relationship of various sedimentary rocks to common plate tectonic settings.

The Origins of Sedimentary Rocks

As we saw in Chapter 5, mechanical weathering breaks down rocks into smaller fragments, without changing their chemical composition; chemical weathering converts unstable minerals to new, more stable compounds through various chemical reactions. Our concern in this chapter is what happens to the products of these reactions—how are they moved from one place to another, deposited as sediment at a new location, and buried by subsequent deposition, and how do they eventually become new rocks?

Geologists classify sediments according to the source of their constituent materials (Fig. 6-2). **Detrital sediment** is composed of transported solid fragments, or *detritus,* of preexisting *igneous, sedimentary,* or *metamorphic rocks.* **Chemical sediment** forms when previously dissolved minerals

(a)

(b)

Figure 6-3 Differential sediment sorting by transport media. **(a)** Because wind is highly selective of the particles it transports, this well-sorted windblown silt near Vicksburg, Mississippi, is limited to very fine sediment particles. **(b)** Glaciers are capable of transporting sediment of all sizes. Their deposits, such as this one in Rocky Mountain National Park, Colorado, are typically very poorly sorted.

either precipitate from solution or are extracted from water by living organisms and converted to shells, skeletons, or other organic substances, which are then deposited as sediment when the organisms die or discard their shells. The different types of sediments and the rocks they form are discussed in more detail later in this chapter.

Sediment Transport and Texture

The raw materials for chemical sediments—the dissolved products of chemical weathering processes—are transported by the water in which they are dissolved. These materials remain in solution until a change in the water's temperature, pressure, or chemical composition causes them to precipitate, or until a living organism extracts them from solution to manufacture some type of biological structure, such as an outer protective shell or an internal skeleton. Precipitated materials are typically deposited where they fall. Extracted materials are deposited (perhaps after further transport by the organism) only after the organism dies or discards its shell.

The vast majority of sediments are detrital, being composed primarily of the solid fragments, or *clasts* (from the Greek *klastos,* meaning "broken"), produced by mechanical weathering or released by erosion from preexisting rocks. Some detrital sediment may also contain the undissolved residual products of chemical weathering. During transport, detrital sediments generally move from high places to low places, largely drawn by the pull of gravity but usually assisted by a transporting medium, such as running water, wind, or glacial ice. When the transporting medium loses its capacity to carry the sediments further, as when a river ceases to flow upon entering relatively still marine water at a coast, the particles are deposited. Each year, an estimated 10 billion tons of detrital sediment, most of it carried by rivers, arrive at the world's oceans.

If you've ever held a handful of beach sand or lake-bottom mud, you can appreciate the variety of textures of detrital sediments—the size, shape, and arrangement of particles—depends on the source of the sediment particles, the medium that transported them, and the environment of deposition. (We distinguish chemical sediments by composition more than texture.)

Grain Size Rocks are commonly broken, crushed, and abraded while being carried by turbulent streams, rasping glaciers, surf crashing against a coast, and violent desert windstorms. Thus, rock fragments are generally smaller when they are finally deposited as sediment. The extent to which these fragments become worn down during transport partly reflects the nature of their parent rocks. Different rocks produce grains of different sizes, shapes, and resistance to weathering. Coarse-grained granite, for example, generally weathers to produce larger grains than does fine volcanic tuff.

The grain size of a sediment also depends on the nature and energy level of the transport medium. For example, a pebble that would be pulverized at the base of a creeping glacier might remain unchanged when transported by an oozing mudflow; the same pebble would be worn down by a white-water river, but might be unaffected by a trickling stream.

The texture of sediment relates not just to the transport medium's ability to weather rock fragments, but also to its ability to carry them at all. A raging, flooding river can transport large boulders along with tiny particles and all sizes in between; in contrast, a gentle wind can carry aloft only minute grains. **Sorting** is the process by which a transport medium "selects" particles of different sizes, shapes, or densities. Wind is the most selective of the transport media; a deposit of windblown silt, which contains particles that are nearly all within a narrow size range, is considered *well sorted* (Fig. 6-3a). At the other extreme, glacial ice and flooding rivers are relatively unselective, transporting particles with a wide range of sizes. Glacial deposits, which may contain the finest particles along with boulders the size of small office buildings, are said to be *poorly sorted* (Fig. 6-3b). Geologists can often determine what medium transported a sediment by a quick visual estimate of its sorting.

Figure 6-4 An alluvial fan at the south end of Death Valley, California. Upon reaching the relatively flat Mojave Desert floor, a Sierra Nevada mountain stream lost the energy it had built up in flowing downslope and deposited its accumulated sediment load.

When a moving current (wind or water) eventually loses its energy and can no longer carry its suspended sediment load, the sediment particles settle out and drop to the Earth's surface. Because their transport requires more energy, the larger, heavier particles are deposited first; the smallest, lightest particles are typically carried farthest and deposited last. When a rushing stream emerges from the rocky confines of a steep mountain gorge onto a broad flat valley plain, it fans out and loses much of its energy, dropping its sediment fairly abruptly where the angle of slope changes sharply. This process produces a wedge-shaped body of poorly sorted sediment, called an *alluvial fan,* in which the coarsest grains are deposited first (near the foot of the slope) and the smaller grains fall to the Earth soon after and slightly farther downstream (Fig. 6-4).

Grain Shape Particles released from rock by mechanical weathering may be jagged and angular, particularly if they originate as irregular grains in a plutonic igneous rock. Abrading during transport, however, wears off a grain's prominent points. Some transport media are particularly efficient in rounding particles. Swiftly flowing rivers, for instance, bounce pebbles and sand grains around vigorously, so that they collide with other particles and with the river bottom, becoming ever smoother and smaller. Sediment particles at the base of a glacier may be ground to a rocky powder, called "glacial flour," while other glacial sediments, embedded in hundreds of meters of ice, are cushioned from such collisions and retain their size and shape. *In general, the more vigorous collisions a particle experiences, and the farther it moves away from its parent rock, the more rounded it becomes* (Fig. 6-5).

Grains of softer minerals, such as gypsum and calcite, become rounded more readily than harder minerals, such as quartz. In a recent field study, fragments of soft sedimentary rock became well rounded after only 11 kilometers (6.6 miles) of stream transport. More durable fragments of granite, carried in the same stream, required 85 to 335 kilometers (53–208 miles) of transport to become comparably rounded.

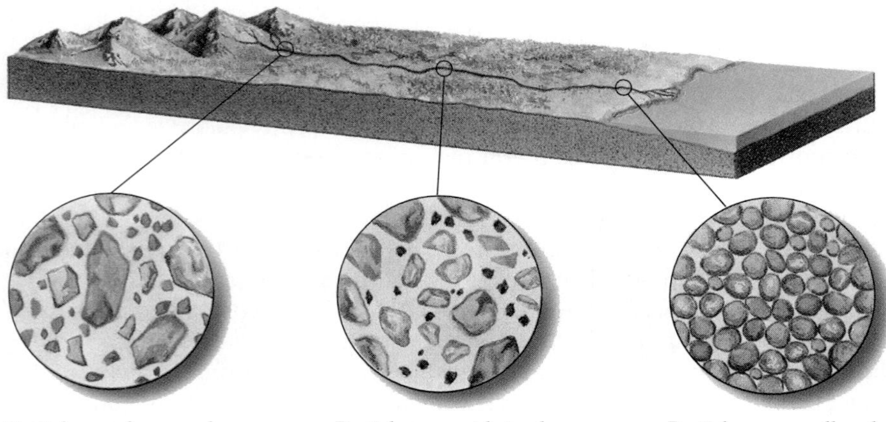

Particles are large and irregular, and consist of a variety of lithologies, including the least resistant.
Particles are mid-sized and of intermediate sphericity, and include resistant and nonresistant lithologies.
Particles are small and nearly spherical, and consist mainly of the most resistant lithologies, such as quartz.

Figure 6-5 As you can see here, a stream's sediment changes along its course; its particles become more rounded as they collide with other particles and the stream's bed. Sediments' compositions change as well, with resistant minerals surviving the journey while the softer ones disintegrate.

Sedimentary Structures

Detrital sediments often contain **sedimentary structures,** physical features that reveal the conditions under which sediment deposition occurred. For example, the upper surface of a layer of sand may display gentle undulating ripples, indicating that the sediments were most likely deposited and shaped by flowing water or strong winds. Let's examine some common sedimentary structures and see how they can be used to interpret past environmental conditions.

Bedding (Stratification) **Bedding,** or **stratification,** is the arrangement of sediment particles into distinct layers *(beds,* or *strata)* of different sediment compositions or grain sizes. A clear break, or *bedding plane,* generally separates adjacent beds, marking the end of one depositional event and the beginning of the next. Bedding planes most often signal a change in the nature of the sediment itself, a shift in the energy with which it was transported, or both. Consider, for example, a typical river bed. Its layers consist principally of the river's usual sediment load of medium-sized particles. Interspersed between these beds, however, you might find occasional layers of heavier, coarse-grained particles. How did these particles arrive there? They were deposited during high-energy flooding episodes, the only times when the river would be capable of carrying the heavier sediment. Similarly, a flooding river typically deposits a particle load of heavy, coarse-grained sediment on top of finer preexisting sediments in the surrounding flood plain; such a difference in grain size would appear as a bedding plane in a cross section of the resulting sediment layers (Fig. 6-6).

Graded Bedding When a sediment load containing a variety of sediment sizes is suddenly dumped into relatively still water, its particles will settle at different rates, depending on their sizes, densities, and shapes. This settling process produces a **graded bed,** a single sediment layer (formed by a single depositional event) in which particle size varies gradually,

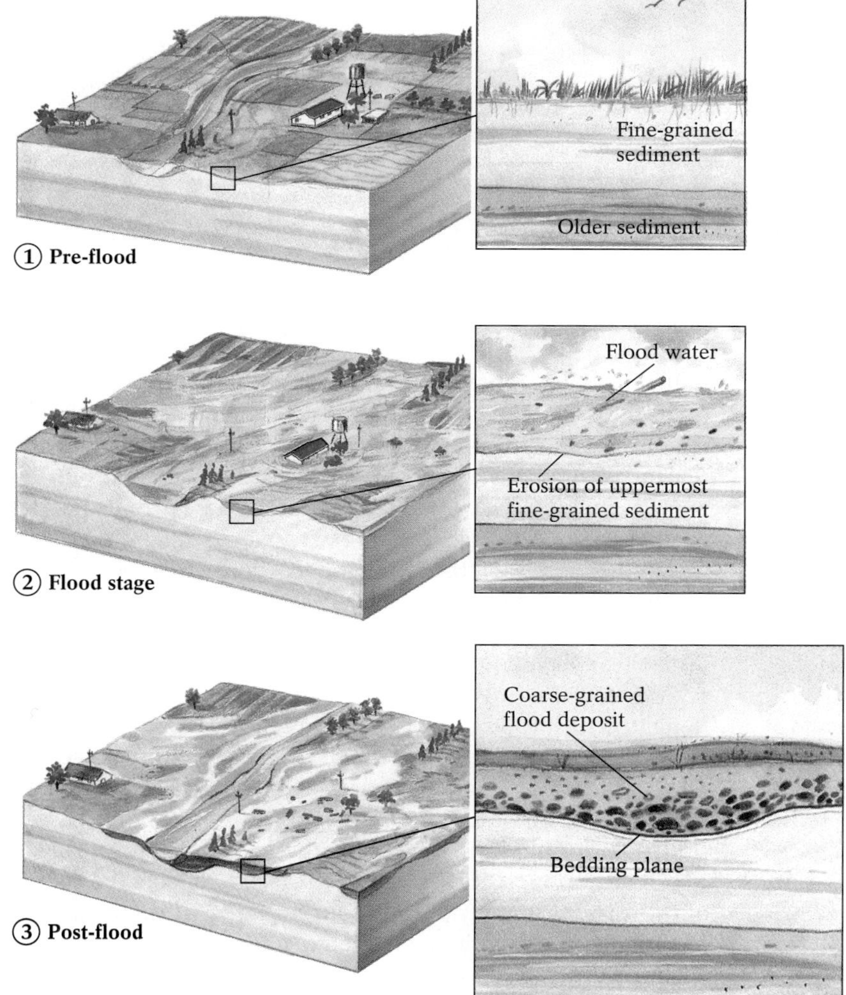

Figure 6-6 Development of a bedding plane due to river flooding. Any depositional event that leaves sediment that differs (either in grain size or composition) from the preexisting sediment leaves a demarcation—that is, a bedding plane—between the resulting sediment layers.

(a)

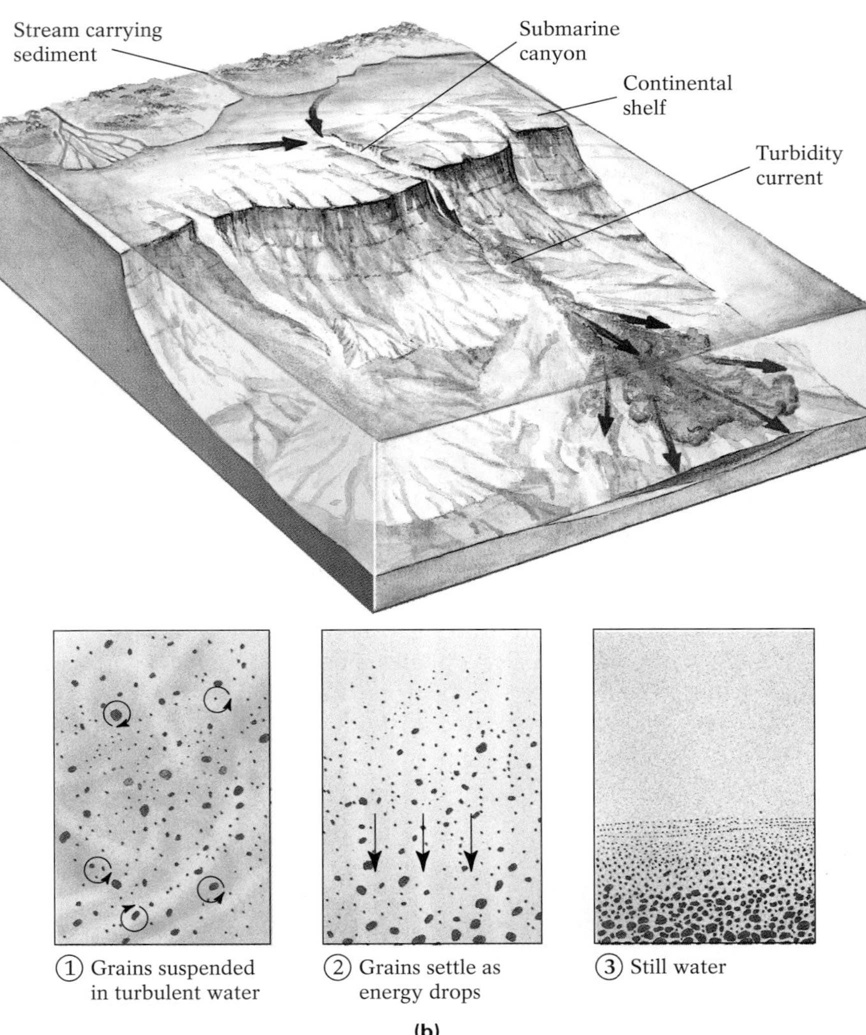

(b)

Figure 6-7 Graded bedding of sediment at Furnace Creek, Death Valley, California. **(a)** Graded beds form as particles of different density, size, and shape settle out of a standing body of water into distinct layers. The larger, heavier particles reach the bottom first and the smaller, lighter particles settle above them. **(b)** Graded sediment is frequently produced by turbidity currents (offshore sediment flows that abruptly lose their energy and drop their particle loads onto the ocean floor).

with the coarsest particles on the bottom and the finest at the top (Fig. 6-7a). To understand how graded beds form, drop a handful of unsorted backyard dirt into a tall glass of water—the largest particles will quickly settle to the bottom, while the finest will settle last and land on top of their larger counterparts.

Graded sediments commonly occur on ocean floors, near shores where muddy streams shed their sediment loads onto the continental shelf (the underwater edge of a continent). Offshore sediment accumulates on a continental shelf as an unstable mass that can easily be dislodged (for example, by an earthquake). When this happens, a dense mixture of sediment and seawater called a *turbidity current* (from the Latin *turbidius,* meaning "disturbed") flows rapidly downslope toward the deepsea floor at 60 kilometers per hour (40 miles per hour) or more (Fig. 6-7b). On reaching the horizontal ocean floor, the turbidity current slows to a virtual standstill, losing its transport energy and dropping its sediment load. The particles from this mixture settle to form a graded bed. Graded beds may also develop when a *lahar,* or volcanic mudflow (discussed in Chapter 4), comes to rest, and its larger, heavier rocks and boulders sink to the bottom of the mud.

Cross-Bedding **Cross-beds** consist of sedimentary layers deposited *at an angle to the underlying set of beds.* They form when particles drop from a moving current, such as wind or a flowing river, rather than settling out of relatively still water or air. We can see cross-beds in wind-deposited sand dunes (Fig. 6-8) and in water-deposited ripples at a river's bottom. Because cross-beds always slope toward the downcurrent direction, they record the flow direction of the current that deposited them. The orientation of cross-beds may also reveal whether a series of rocks has been overturned by tectonic forces (discussed in Chapter 9).

Surface Sedimentary Features The top of a layer of detrital sediment often provides clues to the environmental conditions to which it was exposed during or after deposition. For instance, a pattern of wavy lines, or **ripple marks,** preserved on top of a sediment bed indicates that wind or water currents shaped the particles into a series of shallow curving ridges during deposition. The configuration of these ridges, which are often visible on sandy surfaces, reflects the nature of the current that produced them (Fig. 6-9). If the rocks that form from such sediments remain undisturbed, their ridge crests will always point upward. If, however, the rocks are overturned tectonically, the crests will point downward. Thus, ripples can be used to determine if a layer of sedimentary rock has been overturned.

Mudcracks are fractures that develop when the surface of fine-grained sediment becomes exposed to the air, dries out, and shrinks (Fig. 6-10). They indicate that the watery environment in which the sediment was deposited dried up at some point (as happens, for example, when shallow lakes evaporate). Because they form only at the *top* of a layer of muddy sediment, mudcracks also give clues about whether a layer of sedimentary rock has been overturned (see Fig. 8-4).

Lithification: Turning Sediment into Sedimentary Rock

When a sediment layer is deposited, it buries all previous layers at that location. In time a sedimentary pile may become thousands of meters deep. Sediments buried several kilometers or more beneath the Earth's surface retain heat (produced largely from the decay of radioactive mineral grains and conducted from the deep interior below). In addition, they are compressed by the accumulation of overlying materials and invaded by circulating underground water, which typically carries dissolved ions. Together, the heat, pressure, and the ions alter the physical and chemical nature of both detrital and chemical sediments by a set of processes known collectively as **diagenesis;** sometimes these processes result

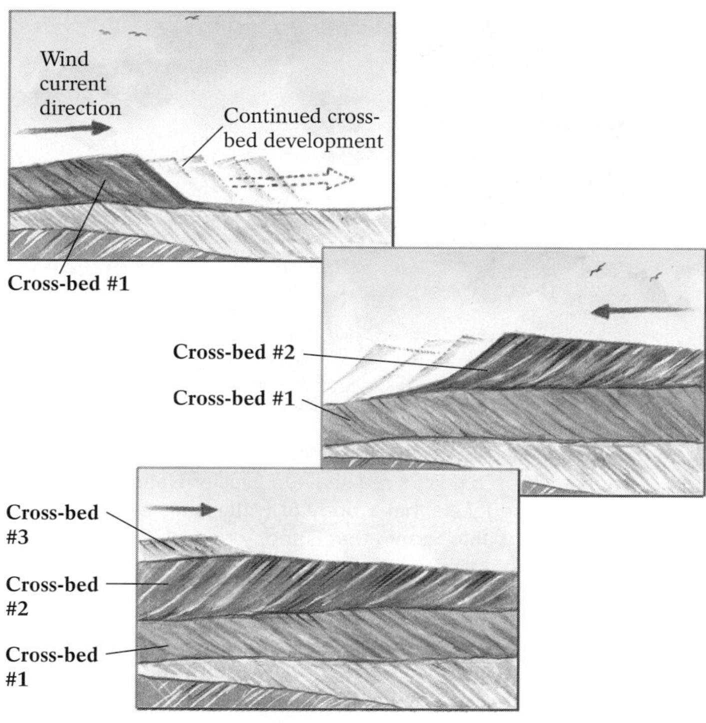

Figure 6-8 The development of cross-bedding in sand dunes. The sets of cross-beds with different orientations that we see here form as wind directions shift through time.

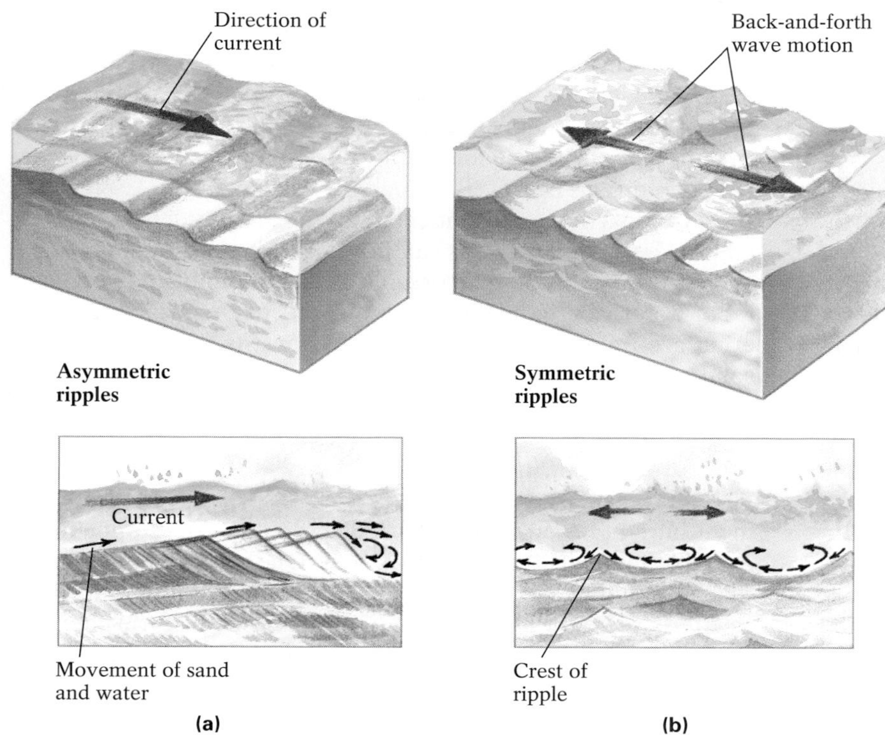

Figure 6-9 Different types of currents produce different ripple patterns. **(a)** A current that generally flows in one direction, such as a stream, produces asymmetric ripples: Sand grains roll up the gently sloping upstream side of each ridge and then cascade down the steeper downstream side. **(b)** Symmetric ripples form from the back-and-forth motion of waves in shallow surf zones at the coast or at the water's edge in a lake. Photo: Exposed rocks show ripple marks, evidence of past current flow (water or wind).

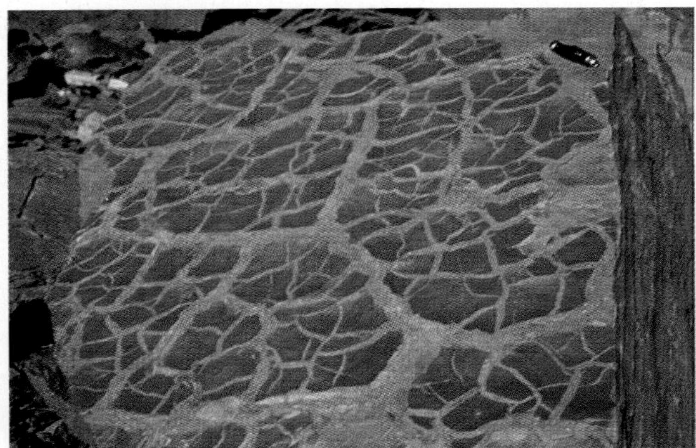

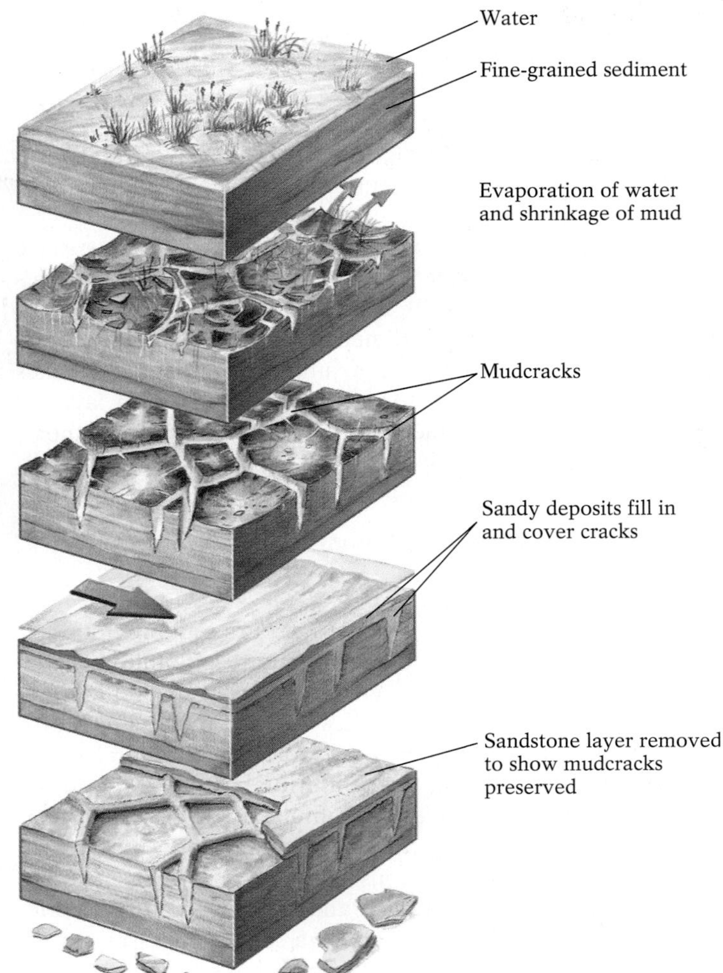

Figure 6-10 The origin of mudcracks. The presence of these mudcracks in oxidized red shale in Glacier National Park, Montana, suggests that a body of water once evaporated to dryness within terrain that is today part of the Montana Rockies.

in **lithification** (from the Greek *lithos,* meaning "rock," and Latin *facere,* meaning "to make"), the conversion of loose sediment into solid sedimentary rock.

Diagenesis generally occurs within the upper few kilometers of the Earth's crust at temperatures less than approximately 200°C (400°F); thus it differs from the intense heat- and pressure-related processes that change rocks deep in the Earth's interior by melting (Chapter 3) or metamorphism (Chapter 7). During lithification, sediment grains are compacted, often cemented, and sometimes recrystallized.

Squeeze a wad of wet clay in your hand, and it will shrink as the water exits. This example illustrates the action of **compaction,** the process by which pressure reduces the volume of a sediment during diagenesis. As sediments accumulate, pressure from the increased weight of overlying material expels water and air from the spaces between deeply buried sediment grains and packs the grains more closely together. When fine-grained muds (particularly those composed of clay minerals) undergo such compaction, weak attractive forces between the grains cause them to adhere to one another, converting loose sediment into more cohesive sedimentary rock.

Cementation takes place when materials originally dissolved during chemical weathering precipitate from water circulating through sediment, creating a chemical cement that binds the sediment grains together (Fig. 6-11). Common cementing agents include calcium carbonate, silica, and several iron compounds. Calcium carbonate forms when calcium ions produced by chemical weathering of calcium-rich minerals, such as calcium plagioclase or calcium-rich pyroxenes and amphiboles, combine with carbon dioxide and water in soil. Silica cements are produced primarily via chemical weathering of the feldspars in igneous rocks. Iron oxides (such as hematite and limonite), iron carbonates (principally siderite), and iron sulfides (such as pyrite) also serve as cements. The Clinton Formation, an extensive sandstone unit in the southern Appalachian Mountains of the southeastern United States, contains so much iron oxide cement that it once represented an important source of iron ore for the steel industry operating in nearby Birmingham, Alabama.

Compaction and cementation do not bind just rock grains. The same processes can lithify shells, shell fragments, and other remains of dead organisms that accumulate in bodies of water where the dissolved products of chemical weathering are transported. Any rock that consists of preexisting solid particles compacted and cemented together—whether preexisting rock fragments or organic debris—is said to have a **clastic** texture. *While clastic sedimentary rocks may be detrital or chemical, they are most often the former.*

The increased heat and pressure associated with sediment burial also promote **recrystallization,** the development of stable minerals from unstable ones. One common mineral that recrystallizes is *aragonite* ($CaCO_3$). This polymorph of calcite is secreted by many marine organisms to form their shells; over time it recrystallizes as stable calcite. This transformation explains why no aragonite exists in ancient carbonate rocks.

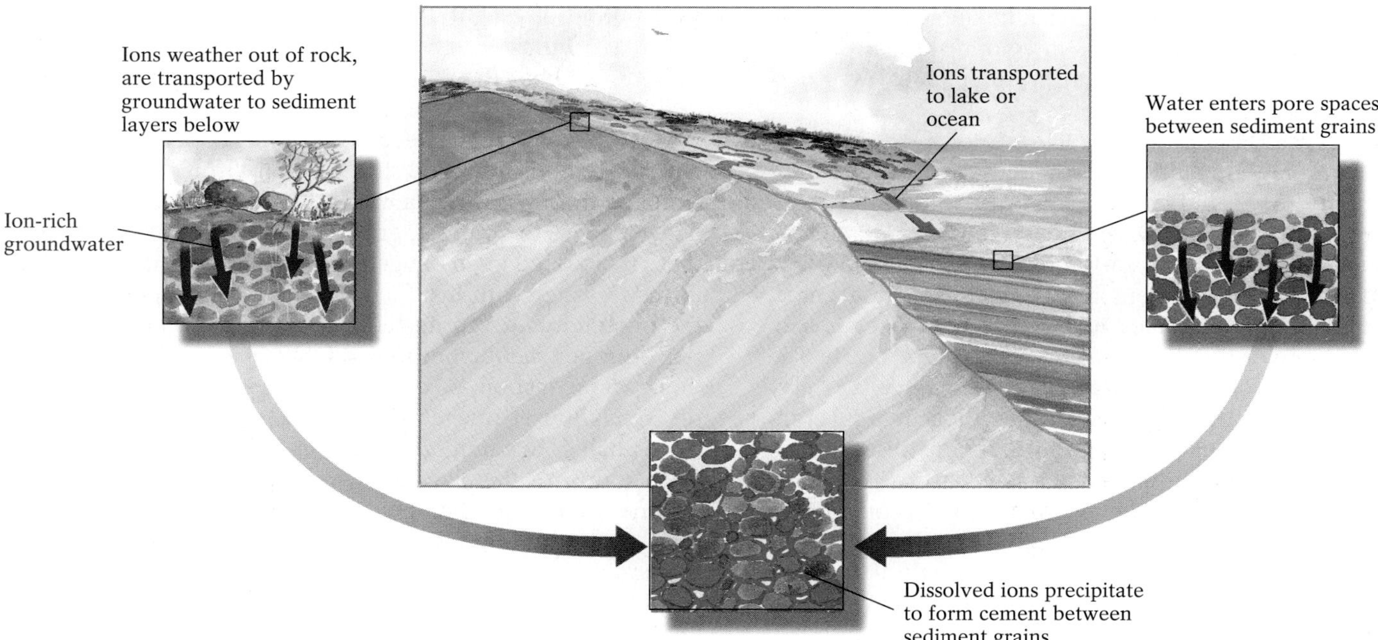

Figure 6-11 Lithification of sediment by cementation. Weathering of source rocks releases ions in solution, which then circulate via groundwater through coarse-grained sediments. When these dissolved materials precipitate as solid compounds in the spaces around sediment grains (called *pore spaces*), they form a cement that binds the grains together.

Classifying Sedimentary Rocks

We generally classify sedimentary rocks as either detrital or chemical, depending on their source material. Each of these broad categories, however, encompasses a wide variety of rock types, each reflecting the diverse transport, deposition, and lithification processes that formed it.

Detrital Sedimentary Rocks

Classification of detrital sedimentary rocks depends on their particle sizes, rather than the composition of these particles (Table 6-1). Shales and mudstones are the finest-grained; sandstones have grains of intermediate size; and conglomerates and breccias contain the largest grains. Note that all detrital rocks are *clastic* because they consist of solid particles cemented together.

Table 6-1 Detrital Sediments and Rocks

Particle Size (mm)	Particle Name		Name of Rock Formed	
<0.004	Clay*	Mud	Shale	Mudstone
0.004–0.063	Silt		Siltstone	
0.064–2	Sand		Sandstone	
2–4	Granule	Gravel	Breccia (if particles are angular)	
4–64	Pebble			
64–256	Cobble		Conglomerate (if particles are rounded)	
>256	Boulder			

1 mm = 0.039 inch

*Note that the term "clay," when used in the context of sediment size, denotes very fine particles of any rock or mineral (as opposed to the term "clay mineral," which refers to a compositionally specific group of minerals); all clay minerals have clay-sized particles, but not all clay-sized particles are composed of clay minerals.

Conglomerates and breccias contain pebble-sized rock fragments, collectively called *gravel;* the other detrital sedimentary rocks consist mainly of smaller grains of the abundant rock-forming silicate minerals, such as quartz, feldspar, and the clay minerals. The sand- and silt-sized grains contain mostly quartz. Because it is relatively hard and chemically stable (and therefore resistant to chemical weathering), quartz is more likely than other common minerals to survive the journey from source rock to deposition site. Clay minerals, which naturally form fine, flat particles, dominate the fine-sized grains; they are also chemically stable at the Earth's surface, so they too can survive the rigors of the Earth's weathering environment. Feldspar minerals, which are less stable at the surface, appear only in sediment subject to minimal chemical weathering, such as in cold, dry regions, or in sediment buried rapidly enough to avoid chemical weathering altogether.

Mudstones More than half of all sedimentary rocks are **mudstones,** the detrital sedimentary rocks containing the smallest particles (less than 0.004 millimeters in diameter). Because such fine particles settle out of only relatively still waters (more energetic waters keep them suspended), most mudstones originate in lakes and lagoons, in deep ocean basins, and on river floodplains after floodwaters recede. The extremely fine particles in mudstones consist largely of clay minerals and micas. They are so small that their mineral composition is best analyzed by X-ray diffraction (discussed in Chapter 2). When these flat or tabular particles become buried beneath hundreds of meters of sediment, compaction flattens them into parallel layers resembling a deck of cards, a characteristic called *fissility* (Fig. 6-12). **Shale** is a fissile mudstone noted for its ability to split easily into very thin parallel surfaces.

A geologist will often place his or her tongue on bits of fine-grained sedimentary rock to distinguish clay-rich mudstones from the slightly coarser quartz-rich siltstones. When moistened by saliva, the particles of the clay minerals feel smooth, whereas the abrasive quartz grains of siltstones feel noticeably gritty.

Shales vary considerably in color, depending on their mineral composition. Red shale contains iron oxides that precipitated from water containing both dissolved iron and abundant oxygen. In contrast, green shale includes iron minerals that precipitated in an oxygen-poor environment. Black shale forms in water with insufficient oxygen to decompose all of the organic matter in its sediment, leaving a black, carbon-rich residue; such conditions might occur in the still waters of a swamp, lagoon, or deep-marine environment, where little circulation of oxygen-rich surface water takes place.

Shales have numerous practical uses. For example, they are a source of the clays used for making bricks and ceramics, such as pottery, fine china, and tile. Mixing shale at various stages with sand, gypsum (hydrous calcium sulfate), and calcium carbonate produces Portland cement, a staple of the construction industry. Oil shale (fine-grained rocks that contain abundant tar) may some day provide a key source of energy. The economic value of these and other sedimentary rocks will be discussed in detail in Chapter 20.

Sandstones **Sandstones,** detrital sedimentary rocks whose grains range from 1/16 millimeter to 2 millimeters in diameter, account for approximately 25% of all sedimentary rocks. Silica or carbonate cement generally surrounds the mineral grains in sandstones. Three major types of sandstones exist, each with its own distinctive composition and appearance. A sandstone composed predominantly of quartz grains (90%), with very little surrounding *matrix* (the finer material that occupies pore spaces between grains), is a *quartz arenite* (from the Latin *arena,* or "sand"). Quartz arenites generally have a light color, varying from white to red depending on the cementing agent (Fig. 6-13a). Their grains are rounded and well sorted, suggesting that they were transported over a long distance.

Arkoses (named for a Greek word that denotes a rock created by consolidation of debris) are distinctive pinkish sandstones containing more than 25% feldspar (Fig. 6-13b). The grains, which are typically derived from feldspar-rich granitic

Still water
Recently deposited particles become partially oriented during settling
Compaction of older, deeper sediment flattens particles to produce thin layers of strongly oriented grains

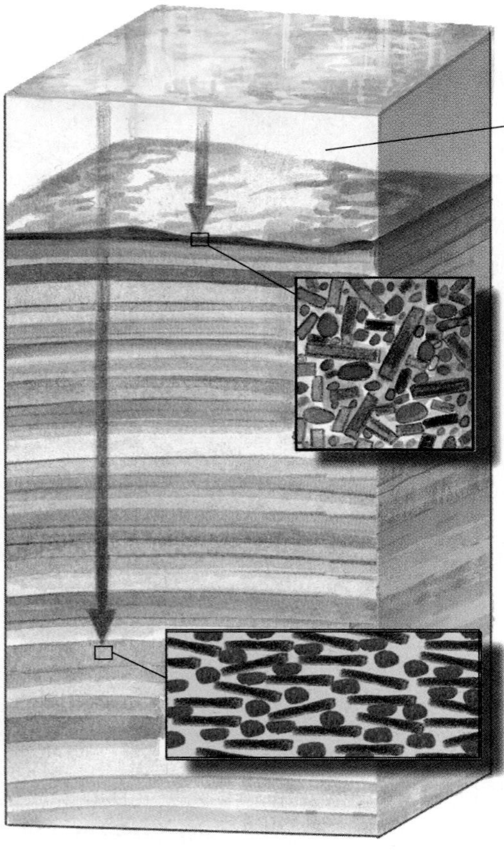

Figure 6-12 The initial deposits of flat or tabular clay and mica grains may be oriented randomly. The weight of subsequent deposits, however, causes these grains to "collapse" into a parallel orientation, producing the typical layered appearance of shales.

source rocks, are generally poorly sorted and angular. This pattern suggests that the grains underwent short-distance transport, minimal chemical weathering under relatively dry climatic conditions, and rapid deposition and burial.

Graywackes (derived from the German *wacken,* or "waste"), also known as *lithic sandstones,* are dark, gray-to-green sandstones that contain a mixture of quartz and feldspar grains, abundant dark rock fragments (commonly of volcanic origin), and a fine-grained clay-and-mica matrix (Fig. 6-13c). The poor sorting, angular grains, and presence of such easily weathered minerals as feldspar suggest that the sediments settled rapidly from turbid water after short-distance transport.

Sandstone's durability has made it a popular building stone. You can see it in the construction of Victorian brownstone houses as well as the Gothic-style edifices found on many college campuses. Sandstones also hold much of the world's crude oil, natural gas, and drinkable groundwater, thanks to the pore space between their sand grains, which is easily saturated with migrating fluids.

Figure 6-13 The three major types of sandstone. **(a)** Quartz arenite, composed predominantly of highly rounded quartz grains. One of North America's most celebrated quartz arenites is the St. Peter Sandstone of Minnesota, Iowa, and Wisconsin. It is most prominently exposed in the Minneapolis–St. Paul area, where the rock is so pure it has been mined, melted, and used in manufacturing glass. Inset: Quartz micrograph. Scale bar = 1 mm. **(b)** Arkose, containing an abundance of angular feldspar grains. Some of North America's classic arkoses can be found in the Red Rocks area along highway I-70, west of Denver, Colorado. Inset: Arkose micrograph. Scale bar = 1 mm. **(c)** Graywacke, distinguished by an abundance of dark volcanic fragments and relatively poor sorting of its particles. We can find good examples of graywacke in the Ouachita Mountains of Oklahoma and Arkansas, and in the coastal mountains of California, Oregon, and Washington. Inset: Graywacke micrograph. Scale bar = 1 mm.

Figure 6-14 Conglomerates and breccias. The grains in these coarse sedimentary rocks reveal much about their history. **(a)** The roundness of the grains in conglomerates suggests long-distance transport by vigorously moving water. **(b)** The angularity of the grains in breccias suggests short-distance transport.

(a)

(b)

Conglomerates and Breccias Conglomerates and **breccias** (BRECH-ahs), the coarsest of detrital sedimentary rocks, contain grains larger than 2 millimeters in diameter. Conglomerates are characterized by rounded grains; breccias include angular grains (Fig. 6-14). Both contain fine matrix material, which is typically cemented by silica, calcium carbonate, or iron oxides. The size of conglomerate and breccia grains makes it relatively easy to identify their parent rocks. Likewise, the shape of the grains provides clues to their transport path: The rounded particles in conglomerates suggest lengthy transport by vigorous currents; the angular grains in breccias suggest brief transport, such as when shattered rock debris accumulates at the base of a cliff.

Chemical Sedimentary Rocks

Whereas detrital sedimentary rocks consist of distinct fragments of preexisting rocks or minerals that have undergone compaction or cementation, chemical sedimentary rocks typically consist of an interlocking mosaic of crystals derived from dissolved compounds. Two kinds of chemical sediments exist: *inorganic* sediments, precipitated directly from solution in water, and *biogenic* sediments, produced by the biological activity of plants and animals. Table 6-2 summarizes the various types of chemical sedimentary rocks that form from lithification of these sediments.

Inorganic Chemical Sedimentary Rocks Inorganic chemical sedimentary rocks form when the dissolved products of chemical weathering precipitate from solution, typically when the water in which they are dissolved evaporates or undergoes a significant temperature change. Four common types of inorganic chemical sedimentary rocks form by varying processes: *inorganic limestones* and *cherts* precipitate directly from both seawater and freshwater; *evaporites* precipitate when ion-rich water evaporates; *dolostone* is a rock whose origin remains the subject of much debate.

Limestones, which are composed largely of calcium carbonate ($CaCO_3$), account for 10% to 15% of all sedimentary rocks. Most limestones forming today are biogenic. Under certain conditions, however, limestone sediments also precipitate inorganically, directly from an aqueous solution. Here's how: Soluble materials usually dissolve more rapidly as water temperature increases—hot coffee, for example, dissolves sugar more quickly than cold. The solubility of calcium carbonate, on the other hand, is directly proportional to the

Table 6-2 Chemical Sedimentary Rocks

	Rock Name	Typical Composition
Inorganic	Inorganic limestone	Calcite ($CaCO_3$)
	Evaporites	Halite ($NaCl$), gypsum ($CaSO_4 \cdot H_2O$)
	Dolostone	Dolomite ($CaMg(CO_3)_2$)
	Inorganic chert	Chemically precipitated silica (SiO_2)
Organic	Organic limestone	Calcium carbonate remains of marine organisms (e.g., algae, foraminifera)
	Organic chert	Silica-based remains of marine organisms (e.g., radiolaria, diatoms, sponges)
	Coal	Compressed remains of terrestrial plants

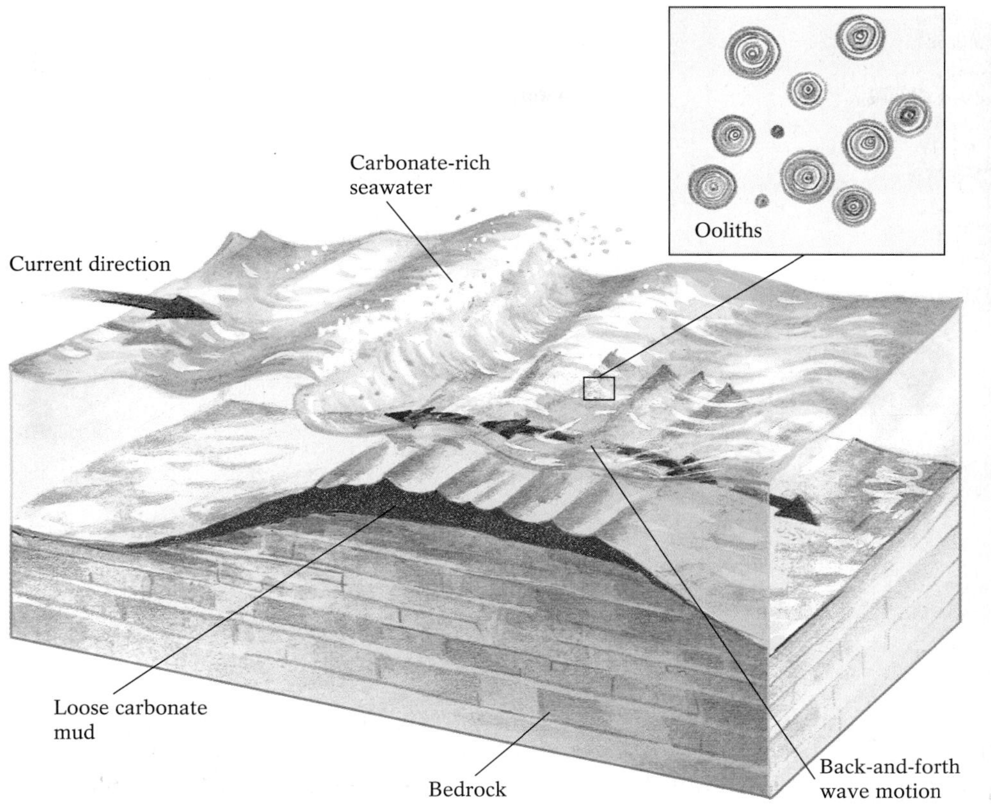

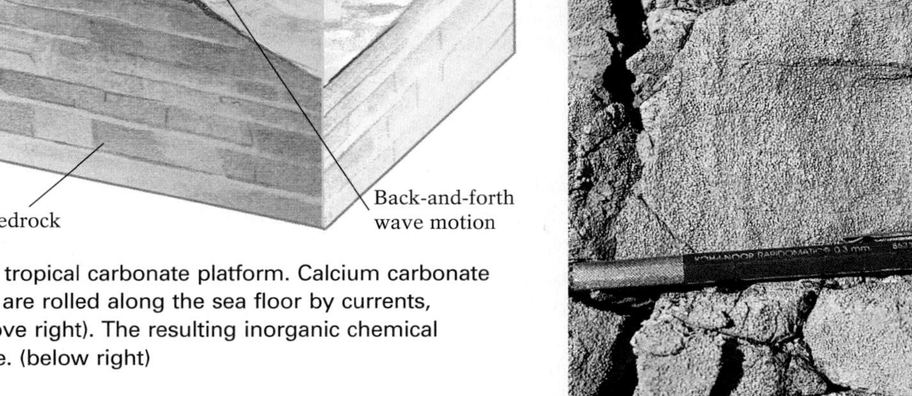

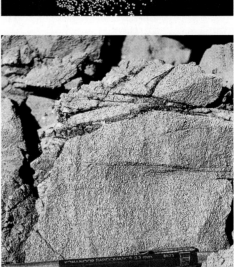

Figure 6-15 The formation of ooliths on a tropical carbonate platform. Calcium carbonate precipitates and coats sand grains as they are rolled along the sea floor by currents, forming spheres called ooliths (shown above right). The resulting inorganic chemical sedimentary rock is called oolitic limestone. (below right)

amount of carbon dioxide (CO_2) in the water, and warm water typically holds less CO_2 in solution than does cold water. In general, as water warms and the proportion of CO_2 present decreases, calcium carbonate in the water becomes *less* soluble; thus *more* calcium carbonate precipitates, taking the form of *inorganic limestone.* Conversely, as water cools and the proportion of CO_2 in it increases, *more* calcium carbonate dissolves and *less* inorganic limestone precipitates.

The amount of $CaCO_3$ that remains in solution is also affected by factors other than temperature: agitation of the water, the presence of photosynthesizing plants, and water depth and pressure. When water is stirred up or agitated, as by wave action, CO_2 escapes to the atmosphere and calcium carbonate therefore tends to precipitate. Aquatic plants remove CO_2 from water during photosynthesis, promoting precipitation of calcium carbonate. A decrease in water pressure allows CO_2 to escape from the water into the atmosphere; because water pressure increases with depth, more calcium carbonate precipitation tends to occur in shallow water than in deep water.

Inorganic limestone precipitates most readily when several of these factors act together. The Grand Bahamas Banks, for example, is a *shallow* submarine platform separated from Florida by the Straits of Florida. Here relatively pure carbonate muds accumulate as carbonate-rich marine waters wash across the broad continental shelves within 30° of the equator. The conditions—a combination of warm water temperatures, active waves and currents, shallow water, and abundant marine plant life—act to remove CO_2 from solution. The resulting calcium carbonate precipitation coats sand grains or other particles on the sea floor; successive coats of the mineral form concentric layers around the growing grains as tidal currents roll them back and forth along the ocean floor (Fig. 6-15). The result is a bed of spheres called *ooliths* (from the Greek *oo,* or "egg," a reference to the sediment's resemblance to fish eggs) about 2 millimeters in diameter. The chemical sedimentary rock formed from this sediment is called *oolitic limestone.*

Inorganic limestone can also form on land. *Tufa* (toofah) is a soft, spongy inorganic limestone that accumulates

Figure 6-16 Evaporite deposits at the Bonneville Salt Flats, west of Salt Lake City, Utah. The modern-day Great Salt Lake is a small remnant of Lake Bonneville, a vast lake that existed in Utah about 15,000 years ago when the local climate was cooler, cloudier, and more humid. A warm, clear, and dry climate evaporated most of Lake Bonneville, leaving these salt deposits.

where underground water emerges at the surface as a natural spring. At the surface, water encounters a low-pressure environment, warms in the sunshine, and bubbles out in an agitated fashion. These factors promote the loss of CO_2, hastening carbonate precipitation. Inorganic limestone, in the form of travertine, also forms in caves when droplets of carbonate-rich water on the ceilings, walls, and floors lose CO_2 to the cave atmosphere and precipitate carbonate rock. (Chapter 16 examines these cave features in more detail.) Public places such as banks, building lobbies, and railroad station waiting rooms feature tufa and travertine as decorative stones.

Evaporites are inorganic chemical sedimentary deposits that accumulate when salty water evaporates (Fig. 6-16). The world's seawater contains, by volume, almost 3.5% dissolved salts. Where marine water is shallow and the climate is warm, evaporation increases the concentration of these salts. When evaporation exceeds the inflow of water into the marine basin, solid crystals precipitate and accumulate at the sea bottom.

Evaporites precipitate from seawater in an orderly sequence: The least soluble crystallize first and the most soluble stay in solution until little water remains. Relatively insoluble gypsum ($CaSO_4 \cdot 2H_2O$), for example, begins to precipitate when roughly two-thirds of a volume of saltwater has evaporated. More soluble salts, such as halite (NaCl), require about 90% evaporation. We rarely see the most soluble salts, such as sylvite (KCl) and magnesium chloride ($MgCl_2$), in evaporites; they precipitate only after more than 99% of the water has dried up.

Because evaporites dissolve so readily in water, they rarely occur at the surface in moist climates, such as in the United States' rainy Northwest or humid Southeast. You can see gypsum at the surface in Colorado, Wyoming, and the arid Southwest, where it forms the dunes of White Sands National Monument, New Mexico. Subsurface salt deposits, however, underlie many areas of the continental United States (for example, in the East, Midwest, Great Plains, Rocky Mountains, and Southwest), wherever shallow inland seas and warmer climates existed in the past (Fig. 6-17). The thickness of salt deposits such as those in central Michigan, exceeding 750 meters (2475 feet), suggests that these ancient seas occupied a vast space.

If, for example, 300 meters (1000 feet) of average seawater evaporated today, only 10.5 meters (35 feet) of salt would remain. To produce the volume of salt now found in Michigan, an ocean about 22 kilometers (13 miles) deep with today's salt content would have had to evaporate completely. Although ancient oceans may have been saltier than oceans today, no *single* ocean is likely to have contained enough salt at one time to produce this thick deposit. Michigan's salt probably resulted from *long-term* precipitation of gypsum and rock salt (a rock composed largely of halite), either from a succession of ancient shallow inland seas that periodically evaporated and then were refilled, or from continuous partial evaporation and refilling of a single long-lasting inland sea. In one case an entire ocean apparently did evaporate completely as recounted in Highlight 6-1 on page 168.

Dolostone is rock composed of the mineral *dolomite,* a calcium-and-magnesium carbonate, $CaMg(CO_3)_2$, similar to limestone in appearance and chemical structure. The origin of dolostone has inspired lively debate for decades. As yet, no strong consensus favors any single mechanism for its origin.

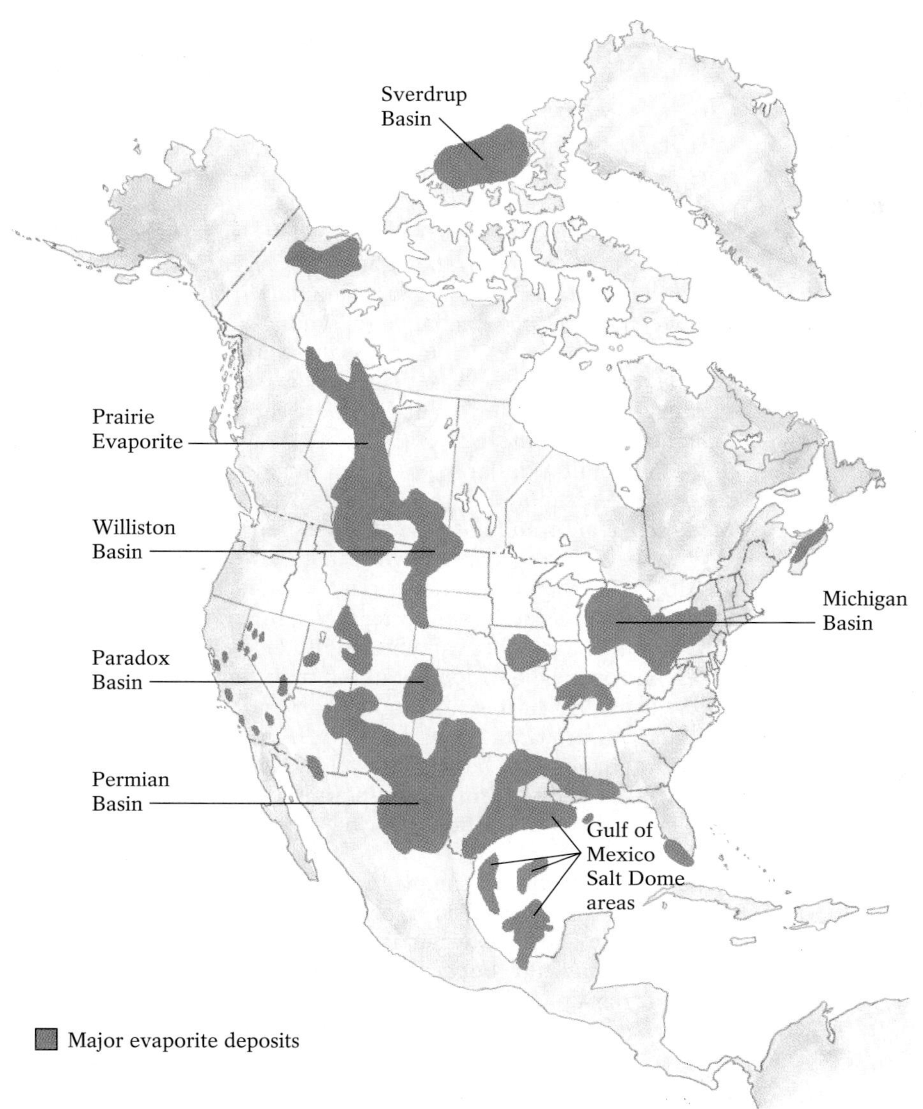

Figure 6-17 The locations of major subsurface evaporite deposits in North America, produced in the ancient past when salty marine water invaded topographic low spots on the North American continent during times of high sea level. The shallow seas produced in this way later evaporated during periods of climatic warming, leaving behind thick evaporite deposits. For example, 400 million years ago, evaporating seas occupying what is now Michigan, Ohio, West Virginia, Pennsylvania, New York, and Ontario produced great deposits of gypsum and rock salt. Younger sedimentary rocks have covered most of these deposits.

For many years, geologists believed that most dolostone formed simply when magnesium ions replaced calcium ions in a preexisting body of limestone. This substitution could happen when ocean water evaporated enough to produce a magnesium-rich solution that then circulated through a limestone bed (Fig. 6-19). This hypothesis predicted dolostone production to occur wherever a tropical or semitropical climate promoted evaporation of salty marine water near a porous body of limestone. Could this process really produce a dolostone mountain range and explain the great thicknesses found today in places such as La Crosse, Wisconsin, Champaign, Illinois, and Bloomington, Indiana? Were these places once located along the shores of tropical inland seas? And why is so little dolostone forming today, even where these ideal environmental conditions apparently exist? Certainly, the rock's ingredients—calcium, magnesium, and carbonate ions—are common enough in seawater. Perhaps other factors, more prevalent in the past, promoted far greater dolostone accumulation than occurs in today's oceans. Highlight 6-2 discusses some of our most recent thoughts about the "dolostone mystery."

Chert is the general name for a group of sedimentary rocks that consist largely of silica (SiO_2) and whose crystals can be seen only through a microscope. Although many cherts have biogenic origins, the most common are inorganic, forming as a chemical precipitate from silica-rich water. (One type, jasper, owes its reddish color to the presence of a small amount of the iron oxide hematite.) *Chert nodules,* fist-sized masses of silica that commonly occur in bodies of limestone and dolostone, are believed to form when portions of these rocks are dissolved by circulating groundwater and replaced by precipitated silica. Because chert is easy to chip and forms

Highlight 6-1 When the Mediterranean Sea Was a Desert

The climate in the lands bordering the Mediterranean Sea, such as Morocco, Algeria, Libya, and Greece, can be oppressively hot. Every year this heat evaporates more than 4000 cubic kilometers (960 cubic miles) of the Mediterranean's water, with only 400 cubic kilometers (96 cubic miles) being replaced by rainfall and inflowing rivers. If these two sources were the only means of maintaining its water level, the Mediterranean would be nearly dry in approximately 1000 years. Today, a massive inflow of Atlantic Ocean water through the Strait of Gibraltar balances the Mediterranean's water deficit. This balance has not always been present, however.

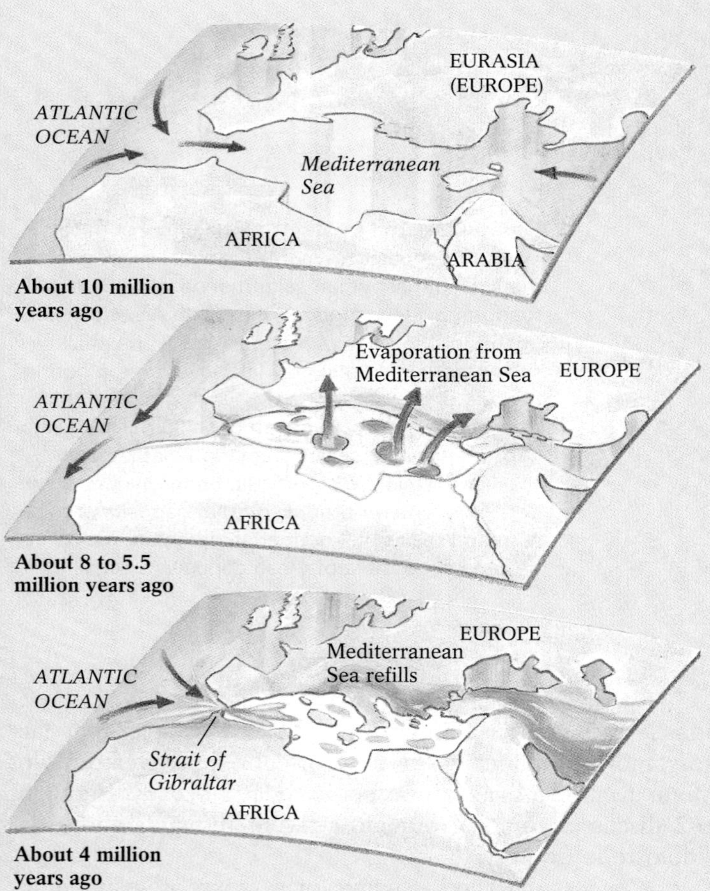

Figure 6-18 The thick evaporite deposits that underlie much of the Mediterranean basin today probably accumulated during a time (8–5.5 million years ago) when a topographic barrier stretched across what is now the Strait of Gibraltar, cutting off the water influx from the Atlantic Ocean. With no water supply to replenish it, the Mediterranean Sea experienced prolonged periods of evaporation lasting several million years, precipitating vast amounts of salt.

Deep-sea exploration in the early 1970s revealed a massive evaporite layer more than 2000 meters (6600 feet) thick beneath the floor of the Mediterranean Sea. This layer is more than 25 times the thickness that would accumulate if the current Mediterranean evaporated to complete dryness. What past conditions could have produced such a great salt accumulation?

One hypothesis proposes that these deposits formed when a barrier restricted water circulation between the Atlantic Ocean and the Mediterranean Sea (Fig. 6-18). According to this hypothesis, about 8 million years ago ongoing convergence between the African and Eurasian plates, which meet beneath the Mediterranean, gradually raised the sea floor in the area of what is now the Strait of Gibraltar. The natural limestone dam created by this movement increasingly blocked the inflow of Atlantic waters. As the replenishing water supply diminished, the Mediterranean rapidly evaporated. With less water to moderate the semitropical heat, temperatures may have risen to 65°C (150°F), further hastening evaporation and producing desert-like conditions. Occasional pulses of Atlantic water over or through the Gibraltar barrier, like a leaky faucet, may have provided the now-dry basin with a periodic supply of salty water that evaporated and added to the thick Mediterranean salt deposits.

Analysis of sediment samples taken from the Mediterranean's floor indicates that deposition of the evaporites ceased 5.5 million years ago, perhaps when a break in the limestone dam expanded dramatically, boosting the trickle of Atlantic water into a spectacular waterfall spanning what is now the Strait of Gibraltar. Such a force would have removed the barrier completely by 4 million years ago. A flow 100 times the volume of Niagara Falls would have been required just to balance regional water loss by evaporation and begin to refill the empty Mediterranean basin. To provide enough water to support the sea life of the time (whose extent and makeup are revealed in the fossil record), a flow in excess of 1000 Niagara Falls would have been required. Even at that flow rate, it would have taken more than a century to refill the basin to its present level. In contributing the water that refilled the Mediterranean, the other oceans of the world probably fell about 10 meters (35 feet) over that time.

The Rock of Gibraltar is a remnant of the ancient limestone dam, a reminder of the time when an intermittently dry Mediterranean basin separated Europe and Africa. Today the Mediterranean never evaporates even to the point of gypsum deposition. If the African and Eurasian plates continue to converge, however, a new rocky barrier to Atlantic water inflow may once again rise from the sea floor and the Mediterranean basin may again dry out and fill with salt.

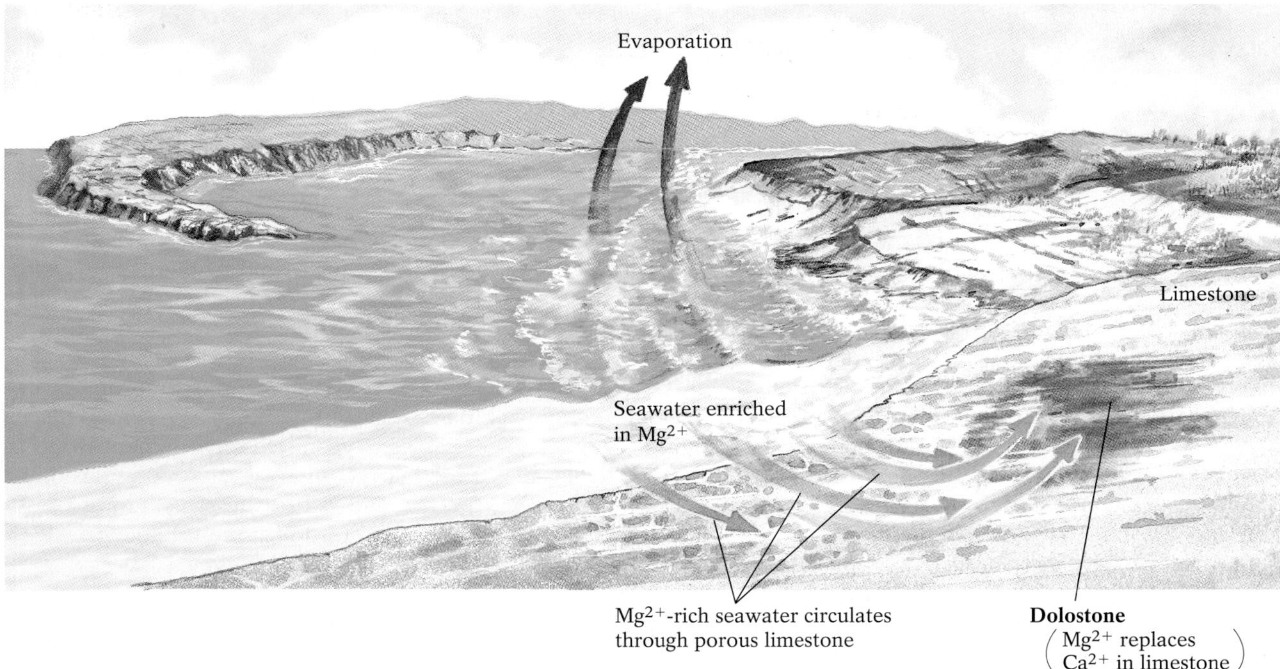

Figure 6-19 One hypothesis for the formation of dolostone states that seawater enriched in magnesium ions (which increase in concentration as the water evaporates) circulates through porous limestone; dolostone then forms as the magnesium ions replace calcium ions in the limestone.

sharp edges, early humans often shaped it into weapons, cutting blades, and other tools (Fig. 6-20).

Biogenic Chemical Sedimentary Rocks Biogenic chemical sedimentary rocks are derived from living organisms. The principal rocks of this type are *biogenic limestones* and *cherts,* which are composed largely of the skeletal remains of marine animals and plants, and *coal,* which formed from the compressed remains of terrestrial plants. In each case, biogenic chemical sediments undergo the same diagenetic processes (compaction, cementation, and recrystallization) that produce detrital sedimentary rocks.

Nearly all biogenic limestones consist of calcite ($CaCO_3$) that formed in a warm marine environment. Warm seawater is nearly saturated with calcium ions that have been chemically weathered from calcium-rich rocks on land and then transported to the sea by rivers and streams. Numerous ocean-dwelling organisms extract the ions to form their calcium carbonate shells and internal skeletons. When these organisms die, their hard parts settle to the sea floor, where they accumulate in great thicknesses and then lithify as clastic biogenic limestone.

Most biogenic limestones form in shallow water along the continental shelves of equatorial landmasses, where the presence of warm water, plentiful sunlight, and abundant nutrients enable marine life to flourish. For example, an abundance of microscopic algae that secrete needle-like calcium carbonate casings inhabit the waters around most Caribbean islands, the Florida Keys, and the east coast of Australia. The deaths of these organisms produce a rain of delicate calcite stalks and branches—so small that 600,000 laid end to end would span this line of type—that descends to the shelf floor and accumulates to produce a calcite-rich mud. Other algae

Figure 6-20 Stone axes made from chert some 150,000 years ago. Because of the sharp edge that forms when chert is chipped, our ancestors constructed many of their tools from this hard, silica-rich rock.

Highlight 6-2 The Dolostone Mystery

Once upon a time in Earth's history, vast quantities of dolostone precipitated from the world's oceans. We see great piles of this carbonate rock in the Dolomite Mountains of the Italian Alps (from which we derive the constituent mineral's name) (Fig. 6-21). Today, however, dolostone is forming in only a few places—in particular at warm coastal tide flats and lagoons where water circulation is restricted and evaporation is extreme. For example, a thin crust of dolomite crystals is developing today above the low-tide level on some limestone bodies in the Persian Gulf states, the Florida Keys, and the Bahamas.

But did all of the Earth's dolostones form in such settings? Can the processes active in these areas today explain the vast layers of dolostone found around the world?

Recent discoveries suggest that the presence of a family of sulfate-consuming bacteria, *today nearly absent from the Earth's oceans,* may be a necessary condition for substantial dolostone accumulation. A trip to the bottom of a foul-smelling, murky, red lagoon near Brazil's scenic Rio de Janeiro may hold the clue that solves the "dolostone mystery." Small dolomite crystals are forming there in the oxygen-poor, sulfate-rich sludge. What is unique about conditions in this lagoon that might account for the dolomite precipitation? The answer apparently lies in the lagoon's reddish murkiness and its smell. Analyses of the lagoon's water reveal the abundant presence of a form of bacteria that acquire life-sustaining oxygen from sulfate ions. In the process, these bacteria manufacture hydrogen sulfide gas, renowned for its rotten-egg stench. Perhaps, researchers thought, a similar community of bacteria flourished in past oceans, producing some of the world's great bodies of dolostone.

To test the relationship between sulfate-consuming bacteria and dolomite precipitation, a research team led by Dr. Judith McKenzie of the Swiss Federal Institute of Technology collected bacteria from the red lagoon's sludge and combined them with carbonate sand, nutrients, and magnesium sulfates—the proposed recipe for dolomite. A year later, when the researchers examined the confection under an electron microscope, they found a white precipitate of dolomite crystals on the sand grains. *A control sample of bacteria-free sludge bore no such crystals.*

Dr. McKenzie has proposed that, as the bacteria ingest sulfates, they use some as nutrients and then excrete the rest, including magnesium ions. Similarly, the bacteria ingest calcium carbonate, use some, and excrete the rest. Thus these bacteria ingest, process, and then excrete all the ingredients for dolomite production. The excreted ions combine chemically on the bacteria's cell walls as dolomite. Once the initial crystals of dolomite have formed, other inorganic processes add calcium, magnesium, and carbonate ions to the bacteria-produced crystal template.

McKenzie's group hypothesizes that hundreds of millions to several billions of years ago, before oxygen-producing trees and grasses added a vast amount of oxygen to the atmosphere, sulfate-consuming bacteria may have thrived in many more places than they do today. (Today's oxygen-rich environment is lethal to such organisms.) If the hypothesis about their relationship to dolomite is correct, the rise and fall of these bacteria may explain the Earth's uneven record of dolostone production.

Figure 6-21 Vast exposures of dolostone in the Dolomite Mountains, a segment of the Italian Alps.

and tiny carbonate-secreting animals living within the mud add to the growing volume of carbonate sediment, which comprises a soft white biogenic deposit called *chalk* (Fig. 6-22).

Biogenic limestone can also form in deep-marine environments. Although few carbonate-secreting organisms live at the cold, lightless, deep-sea floor, microscopic animals called *foraminifera,* or "forams," swim and float in the upper 50 meters (160 feet) of the deep sea (Fig. 6-23). When the forams die, their protective shells accumulate on the sea floor, where they mix with marine mud to form a deep-sea *ooze.* Burial of an ooze by subsequent sedimentation causes its clays and carbonates to become cemented and recrystallized as biogenic limestone.

Biogenic chert is a chemical sedimentary rock composed mainly of silica derived from the remains of a variety of marine organisms. Chert that takes the form of layered

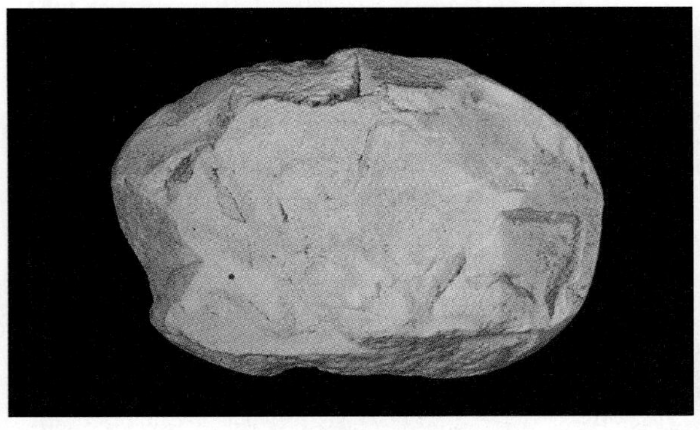

(a)

(b)

Figure 6-22 (a) Chalk is a shallow-marine limestone formed from the carbonate secretions of countless microscopic organisms. **(b)** Accumulation of chalk can eventually produce deposits of impressive size, such as the famed White Cliffs of Dover in England. The cliffs are composed mainly of the skeletons of microscopic marine plants and animals that accumulated some 100 million years ago, when the global sea level was apparently higher and coastal England was under water.

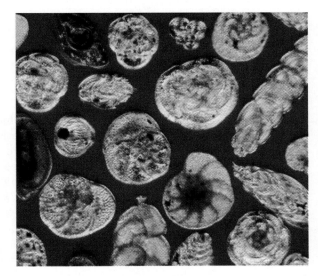

Figure 6-23 Foraminifera, microscopic marine animals whose calcium carbonate shells are an important component of deep-marine limestone. (Magnified 115×.)

beds (rather than the nodules characteristic of inorganic cherts) is believed to be of biogenic origin (Fig. 6-24a). Microscopic examination of such biogenic cherts typically reveals silica-based organic debris, such as the shells of single-celled animals called *radiolaria,* the internal structures of single-celled plants called *diatoms,* and the skeletons of larger, more complex animals such as marine sponges (Fig. 6-24b). One variety of chert, known as *flint,* gets its dark gray-to-black color from the presence of carbon-rich matter.

Coal is a biogenic sedimentary rock composed largely of plant remains. You can often see original plant structures,

(a)

(b)

Figure 6-24 Biogenic chert, found in the form of layered beds **(a),** develops from accumulation and lithification of the silica-based remains of marine organisms (shown here magnified 20×) **(b)**

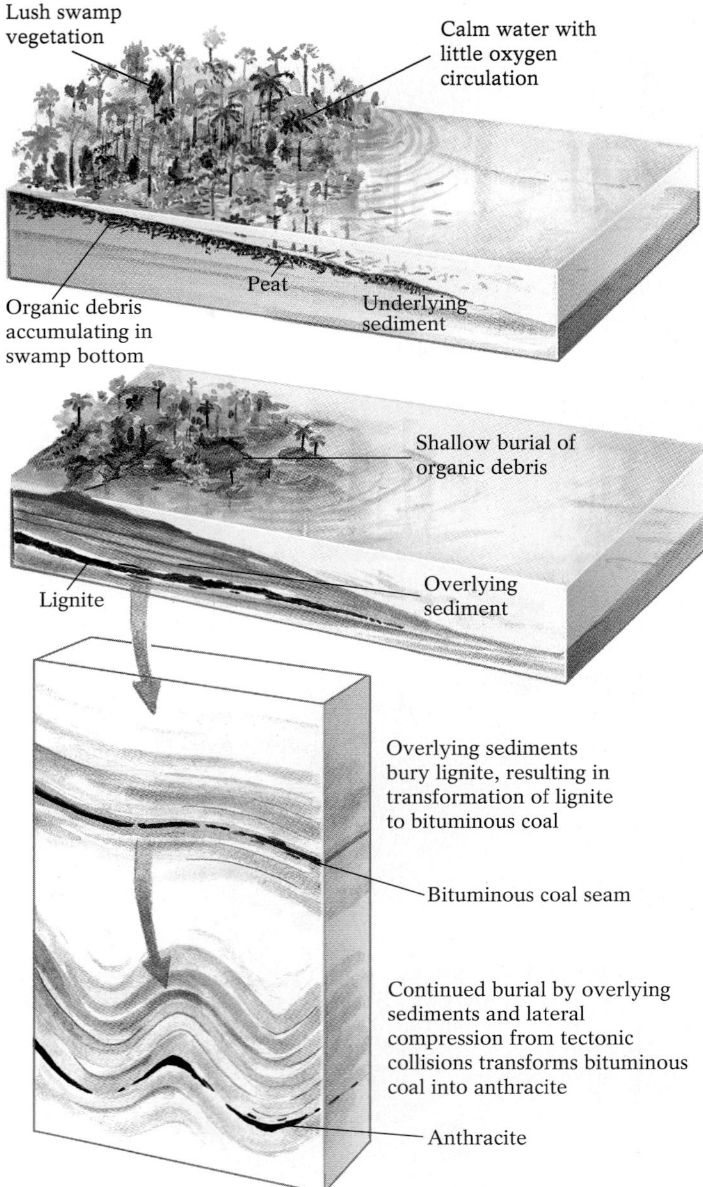

Figure 6-25 The formation of coal from swamp deposits. Abundant organic debris accumulates on the swamp floor and becomes buried before it can decay in oxygen-poor swamp water; the weight of subsequent deposits and the increased temperatures at greater depths transform the debris to progressively harder forms of coal. Above: A body of coal (a "coal seam") in central Utah.

such as fragments of leaves, bark, wood, and pollen, in a lump of coal under magnification, and sometimes even with the naked eye. Although vegetation generally decomposes quickly at the Earth's surface, in an environment characterized by little oxygen and rapid sedimentation, it may be preserved until buried and converted to coal. The stagnant water of a warm, lushly vegetated swamp is ideal for the production of coal. The typically windless conditions found there produce calm water surfaces, and relatively little atmospheric oxygen circulates down to reach the organic debris accumulating at the swamp's bottom. Plant decay quickly depletes any oxygen dissolved in the swamp water, leaving the remaining vegetation to accumulate undecayed.

With the passing millennia, increasing pressure from the weight of overlying sediments expels water, CO_2, and other gases from the accumulating mass of vegetation, increasing the proportion of carbon in the plant residue. Early in the process, when much of the original plant structure remains intact, a soft brown material called *peat* is produced. Overtime, increased heat and pressure create ever-harder and more compact forms of coal, ranging from soft brown *lignite* to moderately hard *bituminous* coal. Ultimately, heat and pressure may convert coal, a sedimentary rock, into dense, lustrous *anthracite,* a metamorphic rock (Fig. 6-25).

Nearby seas generally create and sustain swamps. When sea levels rise, the swamps may be submerged; plant debris then ceases to accumulate and is replaced with sediments typical of marine environments. Falling sea levels restore the swamp, however, allowing accumulation of plant debris to resume. The cyclical nature of deposition in such a changeable environment typically produces alternating coal beds and detrital sedimentary rock layers.

Much of North America's coal developed during two periods in its geologic past: approximately 280 million years ago, in today's coal-mining regions of the Appalachians in Pennsylvania and West Virginia, and nearly 75 million years ago, along the present-day bituminous-producing formations of the Rockies and throughout the plains of Montana, Wyoming, North Dakota, and Saskatchewan. Where are North America's future coal deposits forming? The warm, richly vegetated candidates include the Dismal Swamp of coastal North Carolina, the barrens of the Florida Everglades, and the colorful bayous of Louisiana.

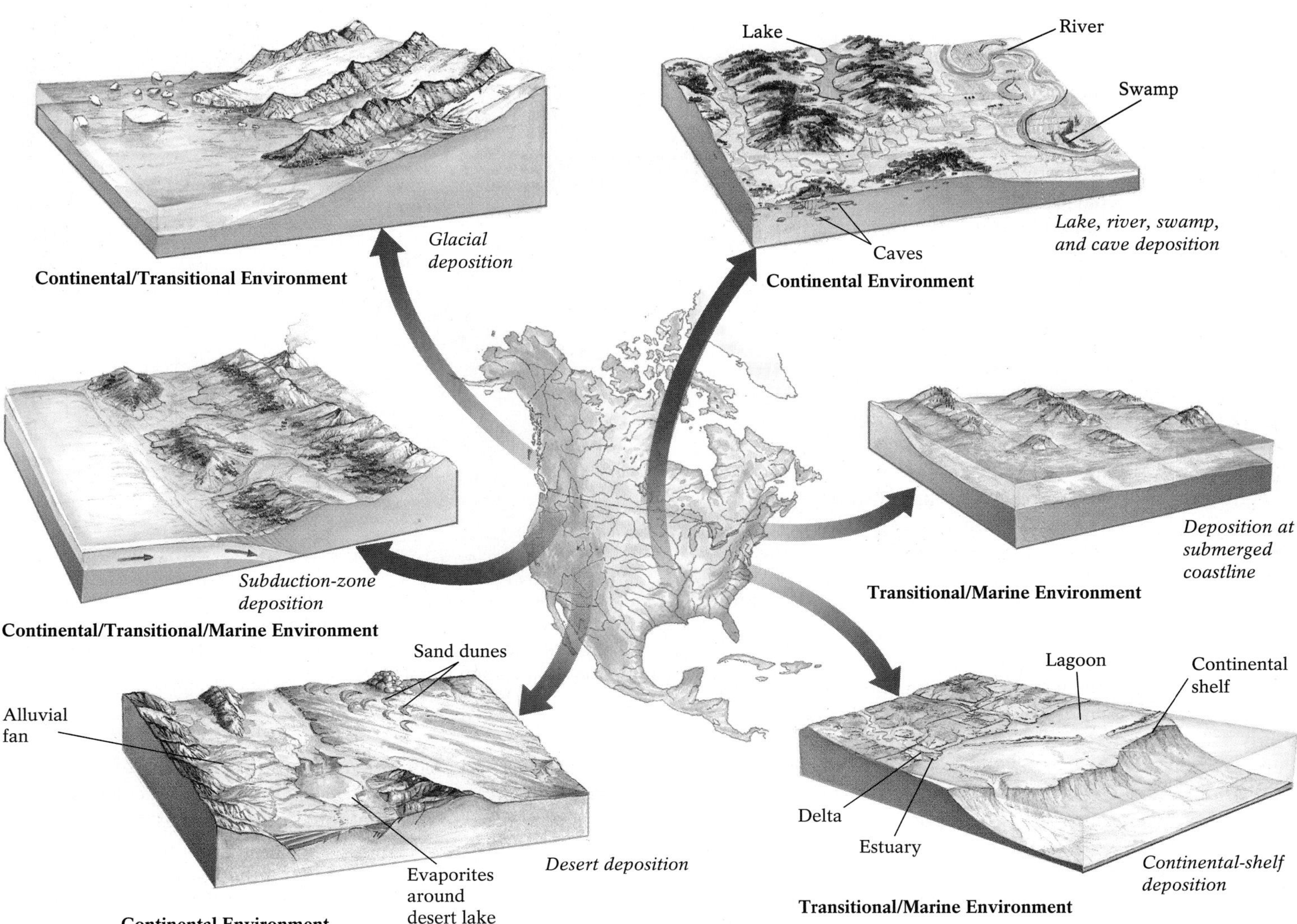

Figure 6-26 Some common geological environments in which sediments accumulate.

"Reading" Sedimentary Rocks

A sedimentologist can analyze sedimentary rock formations, study their fossils and sedimentary structures, and determine the depositional environment that produced each formation. Ultimately, it is possible to deduce a region's unique geologic history—including the sequence of its major geological events, such as the rise and fall of sea levels, and possibly even its past plate-tectonic settings.

Depositional Environments

Where can we find sediment deposits? At virtually any spot on the Earth's surface—from atop the highest peak to the depths of the ocean. Figure 6-26 shows just a small sample of the diversity of depositional environments. **Sedimentary environments** may be *continental* (on a landmass), *marine* (at sea), or *transitional* (in the zone in-between). In this section, we will discuss the principal characteristics of these three categories; in later chapters, we will examine individual environments in greater detail.

At any given time, a sedimentary environment's geological, geographical, biological, and climatic conditions dictate the properties of its sediments, leaving behind tell-tale features that reveal their presence. These properties enable geologists to learn about past environments using the clues found in ancient rocks.

Continental Environments Sedimentation in continental environments is mostly detrital. Sediment layers contain numerous indicators of past water-, wind-, and glacier-flow directions, and plant and freshwater fossils abound. Rivers, lakes, caves, deserts, and glaciers are all continental depositional environments.

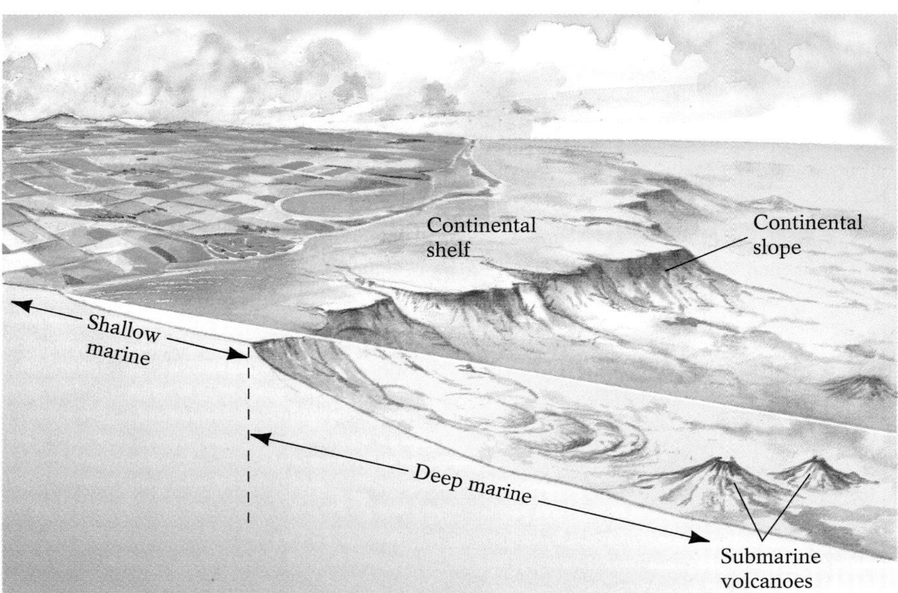

Figure 6-27 Marine sedimentary environments are described as being either shallow marine or deep marine. Shallow-marine environments lie over continental shelves and the shallow platforms that surround many oceanic islands. Deep-marine environments begin at the foot of the continental slope.

Swift river currents carry and then deposit coarse-grained sediments (sand and gravel), forming rippled and cross-bedded structures. Sediments that settle from "standing" floodwaters (after they've dropped their initial, high-energy load of coarse sediment) tend to be well-sorted, fine-grained, and graded. Deposits in lake and swamp environments, where sediments also settle from standing bodies of water, share the same characteristics.

Lake deposits often contain diatoms (siliceous algae) and other organic matter. Reflecting the lush vegetation found in swamps, the sediment deposits there may include peat and organic-rich mud.

In caves, calcite precipitated from underground water is deposited as protrusions from ceilings, walls, and floors. Cave sediments may also include the bones and droppings of cave-dwelling bats, birds, and other animals.

In desert environments, wind serves as a significant transport mechanism. Incapable of moving large particles (except in unusual circumstances), winds typically carry aloft only silt- and, occasionally, clay-sized particles; they transport larger sand particles either in a series of short jumps or by rolling them along the surface. Thus, desert deposits are generally well-sorted. But where steep mountain streams abruptly drop their loads onto flat desert floors, coarse, poorly sorted, poorly stratified alluvial fan deposits form. Evaporites occur where the extreme heat in desert environments has evaporated temporary water bodies.

In a glaciated landscape, deposits vary in composition, texture, and structure. Because flowing ice transports particles of all sizes, its deposits are typically poorly sorted. As the ice melts, it drops its load in an unstratified heap. Sediments carried by meltwater streams and deposited beyond the glacier's margin are generally coarse and well-rounded, because these streams' swift currents can carry large particles and their turbulent flow promotes the forceful collisions that round sediment grains. Glacial lakes are still, however, so deposits there are characteristically fine-grained, graded, and well-sorted.

Transitional (Coastal) Environments Along ocean shores, continental and marine sedimentary processes merge. Breaking waves, tides, and ocean currents pulverize soft mineral grains and shells and sweep fine particles out to sea, leaving behind well-sorted, rounded, sand-sized deposits made principally of durable mineral grains, such as quartz.

When a river joins with an ocean, its freshwater mixes with the salty seawater; this action forms a body of *brackish* (somewhat salty) water called an *estuary*. If the estuary water is not too salty, it may support marine, brackish, and freshwater organisms that contribute biogenic debris to its sediment. Estuaries sometimes contain *deltas*, fan-shaped accumulations of well-sorted sediment formed when river currents slow suddenly upon entering the sea (similar to alluvial fans, although deltas form primarily under water rather than on land). Coastal deltas grow seaward if sediment is deposited at a faster rate than it is removed by waves, tides, and coastal currents.

Sometimes sediment of continental origin is deposited offshore to form narrow islands that lie parallel to the coastline. The shallow body of water between the coast and an offshore island is called a *lagoon*. Fine continental sediments from inflowing streams as well as peat and organic mud may accumulate there.

Continental and marine environments also meet at subduction zones, where explosive volcanism, powerful earthquakes, and mountain building produce a great volume of sediment; some of this sediment is deposited on shore, while other materials are transported offshore by turbidity currents. Transitional deposition also occurs when rising sea levels submerge coastlines, depositing marine sediments directly on top of continental sediments.

Marine Environments Marine environments vary according to their depth (Fig. 6-27). The *shallow-marine* environment lies above the continental shelf, at depths of 200 meters (700 feet) or less. Although a continental shelf may extend sev-

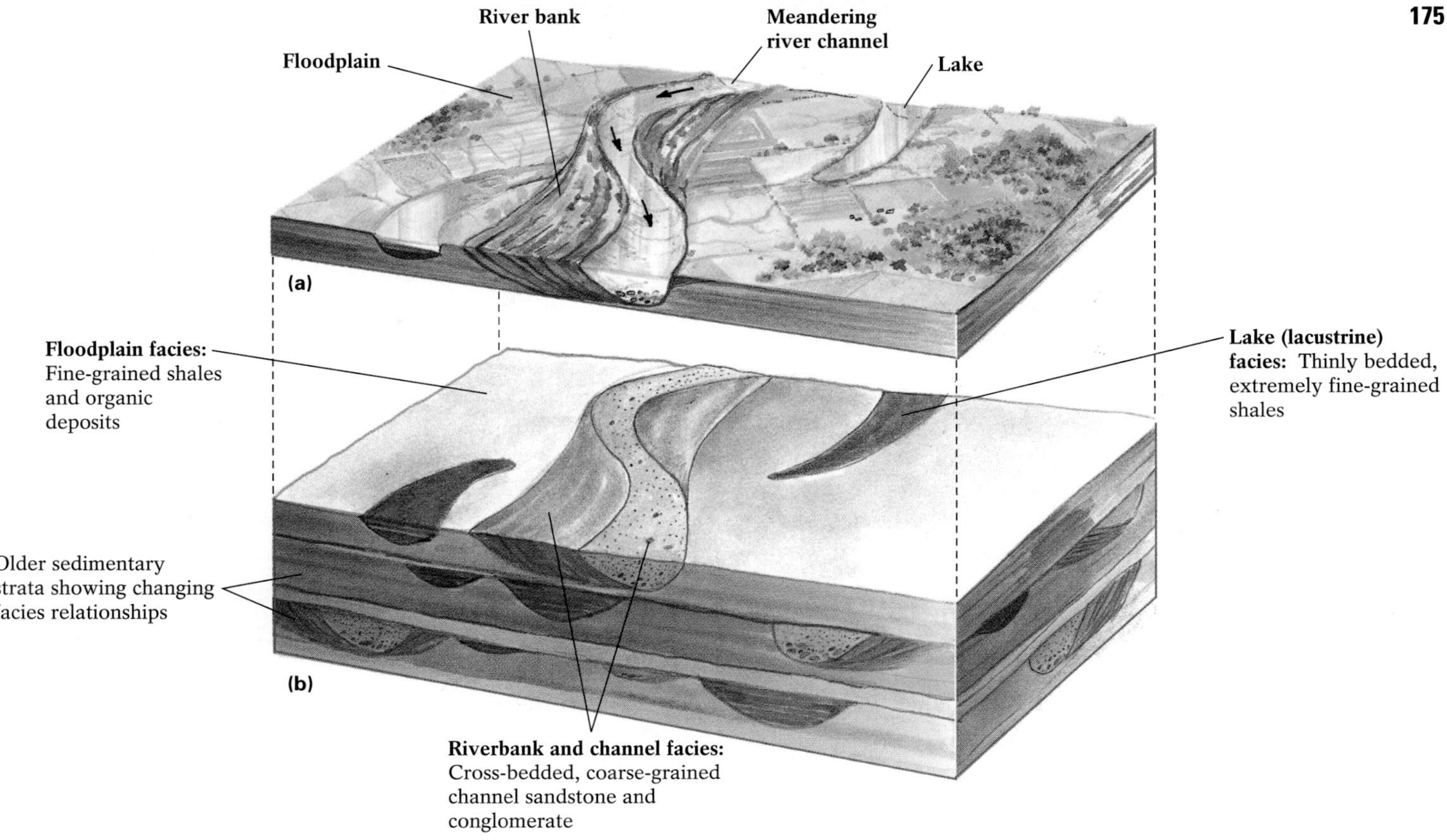

Figure 6-28 An example of sedimentary facies formation. The hypothetical modern river shown in **(a)** is surrounded by a lake-studded floodplain. If we could freeze this scene and convert the sediments to their future horizontal sedimentary rock layer **(b)**, different rocks would represent the river channel, the river bank, the floodplain, and the lakes. Although these rocks are closely spaced geographically and formed at the same time, they differ in composition and appearance because they were deposited in four different sedimentary settings. The changing relationships of facies over time, such as when a meandering river alters its course, can be seen by examining the *vertical* succession of rock layers over a broad area.

eral hundred kilometers into the sea, most are much narrower, with some being less than 1 kilometer (0.6 miles) wide. This narrow zone bordering all of the world's continents receives land-derived sediment carried seaward by waves and tides. Because sunlight penetrates approximately 50 meters (165 feet) below the water's surface, the upper part of the shallow-marine environment abounds in plant and animal life. Thus, its sediments often consist of carbonate-rich sands and mud rife with the remains of diverse marine life forms.

The *deep-marine* environment lies beyond the continental shelf, beginning at the foot of the steep continental slope. In this region sediments consist largely of the remains of calcium carbonate- and silica-secreting microorganisms that have died and fallen from the upper 50 meters (165 feet) of the ocean, red and brown clays derived from weathering of tephra from continental and submarine volcanoes, land-derived deposits carried to the deep-sea floor by submarine landslides, and meteoritic fragments from outer space. These sediments contain few large fossils, as too little sunlight penetrates this region to support many bottom-dwelling organisms (at least few having hard, fossilizable parts).

Sedimentary Facies

Just as sediments deposited in the same place but at different times are recorded as a *vertical* succession of distinct rock layers, sediments deposited at the same time but in different places appear as a *horizontal* continuum of distinct rock types. The set of characteristics (such as mineral content and particle size, shape, and sorting) that distinguish a sedimentary-rock deposit from nearby units *deposited at the same time* is called a **sedimentary facies** (*facies* comes from the Latin for "form" or "aspect"). The nature of a sedimentary facies depends on the particular conditions under which it was deposited. Thus the pattern of facies in a given rock layer reflects the different environments that existed when the sediments were deposited.

Take, for example, the different sediments associated with a modern coastal river. If they were instantly lithified, each distinct depositional setting would appear as a separate facies (Fig. 6-28a): Coarse, cross-bedded sandstones and conglomerates would record the flowing water in the river's channel; fine shales and organic deposits would mark the river's floodplain; and thinly bedded, graded shales, perhaps containing fossils, would be left behind by floodplain lakes. All of these deposits accumulated at the same time, but under different *local* conditions; thus the geological record of this modern coastal river would include channel facies, floodplain facies, and lacustrine ("lake") facies.

As we saw earlier, the vertical succession of rock layers at a specific location shows how depositional conditions at that spot changed over time. Similarly, the vertical succession of rock layers encompassing various sedimentary environments would reveal how these environments changed *in relation to one another* over time. For example, if the course of a river shifted over time, its channel facies might be succeeded by its floodplain facies; the rock record would include a noticeable change in the nature of the sediment—from coarse and cross-bedded to fine and thinly bedded (Fig. 6-28b).

Shifting Sedimentary Facies Sedimentary environments at any location will change over time, as bodies of water dry up, climate warms or cools, sea levels rise and fall, and so on. These changes alter the particle size and composition of the deposited sediments. After lithification, these sediments then appear as distinctive facies in the rock record.

In Figure 6-29, for instance, note the position of the river, beach, shallow-marine, and deep-marine environments at time A. If sea level were to rise, however, the shoreline would migrate inland and a marine environment would replace the river environment. Gradually, a coastal beach would replace the river's original floodplain, a shallow-marine environment would replace the old beach, and a deep-marine environment would replace the former shallow-marine environment. By time B, these environments would in turn be replaced by even deeper marine environments. The sediments deposited at any given location at time B would differ from those deposited at the same location at time A, reflecting the new environment of deposition. Such shifting sedimentary facies can provide clues to changes in ancient environmental conditions; we discover them by comparing vertical sequences of rock from neighboring areas.

Sedimentary Rocks and Plate Tectonics

Some of the Earth's most majestic mountains—the Northern Rockies, the Appalachians, the European Alps, the Urals of Russia, and the Himalayas of China, Nepal, and Tibet—contain sedimentary rock strata that were clearly deposited under water. We can easily see the ripple marks that record rising and falling tides and coastal waves, evaporites and mudcracks from past sea-level fluctuations, and thousands of meters of limestone. These rocks, all derived from sediments that accumulated in shallow-marine environments, ended up in mountains when plate tectonic forces uplifted them in zones of past plate collisions.

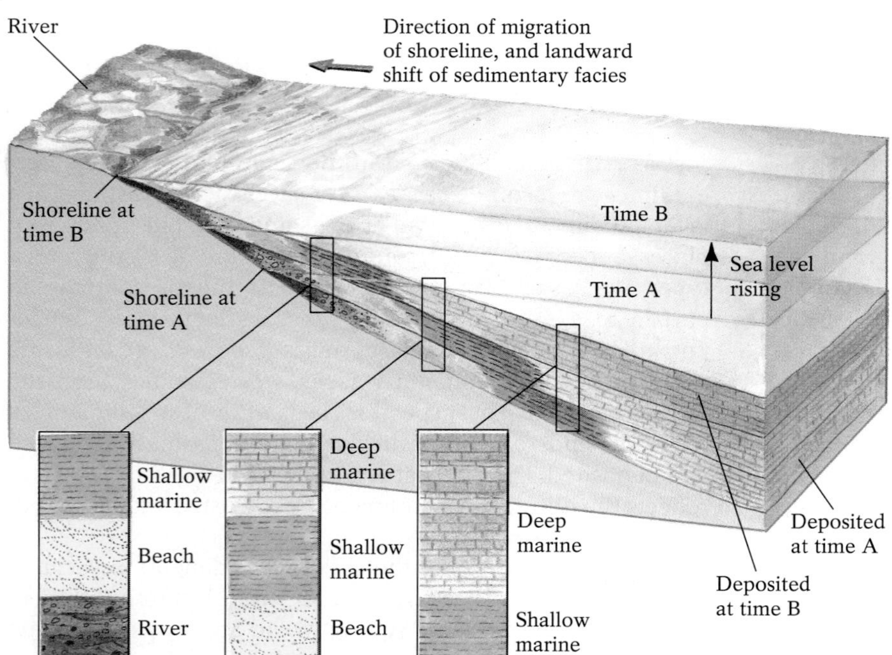

Figure 6-29 Landward migration of sedimentary environments associated with rising sea level.

Sedimentary rocks also provide evidence that lofty mountains once stood where no such range exists today. The fairly squat Taconic Hills of western Connecticut and Massachusetts and eastern New York, for example, are thought to be the remnants of mountains that rose when the plates containing what are now Europe, Africa, and North America collided nearly 375 million years ago. Rivers carried a great volume of coarse detrital sediment from those now-departed mountains into shallow seas to the west; the cross-bedded sandstones derived from these sediments were themselves subsequently uplifted and then cut through by streams, forming the Catskill Mountains of east central New York.

Specific plate tectonic settings are often associated with certain characteristic sedimentary facies. Recently rifted plate margins, such as the East African rift zone, commonly contain large alluvial fan deposits, volcanic graywackes, and extensive lake deposits and associated evaporites—all typical of volcanically active areas that have been faulted downward during rifting (see Chapter 9). Transform boundaries, such as the San Andreas fault system in California, are noteworthy for their rapid sedimentation of angular, feldspar-rich arkosic sands (formed when igneous rock was crushed at the plate boundaries). Rapid sedimentation also takes place near sites of continental collisions, such as in the shadow of the still-rising Himalayas, where a 15-kilometer (10-mile)-thick tongue of sediment extends 2500 kilometers (1500 miles) into the Indian Ocean south of Calcutta, India. Volcanoes that rise above a subducting plate supply sediment that tends to be poorly sorted and rich with angular fragments of volcanic rock, providing testimony about the rapid burial and existence of an active sediment source. By studying these sedimentary rock records, geologists can infer much about past plate margins.

Sedimentary Rocks from a Distance

Speeding down a highway (of course, at or below the legal limit), we may catch glimpses of great outcrops of layered sedimentary rocks off in the distance. Even at highway speeds and from many miles away, these glimpses can tell us much about a sedimentary rock's composition and history. For instance, in moist temperate regions such as the Eastern seaboard of North America, accelerated chemical weathering and erosion remove much of the soft, erodible shale and soluble limestone, leaving prominent ridges of durable, well-cemented sandstone, such as those in the Appalachians of central Pennsylvania. The less resistant rocks usually form the region's long gentle slopes and underlie its broad valley bottoms (Fig. 6-30a).

In the Southwest, durable sandstones form the cliffs and soft shales constitute the slopes. Thick limestone beds may form steep cliffs as well, because the arid climate permits relatively little weathering and erosion (Fig. 6-30b). From the scenic overlooks along the south-rim road of the Grand Canyon, for example, we can readily distinguish the slope-

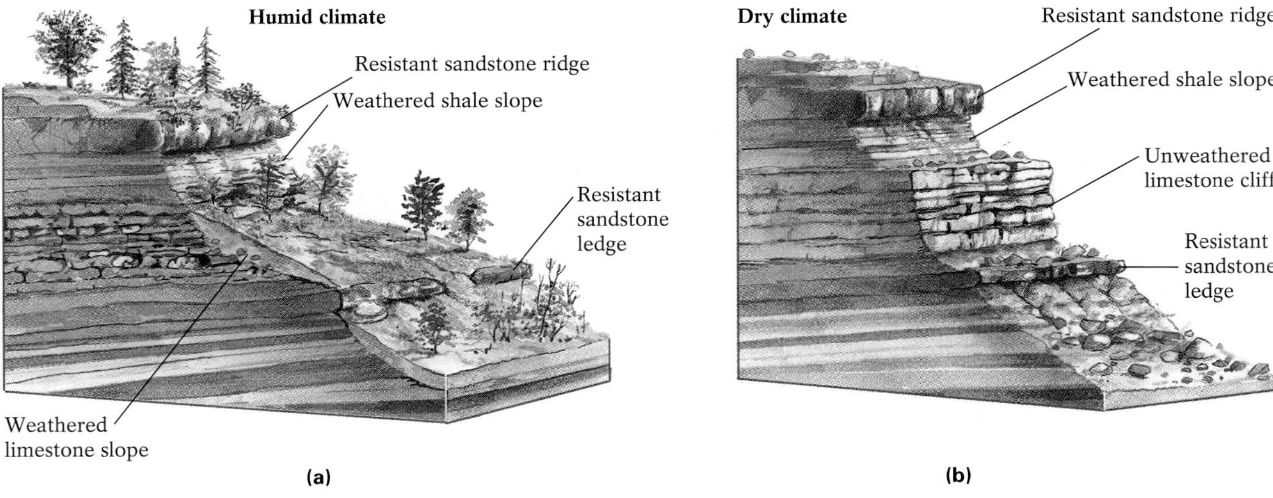

Figure 6-30 Rock types form predictable topographies, often depending on the climate to which they are exposed. In both humid and dry climates, well-cemented, weather-resistant sandstones form prominent ridges and poorly cemented, less-resistant shales form slopes and valleys. Limestones form ridges in dry climates but dissolve away to form slopes in humid climates. Thus the same sequence of rocks that would form a gently sloping, soil-rich topography in a humid climate **(a)**, such as that found in the northeastern United States, might form soil-poor ridges and cliffs in a dry climate **(b)**, such as that found in the southwestern United States.

Figure 6-31 The American Southwest affords numerous vistas of cliffs and canyons in brilliant reds, oranges, pinks, greens, and purples. Here, at Little Painted Desert Park, north of Winslow, Arizona, the sedimentary rocks owe their vivid colors to their mineral content, the materials that cement their grains, the nature of their iron compounds, and the presence or absence of organic matter.

forming shales from the cliff-forming sandstones and limestones, even from great distances.

Sedimentary rocks are typically colorful, more so than other types of rocks. Almost every sedimentary rock contains some iron, and even a 0.1% iron content can lend a deep-red hue to an oxidized sandstone. Red, pink, orange, and brown colors typically denote a continental origin for sedimentary rocks; these colors develop when the rocks are exposed to air during transport or shortly after deposition, causing their iron to oxidize. In oxygen-poor settings, such as organic-rich lagoons and swamps and in deep-marine settings, the iron minerals in sediment take on a characteristic green or purple hue. Low-oxygen, aqueous conditions promote the formation of black and dark gray sediments containing undecomposed organic matter. In many sedimentary rock formations, such as some in the Southwest (Fig. 6-31), color alone tells us much about the environments in which the respective layers formed.

The remainder of this book will revisit sedimentary processes and rocks in a variety of ways. In Chapter 8, we will see how sedimentary rocks record local, regional, and even global geological events, as well as the evolution of the Earth's animals and plants. In Chapter 9, we will discuss how the motion of the Earth's plates crumples and fractures horizontal beds of sedimentary rock into continent-long mountain ranges such as the Rockies and Appalachians. In Chapters 13 through 19, we will explore how gravity, rivers and streams, underground water, glaciers, desert winds, and crashing surf move sediment from place to place. Finally, in Chapter 20, we will discuss some sedimentary processes that produce many of the Earth's valuable natural resources—from coal, oil, and natural gas to the sand and gravel with which we build our cities and highways. First, however, Chapter 7 will examine how all kinds of rocks are heated and pressed to form the remaining major rock group of the rock cycle (discussed in Chapter 1)—metamorphic rocks.

Chapter Summary

Sediment consists of fragments of solid material derived from preexisting rock, the remains of organisms, or the direct precipitation of minerals out of solution in water. A vast amount of sediment accumulates continuously at the Earth's surface, most of which is eventually converted to **sedimentary rocks.** The thin layer of sedimentary rocks in the Earth's crust accounts for about 75% of all the rocks exposed on land.

Geologists classify sediments according to the source of their constituent materials. **Detrital sediment** is composed principally of fragments of preexisting igneous, sedimentary, or metamorphic rock. **Chemical sediment** consists of minerals that dissolved in the process of chemical weathering and that then either precipitated directly out of solution or were extracted from solution by organisms and deposited later in the form of shells, skeletons, and other organically derived materials.

Most sediments are ultimately deposited at some distance from their point of origin. The transport process influences detrital sediments in particular, via such agents as flowing surface water, circulating underground water, wind, glaciers, or coastal waves.

Detrital sediment texture—the size, shape, and arrangement of particles in a deposit—is determined largely by the transport and depositional processes involved (in addition to the weathering resistance of the parent material). During transport, sediment grains undergo **sorting,** a process by which they are carried or deposited selectively, based on the energy of their transport medium and the grain's size, density, and shape. A well-sorted deposit (typically associated with wind transport) consists of particles of one size; a poorly sorted deposit (like those seen with glacial transport) contains particles of varying sizes.

Detrital sediments often display **sedimentary structures,** physical features that reflect the conditions of deposition. **Bedding,** or **stratification,** is the arrangement of sediment particles into distinct layers marking separate depositional events. **Graded bedding** forms when sediment settles through relatively still water. The coarsest grains settle to the bottom first, and grain size decreases gradually toward the top of a layer. **Cross-bedding** refers to sediment layers that are oriented at an angle to the underlying sets of beds, as is typical of sediments deposited by wind and moving currents of water. **Ripple marks** are small surface ridges produced when water or wind flows over sediment after it is deposited. **Mudcracks** occur in the top of a sediment layer when muddy sediment dries and contracts.

After a body of sediment has been buried by subsequent deposits, the increased heat, pressure, and circulating groundwater to which it is exposed produce a number of changes, collectively known as **diagenesis.** The end result of diagenesis is often **lithification,** the conversion of loose sediment into solid sedimentary rock. **Compaction** is the diagenetic process by which the weight of overlying materials reduces the volume of a sedimentary body. **Cementation** of sediment grains occurs when dissolved ions are precipitated as a sort of "cement" in the pore spaces within the sediment. A rock that is formed by compaction and cementation of sediment particles (usually rock or mineral fragments, but sometimes organic debris such as shell fragments) has a **clastic** texture. **Recrystallization** converts certain unstable minerals in sediment into new, more stable minerals.

Detrital sedimentary rocks are classified by grain size. They include the fine-grained **mudstones** (such as **shale**), intermediate-grained **sandstones,** and coarse-grained **conglomerates** (which contain large *rounded* grains) and **breccias** (which contain large *angular* grains). Sandstones are classified further, based on their composition, into quartz-rich arenites, feldspar-rich arkoses, and rock-fragment-rich graywackes.

We base classification of chemical sedimentary rocks not on grain size, but on the composition of the sediment. These materials are further classified as being inorganic or biogenic, depending on how their mineral components were converted from their original dissolved state into solid form. The inorganic chemical sedimentary rocks precipitate directly from water (usually when the water evaporates or undergoes a significant temperature change). They include inorganic limestone, consisting primarily of calcium carbonate that precipitates from either seawater or freshwater; **evaporites,** salts (such as gypsum and halite) that accumulate when seawater evaporates; **dolostone,** a calcium-and-magnesium carbonate believed to form when magnesium ions in groundwater replace some of the calcium ions in limestone; and inorganic chert, composed largely of silica precipitated from seawater or freshwater.

Biogenic chemical sedimentary rocks form when organisms extract dissolved compounds from water, convert them into biological hard parts (such as shells and skeletons), and subsequently deposit them as sediment. They include biogenic **limestone,** composed of calcium carbonate-based remains of marine organisms; biogenic **chert,** composed of silica-based remains of marine organisms; and **coal,** composed of the carbon-rich remains of terrestrial plants.

Sediment accumulates in numerous **sedimentary environments,** which may be continental, transitional (coastal), or marine. Because the properties of any sedimentary rock stem from the specific conditions under which it develops, geologists can distinguish rocks formed in one depositional setting from rocks formed in another setting. A **sedimentary facies** is the set of unique properties that distinguish a rock in a given layer from surrounding rocks formed in different depositional settings. By noting how sedimentary facies change over distance as well as over time, we can interpret not only individual sedimentary environments of the past but also their changing relationships to one another.

Key Terms

sediment (p. 153)
sedimentary rock (p. 153)
detrital sediment (p. 154)
chemical sediment (p. 155)
sorting (p. 155)
sedimentary structures (p. 157)
bedding (stratification) (p. 157)
graded bed (p. 157)
cross-beds (p. 158)
ripple marks (p. 158)
mudcracks (p. 159)
diagenesis (p. 159)
lithification (p. 160)
compaction (p. 160)
cementation (p. 160)
clastic (p. 160)
recrystallization (p. 160)
mudstones (p. 162)
shale (p. 162)
sandstones (p. 162)
conglomerates (p. 164)
breccias (p. 164)
limestones (p. 164)
evaporites (p. 166)
dolostone (p. 166)
chert (p. 167)
coal (p. 171)
sedimentary environments (p. 173)
sedimentary facies (p. 175)

Questions for Review

1. What are the major types of sedimentary rocks? On what bases are they classified?

2. Briefly describe how sorting and rounding of detrital sediment vary with different transport media, such as wind, rivers, and glaciers.

3. Describe the differences between graded beds and cross-beds. Which indicate the flow direction of ancient currents? How can both be used to determine whether a sedimentary bed is upside down or right-side up? Give an example of a setting in which each forms.

4. Explain the processes involved in diagenesis, and describe how each affects the physical properties of sediment.

5. Describe the composition and texture of quartz arenites, arkoses, and graywackes.

6. Explain the relationship between carbon dioxide and precipitation of inorganic limestone. Briefly describe the origin of the following types of limestone: oolitic limestone, tufa, and travertine.

7. What are evaporites? Describe two sedimentary environments where evaporites form.

8. Name and describe the origin of three different types of biogenic chemical sedimentary rocks.

9. Describe how deposition occurs in each of three different *continental* sedimentary environments. How do sediments in deep-marine and shallow-marine environments differ?

10. Name some types of sedimentary facies associated with each of two different plate tectonic boundaries.

For Further Thought

1. Why are the evaporites halite and gypsum much more common than the evaporite sylvite? Why are dolostone sites often associated with evaporite deposits?

2. Under what circumstances might poorly sorted, angular, arkosic sediments appear in a coastal environment?

3. Study the photo of a modern salt flat below and speculate about the environmental conditions that existed when these sediments were first deposited. How would the appearance of the present sediments change if the climate in this area became very moist?

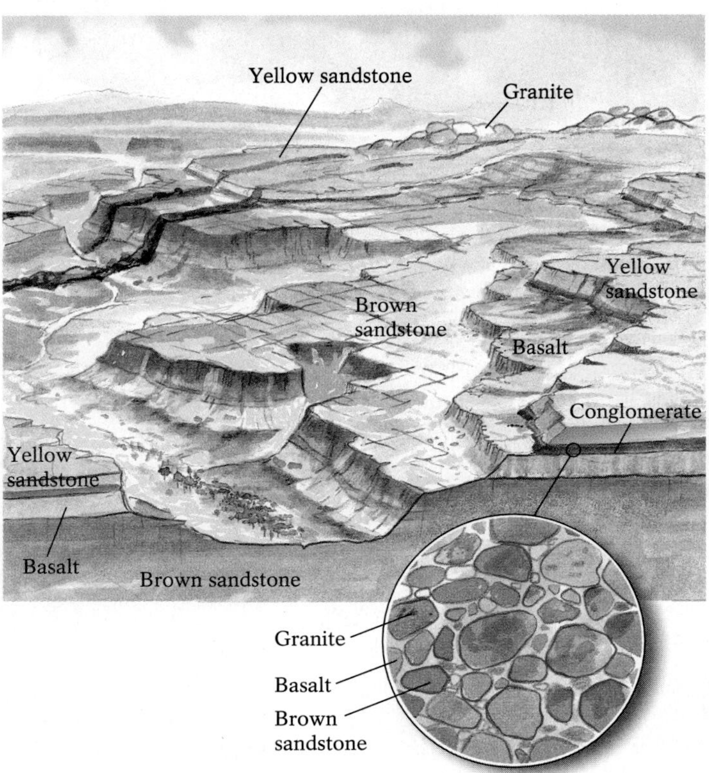

Section of conglomerate

4. In the figure at left, what can you tell about the path taken by the stream that deposited the conglomerate?

5. Why are oolitic limestones common among the rocks of southern Indiana?

6. Why do we believe that few sedimentary rocks are *currently* forming on the Moon?

7

Metamorphism and Metamorphic Rocks

Figure 7-1 These picturesque rocks at Pemaquid Point in coastal Maine have been converted to metamorphic rocks by extreme heat and pressure.

Recalling the rock cycle diagram in Chapter 1, we are reminded that all rocks—even those that have existed in more or less the same form for millions of years—are changeable, evolving in response to changes in the conditions around them (Fig. 7-1). We saw in Chapter 3 that high temperatures deep in the Earth's interior create the magmas that then cool to form igneous rocks. We saw in Chapter 6 that sediments lithify to become sedimentary rocks in the relatively low-temperature environment near the Earth's surface. The third major type of rock, **metamorphic rocks,** generally forms at conditions *between* those that produce igneous and sedimentary rocks (Fig. 7-2). **Metamorphism** (from the Greek *meta,* meaning "change," and *morphe,* meaning "form") is the process by which heat, pressure, and chemical reactions deep within the Earth alter the mineral content and/or structure of preexisting rock *without melting it.* Any rock—whether igneous, sedimentary, or metamorphic—is a candidate for metamorphism. Wherever you are on Earth, metamorphic rocks lie at some depth beneath your feet.

Most metamorphic rocks are buried beneath thousands of meters of sedimentary rock. Near Topeka, Kansas, for example, the rocks extending for thousands of meters below the surface formed from shallow-marine sediments left behind by the inland seas that invaded North America more than 100 million years ago. The ancient rocks that underlie such sedimentary rocks throughout the continent, however, are predominantly metamorphic. In some regions glacial erosion or energetic rivers carving deeply into rising ranges have stripped away the surface rocks, exposing the metamorphic rocks beneath. A geologic map of North America reveals exposed metamorphic rock throughout much of Canada, the northern portions of the midwestern Great Lakes states, and many of the continent's mountainous regions (Fig. 7-3).

Metamorphic processes presumably go on continuously beneath the Earth's surface, but, for obvious reasons, no one has ever seen metamorphism in action. We can observe the destructiveness of lava flowing from Hawai'i's K'ilauea volcano and watch it congeal into another acre of Hawai'ian

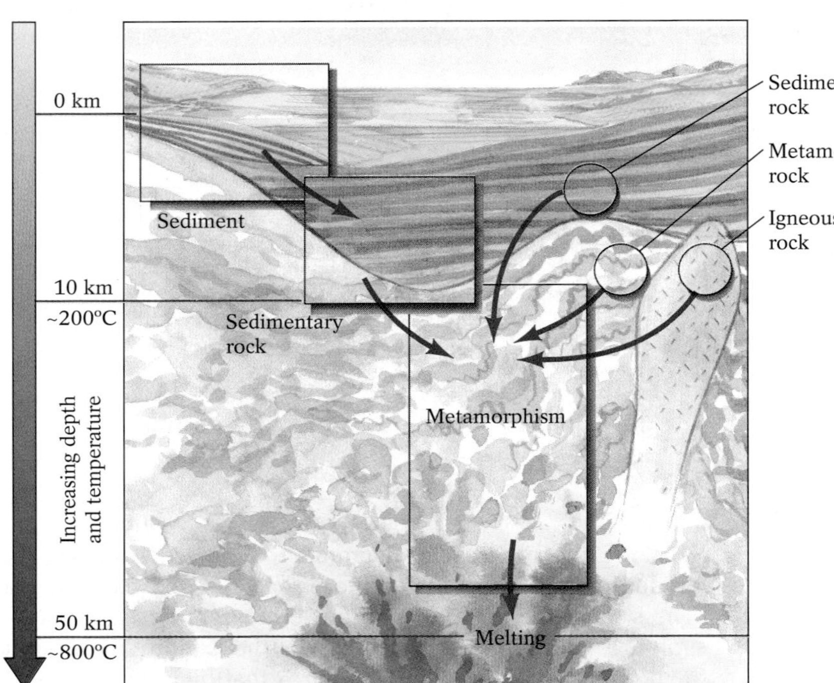

Figure 7-2 Sedimentary, metamorphic, and igneous rocks are metamorphosed at temperatures and pressures greater than those that lithify sediment into sedimentary rock, but less than those that melt rock into magma.

basalt. We can observe a river carrying a load of sand that may someday form a layer of sandstone. But we see metamorphic rocks only after uplift and erosion have stripped away the overlying rocks. Thus we must be geological "detectives," to infer, from surface clues, the deep interior processes that produced them. Much of our knowledge about the causes of metamorphism has come from laboratory experiments and theoretical models that replicate conditions in the Earth's interior.

Metamorphic rocks find many uses—from building and road construction to firefighting material to the marble that Michelangelo used to sculpt his famous statue of David. We'll touch on these uses (and more) later in this chapter. First we will look at the conditions that promote metamorphic processes, the various types of metamorphism, and the common rocks produced. We will then examine how metamorphic rocks vary in grade (the degree to which they differ from parent material) and in the minerals they contain. And we'll see that plate tectonics play a major role in the production of metamorphic rock.

Conditions Promoting Metamorphism

It is a basic geologic principle that rocks and their constituent minerals are most stable in the environment in which they form and least stable in markedly different environments. As we saw in Chapter 5, rocks are more likely to be altered by weathering processes when they become exposed to an environment that differs from the one in which they formed. For example, feldspars created by relatively high-temperature igneous processes deep in the Earth rapidly break down chemically at the planet's surface, an environment characterized by relatively low temperatures, low pressure, and abundant water and atmospheric gases. They are replaced by clay minerals, which are more stable than the feldspars under surface conditions. If the clays then became buried beneath tens of kilometers of overlying sediments and heated to 600°C (1110°F), their new environment would be characterized by higher temperature and pressure. Under these conditions, the clays would react to form new minerals that would be stable under the new conditions. The resulting materials might even include some feldspars.

Rocks undergoing metamorphism remain in a solid state. Their unstable minerals either recrystallize into new, more stable forms or react with other unstable minerals to produce new minerals with stable atomic arrangements. Such rearrangement is possible because heat and pressure break the bonds between *some* atoms or ions in an unstable mineral. Once freed, these atoms and ions may migrate to other sites within that mineral, or to another mineral, and rebond. In this way, minute grains in a fine-grained rock usually give way to an interlocking mosaic of large, visible grains.

Metamorphic processes never break all bonds in a rock's minerals—if all its bonds were broken, a rock would melt (an *igneous* process) and become a magma. Instead, metamorphism occurs when heat and pressure destabilize the minerals in rocks, but do not become high enough to cause melting.

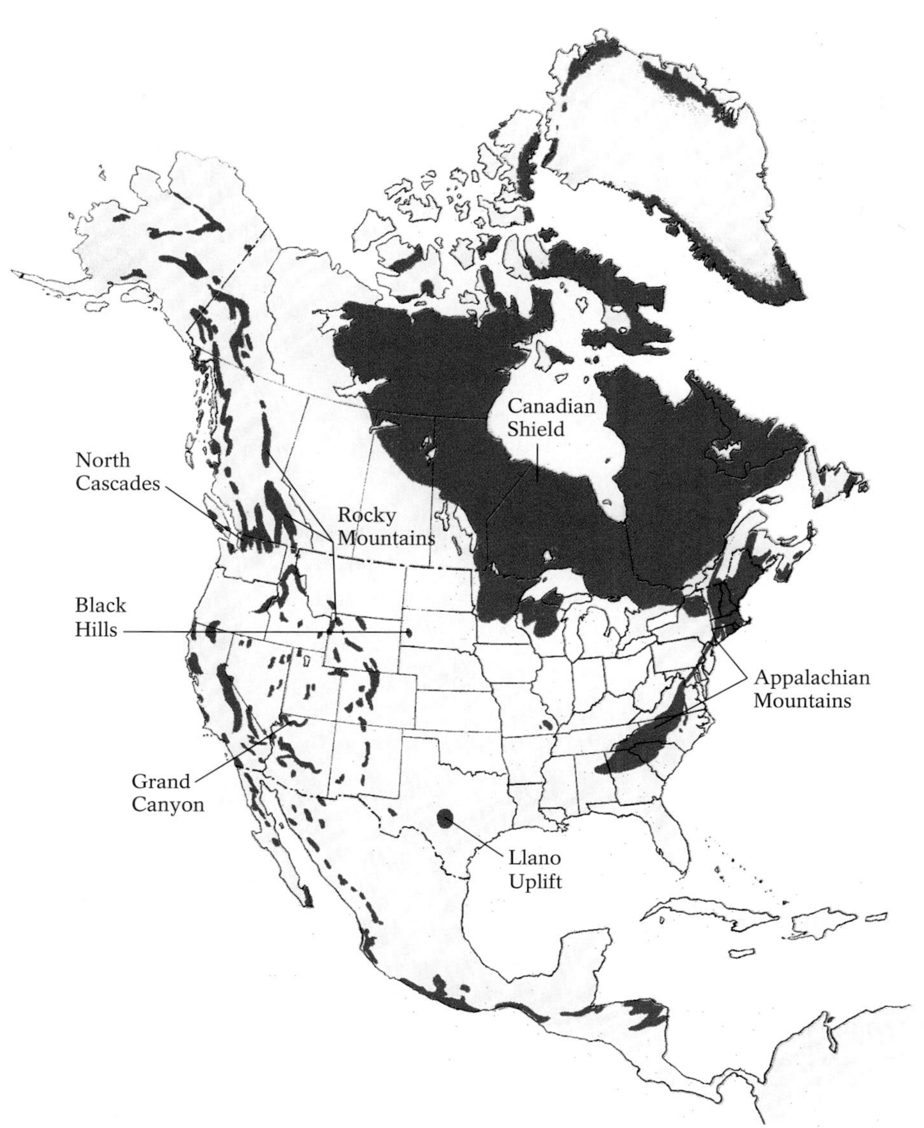

Figure 7-3 Exposed and near-surface metamorphic rocks of North America. Generally buried beneath kilometers of sedimentary rocks, the continent's metamorphic rocks can be seen only where erosion has removed the overlying rock—principally in very deep river valleys, such as the Grand Canyon, in the cores of mountain ranges, such as the Colorado Rockies, the North Cascades of Washington state, the Appalachians of eastern North America, and in the glacially scoured "Canadian Shield" region of Ontario, Quebec, and the adjacent Great Lakes states of Minnesota, Wisconsin, and Michigan.

The presence of circulating ion-rich fluids also promotes metamorphism. The composition of the parent rock largely determines which metamorphic rocks and minerals form.

Heat

Mix the appropriate proportions of flour, yeast, water, and a pinch of salt, and you can start to make bread. Unless you turn on the oven and heat these ingredients to a specific temperature, however, the bread would not rise. Similarly, little happens to the ingredients for making metamorphic minerals and rocks until they encounter a source of heat.

Heat, which accelerates the pace of nearly all chemical reactions, is perhaps the most important factor contributing to metamorphism. Beneath the Earth's surface, temperature generally increases with depth (see the discussion of the geothermal gradient in Chapter 3); in the Earth's crust and upper mantle, temperature increases at an average rate of 20° to 30°C per kilometer of depth (approximately 72°F per mile). Temperatures sufficient to metamorphose rocks—greater than about 200°C (400°F)—are generally reached at about 10 kilometers (6 miles).

As we saw in Chapters 1 and 2, some of the heat in the Earth's crust is *conducted* upward through solid rock from the deep interior. Some heat is also *convected* from deeper within the Earth by rising magma, or is introduced to the rocky crust and upper mantle as radioactive isotopes decay and emit heat energy. Another minor source of heat is the friction created between two bodies of rock as they grind past one another along a fault or at a plate boundary. Together all of these sources produce enough heat to promote metamorphism. Heat does not act alone, however.

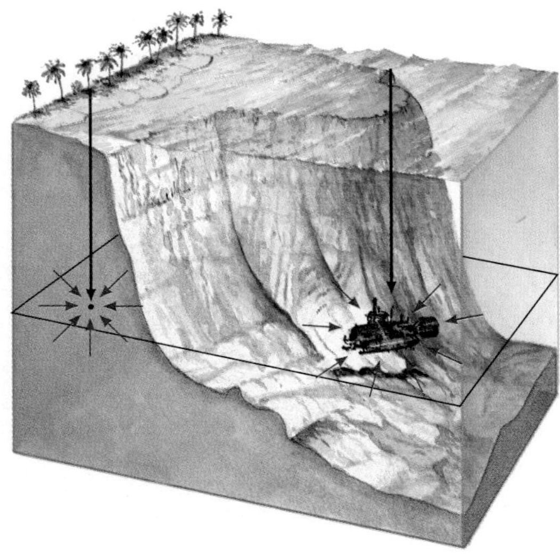

Figure 7-4 Deep burial of rocks generally subjects them to lithostatic, or confining, pressure, an inward-pressing force that acts equally from all directions. (Any object immersed in water—such as this submarine—incurs similar pressure.)

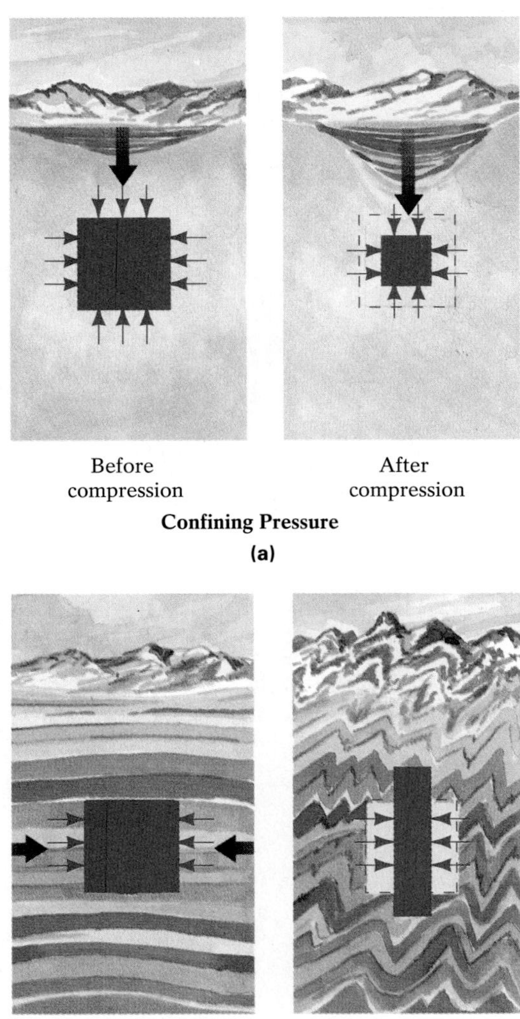

Figure 7-5 (a) A rock subjected to confining pressure becomes compressed without changing shape. (b) A rock subjected to directed pressure has its shape distorted, becoming thinner in the direction of the greatest pushing, or *stress,* and elongated in the direction perpendicular to this stress.

Pressure

Pressure on rocks increases with depth as the thickness of the overlying rock increases (Fig. 7-4). The type of pressure associated with deep burial, called **lithostatic** (*litho,* meaning "rock," and *static* from the Greek *statikos,* or "causing to stand in place") or **confining pressure,** pushes on rocks equally from all sides. As a result of lithostatic pressure, a deeply buried rock becomes compressed into a smaller, denser form, but retains the same shape (Fig. 7-5a).

In some geological settings, pressure does not push equally from all directions, but instead acts in one principal plane. Such **directed pressure,** which commonly occurs where tectonic plates collide to form mountains and along various types of faults (discussed in Chapter 9), distorts the shape of rock (Fig. 7-5b). Directed pressure typically flattens a rock in the plane on which the highest pressure is applied, and lengthens it in the plane *perpendicular* to the principal direction of pressure. It may also deform individual components within rocks, such as fossils or layers of minerals, by stretching and folding them (Fig. 7-6).

Rocks subjected to either confining or directed pressure change in a number of ways. Both types of pressure close pore spaces between mineral grains, producing a more compact, denser rock. Pressure at the contact points between the compressed grains may break some bonds there. Unbonded ions then migrate to other sites with lower pressure where they rebond. This process is especially active when the grains are separated by water, which fosters ion migration. The new, more compact mineral structure that results is more stable under high pressure (Fig. 7-7).

Directed pressure accounts for one particular change in rocks. Because low-pressure sites in rocks under pressure usually occur *perpendicular* to the direction of the pressure, the recrystallized minerals that bond in these sites typically develop a preferred alignment *perpendicular to the directed pressure.* This preferred alignment is known as **foliation** (from the Latin *foliatus,* or "leaf-like"). Foliation also develops when directed pressure mechanically rotates mineral grains into a preferred orientation, again typically perpendicular to the directed pressure (Fig. 7-8). When the mineral grains in a

Figure 7-6 Folded rock in the Blue Ridge Mountains of North Carolina, showing the deformation typical of rocks subjected to directed pressure.

rock are platy (thin and planar) or needle-like, their systematic orientation may give the foliated rock a distinctive layered look.

Geologists measure pressure with various units, including bars and pascals. A *bar* is equal to the pressure applied to the Earth's surface at sea level by the atmosphere (1 bar = 1.02 kilograms per square centimeter or 14.7 pounds per square inch; 1 bar also equals 10^5 pascals). Metamorphism requires more than 1 kilobar (1000 bars, or 10^8 pascals) of pressure, which is approximately equal to the pressure prevailing about 3 kilometers (2 miles) beneath the Earth's surface. As the *temperatures* needed for metamorphism do not normally occur above 10 kilometers (6 miles), metamorphism rarely takes place in such shallow depths unless heat is carried up from greater depths by rising magma or induced by friction along faults or at plate boundaries.

Circulating Fluids

Put two dry chemicals at room temperature into a test tube, and typically nothing happens. Heat the test tube with a Bunsen burner, and still no activity results. But add one more ingredient—a fluid, such as water—and a significant reaction will probably occur. Because dissolved ions move about easily in fluids, the presence of a liquid or gas in or around hot rock under pressure facilitates the migration of unbonded atoms and ions. The presence of a fluid also dramatically increases the potential for metamorphic reactions. For example, water surrounding a mineral grain promotes the exchange of atoms and ions with adjacent grains. Without its presence, unbonded ions migrate very slowly along grain boundaries or through minute pathways within a mineral's atomic structure.

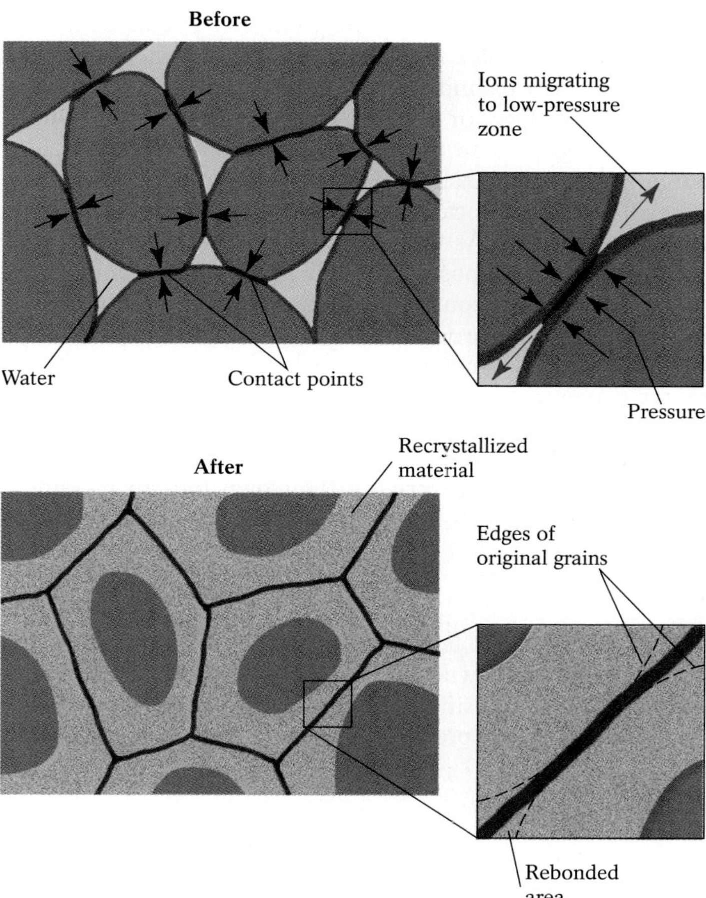

Figure 7-7 Mineral grains may dissolve under the high pressure that exists at the contact points between compressed grains. The pressure strains and then breaks some bonds between ions. The unbonded ions migrate away from the high-pressure contact points to low-pressure areas in the spaces between grains, where they rebond to form more compact, stable structures.

Figure 7-8 Directed pressure creates foliation in rock.

Water, the most important fluid that circulates underground and promotes metamorphism, may come from various sources. For example, it may have percolated down from the Earth's surface, or become trapped in the spaces within sediment layers or in the cracks of a subducting plate. The cooling and crystallization of magmas may release water and other fluids (such as carbon dioxide or sulfur dioxide gas) to surrounding rocks. As water-rich minerals, such as clays and amphiboles, decompose, they too may release water. Most underground fluids contain a variety of ions and acquire more ions as they move through fissures in heated rocks (see Chapter 3). These hot fluids may contribute ions to rocks during metamorphism, changing the chemistry of the rocks and producing a host of *new* minerals. Thus fluids not only serve as the medium through which a rock's own unbonded ions migrate, but may also contribute their own ions to metamorphic reactions.

Parent Rock

The most important factor determining which metamorphic rocks and minerals will form under new environmental conditions is the composition of the parent rock. No matter what the temperature or pressure, parent rock devoid of calcium generally cannot produce a calcium-rich metamorphic rock. In a parent rock that contains only one mineral, metamorphism tends to produce a rock composed predominantly of that same mineral; the key change is its recrystallization into a more stable configuration. For example, the calcite grains in a pure limestone will recrystallize into calcite-rich metamorphic limestone, or *marble.* In a parent rock that contains several minerals, the components typically combine to form a number of new minerals. For example, metamorphism of various clay minerals, quartz grains, mica flakes, and volcanic fragments in graywacke sandstones may release a variety of ions and produce a group of new minerals that were *not* part of the parent rock.

Types of Metamorphism

Heat, pressure, and chemically active fluids interact differently in different geological settings to produce metamorphic rocks. These settings yield distinctive types of metamorphism. Under some conditions, only a narrow area of rock is affected; in other settings, rocks may be changed over a vast region.

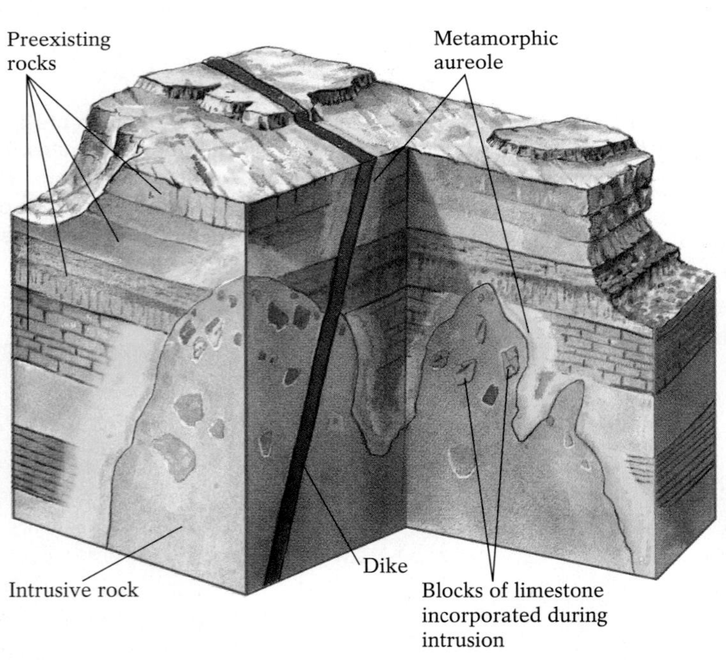

(a)

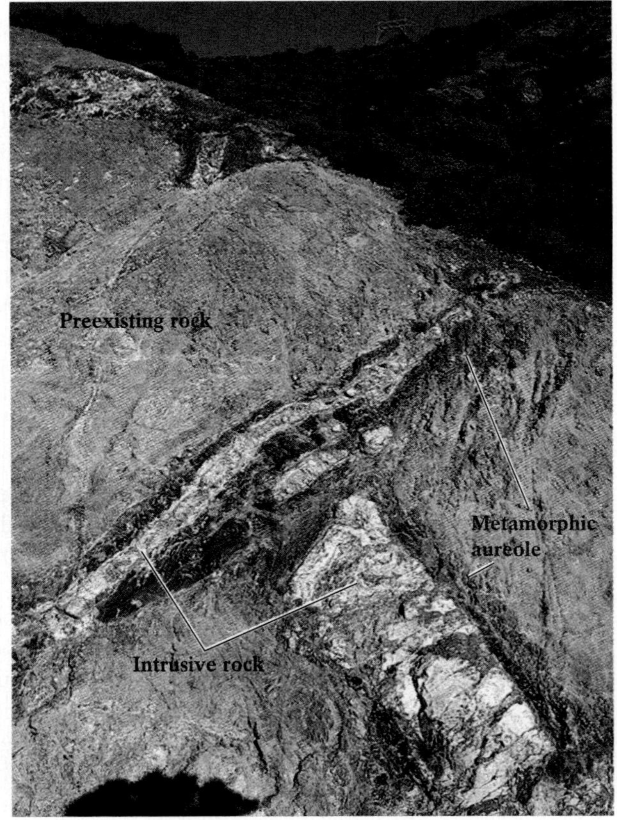

(b)

Figure 7-9 (a) Contact metamorphism occurs when magma intrudes preexisting rocks. The rocks adjacent to the magma become strongly metamorphosed; those farther from the region of direct contact are less affected, although they still receive metamorphic heat from circulating ion-rich fluids. **(b)** An igneous dike in preexisting rock, with an obvious metamorphic aureole.

Contact Metamorphism

Preexisting rock touched by the intense heat of migrating magma may undergo **contact metamorphism.** Metamorphic change within the rock develops entirely from the magma's heat and from reaction with hot circulating fluids; pressure is not a significant factor. Because most rocks conduct heat only poorly, the effects of contact metamorphism *decrease with increasing distance from the magma.* Rocks in direct contact with the intruding magma will be highly metamorphosed; those farther away, which receive little magmatic heat, will be altered only slightly (Fig. 7-9a). The amount of water in the mineral structure of contact-metamorphic rock sometimes indicates how close the rock was to a magma: The closer it was to the heat source, the more water that was driven off.

The effects of contact metamorphism typically vary with the size of the intruded pluton. Contact metamorphism may extend only a few centimeters or meters around a small igneous dike or sill, or it may affect an area of several kilometers around a major batholith. The entire zone of contact metamorphism surrounding an igneous intrusion, marked by changes in mineral content and grain texture, is referred to as a metamorphic **aureole** (Fig. 7-9b).

Regional Metamorphism

Contact metamorphism has relatively local effects, whereas **regional metamorphism** alters rocks for thousands of square kilometers. The latter process produced the vast regions of exposed metamorphic rock in central Canada and the tracts of metamorphic rock found in the Appalachians of New England, the Rockies, and the North Cascades of northwestern Washington and southwestern British Columbia. Two types of regional metamorphism exist: burial metamorphism and dynamothermal metamorphism.

Burial Metamorphism **Burial metamorphism** occurs when rocks are overlain by more than 10 kilometers (6 miles) of rock or sediment. At such depths, the combination of confining pressure and geothermal heat recrystallizes a rock's component minerals. Because the process does not involve directed pressure, burial metamorphic rocks are generally nonfoliated.

Today, burial metamorphism is occurring in such locations as the Gulf Coast of Louisiana, where clay minerals lie beneath 12 kilometers (8 miles) of sediment near the bottom of the Mississippi River's deltaic deposits. Samples collected from deep drill holes in the delta indicate that these clay minerals have already begun to metamorphose into minerals that are more stable at that depth, such as mica and chlorite.

Dynamothermal Metamorphism **Dynamothermal metamorphism** occurs when converging plates trap rocks between them during mountain building (Fig. 7-10). In such a setting, lateral compression—a form of directed pressure—forces some rocks upward, forming mountains, and some rocks downward to depths of several to tens of kilometers. The latter rocks are then subjected to great heat and confining pressure. "Dynamothermal" refers to both the *dynamic* pressures on the rocks and the *thermal* energy applied by geothermal heat and rising, mantle-derived magma. Because of the directed pressure on rocks at convergent boundaries, dynamothermal metamorphic rocks typically have a foliated appearance.

Dynamothermal metamorphism created the vast regions of metamorphic rock that lie

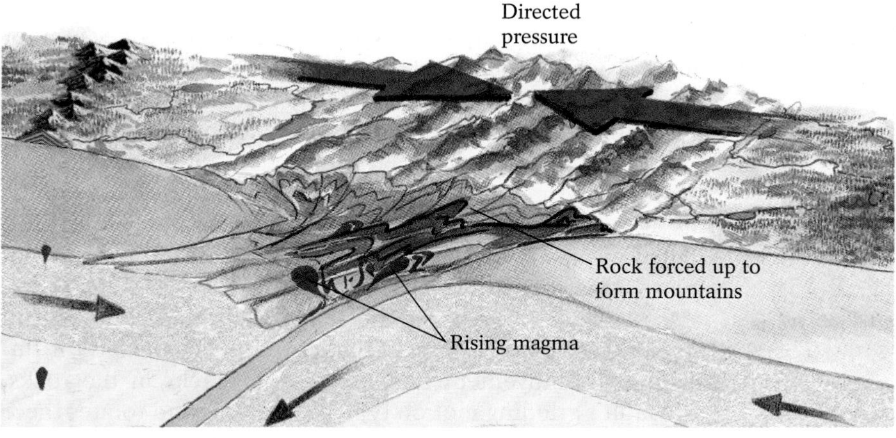

Figure 7-10 Dynamothermal metamorphism is generally associated with the directed pressures and magmatic heat of convergent plate boundaries.

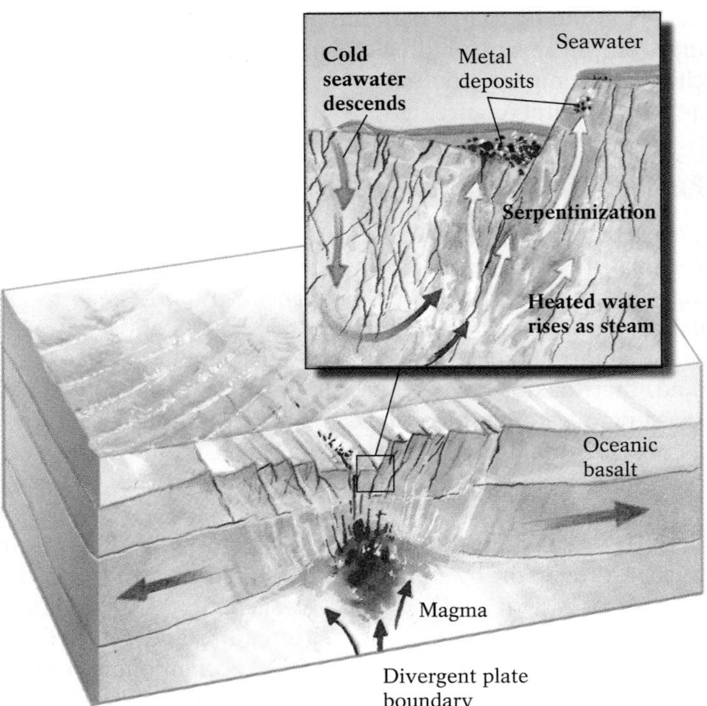

Figure 7-11 Hydrothermal metamorphism occurs most often at mid-ocean ridges, where tectonic plates diverge. Cold seawater comes into contact with hot basaltic magma, heats up, and rises as hot water and steam. Along the way, it dissolves metals and other ions from mantle peridotites, basalts, and oceanic sediments. This process converts olivine and pyroxene minerals to the magnesium silicate serpentine and several other iron-magnesium–rich silicates.

Figure 7-12 Augen gneiss, a common product of fault metamorphism. The augen ("eyes") are the areas of lighter colored rock, which have been flattened and rotated by high pressure.

at the cores of many mountain ranges, including the Alps, Himalayas, and Appalachians. It also formed the roots of more ancient, now-vanished mountains, including those exposed by billions of years of erosion in the northern Great Lakes states of Minnesota, Wisconsin, and Michigan and adjacent parts of southern Ontario and Quebec.

Other Types of Metamorphism

Regional increases in heat and pressure, and contact with intruding hot magmas are not only the only ways in which rock may be metamorphosed. Metamorphism can also occur at sites where rocks are invaded by hot water, at fault zones, and at meteorite impact sites.

Hydrothermal Metamorphism **Hydrothermal metamorphism** is the chemical alteration of preexisting rocks by hot water. This water may come directly from magma, be driven off from rocks undergoing metamorphism, or consist of groundwater that has percolated deep down from the surface and been heated by contact with hot rock. Most hydrothermal metamorphism, however, occurs beneath ocean floors, where seawater penetrates cracks near a divergent plate boundary. Descending until it encounters hot basaltic magma or warm young oceanic rocks, the seawater heats up to about 300°C (572°F). Rising as steam, it passes through the overlying ultramafic mantle rocks. As it moves through the mantle, it dissolves soluble materials and transforms mafic minerals such as olivine and pyroxene into other, more stable silicates—serpentine, chlorite, epidote, and talc (Fig. 7-11). This process, known as *serpentinization,* is responsible for much of the metamorphism in ocean basins. When the hot, mineral-rich steam ultimately reaches the sea floor, it cools and precipitates its dissolved load, often in the form of valuable concentrations of copper, nickel, iron, and lead.

Fault-Zone Metamorphism Rocks grinding past one another along a fault generate a large amount of directed pressure and considerable frictional heat. The pressure and heat in the immediate vicinity of an active fault can be intense enough to produce **fault-zone metamorphism,** although the effects of foliation and metamorphic recrystallization taper off a short distance from the fault. The grinding of rocks in zones of fault metamorphism typically reduces grain sizes, producing a fine-grained material called *mylonite*. In certain cases, however, recrystallization along the fault may also produce unusually large grains (some as large as grapefruits), particularly if fluids aid the movement of ions through cracks in the rocks. The fault's grinding motion typically flattens and rotates these large grains into almond-shaped structures known as *augen* (German for "eyes") (Fig. 7-12).

Shock Metamorphism A dramatic form of metamorphism, called **shock metamorphism,** occurs where a meteorite strikes rocks at the Earth's surface. The tremendous pressures and temperatures generated at meteorite impact sites "shock" minerals in the stricken rocks, causing them to shatter and recrystallize. The minerals produced commonly include stishovite and coesite, two types of SiO_2 that are found within the vicinities of impact craters. Some minerals created at meteorite impact sites, such as stishovite, occur in no other geological setting.

The search for impact-site metamorphic rocks, both on land and undersea, has produced extensive information on shock metamorphism. Such study has led researchers to hypothesize that the Earth's dinosaurs were driven to extinction by an enormous meteorite impact, perhaps on the Gulf Coast of Mexico, about 65 million years ago. (See Highlight 1-1, "What Caused the Extinction of the Dinosaurs?") There is also speculation that meteorites "ferried" the Earth's first organic molecules to our planet, perhaps sparking the creation of life on Earth. (See Highlight 7-1.)

Pyrometamorphism Pyrometamorphism is caused by ultrahigh temperatures that occur occasionally in relatively low-pressure environments. We can see the principal examples of this type of metamorphism where bolts of lightning have recrystallized surface rocks and sediments and where burning subterranean coal seams recrystallized adjacent rocks.

Common Metamorphic Rocks

The first criterion used to distinguish any metamorphic rock is whether it is foliated or nonfoliated. When the parent rock is a simple single-mineral sedimentary rock, such as quartz sandstone, the resulting metamorphic rock is usually nonfoliated or so slightly foliated that the foliation remains invisible to the naked eye. More dramatic, readily visible foliation tends to develop when a multimineral, mica-rich rock is subjected to progressively greater heat and directed pressure. For

Highlight 7-1 Shock Metamorphism, Buckyballs, . . . and the Origin of Life on Earth?

For many years, a host of geoscientists and paleobiologists clung to the belief that life on Earth originated during the planet's infancy, when a random bolt of lightning struck the primordial stew of inorganic molecules at the Earth's surface. This energy was thought to have sparked a chemical reaction that converted simple inorganic molecules into the first amino acids—the building blocks of complex organic molecules. In more recent years, other scientists have proposed an alternative hypothesis: The organic molecules from which all living things evolved were deposited here unceremoniously by huge meteorites that collided with our planet during its earliest days.

How could those molecules survive the searing heat and unimaginable shock of a high-speed meteorite impact? Recent discoveries suggest that such molecules need not have perished upon impact. Perhaps they were encased in some type of protective structure. A group of researchers from the University of California, San Diego, walking the hills of an impact site at Sudbury, Ontario, collected samples of shocked rocks that contained countless *fullerenes*—or buckyballs—minute carbon-based structures named for the great twentieth-century thinker Buckminster Fuller. Some scientists believe that these microscopic yet sturdy molecular cages, composed of 60 carbon atoms arranged in a manner similar to Fuller's geodesic domes (or a soccer ball; Figure 7-13), reached the Earth's surface via the meteorite that struck what is now Sudbury 2 billion years ago. This conclusion stems from the unique chemical makeup of helium gas found trapped within the buckyballs.

The helium in Sudbury's buckyballs is believed to be of extraterrestrial origin, quite possibly manufactured, along with the buckyballs, in the core of a dying "Red Giant" star. According to this fascinating hypothesis, gases from the bloated, carbon-rich star—including helium—were sprayed along with the buckyballs into interstellar space, joining the gas and dust from which our solar system and its meteorites ultimately formed. If we accept that one such meteorite ferried stellar gases and fullerenes to Sudbury, we may also speculate that other kinds of complex organic molecules—such as those that formed the basic building blocks of life—may have taken the same route to our planet. Undoubtedly, their arrival here would have been smashing, as the metamorphically shocked rocks at various impact sites attest.

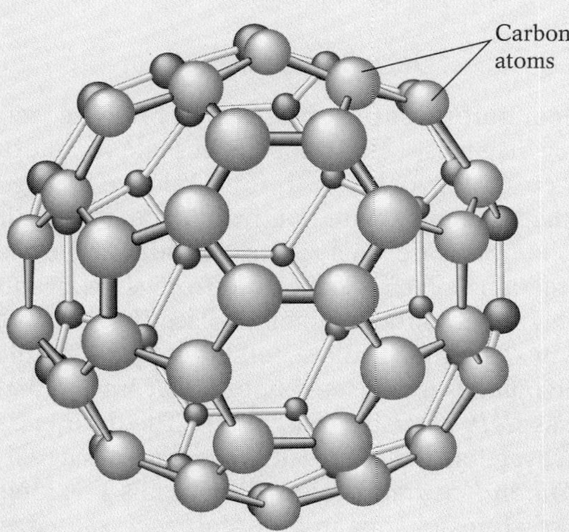

Figure 7-13 This buckyball molecule is composed of 60 covalently bonded atoms of carbon.

Table 7-1 Classification and Derivation of Some Common Metamorphic Rocks

	Rock	Parent Rock(s)	Key Minerals	Metamorphic Conditions
Foliated	Slate	Shale, mudstone	Clay minerals, micas, chlorite	Relatively low temperature and directed pressure
	Phyllite	Shale, mudstone	Mica, chlorite	Low–intermediate temperature and directed pressure
	Schist	Shale, mudstone, basalt, graywacke sandstone, impure limestone	Mica, chlorite, epidote, garnet, talc, hornblende, graphite	Intermediate–high temperature and directed pressure
	Gneiss	Shale, felsic igneous rocks, graywacke sandstone	Quartz, feldspars, garnet, mica, augite, hornblende, staurolite, kyanite	High temperature and directed pressure
Nonfoliated	Marble	Pure limestone or dolostone	Calcite, dolomite	Contact with hot magma, or confining pressure from deep burial
	Quartzite	Pure sandstone	Quartz	Contact with hot magma, or confining pressure from deep burial
	Hornfels	Shale, mudstone, basalt	Andalusite, mica, quartz	Contact with hot magma; little pressure

example, the sequence of metamorphic rocks that forms from a typical marine shale exhibits vivid foliation.

Table 7-1 summarizes the classification of metamorphic rocks. Note that some metamorphic rocks can be produced from a number of different parent rocks; such metamorphic rocks are distinguished *by their appearance and by the conditions under which they formed, rather than by their composition.* Other metamorphic rocks are produced by a specific type of parent rock and thus are classified according to their composition.

Foliated Rocks Derived from Shale or Mudstone

As temperatures and pressures rise beyond the range of sedimentary diagenesis, clay minerals in shale or mudstone gradually recrystallize or react to form relatively long, flat mica flakes that become aligned perpendicular to the direction of the applied pressure. This reaction produces the first in a series of metamorphic rocks derived from shale or mudstone—a strongly foliated rock known as **slate** (from the Old French *esclat,* or "splinter"). Slate tends to break parallel to the mica-rich planes into relatively thin, flat fragments, a pattern known as *slaty cleavage* (Fig. 7-14). (The difference between rock cleavage and mineral cleavage was discussed in Chapter 2. *Mineral cleavage* refers to minerals breaking between planes of atoms or ions in a crystal; *rock cleavage* refers to rocks breaking between planes of minerals.)

Slate's color depends on the chemical composition of its parent shale or mudstone. Red slates are rich in hematite, an iron oxide; green slates contain a significant amount of chlorite; purple slates are stained by manganese oxides; and black slates are rich in carbon-rich organic material. Whatever their hue, the fine-grained slates exhibit slaty cleavage, producing smooth slabs of rock that are commonly used as blackboards, roofing materials, floor tiles, and pool-table tops.

As temperatures increase to about 300°C (575°F), the sizes of the microscopic mica, chlorite, and graphite flakes in slate increase, producing the foliated metamorphic rock **phyllite** (from the Greek *phyllon,* or "leaf-like," so-named for its pronounced thin, wavy foliation). Light reflected off the surfaces of these mica, chlorite, and graphite grains gives phyllite its characteristic sheen. The fine flakes in phyllites continue to grow as heat and pressure increase. When the flakes become visible to the unaided eye, the metamorphic rock is classified as a **schist** (from Greek *schistos,* or "split"), a medium-to-coarse-grained, strongly foliated rock (Fig. 7-15). By this point, no sedimentary structures in the original rock remain.

As temperatures approach 500°C to 700°C (950°–1300°F), the minerals in schists may separate into light- and dark-colored layers. During metamorphic reactions, the liberated ions in unstable felsic minerals migrate and recrystallize in distinct light-colored, single-mineral layers containing larger grains of new, stable minerals such as quartz or feldspar. Meanwhile, ions freed from other minerals, such as biotite and the dark-colored amphiboles and pyroxenes, form intervening layers of dark mafic minerals. This process of mineral separation, known as **metamorphic differentiation,** creates the distinctively layered metamorphic rock **gneiss** (from the German *gneisto,* meaning "sparkle," and pronounced "NICE").

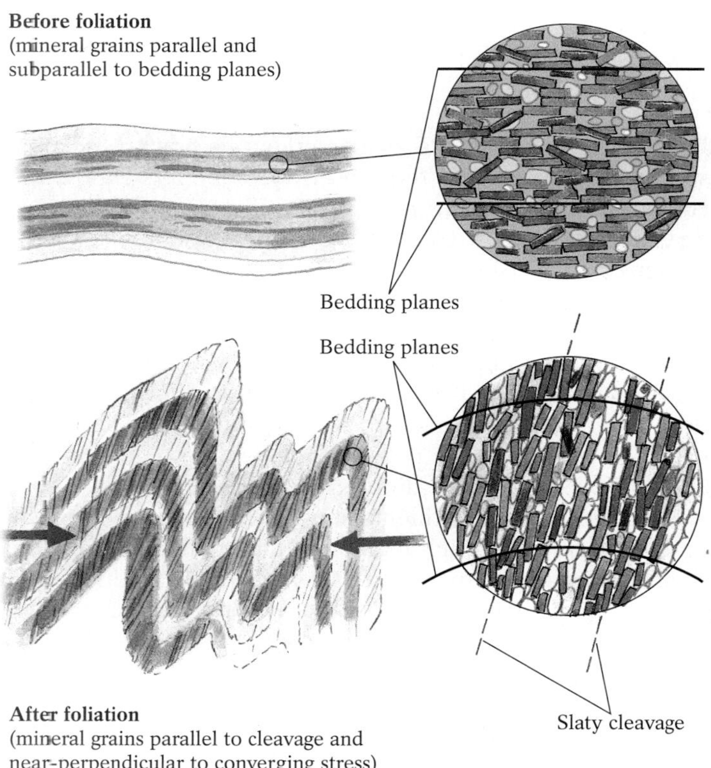

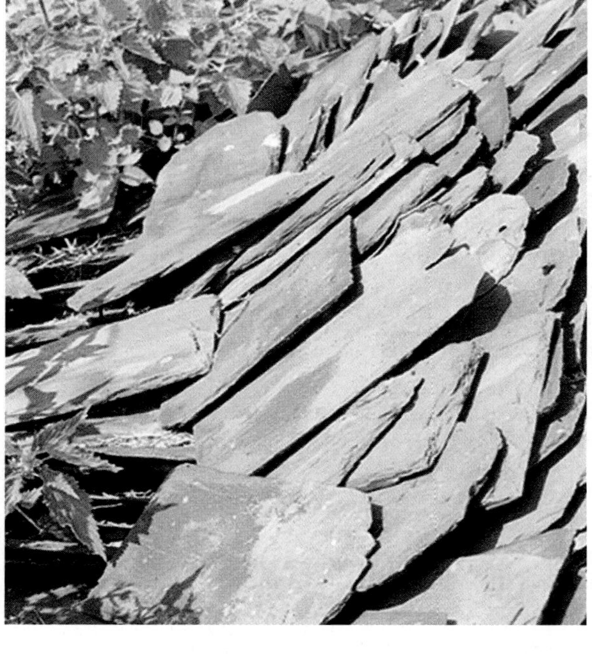

Figure 7-14 The parallel cleavage characteristic of slate (slaty cleavage) is due to foliation. The orientation of the clay and mica flakes within slate, which is perpendicular to the directed pressure applied during metamorphism, determines the direction of the cleavage.

Gneisses are typically found in the cores of ancient mountain ranges that were uplifted during mountain building and then lost their overlying rocks through erosion. For example, gneisses are exposed in the Front Range of the Colorado Rockies and in the North Cascades of Washington. These rocks also occur in currently nonmountainous areas where lofty mountains once stood. The 3.7-billion-year-old Morton Gneiss (Fig. 7-16), exposed in the Minnesota River valley of southwestern Minnesota, and the 600- to 700-million-year-old Fordham Gneiss, exposed in New York's Central Park, are remnants of such long-departed mountain ranges.

Figure 7-15 Mica-rich schist. Large chlorite and mica grains give a glittery appearance to schistose rocks.

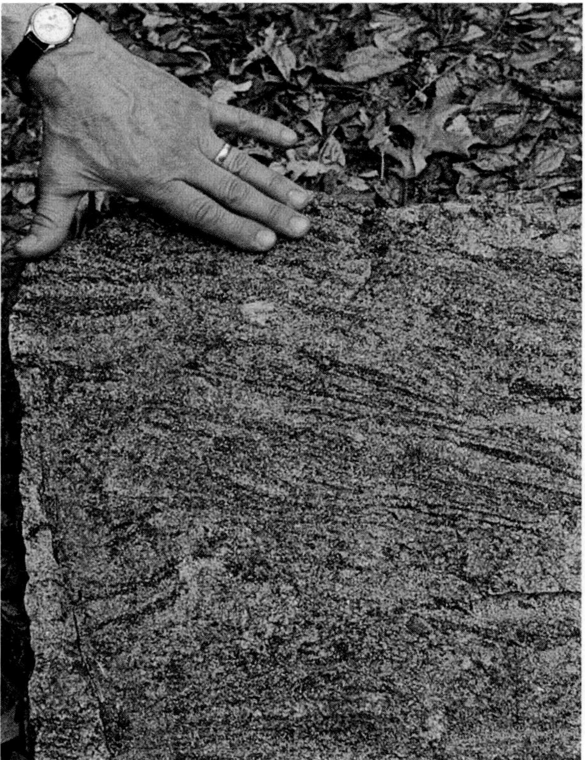

Figure 7-16 A block of Morton Gneiss, photographed on the Minneapolis campus of the University of Minnesota. The rock's characteristic gneissic layering formed under the high directed pressures and temperatures typical of mountain building at convergent plate boundaries.

When they are exposed to even greater heat, gneisses will melt, marking the transition to igneous activity and the end of metamorphism. Sometimes a metamorphic rock that has *partially* melted cools at that stage, producing a rock that is part metamorphic and part igneous. This type of rock is known as a **migmatite** (Greek for "mixed rock"). When gneiss heats to 600° to 800°C (1110°–1470°F), its felsic layers begin to melt. In contrast, the mafic layers, which have *higher melting points,* remain solid. Because the rocks deform at such high temperatures, they typically create intricately contorted, swirled patterns. The melted felsic components cool and crystallize into a granular texture resembling that of plutonic igneous rock (Fig. 7-17).

Foliated Rocks Derived from Igneous Rocks

Like multimineral sedimentary rocks, igneous rocks may also react to form schists and differentiate to form gneisses when exposed to a new set of temperature and pressure conditions. As we saw in Chapter 3, basalt, the most common *volcanic* rock of the Earth's crust, crystallizes at high temperatures. Thus all of its minerals are *anhydrous* ("water-free"). During its time on the sea floor, however, a slab of ocean basalt hydrates somewhat, as seawater enters its cracks and pore spaces. Furthermore, in the subduction zone it is exposed to cooler conditions than those of its initial crystallization. As a result of these new conditions, basalt's calcium plagioclase is converted to a lower-temperature feldspar—sodium plagioclase—and its anhydrous olivine and pyroxene are transformed into hydrous minerals, such as chlorite (a mica-like iron-and-magnesium silicate), epidote (an aluminum-iron-calcium silicate), and amphibole. Because chlorite and epidote are green, the resulting foliated metamorphic rock is called *greenschist.*

When granite and diorite, the most common *plutonic* rocks of the Earth's continental crust, undergo metamorphism under conditions of high temperature and directed pressure, their multimineral compositions react to form gneisses. Because their parent rocks are felsic, however, these gneisses are more felsic and have less-defined foliations than do gneisses derived from shale and mudstone.

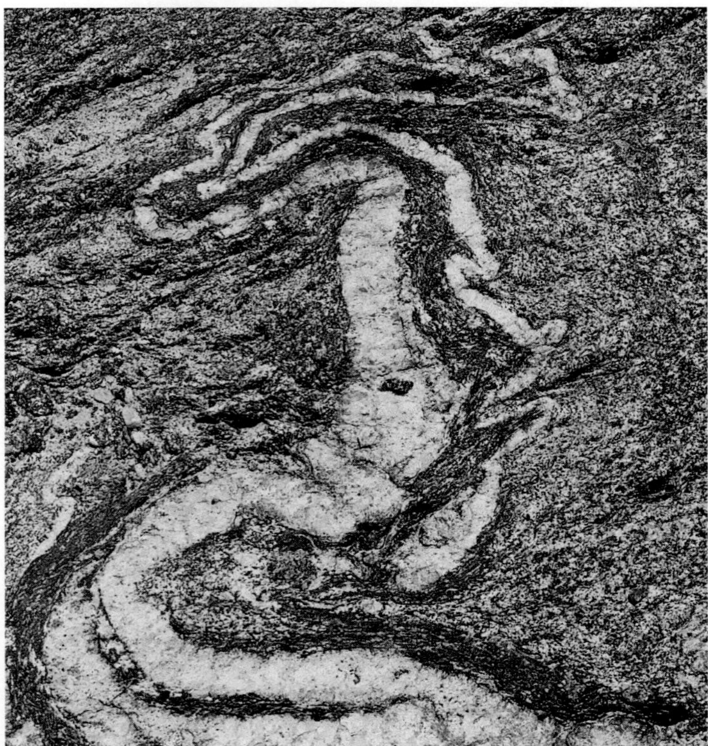

Figure 7-17 Migmatite from the Colorado Rockies. Migmatites are foliated like metamorphic rocks, but their felsic segments contain phaneritic textures like those of plutonic igneous rocks.

Nonfoliated Rocks

Nonfoliated metamorphic rocks may be produced by the increased heat and high confining pressure ensuing from deep burial, from contact with heat associated with an intruding body of magma, from parent rocks that are rich in carbonate minerals or quartz, and even, under certain circumstances, from directed pressures at plate margins.

Sedimentary rocks composed of primarily one mineral—such as quartz or calcite—typically recrystallize as coarse-grained rocks consisting of interlocking *equant* grains (that is, grains having sides of roughly equal length). Strong foliation is less likely to develop in these rocks because they lack *platy* mineral grains (such as those of mica or clay), which can be easily rotated by directed pressure into the preferred orientations. When relatively pure limestones and dolostones are metamorphosed, even by directed pressure, they tend to form a coarsely granular *nonfoliated* mosaic of calcite and/or dolomite grains—a rock called **marble.**

Similarly, pure quartz sandstone metamorphoses to **quartzite,** a very durable nonfoliated rock. To distinguish a hard, well-cemented sandstone from a quartzite, strike both rocks with a rock hammer and then study their fragments under a binocular microscope. When sandstone is shattered, it tends to break *around* its sand grains, because quartz grains are generally stronger than their surrounding cement. During metamorphism, however, the cement recrystallizes. Thus, when a quartzite is shattered, the rock may break directly *through* the grains (Fig. 7-18).

When hot magma intrudes a shale, slate, or basalt parent rock, the accompanying heat drives off virtually all mineral-bound water, promoting recrystallization and metamorphic reactions that produce anhydrous minerals with more compact structures. The resulting nonfoliated rock, **hornfels,** is dark in color, dense, and hard.

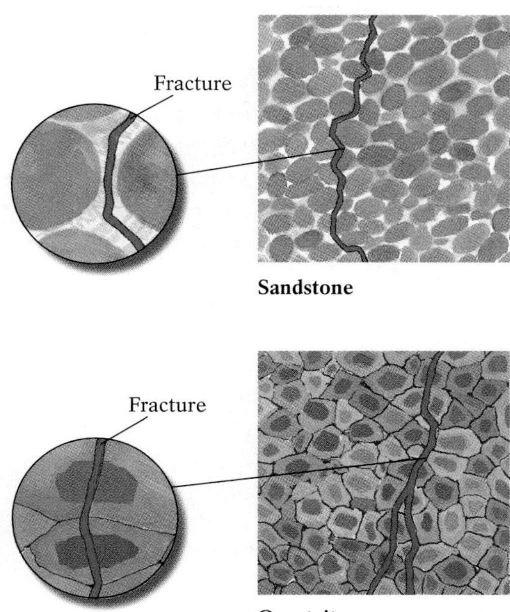

Figure 7-18 When a quartz sandstone is fractured, it tends to break within the cement that surrounds the sand grains. After undergoing metamorphism to quartzite, the rock breaks right through the grains and the recrystallized cement.

Metamorphic Grade and Index Minerals

Because they cannot study directly the environments in which metamorphic reactions naturally occur, geologists must simulate these conditions in the laboratory by subjecting minerals to increasing temperatures and pressures. Also in the laboratory, the chemical compounds that make up common rocks can be combined, and the resulting mixtures then heated and pressurized to varying degrees. In this way, geologists can determine which minerals crystallize under specific temperature-pressure combinations. From such experiments, the conditions that cause individual rocks to metamorphose have been ascertained to within about 40°C (104°F) and 0.5 kilobar of pressure.

Information gained both in the laboratory and in the field has enabled geologists to assign a **metamorphic grade** to each metamorphic rock. This grade indicates the severity of the metamorphic conditions the rock experienced. *Low-grade* metamorphic rocks retain enough of their original character—such as bedding and other sedimentary structures, fossils, and original minerals—to enable geologists to readily identify their parent rock. Low-grade metamorphism occurs at relatively low temperatures (200°–400°C, or 400°–750°F) and pressures (about 1–6 kilobars). In contrast, *high-grade* metamorphic rocks, which have been subjected to much higher temperatures and pressures, lack virtually all of their original structures, fossils, and minerals. High-grade metamorphism occurs at temperatures between 500° and 1000°C (950°–1825°F) and pressures of 12 to 40 kilobars. A variety of intermediate metamorphic grades exist as well.

Index Minerals and Mineral Zones

While working in the field, geologists may not have the means to identify tell-tale minerals. Consequently they may initially classify metamorphic rocks simply by their parent rocks, describing them as *metasedimentary, metaigneous,* or perhaps, *metavolcanic.* But once they know the specific mineral content, typically by microscopic analysis, geologists have clues as to the temperature and pressure conditions under which the rocks formed. Some minerals, because they remain stable only within a *narrow* range of temperatures and pressures, serve as indicators of specific metamorphic environments. These materials are called **metamorphic index minerals,** and areas of rock containing them are designated as specific **mineral zones.**

The presence of a given index mineral throughout a mineral zone signals that the same specific range of temperature and pressure conditions metamorphosed the entire zone, distinguishing it from other zones that formed under different conditions. For example, the aluminum silicates kyanite, andalusite, and sillimanite—three common index minerals that form *only* during metamorphism—are typically associated with separate mineral zones. Each of these minerals has the chemical formula Al_2SiO_5, but each forms under different temperature and pressure conditions, and therefore exhibits a different crystal structure (Fig. 7-19). Andalusite forms at relatively low temperatures and pressures, so metamorphic rocks in an andalusite mineral zone are designated as low-

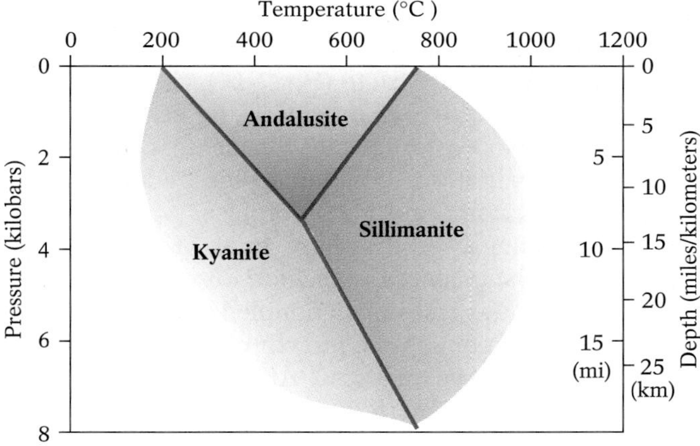

Figure 7-19 The depth, pressure, and temperature relationships that are responsible for the crystallization of the aluminum silicates kyanite, andalusite, and sillimanite. Andalusite is generally found only in rocks that form at relatively low pressures.

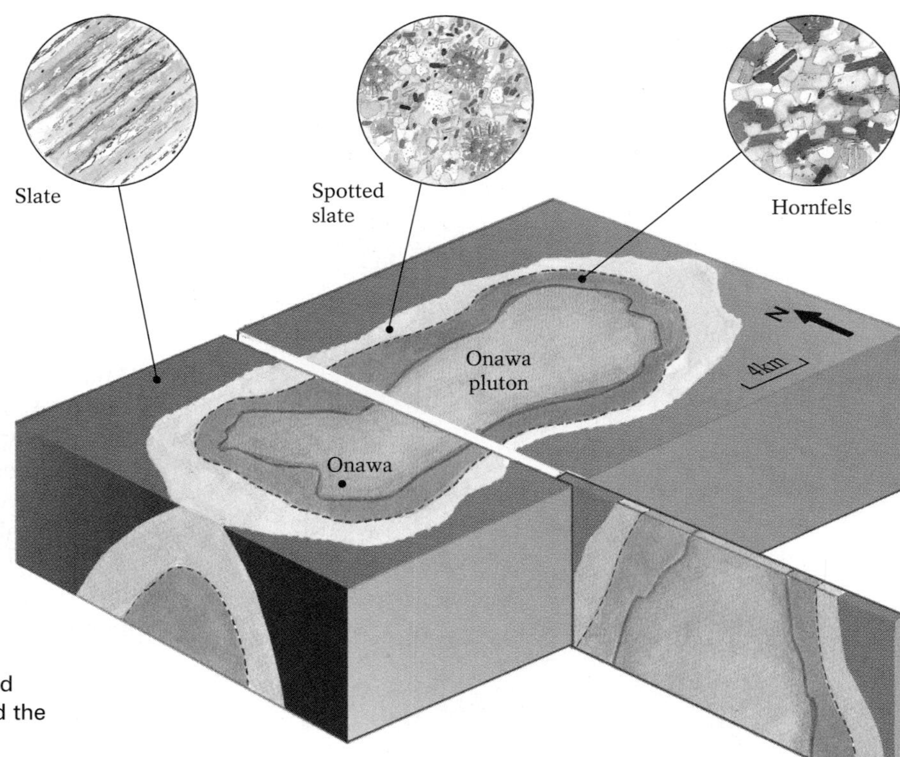

Figure 7-20 Heat from the intrusion of the Onawa pluton in central Maine created a series of metamorphic mineral zones typical of large-scale contact metamorphism. The metamorphic rocks associated with this pluton range from high-grade hornfels to intermediate-grade "spotted slate" (slate containing a number of large, recrystallized grains of andalusite) to low-grade slate located beyond the limits of the metamorphic aureole.

grade. Because kyanite and sillimanite form under higher pressures and temperatures, metamorphic rocks in mineral zones containing these index minerals are considered to be of higher grade than those in an andalusite zone.

We can locate mineral zones in rocks produced by both contact metamorphism and regional dynamothermal metamorphism. In regions of contact metamorphism, the highest-grade metamorphic rocks are those closest to the igneous intrusion; they contain index minerals associated with high temperatures. Lower-grade metamorphic rocks are found at progressively greater distances from the intrusion; they contain index minerals reflecting lower temperatures. A clear example of mineral zones produced by contact metamorphism can be found at Onawa, Maine, where, 365 million years ago, a granitic intrusion invaded and metamorphosed low-grade metamorphic rocks such as slate (Fig. 7-20).

Mineral zones within the rocks of the eastern United States, located from eastern Pennsylvania to coastal New England, indicate variations in metamorphic grade associated with regional dynamothermal metamorphism. Approximately 390 million years ago, the first in a series of collisions occurred between plates that now include Europe, Africa, and North America. These collisions crumpled the eastern edge of North America, initiating the growth of the Appalachians, which extend from Alabama into Atlantic Canada. Compression of the North American plate edge heated the region's sedimentary parent rocks, pressing them into a sequence of metamorphic rocks that increase in grade, from the largely unmetamorphosed rocks of western Pennsylvania to the intensely metamorphosed zones of Massachusetts and New Hampshire (Fig. 7-21).

Metamorphic Facies

One major drawback limits the use of index minerals as universal indicators of metamorphic conditions: If the parent rock did not contain an index mineral's elements in the proper proportions, that mineral will be absent from the metamorphic rocks. For example, the high-grade metamorphic minerals kyanite or sillimanite rarely appear in most metamorphosed limestones, because the limestone usually doesn't provide enough aluminum to form these minerals.

To identify the conditions of temperature and pressure present in the past, *regardless of the composition of the parent rock,* geologists have formulated the concept of metamorphic facies. **Metamorphic facies** are assemblages of different minerals customarily found together in metamorphic rocks. Like an index mineral, a facies indicates the specific temperature and pressure conditions that characterized the development of its component rocks. Moreover, the presence of such a *group* of minerals, all associated with a given set of metamorphic conditions, allows us to infer these conditions even when a particular index mineral is absent. As noted in our discussion of sedimentary facies in Chapter 6, changes in environmental conditions gradually produce different facies. Likewise, the facies in metamorphic rocks reveal information about changing metamorphic environments.

To understand the concept of metamorphic facies, consider the different ways rocks may be metamorphosed in a subduction zone (Fig. 7-22). A slab of basaltic oceanic crust and its overlying oceanic sediment layer are both subjected to regional dynamothermal metamorphism. The layers of continental sediment overlying the subducting slab are subjected

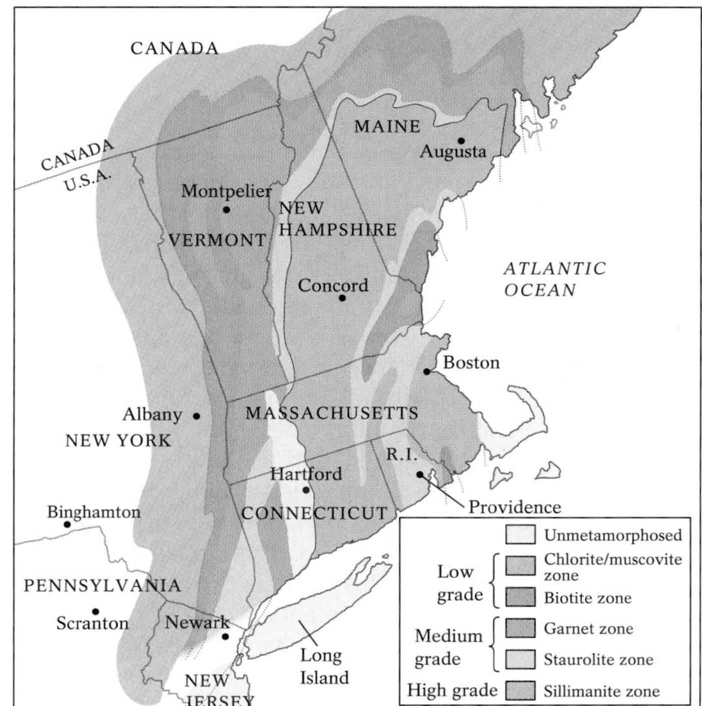

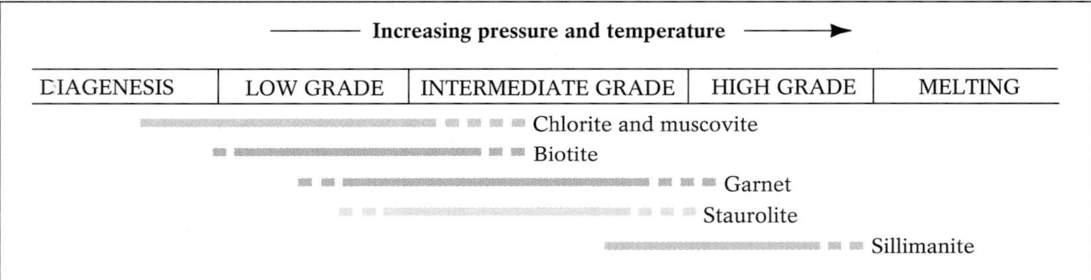

Figure 7-21 (top) The metamorphic mineral zones of the northeastern United States. These zones resulted from the regional dynamothermal metamorphism associated with several episodes of plate convergence between 460 and 250 million years ago. (bottom) General metamorphic mineral zones typical of regional dynamothermal metamorphism.

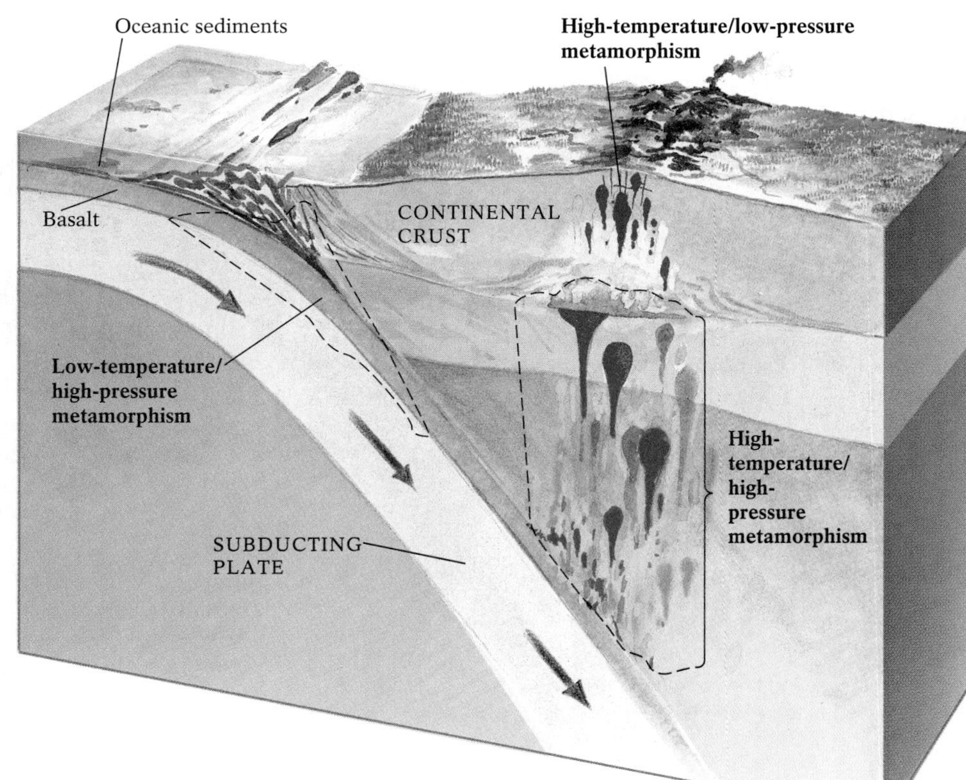

Figure 7-22 The three main metamorphic environments associated with subduction zones: low temperature/high pressure, high temperature/high pressure, and high temperature/low pressure. Each set of conditions produces a different metamorphic facies, regardless of the composition of the parent rock.

to a wholly different type of metamorphism—contact metamorphism from the rising magmas. The descending basaltic slab and its overlying sediments experience the same temperatures and pressures at each location in the subduction zone. Therefore, although the basalt and the oceanic sediments will metamorphose to different minerals because the composition of the parent rocks differ, the new rocks—metamorphosed basalt and metamorphosed sediments—formed at any given location will constitute a single metamorphic facies associated with those particular temperature and pressure conditions. The continental sediments, on the other hand, although they may contain the same constituent minerals as the oceanic sediments, will produce a different facies—one associated with low-pressure, high-temperature metamorphism.

Because much of the Earth's crust is composed of basalt, the common types of metamorphosed basalts provide a convenient way to categorize metamorphic facies (Fig. 7-23). *Hornfels facies,* for example, form under the high-temperature, low-pressure conditions generally found around intruding igneous rocks such as batholiths, plutons, and dikes. *Greenstone facies* form under the relatively low temperatures and pressures associated with regional burial metamorphism, which converts basalts to a greenish, nonfoliated, low-grade metamorphic rock called greenstone. *Greenschist facies* form when the same rocks are subjected to directed pressures, which produce chlorite-rich greenschist.

Amphibolite facies form at intermediate-grade temperatures and pressures, between 450° and 650°C (850°–1200°F), and at pressures of 4 to 10 kilobars. The resulting rock contains a large amount of the amphibole hornblende and some accessory garnet and plagioclase feldspar. If the parent rock includes aluminum-rich micas, clay minerals, and feldspars (as, for example, in the case of a shale or mudstone), then amphibolite-facies rocks will also contain large grains of intermediate-grade aluminous minerals, such as staurolite, kyanite, and sillimanite.

The *granulite facies* of high-grade metamorphic rocks develop at higher temperatures and pressures, which drive out virtually all water in a rock's minerals. These ultradry conditions inhibit melting, producing minerals that are stable beyond 700° to 800°C (1300°–1475°F); these minerals include quartz and feldspar, high-temperature pyroxenes, garnets, and aluminum silicates, such as sillimanite or kyanite.

The *eclogite facies* is a relatively rare assemblage of minerals that forms only in *extremely* high-pressure, high-temperature environments. Found as inclusions in mantle-derived diamond pipes, basalts, and ultramafic intrusions and in the deeper portion of subduction zones, eclogite-facies rocks consist largely of pyroxenes and magnesium-rich garnets. These minerals are stable at pressures exceeding 10 kilobars (corresponding to a depth of about 35 kilometers, or 20 miles) and at temperatures of about 800°C (1475°F). Such environmental conditions suggest that eclogite-facies rocks probably originate in the upper mantle or in the deeper parts of subducting plates.

Blueschist facies contain minerals that are stable only at high pressures and relatively low temperatures. The most prominent is a sodium-rich amphibole called glaucophane, which imparts a distinctive bluish tint to the foliated rocks of this facies. Because high pressures exist principally beneath great thicknesses of overlying rock, *where high temperatures*

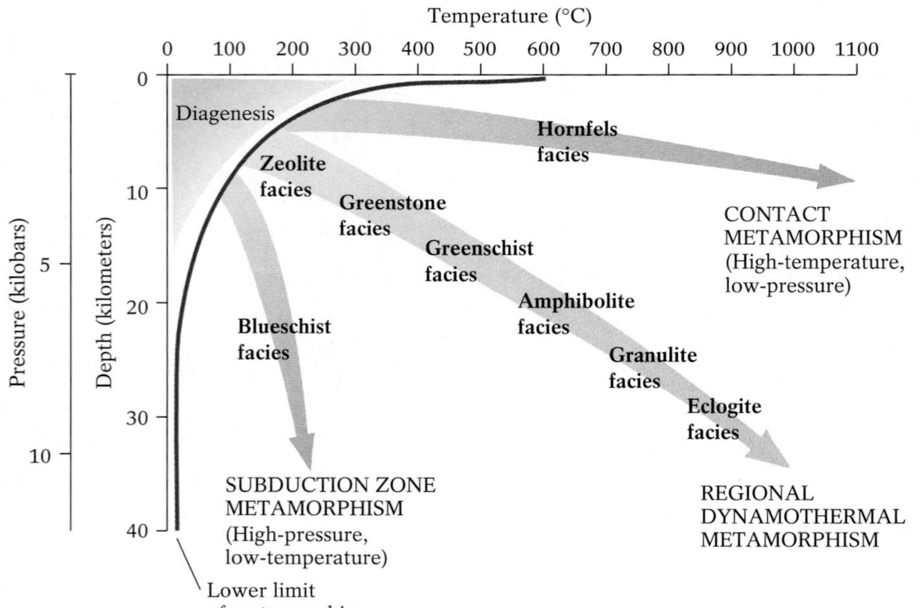

Figure 7-23 The depth, pressure, and temperature relationships that produce the common metamorphic basalts, which geologists use to identify the various metamorphic facies. With such a chart, one should be able to extract such information as: What temperature and pressure conditions create the eclogite facies? To which facies would a rock that is stable at 500°C and 3 kilobars of pressure belong?

should also prevail, the conditions that produce blueschist-facies rocks are quite rare. As we shall see in the next section, such conditions typify subduction zones, where high pressures associated with plate convergence develop at relatively shallow depths and where cold subducting oceanic lithosphere has not been submerged long enough (or deep enough) to have been warmed significantly by geothermal heat.

Plate Tectonics and Metamorphic Rocks

Plate movements create the heat, pressure, and circulating hot fluids that produce much of the Earth's metamorphic rock. Directed pressure at plate boundaries causes virtually all regional dynamothermal metamorphism, and new magmas generated at subduction and divergent zones are responsible for most contact metamorphism.

The existence of vast regions of exposed foliated rock suggests that directed pressures played a hand in the rock's metamorphism, with the foliation pattern itself indicating the direction of ancient tectonic compression. Their high metamorphic grade tells us that some of these rocks formed at great depths, so we can conclude that they have experienced considerable uplift and subsequent erosion.

Certain metamorphic rocks indicate specific plate tectonic settings. In subduction zones, for example, oceanic plates converge on a regional scale, creating several distinct metamorphic environments (Fig. 7-24). In a relatively shallow high-pressure, low-temperature zone in the subduction zone, the subducting rocks are subjected to directed pressure and undergo blueschist-facies metamorphism. In a low-pressure, low-temperature zone above the subducting plate, close to the Earth's surface, heat from rising magmas, the relatively low burial pressures associated with shallow depths, and the directed pressures at convergent plate boundaries combine to

Figure 7-24 Metamorphic facies characteristic of subduction zones. The high-pressure, low-temperature environment produces blueschist-facies rocks; the high-temperature, high-pressure environment produces granulite-facies rocks; and the low-pressure, high-temperature environment produces greenschist-facies rocks.

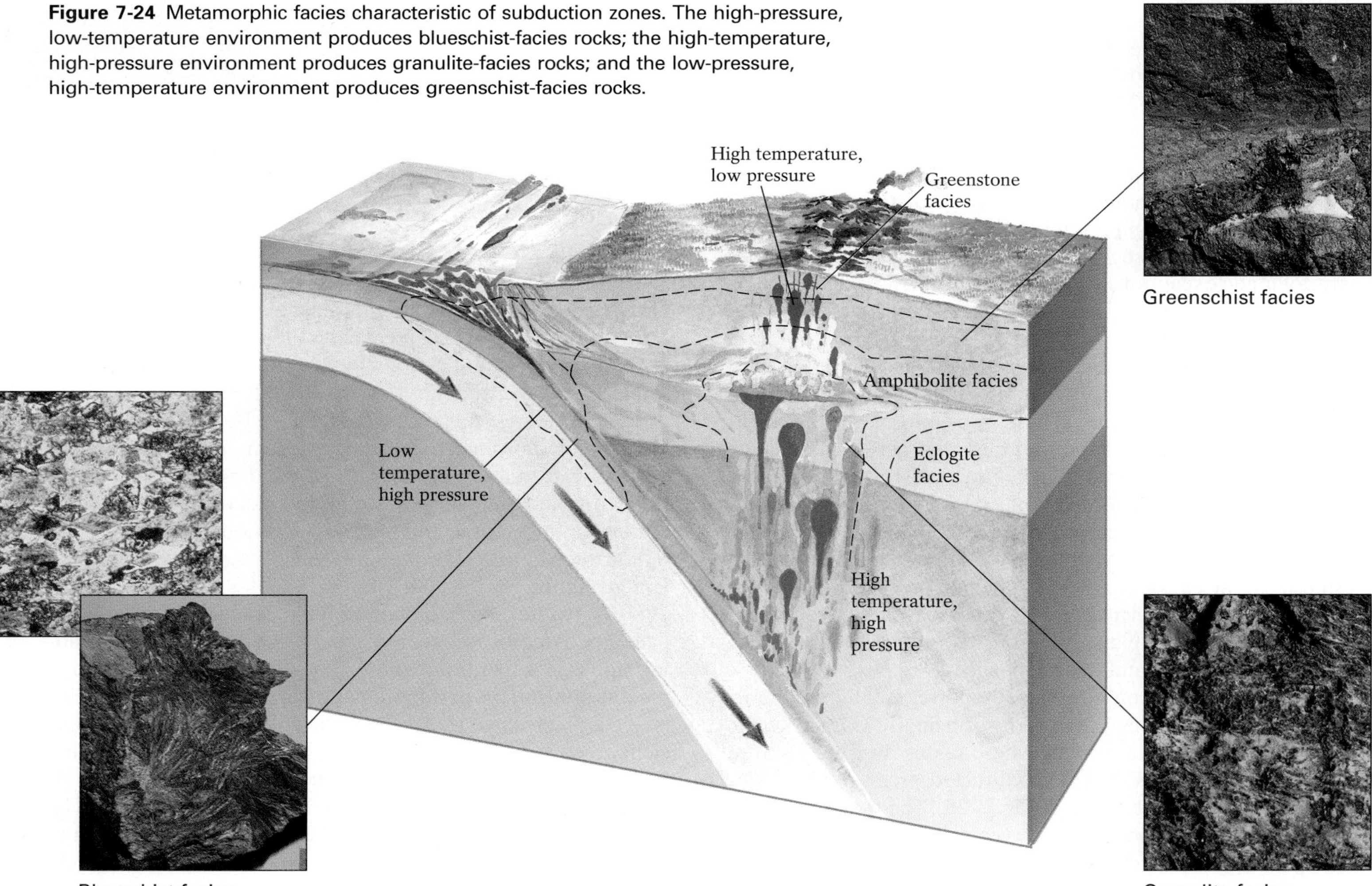

Figure 7-25 Three-billion-year-old greenstone from Ely, Minnesota. The pillow structures in this rock suggest that the parent rock originated as submarine basalts; the fact that the rock is nonfoliated and its structures relatively undeformed indicates that the rock was not subjected to directed pressure.

Figure 7-26 Verd antique, a popular decorative stone, is composed of the hydrothermally metamorphosed rock, serpentinite.

produce greenschist-facies rocks. In a high-pressure, high-temperature zone above the subducting plate but at greater depth, granulite-, amphibolite-, and eclogite-facies rocks form. Most metamorphic rocks associated with subduction zones are foliated (exceptions include single-mineral rocks such as quartzite and marble), because the entire region is subjected to the directed pressures of convergence.

Blueschist-facies rocks represent a particularly useful clue to the location of ancient subduction zones and convergent plate boundaries: These rocks form at the high-pressure, low-temperature conditions that geologists believe occur only when plates subduct. The blueschist-facies rocks in western California, for example, most likely mark the subduction zone that produced the Sierra Nevada batholiths, some of which are visible in Yosemite National Park. Blueschist-facies rocks have also been discovered in the Appalachians of northern Vermont; this area may have been a zone of converging plates, volcanism, and great earthquake activity 300 to 400 million years ago.

Ocean basins contain virtually no regionally metamorphosed rocks. Oceanic rocks form principally from volcanic eruptions at mid-ocean ridges where plates diverge; because these rocks are not subjected to directed pressure, they generally do not become foliated (Fig. 7-25). The metamorphism that does occur in ocean basins typically involves relatively low-grade hydrothermal serpentinization within the fractured mantle rocks beneath mid-ocean ridges.

Metamorphic Rocks in Daily Life

Metamorphic rocks are generally strong and durable, for several reasons. The combination of heat and pressure eliminates pore spaces in the rocks, increasing their density. At the same time, metamorphic reactions replace unstable minerals with stable ones, and sediment grains and their binding cement become interlocked by recrystallization. Because of its durability, metamorphic rock is a popular material for the weather-resistant exteriors of office buildings and as foundation stones for such large-scale projects as bridges and dams. For example, building and road construction uses roughly 1.6 billion metric tons of slate and quartzite every year in the United States.

We value certain metamorphic rocks highly for their appearance. Among architecture's most prized decorative building stones is serpentinite, the rock produced by hydrothermal metamorphism of peridotite at divergent plate boundaries (see Fig. 7-11 on p. 190). This rock, called *verd antique* ("ancient green") by architects, adorns the interiors of such stately buildings as the United Nations and the National Gallery of Art, as well as countless office towers, banks, and libraries (Fig. 7-26). When marine limestone and dolostone are deposited in the cracks that often form in serpentinite, an attractively streaked green-serpentine, white-carbonate rock results.

Sculptors prize the metamorphic rock marble for its appearance, softness, and texture. Because pure marble is snow-white, it shows the details of a sculptor's carving; because it is soft, it responds perfectly to the sculptor's tools. (Calcite, marble's main constituent, measures only 3 on the Mohs hardness scale.) The stone from which Michelangelo fashioned his renowned sculptures resulted from metamorphism when the Apennine mountains of Italy, which contained a layer of pure marine limestone, were squeezed between colliding plates (Fig. 7-27). In contrast, metamorphism of impure limestones yields marbles colored by their impurities—pink, red, or brown from iron oxides; gray or black from carbon-rich material; or green from the calcium-magnesium pyroxene diopside or the common amphibole hornblende.

Soapstone, a talc-rich serpentinite, is another popular material for sculpture (and for laboratory work counters). Soapstone and many other useful everyday commodities are the results of regional dynamothermal metamorphism. Low-grade regional metamorphism of ultramafic rocks produces talc, used in talcum powder, as well as flame-resistant asbestos (another variety of serpentinite), formerly used in automobile brake linings, the safety garb favored by firefighters, and toddler's pajamas (though the latter use has ended because of asbestos' toxicity). Garnet, an industrial abrasive valued for its hardness (and also January's gem-like birthstone), results from intermediate-to-high–grade metamorphism. The high-grade metamorphic minerals kyanite and sillimanite are key components of porcelain casings for spark plugs, because the high temperatures at which they crystallize render them capable of withstanding intense heat.

Metamorphic rocks also occasionally contain important mineral deposits, enabling the countries in which they are found to prosper in the world's marketplace (see Chapter 20). Zinc (in the mineral sphalerite), lead (in galena), copper (in chalcopyrite and bornite), iron (in magnetite), and gold are a few of the valuable minerals found in metamorphic rocks.

Figure 7-27 (left) The Carrara marble quarry in northern Italy. The purity of Carrara marble, which stems from the purity of its parent limestone, has long made it a favored source of sculpting material (below).

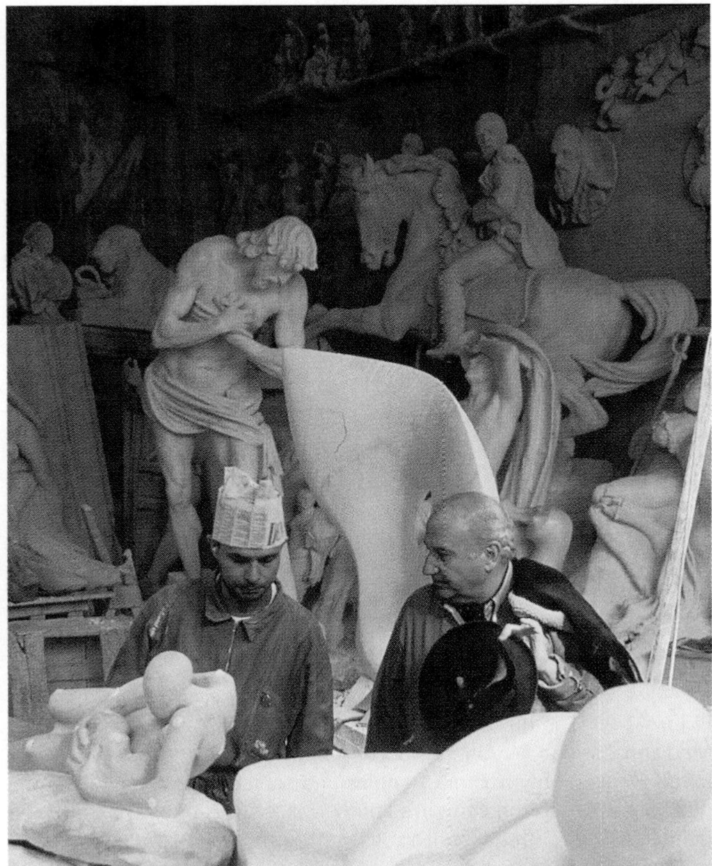

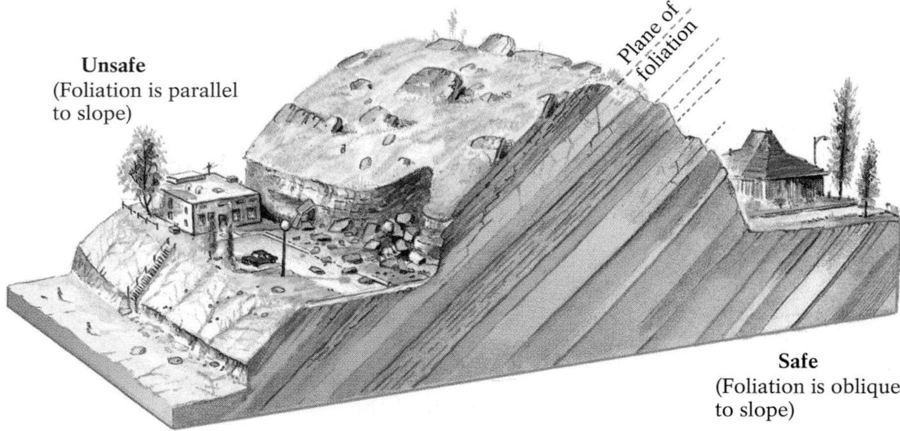

Figure 7-28 Building in areas with metamorphic surface rocks requires consideration of the direction of the rocks' foliation planes. The slope on the left side of this figure is unstable, because the rock's foliation lies parallel to the slope and can readily fail, causing a landslide of thick slabs of rock. The slope on the right is stable, because the foliation is oriented obliquely to the slope.

Potential Hazards from Metamorphic Rocks

While some characteristics of metamorphic rocks have great aesthetic or practical value, others can produce hazardous conditions in natural settings. Foliation, in particular, can prove dangerous. Because it reduces the strength of metamorphic rocks, slaty cleavage can weaken slopes and trigger landslides, particularly where the cleavage planes run parallel to steep slopes (Fig. 7-28). Building in foliated metamorphic terrains requires careful planning and thorough geological investigation to determine the orientation of the foliation.

The weak planes of metamorphic foliation are especially susceptible to earthquake damage. Such was the case at midnight on August 17, 1959, near Yellowstone Park in Montana, when a powerful earthquake shook the Madison Canyon area. A deeply weathered body of gneiss and schist, the foliations of which paralleled the canyon's south wall (Fig. 7-29), slid away from adjacent rock during the shaking. The landslide accumulated debris as it raced downslope at about 100 kilometers per hour (60 miles per hour), burying a campground and its 28 visitors with some 45 meters (150 feet) of rock, soil, and vegetation.

Metamorphic foliation must also be analyzed when designing large-scale engineering projects such as dams. Today, teams of geologists and engineers routinely study local and

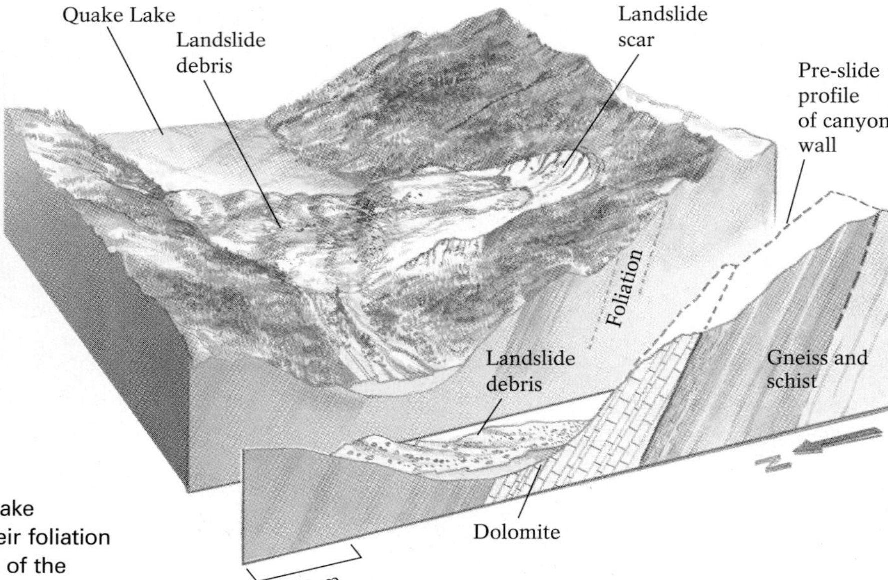

Figure 7-29 The disastrous rock slide of 1959 at Madison Canyon, Wyoming, occurred when an earthquake jolted weak, highly fractured gneiss and schist along their foliation planes parallel to the canyon's south wall. Left: A photo of the landslide aftermath, taken from the northern canyon wall.

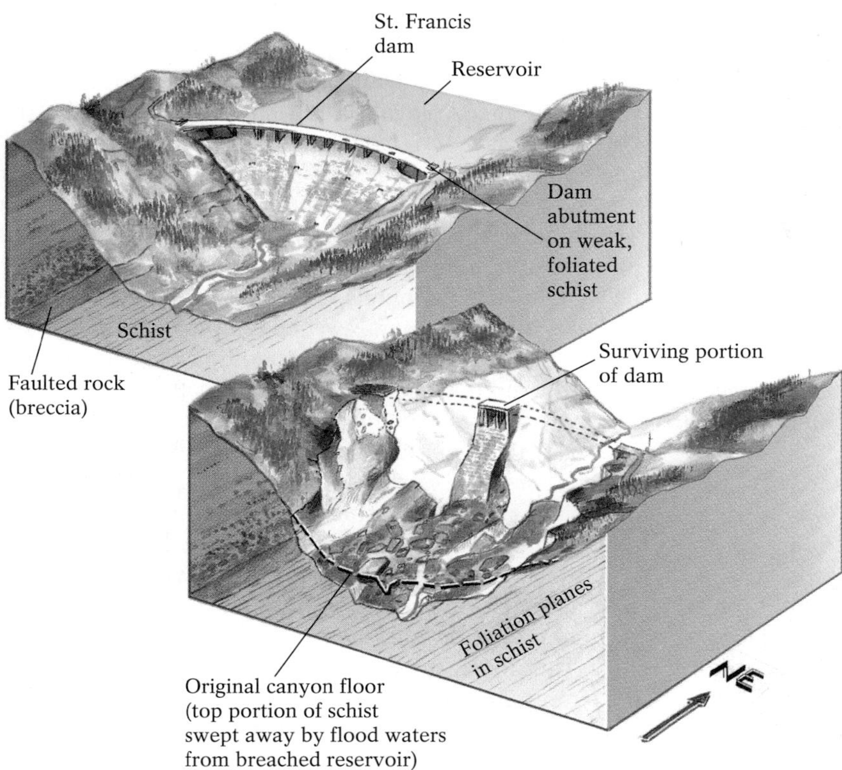

Figure 7-30 The geology of the St. Francis dam, which collapsed in 1928. The weak, foliated rocks that supported one side of the dam and the faulted bedrock below the dam led to the dam's failure.

regional rock formations for evidence of inherent weakness before beginning any major construction. In the past, however, this step was seldom taken—sometimes with disastrous consequences. On the night of March 12, 1928, the St. Francis dam in Saugus, California, failed, claiming 420 lives in the ensuing flood. Its eastern side adjoined loose landslide debris from a weak schist whose foliation planes lay parallel to the canyon wall. This structure, combined with a fault that traversed the dam site, was largely responsible for the dam's failure (Fig. 7-30). The disaster led to subsequent improvements in the way we investigate sites for prospective dams, although community opinion and political pressure may still cause geologists' advice to go unheeded.

The Rock Cycle Revisited

In our preview of the rock cycle in Chapter 1, we saw the principal evolutionary paths of the Earth's rocks and the variety of processes that create them. We have now examined in detail the three basic types of rocks—igneous, sedimentary, and metamorphic—and the major processes (melting, recrystallization, weathering, lithification, and metamorphism) that form and alter them. We can now fully appreciate the relationships between these processes. Any rock, including an existing igneous rock, can be melted and form a new igneous rock. Any rock, including an existing sedimentary rock, can be weathered, eroded, transported, deposited, and lithified to form a new sedimentary rock. Any rock, including an existing metamorphic rock, can be heated and pressed to form a new metamorphic rock. The cycle has no end. Every body of rock on Earth is constantly changing—although, as we have seen, some processes change rocks more slowly than others.

We have also seen how plate tectonic processes contribute to the cycling of the Earth's rocks. The igneous processes associated with rifting and divergent zones produce new oceanic lithosphere, and subduction recycles mafic oceanic lithosphere and its overlying sediments, producing plutons of intermediate and felsic rock. Rocks uplifted by tectonic forces and exposed to the Earth's surface environment become weathered, eroded, transported, deposited, and lithified to form sedimentary rocks. In this chapter, we have seen how the heat and directed pressure at convergent plate boundaries can convert a soft sedimentary rock (such as shale) to a hard metamorphic rock (such as gneiss). Clearly, plate tectonics plays a pivotal role in the evolution of the Earth's rocks.

In the next chapter, we will examine what the Earth's rocks tell us about the past. In succeeding chapters, we will look more closely at how some of the processes that create rocks also shape the Earth's mountains, continents, and ocean floors.

Chapter Summary

Metamorphic rocks, the third major type of rock in the Earth's rock cycle, are formed at temperatures and pressures intermediate between those that form igneous and sedimentary rocks. They are created by **metamorphism** of preexisting rock, when the mineral content and structure of solid rock are altered by heat, pressure, and chemically active fluids but the rock does not melt. During metamorphism the rock may lose many of its original features; its mineral grains will generally grow larger and develop an orderly parallel arrangement, and they may even recrystallize into new minerals that are more stable under the prevailing conditions.

Most metamorphism occurs at depths greater than 10 kilometers (6 miles) below the Earth's surface, where the geothermal gradient and the weight of the overlying rock contribute the heat and pressure necessary to alter solid rock. Water and magmatic fluids also circulate at these depths, enhancing migration of unbonded ions in rocks undergoing metamorphism. Under any given set of conditions, the composition of the parent rock determines which metamorphic rocks can form.

Rocks at great depths are subjected to **lithostatic**, or **confining, pressure,** which pushes at the rock equally from all sides, making it smaller and denser but maintaining its general shape. Some rocks—particularly those at convergent plate boundaries—are also subjected to **directed pressure,** which distorts their shapes, flattening them in a single plane across which the pressure is predominantly applied. The minerals in a rock subjected to directed pressure tend to align in parallel planes perpendicular to the direction of the pressure, giving the rock a preferred mineral orientation, or **foliation.** All metamorphic rocks are categorized as either foliated or nonfoliated.

Metamorphism varies according to the prevailing geological conditions. **Contact metamorphism** occurs where rocks are heated by direct or close contact with magma; the zone of metamorphosed rock surrounding a magma is called a metamorphic **aureole. Regional metamorphism** occurs over a broad area of rock, and may be one of two types: **burial metamorphism,** which results from the application of geothermal heat and confining pressure to rocks buried beneath more than 10 kilometers of overlying rock, or **dynamothermal metamorphism,** which develops from increased heat and pressure (both directed and confining) in an area of plate convergence. **Hydrothermal metamorphism** involves hot circulating fluids. **Fault-zone metamorphism** occurs where rocks grind against one another at active faults. **Shock metamorphism** occurs where rocks become violently heated and compressed by the impact of a meteorite.

Increased heat and directed pressure are responsible for the sequence of foliated rocks that forms from a typical marine shale, for example. This sequence begins with the fine-grained **slate** and **phyllite,** and progresses to coarse-grained **schist** and **gneiss.** Foliation in slates, phyllites, and schists develops as clay and mica grains align, whereas foliation in gneisses is marked by the formation of distinct, single-mineral layers of feldspars, micas, quartz, and mafic minerals. The mineral bands in gneisses result, in part, from **metamorphic differentiation,** in which unstable minerals recrystallize as new, more stable minerals in alternating bands of light (felsic) and dark (mafic) minerals. When temperature and pressure increase beyond metamorphic conditions to the threshold of igneous conditions, a "mixed" rock—called **a migmatite**—develops with both metamorphic and igneous parts.

Nonfoliated rocks commonly develop in the absence of directed pressure or when the parent rock is composed largely of a single mineral. **Marble** forms from the metamorphism of relatively pure calcite-rich limestone or dolostone. **Quartzite** is produced by metamorphism of relatively pure quartz-rich sandstone. **Hornfels** forms where shale or basalt comes in direct contact with hot magma.

The degree to which a metamorphic rock differs from its parent material is described in terms of **metamorphic grade.** Low-grade metamorphism occurs when rocks are heated to temperatures of approximately 200° to 400°C; high-grade metamorphism occurs at temperatures of about 500° to 800°C. The more pronounced the metamorphic conditions, the higher the grade—and the more a metamorphic rock differs from its parent rock.

The minerals in metamorphic rocks can provide clues to the conditions under which the metamorphism occurred. Some minerals, which remain stable only within a narrow range of temperatures and pressures, indicate specific metamorphic conditions; they are called **metamorphic index minerals.** Areas of rock containing these minerals are designated as specific **mineral zones.** Metamorphic environments may be associated not only with individual index minerals, but also with diagnostic assemblages of minerals called **metamorphic facies.** Mineral zones and metamorphic facies delineate areas of rock subjected to the same metamorphic conditions of temperature and pressure.

Different types of metamorphic rocks and facies form in different plate tectonic settings. A variety of metamorphic rocks develop at different venues within a convergent plate boundary. The high-pressure, low-temperature environment in a subduction trench produces blueschists; the low-pressure, high-temperature environment in the continental crust above the subducting plate yields greenschists; and the high-pressure, high-temperature environment in the mantle directly over the subducting plate leads to granulites, amphibolites, and eclogites. Divergent plate boundaries are characterized by hydrothermal serpentization, relatively low-temperature reactions that involve interaction between warm ocean-ridge basalts and circulating seawater.

Key Terms

metamorphic rocks (p. 183)
metamorphism (p. 183)
lithostatic (confining) pressure (p. 186)
directed pressure (p. 186)
foliation (p. 186)
contact metamorphism (p. 189)
aureole (p. 189)
regional metamorphism (p. 189)
burial metamorphism (p. 189)
dynamothermal metamorphism (p. 189)
hydrothermal metamorphism (p. 190)
fault-zone metamorphism (p. 190)
shock metamorphism (p. 190)
slate (p. 192)
phyllite (p. 192)
schist (p. 192)
metamorphic differentiation (p. 192)
gneiss (p. 192)
migmatite (p. 194)
marble (p. 194)
quartzite (p. 194)
hornfels (p. 194)
metamorphic grade (p. 195)
metamorphic index minerals (p. 195)
mineral zones (p. 195)
metamorphic facies (p. 196)

Questions for Review

1. What are the four major factors that induce or influence metamorphism?

2. Describe four ways that rocks may change during metamorphism.

3. Define metamorphic foliation, explain how it develops, and list three metamorphic rocks that are foliated.

4. List the sequence of metamorphic rocks that develops during the progression of shale to migmatite. Describe the successive changes in mineralogy and structure that the rocks undergo.

5. Under what conditions do nonfoliated metamorphic rocks generally form? Give examples.

6. List the six common types of metamorphism, and discuss the geological conditions under which they occur.

7. Kyanite, andalusite, and sillimanite are considered metamorphic index minerals. What is a metamorphic index mineral, and why do these three minerals qualify?

8. List four different metamorphic facies associated with regional dynamothermal metamorphism, and describe the metamorphic conditions that produce them.

9. What is the plate tectonic significance attached to the presence of the blueschist-facies mineral glaucophane in a rock? What type of metamorphism generally occurs at mid-ocean ridges?

10. Under what circumstances might the presence of metamorphic rocks lead to landslides?

For Further Thought

1. Why do quartzites appear to be essentially unfoliated, even though they have been subjected to the directed pressures of regional metamorphism?

2. Why do we rarely find minerals such as chlorite directly next to an igneous intrusion?

3. Referring back to Figure 7-9a (p. 188), suggest why some of the rocks crosscut by the intrusive rocks have not undergone contact metamorphism.

4. The Earth's internal temperature was apparently much higher during the first billion years of its history. Speculate about how metamorphism, metamorphic environments, and the distribution of metamorphic rocks would have differed during that period.

5. From the photo below, estimate the orientation of the directed pressures that metamorphosed these rocks.

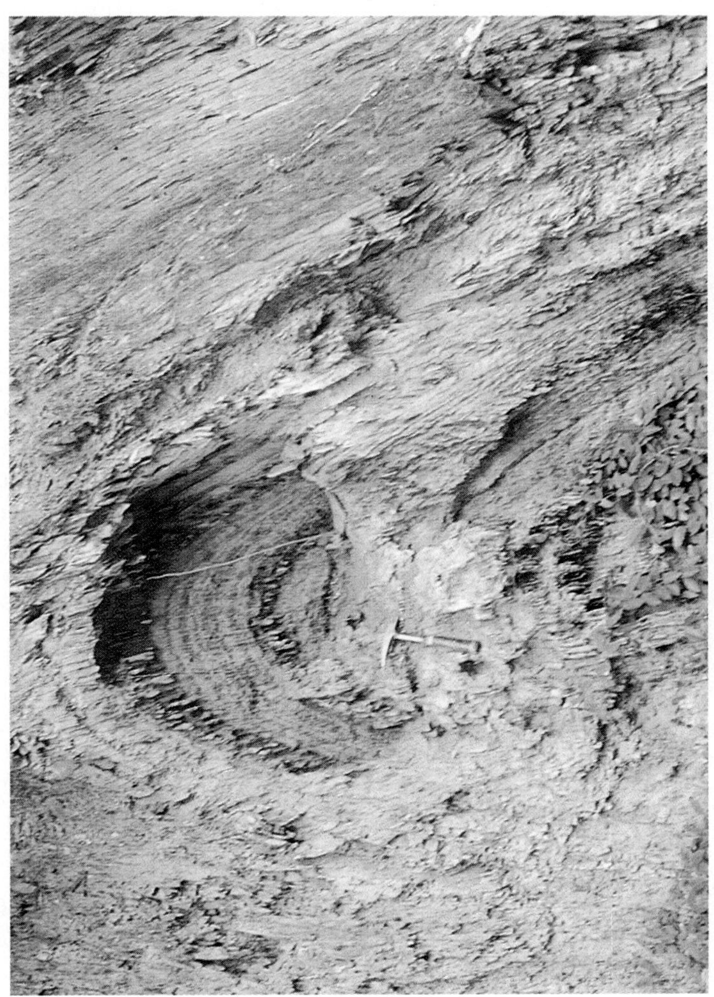

8

Telling Time Geologically

Figure 8-1 This rich cache of dinosaur bones is exposed at Dinosaur National Monument in northeastern Utah.

The first seven chapters of this text explored some basic geological processes—how plates move; how underground molten rock solidifies into igneous rocks and erupts to produce volcanoes; how solid surface rocks weather to form loose grains of sand that, in turn, lithify to form sedimentary rocks; how heat and pressure convert rocks underground into hard, durable metamorphic rocks. One particular factor is vital to all of these and many other geological processes—time.

When thinking about geologic time, we are often overwhelmed by its vastness. For example, mountain building by plate convergence generally takes place at the rate of about 1 meter (a little more than 3 feet) per 5000 years. At that rate, peaks as high as those of the Rockies or Alps (roughly 3000 meters, or 10,000 feet) would have taken 15 million years to form. The Atlantic Ocean, which widens in places at a rate approaching 4 centimeters (1.5 inches) per year as a result of plate divergence will probably take about 50 million years to become 2000 kilometers (1200 miles) wider.

In terms of Earth history, however, the creation of an ocean and the rise of a mountain are relatively rapid events. Geological processes have continuously affected the Earth since the planet's birth *4.6 billion years ago.* Many of the "great moments of Earth history," such as the extinction of the dinosaurs (Fig. 8-1), were virtually instantaneous in a geological sense, requiring "only" 1 million or fewer years to occur. Others, such as the breakup of the supercontinent Pangaea, required tens or hundreds of millions of years to unfold. Virtually all of the major events that have shaped the Earth culminated long before the first humans evolved.

The timing of geologic events and processes is the subject of **geochronology,** the study of "Earth time." Geochronology helps us comprehend the exceedingly slow pace of certain geological processes and the relatively rapid pace of others. Geochronologists attempt to date and understand the geological development of every region on Earth as part of their quest to reconstruct the entire sequence of events that makes up the Earth's geologic record. This pursuit is an important component of the larger field of **historical geology,** which focuses on the origin and evolution of all the Earth's life forms and geologic structures.

To tell time geologically, geologists use a variety of ingenious geologic "clocks." Much of this chapter will explain how trees, radioactive elements, fossils, and other materials function as geologic clocks. But first we must put geologic time into a human perspective, so that we—with our life-spans of 80 or 90 years—might begin to grasp the magnitude of the Earth's longevity.

Geologic Time in Perspective

Here are three statements that represent different human views of the age of the Earth:

> "Heaven and Earth, Centre and substance were made in the same instant of time and clouds full of water and man were created by the Trinity on the 26th of October 4004 B.C. at 9:00 in the Morning."

> "High up in the North in the land called Svithjöd, there stands a rock. It is a hundred miles high and a hundred miles wide. Once every thousand years a little bird comes to this rock to sharpen its beak. When the rock has been thus worn away, then a single day of eternity will have gone by."

> "The Earth is approximately 4.6 billion years old."

There's something here for everyone. The first pronouncement records the thoughts of Archbishop James Ussher of the Irish Protestant Church. Using the genealogical record of the Old Testament to work back to the biblical account of Creation, Ussher, in 1664, dated the Earth as being precisely 5668 years old. This view, which appealed to those who value precision and those who seek theological dominion over science, dominated scientific thought for several centuries. To the naturalists of the eighteenth and nineteenth centuries, however, it proved somewhat "confining." Their observations led them to propose that the Earth would have needed much more time to evolve into its current condition.

The second view comes from a Norwegian folktale recounted by Hendrik Van Loon in his *Story of Mankind.* It should appeal to those among us who fancy ourselves to be deep thinkers. Although imprecise, it successfully conveys to our finite minds some sense of the great extent of Earth history.

The third view—less precise than the first view and less poetic than the second—is generally accepted by scientists today. Yet, how can we begin to grasp the magnitude of 4,600,000,000 years? In human terms, that inconceivable length of time might as well be the eternity of the Norwegian folktale.

Geology professors will never rest until they find a dramatic way to illustrate to their students the great span of Earth history. One way is to picture a timeline stretching across the United States, 5000 kilometers (3000 miles) long, on which each kilometer represents about 1 million years (Fig. 8-2). No rocks from the Earth's first 600 million years have ever been identified, so no direct information is available for this period; the first 600 kilometers (400 miles) of our continent-wide timeline is therefore barren.

The oldest rocks known, recently discovered in Canada's Northwest Territories, are dated at 3.96 billion years. Few of our most ancient rocks contain any hint of life, probably because conditions on Earth during the planet's first billion years were too harsh to support most forms of life. (It is possible that very early life forms might have been extinguished if large meteorites colliding with the Earth generated enough heat to vaporize the planet's first bodies of water and kill everything in them.) The earliest known life forms on Earth were microscopic single-celled organisms believed to have lived 3.77 billion years ago. Scientists have identified fossils of these organisms—so primitive that they had not yet developed a cell nucleus—in the lithified silt that collected in the bottoms of muddy ponds in what is now northwestern Australia.

Over the next 3 billion years, more complex forms of life began to develop and diversify. What allowed these new forms to arise and flourish? Many scientists believe that by about 545 million years ago (4500 kilometers or 2800 miles on our timeline) a more hospitable environment had developed on Earth. The atmosphere had accumulated sufficient oxygen to absorb a significant portion of the Sun's ultraviolet radiation, and this protective atmosphere facilitated an explosion of life in the world's ancestral oceans.

Many momentous events in Earth history are concentrated in the last few hundred kilometers of our timeline: the rise and fall of the dinosaurs, the onset of the Earth's most recent ice ages, the origin of mammals (including humans), and the arrival of humans on the North American continent. Indeed, all of the material taught in a typical history course represents only 0.0000011% of the Earth's total history!

How did geologists pinpoint the timing of the Earth's great events? Why do they believe that the planet is 4.6 billion years old? How did they deduce the ages of the oldest known rock, the oldest fossil, and the landmark fossils in the evolution of modern species? How do they date the rocks and fossils that continue to be uncovered?

Geologists examine time in two ways. First, they use **relative dating,** which compares two or more entities to determine which is older and which is younger. Just as you can generally estimate the relative ages of people by assessing the basic features of the bodies and faces of infants, toddlers, children, adolescents, adults, and senior citizens, geologists have learned certain characteristics that help them determine the relative ages of rocks and other geological features.

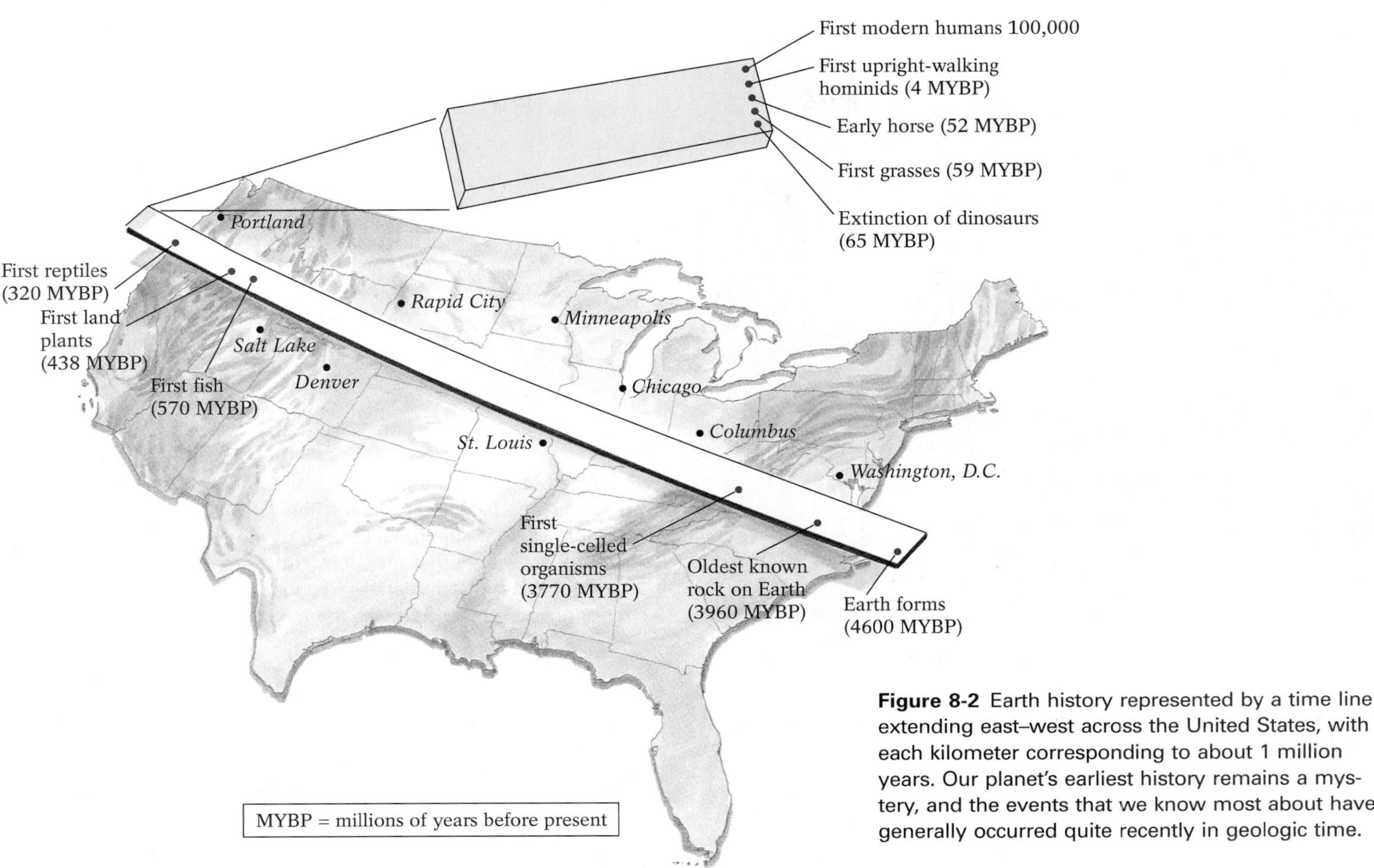

Figure 8-2 Earth history represented by a time line extending east–west across the United States, with each kilometer corresponding to about 1 million years. Our planet's earliest history remains a mystery, and the events that we know most about have generally occurred quite recently in geologic time.

Second, geologists analyze time by **absolute dating,** which establishes how many years ago a specific feature developed or a specific event occurred. To determine the absolute age of any individual, you would have to obtain quantitative information, such as that found on a birth certificate or driver's license. But the Earth has no such written record. Consequently, to fix the absolute age of geological phenomena, geologists use a variety of sources that yield quantitative information.

By combining information from both relative and absolute dating, geologists can date with some assurance a wide variety of geological materials virtually everywhere on Earth.

Determining Relative Age

We can determine the sequence of geological events that has produced an area's rocks by using certain common-sense principles and assumptions. It helps to have knowledge of such fundamental processes as sedimentation, volcanism, and erosion.

Principles of Relative Dating

Using basic scientific logic, an understanding of spatial relationships (such as where rock layers form), and an understanding of fossils, we can usually answer the questions "Which is older?" and "Which is younger?", compare the ages of rocks from different geographic regions, and identify gaps in the geologic record.

Uniformitarianism The most basic principle used to interpret Earth history is James Hutton's **principle of uniformitarianism** (discussed in Chapter 1). It states that the geological processes taking place in the present operated similarly in the past. This cumbersome term can be summed up with a simple, semi-memorable catch-phrase: "The present is the key to the past." From this principle, we can assume that the physical laws that govern nature today—such as gravity—have been constant over time. Thus, even a billion years ago, rivers flowed *downhill,* landslides slid *downhill,* and volcanic ash fell from the sky—just as they do today.

We must temper our acceptance of the principle of uniformitarianism, however, by realizing that the Earth has not

Figure 8-3 Horizontal beds of sedimentary rock outside of Salt Lake City, Utah. Using the principle of superposition, we can instantly determine that the rocks at the peak of this formation are the youngest and those at the bottom are the oldest.

always been the way we see it today. Early in its history, the Earth lacked a well-developed atmosphere; the atmosphere it did have differed dramatically in composition, too. The surface lacked its present-day vast ocean cover, volcanic processes may have been more intense (the Earth's interior was probably hotter), and the planet rotated considerably faster (300 million years ago, for example, our Earth spun more quickly and our days were shorter, with 440 days in a year). Thus the *rates* of geological processes may have differed significantly, even though the basic forces that control those processes—such as gravity and heat flow—were essentially the same.

Armed with our qualified acceptance of the principle of uniformitarianism, we can begin to examine specific aspects of rocks that enable us to deduce their relative ages. First we will look at how they are "positioned."

Horizontality and Superposition As we saw in Chapter 6, most sediments settle out from bodies of water and are deposited as horizontal or nearly horizontal layers. Lava flows also tend to solidify as horizontal layers. This **principle of original horizontality** is fundamental because rock layers are often found in nonhorizontal positions, indicating that they have been tilted or deformed by tectonic forces.

The **principle of superposition** states that rock materials generally get deposited on top of earlier, older deposits. Consequently, in any *undeformed* sequence of rock strata, the *youngest stratum will be at the top and the oldest at the bottom.* Here's an analogy. Suppose you're something of a slob around the house, but you're a well-informed slob: You read the newspaper every day. When you finish reading the paper, you cast it into the corner of your room. Over time, the pile grows. Obviously, unless it grows too high and topples over, the recent history of world events is preserved in the corner—with today's paper on top and progressively older editions below.

The principle of superposition, like the principle of original horizontality, commonly applies to sedimentary rocks and lava flows. In undisturbed horizontal layers of sedimentary or volcanic rock, any given rock bed will be younger than the beds below it, but older than the layers above it (Fig. 8-3).

When tectonic forces have tilted or even overturned a sequence of rock layers (Fig. 8-4), we must look for identifiable sedimentary structures, such as ripple marks, mudcracks, graded beds, or cross-bedding (all discussed in Chapter 6), to identify the upper surface of any one sedimentary layer. Similarly, vesicles—the pea-sized dimples left behind as gas bubbles burst at the tops of some lava flows (see Chapter 3)—may indicate the location of the flows' upper surfaces. Once we have identified the top of a rock layer, we can apply the principle of superposition to date relatively the layers below and above it.

Cross-Cutting Relationships Bodies of igneous rock often appear *within* other rock types, indicating that the latter were intruded, or cut across, by flowing magma. Because the other rock had to exist first before it could be intruded, it must be older than the rock formed by the intruding magma. The **principle of cross-cutting relationships** states that any intrusive formation, such as a dike or sill (Chapter 3), must be younger

(a)

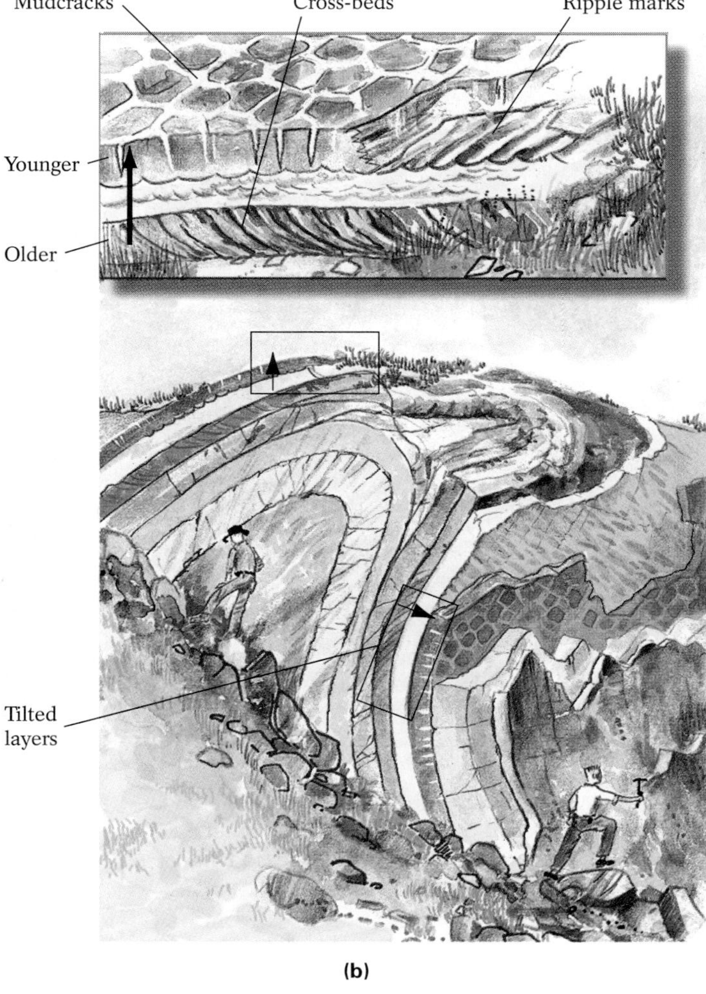

(b)

Figure 8-4 **(a)** Tilted turbidite beds in Zumaya, Spain. **(b)** When sedimentary rocks have been displaced from their initial horizontal orientation, we must use sedimentary structures to identify the top of any individual bed. Once the top of a single bed has been determined, the initial orientation of the beds can be reconstructed using the principle of superposition.

Figure 8-5 A cross-cutting relationship: A mafic dike cuts across these layered sedimentary rocks in the Grand Canyon, Arizona.

than the rock it cuts across (Fig. 8-5). Cross-cutting relationships also provide relative dates for faults—fractures in rocks that have been displaced (discussed in Chapter 9). *Faults must be younger than the rocks they displace* (Fig. 8-6).

Inclusions The **principle of inclusions** states that the fragments of other rocks contained within a body of rock must be older than the host rock. Sedimentary conglomerates, for example, contain pieces of preexisting rock, ranging in size from granules to boulders. These fragments were deposited along with finer sediment, with both types being incorporated into the conglomerate's matrix. Many igneous rocks

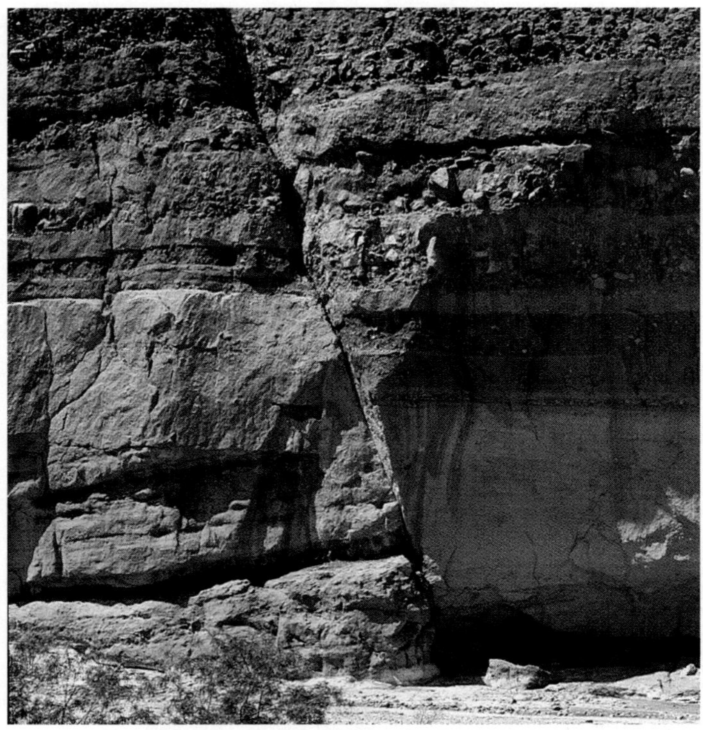

Figure 8-6 The fault that cuts through this sedimentary rock, found in California's Anza Borrego Desert State Park, must be younger than the deposition of any of the rock layers.

also contain pieces of preexisting rock that were broken by intruding magma and then incorporated into the new rock when it solidified. Thus a conglomerate bed must be younger than its included particles, and a body of granite must be younger than its xenoliths ("strange rocks") that were broken but not melted by intruding magma (discussed in Chapter 3) (Fig. 8-7).

Fossils and Faunal Succession Fossils (from the Latin *fossilus,* meaning "something dug up") consist of the remains of ancient organisms, or other evidence of their existence, that became preserved in geologic material. *Paleontologists* study them in part to reconstruct the evolution of life on Earth. When we apply the principles of superposition, cross-cutting relationships, and inclusions to fossil-bearing rocks, we can determine the order in which life forms developed and perhaps became extinct.

Sedimentary rocks, because they form at the Earth's surface, are more likely than other kinds of rocks to contain the remains of organisms. Even in sedimentary rocks, however, fossils are scarce: Only some 1% of all species that ever existed are believed to have been preserved as fossils. As is shown in Highlight 8-1 on page 214, organisms fossilize only under very special conditions. Many extinct organisms composed exclusively of soft tissue contained no preservable parts. Others, however, found their way into the fossil record, even without such hard parts.

Knowing the spatial relationships between rock strata, then, helps geologists estimate the relative ages of the fossils contained in them—and vice versa. Knowledge of a fossil organism's place in the geological record can help date a rock unit in which the fossil appears. The **principle of faunal succession** states that specific groups of fossils follow, or *succeed,* one another in the rock record in a definite order. Thus rocks containing identical fossils are thought to be identical in age. For example, when a geologist in Montana finds a specific assemblage of dinosaur fossils, the surrounding rock is most likely similar in age to rocks containing the same fossils in China.

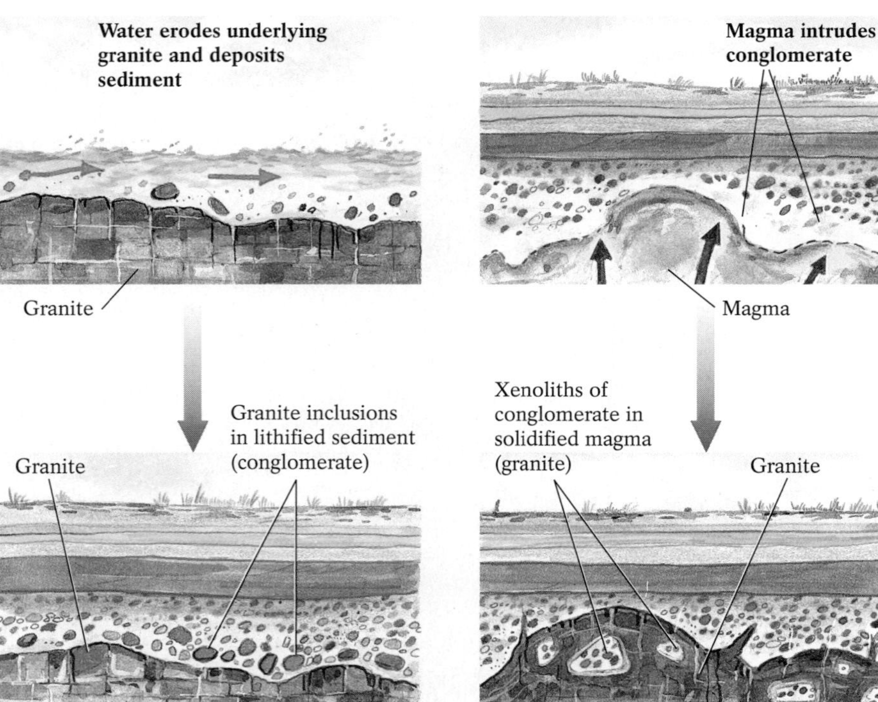

Figure 8-7 Inclusions are always older than the bodies of rock in which they are found. Part **(a)** shows one way in which fragments of preexisting granite can be incorporated into conglomerate sandstone. Conversely, part **(b)** shows how fragments of preexisting conglomerate may become incorporated into granite.

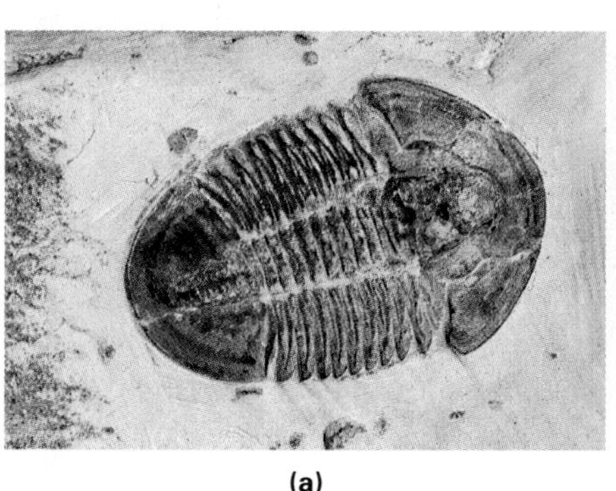

(a)

(b)

(c)

Figure 8-8 Index fossils. Each of these organisms lived during a specific period of Earth history, which enables us to date the rocks in which they are found. **(a)** A 500-million-year-old trilobite; **(b)** 300-million-year-old brachiopods; **(c)** a 200-million-year-old ammonite.

Certain varieties of organisms have existed (or did exist) for hundreds of millions of years. Sharks, for example, have inhabited the oceans for more than 400 million years. Other organisms lived only during relatively brief, specific periods of Earth history. When paleontologists find fossils of such short-lived organisms, known as **index fossils,** they know the age of the rocks to a fairly specific point (Fig. 8-8). The most useful index fossils include those species that were geographically widespread during their short time on Earth. They can be used to date many different—even far-flung—rock formations.

Unconformities

Given all these tools for relative dating, one would expect that geologists are well prepared to unravel all the secrets of Earth history. But several large problems frustrate our efforts. First, no single place on Earth contains all of the rock strata composing the entire geologic record. Thus we must travel from place to place, piecing together a region's geological history from numerous outcroppings of rock. Second, the geologic record at most localities is marked by **unconformities,** boundaries separating rocks of different ages that typically signal great gaps in time.

Unconformities occur where erosion has removed rock layers, or where layers were deposited during certain geologic periods. Many different types of unconformities exist. A *nonconformity* is the boundary between an unlayered body of plutonic igneous or metamorphic rock and an overlying layered sequence of sedimentary rock layers (Fig. 8-9a); rock underlying a nonconformity usually shows evidence of erosion that took place before the overlying rock was deposited.

An *angular unconformity* is the boundary between a sedimentary rock layer that has been deformed and eroded and a later, horizontal rock deposit that overlies it (Fig. 8-9b). One angular unconformity—that found at Siccar Point

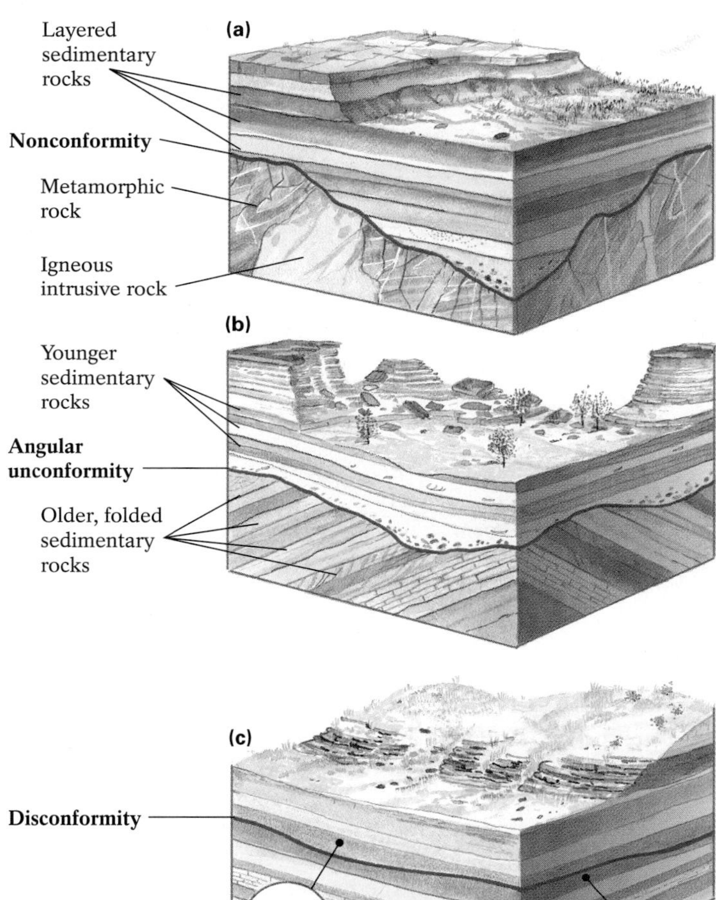

Figure 8-9 The three major types of unconformities between rock layers. **(a)** A nonconformity, between metamorphic or igneous rock and overlying sedimentary rock; **(b)** an angular unconformity, between older, deformed sedimentary rock and younger, undeformed sedimentary rock; **(c)** a disconformity, between parallel layers of sedimentary rock.

Highlight 8-1 How Fossils Form

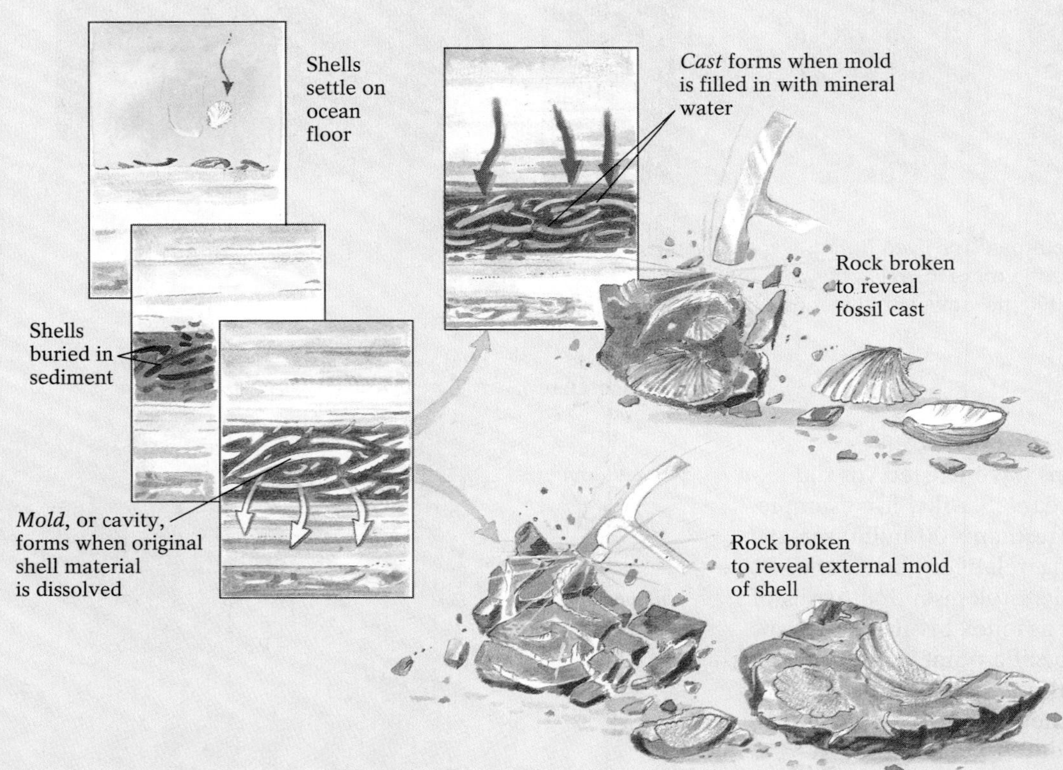
Shells
settle on
ocean
floor
Shells
buried in
sediment
Mold, or cavity,
forms when original
shell material
is dissolved
Cast forms when mold
is filled in with mineral
water
Rock broken
to reveal
fossil cast
Rock broken
to reveal external mold
of shell

Figure 8-10 An organism that has been buried in sediment may be preserved in the form of a mold or a cast in the resulting rock. A *mold* is an imprint of the organism's external form that remains behind in the surrounding rock long after the organism itself has decayed. After an organism's biological substance has dissolved away, circulating groundwater may precipitate mineral matter into the cavity left in the rock, forming a *cast* of the organism.

Most fossils form when the hard parts of organisms become buried in layers of sediment, where they decompose so slowly that the rock lithifying around them preserves their shapes. Geologists have unearthed countless fossils of ancient clam shells, many fish skeletons, and a fair number of dinosaur bones. They have found far fewer remains of worms, slugs, and jellyfish. Organisms with hard parts such as shells and bony skeletons are more likely to remain intact long enough to become preserved in sediment; in contrast, those composed solely of soft tissue would likely decompose rapidly at the surface in most environments. (For the same reason, hard, woody trees are more likely to be fossilized than are plants with no hard parts, such as mosses.) The hard parts of organisms also decompose or dissolve after burial if they come into contact with underground water. For that reason, a fossil is typically composed of some type of replacement material rather than an organism's original biological substance.

After the original bones, shells, and other organic materials have decomposed, a *mold*—an impression of the organism's form—remains in the enveloping sediment. At a later time, the mold may be filled with other materials. For example, circulating ion-rich groundwater may enter a mold and precipitate mineral matter in it. This process creates a solid *cast* of the mold, which is what we may find when we go fossil-hunting. The cast reveals only the organism's *external* form, but in certain cases, the form may be a remarkably faithful replication. Some deceased organisms may decompose very slowly and undergo cell-by-cell replacement of their original material by such secondary substances as calcium carbonate, silica, pyrite, or other substances dissolved in groundwater. This replacement process may retain the *internal* structure of the organism, which stirs great interest in paleontologists. Figure 8-10 depicts the processes that lead to the creation of molds and casts and replacement.

Preservation of an organism's original parts occurs only under unusual environmental conditions. For instance, the skeletons of thousands of mammoths, mastadons, saber-toothed cats, and other late Ice Age mammals were preserved in the viscous tar of the La Brea tar pits, in what is now downtown Los Angeles. Intact woolly mammoths have been dug up from the tundras of Alaska and Siberia, where they had become frozen in arctic *permafrost* (permanently frozen ground). (Paleontologists and others have even dined on thawed, yet still-edible, mammoth

(a)

(b)

(c)

Figure 8-11 Various types of fossils. **(a)** A 3.5-million-year-old fly in amber; **(b)** a saber-toothed cat skeleton from the La Brea tar pits; **(c)** ancient worm burrows. Only the fossils in parts **(a)** and **(b)** contain actual remains of the organisms.

meat.) The original carcasses of entire insects have been recovered from hardened tree sap, or *amber,* in which they became stuck millions of years ago. In each of these cases, the enveloping material did not permit the circulation of water or air, preventing the organism's original material from decomposing.

Even when we don't find fossils showing an organism's physical form, we sometimes find other evidence of its presence through *trace fossils.* If an animal burrowed through sediment, it may have left "tunnels"; if it crawled across soft mud, it may have etched delicate tracks into the sediment. Larger animals may have left behind footprints as they walked around. We have even discovered rounded and polished rocks, called *gastroliths,* that dinosaurs swallowed to help grind their coarse food and aid digestion and lithified "droppings," or *coprolites,* from a variety of creatures. A few of the many types of fossils are depicted in Figure 8-11.

A final note about fossils: Be careful where you dig them up. Recently, a commercial fossil dealer was convicted of stealing fossils from federal land. He is currently appealing his two-year sentence and $5000 fine for illegal excavation of "Suzie," a nearly complete *Tyrannosaurus rex* specimen taken from public and private lands in South Dakota. Apparently, given their considerable value on the museum market, dinosaur fossils are big business. Better ask before you walk off with one.

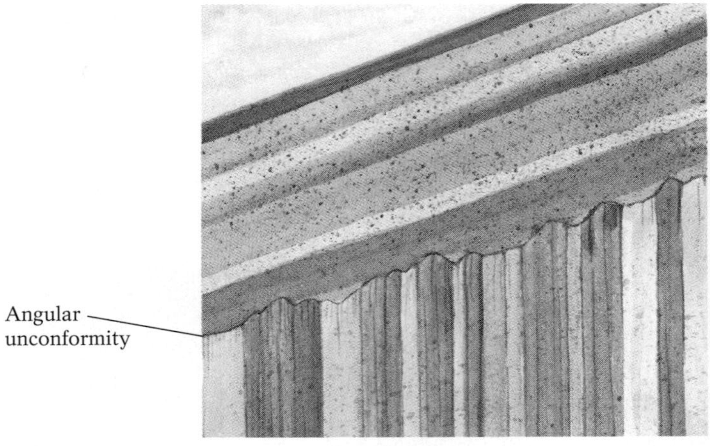

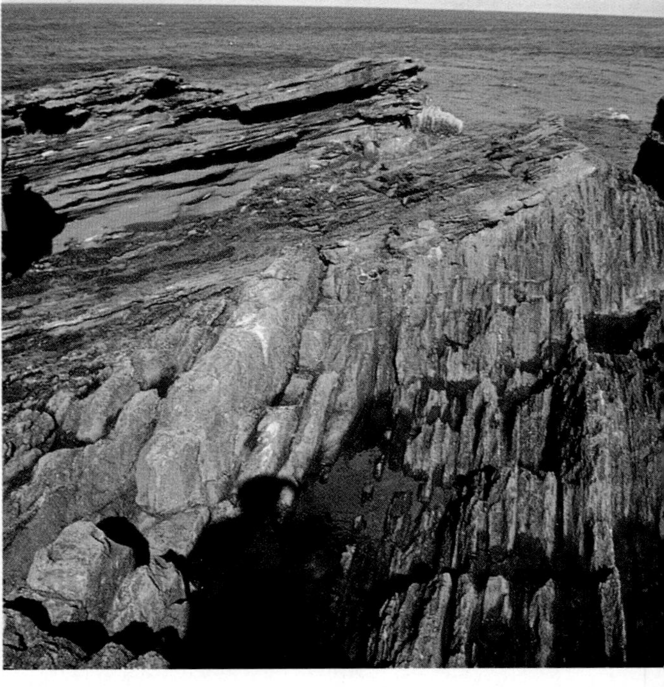

Figure 8-12 The angular unconformity at Siccar Point, Scotland. James Hutton, the founder of modern geology (Chapter 1), correctly interpreted this formation as showing layers of rock that had been deposited horizontally, lithified, tilted to a nearly vertical position, eroded smooth at the surface, and then overlain with subsequent horizontally deposited layers that were also later deformed—a graphic example of the dynamics that shape the Earth's features, and the amount of time necessary to achieve them.

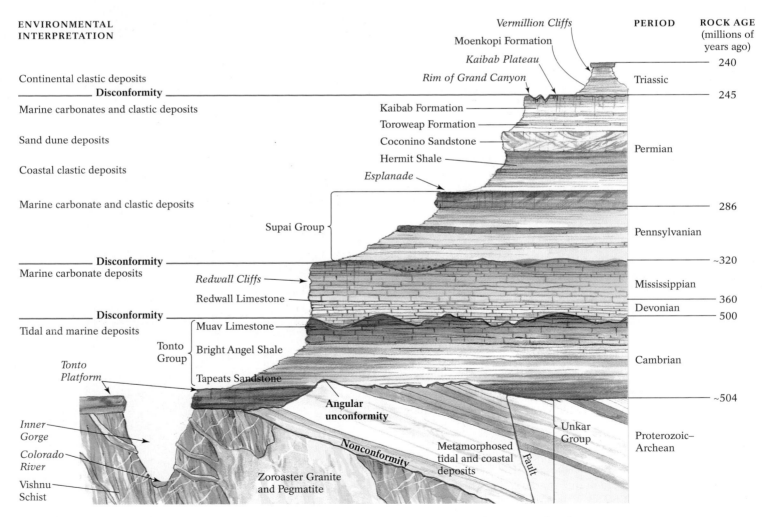

ENVIRONMENTAL INTERPRETATION
Continental clastic deposits
Disconformity
Marine carbonates and clastic deposits
Sand dune deposits
Coastal clastic deposits
Marine carbonate and clastic deposits
Disconformity
Marine carbonate deposits
Disconformity
Tidal and marine deposits

Vermillion Cliffs
Moenkopi Formation
Kaibab Plateau
Rim of Grand Canyon
Kaibab Formation
Toroweap Formation
Coconino Sandstone
Hermit Shale
Esplanade
Supai Group
Redwall Cliffs
Redwall Limestone
Tonto Group: Muav Limestone, Bright Angel Shale, Tapeats Sandstone
Tonto Platform
Angular unconformity
Inner Gorge
Colorado River
Vishnu Schist
Zoroaster Granite and Pegmatite
Nonconformity
Metamorphosed tidal and coastal deposits
Fault
Unkar Group

PERIOD	ROCK AGE (millions of years ago)
	240
Triassic	
	245
Permian	
	286
Pennsylvanian	
	~320
Mississippian	
	360
Devonian	
	500
Cambrian	
	~504
Proterozoic–Archean	

Figure 8-13 The Grand Canyon in northern Arizona affords one of the Earth's most complete exposed rock sequences—almost 2 vertical kilometers (about 1.2 miles) of rock whose age ranges as high as 3 billion years. Even here, however, unconformities represent large gaps in the local rock record.

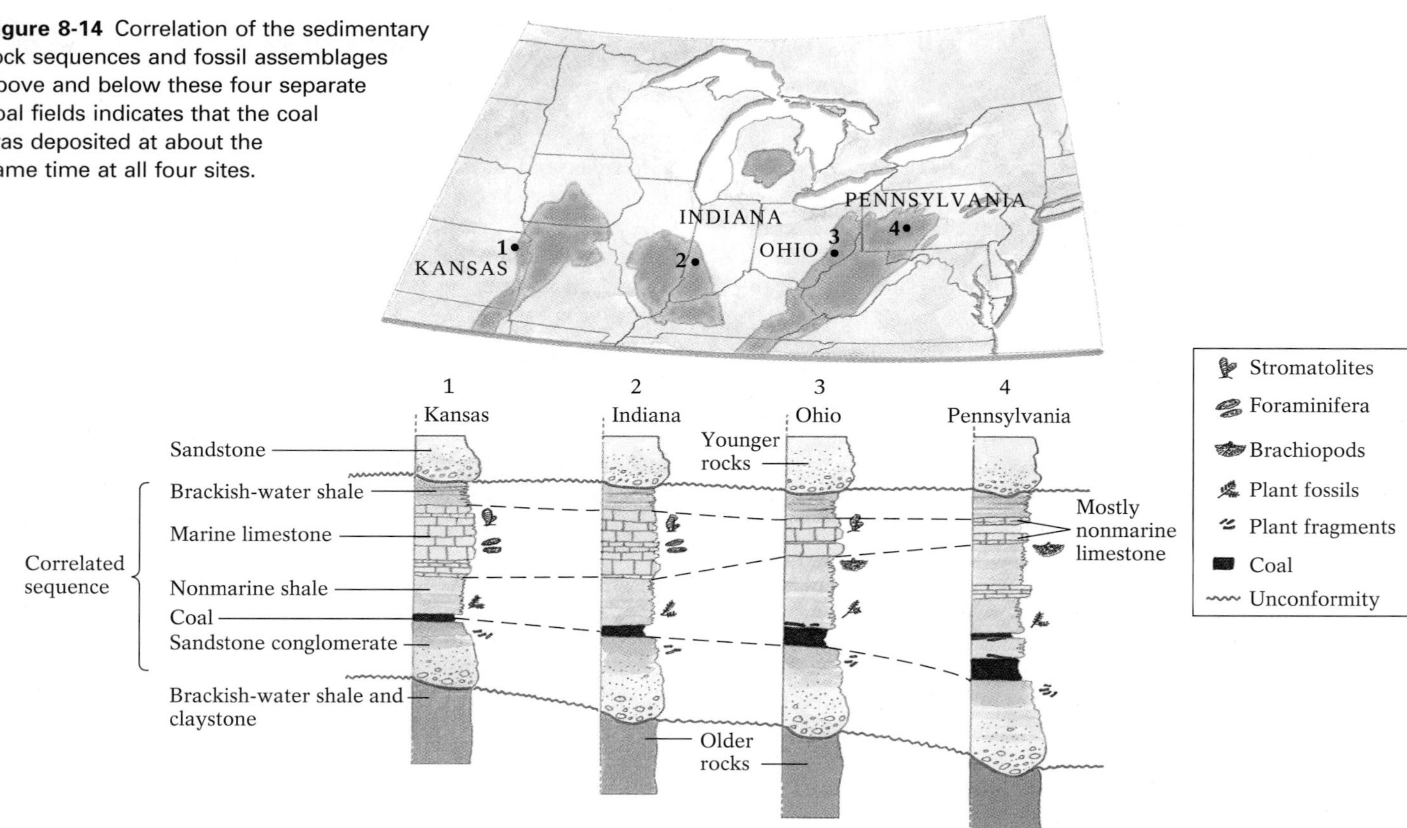

Figure 8-14 Correlation of the sedimentary rock sequences and fossil assemblages above and below these four separate coal fields indicates that the coal was deposited at about the same time at all four sites.

in Scotland—has particular historical significance in geology (see Figure 8-12).

A more subtle type of unconformity, a *disconformity,* occurs between *parallel* layers of sedimentary rock (Fig. 8-9c). We know such a boundary is a disconformity if fossil groupings above and below it differ substantially in age, or if the surface of the lower layer shows evidence of significant erosion. Disconformities can also be present in a succession of lava flows. A recently discovered unconformity in northwest Australia—a relic of perhaps the oldest continental landscape in the world—is the subject of Highlight 8-2.

Virtually every sequence of sedimentary rock layers contains one or more unconformities. In fact, the sedimentary rock record of most regions represents only 1% to 5% of the Earth's existence. In other words, much more of the record is missing than is present. The most complete rock sequences, such as the one exposed in Arizona's Grand Canyon, span only some 20% of the Earth's existence (Fig. 8-13). The following analogy will help show how geologists deal with such incomplete records. Suppose despicable vandals broke into your college bookstore and removed 80% of the pages in this text randomly before you bought it. How would you be able to take a comprehensive final exam in geology without those missing pages, and without the benefit of your professor's thoughtful, informative lectures? One possible solution might be to take out a loan and buy a lot of geology texts—to be sure that every page is available. Another option might be to team up and share texts with your classmates to accomplish the same goal. Earth historians employ similar strategies to piece together isolated glimpses of Earth history from often widely separated localities.

Correlation

To determine equivalence in age between geographically distant rock units, geologists look for paleontological or mineralogical similarities between the units, a process known as **correlation** (Fig. 8-14). In general, because specific environmental factors influence the origin of a region's rocks and their subsequent erosion, the farther apart two rock formations are in terms of geography, the less likely they will possess identical rock sequences. Yet, even formations thousands of kilometers apart may contain individual layers whose similarities suggest that they originated at the same time. The presence of markedly similar fossil assemblages in different rock units, for example, suggests that the layers are the same age. Geologists also base their correlations on individual index fossils; when we find them—even in widely separated rocks—we may conclude that the rocks formed during the same time period.

Highlight 8-2 The World's Oldest Landscape

By most accounts, the birth of the Earth's continents was a violent affair. Lighter minerals rising from the hot convecting mantle first coalesced as volcanic islands surrounded by lava seas (Fig. 8-15). These islands, the planet's first parcels of continental real estate, then collided over and over as they bobbed on the churning mantle below. Some probably plunged back to the Earth's interior, only to remelt and erupt again.

With so much planetary remodeling taking place, it is not surprising that very little of the Earth's early continental rock survived. Just recently, however, geologists searching for zinc deposits in northwestern Australia may have found a small piece of the early Earth's continental puzzle. Two layers of very old continental rock exposed over a distance of some 75 kilometers (46 miles) in Australia's remote, arid Pilbara region are separated by a discernible unconformity. The lower rocks are 3.52-billion-year-old greenstones and granites—a metamorphosed block of old continental crust. These rocks were tilted and eroded before being covered by 3.46-billion-year-old sedimentary rocks, some containing shallow-water ripple marks. The overlying sedimentary rocks contain the Warrawoona cherts, a marine layer in which fossil cells providing some of the oldest evidence of life on Earth have been found.

The unconformable surface between these rocks represents a time gap of some 60 million years. It shows tell-tale signs of weathering and erosion, indicating that wind and water must have scoured the rocks *above* sea level at some point. Thus these 3.52-billion-year-old water-worn continental rocks may constitute the Earth's oldest known *exposed* continental landscape. (Older rocks, such as the 3.96-billion-year-old gneisses of northern Canada, are highly metamorphosed rocks that originated deep in the Earth's interior and probably remained there until more recently exposed.) The rocks of Pilbara (Fig. 8-16) may contain the first direct evidence of continental lands that sat high and dry above ancient Precambrian seas. What's more, they're a half-billion years older than the next-oldest known continental rocks.

The rocks of the Pilbara unconformity may eventually tell us much about our planet's early atmosphere. During those days, the young Sun's rays were probably only 70% as strong as they are today. No evidence suggests that the planet was sharply colder back then, so scientists speculate that the Earth's surface may have been warmed by a "super" greenhouse effect, perhaps caused by a very high carbon dioxide content in the atmosphere. To find traces of the ancient atmosphere, geologists study ancient soils, or paleosols (discussed in Chapter 5). The Pilbara rocks appear to contain such paleosols, and initial analyses of iron carbonate minerals in them indicate that CO_2 levels at the time may have been 100 to 1000 times higher than those found today. Geologists are also searching these rocks for chemical signs of early life—perhaps the oldest unequivocal evidence of life on Earth.

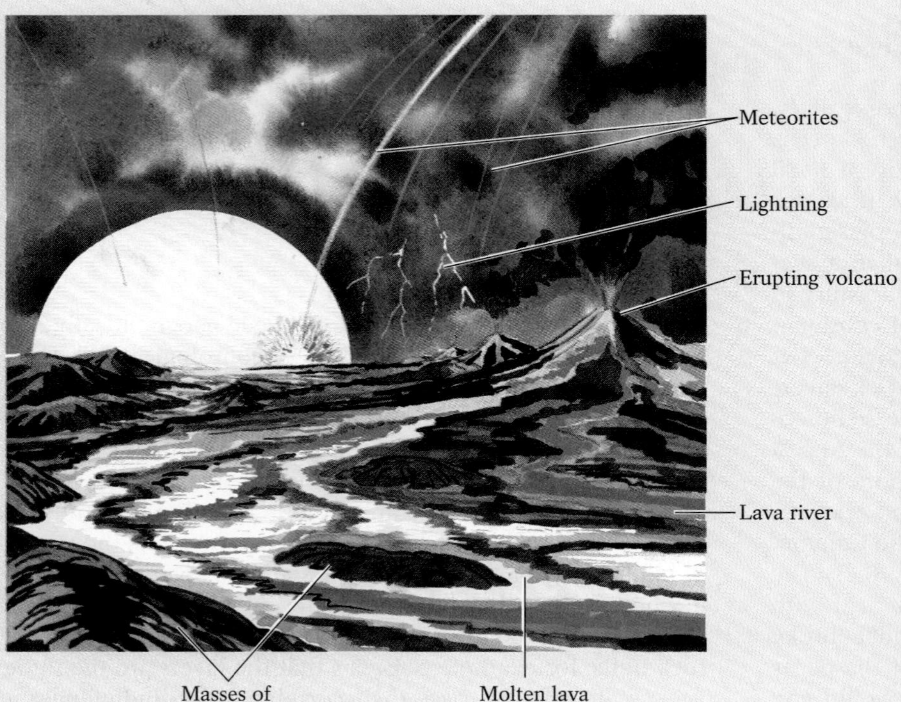

Figure 8-15 A hypothetical view of the early Earth—about 4 billion years ago. The Earth's greater internal heat at that time would have fueled more extensive volcanism; more frequent meteorite strikes would have periodically ruptured the planet's earliest crust.

Figure 8-16 The 3.5-billion-year-old sedimentary rocks of Australia's Pilbara region are some of the oldest surviving remnants of Earth's earliest crust.

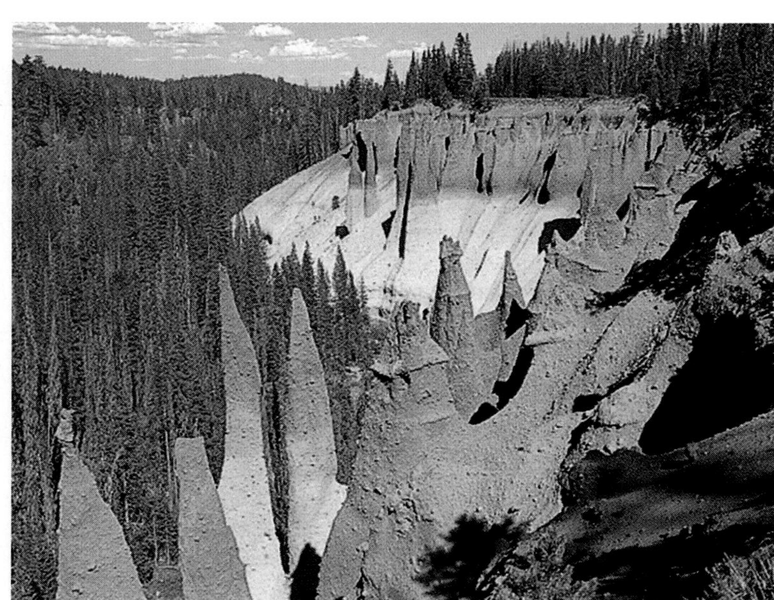

Figure 8-17 Ash from the eruption of Mount Mazama 6700 years ago, at Crater Lake in Oregon. The lighter-colored ash consists of the frothy, highly felsic material that first erupted from the volcano; it was deposited first. The overlying, darker-colored layer is from a lower position in the magma chamber where slightly denser, less felsic material may have settled during differentiation; it erupted later and thus was deposited on top of the lighter-colored material. Rain and snow then eroded the ash beds to produce the pinnacles. The Mazama ash is a key bed found in same-aged rock units throughout the Pacific Northwest.

Distant rock sequences can also be correlated if they contain the same *key bed,* a distinctive stratum that appears at several locations. A key bed records a geological event *of short duration that affected a wide area.* The eruption of Mount Mazama in the Pacific Northwest some 6700 years ago, for example, produced a whitish-tan, highly felsic ash that was scattered so widely that it can be used to correlate the geologic strata of Oregon, Washington, British Columbia, and Alberta (Fig. 8-17). Grab a shovel and dig in much of this region and you would have a good chance of finding Mazama ash.

Distinctive sedimentary facies of the same age in a particularly wide range of formations could indicate global environmental conditions during deposition, such as a widespread climate change or a worldwide rise or fall in sea level. The extensive coal beds on both the North American and Eurasian continents, for example, indicate the occurrence of a distinctively warm episode about 300 million years ago. This episode produced substantial vegetation that later became covered by sediments and other deposits and eventually lithified as coal.

Relative Dating by Weathering Features

We saw in Chapter 5 how the processes of mechanical and chemical weathering convert solid rock to loose regolith and soil. Because rock tends to become increasingly weathered over time, comparing the extent of weathering in different bodies of rock can yield clues to their relative ages. This method of dating generally is limited to geologic materials that are less than a few million years old.

Crack open a rock that has been sitting at the Earth's surface and you will often see a *rind* of weathered material at or near the rock's surface. Such rinds arise when water infiltrates rocks, chemically altering their original composition into some type of weathering product. In general, the longer a rock is exposed to weathering, the thicker its weathering rind will be. Thus, a comparison of weathering rinds (among rocks having the same initial composition) can distinguish, for example, between deposits of different ages. This relative dating technique is most accurate for basalt, whose ferromagnesian minerals (olivine, pyroxenes) weather quickly at the Earth's surface.

A *hydration rind,* similar to a weathering rind, develops when fresh obsidian (volcanic glass), formed from the rapid cooling of felsic magma, is exposed to atmospheric moisture. Hydration incorporates water from the air into obsidian's structure, forming a rind that increases in thickness over time. Thus, we can establish the relative ages of obsidian-bearing materials by comparing their hydration rinds. Because obsidian projectile points and cutting blades are often found at ancient sites of human habitation, archaeologists use hydration rinds to date finds at such sites.

Knowing the degree to which an area's landscape features, from its boulders to its soils, have weathered may enable geologists to date relatively different landscapes (Fig. 8-18). Of course, age is not the only factor that influences weathering; we must also consider such factors as climate and rock composition. After accounting for such variations, however, it is possible to conclude that, in general, an old, deeply weathered landscape will contain weathered and broken-down boulders, whereas young landscapes will more likely display a high concentration of intact surface boulders. An experienced geologist can even estimate the relative age of an intact boulder with a single hearty whack of a geologic hammer: The resounding ring of an unweathered boulder sounds distinctly different from the dull thud of an aged, weathered boulder. A geologist's trained eye also notes the extent to which weathering and other surface processes

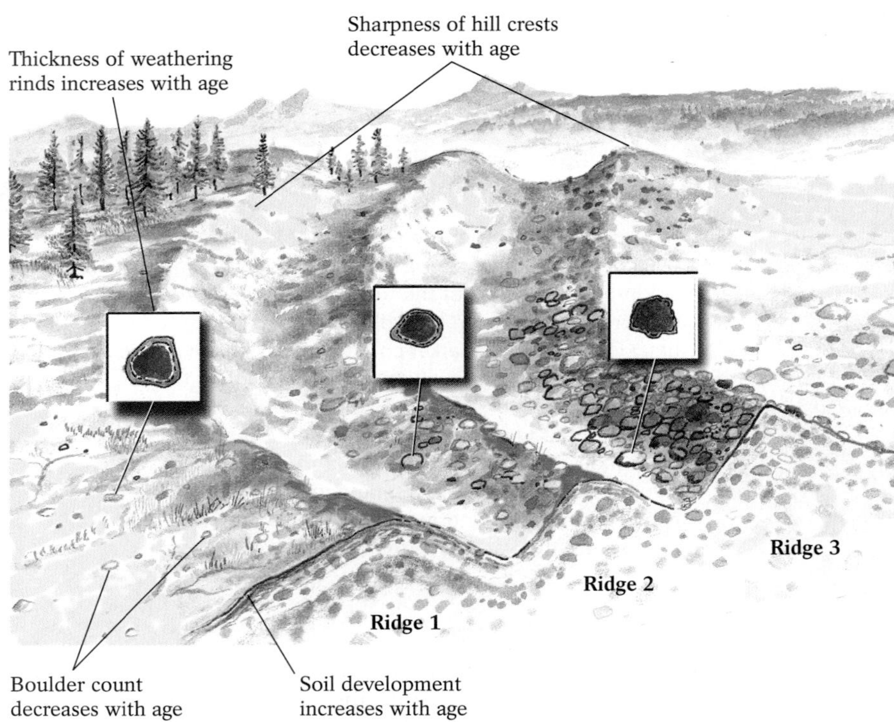
Sharpness of hill crests
decreases with age
Thickness of weathering
rinds increases with age
Ridge 3
Ridge 2
Ridge 1
Boulder count
decreases with age
Soil development
increases with age

Figure 8-18 This hypothetical landscape shows the variety of weathering phenomena that can help determine the relative age of an area and its constituent materials. Ridges 1, 2, and 3 are deposits left behind by a retreating glacier. The sharp hill crests and large number of boulders on ridge 3 indicate relatively little exposure to weathering; the thickness of the weathering rinds on the boulders in ridge 1 indicates that they are fairly old; the soils in ridges 1 through 3 are progressively less developed. These clues suggest that ridge 1 is the oldest and ridge 3 the most recent of the glacial deposits.

alter the shape of hills. An older landscape tends to have gentler slopes and rounded, lower, hill crests; younger landscapes tend to have steeper slopes and sharper hill crests.

Soil, as we saw in Chapter 5, develops over time, and its maturity reflects the climate, topography, biological activity, and parent material that produced it. We can estimate the amount of time that a particular soil needed to develop from the thickness of the overall soil profile, the depth to which its soluble elements have been dissolved away, and the number and degree of development of its horizons. Thus we can determine the relative ages of soils in topographically and climatically similar landscapes having the same rock types by comparing soil-horizon development.

Determining Absolute Age

Relative-dating principles and techniques often tell us only that rock A is older or younger than rock B. Understandably, geologists prefer the more specific "years ago" information yielded by absolute-dating methods. All absolute-dating methods depend on some type of "natural clock"—a process that operates at a constant quantifiable rate over long periods of time. For example, as a tree grows it adds rings to its trunk every year. We can determine a tree's absolute age by counting its rings. Nature has provided a substantial number of such "clocks" that help us determine the ages of a range of geological materials.

Radiometric Dating

One way of dating relies on a form of nuclear clock—the rate of decay of radioactive isotopes within rocks. We saw in Chapter 2 that the atoms of certain chemical elements exist as *isotopes,* containing different numbers of neutrons in their nuclei. *Radioactive* isotopes have nuclei that spontaneously decay by emitting or capturing a variety of subatomic particles. The decaying radioactive isotope, known as the **parent isotope,** evolves into a decay product, called the **daughter isotope.** Loss or gain of *neutrons* converts a parent isotope into a daughter isotope of the same element. Loss or gain of *protons* changes the parent element into an entirely different daughter element having a new set of chemical and physical properties. After a number of intermediate steps, an unstable radioactive parent isotope eventually becomes a stable, nonradioactive daughter isotope.

Unstable nuclei decay in three primary ways:

- *Alpha Decay* Alpha particles are composed of two protons and two neutrons (essentially they are helium nuclei). With the expulsion of an alpha particle, the atomic weight of an atom decreases by 4 and the atomic number decreases by 2. Of course, this change in atomic number produces an atom of an entirely different element. For example, uranium-238, an isotope of the element uranium, has an atomic number of 92 (that is, 92 protons in its nucleus). When an atom of uranium-238 undergoes alpha decay, it loses 2 protons and 2 neutrons and is converted to an atom of an element

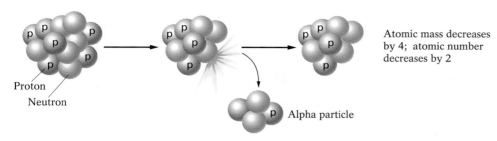

Radioactive parent nucleus

Decay process

Daughter nucleus

Proton

Neutron

Atomic mass decreases by 4; atomic number decreases by 2

Alpha particle

(a) Alpha decay

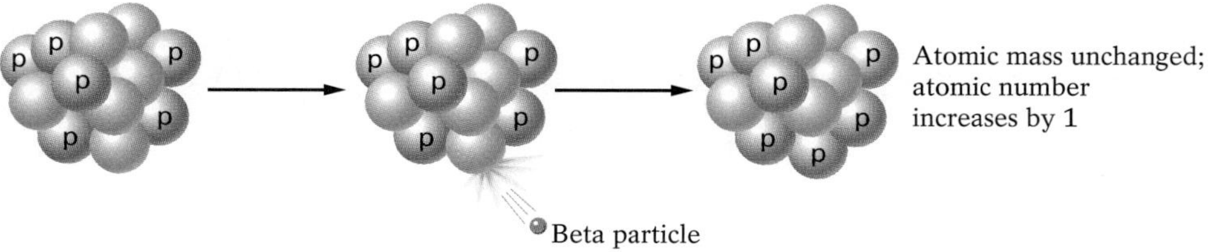

(b) Beta decay

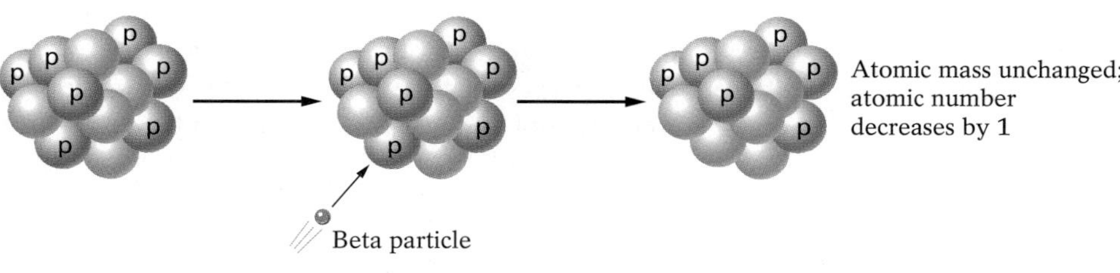

(c) Electron capture

Figure 8-19 Various processes by which unstable atomic nuclei decay. **(a)** alpha decay, which involves expulsion of a particle equivalent to a helium nucleus; **(b)** beta decay, which involves expulsion of an electron; **(c)** electron capture, which involves attraction of one of an atom's orbiting electrons to an atomic nucleus where it combines with a proton to form an additional neutron.

with 90 protons—that is, thorium. After the loss of the 4 nuclear particles, its atomic weight is 234 (Fig. 8-19a).

- *Beta Decay* Beta particles are essentially electrons. Ordinarily, we picture electrons as minute negatively charged particles that orbit the nucleus of an atom and are gained, lost, and shared in chemical bonding. The electrons that make up beta particles, however, do not originate from the orbiting cloud, but instead result from the breakdown of a neutron. A neutron is composed of a proton and an electron. Thus, when a neutron decays and its component electron is expelled, an additional proton is left in the nucleus. The addition of this proton increases the atomic number, creating an atom of a different element. An example: radioactive potassium-40 (atomic number 19) decays by beta decay to calcium-40 (atomic number 20). Note that while the atomic number changes, the atomic weight remains the same; the loss of the neutron is offset by the addition of the proton, and the departure of the nearly weightless electron leaves the atomic weight unchanged (Fig. 8-19b).

- *Electron (Beta) Capture* In electron capture (also known as beta capture), the nucleus abducts an electron from the atom's own orbiting cloud. The newly captured electron joins with a proton—remember, opposites attract—to manufacture an additional neutron. In this case, because a proton is lost by conversion to a neutron, the atomic number decreases by 1 but the atomic weight remains the same. For example, when potassium-40 emits beta particles to form calcium-40, a percentage of the unstable potassium atoms capture electrons. The change in the number of protons creates a new element, argon-40 (atomic number 18). Note that the atomic weight remains the same (Fig. 8-19c).

Radioactive isotopes are incorporated into rocks and minerals in various ways. For example, as minerals crystallize in cooling magma, they sometimes trap atoms of such isotopes within their crystal structure. These radioactive isotopes begin decaying immediately in a continuous process. **Radiometric dating** uses this continuous decay to measure the amount of time elapsed since the rock's formation. As time passes, a rock will contain less and less of its initial radio-

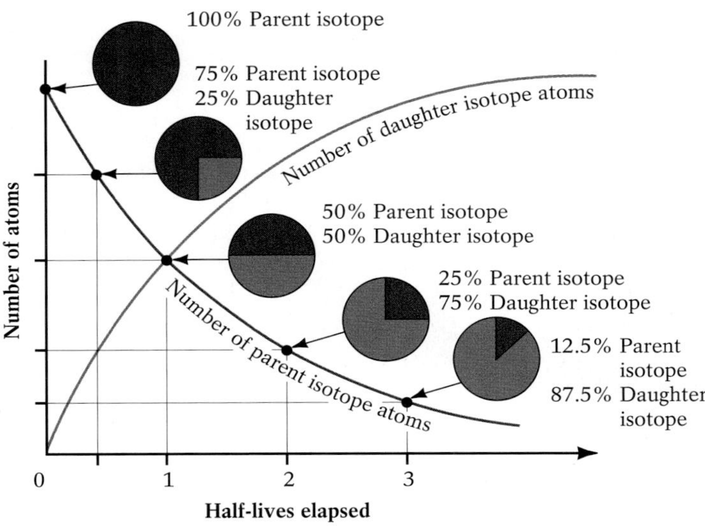

Figure 8-20 Radioactive decay converts a radioactive parent isotope to a stable daughter isotope. With the passage of each half-life, the number of atoms of the radioactive isotope is reduced by half, whereas the number of atoms of the daughter isotope increases by the same quantity. If they know the half-life of the isotope, geologists can measure the ratio of parent to daughter isotopes to determine the absolute ages of some rocks.

active parent isotopes and more and more of their daughter products. Uranium-238, for example, commonly found in zircon crystals in granitic rocks, decays to form a stable daughter isotope, lead-206. Over time, a uranium-rich zircon crystal gains lead-206 as it loses uranium-238. Radioactive decay rates are constant, just like the mechanisms of a clock. The decay rates are unaffected by changes in temperature or pressure or by chemical reactions involving the parent isotope. By measuring the ratio of parent to daughter isotopes and comparing it with the parent element's known rate of radioactive decay, we can determine the absolute age of a rock.

The time it takes for *half* the atoms of the parent isotope to decay is known as the isotope's **half-life.** For example, if a rock has 12 parent and 12 daughter atoms—a parent-to-daughter ratio of 1:1—the original rock must have had 24 radioactive parent atoms, and one half-life has elapsed. After another half-life, the number of remaining parent atoms would halve again, leaving 6 unstable parent atoms and a total of 18 stable daughters—a parent-to-daughter ratio of 1:3. Note that the total number of parent and daughter atoms remains the same (in this example, 24) throughout the progression. Regardless of which radioactive isotope decays, the proportion of parent to daughter atoms exists as a predictable ratio at each half-life (Fig. 8-20).

The half-lives of some isotopes span billions of years. Certainly no one sits in a laboratory waiting for that time to elapse to determine the half-life of a given parent–daughter system. Instead, we place samples containing measured quantities of the radioactive isotope in a device that counts the number of nuclear particles emitted per unit of time; we can then extrapolate the decay rate in years.

We have to know the specific decay rates of individual isotopes to determine the age of a given rock. A *mass spectrometer* measures the precise quantities of parent and daughter isotopes in a substance such as a mineral, rock, bone, tree, or shell to determine their ratio. From the ratio, we can establish how many half-lives (or fractions of half-lives) have elapsed since the radioactive parent isotope was incorporated into the substance. We then multiply the number of half-lives by the known length of one half-life of that isotope to calculate the substance's absolute age.

Factors Affecting Radiometric Results Radiometric dating is typically more accurate and useful for igneous rocks than for other types of rocks. As an igneous rock crystallizes, it locks radioactive isotopes into the crystal lattices of its minerals. That action sets the radiometric "clock" within the rock. It is more difficult to determine the radiometric age of a clastic sedimentary rock, simply because such rocks typically contain mineral fragments of countless preexisting rocks of different ages. Dating these fragments yields the age of their source rocks, but not the age of the sedimentary rock itself. To determine the absolute age of a sedimentary rock, we first apply the stratigraphic principles of relative dating, particularly those involving fossils. We then correlate the rock formation with others containing the same fossils, but with known ages derived from radiometric dating of igneous rocks that may cross-cut, underlie, or overlie the sedimentary rock formation (Fig. 8-21).

Radiometric dating is most reliable when a rock or mineral is a "closed" system—one in which atoms of the parent and daughter isotopes have not migrated into or out of the mineral or rock being dated. Metamorphic rocks are often not closed systems. Heat and circulating liquids and gases during metamorphism may affect the parent–daughter isotopic ratios, essentially "resetting" the isotopic clock. Although these factors do not alter the *rate* at which isotopes decay, they can affect the *relative number* of parent and daughter isotopes ultimately found in a rock or mineral. They may enable atoms to *diffuse,* or migrate away, from their point of origin or, conversely, they may promote incorporation of additional parent or daughter isotopes from elsewhere. Heating, for example, enhances diffusion for two reasons: (1) Increased thermal energy increases the atoms' mobility, and (2) the structure of heated mineral crystals expands, allowing trapped atoms of the daughter isotope to escape into the surrounding rock. In both cases, the loss of diffused daughter isotopes makes a crystal appear younger than it really is. When a rock that has been heated eventually experiences cooler conditions, like those seen after mountain building, and metamorphism ceases, the mineral structure contracts, and its crystals once again begin to trap the daughter isotope (Fig. 8-22). The parent–daughter isotopic ratios of minerals

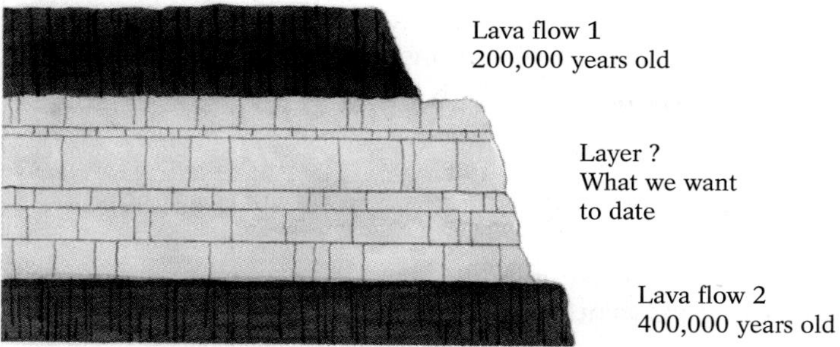
Lava flow 1
200,000 years old
Layer ?
What we want
to date
Lava flow 2
400,000 years old

Figure 8-21 Because the absolute age of sedimentary rocks is typically difficult to measure, geologists commonly rely on bracketing ages. Here, the ages of the surrounding layers of datable igneous rocks provide an age range for the intervening limestone layer.

in metamorphic rocks may not reveal the ages of the *original* pre-metamorphic minerals or rocks. But they can tell us when the metamorphism ended, yielding approximate dates for uplift and mountain-building events.

The reliability of radiometric dating depends, in part, on the condition of the dated materials. In fractured or highly weathered rock material, for example, migrating groundwater may have dissolved and carried away some of the parent or daughter isotopes, resulting in inaccurate parent–daughter ratios. Several recently developed devices, such as the *s*ensitive *h*igh-*r*esolution *i*on *m*icro*p*robe, or SHRIMP, help geologists analyze minute fragments from deep within a crystal, to ensure dating of unfractured, unweathered material.

The age of a substance may also limit the effectiveness of radiometric dating. Young rocks and minerals, for example, may not have produced enough daughter isotopes to permit an accurate measurement. Even when radioactive isotopes with half-lives on the order of hundreds of millions of years are involved, a rock must be roughly 100,000 years old to produce a measurable quantity of daughter isotope. For instance, rocks that contain uranium-238 (half-life of 4.5 billion years) must be at least 10 million years old to yield a measurable amount of decay-produced lead-206.

Very old rocks and minerals may have exhausted virtually all of their parent isotopes. For radioactive isotopes with half-lives of 50 million years, very few parent isotopes will remain after roughly 500 million years, or 10 half-lives. (After 10 half-lives, only 1/1024 of the parent isotope would remain.) With so few parent atoms, the reliability of radiometric dating diminishes substantially. The radioactive isotope carbon-14, for example, has a half-life of only 5730 years; after 12 half-lives, or approximately 70,000 years, the amount remaining is too small to be measured accurately.

We'll now look at some of the most common radiometric-dating systems. Each of these systems has its own strengths and weaknesses. Some can be applied only to very old rocks; others work with only very young rocks and materials. Some are subject to various uncertainties, such as those described earlier in this section. Geologists always run tests on multiple samples of a rock or mineral to eliminate experimental error and the possibility of sample

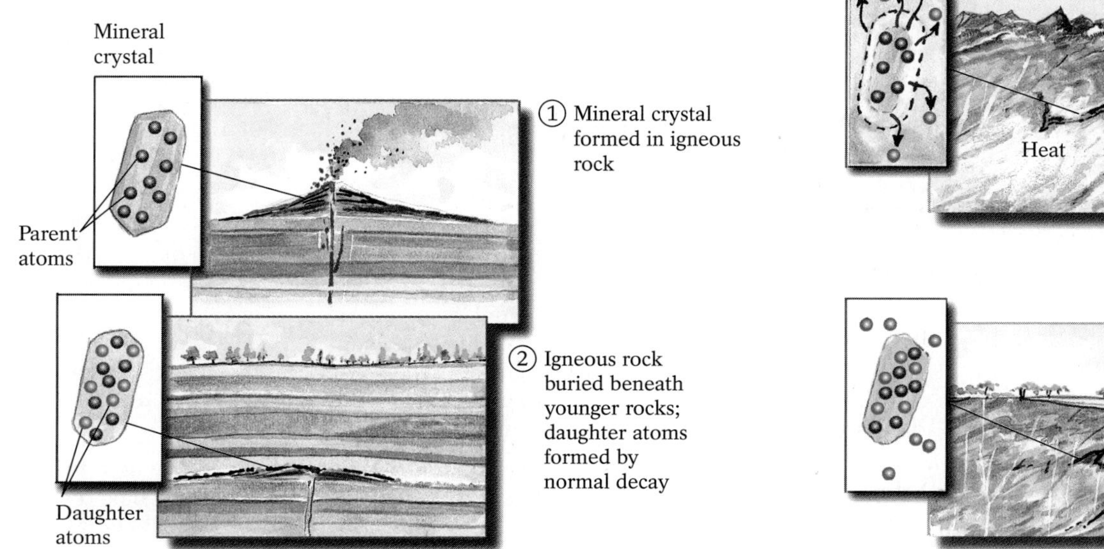
Mineral
crystal
Parent
atoms
① Mineral crystal
formed in igneous
rock
② Igneous rock
buried beneath
younger rocks;
daughter atoms
formed by
normal decay
Daughter
atoms

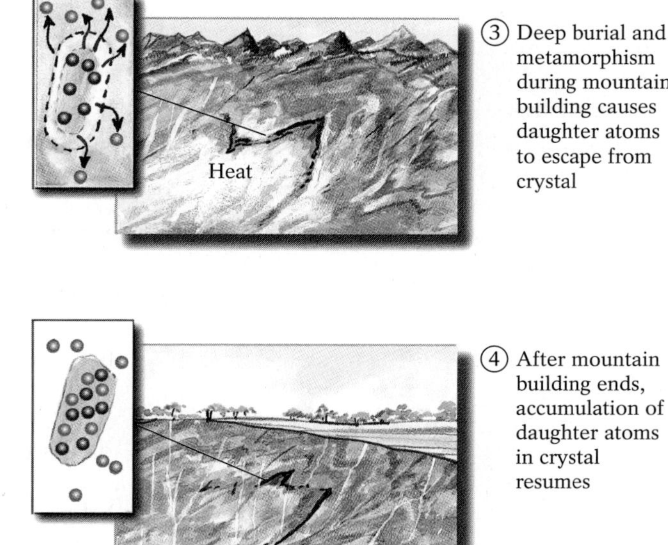
Heat
③ Deep burial and
metamorphism
during mountain
building causes
daughter atoms
to escape from
crystal
④ After mountain
building ends,
accumulation of
daughter atoms
in crystal
resumes

Figure 8-22 Loss of daughter isotopes from rock metamorphosed during mountain building. The heat produced in this process causes mineral crystals to expand, allowing daughter atoms to escape and migrate elsewhere, even as the decay of the parent isotope proceeds normally. The loss of these daughter isotopes during this period skews the resulting parent-to-daughter ratio so that the rock appears younger than it is. After metamorphism ceases and the crystals contract, the daughter isotopes start accumulating again.

contamination. Wherever possible, they also use several dating systems to date the same sample, checking to see whether the results of the different systems converge. Rarely, if ever, is a rock or mineral dated by a single analysis of a single sample. Dating geological materials is painstaking business, taken very seriously by its practitioners. The validity of our interpretations of past events relies heavily on the accuracy of these dates.

Uranium, Thorium, Potassium, and Rubidium Dating Systems

Although scores of radioactive isotopes exist, only a few are useful for dating purposes. Some decay too rapidly, on the order of thousandths of a second; others decay too slowly, on the order of tens of billions of years. Some migrate into and out of rocks and minerals too readily under ordinary environmental conditions, and still others are simply too rare. To satisfy the needs of dating systems, isotopes must be present in common Earth materials, not be overly susceptible to gains or losses of parent and daughter isotopes, and have half-lives of at least several thousand years. Sometimes rocks contain more than one radioactive-isotope system, allowing us to cross-check the dates yielded by each system.

Table 8-1 shows the most frequently used radioactive isotopes, along with their daughter isotopes, half-lives, and

Table 8-1 The Major Isotopes Used for Radiometric Dating

Method	Parent Isotope	Daughter Isotope	Half-Life of Parent (years)	Effective Dating Range (years)	Materials Commonly Dated	Comments
Rubidium–strontium	Rb-87	Sr-87	47 billion	10 million–4.6 billion	Potassium-rich minerals such as biotite, potassium muscovite, feldspar, and hornblende; volcanic and metamorphic rocks (whole-rock analysis)	Useful for dating the Earth's oldest metamorphic and plutonic rocks.
Uranium–lead	U-238	Pb-206	4.5 billion	10 million–4.6 billion	Zircons, uraninite, and uranium ore such as pitchblende; igneous and metamorphic rock (whole-rock analysis)	Uranium isotopes usually coexist in minerals such as zircon. Multiple dating schemes enable geologists to cross-check dating results.
Uranium–lead	U-235	Pb-207	713 million	10 million–4.6 billion		
Thorium–lead	Th-232	Pb-208	14.1 billion	10 million–4.6 billion	Zircons, uraninite	Thorium coexists with uranium isotopes in minerals such as zircon.
Potassium–argon	K-40	Ar-40	1.3 billion	100,000–4.6 billion	Potassium-rich minerals such as amphibole, biotite, muscovite, and potassium feldspar; volcanic rocks (whole-rock analysis)	High-grade metamorphic and plutonic igneous rocks may have been heated sufficiently to allow Ar-40 gas to escape.
Carbon-14	C-14	N-14	5730	100–70,000	Any carbon-bearing material, such as bones, wood, shells, charcoal, cloth, paper, animal droppings; also water, ice, cave deposits	Commonly used to date archaeological sites, recent glacial events, evidence of recent climate change, environmental effects of human activity.

the particular minerals or rocks in which they are commonly found.

1. **Uranium–thorium–lead dating.** Find a rock containing uranium, such as a granite or a marine limestone, and you have *three* radiometric dating systems with which to work. Virtually all uranium deposits contain three radioactive isotopes—uranium-238, uranium-235, and thorium-232. Each isotope decays through a series of steps to form a different isotope of another element—lead. Uranium and thorium isotopes have extremely long half-lives, which makes them invaluable in dating some of the Earth's oldest rocks (Fig. 8-23).

2. **Potassium–argon dating.** Potassium is one of the Earth's most abundant elements, found in such common rock-forming minerals as biotite and muscovite mica, potassium feldspars, and glauconite (a common mineral in sedimentary rocks). All naturally occurring potassium contains some radioactive potassium-40, which decays to argon-40, giving us the potassium–argon system. A gas that does not readily bond with other elements, argon-40 exists in minerals only as the daughter product of potassium-40 decay. When trapped within the crystal structures of certain minerals, it can be measured to yield a reliable date. We must be cautious, however, when dating rocks with this system. Argon-40 gas may escape from a rock that is heated during metamorphism or from one that is very old or extensively weathered. In such cases, measurement of the remaining argon-40 could suggest that the rock is younger than it actually is, because less daughter product would be present in the rock than was actually produced.

   Because potassium is more common than uranium and has a long half-life, the potassium–argon system has proved very useful for dating the Earth's oldest rocks in a wide range of locations. We can also use even extremely small amounts of argon-40 to determine the ages of potassium-rich rocks as young as 100,000 years. In East Africa, biotite grains in volcanic deposits above and below sedimentary layers containing fossils of our earliest human ancestors have yielded potassium–argon dates ranging from 3.6 to 4.0 million years.

3. **Rubidium–strontium dating.** Radioactive rubidium-87, which occurs along with potassium in numerous minerals and rocks, produces a daughter isotope, strontium-87, that does not escape as a gas. The rubidium–strontium system can therefore reliably date both metamorphic and igneous rocks. With its 47-billion-year half-life, this system is most effective for rocks at least 10 million years old. It is used primarily to date very ancient rocks, deep-Earth plutonic rocks, and Moon rocks. It also provides a good cross-check of potassium–argon dates.

Figure 8-23 The oldest rocks yet discovered—3.96-billion-year-old Acasta Gneiss from the Yellowknife area of Canada's Northwest Territories. Uranium and lead isotopes, measured with a sensitive, high-resolution ion microprobe (SHRIMP), were used to date these rocks.

Carbon-14 Dating The radiometric systems discussed previously have such long half-lives that they are generally useless for dating Earth materials that are less than 100,000 years old. Radioactive carbon-14, however, has a relatively brief half-life of 5730 years; **carbon-14 dating,** then, allows us to date materials from 100 to 70,000 years old. This time span encompasses the most recent glaciation and climate change, the latest rise and fall of worldwide sea levels, and the rise of our own species, *Homo sapiens.*

Carbon-14 atoms occur naturally in the atmosphere, where they are produced by cosmic-ray bombardment of nitrogen atoms. These carbon-14 atoms, along with atoms of carbon's stable isotope, carbon-12, then combine with oxygen to form carbon dioxide (CO_2). Some CO_2 that contains carbon-14 dissolves in oceans, lakes, rivers, glaciers, and underground waters, where organisms living in or drinking the water take it up. Plants incorporate atmospheric CO_2 during photosynthesis, using it to produce sugars and starches. When animals eat the plants, they ingest the carbon-14 as well. Thus carbon-

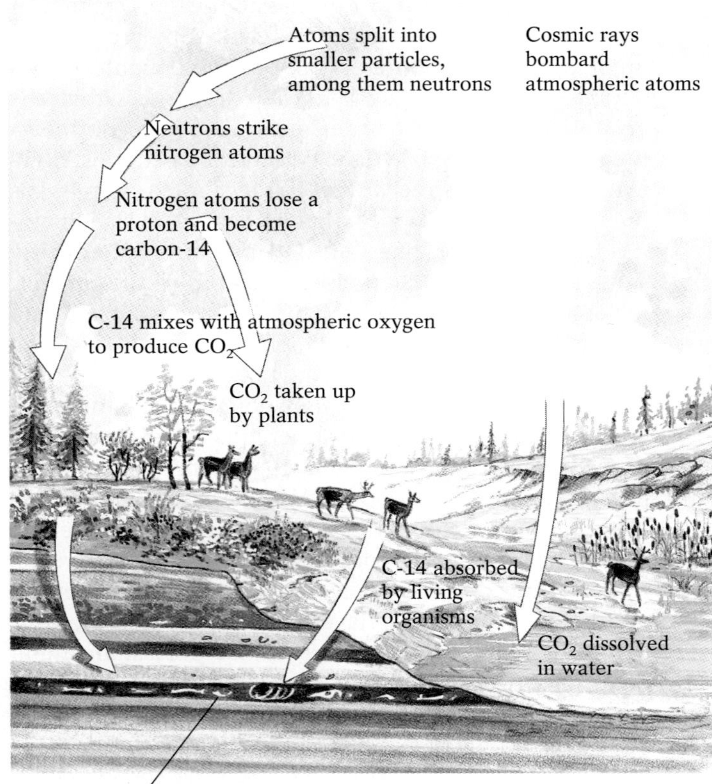

Figure 8-24 The carbon-14 used to date long-dead organisms and ancient artifacts originates in the Earth's atmosphere. Here, it combines with O_2 to produce CO_2, much of which dissolves in water or is taken up by plants, and is ultimately ingested by animals. While alive, an organism constantly replenishes the carbon-14 supply in its body. When it dies, its intake of carbon-14 ceases, and the carbon-14 begins to decay to its daughter isotope, nitrogen-14. Thus, the less carbon-14 left in the remains of an organism, the more time has elapsed since it died.

14 finds its way into the cells of most living organisms, where it continuously decays to its daughter isotope, nitrogen-14. An organism that is alive and taking in nutrients continuously replenishes its supply of carbon-14. As soon as it dies, however, its supply of carbon-14 immediately begins to diminish, and its isotopic clock begins to tick. *The greater the time elapsed since the death of the organism, the less carbon-14 (and the more nitrogen-14) it will contain* (Fig. 8-24).

As long as it's not too old, anything that contains carbon—such as bones, shells, wood, charcoal, plants, peat, paper, cloth, pollen, and seeds—is amenable to carbon-14 dating. We can date charcoal in ancient campfires, house posts of Native American dwellings, seeds and seafood shells in archaeological refuse dumps, and the bones of our predecessors and their prey. Carbon-14 dating has estimated the age of the linen cloth that was wrapped around the Dead Sea

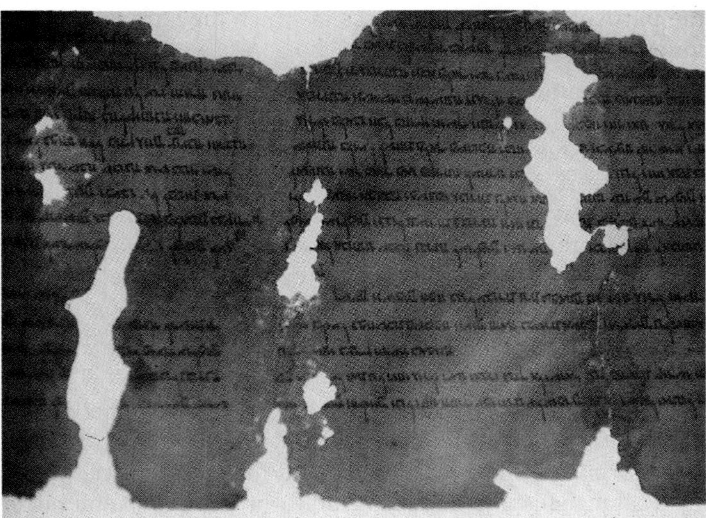

Figure 8-25 The papyrus that comprises the Dead Sea Scrolls, a recorded history of ancient Egypt, has been dated by C-14 dating. Its age—about 2100 years.

Scrolls (2000 to 2200 years old; Fig. 8-25), the papyrus on which the history of ancient Egypt was written (about 2100 years old), and even a fragment of the Shroud of Turin, purported to be the burial cloth of Jesus of Nazareth, although it registers an age of only 700 years. These items all have one thing in common—they were once part of a living organism or are composed of materials from once-living organisms. During life, such organisms take up the radioactive carbon that will eventually be used to date their remains.

One caution is warranted, however. When scientists compare radiocarbon dates with those obtained via other independent dating techniques, they see systematic patterns of "error" in the carbon dates. These discrepancies are presumably due to past variations in cosmic-ray bombardment, which, in turn, affects rates of carbon-14 production and dates. (Cosmic-ray bombardment varies, among other things, with solar-energy output and the strength of the Earth's protective magnetic field.) Thus, scientists either calibrate carbon-14 dates, where possible, with other dating schemes (such as tree-ring dates, discussed later in this chapter), or they simply report them in carbon-14 years, acknowledging that this age may not precisely equal calendar years.

Other Absolute-Dating Techniques

Some absolute-dating methods do not rely *directly* on the measurement of the relative proportions of parent and daughter isotopes. Some, for example, measure other physical changes within a mineral crystal brought on by nuclear decay. Other methods derive data from a process that produces countable annual layers (such as tree rings and lake sediments) or an organism that grows at a fairly constant rate (such as lichen-botanical organisms that grow on rocks).

Fission-Track Dating *Fission* is the division of a radioactive atom's nucleus into two fragments of approximately equal size, a process that releases several subatomic particles (such as alpha and beta particles). When a radioactive atom that is trapped in a mineral undergoes fission, these subatomic particles move at high speed across the orderly rows of atoms in the mineral's crystal lattice, leaving behind tears within the crystal called **fission tracks** (Fig. 8-26). Because fission occurs continuously in minerals containing radioactive substances, *the older the mineral, the more fission tracks per unit area it contains.* Fission tracks can date minerals that range from 50,000 to billions of years old. This method is particularly useful for dating volcanic glasses and other uranium-trapping mineral grains, such as zircons. Fission tracks are usually erased when their host rocks are heated above 250°C (475°F); consequently, they cannot be used to date medium- and high-grade metamorphic rocks.

Dendrochronology (Tree-Ring Dating) In temperate climates, the annual growth of most trees produces concentric sets of dark and light rings within the cross sections of their trunks and larger branches. By counting its rings, we can determine a tree's age—an absolute-dating method known as **dendrochronology.** This method can help us date relatively recent geological events such as landslides or mudflows—wherever trees have become established on new surfaces. Because tree rings are preserved even in long-dead trees, dendrochronology can also date much older geological events and, in certain cases, archaeological artifacts. Changes in climatic conditions, such as prolonged droughts, create a distinctive pattern of tree-ring variations in all trees living within a given geographical region. By comparing and correlating the ring patterns in these trees (even dead ones), we can see where their life spans overlapped. Dates ranging as far back as 9000 years have been established for some areas by tracing ring patterns from progressively older live trees to even older dead trees with overlapping life spans (Fig. 8-27). Archaeologists use dendrochronology to date wooden artifacts, such as house posts and digging sticks, by comparing their distinctive ring patterns with an identical pattern of known age.

Varve Chronology Lakes, particularly those that freeze in the winter and are fed by a large inflow of glacial meltwater in summer, can produce countable annual layers of sediment. **Varves** are paired layers of sediment, typically consisting of a thick, coarse, light-colored layer deposited in summer during periods of high inflow, and a thin, fine, dark-colored layer deposited in winter during periods of low or virtually no inflow (especially if ice is frozen) (Fig. 8-28).

To count the varves underlying a lake, geologists drill through the lake mud to extract a *core,* a continuous sequence of lake sediment. The number and nature of the varves in a core reveal how long ago the lake formed and identify events, such as landslides, that affected its existence. For example, a lake produced when a landslide blocks a river valley may immediately begin to produce varves. Counting the varves tells us when the landslide occurred. Varves can

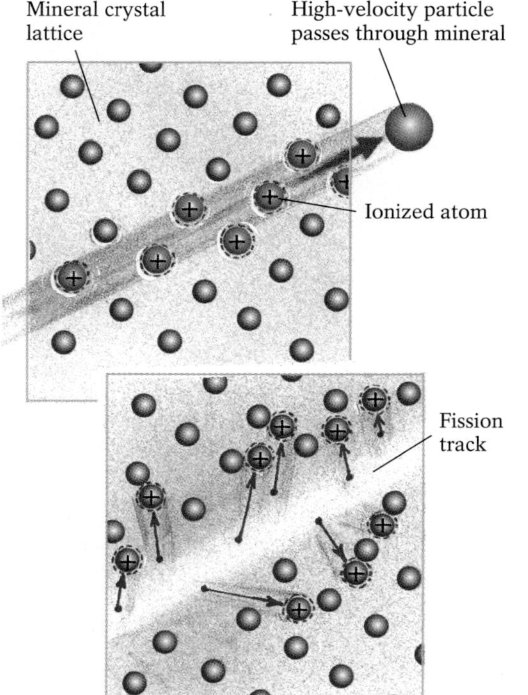

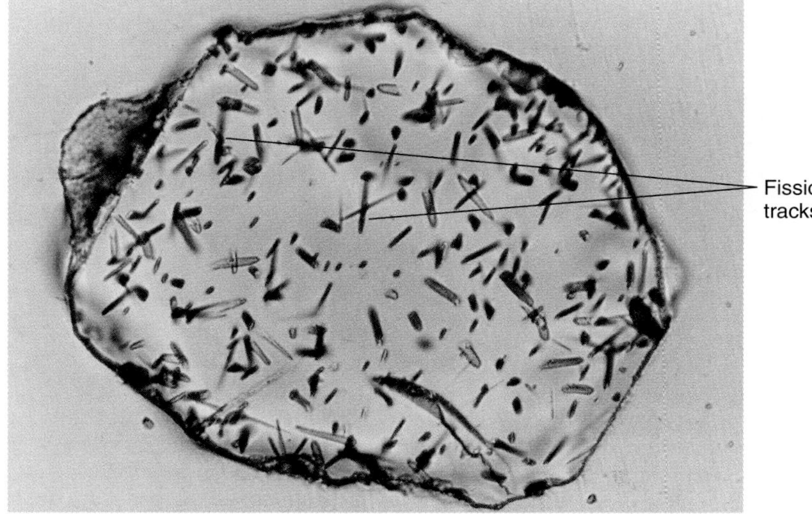

Figure 8-26 Fission tracks are produced by the decay of radioactive atoms trapped in a mineral's crystal structure. High-velocity decay particles emitted by the radioactive atoms disrupt the orderly arrangement of atoms in the crystal lattice, leaving tracks only a few atoms wide. These tracks are typically flooded with a strong solvent and enlarged so that they may be seen (and counted) with an ordinary microscope. The more fission tracks, and the greater the total track length, the older the mineral.

Years of age
500
400
200
150
100
50
Current year
Annual-ring similarities show correlation
Tree growth rings
A
B
C
D
Sediment layers with tree logs to be collected for dendrochronology
A
B
C
D
Buried tree logs

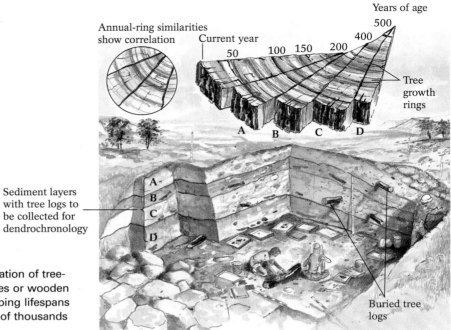

Figure 8-27 Correlation of tree-ring sections in trees or wooden artifacts of overlapping lifespans can establish ages of thousands of years.

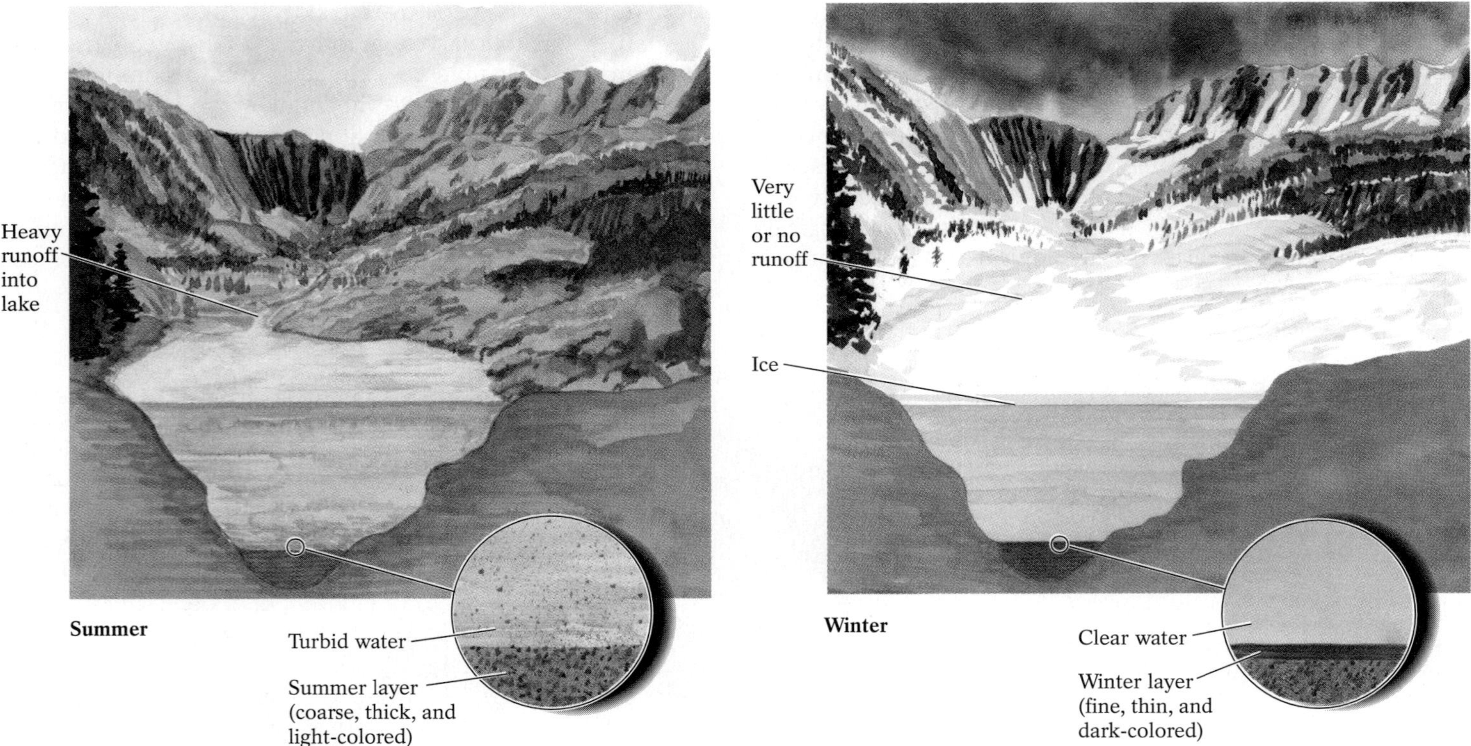

Figure 8-28 The origin of lake varves. A typical varve includes a thick, coarse, light-colored summertime layer produced during high runoff (from snowmelt and spring storms) and high sediment influx, and a thin, fine, dark-colored wintertime layer produced during low runoff and low sediment influx (or no influx if the lake is frozen). Note the varves in the photo. Why do you think the varves vary so noticeably in thickness?

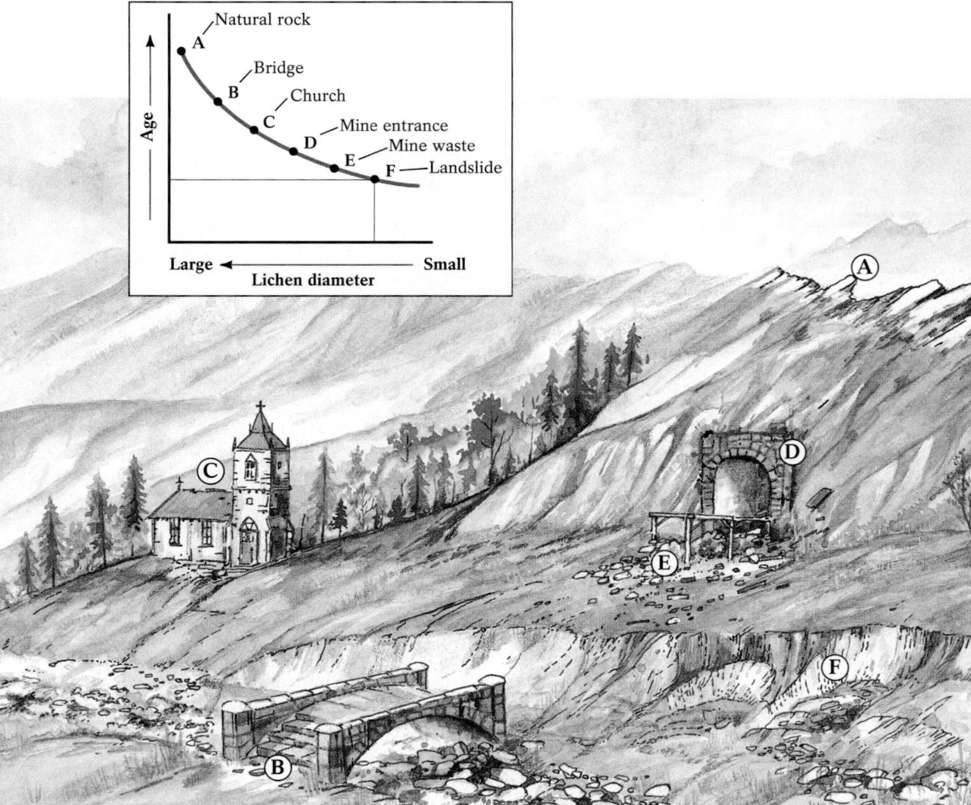

Figure 8-29 Estimating the absolute age of a material from the lichen growing on its exposed surface first requires that we determine the rate of lichen growth in the area. After measuring lichen on surfaces of known age, such as the bridge, the church, and the mine entrance in (left), geologists plot a growth curve that relates lichen diameters to time. Then they can estimate the age of a surface of unknown age by measuring the lichen on it and finding its position on the growth curve. (right) Lichen colonies on a granite boulder. (Chemicals in the lichen have bleached the light-colored areas of rock).

also indicate the presence of repeated landslides in an area, information useful for planning safe development in a community marked by chronically unstable slopes.

Lichenometry **Lichen,** colonies of simple, plant-like organisms that grow directly on exposed rock surfaces, are the basis of a dating method known as **lichenometry.** Given similar rocks and climatic conditions, *the larger the lichen colony, the longer the time since the growth surface became exposed.* Larger lichen colonies are found on old tombstones, for example, than on newer ones. Using tombstones and other stone surfaces of known age on which lichens grow (such as buildings, bridge supports, and old mine tailings), we can develop a growth curve, showing the rate at which lichen have grown in a given area (Fig. 8-29). Then we can compare lichen on a surface of unknown age (within the same area) with the growth curve to determine the surface's age. Because lichen colonies eventually grow together and can no longer be measured individually, lichenometry is useful only for surfaces less than about 9000 years old. These organisms provide accurate dates for young glacial deposits, rockfalls, and mudflows—all events that expose new rock surfaces on which lichen can grow.

Surface-exposure Dating For decades geologists have been forced to accept one major limitation of absolute dating—the availability of very few methods for dating geological materials whose ages fall between the ranges of carbon-14 dating (about 70,000 years) and other isotopic-dating techniques, such as potassium–argon dating (the range of which begins at about 100,000 years). In addition, most of these techniques apply to the dating of a mineral, rock, bone, or shell, rather than a specific landform or geological feature of interest.

Some geologists, less preoccupied with rocks and minerals, prefer to study the origin of landforms—such as glacial *moraines* (the piles of rocks and soil left behind by now-vanished glaciers, discussed in Chapter 17), alluvial fans (the wedges of sediment dropped by rivers when they enter the relatively still waters of oceans or lakes, discussed in Chapter 14), or newly emerging lands rising from the sea, either from plate-tectonic uplift or a relative drop in sea level. In other words, such geologists focus on dating newly exposed surfaces rather than a single rock collected from an outcrop.

In recent years, a new variety of geological dating—one designed to identify the time when a portion of the Earth's surface was first exposed—has become increasingly popular.

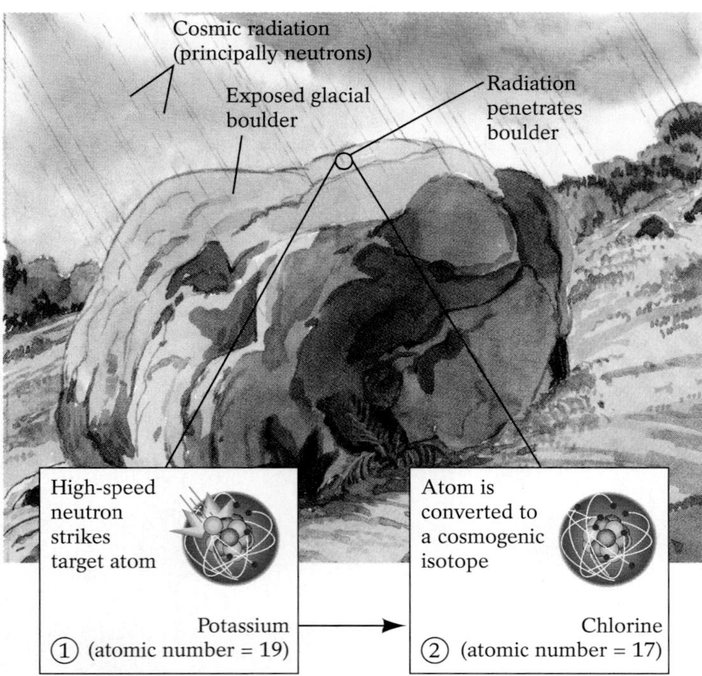

Figure 8-30 As soon as it is exposed, any surface rock begins to be bombarded with cosmic radiation. This converts target atoms into radioactive "cosmogenic" isotopes whose decay rate can then be measured to determine the time of initial exposure.

This method measures *cosmogenic* isotopes, those produced in extremely small quantities at the surfaces of newly exposed rocks and other surface materials from cosmic-ray bombardment.

Intergalactic cosmic radiation, predominantly in the form of neutrons, constantly bombards the Earth at phenomenal speeds. (Countless numbers of these particles are passing through your body as you read this passage.) These neutrons strike the rocks and soils at the Earth's surface, penetrating to a maximum depth of 2 to 3 meters (7 to 10 feet) before being absorbed. As these high-speed neutrons hit target elements in rocks and soils—such as oxygen, silicon, aluminum, iron, magnesium, calcium, and potassium—they knock other atomic particles from the nuclei of these elements, converting these elements into new "cosmogenic" radioactive isotopes (Fig. 8-30). Silicon, magnesium, iron, and aluminum, for example, are typically converted to *beryllium-10;* potassium, calcium, and chlorine become *chlorine-36.*

The atoms of these radioactive substances, like other radioactive atoms, immediately begin to decay. Unlike in the case of non-cosmogenic isotopes, whose number of parent atoms is fixed (such as uranium in the Earth's crust), new atoms of these cosmogenic isotopes are constantly being produced, even as others decay. Thus a simple parent–daughter decay relationship does not apply to these isotopes. Instead, we must know and then subtract their production rates and accurately determine their decay rate, or half-life. Much rigorous testing and study have revealed that chlorine-36 has a half-life of roughly 300,000 years; beryllium-10's half-life is approximately 1.5 million years. Knowing these half-lives enables geologists to date newly exposed surface boulders (Fig. 8-31) whose ages fall between the ranges of carbon-14 dating and other isotope-dating schemes.

Figure 8-31 This massive basaltic boulder was plucked from the Columbia River basalt flows of eastern Washington by a moving glacier about 15,000 years ago. Since the moment it melted from the ice, it has been bombarded by cosmic radiation; its time of initial exposure can be determined by surface-exposure dating.

Of course, any new dating method faces technical challenges. The first involves the extremely small amounts of these cosmogenic isotopes present in the rocks. During the past decade, geologists have developed powerful new devices capable of measuring minute numbers of atoms with remarkable accuracy. The accelerated mass spectrometer (AMS), for example, can detect 5 radioactive atoms in a pool of 1,000,000,000,000,000 atoms. This capability would be akin to finding 5 red garnet grains amidst an entire houseful of white quartz grains. The technology has also been applied to carbon-14 dating. Today geologists can measure very small samples of an organic substance, enabling them to date an object without completely consuming it (as was often the case with past techniques). Thus a priceless item, such as the Shroud of Turin or the Dead Sea scrolls, can be dated through only a tiny morsel of its substance.

Unlike other isotopic-dating schemes, in which the daughter products essentially persist in the rock, the isotopes produced by surface exposure may themselves erode away. Once the isotopic "clock" is set upon exposure to cosmic radiation, a variety of surface events may "reset" it. Ongoing erosion of a boulder's surface, for example, may remove its outer portion and hence some of the exposure-produced isotopes. The boulder itself may become dislodged at some time in its history, exposing a new surface to cosmic radiation. Even the intense heat of a forest fire may cause the outer segment of a boulder to expand and fall off. Geologists who use this exciting new dating method must therefore be on the lookout for the tell-tale signs of surface erosion and disturbance. Nevertheless, surface-exposure dating holds much promise for closing the gap between other isotope-dating schemes.

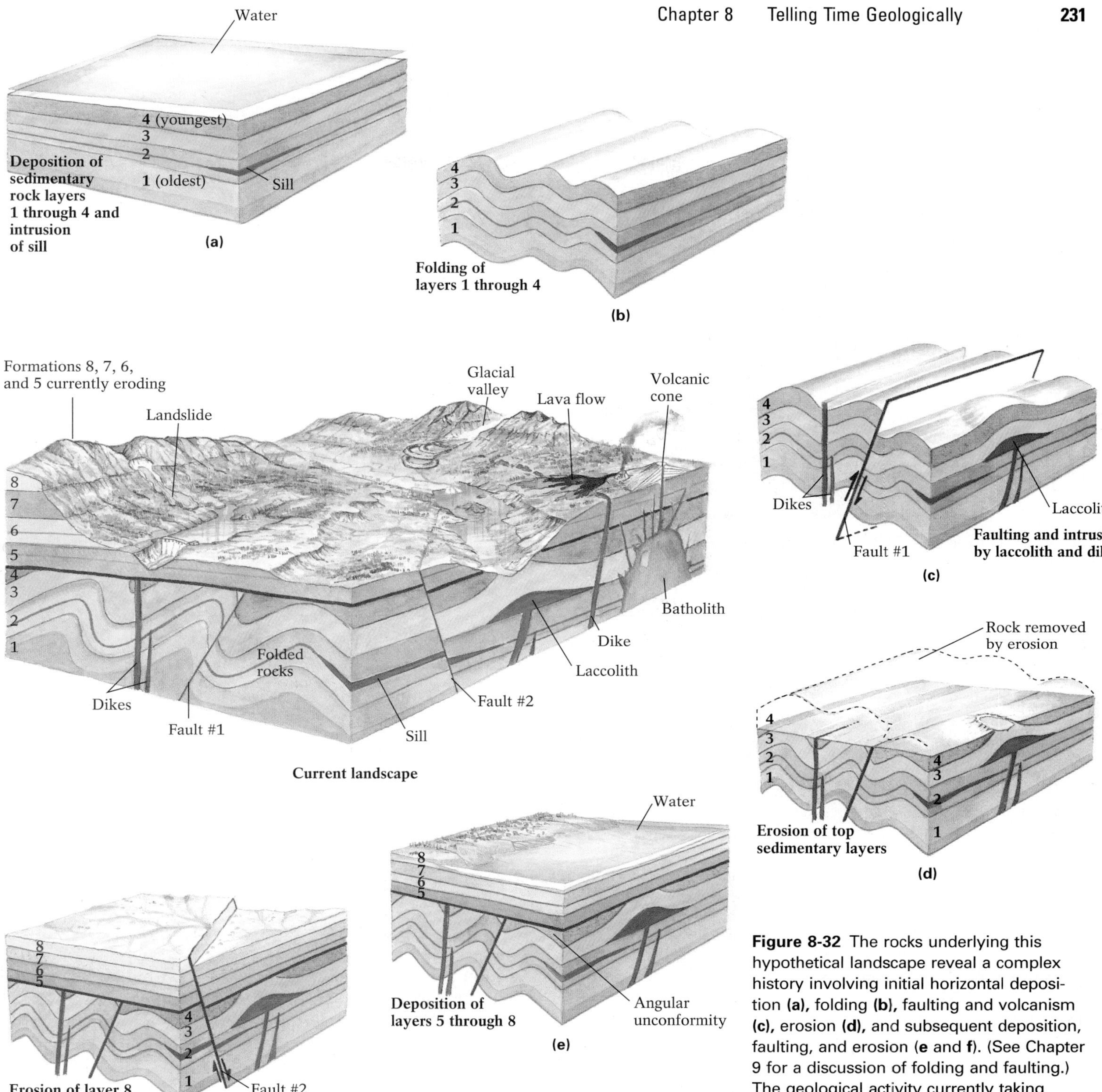

Figure 8-32 The rocks underlying this hypothetical landscape reveal a complex history involving initial horizontal deposition **(a)**, folding **(b)**, faulting and volcanism **(c)**, erosion **(d)**, and subsequent deposition, faulting, and erosion (**e** and **f**). (See Chapter 9 for a discussion of folding and faulting.) The geological activity currently taking place at the surface will be similarly recorded later in the area's stratigraphy.

Combining Relative and Absolute Dating

When geologists and paleontologists study a region for the first time, they use all dating methods at their disposal—both relative and absolute—to reconstruct its geological history. They can use absolute methods on a region's rocks that contain isotopically datable key beds; other materials, such as fossil-free sedimentary rocks, may be dated only relatively. Scientists may write a geological "biography" of some regions, albeit with some chapters missing, by absolutely dating every layer that contains radioactive isotopes and then applying the principles of relative dating to supplement this information. Figure 8-32 shows how the complex history concealed in a region's rocks can be reconstructed, revealing a series of geological events spanning billions of years of Earth history.

Eon	Era	Period	Epoch			Life Forms	Major Events
Phanerozoic (*Phaneros* = "evident"; *zoic* = "life")	Cenozoic	Quaternary	Recent, or Holocene	10,000	Age of Mammals	Spread of modern humans	Eruption of volcanoes in the Cascades
			Pleistocene	1.6		Extinction of many large mammals and birds; *Homo erectus*	Worldwide glaciation; Fluctuating cold to mild in the "Ice Age"
		Tertiary – Neogene	Pliocene	5.3		Large carnivores; Earliest hominid fossils (3.4–3.8 mya)	Uplift of the Sierra Nevada; Linking of North and South America; Beginning of the Cascade volcanic arc
			Miocene	23.7		Whales and apes; Large browsing mammals; monkey-like primates; flowering plants begin	Beginning of Antarctic ice caps; Opening of Red Sea
		Tertiary – Paleogene	Oligocene	36.6			Rise of the Alps; Himalaya Mountains begin to form
			Eocene	57.8		Primitive horse and camel; giant birds; formation of grasslands	Volcanic activity in Yellowstone region and Rockies; Ice begins to form at the poles
			Paleocene	66.4		Early primates; Extinction of dinosaurs and many other species	Collision of India with Eurasia begins; Eruption of Deccan basalts
	Mesozoic	Cretaceous		144	Age of Reptiles	Placental mammals appear (90 mya); Early flowering plants	Formation of Rocky Mountains
		Jurassic		208		Flying reptiles; Early birds and mammals	Opening of Atlantic Ocean
		Triassic		245		First dinosaurs	Breakup of Pangaea begins
	Paleozoic	Permian		286	Age of Amphibians	Coal-forming forests diminish	Supercontinent Pangaea intact; Culmination of mountain building in eastern North America (Appalachian Mountains); extensive glaciation of southern continents
		Carboniferous – Pennsylvanian		320		Coal-forming swamps abundant; Sharks abundant	Warm conditions, little seasonal variations; most of North America under inland seas
		Carboniferous – Mississippian		360		Variety of insects; First amphibians; First reptiles	
		Devonian		408	Age of Fishes	First forests (evergreens)	Mountain building in Europe (Urals, Carpathians)
		Silurian		438		Early land plants	
		Ordovician		505	Age of Marine Invertebrates	Invertebrates dominant; First primitive fishes	Beginning of mountain building in eastern North America (rest of North America low and flat)
		Cambrian		545		Multicelled organisms diversify; Early shelled organisms	Extensive oceans cover most of North America
Proterozoic ("Early Life")	Precambrian			2500		First multicelled organisms; Jellyfish fossil (~670 mya)	Formation of early supercontinent (~1.5 billion years ago); Abundant carbonate rocks being deposited; first iron ore deposits
Archean ("Ancient")				~3800		Early bacteria and algae; Oldest evidence of life	Primitive atmosphere begins to form (accumulation of free oxygen); Oldest known sedimentary rocks
Hadean ("Beneath the Earth")				~4600		Formation of the Earth	Oldest known rocks on Earth (~3.96 billion years ago); Oldest Moon rocks (~4 billion years ago) (4.55 billion years ago); Earth's crust being formed

Figure 8-33 The geologic time scale. Absolute ages shown are in millions of years.

The Geologic Time Scale

The **geologic time scale** (Fig. 8-33) organizes all of Earth history into blocks of time during which important events occurred. The basic divisions of the geologic time scale were established during the eighteenth, nineteenth, and early twentieth centuries by geologists and paleontologists who identified changing fossil assemblages in rocks and then applied the principles of superposition, faunal succession, and cross-cutting relationships to establish their age sequence. In the last half-century, scientists have used radiometric dating to assign absolute-age ranges to these divisions.

The largest time spans shown on the geologic time scale are *eons*. They mark major developments, such as the first occurrence of life on Earth and the worldwide expansion of multicelled life forms. The earliest is the Hadean ("beneath the Earth") Eon, ranging from the time of the Earth's formation some 4.6 billion years ago to 3.8 billion years ago. With the exception of the 3.96-billion-year-old rocks of Canada's Northwest Territories, geologists have found virtually no other Hadean rocks or fossils. The earliest known life forms appear in rock from the Archean ("ancient") Eon, about 3.8 to 2.5 billion years ago (Fig. 8-34).

Eons are subdivided into *eras*, each of which is defined by its dominant life forms. The Mesozoic Era, for instance, is the "Age of Reptiles"—the time of the dinosaurs. Eras are subdivided into *periods*, which record less dramatic biological changes. Periods, in turn, are subdivided into *epochs*, based on even more subtle changes in the assemblage of living creatures and on some nonbiological criteria. Each successive subdivision marks a progressively smaller unit of time. The length of time for each division varies, even for subdivisions at the same level, because the key characteristics and geological conditions that define them are not of uniform du-

Figure 8-34 (top) The fossils in this rock, from the Precambrian Period of the Archean Eon, are *stromatolites,* structures created by a very early form of alga. (bottom) Modern stromatolites in Shark Bay, Western Australia.

ration. Thus the Eocene Epoch of the Tertiary Period of the Cenozoic Era lasted for 21 million years, whereas the Oligocene Epoch of the same period and era lasted for only 13 million years.

Life on Earth

Many geologists believe that simple life forms may have evolved *here on Earth* during the planet's first billion years. Proponents of this view suggest that all of the ingredients necessary to produce amino acids (carbon, hydrogen, oxygen, and nitrogen)—the precursors to life—existed in the primordial stew of the Earth's early atmosphere and oceans. In their minds, a spark of Precambrian volcanic lightning may have energized the reactions that produced organic life on Earth. Others believe that the seeds of life on Earth were *imported*—that is, ferried here by meteorites or other incoming materials from around the solar system. Highlight 8-3 (see pages 236-237) discusses some of these recent findings and hypotheses.

Regardless of how the first spark of life was ignited on Earth, it was most likely extinguished shortly thereafter, either by the impact of one of the massive meteors that littered the developing solar system, or in the wake of one of the early Earth's protracted periods of widespread explosive volcanism. Picture a meteor 400 kilometers (250 miles) in diameter and traveling at a speed of 15 kilometers (10 miles) per second. Its impact could have raised the Earth's surface temperature as high as 1870°C (3120°F). This heat would have rapidly vaporized the early oceans and any life they supported.

It seems likely that life could begin to develop and flourish on Earth without such catastrophic interruptions only after all planets and moons of the solar system had finally formed, clearing the surrounding space of most of its remaining debris. In 1980 geologists studying in northwestern Australia discovered the first evidence of what is believed to be Earth's earliest "inhabitants." They found simple bacteria-like cells in a layer of sedimentary chert that was dated (by fission-track analysis of the surrounding igneous rocks) at about 3.77 billion years—the beginning of the Archean Eon. In South Africa, primitive single-celled organisms have shown up in rocks dated at more than 3.3 billion years.

The Archean Eon gave way to the Proterozoic ("early life") Eon 2.5 billion years ago. The first multicelled organisms arose during this time of increasing biological diversity. Some 670-million-year-old rocks in Australia's Ediacara Hills, for instance, contain fossil evidence of jellyfish, soft marine worms, and a variety of other simple creatures composed entirely of soft tissue. These creatures and their descendants flourished in the world's oceans until some 545 million years ago.

The beginning of the Phanerozoic ("visible life") Eon 540 million years ago marks the first point at which we find abundant fossil evidence of marine creatures possessing hard shells, prominent external spines, and internal skeletons. A number of competing hypotheses attempt to account for this development (which actually had its start in the late Proterozoic Eon). One claims that algae, which rely on photosynthesis for basic life processes, released sufficient oxygen (a product of photosynthesis) to form the ozone layer in the atmosphere; this layer protects life forms on Earth from deadly ultraviolet radiation from the Sun. As atmospheric oxygen fueled their metabolism, animals may have grown larger and more complex, requiring a rigid structure to support their larger mass. Another hypothesis suggests that the oceans—and the creatures in them—contained so much calcium at this time that marine creatures were forced to excrete calcium to avoid calcium poisoning. This excreted calcium ultimately became their hard outer shells. Yet another hypothesis holds that shells and spiny growths developed as protection against attacks from predators.

The Phanerozoic Eon extends all the way to the present, marked by the development of a remarkable succession of plant and animal life. It is divided into three eras: the Paleozoic, Mesozoic, and Cenozoic. The first part of the **Paleozoic** ("ancient life") **Era** was dominated by marine invertebrates (sea creatures without backbones, such as corals, clams, and other shelled organisms); fish, amphibians, insects, and land plants rose to prominence during the latter part of this era. The **Mesozoic** ("middle life") **Era** was dominated by marine and terrestrial reptiles, including dinosaurs; the first birds, mammals, and flowering plants appeared during this era. The **Cenozoic** ("recent life") **Era,** which persists through today, is distinguished by a rich variety of mammals.

When and where did humans arise in the course of Earth history? East Africa may have been the cradle of our species. In what are now Ethiopia and Tanzania, the bones and footprints of the earliest-known hominids (members of the human family, Hominidae) have been found embedded in sedimentary layers that also contain datable volcanic ash. It appears from potassium–argon analysis of these hominid fossils that the earliest hominids walked the Earth from 3.4 to 4.0 million years ago (Fig. 8-35). Eastern Africa provided not only a hospitable climate and abundant resources that enabled our species to evolve, but also the volcanism and sedimentation that preserved the evidence and facilitates its dating.

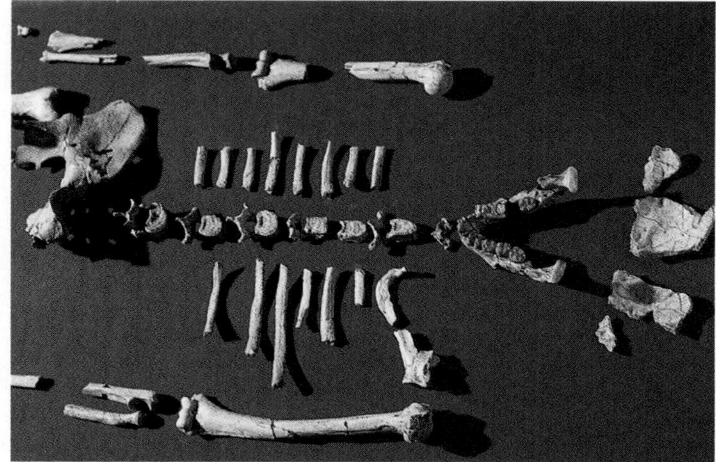

Figure 8-35 The fossil hominid called Lucy, collected in Hadar, Ethiopia, in 1974 by Donald Johanson. Lucy, one of the most complete hominid fossils yet discovered, is believed to be more than 3 million years old.

The Age of the Earth

Geologists have a powerful craving to date things, as can be seen from numerous dating methods mentioned throughout this chapter. Nevertheless, the ability to date the age of the Earth itself—directly and absolutely—continues to elude us. This chapter began with the claim that the Earth is about 4.6 billion years old. But the oldest Earth rocks discovered thus far fall short of that age. We may never find a rock that dates from the very origin of the planet. Many of the Earth's oldest rocks are metamorphic, changed from even older, preexisting rocks. The planet's fierce internal heat, recycling of tectonic plates, catastrophic events such as meteor impacts during the early evolution of our planet, and a host of other geological processes have worked together over these billions of years to change much of the Earth's original material, thereby concealing direct evidence of its true age.

Geologists have had to look elsewhere to tell how old the Earth is. Because all the components of our solar system formed at roughly the same time, the ages of extraterrestrial rocks provide *indirect* evidence of the age of our own planet. Some exciting clues about the Earth's antiquity have come from Moon rocks collected by American astronauts within just a few hundred meters of their landing craft (Fig. 8-36). These rocks were dated by both uranium–lead and potassium–argon systems as being 4.53 billion years old. Remember, the astronauts did not conduct a Moon-wide search for rocks—these rocks just happened to be on the ground where the spacecraft landed. Since, as far as we know, the Moon's rocks were never recycled by plate tectonic and surface processes, any one of them can represent the Moon's earliest development. In addition, radiometric dating of meteorites that have struck the Earth, such as the one shown in Figure 8-37, has yielded virtually the same age—4.6 billion years. Most meteorites are remnants of the early solar system and hence chunks of rock that have not been recycled by planetary tectonics or undergone heating, pressing, or other planetary forces. In all likelihood, they faithfully record the age of the solar system—and of the Earth.

Some clever scientific detective work also enables us to learn something about our planet's date of birth from the Earth itself. We have seen that uranium and thorium decay to produce three stable lead isotopes: lead-206, lead-207, and lead-208. Lead has a fourth stable isotope, lead-204, that is *not* a product of radioactive decay. As lead-204 is not created by the decay of other elements, its concentration in the Earth has remained constant from the beginning, while the concentrations of the other lead isotopes have been increasing through radioactive decay since the origin of the Earth. The relative proportions of the four lead isotopes are roughly the same in every meteorite that does not also contain a radioactive parent, suggesting that these proportions resemble the mixture of lead originally found in our solar system. By comparing these proportions to those found in Earth rocks, we can determine how much lead of radioactive origin has been added.* Scientists have calculated that today's proportions of radioactively produced lead indicate the occurrence of about 4.5 billion years of radioactive decay—a good fit with other data gleaned from Moon rocks and meteorites.

Lunar rocks, meteorites, and the lead isotopes on Earth all provide convincing evidence that the Earth and the rest of our solar system formed some 4.6 billion years ago. Since that distant time, geological processes that appear to be excruciatingly slow *from a human perspective* have accomplished incredible feats. It is humbling to realize, standing before Pike's Peak in Colorado or Mount Katahdin in Maine, that the process of erosion, which seems imperceptibly slow by human time standards, could remove those imposing mountains in a geological instant. The next time you sit by a stream and glimpse the current carrying away sand grains,

*An interesting, practical aside: To develop U-Pb dating techniques, geologists had to create a lead-free lab environment, so that they wouldn't analyze any lead that didn't come from the rocks. Back in the 1950s, such a technological task was extremely difficult because of the volume of leaded-gasoline pollution in the air. Clair Patterson, the geologist who developed U-Pb analysis, used his knowledge of the high concentration of toxic lead in the environment to lead the fight against leaded gasoline.

Figure 8-36 A lunar boulder, as seen from the Apollo 17 spacecraft in December 1972. (The Earth is visible in the far distant background.) Fragments of such Moon rocks taken back to Earth for analysis have been dated as about 4.5 billion years old.

Figure 8-37 The Hoba meteorite, in Namibia. Almost every meteorite that has struck the Earth and been subsequently radiometrically dated yields an age of about 4.6 billion years.

consider how little geologic time it would take for the grains to accumulate and lithify to form a hill or a mountain, and how easily geological change is effected when there's "all the time in the world."

This chapter marks the end of Part 1 in this book, the section devoted to building your foundation of geological basics. In this section we described the basic structure of minerals and rocks, and considered how the major rock groups form: by cooling from a melted state, by lithification of broken-down and dissolved rock fragments into sedimentary rocks, and by metamorphic processes involving the heat and pressure. We discussed how plate tectonic movement affects each of these processes. In this chapter, we examined the vast reaches of geologic time and our methods for measuring it. In Part 2, we will scrutinize the dynamic tectonic processes that have combined to create the Earth's most massive features, such as its immense interior layers, its continents and ocean basins, and its lofty mountain ranges. We will also focus on earthquakes, one of the most dramatic effects of these dynamic processes.

Chapter Summary

When geologists determine the ages of rocks and other materials, they are engaged in **geochronology,** the study of time in relation to the Earth's existence. This subject is part of **historical geology,** the branch of the geosciences that studies the Earth's past. To place the origin of an item such as a rock or sediment in the correct historical perspective, geologists apply either **relative dating** (determining how old it is in relation to its surroundings) or **absolute dating** (determining its actual age in years).

Relative dating relies on several key principles. The **principle of uniformitarianism** states that modern geological processes resemble those that operated in the past. The **principle of original horizontality** states that most sedimentary rocks and lava flows are initially deposited in horizontal layers. The **principle of superposition** states that for tectonically undisturbed sedimentary rocks and lava flows, the uppermost layer in a sequence of rocks is the youngest, with those below it being successively older. The **principle of cross-cutting relationships** states that layers cut across by other layers and features, such as igneous dikes and faults, must be older than the features that cut them. Similarly, rocks that are incorporated within other rocks must be older than the rocks that surround them—an idea known as the **principle of inclusions.** Finally, the **principle of faunal succession** states that organisms have succeeded one another in a recognizable, reproducible sequence in the geological record. **Fossils** are evidence of past life preserved in rock; the most helpful fossils, called **index fossils,** come from species that had wide geographical distribution but lived for only a relatively brief period of time.

When dating rocks and sediments relatively, we often find that two physically adjacent layers were actually deposited at distinctly separate periods of time. The boundary between such layers, called an **unconformity,** represents a gap in the rock record (resulting from either erosion of entire rock layers or periods of nondeposition). To determine the relative histories of geographically distant rock sequences, geologists establish age equivalence between similar individual rock layers through a process known as **correlation.** Correlation involves comparing paleontological and mineralogical similarities between separate rock exposures.

Whenever possible, geologists prefer to date a rock unit absolutely. In some cases, the rock or a mineral in the rock may contain a radioactive **parent isotope** that decays at a constant rate to produce a measurable amount of a nonradioactive **daughter isotope.** The time it takes for half the

Highlight 8-3 Life on Earth—A Foreign Import?

In the 1996 Hollywood hit "Independence Day," Earth is visited and almost conquered by an armada of technologically advanced, malevolent aliens. Exciting stuff. At about the same time of the movie's premiere, a group of planetary geologists and chemists from NASA's Johnson Space Center were publishing their own findings of a proposed alien "invasion." But their alleged aliens were not the stuff of summertime blockbusters; they were purported to be microscopic specks of organic material—1/1000 the width of a human hair. Despite their tiny size, these specks have ignited a firestorm of debate about life on Mars—and Earth—that is destined to persist for years.

The center of attraction in this debate is ALH84001—not a very memorable name for what may be the most exciting clue to emerge in the as-yet unsolved mystery of extraterrestrial life and the origin of life on Earth. This designation has been assigned to a 2-kilogram (4.3-pound) potato-sized meteorite retrieved in 1984 from the Allan Hills, a desolate spot on the vast icy wasteland of Antarctica (Fig. 8-38). For eight years this unspectacular chunk of rock gathered dust in a storage cabinet in Houston until analysis of the gases emitted during its heating revealed the rock's extraordinary origins.

This rock's life story apparently began about 4 billion years ago on the surface of Mars, 78 million kilometers (49 million miles) from Earth. We believe this idea about its origin because the rock's gas content—particularly its mix of oxygen isotopes—is unique to the Martian atmosphere. The great length of ALH84001's radiometrically dated age has generated great interest, because the rock dates from a time when the Martian surface may have been considerably warmer, wetter, and more hospitable to life than the frozen, barren surface of modern-day Mars. (Look ahead to Figure 14-35—Mars' ancient stream beds—for proof of the presence of flowing water there in the distant past.)

Figure 8-38 The meteorite ALH84001.

Over time, Mars lost its atmosphere and, as a consequence, its liquid surface water vaporized. Perhaps 15 million years ago, an asteroid or comet smashed into the Martian surface, hurling great masses of the planet's crust skyward, including ALH84001. This detritus entered into orbit around the Sun, where it remained until about 13,000 years ago, when Earth's gravity caused ALH84001 to plummet to our planet's surface, landing on the South Pole's Antarctic ice sheet.

What microscopic clues have led some scientists to propose that seeds of life once germinated, if only briefly, within ALH84001?

- The rock contains microscopic globules of carbonates formed from a carbon-rich liquid, such as carbon dioxide *dissolved in water.* The presence of water is generally believed to be a prerequisite for life.
- The black rims surrounding the globules contain iron sulfide and magnetite, which, although nonbiological, are known to be produced by primitive terrestrial bacteria.
- The rock contains polycyclic aromatic hydrocarbons (PAHs). Although these compounds may form from nonbiological chemical reactions, they typically form here on Earth from organic decay.
- Electron microscopy reveals minute worm-shaped and ovoid structures similar to the fossils of primitive bacteria found in some of our planet's oldest rocks (Fig. 8-39).

Taken individually, none of these findings would amount to absolute proof of past life. Taken collectively, the evidence is quite suggestive, *but still by no means conclusive.* Shortly after publication of these remarkable findings, other groups of scientists set about refuting them. They argued that nonbiological processes could have created virtually all of the evidence. Carbonates, sulfides, and magnetites have also formed here on Earth by volcanic

atoms of the parent isotope to decay is called the **half-life.** The use of radioactive isotopes to date rocks, some up to 4.6 billion years old, is called **radiometric dating.** Most igneous rocks contain some measurable proportion of radioactive isotopes, such as uranium-238, which decays into lead-206 **(uranium–lead dating);** potassium-40, which decays to form argon-40 **(potassium–argon dating);** and rubidium-87, which decays into strontium-87 **(rubidium–strontium dating). Carbon-14 dating** allows us to date organic substances as old as 70,000 years.

Fission tracks are produced in a crystal's structure as high-speed particles emitted during radioactive decay pass through the crystal; some minerals and their host rocks can be dated by measuring the number and length of such fission tracks in their structures. A variety of other absolute-dating techniques principally date rocks and sediments that

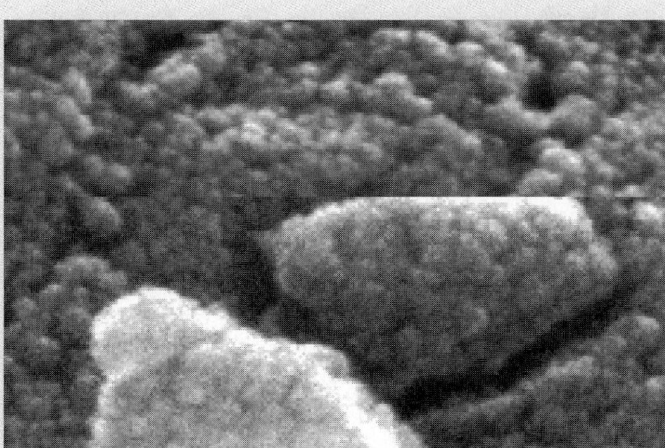

Figure 8-39 Minute structures believed by some geoscientists to be evidence of primitive bacteria.

processes; they do not necessarily require the presence of life-supporting surface water. Similarly, the fossil-like structures may also be inorganic and volcanically produced. Until its batteries petered out in late 1997, NASA's Pathfinder probe continued its exploration of the Martian surface, examining rocks and soil samples. Perhaps we will soon have an answer to the question of whether organic life has ever existed on Mars.

In 1997, another remarkable development gave credence to the idea that life on Earth may have been imported. Louis Frank, a physicist from the University of Iowa, presented convincing evidence that a barrage of cometary "snowballs" reaches the Earth's atmosphere every day (Fig. 8-40). According to Frank, thousands of these icy objects—some the size of houses—vaporize in the Earth's upper atmosphere daily. Each adds an estimated 20 to 40 tons of water vapor to the Earth's water cycle and a smattering of carbon-rich cosmic dust. (Perhaps all of the carbon in our trees and even in our bones originated *out in space,* only to be delivered here by these objects.) If this quantity of extraterrestrial water was imported at its present rate throughout Earth history, it could easily account for the Earth's surface waters, from the vast reservoir of water that fills our oceans to the water that flows from your kitchen tap and occupies each and every one of your body's cells. Such a suggestion represents a radical departure from previous thoughts that virtually all of

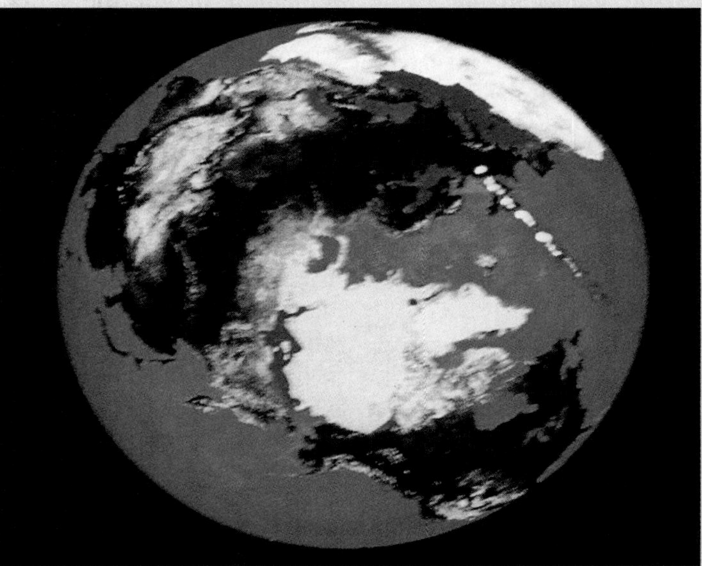

Figure 8-40 The linear vapor trail in the right-hand side of this global image is believed by some scientists to be the remains of an icy comet that vaporized in the Earth's atmosphere, leaving behind a significant parcel of water and cosmic dust. Some scientists suggest that most, if not all, of the Earth's water (and perhaps its seeds of life) was transported here by such comets.

the planet's water originated by degassing of the Earth's interior during the early period of intense volcanism. What's more, these cometary snowballs may have delivered organic compounds that provided the original basis for life on Earth. Analysis of the icy tail of the well-publicized Hale-Bopp comet, for example, revealed the presence of methanol, formaldehyde, carbon monoxide, hydrogen cyanide, hydrogen sulfide, and many other compounds rich in carbon, hydrogen, and oxygen. Such ingredients may have been the recipe for the Earth's earliest life.

Certainly these landmark discoveries will generate years, even decades, of claims and counterclaims, the stuff of which scientific progress is made. Perhaps answers to some of humankind's greatest questions—about the origin of life on Earth and whether life has existed on other planets—will emerge at the other end of this exciting, spirited debate.

are less than a few tens of thousands of years old. Some involve counting periodically accumulated layers, such as the annual growth rings in trees (a dating method known as **dendrochronology**) or the seasonally deposited layers of lake-bottom sediment called **varves.** Other methods measure the accumulation of something that grows at a fairly constant rate, such as the lichen that colonize exposed rock surfaces; dating using lichen is known as **lichenometry.**

Scientists have combined these and other techniques to produce the **geologic time scale,** a chronicle of all of Earth history. Use of these dating methods has also enabled us to determine how old the Earth is, how long ago life on Earth originated, and when our early human ancestors lived. The geologic time scale is divided into eons, eras, periods, and epochs. Most of the evidence used to interpret Earth history comes from the Phanerozoic Eon, a segment of time during which evidence of life began to be abundantly preserved as fossils in rocks. The Phanerozoic Eon is divided into the **Paleozoic** ("ancient life") **Era, Mesozoic** ("middle life") **Era,** and **Cenozoic** ("recent life") **Era;** the Paleozoic was dominated by marine invertebrates (such as primitive clams, snails, and corals), and later fish and amphibians, and the Mesozoic and Cenozoic were dominated by reptiles (such as dinosaurs) and mammals, respectively.

Key Terms

geochronology (p. 207)
historical geology (p. 208)
relative dating (p. 208)
absolute dating (p. 209)
principle of uniformitarianism (p. 209)
principle of original horizontality (p. 210)
principle of superposition (p. 210)
principle of cross-cutting relationships (p. 210)
principle of inclusions (p. 210)
fossils (p. 212)
principle of faunal succession (p. 212)
index fossils (p. 213)
unconformities (p. 213)
correlation (p. 217)
parent isotope (p. 220)
daughter isotope (p. 220)
radiometric dating (p. 221)
half-life (p. 222)
uranium–thorium–lead dating (p. 225)
potassium–argon dating (p. 225)
rubidium–strontium dating (p. 225)
carbon-14 dating (p. 225)
fission tracks (p. 227)
dendrochronology (p. 227)
varves (p. 227)
lichenometry (p. 229)
geologic time scale (p. 232)
Paleozoic Era (p. 233)
Mesozoic Era (p. 233)
Cenozoic Era (p. 234)

Questions for Review

1. Briefly explain the difference between relative and absolute dating.

2. Discuss three of the basic principles that serve as the foundation of relative dating.

3. What qualifies a species to become an index fossil? How are index fossils used in the correlation of sedimentary rock strata?

4. Sketch and label two different types of unconformities.

5. Name three parent–daughter radiometric dating systems, and give the half-lives of each parent isotope, as well as the rocks or sediments that are most likely to be dated by each.

6. Briefly discuss two potential problems that may diminish the reliability of an isotopically derived date.

7. Briefly explain how carbon-14 enters the cells of living organisms.

8. Select three absolute-dating methods. Describe their basic principles, and the materials that can be dated by each technique.

9. Using the geologic time scale, state when each of the following great events in Earth history occurred: the origin of the world's iron ores; the first appearance of a protective atmosphere; the origin of flowering plants; the origin of birds; the age of reptiles; the current ice age.

10. If the oldest rocks found on Earth are less than 4.0 billion years old, what evidence suggests that the Earth is actually 4.6 billion years old?

For Further Thought

1. Why are obsidian and basalt more susceptible to the development of hydration and weathering rinds than granite and andesites?

2. Look at Figure 8-13 on page 216, the geologic profile of the Grand Canyon. The Cambrian Tapeats Sandstone lies unconformably above several different bodies of rock. Identify two different types of unconformities that separate the Tapeats Sandstone from the underlying rocks.

3. Using the various dating methods discussed in Chapter 8, derive the sequence of events that produced the hypothetical landscape below. (Go slowly, and don't jump to premature conclusions. Consider all the principles that we've discussed.) Which of the layers might be dated absolutely?

4. Although geologists claim that "the present is the key to the past" (the principle of uniformitarianism), the Earth has certainly changed throughout its 4.6-billion-year history. Think of two geological processes that operate differently today than they did in the past, and discuss how they vary.

5. Suppose you decided not to accept the 4.6-billion-year age of the Earth that geologists have determined (primarily from the ages of Moon rocks and meteorites and the evolution of lead isotopes on Earth). Devise an alternative strategy for determining the age of the Earth, assuming that you have unlimited funds.

Part 2 Shaping the Earth's Crust

9

Folds, Faults, and Mountains

Figure 9-1 These rocks in the Lower Ugab Valley, Namibia, have been tilted and bent by powerful plate-tectonic forces.

The Rockies, the Alps, the Andes, the Himalayas—something about these peaks draws us to them. Many of us are simply inspired by their beauty, and some even risk life and limb to climb them. What enormous forces could have shaped common rocks into such massive mountain ranges? Elsewhere, we can see rocks that look twisted and bent (Fig. 9-1). What powerful forces could so dramatically distort such seemingly resistant material, shaking our common belief in the hardness of rock?

Around North America, we can visit young mountains, such as the Cascades of the Pacific Northwest, that continue to rise higher even as you read this book. We can also visit older mountains, such as the Appalachians of eastern North America, that may have been much loftier once but have long since reached their maximum heights and begun eroding away. We can even visit the eroded cores of ancient, now-departed mountains, like those widely exposed in Minnesota, Wisconsin, and northern Michigan and throughout the provinces of eastern Canada.

We can even see future mountains growing. In the rocks along Sagami Bay near Yokohama, Japan, lives a colony of clams called *Lithophaga*, or "rock eaters." These creatures scoop out small shelters for themselves from the soft rocks at sea level, and wait there for high tide to flood their homes and bring their meals of marine algae. At the moment of Japan's great earthquake of 1923, the land at Sagami Bay shifted upward, leaving rows of *Lithophaga* to starve 5 meters (16 feet) *above* sea level. Even higher rows of abandoned *Lithophaga* dwellings populate the cliffs at Sagami Bay, including one that correlates with the area's 1703 tremor and another that correlates with its earthquake of 818. The rocks adjacent to the bay have risen roughly 15 meters (50 feet) during the past 2000 years, offering a stark example of one way to build mountains.

Several geological phenomena can cause a landscape to move in such a dramatic manner. In Part 1 of this text, we saw how the movement of the Earth's tectonic plates produces most of the large-scale features on the planet's surface, such as its ocean basins and continents. With few exceptions, mountains also owe their existence to plate tectonics. The same forces that carry the Earth's plates for thousands of kilometers tear the edges of those plates apart, rupturing them into huge displaced blocks, or squeeze plate edges together, crumpling and uplifting them into great folds of rock.

In this chapter and throughout the rest of Part 2, we will focus on how plate movements invoke the vast geological processes that deform the Earth's crust, creating many of its structures. The branch of geology devoted to crustal deformation and the creation of mountains is known as **structural geology.**

Stressing and Straining Rocks

For a graphic view of what happens to horizontal marine sediments that are subjected to powerful tectonic forces, take a look at Figure 9-2. The sandstone layers of California's Borrego Desert were crumpled and folded during a plate collision; the rocks in California's eastern Sierra Nevada mountains were broken and shifted by a pulling-apart motion that typifies the tectonics of southwestern North America. Compare these rocks with marine sedimentary rocks found near Salt Lake City (see Figure 8-3 on page 210), which have lain relatively undisturbed for hundreds of millions of years.

Plates that converge, diverge, or slide past one another by transform motion subject their crustal rocks to powerful stresses. **Stress** is the force applied to a rock per unit area, usually expressed in *pascals* (1 pascal = 1 newton per square meter) or *kilopascals* (1000 pascals). (A *newton* is the force required to accelerate a mass of 1 kilogram at a rate of 1 meter per second squared. If a rock is sufficiently stressed, it becomes deformed, changing in shape and often volume. This change in shape is called **strain.**

Rocks may be stressed in three ways, corresponding to the three basic types of plate-boundary movements. Rocks at converging plate margins are pushed together, in a type of stress called **compression** (Fig. 9-3a). Compressed rocks are generally crumpled, causing them to thicken vertically and shorten laterally. Rocks at diverging or rifting plate margins are pulled apart; this type of stress is known as **tension** (Fig. 9-3b). Tensional stress stretches, or extends, rocks so that they thin vertically and lengthen laterally. Rocks at transform plate margins are forced past one another in parallel but opposite directions, in a process known as **shearing stress** (Fig. 9-3c). Shearing stress slices rocks into parallel blocks, breaking and displacing preexisting rocks and structures.

Types of Deformation

When subjected to stress, rocks strain in different ways (Fig. 9-4). When the stress—whether compression, tension, or shear—is minor, a rock may return to its original shape and volume after the stress is removed, much as a stretched rubber band regains its original shape after use. Such a temporary change is described as **elastic deformation.** A rock that is strained elastically is not permanently deformed. A greater amount of stress, however, may deform a rock to the point it cannot return to its original shape after removal of the stress. The maximum stress a rock can withstand before becoming deformed permanently is called its **yield point,** or **elastic limit.**

(a)

(b)

Figure 9-2 **(a)** The person in this photo is standing next to a fold in layers of strong, well-cemented sandstone rock in California's Borrego Desert. **(b)** These rocks, from the Sierra Nevada of California, have been broken and shifted by powerful rifting-type forces.

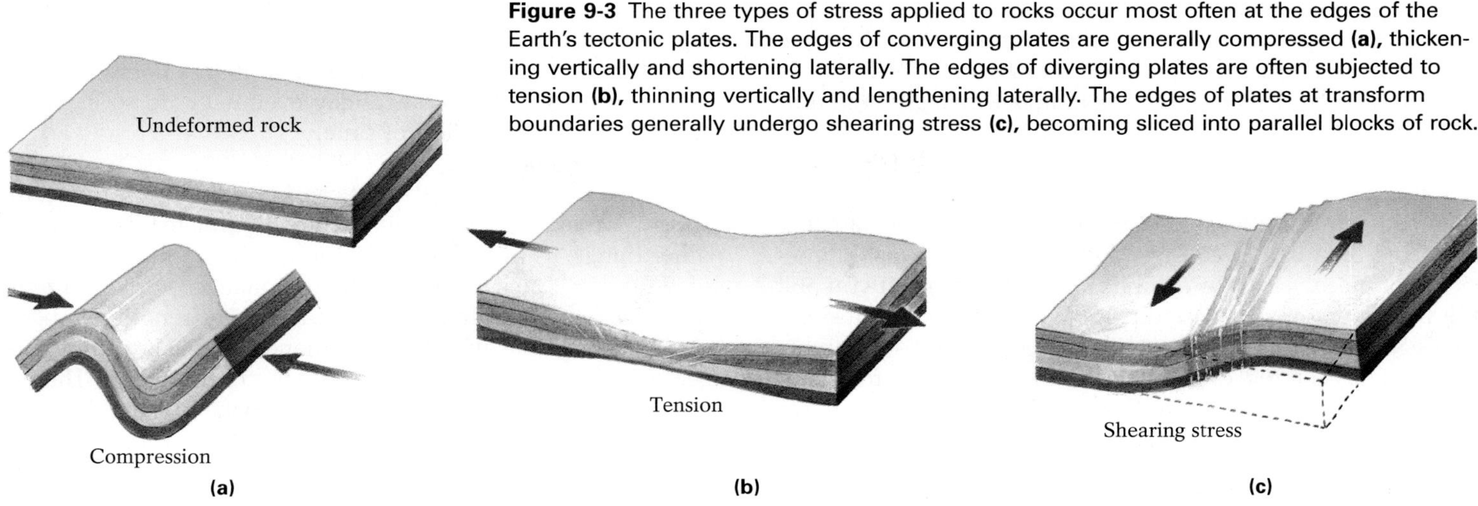

Figure 9-3 The three types of stress applied to rocks occur most often at the edges of the Earth's tectonic plates. The edges of converging plates are generally compressed **(a)**, thickening vertically and shortening laterally. The edges of diverging plates are often subjected to tension **(b)**, thinning vertically and lengthening laterally. The edges of plates at transform boundaries generally undergo shearing stress **(c)**, becoming sliced into parallel blocks of rock.

The pressure and temperature conditions when the stress is applied determine in large part how the rocks deform at the yield point. Rocks subjected to great stress under conditions of low temperature and pressure, or sudden stress, may crack or rupture when they exceed the yield point because their mineral bonds break outright all along the zone of maximum stress. Such large-scale bond breaking results in a type of permanent deformation known as **brittle failure.** When struck sharply with a rock hammer, even a rock as hard as granite may crack.

Sometimes rocks undergo stress under conditions of high temperatures and pressures, such as when they lie buried several kilometers beneath the Earth's surface, or sometimes stress is applied very slowly. Under these conditions, a rock may also exceed its yield point but instead undergoes **plastic deformation,** an *irreversible* change in shape or volume that occurs *without the rock breaking.* During plastic deformation, atoms move about and adjust to stress by breaking and reforming their bonds on a microscopic scale. Atoms in the stressed material generally move from areas of maximum

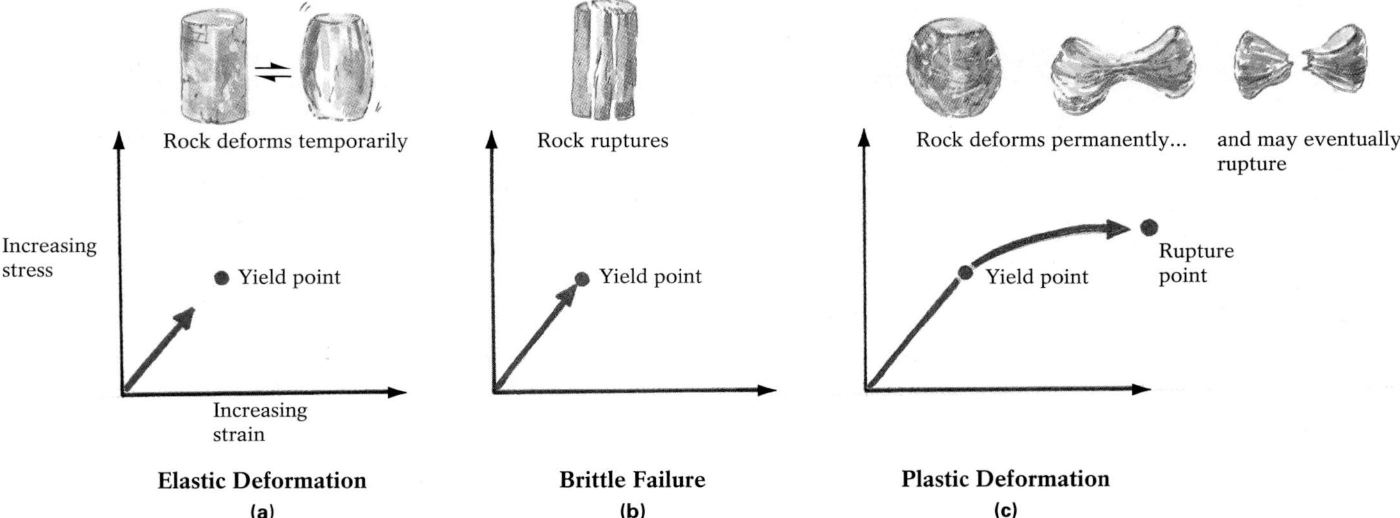

Figure 9-4 The relationship between stress and strain in different types of rock deformation. Within the range of elastic deformation **(a)**, application of stress produces a proportionate amount of reversible strain. Stress that exceeds the yield point of a rock causes it to either rupture due to brittle failure **(b)** or deform plastically **(c)**. Within the range of plastic deformation, even small amounts of additional stress create a great deal of irreversible strain.

stress and re-form in areas of lower stress, much as toothpaste moves upward and away from your clenched fist when you squeeze the tube. Sometimes rocks initially respond to stress by elastic or plastic deformation, but then undergo brittle failure if the stress suddenly increases.

Silly Putty is the perfect medium for demonstrating different types of rock deformation. When rolled into a ball and bounced gently on a table top, Silly Putty is subjected to a low level of stress. Because it does not change shape appreciably or permanently, it exhibits *elastic behavior.* If you pull the ball slowly in opposite directions until the stress just exceeds Silly Putty's yield point, its shape changes considerably and remains altered even after the stress ceases—that is, it deforms *plastically.* Beyond this point, even small increases in tensional stress cause a great deal of plastic deformation, and you can easily pull the putty into long, wispy filaments. Finally, when the original ball of Silly Putty is pulled firmly and abruptly, it fractures into two sharp-edged pieces, exhibiting *brittle failure.*

Factors Affecting Rock Deformation

The intensity of the applied stress is not the only factor that determines how rocks will deform. Laboratory experiments that simulate "real-world" conditions have shown that lithostatic pressure, heat, the amount of time that a stress is applied, and a rock's composition all affect its response to stress.

Lithostatic Pressure Lithostatic pressure from overlying rocks (discussed in Chapter 7) increases with depth. Near the Earth's surface, relatively low lithostatic pressure stresses rocks only moderately. Thus their bonds may stretch but do not break, and the rocks do not reach their elastic limit. Such shallow rocks tend to deform *elastically.* The same rocks, if buried several kilometers beneath the surface, would undergo greater stress and would be more likely to exceed their elastic limit. Their bonds would tend to break and reform, and the rocks would deform *plastically.*

Heat Increased heat excites the atoms in a rock's minerals, weakening the bonds between them until they break. Thus, the bonds in a rock that is stressed at low temperatures—like those found near the Earth's surface—may stretch somewhat without breaking. Like low lithostatic pressures, relatively low temperatures promote elastic deformation. If, however, a rock is stressed at high temperatures—generally deeper beneath the Earth's surface—its minerals' bonds will tend to break and re-form, promoting plastic deformation. Just as a blacksmith's furnace heats a metal bar so it can be fashioned into a horseshoe, so the Earth's internal heat renders rocks more plastic.

Time We can simulate the heat and pressure exerted on rocks deep in the Earth's interior in laboratories, but we cannot simulate the passage of eons of time. Clues in real-world settings, however, hint at how time affects deformation processes. Consider, for instance, how sturdy wooden bookshelves gradually sag over time from sustaining the weight of heavy volumes. This development suggests that a force that is initially insufficient to deform a substance may eventually do so if maintained over a long period of time.

Composition Rocks vary significantly in their responses to stress. A rock's elastic limit is directly related to, among other things, its mineral composition, texture, internal planes of weakness (such as metamorphic foliation), fluid content (high fluid content tends to reduce a rock's strength), and geologic history (has it been strengthened or weakened by the geological forces and processes it has experienced?). Thus a relatively unweathered, unfaulted block of dry, coarse-grained granite will deform differently from a thinly bedded layer of limestone riven with water-filled solution cavities.

Deformed Rocks in the Field

In the field, virtually every rock outcrop contains visible evidence of strain. We can see deformation most readily in sedimentary rocks, largely because most were originally horizontal and laterally continuous. When sedimentary layers deform plastically, their crumpled appearance makes them easier to distinguish than plastic deformation in a mass of plutonic igneous rock, such as granite. Likewise, fractures and displacements from brittle failure are more clearly visible in sedimentary rock, thanks to the obvious offset of the rock layers.

When geologists find deformed rocks, they are curious about the type of stress (compression, tension, or shearing) that produced the deformation and about the distance and direction that the deformed rocks moved. This information enables them to identify past plate motions and other geological events. They also want to understand how the deformation continues underground, because this knowledge helps locate subsurface resources, such as oil, coal, natural gas, and valuable metals and other minerals. Geologists estimate the subsurface deformation of rocks by extrapolating from the orientation of their surface features.

The orientation of a geological plane—either a structure or rock layer—in three-dimensional space is determined in relation to the four principal compass directions (north, south, east, and west). One key component of a geological plane's orientation in space is its **strike,** defined as the compass direction of the line that forms at the intersection between the plane (such as a layer of sedimentary rock) and an imaginary *horizontal* plane (Fig. 9-5a). We can determine strike most easily where a geological plane, such as a layer of sedimentary rock, intersects the horizontal plane of a body of water (Fig. 9-5b). A compass tells us the direction of the horizontal water line crossing the geological plane—the rock's strike.

The **dip** of any geological plane is the angle at which it is inclined relative to the horizontal (Fig. 9-5c). It may be

Figure 9-5 Because the surface of a body of water is horizontal, it can help determine the orientation of rock outcrops relative to the Earth's surface. The plane of the water traces a strike line across the exposed geologic surface of a partially submerged outcrop, and provides a horizontal reference against which to measure a plane's angle of dip (using an inclinometer). Note that a plane's strike and its dip are perpendicular to each other.

measured with an *inclinometer,* a device much like a carpenter's bubble level contained within a geologic compass. Whereas strike is expressed only as a direction, dip requires a measured angle and a *direction.* The directions of strike and dip are always perpendicular to one another. Measuring the strike, dip angle, and dip direction of a geological plane establishes its orientation in three-dimensional space. On geologic maps, these measurements are shown by a *strike-and-dip symbol.* By making hundreds of strike-and-dip measurements at numerous outcrops and recording the data on a base map or aerial photograph, geologists can determine the three-dimensional *subsurface* form of structures barely visible at the surface.

Folds

Folds are bends in strata that develop when layers of rock deform plastically. The most commonly observed folded rocks comprise sedimentary rock sequences. When deeply buried sedimentary rocks are compressed, generally at the edges of converging plates, they fold down and up, forming a series of troughs and arches—much like a rug that bunches up when pushed on a polished hardwood floor. Folds in rock vary in size from the small crinkles visible in some fist-sized hand specimens (Fig. 9-6a), through the obvious crenulations seen in the roadcuts that border many major highways (such as those through the Appalachian Mountains of eastern North America) (Fig. 9-6b), to the mountain-sized folds of Stair Hole at Lulworth Cove, Dorset, England (Fig. 9-6c).

(a)

(b)

(c)

Figure 9-6 Folded rocks of various sizes. **(a)** A hand specimen of folded gneiss; **(b)** folded rock strata along Route 23 in Newfoundland, New Jersey; **(c)** folded sedimentary rocks in Dorset, England, which result from strong compressional forces at a former convergent plate margin.

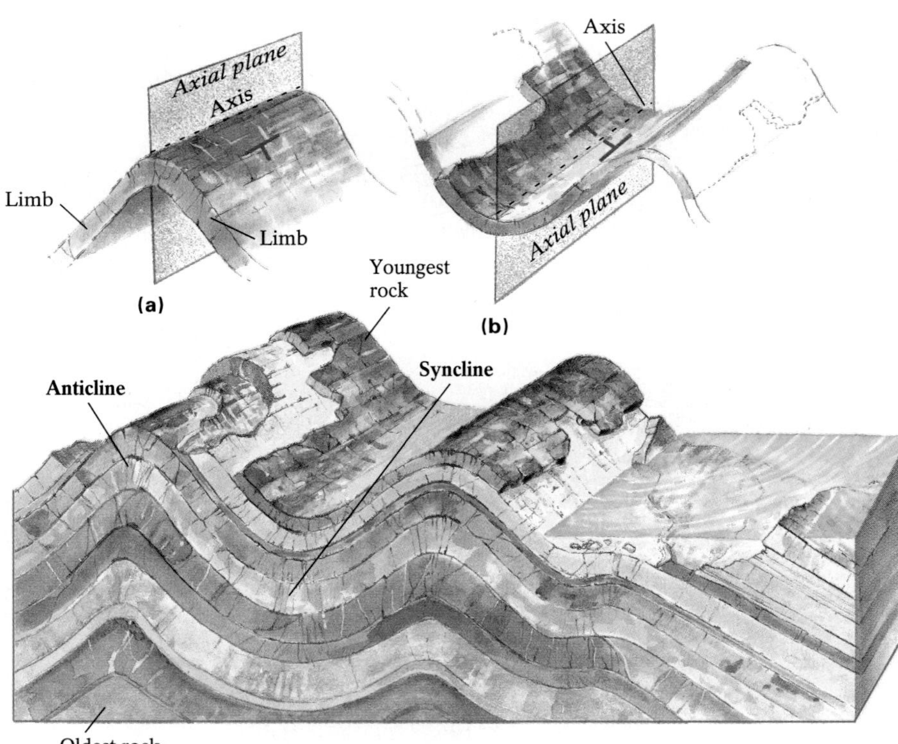

Figure 9-7 The geometry of anticlines and synclines.

Synclines and Anticlines

The most common types of folded rocks are the tightly folded arches and troughs that occur most often from compression at active convergent plate margins. Concave-up, trough-like folds, which typically expose a region's *youngest* rocks at their cores after erosion, are called **synclines** (from the Greek for "inclined together"). Convex-up, arch-like folds, which when eroded typically expose a region's *oldest* rocks at their cores, are called **anticlines** (from the Greek for "inclined against") (Fig. 9-7).

A fold's sides are referred to as its *limbs.* Folding typically produces a series of alternating synclines and anticlines, with adjacent folds sharing a common limb. Each syncline and anticline has an *axial plane,* an imaginary plane that divides the fold into two approximately equal parts. The *axis* of the fold—its line of maximum curvature—appears where the axial plane intersects the fold's crest. An anticline's axis resembles the peaked roofline of an A-frame structure, whereas a syncline's axis is akin to the keel of a sailboat.

Although in diagrams anticlines may look like *topographic* hills and synclines like *topographic* valleys, they are actually structures in the rocks, not surface landforms. Anticlines do not always form hills or ridges in the landscape, nor do synclines invariably underlie valleys. The layers in the folded rock resist erosion differently. Thus, if a resistant sandstone were exposed in the trough of a syncline, and readily eroded rocks formed the crests of adjacent anticlines, structural anticlines would tend to erode so as to form topographic valleys and structural synclines would eventually stand out prominently as topographic ridges (Fig. 9-8). The rocks across a region are not necessarily continuous. Thus a syncline 10 kilometers (6 miles) away from an anticline may have resistant sandstone exposed at the surface and erode in a different manner.

In the field, vegetation may conceal folds, especially in moist, continuously vegetated regions such as those of eastern North America. Folds may also be covered by recent sediments or be so weathered and eroded that only scattered small outcrops remain. Strikes and dips can establish the orientation of these scattered outcrops; many of them are limb remnants of partially eroded folds. They enable geologists to construct an accurate picture of the regional fold pattern.

Fold Symmetry Folds vary from simple to complex (Fig. 9-9). Under relatively gentle compression, anticlines and synclines appear as broad *symmetrical,* or *open,* structures with near-vertical axial planes and gently dipping limbs inclined at roughly the same angle. These forms occur most commonly at the edges of mountain belts, in areas that are relatively quiet tectonically. Tightly folded anticlines and synclines are more common. They typically form at active or formerly active convergent plate margins, where compression may range from moderate to intense. If compression forces one limb to

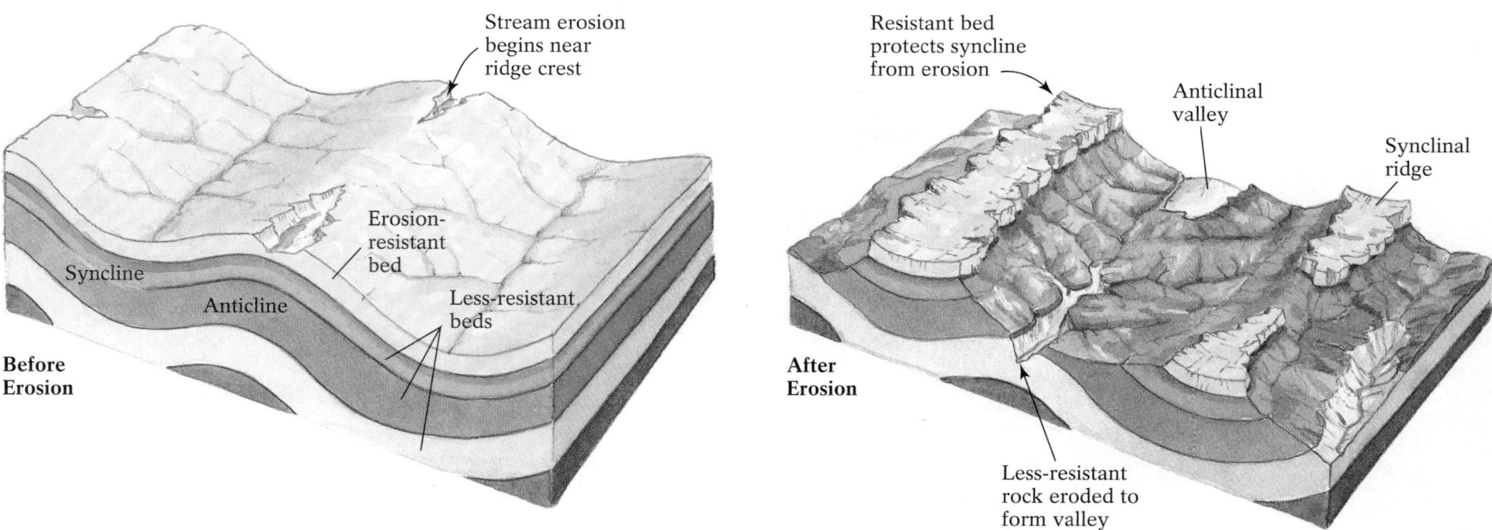

Figure 9-8 Anticlinal valleys and synclinal ridges. The composition of exposed bedrock structures strongly controls their susceptibility to weathering and erosion. If weak, erodible rock crops out along the crest of an anticline, its erosion will produce a valley at that site. If the axis of a syncline consists of strong, resistant rock, its surface expression will be a ridge.

move more than the other, folds may become *asymmetrical.* Prolonged directed pressure may cause the limb of an asymmetrical fold to rotate past vertical, producing an *overturned* fold. Under conditions of extreme compression, a fold may even rotate until its axial plane becomes essentially horizontal, paralleling the Earth's surface. Such *recumbent* folds typify highly deformed mountain belts, such as the northern Rockies, Appalachians, Himalayas, and Alps.

More complex structures develop when an entire sequence of anticlines and synclines tilts so that the structures' axes actually intersect the Earth's surface. We identify such *plunging* folds by the characteristic *zigzag* pattern that emerges as they erode. The Valley and Ridge province of the folded Appalachians of Pennsylvania provides some of North America's best examples of plunging folds (Fig. 9-10).

Plate Tectonics and Folding Because folds are generally associated with compression, they occur primarily at convergent plate boundaries, such as subduction zones and zones of continental collision (Fig. 9-11). At subduction zones, loose sediments caught between the converging plates experience intense compression and commonly become folded into complex patterns, both within the oceanic trench and in the area between the trench and the growing volcanic arc. Continental plate collisions produce the most extensive folding. Generally too buoyant to subduct, the upper portion of the continental lithosphere gets pushed up and folded spectacularly, whereas the lower crust and the plate's underlying mantle component are pushed down. Compressed sediments in zones of past or ongoing plate collisions form such major mountain belts as the Alps, Himalayas, Appalachians, and Canadian Rockies.

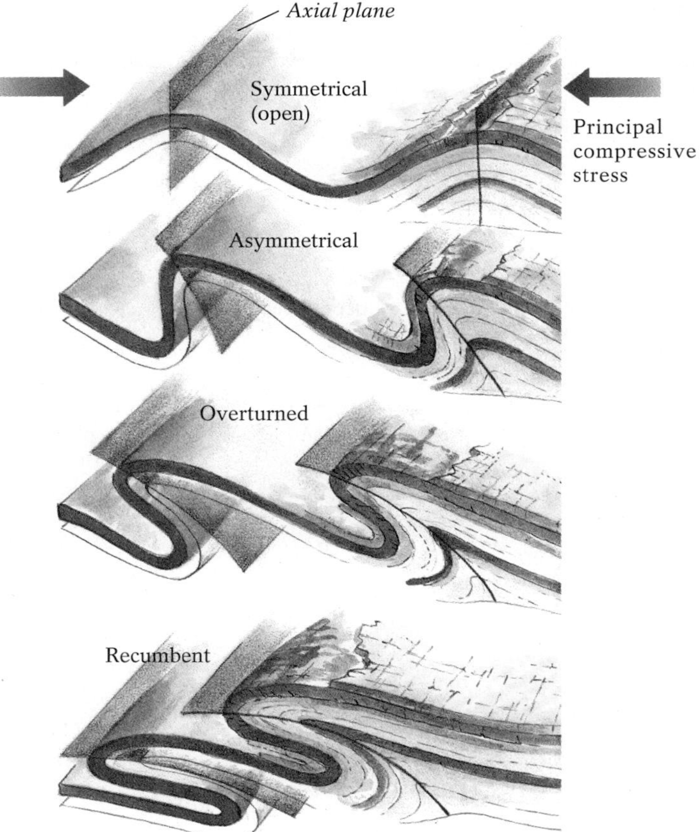

Figure 9-9 Folds vary from broad, open structures, with limbs dipping at roughly the same angles, to overturned and recumbent folds. Such evidence of compression generally occurs where continental plates have collided.

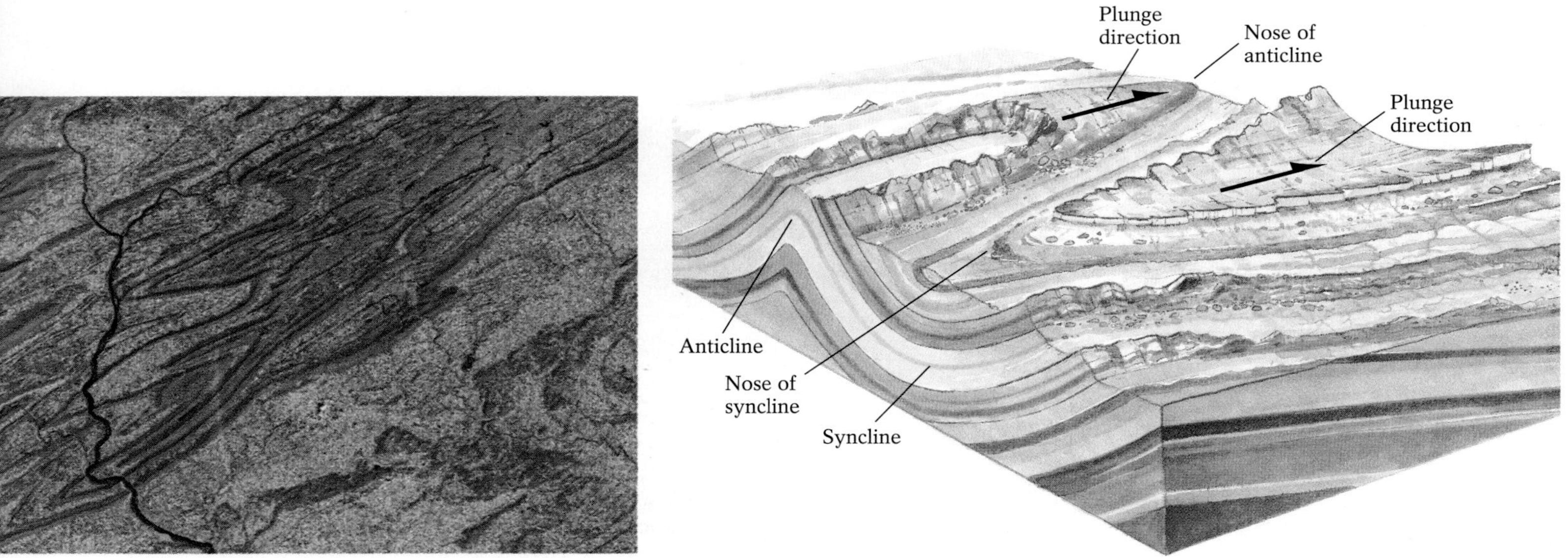

Figure 9-10 (left) Plunging folds in the Valley and Ridge province of central Pennsylvania, near Harrisburg. These structures consist of sedimentary beds that were folded and tilted by the tectonic plate collisions that produced the Appalachian mountains. (right) We can determine the direction in which the folds are plunging by identifying the anticlines and synclines, and then noting the direction in which the *nose* (the area of maximum curvature of the fold, seen on the zigzag pattern of plunging folds) points. The nose of an anticline points toward the plunge direction of a fold, whereas the nose of a syncline points away from its plunge direction.

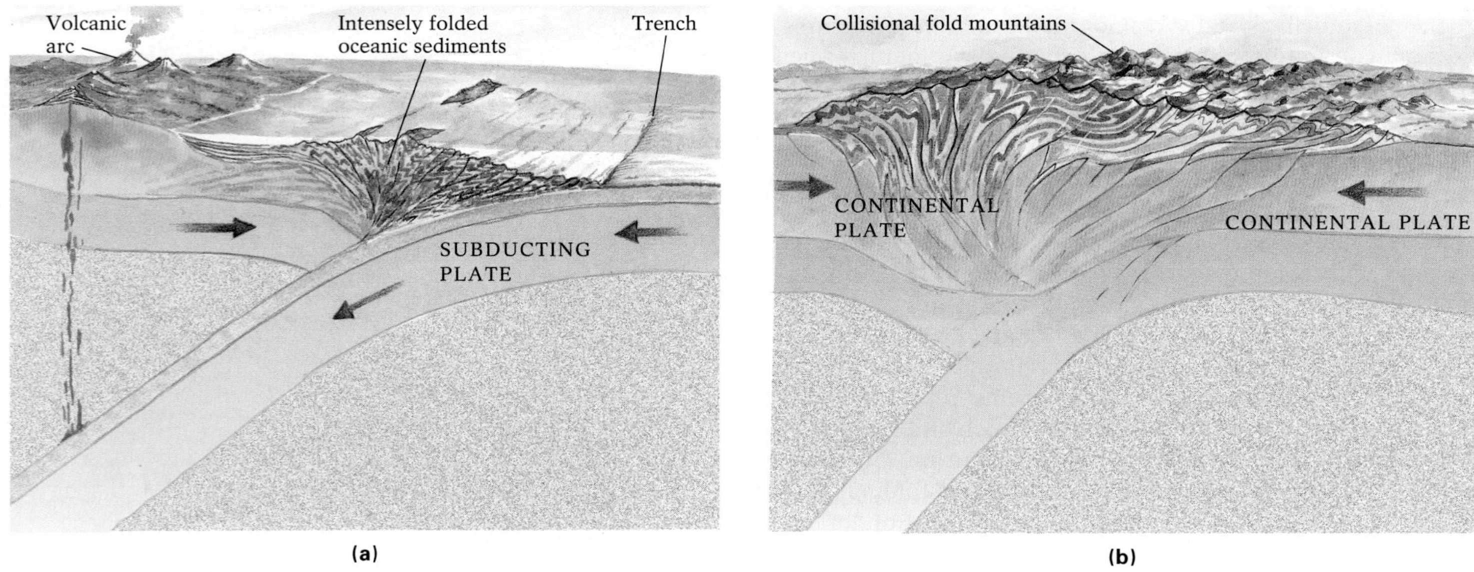

Figure 9-11 Folding occurs most often at convergent plate boundaries. **(a)** At subduction zones, soft marine sediments are compressed within the oceanic trench and between the trench and its associated volcanic arc, creating extensive folding. **(b)** At continental collision zones, sediments between the plates are intensely folded and metamorphosed.

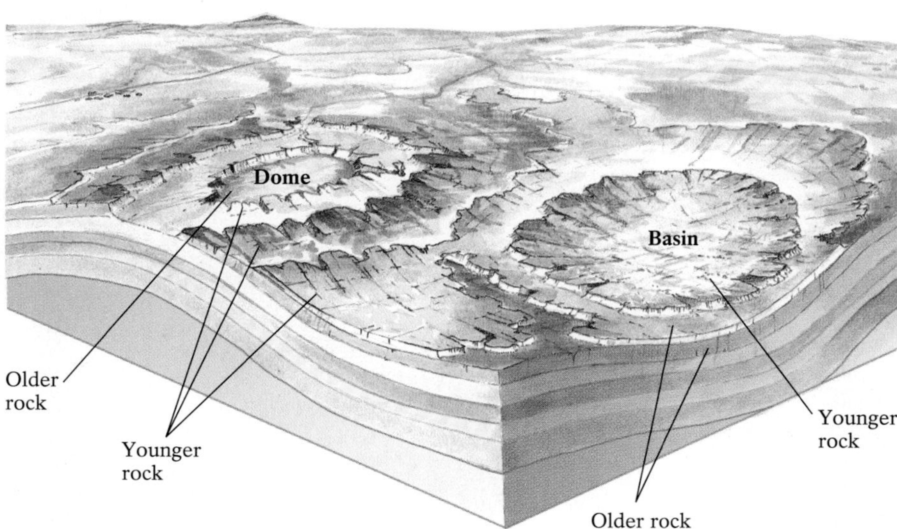

Figure 9-12 Structural domes and basins resemble anticlines and synclines, respectively, especially with regard to the ages of rocks within their cores. Because these structures often form far from plate boundaries, they may reflect the effects of vertical motion related to crustal density variations rather than plate-edge activity.

Domes and Basins

A structural **dome** is a subtle, oval-shaped bulge of rock layers that, in cross section, resembles an anticline. As in an anticline, when the surface of a dome is breached, the oldest rocks lie at its center, and all other layers dip away from them (Fig. 9-12). A structural **basin** is an oval-shaped bedrock depression that, in cross section, resembles a syncline. As in a syncline, the youngest rocks in a basin are found at its center, and the other layers dip toward them.

Unlike anticlines and synclines, domes and basins sometimes form in mid-continent, mid-plate locations. Some geologists have therefore concluded that whereas anticlines and synclines result primarily from the *lateral* forces associated with plate movements, domes and basins more likely result from the *vertical* forces associated with variations in crustal density. For example, some domes appear to form where low-density materials such as salt, upwelling magma, or warm mantle currents rise toward the surface, pushing the overlying sedimentary strata upward. Some basins appear to form where materials near and just below the surface are dense enough to depress and deform the underlying materials.

We see domes more readily than basins. They stand out prominently at the surface, whereas basins usually become filled with sediment and are thus obscured. Noteworthy in North America are the Adirondack dome of northern New York, the Ozark dome of the southern Mississippi River valley, the Nashville dome of the Tennessee River valley, and the oil-producing domes of eastern Wyoming. The scenic Black Hills of South Dakota are actually a large oval dome, whose 2208-meter (7242-foot)-high Harney Peak is the highest point in the state. There, the region's sedimentary rocks dip away from an exposed core of Precambrian igneous and metamorphic rocks, which serve as the medium for the prominent sculpture atop Mount Rushmore. Notable structural basins include the oil- and coal-bearing Williston basin of North Dakota, the Illinois basin of southern Illinois, and the Michigan basin, which forms virtually all of lower Michigan (with the campus of Michigan State University lying at its center).

Faults

Most rocks are brittle at low temperature and low lithostatic pressure. Those conditions prevail at or near the Earth's surface, so virtually every solid surface rock and nearly all layers of unconsolidated sediment contain evidence of brittle failure in the form of cracks, or *fractures*. The orientation of these fractures provides clues to the stresses applied to the rocks in the past. Some fractures seem to be random, and show no evidence of relative movement (Fig. 9-13a). A variety of different stresses may produce these fractures, such as stresses associated with cooling igneous rocks or freezing surface rocks. Fractures that are oriented systematically (for example, at right angles to one another), again

(a)

(b)

(c)

Figure 9-13 Products of brittle failure in surface rocks. **(a)** These rocks from Acadia National Park, Maine, have been fractured by the application of a variety of stresses, including the weight of the glaciers that once sat upon them, the cracking that occurs during mechanical exfoliation (Chapter 5), the contraction that accompanies the cooling of plutonic igneous rock, and the stress from past plate-edge interactions. **(b)** The patterned appearance of these joints in Arches National Park, Utah, indicates that these rocks experienced some type of regional stress, perhaps related to plate interactions. **(c)** These faulted rocks have clearly moved relative to one another, as can be seen by the displacement of the various colored rock layers.

with no evidence of significant relative movement, are called **joints** (Fig. 9-13b). Joints typically form in response to a specific set of stress conditions, such as those prevailing at a plate boundary.

The most dramatic examples of cracks in rock are **faults,** fractures that show noticeable relative movement (Fig. 9-13c). Fault blocks are the rock masses that lie on either side of the *fault plane,* the approximately planar (flat) surface along which the movement occurs. Strike-and-dip measurements help determine the orientation of a fault plane and each fault block's relative direction of movement (Fig. 9-14).

An area's rock outcrops provide both obvious and subtle clues that help geologists identify faulted rocks. The most obvious expression of faulting is a visible displacement of rocks, such as that shown in Figure 9-13c. In a rock outcrop that contains only one type or layer of rock, the clue may take the form of pulverized rock along a fracture, produced by grinding as adjacent rock masses moved. In addition, rocks in contact at a fault plane often have a polished surface called *slickensides,* the result of being abraded and smoothed by pulverized rock acting like jeweler's grit. In more subtle cases, faults may be identified where a local rock sequence is discontinuous with the adjacent stratigraphy. For example, if a distinctive rock bed disappears abruptly, sometimes reappearing at a greater depth some distance away, faulting has very likely disrupted the lateral continuity of the rock layers. As you can see in Figure 9-15, such discoveries may be associated with significant economic benefits.

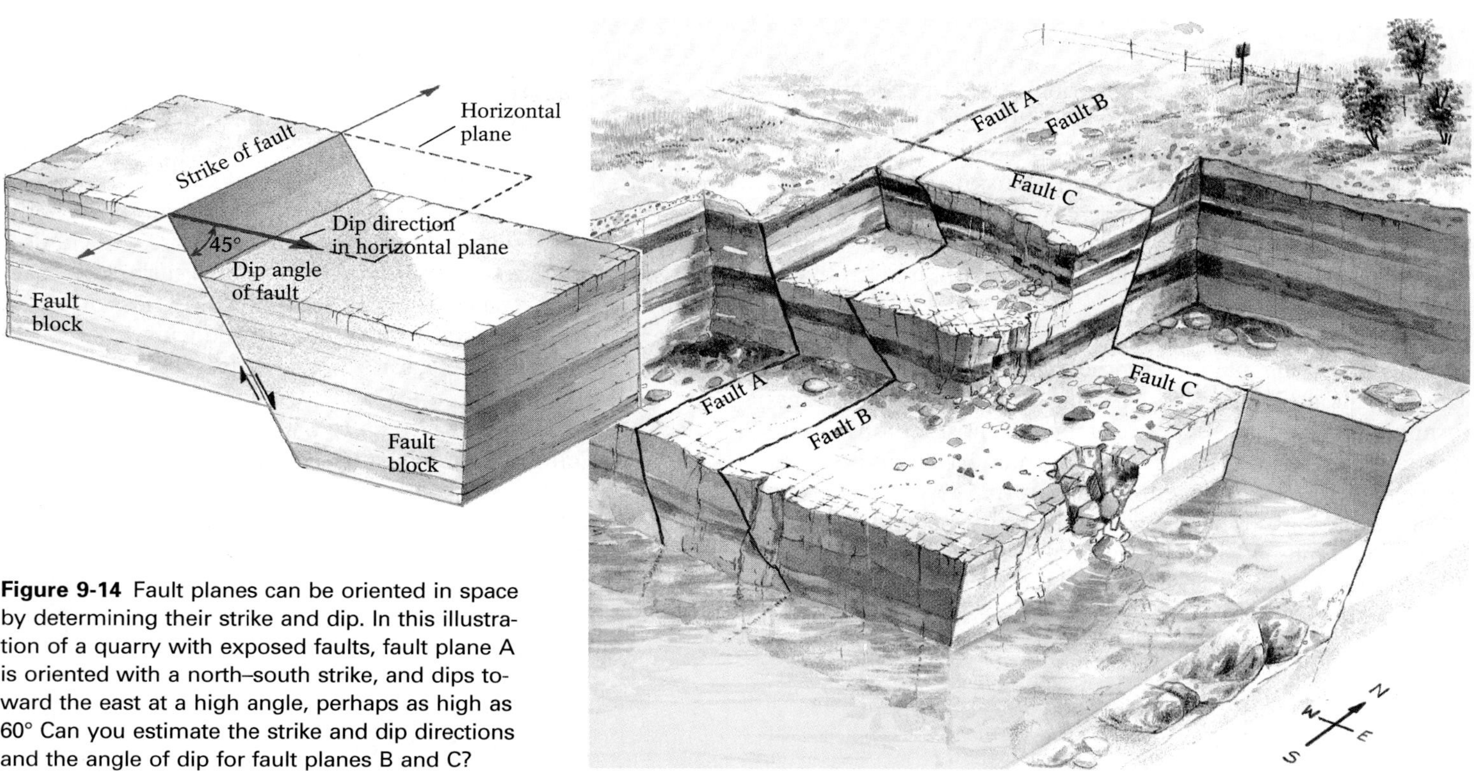

Figure 9-14 Fault planes can be oriented in space by determining their strike and dip. In this illustration of a quarry with exposed faults, fault plane A is oriented with a north–south strike, and dips toward the east at a high angle, perhaps as high as 60° Can you estimate the strike and dip directions and the angle of dip for fault planes B and C?

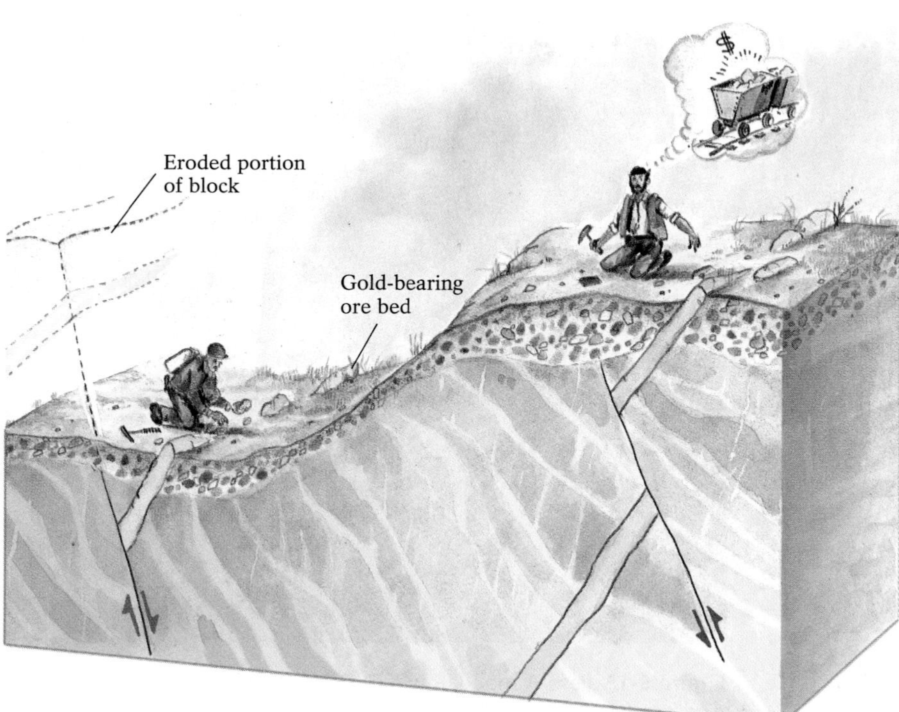

Figure 9-15 By offsetting adjacent blocks of rock, faults disrupt the continuity of individual rock layers. One economic implication of faulting is shown in this illustration, in which the relative movement of the blocks is indicated by arrows. At left, the valuable ore bed that was present in the higher fault block has been eroded away, leaving none on that side of the fault. At right, the ore body continues at a slightly greater depth in the lower fault block. If he digs deep enough, the miner at right will find a significant lode of gold, whereas the miner at left will find only a limited amount.

Types of Faults

All faults form as a result of shearing motion between adjacent blocks of rock, during which one or both blocks move, or *slip,* relative to their original position. Such slippage is usually caused by the compressional, tensional, or shearing stresses associated with plate-tectonic movement. We generally classify faults according to the direction of the relative movement between fault blocks, which, in turn, relates to the type of stress causing the fault.

In **strike-slip faults,** fault-block movement is largely horizontal, occurring parallel to the strike of the fault plane. These faults are caused by shearing stress (usually at transform plate boundaries). North America's most famous strike-slip fault, California's San Andreas, underwent one of its greatest recorded displacements on April 18, 1906, the day of the great San Francisco earthquake. If you and a friend had been standing on the shores of Tomales Bay, about 30 kilometers (20 miles) northwest of San Francisco, staring at one another across the San Andreas fault at precisely 5:12 A.M., within seconds you would have moved 7 meters (25 feet) to one another's right.

The San Andreas fault, an active fault 1000 kilometers (600 miles) long, extends from the Pacific Ocean beyond Cape Mendocino, northwest of San Francisco, southward beyond California's border into the Gulf of California in western Mexico (Fig. 9-16). Parts of the fault creep along daily, causing relatively little damage (except to the creaking houses directly over the fault and to sidewalks and curbs, which require frequent repaving). Other segments move less often but in large rapid displacements, typically producing great earthquakes. The 1906 quake, for example, offset fences and other linear features along more than 400 kilometers (250 miles) of the fault, from Point Arena to San Juan Bautista. By comparison, California's Loma Prieta earthquake of 1989 produced far less displacement; a maximum of 2.2 meters (7 feet) of motion was measured east of Santa Cruz, and the surface rupture stretched for only a few tens of kilometers.

The surface expression of a strike-slip fault may stretch for hundreds of kilometers as low linear features—both ridges

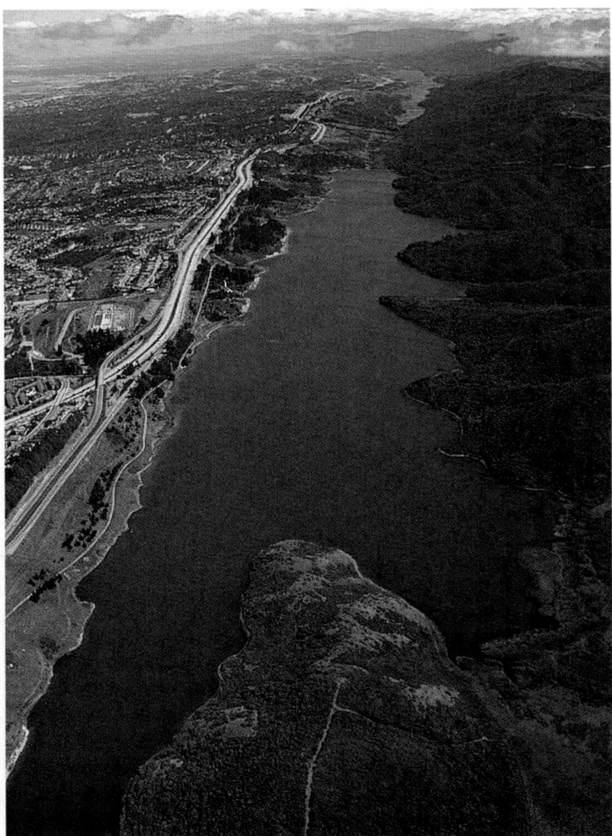

Figure 9-16 (left) A map showing the length of the San Andreas fault. (above) A section of the fault just south of San Francisco, containing San Andreas Lake and Crystal Spring Reservoir—examples of depressions and erosional basins that commonly occur along strike-slip faults.

and depressions—that disrupt the works of both nature and humanity (see Figure 1-25 on page 24). As the pulverized rock of a strike-slip fault zone erodes, several distinctive landforms may be produced: linear valleys, chains of lakes (such as Loch Ness, in Scotland's Great Glen fault zone), sag ponds (depressions) caused by uneven ground settling within the fault zone, and topographic saddles (notches in the skyline of faulted hills; Fig. 9-17).

Fault movements are more often vertical than horizontal. In **dip-slip faults,** the fault blocks move up and down, parallel to the dip of the fault plane (Fig. 9-18). Such faults are typically marked by steps in the landscape that form where the land surface shifts upward or downward. You will see many old mineshafts along dip-slip faults, because that's where precious metals became concentrated when hot, mineral-rich solutions migrated along faulted rocks. We identify dip-slip fault blocks by their position relative to the fault plane, using terms derived from mining traditions. The block above the fault plane, from which the miners hung their lanterns, is known as the *hanging wall;* the block below the fault plane, on which the miners stood, is called the *footwall.*

Various types of dip-slip faults can be distinguished by the relative movement of their fault blocks. In a **normal fault,** the hanging wall has moved downward. This type of fault is generally associated with tensional stress, which typically occurs where plates are rifting or diverging. Tensional stresses initially stretch and thin the Earth's crust, then eventually fracture rocks into faulted blocks. A fault block that drops during normal faulting produces a depression called a **graben** (from the German for "grave"). The blocks that remain stand-

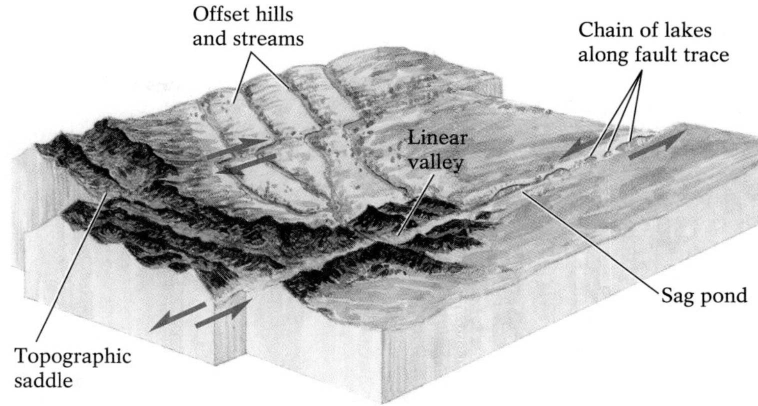

Figure 9-17 Horizontal movement along strike-slip faults creates offset topographic features and distinctive erosional landforms, such as linear valleys, lake chains, and sag ponds.

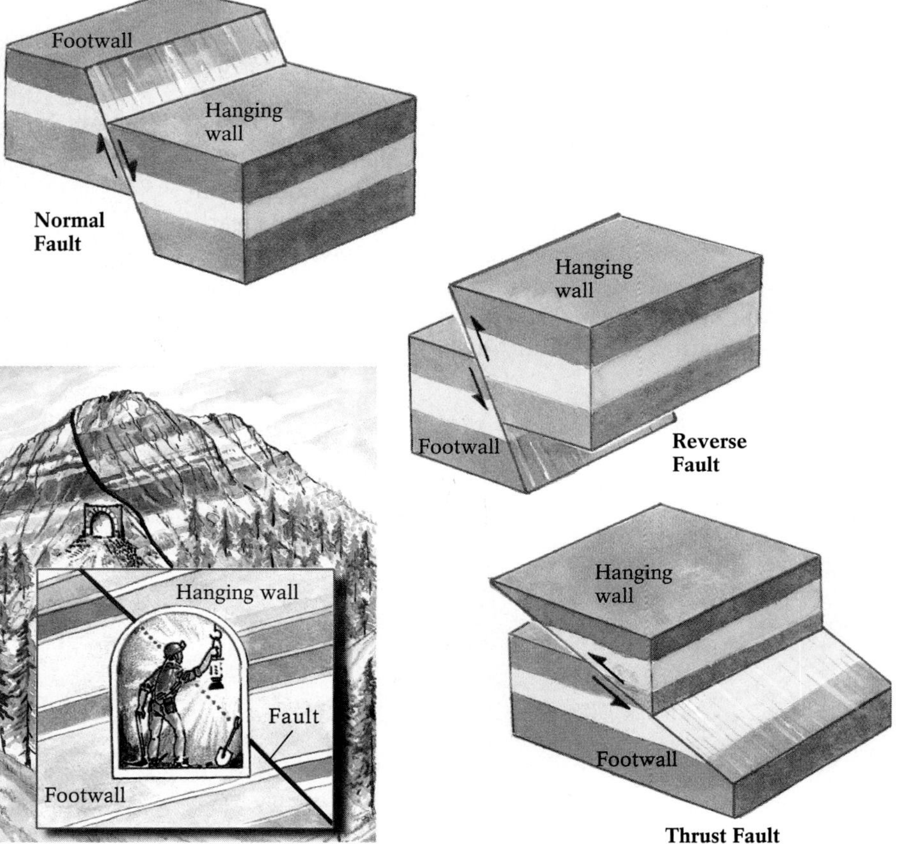

Figure 9-18 Dip-slip faults. (above) A dip-slip fault along McGee Creek, south of Mammoth Lakes, California. The rocks in the foreground (in front of the "step" in the landscape) have slipped downward relative to the rocks in the background. (right) Dip-slip faults are classified by the relative movement of their fault blocks. In normal faults, the hanging wall moves downward relative to the footwall; in reverse faults, the hanging wall moves upward relative to the footwall; thrust faults are low-angle reverse faults.

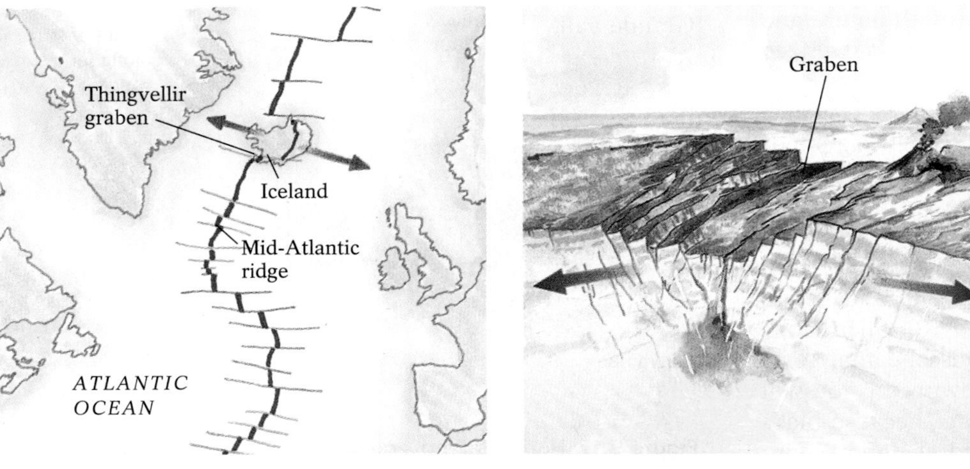

Figure 9-19 Iceland's central valley, the Thingvellir, formed when tensional stress and normal faulting within the diverging mid-Atlantic ridge system created a graben.

ing above and on either side of a graben are referred to as **horsts** (from the German for "height"). These features characterize the normal faults of East Africa's rift valleys, as well as Earth's 60,000 kilometers (40,000 miles) of mid-ocean divergence zones (Fig. 9-19).

In a **reverse fault,** the hanging wall has moved upward relative to the footwall. This type of fault is generally associated with the powerful horizontal compression that develops where plates converge. As the hanging wall moves over the footwall, it carries along deeper, older rocks, transporting them over younger rocks to create a rock sequence that violates the principle of superposition (described in Chapter 8).

A **thrust fault** is a type of reverse fault in which the blocks move at a relatively low angle (less than 45°). A very-low-angle thrust fault (perhaps as low as 5–10°) produces an *overthrust,* in which enormous slabs of rock move nearly horizontally for tens of kilometers. In the Lewis Overthrust in Glacier National Park, Montana, and Waterton-Lakes National Park in Alberta, a 3-kilometer (2-mile)-thick slab of Precambrian marine sedimentary rock moved more than 50 kilometers (30 miles) eastward, coming to rest atop much younger continental sedimentary rock (Fig. 9-20).

In many faults, the slip is not solely vertical (up or down along the dip of the fault plane) or horizontal (along the fault plane's strike). Such fault motions—called **oblique slip**—combine both strike-slip and dip-slip motion (Fig. 9-21).

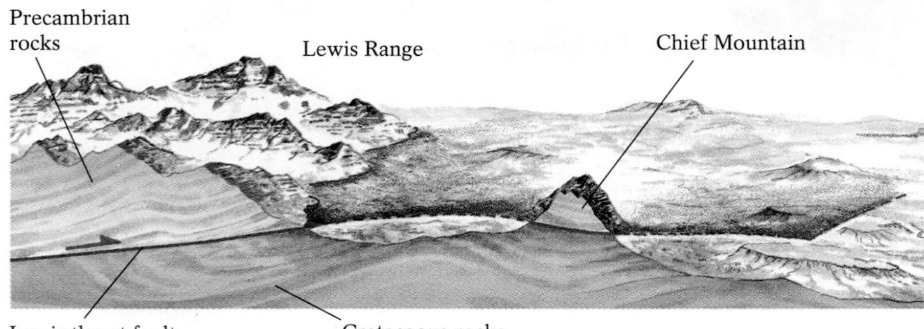

Figure 9-20 The Lewis Overthrust, in Glacier National Park, Montana, formed when compression moved a slab of 800-million–1.1-billion-year-old Precambrian marine sedimentary rock onto the top of a layer of 150-million-year-old continental sedimentary rock. Photo: Chief Mountain, formed by subsequent erosion of both layers, contains Precambrian rock at its top and the younger Mesozoic rock at its base.

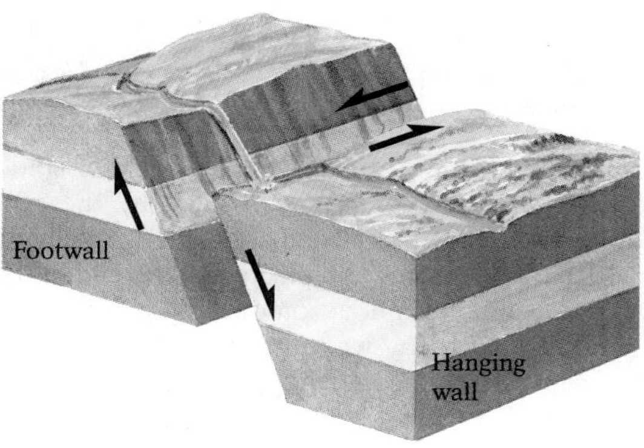

Figure 9-21 Oblique slip is fault motion that involves both dip-slip and strike-slip movement of fault blocks.

Plate Tectonics and Faulting

Each of the three major types of plate margins is associated with a particular type of stress, and thus with a particular type of fault (Fig. 9-22). Normal faults commonly occur where tensional stresses pull apart the Earth's lithosphere, such as along the axes of the world's mid-ocean ridges and at sites where continents are rifting, such as in the north–south Rio Grande valley in New Mexico, where North America may be splitting apart.

Reverse and thrust faults, which are produced by compressive stresses, are concentrated along convergent plate boundaries, both where oceanic plates subduct and where continental plates collide. The major faults in Japan, the Philippines, and other western Pacific islands are produced largely by convergence of two ocean plates (with one subducting beneath the other); in contrast, the faults in the Andes of South America and Cascades of Washington, Oregon, and northern California have developed where an ocean plate subducts beneath an adjacent continental plate. Faults in the Alps and Himalayas owe their existence to collisions between continental plates—those in the Alps from past collisions of Africa and Europe, and those in the Himalayas from the ongoing collision of Asia and India.

Some of the world's longest continuous faults, such as the San Andreas fault in California and the Anatolian fault in Turkey, are strike-slip faults that coincide with transform boundaries. At these boundaries, plates slide past one another, producing shearing stress.

The plate-tectonic forces that produce large dip-slip faults in thick sequences of marine sedimentary rocks, as well

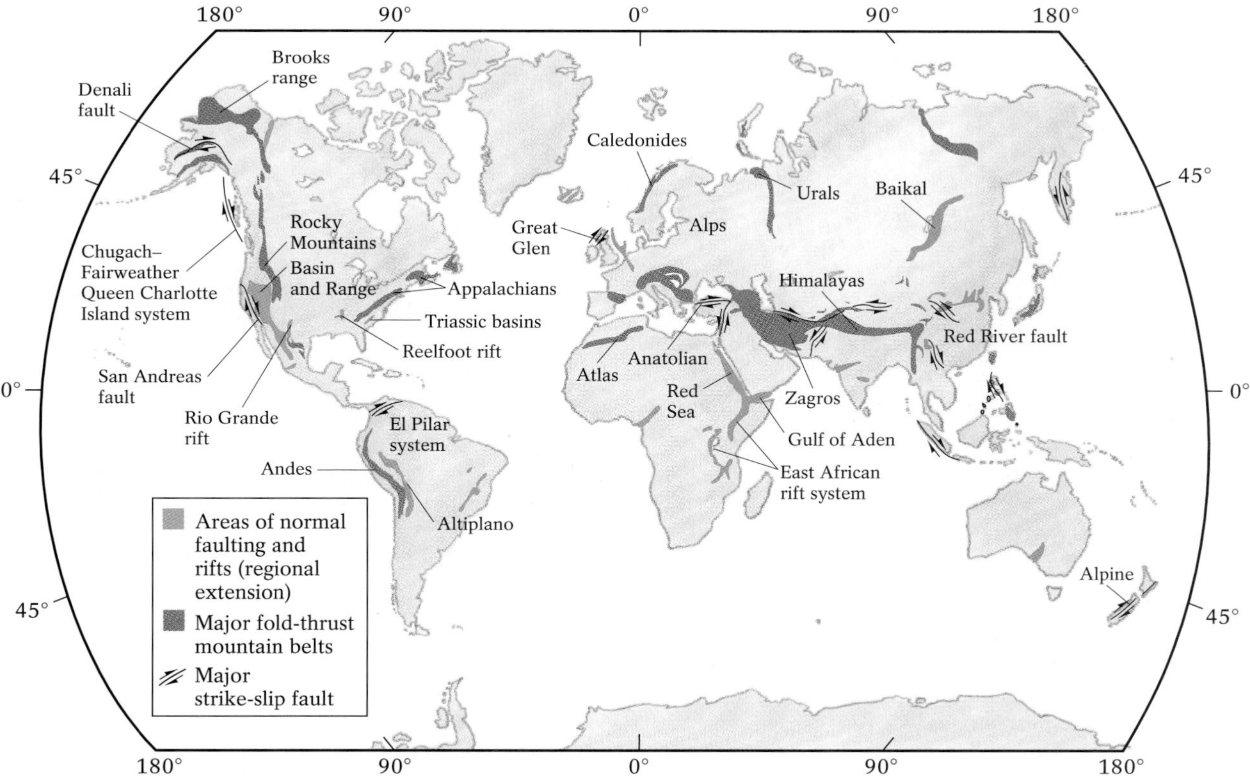

Figure 9-22 The correlation of different fault types with the Earth's plate boundaries. Normal faults typically occur at rifting and divergent zones. Reverse and thrust faults commonly occur at convergent zones, including subduction zones and zones of continental collisions. Strike-slip faults occur principally at transform boundaries.

Highlight 9-1 Folds, Faults, and the Search for Fossil Fuels

The search for the Earth's dwindling fossil fuel reserves depends on the identification of structural features that trap oil and natural gas in recoverable quantities. Heating of organic material in marine sediments (the source rocks) produces oil underground. Less dense than water and the surrounding rocks, oil and gas tend to rise toward the surface, migrating through permeable rocks and fractures until their passage is blocked. At these locations, the fluids accumulate in the pores and cracks of other rocks (the reservoir rocks) from which they can be readily collected. In folded terrain, the passage of oil and gas may be blocked where impermeable rocks cap the crests of anticlines (Fig. 9-23). The fluids then accumulate in the pore spaces of permeable sedimentary rock layers. For this reason, searches for oil often concentrate on known oil-producing marine sediments beneath anticlinal crests. Indeed, several oil deposits may appear at different stratigraphic levels within a single anticline.

Unfortunately for motorists and for residents of cold regions, most large, exposed anticlines were drilled by the 1920s, and most of the recoverable oil in buried anticlines (probably between 20% and 40% of the total) was collected by the end of the 1950s. New recovery techniques allow some of these anticlines to continue to produce, but today we must seek less-obvious oil traps, including some in isolated, relatively unspoiled arctic regions.

Faulting sometimes places an impermeable bed next to oil-bearing permeable ones, creating a barrier to oil flow (Fig. 9-24). Thus a petroleum geologist also looks at faults to identify sites where expensive test drilling is most likely to be productive. Faults and their associated earthquakes may wreak expensive havoc on property and safety, but they also contribute one important economic benefit: They trap oil. In Chapter 20, we will consider the future of fossil fuels and evaluate the potential for some alternative energy sources.

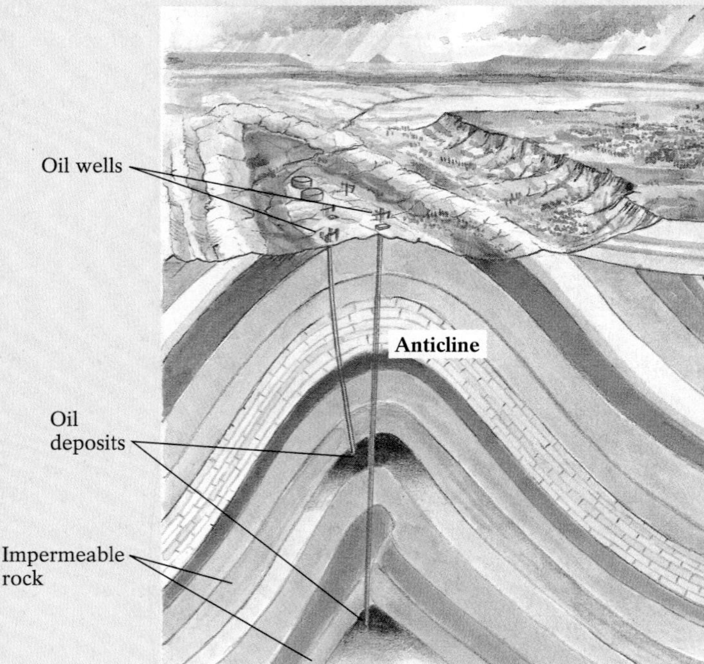

Figure 9-23 The accumulation of oil and natural gas at the crests of anticlines. The low densities of these fluids allow them to migrate readily through permeable sedimentary bedrock until they encounter an overlying cap of impermeable bedrock. The fluids become concentrated at the crests of anticlines, which are the highest points within the permeable beds.

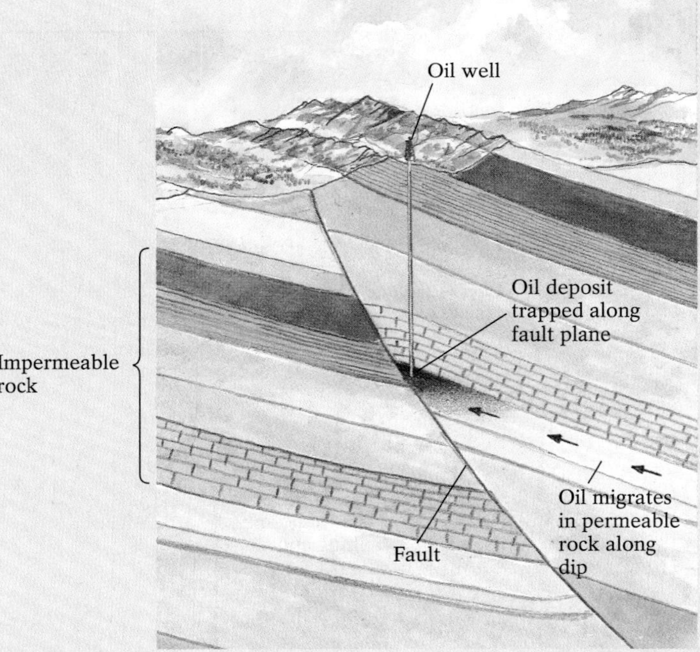

Figure 9-24 The accumulation of oil along fault planes. Oil migrates upward through permeable sedimentary rock because of its low density, but its path to the surface may become blocked by an impermeable bed moved to that location by faulting.

as those that yield vast folds in marine rocks, sometimes trap oil and natural gas within rock bodies. Highlight 9-1 describes how knowledge of bedrock structures can aid in the search for these fuels.

Building Mountains

A drive down Route 1 along the southern California coast affords an opportunity to see mountain building in action. Near San Clemente we find a giant stairway of stepped terraces, each of which was cut by waves at sea level and subsequently uplifted (Fig. 9-25). Each terrace remained at sea level during a temporary pause in the uplift of California's coast, long enough for the pounding surf to erode the surface before the next pulse of uplift carried it safely above the wave action. The highest terrace, 400 meters (1300 feet) above the present sea level, is the oldest.

We all know a mountain when we see one, but geologists have a precise way of defining the structures that rise above the Earth's surface. A *mountain* is a part of the Earth's crust that stands more than 300 meters (1000 feet) above the surrounding landscape, has a discernible top or *summit,* and possesses sloping sides. Every continent has mountains, as does every ocean basin. Some mountains, such as Georgia's Stone Mountain, stand alone, isolated and towering above their surroundings. Some group together in *ranges,* a succession of high peaked structures, such as the volcanic Cascades of the Pacific Northwest. Some ranges form continent-long mountain groups, or *systems,* such as the Appalachian mountain system, which comprises the Great Smokies of Tennessee and North Carolina, the Blue Ridge of Virginia, the Catoctin Mountains of Maryland, the Poconos of Pennsylvania, the Catskills of New York, the Taconics of Connecticut, the Berkshires of Massachusetts, and the Green and White Mountains of Vermont and New Hampshire. As a consequence of their diverse origins, mountains vary in composition and shape,

Figure 9-25 (left) Uplifted marine terraces at San Clemente Island, California. In time, these structures may become mountains. (below) A terrace-producing scenario: Terrace 1 forms at sea level, where crashing surf shapes its surface; subsequent uplift then elevates this terrace beyond the reach of the waves, preserving it and enabling terrace 2 to form at sea level.

even within a system. Indeed, mountain building may vary in so many ways that even adjacent ranges may have been formed by completely different processes at different times (Fig. 9-26).

Some mountains, such as the Catskills of upstate New York, are actually just the uplands that remain after streams have cut deep valleys into a plateau. Some, such as Mauna Loa and Mauna Kea of Hawai'i, are undeformed accumulations of basalt that erupted initially from sea-floor hot spots. Complex systems, such as the Appalachians of eastern North America and the Alps of southern Europe, may have formed by multiple episodes of sedimentation, intense folding, thrust faulting, plutonism, volcanism, and metamorphism during a series of plate collisions.

When we gaze upon mountains, we are truly seeing eons of the planet's past. A mountain and its rocks represent several periods of geologic and tectonic activity. Except in the case of volcanic mountains, the rocks within a mountain are usually much older than the mountain itself. That is, they existed for a long time before being uplifted. Thus, while today's mountains may have formed from one tectonic process, their component rocks generally formed much earlier, perhaps from entirely different processes.

Mountains formed during most of the Earth's past, and they continue to form today. The Himalayas of India, China, and Tibet are young and still rising. The Appalachians and the Urals of central Europe are so old that the primary tectonic forces that created them ceased to operate hundreds of millions of years ago; the principal processes affecting them today are weathering and erosion. Some mountains are so old that they have completely worn away, and are now just flat regions in continental interiors. These remnants, called *continental shields,* consist of the deeply eroded cores of some of the Earth's earliest mountains; in some cases, the mountains were originally uplifted more than 3 billion years ago. Most shields are composed largely of highly metamorphosed, highly deformed rocks.

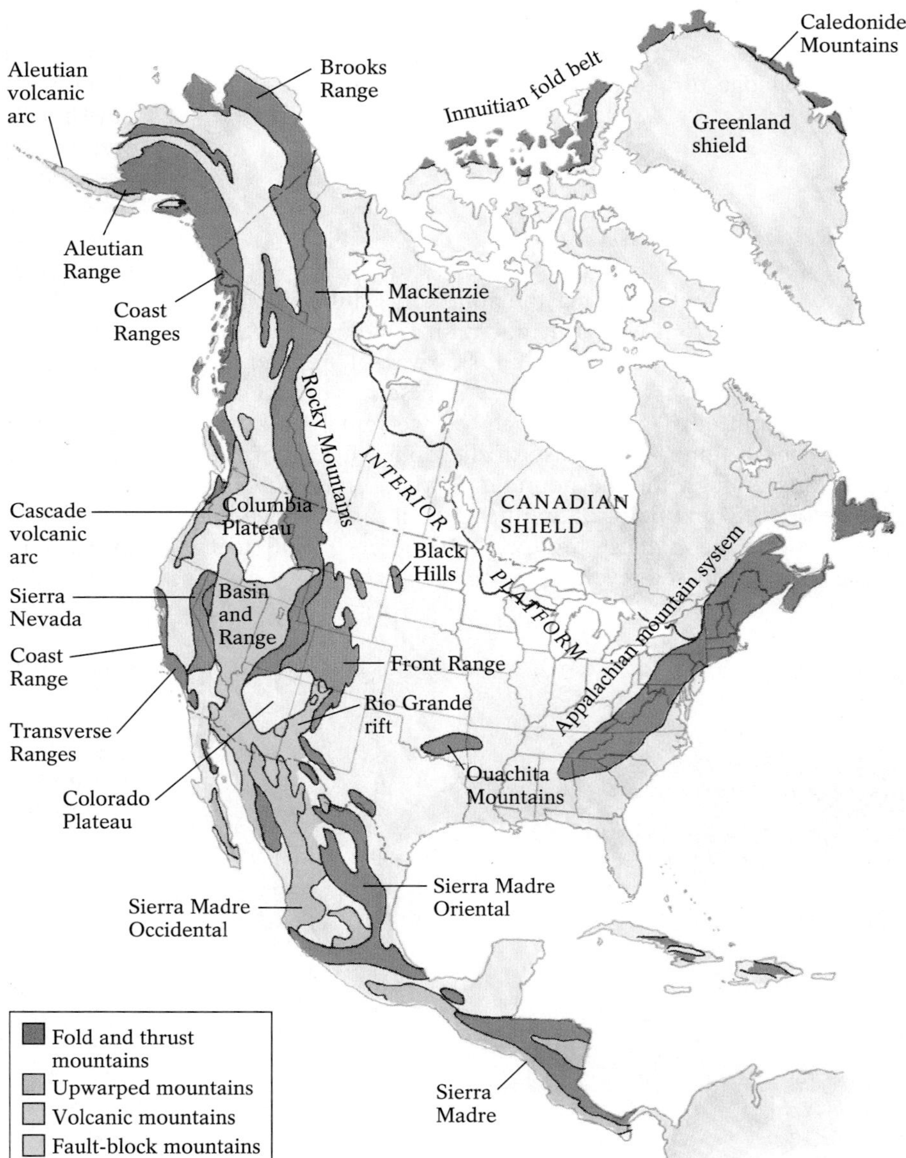

Figure 9-26 The mountain ranges of North America. Some were formed by volcanism, fold-and-thrust mountains developed at the convergent plate boundaries associated with past continental collisions, fault-block mountains resulted from normal faulting due to recent tensional stresses in western North America, and upwarped mountains represent the outcome of uplift within the interior of the North American plate.

Types and Processes of Mountain Building

Geologists refer to the processes of mountain building as **orogenesis** (from the Greek *oro,* meaning "mountain," and *genesis,* meaning "birth"). We examined one of the Earth's principal orogenic processes—volcanism—in Chapter 4. Volcanic mountains form around volcanic vents through the accumulation of lava flows and pyroclastic materials. Some, such as Hawai'i's shield volcanoes, rise atop intraplate hot spots. Others, such as the submarine peaks of the mid-Atlantic ridge, erupt at divergent plate boundaries. Still others emerge in subduction zones: South America's Andes and North America's Cascades were built where oceanic plates subducted beneath continental plates. Volcanic island peaks, such as Alaska's Aleutians and the Pacific's Philippines, formed where one oceanic plate subducted beneath another.

In this chapter we focus on the mountains formed from processes that deform and uplift materials in the Earth's crust. Three principal nonvolcanic types of mountains exist, each originating in a distinctive tectonic setting and each identified by its geological structure: fold-and-thrust mountains, fault-block mountains, and upwarped mountains.

Fold-and-Thrust Mountains **Fold-and-thrust mountains** develop where continental plates collide. In places, they produce exceptionally high mountain systems (3 kilometers [2 miles] or taller). Typically, these rocks consist predominantly of marine sediments that have been intensely folded, thrust-faulted, and, sometimes, metamorphosed and intruded by large plutons. A succession of mountain-building events have deformed most of them more than once. Fold-and-thrust systems include the European Alps, the Himalayas of India, Tibet, and China, the Urals of eastern Europe, and the northern and Canadian Rockies. Highlight 9-2 discusses the complex origin of the Appalachians of eastern North America, a series of ancient fold-and-thrust mountains that have been largely subdued by erosion.

Fault-Block Mountains **Fault-block mountains** are bounded on at least one side by high-angle normal faults. They generally form where the Earth's crust has been stretched, thinned, and ultimately fractured by tensional stresses. The crust is broken into a number of high, tilted horsts interspersed with much lower grabens. Fault-block formations can be seen across the western United States (see Figure 9-27), from the Sierra Nevada of California, eastward through Nevada, Utah, and Arizona, to Colorado and New Mexico, and from Mexico north to the Tetons of northwestern Wyoming. This Basin and Range province is characterized by normally faulted grabens and horsts. Picturesque Jackson Hole is a graben at the foot of a fault block that rises 2000 meters (7000 feet) to the Grand Tetons.

The Basin and Range province, like most mountain systems, has experienced several mountain-building periods. Most rocks in this region probably originated as sediments and oceanic crust that accumulated off the west coast of North America hundreds of millions of years ago. During this time, the Pacific and North American plates periodically converged, compressed, folded, and thrust-faulted these rocks against what was then North America's western coast, producing the Sierra Nevada and Rocky Mountain ranges. Compression of the continental margin ended 30 million years ago, when both the eastward-moving oceanic plate and the diverging oceanic ridge farther west that produced it became completely subducted. Subsequently, the southwestern landscape between the Sierra Nevada and Rocky Mountain ranges became stretched, and steep normal faults broke it into numerous individual fault-block mountain ranges and intervening sediment-laden basins (Fig. 9-27). Extensive volcanism accompanied the faulting, as basaltic magma rose to the surface from the mantle.

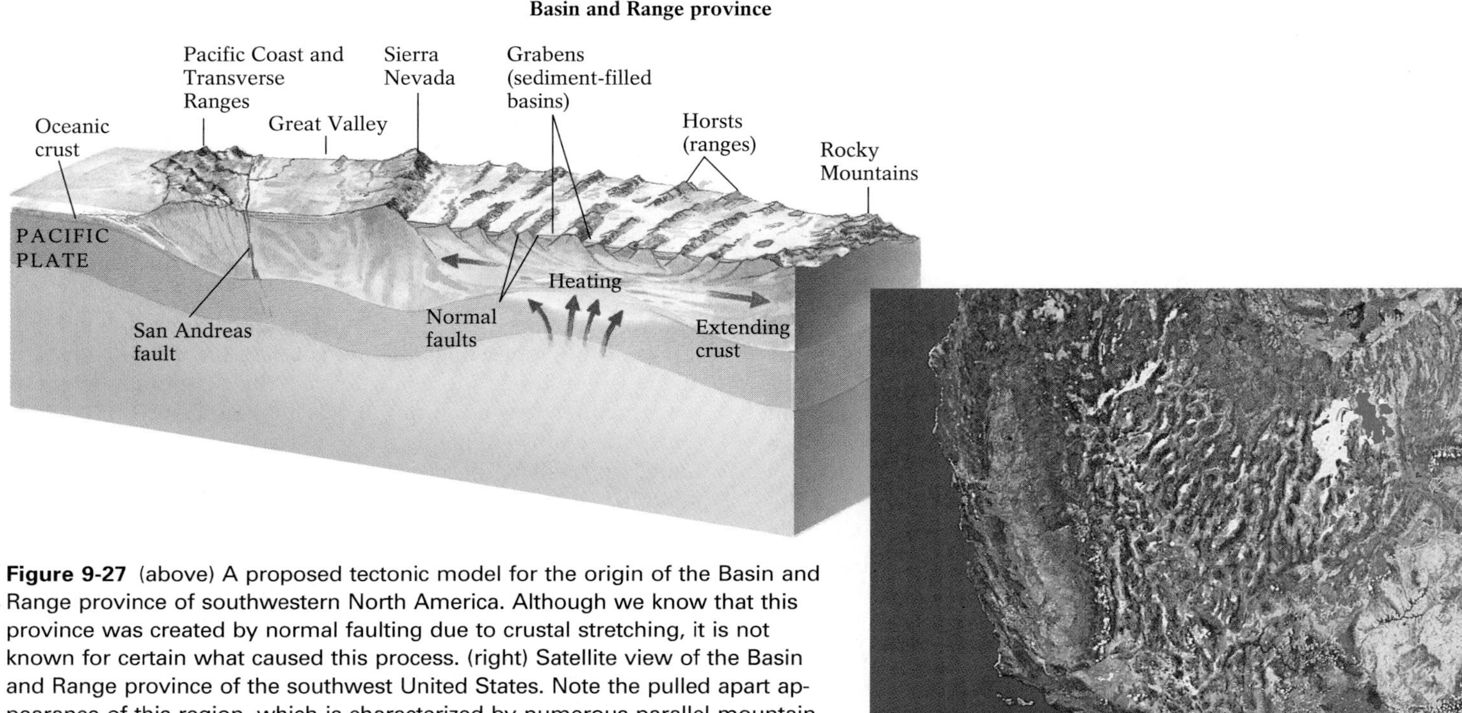

Figure 9-27 (above) A proposed tectonic model for the origin of the Basin and Range province of southwestern North America. Although we know that this province was created by normal faulting due to crustal stretching, it is not known for certain what caused this process. (right) Satellite view of the Basin and Range province of the southwest United States. Note the pulled apart appearance of this region, which is characterized by numerous parallel mountain ranges separated by intervening depressions (the basins).

Highlight 9-2 The Appalachians: North America's Geologic Jigsaw Puzzle

The Appalachians of eastern North America are a mosaic of folded, thrusted, and metamorphosed provinces that evolved over nearly a billion years of Earth history. The North American segment of the system extends 3000 kilometers (2000 miles) with a northeast–southwest orientation, from Lewis Hill (810 meters [2672 feet] high) in eastern Newfoundland to Cheaha Mountain (730 meters [2407 feet] high), some 75 kilometers (40 miles) east of Birmingham, the highest point in Alabama. When we reconstruct the Northern Hemisphere's plate boundaries, however, we find that the Appalachians actually extend a good deal farther—to the Caledonides of western Europe and on to Norway. To the west of North America's Appalachian Mountains lies the Appalachian plateau, a region of relatively unfolded, unmetamorphosed, coal-bearing rocks. During Paleozoic time, this plateau consisted of lushly vegetated wetlands located adjacent to the rising Appalachians. To the east lie the more recent deposits of the Atlantic coastal plain.

The Appalachians' width of 600 kilometers (400 miles) covers three principal provinces, separated by major thrust faults. Each province developed at a different time and has distinct rock formations, but all were affected by the same series of mountain-building episodes. West to east, the Valley and Ridge province consists of a thick sequence of relatively unmetamorphosed Paleozoic sediments that were folded and thrust to the northwest by compression from the southeast. The Blue Ridge province includes highly metamorphosed Precambrian and Cambrian crystalline rocks. The Piedmont province consists of metamorphosed Precambrian and Paleozoic sediments and volcanic rocks that were intruded by granitic plutons (Fig. 9-28).

For the last century, a legion of geologists have studied these mountains in an effort to decipher their complex history. One popular model for the evolution of the Appalachians begins with the late Precambrian plate collisions (about 1.1 billion years ago) that assembled a pre-Pangaea supercontinent. Approximately 800 to 700 million years ago, this Precambrian supercontinent began to break up, with North America rifting from Eurasia and Africa. This movement created an ancestral Atlantic Ocean and an additional continental fragment that was separated from the North American plate by a marginal sea (Fig. 9-29a). Then, about 700 to 600 million years ago, subduction began within the ancestral Atlantic Ocean, and an island arc developed above this subduction zone (Fig. 9-29b). The oceanic crust in the marginal sea began to subduct between 600 and 500 million years ago, and a volcanic arc developed within the continental fragment (Fig. 9-29c). Meanwhile, subduction continued beneath the island arc, and the ancestral Atlantic Ocean began to shrink.

By 500 million years ago, the marginal sea had entirely subducted, allowing the continental fragment to collide with the eastern edge of the North American plate (Fig. 9-29d). In this collision, the "foreign" rocks of the continental fragment were thrust-faulted northwesterly over younger continental and marine sediments along the continental shelf of the North American plate.

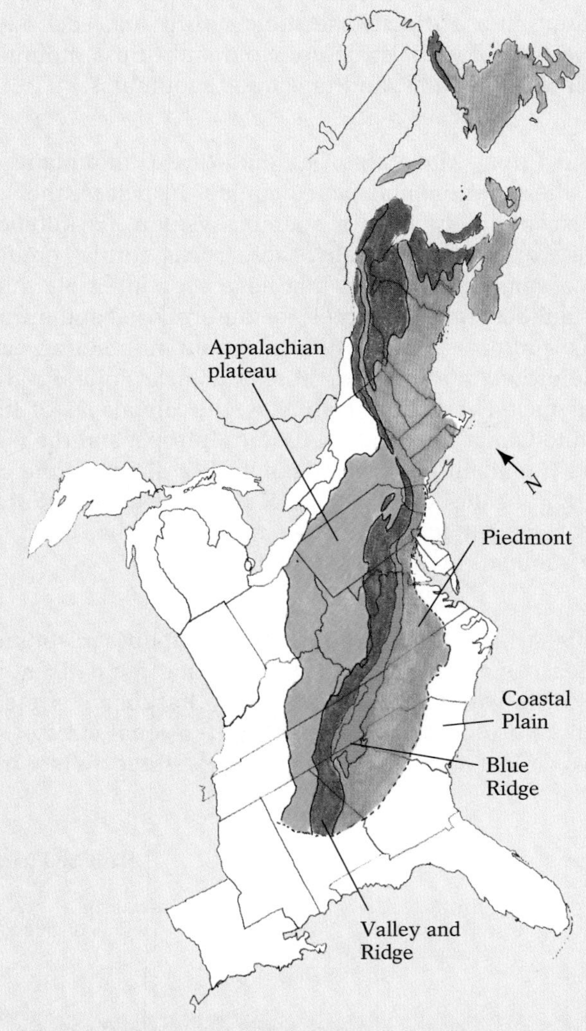

Figure 9-28 The provinces of the Appalachian mountain system in eastern North America.

This first mountain-building episode in the long geologic evolution of the Appalachian system is called the *Taconic orogeny.* Today's Blue Ridge Mountains eventually developed on top of the structures formed during the Taconic orogeny. Preserved in these mountains are highly deformed masses of gabbro, serpentinite, and pillow basalts from the oceanic crust that once underlay the marginal sea, as well as metamorphosed Precambrian–Cambrian marine sediments from deeper parts of the marginal sea. These rocks belong to what is now the western or inner Piedmont province in eastern North America. Excellent views of these exposed rocks may be seen in Newfoundland, metropolitan Baltimore and Washington, D.C., and east of the Blue Ridge Parkway in Virginia and North Carolina.

The second mountain-building episode, the *Acadian orogeny*, occurred 400 to 350 million years ago, at the end of the Devonian Period (Fig. 9-29e). The ancestral Atlantic Ocean continued to diminish in size as Africa began its northwestward convergence. The island arc that developed earlier (see Figure 9-29b) now collided with the eastern margin of the North American plate, pushing the Blue Ridge and western Piedmont on top of the plate and farther to the northwest. These rocks form the eastern Piedmont province. Much of New England's widespread metamorphism (discussed in Chapter 7) and many of the East Coast's granite plutons (such as those in Acadia National Park in Maine) resulted from this collision and the subduction of the ancestral Atlantic Ocean crust beneath the eastern edge of the North American plate.

The third and final mountain-building collision, the *Allegheny orogeny*, occurred between 350 and 270 million years ago (Fig. 9-29f). As the ancestral Atlantic Ocean continued to subduct, the continental edge of Africa approached North America—an episode recorded in the plutonic intrusive rocks found today along the eastern edge of Piedmont province. The orogeny culminated when the continental crust of Africa collided with North America, finally closing the ancestral Atlantic Ocean and forming the supercontinent Pangaea. This collision produced the fold-and-thrust structures of the Valley and Ridge province, and thrust the Blue Ridge and Piedmont provinces farther inland. In Africa, a corresponding orogenic belt is marked by the Mauritanide Mountains.

The last great East Coast tectonic event involved the opening of the modern Atlantic, some 200 million years ago (Fig. 9-29g). At this time the African plate once again separated from the North American plate, but a healthy chunk of the African plate remained attached to eastern North America from New York City to Florida. Thus, the three plate collisions that produced the Appalachians added to the East Coast an ancient (Precambrian) fragment of North America that had earlier been rifted away, an island arc that originated on the African–European side of the ancestral Atlantic basin, and a massive slice of the African plate.

The Appalachians' complex geology and long history of multiple orogenic events are characteristic of fold-and-thrust mountain systems. Despite geologists' extensive studies of the Appalachians, many unresolved questions remain. New discoveries also continue to be made, such as the recent finding of eclogites (extremely high-pressure, high-temperature metamorphic rocks discussed in Chapter 7) in the Blue Ridge Mountains.

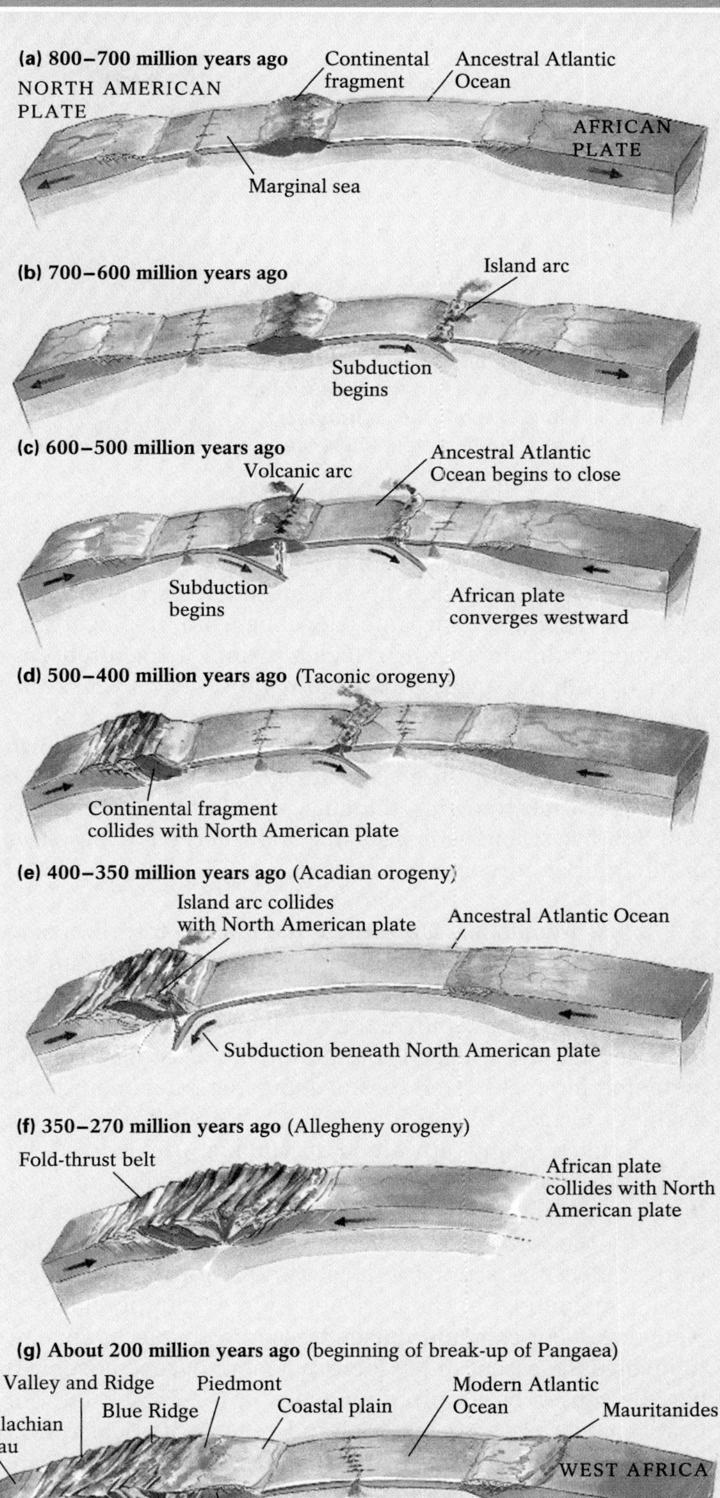

Figure 9-29 A model for the evolution of the southern Appalachians. The northern Appalachians developed in part from collision with Europe.

Figure 9-30 The Adirondack Mountains of northern New York. The center peak in the background is Mount Marcy, the Adirondacks' tallest at 1630 meters (5344 feet).

What caused the normal faulting of the Basin and Range province? One hypothesis proposes that the faulting occurred when the warm, slow-flowing asthenosphere spread laterally from the subducted oceanic ridge beneath the overlying lithosphere, thinning and stretching it until it fractured. Another hypothesis suggests that the region's volcanism, faulting, and crustal thinning arose when a new independent hot spot developed within the southwest portion of the North American plate. The Basin and Range province is clearly a result of crustal stretching, thinning, and faulting in the American Southwest, but geologists have not reached a consensus on the precise mechanism involved.

Upwarped Mountains **Upwarped mountains** form when a large area of the Earth's crust bends gently into broad regional uplifts without much apparent deformation of the rocks. After erosion removes the overlying sedimentary strata from this feature, a rugged core of durable older igneous and metamorphic rocks often towers above the surrounding sedimentary terrain.

Because many upwarped mountains are located far from plate boundaries, our current understanding of plate tectonics does not explain their origin. Geologists do not yet agree on the forces that drive upwarping. Some believe upwarps reflect the actions of local vertical forces at plate interiors, in contrast to the horizontal forces that dominate at plate edges. For example, crustal upwarping in places appears to involve the ascent of low-density material. This movement may be caused by localized increases in the Earth's interior heat (heating expands material and reduces its density) or, alternatively, by lower-density mantle materials that are forced upward as heavier materials sink toward the Earth's core. Either way, older buoyant rock probably rises, arching overlying layers of sedimentary rock.

In many places, upwarped mountains consist of ancient Precambrian rocks that became exposed following the erosion of younger sediments. We can see the results of this process in the Adirondack Mountains of northern New York, which may have formed from a combination of upwarping and plate-tectonic forces. Approximately 1 billion years ago, the rocks now at the core of the Adirondacks were compressed, folded, thrusted, and metamorphosed at a collisional plate boundary, and then intruded by igneous plutons. By the end of the Precambrian era, some 600 million years ago, these mountains had been eroded down to a fairly flat surface on which younger sedimentary rocks formed. Then, for unknown reasons, upwarping began during the Cenozoic Era (the last 65 million years) after hundreds of millions of years of relative stability. During the last 2 to 3 million years, streams and glaciers eroded the uplifted Paleozoic and Precambrian rocks into deep valleys and prominent mountains, creating the 1000 meters (3500 feet) of *relief* (the difference between an area's highest and lowest points) that now typifies the region (Fig. 9-30). Local earthquake activity that occurred as recently as 1989 suggests that uplift continues in the Adirondacks today, even though the area is thousands of kilometers from a plate boundary.

Mountain Building on Our Planetary Neighbors

Some of our planetary neighbors have mountains that resemble those found on Earth. Others have mountains that appear to have formed from processes not observed on our planet. We have studied the three bodies nearest to the Earth—the Moon, Mars, and Venus—by robot satellites, robot landers, telescopic inspection, and, in the case of the Moon, human visits. The data collected suggest that Earth-like plate tectonics is absent on the Moon and Mars; thus orogenic processes on these two bodies must differ from those on Earth. Plate tectonics may operate on Venus, suggesting a more Earth-like orogenesis may be at work on that planet.

Mountains of the Moon The Moon's major highlands, some as lofty as our Mount Everest, call for an orogenic explanation. Nevertheless, geologists have concluded that the Moon is tectonically dead, because highly sensitive earthquake-detecting devices have documented relatively few and very small quakes there. Without stress and strain, and faulting and folding, how did the lunar mountains form?

The Moon's highlands consist of rocks that apparently formed some 4.5 billion years ago (the date consistently yielded by rocks collected by Apollo astronauts). During the earliest period of its existence, the Moon was beginning to cool down after being heated by meteorite bombardment over its first few hundred million years. The impact and accretion of these high-speed space fragments may have melted the Moon's outer 100 to 150 kilometers (60–90 miles). As the surrounding region of space cleared and meteorite impacts dwindled, the Moon's magma sea gradually cooled.

The first components to crystallize were low-density calcium feldspars, which rose to the Moon's surface. This primordial lunar crust consisted of anorthosite, a rock that is almost completely composed of the calcium feldspar anorthite. Meteoroid impacts over the last 4 billion years have apparently dislodged chunks of anorthosite and hurled them about, forming mountains of rock debris. Thus, the Moon's mountains appear to have developed largely through a surface process of fragmentation and accumulation. Some lunar geologists suggest that meteorite impacts also generated local magmas that fueled volcanic eruptions. In that case some lunar mountains may have a volcanic origin.

Mountains of Mars In the apparent absence of Earth-like plate tectonics on Mars, the principal mountain-building process on this planet seems to be volcanism—on a grand scale. Olympus Mons, a single mountain in Mars' Tharsis region, is nearly 600 kilometers (400 miles) in diameter (about the size of Ohio) and more than 23 kilometers (14 miles) high. As we learned in Chapter 4, its huge size suggests that Olympus Mons is a shield volcano that has continued to grow atop its magma source in the stationary Martian lithosphere. Although the 1997 Pioneer mission has begun to reveal much about the composition of Mars' surface rocks, their relationship to the planet's orogenic history remains unclear. The presence of andesitic rocks, which on Earth strongly indicates past subduction, may someday tell us much about Mars' tectonic history. Stay tuned.

Mountains of Venus Venus may well be tectonically active. If so, its mountain-building processes may more closely resemble those on Earth than the processes at work on the Moon and Mars. Until recently, however, Venus' 25-kilometer (15-mile)-thick cloud cover of sulfuric acid and CO_2 blocked the view of conventional telescopes and prevented us from seeing Venutian orogenesis in action. This CO_2 cover, which begins approximately 70 kilometers (45 miles) above the planet's surface, creates the solar system's ultimate greenhouse effect by trapping heat radiated from the planet's surface. Even when Earth probes penetrated the cloud, the surface temperatures of 500°C (900°F) and the overwhelming atmospheric pressure (90 times greater than Earth's atmospheric pressure) caused the spacecrafts' cameras to malfunction. The Soviet probe Venera survived for about an hour in 1975, sending back a few fuzzy photographs of the surface. In 1990, the U.S.-launched Magellan probes used radar, which can pierce clouds, to provide scientists with exciting images of the mountainous topography of Venus.

The Magellan images show a pattern of lowlands and highlands that have led some geologists to speculate about ongoing plate-tectonic activity on Venus. The highlands may be analogous to the Earth's prominent continents, and the plains analogous to the Earth's low-lying oceanic crust. The highlands, which constitute only 8% of Venus' surface, are mostly plateau-like structures, although a few discrete mountain ranges exist (Fig. 9-31). Maxwell Montes on the Ishtar highlands, for example, rises to a height of 11 kilometers

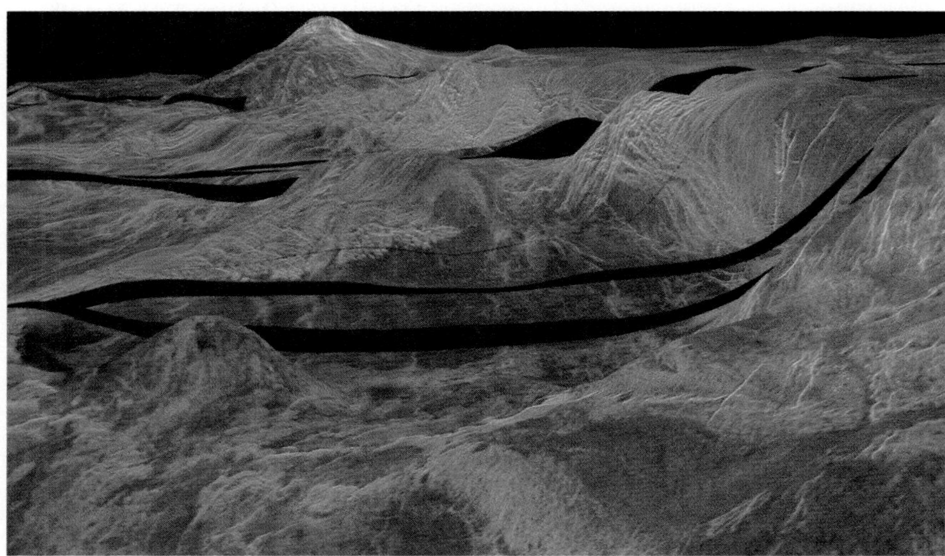

Figure 9-31 Venus' Ishtar Mountains and adjacent lowlands, as constructed from the high-resolution radar images provided by the Magellan probe in 1990. The black swaths that traverse the image are caused by missing radar data.

(7 miles) above the surrounding lowlands, or 2 kilometers higher than Mount Everest. Venus also contains structures that have been tentatively identified as 5-kilometer (3-mile)-high shield volcanoes and enlarged tectonic valleys, possibly grabens. Its surface may contain folded and faulted mountain belts as well. Of course, we cannot confirm these speculations until our ability to study the Venutian surface improves.

This chapter has often referred to the connection between tectonic stress, faults, mountain building, and earthquakes. The next chapter focuses specifically on earthquakes: their causes, their often-catastrophic effects, and the ways in which humans attempt to cope with them.

Chapter Summary

The rocks that make up the Earth's lithosphere are often subjected to great forces—particularly at the edges of tectonic plates—to which they respond by bending or breaking. **Structural geology** is the study of such deformed rocks. When a sufficient force, or **stress** (force per unit area of rock), is applied to rocks, they may **strain,** changing in shape or volume. Three principal types of stress exist: **compression,** which squeezes rocks together; **tension,** which stretches and pulls rocks apart; and **shearing stress,** which grinds rocks by forcing them past one another in opposite but parallel movement.

Subjected to only a minor amount of stress, rocks undergo **elastic deformation.** A greater amount of stress may cause a rock to exceed its **yield point,** or **elastic limit,** and experience permanent deformation. Under certain conditions, particularly when a great amount of stress is applied rapidly to relatively cool and shallow rocks, the rocks may break, in a process known as **brittle failure.** Under other conditions, particularly when stress is applied gradually to relatively warm, deep rocks, the rocks undergo **plastic deformation** without breaking.

When geologists find deformed rocks at the surface, they determine the rocks' orientation in space so as to estimate their subsurface structure. To accomplish this goal, they measure the rocks' **strike** (the compass direction of the line forming the intersection between a geologic plane, such as a bed of sedimentary rock, and the horizontal plane) and its **dip** (the angle at which the rocks are inclined relative to the horizontal).

When rock layers deform plastically, they form a series of **folds,** including trough-like **synclines** and arch-like **anticlines.** Folds occur most often at convergent plate boundaries, such as within the trenches of subduction zones and at the collision zones between masses of converging continental lithosphere. Plastic deformation of rocks also produces oval-shaped bulges, or **domes,** in the Earth's surface as well as bedrock depressions, or **basins.**

Brittle failure may produce randomly distributed fractures, systematically distributed **joints,** or **faults.** Faults are cracks along which marked movement has occurred. Geologists classify faults according to the relative direction in which the affected rocks, or fault blocks, have slipped. **Strike-slip faults** are marked by horizontal movement, whereas **dip-slip faults** are characterized by vertical movement.

The most common types of dip-slip faults are **normal faults,** which develop under tensional stress. In a normal fault, the fault block that moves downward produces a depression called a **graben;** the blocks that remain standing above and on either side of a graben are known as **horsts. Reverse faults** are dip-slip faults that develop under compressional stress. **Thrust faults** are low-angle reverse faults. In normal faults, the fault block above the fault surface (the hanging wall) moves downward relative to the fault block below the fault surface (the footwall). In reverse faults, the hanging wall moves upward relative to the footwall.

Strike-slip faults generally occur at transform plate boundaries, normal faults at divergent plate boundaries, and reverse faults at convergent plate boundaries. In many faults, the slip is not solely vertical (up or down the dip of the fault plane) or horizontal (along the fault plane's strike). Such mixed fault motions—called **oblique slip**—combine both strike-slip and dip-slip motion.

The most dramatic effect of stress-induced rock deformation is the creation of mountains, or **orogenesis.** Compressional forces produce combination **fold-and-thrust mountains.** Tensional forces may cause regional normal faulting, which produces uplifted ranges, or **fault-block mountains,** separated by basins. **Upwarped mountains** are broad regional uplifts that may form by the rise of low-density material below the Earth's surface.

Mountain-building also occurs on some of our planetary neighbors. Because the Moon and Mars appear to be tectonically inactive, geologists believe that their mountains formed by entirely different orogenic processes than those operative on Earth. Lunar mountains were probably produced mainly by meteorite impacts, whereas Martian mountains seem to be largely of volcanic origin. The mountains on Venus, however, may have risen in response to somewhat Earth-like tectonic forces.

Key Terms

structural geology (p. 242)
stress (p. 242)
strain (p. 242)
compression (p. 242)
tension (p. 242)
shearing stress (p. 242)
elastic deformation (p. 242)
yield point (p. 242)
elastic limit (p. 242)
brittle failure (p. 243)
plastic deformation (p. 243)
strike (p. 244)

dip (p. 244)
folds (p. 245)
synclines (p. 246)
anticlines (p. 246)
dome (p. 249)
basin (p. 249)
joints (p. 250)
faults (p. 250)
strike-slip faults (p. 252)
dip-slip faults (p. 253)
normal fault (p. 253)
graben (p. 253)
horsts (p. 254)
reverse fault (p. 254)
thrust fault (p. 254)
oblique slip (p. 254)
orogenesis (p. 258)
fold-and-thrust mountains (p. 259)
fault-block mountains (p. 259)
upwarped mountains (p. 262)

Questions for Review

1. What are the three principal types of tectonic stress, and at which type of plate boundary does each develop?

2. Compare brittle failure and plastic deformation of rocks. Where does each type of deformation commonly occur within the Earth?

3. List three factors that affect rocks' deformation and describe the effects of each.

4. Determine which way is north and then hold this book so that its direction of strike is 45° west of north, and it dips 45° and toward the northeast; toward the southwest.

5. Draw an anticline and syncline, and label the limbs, axial plane, and axis of each. Why does the core of an eroded anticline contain the formation's oldest rocks, and the core of an eroded syncline hold its youngest rocks?

6. Distinguish between fractures, joints, and faults.

7. What are four ways to identify faults in the field?

8. Draw two simple sketches illustrating the difference between normal faults and reverse faults. Label the hanging wall and footwall; use arrows to show their relative direction of movement.

9. Within deformed geological terrain, where might you look for a subsurface accumulation of oil?

10. Using North American examples, briefly discuss two different styles of orogenesis.

For Further Thought

1. In what types of rocks might you expect to find fractures but no faults?

2. Identify the type of fold and the type of fracture that appear in the photo below.

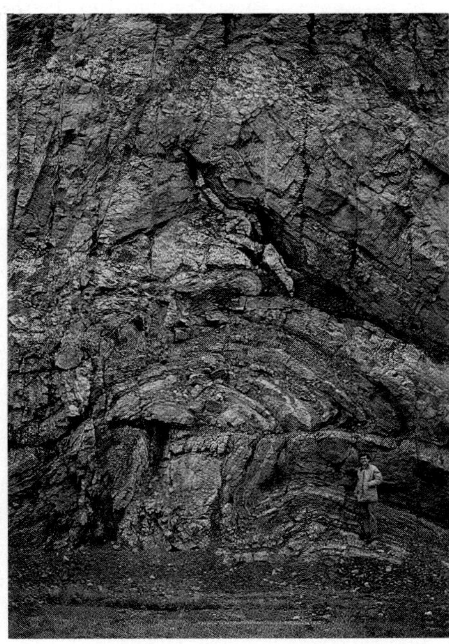

3. Speculate about how the sedimentary layers in the diagram below became arranged in their present stratigraphic order.

4. Suggest future plate-tectonic scenarios that would result in development of normally faulted grabens in Minnesota and the creation of Appalachian-type fold-and-thrust mountains along the Louisiana coast.

5. How might the Moon's surface appear if plate tectonics had been active there for its entire history? How might the Earth's surface look if plate tectonics had never developed here?

10

Earthquakes

From 1985 to 1996, Los Angeles, Oakland–San Francisco, Tokyo, Mexico City, Kobe (Japan), and many other regions have experienced cataclysmic earthquakes (Fig. 10-1). In such heavily populated and industrialized areas, the damage caused when powerful tremors shake vast tracts of land can be devastating. Skyscrapers, highways, hospitals, and homes collapse. Thousands of people die in the wreckage, and many more are displaced from their homes or lose their most treasured possessions. Governments shift abruptly into emergency mode to repair cut-off water supplies, restore snapped electrical lines, and douse the fires produced when gas from severed gas lines ignite. Ironically, the forces that cause such devastation are, in most cases, the very same forces that created our planet's most impressive mountain ranges and some of its most beautiful scenery.

During the 1989 San Francisco Bay area's Loma Prieta earthquake (named for a mountain peak near the quake's point of origin), tremors sped through the rocks and soils of northern California at 11,000 kilometers per hour (7000 miles per hour), toppling buildings, freeways, and bridges, rupturing gas mains and igniting fires, and triggering landslides throughout the region. The quake emanated from the 35-kilometer (20-mile)–long South Santa Cruz mountain segment of the San Andreas fault, approximately 80 kilometers (50 miles) southeast of San Francisco and 16 kilometers (10 miles) northeast of Santa Cruz. Although public attention was riveted on the fires in the Marina district along San Francisco's waterfront and on the rescue efforts on Oakland's collapsed Nimitz Freeway, the most extensive damage occurred closer to the earthquake's point of origin, in the towns of Santa Cruz, Los Gatos, and Watsonville. In all, the Loma Prieta earthquake caused more than $5 billion in damage and claimed 65 lives.

Earthquakes are among the greatest catastrophes that geological forces can produce, and have devastated numerous civilizations throughout history. Virtually every culture that has experienced these events has sought to explain their cause. Some cultures invoke images of giant creatures that carry the Earth through the sky, occasionally stumbling and

Figure 10-1 This enormous rent in the Earth's surface appeared during the powerful February 4, 1998, earthquake in Afghanistan that took thousands of lives.

jostling the planet. In Japanese tradition, the source of all tremors is a giant catfish; in many Native American traditions, it is a giant tortoise; and for the farmers of Mongolia, the source is an enormous frog. The Wanyamwasi of western Africa believe that the Earth is supported on one side by mountains and on the other by a giant, who causes the Earth to tremble whenever he relinquishes his grip to embrace his wife.

In this chapter, we will discuss the causes of earthquakes and how geologists study and evaluate them. We will also examine how humans have learned to cope with their effects, and how we are trying to predict and prepare for them.

The Causes of Earthquakes

The sudden release of energy accumulated in deformed rocks causes the ground to tremble during an **earthquake.** In Chapter 9 we saw that rock becomes deformed, or strained, when subjected to stress, usually at or near plate boundaries (Fig. 10-2). Deep within the Earth, when rock is exposed to stresses that exceed its yield point, it deforms plastically to form folds. At shallower underground levels closer to the Earth's surface, where rock is relatively cool and experiences less lithostatic pressure, it deforms elastically until—if the stress persists—it ruptures. At these shallow depths, stressed rocks accumulate *strain energy.* Strain energy builds until the rocks either return to their original shape or rupture, much as a continuously stretched rubber band, pulled with increasing force, eventually breaks if it does not snap back to its original shape. Rocks that rupture create new faults; rocks at preexisting faults, although temporarily held in place by friction between the fault blocks, eventually break free and shift under stress. In both cases—new ruptures or slippage along old ones—rocks move suddenly, releasing pent-up strain energy. The movement occurs *as* the Earth quakes.

The precise subterranean spot at which rocks begin to rupture or shift is known as earthquake's **focus** (or *hypocen-*

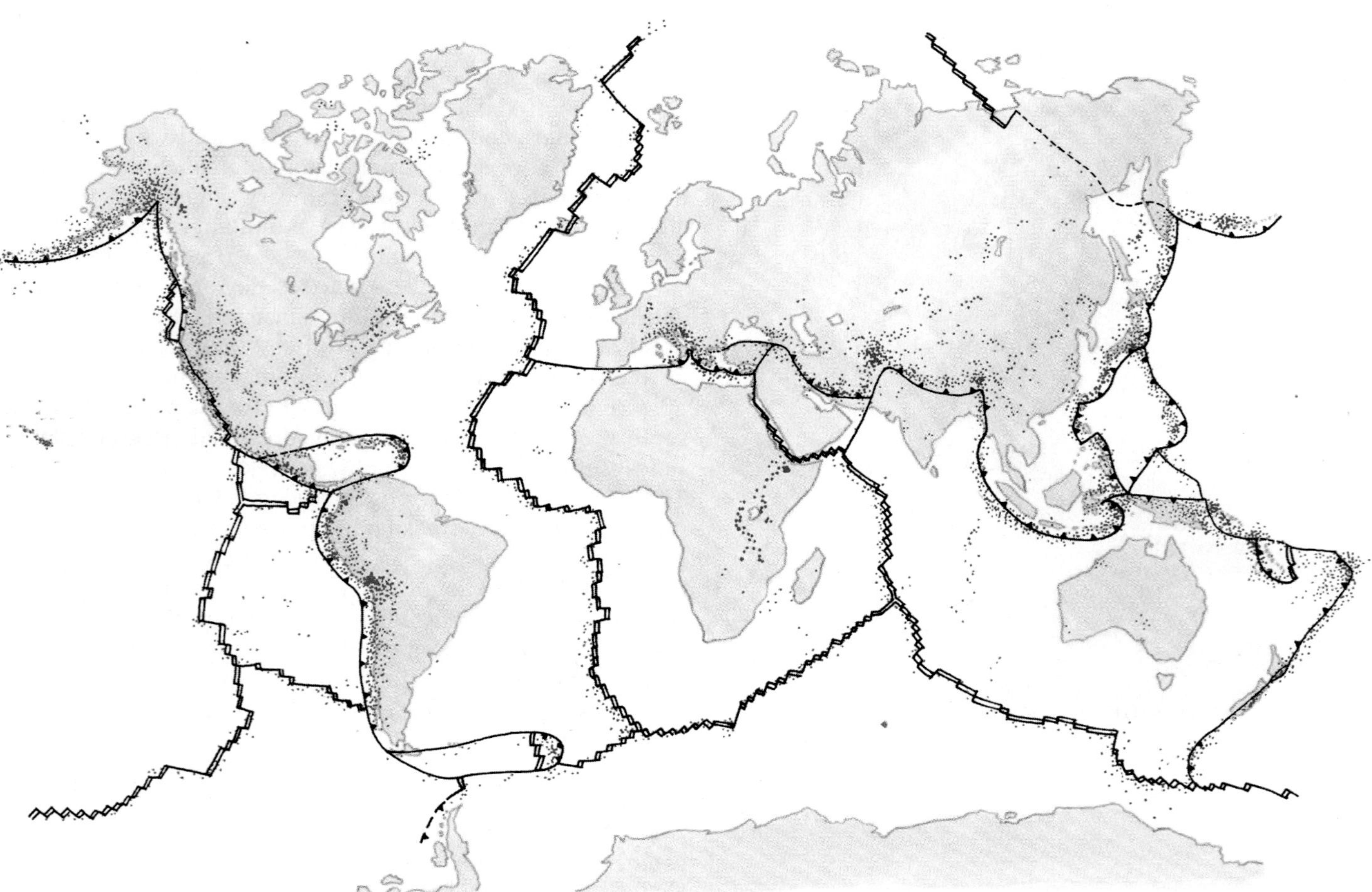

Figure 10-2 The worldwide distribution of earthquakes occurring during the last 100 years. (The dots represent significant earthquakes—greater than magnitude 4.) Notice that earthquakes essentially define current, ancient, and incipient plate boundaries.

ter). The quake's **epicenter** is the point on the Earth's surface that lies directly above the focus; thus it generally occurs where the quake's greatest impact is felt. Figure 10-3 shows the relationship between the earthquake's focus and its epicenter. (A quake's surface effects generally decrease with increasing distance from its epicenter.) After a major earthquake, the rocks in the vicinity of the focus continue to reverberate as they adjust to their new positions, producing numerous, generally smaller, earthquakes called **aftershocks.** Aftershocks, which may continue to occur for as long as one to two years after the initial quake, can produce significant damage by shaking already-weakened structures. In the aftermath of the main shock from southern California's 1994 Northridge earthquake, thousands of aftershocks caused new damage and added significantly to the quake's $15 billion cleanup and rebuilding costs.

After releasing their stored strain energy during an earthquake, the rocks along a fault cease to move, and friction between the fault blocks temporarily locks them in place. As the rocks experience subsequent stress, strain energy accumulates again. The friction holding the fault blocks is eventually overcome, and the rocks then lurch, releasing newly accumulated energy in the form of another earthquake. This series of events explains why movement along faults is generally sporadic, and earthquakes occur periodically rather than continuously.

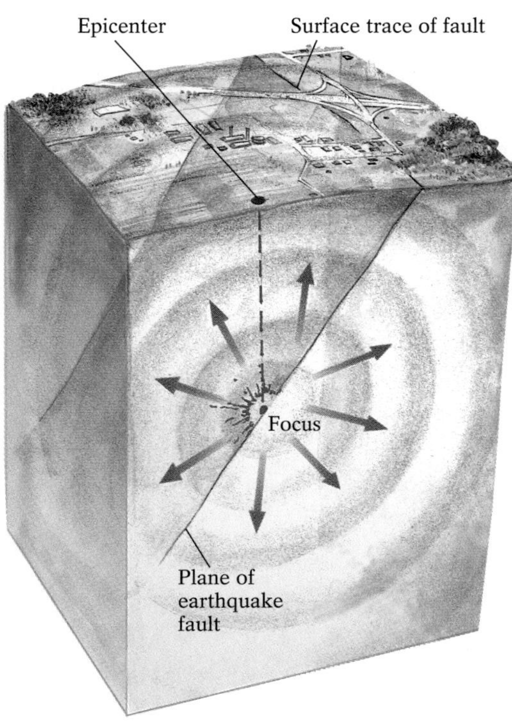

Figure 10-3 At the moment an earthquake occurs, pent-up energy is released at the quake's focus and transmitted through the Earth. The point on the Earth's surface directly above a quake's focus is its epicenter.

Unfortunately for the citizens of California, seismologists agree that neither the Loma Prieta nor the Northridge quake was the long-awaited "Big One." California is long overdue for an earthquake of even greater magnitude than either of those events—one powerful enough to relieve the energy that has been building along some segments of the San Andreas and nearby faults for as long as 150 years.

We will now describe the transmission of the strain energy from an earthquake's focus to the Earth's surface, which causes the vibrations and faulting associated with an earthquake.

Seismic Waves

When you toss a large stone into a placid lake, the energy of the falling stone is transferred on impact to the water. Ripples pass through the water in all directions, eventually dying out at some distance from the point of impact. Likewise, when you speak, sound energy produced by the vibration of your vocal cords is transmitted through the air in the form of sound waves. These too die out with greater distance from their source.

When rocks break or a locked fault lurches free, the released energy is similarly transmitted at great speed through the surrounding rocks in all directions. Earthquake energy, like all types of energy, moves from one place to another in the form of waves. At the instant that an earthquake occurs, the released energy is transmitted through the Earth as **seismic waves.** (The term *seismic,* from the Greek for "shaking," refers to anything related to earthquakes.)

Seismology is the study of earthquakes and the Earth's interior. It analyzes information yielded by two main types of seismic waves: **body waves,** which transmit energy through the Earth's interior in all directions from an earthquake's focus, and **surface waves,** which transmit energy along the Earth's surface, moving outward from a quake's epicenter.

Body Waves Two types of body waves are distinguished by the speed and type of motion with which they travel through the Earth. Primary waves, or **P waves,** are the fastest seismic waves; they pass through the Earth's crust at a velocity of 6 to 7 kilometers (about 4 miles) per second. Thus they are the first signal to arrive at an earthquake-recording station after a quake occurs. The energy released when rocks break or slip near the earthquake's focus, compressing (reducing) the rock's volume, initiates the P waves. These rocks then expand elastically, or *dilate,* past their original volume after generating the wave, thereby compressing adjacent rocks. The rocks are alternately compressed and dilated again as the next wave of seismic energy passes . . . and so on. In this way, rocks in the path of P waves in turn compress and expand in the same

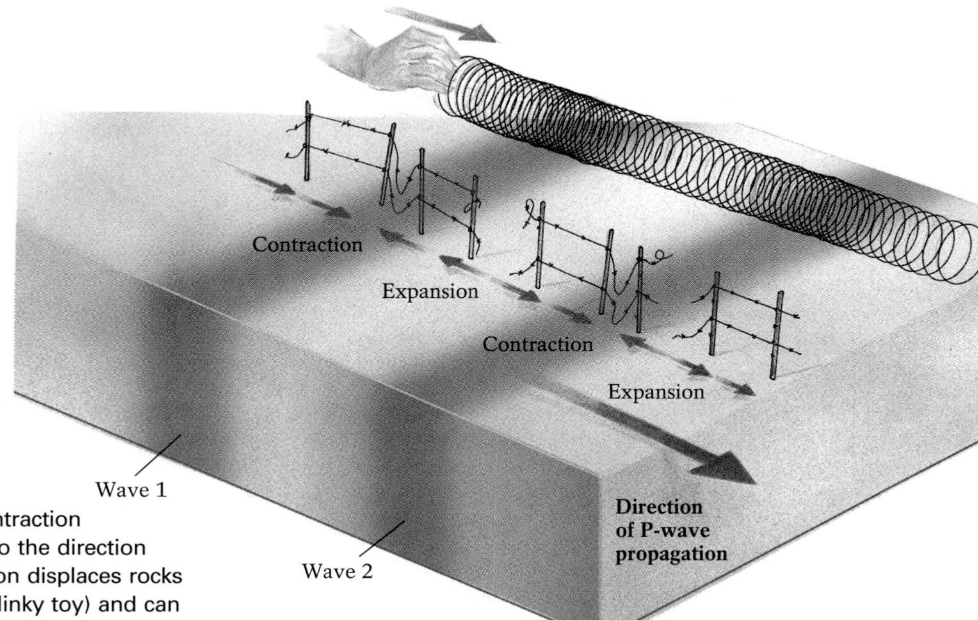

Figure 10-4 Primary waves, or P waves, are compressional. They produce the alternate contraction and expansion of rocks in a direction parallel to the direction of wave propagation. The back-and-forth motion displaces rocks as P waves pass (like the vibrating coils of a Slinky toy) and can cause powerlines and fences to snap.

direction as the waves are traveling (Fig. 10-4), much as the coils in a stretched Slinky toy vibrate back-and-forth when struck. The energy imparted when you strike one end of the Slinky is transmitted by a compressional wave, like a P wave, that moves from one end of the toy to the other. As you can see, vibration from P waves is *parallel* to the direction in which the wave travels. During an earthquake, we feel the arrival of P waves at the Earth's surface as a series of sharp jolts that occurs as surface rocks alternately compress and expand with passing waves of energy.

Secondary waves, or **S waves,** move much more slowly than P waves, passing through the Earth's crust at a velocity of about 3.5 kilometers (2 miles) per second. Consequently, they take longer than P waves to arrive at earthquake-recording stations. As shown in Figure 10-5, the motion of rock in an S wave is similar to that of a rope fixed at one end and flicked sharply up and down at the other end: The rocks move up and down, *perpendicular* to the direction in which the S wave travels. An S wave is generated in much the same way as the once-popular stadium "wave" seen at sporting events. As successive individuals stand and then sit, the motion of the components of the wave—the fans—occurs perpendicular (up and down) to the direction of the wave as it travels around the stadium (sideways). Likewise, as an area of rock

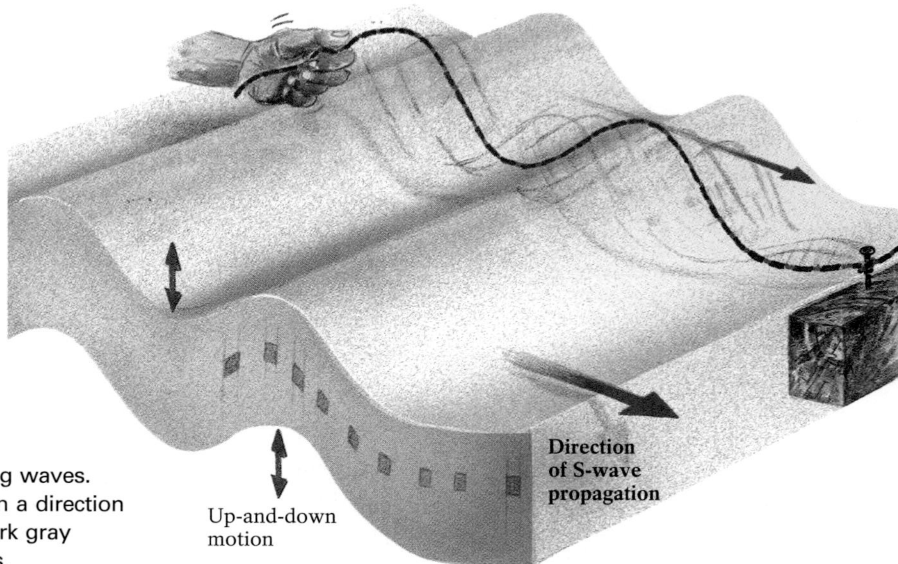

Figure 10-5 S waves, or secondary waves, are shearing waves. They generate a motion that displaces adjacent rocks in a direction perpendicular to the direction of wave propagation. Dark gray rectangles show how rocks deform as the wave passes.

moves up or down when an S wave passes, the wave creates a drag on neighboring rocks, pulling them along as well. This movement has a shearing effect on the rocks (see Chapter 9), changing their shape but not their volume. During an earthquake, P waves strike with a succession of compressional jolts, but S waves impart a continuous wriggling motion.

Surface Waves When an earthquake occurs, some body waves move outward from the focus and spread up toward the epicenter, where they cause the Earth's surface to vibrate. This vibration generates surface waves, which travel within the upper few kilometers of the Earth's crust. The slowest of the seismic waves, surface waves arrive last at earthquake-recording stations; they travel through the crust at a velocity of about 2.5 kilometers (1.5 miles) per second. Two main types of surface waves exist: one having a side-to-side whipping motion like a writhing snake, and the other having a rolling motion that resembles ocean swells (Fig. 10-6). People who have experienced the latter type of surface wave have compared it with a brisk walk across a waterbed—some have even become "seasick." When both types of surface waves occur together, they cause objects to rise and fall while being whipped from side to side; this combination is responsible for most earthquake damage to rigid structures.

Measuring Earthquakes—Which Scale to Use?

When an earthquake occurs, citizens and the news media clamor for information about the tremor's strength. City planners and engineers, who need to know how human-made structures fare in quakes of different sizes, are also eager to classify a quake's strength. So too are seismologists, who must estimate how much of a fault's accumulated strain has been released so as to help predict future quakes. Calculating an earthquake's strength, however, has proved both challenging for scientists and confusing to the general public. Over the past century or so, geologists have used several different scales to classify the strength of earthquakes.

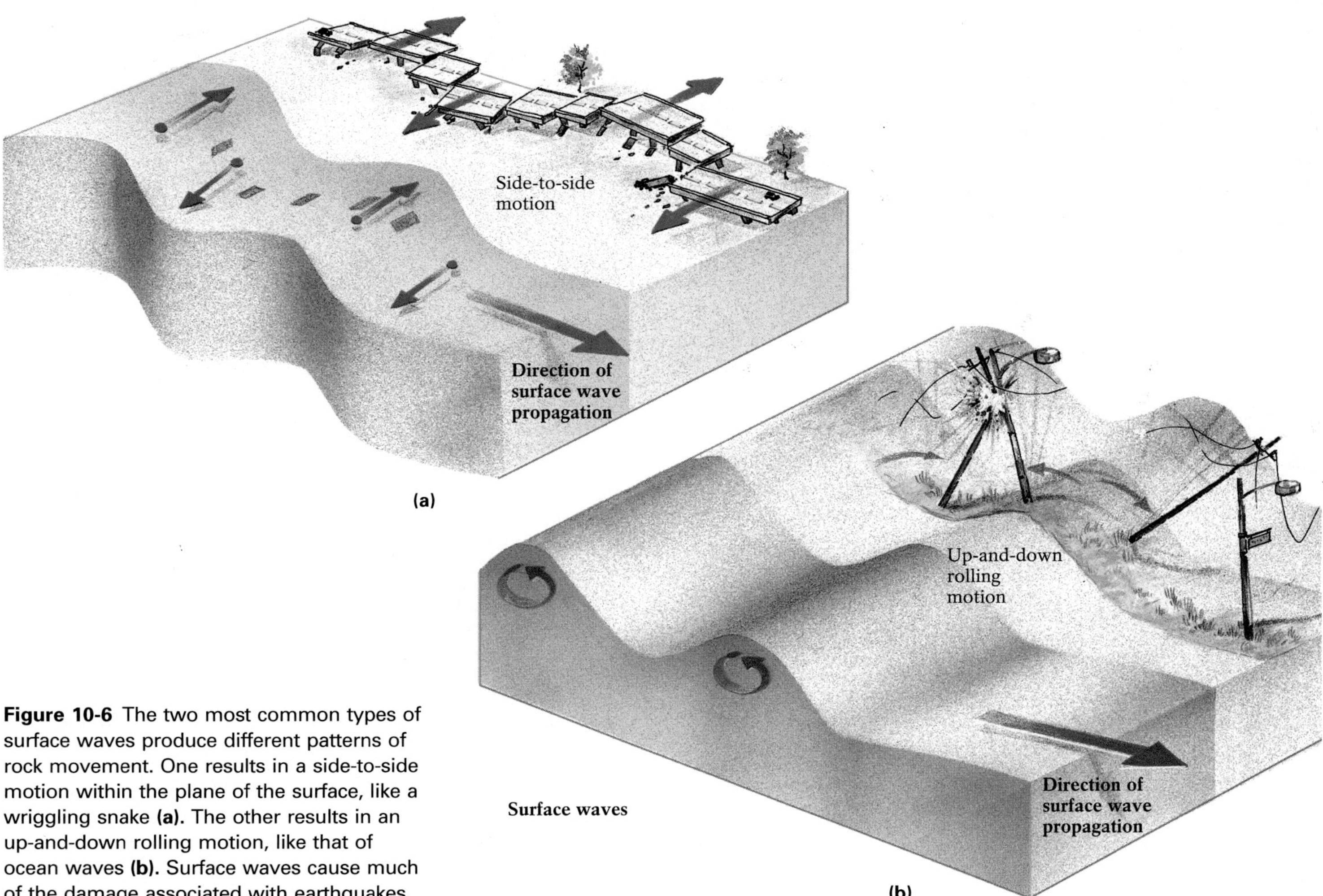

Figure 10-6 The two most common types of surface waves produce different patterns of rock movement. One results in a side-to-side motion within the plane of the surface, like a wriggling snake **(a)**. The other results in an up-and-down rolling motion, like that of ocean waves **(b)**. Surface waves cause much of the damage associated with earthquakes.

The Mercalli Intensity Scale An earthquake's strength depends on how much of the energy stored in the rocks is released. Early efforts to classify earthquakes focused on their destructiveness, or *intensity*, as estimated from observable damage to property and from survivors' descriptions of ground shaking. In 1902, the Italian seismologist Giuseppe Mercalli developed the **Mercalli intensity scale,** which correlates quake intensity to increasing levels of damage to human-made structures and to eyewitness descriptions of the seismic event. His classification scheme is listed in Table 10.1.

Although the Mercalli and similar intensity scales are sometimes used in earthquake planning and earthquake-hazard mapping, this type of descriptive scale cannot locate an earthquake's epicenter accurately, nor can it distinguish an extremely strong earthquake that occurs at a far distance from a less severe event that occurs nearby. In addition, the Mercalli scale cannot distinguish damage to structures that results from the actual intensity of ground motion from damage resulting from poor construction quality or related to the nature and stability of the local soil. Moreover, this scale is obviously ineffective in uninhabited areas, where no human-made structures will sustain damage and no people will be present to describe ground motion.

Yet, because it puts earthquakes into a human perspective, the Mercalli scale offers a tangible way to characterize the local effects of an earthquake. Several other scales that use *quantitative* measures of the amount of energy released by a quake, however, have proven more useful. Most rely at least in part on the use of a seismograph.

Table 10-1 Modified Mercalli Intensity Scale

Intensity Value	Intensity Description
I	Not felt, except by a very few persons under especially favorable circumstances.
II	Felt by only a few persons at rest, especially on upper floors of buildings. Delicately suspended objects may swing.
III	Felt quite noticeably indoors, especially on upper floors of buildings, but many people do not recognize it as an earthquake. Standing automobiles may rock slightly. Vibration resembles a passing truck. Duration estimated.
IV	During the day felt indoors by many people, outdoors by few. At night some awakened. Dishes, windows, doors disturbed; walls make creaking sound. Sensation like heavy truck striking building. Standing automobiles rocked noticeably.
V	Felt by nearly everyone; many awakened. Some dishes, windows, and other objects broken; cracked plaster in a few places; unstable objects overturned. Disturbance of trees, poles, and other tall objects sometimes noticed. Pendulum clocks may stop.
VI	Felt by all; many become frightened and run outdoors. Some heavy furniture moved; a few instances of fallen plaster and damaged chimneys. Damage slight.
VII	Everyone runs outdoors. Negligible damage in buildings of good design and construction; slight to moderate damage in well-built ordinary structures; considerable damage in poorly constructed or badly designed structures; some chimneys broken. Noticed by persons driving cars.
VIII	Slight damage in specially designed structures; considerable damage in ordinary substantial buildings, with partial collapse; great damage in poorly built structures. Panel walls thrown out of frame structures. Chimneys, factory stacks, columns, monuments, walls collapse. Heavy furniture overturned. Sand and mud ejected in small amounts. Changes in well water. Persons driving cars disturbed.
IX	Damage considerable in specially designed structures; well-designed frame structures thrown out of plumb; great damage in substantial buildings, with partial collapse. Buildings shifted off foundations. Ground cracked conspicuously. Underground pipes broken.
X	Some well-built wooden structures destroyed; most masonry and frame structures with foundations destroyed; ground badly cracked. Rails bent. Landslides considerable from river banks and steep slopes. Shifted sand and mud. Water splashed, slopped over banks.
XI	Few (masonry) structures remain standing. Bridges destroyed. Broad fissures in ground. Underground pipelines completely out of service. Earth slumps and land slips in soft ground. Rails bent dramatically.
XII	Damage total. Waves seen on ground surface. Lines of sight and level distorted. Objects thrown into the air.

Source: U.S. Federal Emergency Management Agency (FEMA).

The Modern Seismograph A **seismograph** is a device that measures the magnitude of an earthquake by first sensing and recording the seismic waves generated, and then producing a **seismogram,** a visual record of the arrival times of the different waves and the magnitude of the shaking associated with them. Figure 10-7 illustrates the components of a seismograph and describes how this device works. Seismographs have been placed at more than 1000 stations around the world. By comparing recordings made of a single earthquake event generated at numerous stations, seismologists can locate an earthquake's epicenter and focus and determine how much energy the earthquake released.

A modern seismograph receives signals from a *seismometer,* a device that amplifies wave motion electronically so that the seismograph can detect even weak or distant disturbances. Seismometers are sensitive enough to detect passing traffic, high winds, and nearby crashing surf. They can even detect earthquakes that occur on the opposite side of the globe. Seismometers can also detect underground nuclear explosions and, as we will discuss later, are used to monitor compliance with nuclear test-ban treaties.

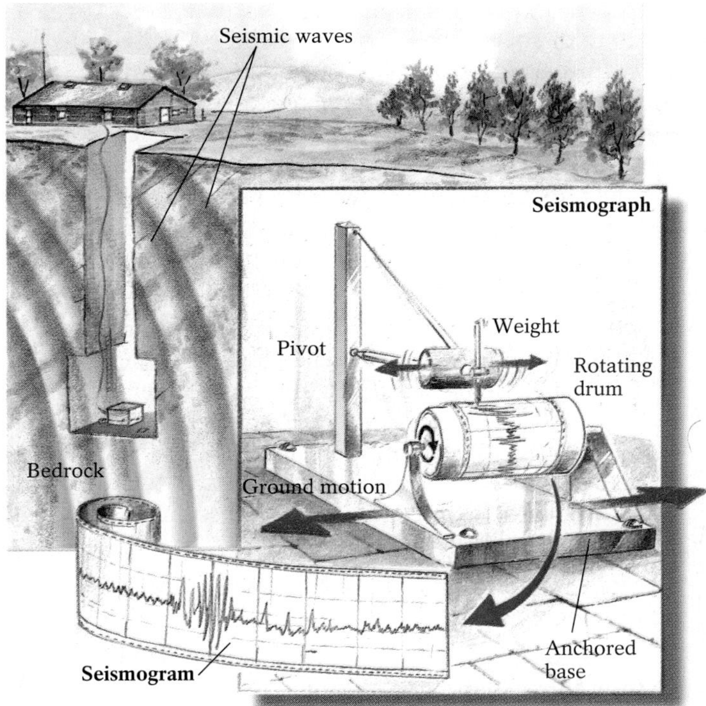

Figure 10-7 The functioning of a traditional seismograph. At its most basic level, a seismograph consists of a mass suspended by a spring or wire from a base that is firmly anchored. (The entire seismograph is usually encased in a protective box and bolted to the Earth's crust meters below the surface.) The base moves with the Earth during an earthquake, while the suspended mass remains motionless. When a seismic wave jostles the seismograph beneath the suspended stylus, the stylus records the shaking on a roll of paper that turns at a steady rate on a rotating drum anchored to the base, producing a tracing called a seismogram. The amplitude of the seismogram's squiggles is proportional to the amount of energy released by the earthquake. (Note that seismographs can also be oriented vertically, to record vertical ground movement.)

The Richter Scale By 1935, Dr. Charles Richter, head seismologist at the California Institute of Technology, had grown weary of local journalists' constant questioning about the relative sizes of California's frequent earthquakes. In response, he devised a way to use seismograms to determine an earthquake's *magnitude,* or the amount of energy released. The **Richter Scale** defined magnitude in terms of the amplitude of the largest peak traced on a seismogram for:

- A Wood-Anderson seismograph,
- Located at a standard distance of 100 kilometers (60 miles) from an earthquake epicenter,
- *In California.* (He calibrated his scale to reflect the properties of the rocks of the Earth's crust in California.)

To accommodate the vast range of earthquake sizes, Richter developed a logarithmic scale, in which each successive unit corresponded to a 10-fold increase in the amplitude of the seismogram's tracings. By Richter's definition, a tremor of magnitude 1.0 would swing the arm of the Wood-Anderson seismograph one-thousandth of a millimeter. A magnitude 2.0 quake would swing the arm 10 times as much, or one-hundredth of a millimeter, and so on. Thus, a magnitude 4.0 quake would trace a peak with a one-millimeter amplitude, a magnitude 5.0 would trace a 10-millimeter, and a peak magnitude 7.0 quake would correspond to a peak of a full meter.

The actual amount of energy released during tremors of different Richter magnitudes varies even more than their graphic peaks. Each one-unit increase in the Richter scale corresponds to an estimated 33-fold increase in energy released. Thus, the energy released by a magnitude 6.0 earthquake is approximately 33 times greater than the energy released by a magnitude 5.0 quake and more than 1000 times greater (33 × 33) than the energy released by a magnitude 4.0 event.

Because the Richter scale uses logarithmic measures of energy, it requires some interpretation to compare earthquakes of different magnitudes. To demonstrate the difference in energy released, let's examine earthquakes that have occurred in metropolitan Los Angeles. Some seismologists have suggested that the area is struck by an earthquake of a magnitude greater than 8.0 every 160 years or so. The last quake of this magnitude occurred there in 1857. You might think that an occasional magnitude 6.0 quake would relieve a significant amount of accumulated strain energy and forestall or even prevent the next "Big One." The energy released by a magnitude 8.0 earthquake, however, is 1000 times (33 × 33) that released by a magnitude 6.0 quake. Thus Los Angeles would need a swarm of a thousand earthquakes of magnitude 6.0 to dissipate the accumulated strain energy and avert the next magnitude 8.0 quake. As southern California experiences on average only one quake of magnitude 6.0 every five years, the region can still expect the occurrence of a seismic event 1000 times as violent as any in recent memory.

Given its virtual universal acceptance among news reporters and anxious citizens, why have seismologists deemphasized the importance of the venerable Richter scale during the last decade? The Richter scale has several drawbacks:

- Although theoretically without an upper limit, the scale cannot effectively measure quakes with a magnitude of 7.0 or greater. (The amplitude of a seismographic peak of an 8.0 quake would be 10 meters, or well beyond the limits of the machine.) Thus the Richter scale generally underestimates the energy released by "major" and "great" earthquakes (magnitudes of 7.0–7.9 and greater than 8.0, respectively).
- More modern devices have replaced Wood-Anderson seismographs.
- The scale is less effective for crustal rocks that differ from those found in California. Because long-traveling seismic waves may pass through numerous regions of different crustal composition, the scale works best for "local" earthquakes within a few hundred kilometers of the seismograph. Seismologists must factor in a number of corrections for earthquakes that occur beyond this distance.

Although the well-engrained Richter scale continues to be widely cited, it seems to work best for *earthquakes of magnitude 7.0 or less or nearby earthquakes in California.* Consequently, the U.S. Geological Survey and other agencies have recently adopted another scale—the moment-magnitude scale—that consistently measures the magnitude of earthquakes (including truly large ones) and the energy they release, but remains unaffected by variations in rock type.

The Moment-Magnitude Scale During the past few decades, seismologists have discovered that they can more accurately gauge a large earthquake's total energy by measuring a quantity called the **seismic moment,** defined as:

$$\begin{aligned} \text{Moment} = &\ (\text{total length of fault rupture}) \\ &\times (\text{depth of fault rupture}) \\ &\times (\text{total amount of slip along rupture}) \\ &\times (\text{strength of rock}) \end{aligned}$$

The last factor—the strength of the rock—is measured by its *elastic modulus,* the ratio of stress to strain in the rock. The values produced through this equation generate the **moment-magnitude scale.**

You can see from this equation that, in general, the longer the fault, the greater its potential for producing a large quake. Similarly, the stronger the rock, the greater its potential to store strain energy and thus generate large quakes. Recent studies in southern California indicate that the subsurface presence of soft, warm rocks, such as schist, may actually somewhat reduce the region's potential for extremely high-magnitude earthquakes.

One principal benefit of the moment-magnitude scale is that seismologists can now *directly* measure the size of the earthquake from its "cause" (the rupture in the rocks and the distance that the rocks have moved) rather than solely from its "effect" (the tracing of seismic waves on a seismograph). To determine length and depth of rupture and total slip requires either field observations (if the rupture reached the surface) or analysis of the precise location of the main shock and its aftershocks (if it failed to break the surface). Because of the time required immediately after an earthquake to complete these analyses, scientists generally first estimate a *preliminary magnitude* based on the seismographic tracings. Later, after more thorough analysis, an updated moment magnitude will be announced that may differ—although usually not substantially—from the preliminary estimate.

Earthquake Magnitudes on a Human Scale In their attempts to quantify the strength of earthquakes for the general public, Richter and his seismological colleagues probably confused more people than they educated with their logarithms and seismic moments. Table 10-2 illustrates various earthquake magnitudes and the energy they release in "human terms." Table 10-3 shows how many of each of these events we can expect annually throughout the world.

Locating an Earthquake's Epicenter

During the last 40 years, increased international cooperation in the field of seismology has led to standardization of equipment and procedures and has encouraged scientists to share data freely. Since the early 1960s, stations from many nations have fed their data directly to the Earthquake Information Center of the U.S. Geological Survey in Golden, Colorado. The Center's high-speed computers continuously analyze incoming signals, quickly locating earthquake epicenters and calculating quake magnitudes.

We can use the different velocities of the various types of seismic waves to determine the location of a quake's epicenter. Just as two sprinters who set out together but run at different speeds will reach the finish line at different times, so do the P and S waves from an earthquake reach a seismograph station at different times. For example, a seismograph located 600 kilometers (375 miles) from an earthquake's epicenter might begin to record P waves about 100 seconds after the quake begins and S waves about 70 seconds later. Because we know the velocity at which each type of wave travels through various rock types, we can use the difference between their arrival times and our knowledge of the rocks through which they traveled to calculate the distance from the seismograph stations to the earthquake's epicenter.

Table 10-2 Earthquakes and Their Energy Equivalents

Moment Magnitude	Approximate Energy Equivalent
−2	100-watt lightbulb left on for a week
0	1-ton car going 40 kilometers (25 miles) per hour
2	Amount of energy in a lightning bolt
4	1 kiloton (1000 tons) of explosives
6	Hiroshima atomic bomb
8	1980 eruption of Mount St. Helens
10	Annual total U.S. energy consumption (largest recorded earthquake—9.5, Chile, 1960)

Table 10-3 Frequency of Earthquake Occurrence Based on Observations Since 1900

Description	Magnitude	Annual Average Number of Events
Great	8.0 or higher	1
Major	7.0–7.9	18
Strong	6.0–6.9	120
Moderate	5.0–5.9	800
Light	4.0–4.9	6200 (estimated)
Minor	3.0–3.9	49,000 (estimated)
Very Minor	2.0–3.0 1.0–2.0	1000/day 8000/day

By comparing the arrival times of seismic waves from earthquakes of known origin, recorded at stations around the world, geologists have developed travel-time charts that convert the lag between P- and S-wave arrival times into the distance to a quake epicenter. For quakes of unknown origin, the gap between the arrival times of the two types of waves is used to calculate the distance from the seismograph stations to the earthquake's epicenter. Figure 10-8a shows how seismologists work with data on P waves and S waves gathered at different locations to find the quake's epicenter.

The linear distance from an earthquake to a single seismic station will not by itself identify the location of the epicenter, as the quake could have occurred at that distance from the station *in any direction.* This distance may be represented on a map as the radius of a circle that has the station at its center. When similar circles derived from the lag times in P- and S-wave arrivals are drawn from three different stations (see Figure 10-8b), they intersect at the quake's epicenter. To pinpoint the precise location of an earthquake epicenter, then, at least three seismic stations must record the

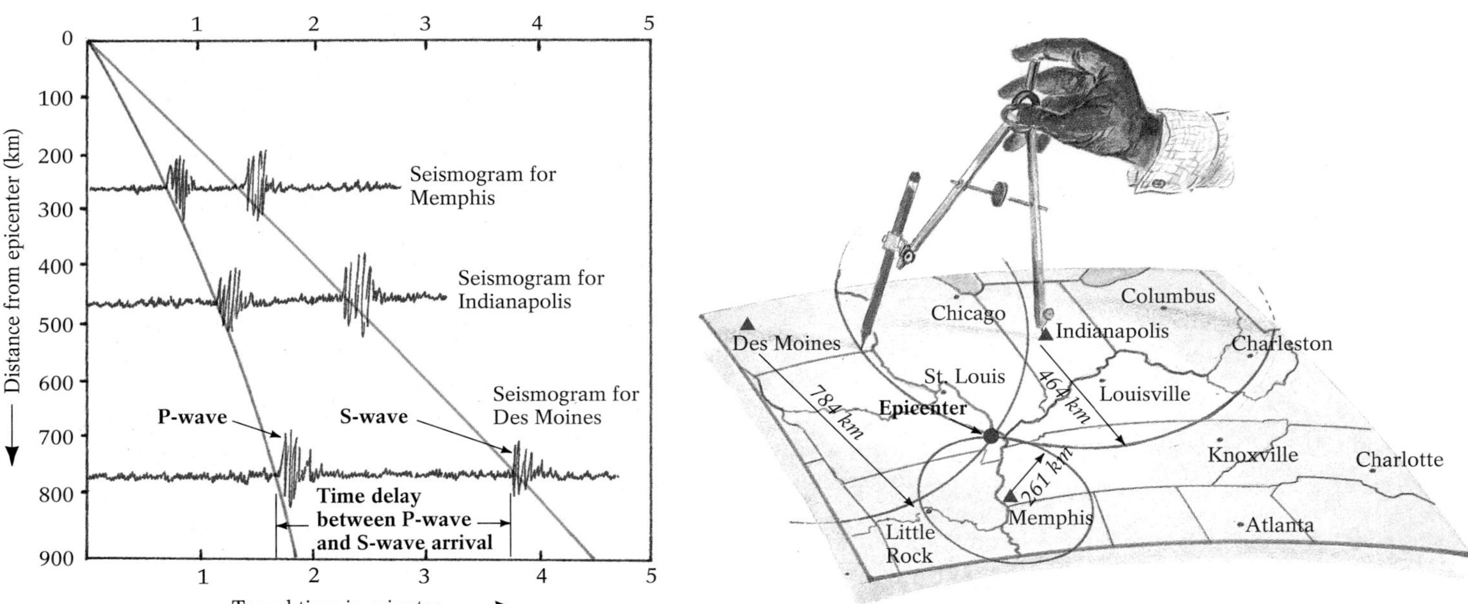

Figure 10-8 **(a)** Seismograms from various seismograph stations, recording the arrival times of P and S waves at each station from an earthquake in the New Madrid, Missouri, area. The closeness of the two wave types at the Memphis, Tennessee, recording station indicates that the earthquake occurred nearby. Using our knowledge of the velocities of P and S waves through the Earth's crust, and the 120-second and 186-second delays at Indianapolis, Indiana, and Des Moines, Iowa, respectively, we can determine each station's distance from the earthquake epicenter. **(b)** By drawing a circle around each of these stations on a map, with each circle's radius proportional to the station's distance from the quake's epicenter, we can determine the location of the epicenter—it is marked by the single point at which the three circles intersect on the Earth's surface.

event. (Note that the three circles will intersect if the focus of the earthquake is very close to the Earth's surface—that is, if the focus lies within a few kilometers. Deeper foci produce a triangle around the quake's epicenter, the size of which is proportional to the earthquake's depth.)

The value of being able to locate an earthquake's epicenter was clearly demonstrated on May 23, 1989, when seismographs around the world detected a "great" earthquake—magnitude 8.3. Moments after the quake, a worldwide network of more than 100 state-of-the-art seismological facilities transmitted data to high-speed computers at the U.S. Geological Survey facility at Golden, Colorado. Within minutes seismologists determined that the earthquake had occurred along the Macquarie Ridge, a submarine mountain chain 800 kilometers (500 miles) southwest of New Zealand. As we will discuss shortly, a powerful quake that strikes the ocean floor often produces enormous sea waves that can devastate coastal regions. A sea-wave alert was issued for coastal communities around the Pacific Ocean basin, allowing the successful evacuation of all people in those areas.

Earthquake Depth and Magnitude

When an earthquake occurs, news accounts generally identify its location by making reference to the nearest town or city. In truth, an earthquake occurs at some depth below such a site, at the quake's focus. We describe an earthquake as "shallow" if its focus lies less than 70 kilometers (45 miles) below the surface, "intermediate" if its focus is 70 to 300 kilometers (45 to 180 miles) below the surface, and "deep" if its focus occurs more than 300 kilometers (180 miles) below the surface. Earthquakes have been known to occur as far as 700 kilometers (435 miles) below the Earth's surface, though more than 90% of all earthquakes emanate from depths of less than 100 kilometers (60 miles) and the vast majority of catastrophic quakes originate within 60 kilometers (40 miles) of the surface.

Few large earthquakes occur at depths greater than 100 kilometers (60 miles), because heat weakens deeper rocks, causing them to lose some of their ability to store strain energy. The rocks within a few tens of kilometers of the surface are typically brittle and elastic enough to accumulate a vast amount of energy; thus they are more likely to produce large quakes. The depth of the focus of the 1964 Good Friday earthquake in Alaska, for instance, was 33 kilometers (20 miles); the focus of the 1994 Northridge earthquake was 21 kilometers (13 miles) deep; that of the 1989 Loma Prieta earthquake was 18 kilometers (12 miles) deep; and that of the Kobe, Japan, earthquake of 1995 was 20 kilometers (12.5 miles) deep.

General limits apply to the magnitudes of quakes at different depths. Shallow earthquakes have been recorded with magnitudes as high as 9.5, and intermediate ones have magnitudes that can reach 7.5. Deep quakes, however, are rarely recorded with magnitudes greater than 6.9. The magnitude 8.2 earthquake that struck Bolivia in June 1994 did occur at a depth of almost 700 kilometers (400 miles); seismologists are analyzing the data from this quake to determine how deep, presumably warm, plastic rocks could store enough energy to produce such a powerful tremor. On a human note, this quake produced body waves that caused downtown buildings in Minneapolis and Toronto—several thousand kilometers to the north—to sway.

All of this attention is paid to seismic waves and to the location, depth, and magnitude of earthquakes because they bear strongly on the quality of life in the Earth's many earthquake zones. Before moving on to look at the wide range of effects from earthquakes, however, take a moment to consider Highlight 10-1, which illustrates how our growing knowledge of seismic waves plays a significant role in human affairs.

The Effects of Earthquakes

During an earthquake, many geological effects occur simultaneously: Large areas of the ground shift position as displacements occur along faults; the passage of surface waves sets off landslides and mudflows; in places, the seemingly solid earth starts to flow as if liquid; surfaces of bodies of water rise and fall; groundwater levels fluctuate as the Earth compresses and expands with the passage of P waves; and enormous sea waves, generated principally by submarine fault displacements, strike coastlines. In this section we describe each of the major earthquake effects.

Ground Displacements

Among the most obvious geological effects of an earthquake is large-scale displacement of the landscape. After fault blocks have unlocked and moved releasing pent-up energy, geological features and geographical landmarks—such as those in Figure 10-9—are often markedly displaced. During the 1906 San Francisco quake, for example, a 432-kilometer (270-mile)-long rupture opened along the strike-slip San Andreas fault. The west side of the fault then lurched laterally northward 7 meters (25 feet), as determined by measuring displacements of fences and other linear features near the epicenter. By comparison, the lateral displacement produced by the 1989 Loma Prieta quake was only 2.2 meters (7 feet). During the last 30 million years of motion along the San Andreas fault, the rocks and landforms on the west side have shifted an estimated 560 kilometers (350 miles) toward the northwest.

Vertical ground displacements occur during movement along dip-slip faults. In southern California's 1994 Northridge earthquake, the Santa Susana Mountains rose about 30 centimeters (1 foot). During the Good Friday earthquake of 1964, reverse faulting thrust upward 500,000 square kilometers (200,000 square miles) of the Alaska mainland and the adjacent sea floor. The maximum upward land shift to-

Highlight 10-1 Seismic Eavesdropping—Monitoring the Peace

In an international climate characterized by increasing anxiety regarding "who has them and who doesn't," the world's community of nations is preparing to adopt the Comprehensive Test Ban Treaty (CTBT), a global attempt to monitor all nuclear testing. The product of nearly 40 years of heated discussion and negotiation, this treaty bans *all* nuclear detonations. The difficult task of monitoring nations' compliance with this treaty falls largely to seismologists and their hypersensitive seismometers.

A new 170-station seismic network forms the backbone of the detection system. At present, the system can detect unmuffled explosions with yields as low as 1 kiloton of energy, roughly equivalent to a magnitude 4.0 earthquake. (A kiloton equals 10^{12} calories of energy; the first U.S. nuclear detonation—the Trinity test in Alamogordo, New Mexico—had a yield of 21 kilotons.) The system can also pinpoint detonations within an area of 1000 square kilometers [equal to a circle with a radius of 18 kilometers (11 miles)]. The main trick is to distinguish nuclear blasts from other seismic signals, such as those sent out by actual earthquakes or conventional explosions such as mining blasts.

At higher energies (greater than magnitude 4.0), the seismograms of earthquakes and nuclear detonations differ noticeably. Consider the seismic record of a recent French nuclear test in the South Pacific, equivalent to a magnitude 5.0 quake. The shape of the largest peak reveals the blast's true nature. An explosion, which releases virtually all of its energy in less than a second, generates all of its energy waves instantaneously, causing the peak to be quite sharp. It then tapers off smoothly and uniformly as the energy dissipates. By comparison, the peaks on the seismograms of earthquakes appear "sloppy." The ground may begin to vibrate before the main shock (foreshocks), and then continue to vibrate for seconds or minutes after the main shock as the rocks continue to slip and adjust. This pattern creates a complex mix of seismic waves moving away from the quake's focus at different velocities. As a result, the seismogram of a magnitude 5.0 earthquake starts off slowly, grows to its peak erratically, and then decays unevenly—a pattern easily distinguished from that of a nuclear explosion.

At present, an average of 60 seismic events per day are analyzed by the International Data Center (IDC) in Arlington, Virginia, the prototype for the treaty's detection facility. One recent event in Lop Nor, the Chinese test site in central Asia, was closely scrutinized by the IDC. The relatively small shock, about magnitude 3.5, could have been a low-yield detonation. The seismic network pinpointed its source at a location several hundred kilometers from Lop Nor, and the seismographic pattern indicated that the event was a small earthquake, not a nuclear test. This demonstrated the value of the system that the world community hopes will maintain nuclear peace. This system was again used in 1998 to determine the number and yield of nuclear detonations by India and Pakistan.

Figure 10-9 Wallace Creek, on the Carrizo Plain in southern California, offset by the San Andreas fault.

taled 12 meters (39 feet). In some places, the sea floor dropped as much as 16 meters (52.5 feet). While coastal locations that fell below sea level suffered extensive flooding, uplift at the coast left harbors, fisheries, and docks high and dry, requiring relocation of those facilities.

Landslides and Liquefaction

A large earthquake's violent shaking often jostles and dislodges large masses of unstable rocks and soils from hillsides, causing them to rush downslope. Such *landslides* may also occur as rock fragments detach from bedrock, thick layers of sedimentary rock slide along weak bedding planes, foliated metamorphic rock shifts along foliation planes, or loose sediment cascades down slopes.

Earthquakes may also cause **liquefaction,** the conversion of unconsolidated sediment with some initial cohesiveness into a mass of water-saturated sediment that flows like a liquid, *although no water has been added.* When fine, moist sediment becomes disturbed during an earthquake, the jostling increases the pressure on the water between sediment grains, forcing them apart. Without frictional contact between adjacent grains, the sediment loses its cohesiveness and strength, and once-firm ground becomes transformed into a slurry of mud. When sand deposits suddenly behave like liquids during violent tremors, they often produce *sandblows,* fountaining geysers of sand and water that become pressurized, burst through overlying dry sediments, and erupt at the surface. These eruptions typically produce a cone of

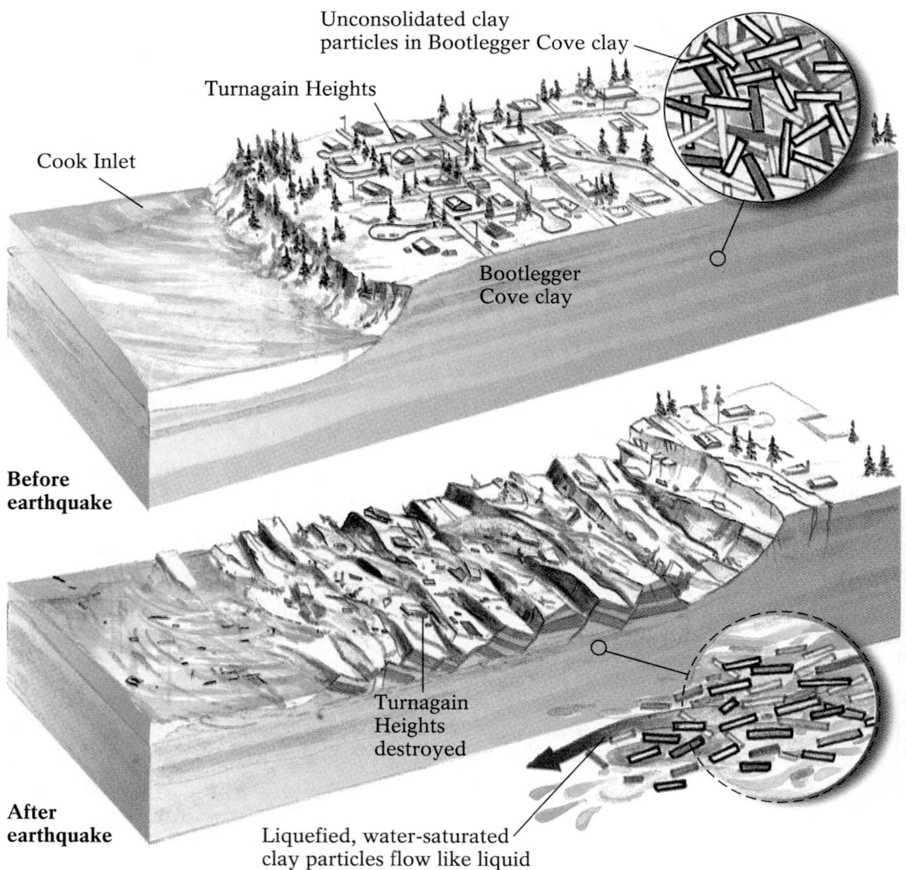

Figure 10-10 Liquefaction occurs when the grains in a layer of wet, fine-grained sediment are shaken during an earthquake. In 1964, the clay layer beneath the entire Turnagain Heights area of Anchorage, Alaska, liquefied, producing a chaotic landscape of jumbled homes and asphalt blocks.

sand shaped like a miniature volcano. Sandblows from great earthquakes may range from 20 to 60 meters (65 to 195 feet) in diameter and rise to heights of 1 meter (3 feet).

During Alaska's 1964 quake, for example, the sediment beneath the Turnagain Heights neighborhood of Anchorage liquefied (Fig. 10-10). The devastated area, which was deemed worthless for future development, was eventually bulldozed into Earthquake Park. During the Loma Prieta quake, liquefaction of the bay muds and landfill soils beneath the Nimitz Freeway in the Oakland area and the Marina district of San Francisco caused many structures to collapse and was responsible for much of that quake's death toll. Liquefaction also contributed significantly to the widespread destruction in the Kobe, Japan, earthquake in 1995 and the Mexico City earthquake in 1985—even though the latter quake's epicenter was 500 kilometers (310 miles) away. (Mexico City is built on the wet, fine-grained sediment of an ancient lake bed.)

The combined effects of liquefaction and landsliding can cause large blocks of land to move significant distances. On June 7, 1692, a relatively moderate quake sent the entire city of Port Royal, on the Caribbean island of Jamaica, sliding into the sea. The town finally came to rest beneath 15 meters (50 feet) of water. The picturesque city—infamous as a base for numerous pirates and other scoundrels—was built on loose, wet, steeply sloping sediment that was prone to liquefaction. When marine archaeologists explored the city in 1959, it lay protected and virtually intact beneath 3 meters (10 feet) of marine silt (Fig. 10-11); in one kitchen, a copper kettle still contained the evening's meal of turtle soup. Eyewitnesses described the scene in the following way:

> Whole streets (With Inhabitants) were swallowed-up by the opening Earth, which then shutting upon them, squeezed the People to Death. And in that Manner several are left buried with their Heads above Ground; only some Heads the dogs have eaten; others are covered with Dust and Earth, by the People who yet remain in the Place, to avoid the Stench.
>
> Many People were swallowed up; some the Earth caught by the middle, and squeezed to Death; the Heads of others only appeared above Ground; some were swallowed quite down, and cast up again by great Quantities of Water; others went down, and were never seen again.

Figure 10-11 Historical account of the disastrous Port Royal earthquake and liquefaction event of June 1692. Liquefaction is the conversion of unconsolidated sediment into a mass of water-saturated sediment that flows like a liquid, although no water has been added.

Seiches

Seismic waves cause the water in an enclosed or partially enclosed body of water, such as a lake or bay, to move back and forth across its basin, rising and falling as it sloshes about. This phenomenon is called a *seiche* (pronounced "SAYSH"). In 1964, the water at one end of Kenai Lake, south of Anchorage, rose 9 meters (30 feet), overflowed its banks, and flooded inland before reversing its direction. The back-and-forth motion of the water stripped the soils around the lake down to bare bedrock. During Montana's Madison Canyon quake of 1959, the Hebgen Lake dam overflowed because of a seiche that oscillated back and forth across the lake every 17 minutes for 11 hours.

Seiches may develop at great distances from powerful earthquakes—sometimes so far from the epicenter that no other effects of the quake are felt. The great Alaska earthquake, for example, set off seiches in reservoirs as far away as Michigan, Arkansas, Texas, and Louisiana. Water rose and fell 2 meters (6.5 feet) in bays along the Texas Gulf Coast, damaging boats by buffeting them with waves. Even some swimming pools in Texas, 6000 kilometers (4000 miles) from the quake's epicenter, developed seiches.

Tsunami

According to a Japanese folktale, a venerable grandfather who owned a rice field at the top of a hill felt the sharp jolt of an Earth tremor one day just before the harvest. From his hilltop vantage point, he saw the sea pull back from the shore. Curious villagers rushed out to explore the exposed tidal flats and collect shellfish. From experience, the old man knew of the grave danger to his neighbors. With his grandson by his side, he dashed about his fields, setting fire to his crop. The villagers saw the smoke, and hurried up the hill to aid their neighbor. As they beat out the flames, they saw the old man scurrying ahead, setting new fires near the hill's crest. Hoping to prevent him from destroying all of his crops, they rushed up the hill to stop him. Moments later, the villagers saw a tremendous wall of water surging onshore, flooding the flats where they had just been standing—and they understood that the old man had sacrificed his harvest to save their lives.

More than any other country, Japan, with its thousands of kilometers of island coastlines nestled precariously within one of the Earth's principal seismic zones, has suffered from **tsunami** (Japanese for "harbor wave"). In open water, tsunami ("tsunami" is both singular and plural) can move at speeds exceeding 800 kilometers per hour (500 miles per hour), making them far larger and faster than ordinary wind-generated waves.

Out at sea, tsunami are barely perceptible, being only about 1 meter (three feet) high. As the crests of successive waves can be separated by as much as 160 kilometers (100 miles), these waves appear to be mere bumps on the sea. [This reality contradicts the Hollywood myth perpetuated in the film *The Poseidon Adventure,* in which a ship is overwhelmed in the open sea by a tsunami 30 meters (100 feet) high.] Only in the shallow waters of enclosed bays and harbors, where drag against the ocean floor causes the fast-moving waves to bunch up, do tsunami become visible as devastating walls of water. In 1739, at Cape Lopatka on Kamchatka Island, one reached a height of 65 meters (210 feet), the equivalent of a 20-story building.

Most tsunami originate from a large, rapid displacement of the sea floor during submarine faulting, when the sea floor either drops through normal faulting or rises via reverse faulting. (These waves may also result from submarine landslides.) When a section of the sea floor drops downward, a trough in the ocean surface appears, temporarily withdrawing water from the coast. A wave is created as water rushes in to fill the trough, overcompensates, and travels to land as a

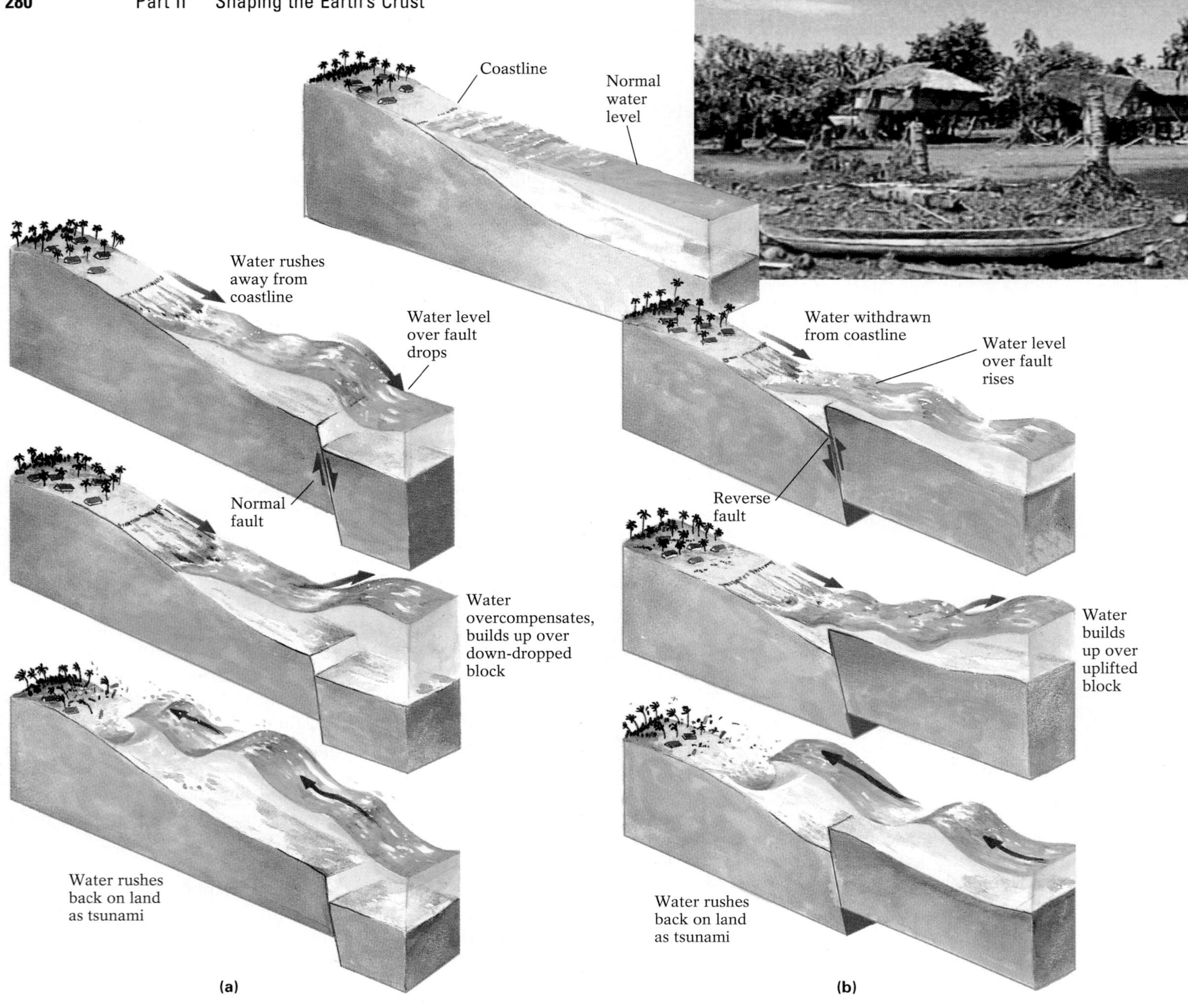

Figure 10-12 **(a)** When a portion of the sea floor is shifted downward from normal faulting, a depression forms in the sea surface. A tsunami is generated as water surrounding the depression rushes in to fill it and overcompensates, forming a "bump" in the sea surface that is propagated in all directions as a large wave. **(b)** When sea-floor uplift occurs from reverse faulting, the sea surface is displaced upward and water is withdrawn from the shore, forming a tsunami. Photo: In July 1998, a magnitude-7.0 earthquake occurred about 12 kilometers (8 miles) off the coast of Papua New Guinea. The close proximity between the quake and the coastline left virtually no time for evacuation before the tsunami—7 meters (23 feet) high—took more than 2000 lives.

potentially catastrophic crest—a tsunami (Fig. 10-12a). When a section of the sea floor is thrust upward, the movement displaces water upward, pulling along water from the shore; the wave later arrives at the coast as a tsunami (Fig. 10-12b).

In most cases, tsunami consist of several waves that arrive at irregular intervals over several hours. Moreover, these giant waves can travel great distances from a quake epicenter and generally go undetected before striking a coast. In 1960, a tsunami generated by a catastrophic quake in southern Chile struck the Hawai'ian shore at Hilo 7 hours later, where it took 61 lives; 22 hours after the quake, the tsunami had traveled 17,000 kilometers (11,000 miles) to reach Honshu and Hokkaido in Japan, where it took 180 lives. Figure 10-13 shows the damage inflicted on the giant sculptures of Easter Island. Highlight 10-2 dramatically illustrates how a far-traveled tsunami that occurred long ago may help prepare

Figure 10-13 A tsunami that originated with Chile's powerful 1960 earthquake upended the renowned sculptures of Easter Island, located in the Pacific Ocean west of South America.

Highlight 10-2 A Mysterious Tsunami Strikes Japan

Figure 10-14 The famous print *Beneath the Waves Off Kanagawa,* by Japanese artist Katsushika Hokusai (1760–1849), depicts the helplessness of a group of boatmen in the face of tsunami-sized waves.

Sometime around midnight on January 27, 1700, a 2-meter (7-foot)-high tsunami struck Japan's Pacific Coast. (Artist Katsushika Hokusai captured this tsunami in a famous print that appears as Figure 10-14.) The tsunami swept the wintry sea inland, swamping coastal homes and their surrounding rice paddies. Yet Japan itself had not been struck by an earthquake that night. The identity of the unknown seismic culprit—a large earthquake somewhere within the Pacific Basin—had, until now, never been identified. Recent research by seismologists at the University of Tokyo and the Geological Survey of Japan points the guilty finger at the Cascadia subduction zone, a fault system that extends from Vancouver Island, British Columbia, to northern California.

Looking for the source of Japan's mysterious 1700 tremor, Japanese seismologists searched the historical records of other Pacific Rim regions. None was found, for example, in South America, where Spanish colonists and explorers kept thorough records. They were left with vast areas of Alaska, the Aleutians and Kuril Islands, and western North America as possibilities. Unfortunately, those areas lack written historical records from the time of the quake. The seismologists then consulted the work of nineteenth-century ethnographers—anthropologists who interview a region's inhabitants and compile their histories through their stories and ancient oral traditions. Although no stories of an earthquake around 1700 come from Alaska or other Northern Pacific lands, dozens of Native American legends describe a quake and tsunami that struck the Pacific Northwest's coast at the appropriate time.

The next piece of evidence came from teams of muck-covered geologists, who combed the marshlands of coastal Washington for physical evidence of a great earthquake that may have struck the Northwest about 300 years ago. In some places, broad areas of the coast have dropped several meters below sea level, swamping stands of red cedar and Sitka spruce. Covering these fallen surfaces is a thin, continuous sheet of sand left behind by what has been estimated as a 10-meter (33-foot)-high tsunami. Carbon-14 dating of organic deposits within these sediments and tree-ring dating of the drowned trees places these events within a few decades of 1700, making them a solid suspect in the Japanese earthquake mystery.

Working on the assumption that the Japanese tsunami and the northwestern quake were caused by the same event, Japanese seismologists have estimated that their relatively small tsunami could have been a faint signal of Northwest's great quake. Using computer models that derive earthquake magnitudes from the size of resulting tsunami, they estimate that a tsunami spawned by a magnitude 8.0 quake, traveling 8000 kilometers (5000 miles) from the Cascadia subduction zone, would arrive as a 40-centimeter (16-inch)-high swell 10 hours after the initial tremor. Although such an arrival time is consistent with Native American stories of an early evening quake, the resulting tsunami would be far too small. If a tsunami generated by a magnitude 9.0 quake traveled the same distance, however, it would measure 2 meters (7 feet) high, precisely the height of the swell that struck Japan that January night in 1700.

In light of the emerging story of Japan's 1700 tsunami, some North American seismologists now speculate that the 1000-kilometer (600-mile)-long Cascadia subduction zone may be capable of rupturing en masse, generating an earthquake of magnitude 9.0 or higher. (A similar event occurred off the coast of Chile in 1960 when the Andean subduction zone ruptured, generating history's largest recorded earthquake—magnitude 9.5. The resulting tsunami sped across the Pacific, striking the Japanese coast and taking 140 lives.) In light of this thoughtful seismological detective work, the Pacific Northwest must accept the possibility that an earthquake of magnitude 9.0 or greater that shakes for three to five minutes, may send 10-meter-high tsunami onshore in the future.

Figure 10-15 In the earthquake of 1906, the buildings and homes along San Francisco's Fisherman's Wharf District that were not destroyed by the city's widespread fires were left listing and severely damaged by liquefaction of the underlying wet, loose sediments.

today's cities on North America's west coast for its next great earthquake.

Tsunami that strike a coast at high tide often exact a greater human toll than the earthquakes that generate them: Of the 131 deaths attributable to the 1964 Alaskan quake, 109 resulted from tsunami that buffeted the southern Alaskan coast. Tsunami from that quake also hit the coasts of Oregon and California, causing 16 deaths there. At Crescent City, California, the first wave arrived at low tide. It crested about 4 meters (14 feet) above sea level, destroying several lumber boats and releasing a morass of giant redwood logs. After three additional small waves had passed, residents returned to the coast, believing the worst was over. A fifth wave struck at high tide, however, cresting at more than 10 meters (33 feet) and washing 12 people out to sea.

Fires

On September 1, 1923, a powerful earthquake struck the Tokyo–Yokohama area of Japan at midday. The tremor overturned thousands of open cooking fires. Fanned by high winds, the fires coalesced and spread until, by evening, the glow from the blaze was bright enough to read by as far away as 15 kilometers (10 miles) from the city. Before residents managed to extinguish the flames, 500,000 buildings in the mostly wood-frame city were destroyed and 120,000 people perished.

Each year, Japan observes a national day of earthquake preparedness in commemoration of this tragedy. After the earthquake, Tokyo was rebuilt with heightened awareness—with broader avenues to facilitate passage of fire-fighting equipment and concrete (rather than wooden) buildings. Millions of liters of water are now stored in underground cisterns and quake-resistant structures, and the city's 30,000 taxis are equipped with fire extinguishers.

Earthquake damage to gas mains, oil tanks, and electrical power lines typically ignites fires in inhabited regions. Because water mains are frequently ruptured as well, the area's fire-fighting capability may be impeded. During the 1906 earthquake, San Francisco's wooden buildings were tinder for a blaze that caused 80% to 90% of the city's damage (Fig. 10-15). The fire raged uncontrolled for three days, largely because water lines had broken into hundreds of segments that subsequently lost pressure. Finally, workers created a firebreak by dynamiting rows of buildings. Sadly, it is believed that the main blaze was initially kindled several hours after the quake by a woman on Hayes Street, who, shaken by the events of the day, attempted to restore some normalcy to life by preparing a ham-and-eggs breakfast. Unfortunately, the flue of her stove had been damaged during the earthquake; the resulting fire consumed more than 500 square blocks of the city's central business district.

The World's Principal Earthquake Zones

As we have seen, the vast majority of earthquakes are concentrated at plate boundaries, whereas plate interiors are generally seismically inactive. Nevertheless, major earthquakes do occasionally strike within a plate interior: Charleston, South Carolina, Boston, Massachusetts, and Memphis, Tennessee—all located far from fault-prone plate boundaries—have all experienced a major quake within the past 250 years. In the following sections, we describe many of the world's most active earthquake zones.

Earthquake Zones at Plate Boundaries

The depths of earthquakes vary significantly across the different types of plate boundaries. As Figure 10-16 shows, in general earthquake foci are relatively shallow at rifting, divergent, and transform plate margins, but are substantially deeper at subduction-type margins. At oceanic divergent zones, most earthquakes are extremely shallow, occurring at depths of less than 20 kilometers (12 miles), as strain energy builds up only within the uppermost, brittle portion of an oceanic plate. Deep earthquakes are virtually nonexistent in these zones, because the asthenosphere—located about 100 kilometers (60 miles) beneath oceanic plates—consists of partially melted, softer material that deforms plastically and therefore produces no earthquakes. The same characteristics apply to continental rift zones and continental collision zones, and along continental transform boundaries. Earthquakes in these zones rarely occur at depths greater than 50

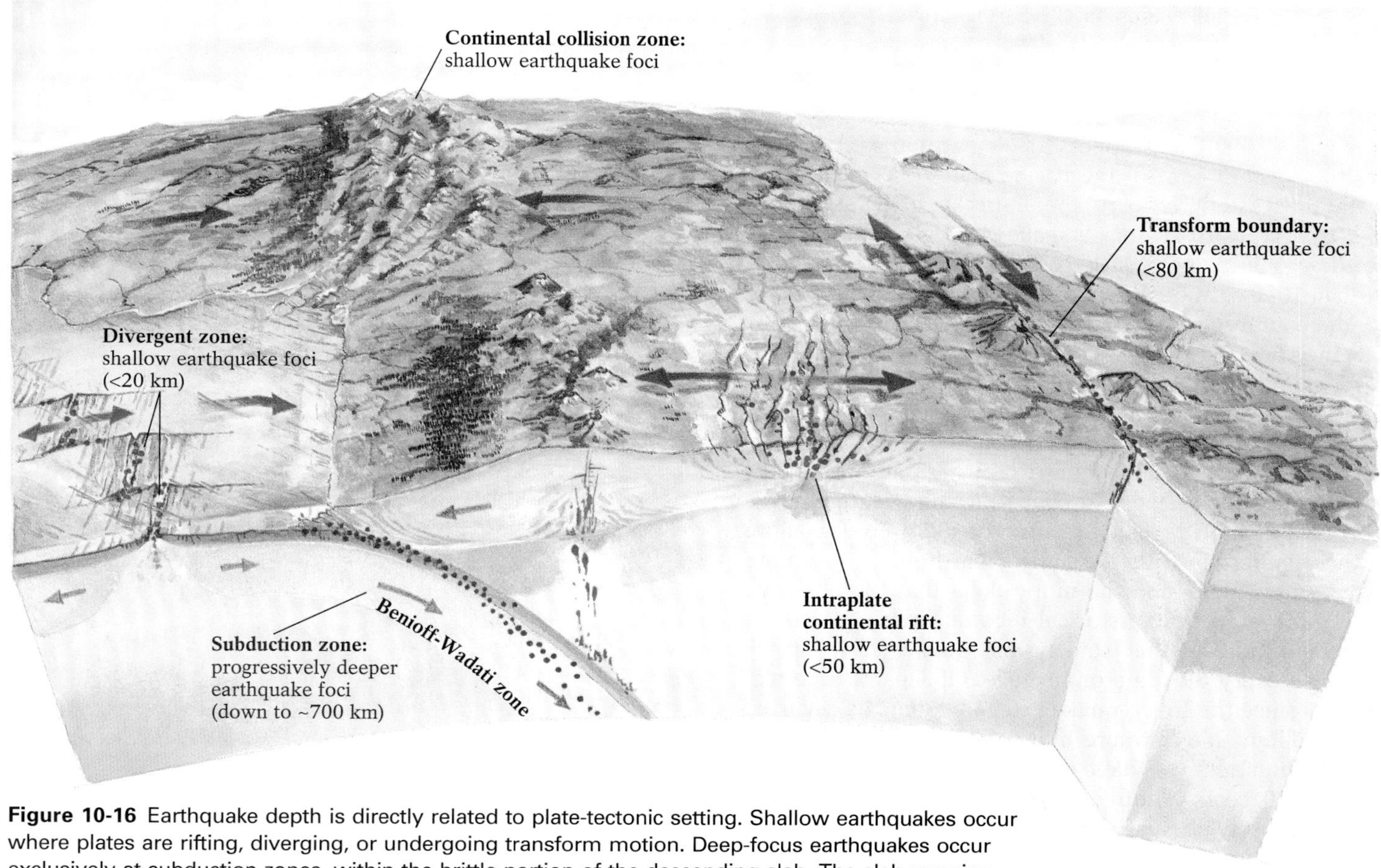

Figure 10-16 Earthquake depth is directly related to plate-tectonic setting. Shallow earthquakes occur where plates are rifting, diverging, or undergoing transform motion. Deep-focus earthquakes occur exclusively at subduction zones, within the brittle portion of the descending slab. The slab remains largely intact and brittle (except for possibly some minor melting at its very top) as it passes through the partial-melting zone. It continues to generate earthquakes until it descends to about 700 kilometers, at which point it has warmed sufficiently to deform plastically.

to 80 kilometers (30 to 65 miles)—the maximum thickness of the brittle portion of continental plates.

At subduction zones, however, progressively deeper quakes occur within the descending slab to a depth of about 700 kilometers (450 miles). The region within which earthquakes arise at subducting oceanic boundaries is called the **Benioff-Wadati zone;** it is named for the American seismologist Hugo Benioff, who first observed that earthquake foci become progressively deeper inland from ocean trenches, and Japanese seismologist Kiyoo Wadati, who pioneered methods of comparing earthquake sizes.

From the onset of subduction at an ocean trench to a depth of approximately 300 kilometers (200 miles), intense friction between the descending slab and the overriding plate causes ruptures along the slab's brittle upper surface, resulting in shallow and intermediate-depth earthquakes. These events are generally the most powerful subduction-zone earthquakes. As the slab descends farther, the center of quake activity shifts to the still-brittle interior of the slab, which has not been in the Earth's interior long enough to warm up and soften. By the time the slab reaches approximately 700 kilometers (450 miles) below the surface, it has been heated sufficiently to soften throughout, and thus deforms plastically instead of undergoing brittle failure. Hence, earthquakes rarely occur at depths greater than 700 kilometers.

The magnitude of earthquakes varies considerably with the type of plate boundary involved. Almost all major earthquakes occur along either convergent or transform boundaries. (Earthquakes at divergent plate boundaries are typically of relatively low magnitude, because rocks being stretched tend to break before accumulating too much strain energy.) Every year, about 80% of the world's earthquake energy is released in the Pacific Rim region, where the oceanic plates of the Pacific Ocean basin subduct beneath adjacent plates along most of the ocean's perimeter (see Figure 10-2). This region includes the earthquake zones of Japan, the Philippines, the Aleutians, and the west coasts of North, Central, and South America. Most of the remaining 20% of the Earth's earthquake energy is released along the collision zone stretching through Turkey, Greece, Iran, India, and Pakistan to the Himalayas, southern China, and Myanamar (Burma). Several of the Pacific Rim's most prominent earthquake zones are discussed next.

Japan On January 17, 1995, an earthquake devastated Kobe, Japan, causing $100 million in damage and earning the dubious distinction as history's most expensive natural disaster. The quake took 5378 lives, leveled 152,000 buildings, and incinerated the equivalent of 70 U.S. city blocks. Yet, this magnitude 6.9 event was classified as only a strong quake, not a "major" or "great" one. Kobe's earthquake occurred where the western edge of the Pacific plate descends steeply beneath the eastern edge of the Eurasian plate at Japan's Nankai trench. By itself, Japan is shaken by 15% of the planet's released seismic energy; each year, its residents experience more than 1000 earthquakes of magnitude 3.5 or greater. Since 1900, the country has suffered 25 earthquakes as powerful as San Francisco's great quake of 1906. Landslides, tsunami, fires—Japan has experienced the worst of each.

Even so, Japan's next quake may be its worst. Like other nations located at other quake-prone plate boundaries, Japan experiences periodic earthquakes whenever one of its faults accumulates enough energy to overcome the frictional resistance between fault blocks and then breaks free. The more time that elapses since the last major earthquake, the more energy accumulates—and the more likely that the next one will be quite powerful. Historical records from A.D. 818 to the present indicate that cataclysmic quakes have struck Tokyo once every 69 years on average. Of Japan's nine major quakes since the Tokyo quake in 1923, none has occurred in metropolitan Tokyo, where more than 20 million people now live. The stress on Tokyo's rocks has been building for more than 70 years. Another great quake is probably overdue in the densely populated Tokai region to the south, where the last major quake released accumulated energy on December 24, 1854.

Figure 10-17 Effects of the Mexico City earthquake of September 19, 1985 (magnitude 8.1). Due to the violent shaking of Mexico City's geological foundation of unconsolidated lake clay, the city—500 kilometers (310 miles) from the quake epicenter—actually suffered greater damage than areas closer to the epicenter.

Mexico and South/Central America Earthquakes have claimed hundreds of thousands of lives in Chile and Peru, along the west coast of South America, where the Nazca plate (see Figure 1-16) flexes downward to form the Peru–Chile trench and the Andes mountains. On May 31, 1970, an earthquake leveled virtually every dwelling along a 70-kilometer (45-mile) strip of the Peruvian coastline, taking 120,000 lives, demolishing 200,000 buildings, and leaving 800,000 people homeless. Similarly, subduction of the Cocos plate produces the tremors that have periodically devastated such cities as Mexico City, Mexico, and Managua, Nicaragua. Figure 10-17 shows some of the tragic aftermath of the Mexico City quake of September 1985.

Western North America Western North America, from southern California to Alaska, owes much of its extensive earthquake history to interaction between the Pacific plate and the North American plate. The southern end of the transform boundary between these plates extends for 1000 kilometers (600 miles), from Baja California to Cape Mendocino northwest of San Francisco. Between Cape Mendocino and Vancouver, British Columbia, the Pacific–North American transform boundary is interrupted by two small plates (the Gorda plate off the coast of northern California and Oregon, and the Juan de Fuca plate off the coast of Washington and southern British Columbia) that are being subducted beneath westward-drifting North America. The Cascade volcanoes, from Lassen Peak and Mount Shasta in northern California to Mount Garibaldi north of Vancouver, are products of this subduction zone. The transform boundary between the Pacific and North American plates resumes north of Vancouver and extends to Alaska, where the Pacific plate subducts beneath the North American and Eurasian plates, producing the volcanic Aleutians. Thus the transform boundaries and subduction zones of the western states and provinces form an almost continuous stripe of earthquake-prone country.

Intraplate Earthquakes

On June 10, 1987, a rare mid-plate earthquake unnerved the citizens of Lawrenceville, Illinois, and 16 surrounding states and Canadian provinces. Registering a 5.0 magnitude, the quake triggered alarms at a nuclear power plant, cut phone service, and shook hospital patients from their beds in Iowa and West Virginia.

Though generally shallow [less than about 50 kilometers (30 miles)], mid-plate earthquakes tend to be of lower magnitude than those at active plate margins, apparently because less strain energy builds at mid-plate locations. On the other hand, the older, colder (and therefore more brittle), and less fractured rocks at mid-plate transmit seismic waves more efficiently than the young, warm rocks at active plate margins. Thus mid-plate earthquakes, such as those occurring in eastern North America, are felt over a significantly larger area than are plate-boundary quakes, such as those arising in the west.

Locating mid-plate faults to determine their history and predict their future seismic activity can prove difficult, particularly in eastern North America, where heavy vegetation conceals rocks and their faults. Because of the infrequent and sporadic nature of this tectonic activity, any surface evidence is usually obscured or eradicated with the passage of time. A few mid-plate faults, such as the network of faults in the English Hills of southeastern Missouri and northeastern Arkansas, actually break the surface—a real "treat" for eastern geologists.

In the East, subsurface faults are typically discovered by seismic-wave studies, which have revealed the crust of eastern North America to be riddled with a deep irregular network of old, near-vertical faults. Those along the East Coast may date from the rifting of North America from Europe and Africa during the initial breakup of the supercontinent Pangaea, 200 million years ago. Faults in the Midwest may date from an older rifting event that occurred some 570 million years ago. At that time, North America apparently rifted, opening the Iapetus Sea, in a manner similar to the rifting that created the present Atlantic Ocean. The so-called Iapetan faults, which are today landlocked because of subsequent plate-tectonic events (principally the formation of the Appalachians, discussed in Chapter 9), run along the west side of the Appalachians from northeastern Canada, along the St. Lawrence River, through Lake Ontario, and then southward through eastern Tennessee into Alabama.

The exact causes of intraplate earthquakes remain a mystery. Some hypotheses cite processes that are unrelated to plate-tectonic motion. One, for example, proposes that after long-term erosion has removed vast quantities of surface materials, the buoyant unloaded crust rises, generating enough stress to arouse old faults. Another suggests that the load of mid-continent sediment deposition weighs down the crust, producing pressure that reactivates its faults. A third hypothesis proposes that these earthquakes may be triggered by excessive rainfall, which infiltrates and saturates the network of subsurface fractures, reducing friction between adjacent blocks of rocks and reactivating long-dormant faults.

Another hypothesis views intraplate seismic activity as a precursor to future plate configurations, and perhaps as an early warning of rifting. According to this scenario, plate rifting is initiated by rising masses of heat-softened mantle material. Such thermal plumes may have set off quake activity near Boston and Charleston, and seismic studies in those areas have identified extremely young plutons that are considered evidence of incipient rifting and future plate-edge locations.

A simpler model holds that plate motion stresses the weakened and faulted crust at some intraplate sites, initiating infrequent but sometimes powerful quakes. The lateral motion of the westward-drifting North American plate, for example, may create frictional stress within the plate as it rides atop the underlying asthenosphere. Strain energy may accumulate over long periods, until it exceeds the strength of the rocks or the resistance along their faults. Today's northeast-to-southwest trend of the Iapetan faults is oriented so that they absorb the stresses generated by the continual growth of the Atlantic Ocean.

Earthquakes in Eastern North America Because relatively few earthquakes have released the pent-up seismic energy in the interior of the North American plate, large quakes may be looming in this region. Seismologists working with earthquake records that date back to 1727 predict that there is a better-than-even chance of a magnitude 6.0 earthquake striking east of the Rocky Mountains before the year 2020.

Past eastern earthquakes have included a sharp tremor, with an estimated magnitude of 5.0, that struck New York City on August 10, 1884. Felt from Washington, D.C., to Maine, the quake hurled horsecars from their tracks on Fifth Avenue and sent a two-ton safe sliding across the lobby of the Manhattan Beach Hotel in Brooklyn. In 1929, a powerful tremor toppled 250 chimneys in Attica, New York; in 1983, the Adirondack Mountains of New York were rocked by a quake of magnitude 5.2. In Massachusetts, earthquakes occurred near Plymouth in 1638, at Cape Ann 80 kilometers (50 miles) northeast of Boston in 1755, and in Cambridge in 1775. The 1755 quake, which was felt from Nova Scotia to South Carolina, shattered roof gables on colonial homes, destroyed 1500 chimneys, and snapped the gilded-cricket weather vane off the roof of Boston's Faneuil Hall. In Canada, an earthquake emanated from fractured rocks 20 kilometers (12 miles) beneath the evergreen forests of the small lumber town of Chicoutimi, 150 kilometers (100 miles) north of Quebec City, on November 25, 1988; measuring a surprising magnitude 6.0, it was felt as far away as Washington, D.C. One of the most powerful eastern earthquakes, estimated at magnitude 7.0, razed much of Charleston, South Carolina, in August 1886 and took the lives of 60 residents.

The most powerful recorded seismic events of eastern North America comprised the New Madrid, Missouri, quakes, which began on December 16, 1811, and lasted for 53 days, ending February 7, 1812. The series of earthquakes included three main shocks and more than 1500 powerful aftershocks. The most violent aftershock was the last—it shook a 2.5-million-square-kilometer (1-million-square-mile) area from Quebec to New Orleans and from the Rocky Mountains to the eastern seaboard. Church bells tolled in Boston, 1600 kilometers (1000 miles) away, windows rattled violently in Washington, D.C., and pendulum clocks stopped in Charleston. Based on observations by fur trappers and an eyewitness account from naturalist John Audubon, geologists believe that each of the three principal tremors may have exceeded a magnitude of 8.5.

Near the estimated epicenter of the New Madrid quake, entire forests were flattened when the violent shaking

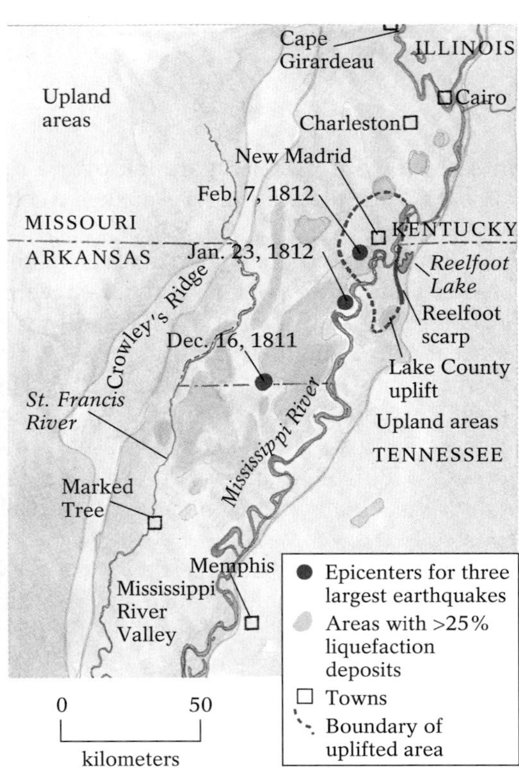

Figure 10-18 The effects of the New Madrid, Missouri, earthquakes of 1811–1812 included liquefaction of sediments, uplift of some areas, and subsidence of other areas to form several new lakes. The lake shown here, Reelfoot Lake in northern Tennessee, formed when a vast area of prairie land subsided during the New Madrid quakes.

snapped tree trunks. Large fissures opened, including some too wide to cross on horseback. Sandblows, water, and sulfurous gas erupted as the shaking ground compacted. Thousands of square kilometers of prairie land subsided, flooding to form the St. Francis and Reelfoot lakes in northwestern Tennessee (shown in Figure 10-18) as well as swamps that today stretch 300 kilometers (190 miles) from Cape Girardeau, Missouri, to northern Arkansas. Uplifting of other areas exposed lake beds that have since dried out. The Mississippi River became dammed in some spots by uplifts and landslides, but flowed wildly elsewhere, with waves that overwhelmed many riverboats. The flow of the river even reversed direction where uplifts changed lowlands into highlands. Whole river-channel islands disappeared, and offsets along the river produced new cliffs and waterfalls.

Because they occurred at a mid-plate location and produced a great variety of landscape changes, the New Madrid events hold considerable interest for geologists. It is critical that geologists understand these events, because the area could again experience seismicity of the same magnitude—a circumstance for which it is now woefully unprepared. Recently, geologists and archaeologists have joined forces to determine the frequency with which great earthquakes occur in the mid-continent. Their studies have focused on ancient sandblows that contain pits dug by prehistoric people that contain datable pottery fragments. From the age of these artifacts, we estimate that the last great quake in Missouri, prior to the events of 1812, occurred about 1100 years ago. This finding suggests that the time between great quakes in that region may be on the order of 1000 years. Residents of that area, however, should not rest easy; their region is probably capable of producing more frequent strong (magnitude 6.0–6.9) and major quakes (magnitude 7.0–7.9). Dating of sandblows by artifacts and carbon-14 dating of buried wood indicate that tremors of magnitude 6.4 or greater have shaken the region at least four times during the last 2000 years (roughly around the years 500, 900, 1300, and 1600 A.D.). Most definitely, many of today's homes and high-rise office buildings are not built to withstand such a shake. In the nineteenth century, the Mississippi valley was sparsely populated; the same area, from St. Louis to Memphis, is currently home to more than 12 million unprepared citizens.

Coping with the Threat of Earthquakes

Their fresh memories of the tragic destruction of Kobe, Japan, have spurred great concern among residents of North America's earthquake zones. Recent studies that examined the likelihood of similar events, occurring along the Hayward fault in California's Oakland–Berkeley area and the Seattle fault, which traverses metropolitan Puget Sound, have raised serious questions about how well we're prepared for such potential urban catastrophes.

Can anything be done to prevent damage from earthquakes and to protect the millions of people in North America and elsewhere who live in vulnerable regions? In this section, we will address this question. Although the U.S. government has prepared a checklist of helpful, logical "do's and don'ts" (Table 10-4), even more can be done.

Table 10-4 Earthquake Do's and Don'ts

What to Do Before an Earthquake

Check for potential fire risks, such as defective wiring and leaky gas connections. Bolt down water heaters and gas appliances.
Know where and how to shut off electricity, gas, and water at main switches and valves.
Place large and heavy objects on lower shelves. Securely fasten shelves to walls. Brace or anchor top-heavy objects.
Do not store bottled goods, glass, china, and other breakables in high places.
Securely anchor all overhead lighting fixtures.

17 Survival Items to Keep on Hand—in a Large, Lockable Plastic Trash Barrel

Portable radio with extra batteries
Portable fire escape ladder for buildings with multiple floors
Flashlight with extra batteries
First-aid kit containing any specific medicines needed by members of the household
First-aid manual
Fire extinguisher
Bottled water—one gallon per person, per day
Adjustable wrench for turning off gas and water
Smoke detector
Matches
Axe (for chopping wood for heating)
Canned and dried foods for one week for each member of household
Nonelectric can opener
Telephone numbers of police, fire, and doctor
Portable stove with propane or charcoal
Sleeping bags for each person/tent or plastic tarps with cord
Cash (banks and automatic teller machines may be closed or inoperable)

Note: The status of these items should be checked every six months.

What to Do During an Earthquake

If you are outdoors, stay outdoors; if you are indoors, stay indoors. During earthquakes, most injuries occur as people enter or leave buildings.
If indoors, take cover under a heavy desk, table, or bench, or in doorways, or halls, or against inside walls. Stay away from glass.
Don't use candles, matches, or other open flames either during or after tremors. Extinguish all fires.
If in a high-rise building, don't dash for exits; stairways may be broken or jammed with people. Never use an elevator.
If outdoors, move away from buildings, utility wires, and trees. Once in the open, stay there until the shaking stops.
If in a moving car, drive away from underpasses and overpasses. Stop as quickly as safety permits, but stay in your vehicle. A car may shake violently on its springs, but it is a good place to stay until tremors stop. When you drive on, watch for fallen objects, downed wires, and broken or undermined roadways.

The Most Common Causes of Earthquake Injuries

Building collapse or damage
Flying glass from broken windows
Falling pieces of furniture, such as bookcases
Fires from broken gas lines, electrical shorts, and other causes, aggravated by a lack of water caused by broken water mains
Fallen power lines

What to Do After an Earthquake

Be prepared for aftershocks.
Check for injuries; do not move seriously injured persons unless they are in danger of sustaining additional injury.
Listen to the radio for the latest emergency bulletins and instructions from local authorities.
Check utilities. If you smell gas, open your windows and shut off the main gas valve. Leave the building and report the gas leakage to the authorities. If electrical wiring is shorting out, shut off the current at the main meter box.
If water pipes are damaged, shut off the supply at the main valve. Emergency water may be obtained from hot-water tanks, toilet tanks, and melted ice cubes.
Check sewage lines before flushing toilets.
Check chimneys for cracks and damage. Undetected damage could lead to fire. Approach chimneys with great caution. The initial check should be from a distance.
Do not touch downed power lines or objects touched by downed lines.
Immediately clean up spilled medicines, drugs, and other potentially harmful materials.
If the power is off, check your freezer and plan meals so as to use foods that will spoil quickly.
Stay out of severely damaged buildings. Aftershocks may shake them down.
If you live along a coast, do not stay in low-lying coastal areas. Do not return to such areas until local authorities tell you that the danger of a tsunami has passed.

Source: U.S. Federal Emergency Management Agency (FEMA).

Limiting Earthquakes Caused by Humans

Any earthquake-defense plan should start with policies to eliminate quakes caused by humans, such as those initiated by the building of dams to create large reservoirs. The additional water accumulated in these reservoirs applies considerable weight over a geologically short time period to underlying rocks. Eventually the water may seep into deep, faulted rocks, reducing the friction between adjacent rock masses and activating old, previously stable faults. In this way, dam building may stimulate seismic activity in areas that have been inactive for centuries.

The Hoover Dam, which impounds Lake Mead, was erected across the Colorado River at the Arizona–Nevada border in an area that had been relatively free of earthquakes. In 1936, shortly after the dam was completed, a series of 600 tremors began that lasted for 10 years. The largest, measuring about magnitude 5, was felt throughout the region. It is believed that the 42 cubic kilometers (10 cubic miles) of water contained in young Lake Mead depressed fractured crustal blocks and lubricated subsurface fractures in the bedrock, causing the blocks to slip. The massive Three Gorges Dam, currently under construction along China's Yangtze River, could also trigger an earthquake of magnitude 6 or greater, and endanger thousands of people living downstream of its vast reservoir if the dam is breached.

Another way that humans may cause earthquakes was revealed during the early 1960s in Denver, Colorado. This community is built over fractured billion-year-old granite that had been quake-free for the 80 years prior to 1962. From 1962 until 1965, however, it experienced 700 tremors measuring from 3.0 to 4.3 in magnitude; the largest produced significant damage in the area. In 1962, engineers at the U.S. Army's Rocky Mountain arsenal had drilled a 3660-meter (12,000-foot)-deep well into faulted bedrock beneath the Denver area. Every month, as much as 16 million liters (4 million gallons) of liquid waste were injected into this well under high pressure. The high fluid pressure widened bedrock joints, and the liquid waste lubricated potential slip surfaces along major faults. When the arsenal was informed of its responsibility for the tremors, it discontinued the project. As shown in Figure 10-19, the level of quake activity dropped dramatically in response.

Determining Seismic Frequency

Seismologists study the earthquake history of an area to determine the statistical probability that future earthquakes will strike there. If you hear that there is a 40% chance of a magnitude 7.5 earthquake occurring along the Palm Springs section of the San Andreas fault within the next 30 years, for instance, that "forecast" is based on the frequency and magnitude of past events in that area. Seismologists use maps such as the one shown in Figure 10-20 to make general predictions by identifying areas that have experienced quakes

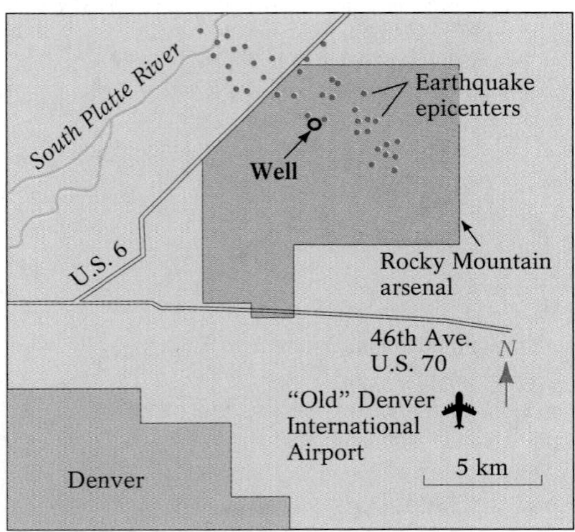

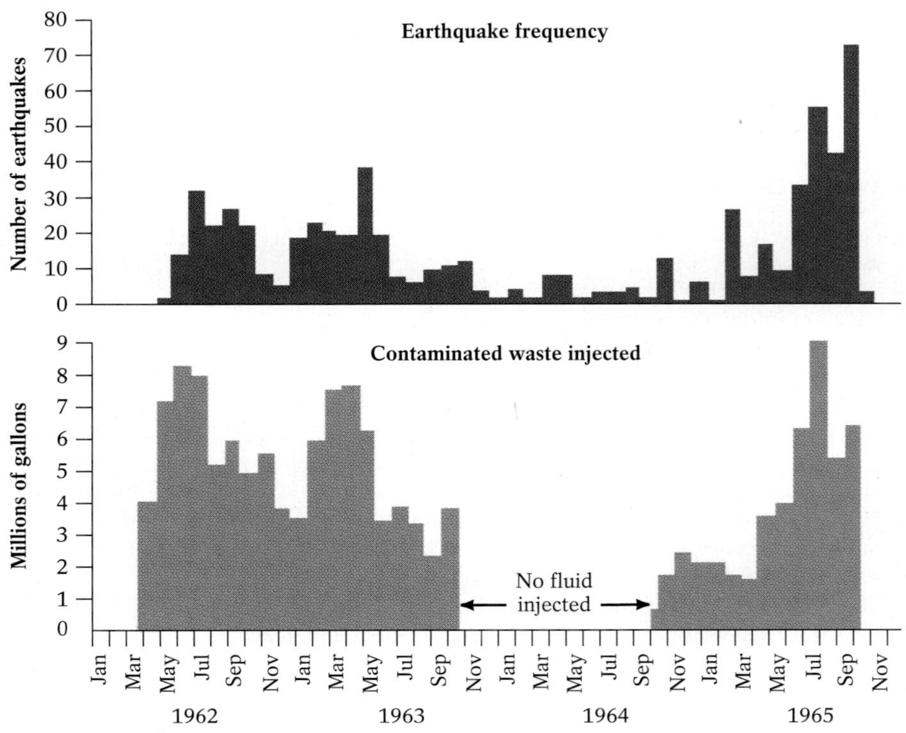

Figure 10-19 Graphs correlating a series of earthquakes that occurred in Denver during the early 1960s with the high-pressure disposal of liquid waste deep in the region's underlying fractured bedrock during those years. Note how the frequency of seismic events coincided with the timing and quantity of waste disposal.

in the past and that should therefore take earthquake precautions, such as adopting stringent building codes. The maps do not, however, indicate the frequency with which destructive tremors arise. For example, Charleston, South Carolina, with its one large 1886 event, is represented on such a map in the same way as southern California, which experiences frequent tremors. Moreover, because such a map relies on historically recorded and therefore relatively recent events, its information tells us little about the pattern of seismic events over thousands of years. To develop a long-term history for North America, we must attempt to date fault and quake activity that preceded the 400 to 500 years of recorded North American history.

In California, for example, substantial written records exist for only the past 150 years, and they constitute too brief a set of earthquake observations to determine past quake

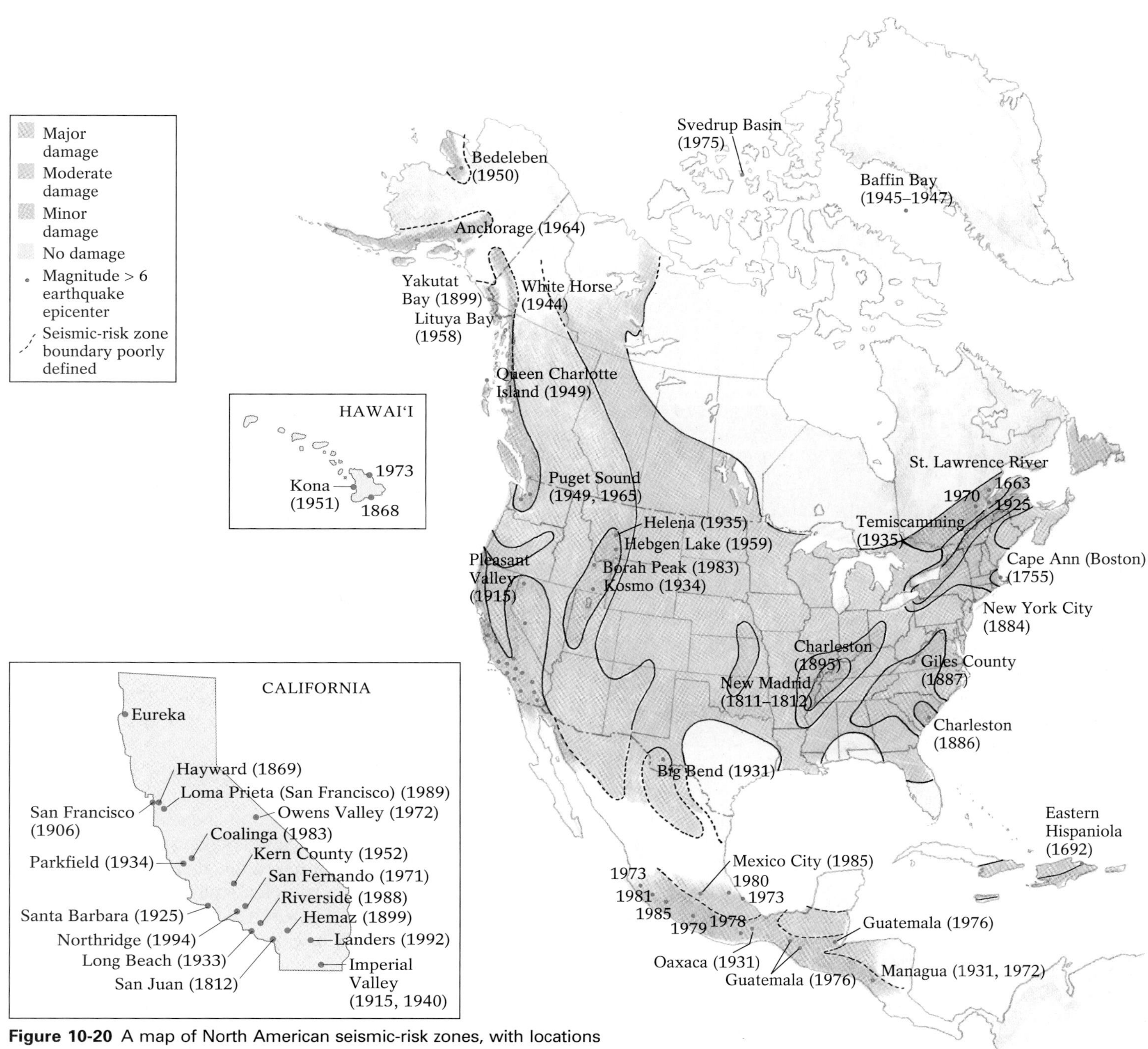

Figure 10-20 A map of North American seismic-risk zones, with locations and dates of some of the continent's major historical earthquakes.

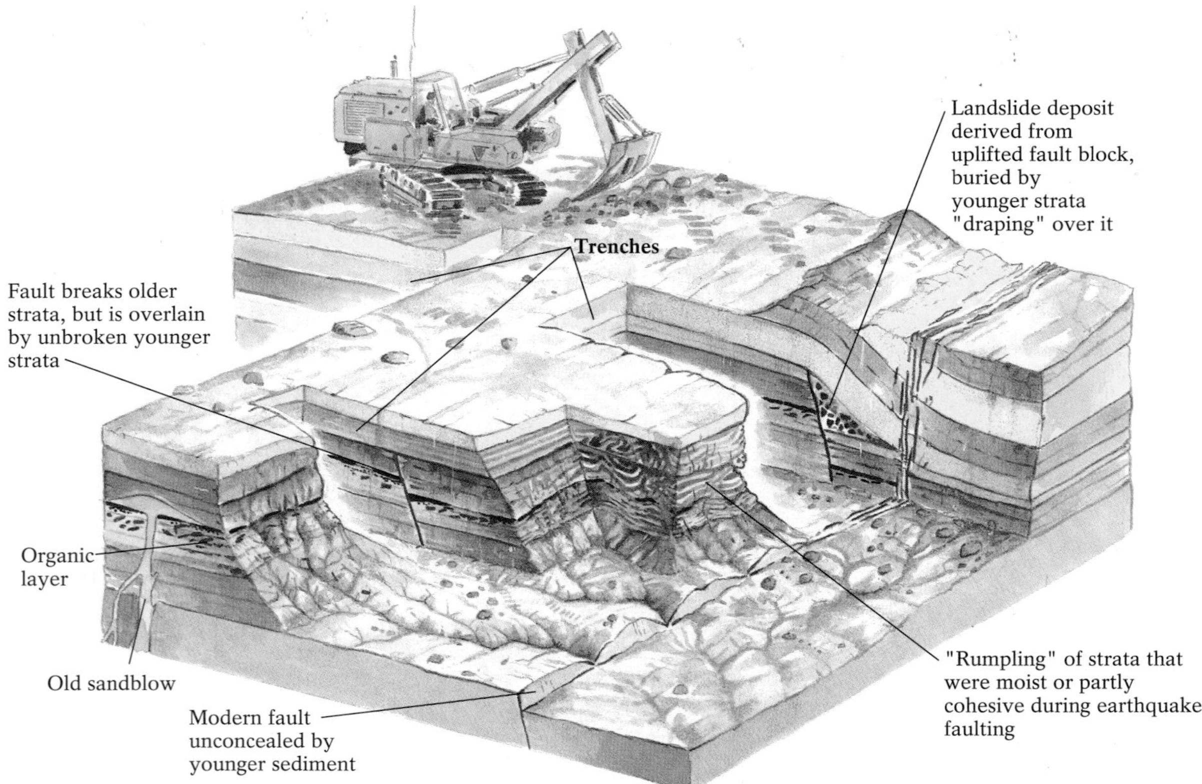

Figure 10-21 (above) Trenches excavated across a fault expose the stratigraphy of the fault zone. According to the principle of cross-cutting relationships, any movement along a fault must be younger than the age of the youngest rock or sediment that it displaced, and older than the oldest rock or sediment that it did not displace. Thus each of the seismic events shown here can be dated relative to one another, and its absolute age estimated in relation to the organic deposit. (left) Vertically offset deposits near Palmdale, California, showing evidence of a previous earthquake there.

frequency. To date earlier seismic events in California, we must look for geological evidence of a fault, such as topographic breaks in the landscape, linear valleys that may follow a fault line, areas of crushed rock, small lakes that may have formed where impermeable crushed rock prevents surface water from seeping underground, or linear features that have been offset by fault movements (such as the orange grove in Figure 1-25 on page 24).

Once geologists find a fault, they then search for evidence of past seismic events related to the fault and try to date them to determine the fault's *recurrence interval*, the average amount of time between its associated quakes. One method is to examine the area's geologic strata. Trenches excavated across the San Andreas fault, such as the one shown in Figure 10-21, reveal a complex stratigraphy that includes vertically offset organic deposits. Geologists apply carbon-14 dating and the principle of cross-cutting relationships to determine when the offsets occurred, enabling them to estimate when the related seismic events may have taken place. Northeast of Los Angeles, for example, evidence suggests that nine major events have taken place during the past 1400 years, with an average recurrence interval of 160 years. This fact

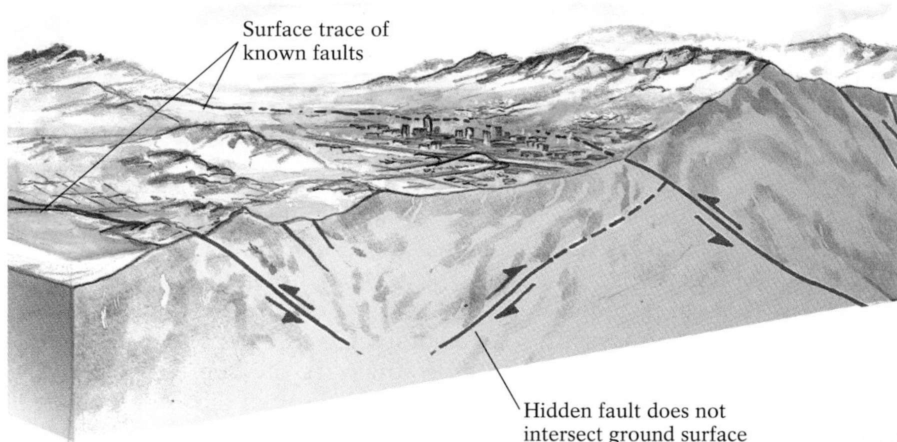

Figure 10-22 Hidden faults, because they are not evident at the surface, often go undetected until an earthquake occurs. One such hidden fault is believed to have caused the Whittier earthquake of 1987, which claimed eight lives and caused $360 million in damage in Pasadena. We know now that Los Angeles is underlain by numerous hidden faults, such as those recently discovered at Echo, MacArthur, and Elysian Parks.

unnerves many southern Californians, given that the last great earthquake in that area occurred in 1857.

Seismologists can estimate the frequency of past earthquakes only for visible faults. Even though we have lasers that can measure fault displacements of hundredths of a millimeter and ultrasensitive seismometers that can detect faint tremors on the opposite side of the Earth, faults may still lie undiscovered beneath downtown Los Angeles and elsewhere. After an earthquake shook Whittier, California, on October 1, 1987, local geologists identified numerous hidden, or "blind," faults in the Los Angeles area (Fig. 10-22); many of these are thrust faults that have uplifted the mountains surrounding the Los Angeles basin. The faults had little or no surface expression until the quake exposed their locations through compressed pavement, uplift, and fractures at the surface.

One such hidden fault, part of the Oak Ridge fault system, produced the magnitude 6.7 quake that devastated the northern San Fernando Valley, 30 kilometers (20 miles) northwest of downtown Los Angeles, in January 1994. This tragedy cost 63 lives, led to thousands of injuries, produced a mass disruption of daily life, and created more than $15 billion in damage. Although geologists had suspected its existence for at least half a century, the hidden fault's extent and depth remained unknown until surface damage and aftershocks gave clues to its location. Another recently discovered fault, the Elysian Park fault, runs along Wilshire Boulevard through Hollywood toward Santa Monica, threatening such landmarks as Dodger Stadium, Beverly Hills, and the world-famous "HOLLYWOOD" sign. The discovery of two additional hidden faults in Los Angeles, one through Echo Park and the other through MacArthur Park, has further heightened the city's awareness of its undiscovered danger zones. Although the San Andreas fault is more likely than these recently discovered faults to produce a powerful earthquake of, for example, 7.5 to 8.0 in magnitude, it is located 50 kilometers (30 miles) east of downtown Los Angeles; consequently, its earthquakes may cause less damage than lower-magnitude tremors generated on hidden faults directly beneath the city.

To make matters worse, seismologists studying a 1957 earthquake in Mongolia discovered that that great quake—with a magnitude exceeding 8.0—involved *two* sets of faults in the same region. During that quake, the land lurched along a 250-kilometer (150-mile) segment of a San Andreas-type strike-slip fault; simultaneously, a nearby 50-kilometer (30-mile)–long network of thrust faults ruptured. This finding suggests that a major event along the San Andreas fault could potentially activate one or more of Los Angeles's hidden faults, such as the one that uplifted the San Gabriel Mountains.

Building in an Earthquake Zone

Safe use of earthquake-prone regions requires that localities plan *where* they construct various structures and then engineer those facilities to survive the region's expected quake magnitudes. Certain structures, such as nuclear power plants and large dams, are considered *critical facilities,* because their destruction by an earthquake could take a heavy toll in human lives. Critical facilities must be sited on stable ground and as far as possible from active faults as well as major population centers and heavily traveled highways. The builders must make every effort to locate hidden faults prior to construction. Extensive geological surveying must also precede the siting of public facilities that would be essential for coping with a quake disaster, such as fire and police departments, hospitals, other civil-defense installations, and the local communications network.

In the following sections, we describe how the stability of the ground and the structures built determine the likelihood that an earthquake will produce damage.

Ground Stability When seismic waves pass through adjacent areas of solid bedrock and soft, wet sediment, the shaking is generally more pronounced in the loose sediment than in the

Figure 10-23 Differential earthquake damage in Varto, Turkey, in 1953. Variations in the destructiveness of earthquake shaking are often controlled by the underlying geologic materials: The collapsed houses shown here were built on the unconsolidated sands of an old river channel, while the surviving houses rest on a solid bedrock bench.

Figure 10-24 The devastation of villages from the May 27, 1995, earthquake on Sakhalin Island cost the lives of 2325 of this village's 3200 inhabitants.

solid bedrock. While the bedrock moves in short sharp jolts, the loose sediment develops a continuous rolling motion that can magnify the quake's intensity. As a result, structures built on unconsolidated materials typically suffer far more damage than comparable ones located on solid bedrock, as illustrated by the collapsed buildings within the old channel deposits shown in Figure 10-23. During the 1906 San Francisco quake, for instance, the buildings constructed on wet bay mud and landfill sustained four times as much damage as those built on nearby bedrock; during the 1989 Loma Prieta earthquake, the waterfront areas again suffered the greatest damage. The destruction seemed to hopscotch around the Bay Area, striking those structures sited on the most unstable materials, including the rubble from the 1906 quake.

Although the epicenter of the 1985 Mexico earthquake was near Acapulco, the damage was greater and the death toll far higher in more-distant Mexico City. Acapulco is built on solid volcanic bedrock, which shook far less violently than the ancient lake clays on which Mexico City is situated. Both the California and Mexico City events demonstrate that, in quake-prone areas, geological surveys are essential to identify solid bedrock upon which communities can be built safely. For their part, potential home buyers should inquire about the ground beneath their prospective property.

Structural Stability The amount of damage sustained by a building during an earthquake reflects the magnitude and duration of the quake, the nature of the ground on which the building stands, and the building's design. The earthquake that destroyed several cities in Armenia in 1988, taking more than 50,000 lives, was of the same magnitude as California's 1989 Loma Prieta quake, which took 65 lives. India's catastrophic earthquake of 1993 was of lower magnitude than either of these events, yet took almost 18,000 lives. The greater damage and loss of life in Armenia and India largely resulted from the poor construction of the buildings there. San Francisco's structures were built to meet strict standards and with the expectation that they would have to survive high-magnitude quakes; constructed using superior designs, methods, and materials, they could better withstand the shaking when it did occur.

The most secure structures are built of strong, flexible, but relatively light materials. Wood, steel, and reinforced concrete (containing steel bars) tend to move *with* the shaking ground. In contrast, unreinforced concrete, heavy brickwork and concrete block, stucco, and adobe (mud brick) are generally inflexible. Their components move independently and in opposition to the shaking, battering one another until the structure collapses. The collapsed houses in Figure 10-24, for example, were made of stone and other inflexible materials. Concrete-frame structures, such as the large parking garage that collapsed during California's 1994 Northridge quake and the I-880 Cypress Expressway that failed disastrously during the 1989 Loma Prieta quake, appear particularly vulnerable during large earthquakes.

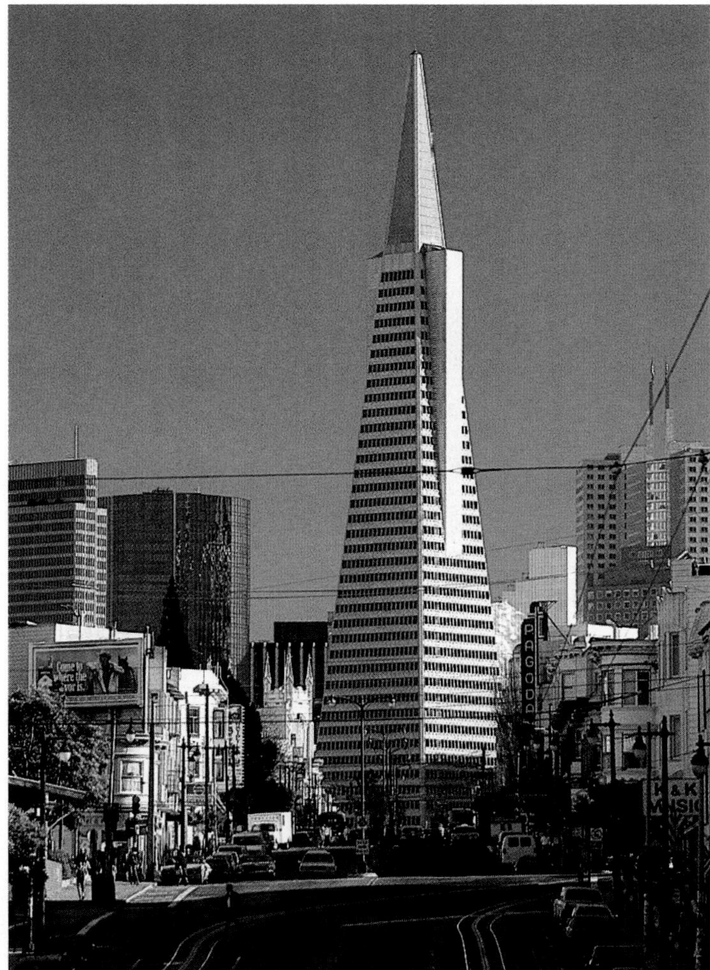

Figure 10-25 The Transamerica Building, in San Francisco, California. The pyramidal shape of this structure makes it less likely than more traditional buildings to suffer extensive earthquake damage.

Well-designed and engineered skyscrapers may be the safest buildings. A building with structural elements (floors, walls, ceilings) that are joined firmly may sway in a quake like a tree in the wind, but it is unlikely to collapse. Distributing a building's weight with the greater proportion at the bottom also enhances safety. The momentum that develops when a top-heavy building swings from side to side may topple it. The pyramidal Transamerica Building in San Francisco (Fig. 10-25), which tapers progressively toward the top and therefore has most of its weight concentrated in its lower floors or below-ground foundation, can sway with the shaking Earth but will not develop damaging momentum near its top. The buildings that suffer most dramatically during powerful tremors include those with unsupported roofs that extend across a broad span, such as bowling alleys, supermarkets, and shopping malls.

Another general rule about building in an earthquake zone can be summarized as "the simpler, the better." Projections or decorative elements that are not securely attached may loosen or fall during a tremor. Many nineteenth-century buildings have projecting decorations that remain secured only by deteriorating mortar. Realizing the danger, the cities of Los Angeles and San Francisco have recently begun requiring owners to secure such embellishments with special braces.

Earthquakes also play havoc with highways, endangering motorists and severing vital transportation lifelines. Through painful experience, we now know that elevated roadways must be strongly connected to their support columns, that the columns themselves must be of equal height so that they flex at the same rates, and that, wherever possible, the roadways should be sited on solid bedrock. (Unfortunately, North America's greatest period of highway building was during the 1950s and 1960s, when less was known about earthquake engineering.) The collapse of Oakland's Nimitz Freeway during the 1989 Loma Prieta earthquake clearly demonstrated that multilevel roadways are particularly inappropriate in quake-prone areas, though many such structures remain in California. Costly efforts to increase the strength and stability of California roadways and bridges after the Loma Prieta earthquake did pay off during the 1994 Northridge earthquake; among the dozens of bridges and overpasses that had been upgraded, only one collapsed in that quake.

Earthquake Prediction: The Best Defense

In March 1966, two powerful tremors rocked a broad area 200 kilometers (120 miles) south of Beijing, the capital of the People's Republic of China. The death toll and property damage were so appalling that Premier Chou En Lai declared the "People's War on Earthquakes." For his army, Chou recruited 10,000 seismologists and 100,000 amateur assistants from the ranks of China's college students, telephone operators, and factory workers. Their mission was to set up 250 regional seismic stations and analyze a huge volume of seismic data so that China might better predict and prepare for future earthquakes. The effort paid off handsomely on February 4, 1975.

In late 1974 and early 1975, residents of Liaoning province in southern Manchuria reported that their well water contained high levels of radon gas; among other precursor signals of an impending earthquake, thousands of snakes crept out of winter hibernation and froze to death on icy roads. Chou's army focused its attention on Haicheng, the region's center, which supported 90,000 inhabitants. On February 1, 1975, a series of seismic spasms struck the city; the largest, on the morning of February 4, had a magnitude of 4.8. Early that afternoon Chinese seismologists issued an earthquake advisory, ordering residents to extinguish their cooking fires and evacuate to the city's parks and fields. At 7:36 P.M., a powerful tremor (magnitude 7.6) destroyed almost 90% of Haicheng's tile-roofed dwellings. A disaster of

Figure 10-26 Citizens of Beijing living in tent cities on the streets, after the earthquake of 1976.

enormous proportions was averted, demonstrating the effectiveness of accurate earthquake prediction.

One year later, however, in May 1976, the same network of professional and amateur seismologists failed to predict the greatest natural disaster of this century. A magnitude 8.0 earthquake struck the coal-mining and industrial center of Tangshan, located about 150 kilometers (90 miles) east of Beijing, taking 650,000 lives and injuring 780,000. Fearing aftershocks, Beijing's residents lived on the streets, in fields, and in parks for three months (Fig. 10-26). These two events—one glorious hit, one tragic miss—illustrate the uneven current state of earthquake prediction.

To justify an evacuation such as the one undertaken in Haicheng, seismologists monitor and analyze certain prequake signals, then attempt to make short-term predictions (on the order of months, weeks, or days). Long-term prediction (on the order of decades or years) is often used to establish building codes and construction standards and to site critical facilities such as nuclear power plants. Recent efforts at long-term prediction have focused on areas where earthquake activity has been unexpectedly absent. Recent short-term predictions have involved monitoring the buildup of stress in fault-zone rocks and other potential precursors.

Long-Term Prediction from Seismic Gaps When two adjacent plates move, strain energy accumulates in the rocks along their shared boundary. Although this energy probably accumulates uniformly throughout the seismically active zone, it may be released irregularly, with earthquakes occurring in different places and at different times. Some fault segments release their strain energy continuously, as fault blocks move without locking up. This nearly constant motion, called **tectonic creep,** prevents the large buildup of strain energy that causes great earthquakes, but it can gradually deform or break any features subjected to it (Fig. 10-27). The creeping Hayward fault in Berkeley, California, runs through the football field at Memorial Stadium on the University of California campus, bending the seats in the bleachers, opening large cracks in the stadium walls, and damaging a large drainage culvert beneath the field.

According to some seismologists, locked fault segments that have not experienced seismic activity for long periods—known as **seismic gaps**—represent prime candidates for major earthquakes. Once geologists identify a seismic gap, they estimate the rate of plate movement and determine the amount of strain energy built up in the fault segments. Some seismologists believe that the longer the period of seismic "silence," the more strain energy is being accumulated, and the more powerful the eventual earthquake will likely be. Figure 10-28 shows the areas surrounding the Pacific Ocean

Figure 10-27 A low wall bent by tectonic creep along the Calaveras fault in California. The creeping of the fault continuously confounds road-construction engineers, who can't keep the curbs straight.

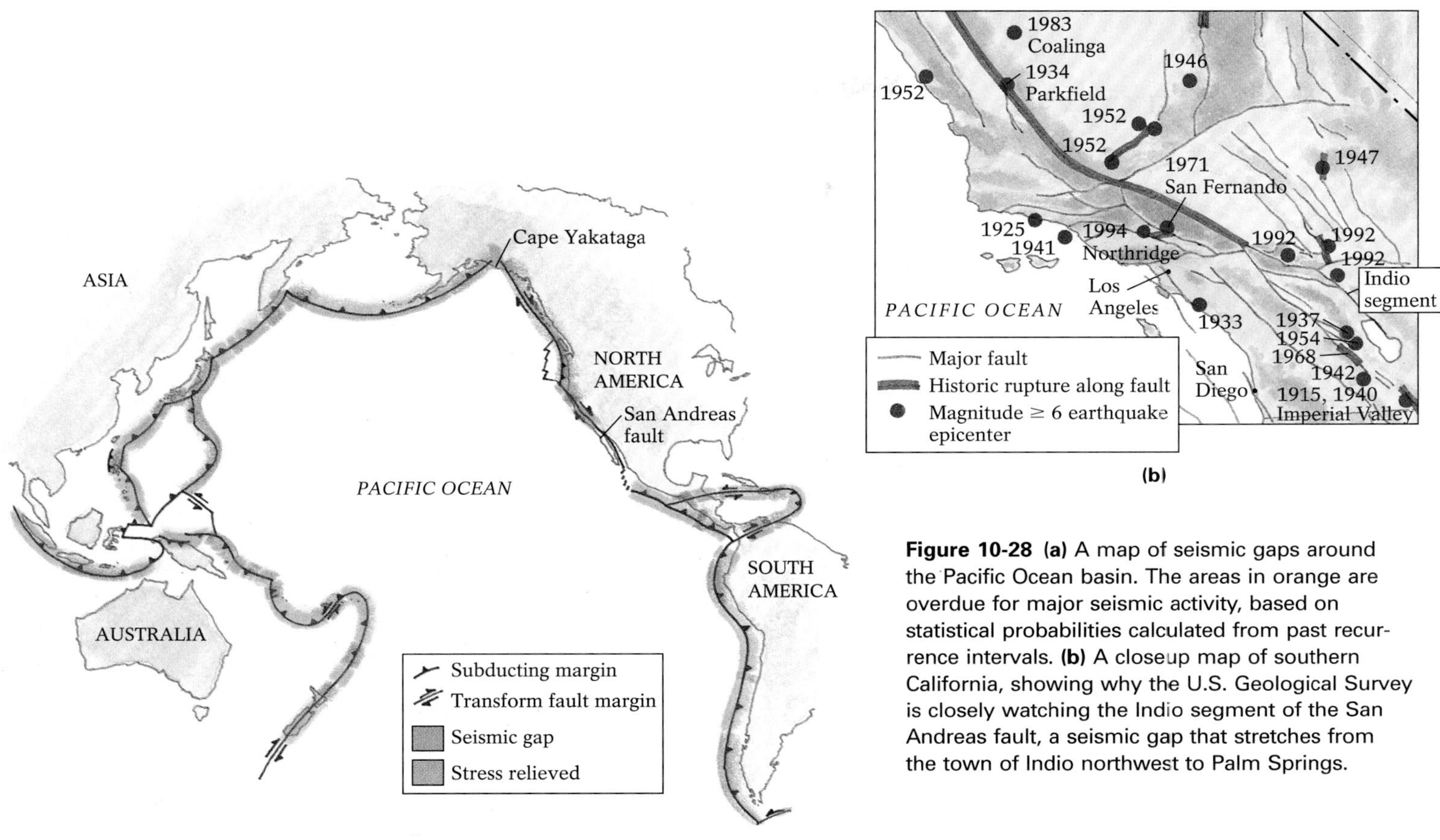

Figure 10-28 **(a)** A map of seismic gaps around the Pacific Ocean basin. The areas in orange are overdue for major seismic activity, based on statistical probabilities calculated from past recurrence intervals. **(b)** A closeup map of southern California, showing why the U.S. Geological Survey is closely watching the Indio segment of the San Andreas fault, a seismic gap that stretches from the town of Indio northwest to Palm Springs.

basin that have not suffered their expected share of earthquakes in recent times.

In southern Alaska, the area around Cape Yakataga has been unnervingly quiet in a seismic sense for more than a century. Based on the amount of accumulated strain in the rocks, geologists predict that a magnitude 8 earthquake is due there soon. Fortunately, the area is sparsely populated.

The entire stretch of the San Andreas fault in densely populated southern California has also been seismically quiet for an unusually long time. In fact, from an historical standpoint, Los Angeles itself has suffered only two strong earthquakes (6.0–6.9) during the last two centuries and thus is markedly overdue for increased earthquake activity. According to the Southern California Earthquake Center's calculations, the city has stored enough strain energy in the last 200 years along its six "known" fault systems (other faults remain hidden) to generate 17 Northridge-size quakes, leaving enough energy for 15 quakes of magnitude 6.7 sitting beneath the city. These studies also indicate that Los Angeles is overdue for a quake of magnitude 7.2–7.6, one that statistically occurs every 140 years or so. The area's last quake of this size occurred 210 years ago. Thus, contrary to the belief that the 1994 quake has bought them some time, Angelenos are actually enjoying a seismic lull that can't last forever.

Recently, however, other seismologists have vigorously challenged the seismic-gap theory. They protest that faults cannot be broken up into discreet segments that build and release strain independently. According to these seismologists, when faults slip during a large earthquake, it increases the stresses on other nearby faults. These "anti-gap" seismologists suggest that the occurrence of a large quake actually *increases* the likelihood of other large earthquakes. In contrast, proponents of the gap theory argue that once a fault segment has released its accumulated strain energy, it quiets down and becomes less dangerous for a period of years until significant strain energy rebuilds. Thus they propose that the likelihood of subsequent quakes *decreases.* Given that the seismic-gap theory has been instrumental in estimating future earthquake potential and has greatly influenced building codes and preparedness strategies, this lively debate seems certain to continue for years.

The role played by seismic-gap theory in structural engineering in earthquake zones is dramatically illustrated in the current construction of the 260-meter (850-foot)-high

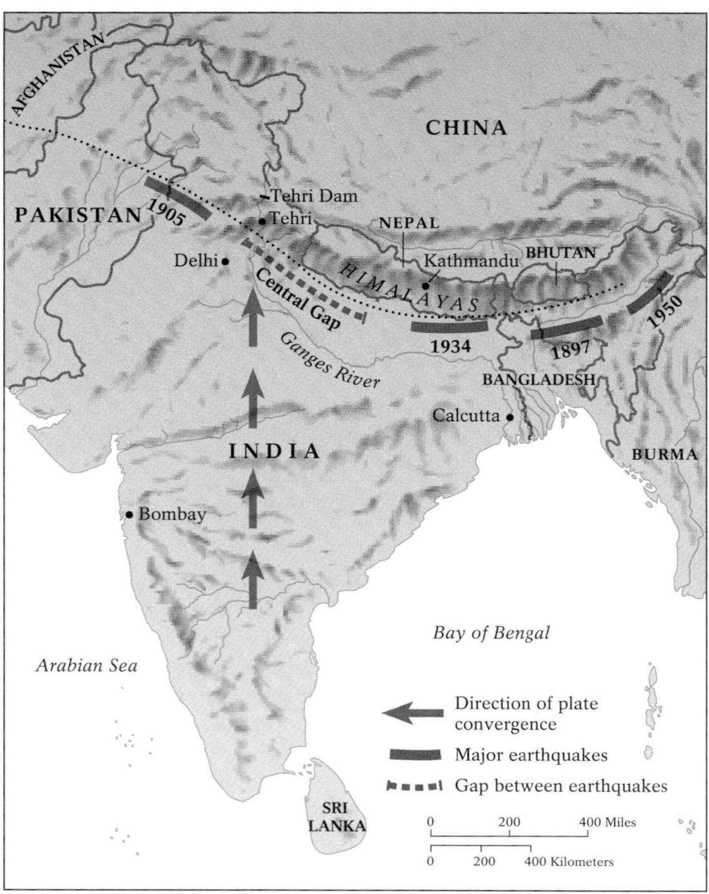

Figure 10-29 Northern India's Tehri Dam project, shown in progress above, is highly controversial, given its location in the Himalayas' Central Gap region. Despite ongoing compression between India and Asia, this region's seismic gap may be overdue for a major earthquake. (See map at left.)

Tehri Dam in northern India (Fig. 10-29a). The dam will occupy the so-called "central gap," shown in Figure 10-29b. According to seismic-gap theory, this area is overdue for the next great Himalayan earthquake, one whose magnitude may exceed 8.0. Such a massive event would undoubtedly endanger the lives of hundreds of thousands of people in the holy cities of Rishikesh and Hardwar, downstream of the dam in the Bhagirathi River valley.

Short-Term Prediction by Detecting Dilatancy When rocks become stressed, they develop countless minute cracks long before they rupture completely. These cracks begin to appear when accumulated strain energy approaches about one-half the amount needed to produce rock failure. As cracking continues, the rocks expand in volume, or *dilate.* **Dilatancy** produces a number of side effects that geologists can monitor to make short-term earthquake forecasts, perhaps on the order of days, weeks, or even hours. Figure 10-30 illustrates how a number of these effects appear before, during, and after an earthquake. The phenomena include the following events:

1. *Swarms of micro-earthquakes.* As hairline cracks open and rocks dilate, countless micro-earthquakes (magnitude less than 1) occur that are detectable only with the most sensitive seismometers. A swarm, or cluster, of these tremors sometimes precedes a major quake. Micro-earthquakes that are followed by a more powerful quake are called **foreshocks.**
2. *Tilt or bulge in rocks.* The surface of a dilated rock may bulge upward. Sensitive tiltmeters can measure changes in surface slope, and frequent land surveys can detect changes in ground elevation. In 1964, a major earthquake in Niigata, Japan, was preceded by 10 years of surface bulging near the quake's epicenter.
3. *Changes in seismic-wave velocity.* When rocks dilate and cracks develop, the velocity of any seismic waves passing through them declines. Later, if groundwater fills the cracks, the seismic waves speed up again, because the water transmits seismic waves more efficiently than air does. Seismologists can determine the extent and rate of fracturing within rocks and the presence of groundwater by periodically setting off small explosions nearby and monitoring any changes in seismic-wave velocities.
4. *Variations in electrical conductivity.* Rock is a relatively poor conductor of electricity, and it becomes an even poorer one when dilated by air-filled cracks. But when

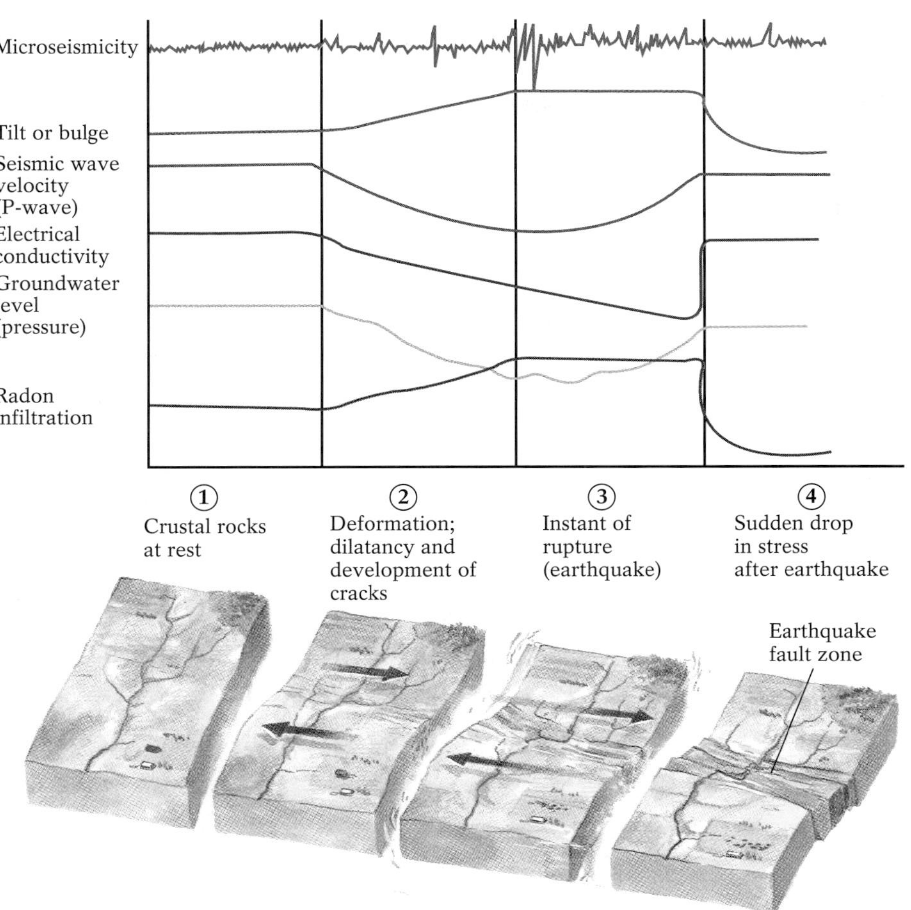

Figure 10-30 Some phenomena caused by the dilatancy of highly stressed rocks, and their relationship to seismic activity.

impure water, a good conductor, enters a network of connected fractures in a stressed rock, the rock's electrical conductivity increases; this increase sometimes precedes an earthquake by days or hours. Thus improved electrical conductivity may indicate that water is filling new fractures in highly stressed rocks that are about to fail and generate an earthquake.

5. *Anomalous radio-wave signals.* Geologists have sometimes noticed that the atmosphere contains fewer radio waves several days before an earthquake. They hypothesize that as underground electrical conductivity increases before an earthquake, the ground absorbs radio waves from the atmosphere. In May 1984, a marked reduction in radio waves preceded a magnitude 6.2 earthquake near Hollister, California, by six days. Similar effects were also noted prior to the Loma Prieta quake of 1989. Proponents of radio-wave monitoring claim that it will someday provide short-term warning—as little as three to five days—of impending earthquakes.

6. *Changes in groundwater level and chemistry.* As rocks dilate and groundwater seeps into the fresh cracks, the water level in nearby wells may drop perceptibly; some wells may even run dry. Groundwater chemistry is also monitored to detect increased proportions of elements that normally reside tens of kilometers beneath the surface, but that rise and infiltrate the groundwater in surface wells after stressed rocks dilate and crack. Prior to recent powerful quakes in China and Armenia, geologists observed a sharp increase in radioactive radon gas dissolved in well water near the epicenters. They believe that radon, produced by the radioactive decay of uranium in deep granitic rocks, migrated through fractures in the dilated rocks and entered the water of near-surface wells. Constant monitoring of the radioactivity of groundwater may therefore provide pre-quake warnings by signaling increased dilatancy in area rocks.

Unusual Animal Behavior When dilatancy occurs, the cracking rocks evidently emit high-pitched sounds and minute vibrations that are imperceptible by humans but audible to many animals. Farmers and scientists alike have often observed unusual animal behavior—perhaps related to the dilatancy events—shortly before earthquakes. Dogs have been known to howl incessantly, as they did in the San Francisco streets the night before the great 1906 earthquake. Cattle, horses, and sheep refuse to enter their corrals, and ordinarily timid rats leave their hideouts and make their way fearlessly through crowded rooms. Shrimp crawl onto dry land, fish jump from water, ants pick up their eggs and migrate, and ground-dwelling birds (such as chickens) roost in trees. In cold weather, snakes leave warm subsurface hibernation dens only to freeze on icy ground. Such episodes of extraordinary animal behavior may serve as pre-quake warnings.

Predicting Parkfield Earthquakes Consider the following series of numbers: 1857 . . . 1881 . . . 1901 . . . 1922 . . . 1934 . . . 1966. Those are the years in which Parkfield, California, was struck by a major earthquake on its segment of the San Andreas fault. The sequence would have looked better if the 1934 was 1944, but the U.S. Geological Survey, overlooking the anomaly, concluded in 1988 that Parkfield was in the midst of a seismic cycle lasting 20 to 22 years. All of the Parkfield earthquakes were remarkably similar in nature and magnitude. Each began with foreshocks that struck within 2 kilometers (1.25 miles) of the epicenter of the main shock, and each had a magnitude in the 5-to-6 range. For these reasons, Parkfield was the subject of the first long-term earthquake prediction made by the U.S. Geological Survey in 1988: There was a 95% statistical probability that a quake of magnitude 5.5 to 6.0 would occur in that area before the end of 1993.

Parkfield, a tiny hamlet of 34 citizens, has been overrun by geophysicists on 24-hour call since the mid-1980s. At the cost of millions of dollars and thousands of research hours, a large concentration of instruments has been installed to detect foreshocks and to document the precursor signals that may possibly accompany a major quake. On November 13, 1993, a magnitude 4.8 earthquake struck Parkfield, prompting the U.S. Geological Survey to issue a high-level alert on November 15 that predicted a magnitude 6 earthquake within 72 hours. After three days passed and the quake failed to materialize, the alert was canceled. As of May 1998, Parkfield has remained seismically silent, although recent stirrings of the San Andreas fault near Parkfield have heightened speculation about the imminent arrival of the overdue tremor.

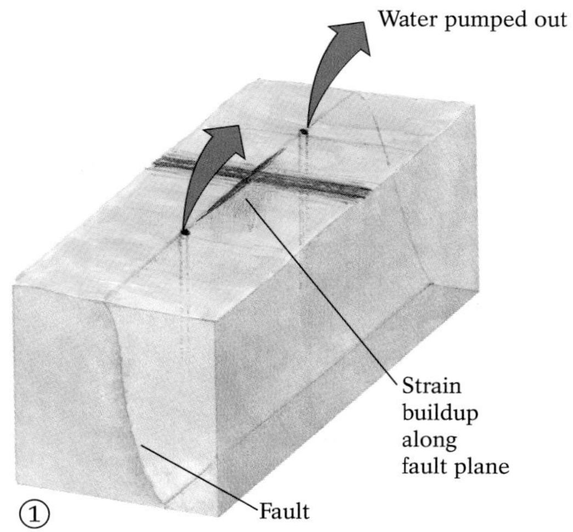

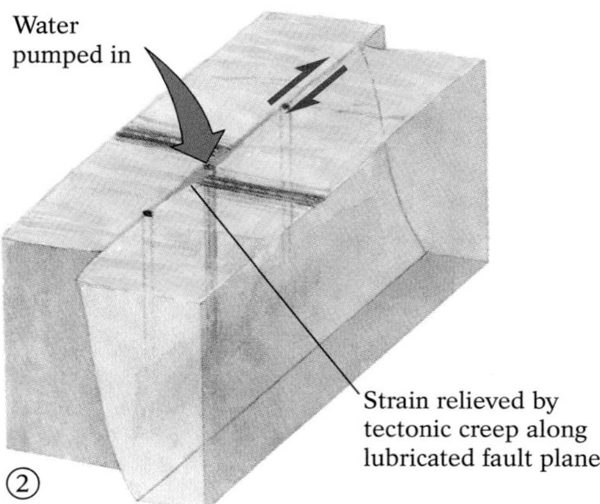

Figure 10-31 One U.S. Geological Survey plan to control future earthquake activity along major transform faults, such as the San Andreas, is to pump water into locked fault segments, facilitating tectonic creep and preventing a major buildup of strain energy.

Can Earthquakes Be Controlled?

Although we cannot slow or stop the motion of the Earth's tectonic plates, we may someday be able to control or prevent earthquakes by reducing the energy buildup along segments of individual faults. The U.S. Geological Survey has drawn up a plan to prevent the San Andreas fault from locking by converting dangerous seismic gaps into zones of less-threatening tectonic creep. Figure 10-31 illustrates the proposed actions. Hypothetically, a stressed segment could be temporarily locked in place by drilling deep wells at each end of the segment (points A and B in panel 1 of Figure 10-31) and withdrawing the lubricating pore water from each end. Then, ever so cautiously, water would be injected into a well in the center of the segment. In theory, this process would lubricate the fault so that its energy would be released gradually by tectonic creep, and only between the locked ends of the controlled segment. By tackling small segments along the entire fault, pumping water into locked areas in turn, it may be possible to dissipate much of the fault's energy before it accumulates to deadly levels.

Is such an earthquake-prevention strategy really plausible? No one knows if injection of water would cause only small displacements. What if the proposed procedure unleashed the accumulated energy of more than 130 years and caused the next great quake in the Los Angeles area? The

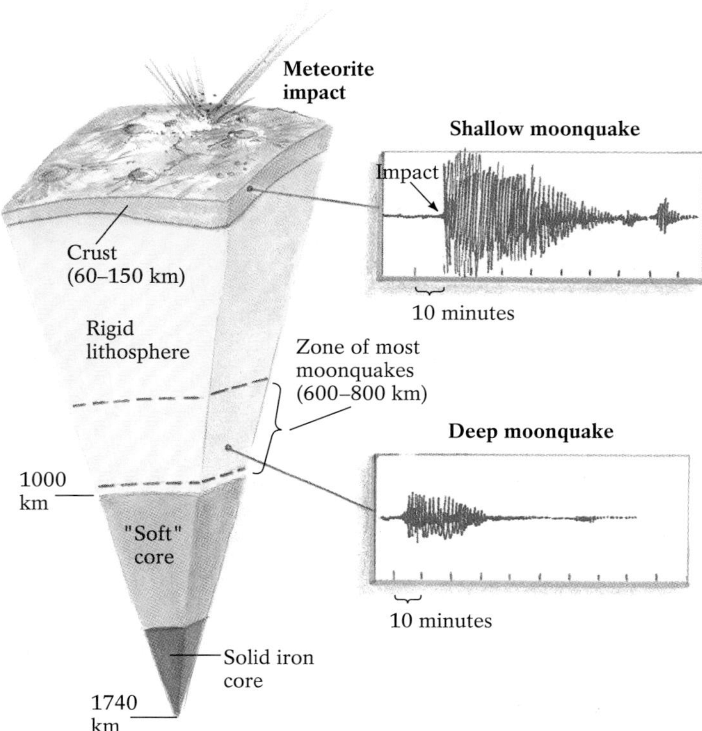

Figure 10-32 The Moon has two centers of seismic activity: one at the surface, where meteorite impacts break rocks and generate small tremors, and one in its mantle, where fractured rocks may be stressed in response to tidal flexing of the Moon as it moves through its elliptical orbit about the Earth.

plan's supporters believe it deserves serious consideration *after* the next San Andreas earthquake, when the injected water would be more likely to unleash only a modest accumulation of energy.

The Moon: A Distant Seismic Outpost

In addition to our burgeoning knowledge of the pattern of terrestrial earthquakes, geologists are compiling data on extraterrestrial quakes. America's Apollo astronauts installed highly sensitive seismometers on the Moon that have been sending data to Earth-bound scientists for more than 20 years. The lunar seismometers are so sensitive that they can detect the impact of a softball-sized meteorite that strikes anywhere on the Moon's surface. Indeed, many of the recorded "moonquakes" are actually meteorite impacts.

Even so, only some 3000 weak moonquakes (magnitude 0.5 to 1.5) are recorded annually, compared with 1 million or more earthquakes that occur each year on this planet. The low magnitude and relative scarcity of moonquakes demonstrate that plate-tectonic processes are inactive on the Moon.

Those moonquakes that are not attributable to meteorite impacts appear to be concentrated in the lunar interior, at depths of 600 to 800 kilometers (400 to 500 miles) (Fig. 10-32). Most deep moonquakes occur when the Moon moves closest to the Earth, suggesting that the Earth's gravitational attraction may induce slight adjustments in deep, fractured lunar rocks, prompting them to release small quantities of strain energy.

This chapter has presented many of the principles underlying the science of seismology. In Chapter 11, we will explore variations in the behavior of seismic body waves, and see what they have taught us about the structure and composition of our planet's otherwise inaccessible interior.

Chapter Summary

Earthquakes are vibrations of the ground that occur when strain energy in rocks is released suddenly, causing rocks to shift along preexisting faults or fractures or to create new faults. The precise subterranean spot where rocks begin to fracture or shift is a quake's **focus,** and the point at the surface directly above the focus is its **epicenter.** A major quake is generally followed by **aftershocks,** tremors that occur as rocks adjust to their new positions.

Earthquake energy is transmitted through the Earth as **seismic waves.** The study of earthquakes is known as **seismology.** The most common seismic waves include **body waves,** which transmit energy through the Earth's interior, and **surface waves,** which transmit energy along the planet's surface. Two types of body waves exist: **P waves,** which cause rocks in their path to be alternately compressed and expanded in a direction parallel to the wave's movement; and **S waves,** which cause rocks to shear and move perpendicularly to the wave's direction.

The intensity of earthquakes, quantified as a function of the damage caused in human terms, is estimated with the **Mercalli intensity scale.** A **seismograph** records the arrival times of seismic waves at a specific location by tracing lines on a **seismogram.** The **Richter scale,** which quantifies earthquake magnitude (a measure of the energy released), is based on the amplitudes of those traced lines. It has recently been largely replaced by the **moment-magnitude scale,** which uses the **seismic moment** (the length of fault rupture and other criteria) to determine earthquake magnitudes. Because seismic waves travel at different velocities, the lag in arrival times between the faster-moving P waves and slower-moving S waves at different (and at least three) seismograph locations can be used to pinpoint the epicenter of an earthquake. Each year hundreds of thousands of earthquakes are measured that have magnitudes of 1.0 to 2.0. Quakes exceeding magnitude 3.5 can be felt by humans, magnitudes of 6.0 denote major quakes, and exceptionally powerful earthquakes occur once every 5 to 10 years that exceed magnitude 8.0.

The effects of high-magnitude earthquakes include large ground shifts along faults, landslides, **liquefaction** (the conversion of formerly stable, fine-grained sediment into a fluid mass), seiches (the back-and-forth sloshing of water in enclosed bays or lakes), large fast-moving sea waves called **tsunami,** and fires.

The world's principal earthquake zones are located at or near plate boundaries, with the deepest quakes generally taking place at subduction zones. The foci of earthquakes at subduction zones exhibit a distinctive pattern—they occur at progressively greater depths inland from ocean trenches. The area in which such a pattern occurs is called a **Benioff-Wadati zone.** The subduction zones surrounding the Pacific Ocean basin (particularly near Japan, Alaska, and the west coasts of North, Central, and South America) and the broad collision zone stretching from the Arabian peninsula to the Himalayas of Asia are the world's most seismically active regions. Powerful earthquakes at such localities as New Madrid, Missouri, and Charleston, South Carolina, indicate that as-yet-unexplained tectonic forces are also at work within plate interiors.

Human attempts to cope with the threat of earthquakes involve limiting quakes caused by humans, assessing local seismic history and future risk, planning land-use carefully, building quake-resistant structures, and, most importantly, developing ways to predict earthquakes. Long-term forecasts, on the order of tens to hundreds of years, focus on the frequency of past earthquake events. They are based on the premise that some areas along an active fault are constantly releasing their energy as they inch along by **tectonic creep,** but other areas within seismic zones remain locked and accumulating a large amount of energy. The latter zones, called **seismic gaps,** are considered to be overdue for an earthquake. Short-term predictions, on the order of months, weeks, or days, center around the phenomenon of **dilatancy,** the volumetric expansion that occurs as minute cracks open in highly stressed rocks. Dilatancy may spawn swarms of earthquakes—called **foreshocks** if they precede a larger quake. It may also produce tilting or bulging of surface rocks; changes in seismic-wave velocities, electrical conductivity, and groundwater levels and chemistry; anomalies in radiowave signals; and unusual behavior in animals. These events are some of the most important signals used to predict earthquakes.

Lunar seismic activity takes the form of "moonquakes," tremors that are much less frequent and smaller in magnitude than our earthquakes. Moonquakes are believed to be produced primarily by meteorite impacts and by the adjustment of lunar rocks to the Earth's changing gravitational pull as the Moon travels in its elliptical orbit around the Earth.

Key Terms

earthquake (p. 268)
focus (p. 268)
epicenter (p. 269)
aftershocks (p. 269)
seismic waves (p. 269)
seismology (p. 269)
body waves (p. 269)
surface waves (p. 269)
P waves (p. 269)
S waves (p. 270)
Mercalli intensity scale (p. 272)
seismograph (p. 273)
seismogram (p. 273)
Richter scale (p. 273)
seismic moment (p. 274)
moment-magnitude scale (p. 274)
liquefaction (p. 277)
tsunami (p. 279)
Benioff-Wadati zone (p. 283)
tectonic creep (p. 294)
seismic gaps (p. 294)
dilatancy (p. 296)
foreshocks (p. 296)

Questions for Review

1. Draw a simple diagram showing the difference between the focus and the epicenter of an earthquake.

2. Describe the differences between P and S waves.

3. What is the fundamental difference between the Mercalli intensity scale and the moment-magnitude scale? How much more powerful than a magnitude 5 earthquake is a magnitude 7 earthquake?

4. Draw a simple sketch to illustrate how geologists use P- and S-wave arrival times to locate the epicenter of an earthquake.

5. Describe how earthquakes cause tsunami and liquefaction.

6. Why are the foci of divergent-zone earthquakes limited to depths of less than 100 kilometers, whereas those of subduction zones extend to 700 kilometers? Why don't earthquakes occur below 700 kilometers? If you heard that an earthquake occurred at a depth of 300 kilometers, what could you conclude about its geological setting?

7. Name three geographical areas that suffer significantly from earthquakes, and specify the plate-tectonic setting in which each is located.

8. Discuss how human activity was responsible for the swarm of earthquakes that struck Denver, Colorado, during the early 1960s.

9. How would you determine whether a given location had experienced earthquakes in the past? How could you predict the probability of an earthquake's occurring there in the future?

10. Before moving into a community, what should you look for to determine if the area is especially vulnerable to earthquake damage?

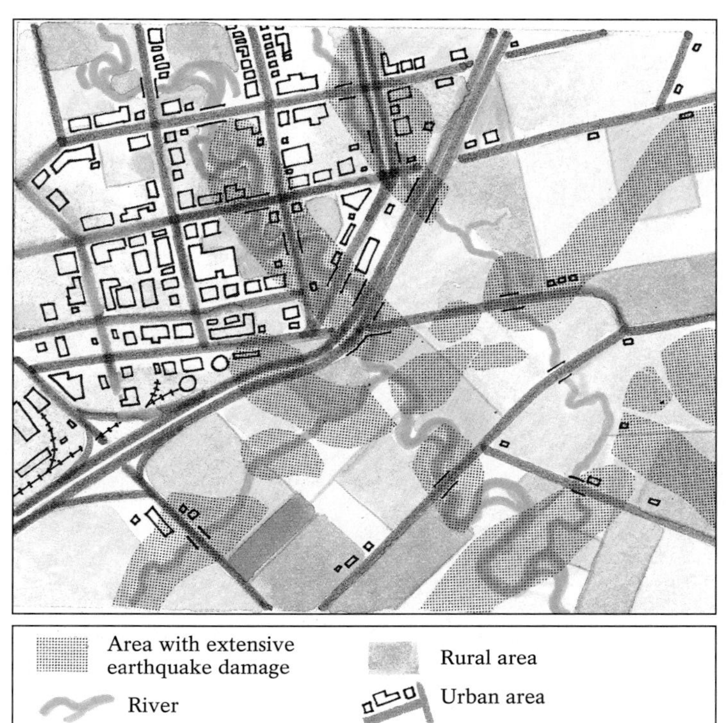

For Further Thought

1. Speculate about the nature of the geologic materials (solid bedrock or loose sediment) found in the map area in the figure at left. Where do you think the major faults lie?

2. If you lived in an area that had suffered no earthquakes during its past recorded history, but which then began to experience occasional tremors, how would you explain the apparently sudden onset of seismicity? Propose at least three hypotheses.

3. What evidence would you need to estimate the magnitude of an earthquake that occurred before the advent of the seismograph?

4. If you were a city planner in an earthquake-prone region and were responsible for coordinating plans for an urban center that will include schools, hospitals, fire and police stations, residential areas, parks, commercial areas, a mass-transit system, power plants, dams, and other facilities, how would you proceed in determining the optimal location of these components of the community?

5. Suppose all earthquake activity on Earth ceased. What set of geological circumstances could explain such a change?

11

Geophysical Properties of Planet Earth

What is the Earth's interior like? In 1864, author Jules Verne, writing *Journey to the Center of the Earth,* depicted the subterranean world as a yawning abyss filled with giant sea serpents and other grotesque creatures. Today we know what really fills the Earth's interior: layers of rocks and metals, subjected to increasing heat and pressure and becoming increasingly dense at progressively deeper levels. For today's scientists, the story of the origin and structure of the Earth's interior is at least as exciting as, if perhaps less fanciful than, Verne's fantastic imaginings.

Humans have lived in space for months at a time, where they used telescopic probes to study both our immediate galactic neighborhood and the distant cosmos. They have walked on the Moon, collecting its rocks and conducting numerous experiments on the lunar surface. They have also dived in submarines to the floors of the Earth's oceans to study the secrets there. The Earth's deep interior, however, remains a frontier that no one is likely ever to visit *in person.* The deepest hole ever drilled (begun in northern Russia in 1970, and still in progress) extends only 0.2% of the distance to the planet's center, piercing about 13 kilometers (8 miles) into the Earth.

Only in rare circumstances do rocks from deep in the Earth appear at or near the surface. Volcanic eruptions that bring up unmelted subterranean rocks within rising deep-origin magmas give us a few indirect glimpses into the makeup of the Earth's interior. Rocks from perhaps as deep as 400 kilometers (250 miles) have been found within diamond-bearing kimberlite pipes (Fig. 11-1), which are products of deep explosive volcanism. [As discussed in Chapter 2, diamonds form when carbon atoms become compressed by the crushing pressures of the Earth's interior, at least 150 kilometers (90 miles) beneath the surface.] The collisions between continental plates have sometimes transported rocks from tens and perhaps hundreds of kilometers deep—even

Figure 11-1 Because they are carried up from deep within the Earth's mantle, diamonds such as this one provide clues to the nature of the planet's interior.

ultramafic rocks from the upper mantle—to the surface along massive reverse faults. Analysis of these relatively few exposures of deep-origin rocks confirms that the composition of the Earth's interior varies with depth, but we are still left with many unanswered questions regarding our planet's internal structure.

This chapter explains how geologists explore the Earth's interior and introduces some remarkable recent discoveries about the deep Earth. We will discuss how geologists use seismic waves to probe the interior, and how they extend the information gained via this method to speculate about the Earth's internal layering. We then turn to some fundamental properties of the Earth: its internal heat and its gravitational and magnetic fields. We describe these properties, explore how variations in the Earth's internal composition and structure affect them, and see how geologists apply their knowledge of these properties to interpret the Earth's evolution. The magnetism in ancient rocks, for example, enables geologists to trace past movements of the Earth's plates. Because study of the fundamental properties of the Earth's interior depends on the physical principles underlying seismic waves, heat flow, gravity, and magnetism, this branch of the geosciences is called **geophysics.**

Investigating the Earth's Interior

Most of what we know about the Earth's interior comes from analysis of variations in the speed of seismic waves. Like all waves, seismic waves tend to travel in a straight line and at an unchanging velocity as long as they continue passing through a homogeneous medium at constant temperature and pressure. Comparisons of data collected at seismograph stations around the world, however, show that seismic waves occasionally slow down or speed up.

These changes in speed indicate that they are passing through materials of varied composition and structure at different temperatures and pressures. From the differing speeds of seismic waves, we conclude that the Earth's interior is not homogeneous, nor do its temperature and pressure remain the same at all depths.

In the next section, we consider the behavior of seismic waves, using that information to build a model of the structure and composition of the Earth's internal layers.

The Behavior of Seismic Waves

In Chapter 10, you learned that P waves compress the materials—whether in solid, liquid, or gaseous form—through which they pass. Different media conduct P waves at different rates. After being compressed by a passing P wave, an elastic medium such as solid granite returns to its original volume more swiftly than does an inelastic medium composed of a liquid or gas. Hence P waves accelerate when they pass through solids of greater elastic strength and slow down when they enter fluids (liquids or gases), which are inelastic.

The inelasticity of fluids also explains why S waves disappear altogether upon entering them. S waves move through material by shearing, a form of deformation in which adjoining material is temporarily displaced, then returns to its original shape after the wave exits (see Chapter 10 for a discussion of deformation). Fluids are not elastic, and thus cannot deform and then return to their original shape—they simply flow away from the shearing stress. Consequently, S waves can pass through only solid media. They die out completely when they encounter an inelastic fluid medium within the Earth's interior. The changing velocities of P and S waves as they pass through media of different properties dictate the paths they take through the Earth's interior and are the basis of many seismological studies.

Because a seismic body wave emanates from an earthquake's focus in all directions, its effect on the surrounding rock takes the shape of an ever-widening sphere traveling out from the focus (see Figure 10-3). The surface of this sphere is known as a *wave front.* Of course, earthquakes emit not just one wave, but a series of waves, so we often illustrate seismic behavior by using a concentric series of wave fronts. The direction of wave-front movement is represented by arrows, or *rays,* drawn perpendicular to the wave front's surface. Visualizing the paths of seismic waves in this way and determining the velocity of the waves as they travel along those paths provide evidence of the composition and structure of the Earth's layers.

When seismic waves move from one medium to another—such as between the major layers of the Earth (the crust, mantle, and core)—their paths suddenly change. If they approach the boundary between the adjoining media at a relatively small angle, they may bounce, or *reflect,* off the surface, much as sunlight reflects off the shimmering surface of a lake. If they approach at a somewhat greater angle, some waves may reflect, while others may enter the new medium. The velocity of the waves in the latter category changes, causing them to bend, or *refract* (Fig. 11-2a).

Seismic-wave velocities also change as waves travel *within* individual layers of the Earth's interior. Seismologists have determined that as seismic waves pass through a layer of the Earth's interior, their velocity increases with depth; as a result, the deeper segment of a wave front travels faster than less-deep segments. This effect causes much of the wave front to curve back toward the Earth's surface (Fig. 11-2b). Using our knowledge that the speed of seismic waves increases with the rigidity of the materials through which they pass, we can infer that any given layer of the Earth becomes more rigid with increasing depth.

By analyzing the great volume of data from seismographic readings around the world, seismologists have estimated the density, thickness, composition, structure, and physical state of each portion of the Earth's interior. This long-distance approach to the study of the Earth's deep

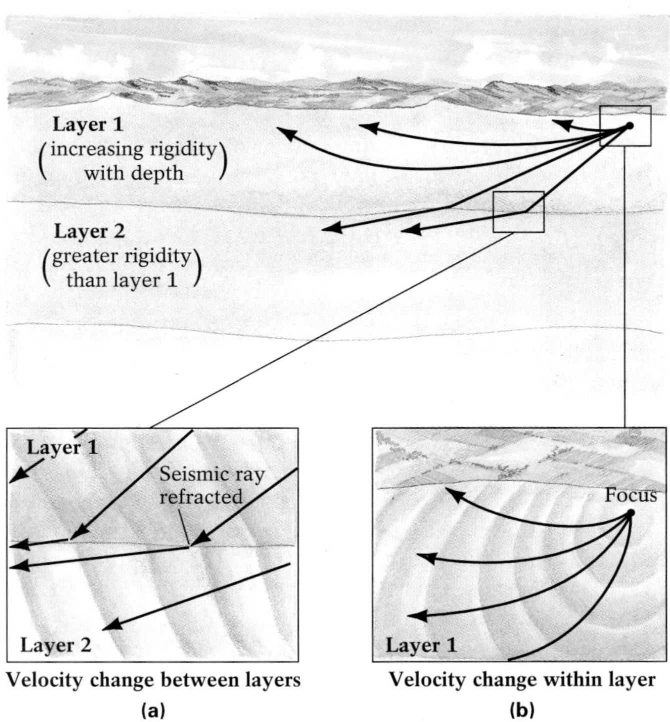

Figure 11-2 The velocity of seismic waves changes as the waves pass through various media in the Earth's interior that have different elasticities. **(a)** Seismic waves speed up or slow down as they pass between two layers, causing the waves to bend, or refract. **(b)** Because rigidity generally increases with increasing depth, seismic waves speed up as they descend within individual layers, causing their paths to curve back toward the surface.

interior has improved dramatically with the advent of seismic tomography—a revolutionary way of looking at the Earth developed during the 1990s. Highlight 11-1 discusses this innovative approach to deep-Earth investigation, which was inspired by a sophisticated medical diagnostic tool.

Based on data produced by seismic tomography and several other new techniques, we have concluded that the Earth's interior is a far more complex place than was once thought. Yes, it contains the three principal concentric layers examined briefly in Chapter 1—a thin outer crust, a large underlying mantle, and a central core. As we shall see, there's a lot more going on down there for scientists to ponder. Figure 11-3 presents an overview of these layers, which we describe in more detail in the next sections.

The Crust

The crust, the Earth's rocky outer segment, is composed principally of silicate-rich igneous rocks. It is the most accessible of the Earth's layers: We can sample it directly from both the continents and the sea floor by drilling. From these investigations, we know that continental crust is typically felsic (average composition—andesite and diorite) with a base of metamorphosed and relatively thick gabbro. On the other hand, oceanic crust is typically mafic (average composition—basalt and gabbro) and relatively thin.

Even more of our knowledge of the crust has come from seismological surveys. Key information is provided by earthquakes that occur within 50 kilometers (30 miles) of the Earth's surface. The direct seismic waves recorded within

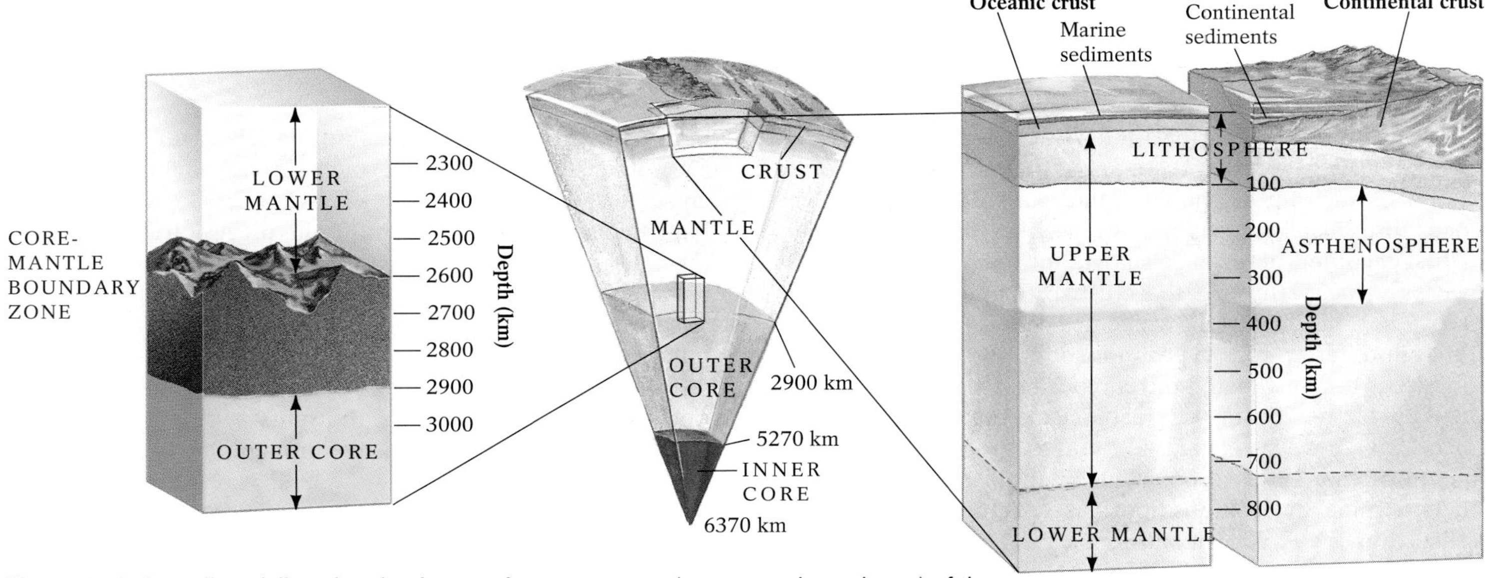

Figure 11-3 As well as delineating the three major components (crust, mantle, and core) of the Earth, seismological studies provide details about the different natures of the layers. For example, the studies reveal that continental crust is significantly thicker and less dense than oceanic crust. These studies also show the rugged topography of the Earth's core-mantle boundary.

Highlight 11-1 Seismic Tomography: A Planetary CAT Scan

How do we study the Earth's interior? To examine the mantle, the mantle–core boundary, and even the core, geophysicists have developed their own type of CAT (computerized *a*xial *t*omography) scan, based in principle on a diagnostic device originally designed for medical purposes. In medical scanning, a patient swallows or is injected with a radioactive (but safe) substance. X-ray–like images are then made from numerous angles along successive planes of a portion of the patient's body. A computer calculates the amount of radiation absorbed by the internal organs at every point along each plane, producing a series of three-dimensional images. The sequence provides a view of every part of every organ within the scanned part of the body.

Inspired by the CAT-scan technique, geophysicists have developed **seismic tomography,** a process that uses seismic waves to provide three-dimensional views of the Earth's interior. Here's how the concept works.

When an earthquake or an underground nuclear blast occurs, seismic body waves race through the Earth's interior on their way to the thousands of seismographs around the world. When seismologists know the precise timing of a tremor, they fully expect its resulting seismic waves to arrive at each of the world's seismographs at a *specific* moment following the quake, based on each seismograph's distance from the seismic event. During the last few decades, however, seismologists have puzzled over seismic waves that seem to arrive measurably "early" or "late." What could be accelerating or impeding the progress of these waves? We now believe that seismic waves slow down when they pass through warmer regions of the interior (where the rocks would be less rigid than surrounding rocks). Conversely, we believe that they speed up as they pass through colder regions (where the rocks would be more rigid than surrounding rocks).

Seismologists using powerful computers have synthesized data from hundreds of thousands of recent earthquakes from around the world to generate a series of cross-sectional views of the Earth. A sequence of such cross sections provides a three-dimensional image of the planet's interior.

Seismic tomography is revolutionizing our view of the Earth's interior. This technique has recently shown, for example, that the outer surface of the Earth's core is not smooth and featureless, as was once believed. Instead, it projects upward into the mantle in some places; in other locations, the mantle extends downward into the liquid outer core. Some of the core's peaks extend higher into the mantle than Mount Everest, the highest mountain at the Earth's surface, and some of its valleys are six times deeper than the Grand Canyon. Seismic tomography has also revealed where heat is concentrated beneath the Earth's surface (such as beneath Hawai'i) and where the interior is anomalously cold (Fig. 11-4a). It has even demonstrated that subducting plates may descend all the way through the mantle to the core–mantle boundary (Fig. 11-4b).

These findings have compelled geologists to reassess their "simple" three-layer model of the Earth. As we shall discuss in Chapter 12, they have also challenged us to rethink our models for the driving mechanisms of plate tectonics.

about 1100 kilometers (680 miles) of the focus pass exclusively through the crust. If we know the precise time and location of such an earthquake, the exact distance from the focus to each seismograph, and the exact time when a wave is first recorded on each seismogram, we can determine the wave's velocity through the crust. By comparing velocities derived in this way to the velocities of seismic waves passed through different rock types in laboratory experiments, geophysicists have estimated the composition of deep, otherwise inaccessible regions of the crust. Seismological measurement of the shock waves generated by underground nuclear testing during the 1960s also provided data that have been used to determine the composition and structure of the Earth's crust; thus the Cold War's arms race did produce some scientific benefit.

When an earthquake occurs in San Francisco, for example, the direct seismic waves that are recorded at Fresno State University in Fresno, California [about 290 kilometers (180 miles) away], have traveled through only continental crust. Because we know the precise timing of some of San Francisco's past earthquakes and the precise arrival times of the P waves recorded in Fresno, we have been able to determine that P waves move through the continental crust between these cities at a velocity of about 6 kilometers (3.7 miles) per second.

Because basaltic oceanic crust is more rigid than continental crust and therefore transmits seismic waves more efficiently, P waves travel faster through oceanic crust. Seismological studies have shown that P waves travel through basalt and its underlying plutonic equivalent, gabbro, at a rate of about 7 kilometers (4.4 miles) per second.

What do these seismic velocities tell us about the composition of continental and oceanic crust? We answer this question next.

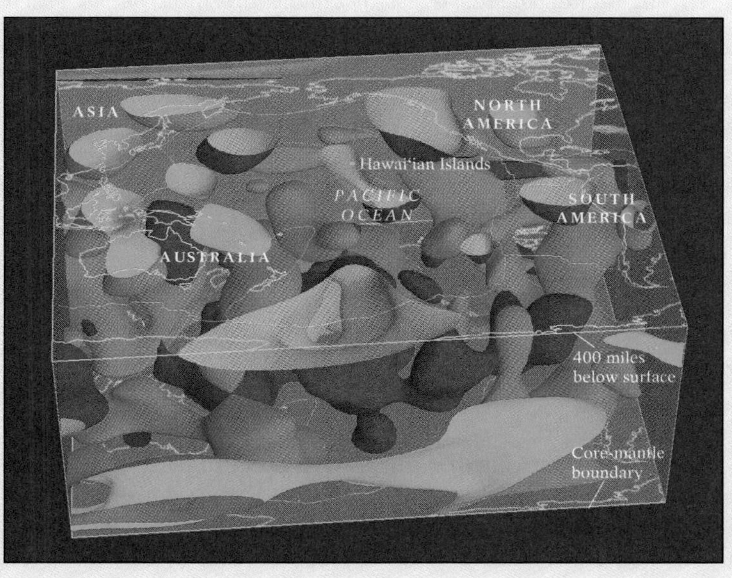

(a)

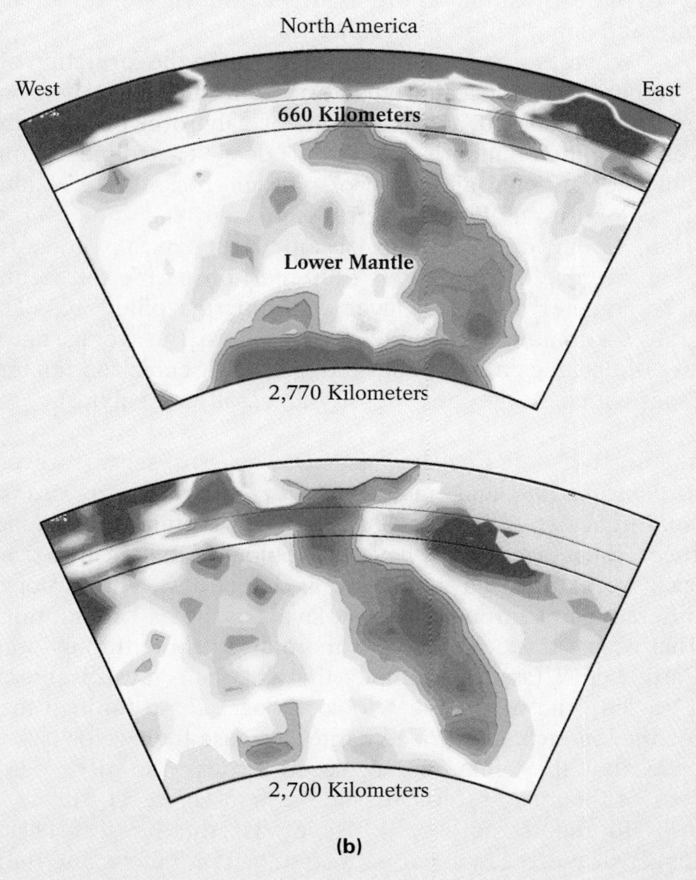

(b)

Figure 11-4 **(a)** This seismic tomographic image of the Earth's mantle shows hot regions (shades of reddish orange) and cold regions (blue). The continents are outlined in white. Note the warm regions in the vicinity of the volcanic Hawai'ian Islands and along the coast of the Pacific Northwest. Geophysicists are analyzing tomographic data to determine the types of relationships that exist among interior heating, tectonic activity and other processes, and surface features. **(b)** Tomographic images of subduction zones. These images indicate that subducting plates—the cold blue regions—descend to the deep mantle.

Continental Crust At its thinnest points, where plates are being stretched or rifted, continental crust is generally less than 20 kilometers (12.5 miles) thick. Where continental plates have collided to form mountains, the crust may span 70 kilometers (45 miles). In North America, the thinnest crust—in the basins between mountain ranges in Nevada, Arizona, and Utah—is only 20 to 30 kilometers (12.5–19 miles) thick. The thickest crust—in the Rocky Mountains of Montana and Alberta and the Sierra Nevada of California—is more than 50 kilometers (30 miles) thick.

Slight variations in seismic-wave velocities indicate that the upper portion of continental crust differs in composition from the lower portion. In upper continental crust, P-wave velocity is about 6 kilometers (3.7 miles) per second; this region comprises a complex jumble of basalt and andesite, numerous granitic intrusions, vast subsurface regions of high-grade metamorphic rocks, and a nearly continuous blanket of sediment and sedimentary rock. When P waves pass through the deepest parts of continental crust, they accelerate to about 7 kilometers (4.4 miles) per second. The increased velocity is believed to reflect either a change in composition to metamorphosed gabbro or a pressure-induced increase in the lower crust's rigidity. The average composition of continental crust, which is characterized by near-surface felsic rocks and deeper mafic rocks, probably matches the composition of andesite or diorite. The density of continental rock varies between 2.7 grams per cubic centimeter (2.7 times as dense as water) near the surface and 3.0 grams per cubic centimeter near its mafic base.

Oceanic Crust Oceanic crust is more difficult to sample than continental crust, because it lies below an average depth of about 4 kilometers (2.5 miles) of seawater. This limitation has been largely overcome by new methods of deep-sea drilling that enable geologists to collect ocean-floor rocks and by seismic studies that help them infer the structure and composition of deeper segments of oceanic crust. Together, these types of investigations afford us an accurate view of the structure and composition of ocean crust.

Whereas many processes contribute to the formation of continental crust, oceanic crust forms almost exclusively from eruptions of basalt and intrusions of gabbro at oceanic divergent plate boundaries, as well as deep-sea sedimentation and hydrothermal alteration of plutonic rocks. As a result, the composition of oceanic crust is virtually the same everywhere. From top to bottom, oceanic crust typically consists of an average of 200 meters (700 feet) of marine sediment, a 2-kilometer (1.2-mile) layer of submarine pillow basalts, and a 6-kilometer (3.7-mile) layer of gabbro. The average density of oceanic crust is about 3.0 grams per cubic centimeter (approximately the same density as basalt and gabbro).

The Crust–Mantle Boundary Often, two sets of seismic waves *of the same type* and emanating *from the same source at the same time* arrive at a single seismograph station *at different times.* Such occurrences have led geologists to conclude that such waves must have traveled at different velocities along different paths through different layers of the Earth's interior. That is, one set took a shorter, more direct route through the Earth's crust (as, for example, the light blue rays in Figure 11-5 did); the other waves took a longer route through the mantle, but actually arrived earlier because their velocity increased as they entered and passed through the more rigid rock of the mantle (the dark blue rays in Figure 11-5).

To understand how a seismic wave that has traveled a longer distance can arrive at a seismograph before one that has moved across a shorter distance, consider the following analogy: a motorist who drives a longer distance on a freeway rather than a shorter distance on a city street characterized by many traffic lights, stop signs, and slower speed limits. Unless snarled by rush-hour traffic, a longer freeway trip at higher speeds will still enable the motorist to reach his or her destination sooner than a shorter, slower journey through downtown traffic would.

In 1910, Croatian seismologist Andrija Mohorovičić attributed these discrepancies in the arrival times of seismic waves to a **seismic discontinuity,** an abrupt change in seismic-wave velocity that reflects a marked change in the composition of materials within the Earth's interior. The boundary between the base of the crust and the top of the more rigid mantle—named for its discoverer—is known as the Mohorovičić discontinuity, or simply the **Moho.**

The Mantle

The mantle, the layer below the crust, constitutes the largest segment of the Earth's interior. Accounting for more than 80% of the planet's volume, it extends to a depth of about 2900 kilometers (1800 miles). At greater depths, as pressure increases, the density of mantle rock increases as well, from about 3.3 grams per cubic centimeter at the Moho to 5.5 grams per cubic centimeter at the mantle–core boundary. Over the long term, the high temperatures in this region cause mantle rocks to deform and actually flow, albeit quite slowly, in convection currents. In the short term, such as the time it takes for seismic waves to pass through them, mantle rocks behave as elastic solids, allowing the passage of both compressional P waves and shearing S waves. Hence, the mantle displays characteristics of both solid and not-so-solid materials.

The minerals in mantle rocks are dense silicates or oxides that contain iron and magnesium, such as olivine and py-

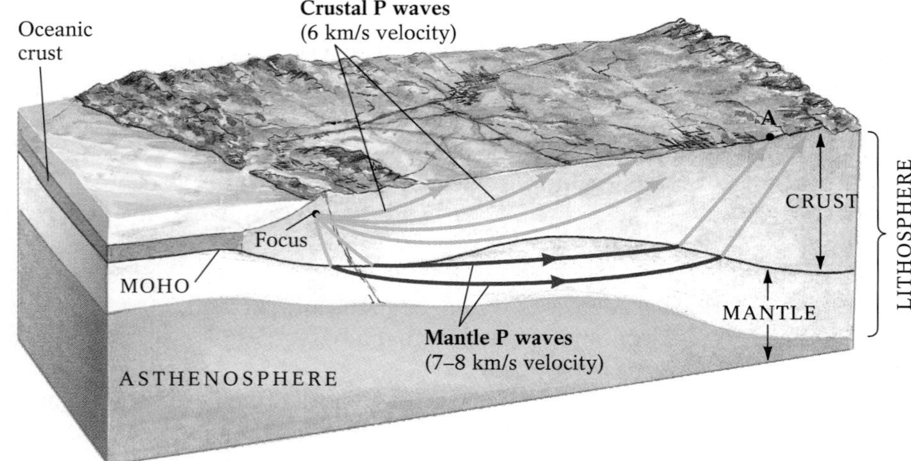

Figure 11-5 A seismic wave that travels through the mantle may arrive at a given location faster than one that travels through only the crust—even though the latter wave travels a shorter distance—because waves move through the mantle more quickly than they do through the crust. Here, the P waves that have traveled through the mantle have already arrived at point A, whereas those that are traveling through only the crust are still en route.

roxene (discussed in Chapter 2). The specific minerals and their internal structures vary as pressure and temperature increase with greater depth. We know these mineralogical variations in the mantle exist because P- and S-wave velocities, which generally increase steadily as the waves penetrate more deeply, change abruptly at certain depths within the mantle. These aberrations in seismic-wave velocity indicate distinct mineralogical changes and mark the boundaries between the sublayers of the mantle: the upper mantle (which contains the asthenosphere), the transition zone, and the lower mantle.

The Upper Mantle Until recently, much of our knowledge about the upper 400 kilometers (250 miles) of the mantle derived from the chunks of it that were incorporated into rising magma or the small amounts exposed in mountain ranges where portions of continental lithosphere collided in the past. In 1993, researchers successfully collected samples of mantle rock *directly* from one of the few places where the mantle lies exposed at the Earth's surface. On the Pacific Ocean floor, west of the South America's Galapagos Islands, the Pacific, Cocos, and Nazca plates are diverging from one another. At the western tip of an east–west spreading ridge between the Nazca plate (heading south) and Cocos plate (heading north) lies the Hess Deep—a rifted valley, 5.1 kilometers (3.2 miles) deep. Within this valley, huge blocks of faulted oceanic lithosphere lie broken and tilted. In January 1993, members of a team aboard the scientific research vessel *Resolution* extracted cores containing mantle rock, the first time the mantle had been pierced directly by drilling.

These mantle-rock samples and other rocks collected from collisional mountain ranges and rising plutons reveal that the upper mantle is dominated by peridotite. Peridotite is a coarse-grained metamorphic rock composed largely of olivine, pyroxene, calcium-rich and magnesium-rich garnet, and other metamorphic minerals that form under high-pressure conditions.

At the Moho, the velocity of descending seismic P waves increases to about 8 kilometers (5 miles) per second as they pass from the crust to the mantle; in contrast, the velocity of P waves continues to increase within the mantle, because its rigidity increases with depth. Seismic-wave velocities, however, do not increase uniformly through the mantle. P waves, for example, accelerate to 8.3 kilometers (5.2 miles) per second by about 100 kilometers (62 miles) beneath the Earth's surface, but then slow to less than 8 kilometers (5 miles) per second. They remain at this lower velocity until they reach a depth of about 350 kilometers (217 miles). This 250-kilometer (155-mile)-thick region—another seismic discontinuity—is known as the **low-velocity zone** (Fig. 11-6). This weak, heat-softened, deformible layer underlies the Earth's plates. S waves also decelerate markedly within the low-velocity zone.

The low-velocity zone is near its melting temperature and, in places, partially melted (most likely containing about 1% liquid). Thus, it is less rigid than the rest of the mantle, and impedes the progress of seismic waves. That the low-velocity zone is the only portion of the mantle containing any liquid reflects the unique interplay of temperature and pressure in this region.

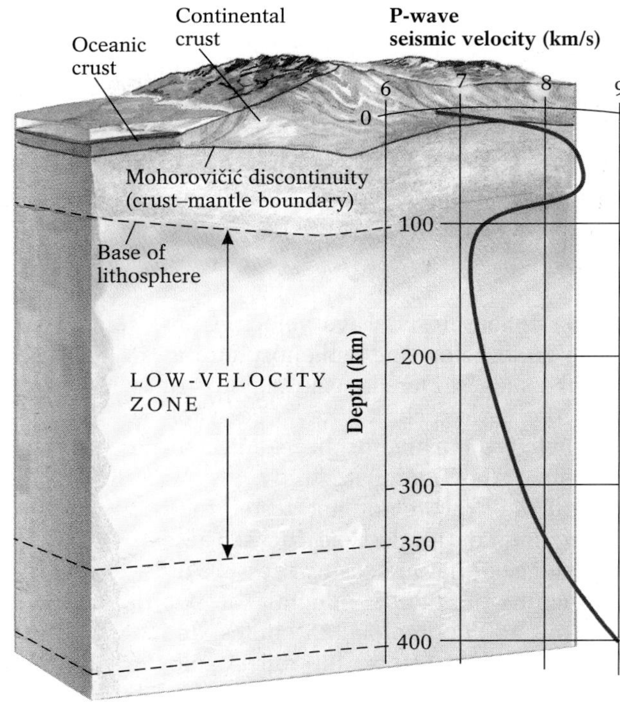

Figure 11-6 The velocities of seismic waves vary across the boundaries of the Earth's interior layers. These seismic discontinuities, such as occur at the Mohorovičić boundary (or Moho) and the low-velocity zone within the upper mantle, result from the changing density, rigidity, and elasticity of the Earth at various depths.

As noted earlier, both temperature and pressure increase with depth in the Earth's interior. Within the low-velocity zone, however, the effect of higher temperatures predominates and mantle rock approaches its melting temperature. Above and below the low-velocity zone, as illustrated in Figure 11-7, the effects of pressure overcome the effects of temperature and maintain the solid state of the mantle. As we discussed briefly in Chapter 1, the crust and the uppermost layer of the mantle, which make up the Earth's plates, constitute the lithosphere. The low-velocity zone beneath the plates constitutes the asthenosphere.

The asthenosphere's most significant effects relate to plate tectonic motion. Many Earth scientists believe that the slow flowing of this low-velocity zone transports the Earth's tectonic plates. (This and other hypotheses proposed to explain plate motion are discussed in Chapter 12.)

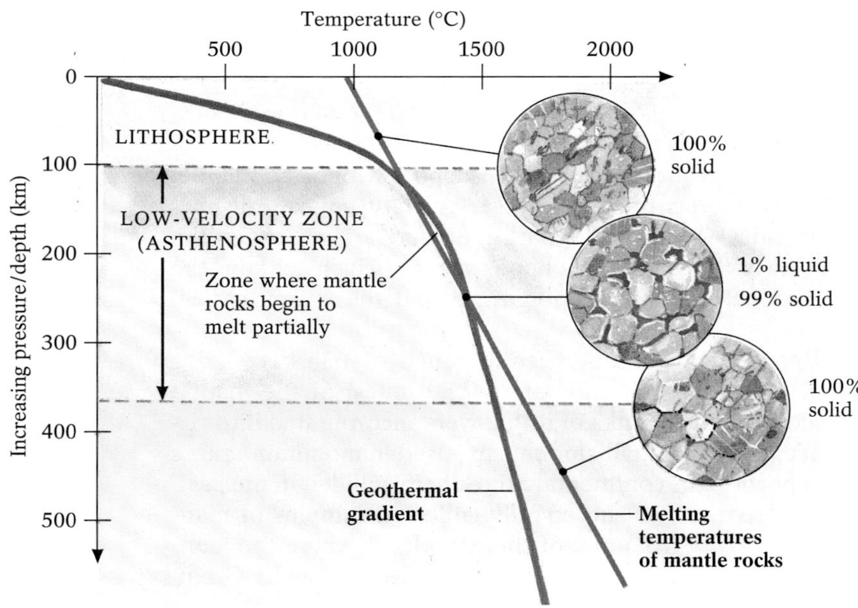

Figure 11-7 The heat-softened, partially melted condition of the low-velocity zone (the asthenosphere) of the upper mantle derives from the temporary dominance of rising temperature over increasing pressure. Above and below this 250-kilometer (155-mile)-thick layer, mantle rock is completely solid.

The Transition Zone Seismic-wave velocities increase in the zone below the asthenosphere as mineralogical changes increase the rigidity of mantle rocks there. No rocks have ever been collected from below the asthenosphere, but laboratory experiments have simulated the pressures and temperatures of this region and tested their effects on ultramafic mantle-type rocks. These experiments suggest that increasing pressure alters the structure of mantle minerals at two points. At about 400 kilometers (250 miles) and again at about 700 kilometers (440 miles) below the Earth's surface, mineral structures become compressed but do not change in composition. In the first phase change, magnesium olivine is compressed to form the structure of the mineral spinel (Fig. 11-8). In the second phase change, spinel and other mantle minerals, such as garnet, are compressed further to form simple metallic oxides such as MgO; the materials become approximately 10% denser and undergo a corresponding increase in rigidity. These effects define the mantle's **transition zone.**

Until recently, most hypotheses about the structure of the mantle suggested that the transition zone was an impenetrable barrier separating the Earth's upper mantle and crust from the deep recesses of the planet's interior. The advent of seismic tomography, however, has begun to change this picture of a two-tiered Earth. Highlight 11-2 discusses how sinking lithospheric plates may periodically breach the barrier between the upper and lower mantle.

The Lower Mantle The lower mantle extends from 700 to 2900 kilometers (from 440 to 1800 miles) below the Earth's surface. Despite extremely high temperatures, the pressure at such depths—produced by the weight of the overlying crust and upper mantle—is great enough to keep lower-mantle materials solid. It also compresses minerals and crystal structures and, with increasing depth, steadily increases the rigidity of lower-mantle rocks. As a result, the velocity of P waves reaches about 13.6 kilometers (8.5 miles) per second at the base of the mantle. Laboratory studies simulating the pressure and temperature conditions of the lower mantle suggest that this region of the Earth probably consists of dense magnesium silicates and oxides, such as *perovskite* ($MgSiO_3$), perhaps the Earth's single most abundant mineral, periclase (MgO).

The Mantle–Core Boundary The boundary between the mantle and the core lies about 2900 kilometers (1800 miles) below the surface, almost halfway to the center of the Earth (or roughly the distance from Boston to Houston). Of all the Earth's internal horizons, the mantle–core boundary—another prominent seismic discontinuity—has the most dramatic effect on seismic waves. In this region, we find the greatest change in the composition and rigidity of Earth materials. At the boundary, the velocity of P waves suddenly decreases from 13.6 kilometers (8.5 miles) per second to 8.1 kilometers (5 miles) per second, and S waves vanish altogether. Because we

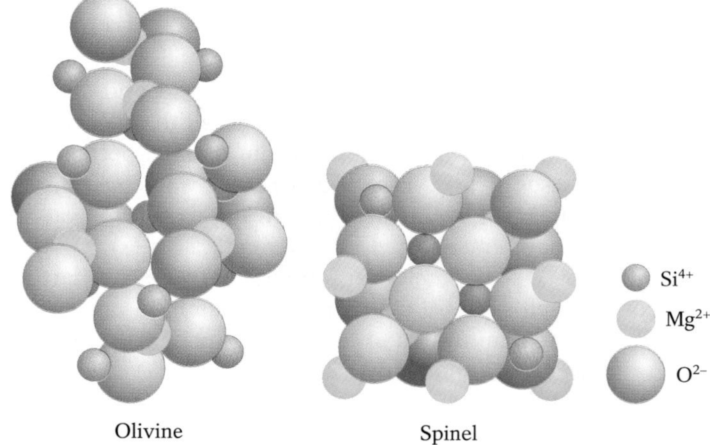

Figure 11-8 Within the mantle's transition zone, increasing pressure causes ultramafic mantle minerals to collapse into denser, more compact structures. In one such phase change, the structure of olivine gives way to that of spinel, another magnesium-rich mineral with a more compressed structure.

Highlight 11-2 The Mantle Flush: Breaching the Barrier Between the Upper and Lower Mantle

Look at the oddly mottled, blue-tinted sphere in Figure 11-9. This image is a supercomputer's hypothetical representation of what happens at the base of the Earth's upper mantle as masses of "cold" subducted lithosphere accumulate there, grown denser by the pressure-induced phase changes of the transition zone.

One feature in particular catches our eye: the blue, near-vertical cylinder in the lower-right quadrant, where a portion of the upper mantle plunges sharply into the deep recesses of the lower mantle toward the Earth's core (depicted here in neon green). Some Earth scientists now believe that every 500 million years or so, great slabs of dense, subducted lithosphere, pooled at the base of the upper mantle, "flush" through the transition zone—once thought to represent an impenetrable barrier between the upper and lower mantle—and then sink toward the Earth's core. According to this hypothesis, such flushes may be partly responsible for the breakup of past supercontinents—such as Pangaea, the most recent conglomeration of continent lithosphere, which began to rift apart about 200 million years ago (discussed in Chapters 1 and 9). According to this hypothesis, when cold rock flushes toward the core, great plumes of hot, buoyant material (perhaps from the core–mantle boundary) rise to take its place. This movement fuels volcanism and divergence, such as that characterizing the rifted margins of disassembled supercontinents such as Pangaea.

To test such computer models (which, of course, are only hypothetical), geologists have studied the transition zone in the "real" mantle, by using seismic tomography for deep-Earth mapping. Researchers at Harvard University, for example, have recently produced tomographic maps that show the vast areas of cold patches of rock pooled above the mantle's transition zone. Researchers at the University of Minnesota have demonstrated that substantial cold patches are also concentrated at the *base* of the lower mantle, immediately above the core–mantle boundary. Logically, if subducting slabs were *continuously* slipping through the transition zone, such cold patches would be distributed uniformly throughout the lower mantle. These two concentrations of cold rock, however, suggest that after a flush occurs, slabs of cold rock descend for millions of years until they reach the core–mantle boundary. There they founder above the denser material of the core and accumulate. Meanwhile, the mantle prepares for its next flush by concentrating dense, subducted slabs above the transition zone until their accumulated mass can eventually push through it.

Figure 11-9 This reconstruction of the Earth's interior results from seismic-tomographic studies. Note the blue, near-vertical cylinder in the lower-right-hand quadrant: this area suggests the presence of a cold slab of lithosphere that has flushed through the mantle's transition zone toward the mantle-core boundary.

Recent tomographic studies have identified several isolated areas where a cold slab appears to be passing through the transition zone to the lower mantle—a sign that such an event is physically possible. Such evidence suggests that subducted slabs may have begun to gather at the base of the upper mantle for another future "flush."

As we shall see in Chapter 12, these deep-mantle churnings may profoundly affect our fundamental understanding of plate-tectonic motion and play a dominant role in proposed supercontinent cycles.

know that P waves slow substantially when they encounter materials of low rigidity and that shearing S waves are not transmitted at all by a liquid medium, we can infer that the outer portion of the core is a liquid. At the mantle–core boundary, highly compressed ultramafic rock with densities of about 5.5 grams per cubic centimeter gives way to a molten iron–nickel mix with densities ranging from 10 to 13 grams per cubic centimeter.

Until recently, geologists thought that the core–mantle boundary, like the other thresholds between the Earth's principal layers, was a distinct, relatively featureless surface. Seismic tomography, however, has given us glimpses into what may be the Earth's most rugged topography—one comprising great *inverted* mountains and deep *inverted* valleys. Today we speak of "anti-continents," consisting of vast slabs of rock that reside at the floor of the mantle. One such continent-sized slab, underlying a portion of Alaska, is roughly 270 kilometers (170 miles) thick and some 1520 kilometers (950 miles) across. Likewise, we now know of "anti-mountains," lofty peaks that hang from the underside of the mantle-like stalactites dripping from the roof of a cave, and "anti-valleys," deep upside-down ravines at the floor of the mantle that lie between the anti-mountains. Using seismic tomography, we have also identified *ultra-low-velocity zones* (ULVZs)—vast heat-softened regions of the lower mantle where seismic velocities drop sharply. Some investigators believe that these zones may be the sources of the great plumes of rising hot mantle material that may be driving the motion of the Earth's plates and fueling the volcanism that occurs at intraplate hot spots, such as the Hawai'ian Islands and Yellowstone National Park.

With each passing year, ongoing refinements of seismic-tomographic studies (and the emergence of more powerful computers) have enabled geophysicists to sharpen the image of the core–mantle boundary, using both natural earthquake waves and occasionally "artificial" ones. On May 21, 1992, a Chinese nuclear detonation at the testing grounds at Lop Nor generated a magnitude 6.5 shock wave that grazed the core–mantle boundary beneath the Arctic Circle near the northwest tip of Alaska. Because we know the timing of this event so precisely, analysis of seismic velocities associated with this event has provided us with the best resolution of the topography of the core–mantle boundary to date. A slab of cold material—128 kilometers (80 miles) thick and 300 kilometers (185 miles) across—hovers above the liquid outer core of the mantle like some sort of "anti-island." More such fascinating discoveries are undoubtedly forthcoming at this horizon, the most remote of the Earth's frontiers.

The Core

The Earth's core, which is slightly larger than the entire planet Mars, accounts for 3486 kilometers (2166 miles) of the planet's total radius of 6370 kilometers (3959 miles). Although it constitutes only one-sixth of the Earth's total volume, the core is so dense that it provides more than one-third of the planet's total mass. Because the core is located at such a great depth, pressures within this region are 3 million times higher than the atmospheric pressure. Core temperatures are believed to exceed 4700°C (8500°F).

The core most likely consists of a mixture of dense metals, such as iron and nickel (at least 90% iron), along with a few other light elements (perhaps sulfur, silicon, and oxygen). Geologists hypothesize that the Earth has an iron–nickel core for several reasons:

1. This composition is consistent with seismic-wave data.
2. The composition of iron meteorites, which are thought to resemble some of the materials that formed the Earth, is predominantly iron.
3. Mathematical analysis of the densities of the whole Earth and of its interior layers predicts a core density similar to that of an iron–nickel mixture (plus some mixed light elements) at core pressures. The average density of the core is estimated to be 10.7 grams per cubic centimeter.

Seismic-wave data suggest that the core has two layers: a liquid outer core 2270 kilometers (1411 miles) thick, surrounding a solid inner core with a diameter of 1216 kilometers (756 miles, or slightly larger than the Moon). Although both layers probably consist of iron and nickel mixtures, the interplay of temperature and pressure within them differs. Heat dominates in the molten outer core, enabling its materials to exceed their melting temperatures; pressure dominates in the solid inner core, keeping its materials below their melting temperatures and thus maintaining them as solids.

Differentiation is also still at work in the core. The entire core may originally have been liquid, but cooling during the last 4 billion years has caused it to solidify into an inner core probably made almost entirely of iron and nickel. Lighter components, perhaps sulfur and silicon, have remained in the molten iron and nickel, forming a liquid outer core where the concentration of elements with lower melting points (such as sulfur and silicon) makes a molten state more likely.

The Liquid Outer Core Geologists monitoring seismic-wave arrival times after earthquakes have discovered that no direct S waves appear within an arc of 154° lying directly across the globe from an earthquake's epicenter. This region, analogous to the shadow cast by an object that intercepts light waves, is called the **S-wave shadow zone** (Fig. 11-10). This seismic shadow must exist because of the S waves' inability to penetrate the intervening materials. As S waves do not travel through a liquid medium, geologists have concluded that a liquid region surrounds the Earth's center, and that its outer boundary is indicated by the border of the S-wave shadow zone.

P waves, which do travel through liquids, cast a different kind of shadow. These seismic waves reach the side of the planet opposite the epicenter of an earthquake, but their arrival is somewhat later than we would expect based on their initial speed; P waves are also missing in the zone between 103° and 143° from the epicenter. These phenomena can be explained by known P-wave behavior at solid–liquid boundaries. When a P wave grazes the mantle–core boundary, it continues on a curved path to the surface, where a seismograph records it at about 103° from the epicenter. When a P wave enters the liquid outer core, it is refracted, emerging 143° or more from the epicenter (Fig. 11-11). The region between 103° and 143° of an epicenter, where no P waves arrive, forms the **P-wave shadow zone.**

Figure 11-10 Because shear waves cannot pass through an inelastic medium, S waves stop when they encounter the Earth's liquid outer core. This obstruction creates the S-wave shadow zone, a broad area on the opposite side of the Earth from an earthquake's epicenter where no direct S waves appear. Note that the position of the shadow zone will vary with the location of the epicenter of the earthquake.

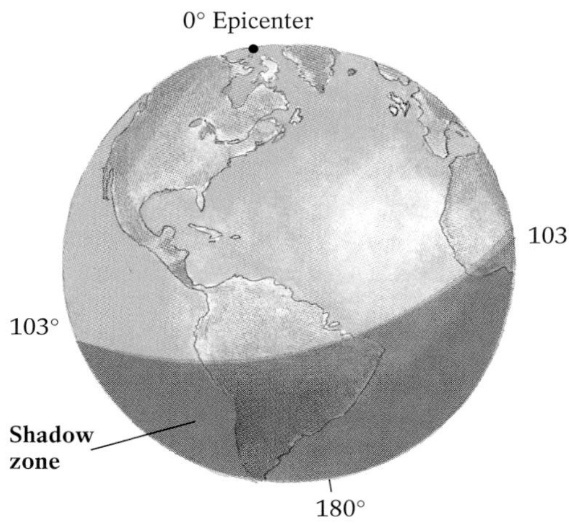

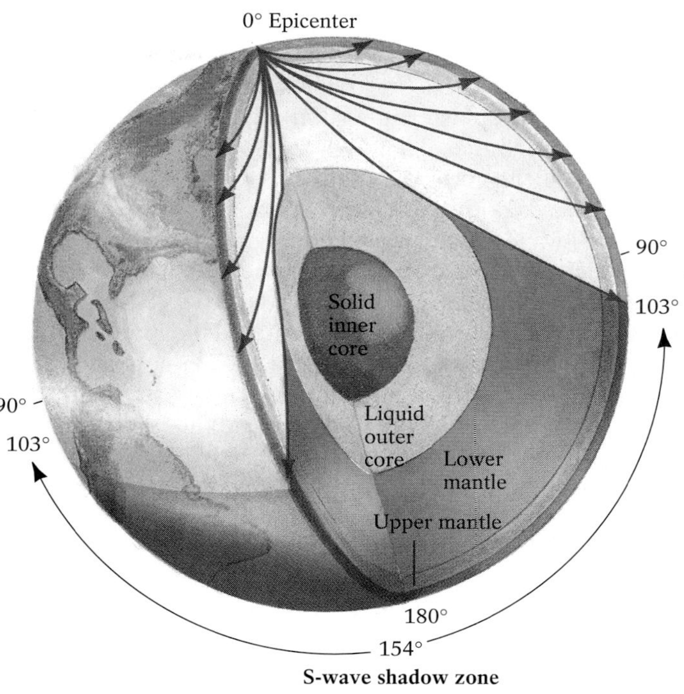

Figure 11-11 Depending on the angle at which they approach the Earth's liquid outer core, P waves either pass through it and refract or reflect off its surface. This development creates a band on the opposite side of the Earth from an earthquake's epicenter in which no direct P waves appear—the P-wave shadow zone. Note that the position of the shadow zone will vary with the location of the epicenter of the earthquake.

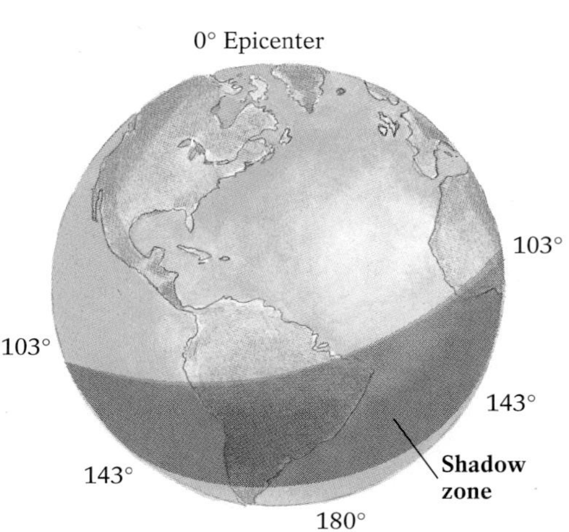

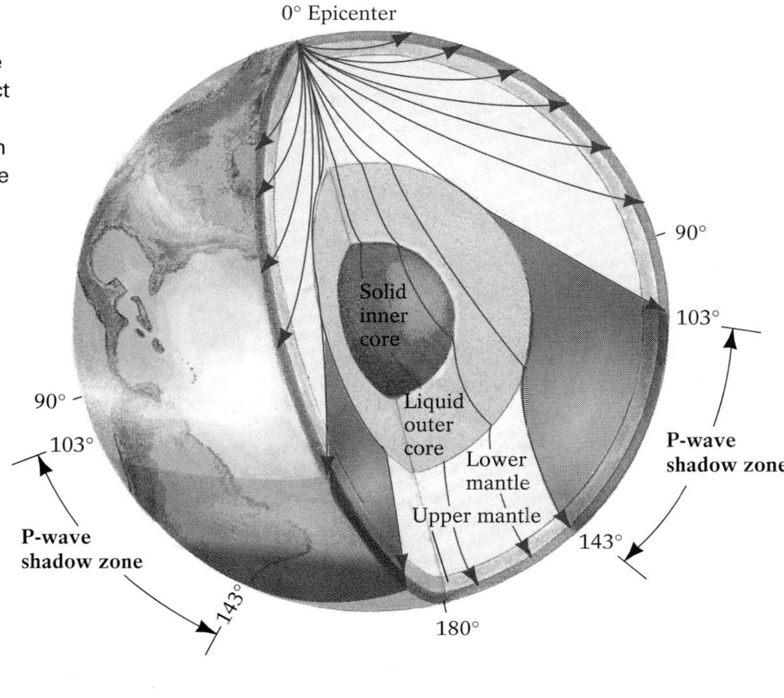

The Solid Inner Core Once they confirmed the existence of the liquid outer core, geophysicists puzzled over another enigma: Why did the P waves that passed directly through the entire planet arrive on the other side *earlier* than expected? P waves travel faster in a solid region, leading geologists to hypothesize that the waves' early arrival indicated that some part of the core was solid.

To test this hypothesis, seismologists analyzed the arrival times of P waves generated by underground nuclear test blasts, for which they knew exactly the time and place of origin. These studies showed conclusively that the velocity of these waves increased abruptly as they entered the inner core, supporting the idea that the inner core is solid. Knowing the velocities of P waves as they pass through the crust, mantle, and outer core made it possible to determine the increase in their velocity as they move through the inner core. The change in velocity, together with the reflection and refraction patterns of P waves, established that the outer core–inner core boundary lies 5100 kilometers (3170 miles) beneath the Earth's surface. Figure 11-12 summarizes how P and S waves behave as they pass through the Earth's interior.

After they demonstrated the existence of a solid inner core, geophysicists began to speculate about how its immersion within the surrounding liquid core might affect its motion. Recently, these scientists have discovered something quite remarkable: Comparisons of earthquake waves that pass directly through the inner core to those that just barely miss the inner core indicate that the inner core spins faster than the surrounding layers of the Earth. Geophysicists studied dozens of records from 1967 to 1995 of seismic waves that traveled between quake epicenters in the South Atlantic's South Sandwich Islands and seismographs at College, Alaska. They observed that seismic-wave arrival times become shorter for waves that travel directly through the inner core, whereas those that miss the inner core arrive at precisely the same time as they have in the past. From their calculations, these scientists have concluded that the inner core spins about 1° faster than the rest of the Earth. Figure 11-13 illustrates the extra degree of inner-core rotation that occurred between 1900 and 1996. Using this rate of additional rotation, we can extrapolate that the inner core would experience a full extra rotation every 400 years or so.

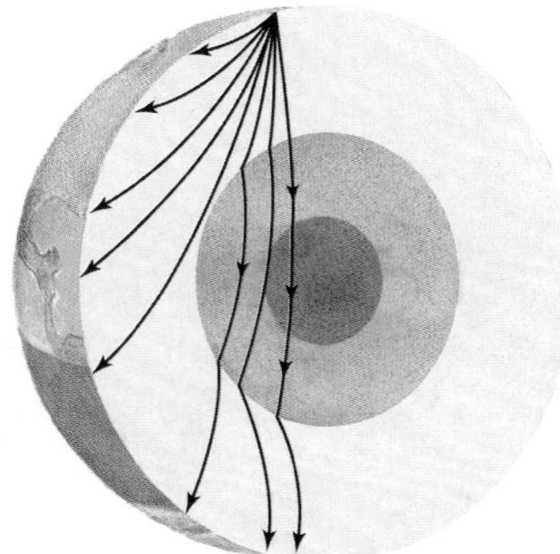

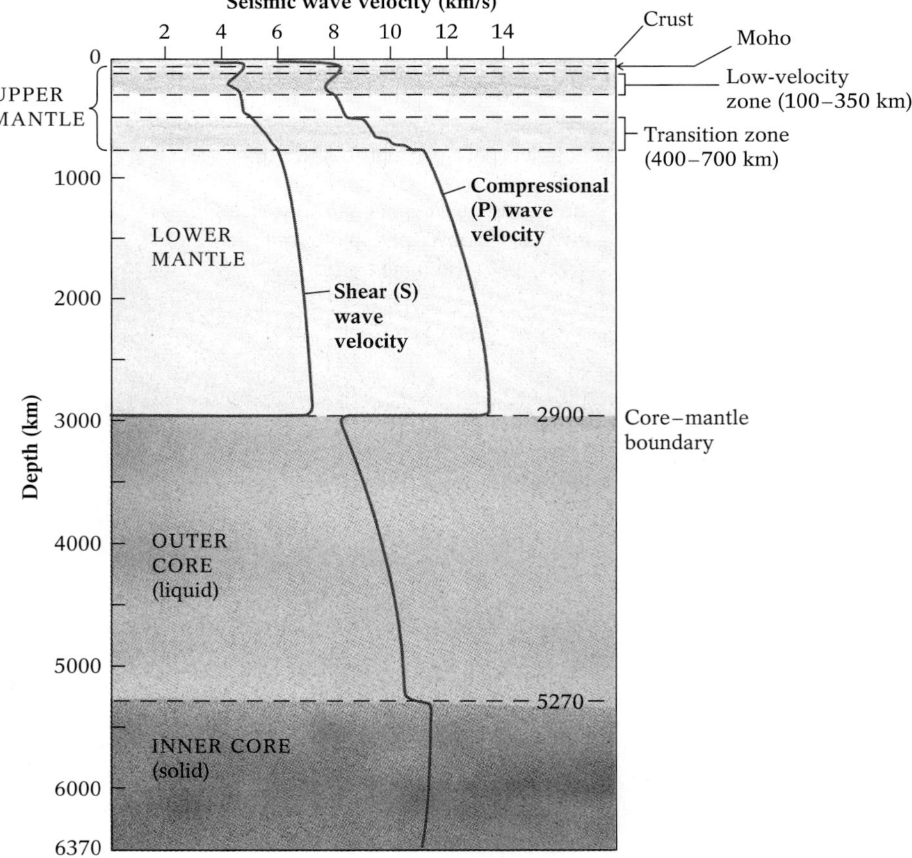

Figure 11-12 The behavior of P and S waves as they pass through the Earth's interior layers. P waves accelerate as they cross the crust–mantle boundary (the Moho), slow down as they enter the less rigid low-velocity zone (or asthenosphere), speed up again as they approach and pass through the lower mantle, slow precipitously as they enter the less rigid liquid outer core, and speed up once more as they pass through the solid inner core. Like P waves, S waves accelerate as they cross the crust–mantle boundary and slow down dramatically as they enter the asthenosphere; these waves speed up again as they enter the lower mantle, but stop altogether when they reach the liquid outer core.

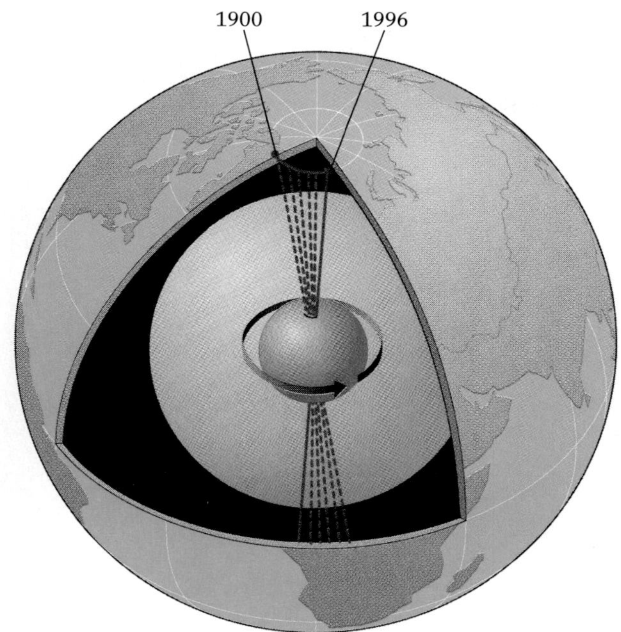

Figure 11-13 The solid inner core (in red) spins independently and faster than the rest of the Earth. Between 1900 and 1996, the inner core rotated an extra one-quarter revolution compared with the other interior layers.

Apparently the liquid surrounding the solid inner core enables this region to spin unfettered, like a beach ball spinning freely in a bathtub. Some geophysicists believe that the combination of the swirling currents in the flowing liquid core and the pull of the outer core's magnetic field tug at the inner core, twirling it faster than the rest of the planet. Much more study is needed to explain fully why the Moon-sized solid inner core spins solo, somewhat like a planet within a planet.

The Earth exhibits some fundamental properties that we can explain with our knowledge of its internal layers. In the sections that follow, we discuss the planet's internal temperatures and the origin of its gravitational and magnetic fields. The study of these physical properties provides us with additional insight into the Earth's internal structure and composition.

Taking the Earth's Temperature

We can estimate the pressure at any depth in the Earth's interior by totaling the pressures applied by each overlying layer. No similar correlation exists between depth and temperature, however. To estimate interior temperatures, we must apply logic to correlate our few direct observations with experimentally derived data and theoretical models.

Direct evidence of the Earth's interior temperatures is, of course, quite limited. Gold miners in South Africa's Western Deep Levels, near Johannesburg, report that conditions become hotter as they descend. Air-conditioning is required to offset temperatures of 40°C (104°F) at the bottom of the mine. As you will recall from Chapter 3, the *geothermal gradient* is the rate at which the Earth's internal temperature increases with depth. Near the surface, the geothermal gradient is about 30°C (86°F) per kilometer, meaning that the temperature at 1 kilometer beneath your geology classroom would be 30°C warmer than the temperature at the base of the building. If the Earth's internal temperature continued to increase at that rate, mantle rock would melt at depths below 100 kilometers (60 miles). From seismological studies, however, we know that the mantle remains a solid at that depth. Two principal reasons explain this behavior. First, the pressure at such depth inhibits melting (see Chapter 3). Second, the high concentration of radioactive isotopes in crustal rocks adds heat at and near the surface; lower concentrations of these isotopes found deeper in the Earth's interior reduce the geothermal gradient.

Although the deep inner Earth may contain fewer radioactive elements and thus has less radioactive decay available for use as a heat source, it maintains a high temperature through other means—the residual heat produced in the Earth's earliest stages of formation, and the heat released as the liquid outer core solidifies.

Determining Interior Temperatures

To remain solid at any given depth, a region's temperature must be below the melting points of its constituent materials at the prevailing pressure. Knowing the composition of a solid portion of the Earth's interior and the pressure at that depth, we can determine the melting temperature of each of its components and thus derive the theoretical maximum temperature possible at that depth (without melting). For example, we know that the crust and upper mantle remain solid down to the low-velocity zone, and we have a good idea of the mineral components of these regions. Using this information, we can determine the maximum temperature of each of these layers at the pressures expected as these depths.

The low-velocity zone is near its melting temperature, and in some places has partially melted; hence its temperature at those locations must exceed the melting points of some of its mineral combinations. For 2500 kilometers (1500 miles) below this zone, down to the mantle–core boundary, the mantle is solid. This knowledge suggests that rising pressures in this zone result in high melting points for its component materials—higher than the zone's slowly rising temperatures. The same logic suggests that the liquid outer core's temperatures must be higher than the melting points

Highlight 11-3 Simulating the Earth's Core in the Laboratory

Until the last few decades, laboratory simulations of the Earth's internal heat and pressure could recreate conditions only to depths of 300 kilometers (200 miles). The jaws of a hydraulic press could produce pressures about 100,000 times atmospheric pressure—only 1/30 of the estimated pressure at the core. A great advance came in the 1960s with the introduction of the diamond-anvil high-pressure cell. This device squeezes a sample substance between two diamonds and heats it with lasers (Fig. 11-14). It can produce pressures equivalent to 3 million times atmospheric pressure and temperatures as high as 3000°C (5400°F). Although these conditions correspond to those found in the lower mantle, they still fall far short of the estimated conditions at the core.

In 1987, using further technological advances and considerable ingenuity, several laboratories successfully simulated the inner core pressures, which are estimated at 3 to 4 million times atmospheric pressure. By detonating chemical explosives wrapped around rocks that had been fitted with internal electrical sensors, scientists at the University of California at Berkeley, the California Institute of Technology, and the University of Illinois generated shock waves that compressed the rocks to within the range of core pressures. Sensors recorded various properties of the compressed rocks in the few millionths of a second before the shock waves destroyed them.

In another recent experiment, metal and plastic "bullets" were shot into a thin film of iron at a velocity of 27,000 kilometers (17,000 miles) per hour. This study simulated the application of core-level pressures to the materials that most likely make up the core. The results of these pioneering experiments yielded new data about the behavior of Earth materials at the core's extremely high pressures. Based on the new information, scientists even revised their melting-point estimates of these materials. Previous estimates of melting points at the mantle–outer core and outer core–inner core boundaries, and hence of temperatures in those regions, had been significantly lower.

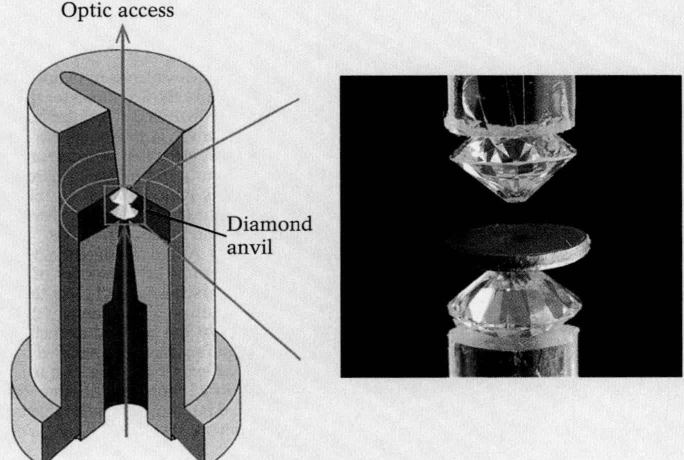

Figure 11-14 A diamond-anvil high-pressure cell, a device that simulates conditions in the Earth's lower mantle by subjecting samples to extreme pressure and heat.

of its component materials at prevailing pressures. From inferences of the outer core's composition and from laboratory experiments, geologists have estimated that its temperature is 4800°C (8650°F) or higher. In the solid inner core, the effects of pressure and composition cause the material to solidify, keeping the melting points of its materials higher than the prevailing temperatures there. From inferences of the inner core's composition and estimates of its components' melting points at the extremely high prevailing pressures, geologists have estimated the temperature at the inner core–outer core boundary to be about 7600°C (13,700°F). The center of the core must be somewhat hotter.

In recent years, geologists have developed new experimental methods for estimating the extreme temperature and pressure conditions of both core regions ever more accurately. Highlight 11-3 describes some of these recent technological advances.

Heat at the Earth's Surface

The amount of internal heat near the Earth's surface varies significantly from one place to another. Because uranium, potassium, and thorium isotopes—the primary sources of radioactivity on Earth—are generally associated with granitic rocks, radioactive decay produces less heat in areas where granite is absent. Oceanic crust is mafic (composed largely of basalt and gabbro) and thus contains virtually no granitic rocks; geologists had long assumed that oceanic rocks were cooler than their granitic continental counterparts. Thousands of temperature measurements taken via deep-sea probes, however, show that in some places even more heat rises from oceanic crust than from continental crust. Along divergent plate boundaries and at intra-oceanic hot spots, for example, rising thermal plumes convect internal heat from great depths beneath the oceans, perhaps from as deep as the mantle–core boundary. Most of the heat of the oceanic crust, therefore, comprises residual heat from the formation of oceanic lithosphere at divergent plate boundaries. This heat does not result from near-surface radioactive decay, unlike in continental crust.

Is the Earth Still Cooling?

The Earth has been cooling continually since its origin 4.6 billion years ago. Like all heat, the Earth's internal heat always flows from warmer to cooler areas. In the Earth's interior this transfer is accomplished by convection and conduction, processes that continually transport heat from deep within the Earth to the planet's surface (or at least near it). The heat then dissipates in the atmosphere, contributing to the planet's cooling. Because of the Earth's substantial size and the extremely low heat conductivity of rock, this cooling process has progressed quite slowly throughout Earth history and continues today. Heat from the origin of the Earth is still making its way to the planet's surface.

The overall pattern of terrestrial cooling, however, is not a simple one. Although the Earth continues to cool, several processes counteract this effect by producing a substantial amount of heat. The decay of radioactive isotopes, for example, has been giving off heat since the Earth's formation. Although the abundance of radioactive isotopes is continually decreasing and the amount of heat they produce is diminishing accordingly, these materials remain a significant source of heat in the Earth's continental crust and mantle. In addition, ongoing solidification of the core releases heat deep within the planet. Friction produced by the motion of the Earth's plates, particularly at subduction zones, also serves as a local heat source. Nevertheless, although additional heat is still being produced, the *net* effect of the Earth's cooling and heating processes is that the Earth continues to cool over time.

The Earth's Gravity

The Earth hurtles through space at a speed of approximately 107,000 kilometers (66,500 miles) per hour on its annual journey around the Sun, and a point on the equator travels around the spinning Earth's axis at a velocity of 1700 kilometers (1060 miles) per hour. Why, then, don't you fly off the chair on which you're sitting? **Gravity,** the force of attraction that all objects exert on one another, pulls you and everything around you toward the Earth's center. This force accounts for every object's *weight* at the Earth's surface. The force of gravity acts in direct proportion to the product of the masses of the attracted objects, and is inversely proportional to the square of the distance from the center of one mass to the center of the other, according to the following expression (the symbol $\propto$ means "is proportional to"):

$$\text{Force of gravity} \propto \frac{\text{mass of A} \times \text{mass of B}}{\text{distance}^2}$$

The Earth's enormous mass provides the strong gravitational attraction that affects everything on or near it, and this force greatly overpowers the attraction between smaller objects in its vicinity (Fig. 11-15). Because the force of gravity decreases sharply as the distance between objects increases, a meteor hurtling through space close to the Earth may be drawn to our planet by a greater gravitational force than that exerted by the much larger, but more distant, Sun.

Gravity measurements help us explore the Earth's interior by giving us clues about what may lie below various features on the planet's surface. To use gravity as a tool to explore the Earth's interior, we must measure this force at various locations and assess our results.

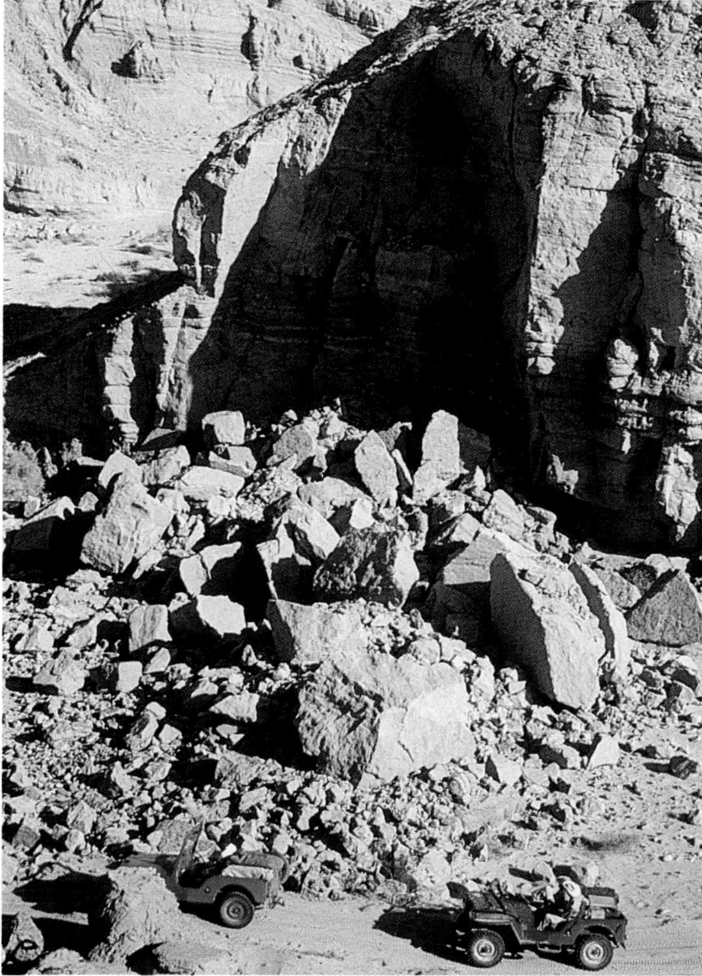

Figure 11-15 Because the mass of the Earth is so much greater than that of any object on its surface, all objects—such as these fallen rocks—are attracted by gravity toward the Earth's center.

Measuring Gravity

A *gravimeter* measures variations in the Earth's gravity. This device contains a mass suspended from a spring so sensitive that it registers even minute differences in gravitational attraction to the Earth's center. For example, a gravimeter placed on a thin rug, instead of directly on the floor, would record a reduction in gravitational attraction because of the *increased* distance to the Earth's center.

Thousands of gravimetric measurements have been taken everywhere from mountain tops to ocean floors. They show that the Earth's gravity varies significantly from place to place and from high elevations to low ones (Fig. 11-16). The pull of gravity is stronger at the poles than at the equator, because the rotation of the Earth makes the planet bulge at the equator and flatten out at the poles; as a result of this bulge, the equator lies 21 kilometers (13 miles) farther from the center of the Earth than do the poles. The pull of gravity is also somewhat lower at the top of a high mountain than at sea level, reflecting the peak's greater distance from the Earth's center. This difference explains why a person would weigh a bit less when standing on top of a high mountain summit near the equator than when standing at sea level near the North or South Pole.

Gravity Anomalies

Even after we adjust them for the effects of altitude and latitude, gravity readings are not the same everywhere on Earth. The remaining differences between actual gravimetric measurements and the expected theoretical values are called **gravity anomalies.** Gravity anomalies result primarily from local variations in the density of Earth materials that, in turn, affect the Earth's mass at a given location and consequently its gravitational pull.

Negative gravity anomalies—attractions lower than the theoretically expected value—arise where relatively low-density rocks or sediments below the surface lie amid denser surrounding rock. They often occur in continental mountain ranges, which generally have deep roots of low-density felsic rock embedded in the higher-density rocks of the upper mantle, or in regions where denser mantle rocks surround subducted basaltic lithosphere. Negative anomalies also occur along oceanic trenches, which are typically filled with water and low-density sediment, and therefore register a lower gravitational attraction than the denser oceanic crust that surrounds them.

Analysis of negative gravity anomalies can have practical applications as well. Along the Gulf Coast of Louisiana

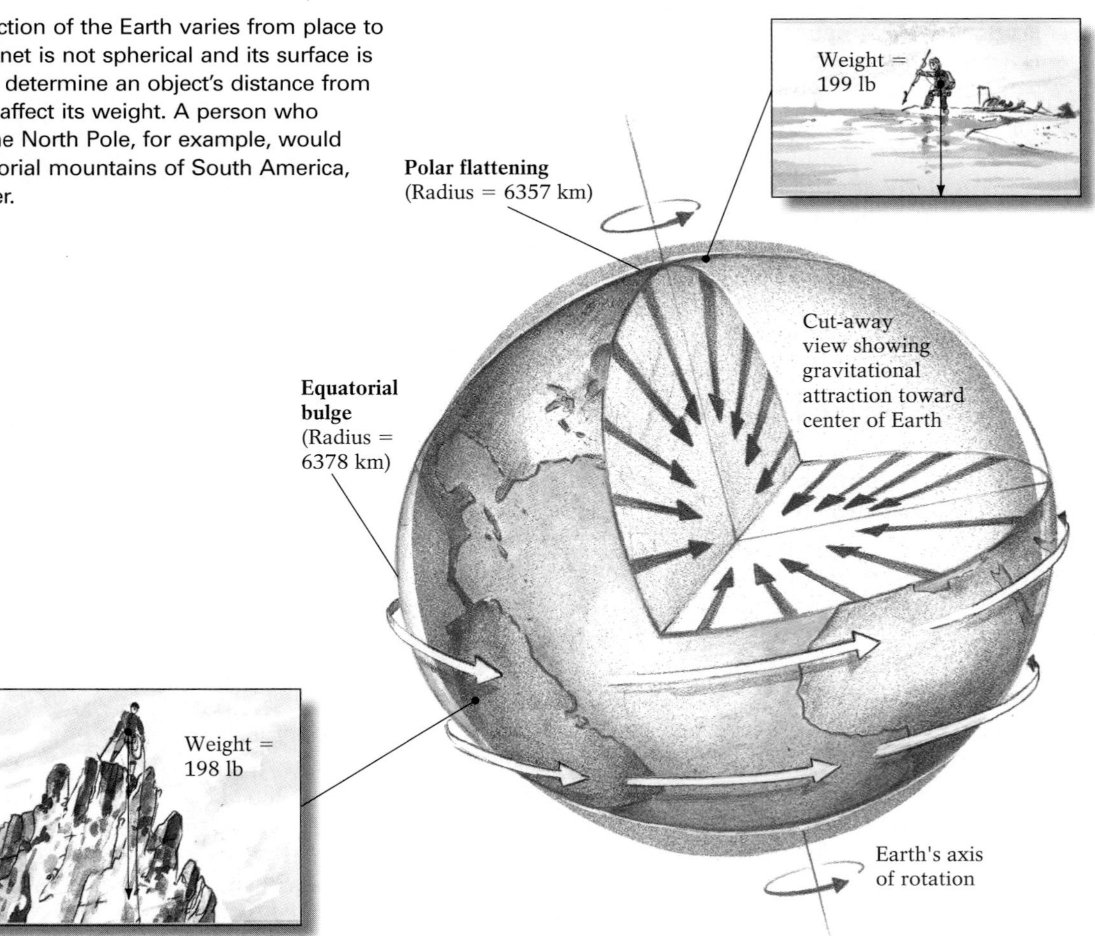

Figure 11-16 The gravitational attraction of the Earth varies from place to place on its surface. Because the planet is not spherical and its surface is irregular, both latitude and elevation determine an object's distance from the Earth's center and consequently affect its weight. A person who weighs 199 pounds at sea level at the North Pole, for example, would weigh only 198 pounds in the equatorial mountains of South America, where the gravitational pull is weaker.

and Texas, anomalously low gravity signals the presence of subsurface low-density salt deposits, which were originally produced by evaporation of ancient oceans and are now surrounded by denser sedimentary rocks (Fig. 11-17). As we shall see in Chapter 20, such impermeable salt deposits often trap migrating crude oil. As a result, gravity studies are sometimes employed to search for oil.

Positive gravity anomalies—attractions higher than the theoretically expected value—result from the presence of high-density rocks below the surface, often where the Earth's crust is thinnest. These anomalies often occur over ocean basins, for example, where high-density mantle rocks lie relatively close to the surface below thin oceanic crust. They also arise where continental crust has thinned substantially, perhaps as a prelude to plate rifting. One such positive anomaly stretches southward along the Minnesota–Wisconsin border from Lake Superior through Iowa and Kansas to Oklahoma (Fig. 11-18a). Known as the "mid-continental gravity high," this anomaly marks the location of an ancient rift in the North American plate into which basaltic magmas rose 1.1 billion years ago. The rifting stopped before the continent divided completely, leaving mid-continental basaltic flows at the surface. Today high gravity readings record the

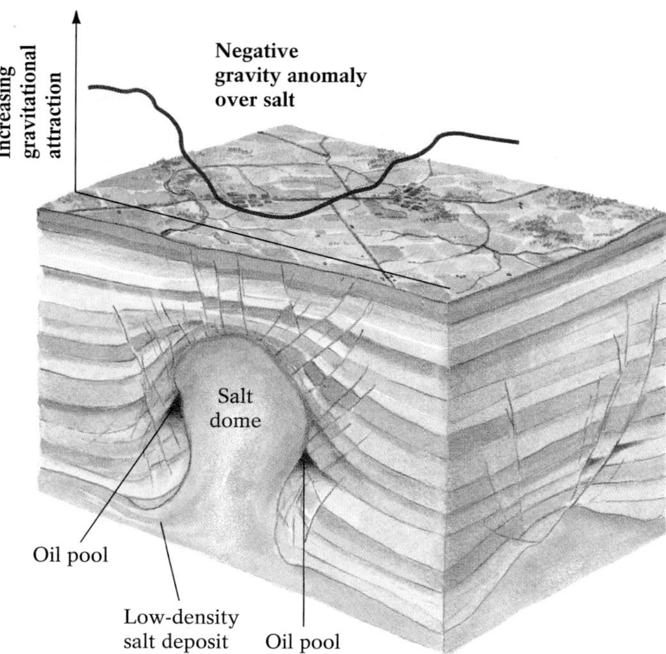

Figure 11-17 A negative gravity anomaly may be caused by a concentration of low-density salt below the surface.

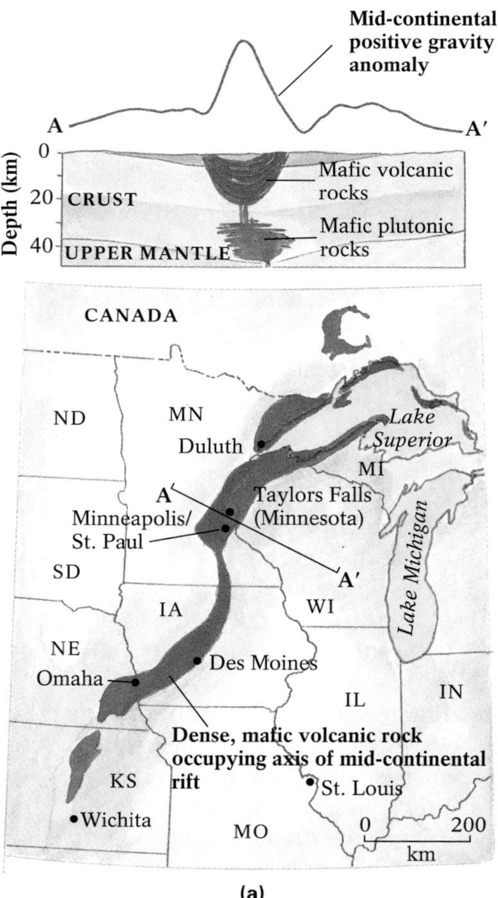

(a)

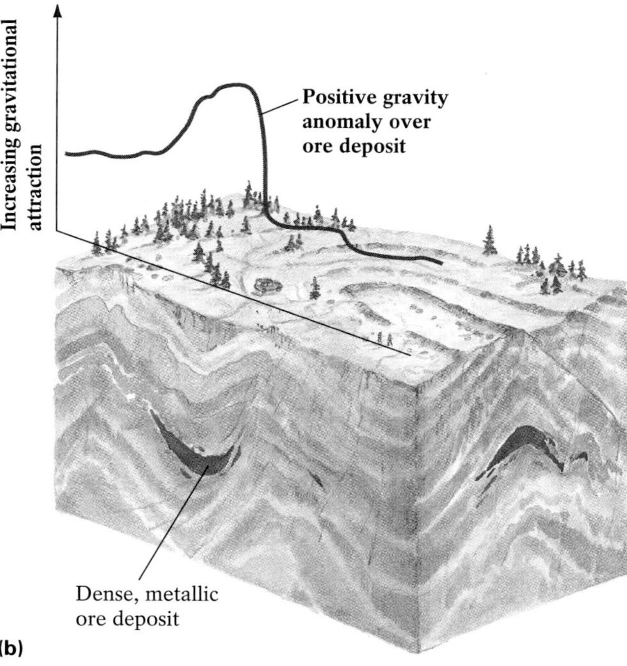

(b)

Figure 11-18 Positive gravity anomalies are caused by the presence of high-density rocks near the surface. **(a)** The mid-continental gravity high, a nearly continent-long positive gravity anomaly found in North America, probably formed when the North American plate began to rift 1.1 billion years ago and a body of mantle-derived mafic igneous rock intruded into the continental rocks of the Midwest. The surface expression of this rifting event can be seen in the basaltic flows at Taylors Falls, Minnesota, along the Minnesota–Wisconsin border. **(b)** A positive gravity anomaly may also be caused by a near-surface concentration of dense metallic ore.

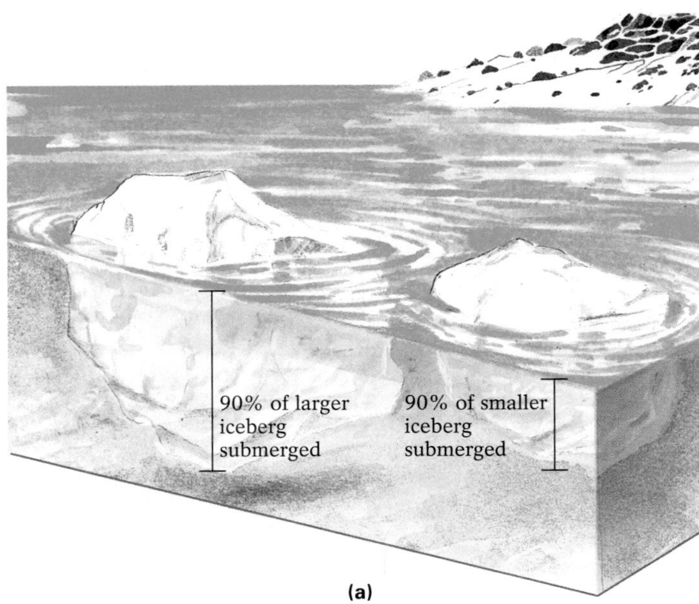

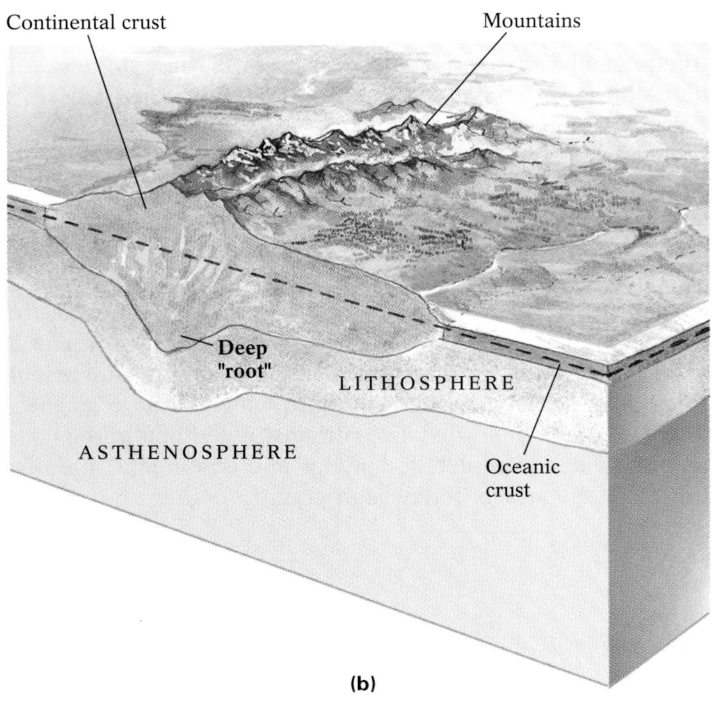

Figure 11-19 The principle of isostasy states that the depth to which a floating object sinks into underlying material depends on the object's density and thickness. **(a)** All floating blocks of ice have the same density; thus they sink in water so that the same proportion of their volume (90%) becomes submerged. The thicker the block of ice, the greater this volume will be. **(b)** Continental lithosphere, because it is of lower density than oceanic lithosphere, is more buoyant and has a larger proportion of its volume above the asthenosphere. Because it is so thick compared with oceanic lithosphere, however, continental lithosphere extends farther into the asthenosphere (that is, it has a deep "root").

presence of a shallow, 50-kilometer (30-mile)-wide zone of dense subsurface rock.

Not all positive anomalies are associated with thinning crust. Some may occur where masses of ultramafic mantle rocks become caught at or near the surface between colliding plates and form part of such continental mountains as the Alps and Appalachians. Higher-than-expected gravity readings may also indicate concentrations of dense metallic ore in the crust (Fig. 11-18b).

Isostasy

Gravity has another effect on the outer layers of the Earth that explains, in large part, why some regions of the lithosphere "ride high" (such as the continents), whereas others "sink low" (such as the oceans). The Earth's lithosphere "floats" on the underlying denser, heat-softened, partially melted asthenosphere of the upper mantle. Because areas of the lithosphere vary significantly in terms of their density and thickness, gravity pulls some of this layer's segments (those with more mass) down farther into the asthenosphere. In addition, because the weight and dimensions of any given segment change over time, that portion's position relative to the mantle changes over time. This equilibrium between lithospheric segments and the asthenosphere beneath them is called **isostasy.**

When an iceberg breaks off a coastal glacier, it may initially plunge below the surface of an adjacent bay. It immediately bobs back up, however. The iceberg is buoyed upward, because ice is less dense than water. Isostasy explains why the iceberg adjusts its position until it displaces a volume of water equal to its own total weight. Water is 10% more dense than ice, so a volume of water equal to the weight of the iceberg will amount to 90% of the iceberg's volume. Thus every iceberg, like every ice cube floating in a water glass, keeps 10% of its volume above the surface of the water and 90% below it (Fig. 11-19a). If some of the ice melts, the iceberg immediately rises, adjusting its position isostatically to maintain the same 10:90 proportion of ice above and below the water.

In the same manner, each segment of the Earth's lithosphere, floating on the denser underlying asthenosphere, rises or sinks to achieve its own isostatic equilibrium (Fig. 11-19b). Because continental lithosphere is less dense than oceanic lithosphere, it is more buoyant—that is, a larger proportion of it "floats" above the asthenosphere. Because continental lithosphere is much thicker than oceanic lithosphere, however, it extends much deeper into the asthenosphere. If we could look below the surface of the lithosphere, we would see that tall continental mountain ranges have proportionately deep "roots" extending far into the asthenosphere; oceanic lithosphere, which is much thinner than continental lithosphere, does not extend nearly as far into this layer.

Each segment of the lithosphere constantly undergoes isostatic adjustment as mass is added to its surface in some places and removed in others. The isostatic response to variations in crustal loads is similar to the reaction of a ship to the loading or unloading of cargo: The more cargo, the deeper in the water the ship will ride. If we unload the cargo, however, the ship rises. As lava flows, accumulating sediments, or advancing glaciers add weight to a given segment of the Earth's crust, that segment gradually sinks deeper into the asthenosphere. Conversely, when materials are removed from a given segment of the crust—through erosion of a hill or melting of a glacier, for example—that segment slowly rises. Figure 11-20 shows some of the isostatic adjustments that occur when mass is added to or removed from the lithosphere.

Magnetism

About 2500 years ago, Greeks noticed that some dark rocks near the Asia Minor city of Magnesia had a strange property: Objects made of iron became attached to them. For centuries magnetic rocks were thought to have wonderful abilities—to comfort the sick, stop hemorrhages, cure toothaches, increase a person's gracefulness, and even reconcile estranged spouses. A truly practical purpose for magnetic rock was discovered 1500 years ago by the Chinese, who observed that a suspended sliver of such rock would turn freely and always come to rest in the same position. This discovery led to the invention of the magnetic compass. In 1600, England's Sir William Gilbert proposed that the Earth itself behaves like a giant magnet, setting the stage for our modern understanding of magnetism.

Magnetism is the force, associated with moving charged particles (such as electrons), that enables certain substances to attract or repel similar materials. In our discussion of minerals in Chapter 2, we examined the role of an electrical charge in attracting charged particles. Magnetism in minerals arises not from the charges on the particles, but rather from the *motion* of charged particles. Electrons, for example, are negatively charged particles that are in constant motion; certain patterns of electron motion create magnetic properties in some substances.

Electrons move not only by orbiting the nucleus of an atom, but also by spinning about their own axes. Each orbiting, spinning electron creates and is surrounded by its own region of magnetic influence, known as its **magnetic field.** Every substance contains vast numbers of spinning electrons, so why then are only certain substances magnetic? When electrons spin randomly, their magnetic fields cancel out one another; consequently, the substance containing them is not magnetic. In certain substances, however, electrons align their magnetic fields and tend to spin in one direction more than in another. The magnetic fields of these electrons enhance one another, giving the substance itself an overall magnetic field. When magnetic substances are subjected to a strong *external* magnetic field, their internal fields become further aligned. They then become more strongly and, in some cases,

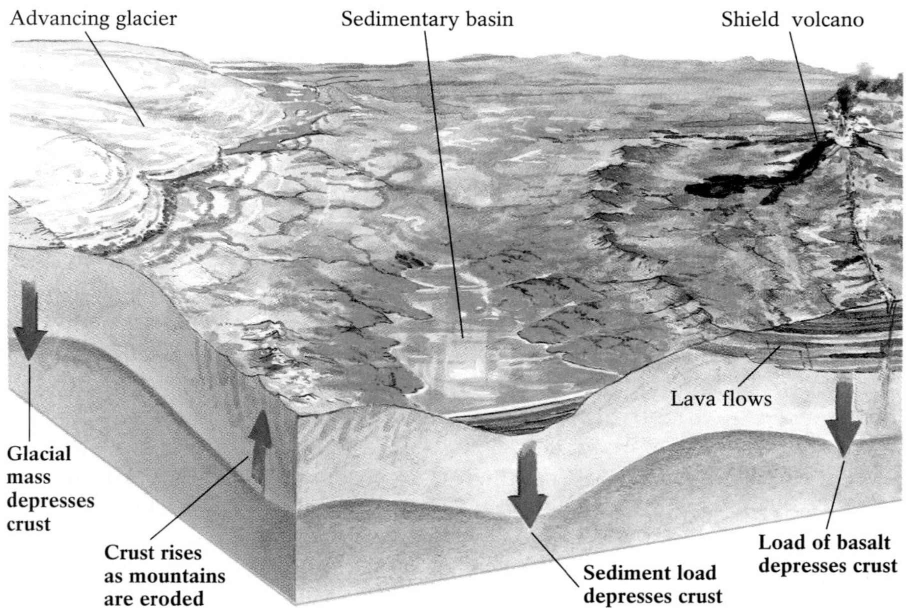

Figure 11-20 Isostatic adjustments in the Earth's crust. As mass is added to a segment of the lithosphere, the segment tends to subside. As mass is subtracted from a segment of the lithosphere, the segment tends to rise.

Figure 11-21 The magnetic field of a simple dipolar bar magnet includes a north pole, from which the lines of magnetic force emerge, and a south pole, at which the lines of magnetic force reenter the magnet. In the same way that electrically charged particles attract opposite charges and repel like charges, magnetic north and south poles attract one another, while like poles repel one another.

permanently magnetized. We can measure the intensity of a magnetic field with a sensitive device known as a *magnetometer.*

Although magnetic fields are invisible, we can see the pattern created by their lines of magnetic force by shaking iron filings onto a sheet of paper placed over a bar magnet. The filings form a pattern that shows numerous closed loops of magnetic force emerging from the magnet's north pole and entering its south pole (Fig. 11-21).

Certain compounds of iron, such as a variety of magnetite called *lodestone* (Fe_3O_4), are strong natural magnets; the rocks of Magnesia, for example, contain a high concentration of magnetite. Mafic rocks, such as basalt and gabbro, usually contain some magnetite. Other iron compounds, such as hematite (Fe_2O_3), become temporary magnets after exposure to a strong external magnetic field; they lose their magnetism following the removal of the external field.

A rock can be *demagnetized* by chemical weathering, lightning strikes, and heat, all of which can alter the orientation of electrons. Heat, for example, causes electrons to vibrate and shift from their aligned positions into random orientations. Magnetite loses its magnetic properties completely when heated to temperatures exceeding 580°C (1075°F), the so-called *Curie point* (named for the French chemist, Pierre Curie).

The Earth's Magnetic Field

The Earth's magnetic field penetrates and surrounds the planet. It extends into space for more than 60,000 kilometers (37,000 miles) beyond the Earth and its atmosphere. Every cubic centimeter of the planet is permeated with this field's invisible lines of magnetic force. Because these lines of force curve, the angle at which they intersect the Earth's surface varies from place to place. As shown in Figure 11-22, they are perpendicular to the surface at the planet's magnetic poles, and parallel to the surface at the magnetic equator. At all points in between, the lines of magnetic force intersect the surface at an angle that increases toward the magnetic poles and decreases toward the magnetic equator. As we will see, the Earth's magnetic north and south poles can move and even switch places—a phenomenon that can help geologists decipher the history of the Earth's rocks.

The Origin of the Earth's Magnetic Field Albert Einstein described the origin of the Earth's magnetic field as one of the planet's most important unresolved mysteries. Although physicists have constructed a conceptually acceptable explanation for these origins, they still struggle to understand many of the "truths" about the magnetic field: Why does the polarity of the field reverse itself from time to time? How long does it take to complete a reversal? How might such a magnetic reversal affect life on Earth?

We know relatively little about what precisely causes the Earth's magnetic field. The high temperature of the Earth's deep interior—far higher than the Curie points of all known materials—represents an argument against a permanent magnet in the Earth's interior. Most naturally occurring magnetic minerals can exist only within the Earth's upper 30 kilometers (20 miles); below that depth, they would suffer demagnetization by the Earth's internal heat. Seismological and gravimetric studies have not found any evidence of a huge concentration of magnetite ore that could act as a permanent magnet in the Earth's interior. The rapid and frequent changes in the polarity of the Earth's magnetic field also argue against the existence of a permanent internal magnet. Thus the Earth's global magnetic field cannot be a product of simple rock magnetism. What, then, could generate an Earth-sized magnetic field?

Our present understanding of the origin of the Earth's magnetic field derives from our knowledge of the movement of electrons. In the nineteenth century, physicists discovered that an electrical current can create a magnetic field. Such a current can be generated spontaneously by moving an electrically conductive substance through an existing magnetic field, in a system known as a *self-exciting dynamo.* Today we have put this knowledge to use to generate electricity in power plants by rotating an electrical conductor in a magnetic field. A self-exciting dynamo stimulates itself to produce more electricity, which in turn produces a stronger magnetic field, which then produces more electricity.

The Earth functions as a gigantic self-exciting dynamo, with its magnetic field being generated by an electrical current (Fig. 11-23). The electrical current is produced by movement of electrons in the molten iron in the liquid outer core. The rotation of the Earth sets the liquid in motion, and these

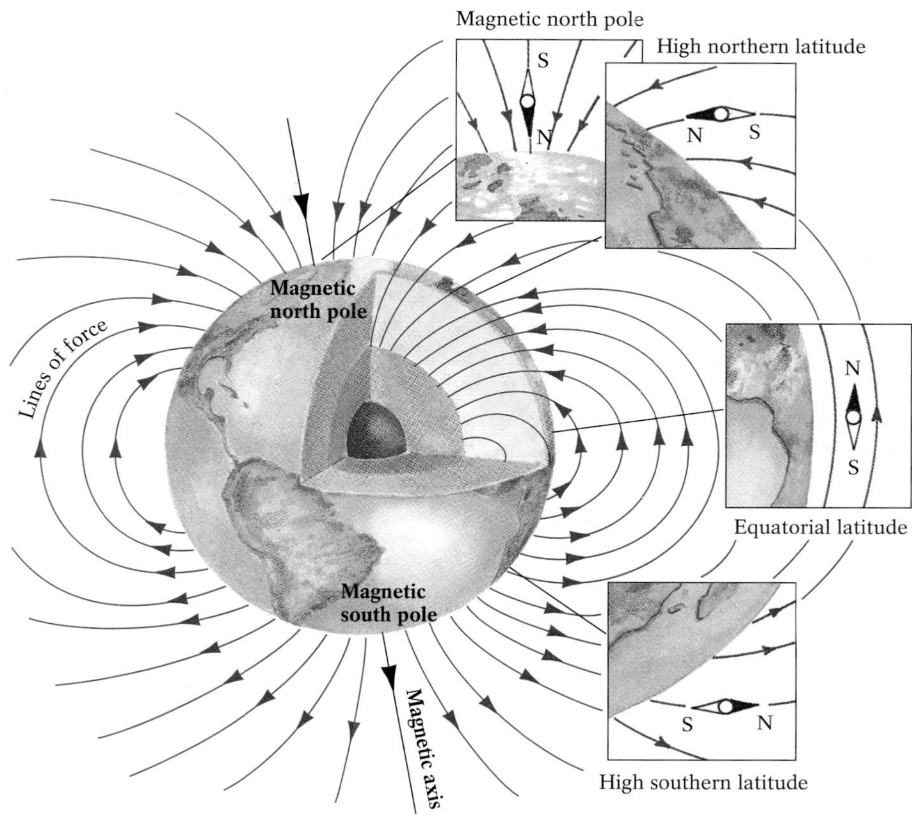

Figure 11-22 The Earth's prevailing magnetic field. Unlike those from the simple bar magnet shown in Figure 11-21, lines of magnetic force emerge from the Earth at the planet's magnetic south pole, located near McMurdo Sound in Antarctica, and reenter at the planet's magnetic north pole, near Prince of Wales Island in the Canadian Arctic. (Note that the planet's magnetic poles do not correspond exactly to its geographic North and South Poles.) In aligning itself along the planet's magnetic force lines, a freely suspended magnetic needle at the Earth's surface would settle perpendicular to the surface at the Earth's magnetic poles, parallel to the surface at the magnetic equator, and at various angles to the surface at points in between.

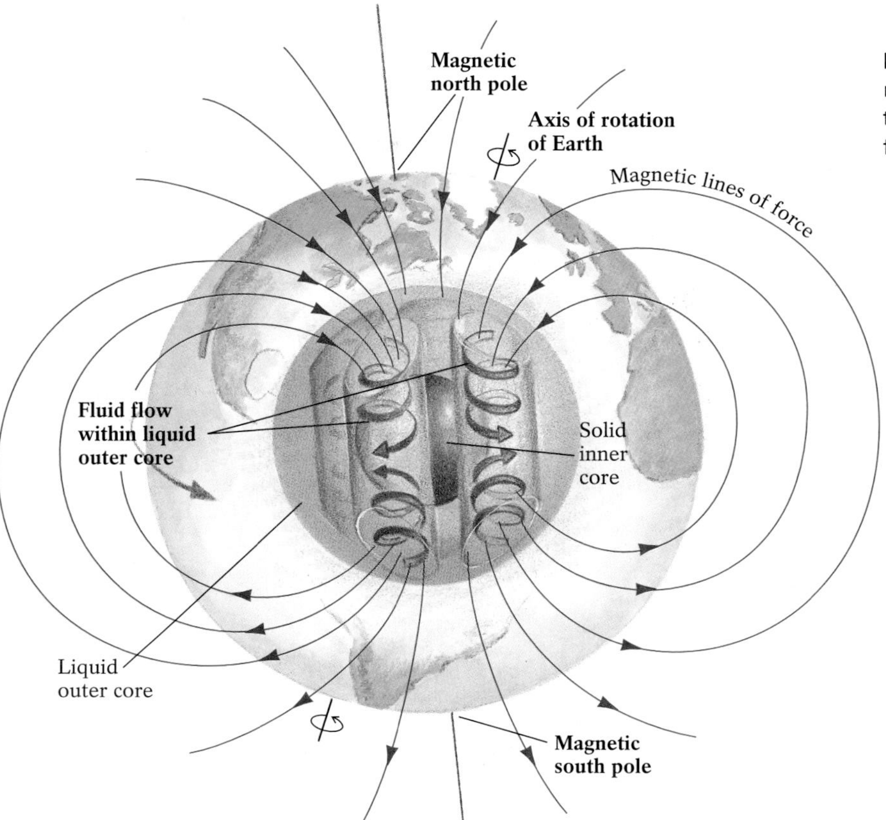

Figure 11-23 The Earth's magnetic field is generated by the flow of electrically conductive fluid in its outer core.

moving currents then generate the Earth's magnetic field, which in turn generates more electrical currents. In this way, the system continues to stimulate itself.

Magnetic Reversals The Earth's magnetic field changes frequently, perhaps more often than any global feature other than climate and sea level. For instance, the location of the magnetic poles changes over time with respect to the geographic poles, although *on average* the magnetic poles do coincide closely with its geographic poles. At present, the magnetic north pole is drifting westward at a rate of about 0.2° per year. In addition, at intervals averaging a half-million years, the Earth's magnetic field actually reverses—magnetic north and magnetic south have exchanged places hundreds, and perhaps even thousands, of times during the Earth's history. The effects of such **magnetic reversals** were first discovered in France at the turn of the century, when scientists noticed that some layers of volcanic rock were magnetized in the opposite direction as other layers.

Magnetic reversals are believed to result from variations in outer-core convection. Turbulence arising from this variable flow may strengthen or weaken the magnetic field. During a magnetic reversal, the Earth's field may weaken substantially for several centuries. As a new reversed field gradually develops, its strength may fluctuate erratically. Eventually it builds until it becomes well established. Most researchers believe that 1000 to 5000 years elapse during a magnetic reversal—that is, between the existence of one stable field and the establishment of a stable field of reversed polarity. A recently discovered reversal in the layered volcanic

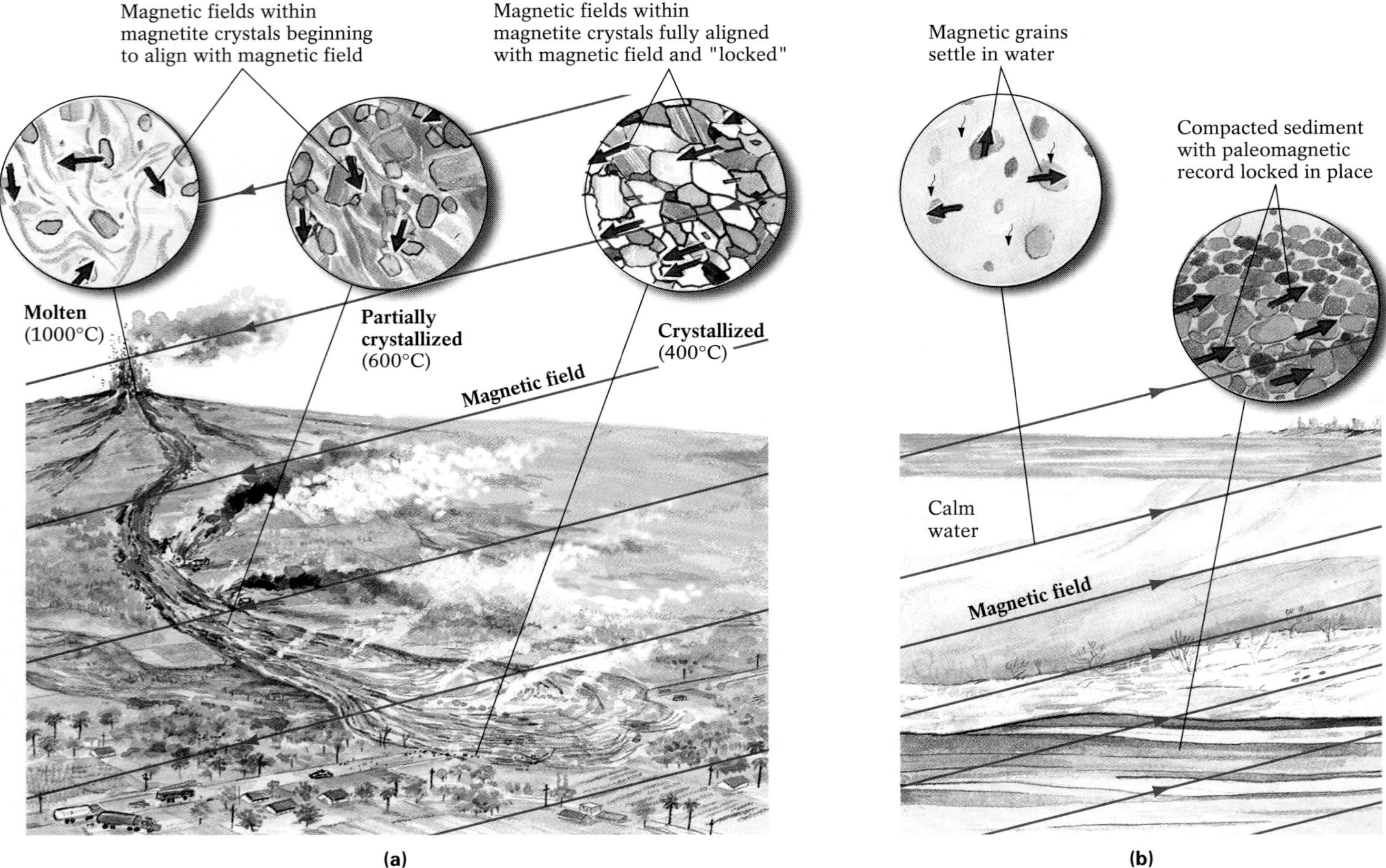

Figure 11-24 **(a)** The magnetic-field polarity within magnetite crystals in molten basaltic flows are free to align with the Earth's prevailing magnetic field. The solidification of the lava fixes the magnetite crystals in place, preserving a record of the Earth's magnetic field at the particular time and place. **(b)** As they fall, magnetic grains of sediment settling through relatively still bodies of water are free to rotate until they become aligned with the Earth's field. Burial of these grains by subsequent deposits locks them in place, creating a paleomagnetic record of the field.

rocks of Steens Mountain in southeastern Oregon, however, suggests that reversals may occur even more swiftly . . . perhaps on the order of tens to hundreds of years.

Today's polarity, which is characterized by lines of magnetic force that emerge from the magnetic south pole and reenter the magnetic north pole (see Figure 11-22), began about 780,000 years ago; this polarity is referred to as *normal.* (The lines of force for a normal magnetic field—as shown in Figure 11-22—are oriented exactly opposite to those in a typical bar magnet—as shown in Figure 11-21.) When the Earth's magnetic field reverses, the lines of force emerge from the magnetic north pole and reenter the magnetic south pole, creating a *reversed* polarity. At present, the needle in a hand-held compass points toward the north; during a period of reversed polarity, the needle would point toward the south.

What effects do these reversals in the Earth's magnetic field have on surface-dwellers? Although little is known for certain, such field reversals may profoundly affect life on Earth. The strength of the reversing field may weaken until virtually no magnetic field exists for a brief time. After the polarity reverses, the strength of the field rebuilds. Because the magnetic field shields the Earth's surface from a significant portion of incoming solar radiation, a reversal may expose life on Earth to a much higher level of ultraviolet radiation for a brief period. Conceivably, people with fair skin might be well advised to use sunblock-1000 during a reversal. A reversal would probably affect the well-being of all surface-dwelling organisms—but exactly how, we do not yet know.

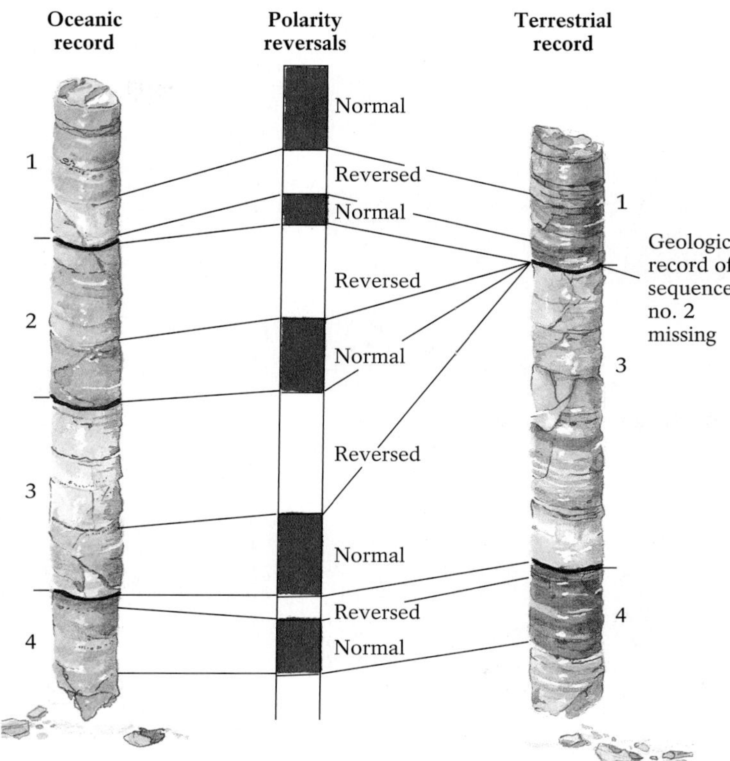

Figure 11-25 The terrestrial record of magnetic reversals has been interrupted by erosion and the intermittent nature of continental volcanism and sedimentation. The oceanic record, which has been generated by uninterrupted marine sedimentation, is more complete. Note that as the rate of sedimentation varies, the thicknesses of the layers corresponding to specific magnetic reversals also vary.

Paleomagnetism The geologic record often contains evidence of **paleomagnetism,** past changes in the Earth's magnetic field. Such changes are most obvious in mafic igneous rocks, such as basalt, and in lake and marine sediments. As mafic lava cools, it forms small crystals of magnetite. The magnetic fields of these crystals, for a while, are free to align themselves with the Earth's prevailing magnetic field (Fig. 11-24a). By the time the entire lava body has solidified, the magnetite crystals and their magnetic fields have become frozen in place, thus preserving the alignment of the Earth's magnetic field at that time. Similarly, when magnetic grains settle through a relatively still body of water, such as a lake or protected bay, the grains themselves are free to rotate like compass needles and become aligned with the Earth's field as they fall (Fig. 11-24b). Their burial by subsequent sedimentation means that the grains can no longer rotate freely, and thus their magnetic orientation is locked in place. In this way, magnetite-rich sediments and lava create a permanent record of the Earth's past magnetic fields.

The quality of the magnetic record varies significantly from place to place, particularly from land to sea floor. The terrestrial ("on land") record of the Earth's magnetic field reversals, for example, is highly susceptible to loss due to weathering and erosion. Consequently, it is—at best—incomplete. The most complete record appears on the ocean floor, especially in areas where uninterrupted marine sedimentation has suffered little or no erosion (Fig. 11-25). In addition, sea-floor spreading provides a unique opportunity to study past magnetic field reversals. These events are recorded in the basaltic lavas that have continuously erupted and cooled at mid-ocean ridges during this spreading (see Figure 12-6).

The ocean-floor record shows that during the past 75 million years alone, the Earth's magnetic field has reversed itself at least 171 times. Periods of stable polarity, either normal or reversed, have lasted from 25,000 to several million years. No definite pattern characterizes the length of time between magnetic field reversals, and geophysicists do not know when the next reversal will occur.

Paleomagnetism is a powerful tool that enables geologists to reconstruct past geographic arrangements of the Earth's landmasses. Because the magnetic fields within a rock's magnetite crystals become aligned with the Earth's magnetic field at the time of the rock's formation, the crystals function

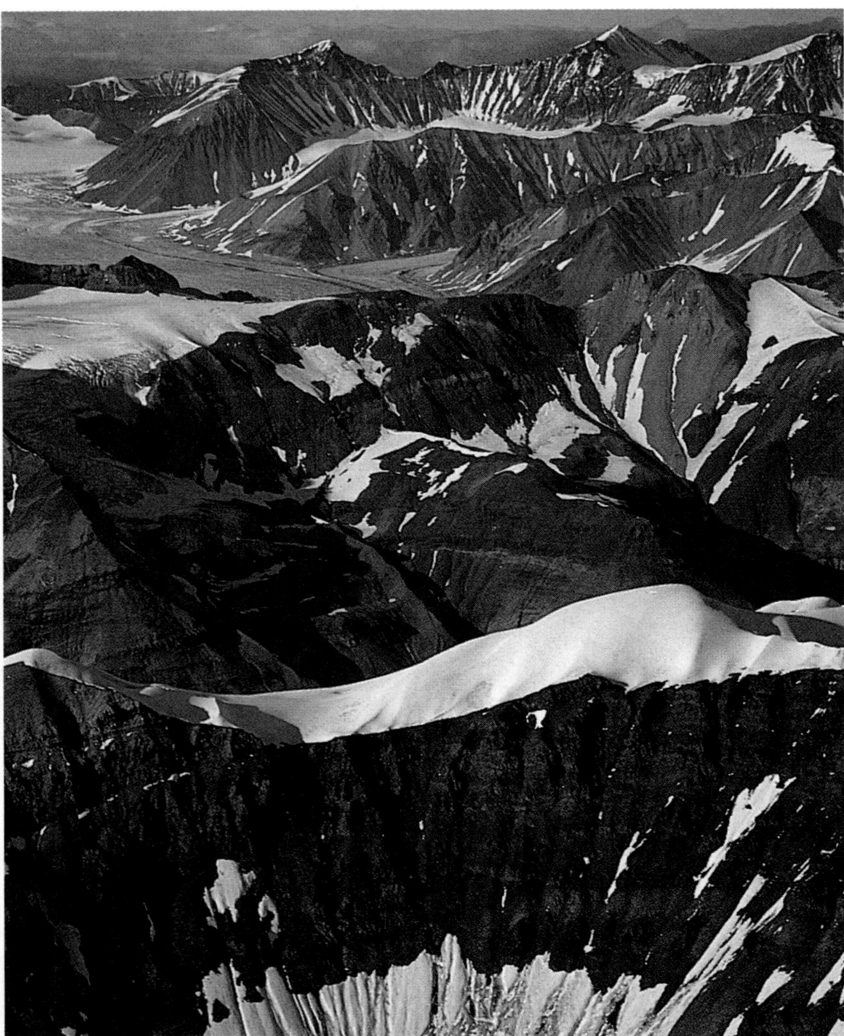

Figure 11-26 The Wrangell Mountains of Alaska. Igneous rocks in these mountains contain magnetite grains that indicate they formed near the equator, suggesting that the mountains were moved a great distance by plate motion. Sedimentary rocks in the mountains contain fossils of warm-water organisms that corroborate the paleogeographic origin of the igneous rocks.

as a kind of "paleo" compass needle, pointing to their original magnetic poles. We can estimate the rock's original geographic *latitude* by comparing the angle between the magnetic polarity of the crystals and the Earth's surface to the angle at which the lines of force of the Earth's magnetic field intersect the surface. (This angle increases with increasing latitude; see Figure 11-22.) The extent to which any magnetite-bearing rock has been moved from its original position *in the north–south direction* presumably reflects movement of the tectonic plate on which it is located. Thus we can use paleomagnetism to estimate how far a rock (and therefore its plate) has moved with respect to latitude. By studying the magnetite grains in the igneous rocks of the Wrangell Mountains of Alaska, shown in Figure 11-26, geologists have been able to determine that these rocks probably formed near the equator and moved to their present polar location by plate motion.

Other Magnetic Fields in Our Solar System

Geophysicists now believe that to generate a magnetic field, a planet must:

- Have a metal core;
- Maintain part of the core as a flowing liquid; and
- Spin on its axis to set the liquid in motion.

All of the planets and moons of our solar system spin. All probably have differentiated interiors as well. But only some appear to contain liquid-metal interior layers; without this feature, the planet or moon cannot produce a self-exciting magnetic field.

The Earth's Moon, for example, does not appear to contain a liquid interior layer, so we would not expect it to generate a magnetic field. Indeed, it does not. The Moon is partially differentiated, however, and its materials could not have migrated into layers according to their different densities unless it had been at least partly melted at one time. Magnetometer-toting astronauts and Earth-bound lunar scientists have found aligned magnetite crystals in lunar rocks, suggesting that the Moon's small, iron-rich core was once molten and generated a magnetic field. These rocks, which became magnetized billions of years ago during their formation, have not been reheated and demagnetized since then. This evidence suggests that the Moon has been tectonically inactive for quite some time.

The existence of a magnetic field on Mars has inspired much debate in recent years. Because Mars displays angular momentum as it rotates on its axis, many geophysicists believe that this planet contains a metallic core, one of the prerequisites for the development of a magnetic field. Magnetic measurements conducted by the Viking probe in 1976 detected no evidence of a magnetic field on Mars, suggesting to some that its metallic core might be completely solid. Later measurements made in 1997 during the Mars Global Surveyor mission, however, revealed that a weak magnetic field does exist.

Although only one eight-hundredth as strong as Earth's magnetic field, Mars' field is still stronger than NASA's scientists had expected. It is not yet known whether the current field is a fading remnant of a much more robust field from a time when Mars had a substantial liquid-metal core. Proof that a more vigorous magnetic field existed on this planet in the past awaits identification of magnetic reversals in Mars' volcanic rocks.

Like Earth, Mercury appears to have undergone extensive planetary differentiation. Because Mercury has an unusually high density, planetary geologists believe that it has an extremely large, iron-rich core, and that all or most of this planet was once molten. Although Mercury rotates so slowly on its axis (one of its days lasts 59 hours) that it probably could not generate the necessary currents in its liquid core to form a self-exciting dynamo, the magnetometer on the 1973 Mariner-10 probe did detect a weak magnetic field on Mercury (about 1% as strong as the Earth's field). The pattern of the lines of force on this planet closely resembles that of the Earth's field.

The existence of a magnetic field on Mercury supports the hypothesis that the planet's core is at least partly molten. Conversely, Venus, with its slow rotation, does not possess a detectable magnetic field, even though it is believed to contain a partly liquid iron core.

Jupiter and Saturn, two large outer planets in our solar system, are composed principally of hydrogen in liquid form. At extremely high pressures, liquid hydrogen exhibits the properties of a metal, providing these planets with a conductive inner layer. Both planets spin quite rapidly, with Jupiter's day lasting 9 hours, 50 minutes, and Saturn's day lasting about 10 hours, 30 minutes. Their rapid spin and their conductive liquid interiors give these planets the strongest planetary magnetic fields in the solar system. The strength of Jupiter's field was measured by Pioneer and Voyager probes as being at least 10 times as strong as the Earth's field; Saturn's field is nearly as strong as that of Jupiter.

Our knowledge of the geophysical properties of planet Earth has grown considerably during recent years, thanks in part to rapid advances in space-satellite technology. In the next chapter, we will see how these and other technological advances are enabling Earth scientists to track the movements of the continents, study the details of the ocean floor, and reconstruct the past positions of the Earth's plates.

Chapter Summary

Geophysics is the study of the fundamental properties of the Earth's interior, especially heat flow, gravity, and magnetism. Much of the data that contribute to this study come from seismological stations' recordings of the seismic body waves (compressional P and shearing S waves) that emanate from the foci of earthquakes. **Seismic tomography,** a research technique adapted in principle from medical CAT scans, uses seismic waves to generate three-dimensional images of the Earth's interior.

By analyzing the waves' arrival times as recorded by seismographs, geophysicists have identified the thickness, density, composition, structure, and physical state of the layers of the Earth's interior. We now believe that the Earth consists of the following layers: continental and oceanic crust, each of differing composition, thickness, and structure; a **seismic discontinuity**—a boundary where seismic waves are refracted as they speed up or slow down because of a marked change in rigidity—at the base of the crust and the top of the mantle, known as the **Moho;** a thick, multilayered mantle whose upper portion contains the **low-velocity zone** (another seismic discontinuity), in which seismic waves slow down temporarily; a distinct **transition zone** separating the upper- and lower-mantle zones; a seismic discontinuity at the mantle–core boundary; and a two-part core consisting of a liquid outer core and a solid inner core.

The **S-wave shadow zone** and **P-wave shadow zone** are segments of the Earth opposite an earthquake's focus where no direct S and P waves are recorded. Because shearing S waves do not travel through a liquid, and P waves are refracted when they enter a region of lower rigidity (such as a liquid), the absence of these seismic waves in the S- and P-wave shadow zones confirms that the Earth's outer core is liquid.

Because we have few direct means of measuring the Earth's deep interior temperatures, scientists must estimate them using logic and laboratory experimentation. If the Earth is solid at a particular depth, the temperature there must be lower than the melting point of the material at that pressure. Recent experiments suggest that the temperature at the mantle–outer core boundary exceeds 4800°C (8650°F) and that the temperature of the outer core–inner core boundary is about 7600°C (13,700°F).

Gravity is the force of attraction that any object exerts on another object; it is proportional to the product of the objects' masses, and inversely proportional to the square of the distance between their centers. The Earth's gravitational attraction is far greater than that of any object on or near it. Gravity at the Earth's surface depends on several factors. The altitude of the land surface, and thus the distance from the Earth's center of gravity, produces a greater attraction in lowlands than on mountain tops. Likewise, the Earth's "flattened" sphere shape gives a higher gravitational attraction at the poles (closer to the Earth's center) than at the equator. The composition of the Earth's underlying interior layers also affects gravity. High-density materials, for example, possess a greater gravitational attraction than low-density materials. **Gravity anomalies** are local deviations in the Earth's gravity caused by variations in density in the planet's interior. Negative gravity anomalies, in which the force of the attraction is lower than expected, occur in mountainous areas where low-density continental crust is particularly thick; they also occur above sediment-filled oceanic trenches and in regions containing rising subterranean bodies of salt. Positive gravity anomalies, in which the force of attraction exceeds the expected level, are commonly found where high-density rocks lie at shallow depths—for example, where dense mantle material rises beneath rifting or diverging plate boundaries, where mantle rocks lie at or near the surface within collisional mountains, and above large deposits of metallic ores.

Isostasy is the continual process of depth-adjustment that the Earth's lithospheric segments undergo as they float on the denser underlying asthenosphere. Surface processes that add mass to the segments, such as volcanism, glaciation, sedimentation, and mountain building, cause them to sink; processes that remove Earth materials from the segments, such as erosion and glacial melting, cause them to rise.

Magnetism is a property of some materials, associated with the movement of electrons, that causes them to attract or repel other materials having this property. The Earth has a **magnetic field,** similar to that of a simple bar magnet. This field probably arises from the flow of an electrical current within the Earth's liquid outer core. The invisible lines of force of the magnetic field emerge from the magnetic south pole and reenter the magnetic north pole. The Earth's magnetic field periodically weakens and changes the direction of its polarity, a phenomenon known as **magnetic reversal. Paleomagnetism** is the preservation of past changes in the Earth's magnetic field within the geologic record. Geologists use this tool to reconstruct past geographic arrangements of the Earth's landmasses.

Other planets in our solar system have a magnetic field if they are differentiated, have a conductive liquid layer, and spin fairly rapidly, keeping the liquid in motion.

Key Terms

geophysics (p. 304)
seismic tomography (p. 306)
seismic discontinuity (p. 308)
Moho (p. 308)
low-velocity zone (p. 309)
transition zone (p. 310)
S-wave shadow zone (p. 312)
P-wave shadow zone (p. 312)
gravity (p. 317)
gravity anomalies (p. 318)
isostasy (p. 320)
magnetism (p. 321)
magnetic field (p. 321)
magnetic reversals (p. 324)
paleomagnetism (p. 325)

Questions for Review

1. Describe the behavior of a seismic wave as it enters a more rigid medium.

2. Explain how oceanic and continental crust differ in composition and thickness.

3. What is the Moho, and how was it discovered?

4. What is the major difference between the upper mantle and the lower mantle?

5. What is the S-wave shadow zone, and what causes it?

6. What are three possible sources of the Earth's internal heat?

7. Discuss how and why the Earth's gravity varies with topography and latitude.

8. Describe two places where you might find positive gravity anomalies and two where you might find negative gravity anomalies.

9. Briefly describe two geological settings where land surfaces rise isostatically and two settings where surfaces subside isostatically.

10. Briefly discuss our current theories of how the Earth's magnetic field originates.

For Further Thought

1. The outer core of the Earth's interior is in a liquid state. Why doesn't the liquid rise to the surface and erupt as lava?

2. In films of the first humans walking on the Moon, the astronauts seemed to be jumping and bounding because of the weak force of gravity. Why is the gravitational attraction of the Moon so much weaker than that of the Earth? What type of gravitational attraction

would we be likely to find on the surface of a planet that is much larger and composed of much denser material than the Earth?

3. What kind of gravity anomaly would exist on an enormous body of dense platinum ore at the bottom of a canyon in Antarctica? Above a large body of subterranean salt at the top of a lofty mountain at the equator?

4. If the Earth's internal heat supply became completely exhausted, what would happen to the asthenosphere, and how would isostatic adjustments at the Earth's surface be affected? How would the Earth's magnetic field change?

5. Has the Earth always had a magnetic field? What evidence supports your answer?

12

Plate Tectonics: Creating Oceans and Continents

Figure 12-1 The Himalayas in Tibet, located at the boundary between India and China, are a spectacular example of the mountain-building capacity of colliding lithospheric plates.

We have discussed plate movement throughout the first 11 chapters because this motion affects virtually every aspect of geology—from the origin of earthquakes, volcanoes and mountain ranges (Fig. 12-1) to the rise and fall of sea level. Since 1957 when the U.S.S.R. launched Sputnik 1, the first space satellite, into orbit around the Earth, satellite technology has expanded our knowledge of plate tectonics. Data from satellites show us how the earth's plates change and move over time.

Every year, ground-based laser beams are bounced off the Laser Geodynamics Satellite (LAGEOS), recording the amount of time required for the beams to make a round-trip journey. With that information, geologists can determine the precise distance between the satellite and the ground station. Because LAGEOS speeds around the Earth at exactly the same velocity as the Earth rotates on its axis, its position over the planet remains fixed. Thus any change in the laser beam's travel time records a change in the geographic position of the landmass where the ground station is located.

Similarly, a constellation of 24 orbiting satellites that constitutes the Global Positioning System (GPS) can locate precise points on the Earth's moving plates. Much like the triangulation used to locate earthquakes (discussed in Chapter 10), the overlapping beams from a trio of satellites can locate accurately a single geographic point on the Earth's surface (Fig. 12-2). Geoscientists can then measure to within a few millimeters the movement of the plate that contains that point. In this way, space-age technology is used to confirm that the Earth's plates do indeed move—and shows that they do not all move at the same rate.

We examined the development of the theory of plate tectonics and reviewed the evidence for it in Chapter 1. In later chapters, we detailed some of the basic assumptions of that theory:

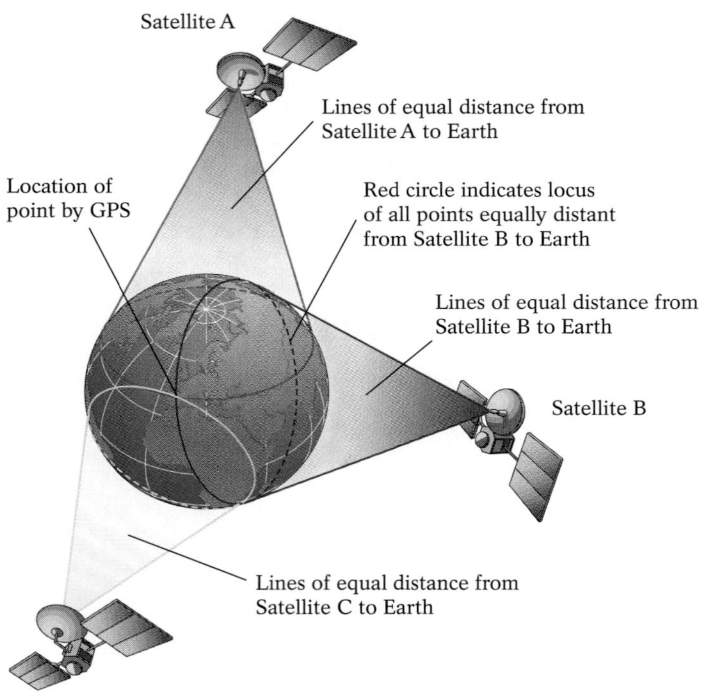

Figure 12-2 The satellite-based Global Positioning System enables geologists to locate changes in precise points on the Earth's surface, confirming that the Earth's plates do indeed move.

- The Earth's lithosphere consists of rigid plates averaging 100 kilometers (60 miles) in thickness and ranging from about 70 kilometers (43 miles) thick for the oceans to 150 kilometers (90 miles) thick for the continents.
- The plates move relative to one another by divergence, convergence, or transform motion.
- Oceanic lithosphere forms at divergent plate boundaries and is consumed at a subduction zone, one type of convergent plate boundary.
- Four basic types of convergent plate boundaries exist: those that occur where an ocean plate subducts beneath another ocean plate; those that occur where a subducting ocean plate tows a continent to a subduction zone; those that occur where an ocean plate subducts beneath an adjacent continent; and those that occur where two continental plates collide.
- Most earthquake activity, volcanism, faulting, and mountain building take place at plate boundaries.
- Plates generally do not deform internally; that is, the centers of plates tend to be geologically stable.

The theory of plate tectonics addresses the origin of both the continents and the oceans. Its predecessor, Alfred Wegener's continental drift hypothesis, had generally ignored the ocean basins—it primarily sought to explain past movements of the continents. When Wegener proposed his hypothesis in 1915, the ocean basins were almost entirely unexplored. Like most geologists of his time, Wegener believed that the ocean floors were very old, featureless plains. He thought that the oceans passively surrounded the drifting continents, which, according to Wegener, moved by plowing through the ocean floors on their way to distant points on the globe. This lack of knowledge about the geology of the sea floor led to most of the conceptual problems that ultimately consigned the continental-drift hypothesis to decades of rejection. Intense deep-sea exploration, which began in the 1950s, brought the discovery of the spreading ocean ridges and revealed the true nature of the ocean's role in plate motion. Only then could continental drift become a useful concept, underlying today's better understanding of plate tectonics.

In this chapter we take a large-scale view of plate motion and its effects. We will examine plate motion rates, the types of landforms created by moving plates, and the mechanisms driving plate motion. We will see that plate movements are largely responsible for the creation of the Earth's ocean basins and the geology of the continents.

Determining Plate Velocity

A plate's velocity is a measure of both its speed *and* its direction of movement. Geologists can estimate a plate's speed in *absolute* terms only when they have a fixed point outside of the plate to use as a reference (such as a distant star or an orbiting satellite). We can measure changes in distance from such a fixed point to any feature on the moving plate. Given two or more moving plates and no fixed reference point, we can determine only each plate's velocity *relative* to the other plates (Fig. 12-3).

Hot spots are localized regions where plumes of hot mantle material rise from great depths—perhaps from as deep as the mantle–core boundary—to the base of the lithosphere. Although these plumes may themselves move as the deep mantle convects, they appear to do so very slowly relative to the overlying plates. The hot spot that fuels Hawai'ian volcanism, for example, appears to be moving southeastward, in the opposite direction from the overlying Pacific plate—but at $^1/_{10}$ the speed. Thus, after accounting for their slow motion, hot spots can be used as "fixed" reference points that allow us to determine both the speed and direction of the plates that move above them. Approximately three dozen hot spots, shown as red dots in Figure 12-4, are active today, and perhaps 100 or so have been active in the past 10 million years. Many appear at or near divergent plate boundaries, but some (such as the one beneath Yellowstone National Park in Wyoming) reside in plate interiors, where they fuel intraplate volcanism.

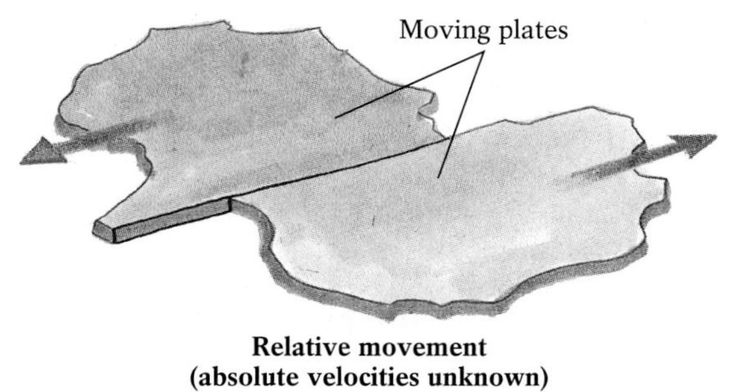

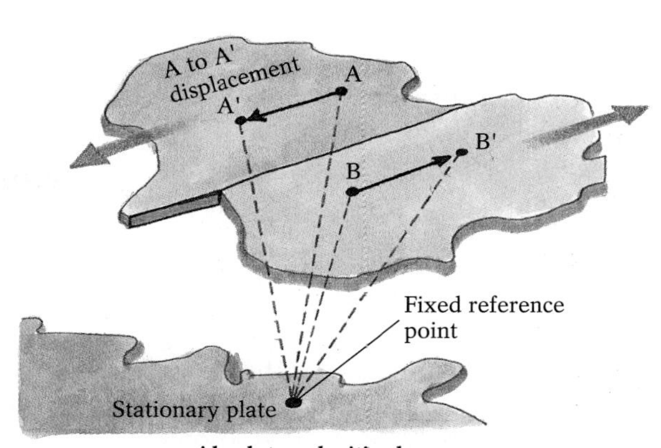

Figure 12-3 Determining relative and absolute plate speeds. If two plates are both moving, and we have no fixed reference point by which to measure their displacement, we cannot establish the absolute speed of either one; the changing distance between the moving plates reflects only their motion *relative* to one another. A plate's *absolute* speed is determined by measuring its changing distance from a fixed point, such as a feature on an adjacent stationary plate.

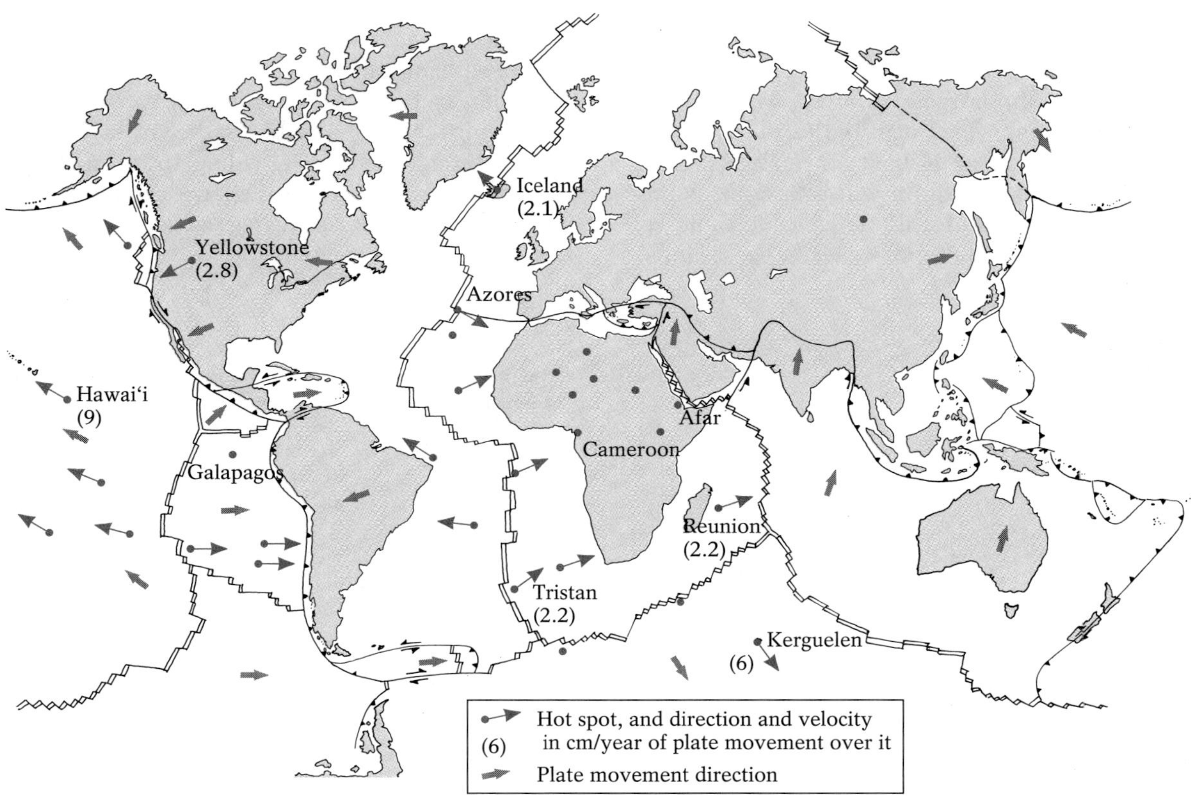

Figure 12-4 Hot spots and absolute plate motion around the world. Hot spots, which may produce volcanoes, usually occur at divergent plate boundaries, but can also be found under plate interiors. Because hot spots are essentially fixed relative to the faster-moving plates above them, geologists can use them as references to determine the absolute velocities of the world's tectonic plates.

Plate Movement over Hot Spots

Hot spots beneath oceanic plates produce volcanoes on the sea floor directly above them. As a plate moves across a hot spot, a chain of submarine volcanoes forms, each of which may eventually extend above sea level and become an island. As they pass beyond the hot spot, these active volcanoes becomes extinct, and new volcanoes form in their place at the end of the trail of volcanic islands. We can deduce the direction of the plate's movement from the locations of the extinct volcanoes, and we can determine the plate's speed from the volcanoes' ages and the distances between each volcano and the hot spot.

Volcanoes that have not grown above sea level form conical submarine mountains called **seamounts.** Some island volcanoes that originally formed over oceanic hot spots have since been worn flat by weathering, wave action, and stream erosion. Called **guyots** (pronounced GHEE-owes), these worn-down forms are found below sea level, principally because the lithosphere that carried them has cooled, become more dense, and subsided deeper into the mantle.

All oceans contain numerous hot-spot islands, seamounts, and guyots, generally arranged in long chains that form as drifting plates trail away from hot spots. Most prominent, shown in Figure 12-5, is the Hawai'ian Island–Emperor Seamount chain, which extends first to the northwest, then north, through more than 6000 kilometers (4000 miles) of the central Pacific, before ending at the Aleutian trench. From potassium–argon dating of basalt from the entire chain, geologists have found that these volcanic features developed over a span of more than 75 million years, indicating that this hot spot is long-lived.

The Hawai'ian Islands are relatively recent features formed by the central Pacific hot spot. Kaua'i, whose rocks date from 5.6 to 3.8 million years ago, is the oldest, most northwestern island in the chain. Hawai'i, the "Big Island," which today experiences nearly continuous eruptions, is the youngest, most southeastern, and only actively volcanic island in the chain. Between Kaua'i and Hawai'i, the islands' rocks are progressively younger toward the southeast. To the southeast of Hawai'i is the Loihi seamount (Loihi means "the long tall one" in a Polynesian dialect), which will likely become the next island in the chain. Loihi's summit rises some 4400 meters (14,500 feet) above the sea floor, still several thousand meters below sea level. If Loihi's eruptions continue at their current rate, however, the volcano will replace the island of Hawai'i as the volcanic centerpiece of the central Pacific sometime during the next few hundred thousand years.

Several other central Pacific seamount chains exist, all of which are marked by a noticeable bend, presumably formed after the Pacific plate changed direction from due north to its present northwesterly trend. Basalt from volcanoes at these bends dates to about 40 million years ago; thus we know that the Pacific plate changed direction at that time. We can estimate the rate of Pacific plate movement by using the Hawai'ian hot spot as a fixed reference point. For example, Midway Island is located about 2700 kilometers (1700 miles) northwest of the hot spot, and its basaltic rocks are 27.2 million years old. Dividing the distance from the hot spot by the age of the rocks, we conclude that the Pacific plate has moved at an average speed of nearly 10 centimeters (4 inches) per year.

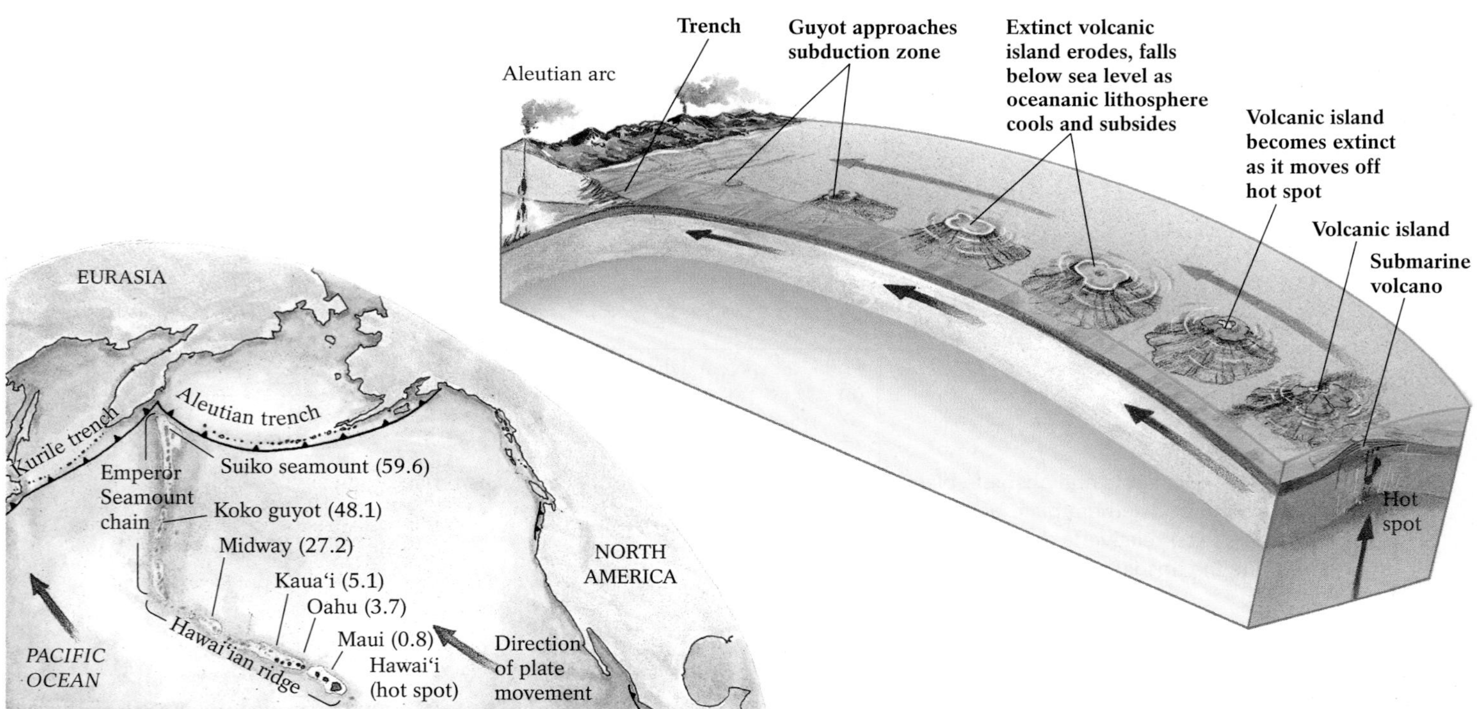

Figure 12-5 The same mid-Pacific hot spot has fueled the eruptions that produced every volcanic island and submarine mountain in the 6000-kilometer (4000-mile)-long Hawai'ian Island–Emperor Seamount chain. As the moving plate carries each island away from the hot spot, the volcanoes lose their heat source and become extinct. Weathering and erosion gradually reduce their heights above sea level. Ultimately they become submerged as the lithosphere cools and subsides isostatically. (Volcano ages, in millions of years, are shown in parentheses.)

Hot spots beneath continental plates do not produce such obvious volcanic chains, probably because much of their magma cools and solidifies within the thick continental lithosphere and never reaches the surface. Nevertheless, some chains of igneous formations may comprise the eroded remains of ancient volcanoes formed over continental hot spots. One such sequence on the North American plate stretches from southwestern Idaho to northwestern Wyoming. Its older features—the basaltic plains of the Snake River—formed largely between 5 million and 1 million years ago; the younger geysers, volcanic cliffs, and uplifted plateaus of Yellowstone National Park formed in the last million years or so.

Tracking Magnetic Field Reversals

We also can use our growing knowledge of the reversals of the Earth's magnetic field (discussed in Chapter 11) to determine the rates of oceanic plate motion. During World War II, military vessels bearing magnetometers traversed the Pacific and Atlantic Oceans seeking metal from wrecks of submarines and other military hardware that might have sunk to the ocean floor. The magnetometers recorded a puzzling pattern of magnetic field variations: Alternating bands of slightly stronger and weaker magnetism along the ocean floor paralleled the mid-ocean ridges.

At first, these bands of variable magnetism perplexed technicians, who assumed that their equipment was malfunctioning. In the early 1960s, however, Fred Vine, a young Cambridge University student, named the magnetic stripes **marine magnetic anomalies** and proposed that they constituted evidence of sea-floor spreading at divergent plate boundaries. Vine and his associate Drummond Matthews explained that as basaltic lava cools at spreading mid-ocean ridges, it becomes magnetized in the direction of the Earth's magnetic field at the time of eruption; basalts that erupt at times of reversed polarity will therefore display evidence of the magnetic reversal (Fig. 12-6). Magnetometers towed above a normally magnetized stripe record a slight strengthening (about 1%) of the magnetic field; when towed above a stripe formed under reversed magnetism, they record a slight weakening of the field. (The Earth's magnetic field is enhanced in the vicinity of rocks of normal polarity and weakened near rocks of reversed polarity.)

Dating any rock along a marine magnetic reversal tells us the age of the basalt along the entire length of the reversal. Because oceanic plates grow continuously at divergent zones and therefore contain an uninterrupted record of the oscillations of the Earth's magnetic field, we can use the sequence of magnetic anomalies in oceanic basalt to date any area of the sea floor. In addition, geologists can use marine magnetic anomalies to estimate rates of plate motion. By measuring an anomaly's distance from the spreading ridge, they can determine the rate of spreading and therefore the rate of plate motion. For example, if an anomaly that is known to be 4.5 million years old is located 90 kilometers (55 miles) from the ridge crest, the spreading rate is 2 centimeters (0.8 inch) per year.

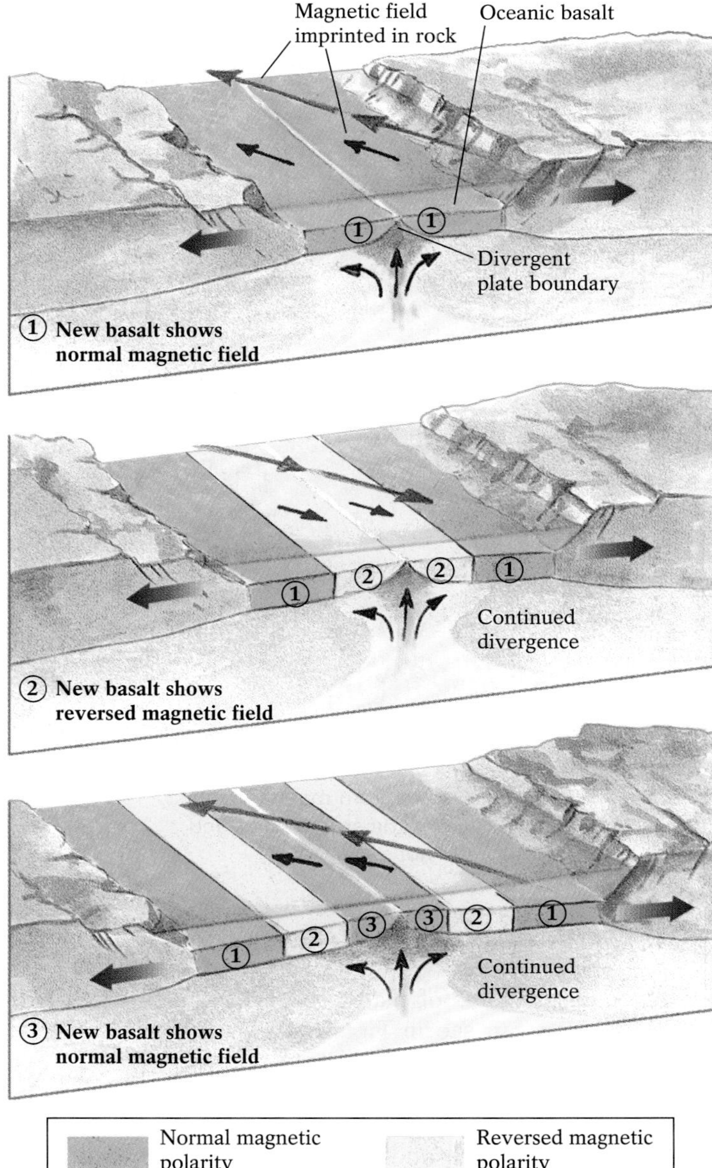

Figure 12-6 Marine magnetic anomalies reflect reversals in the Earth's magnetic field. As basaltic lavas cool and solidify at mid-ocean ridges, the magnetic fields of their magnetite crystals become aligned with the prevailing direction of the Earth's magnetic field. Each resulting stripe of basalt has either normal magnetism (like today's field) or reversed magnetism (opposite to today's field).

How fast, then, do plates move? Using information from orbiting satellites, oceanic hot spots, and magnetic anomalies,

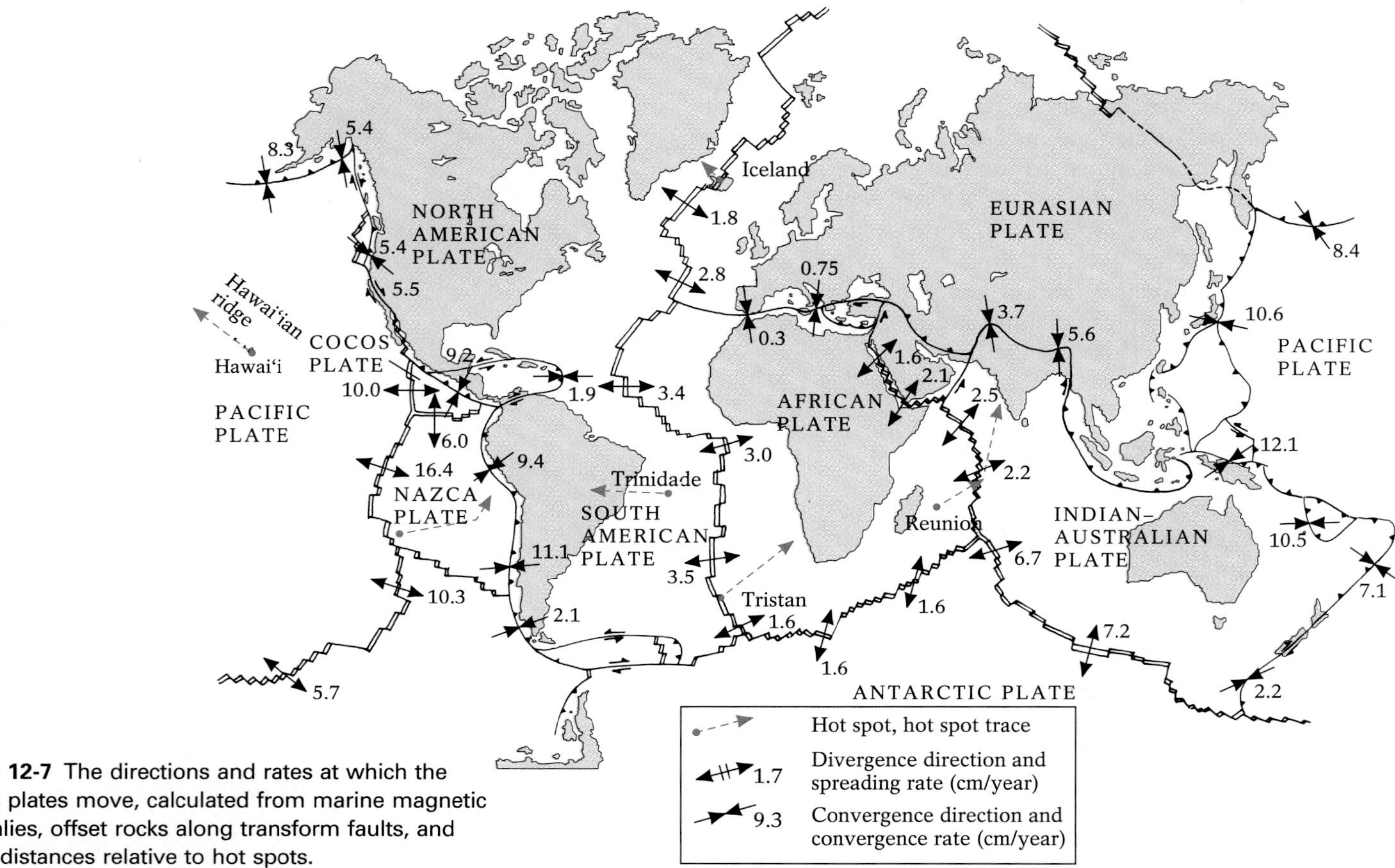

Figure 12-7 The directions and rates at which the Earth's plates move, calculated from marine magnetic anomalies, offset rocks along transform faults, and island distances relative to hot spots.

we can measure rates of plate motion and verify their accuracy. As you can see in Figure 12-7, the Pacific, Nazca, Cocos, and Australian-Indian plates, among the Earth's fastest-moving travel more than 10 centimeters (4 inches) per year. All these plates are composed principally of oceanic lithosphere. The North and South American, Eurasian, and Antarctic plates, among the slowest-moving, shift their positions by 1 to 3 centimeters (0.4–1.2 inches) per year. Each of these plates largely consists of continental lithosphere. As we shall discuss later in this chapter, oceanic lithosphere generally moves more swiftly than continental lithosphere.

The Nature and Origin of the Ocean Floor

If the water was drained from ocean basins, we would see chasms as deep as 11,000 meters (36,000 feet), volcanic ranges thousands of kilometers long, chains of flat-topped mountains, long linear fractures oozing lava, and faulted cliffs stretching like walls for great distances. Most of these features were discovered and mapped in the early 1940s, a by-product of the U.S. Navy's World War II search for enemy submarines and safe passages. The principal method of mapping at that time involved **echo-sounding sonar.** A ship using this technology emits a sharp pinging noise. The sound waves travel at a speed of about 1500 meters (5000 feet) per second, the speed of sound in seawater. The time that elapses while the sound waves bounce off the sea floor and return to a listening device is used to calculate the depth of the ocean bottom and map its topographic features.

After World War II, devices using more sophisticated technology were developed. Today, we can map subsurface ocean-floor rocks by means of **seismic profiling,** which employs more powerful energy waves generated by small targeted explosions. Some of these waves reflect off the sea floor; others penetrate the surface and reflect off layers of underlying sediment and rock, enabling geologists to map the subsurface as well.

Deep-sea drilling and submersible vessels photograph and take samples from the sea floor. Between 1965 and 1980, a group of universities, research laboratories, and government agencies jointly funded and staffed the Deep Sea Drilling Project (DSDP), currently known as the Ocean Drilling

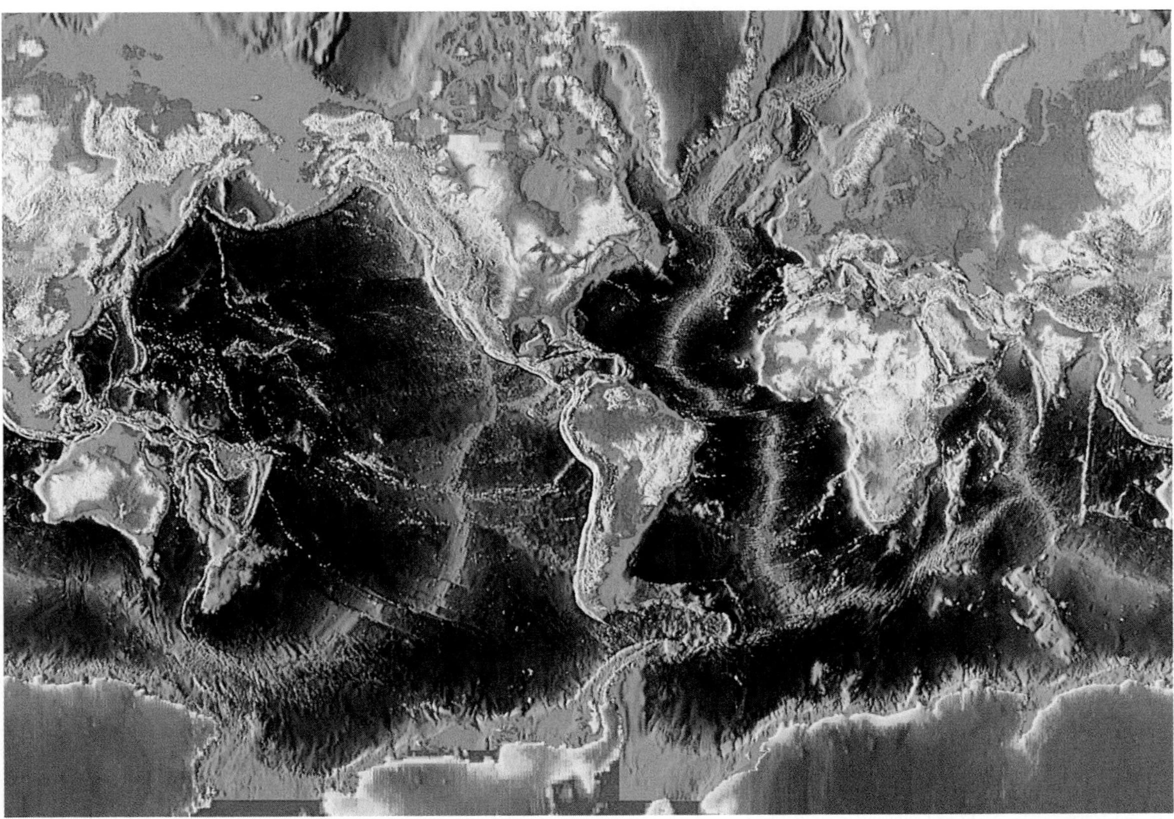

Figure 12-8 This map, which shows the significant variations in the morphology of the sea floor, was created using satellite-gathered data on the elevation of the seawater surface. Positive gravity anomalies produced by masses such as submarine mountain ranges attract large quantities of seawater, causing an upward expansion of the ocean surface (light blue areas); conversely, negative gravity anomalies produced by bedrock depressions lower the ocean surface (dark blue areas).

Project (ODP). During the voyages of its *Glomar Challenger* research vessel, scientists collected more than 96 kilometers (60 miles) of deep-sea cores that have been studied for clues to the origin of the Earth's ocean basins. In addition, oceanographers on ALVIN, the best-known deep submersible, continue to do field work at the sea floor; along the way, they have discovered new life forms and previously unknown geological processes. More recently, remote-controlled video cameras have permitted even more extensive deep-sea exploration (and studied in detail such deep-sea curiosities as the *HMS Titanic*).

Exciting new findings have come from space satellites that can indirectly image the entire ocean floor by bouncing a beam of microwaves between the satellite and the sea surface. NASA's SEASAT, a satellite dedicated to studying the ocean floor, has confirmed that the ocean's surface contains bulges and depressions that correspond to variations in sea-floor topography (Fig. 12-8). The added mass of a chain of submarine mountains, for example, produces a positive gravity anomaly (see Chapter 11 for a discussion of gravity anomalies) that attracts water, creating a sea-surface bulge as high as 30 meters (100 feet). Likewise, the lesser mass of a deep-sea trench produces a negative gravity anomaly and a corresponding sea-surface depression as large as 60 meters (200 feet) deep. In this way, satellites have charted vast unexplored areas of the world's ocean floors. In the remainder of this section, we describe how rifting forms ocean basins and how subsequent movement of ocean plates shape the features of the ocean floor.

Rifting and the Origin of Ocean Basins

Studies of the topography and bedrock composition of the ocean floor have enabled us to speculate about the origin and evolution of ocean basins. We have learned that oceans form where an existing continent rifts into two or more smaller continents, which then diverge. Rifting may begin when a warm current of mantle material, rising at one or more hot spots, stretches and thins the overlying continental lithosphere, causing its surface to break into a three-branched

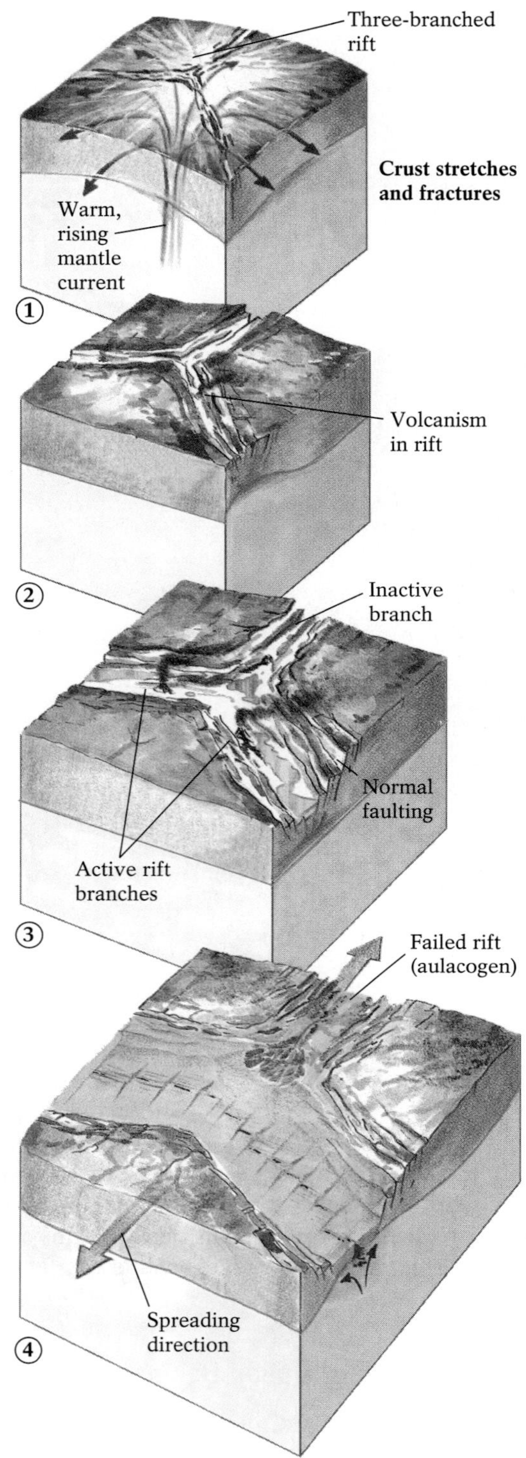

Figure 12-9 Active rifting of a continental plate. Rising warm currents from the Earth's interior flow beneath the continental litho-sphere, stretching and then tearing it. The initial branched rift typically consists of three fractures, two of which continue to diverge. The third, the aulacogen, becomes inactive.

fracture (Fig. 12-9). Such a rift, resulting from the uplift and breaching of the lithosphere above a hot spot, is called an *active rift*. Typically, as the mantle current flows beneath the rifting lithosphere, two of the fractures separate further, while the third becomes inactive. This inactive fracture forms a linear depression known as an *aulacogen* (from the Greek, meaning "furrow"), or *failed arm*.

A good modern example of an active rift, seen in Figure 12-10, is the East African rift zone. There, a three-branched fracture has already separated formerly contiguous Africa and Arabia. The rift's two active arms are the Red Sea and the Gulf of Aden, where divergence continues today. The East African branch of the three-part rift, which is currently occupied by the large lakes near which much fossil evidence of early humans has been found, is probably an aulacogen. The Great Rift Valley of East Africa has not grown wider for several million years, and it appears that rifting there may have either halted altogether or slowed dramatically. The valley may never evolve into an ocean, but instead may remain a sediment-filled topographic depression that will someday accommodate Africa's next great river system. The Red Sea, on the other hand, hovers on the threshold of becoming a true ocean. When rifting fully breaches the Sinai and Somali peninsulas, the Red Sea will connect the Mediterranean and the Indian Ocean. It has already begun to develop an area

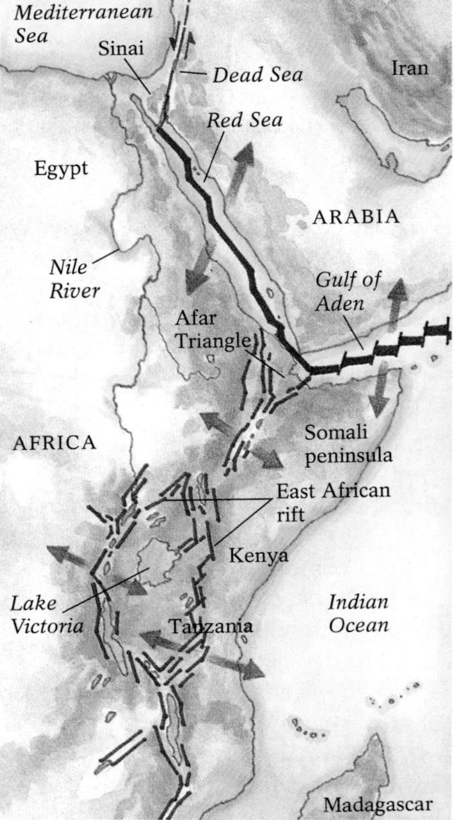

Figure 12-10 The East Africa rift zone has two actively expanding fractures—the Red Sea and the Gulf of Aden. The valleys and gorges of Kenya and Tanzania are the failed arm, or aulacogen, of the rift.

of true oceanic basaltic crust characterized by marine magnetic anomalies.

Many of the world's great river valleys, such as South America's Amazon River, Africa's Niger River, and North America's Mississippi River, may occupy the failed arms of ancient rifts. Intraplate earthquakes in the upper Mississippi valley, such as those recorded in the New Madrid area of Missouri and Tennessee, may have resulted from the release of stresses built up along such an ancient fracture. Another example of a modern river located in a possible aulacogen is the Connecticut River, which occupies a 180-million-year-old rift valley that dates from the breakup of Pangaea. Note that these rivers did not in any way cause the rifting; they are much younger than the rifts. Their modern courses simply follow the lowland topography of these ancient rift zones.

The two actively rifting fractures of a rift system are marked by high heat flow, normal faulting, frequent shallow earthquakes, and widespread basaltic volcanism. Normal faulting at rift fractures produces steep-walled rift valleys. Sedimentation in these valleys is particularly vigorous because of the sharp topographic contrast between the valley floors (graben) and the high-standing horsts marking the edges of the rifted plate. Figure 12-11 shows the stages of the growth of an ocean basin in a rift valley.

As rifting progresses, the rift valley lengthens and widens, possibly expanding until it reaches an adjacent ocean; the waters of the ocean may then flood the low-lying graben. If the new seaway later becomes isolated from the adjacent ocean, perhaps due to a drop in sea level, its water will likely evaporate, particularly in hot, arid climates. At this stage in the rifting process, rising and falling sea levels in rift valleys may produce thick layers of salty evaporites that alternate with shallow-marine clastic sediments.

Further rifting deepens the valley, and the thick sediment load subsides under its own weight. More seawater then floods in, followed by more evaporite deposition. Sediments accumulating in the seaway are deposited directly on rifted continental rock at first, and then on the new oceanic crust that forms as basaltic magma rises and cools in the rift. Sediments from the raised margins of the rift accumulate offshore. The high-standing cliffs above a rifted valley erode; eventually they may be reduced to sea level. As the rocks at the rift margins cool, the land surface subsides and the one-time rift edges become the foundations for continental shelves. After rifting ends, the original rift edges are no longer plate margins, nor are they volcanically or seismically active. Instead, they are now **passive continental margins.** The actual plate margins are half an ocean away, at the still-active spreading center.

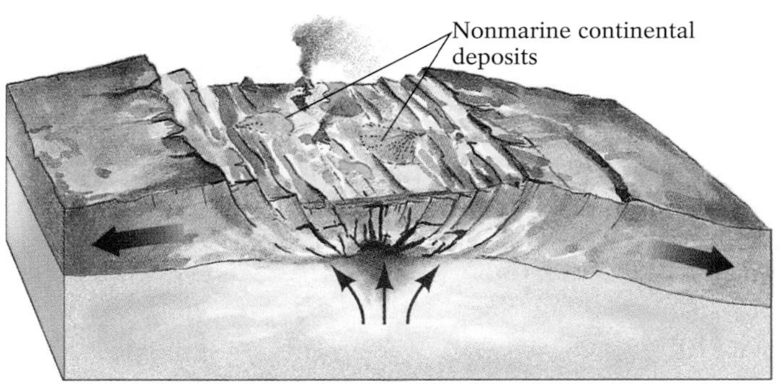

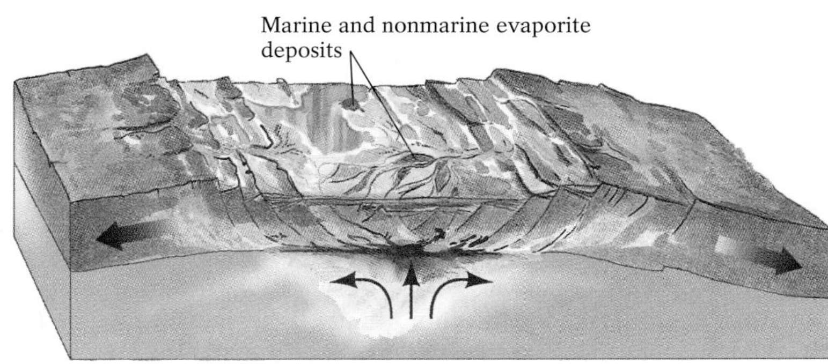

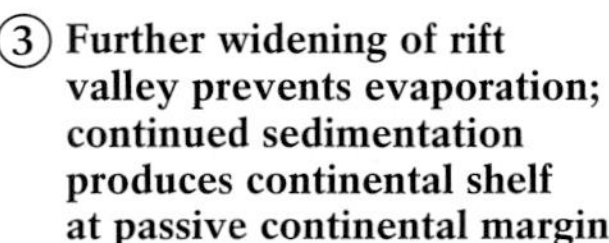

① Erosion of steep rift-valley walls causes rapid sedimentation

② Rift valley widens; water alternately fills basin and evaporates with changing sea levels

③ Further widening of rift valley prevents evaporation; continued sedimentation produces continental shelf at passive continental margin

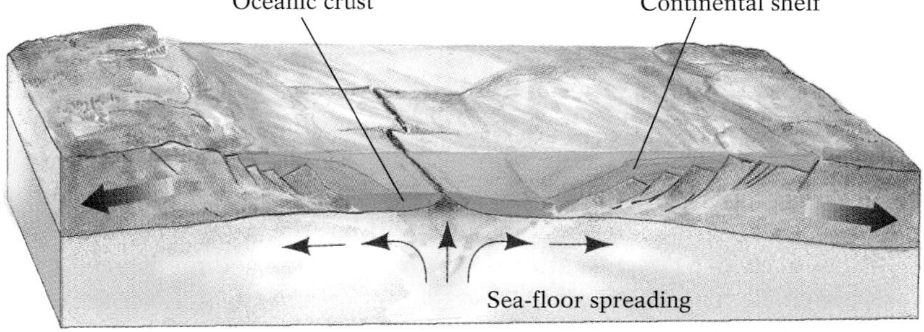

Figure 12-11 The growth of an oceanic basin in a rift valley, and the types of sedimentation that occur during its various stages.

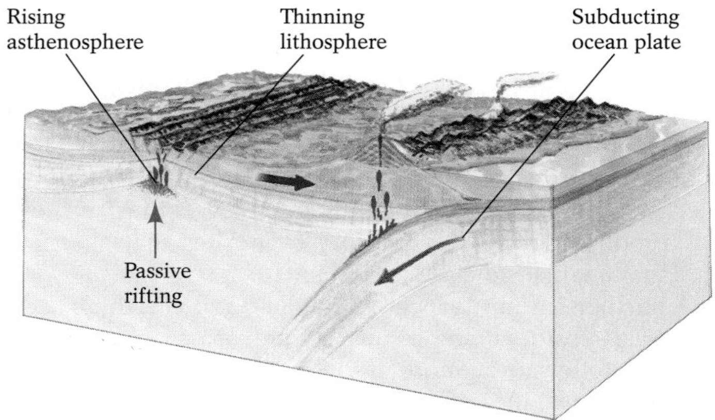

Figure 12-12 Passive rifting occurs where subducting oceanic lithosphere tugs at the overlying plate, causing it to stretch, thin, and then rift at some distance landward from the subduction zone's volcanic arc.

A second type of rift zone, a *passive rift,* is similarly marked by continental stretching, thinning, and rifting. Such rifts may develop not above a hot spot, but rather in association with subduction zones. As oceanic lithosphere subducts, the "suction" produced by its descent may tug powerfully on the adjacent continent, stretching and ultimately rifting it (Fig. 12-12). Uplift, high heat flow, and stretching and thinning of continental crust from *passive* rifting appear to be occurring in the Rio Grande area of Colorado and New Mexico, and in the Basin and Range province of Arizona, Nevada, and Utah. In the future, these areas may become rifts that cleave North America into two or more smaller continents, perhaps separated by a major seaway.

The final stage in the evolution of rifted continental margins can be observed today on opposite sides of vast oceans, such as the Atlantic Ocean along the east coasts of North and South America and the west coasts of Europe and Africa. Some evidence of the Atlantic's initial rifting lies buried beneath the sediment of these continental shelves. On land, much of the rifting record is evident in the fault-bound, 200-million-year-old sediments of the northeastern United States and in the numerous outcrops of basalt found along the Eastern seaboard, from the Canadian Maritime provinces and New England to North Carolina. Faulted basins and rocks of the same age and composition occur throughout the British Isles and northwestern Africa. We now know that, about 180 million years ago, these British and African rocks coexisted with today's North American rocks on the supercontinent of Pangaea. Their journey to their present locations began when a rift separated the Americas from Europe and Africa. It may have taken as many as 20 hot spots to rift Pangaea into today's major continents, most of which had taken shape by about 65 million years ago. Continents are most likely to rift when numerous hot spots are aligned, like the dotted perforations between two postage stamps.

Rifting sometimes ceases following an initial period of faulting and volcanism. We do not yet understand why it terminates. One such aborted rift is believed to have begun about 1.1 billion years ago, as an outpouring of mafic lava reached the surface in the middle of the North American continent. Basalts and gabbros of this age are found along the north and south shores of Lake Superior in Minnesota and Wisconsin. Farther south, similar rocks form a linear zone of anomalously high gravity buried beneath Cambrian and younger rocks. This mid-continent gravity high, tens of kilometers wide, stretches through Iowa and Missouri to the Oklahoma panhandle (see Figure 11-18). Had this rifting continued, Midwesterners today might be taking transoceanic flights to travel from Milwaukee to Minneapolis.

Divergent Plate Boundaries and the Development of the Ocean Floor

As rifted plates diverge, oceanic crust forms at the spreading center, which lies between the two new plates of oceanic lithosphere. As the two plates grow, an underwater mountain range, or **mid-ocean ridge,** uplifted by the hot, low-density material rising from the underlying mantle develops between the plates. Today, mid-ocean ridges stretch continuously for about 65,000 kilometers (40,000 miles) across all major basins (see Figure 1-15). They stretch as wide as 1500 kilometers (900 miles), and in some places their peaks rise more than 3 kilometers (2 miles) from the ocean bottom. Mid-ocean ridges are the largest raised topographic features on the Earth's surface, and their length and breadth account for approximately 23% of the Earth's total surface area.

Whereas most continental mountain systems consist principally of folded metamorphic and sedimentary rocks and massive batholiths of felsic igneous rock, ocean-ridge chains contain only relatively undeformed basalt. Most ridges are split down the middle by an *axial rift valley.* Such a valley forms as normal faulting along the *axial ridge crest* causes a large central block to drop downward. Some axial rift valleys are deeper than the Grand Canyon and three times as wide. Recent dives by submersibles directly into these valleys have yielded evidence of remarkable, previously unknown, biological and geological processes, which are described in Highlight 12-1 (p. 342).

Oceanic lithosphere everywhere has a similar structure (Fig. 12-13). Its upper surface to approximately 200 meters (700 feet) below the sea floor consists of unconsolidated sediment of siliceous or carbonate ooze from the remains of microscopic marine organisms, fine reddish-brown clay from weathering of iron-rich marine lavas, or both. In general, the older the oceanic plate, the thicker its sediment layer becomes, as sediment accumulates with time. Below this sediment layer is a 2-kilometer (1.2-mile)-thick layer of oceanic basalt, its top characteristically pillowed from the underwater eruption of lava. Under the basalt lies 5 to 6 kilometers (3–4 miles) of gabbro, formed from slow plutonic crystal-

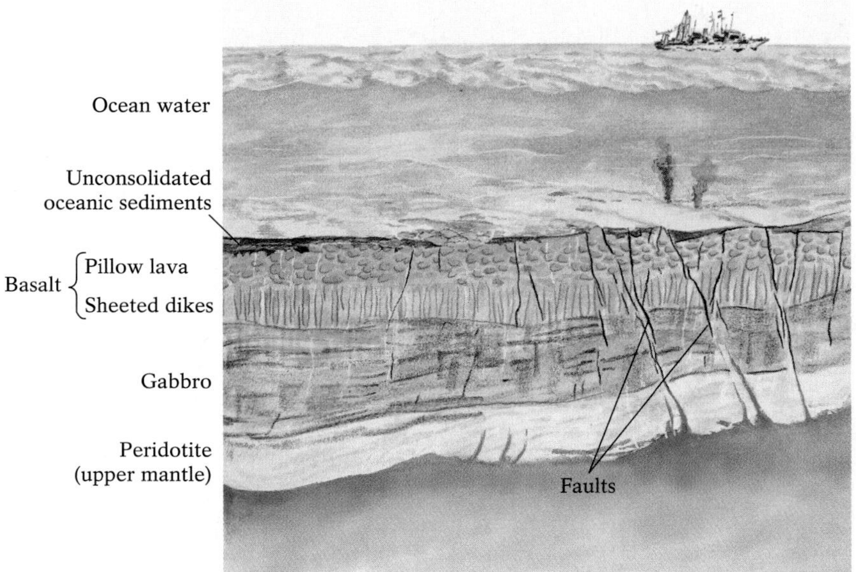

Figure 12-13 The layers of the ophiolite suite, which make up oceanic lithosphere.

lization and crystal settling within mafic magma. A layer of the ultramafic mantle rock peridotite resides at the base of the typical oceanic lithosphere.

The entire sequence of ocean-floor rock may be altered chemically as seawater penetrates its faults and fissures. Water reacts with the pyroxene in the basalt and gabbro to form a green mineral, chlorite; it also reacts with the magnesium olivine in the ultramafic peridotite to form the magnesium silicate mineral serpentine. These reactions may eventually produce *serpentinite,* a soft, greenish rock. The geological name for the group of rocks that make up oceanic lithosphere is the **ophiolite suite** (from the Greek *ophis,* meaning "serpent," and *lithos,* meaning "rock"—a reference to the green snake-like swirls within serpentinites).

Transform Boundaries and Offset Mid-ocean Ridges

Transform boundaries occur where plates slide past one another in opposite directions, like two passing trains headed for opposing destinations. They account for approximately 15% of the Earth's total length of plate margins. Much of this length occurs as distinct perpendicular offsets of the Earth's oceanic ridge system. (See Figure 1-15.) In large part, these offsets represent an artifact of the original rifting event that produced the ocean basin and its mid-ocean ridge. They are further accentuated by the variable rates of divergence along different segments of the ridge.

As we saw in Chapter 9, transform boundaries contain faulted plate blocks that move in opposite directions, producing stresses that typically cause powerful earthquakes. Earthquakes associated with oceanic transform boundaries almost always occur *between* the offset ridge segments, where the opposing motion of the two spreading ridges generates enough shearing stress to break rocks (Fig. 12-14). Beyond

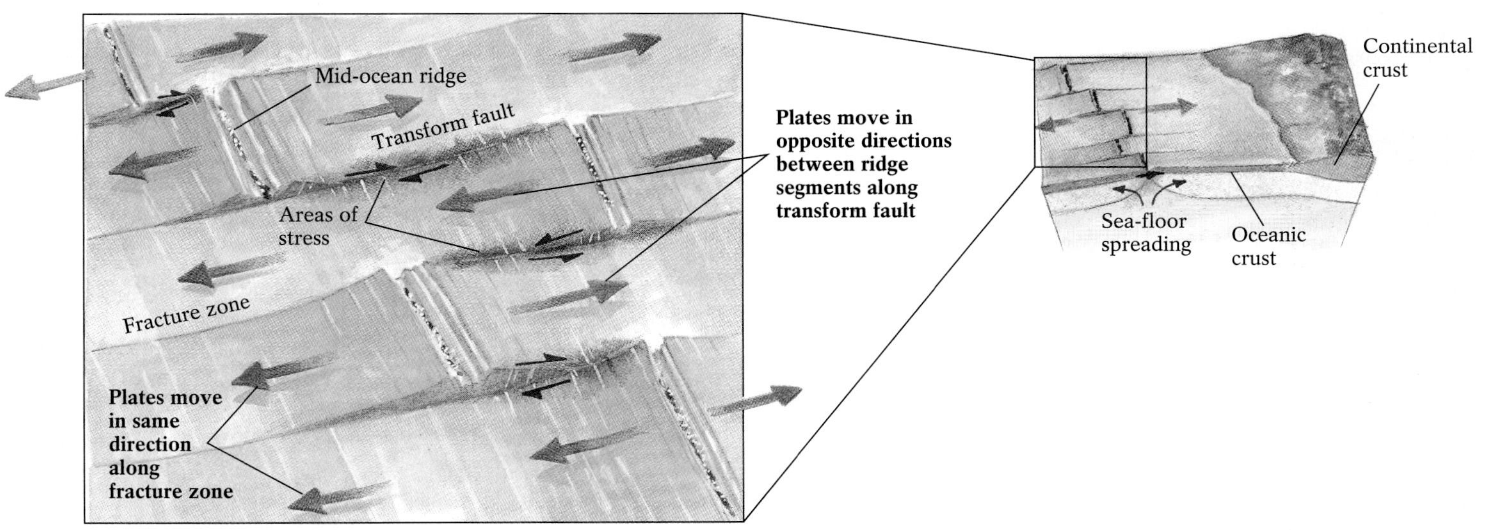

Figure 12-14 The direction of motion of the plates on opposite sides of an oceanic transform boundary is determined by spreading from mid-ocean ridge segments. The plates move in opposite directions only *between* these segments; consequently, the stresses associated with transform movements build up, and earthquakes occur only here. Fracture zones are the seismically inactive portions of transform boundaries beyond offset ridge segments, where adjacent plates move in the same direction. Note the high relief on either side of the fracture zones—a result of the differing age of the lithosphere and the effect of differential subsidence of rock having a differing density. (The younger, warmer, buoyant lithosphere at the divergent zone stands prominently above older, colder, denser, subsiding lithosphere across the fracture zone.)

Highlight 12-1 The Unseen World of Divergent Zones

Since the early 1970s, oceanographers using deep-sea submersibles have studied the rift valleys of the mid-Atlantic ridge, the Galápagos ridge off the coast of Ecuador and Peru, and the East Pacific rise south of Baja California. They have photographed the eruption of pillow lavas and the chemical interaction of cold seawater and warm basalt. During recent dives into the valleys of the East Pacific rise and the Juan de Fuca ridge (off the coast of Washington state), oceanographers returned with videos showing plumes of black sulfurous "clouds" of mineral-laden water rising from vertical chimney-like structures. These hydrothermal vents emit extremely hot water carrying gases, such as hydrogen sulfide, and numerous metallic sulfides. The water nourishes a complex community of hitherto unknown life forms, including giant clams and exotic tube worms (Fig. 12-15) that thrive in this high-temperature world untouched by sunlight.

Vast quantities of valuable minerals are accumulating within axial rift valleys. Seawater seeps down into newly formed oceanic crust, becomes heated by the underlying magma reservoir, and then dissolves copper, iron, zinc, cobalt, silver, and cadmium out of the warm mafic rocks. These 400°C (750°F) mineral-rich waters rise and erupt at the sea floor, creating dark plumes called *black smokers* (Fig. 12-16). Contact with cold seawater precipitates minerals from the plumes, encrusting the basaltic flows around each plume vent to form a chimney-like structure. Similar ore combinations occur within many of the Earth's folded mountains; many are likely ancient slabs of oceanic crust that originated in this manner.

Figure 12-15 Bacteria around volcanic vents at a mid-ocean ridge derive energy from heat-generated chemical reactions involving compounds such as hydrogen sulfide (H_2S). More complex creatures, such as these giant tube worms, subsist on these bacteria.

Figure 12-16 Black smokers consist of hot plumes of mineral-rich water vented at volcanically active regions of the sea floor. Accumulations of precipitated minerals form chimney-like structures around the plumes. The "smoke" here consists of hot water and particles of iron, copper, and other sulfide minerals.

Figure 12-17 The process that created the San Andreas fault began about 30 million years ago, as the westward-drifting North American plate made contact with a segment of the spreading center separating the Farallon and Pacific plates. The North American plate had not yet reached a more northern segment of the spreading center, which was offset to the west by a transform boundary; subduction of the southern portion of the spreading center under North America effectively split the Farallon plate into a separate northern plate (now called the Juan de Fuca and Gorda plates) and a southern plate (now called the Cocos). As the southern Farallon subducted, the transform boundary between the northern and southern plates elongated. About 3 million years ago, it "moved" onshore as the sea floor was uplifted and exposed above sea level, becoming the San Andreas fault. At roughly the same time, the Cocos plate's spreading center (the East Pacific rise) met Baja California, separating it from the rest of the continent and forming the Gulf of California. Today, the North American plate overrides what remains of both the northern and southern Farallon plates; in a few million years, the San Andreas fault will most likely extend for the full length of the West Coast.

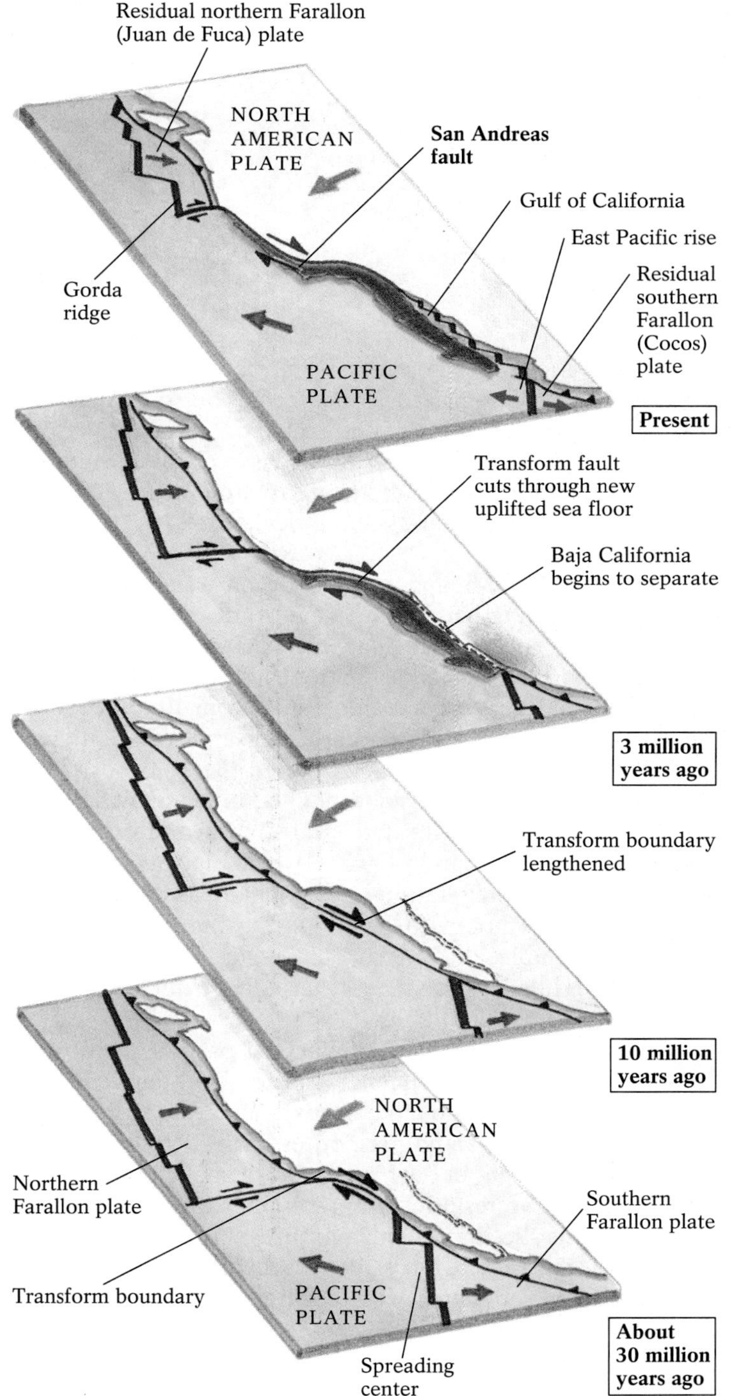

this inter-ridge area, plate movement occurs in the same direction on both sides of the transform boundary, mitigating the shearing stress and minimizing earthquake activity.

California's San Andreas fault zone is a transform boundary—part oceanic, part continental—that extends from the northern end of the East Pacific rise in the Gulf of California to the southern end of the Gorda ridge off the coast of southern Oregon and northern California. It was created by a complex sequence of events involving both oceanic transform movement and subduction. Figure 12-17 summarizes the main stages in the evolution of the San Andreas transform boundary.

Convergence and Subducting Plate Margins

Oceanic plates diverging from mid-ocean ridges eventually encounter less-dense plates and are subducted under them, descending into the Earth's mantle. Large segments of the ocean floor have been swallowed in this way. The Pacific, for instance, is a shrinking ocean, with subduction occurring along most of its margins. The spreading ridge of the East Pacific rise is no longer in the ocean's center, because subduction along the west coasts of the Americas has consumed the eastern part of the Pacific. As oceanic plates descend, they elevate the overriding continental plates; the large-scale destruction of Pacific oceanic lithosphere has produced the great western mountain ranges of the Andes, Sierra Nevada, Cascades, and Rockies.

Pacific plate subduction also explains why we observe an asymmetrical magnetic anomaly pattern off the coast of Washington and British Columbia, while a beautifully symmetrical pattern surrounds the mid-Atlantic ridge at Iceland

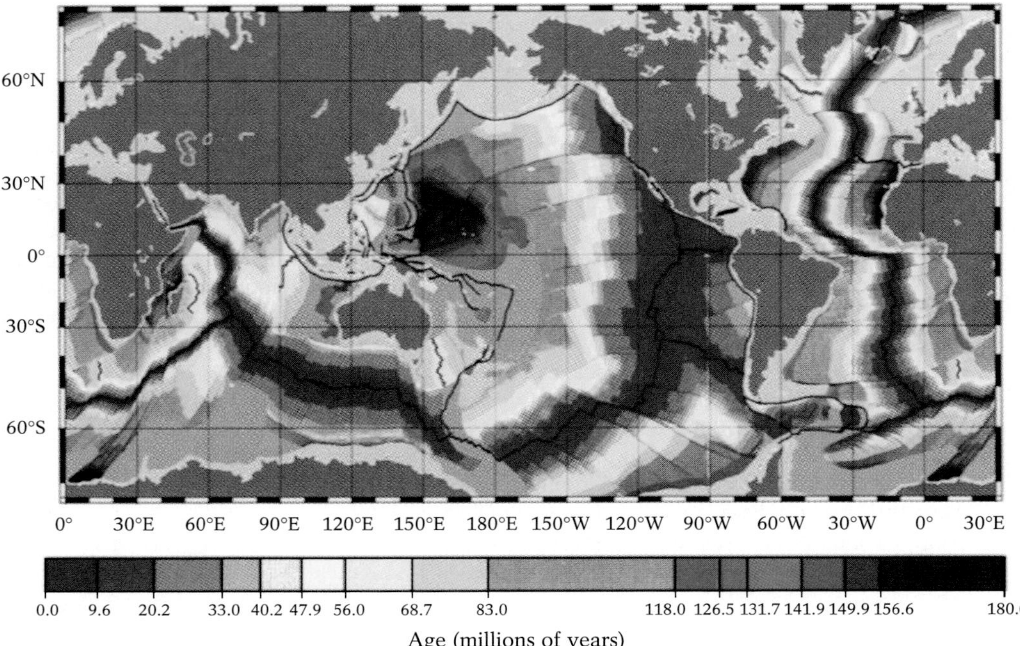

Figure 12-18 The ages of oceanic lithosphere segments around the world, as determined by dating marine magnetic anomalies. The colored stripes represent oceanic lithosphere of the ages indicated in the key. The width of each stripe is proportional to the spreading rate at the mid-ocean divergent plate boundaries. A symmetrical anomaly pattern (upper right) characterizes a mid-ocean spreading center, such as can be seen within the Atlantic Ocean basin. An asymmetrical pattern (center) shows that subduction has consumed part of an oceanic plate along the northwestern coast of North America.

(Fig. 12-18). Most of the oceanic plate east of the Gorda and Juan de Fuca spreading ridges has been consumed by subduction, eliminating most of the magnetic anomaly pattern east of the ridges. Neither of the oceanic plates forming at the mid-Atlantic ridge is subducting; therefore the anomaly patterns of both places are symmetrical.

Ocean Trenches and Sedimentation **Ocean trenches** develop where a subducting plate flexes downward into the mantle, forming deep, curvilinear, relatively narrow depressions in the Earth's surface. Some Pacific trenches, particularly in the tectonically active western region, range from 40 to 120 kilometers (25–75 miles) wide and are thousands of kilometers long. The deepest is the 11,022-meter (36,161-foot)-deep Marianas trench near the island of Guam. Sediment scraped from the subducting plate and eroded from the overriding plate accumulates in these trenches, which may become completely filled if sedimentation rates are high, as they are next to rapidly rising mountain belts. During the last 3 million years, for example, the trench off Oregon and Washington has become filled by sediments eroding from the rising, glaciated Cascade Mountains.

As a cold oceanic plate subducts, it packs a mixture of fine-grained deep-sea sediments, coarser land-derived sediments, siliceous and carbonate oozes, submarine basalts, and serpentinized gabbros and peridotites against the inner wall of the trench. Caught in the high-pressure zone between converging plates, the resulting mass solidifies to become rocky material that is sliced, crushed, and thrust into a chaotic jumble called a **mélange.** Cold slabs of oceanic lithosphere and their associated sediments are subducted relatively rapidly to depths of 30 kilometers (20 miles) or more, where they metamorphose under the unique combination of low temperature and high pressure. The blueschist minerals that commonly occur within a mélange signal the occurrence of metamorphism within a subduction zone. (See Chapter 7 to review subduction-zone metamorphism.)

In North America, mélanges that have been uplifted subsequent to their formation punctuate the Franciscan rocks of the coast ranges of California plate (Fig. 12-19), a prod-

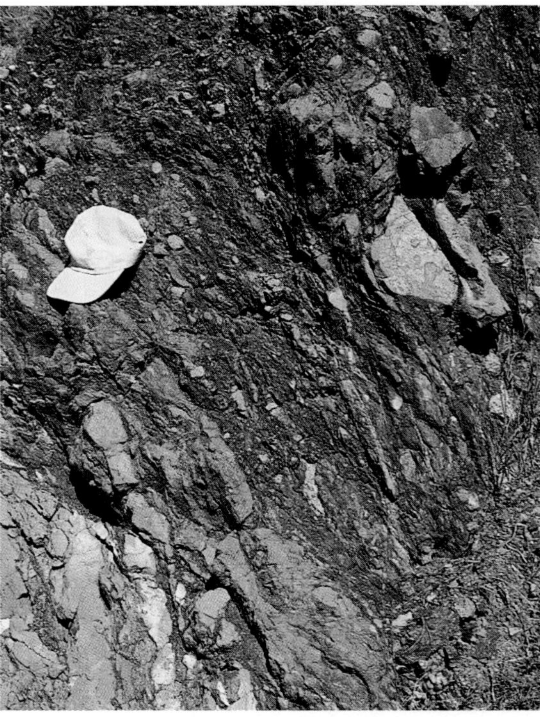

Figure 12-19 A mélange along the Sonoma County coast in California. These rocks—a jumbled mixture of sea-floor and land-derived materials that have undergone high-pressure/low-temperature metamorphism—formed in the ocean trench associated with the subduction of the Farallon plate under North America.

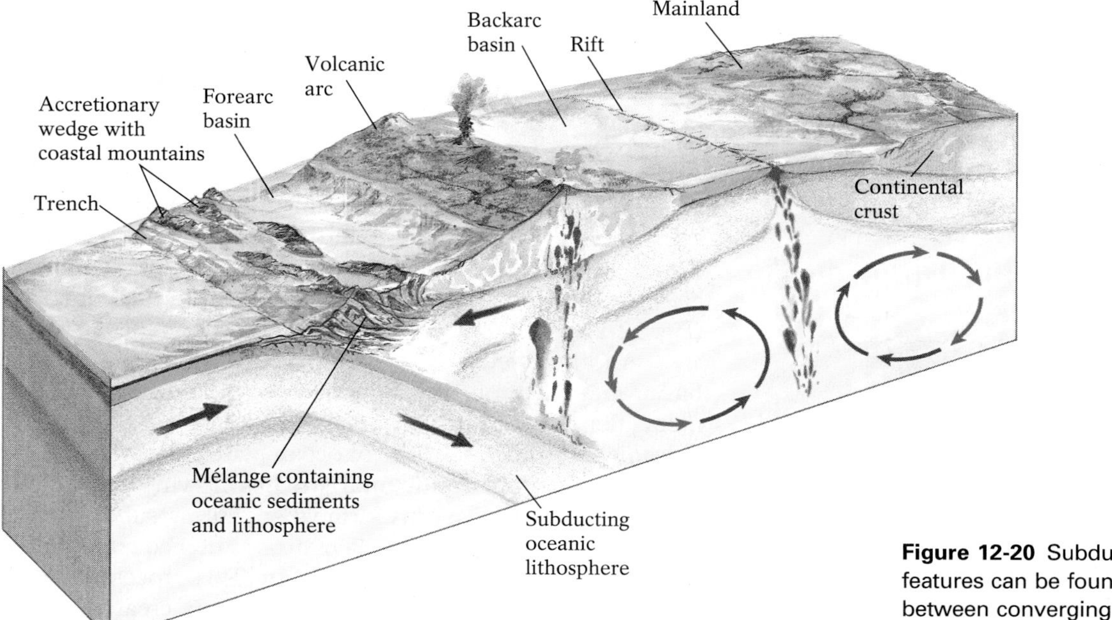

Figure 12-20 Subduction-zone features. These features can be found at the subduction zones between converging oceanic and continental plates or between two plates of oceanic lithosphere.

uct of the consumption of the Farallon plate. They are also found in the Klamath Mountains of southern Oregon, the Kootenay Mountains of eastern British Columbia, the coastal mountains of south-central Alaska, the Blue and Wallowa Mountains of eastern Oregon, and the Appalachians of New England and the Canadian maritime provinces. Each mélange represents strong evidence of past subduction.

Other Features of Subduction Zones The process of subduction forms various characteristic structures, many of which are illustrated in Figure 12-20. Mélanges and other thrust-faulted rocks pile up within an ocean trench to form an **accretionary wedge,** a mass of sediments and oceanic lithosphere scraped from the subducting plate and plastered onto the edge of the overriding plate. As rocks accumulate over time, the wedge thickens and is lifted isostatically, forming a linear range of mountains just inland of the trench. The coastal ranges of California, such as those found near Big Sur (Fig. 12-21), originated in this way. Farther inland is the **volcanic arc,** a chain of volcanoes fueled by magmas rising from above the subducting plate.

Between the accretionary wedge and volcanic arc is a sediment-trapping depression called the **forearc basin.** Sediment eroded from both the uplifted accretionary wedge and the volcanic arc accumulates in the forearc basin. Thus this sediment may be of either marine or terrestrial origin.

The distance between the ocean trench and the volcanic arc is termed the *arc-trench gap.* The breadth of the arc-trench gap is determined by the angle at which the subducting plate descends. Steep subduction—typical of the relatively rapid descent of old, dense oceanic lithosphere—transports a subducting plate relatively quickly to greater depths and higher

Figure 12-21 Coastal mountains typically begin as an accretionary wedge at a subduction zone. These coastal mountains near Big Sur, California, originated as offshore turbidites of continental origin that became packed in the Farallon trench and were dragged downward tens of kilometers before being uplifted isostatically.

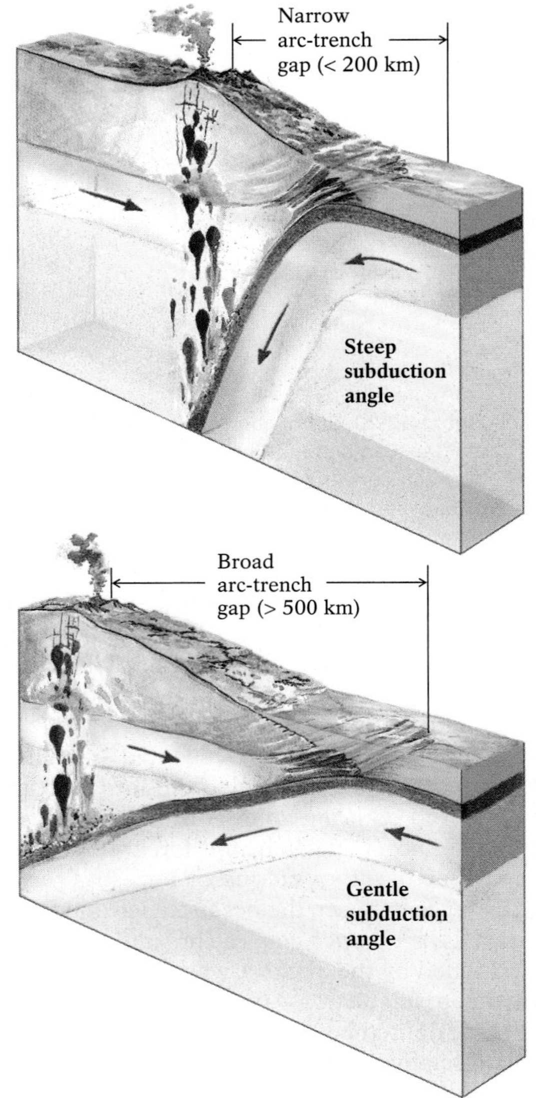

Figure 12-22 The breadth of an arc-trench gap is proportional to the size of the angle of subduction. A steep angle produces a narrow arc-trench gap; a gentle angle produces a broad arc-trench gap.

temperatures. This typically produces a narrow arc-trench gap and a narrow forearc basin. Where the angle of subduction is relatively gentle—typical of young, warm, buoyant oceanic lithosphere—a subducting plate must traverse more lateral distance to reach greater depths. Such gentle subduction angles produce broad arc-trench gaps and wide forearc basins (Fig. 12-22).

Another sediment-trapping depression, or **backarc basin,** may form on the inland side of a volcanic arc. Sediments deposited in this region are derived from the eroding volcanic arc, and in the case of ocean plate–continental plate convergence, from continental streams flowing toward the sea as well. Backarc basins form when, during subduction at an ocean trench, the overriding plate is stretched and thinned on the "back" side of the volcanic arc. This process may occur because of either local mantle currents created by the subduction or tensional stresses in the overriding plate that result from the gravitational pull ("slab suction") generated by the subducting plate (Fig. 12-23). In either case, the crust of the overriding plate may stretch so much that it eventually rifts, which in turn reduces the pressure on the underlying mantle and lowers the melting point of the mantle's rocks at that point. Currents of basaltic magma may then rise and solidify into new oceanic crust, a process called **backarc spreading.** Backarc spreading beneath the Sea of Japan, for instance, is causing the volcanic island arc of Japan to move eastward, as the backarc region expands from the creation of new sea-floor crust between Japan and China. Such spreading is marked by thinning plates, high heat flow, normal faulting, frequent earthquakes, and basaltic volcanism.

Convergence and Continental Collisions

Given enough time, the entire oceanic portion of any subducting plate will be reabsorbed into the Earth's mantle. At this point, the plate's continental portion (if it has one) reaches the subduction zone, where it may encounter another continental plate. This process results in the third principal type of

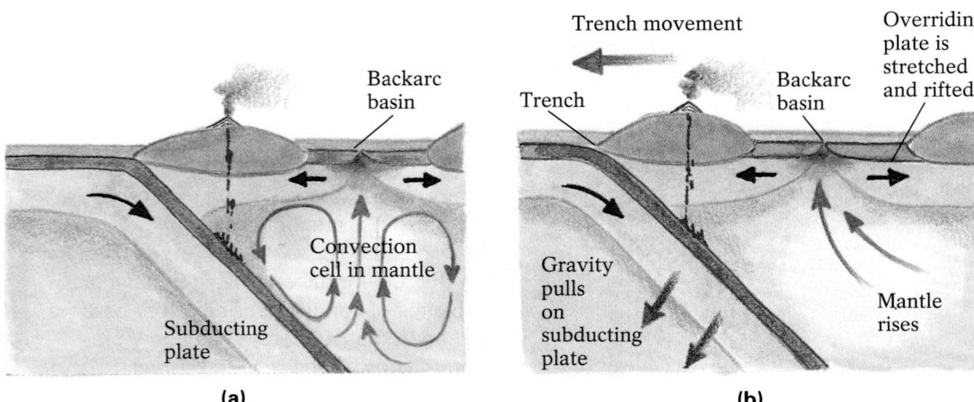

Figure 12-23 Two proposed mechanisms of backarc spreading. **(a)** Subduction-induced mantle currents. **(b)** Tension from gravitational pull on the subducting plate.

Highlight 12-2 Do Continents Subduct?

Since the inception of plate tectonics roughly 30 years ago, conventional wisdom has held that only oceanic plates are dense enough to subduct. Once created, continental lithosphere may rift, drift, or collide with other continents, but it supposedly remains essentially stuck at the Earth's surface because of its buoyancy relative to the underlying mantle. Recent discoveries in southern Africa, southwestern Norway, and the southern Swiss Alps, however, are compelling today's geologists to rethink this conventional wisdom and ponder two questions: Do continents subduct? Have slabs of continental lithosphere been recycled within the Earth's mantle?

The first solid clues in support of continental-lithospheric subduction have been found—of all places—within crystals of South African diamonds. As we discussed in Chapter 2, diamonds form when carbon atoms are squeezed under enormous pressure at depths of at least 120 kilometers (75 miles) within the Earth's mantle. During their formation, minute specks of mantle materials are sometimes included within a diamond's structure. Ultimately, spectacular gas-driven volcanic eruptions dispatch these gems to the surface, with these mantle-derived minerals hitching a ride.

In light of this scenario, how do geologists explain the presence of minute *staurolite* crystals—minerals that do not typically occur in the mantle—in some South African diamonds—minerals that can form *only* in the mantle? Staurolite is a relatively common medium-grade metamorphic mineral that forms in the Earth's continental crust, typically at a pressure of about 10 kilobars (10,000 times atmospheric pressure) and a temperature of 575°C (1070°F). Diamonds, on the other hand, require about 40 kilobars of pressure (40,000 times atmospheric pressure) and temperatures in the neighborhood of 900°C (1650°F). Thus the presence of specks of staurolite in diamonds suggests that metamorphosed continental lithosphere was subducted to diamond-forming depths and then returned to the surface by explosive volcanism.

Now the scene shifts, and the mystery continues on Fjortoft Island, off the southwestern coast of Norway. There, tiny greenish diamonds (Fig. 12-24) have been found embedded in a mass of gneiss, a high-grade metamorphic rock that forms within continental crust at a pressure of about 15 kilobars and a temperature of 650°C (1205°F). Their presence suggests that a mass of continental crust—thought to consist of carbonaceous ancient sedimentary rocks—was pushed down into the mantle to diamond-forming depths before being thrust back to surface, perhaps in some type of mountain-building event. (Carbon in the sedimentary rocks may have provided the necessary ingredient for diamond-making.) Diamonds have also been found in similar metamorphic rocks in northern Kazakhstan and eastern China.

Figure 12-24 Arrows point to minute diamonds embedded within high-grade gneiss in coastal Norway. (inset) Photomicrograph of a diamond grain from the rocks of Kazakhstan.

Finally, whereas the other clues comprise specks of staurolite and scattered microscopic crystals of diamond, the existence of a block of continental lithosphere in the southern Swiss Alps—a kilometer long and a half-kilometer wide—further supports the premise that continental rocks can subduct. The rocks of Arami Massif, near the Italian border, contain deep-red crystals of iron-rich garnet mixed with microscopic crystalline rods of iron titanium oxide. This unique mineral assemblage is believed to form deep within the upper mantle—perhaps at depths as great as 400 to 650 kilometers (250–415 miles). Here again, rocks of the Earth's continental crust contain minerals that could have crystallized only under the enormous pressures and temperatures of the mantle—further proof that tectonic forces may have recycled continental lithosphere within the Earth's mantle.

plate convergence—continental collision. As we have seen, continental plates have relatively low density and, compared with the underlying mantle, are buoyant. For decades, conventional plate-tectonic wisdom has suggested that when the forward edges of two continental plates converge, neither subducts. The lightweight continents simply collide and compress each other's rocky boundaries, becoming welded together into a single larger block of continental rock. Although this process has probably occurred in many areas, recent studies have begun to suggest a different story—one involving subducting continental edges that are dragged down to the mantle, perhaps only temporarily, and then bob back to the surface. Highlight 12-2, "Do Continents Subduct?", introduces some recent hypotheses about this issue.

The boundary between collided continents is known as a **suture zone.** During a collision, continental crust within a suture zone thickens, because one plate thrusts slightly beneath the other. The crust also thickens there, because slices faulted from the colliding edges of both plates become swept into fold-and-thrust mountains. Fragments of an oceanic plate often become trapped between colliding plates; consequently mountains that rise high above sea level sometimes contain large masses of ophiolite rocks. The complex folded and thrust-faulted mountains in suture zones may also incorporate any other material that lies in the path of the colliding plates. This development explains why remnants of stratovolcanoes and felsic batholiths from collided volcanic arcs, or metamorphosed mélanges from the subduction trenches of departed oceanic plates, can sometimes be identified within a mountain's rocky structure.

Such collisions produced many of the Earth's mountain ranges, including the Alps in southern Europe (Fig. 12-25), the Appalachians in eastern North America, and the north–south Urals that formed during the collision of Europe and Asia. Today, continental plates are actively colliding in only one area: from the northern shore of the eastern Mediterranean, where Africa impinges on southern Europe, across to the Himalayas, where India butts against southern Asia. The story of the ongoing creation of the Himalayas, the Earth's loftiest mountains, is summarized in Highlight 12-3.

Based on evidence from the past, ancient suture zones are likely sites for future rifting. When a plate is stretched, either actively or passively, it tends to rift at its weakest points, which are typically old sutures. One convincing piece of evidence is the rift that currently exists along the ancient African–European suture zone, in which the Mediterranean Sea has formed and re-formed several times. Today the Mediterranean lies between the African and Eurasian plates, whose multiple collisions have in the past intermittently uplifted the European Alps and northern Africa's Atlas Mountains. After a lengthy period of rifting, divergence, and ocean building, subduction and convergence may resume, reuniting formerly contiguous landmasses. Along the northern shore of the present-day Mediterranean, subduction is again closing the gap between Africa and Europe. If this convergence continues, a collision will eventually occur and form a brand-new set of Alps. Italian volcanism, devastating Greek and Turkish earthquakes, and uplift of the island of Cyprus from the sea floor offer some hints of the impending collision.

Similarly, the modern Atlantic Ocean opened 180 million years ago along a suture zone that had formed some 40 million years earlier, when North America, Europe, and Africa collided to form the supercontinent Pangaea, creating the Appalachian mountain system (see Highlight 9-2). Rifting, however, does not always follow the line of an old suture zone precisely: When Pangaea rifted apart, for example, a large block of Africa, hundreds of kilometers wide, stretching from New York to Florida, remained attached to North America east of the Appalachians.

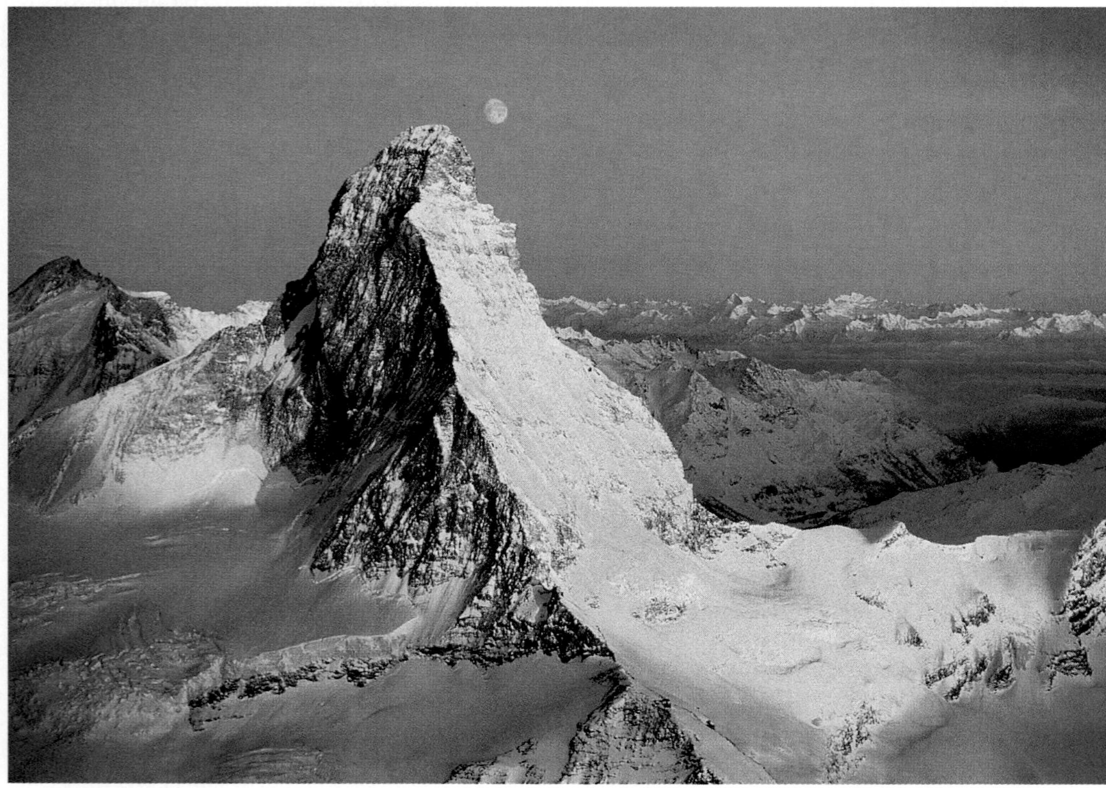

Figure 12-25 The Matterhorn in the Swiss Alps formed when the southern edge of the Eurasian plate collided with the northern edge of the African plate.

Highlight 12-3 Convergence and the Birth of the Himalayas

The origin of the Himalaya Mountains, depicted in Figure 12-26, involved convergence, collision, and suturing of India to Asia that took place over millions of years. These events have had far-reaching consequences for the entire Earth, perhaps even contributing significantly to the triggering of our most recent ice age. The uplift of this mountain chain certainly changed the climate of Asia by isolating it from the southern oceans.

The process leading to the Himalayan orogeny began during the Mesozoic Era, when India separated from the Pangaean assembly of continents (see Chapter 1). About 180 million years ago, India and Madagascar broke away from Pangaea and began to drift northward. (Soon after, Madagascar rifted away from India.) The Tethys Ocean, which separated the Indian continent from Asia, was being consumed along a subduction zone south of Asia. The passage of the drifting Indian lithosphere over the stationary Reunion hot spot some 65 million years ago brought a period of intense volcanism, marked by the eruption of the extensive Deccan basaltic flows. Today, the Reunion hot spot lies about 5000 kilometers (3000 miles) southwest of India in the Indian Ocean.

(An interesting aside: Because of the 65-million-year age of this volcanic event, some geologists believe that the eruption of the Deccan basalts may have caused—or at least contributed to—the demise of the Earth's dinosaurs. These scientists suggest that high concentrations of CO_2 and other gases emitted during these eruptions may have produced a substantial greenhouse effect that altered the Earth's climate sufficiently to disrupt the planet's food chain and initiate widespread extinctions. This idea represents an alternative view to the meteorite hypothesis discussed in Chapter 1.)

For the next 15 million years, India continued to move rapidly northward as subduction along the southern edge of Asia continued. The collision of India with Asia began about 50 million years ago, when the northern margin of India collided with the Tibetan microcontinent, first in the northwest, then about 10 million years later in the east. For the past 40 to 50 million years, sediments on the leading edge of the Indian plate and accretionary wedges, volcanic arcs, and batholiths along the southern edge of the Asian plate have been folded and thrust-faulted up onto the continents to form the modern Himalayan mountain belt.

Continued convergence of India into the Asian continent has resulted in repeated thrust faulting of the leading edge of the Indian plate, increasing the thickness of the continental crust beneath the emerging Himalayas. This increased thickness appears today as the Tibetan Plateau—the highest plateau on Earth. Several suture zones mark the boundary between the colliding continents and continental fragments trapped within the collisional zone. India is still moving into Asia today at nearly 5 centimeters (2 inches) per year. Evidence of this ongoing plate convergence can be seen in this region's frequent earthquakes.

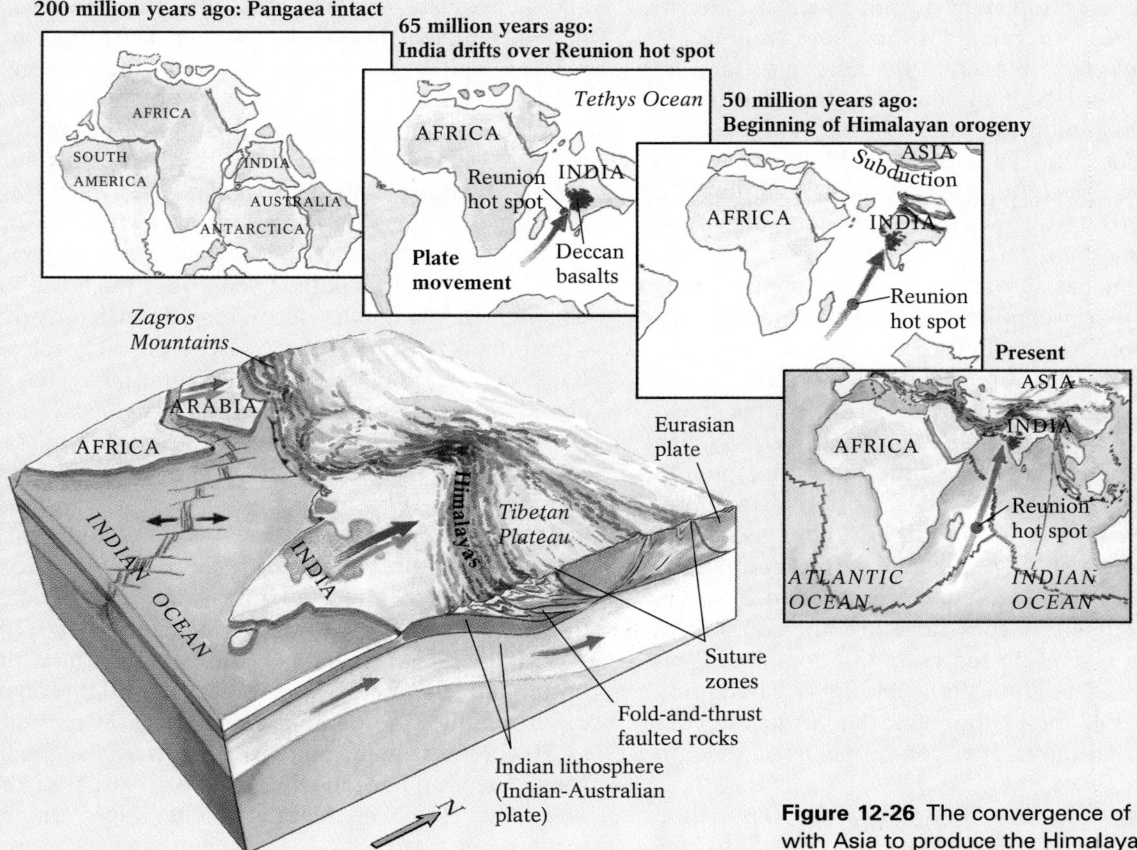

Figure 12-26 The convergence of India with Asia to produce the Himalayas.

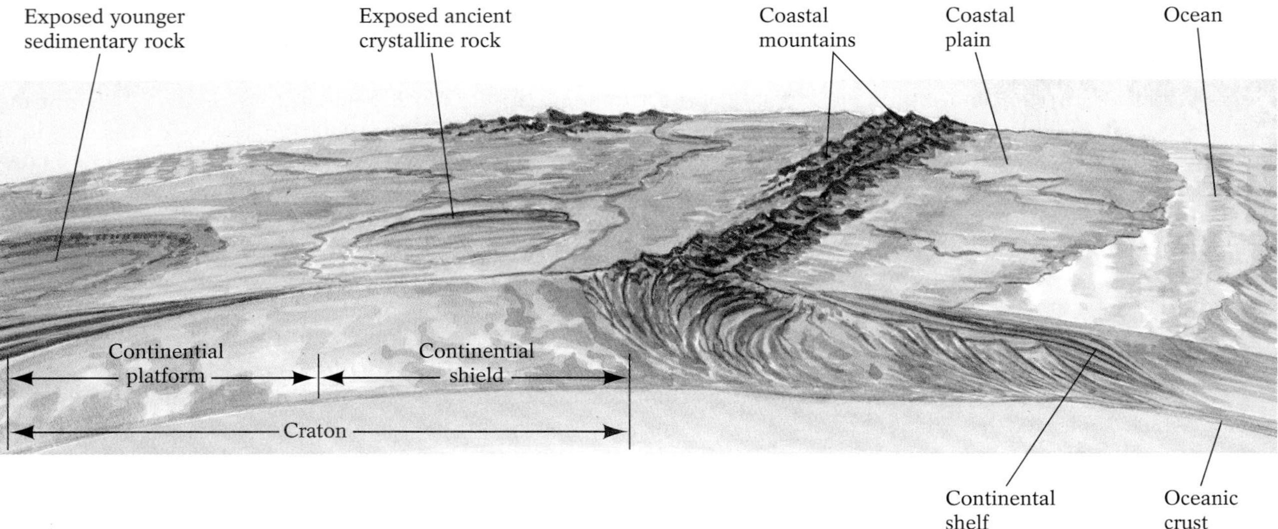

Figure 12-27 The anatomy of a continent. Every continent contains a tectonically stable nucleus, or craton, that consists of one or more ancient, crystalline continental shields and a surrounding continental platform of sedimentary rock. The outer edge of the continent is typically marked by coastal mountains and plains and a submarine continental shelf.

The Origin and Shaping of Continents

Because most continents typically do not subduct (except, perhaps, at their edges), continental lithosphere is more apt than oceanic lithosphere to remain at the Earth's surface. As a result, whereas the rocks of the current ocean basins are rarely more than 200 million years old (older rocks have been recycled by subduction), rocks dated at 4 billion years old have been found on the continents. For clues regarding the earliest stages in Earth history we must look here, not to the younger ocean floors.

Every continent has the same basic components (Fig. 12-27). The oldest parts comprise the **continental shields,** broad areas of exposed crystalline rock in continental interiors that have not changed appreciably for more than 1 billion years. Every continent contains at least one large shield area. North America's Canadian Shield extends across much of Canada from Manitoba to the Atlantic coast, dipping into the northern United States from northern Minnesota and Wisconsin to upstate New York's Adirondack Mountains.

Surrounding the continental shield is the **continental platform,** where the continental shield is covered by a veneer of younger sedimentary rock. Together, the continental shield and platform constitute the **craton,** a continental region that has remained tectonically stable for a vast period of time. At the edges of the craton, near the borders of continents, we find coastal mountains, coastal plains, and continental shelves.

In this section we discuss the origin of the continental lithosphere and the movement of plates that caused the formation of the continents.

The Origin of Continental Lithosphere

In the Earth's first few hundred million years of existence, the planet had no continents, no oceans, no atmosphere, and—obviously—no land or sea life. The planet's surface may have resembled that of the present-day Moon, being pockmarked with craters caused by the impact of countless fragments of interplanetary debris. Four billion years later (by about 600 million years ago), continental landmasses with primitive plants covered 30% of the Earth's surface; vast oceans, teeming with diverse plant and animal life, constituted the remaining 70%. A thick protective atmosphere covered the entire planet. How did such dramatic changes arise?

A few hundred million years after the Earth formed (see Chapter 1), the impact of infalling debris converted gravitational energy to heat energy. Heat was also released as the Earth developed its differentiated internal layers. The Earth's abundant supply of radioactive isotopes decayed to produce even more heat. Geologists hypothesize that heat from these three sources melted materials within the primordial Earth, causing them to become differentiated into distinct layers as gravity gradually separated materials of different densities. The densest substances migrated downward to form the iron–nickel core, and the lightest substances floated upward to form the rocks of the Earth's initial crust (see Chapter 11).

Four billion years ago, the newly formed mantle was hotter and less viscous than it is today, and it probably flowed more rapidly. Vigorous mantle convection currents brought great volumes of magma to the surface, and basaltic lava spewed from numerous hot spots across the Earth's surface. Enormous clouds of steam and other gases erupted as well, forming the Earth's first atmosphere and condensing to create the first global ocean. Evidence from recently discovered

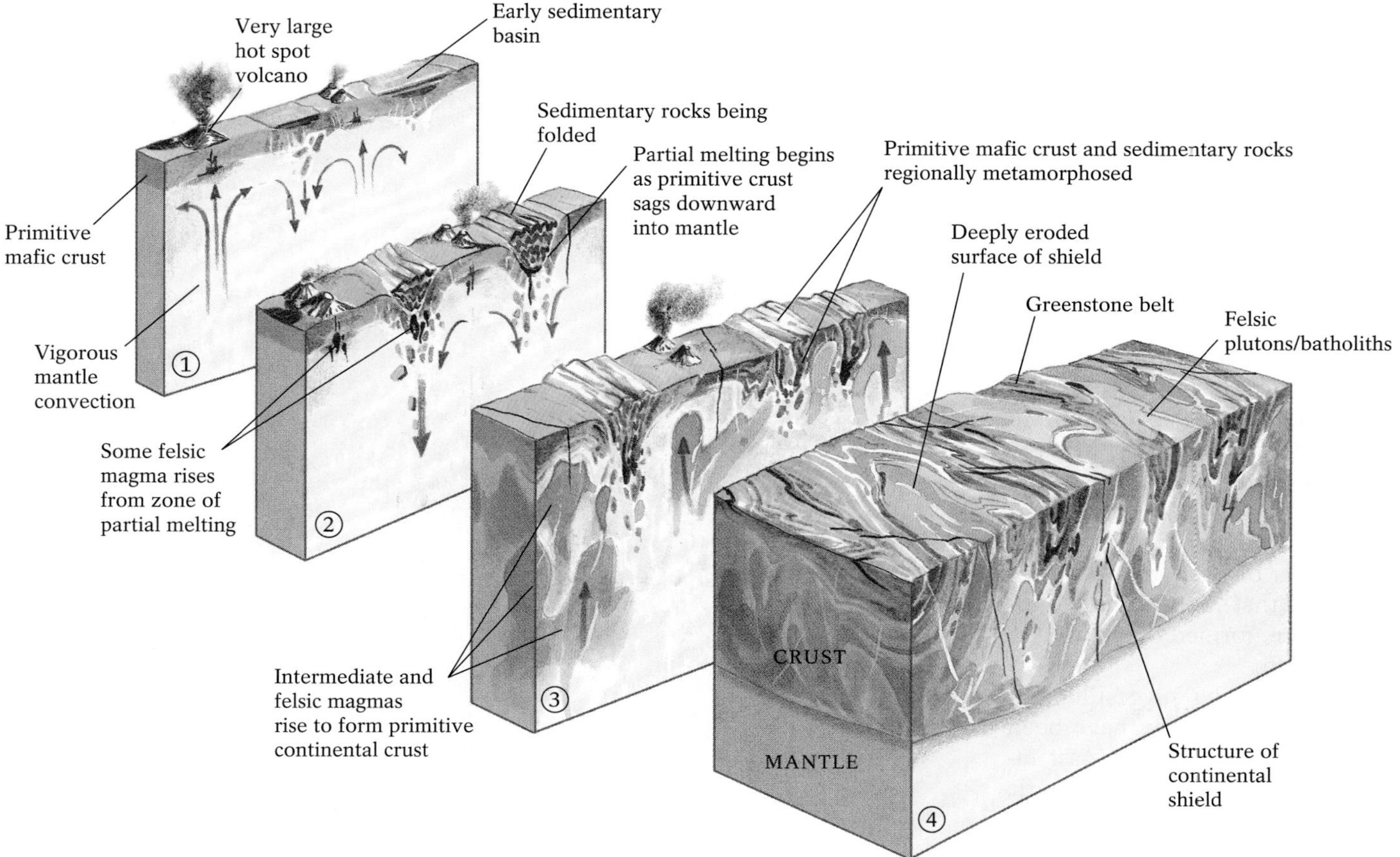

Figure 12-28 Continental shields formed when the Earth's interior was considerably hotter than it is today. Vigorous convection brought warm mantle material to the surface, where it erupted as mafic and ultramafic lavas. As the flows cooled and became denser, they subsided into deeper regions. There, they partially melted and differentiated, eventually solidifying as intermediate and felsic plutons. These ancient, low-density plutons are the oldest rocks on Earth and, together with metamorphic greenstone belts, make up the continental shields.

rocks in Australia, Greenland, and North America supports these hypotheses. Although they are now metamorphic, these ancient rocks (dating from 3.8 to 3.96 billion years ago) were originally sedimentary, suggesting that some type of atmosphere must have existed to weather sediment from preexisting rocks. Some rocks even appear to have been transported by and deposited in bodies of water. We can therefore conclude that as early as 4 billion years ago, the Earth had both an atmosphere and bodies of water.

The Earth's early mafic crust was probably thinner, hotter, and thus more buoyant than the present-day mafic oceanic lithosphere. Consequently, it was less likely to be subducted. The first felsic *continental* lithosphere probably formed as a result of plate-tectonic processes that differed somewhat from those of today. As can be seen in Figure 12-28, before distinct continents existed, early landmasses probably consisted of huge volcanic islands that became enlarged as great volumes of mafic (basaltic) and ultramafic lava erupted from the mantle. As the eruptions continued, the islands grew and thickened, and early sedimentary basins developed around them. Older sections of the early lithosphere cooled and became more dense, until some parts sagged downward into the hotter interior, folding and dragging the sedimentary basins along with them. The basalts, ultramafics, and sediments partially melted and then differentiated, producing the first intermediate and felsic magmas. Continued partial melting and differentiation of these rock materials eventually created new sets of volcanic islands that consisted largely of metamorphic basalts, called *greenstones*, and felsic plutons of varying compositions. This combination of rock types characterizes most ancient continental shields.

After small, lower-density felsic landmasses developed, subduction began to occur where the dense oceanic lithosphere and the light continental lithosphere converged. Recycling of primitive oceanic lithosphere increased the volume of andesitic rocks and their characteristic stratovolcanoes (Chapter 4). In time, the continental nuclei became enlarged as drifting continental plates collided, fold-and-thrust mountains were uplifted, and forearc mélanges accumulated. These masses of ancient continental lithosphere have since been weathered and eroded over the ages, a process that removed the overlying sediments to expose their deep plutonic and metamorphic roots. The landmasses subsequently rifted, creating numerous centers of continental growth around which additional subduction and accretion took place, further increasing the extent of continental lithosphere. Today, these centers of growth form the Earth's stable shields.

Nearly all of the continental shields developed during the Precambrian Era, with much of the early continental lithosphere forming within the Earth's first billion years. Some remnants of this early lithosphere still remain, such as the 3.96-billion-year-old Acasta Gneiss of the Yellowknife Lake in the Canadian Arctic. The noticeable scarcity of rocks of such great age initially suggested that most of the continental lithosphere formed long after the birth of the Earth.

Recent discoveries suggesting that continental lithosphere may have been subducted and recycled in the past indicate that most continental lithospheric growth had occurred by about 3.5 billion years ago. This view is increasingly supported by recent studies of the isotope geochemistry of shield rocks, mantle rocks, and the solar system's primordial materials in meteorites. In the last billion years or so, plate-tectonic activity appears to have slowed considerably, probably because the Earth's internal heat supply has diminished. Apparently little continental material has been added during the last billion years, even though the geographic configuration of the continental lithosphere has changed radically.

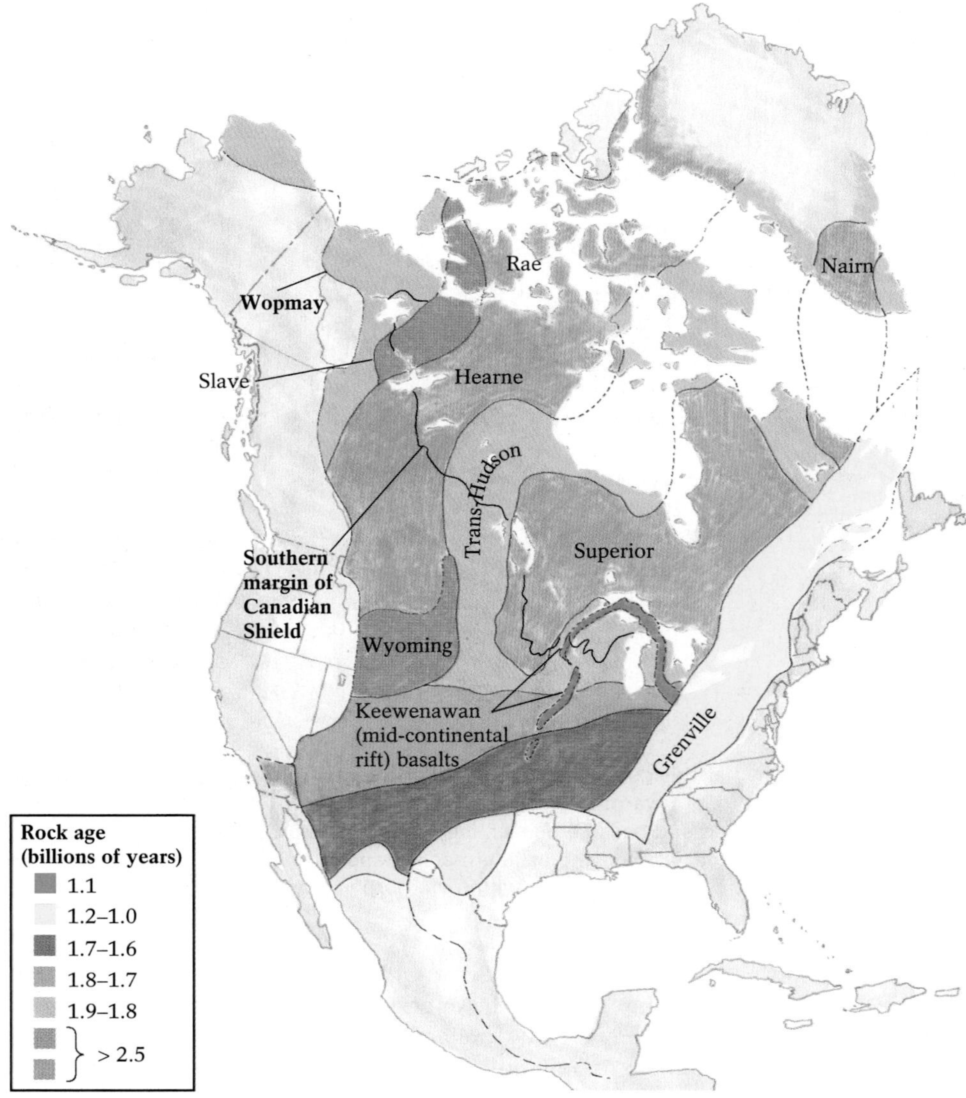

Figure 12-29 The geological provinces of North America. Rocks of the Canadian Shield, North America's ancient crystalline nucleus, are exposed throughout central and southern Ontario and Quebec as well as in northern Michigan, Wisconsin, and Minnesota. In most other places, however, younger rock covers the provinces.

While some continental lithosphere continues to return to the mantle through the continental erosion–marine deposition–subduction cycle, and, as we discussed in Highlight 12-2, some may have even undergone subduction, an approximately equal amount has presumably been produced by the cycle of subduction–partial melting–volcanic arc eruption. Hence, the Earth's overall volume of continental lithosphere has probably remained fairly constant for roughly the past 3 billion years.

The core of the Canadian Shield, which has become exposed in Ontario, western Quebec, and the northern Great Lakes states, dates between 3.8 and 2.5 billion years ago. Progressively younger regions—added on during subsequent collisions between landmasses—surround these ancient rocks (Fig. 12-29). Like the cores of other continents, the North American nucleus is composed of large granitic intrusions, complexly deformed metamorphic rocks, and greenstone belts containing relatively undeformed, metamorphosed basalts.

Greenstone belts, so-named for their abundance of green rocks colored by the metamorphic mineral chlorite, are linear masses of metamorphosed pillow basalts, tens to hundreds of kilometers long, that are surrounded by metamorphosed oceanic mudstones. These materials probably comprise ancient ocean-floor rocks that became incorporated into a continental shield, perhaps during early episodes of plate subduction and collision. Almost all greenstone belts lie in rocks dated at 2.5 billion years or older.

The Changing Shape of the Earth's Continents

Rocks that are no more than 600 million years old border North America on both the Atlantic and Pacific coasts. Recent evidence indicates that many of these younger rocks are geological immigrants—fragments of distant plates that were transported here by convergent plate motion and were attached to our continent by collisions. Such fault-bounded rock bodies, which originated elsewhere, are called **displaced** (or *exotic*) **terranes.** The hunt for such terranes is an exciting aspect of continental geology.

We can distinguish displaced terranes from their surrounding rocks by their ages, geological structures, stratigraphies, fossil assemblages, and magnetic properties. Some may have been island arcs that formed in an ocean basin and then were towed to a continental margin by subduction of the intervening ocean plate. Others are **microcontinents,** or pieces of continental lithosphere broken from larger, distant continents by rifting or transform faulting. (For example, Baja, California and the portion of California west of the San Andreas fault may eventually separate from the rest of western North America and become an independently moving microcontinent.) Geologists may locate and map displaced terranes by searching for geological evidence of former plate margins. As we have seen, former convergent boundaries may contain evidence of subduction and suturing in the form of mélanges, blueschist minerals, and slices of ophiolite rocks.

More than 100 fault-bounded regions of various sizes, accreted to western North America from Alaska to California, have been identified as displaced terranes (Fig. 12-30). Most were swept against the western boundary of the westward-drifting North American plate after about 200 million years ago, when Pangaea rifted apart and the mid-Atlantic ridge began to develop. The gradual subduction of an ancient plate of the Pacific basin beneath the continent's western edge allowed dozens of buoyant fragments to become accreted to North America's western shore. The continent's west coast 200 million years ago probably extended south from present-day central British Columbia through Idaho to Utah or Arizona. Virtually all lands farther west (except for some younger volcanics and sediments) probably comprise displaced terranes.

One of the most extensive displaced terranes of western North America is Wrangellia, stretching more than 3000 kilometers (1900 miles) from the Wrangell Mountains east of Anchorage, Alaska (see Figure 11-26), through Vancouver Island, British Columbia, to Hells Canyon, Idaho. Wrangellia's rocks and fossils are distinctive: They boast thick sequences of basalts, shallow marine limestones, and the fossil clam *Daonella,* a species found in rocks native to Asia but nowhere else in North America. The magnetic record of Wrangellian basalts suggests that they formed near the equator. About 100 million years ago, this microcontinent may have drifted northeast from equatorial Asia for 4000 kilometers (2500 miles) or more before colliding with North America.

But where did Wrangellia first collide along the western margin of North America? Geoscientists who study the displaced terranes of western North America are engaged in a lively debate over this question. One view—based largely on the paleomagnetism of 90-million-year-old rocks in southwestern British Columbia—proposes that Wrangellia collided with North America at about 25°N latitude, near what is now Baja, California. Between about 83 and 45 million years ago, this large block of land—nicknamed "Baja British Columbia"—journeyed north for roughly 3000 kilometers (2000 miles) along a continent-long San Andreas-type transform fault. Similar long-traveled rocks in southeastern Alaska tell the same story. An opposing view—which states that Wrangellia collided with North America where we find it today—cites the absence of field evidence to support the existence of the proposed fault zone as the principal grounds for rejecting the Baja B.C. hypothesis.

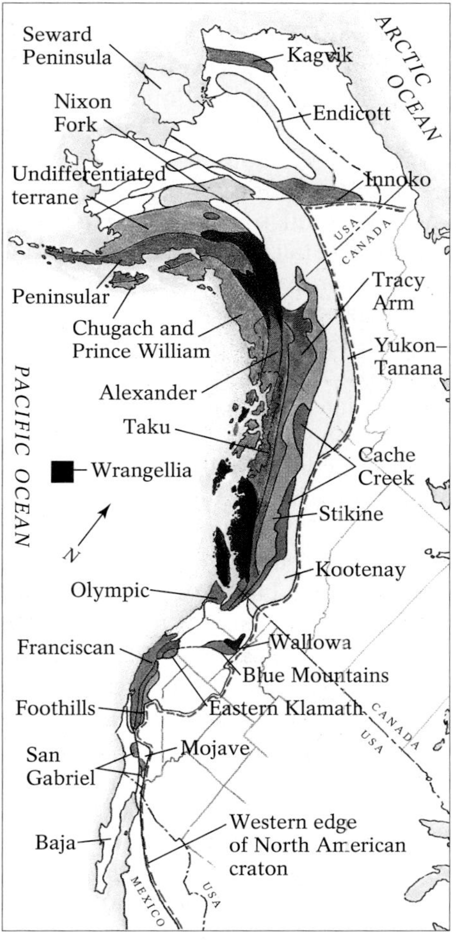

Figure 12-30 The displaced terranes attached to western North America. They are believed to be the remnants of assorted microcontinents, islands, and other unsubductable landmasses that became attached to the continent during subduction of earlier Pacific plates.

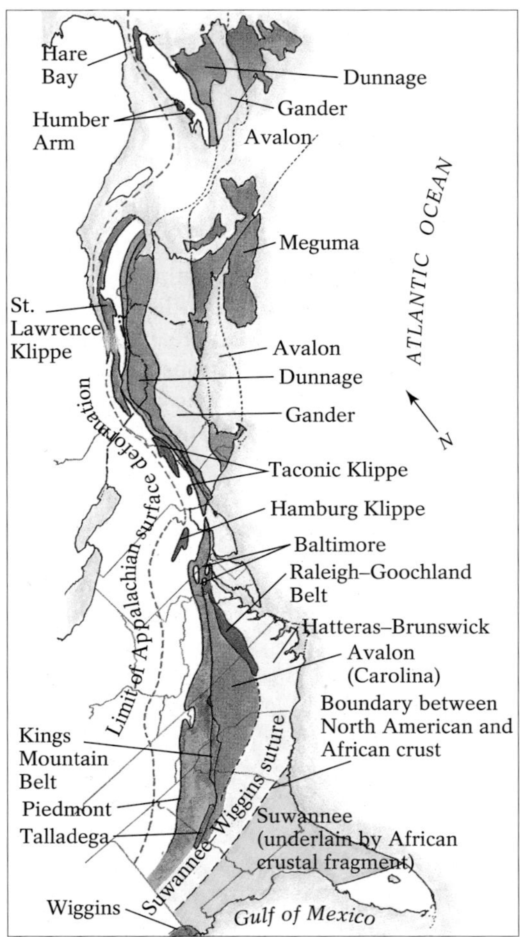

Figure 12-31 The displaced terranes of eastern North America, many dating from the assembly of the supercontinent of Pangaea. They remained attached to North America after Pangaea rifted to form the modern Atlantic Ocean.

The displaced terranes of eastern North America (Fig. 12-31) have a longer and more complex tectonic history. The Appalachian mountain system, for example, is believed to contain foreign island arcs and pieces of Africa that became added to North America when an ancient Atlantic Ocean closed in several stages between 500 and 250 million years ago (see Chapter 9). In the eastern Canadian maritime provinces of New Brunswick, Newfoundland, Prince Edward Island, and Nova Scotia, in adjacent parts of eastern Maine, Massachusetts, and Rhode Island, and further south, displaced terranes record closing oceans, accreted island arcs, continental collisions, transform faulting, and rifting, all dating from 600 to 150 million years ago.

The Earth's Plates Before Pangaea

The geologic record becomes progressively more difficult to read as we go back in time; much of what we know about the Earth's plate-tectonic history occurred within the last 200 million years, after the breakup of the supercontinent Pangaea. Paleomagnetic and other geologic studies, however, are beginning to yield more precise knowledge of the geographic origins of ancient rocks, and continuing efforts to identify displaced terranes have given us a clearer picture of ancient landmasses. As we try—cautiously—to reconstruct plate configurations that predate Pangaea's breakup, we must always remember that, even though rifting often takes place at suture zones, the continental blocks that collided to assemble past supercontinents probably did not have the same shapes as those we recognize today.

For example, exciting new research suggests that about 700 million years ago, a proposed supercontinent named *Rodinia* (from the Russian for "motherland") occupied much of the Southern Hemisphere. And growing evidence—particularly paleomagnetic records—indicates that 700 million years ago, North America may have sat near the South Pole *wedged between Antarctica and Australia on the west and South America on the east.* Thus the Precambrian rocks of the American Southwest may have been linked to similar rocks that today dot the icy wastelands of Antarctica's Transantarctic Mountains. This idea is known as the "SWEAT" (Southwest-East Antarctic) hypothesis. The breakup of Rodinia, which apparently occurred about 650 million years ago, may have left a 1200-km^2 (500-square-mile) chunk of what is now Alabama and Arkansas attached to present-day western Peru. At the time, Peru may have been North America's neighbor to the east. The fossils found in the western foothills of the Peruvian Andes are identical to those found in Alabama, but appear completely unrelated to those found throughout the rest of South America.

Five hundred million years ago (well before Pangaea formed), Gondwana, another proposed supercontinent that may have been a remnant of Rodinia, may have occupied the Southern Hemisphere somewhere near the South Pole (Fig. 12-32). Containing all of the Southern Hemisphere landmasses, Gondwana would eventually become the southern part of Pangaea. At that time, three northern landmasses also existed, each probably separated from the others by a sizable ocean. These independent continents—which would become Laurasia, the northern half of Pangaea—included most of what is now North America, northern Europe, and a combination of southern Europe and parts of Africa and Siberia.

The pre-Laurasian landmasses and Gondwana began converging about 420 million years ago. The partially subducted sea floor from this event appears today as ophiolite rocks in Nova Scotia and Newfoundland. This collision produced the northern Appalachians and corresponding ranges in the British Isles and Norway; it also contributed some displaced terranes to northeastern North America. The Avalon terrane of Nova Scotia, for example, may have been an island arc trapped between the colliding pre-Laurasian continents. Meanwhile, Siberia collided with northern Europe, completing the formation of Laurasia and producing the Ural Mountains of central Russia.

During the next 100 million years, Laurasia and Gondwana collided and formed Pangaea. The final collision in the assembly of the supercontinent involved Africa, a separate

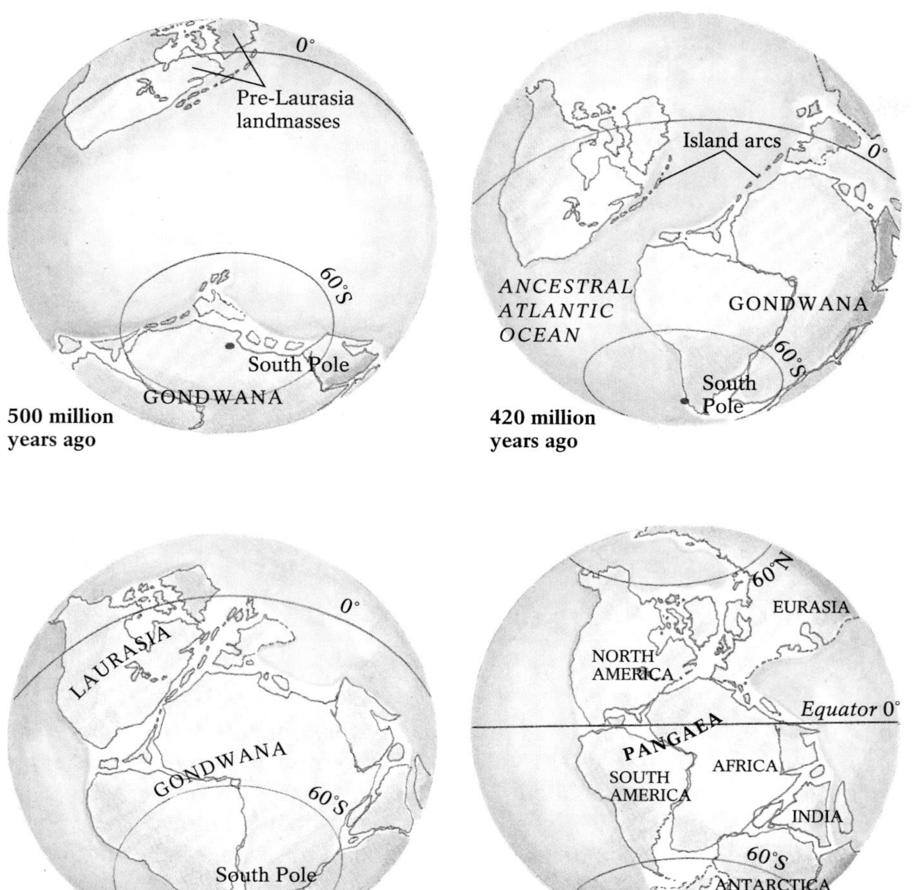

Figure 12-32 The origin of the supercontinent Pangaea.

continent that is now southeastern North America, and North America. This last stage in the formation of Pangaea created the southern Appalachians. By about 225 million years ago, at the close of the Paleozoic Era, Pangaea was a single vast landmass that stretched from pole to pole. Virtually all of the Earth's continental lithosphere remained joined together for the next 50 million years or more, until Pangaea began to rift and water flowed in to form the modern Atlantic, Indian, and Antarctic oceans.

Looking into the Plate-Tectonic Future

By extrapolating current plate motions into the future and speculating about the location of the future rifts and subduction zones, geologists have developed a probable view of the plate-tectonic world 50 to 100 million years from now (see Figure 1-34). Australia will likely collide with southeast Asia. Transform motion along the San Andreas fault will cleave western California completely from the North American continent, allowing it to move northward as a separate microcontinent. The glamorous beaches of the Mediterranean's Riviera may evolve into rugged landlocked deserts, wedged between the colliding edges of the African and Eurasian plates. Rifting may continue in East Africa and begin in the American Southwest. Subduction may resume along the east coast of North America, bringing explosive volcanism and powerful earthquakes to such cities as New York City, Boston, and Philadelphia. Scientists have constructed computer models that predict that within the next few hundred million years, all of the Earth's landmasses will reunite as another Pangaea-like supercontinent.

The Driving Forces of Plate Motion

We have documented conclusively that plates move, and we now know a great deal about the plate-tectonic history of the Earth. Nevertheless, we cannot explain with certainty what drives plate motion. This situation is akin to knowing what a car looks like and how fast it goes, but having no idea what makes it run. Fortunately, because excellent evidence proves the fact of plate motion and we can identify its effects, we can accept the general theory of plate tectonics, even though we do not yet understand all of the forces behind it.

As we stated in Chapters 1 and 11, heat from the Earth's interior probably sets plates in motion. Heat rises through the mantle by *convection*—the flow of currents in a fluid caused by variations in their temperature. (Hot materials are less dense and therefore rise; cold ones are more dense and therefore sink.) Slow convection currents in the warm asthenosphere produce movement of the cold, brittle lithosphere directly above it. When the plate-tectonic theory was initially proposed three decades ago, the plates were viewed as passive hitchhikers on the flowing asthenosphere. Recent findings, however, suggest that some plates may contribute actively to their own mobility.

In this section we explore some explanations for plate movement.

Ridge Push, Slab Pull, or Plate Sliding?

Several factors may affect the movement of the Earth's plates—in addition to the drag created by convection of the warm, soft, flowing asthenosphere. Three of these factors are

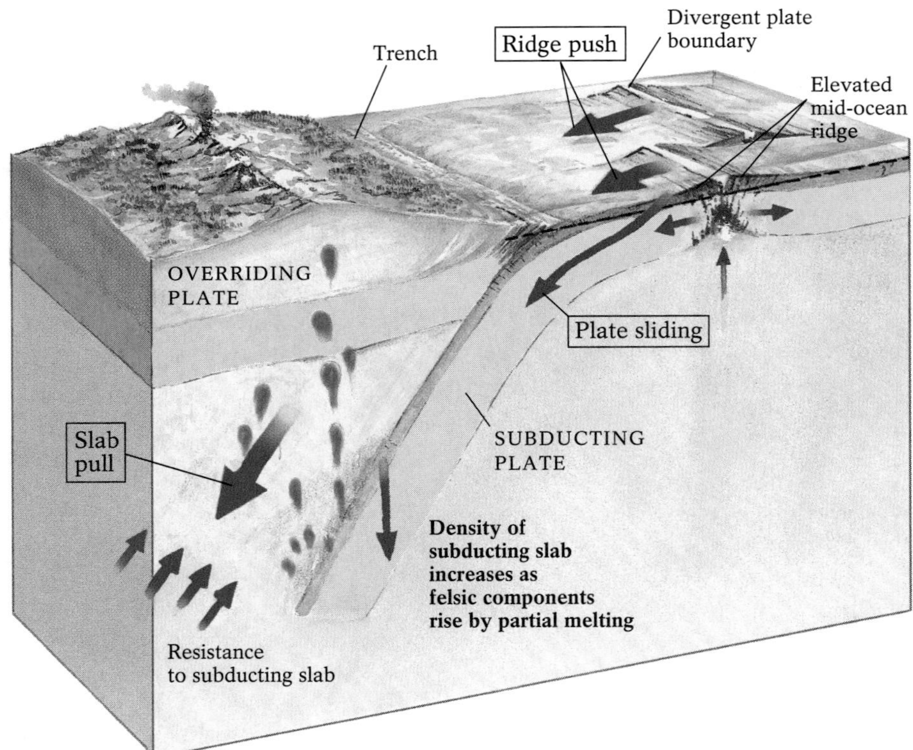

Figure 12-33 Three factors that may help drive plate tectonics: *ridge push* at divergent plate boundaries; *slab pull* as old, dense plates subduct into the mantle; and gravity-induced *plate sliding* at ocean ridges.

illustrated in Figure 12-33. When an ocean ridge grows at a divergent plate boundary, does the rising magma wedge itself between adjoining oceanic plates and actively push them apart? Would this force be powerful enough to move a 5000-kilometer (3000-mile)-wide plate and drive its distant edge down into the mantle at a subduction zone? If plates are being actively pushed, we might expect them to fold up accordion-style. But this case doesn't happen.

In fact, ocean-exploring submersibles, diving deep into axial rift valleys, have identified thousands of tears and fissures within oceanic lithosphere. This evidence strongly suggests that plates are being pulled and stretched at divergent boundaries, rather than being pushed and compressed. Thus ridge pushing probably does not represent a major driving force of plate motion. So does this stretching and tearing indicate that oceanic plates are being actively pulled? This explanation seems theoretically plausible. Old, cold oceanic lithosphere is denser than the warm asthenosphere and therefore sinks. The lower portion of a descending slab becomes even more dense as oceanic crust metamorphoses to denser eclogite-facies rocks (discussed in Chapter 7). The increased density may actively drag oceanic plates into their subduction-zone graves.

Does any evidence exist to prove that this process actually occurs? In some places, where two continents have collided and then joined, the subducting oceanic plate that once lay between them breaks off and continues to descend deeper into the mantle. These broken lithospheric slabs are clearly detectable by their relative coolness as compared with surrounding mantle rocks using seismic tomography. (This relationship is discussed in more detail in Chapter 11.)

Recent studies also show that rates of plate motion are proportionate to the amount of a plate's margin undergoing subduction: The fastest-moving plates—typically oceanic plates—subduct along a significant proportion of their margins, whereas the slowest-moving plates—typically continental plates—tend to lack subducting margins. This finding supports the hypothesis that gravitational pull on a descending plate contributes to general plate motion. It seems unlikely, however, that this force is powerful enough to influence divergence at the opposite side of the plate, thousands of kilometers away. Moreover, lithospheric plates would most likely break if such large-scale pulling forces tugged at them for thousands of kilometers. Furthermore, the Atlantic Ocean's floor diverges from the mid-Atlantic ridge at a rate of about 2 centimeters (nearly 1 inch) per year, even though no slabs sink along any of its margins. Some force other than slab pull must therefore produce the motion of the Atlantic Ocean segments of the North and South American, Eurasian, and African plates.

Another hypothesis suggests that gravity causes oceanic lithosphere to move laterally away from elevated mid-ocean ridges. Newly formed oceanic lithosphere may be literally sliding down the slope of the uplifted asthenosphere beneath ridge crests. Mathematical calculations indicate that oceanic plates could slide down even gentler slopes at a rate of several centimeters per year, which matches the observed rate for the mid-Atlantic ridge. In the initial stages of rifting, however, plates begin to diverge before an oceanic ridge forms. Hence, gravity alone cannot account for plate motion.

Geologists do not yet agree on the relative importance of these possible tectonic influences, although many believe that slab pull is the most powerful. Are oceanic plates—the prime candidates for slab pull—driven by different forces or combinations of forces than those that drive continental plates? Do these forces, perhaps acting together, play a more important role in driving plates than the convection cells discussed in Chapter 1?

Convection Cells, Revised

Although many geologists accept that convective flow in the mantle is an important mechanism of plate movement, some question whether convection cells can exist within the solid lower mantle. If they do, does the mantle flow rapidly enough to drive the Earth's plates at their observed velocities? Advocates of mantle convection continue to question the dimensions of convection cells: Figure 12-34 illustrates several proposed configurations for mantle convection cells. Do they arise at shallow depths, remaining largely confined to the soft, mobile asthenosphere? Do they extend down to 700 kilometers (450 miles), the depth to which subducting plates are known to penetrate? Do they fill the entire mantle, being energized by heat that rises from the Earth's liquid outer core? One model proposes that convection cells are stacked in two tiers. The first tier—large, deep, very-slow-moving cells—transmits heat from below to power the second tier—small, shallow, relatively fast-moving cells—above it; this second tier carries the Earth's lithospheric plates.

Seismic tomography (discussed in Chapter 11) can generate three-dimensional images delineating warm and cold regions of the mantle. (Recall that warmer, less rigid rocks slow the progress of seismic waves, whereas colder, more rigid rocks transmit seismic waves more quickly.) In one seismic-tomographic image, cold continental rocks appear to extend beneath the North American and Eurasian plates to depths of 400 to 600 kilometers (250–370 miles), uninterrupted by a warm asthenosphere that rises up at ocean ridges and hot spots. This temperature pattern casts doubt upon the simple convection-cell model proposed during the early years of the plate-tectonic revolution. This model requires cold continental lithosphere to move as a passenger on warm currents in a *shallow* asthenosphere.

The emerging field of seismic tomography can help us to "view" what's happening 600 kilometers (400 miles) beneath our feet. Consider, for example, South America, which drifts westward at a rate of 3.5 centimeters (1.4 inches) per year—relatively rapid motion for continental lithosphere. South America is not subducting at its western margin, however, and thus no slab pull is at work. Clearly South America defies the conventional wisdom that continents are simply swept along passively by oceanic slab pull, like so many soup cans on the conveyor belt at a supermarket checkout line.

Recent tomographic studies of the Paraña basalt province of southeastern Brazil may hold the key to understanding continental-plate motion. The Paraña basalts erupted 130 million years ago during the rifting of South America from Africa. These flood-basalt eruptions were fueled by a large hot spot—300 kilometers (200 miles) in diameter—that rose from deep within the Earth's mantle, perhaps from the core–mantle boundary. Like the hot spot that currently fuels Hawai'ian volcanism, this plume most likely remained stationary relative to plates that moved above it.

Since the time of the Paraña eruptions, South America has drifted roughly 3000 kilometers (1800 miles) to the west, presumably *away* from the hot spot. Nevertheless, seismic tomography indicates that, despite the movement of the plate, a large remnant of the hot spot—in the form of a cylindrical body of hotter-than-expected rock—still underlies the Paraña region, reaching to a depth of about 600 kilometers (370 miles). To confuse matters even more, at the Paraña basalt's original location near what is now the mid-Atlantic ridge, hot spot volcanic activity continues; its most recent creation is the volcanic island of Tristan de Cunha. How can the hot spot be in two places at one time—under Brazil and under Tristan de Cunha?

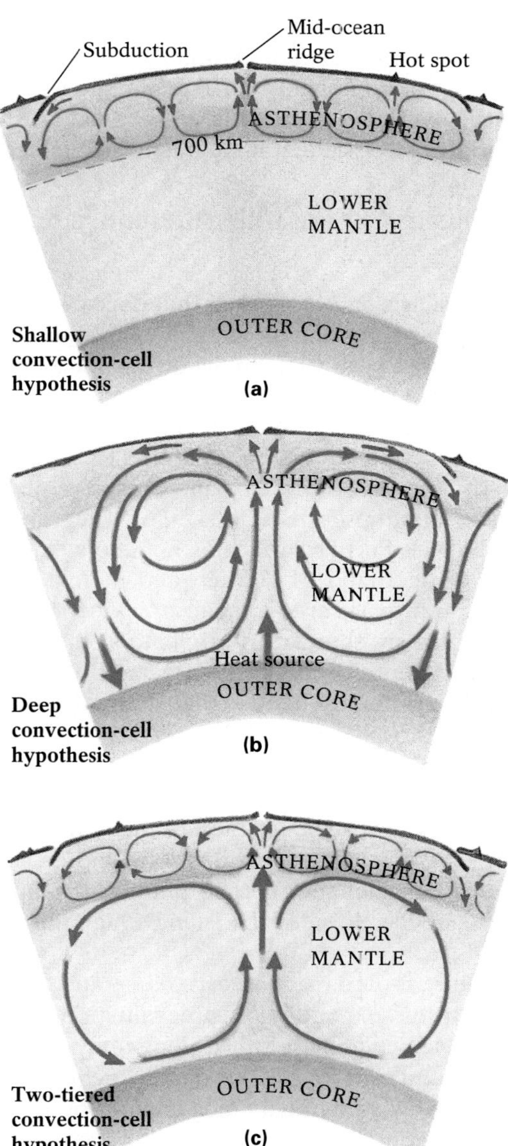

Figure 12-34 Several hypotheses have been proposed to explain the configuration of the Earth's convection cells. **(a)** The cells may be found at shallow depths, remaining confined to the asthenosphere. **(b)** The cells may extend through the entire mantle to the outer core. **(c)** Some hypotheses propose two sets of convection cells that meet at a depth of 670 kilometers (420 miles), the zone within the mantle where seismic waves are known to accelerate in response to changes in the chemistry and structure of mantle materials (see Chapter 11).

The answer lies in the upper mantle beneath Brazil, where a portion of the hot spot appears to be affixed to the overlying continental lithosphere, beheaded by South America's westward drift. As you can see in Figure 12-35, South America appears to be moving by flow *within* the underlying mantle, contradicting the long-held tenet that continents move independently atop the flowing asthenosphere. Although the upper-mantle portion of the Paraña basalt's hot spot is moving, the deeper, lower-mantle portion has remained behind, fueling the volcanism at Tristan de Cunha. Thus a new model for continental-plate motion has emerged —one that invokes flowing currents in an upper mantle bound to the overlying continents, rather than attributing all plate motion to slab pulling at subduction zones.

Thermal Plumes—A Possible Alternative to Convection Cells

Some geophysicists have proposed that deep-Earth heat rises to the asthenosphere not as distinct convection cells, but rather as scattered vertical columns of warm upwelling mantle material called **thermal plumes.** Perhaps resembling the rising blobs in a 1960s lava lamp, these relatively narrow, stationary plumes, 200 to 400 kilometers (125–250 miles) in diameter, rise beneath continents and oceans, at plate boundaries, and in plate interiors (Fig. 12-36). They are typically manifested at the Earth's surface as hot spots (such as those found at the Big Island of Hawai'i and in Yellowstone National Park) and may carry enough energy to move plates. Thermal plumes are thought to originate at the mantle–core boundary, at a depth of 2900 kilometers (1800 miles).

Scientists generally believe that thermal plumes lift up the overlying lithosphere, which becomes domed. The plumes then spread laterally beneath the lithosphere, applying a dragging force to the base of the lithosphere and causing it to rift. Geologists view the location of many thermal plumes at or near diverging plate boundaries as more than coincidental. Some consider the thermal plumes beneath Iceland, for example, as the cause of the island's high altitude above sea level.

The thermal-plume hypothesis, like many other aspects of deep-Earth study, is undergoing reevaluation in light of recent evidence revealed by seismic tomography. Although conventional wisdom calls for relative narrow, cylindrical plumes, geologists are now studying a "megaplume," a rising current of mantle material 2500 to 4000 kilometers (1500–2500 miles) wide. Flowing eastward from beneath the eastern Atlantic, this broad region of warmer-than-expected mantle extends across Africa and southern Europe.

Studies of the geochemistry of the volcanic rocks from the eastern Atlantic's Canary Islands off the coast of Morocco, across the volcanoes of Italy, to the ancient volcanoes of the Rhine Valley of northwestern Germany indicate clear similarities in the chemical compositions of these rocks. Some geologists have suggested that northern Africa and southern

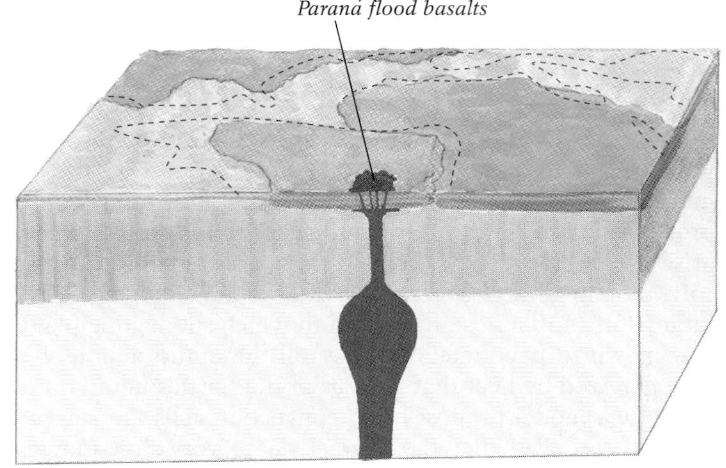

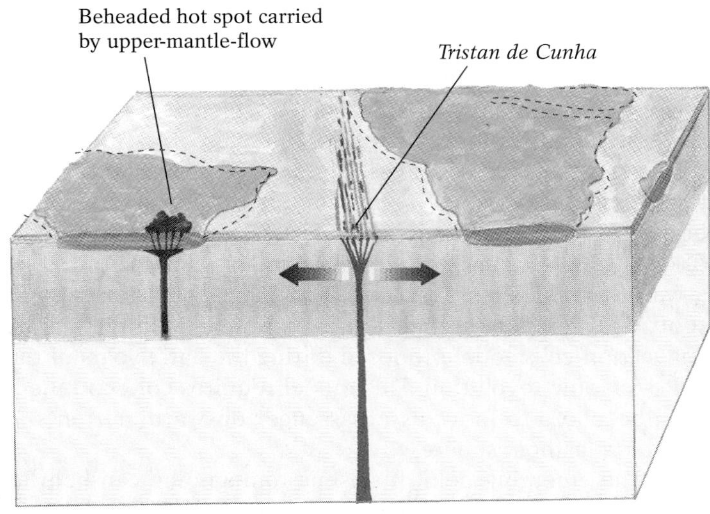

Figure 12-35 Seismic tomography indicates that a detached portion of a hot-spot plume underlies the Paraña basalts of Brazil. This finding suggests that the upper mantle is wedded to the overlying continental plate, which moves along with it as the upper mantle flows. The deeper portion of the hot spot remains 3000 kilometers (1800 miles) to the east, where it continues to fuel the volcanism at Tristan de Cunha.

Europe may sit above the same massive upwelling of hot mantle material. This phenomenon, unknown before the advent of seismic tomography, could explain why parts of the continental lithosphere from the western Mediterranean to northwestern Germany appears to be thinned and stretched. The Rhine Valley graben, a normally faulted region marked by crustal extension, is one example of this continental stretch.

Plate-Driving Mechanisms: A Combined Model

A single model is unlikely to explain all plate movements. More probably, a variety of driving mechanisms—including some combination of slab pull, gravity-induced sliding, upper-mantle convection, and thermal plumes—control the motions of oceanic plates, whereas other combinations may control the motion of continental plates. Some form of convection

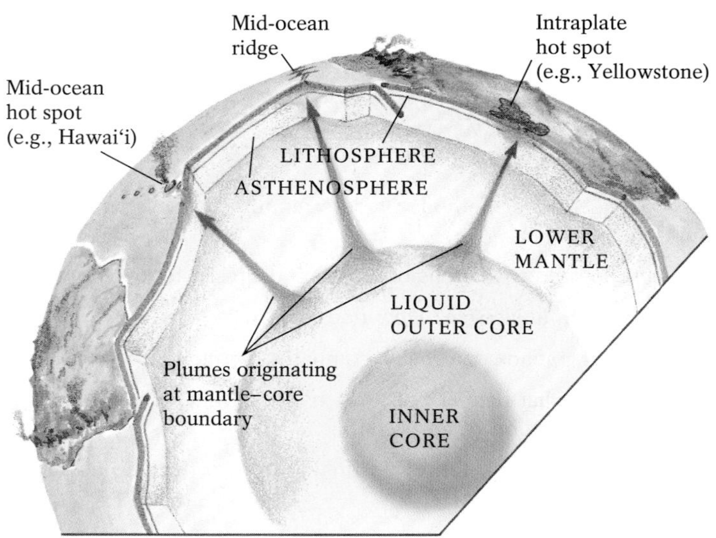

Figure 12-36 Thermal plumes are an alternative explanation to convection cells for the force that drives plate motion. They are also believed to fuel surface hot spots such as those that created the Hawai'ian Islands and Yellowstone's volcanoes and hot springs. The source of the heat that creates plumes remains uncertain, although some geologists believe that the plumes rise from the mantle–core boundary, heated by the liquid outer core.

beneath the lithosphere apparently transports mantle heat upward to drive the mid-ocean plate divergence of Atlantic-style tectonics. Once mantle convection initiates plate motion and oceanic ridge crests develop, gravity may induce the lateral motion of oceanic plates as they slide from topographically high crests over the underlying soft, weak asthenosphere. In addition, gravity enhances plate subduction by returning cold, dense oceanic lithosphere to the Earth's interior; this mechanism may drive Pacific-style tectonics. The motions of continental lithosphere remain the most problematic and warrant further study, particularly with the aid of seismic tomography. In the future, of course, we may even learn about other driving forces of which we are now unaware.

In the mid-1960s, plate tectonics was an exciting new hypothesis. Since then, it has graduated to full-fledged theory as Earth scientists have gathered evidence confirming most of its precepts. Although most now believe that tectonic plates exist and move, the oceans and continents continue to be explored intensively in an effort to address the numerous unanswered questions.

Chapter Summary

New technologies, such as satellite-based studies of plate motion and geomagnetic research, have enabled Earth scientists to study the mechanisms of plate tectonics. The actual velocity of plates can now be measured directly by reflecting ground-based laser beams off reflectors on satellites in known orbits. **Hot spots,** areas of particularly high heat flow in the mantle that may fuel intraplate volcanism, can serve as fixed reference points to permit the measurement of absolute plate velocities. The structures formed as a plate moves over a hot spot can reveal both plate speed and direction. Such structures include volcanic islands such as the Hawai'ian Islands; submarine mountains, or **seamounts;** and older, eroded **guyots,** which initially extended above sea level but later subsided below sea level. **Marine magnetic anomalies,** the stripes of alternating magnetic polarity in oceanic lithosphere, are also used to estimate rates of divergence and, hence, of plate motion.

Our understanding of the Earth's tectonic past comes largely from our growing knowledge of the evolution of the world's ocean floors. In the last four decades, marine geologists and oceanographers have used **echo-sounding sonar** to study the topography of the sea floor and **seismic profiling** to study its underlying layers.

The accumulation of geophysical data has clarified the sequence of events that produces ocean basins. Rifting begins when a warm current of mantle material rises under a continental plate, stretching and thinning the plate until it tears. Three radiating rift valleys typically form simultaneously; two of these eventually diverge, while the third—the failed arm, or aulacogen—commonly becomes inactive and fills with sediment or a stream network. After rifting between two new plates ends, the rift edges become inactive tectonically, forming **passive continental margins.**

As divergence continues, a full-blown seaway develops; new oceanic lithosphere forms at a **mid-ocean ridge,** where mantle-derived basalts erupt to produce a linear chain of volcanic mountains. The typical rock sequence within oceanic lithosphere, called the **ophiolite suite,** consists of a veneer of deep-marine sediment, underlying layers of pillow basalts and related intrusive gabbros, and a bottom layer of ultramafic mantle peridotite. The entire suite may be chemically altered by reaction with infiltrating seawater. The mid-ocean ridge generally becomes divided into short, offset segments that are cut by oceanic transform boundaries.

A diverging oceanic plate gradually cools and becomes more dense. It may eventually sink back into the Earth's interior by subducting under a less dense plate far from the mid-ocean ridge. An **ocean trench** develops where a subducting plate flexes downward into the mantle, forming a depression in the Earth's surface. Subduction produces several rock types and geologic structures: Chaotic mixtures of oceanic sediments and ophiolite rocks called **mélanges** pile up within the trench, forming masses of rock called **accretionary wedges** that become attached to the edge of the overriding plate. As the subducting plate sinks into the warm mantle, its expelled water promotes partial melting of the overlying mantle. When these magmas erupt, they form a chain of volcanoes called a **volcanic arc.** Sediments accumulate on either side of a volcanic arc in topographic depressions called **forearc** and **backarc basins. Backarc spreading,** divergence within the backarc basin, sometimes separates these coastal features from the mainland.

Once subduction of an oceanic plate is complete, the continental blocks on either side of the plate undergo continental collision and connect to form a larger continent. The boundary between them, the **suture zone,** is typically marked by a collisional fold-and-thrust mountain range such as the Himalayas. Because they represent weak spots in the Earth's lithosphere, suture zones are candidate sites for subsequent rifts.

Because the oldest oceanic rocks date from only 200 million years ago, we must rely on the older, more complex geology of the continents to reveal the Earth's earlier tectonic history. The nucleus of a continent consists of ancient bodies of rock called **continental shields;** these shields are generally surrounded by younger sedimentary rocks, which form the **continental platform.** Together the shield and platform constitute the continental **craton,** the tectonically stable portion of a continent. The rocks exposed at the outer edges of the continents may be **displaced terranes;** their fossils, stratigraphic sequences, rock types, and magnetic signatures suggest that they originated elsewhere. **Microcontinents** are pieces of lithosphere broken from larger continents that may eventually become displaced terranes. Information from these types of rocks has allowed geologists to speculate (cautiously) about the past and future configurations of the Earth's landmasses.

Although geologists have learned much about the origin of both oceanic and continental rocks, they remain uncertain about the precise mechanisms that drive the motion of specific individual plates. New data—particularly from seismic tomography—have cast some doubt on the simple convection-cell, slab-pull model for plate motion. Plate motion may instead be driven by the combined effects of **thermal plumes,** rising currents of hot mantle rock, and gravity, which pulls newly formed plates down the slopes of mid-ocean ridges and older plates back into the mantle at subduction zones.

Key Terms

hot spots (p. 332)
seamounts (p. 333)
guyots (p. 334)
marine magnetic anomalies (p. 335)
echo-sounding sonar (p. 336)
seismic profiling (p. 336)
passive continental margins (p. 339)
mid-ocean ridge (p. 340)
ophiolite suite (p. 341)
ocean trenches (p. 344)
mélange (p. 344)
accretionary wedge (p. 345)
volcanic arc (p. 345)
forearc basin (p. 345)
backarc basin (p. 346)
backarc spreading (p. 346)
suture zone (p. 348)
continental shields (p. 350)
continental platform (p. 350)
craton (p. 350)
displaced terranes (p. 353)
microcontinents (p. 353)
thermal plumes (p. 358)

Questions for Review

1. Describe three methods of determining the velocity of plate motion.

2. Describe three methods of mapping deep-sea topography.

3. What is an aulacogen? How does it form?

4. List four phenomena that accompany continental rifting.

5. Describe what happens at a rift margin, from the onset of rifting to the formation of an ocean.

6. Draw a simple sketch of an ophiolite suite (include all of the rock types and structures).

7. Why do the earthquakes associated with oceanic transform boundaries generally occur only between offset oceanic ridge segments?

8. List the three common types of convergent plate boundaries, and name a geographic example of at least two types.

9. Sketch a convergent boundary between an oceanic plate and a continental plate. Be sure to include all of the important associated landforms.

10. Briefly describe the fundamental differences between the convection-cell hypothesis and the thermal-plume hypothesis of plate tectonics.

For Further Thought

1. Of the two patterns of marine magnetic anomalies shown below, which shows a faster rate of sea-floor spreading? Explain.

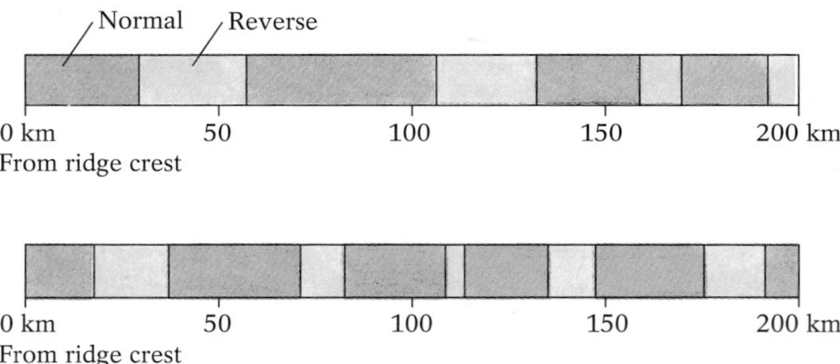

2. What would be some of the major geologic repercussions if a new rift opened between Ohio and Indiana?

3. How would the east coast of North America change if the oceanic segments of the plates that make up the Atlantic Ocean basin began to subduct? How would it change if the Atlantic Ocean lithosphere subducted completely?

4. How might the development of a new subduction zone along the East Coast affect plate interactions on the West Coast?

5. If you had unlimited financial resources, how would you determine the real mechanisms driving plate-tectonic motion?

6. How would you explain the absence of plate tectonics on most of the other planets in our solar system? If a planet had been active tectonically, why would its activity stop?

Part 3 Sculpting the Earth's Surface

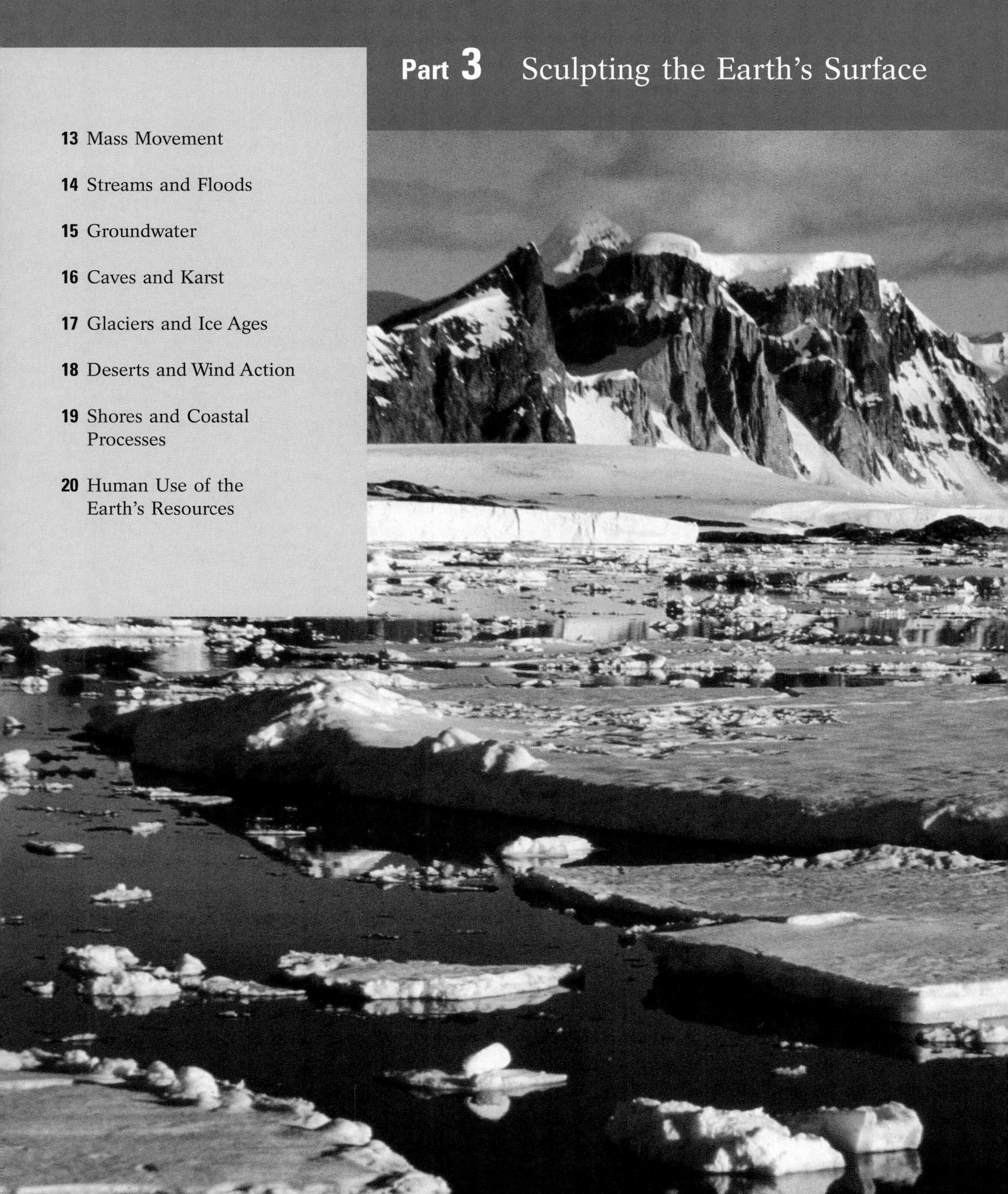

13

Mass Movement

Figure 13-1 This massive rockfall occurred on January 1997 along Highway 140, several kilometers east of the Arch Rock entrance to Yosemite National Park, California.

At 11:37 P.M. on August 17, 1959, a magnitude 7.1 earthquake shook thousands of square kilometers near West Yellowstone, Montana. The soil surface rolled like sea waves, and lakes and rivers sloshed back and forth. A wall of water rushed across Hebgen Lake and over its dam, sweeping away campgrounds on its way downstream. The quake also triggered a landslide encompassing more than 80 million tons of rock and weathered regolith that hurtled downslope at speeds reaching 150 kilometers (90 miles) per hour. Some of the boulders swept along measured 9 meters (30 feet) in diameter. As the slide tumbled into Madison Canyon, it compressed and forcefully expelled the air in its path, producing hurricane-force winds that battered the valley. Two-ton cars were blown into the air; one flew more than 10 meters (33 feet) before being dashed against a tree. By the time the slide mass finally came to rest, it had covered the valley floor with 45 meters (150 feet) of bouldery rubble, buried U.S. Highway 287, and taken the lives of 28 campers.

The Madison Canyon disaster represents an extreme example of **mass movement**, the process that transports Earth material (such as bedrock, loose sediment, and soil) down slopes by the pull of gravity (Fig. 13-1). *Every* slope is susceptible to mass movement. Sometimes, as in Madison Canyon, the movement is as fast as 160 kilometers per hour (100 miles per hour) and involves a great mass of material that may travel more than 80 kilometers (50 miles) from its source. More often the movement is slow, even imperceptible, and involves just the upper few centimeters of loose soil on a gentle hillside.

Of the 20,000 lives lost as a result of all natural disasters in the United States from 1925 to 1975, less than 1000 deaths were a direct result of mass movement. Mass movement, however, accounted for a staggering $75 billion in damage, compared with only $20 billion in damage for all other natural catastrophes. Thus, while few of us are likely to perish in a mass-movement event, we are much more prone to incur some costs—perhaps because of a cracked home foundation or a living room filled with flowing mud—as a result of this geological process. In this chapter, we will examine

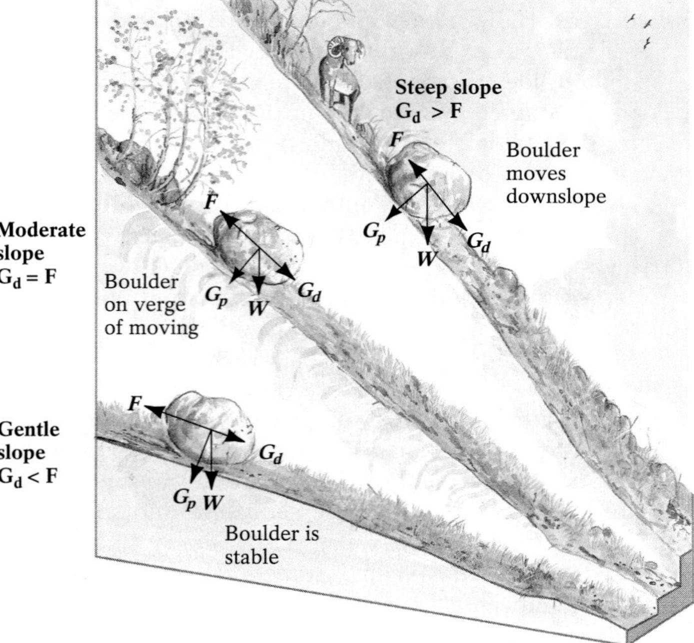

Figure 13-2 (top) This boulder sits on a steep hillside in Acadia National Park, Maine, without sliding off because the friction holding it in place exceeds the gravitational force that would drive it down-slope. (bottom) The principal force tending to hold materials in place on a slope is friction (*F*); the principal force tending to drive materials downslope is the parallel component of gravity (G_d). These forces are affected by several factors, including the steepness and slipperiness of the slope as well as the weight (*W*) and water content of the material on the slope.

both the underlying and immediate causes of mass movement, various types of mass movements, and some ways to prevent these disasters.

What Causes Mass Movement?

The underlying cause of all mass movement is the same: *gravity* coaxes materials on a slope downslope. Some factors loosen materials and promote downslope movement; others resist such movement. In a general sense, *mass movement occurs when the factors that drive materials downslope overcome the factors that resist downslope movement.*

Two principal factors provide resistance to mass movement: the friction between a slope and the loose material at its surface; and the strength and cohesiveness of the material composing the slope, which prevents it from breaking apart and slipping at its surface. Counteracting these forces are steep slopes, the materials' water content, a lack of vegetative cover, and a slope's history of human and other animal disturbances.

Gravity, Friction, and Slope Angle

Look at the boulder in Figure 13-2. Why does it just sit there so precariously? Two main factors determine whether the boulder will stay put or hurtle down the slope: gravity and friction. Gravity promotes the boulder's downslope movement; friction resists it. Gravity pulls on the weight (W) of the boulder, drawing it downward toward the Earth's center. On a slope, the force of gravity on the boulder actually has two components: one parallel to the slope (G_d, for $\text{gravity}_{\text{downslope}}$), which works to pull surface materials downslope, and one perpendicular, or *normal,* to the slope (G_p, for $\text{gravity}_{\text{perpendicular}}$). The normal force, G_p, contributes significantly to the effects of friction (*F*), the force that opposes motion between two bodies in contact. The amount of friction between a boulder and a slope depends on the magnitude of G_p and the slipperiness of the surfaces in contact.

The boulder in Figure 13-2 is likely to stay put if G_d is relatively low and/or G_p and its related factor, *F*, are relatively high. This scenario typically occurs when a slope is gentle and/or its surface is very rough. The boulder is likely to roll down if G_d is relatively high and/or G_p and *F* are relatively low. This situation typically arises when the slope is steep and/or its surface is very smooth.

Several natural and artificial processes can create steeper slopes. Faulting, folding, and tilting of strata, river cutting, glacial erosion, and coastal wave cutting are all natural processes; quarrying, road cutting, and waste dumping are some human activities that can steepen and destabilize slopes (Fig. 13-3).

Slope Composition

The materials that make up the slope also influence the likelihood that mass movement will occur. A slope may be composed of any combination of solid bedrock, weathered bedrock, soil, vegetation, and a variable amount of water. Under certain conditions, some of these materials promote slope stability; in other situations, they promote instability and mass movement.

Solid Bedrock Solid rock tends to be completely stable, even when it forms a vertical cliff. Its degree of cementation (if clastic sedimentary rock) or the interlocking pattern of its crystals (if igneous, metamorphic, or chemical sedimentary rock) imparts a strength to the rock that usually exceeds the downslope force trying to pull it apart. The stability of solid rock diminishes, however, if any of the following geological events or circumstances leave it vulnerable to breaking apart under gravity:

- Tectonic deformation shatters the rock into a network of joints, fractures, or faults.
- The mechanical weathering processes of freeze–thaw, thermal expansion and contraction, exfoliation, or root penetration open significant cracks in the rock.
- The rock is sedimentary (such as sandstone) and developed bedding planes during deposition.
- The rock is soluble (such as limestone) and formed large cavities through dissolution.
- The rock is igneous and developed a joint pattern during cooling (such as columnar basalt).
- The rock is a foliated metamorphic rock with marked rock cleavage or schistosity (such as slate or schist).

When a plane of weakness—whether a fault, crack, joint, or bedding plane—lies roughly parallel to the surface of a steep slope, the rock is even more likely to break along that plane and slide downslope (Fig. 13-4).

Unconsolidated Materials A slope composed of loose, dry material remains stable only if the friction between its components and underlying materials exceeds the downslope component of gravity (G_d), which increases with the steepness of the slope. Different materials form stable slopes at different slope angles. If you pile up dry sand, for example, it forms a small conical hill; when the hill attains a certain slope, the slope angle remains constant. When you add more sand, the extra material simply cascades down the slope, gradually piling up around its base. The maximum angle at which an unconsolidated material is stable is known as that material's **angle of repose.** For dry sand, like that found in sand

Figure 13-3 Some common processes that oversteepen slopes.

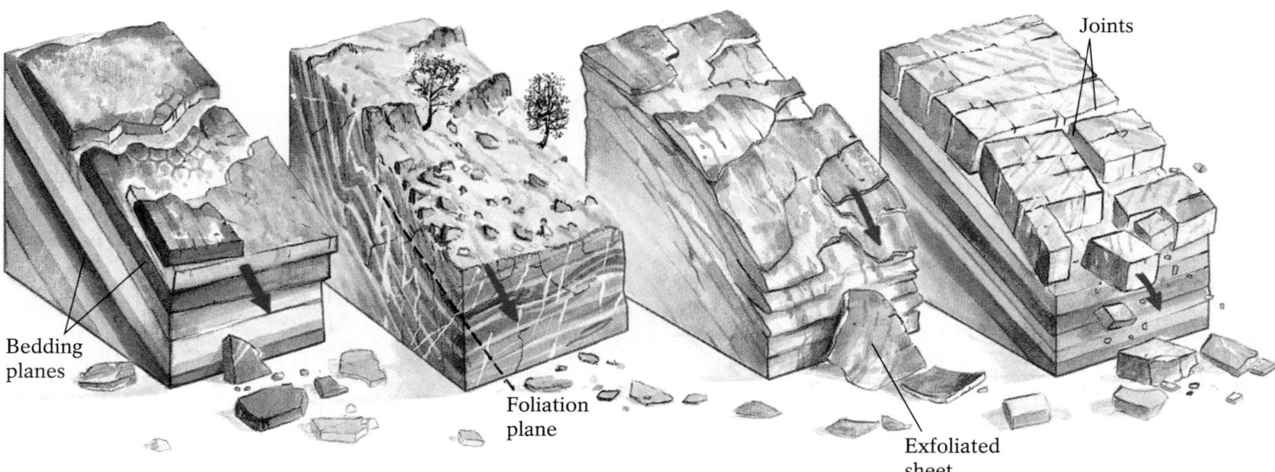

Figure 13-4 Slopes such as these, with planes of weakness parallel to their surface, are especially susceptible to mass movement.

dunes, the angle of repose generally ranges between 30° and 35° (Fig. 13-5a).

The angle of repose of any loose material depends partly on the size and shape of its particles. Large, flat, angular grains with rough, textured surfaces generally have steeper angles of repose than smaller, rounded, smooth grains. Imagine stacking a pile of highly polished marbles, and another pile of chocolate-chip cookies (with big chips). The flatter shape and bumpiness of each cookie's surface enables you to stack them more steeply than the marbles. In nature, the steepest slopes of loose materials, with angles of repose greater than 40°, consist of large, highly angular boulders. These large boulder piles, called **talus slopes,** typically form at the foot of cliffs that have been weathered by frost wedging and other mechanisms (Chapter 5) (Fig. 13-5b).

The tendency of loose material to move also depends on the arrangement of its particles. Rapidly deposited parti-

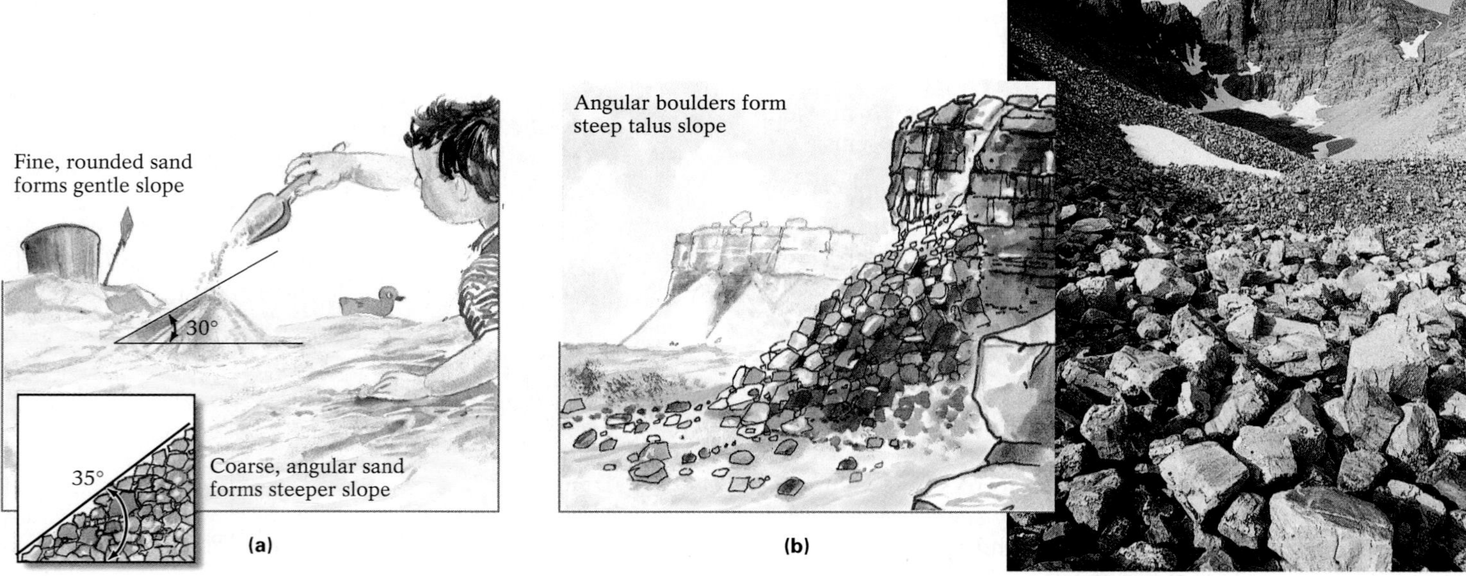

Figure 13-5 The angle of repose of unconsolidated materials depends largely on particle size and shape. **(a)** Coarse, angular sand forms a steeper slope than does fine, rounded sand. **(b)** A talus slope, composed of large and irregularly shaped boulders, can form slopes greater than 40°. (photo) Talus slope at Wheeler Peak, Great Basin National Park, Nevada.

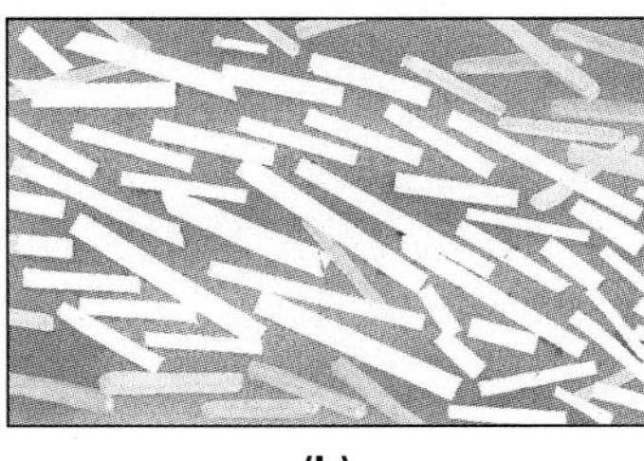

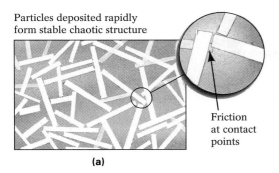

Figure 13-6 The effect of particle arrangement on stability. **(a)** When loose material is deposited rapidly, its particles become oriented at angles to one another, producing friction between the grains and imparting strength to the material. **(b)** When particles are deposited slowly and in parallel planes, both the friction and the strength of the material are reduced.

Figure 13-7 A little water can initially make sediments more cohesive. Partially saturated sand, for example, can be molded into intricate sand sculptures, such as this one at Cannon Beach, Oregon.

cles are generally arranged somewhat chaotically, in a structure that resembles a house of cards (Fig. 13-6a). Although this structure appears unstable, the friction between the particles at their contact points can impart enough strength to resist mass movement for centuries, even in the case of fine-grained particles. Loose particles that are deposited slowly, however, generally arrange themselves in a more organized fashion, in a structure that resembles a stack of pancakes (Fig. 13-6b). In this configuration, the particles tend to slide past one another, promoting mass movement.

Vegetation Vegetation, especially the deep root networks of large shrubs and trees, binds and stabilizes loose, unconsolidated material. Removal of this flora by forest fire or clear-cutting for timber or farming allows loose material to move downslope, especially shortly after a rainstorm. Several decades ago in the town of Menton, France, for example, farmers decided to remove deeply rooted olive trees from the area's steep slopes to plant more profitable, shallow-rooted carnations. This horticultural miscalculation contributed to landslides that took 11 lives. Similarly, widespread tree cutting in Philippine forests was a primary cause of the tragic landslides that claimed 3400 lives in November 1991. More recently, vegetation loss from wildfires in the hills of the Oakland–Berkeley area and along southern California's coast has triggered hundreds of mass-movement events; likewise, fires removed the vegetation from the slopes of Sarno, Italy (near Naples), causing landslides in May 1998, that took more than 100 lives.

Water Water, *more than any other factor,* is likely to cause previously stable slopes to fail and slide. Initially, the presence of a small amount of water actually increases the cohesiveness of loose material: It binds adjacent particles by surface tension, a weak electrical force on the surface of a drop of water that attracts and holds other particles, such as sand grains (Fig. 13-7). Water also fosters the growth of stabilizing vegetation. On the other hand, excessive water can promote slope failure:

- It may reduce the friction between surface materials and underlying rocks, or between adjacent grains of unconsolidated sediment.
- It may counteract some or all of G_p—the perpendicular component of gravity—by buoying upward the weight of slope materials. (*Pressurized* water counteracts the normal component of gravity even more effectively.)

If you have ever slid down a giant water-park slide, you have experienced firsthand the role water plays in reducing friction between an object and a slope. A thin film of water on the slide reduces friction dramatically, allowing you to hurtle down at a high speed. Even solid bedrock can be moved with the aid of water: Infiltrating water reduces the friction between adjacent rock masses that are separated by

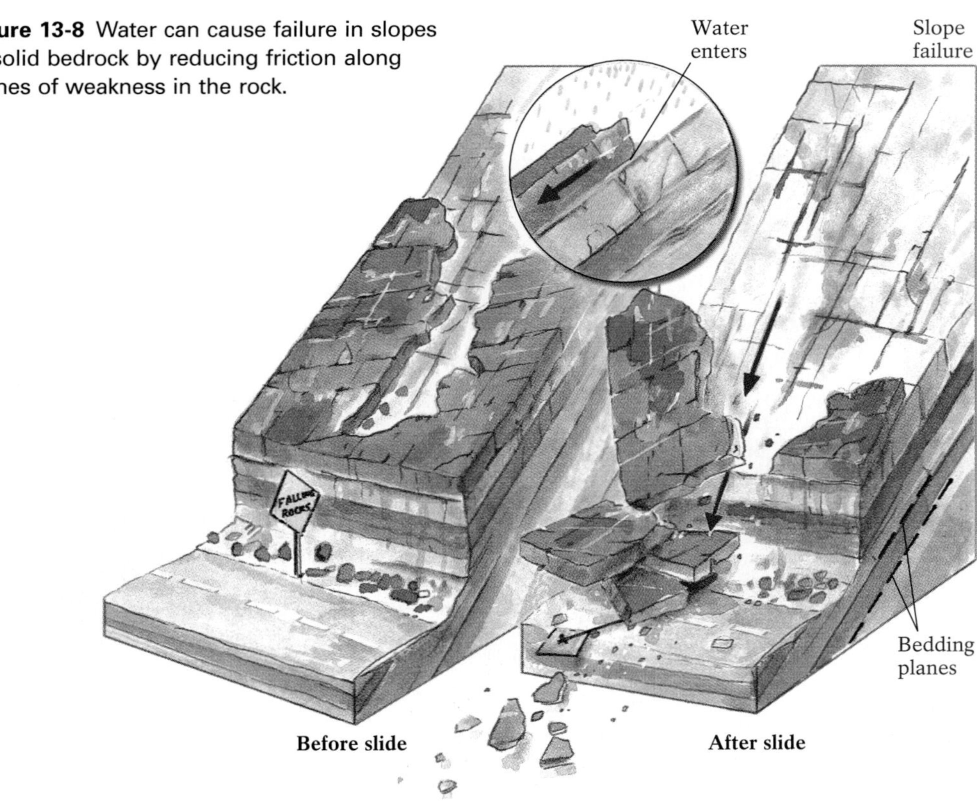

Figure 13-8 Water can cause failure in slopes of solid bedrock by reducing friction along planes of weakness in the rock.

some type of plane of weakness, such as a bedding plane, fault, or metamorphic foliation (Fig. 13-8).

In addition, saturation with water reduces cohesiveness *between individual sediment grains.* In general, an unconsolidated sediment's cohesiveness derives from the internal friction between grains in contact and the weak attractive forces (surface tension) between grains. When water completely surrounds sediment grains, it isolates them from adjacent grains and effectively eliminates the friction and the surface tension between them. Thus, although damp soil may be more cohesive than dry soil, *saturated* soil may become a formless mass of flowing mud as water pressure forces apart individual grains (Fig. 13-9).

Figure 13-9 Saturation with water promotes mass movement in unconsolidated slope materials by decreasing the internal friction and surface tension between particles.

Setting Off a Mass-Movement Event

Before a stable slope becomes unstable and fails, it may develop a fragile balance, or equilibrium, between the forces that tend to drive movement downslope and the forces that tend to resist such movement. Some event may then tip the delicate balance in favor of the driving forces, setting off downslope movement. The trigger mechanisms, or immediate causes, of mass movement can be either natural or human-induced.

Natural Triggers A number of climatic and geological events can set off mass movements. Torrential rains, earthquakes, and volcanic eruptions can all send loose materials downslope. Heavy rainfalls, for example, can quickly saturate a thick, clay-rich regolith that may have taken millennia to accumulate. At some point, the slope's sodden material can no longer withstand the force of gravity, and it slides downhill. In 1967, for instance, a three-hour downpour in central Brazil triggered hundreds of slope failures, taking more than 1700 lives.

An earthquake can dislodge massive blocks of bedrock and enormous volumes of unconsolidated material. The series of powerful earthquakes that struck New Madrid,

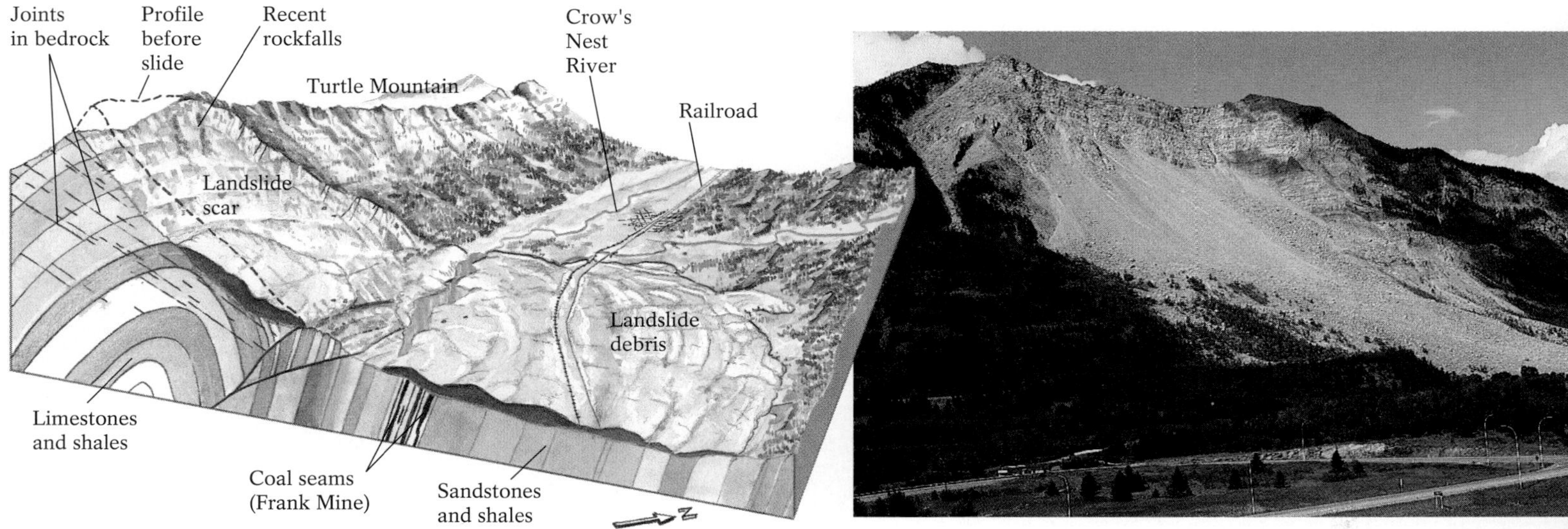

Figure 13-10 The 1903 Turtle Mountain landslide in Frank, Alberta. (left) The disaster began when an enormous mass of limestone broke free and moved rapidly downslope to a protruding rock ledge, where it was launched airborne toward the valley below. After a 900-meter (3000-foot) drop, the rock struck the weak shales and coal seams in the valley, shattering into a great avalanche of crushed rock that spread at speeds as high as 100 kilometers (60 miles) per hour to a distance of 3 kilometers (2 miles) across the valley. The mass of crushed rock had such momentum that it actually ascended 120 meters (400 feet) up the opposite side of the valley. Ironically, 16 men working in the mines survived by digging their way out through a soft coal seam. (right) Turtle Mountain as it looks today, showing evidence of more recent rockfalls within the scar from the 1903 slide.

Missouri, between December 1811 and February 1812 initiated hundreds of mass-movement events along a 13,000-square-kilometer (5000-square-mile) area of the Mississippi Valley. In the most tragic event of its kind in recent memory, a 1970 Peruvian earthquake of magnitude 7.7 shook the Andes so violently that one landslide sent more than 150 million cubic meters (5 billion cubic feet) of ice, rock, and soil tumbling, burying several mountainside towns and their 30,000 inhabitants (see Figure 13-23). Earthquakes may also initiate liquefaction of wet, unconsolidated materials, as we saw in the 1964 quake in Anchorage's Turnagain Heights neighborhood (see page 278 in Chapter 10).

When volcanic eruptions propel large quantities of hot ash onto snow- or glacier-covered slopes, they produce enormous volumes of meltwater. This water may then mix with fresh ash and other surface debris to form a muddy slurry, or *lahar* (see Chapter 4), that flows rapidly down steep volcanic slopes, burying the landscapes and communities below.

Human-Induced Triggers A variety of human activities can also set off mass movements. Mismanagement of water and vegetation, oversteepening and overloading of slopes, mining miscalculations, and even loud sounds can all trigger slope failure.

When we overirrigate slopes for farming, install septic fields that leak sewage, divert surface water onto sensitive slopes at construction sites, or overwater our sloping lawns, we introduce liquids that destabilize slopes by reducing friction. One Los Angeles family went on vacation and inadvertently left on their lawn sprinklers. When they returned, they found their hillside lawn *and their home* sitting on the highway in the valley below.

When we clear-cut forested slopes or accidentally set forest fires, we eliminate the deep, extensive root networks that bind loose materials together. When we cut into the bases of sensitive slopes to clear land for homes or roads—especially if we dump the removed or other material at the tops of the slopes—we oversteepen the slopes and jeopardize their equilibrium. Building housing developments on hillsides or cliffsides—a common practice in California—can initiate mass movement when slopes bearing the extra weight of roads and buildings later become weakened by heavy rainfall.

The rumble of a passing train, the crack of an aircraft's sonic boom, and the blasting that accompanies mining, quarrying, and road construction can trigger mass movement as well. Vibrations from these activities can force apart grains of loose sediment, eliminating the friction between them.

Sometimes human activities, such as mining, may combine with natural factors to increase the probability of mass movement. Such a situation led to the Turtle Mountain landslide of 1903, near the Canadian Rockies town of Frank, Alberta, which claimed 70 lives (Fig. 13-10). Removal of a large volume of coal near the foot of the mountain apparently undermined its steep slope, weakening its already unstable fractured structure. In addition, the month preceding the slide had been a particularly wet one in the southern Alberta Rockies, and water from copious snowmelt probably entered the fractures in Turtle Mountain. This water loaded the slope and reduced friction between fracture surfaces, further increasing the risk of a mass movement.

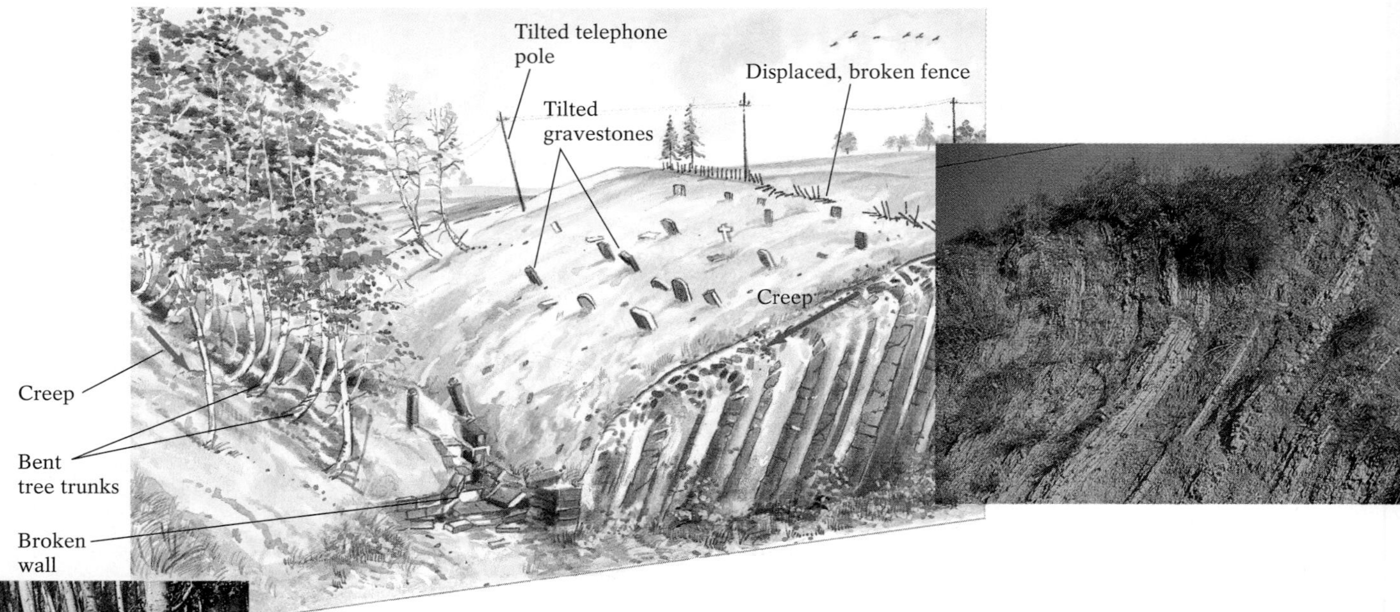

Figure 13-11 Creep affects every near-surface feature, including soil, rock, vegetation, and a variety of shallow-rooted, human-made items such as fences, gravestones, and utility poles.

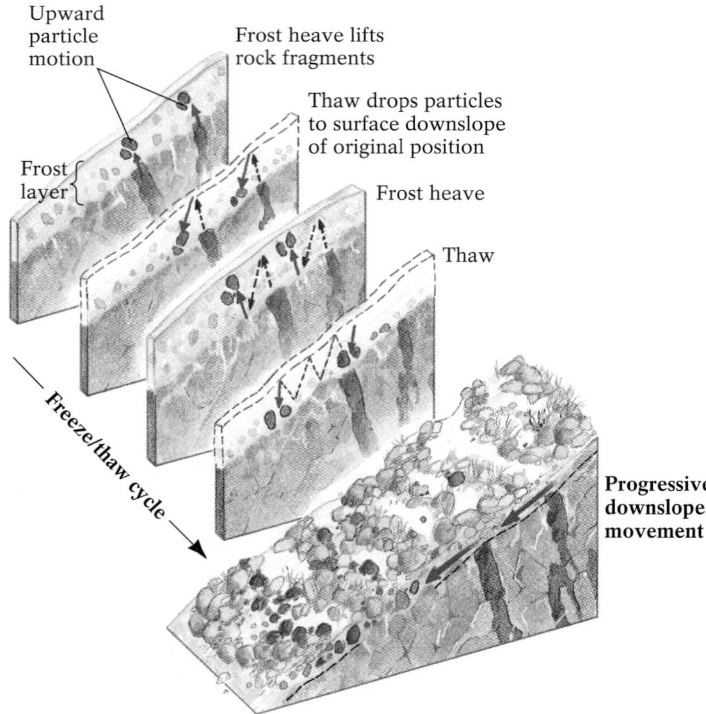

Figure 13-12 Frost-induced creep. Ice that forms beneath soil particles displaces the particles, pushing them toward the surface. Gravity carries the displaced particles a short distance downslope as the frost thaws, and the particles sink back down.

Types of Mass Movement

No universally accepted classification scheme for mass movements exists. Ask a soil scientist, a geologist, and a civil engineer to classify a mudflow (a rapidly flowing slurry of mud and water), and you'll likely receive three different answers. Geologists, however, generally classify mass movements based on the speed and the manner with which the materials move downslope.

Slow Mass Movement

Creep, the slowest mass movement, is measured in millimeters or centimeters per year. It occurs virtually everywhere, even on the gentlest slopes. Creep generally affects unconsolidated materials, such as soil or regolith, that rarely exceed a few meters in depth.

Because of friction with underlying slope material, a mass of creeping material tends to move faster at its surface and more slowly at deeper layers. This disparity in movement causes most originally upright implanted objects, such as lamp posts and telephone poles, to slant over time. Telephone poles inserted vertically in a creeping slope near Yellowstone National Park inclined 8° after 10 years. In cemeteries, older gravestones are generally more slanted than younger ones because they have been creeping for a longer period. Trees on creeping slopes continually adjust their shape to remain vertical and absorb maximum sunlight; as a result they often display contorted trunks (Fig. 13-11).

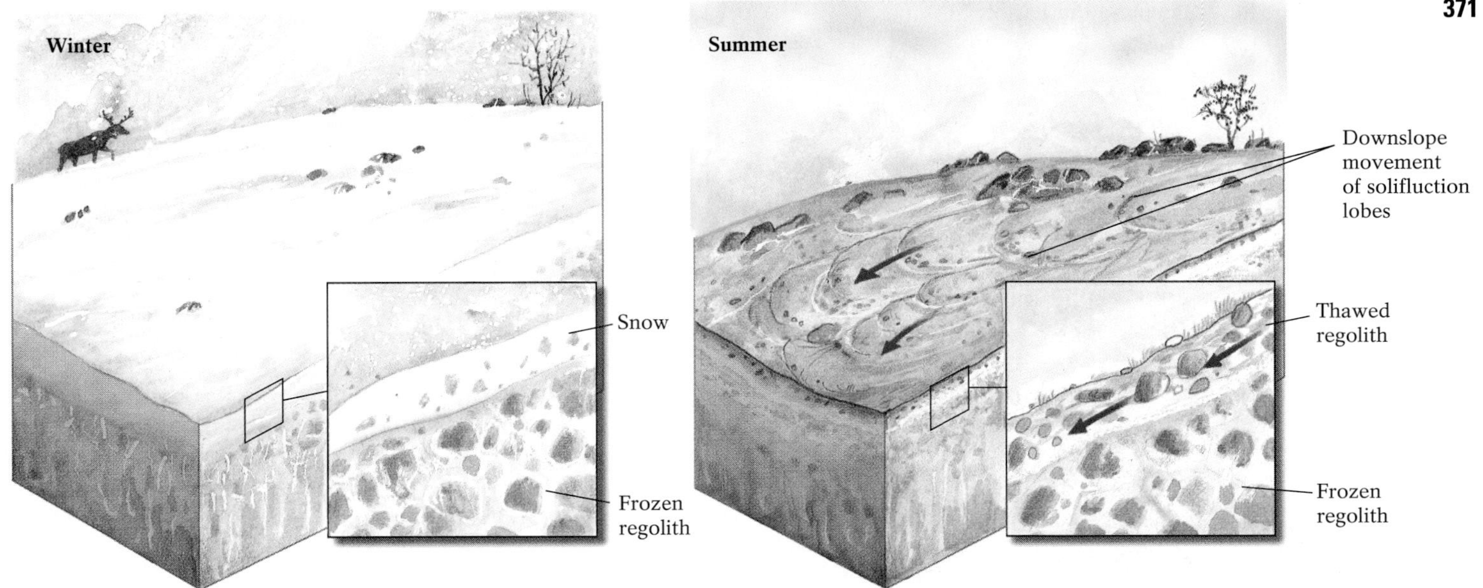

Figure 13-13 Solifluction occurs where the soil and regolith are frozen to depths of hundreds of meters. During the brief summer season, the surface materials thaw out to a depth of several meters and flow downslope with the aid of the water released by melting. Solifluction typically creates lobe-shaped masses of slowly moving sediment.

Loose creeping material gets continuously rearranged as each individual particle responds to the influence of gravity. A particle can be dislodged and set in motion in any number of ways: the burrowing of an insect or larger animal, surface trampling of passing animals, the splash of raindrops. Even the swaying of plants and trees in the breeze can cause their roots to move enough to jostle loose soil cover and send it creeping downslope.

In sufficiently cold environments, where soil water freezes periodically and ice develops in the ground, the volume of the near-surface materials expands upon freezing and displaces soil particles at the slope surface. When thawing causes the near-surface materials to contract, gravity pulls the displaced surface particles slightly downslope before they settle back into a stable position (Fig. 13-12). In clay-rich soils, periodic wetting and drying swells and shrinks materials, moving soil particles in a similar manner. Each of the countless particles in such soils or regolith is gradually but constantly nudged downhill, even on a gentle slope.

A special variety of creep, seen principally in very cold regions, such as Alaska's tundra, is **solifluction** ("soil flow"). This relatively fast form of creep develops where the warm sun of the brief summer season thaws only the upper meter or two of soil or regolith. Because the underlying *permafrost* (permanently frozen ground) remains impermeable to water, the thawed soil becomes waterlogged and flows downslope at rates of 5 to 15 centimeters (2–6 inches) per year (Fig. 13-13).

Rapid Mass Movement

Rapid mass movements can send materials rushing down-slope at speeds measured in kilometers per hour or even meters per second. They are further classified according to their type of motion—falls, slides, slumps, and flows (Fig. 13-14). Each type of movement can occur at different speeds and include a variety of materials. Often one type of rapid mass movement evolves into another.

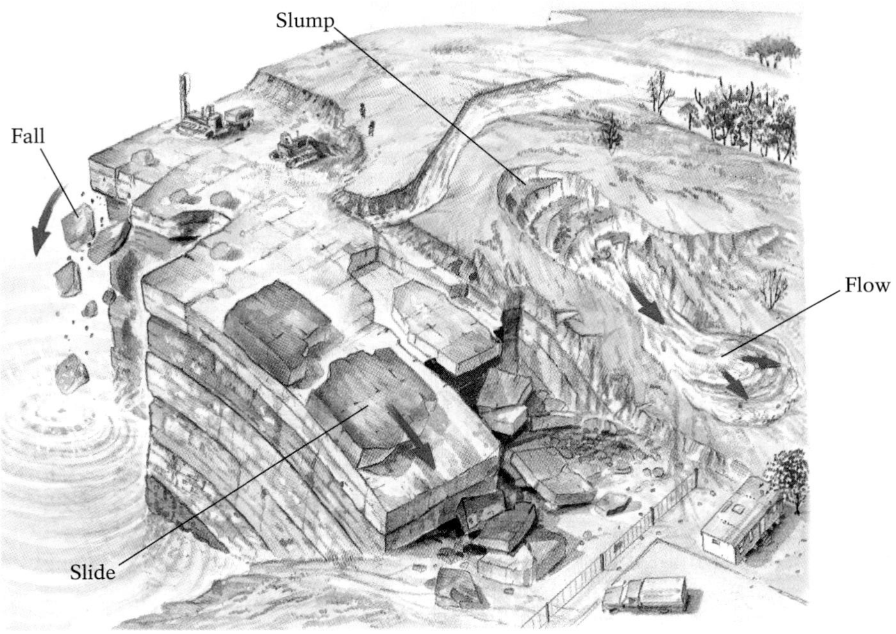

Figure 13-14 Rapid mass movements include falls, slides, slumps, and flows. (See text.)

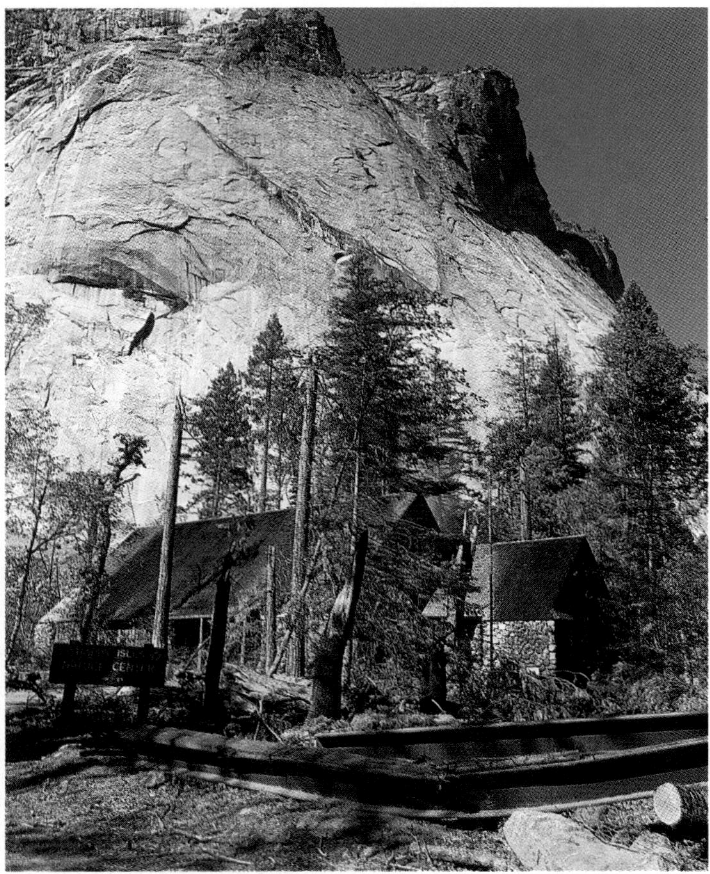

Figure 13-15 Rockfalls, such the one at Happy Isles in July 1996, occur when large masses of diorite fall from steep, exfoliating slopes in Yosemite National Park.

Falls A **fall,** the fastest type of rapid mass movement, occurs when loose rock or sediment breaks free, often at a plane of weakness, and drops from very steep or vertical slopes. The falling material plummets through the air to the ground or water below, traveling at the acceleration of gravity—9.8 meters (32 feet) per second squared (Fig. 13-15). Even a relatively small rockfall can prove deadly; large falls are often disastrous. On July 10, 1996, in the Happy Isles area of Yosemite National Park, a 162,000-ton mass of diorite fell from the Yosemite Valley's steep canyon wall and accelerated to a speed of 260 kilometers per hour (160 miles per hour), killing one hiker and injuring 20 others. The powerful blast of air that preceded the falling mass knocked down more than 2500 trees in its path. Fortunately the fall occurred at 6:52 P.M., after the Happy Isles nature center and snack shop had closed for the day. Most day hikers had already come off the Mist Trail, the site of the rockfall.

A fall that occurs at the coast may create a large sea wave. In 1958, a spectacular coastal rockfall, set off by an earthquake, sent 31 million cubic meters (1 billion cubic feet) of bedrock plummeting 900 meters (3000 feet) into Lituya Bay near Anchorage, Alaska. An enormous wave, hundreds of meters high, raced across the bay; it lifted boats anchored immediately inside the bay mouth and transported them out to sea.

Slides and Slumps A **slide** occurs when rock, sediment, or unconsolidated material breaks loose and moves along a preexisting plane of weakness, such as a fault, fracture, or bedding plane, *as a single, intact mass.* A slide may involve a small displacement of soil over solid bedrock, amounting to just a few meters of movement, or a huge displacement of an entire mountainside. Large slides may be preceded by days, months, or even years of detectable accelerated creep.

In the largest events, a great mass of material detaches along an identifiable plane of weakness, or **slip plane.** The nature of a slip plane varies with the local geology. Unconsolidated sediments often develop slip planes where sedimentary material changes—for example, along the boundary between a layer of sand and a layer of clay. Slip planes also appear where unconsolidated sediments make contact with the underlying bedrock. Highlight 13-1 describes a recent damaging slide in southern California that occurred along the contact between loose and solid slope materials.

In solid bedrock, slip planes develop where planes of weakness exist within the rock itself (see Figure 13-4). In sedimentary rocks, the slip plane often comprises a bedding plane within a weak rock layer, such as thinly bedded shale that inclines toward an open space, such as a river valley (Fig. 13-16). In plutonic igneous rocks, the slip plane may consist of a large joint produced by exfoliation. In metamorphic rocks, slip is likely to occur along foliation surfaces.

A slide's slip plane is generally flat, or planar. Where the plane is irregular or nonplanar, it is called a slip *surface.* A **slump** is a slide that separates along a *concave* slip surface, moving in a downward and outward motion. Whereas slides typically move along a preexisting slip plane, slumps generally create their own slip plane within a body of unconsolidated material. They are distinguished by the crescent-shaped scar that forms in the landscape where the slump detaches. The steep, exposed cliff face that forms where the slump mass pulls away from the rest of the slope is known as a **scarp.** Slumps do not usually travel far from their point of origin, nor do they typically move at high velocities. That explains why the slump block itself usually remains intact (Fig. 13-17).

Flows A **flow** occurs when a mixture of rock fragments, soil, and/or sediment moves downslope as a highly viscous fluid. Flows typically move more swiftly and are more dangerous when they have a high water content. In February 1969, for example, drenching rainstorms moved into southern California from the Pacific, dumping 25 centimeters (10 inches) of rain in nine days. In the next six weeks, the Los Angeles area received an additional 60 centimeters (25 inches) of rain,

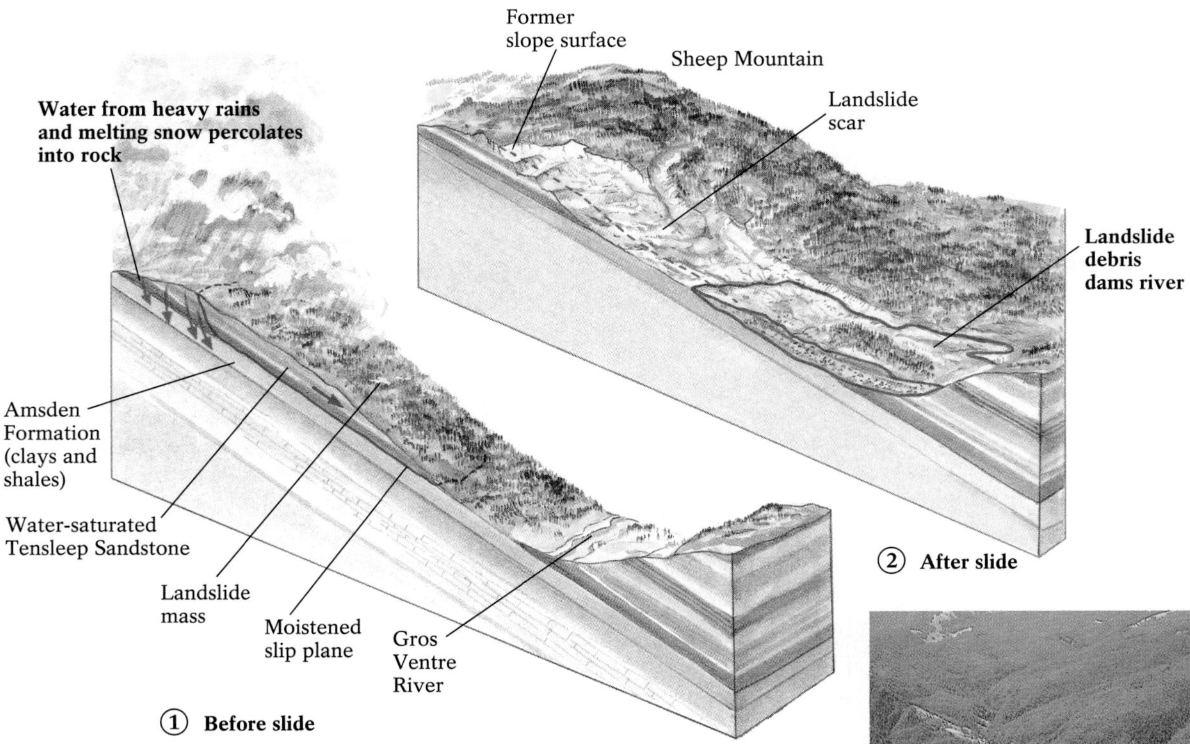

Former slope surface
Sheep Mountain
Landslide scar
Landslide debris dams river
Water from heavy rains and melting snow percolates into rock
Amsden Formation (clays and shales)
Water-saturated Tensleep Sandstone
Landslide mass
Moistened slip plane
Gros Ventre River
① Before slide
② After slide

Figure 13-16 In 1925, Sheep Mountain, along the Gros Ventre River in Wyoming, was the site of one of North America's greatest mass-movement events. (above) The geology of the Sheep Mountain slide. Layers of clay-rich shales alternate with massive beds of limestone and sandstone. Water could not drain into an impermeable shale layer below the overlying sandstone, causing the overlying rock to become saturated and slide downhill. (photo) Landslide debris dams the Gros Ventre River.

more than twice the normal amount for its "rainy season." Los Angeles' precipitation patterns are notoriously erratic: The previous summer had seen a prolonged drought, when brush fires removed the vegetation from hundreds of thousands of acres in the Santa Monica Mountains, which consist largely of clay-rich marine shales. The torrential rains in 1969 triggered numerous flows as saturated clayey soils, denuded of binding vegetation by fire and drought, flowed over impermeable bedrock. One hundred lives were lost, 15,000 people were forced out of their mud-covered homes, and $1 billion in property damage was incurred.

Like other mass movements, flows are classified by their composition and velocity. A flow may contain a wide variety of materials, such as loose rocks, soil, trees and other vegetation, water, snow, and ice. Its velocity is determined by its water content, the nature of its component materials, the nature of the underlying materials, the angle of the underlying slope, and the vertical distance it travels.

Earthflows are relatively dry masses of clayey or silty regolith. Because of their high viscosity, they typically move as slowly as a meter or two per hour (though they may move as rapidly as several meters per minute). Their slow motion ensures that they rarely threaten life, although they may badly damage structures in their paths.

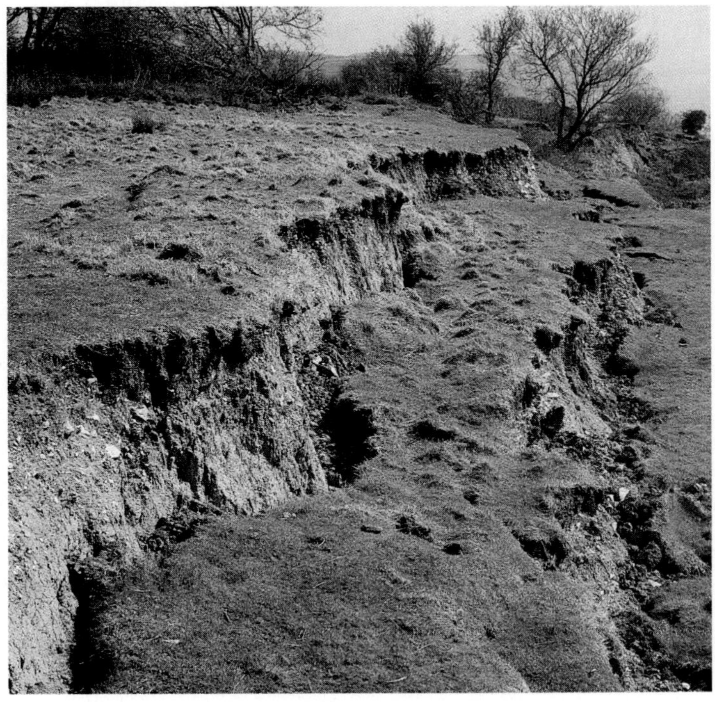

Figure 13-17 Slumping on cliffs in Dorset, England.

Highlight 13-1 The La Conchita Mudslide

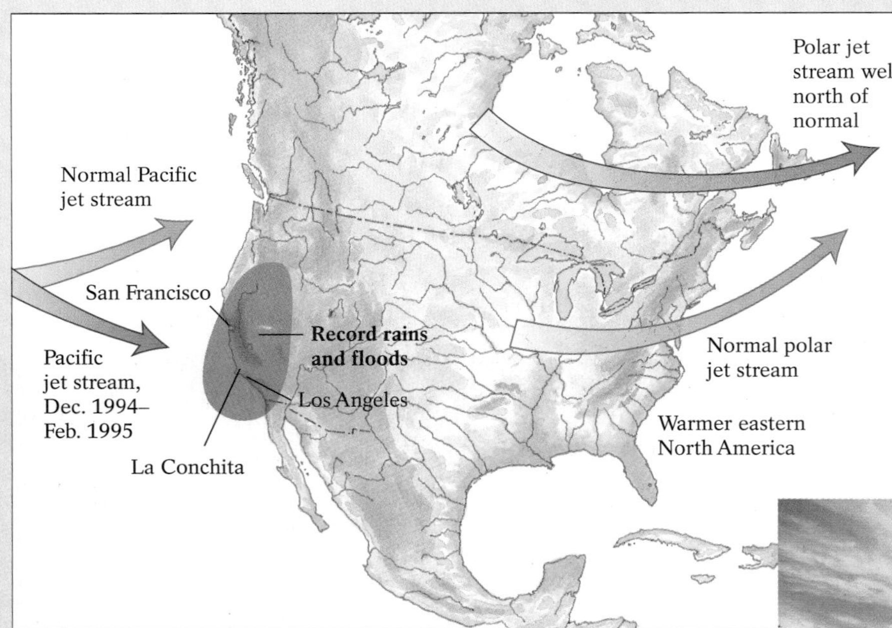

Figure 13-18 **(a)** Storm tracks during the winter of 1994–1995 were diverted to southern California from their normal path across the Pacific Northwest by that year's El Niño effect. At the same time, the polar jet stream in eastern North America was displaced northward; as a result, cold arctic air failed to penetrate southward, allowing the east coast to experience a significantly warmer-than-average winter. **(b)** Damage from the La Conchita mudflow.

On September 29, 1994, residents of La Conchita, California, an unincorporated community located along the Pacific Coast Highway near Santa Barbara, met with a team of local geologists and engineers. The scientists informed the community that cracks had developed in the unconsolidated materials that form the steep slopes surrounding the town. They warned that the slopes could fail at any time, as they had several times during the last decade.

La Conchita sits precariously atop a 10,000-year-old landslide that overlies westward-dipping sedimentary rocks. Extensive irrigation of the avocado and citrus groves at the head of the slope enhances the region's slide potential.

In January and February of 1995, a succession of drenching Pacific storms saturated the unconsolidated material and created pools of water at the heads of the slope's numerous cracks and fissures (see Fig. 13-18a). On February 9, 1995, geologists returned to warn the residents of La Conchita that creep on the surrounding slopes had accelerated significantly and the fissures had become enlarged. According to scientists, there was a 50-50 chance that the slopes would fail soon.

At 2:10 P.M. on Saturday, March 4, 1995, a 600-ton wall of mud—more than 80 meters (260 feet) wide and 10 meters (33 feet) high—slid a distance of 300 meters (1000 feet), roaring down upon La Conchita (Fig. 13-18b). Nine homes were destroyed, some were engulfed in mud, others were shoved from their foundations. Two hundred residents in 60 other homes were evacuated.

Local authorities continue to monitor the slopes, because additional, perhaps larger slides are likely along new fissures. Other communities along the Pacific Coast Highway (U.S. 101) remain on slide alert as well, particularly those in areas that lost their binding vegetation during the tragic wildfires of 1993 and 1994. The combination of earthquakes, fires, drenching rains, steep slopes, dipping rocks, and loose soils in this part of California makes it exceedingly vulnerable to mass movement. Stay tuned—you'll be hearing more about such events in the region sometime soon.

A typical earthflow moves more rapidly than creep but more slowly than a *mudflow,* which is a swifter-flowing slurry of regolith mixed with water. Heavy rainstorms that soak into a thick blanket of regolith generally trigger mudflows. They are fine-grained, comprising 80% or more of sand-sized and finer grains of soil and regolith. The consistency of mudflows can vary from wet concrete to muddy water. Because they are typically saturated with water, these flows tend to travel through topographic low spots, such as valleys and canyons, just as streams do, and they can move considerable distances on slopes as gentle as 1° to 2° (Fig. 13-19).

Mudflows are most likely to develop after heavy rainfalls on sparsely vegetated slopes with abundant loose regolith. Such conditions are common in the canyons and gullies of semi-arid mountains, where loose sediment accumulates between infrequent storms and few roots bind the loose materials. The occasional cloudburst washes large quantities of sediment into arroyos (dry stream channels) and canyons. The saturated sediment then rushes rapidly down the channel.

Occasionally, a quasi-solid body of partly waterlogged clayey sediment almost instantaneously transforms into a highly fluid mudflow, or **quick clay** (Fig. 13-20). Quick clays occur when ground vibrations, caused by an earthquake, explosion, thunder, or even the passing of heavy vehicles, increase the water pressure between sediment particles. The particles separate from one another, reducing the friction between them. Quick clays may even occur spontaneously, without triggering vibrations, when groundwater dissolves away the salts that typically bind *marine*-clay deposits.

Certain regions of North America appear vulnerable to quick-clay flows. Fifteen thousand years ago, much of eastern Canada and New England was covered by an ice sheet several kilometers thick, whose weight depressed the land beneath it. When the ice finally melted, the land was below sea level. The salty seawater that was able to flow inland left behind marine-clay deposits—a loose, disorganized framework of platy clay particles held together in part by salt. On November 12, 1955, near the town of Nicolet, Quebec, along the Gulf of St. Lawrence, 165,000 cubic meters (5.8 million cubic feet) of this marine clay, its binding salt dissolved by thousands of years of groundwater flow, liquefied spontaneously into quick clay that flowed toward the Nicolet River valley. The flow stopped just meters before engulfing the crowded local cathedral, and only three fatalities occurred. A similar, but much larger and more catastrophic quick-clay flow occurred on May 4, 1971, in the Quebec

Figure 13-19 Tragic mudflow in Sarno, Italy, on May 6, 1998. This mudflow, which took more than 100 lives, formed when water from heavy spring rainfall mixed with rock and soil on nearby slopes, producing a muddy slurry that overwhelmed the low-lying sections of the village.

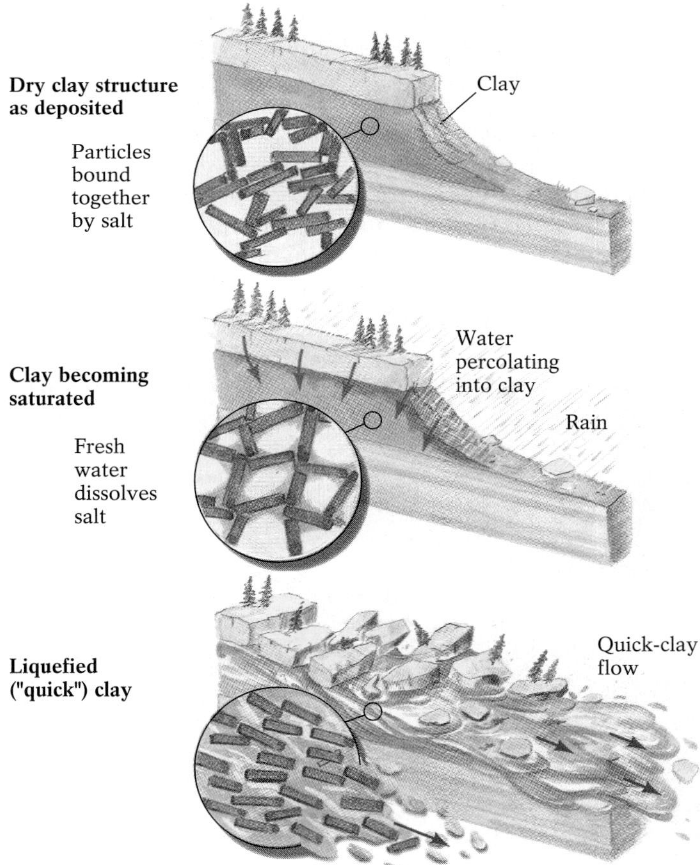

Figure 13-20 The development of quick-clay flows.

Figure 13-21 Rio Nido, in Sonoma County, California, after the El Niño–induced mudslide of February 7, 1998.

village of Saint-Jean-Vianney, where a layer of marine clay liquefied instantaneously: 6.9 million cubic meters (240 million cubic feet) of clay buried 40 homes and took 31 lives. Fig. 13-21 shows the aftermath of a mudflow in Southern California.

Volcanic eruptions can also produce catastrophic mudflows, known as *lahars.* Lahars develop when hot volcanic ash melts snow or glacial ice or when the rainfall that often accompanies an eruption saturates fresh volcanic ash and the accumulated soil and ash layers from previous eruptions. This type of flow may rush downslope at very high speeds. The 1980 eruption of Mount St. Helens in Washington state, for example, set off a lahar that traveled between 29 and 55 kilometers (18–35 miles) per hour, sweeping away homes, bridges, and everything else in its path.

Figure 13-22 A debris flow caused by a one-night flood in a small stream, in Britannia Beach, British Columbia.

Lahars can also occur during a volcano's noneruptive periods. Washington's Mount Rainier has generated more than 60 lahars during the last 10,000 years, many of which occurred during volcanically quiet periods. A major one took place about 5000 years ago, when the summit of the mountain became unstable after millennia of steam emissions had converted solid volcanic andesite to loose clay. More than 450 meters (1500 feet) of the mountain, which stood at an estimated 4850 meters (16,000 feet) at the time, mixed with existing snow and ice and then flowed to the lowland below. We can infer that this event was probably a noneruptive lahar, because no ash layers record an eruption at that time. Today Mount Rainier continues to vent steam and convert its andesitic summit to clay, raising concerns about the possibility of another noneruptive lahar in the near future. For this reason, the U.S. Geological Survey monitors this volcano vigilantly and has drafted an evacuation plan for nearby communities, where more than 50,000 people live today.

Debris flows, like mudflows, are common in sparsely vegetated mountains in semi-arid climates, are typically triggered by the sudden introduction of large amounts of water, and tend to follow topographic low spots. These flows, however, consist of particles that are generally *coarser than sand-size;* often they contain boulders one meter or more in diameter (Fig. 13-22). Hence, debris flows require a steeper slope than other types of flow to trigger their downward movement. A mass movement that begins as a fall or slide can eventually break into a mass of rubble that continues

(a)

(b)

Figure 13-23 Triggered by a major earthquake, the Yungay debris avalanche of May 1970 buried several towns, taking 30,000 lives. **(a)** Before the avalanche. **(b)** After the avalanche.

downslope as a debris flow. Debris flow velocities range from 2 to 40 kilometers (1–25 miles) per hour.

The swiftest and most dangerous debris flows are *debris avalanches,* which are common in the Appalachians, New England's Green and White Mountains, and the Pacific Northwest's Cascades and Olympic Mountains. These mass movements occur on very steep slopes, are triggered during and after heavy rains, and become enhanced by the removal of vegetative cover (by fire or logging, for example). During an avalanche, the entire thickness of soil and regolith may pull away from the underlying bedrock and rush downslope through narrow valleys. In 1973, when Hurricane Camille dropped 60 centimeters (25 inches) of rain in Virginia's Blue Ridge Mountains, debris avalanches claimed 150 lives.

The high velocity of debris avalanches accounts for their great damage. Sometimes they move so fast that they can propel debris through the air. In Yungay, Peru, a powerful earthquake on May 31, 1970, dislodged an avalanche of debris that hurtled downward at speeds exceeding 200 kilometers (120 miles) per hour (Fig. 13-23). On patches of the slope in the avalanche's path, grass and flowers—including shrubs more than 1 meter (3.3 feet) high—were undisturbed, suggesting that the debris flew through the air above them. Some geologists believe that compressed air, trapped below the rapidly moving debris, formed a supporting air cushion that reduced friction between the debris mass and the underlying surface. Eventually, as gravity pulled the flying debris back down to the slope's surface, a powerful wind gust expelled this air cushion. Tangles of downed trees that appear to have been uprooted by a great blast of air often dot the edges of large debris avalanches.

Avoiding and Preventing Mass-Movement Disasters

What can we do to avoid and prevent mass-movement disasters? We can begin by assessing the likelihood for slope failure in a given area. That process requires analyzing both the local geology and the historical records of past mass-movement events. The next step is to avoid human activities that contribute to slope failure. Finally, a mass-movement defense plan calls for preventive measures—stabilizing slopes at risk of failure.

Predicting Mass Movements

Mass movements may well be the most readily predicted of all geological hazards. Simple eyewitness accounts of past and ongoing mass movements provide excellent clues to their future behavior. In the spring of 1935, when the autobahn was being laid through some clay deposits between Munich, Germany, and Salzburg, Austria, a series of slides caught German engineers by surprise. Had they heeded the construction crews' observation that the slope "wird lebendig" (was "becoming alive"), they might not have been so shocked.

The next step in prediction involves thorough terrain analysis. Geologists study the composition, layering, and structure of slope materials, determine their water content and drainage properties, measure slope angles, search for field evidence of past mass-movement events, and implant instruments in boreholes on slopes to monitor the movement

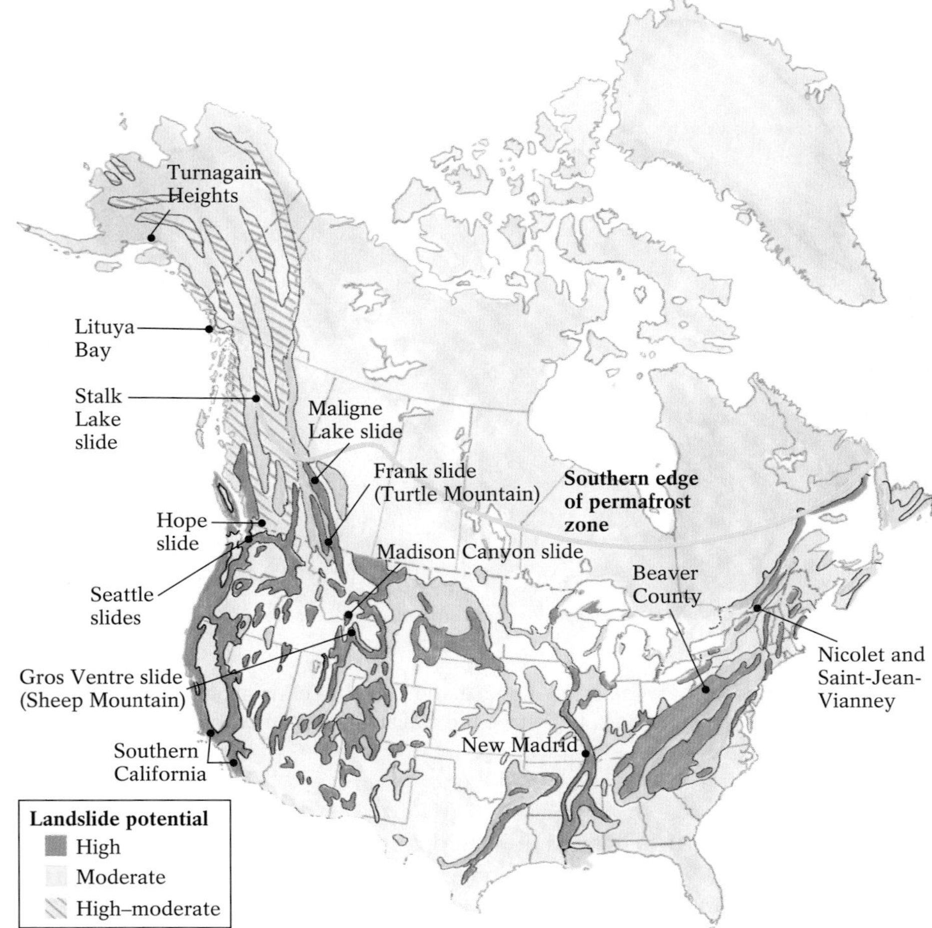

Figure 13-24 A mass-movement hazard map for the United States and Canada. Mass-movement hazards concentrate where slopes are steep, such as in western mountains and the Appalachians; where annual rainfall is high, such as in the humid Southeast and misty Northwest; where droughts and human activities destabilize sensitive slopes, such as in southern California; and where repeated freezing and thawing contribute to slope instability, as in the permafrost regions of Canada.

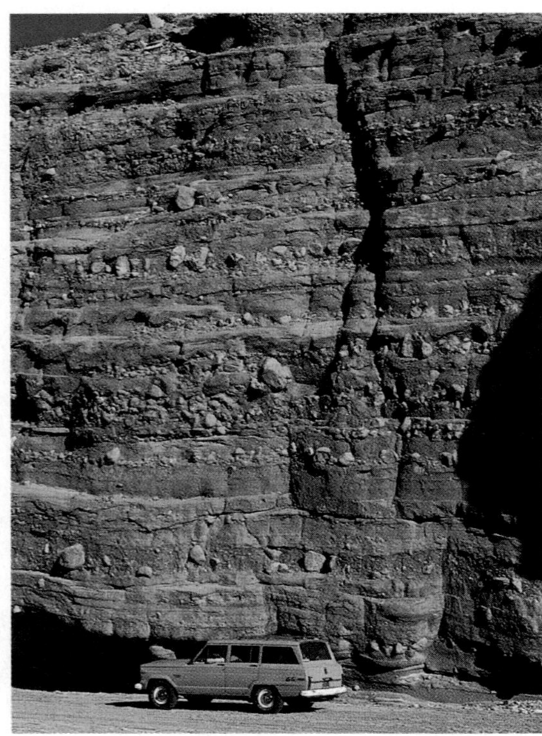

Figure 13-25 Lithified mudflow deposits from the Pliocene Epoch, in Borrego, California.

and deformation of slope materials. U.S. and Canadian geological surveys have issued slide-hazard maps based on such analyses (Fig. 13-24). The U.S. Office of Emergency Preparedness also maintains an inventory of slide-prone areas.

Evidence of earlier slope failures includes talus slopes, landslide scars, and jumbles of slide and flow debris. Mass-movement deposits are usually quite coarse and bouldery, and very poorly sorted. A few mass-movement deposits (such as the remnants of mudflows) appear crudely stratified, a consequence of large, weighty boulders settling toward the bottom of the flow (Fig. 13-25). Their protective muddy matrix and the short distances they travel tend to keep the boulders in mass-movement deposits quite angular. (When these mass-movement deposits become lithified, they typically form breccia, a sedimentary rock.) Because they don't travel far, nearly all the rock fragments in such deposits are locally derived.

It's not always easy to detect earlier slope failures, however. Many very large slumps and slides are difficult to identify after creep and revegetation have modified them. Sometimes the crescent-shaped scarp at the head of a slump or the hilly chaotic terrain that marks a past slide or flow is so large that it cannot be detected up close; only aerial photographs can reveal it. Other slides now lie beneath hundreds or thousands of meters of seawater, as is the case along the southeastern flank of the picturesque island of Hawai'i (see Highlight 13-2).

Highlight 13-2 Crumbling K'ilauea

The sea floor along the southeastern shore of the Big Island of Hawai'i is littered with the lost remnants of the K'ilauea volcano. Even as geologists map K'ilauea's surface lava flows and crater, the mountain itself is tearing away from the rest of the island and sliding toward the ocean at a pace of roughly 10 to 20 centimeters (4 to 8 inches) per year. Although seemingly slow by human standards, this rate constitutes a brisk gallop by geologic-time reckoning; the southeastern flank of K'ilauea is probably the fastest-moving piece of real estate on Earth.

A check of the sea floor off the coast confirms that the crumbling of this volcano has been ongoing for thousands of years (Fig. 13-26). A side-scanning sonar-produced map reveals a dump site for slides of monumental proportions. Slices of past volcanoes, some larger than Manhattan Island, have come to rest after massive slides—some as far as 200 kilometers (120 miles) from their starting point. Slides of such magnitudes apparently spawned tsunami that may have reached heights equal to that of the Empire State Building.

The process that appears intent on dismembering the volcano is born of the very magma that creates it. As magma fills the volcano's subterranean reservoir, K'ilauea inflates like a balloon and pushes outward against its faulted southeastern flank. The flank detaches along the fault, destabilizes, and slides.

At K'ilauea, geologists from the Hawai'ian Volcano Observatory are using recent earthquake data to map the hidden fault that may trigger the next great landslide. In addition, they continue to monitor the flank's inexorable march toward the sea by using the Global Positioning System (GPS) to measure the precise locations of selected benchmarks on the volcano's slopes. As the flank slides southeastward, its bottom portion is apparently bulldozing the sea floor, piling up rubble like a mound of snow in front of a snowplow.

When the current flank eventually gives way, Hawai'i will undoubtedly export some of the disaster to other Pacific Rim locales. A massive Pacific-wide tsunami is a strong likelihood. To understand the potential magnitude of such a wave, consider that scientists have retrieved fragments of coral, which grow below sea level, from as high up as 365 meters (1200 feet) on the slopes of a mountain on the Hawai'ian island of Lanai.

When will this great catastrophe take place? The best estimates suggest that such events occur every 100,000 to 200,000 years. Researchers are now attempting to date past slides. Yet, even if Hawai'i's next great slide is a long way off, movement along K'ilauea's massive southeastern fault has apparently triggered the islands' largest historic earthquakes, some with magnitudes as great as 8.0. Such events spawn life-threatening tsunami, reason enough to pursue the research that will let us better understand how the Hawai'ian islands have evolved.

Figure 13-26 (left) Large fractures on the slopes of Hawai'i's K'ilauea volcano indicate that its southeastern slopes are being pulled apart by large-scale mass movements. **(right)** parcel of forested land that has slid into the Pacific Ocean from the island of Hawai'i's southeastern coast.

Figure 13-27 Methods of dating mass-movement events. Carbon-14 dating is helpful if the moving mass incorporated trees or other types of vegetation. Lichenometry can be used if lichens have colonized bedrock exposed or boulders shattered by the event. Dendrochronology is useful if the moving mass encroached upon living trees that survived the event: The rings of an affected tree grow asymmetrically from the moment the mudflow bends it, so we can determine how long ago the mudflow occurred by counting the asymmetrical rings. Newly exposed boulders begin to accumulate cosmogenic isotopes at the surfaces immediately after the mass-movement event.

We can establish a pattern of recurrence for mass-movement events in a particular area by dating prior mass-movement events and determining their trigger mechanisms. Vegetation is usually the key to dating. A flow on an arid, treeless slope provides little evidence as to its timing. A flow on a vegetated slope, however, often incorporates organic material that can be dated with carbon-14. Dendrochronology, lichenometry, and surface-exposure studies (Chapter 8) can also help date a mass-movement event. A tree engulfed by a flow or slide may become bent by the flowing mass and reveal a datable disruption in its tree-ring pattern; the extent of lichen growth on a shattered rock can reveal how long ago the rock was broken; the buildup of cosmogenic isotopes on the surface of a freshly exposed boulder in a landslide or rockfall can also help geologists date mass-movement events (Fig. 13-27).

After we establish the dates of mass-movement events, we can compare them to the dates of other events that might have triggered the mass movement. If heavy rain is a suspected trigger, mass-movement dates are compared with climatic records; if the records confirm a correspondence, we can predict how an area's slopes will respond to various amounts of rainfall in the future. We know, for example, that most slides in the San Francisco Bay area occur after storms drop more than 15 centimeters (6 inches) of rain on steep, already saturated slopes.

Although we can never predict precisely when a mass-movement event will occur, we can stay alert to the potential dangers of specific slopes. Where property is valuable, or where mass movement poses a particular threat to safety, it makes sense to install expensive landslide-warning devices that signal a slide or flow in progress. In the slide-prone southern Rockies the Denver and Rio Grande Railroad has installed electric slide detectors—wired to respond to the high pressure associated with an encroaching slide mass—upslope from certain bridges. Contact with a moving mass breaks an electrical circuit in the detectors, sending a "stop" signal to approaching trains.

A less expensive early-warning system calls for observing changes in animal behavior. Shortly before the entire Swiss village of Goldau was destroyed in 1806, the town's livestock acted nervously and the beekeeper's bees abandoned their hives. Within hours, a block of rock 2 kilometers (1.2 miles) long and 300 meters (1000 feet) wide broke loose from a steep valley wall, burying the town and its 457 inhabitants in a massive rockfall.

Avoiding Mass Movements

It may seem obvious that it is imprudent to build on sensitive slopes, especially because few insurance companies cover homes and property for mass-movement losses. But people

Highlight 13-3 How to Choose a Stable Home Site

At some point in the future, you may wish to own your own home (if you do not already do so). Having studied physical geology, you will want to ensure that your dream house doesn't fall victim to a geological nightmare.

Suppose you're exploring southern California's scenic beachfront locales. You happen upon the mosaic sign for the town of Portuguese Bend, but notice that the beautiful ceramic signpost is cracked in two. That find should raise your newly grown geologic antenna. You check the local real-estate listings and note a house priced at $50,000 that should be worth $500,000. Your intuition warns you that something is amiss. What else should you look for? How can you tell—in southern California or anywhere else in the world—if you're in mass-movement country?

First, examine the property itself. Do you see any signs of an old mud or debris flow? Look for exposed rocks or fresh breaks in the landscape that reveal the underlying geologic material. Next, investigate the entire neighborhood. Is there evidence of a slump scar upslope where a block may have broken away? Is the site of your dream house in the middle of a large potential slide mass? Evidence of current or past activity may be hundreds of meters or even several kilometers away. As you drive through the community and its environs, look for fences that are out of alignment, and for power and telephone lines that seem too slack in some places and too taut in others (Fig. 13-28).

Look carefully at the house itself, and if possible at neighboring ones. Do you see large cracks in the foundation (small cracks may form during the initial drying and settling of the concrete)? Doors and windows that stick may indicate that once-linear structural features are now out of line, although poor craftsmanship or high moisture content may be the culprits. A cracked pool lining might explain why a swimming pool doesn't retain water. Finding only one such problem may not indicate danger, but the presence of several problems should send a strong warning.

If the geology, topography, and hydrology of a home site all raise questions about slope stability, the location may well be subject to progressive slope failure. You might confirm your suspicions by checking newspaper accounts and the records of the local housing authority, by contacting the state geological survey, and by interviewing the property's neighbors. Of course, anyone purchasing property on a significant slope would be well advised to seek a professional geological inspection before finalizing the deal.

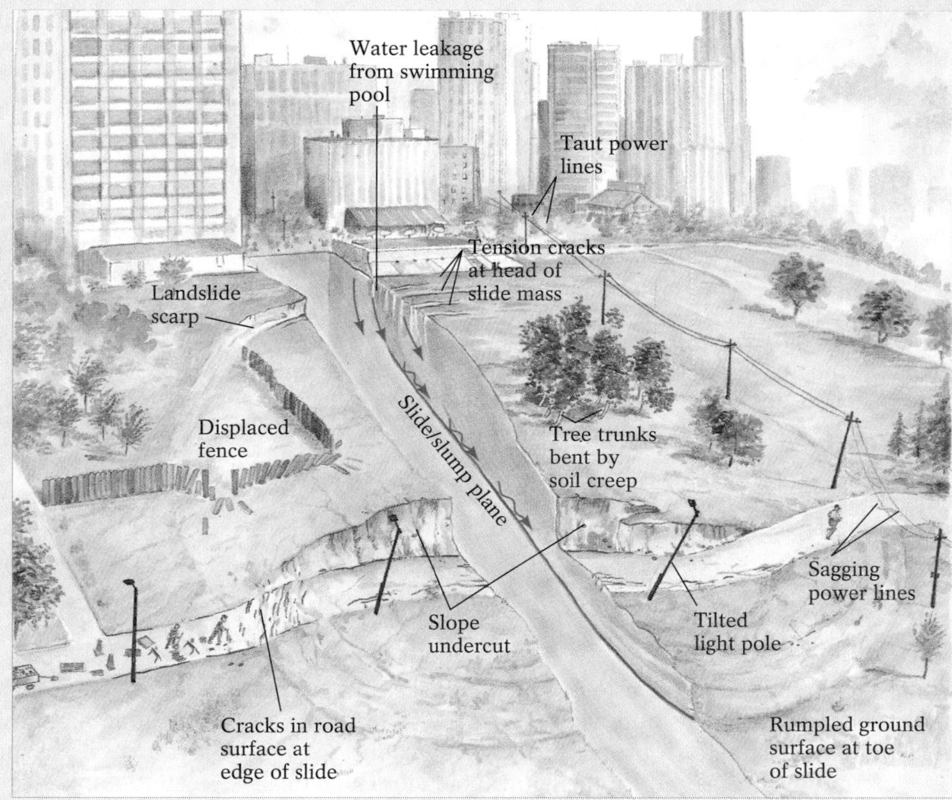

Figure 13-28 Various signs of past, current and potential future mass movement in an urban area. If the power lines in a neighborhood are very loose, the poles that hold them may have moved closer together, as one might expect at the toe of a slide where the slope is bunching up like a rumpled carpet. At the head of the slide, taut lines may indicate that the poles on the slide mass are moving away from those upslope. The poles in the middle of a slide mass may keep their original spacing because the mass may not be deforming much internally.

in designated susceptible areas are often reluctant to abandon ancestral homelands, commercially or agriculturally valuable properties, and scenic hillsides. Thus they rarely practice avoidance of mass movement, as the overdevelopment of southern California's steep canyon slopes and unstable oceanfront cliffs attests. Highlight 13-3 examines the importance of mass-movement avoidance to prospective home buyers.

People in developing nations often overcultivate unstable slopes to feed the hungry, overgraze already sparse veg-

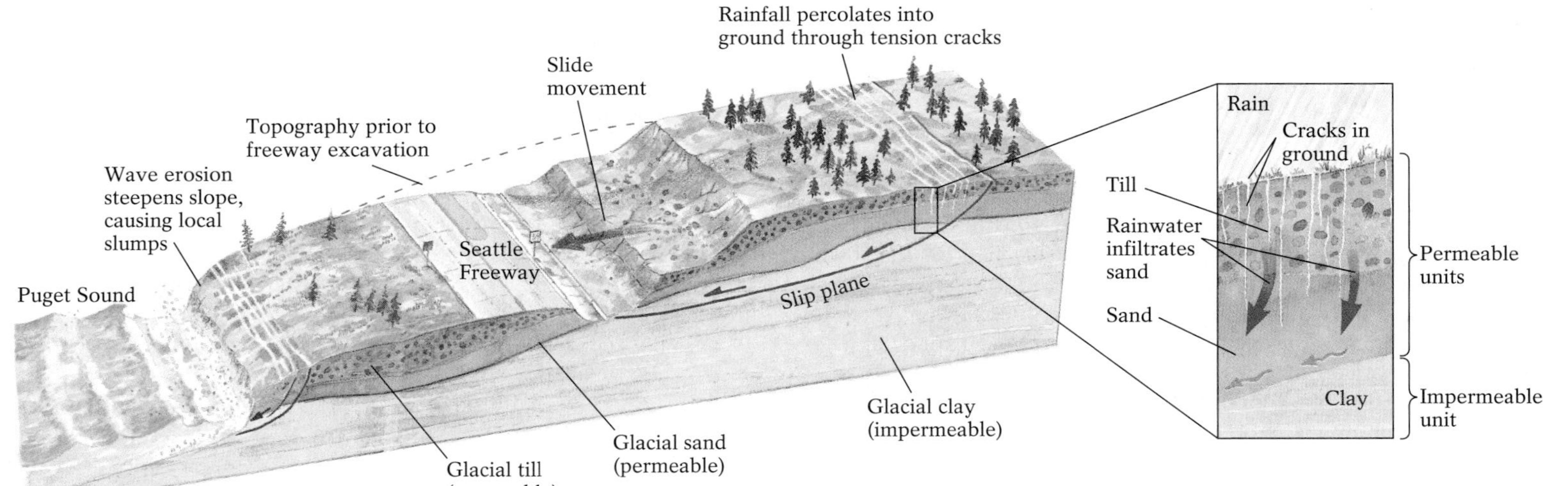

Figure 13-29 The slide-prone geology of Seattle, Washington. (above) Virtually all of this city is built on 15,000-year-old glacial deposits consisting, in part, of a permeable layer of sand overlying an impermeable layer of clay. Water from Seattle's well-known rainfall infiltrates the sand and filters down until it encounters the clay. There the water becomes trapped, high water pressures build up, and the overlying sand layer is buoyed upward. The sand layer is then likely to slip—especially during the rainy months of January, February, and March—taking some of the underlying clay along with it. In addition to the effect of water pressure, Seattle's oversteepened hills, occasional earthquakes, and high level of human activity on slopes increase the city's potential for slope failure. (photo) Tragic slide on Bainbridge Island, Washington, January 1997.

etation to maintain their livestock, and deforest slopes to provide wood for dwellings and heating. They act out of necessity, but all these actions promote mass movement. As long as we continue to use and inhabit unstable slopes in these ways, *prevention* will remain the most practical way to deal with the threat of mass movement.

Preventing Mass Movements

It is far less expensive and much safer to stabilize a slope before it fails than to clean up after the fact. Unfortunately, the dollars lost to mass movements continue to exceed by a factor of 50 the money spent to prevent them. Civil engineers estimate that we could avoid more than 95% of all slope failures by undertaking regional, local, and site surveys and aggressive remedial action.

The first step in developing a prevention plan for any given site is to acquire a clear understanding of its subsurface geology and identify its potential trigger mechanisms. Unfortunately, detailed studies require costly drilling and sophisticated analytical techniques. In addition, some settings include highly variable geological features, requiring tests at a number of locations. In other settings, however, the local stratigraphy is similar virtually everywhere within the region. For example, the entire city of Seattle, one of the urban landslide capitals, is highly susceptible to slope failure. The Puget Sound region of Washington state, including Seattle, typically contains a layer of permeable sand overlying a layer of impermeable clay, which promotes the buildup of excess water near the surface and consequently damaging landslides and mudflows (Fig. 13-29). These geological circumstances proved particularly costly and tragic during the intense winter storms of 1997 (see Figure 13-29 inset).

The second step in preventing slope failure is to determine how best to enhance the forces that resist mass movement or reduce the forces that promote it. Geologists use several methods, often simultaneously, to increase slope stability. Nonstructural methods, which do not require costly engineering efforts, include management of vegetation and introduction of soil-strengthening agents. Structural methods, usually very expensive, involve building structures such as retaining walls and drains or actively modifying the terrain.

Nonstructural Approaches One nonstructural method of preventing mass movement is the planting of fast-growing trees and other vegetation that will develop extensive, deep-root networks that bind near-surface soils. In addition, broad-leaved species, such as elms, maples, and oaks, protect slopes via their large leafy canopies; some falling rain lands on the leaves and evaporates, reducing the likelihood of soil saturation.

Another nonstructural approach adds chemical solutions to soil or regolith to bind clay-mineral particles and increase the strength of loose slope materials. This step can even be taken after a flow has begun: In Norway, a quick-clay flow was halted by bulldozing sodium- and magnesium-chloride salts into the moving mass. Cement injected into a slide or flow mass also lends strength and cohesiveness to unconsolidated sediment.

Structural Approaches Structural mass-movement prevention efforts aim at modifying a slope, reducing its water content, or building supports for it. One slope-modification tactic is to unload it by removing the excess weight of buildings and loose soil. An overly steep slope can be graded by moving material from the top and spreading it at the toe to reduce the slope angle (Fig. 13-30a). More aggressive approaches involve removing all material overlying a potential slip plane (Fig. 13-30b) or cutting flat terraces into a slope to reduce weight and create areas where small slide masses can come to rest (Fig. 13-30c).

Perhaps the best way to prevent mass movement is through a reduction of the water content of a slope, either by not allowing water to enter or by removing water after it has entered. Erecting small earthen barriers to divert water away from a sensitive slope or covering a slope with thin sheets of plastic or a layer of concrete can prevent infiltration. In already-saturated slopes, drains, such as perforated pipes or a layer of coarse permeable gravel, can collect water and channel it away. It is also possible to pump water from a slope to reduce the weight overlying potential slip planes, increase friction along the planes, and decrease swelling and weakening of soil clay. In Pacific Palisades, California, engineers excavated tunnels through an unstable slope and circulated hot air through them to dry the subterranean pore water.

Slopes can also be frozen to stabilize them temporarily. During construction of the Grand Coulee Dam in Washington, the circulation of liquid nitrogen froze 377 points in the damsite's slope. A section of slope 12 meters (40 feet) high, 30 meters (100 feet) long, and 6 meters (20 feet) thick was frozen every day.

Bolts or pins can hold a potential rockslide in place (Fig. 13-31a), but their installation is a costly procedure used only in particularly dangerous situations or in cases where valuable property is at risk. Retaining walls, like those used to

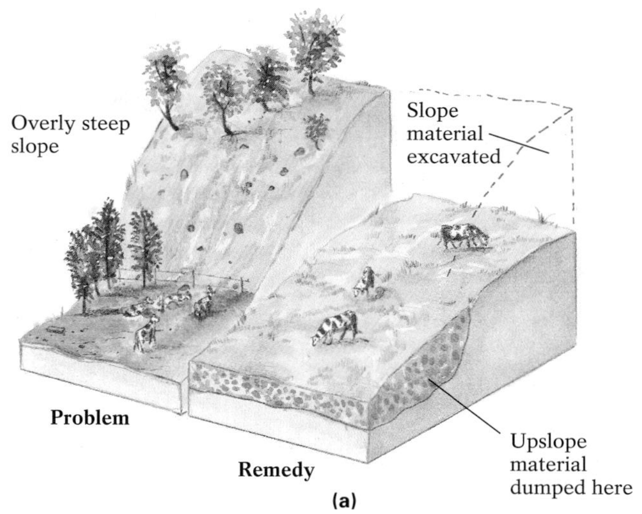

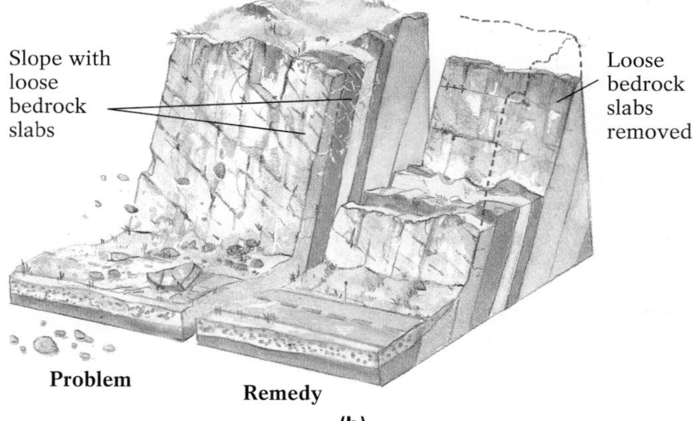

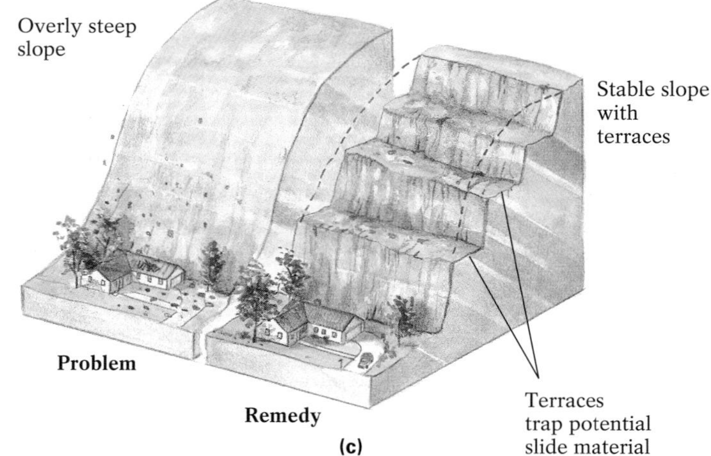

Figure 13-30 Structural methods of slide prevention attempt to diminish the forces driving downslope movement or enhance the forces resisting movement. **(a)** Grading a slope requires moving material from its top to its base to decrease the slope angle. **(b)** Removing loose slope material helps prevent unexpected falls. **(c)** Terraces cut into a slope reduce a steep slope angle and catch any falling debris.

Figure 13-31 Structural supports to prevent slope failure include rock bolts and retaining walls. **(a)** Rock bolts stabilize loose jointed granite on Storm King Mountain, overlooking Storm King Highway in Westchester County, New York. **(b)** A retaining wall supports the ground underlying Route 659, near Blacksburg, Virginia.

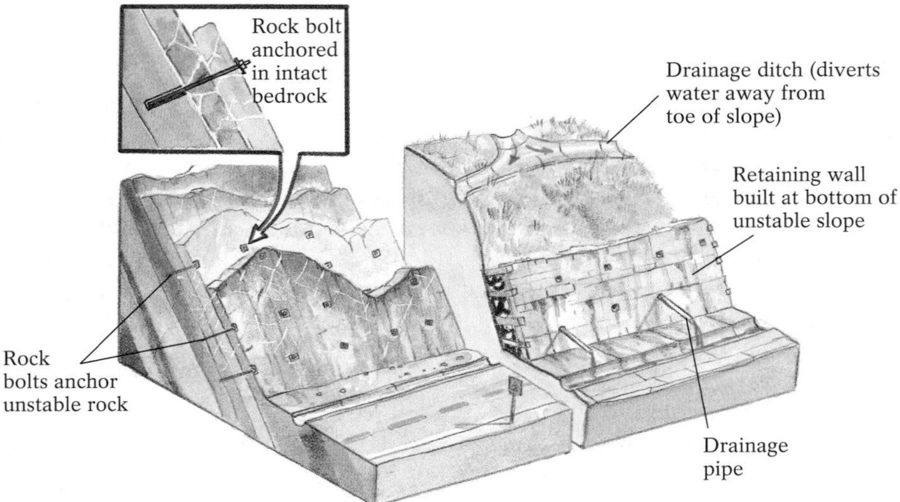

(a)

(b)

shore up excavation walls at construction sites, can provide support to the base of a slope; their installation requires driving steel or wooden piles through loose surface materials into the underlying stable materials (Fig. 13-31b).

Extraterrestrial Mass Movement

Images from NASA's space probes reveal that mass movement occurs on two of Earth's nearest neighbors—the Moon and Mars. Even though the Moon's gravity is only one-sixth that of Earth, mass movement is one of the few ways in which its surface has been modified. With the exception of recently discovered ice patches in some of the Moon's sun-shaded craters, the Moon's surface is a very dry place, and thus mass movement remains largely limited to dry processes. Meteorites are the prime trigger; numerous daily impacts set off avalanches of rock and regolith. After an impact crater forms, slumps occur on its inner walls, which are generally steeper than the angle of repose of loose lunar regolith. Marks left by rolling and sliding boulders may remain unchanged for millions of years on the Moon's surface, virtually free of weathering (Fig. 13-32).

The surface of Mars exhibits a wider range of mass-movement features. Trigger mechanisms for Martian slides and flows include meteorite impacts, "marsquakes," and related faulting. In the past, periodic melting of the planet's subsurface permafrost may have released large amounts of water that oversteepened slopes and promoted development of massive slides, slumps, and debris flows. Such melting may

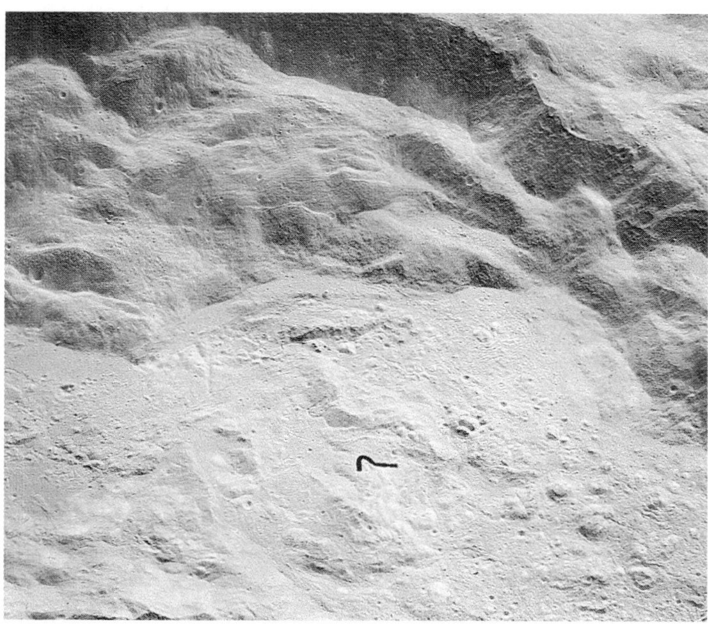

Figure 13-32 A lunar impact crater showing landslide scarps and slumps, as photographed by NASA's Apollo 16 mission. The small hook-shaped mark in the center of the photo is a dust particle on the camera lens at the time of transmission.

have formed rapid mudflows that eroded all or parts of the Valles Marineris, one of Mars' largest canyons (Fig. 13-33).

In this chapter we have explored the many ways that gravity drives erosion and modifies the Earth's surface. In Chapters 14 through 17, we will examine the role that water—as stream water, unseen groundwater, and frozen glacial water—plays in transporting weathered materials and modifying the Earth's landscapes.

Chapter Summary

Mass movement is the process in which the pull of gravity transports masses of Earth materials down slopes. It occurs when the factors that drive material downslope overcome the factors that tend to resist this movement. Gravity is the principal driving force behind mass movement, and the strength and cohesiveness of slope materials and the friction between them and the underlying slope are the principal factors resisting it.

The steepness of the slope, the nature of its materials (for example, unconsolidated sediment or solid bedrock), the

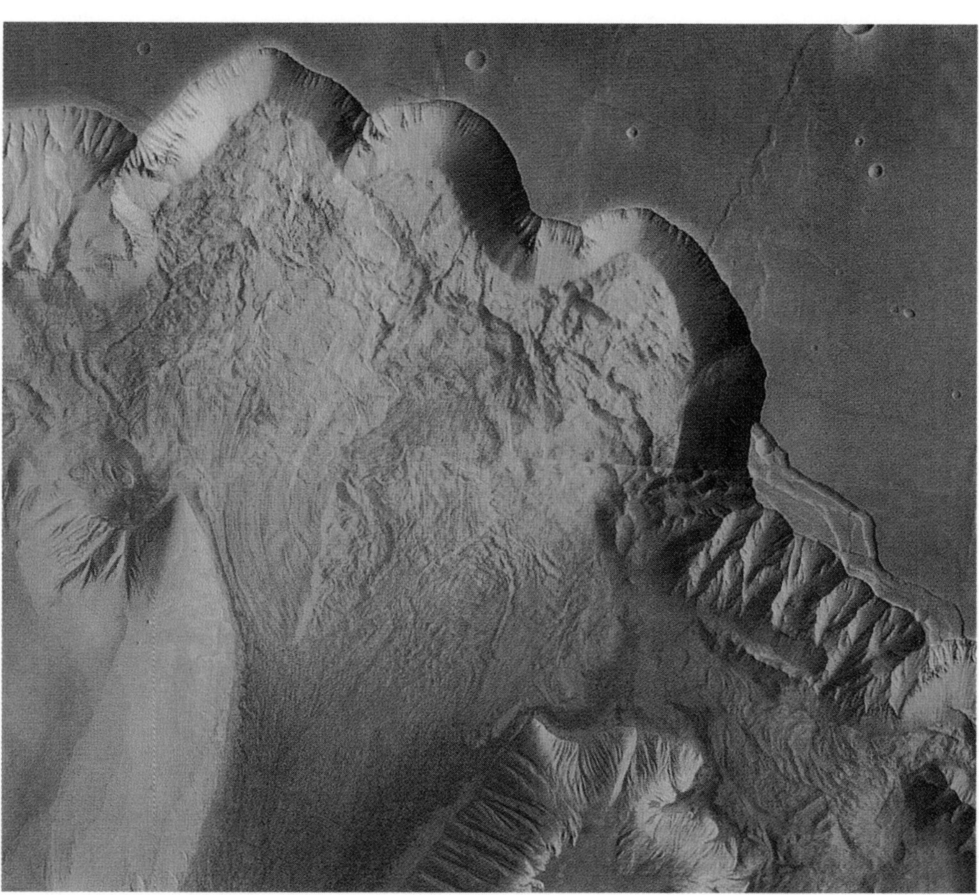

Figure 13-33 Mass movement on the Martian surface is evident in the Valles Marineris, a 2-kilometer (1.2-mile)-deep canyon. Satellite images show that some of the flows are similar in form to those on Earth, and may have traveled as far as 30 kilometers (20 miles) across the canyon floor. Slumps that closely resemble their counterparts on Earth are visible along the margin of the canyon.

amount of water within the materials, the slope's vegetative cover, and the slope's history of human and other disturbance determine whether mass movement will occur. Of these factors, slope steepness (and its effect on the downslope component of gravity) and water content (and its effect on friction) are the primary determinants of mass-movement potential. If a slope is composed of solid bedrock, mass movement is enhanced when the bedrock is jointed, faulted, foliated, bedded, or contains cavities. The effect of slope steepness is particularly pronounced for loose, unconsolidated material, which is stable only until it reaches its **angle of repose,** the maximum angle that it can maintain without initiating downslope particle motion. In general, large, irregularly shaped particles have a higher angle of repose. This relationship is readily apparent on **talus slopes,** the large boulder piles that accumulate at the base of mountains, which often maintain slopes of 40° or more. Water contributes to mass movement by increasing the weight of slope materials, reducing friction between planes of weakness, and reducing internal friction between loose grains in unconsolidated sediment.

A variety of natural and artificial causes trigger mass movement. Earthquakes, volcanic eruptions, heavy rains, and snowmelt are common natural triggers. Ill-advised excavation of unstable slopes and overloading of unstable slopes by dumping or overbuilding are just a few human activities that set off mass movement.

Mass-movement processes are classified by their velocity and composition. In general, slow mass movement, called **creep,** occurs on virtually all slopes composed of unconsolidated soil or regolith. Creep may be enhanced by burrowing organisms, frost action, the impact of raindrops, and trampling by passing animals. A special type of creep occurring principally in very cold regions is **solifluction,** the movement of waterlogged soil over frozen ground.

Several types of rapid mass movement have been identified. In a **fall,** materials become dislodged and fall from a steep or vertical slope without contacting the slope. In a **slide,** rock or regolith detaches along a plane of weakness, or **slip plane,** and moves as an intact mass down slope. A **slump** is a slide that separates along a concave slip plane, leaving behind a crescent-shaped **scarp,** the steep, exposed cliff face that remains where the slide mass pulls away. A **flow** is a slurry of loose material and varying amounts of water.

Flows may contain a wide variety of materials, such as loose rocks, soil, trees and other vegetation, water, and ice. Earthflows are relatively dry and quite slow-moving; mudflows are faster-moving slurries composed of regolith mixed with water. Occasionally, a partly waterlogged mass of solid clayey sediment becomes transformed into a highly fluid flow in an instant, when water separates its sediment particles and reduces the friction between them. Such sediments, called **quick clays,** can be set in motion by an earthquake, explosion, or any other source of vibration. **Debris flows** consist of particles coarser than sand-size, and may contain boulders one meter or more in diameter.

We can predict a slope's potential for mass movement by analyzing its stability, examining evidence of past events and historical records, using landslide-warning devices, and observing changes in animal behavior. Avoiding mass-movement events requires people to refrain from building on sensitive slopes. Prevention of mass movement may follow a nonstructural strategy, such as increased planting of vegetation with extensive root systems and adding chemicals to soil to bind particles. Structural approaches include unloading slopes, grading slopes, building retaining walls, sealing slopes from water infiltration, and draining water from slopes.

Mass movement also occurs on the Moon and Mars. Lunar mass movement primarily involves dry materials and is typically triggered by meteorites; Martian mass movement may be caused by periodic melting of its permafrost, and is triggered by meteorite impacts, "marsquakes," and related faulting.

Key Terms

mass movement (p. 363)
angle of repose (p. 365)
talus slopes (p. 366)
creep (p. 370)
solifluction (p. 371)
fall (p. 372)
slide (p. 372)
slip plane (p. 372)
slump (p. 372)
scarp (p. 372)
flow (p. 372)
quick clay (p. 375)
debris flows (p. 376)

Questions for Review

1. Draw a simple sketch that illustrates the components of the force of gravity that act on an object located on a slope. Include the force of friction.

2. List at least one type of weakness plane that can be found within each of the three common types of rock (igneous, metamorphic, and sedimentary).

3. What does the "angle of repose" of an unconsolidated sediment describe? What is the general particle size and shape of a sediment that would lie at a steep angle of repose?

4. Describe three ways that water contributes to mass movement.

5. List three different mass-movement trigger mechanisms.

6. What is the fundamental difference between a flow and a slide?

7. Describe how solid regolith changes instantaneously to a liquid to produce quick clay.

8. Describe two ways that you might date a landslide.

9. If you were thinking of moving into a slide-prone region, how would you determine if your prospective home was located on a stable slope?

10. Draw a simple sketch illustrating how slopes can be graded.

For Further Thought

1. How would the prospects for mass movement change if a completely clear-cut slope was replanted and became densely reforested? How might the recent forest fires in Yellowstone National Park have affected the stability of the park's slopes?

2. In which North American states or provinces would you expect to find the most mass movement? Explain your reasoning.

3. Discuss the possible ways that plate-tectonic activity could increase the likelihood of mass movement in a given area.

4. What type of mass movement is likely to happen in the photo at right? What steps would you take to stabilize this material?

5. Why do a greater quantity and wider variety of mass-movement events take place on the Earth than on other planets in our solar system?

14

Streams and Floods

Look down from an airplane on a clear day and you will almost certainly see a river. Rivers are perhaps the most common geological feature on the Earth's surface. They flow in virtually every geological and geographical setting, with the exception of large portions of the Arctic and the entire Antarctic. They rush down mountainsides, meander across farmlands, and bisect or border cities and towns. Most major cities are located on the banks of rivers, having been founded at convenient river crossings or sites where waterfalls or rapids prevented boats from proceeding upstream. Historically, a city's success and prosperity has been proportional to the size of its rivers.

At best, life along the banks of a river is a "good news–bad news" proposition. *The good news?* Rivers provide a steady supply of water for home and industrial use and, through irrigation, enable us to make a desert bloom. Crops grown in the fertile soils near rivers provide nourishment for more than one-third of the Earth's human population. Rivers enable cities and towns to transport goods inexpensively and, in modern times, represent a source of clean "hydroelectric" power. Of course, we also fish, swim, sail, windsurf, and waterski in rivers, and even the most sedentary individuals enjoy just sitting on their banks and gazing at them. *The bad news?* Rivers bring floods, the most universal of natural disasters (Fig. 14-1). Many streams flood every two or three years, causing injuries, drowning deaths, and starvation due to loss of crops and livestock. Floods damage property, disrupt transportation networks, and spread diseases caused by insects and microorganisms in contaminated floodwaters.

The health and welfare of much of humankind are intimately tied to the rivers, streams, and creeks that flow by and through our farms, towns, and cities. Thus we should learn about how they form, work, and affect our lives. We will begin by considering the origin of the water that keeps these Earth-surface arteries flowing.

Figure 14-1 Flood damage from the April 1997 flood in Grand Forks, North Dakota.

The Earth's Water

The amount of water *in, on,* and *above* the Earth totals an estimated 1.36 billion cubic kilometers (326 million cubic miles). This volume has apparently remained fairly constant for more than a billion years, although some scientists have recently suggested that incoming icy meteorites continue to contribute significantly to the Earth's water supply. Approximately 97.2% of this water resides in the Earth's oceans; 2.15% is frozen in ice caps and glaciers; and the remaining 0.65% occurs in lakes, streams, groundwater, and the atmosphere. Figure 14-2 illustrates an analogy to help you visualize these proportions. Most of the Earth's water originated in its mantle and came to the surface as **juvenile water,** the hot magma-derived water vapor that accompanies volcanic eruptions. Some water may eventually return to the interior as water-rich oceanic plates subduct.

Water heated by solar energy evaporates from the Earth's surface and enters the atmosphere. Because air temperature decreases with altitude, water vapor cools as it rises and condenses around microscopic ice particles, creating the water droplets that form clouds. When the droplets become too heavy to remain suspended, the water precipitates as rain, snow, sleet, or hail (depending on air temperatures aloft). Some of the water falls back into bodies of water—primarily

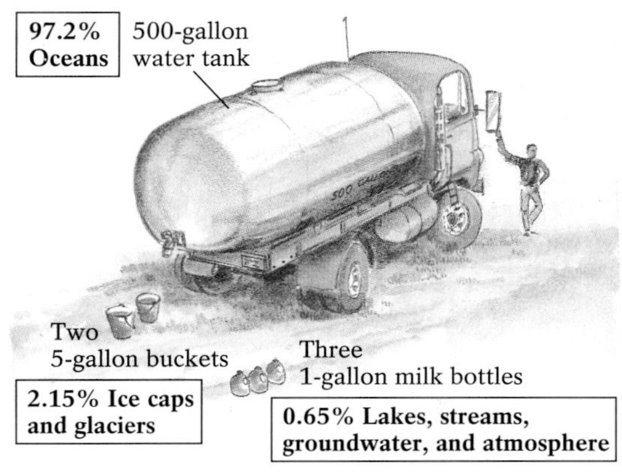

Figure 14-2 Distribution of the Earth's water, by relative volume.

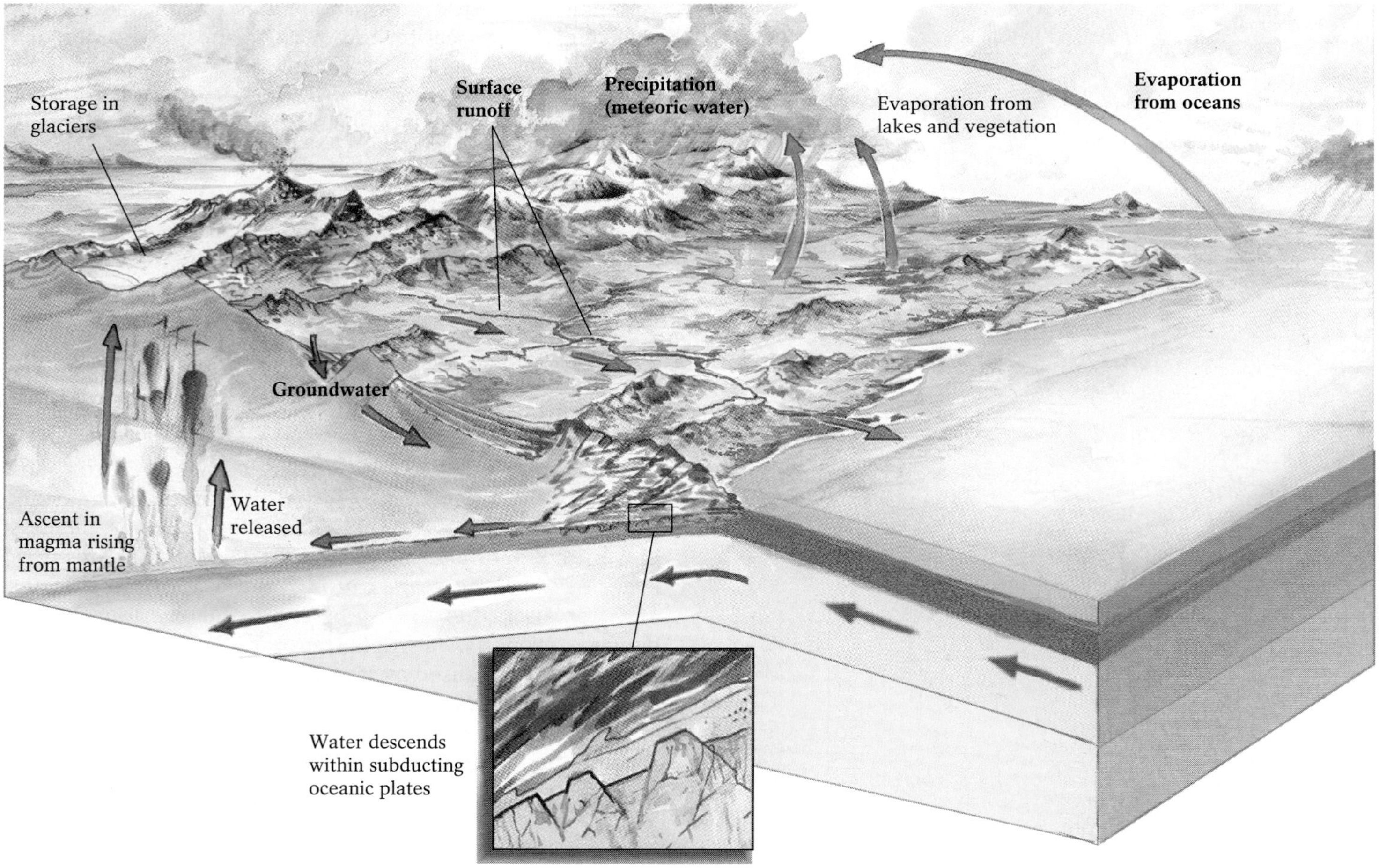

Figure 14-3 The hydrologic cycle. All the water that falls from the atmosphere onto the Earth's surface eventually enters the vast oceanic reservoir through one or more of the pathways of the cycle.

the Earth's oceans—and the rest onto land. This precipitation is known as **meteoric water.**

Water that falls on land may experience one of a number of possible fates. Snow on a high peak may be stored in the snowpack or in a glacier for tens, hundreds, or even thousands of years. Likewise, rain that falls into a lake may remain stored for many years. Rainfall may evaporate back into the atmosphere, run across the landscape as a sheet of water or in distinct channels, or become absorbed into the ground. Within the ground it may be picked up by plant roots, be absorbed by soil particles, or pass into the underground water system. Ultimately, much of the water that strikes the Earth's surface—even if it is stored for many years—will eventually enter a stream and flow back to the sea. Thus the Earth's water moves perpetually among its oceans, land, and atmosphere and, via plate subduction and volcanism, between its mantle and its surface. This movement of water is referred to as the **hydrologic cycle** (Fig. 14-3).

In this chapter, we will explore the *surface-water segment* of the hydrologic cycle and learn how rivers flow and create distinctive landforms by erosion and deposition. We will also examine why they flood and discuss efforts to alleviate flooding.

Streams

Any surface water whose flow is confined to a narrow topographic depression, or channel, is a **stream**—whether it is known locally as a river, bayou, creek, brook, or run. (Geologists commonly use the terms *stream* and *river* interchangeably.) The United States has an estimated 2 million streams, ranging in size from the smallest creeks to the mighty Mississippi River. The flat land immediately surrounding a stream channel, which would be submerged if the stream were to overflow its banks, is called its **floodplain.**

Every stream is sustained by a water-collecting area known as a **watershed** or **drainage basin;** it consists of the total area from which overland flow of precipitation (rain and snowmelt) reaches a stream. Some streams receive water from a network of smaller streams, called **tributaries,** in their drainage basin. In North America, watersheds range from as small as a square kilometer (the area drained by a small tributary) to as large as 3.2 million square kilometers (1.25 million square miles; the area drained by the Mississippi River system). Every drainage basin is bounded by an area of higher topography, called a **drainage divide,** that separates it from an adjacent drainage basin. Divides may comprise low ridges between two small tributaries or continent-spanning mountain ranges (Fig. 14-4). The Rocky Mountain portion of North America's continental divide, for example, stretches from the Canadian Yukon to New Mexico. Rain falling anywhere along this divide flows westward toward the Pacific Ocean, eastward toward the Atlantic Ocean (via the Gulf of Mexico), or northward toward the Arctic Ocean.

At a drainage divide, rainfall initially flows downslope as *overland flow*—broad sheets of unconfined water only a fraction of a centimeter thick. These flows represent a stream's *headwaters.* Wherever the headwaters encounter slight surface depressions or changes in surface composition, they erode narrow depressions called *rills.* Throughout the stream's *transport area,* flowing water shapes and travels through progressively larger channels. Downslope, where the

Figure 14-4 North America's drainage divides and major drainage basins. The drainage basin of the Mississippi River encompasses roughly 40% of the lower 48 states in the United States.

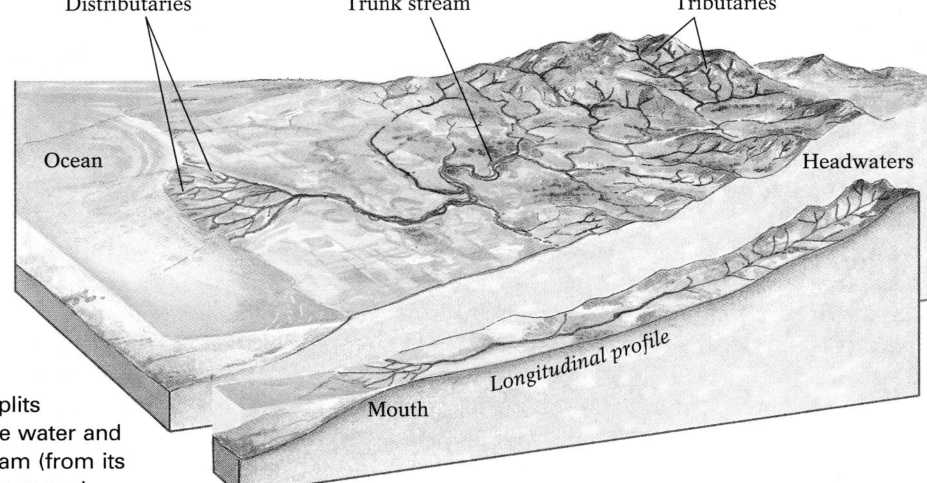

Figure 14-5 A stream system network. All major stream systems consist of tributaries that coalesce to form a trunk stream. The trunk stream transports water and sediment downstream, until it splits into a network of smaller distributaries that deliver the water and sediment to the sea. The longitudinal profile of a stream (from its headwaters to its mouth) is characteristically concave upward.

rills carry more water, they merge into larger, branching channels. Like the branches of a tree, these tributaries converge to form the main stream, known as the **trunk stream.** The trunk stream traverses the greatest part of the transport area and carries the most water and sediment. At the *mouth* of the trunk stream, at sea level, its water and sediment load begin to disperse. Numerous small channels, or **distributaries,** may branch off, carrying water and sediment across lowlands to an ocean (Fig. 14-5).

Stream Flow and Discharge

The flow of a stream is driven by its **gradient,** or slope—its vertical drop in elevation over a given horizontal distance. The steeper its gradient, the more rapidly a stream flows. The gradient, which is expressed in terms of meters per kilometer (or feet per mile), depends on the topography over which the stream flows, but generally decreases from its headwaters to its mouth. In North America, gradients range from 66 meters per kilometer (350 feet per mile) for the upper part of the Uncompahgre River in the Colorado Rockies to 0.1 meter per kilometer (0.5 foot per mile) for the lower Mississippi River downstream of Cairo, Illinois. All streams—even those that appear to be calm—flow *turbulently* (from the Latin for "turmoil"), with their water moving in swirls and eddies. Although the overall direction of flow is downstream, some water may swirl upward, like dry leaves caught in autumn gusts, or descend violently, like the vortex produced as the last of your bathwater disappears down the drain.

Stream velocity measures the distance that a stream's water travels in a given amount of time. It is usually expressed in meters (or feet) per second. Slow-moving streams have velocities of less than 0.27 meter per second (1.0 foot per second); swift-flowing streams can have velocities exceeding 10 meters per second (33 feet per second).

The velocity of the water varies from place to place within a stream. If no rocks and other barriers disturb the stream's course, its *local* velocity will depend partly on the gradient of the channel and partly on the portion of the channel in which the water flows. Velocity is slowest at the sides and bottom of the stream, for example, because of friction between the water and the channel. It is greatest in the center of a stream, in a straight segment of the channel, equidistant from the banks and just below the surface; here the water encounters no friction from either the bed or the atmosphere. Where a stream curves, velocity is greatest at the outside of the curve and slowest at the inside.

Velocity increases where a stream's channel narrows, forcing the same amount of water that was held by a wider channel through a smaller area. This situation resembles what happens when you try to spray an elusive friend with a hose; you can increase the force of the water spray and the distance it travels by putting your thumb across part of the nozzle, thereby narrowing the opening through which the water passes.

The texture of a stream's bed helps to determine its velocity as well. Upstream in the headwaters, boulders in the channel bed create significant frictional drag, sending turbulent waters swirling upward, downward, and sideways. Because the stream expends most of its energy against its bed, its actual *downstream* velocity is relatively low. Further down the stream, velocity generally increases *despite a decrease in gradient,* for two reasons; The volume of water grows as tributaries join the trunk stream, and the stream flows over a smoother bed of sand, silt, and clay, which minimizes frictional resistance to flow.

Related to stream velocity is **stream discharge,** the volume of water passing a given point on the stream bank per unit of time. Discharge is usually expressed in cubic meters (or feet) of water per second. It can be calculated when the stream's width, depth, and velocity are known. The formula for a *hypothetical* square or rectilinear stream is

$$\text{Discharge} = \text{width} \times \text{depth} \times \text{velocity}$$

Because stream channels are rarely square or rectilinear, a corrected estimate for a more realistic stream cross section is

$$\text{Discharge} = (\text{width} \times \text{depth} \times \text{velocity})/2$$

A stream's discharge is dictated principally by the size of its drainage basin and the amount of precipitation that basin receives. The discharge of minor tributaries can be as small as 5 to 10 cubic meters (180 to 360 cubic feet) per second. In contrast, the discharge of the largest river in North America, the Mississippi, is about 18,000 cubic meters (600,000 cubic feet) per second. The discharge of the largest river on Earth, the Amazon of South America—which drains an area equal to nearly 75% of the continental United States—is roughly 200,000 cubic meters (7 million cubic feet) per second, or one-fifth of the Earth's entire freshwater stream flow. To appreciate the size of this massive amount of water, consider that one day's discharge from the Amazon would satisfy New York City's freshwater needs for more than five years.

A stream's daily discharge is also influenced by such climatic conditions as the amount and timing of precipitation and the volume of snowmelt. The ability of local soils to absorb water affects discharge levels as well.

The Geological Work of Streams

Streams carry only about one-millionth of the Earth's water. Nevertheless, they are the planet's most important geological agent of surface change, because they also erode, carry, and deposit sediment. A stream cuts down through uplifted land toward its **base level,** the lowest level to which it can erode its channel. For most streams, the *ultimate* base level is sea level; streams that flow across land areas below sea level—such as the Jordan River in the Middle East—may, however, have base levels *below* sea level. Streams may also reach *temporary* base levels, such as when they encounter extremely durable layers of bedrock that halt their downcutting for a considerable length of time.

When global sea levels fall, coastal streams respond quickly by cutting downward to the newly lowered base level. This situation typically arises during periods of glaciation, when a great volume of seawater converts to continental ice. Conversely, during periods of warming and glacial melting, sea level rises and streams deposit their sediments further inland, where their discharges meet the rising sea.

Figure 14-6 Stream flowing over and eroding a newly formed fault scarp that cuts across some glacier-produced hills.

Graded Streams

Streams, which are among the most dynamic geological agents, respond *immediately* to changes in their environment. When a torrential storm dumps 25 centimeters (10 inches) of rain on a drainage basin, the streams in that region rise quickly, flow more rapidly, erode greater volumes of sediment, and deposit that sediment downstream wherever stream flow becomes blocked or slowed. In a similar fashion, streams respond immediately to local tectonic events. If, for example, normal faulting occurs across a stream's channel and the *downstream* fault block drops, a waterfall immediately forms; the stream then uses its increased energy to erode vigorously the newly uplifted land (Fig. 14-6).

In an unchanging environment, a stream would maintain its gradient at such an angle that it would flow just swiftly enough to transport all the sediment supplied to it from the drainage basin, with little net erosion or deposition. In reality, however, a stream's environment is always changing. Thus the stream's gradient must constantly adjust to achieve a balance between erosion and deposition. A stream in such a state of *dynamic* equilibrium is called a **graded stream.** The equilibrium of any graded stream is merely temporary, lasting only until the next change in its environment. For example, a sudden increase in sediment load, such as from a volcanic mudflow or the collapse of a stream bank, may disturb a stream's equilibrium. Similarly, a rapid increase in discharge, such as from a heavy rainstorm, will force the stream to adjust. A graded stream responds *instantaneously* to any such change in sediment load, discharge, or base level so as to reduce that change's effects and to establish a new graded state.

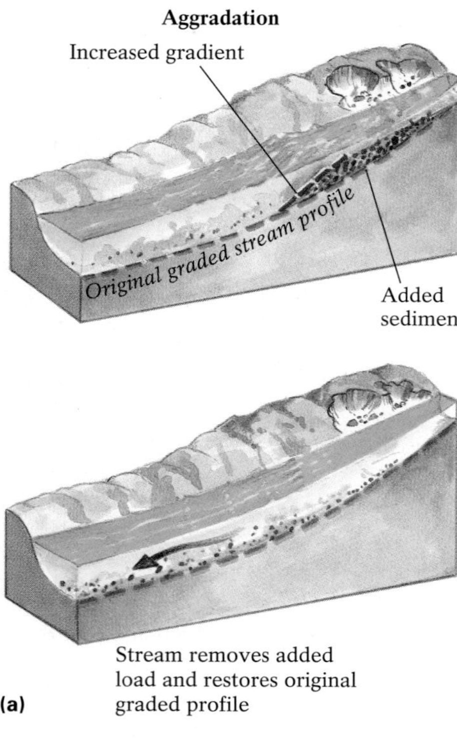

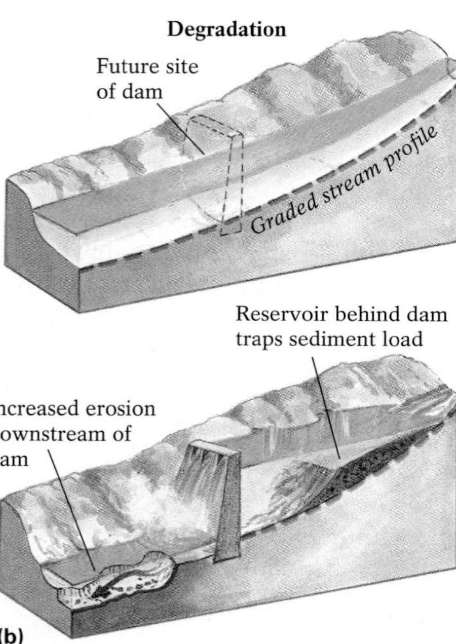

Figure 14-7 Aggradation and degradation of graded streams. **(a)** When a new load of sediment enters a graded stream, the stream becomes aggraded; its gradient instantaneously increases, causing the stream to flow faster and begin to remove the added sediment load. In time, the stream returns to its original gradient. **(b)** When a stream's sediment load is reduced or eliminated (by settling in a new reservoir, for example), the stream bed degrades downstream. It scours its bed, eroding and transporting more sediment, and thereby decreasing its gradient.

Let's say that several loads of coarse gravel are dumped into a graded stream that previously transported only fine sands and silt. The stream initially cannot transport this new material. The coarse sediment settles onto the stream bed, immediately steepening the channel gradient, in a process known as **aggradation** (Fig. 14-7a). The new, steeper gradient increases the stream's velocity; the added sediment is eroded until the graded equilibrium is restored. The streams of central California aggraded in this fashion during the Gold Rush of the nineteenth century, when mining operations dumped millions of tons of debris into them.

What happens when the sediment load of a graded stream suddenly decreases? The stream now has a steeper gradient and a swifter flow than needed to carry the reduced load. As a result, it will scour its bed, eroding more sediment and reducing its gradient, in a process called **degradation** (Fig. 14-7b). Degradation typically occurs downstream from a dam. For example, construction of Hoover Dam on the Lower Colorado River caused the river to degrade its slope downstream as far south as Yuma, Arizona, a distance of 560 kilometers (350 miles) from the dam. To prevent scoured slopes from undermining dams, engineers must design and build structures that protect the dam from degradation.

Stream Erosion

The principal result of streams' erosive power is the creation and deepening of valleys, the Earth's most common continental landform. The composition of a valley's rocks and sediments strongly affects the rate at which it erodes. It could take thousands or millions of years for a stream to cut a valley through a granitic batholith, whereas a single powerful flood can erode a sizable valley in unconsolidated sands.

Left to its own devices, a stream would cut vertically through uplifted terrain like a saw cutting through a board, forming a chasm with near-vertical walls. Why, then, do most river valleys have a distinctive V-shape (Fig. 14-8)? As the river carves out the valley, overland flow and mass movement remove loosened material from its slopes. For example, although downcutting by the Colorado River largely produced the 1810 meters (5940 feet) of *vertical* relief that characterize the Grand Canyon (*relief* is the difference in elevation between its highest and lowest points), overland flow and mass movement have eroded the 21-kilometer (13-mile) width *across* the canyon.

In temperate and humid regions, chemical weathering contributes to V-shaped valley development by breaking down exposed bedrock; the remnants are more easily removed by overland flow and mass movement. In such climates, stream valleys typically have gradually sloping walls. In arid regions, however, only negligible chemical weathering takes place; with little loose sediment to remove, valleys in those areas typically assume the form of steep-walled canyons.

Figure 14-8 The Yellowstone River in Wyoming cuts down rapidly and vertically through the uplifted rocks of its canyon. Mass movement in the upper portion of the canyon helps produce the valley's characteristic V shape.

A stream's erosive power grows as its velocity rises; the increase in rate of erosion is approximately equal to the square of the increase in velocity. Thus, if a stream's velocity doubles, its erosive power increases by a factor of four. For this reason, a flooding stream—which features sharp rises in velocity and discharge—increases dramatically in erosive power.

Processes of Stream Erosion Streams erode their channels by abrasion, hydraulic lifting, and dissolution. **Abrasion** is the scouring of a stream bed by transported particles. Fine particles suspended in the water constantly burnish the bed's surface. Some large pebbles rotate in swirling eddies, carving circular depressions called *potholes* into the bedrock (Fig. 14-9); others bounce against the underlying bedrock. Larger rocks may roll along the bed floor. Erosion by abrasion is most efficient in swift-flowing floodwaters laden with coarse sediment and rocks.

Hydraulic lifting, or erosion by water pressure, occurs when turbulent streamflow through fractures in the bedrock dislodges sediment grains and loosens large chunks of rock. Like abrasion, hydraulic lifting is most active during high-velocity floods; a 1923 flood along the Wasatch Mountain front of central Utah, for example, lifted 90-ton boulders and transported them more than 8 kilometers (5 miles) downstream.

When a stream crosses soluble bedrock (such as limestone, dolostone, and evaporite) some of the bedrock may dissolve, contributing to stream erosion. Every year, *dissolution* removes an estimated 3.5 million metric tons of rock from the continents and carries it out to sea. The Niagara River alone carries 60 tons of dissolved rock over Niagara Falls *every minute.*

Drainage Patterns Streams erode their networks of tributary valleys in distinctive **drainage patterns,** whose shapes are determined largely by topography, composition, and structure of the terrain. A geologist can deduce much about a landscape simply by looking at its stream drainage pattern from an airplane or on aerial photographs or by examining the web of fine blue stream lines that form the pattern on a topographic map.

Figure 14-9 A stream and potholes in Jasper National Park, Canada. The potholes were abraded in the solid bedrock by stones swirling in the stream. Potholes can be as deep as 5 meters (17 feet) and as large as 2 meters (7 feet) in diameter. The stones that ground the potholes typically lie at their bottoms.

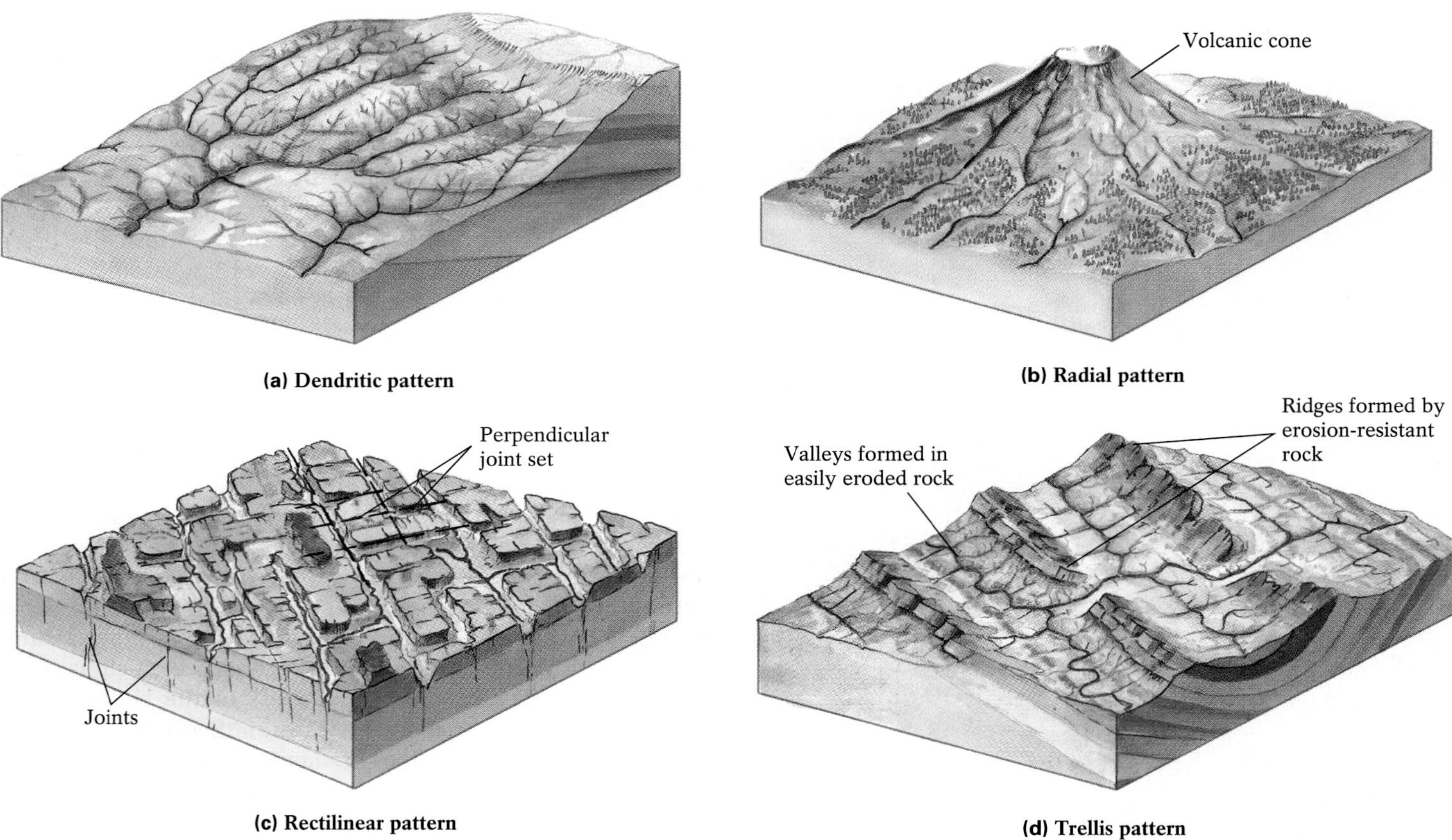

Figure 14-10 Four types of drainage patterns, produced by stream erosion of various bedrock types and under various structural conditions. **(a)** Dendritic drainage patterns often develop on undeformed sedimentary rock of uniform composition. **(b)** Radial drainage patterns usually form on volcanic cones of homogeneous composition. **(c)** Rectilinear drainage patterns often appear where bedrock is cut by perpendicular joints. **(d)** The best place to look for trellis drainage is in an area with long parallel folds of sedimentary rock, such as the Ridge and Valley province of the Appalachians.

When the rocks or sediments that underlie a drainage basin are uniform in composition and undeformed, they tend to erode into a branching, *dendritic* (from the Greek *dendron,* or "tree") drainage pattern. In general, dendritic patterns develop on relatively flat sedimentary rocks, newly exposed coastal muds, and newly exposed unjointed, massive plutonic igneous or metamorphic rocks (Fig. 14-10a).

In a landscape with topographic peaks, such as young volcanic cones, lava domes, and structural domes (discussed in Chapter 9), a network of streams typically drains from a central high area in a pattern resembling the spokes of a wheel. This pattern is called *radial* drainage (Fig. 14-10b).

A *rectilinear* drainage pattern, which looks like a grid of city streets (Fig. 14-10c), forms when stream erosion enlarges systematic joints, fractures, and faults in bedrock. Because these features usually develop in perpendicular sets, the stream pattern shows many right-angle bends.

A *trellis* drainage pattern develops where narrow valleys, underlain by soft, easily eroded rocks, are separated by parallel ridges that consist of resistant durable rocks. Short, steep tributaries, resembling the parallel slats of a garden trellis, drain down the ridge slopes perpendicular to the main stream valleys (Fig. 14-10d).

Stream Piracy The slopes of the two sides of a ridge usually differ. A stream on the steeper slope of a divide tends to flow more swiftly and erosively than one on the opposing slope. One side of a divide may also be eroded more vigorously if it contains less-resistant surface rocks, sediments, or soils. Similarly, if one side of a divide receives more rain than the other, its streams will have greater, more erosive discharges.

When two streams on opposite slopes erode headward toward a common drainage divide, the stream that erodes more vigorously may erode a channel *through the divide itself,* capturing the headwaters of the stream on the other side. This process, colorfully named *stream piracy,* is illustrated in Figure 14-11. A stream that has lost its headwaters in this fashion, described as *beheaded,* simply stops flowing near

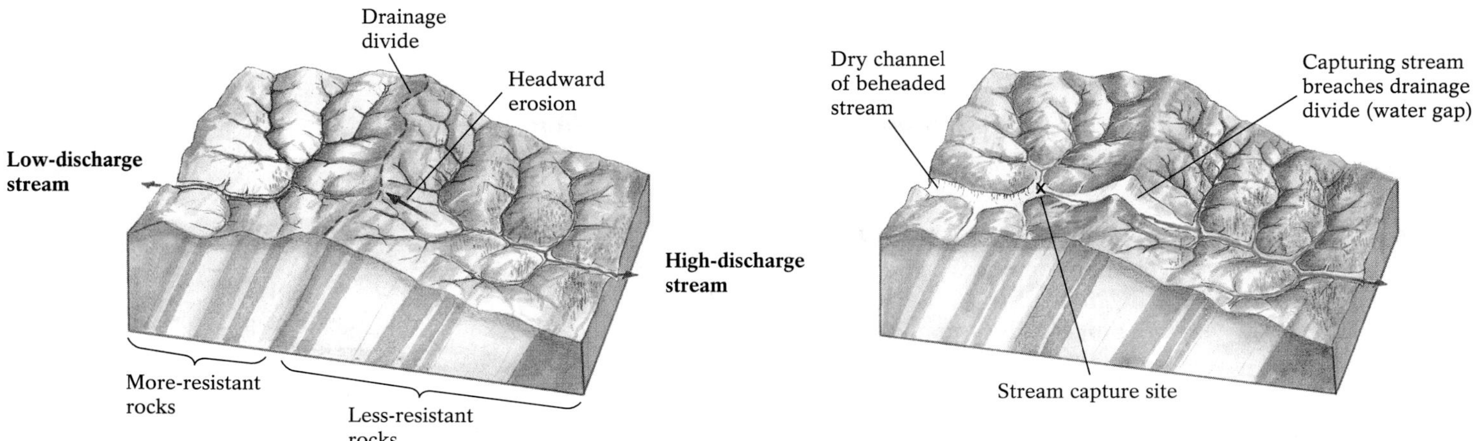

Figure 14-11 Stream piracy. When two streams erode headward toward a drainage divide, the stream that erodes more vigorously may breach the divide and capture, or "pirate," the headwaters of the other. The stream that loses its headwaters is "beheaded."

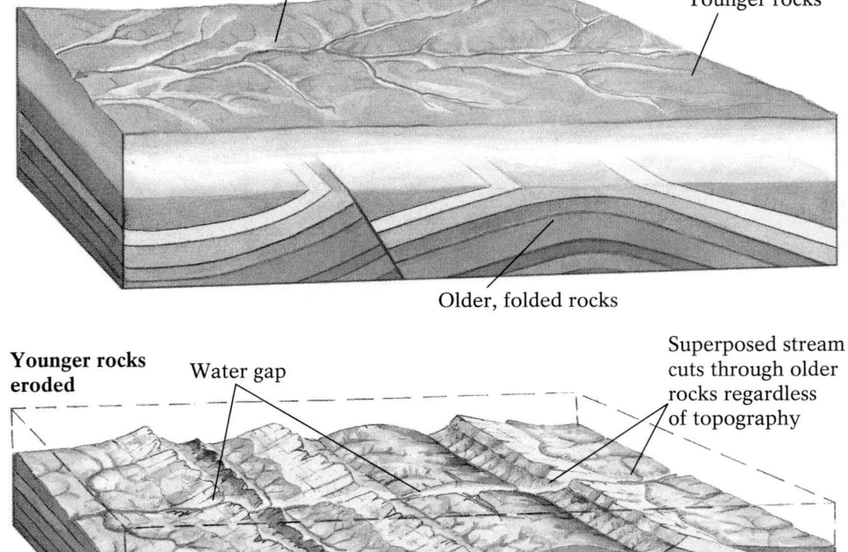

Figure 14-12 Superposed streams. A drainage pattern that seems inconsistent with the surrounding topography—such as a dendritic pattern cutting across a folded landscape—may indicate the stream was established on a different, overlying layer of rock that has since eroded away. Photo: The Delaware River, a superposed stream that forms the border between Pennsylvania and New Jersey. The Delaware Water Gap (near top of photo) was created as the river cut down through a formerly buried ridge, carving out a steep-walled gorge.

the new divide. A pirating stream can create a *water gap,* a steep-walled rocky gorge cut through a ridge or mountain range. (If the pirating stream later becomes diverted from its course and abandons this gorge, the water gap changes into a dry notch in the ridge, known as a *wind gap.*)

Superposed and Antecedent Streams A *superposed stream* is a flow that cuts downward through a number of rock layers of different compositions and structures, yet maintains the drainage pattern established at the original surface. Imagine a dendritic pattern of streams that becomes established on top of flat, undeformed sedimentary rock overlying folded subsurface rocks or masses of plutonic igneous rock. As the sedimentary rock erodes away, its dendritic drainage pattern may be superposed on the layers underneath, which are quite different, such as the folded rocks shown in Figure 14-12.

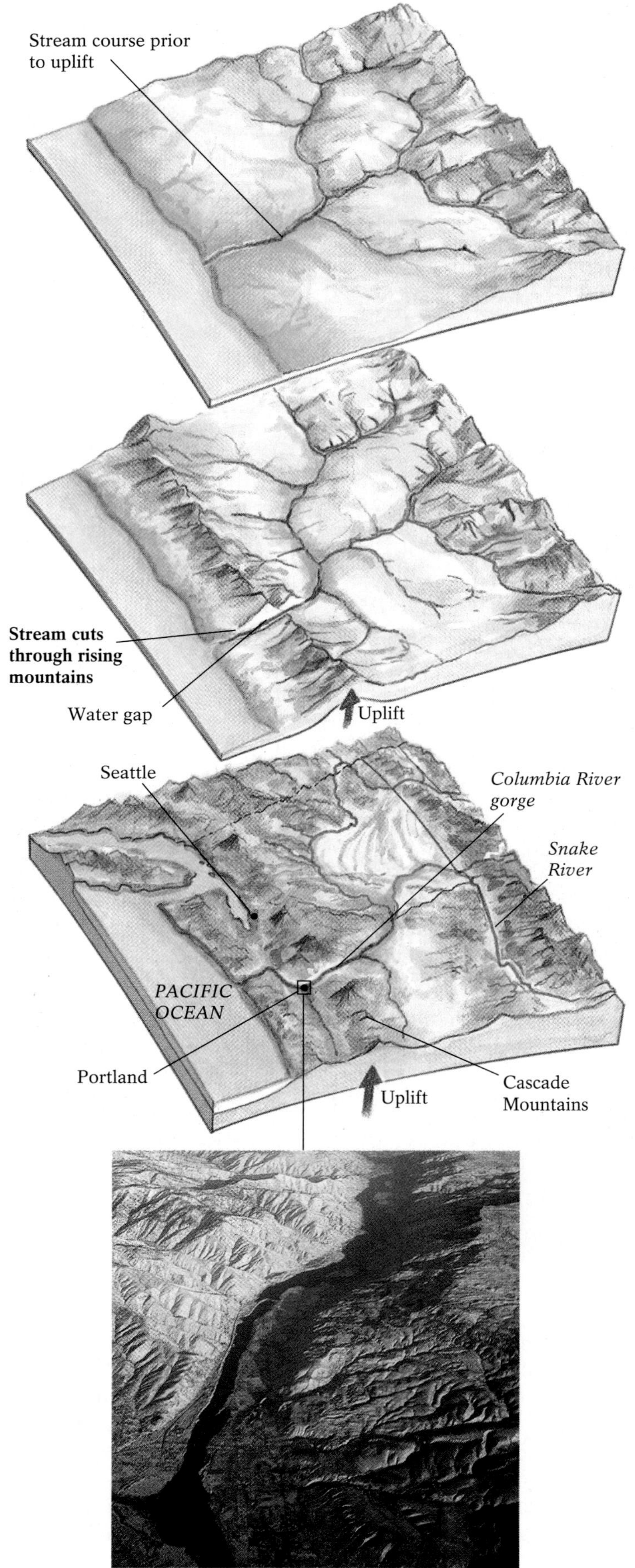

Figure 14-13 Antecedent streams. The Columbia River gorge, which cuts across the Cascades mountain range in Washington state, is a typical antecedent stream valley.

Antecedent streams are older streams that cut through recently uplifted landscapes. If uplift outpaces a stream's ability to cut down through the heightened rock, the emergence of a topographic barrier simply diverts the stream to another course around the barrier. If uplift is slow and steady, however, stream-cutting may match its pace, allowing the stream to maintain its original course and cut a water gap through the rising ridge or mountain range (Fig. 14-13).

Channel Patterns Stream channels can by *straight, braided,* or *meandering.* A single stream may contain segments of all three shapes between its headwaters and its mouth.

Straight channels are rare, but can occur where linear fractures, joints, and faults in resistant bedrock direct a stream's path. Straight segments also arise in areas of active uplift, where the headwaters of a stream flow rapidly down steep slopes.

Networks of converging and diverging stream channels separated by narrow sand and gravel bars are known as **braided streams** (Fig. 14-14). They develop where streams with an exceptionally high sediment loads deposit excess sediment on their channel beds. These deposits build up until they break through the water surface, forming islands that divert the streamflow. Braided streams often occur where a stream's banks consist of loose, highly erodible sediment. They also form where the slope of a stream bed decreases abruptly (such as at the foot of a mountain), reducing the stream's ability to carry its sediment, or where the stream's discharge drops sharply (such as in deserts, where much of a stream's water evaporates or seeps into the sand).

A **meandering stream** (from the Latin *maendere,* meaning "to wander") winds its way across relatively flat landscapes in evenly spaced loops. The loops, or *meanders,* form as the stream erodes its banks; erosion and deposition then combine to exaggerate these curves (Fig. 14-15). Because flow velocity increases toward the outside bank of a curve, the water is most turbulent and erodes the most sediment in that portion of the stream bed. The sediment eroded from the outside bank is carried downstream and deposited on the *inside* bank of the next curve, where velocity and turbulence are minimal. In this way, a meandering stream transfers sediment from its erosional outer banks to its depositional inner banks, with very little net erosion or deposition characterizing the stream as a whole.

Over time, meander curves typically become more pronounced and closer together. Eventually they form a series of loops separated only by thin strips of floodplain. During a flood, a stream may cut through a separating strip, bypassing

Figure 14-14 A braided channel of the Toklat River, in Denali National Park, Alaska.

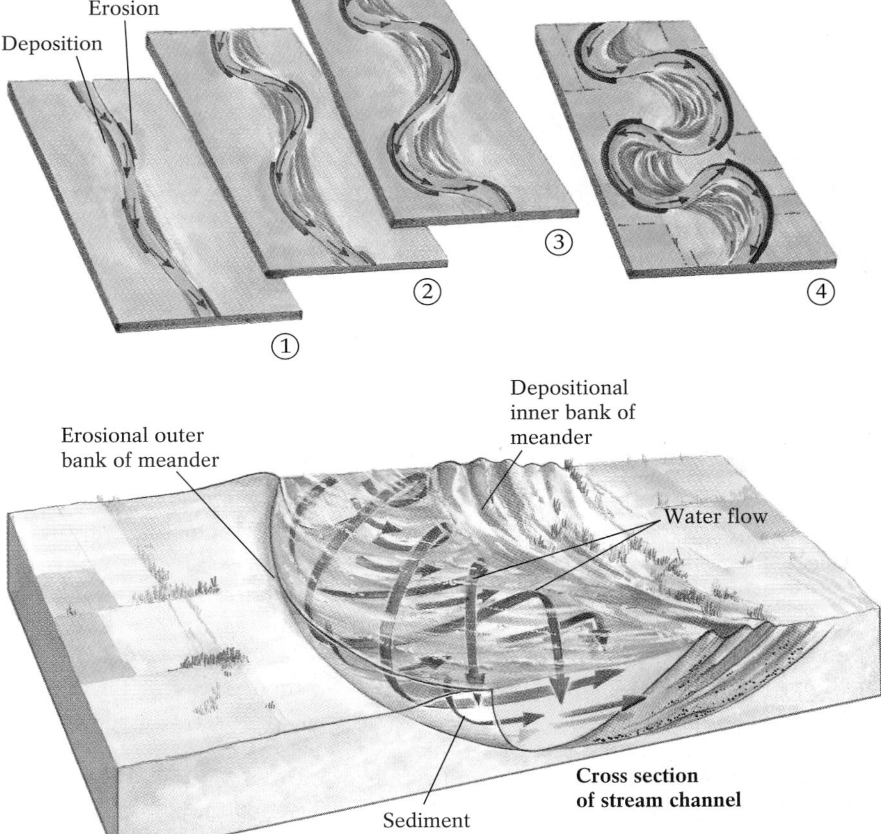

Figure 14-15 The evolution of meandering streams. As a somewhat curving stream flows, the high-velocity water at the outside of each bend produces a cross-channel, corkscrew-like flow that erodes the outer bank. The eroded material is deposited downstream, at the inner bank of the next bend. Meander bends typically grow more pronounced as progressively more sediment is removed from their outside banks and added to their inside banks. Photo: The meandering Little Bear River, in Utah.

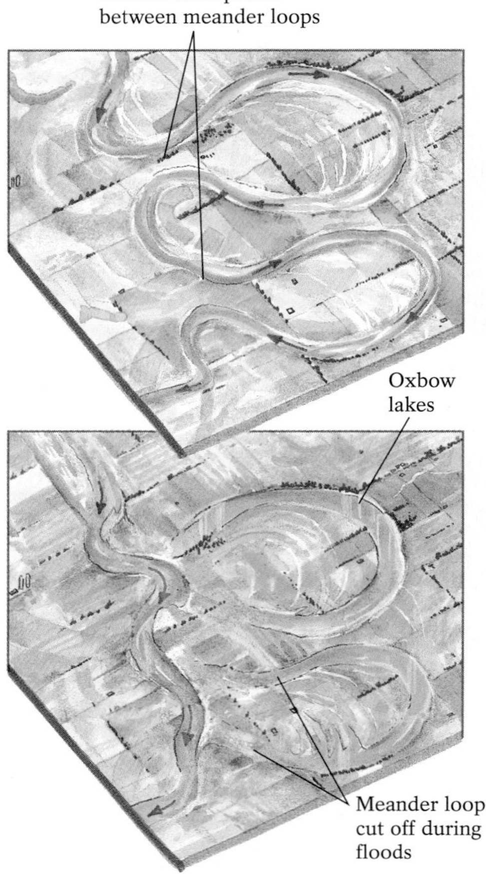

Figure 14-16 (left) When a stream's meander bends become so pronounced that the land between successive loops narrows to thin strips, the stream may completely bypass or cut off the bend (especially during floods), forming oxbow lakes. (right) Failed attempt by the Union Army to cut off a segment of the Mississippi River near Vicksburg, Mississippi, during the American Civil War.

an entire loop (Fig. 14-16). The isolated meander loop evolves into a crescent-shaped body of standing water called an **oxbow lake,** so named because its shape resembles the curved collars worn by farmers' oxen. Subsequent flood deposits may fill the oxbow lakes.

From 1765 to 1932, the Mississippi River cut 19 meanders between Cairo, Illinois, and Baton Rouge, Louisiana—equal to about 400 kilometers (250 miles) of the river's course. During the Civil War, General Ulysses S. Grant, commander of the Union Army, tried unsuccessfully to *create* a cutoff artificially so that Union forces traveling by river could elude the Confederate guns at Vicksburg, Mississippi. Nature finally achieved that task in 1876, the year of the United States' 100th birthday, when the "Centennial" cut off separated Vicksburg from the river's active channel. Despite the presence of so many cutoffs, the Mississippi course has *not* been shortened appreciably. Existing meander curves have expanded, and once-straight cutoff sections have themselves begun to meander.

Waterfalls and Rapids Waterfalls and rapids crop up at sudden drops in topography along a stream's course. Usually they occur where erosion removes softer sections of bedrock, leaving more resistant rock as a "step" in the stream's profile, or where faulting lowers or raises a portion of the stream's profile. Whitewater rapids often represent relics of waterfalls whose "steps" were eroded into irregular rocky beds over which water rushes turbulently.

A waterfall lasts only as long as the conditions that create it persist. The energetic plunging water of a waterfall erodes a deep pool at its base, called a *plunge pool,* that undermines the step from which the water falls. As the step is progressively eroded, the waterfall migrates upstream. Ultimately, as shown in Figure 14-17, when the step disappears, the stream reestablishes a graded profile. One Montana waterfall, created by fault motion during Yellowstone's Madison Canyon earthquake of August 1959, lasted barely a year. The quake forced a tributary to the Madison River—Cabin Creek—to flow over a fresh 3-meter (10-foot)-high fault scarp. By June 1960, the scarp had almost completely eroded, and a small set of rapids replaced the waterfall. Five years later no trace of the rapids or the waterfall could be found.

Thousands of waterfalls and rapids are scattered across North America, from picture-postcard falls such as Yosemite Falls in California's Sierra Nevada, to countless small falls and rapids that dot the springtime streams of the eastern

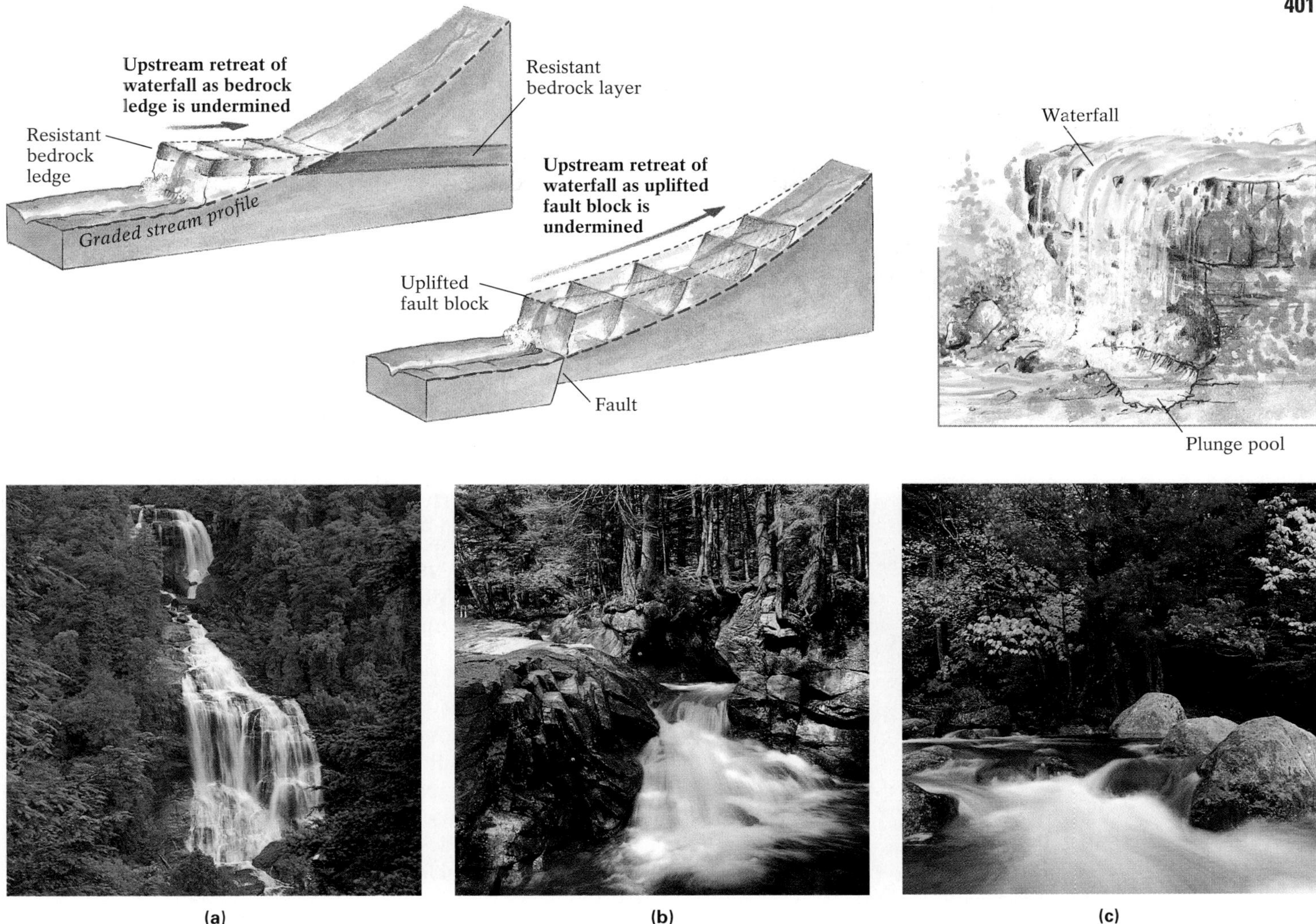

Figure 14-17 The evolution of waterfalls and rapids. When resistant rock or faulting (or relative uplift of any kind) interrupts the graded profile of a stream, the stream initially cascades over the "step" in the landscape. Eventually, the force of the free-falling water excavates a plunge pool at the foot of the falls, undermining the cliff face and causing the upstream retreat of the falls. The falls are gradually eroded down to rapids, which are in turn eventually completely eroded, returning the stream to its graded state. **(a)** Whitewater falls in Nantahala National Forest, North Carolina; this water drops 125 meters (411 feet) in a distance of 150 meters (500 feet). **(b)** A significantly smaller waterfall in Franconia Notch State Park, White Mountain National Forest, New Hampshire. **(c)** Rapids along the Ellis River in White Mountain National Forest.

United States. Many of the latter falls occur where streams descend from the durable igneous and metamorphic rocks of the Appalachians into the soft sediments of the Atlantic coastal plain.

Best known of the eastern falls is, of course, Niagara. This favorite of honeymooners and tourists alike is roughly 55 meters (176 feet) high and 670 meters (2200 feet) wide. It began to form about 12,000 years ago, when the last great North American ice sheet retreated north of the Great Lakes region; its withdrawal uncovered a ridge of resistant dolostone. Since then, the falls has retreated southward 11 kilometers (7 miles) from its point of origin at Lake Ontario. Billions of liters of water annually thunder through the 50-kilometer (30-mile)-long gorge of the Niagara River between Lake Erie and Lake Ontario, plunge over the falls, and erode the shale bed below the resistant cap of dolostone, undercutting the falls.

The future of Niagara Falls is uncertain. At its past rate of retreat [averaging about 1 meter (3 feet) per year], Niagara Falls would have eroded the remaining 30 kilometers (20 miles) of dolostone all the way to Lake Erie in 30,000 years. The United States and Canada, however, divert approximately 75% of the river's discharge into four large tunnels to generate thousands of megawatts of hydroelectric power (enough to supply more than 2 million homes). The diversion has slowed the retreat of the falls to less than 0.5 meter (less than 2 feet) per year, possibly doubling its remaining life.

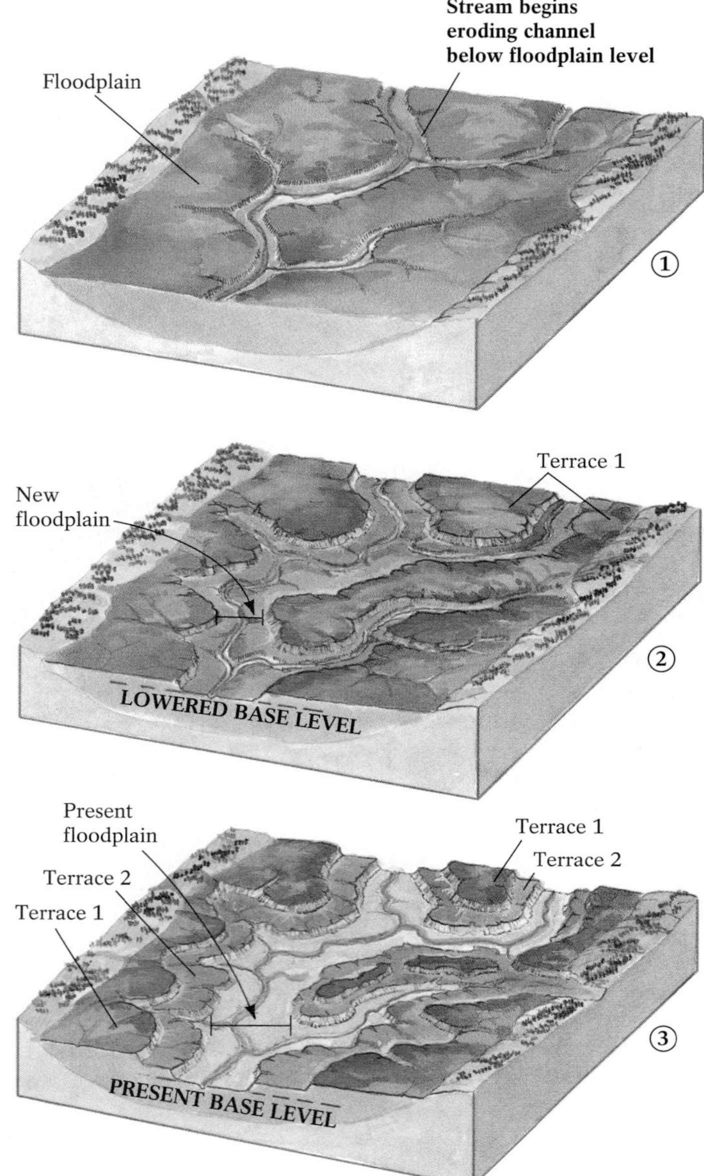

Stream Terraces When a stream's discharge increases over a long period of time (perhaps because of climate change) or when its base level drops (perhaps because of a global sea-level change or local tectonic movement), it typically erodes its channel to a lower depth. Its floodplain is left high and dry above the new channel, forming a gently sloping topographic "bench" known as a **stream terrace.** A terrace usually stands high enough above a stream that even the most extreme flood cannot touch it. You can often see tiers of terraces within a single stream valley—remnants of a once-continuous floodplain surface that spanned the entire valley before being cut down to progressively lower levels (Fig. 14-18).

Stream Transport

The world's streams deliver about 45 trillion cubic meters of water to the sea every year, along with 9 to 10 billion tons of sediment. The Mississippi River alone carries more than 1 million tons of sediment *daily* to the Gulf of Mexico. The amount and type of sediment that a stream transports are largely determined by the flow's velocity, the composition and texture of its sediment, and the characteristics of the bedrock that it crosses. High-velocity streams can erode and transport large boulders. High velocity is also required to pick up small, flat clay particles and mica flakes, because electrical charges on these sediments' surfaces cause them to cling to the stream bed. Once in the streamflow, however, these particles can travel long distances, often remaining suspended in the water all the way to the sea.

The maximum load of sediment that a stream can transport is its **capacity.** Capacity is expressed as the *volume* of sediment passing a certain point on the channel bank in a given amount of time. It is proportional to discharge: The more water flowing in the channel per second, the greater the volume of sediment that can be transported in that time.

The diameter of the largest particle that a stream can transport provides a measure of a stream's **competence.** Competence is proportional to the *square of a stream's velocity:* The greater a stream's velocity, the larger the particles that it can transport. Thus, when velocity doubles—for example, in a flood—competence increases by a factor of four. This relationship explains why a stream that ordinarily carries only fine gravel can sweep large boulders downstream during a flood. During the 1933 flood in California, for example, the Tehachapi River carried steam locomotives from the Santa Fe Railroad hundreds of meters downstream and then buried them under tons of transported stream gravel.

Figure 14-18 The creation of stream terraces. (1) A terrace forms when a stream erodes its channel below the level of its floodplain, either because its discharge has increased or because its base level has fallen. (2, 3) When a stream undergoes a series of such changes, it may create a flight of tiered terraces in a single valley. Photo: Terraces above the Cave River at Arthur's Pass, New Zealand.

Depending on a stream's discharge, its velocity, and local geological conditions (such as the composition of the stream banks), four combinations of capacity and competence are possible:

- *High capacity, high competence.* A rapidly flowing, sediment-laden stream transporting large particles—such as in a flood.
- *High capacity, low competence.* A slow-moving, sediment-laden stream—such as one flowing in a broad, downstream valley through a channel composed of easily eroded sediment.
- *Low capacity, high competence.* A swift-flowing, relatively sediment-free stream transporting large boulders along its bed—such as in a stream's mountainous headwaters, particularly in a region of exposed durable bedrock.
- *Low capacity, low competence.* A slow-moving stream on a relatively gentle slope flowing through resistant, relatively unweathered bedrock.

Sediment Load Streams transport sediment in different ways, depending on the particle size involved (Fig. 14-19). Very fine solid particles are usually distributed within stream water as a **suspended load.** Coarse particles that move along the stream bottom form the **bed load.** Sediment carried invisibly as dissolved ions in the water constitutes the **dissolved load.**

Most of the world's stream-borne sediment—about 7 billion tons annually—travels as suspended load. These fine particles generally drift along in a turbulent, fast-moving stream, like confetti caught in swirling air currents. Approximately 70% of the Mississippi River's annual load of 500 million tons of sediment takes the form of suspended particles. The Colorado River derives its characteristic red color from suspended particles of silt and sand eroded from red formations in the Grand Canyon.

More than 10% of the world's stream-borne sediment is transported by the remarkable Huang Ho (Yellow River) of central China. In fact, half of the Huang Ho's flow volume consists of sediment. During floods, sediment may represent as much as 90% of the river, reducing its flow rate to a languid pace mimicking the flow of molasses. Much of this sediment comes from the 200-meter (650-foot)-thick blanket of yellow, windblown silt through which the river flows. Little vegetation anchors the loose silt in this extremely arid region, allowing periodic storms to wash huge volumes of sediment from the adjacent slopes. The Huang Ho's annual sediment load would suffice to build a wall 25 meters (80 feet) high and 25 meters wide that would encircle the planet.

Although a stream's velocity and discharge largely determine the volume of its suspended load, particle shape and density also play a role. Both flat and low-density particles tend to remain suspended, somewhat like a Frisbee floating on a cushion of air. A flat particle of low-density muscovite mica, for example, resists settling through the water column; it can travel thousands of kilometers without dropping onto the stream bed. Denser particles settle at a faster rate. A gold particle, whose density is 19 times that of water (specific gravity = 19.3), falls to the stream bed much more rapidly than a grain of quartz sand (specific gravity = 2.65).

Unlike its suspended load, which moves constantly, a stream's *bed load* moves only when the water's velocity becomes rapid enough to bounce, lift, or roll particles. The coarse material that makes up a stream's bed load varies in terms of particle size and movement. The sandy portion skips along the bed with a bouncing motion known as *saltation* (from the Latin *saltare,* meaning "to jump"). Saltating sand grains are lifted from the bed by hydraulic lifting or collisions with other saltating grains. Once bounced into the stream, these grains are carried upward and forward by the stream's turbulence and velocity until the pull of gravity returns them to the bed. Larger, heavier cobbles and boulders usually move by a rolling motion called *traction.* The largest boulders can be moved only during major floods. These rocks often feel slippery because they settle in and become covered with algae; few algae would grow if the boulders were in constant motion.

Natural waters are never pure H_2O; instead, they contain ions dissolved from surrounding bedrock. Streams carry these ions as their *dissolved* load. (Precipitation from a water supply's dissolved load coats the inner walls of tea kettles, for example.) The most common dissolved ions are

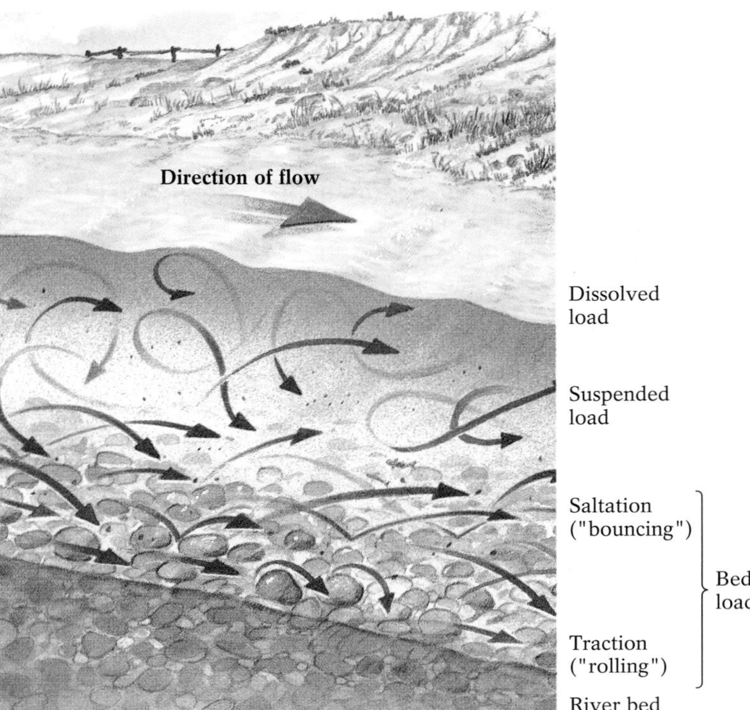

Figure 14-19 Sediment distribution and movement within a stream.

bicarbonate, calcium, sodium, and magnesium, though smaller amounts of chlorides, sulfates, nitrates, silica, and organic acids may be present as well. The amounts and kinds of dissolved ions in stream water depend on the factors that control chemical weathering—such as climate, vegetation, and bedrock composition—and the acidity of the water. Velocity has no effect on a stream's ability to carry dissolved load, although it might influence the rate at which material enters solution. In general, higher dissolved loads are found in streams that flow through warm, moist regions of low relief, lush vegetation, and outcrops of soluble bedrock, such as limestones and evaporites.

In addition, the volume of a stream's dissolved load depends on the drainage basin's topography, which determines whether precipitation runs off rapidly as sheet flow and stream flow or instead infiltrates the groundwater system. Dissolved ions are most abundant in local groundwater, whose slow movement through soluble bedrock promotes dissolution more effectively than contact of a flowing stream with its bed.

Stream Deposition

When the downward pull of gravity on stream particles overcomes the buoyant effects of stream velocity and turbulence, the particles become deposited on the stream bed, with the heaviest particles settling out first. Deposited stream sediments are described collectively as **alluvium** (a word of French and Latin derivation, meaning "washed over"). As much as 75% of a stream's alluvium drops "on land"—into its channel, onto its floodplain, onto flat valley floors at the feet of steep gradients, and into standing bodies of water along its course. The remainder settles where the stream enters the sea. In this section we describe the types of deposition exhibited by streams.

In-Channel Deposition During normal (nonflooding) flow, a stream with a large amount of coarse bed-load sediment typically deposits some of its load *in its channel* as mid-channel bars. As sand and gravel accumulate, the stream is diverted around the channel bars and assumes a braided form (see Figure 14-15). The bars become increasingly resistant to erosion as vegetation begins to grow on them.

In-channel deposits may also take the form of **point bars,** in which sediments scoured from the outside banks of meandering streams are dropped on the streams' less turbulent inner banks (see Figure 14-16). As a point bar grows, the water flowing over it becomes shallower; this development increases the friction between the water and the surface of the bar and by decreasing stream velocity even further, promotes even more deposition. Figure 14-20 shows an example of growing point bars in Montana. Because meandering streams drop their heaviest particles there, point bars are excellent places to search for such "heavies" as gold, platinum,

Figure 14-20 Growing point bars on the Madison River, Montana.

and silver. The gold deposits in the rivers of the Yukon and Alaska, which beckoned thousands during the gold rushes of the nineteenth century, were formed in this manner.

Floodplain Deposition A rising stream may overflow its banks and spill onto the surrounding floodplain. As noted earlier, because of its increased velocity, floodwater can carry a large volume of sediment. As this water escapes from the channel and spreads over the relatively flat floodplain, however, its velocity decreases and the stream drops its sediment load. As its name suggests, a floodplain is a major repository of flood-borne stream sediment. Some typical features associated with floodplains are illustrated in Figure 14-21.

Levees (from the French *lever,* meaning "to raise") are ridges of sediment deposited on both banks of a stream as the decrease in velocity, turbulence, and competence at the banks prompts floodwater to begin dropping its sediment load. The coarser sediment accumulates adjacent to the channel and forms the levees; the finer sediment settles farther out on the floodplain, enriching agriculture for miles around. The levees, whose heights increase with each flood, tend to be the highest points on a floodplain. If they reach sufficient heights, they can sometimes prevent lower-volume flows from overflowing the channel banks. Between floods, trees that prefer well-drained growth sites may take root on and stabilize the coarse-grained levee deposits, further enhancing flood prevention (Fig. 14-22). Towns and farms tend to prosper along the gentle, relatively dry, landward slopes of levees, that is, until the next levee-topping flood.

Growing levees can divert tributaries away from the main stream channel, either on a temporary basis or over the long term. Diverted tributaries may be forced to flow parallel to the main channel for tens or hundreds of kilometers before they can cross a low spot in the levee and join the

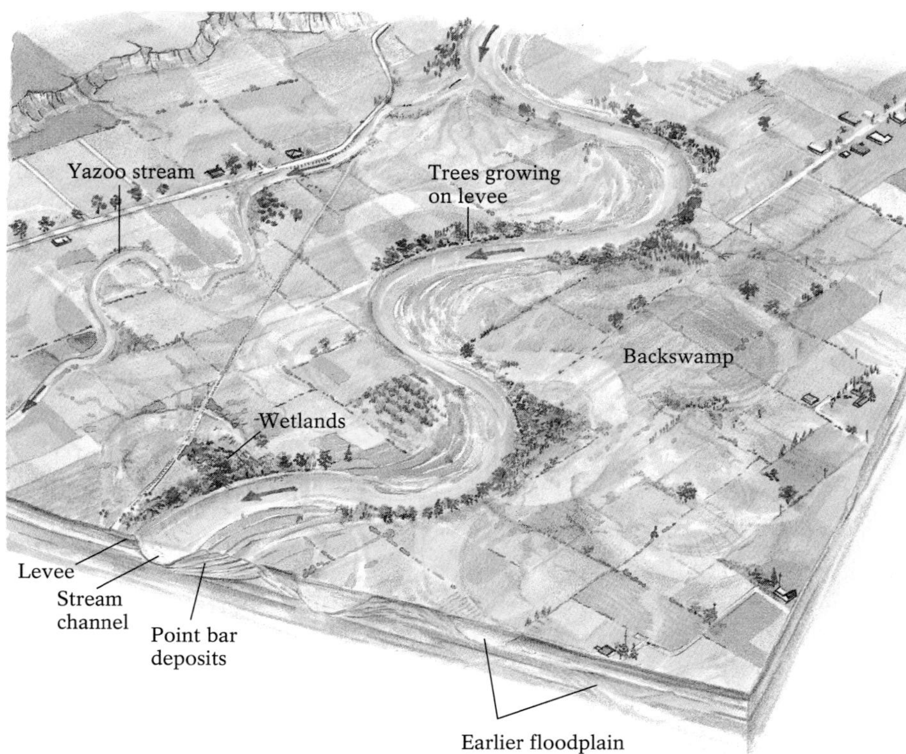

Figure 14-21 Floodplain features, produced when streams overflow their banks and deposit their sediment on the surrounding land.

trunk stream. Such parallel streams, which remain isolated at the margin of a floodplain, are called *yazoo streams*—named for the Yazoo River, which shadows the Mississippi River for 320 kilometers (200 miles) before finally connecting with it near Vicksburg, Mississippi.

Beyond the levee slopes lies the **backswamp,** that portion of the floodplain where deposits of fine silts and clays settle from standing waters following a flood. After flooding subsides, the water left behind on the floodplain typically either evaporates or slowly infiltrates the groundwater system. If the region is marked by surface depressions, however, water may remain at the surface as environmentally valuable **wetlands**—lakes, marshes, and swamps that help to replenish the groundwater system. Wetlands may also serve as habitats for migrating birds and other wildlife. In the past, these areas were routinely drained to allow farming or other land uses, such as industry or housing. Today, conversion of environmentally valuable wetlands to such uses often spurs hot debate and may be subject to environmental restrictions.

Flooding is not the only force that reshapes floodplain landscapes. The point bars deposited within a stream's channel, for example, become major parts of a floodplain's geological structure. As a meandering stream erodes the outside curves and deposits sediment on the inside curves of its meanders, its channel migrates snake-like through its own previous floodplain deposits of sand and gravel. A mature floodplain may consist of a succession of such migrated, partially eroded point bars (see Figure 14-21). As their positions shift in the floodplain, meanders may eventually threaten communities. The town of New Harmony, in southwestern Indiana, waited anxiously for years as the meandering Wabash River approached it. The river would have reached the town in 1994 if civil engineers had not halted its migration, less than a kilometer to the north, by erecting a concrete-block barrier that stabilized its eroding banks and prevented further erosion.

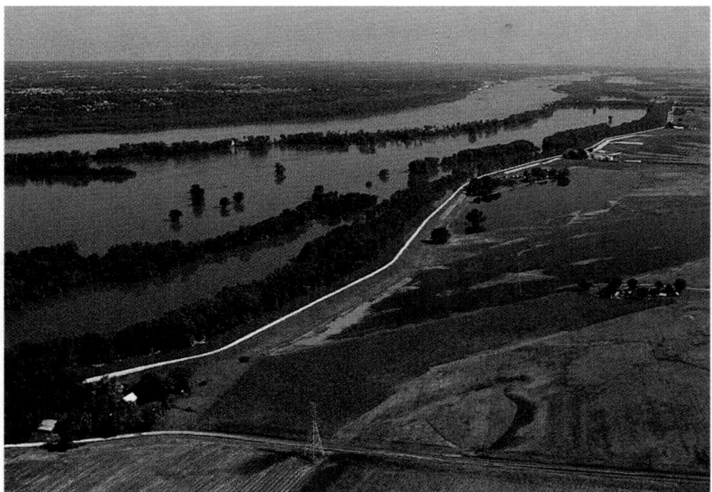

Figure 14-22 A vegetated levee on the bank of the Mississippi River. Natural levees, composed of coarse sediment, are generally the best-drained parts of the floodplain. Consequently, they are most capable of supporting trees that cannot tolerate saturated soil conditions.

Figure 14-23 An alluvial fan in Death Valley, California. Alluvial fans form when streams flowing from steep mountain slopes encounter a sharp reduction in slope at the foot of the mountain and a substantial widening of their valleys. They flow over this plain more slowly and less energetically, dropping their sediment loads as fan-shaped deposits.

Deposition Where Stream Valleys Widen When a stream leaves a narrow, high-gradient mountain valley and encounters a wide valley floor, it loses velocity quickly. In turn, its transport energy declines sharply. Free from the confinement of the narrow valley, the stream spreads out across the lowland plain and deposits much of its coarse sediment load, forming a braided, fan-shaped deposit known as an **alluvial fan** (Fig. 14-23).

Deposition into Standing Water Where a stream enters a standing body of water, such as a lake or ocean, its flow decreases abruptly. It typically deposits its suspended load of fine sand, silt, and clay in the standing water as a **delta,** a roughly triangle-shaped alluvial deposit that fans outward from its apex at the mouth of a stream (Fig. 14-24). [The Greek historian Herodotus named this formation in the fifth century B.C., after noting that the sediment deposit at the mouth of the Nile River resembled the uppercase Greek letter delta (Δ).]

The coarser stream sediment is deposited first, close to the mouth of the stream, producing the delta's angled *foreset beds.* Finer sediments settle farther from the mouth, in horizontal layers known as *bottomset beds.* The greater

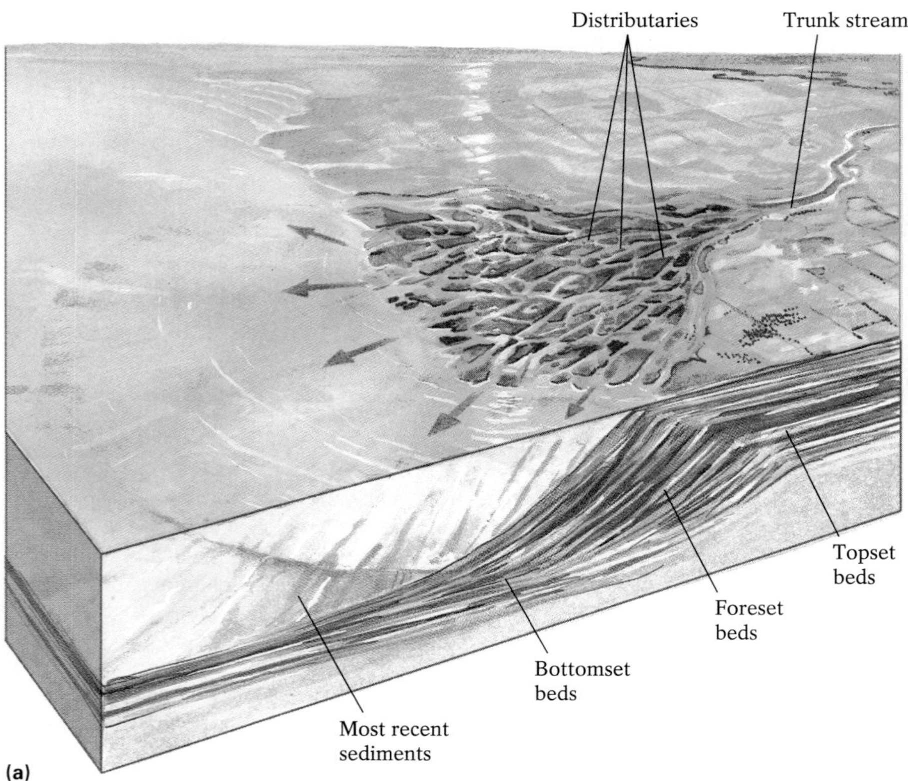

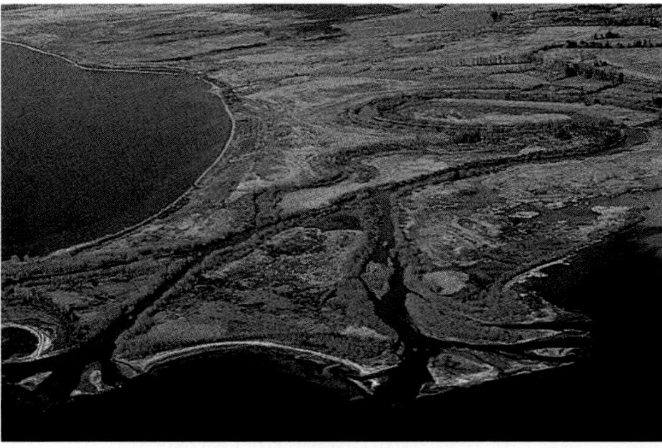

Figure 14-24 **(a)** The anatomy of a delta. **(b)** The Tengarito River delta, New Zealand. **(c)** A cross-bedded deltaic deposit from the Cretaceous Period, in Colorado.

volume of the nearshore sediment gradually expands the foreset beds outward, extending the delta and burying previously deposited bottomset beds. Flooding at the surface of the expanding delta spreads thin, horizontally layered sediments on top of the inclined foresets, forming *topset beds.*

A delta continues to grow as long as its stream deposits more sediment than is eroded from the delta by waves and shoreline currents. The Earth's great deltas typically occur where major rivers—such as the Mississippi, the Ganges in Bangladesh, the Indus in Pakistan, and the Nile in northern Africa—deliver large sediment loads to relatively quiet coastal waters. In contrast, deltas do not form at the mouths of large rivers, such as the Amazon in South America and the Niger in Africa, where vigorous waves and currents immediately sweep away discharged sediments. The largest river in North America's Pacific Northwest, the Columbia River, has no delta, because the Pacific's powerful winter storm waves carry away virtually all of its sediment. The St. Lawrence River, which rises in Lake Ontario and flows past Montreal and Quebec, has no delta for two reasons: (1) Much of the region's sediment settles to the bottom of the Great Lakes and never enters the river, and (2) the river does not accumulate a delta-building load on its relatively short, 600-kilometer (380-mile) journey from the Great Lakes to the Gulf of St. Lawrence and the northern Atlantic.

As a stream's delta grows outward, its surface gradient decreases. The stream flows more slowly and gradually loses the capacity to carry its sediment load. As the stream drops more sediment in its channel, it becomes clogged. The pent-up flow subsequently branches into a new network of *distributaries.* The channel also becomes shallower, increasing the likelihood that the stream will flood and break through its levees. Freed in these ways from its original channel, the stream may discover a more direct route to the sea (usually one with a steeper gradient) and begin to build a new delta lobe at its new outlet (Fig. 14-25). As the younger distributaries capture more of the sediment load, the new delta grows. The older, abandoned delta segments, which lack a replenishing sediment supply, may begin to erode away. They may also subside as their underlying sediments undergo compaction. Thus the shapes and locations of a stream's active and abandoned deltas change continuously.

Offering fertile soils of fresh floodplain silt and intricate networks of navigable waterways, deltas have served for thousands of years as coastal centers of agriculture and commerce. Ancient deltas represent valuable sources of oil and natural gas (from the decomposition of marine fauna) and coal (from accumulation of plant remains in stagnant deltaic swamps). Highlight 14-1 focuses on the history and economic significance of North America's greatest delta—that of the Mississippi River.

Controlling Floods

Many streams overflow their banks every two or three years, causing flooding. Floods usually result from local weather conditions—simply too much rain falls or snow melts, producing more water than the channel can carry. The occurrence of a succession of storms typically increases the severity of the ensuing floods. If rain falls on unsaturated ground, the

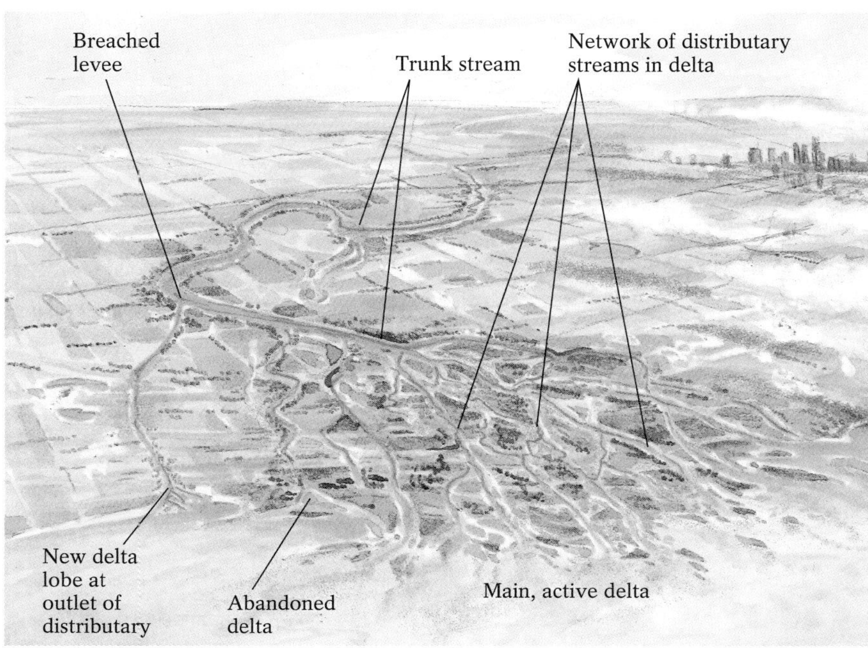

Figure 14-25 The large sediment loads and gentle gradients of deltaic streams promote in-channel deposition. The streams must then either diverge around these deposits or overflow their channel walls, forming new distributaries and delta lobes.

Highlight 14-1 The Mississippi River and Its Delta

The Mississippi River begins its 3750-kilometer (2350-mile) journey as a creek a few meters wide and 10 centimeters (4 inches) deep that trickles from Lake Itasca in the pine forests of northern Minnesota. In its first 2100 kilometers (1300 miles), the *upper* Mississippi cascades over bedrock falls and rapids, wherever it is not dammed or otherwise restrained. South of Cape Girardeau, Missouri, the Ohio joins the Mississippi. Together, the two rivers form a majestic alluvial valley—the much-tamer *lower* Mississippi valley—that stretches for more than 1600 kilometers (1000 miles) and reaches widths of 48 to 200 kilometers (30–125 miles). By the time the Mississippi empties into the Gulf of Mexico 200 kilometers (120 miles) south of New Orleans, it carries water and sediment from the Missouri, the Ohio, the Arkansas, and more than 100,000 other streams, and has deposited fertile silt over more than 770,000 square kilometers (300,000 square miles) of floodplain farmland (and, occasionally, over cities).

Much of the lower Mississippi valley may have once been a shallow marine bay located in an aulacogen (see Chapter 12), a failed rift arm formed approximately 200 million years ago when North America split from Europe and Africa. The Mississippi River's first delta grew southward into this bay, filling the aulacogen with a succession of southward-growing deltas. As a series of ice ages caused the Earth's water to accumulate in glaciers during the last 3 million years (discussed in Chapter 17), global sea levels fell and the Mississippi River's marine bay receded. During this period, the Mississippi River transported the enormous supply of sediment left by the vast ice sheets that blanketed Canada and the Great Lakes region of the United States, and its delta grew steadily. Beginning 12,000 years ago, after the last period of worldwide glaciation, the Earth's climate warmed and sea levels rose. Approximately 9000 years ago, the Mississippi River began to construct its present-day delta (Fig. 14-26).

Today the Mississippi delta consists of at least seven distinct lobes, each of which was once the river's active delta. Together, these lobes constitute about 40,000 square kilometers (15,000 square miles) of real estate added to southern Louisiana (Fig. 14-27). The active portion of today's delta, known as the Balize delta, dates from 1000 years ago. Its bird's-foot shape has been built by three major distributaries, whose gradients are now too gentle to transport sediment effectively. The inactive lobes of the Balize delta were each abandoned in turn as the river sought out shorter, steeper routes to the gulf. Wave action has eroded away many more abandoned lobes, and the weight of overlying sediments has depressed others below sea level. Most of the very large Balize delta now lies under water. The Mississippi River *bayous,* the colorful network of lakes and minor streams in southern Louisiana, consist of abandoned channels that crisscross inactive lobes.

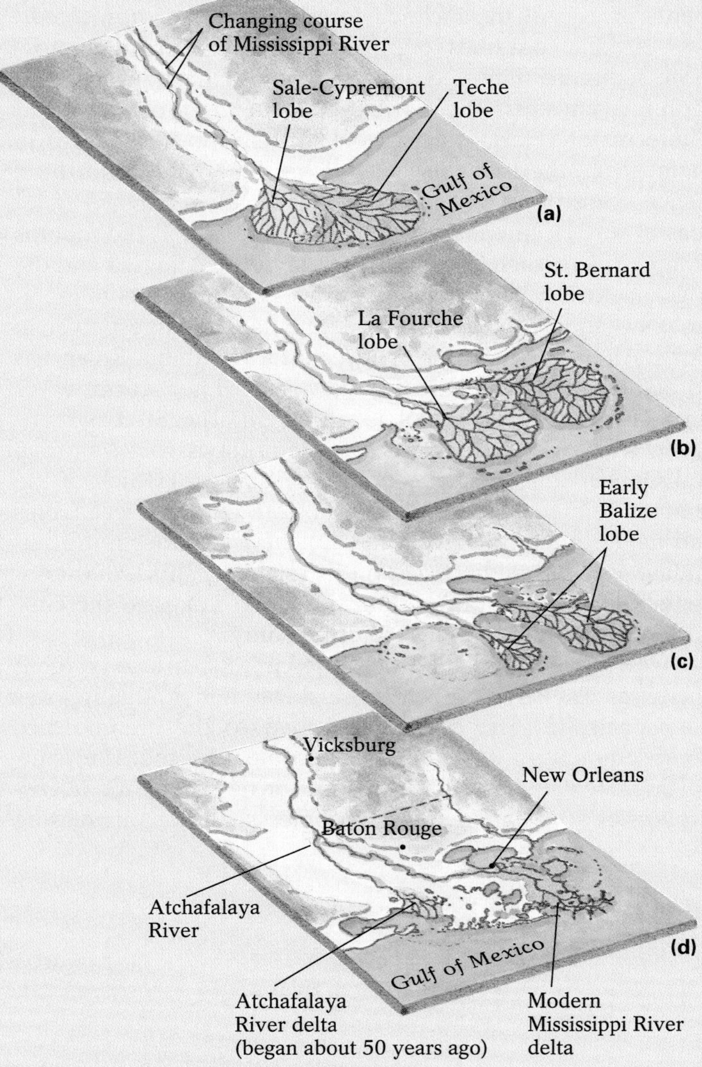

Figure 14-26 The evolution of the Mississippi River delta plain. **(a)** The Sale-Cypremont delta (active 7500 through 5000 years ago), one of the earliest known deltas of the Mississippi, was abandoned when the river diverted its sediment to the Teche lobe (active 5500–3800 years ago). **(b)** Delta growth shifted farther east when the St. Bernard lobe (active 4000–2000 years ago) was deposited. Delta building moved west again, forming the La Fourche lobe (active 2500–1500 years ago). **(c)** By some 1000 years ago, the modern bird-foot (Balize) lobe had begun to be constructed. **(d)** In the last 100 years or so, the Atchafalaya River, a distributary, has begun to carry more of the Mississippi's flow. Were it not for flood-control structures built by the U.S. Army Corps of Engineers, the Atchafalaya would probably become the main channel of the Mississippi, as it presents a shorter route to the Gulf of Mexico.

For several decades, the Mississippi River has been trying to cut a steeper, far shorter course along the Atchafalaya River, located 100 kilometers (60 miles) to the west of its present channel. If and when it succeeds, the lower 500 kilometers (300 miles) of the present river will be cut off, imperiling the livelihood of Baton Rouge and New Orleans and rendering worthless the great investments made in navigational improvements along its course.

Thus far human intervention has managed to preserve the river's course. In 1973, a winter of unusually high precipitation produced raging floods in the lower Mississippi valley, which might have opened the Atchafalaya channel and caused the river to bypass New Orleans, leaving it a sleepy bayou town instead of a center of river commerce. Floodgates diverted water from the main channel to Lake Pontchartrain, however, preventing the river from breaching its levee and taking the Atchafalaya course. Continued dredging to deepen the lower Mississippi channel enough to accommodate the river's flow has saved Baton Rouge and New Orleans for now, but the river may yet have its way.

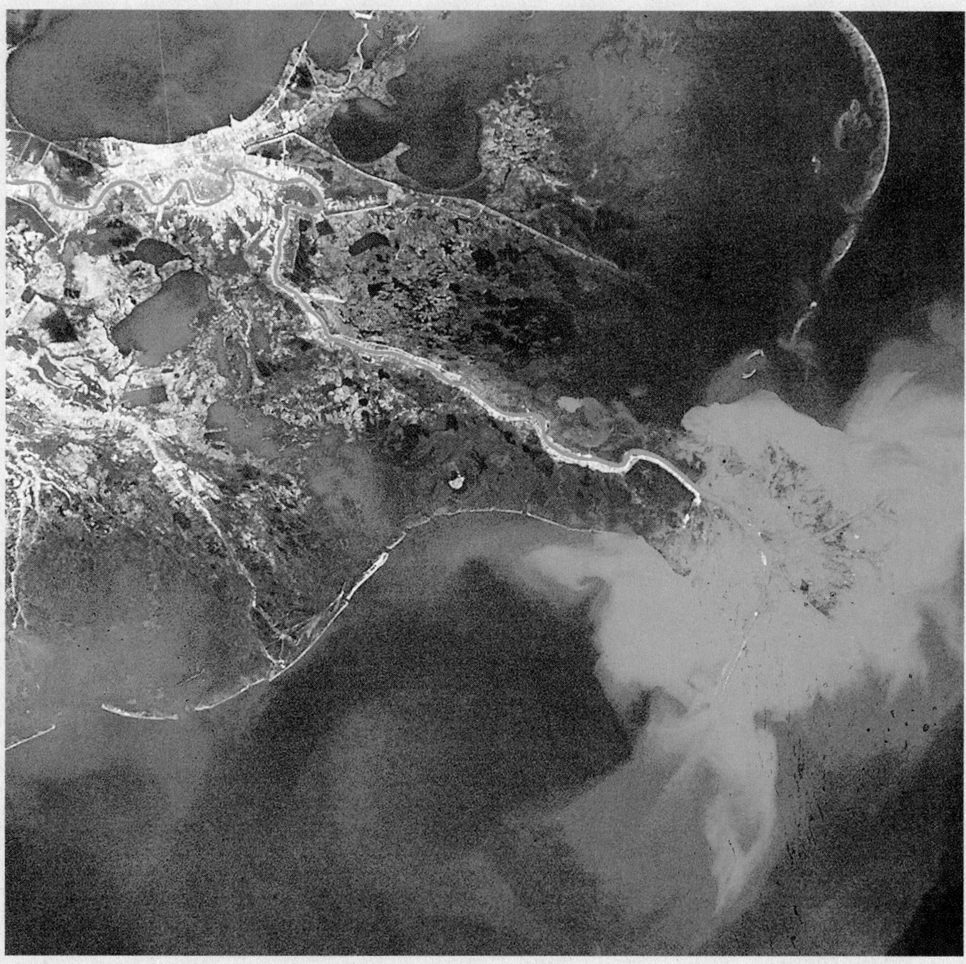

Figure 14-27 The modern delta of the Mississippi River. Today's delta, shaped like a bird's foot, consists of the modern segment centered at Head of Passes, Louisiana, and a number of inactive, abandoned lobes that no longer receive sediment. During the past century, the river has attempted to shift away from this lobe and establish a new course along the Atchafalaya River. Unless humans intervene, it will eventually do so.

soil may soak up the water and delay its entry into streams. After the ground becomes saturated, however, it cannot absorb the next downpour. The excess water then runs off the surface without delay and *directly* into local streams, causing them to rise rapidly.

The geology and topography of the surface help determine whether water runs rapidly into streams or infiltrates the groundwater system, where it could remain long enough to avert flooding. Is the bedrock near the surface *permeable* (that is, easily infiltrated, allowing water to pass through), like sandstone? Or is it relatively impermeable, like granite? Is the rock highly fractured and faulted? (Fractured rocks can more readily absorb water.) Are the surface soils clay-rich types, which swell when they become wet and seal the surface against infiltration? Is the local topography steep or gently sloping? (Steep slopes promote runoff; gentle slopes enhance infiltration.) The steepness of the Rocky Mountain slopes were largely responsible for the catastrophic Big Thompson Creek flood of 1976. Big Thompson Creek, in the Colorado Rockies northwest of Denver, received more than 30 centimeters (12 inches) of rain on the evening of July 31, 1976. This volume of precipitation, which nearly equaled the average annual precipitation for the area, could not be absorbed by the steep exposed bedrock of the canyon. As a result, discharge in the creek rose rapidly to more than four times its previous record. The surface of the creek rose from two-thirds of a meter (about 2 feet) to more than 6 meters (20 feet). In all, the flood took 144 lives and caused $35 million in damage. (Many victims would have survived had they climbed upslope instead of trying to outrun the wall of water that thundered down the narrow gorge.)

Flooding becomes more likely when heavy rains fall on an area where extensive forest fires, droughts, or widespread clear-cutting has thinned vegetation. Trees and their root systems tend to keep soils open, enhancing water infiltration and thus reducing the chance of floods. As communities expand into unpaved, undeveloped areas, new roads, buildings, and parking lots replace vegetated grounds. Their existence prevents surface water from infiltrating the groundwater system, increasing the potential for flooding.

Flood Prediction

Floods are usually seasonal, occurring during heavy spring rains and snowmelt in temperate climates and during rainy seasons elsewhere. Because most streams flood annually, the residents of a flood-prone region may choose to be eternally prepared—by keeping rowboats in basements, building structures on stilts or cinderblocks, and maintaining supplies of sandbags, for example. *Unusually large* floods, however, cause the greatest damage. Can such events be predicted?

Because the vagaries of weather remain largely a mystery to us, flood prediction is necessarily imprecise. Our best hope lies in analyzing the frequency of past floods and using those data to determine the *statistical* probability of a major flood occurring within a given period of time. Stream discharge, for example, is monitored over the long term by creating *hydrographs,* graphical plots that show day-to-day and year-to-year variations in discharge. Hydrographs also illustrate how human activities such as urbanization and forest clear-cutting affect discharge (Fig. 14-28). Hydrograph data for a stream are then used to plot its flood-frequency curve (Fig. 14-29), which allows us to estimate the probability of a particular flooding discharge being equaled or exceeded in that stream in any given year.

Consider the following example. After recording a stream's discharge over a 100-year period, we determine that a peak discharge as great as 2000 cfs (cubic feet per second) occurs once every 10 years. Thus, the chance of such a discharge arising in any particular year is 1 in 10 (a 10% probability). When that discharge does take place, we call the event a 10-year flood. Similarly, recorded data may show that the discharge of the same river reaches 3500 cfs once every 100 years. When there is a 1-in-100 chance (a 1% probability) of such a discharge occurring in any particular year, the event is termed a 100-year flood. These probabilities are updated as the discharge record grows. Several cautions must be applied to flood predictions based on this method, however:

- Few places in North America have peak-discharge records spanning more than 100 years. Thus our calculation of a stream's 100-year flood, extrapolated from a shorter record, is imprecise and typically *underestimates* its size. A community whose structures are engineered based on such flood estimates may be rudely awakened by much larger flood events.
- Substantial changes in a watershed throughout the period of stream monitoring, *primarily from human development,* typically increase stream discharge substantially. Thus the discharge of a stream today during its so-called 100-year flood *may be significantly higher* than the peak discharge encountered decades ago, before extensive development of the watershed.

The discharge associated with the 100-year flood of one stream is not necessarily the same as that of any other stream. Every stream possesses a unique flood-frequency curve, with specific discharges being associated with floods of particular magnitudes.

Nature seldom conforms precisely to our statistical predictions. An area that endures one 100-year flood will not necessarily enjoy a 99-year respite before the next one. Residents living along the banks of the Mississippi River, for example, had to cope with 100-year floods in 1943, 1944, 1947, and 1951. Moreover, because of inadequate early record-keeping, flood frequency is sometimes hard to define. Some agencies have classified the 1993 Mississippi flood (see Highlight 14-2) as a 200-year flood; others have deemed it a 500-year event.

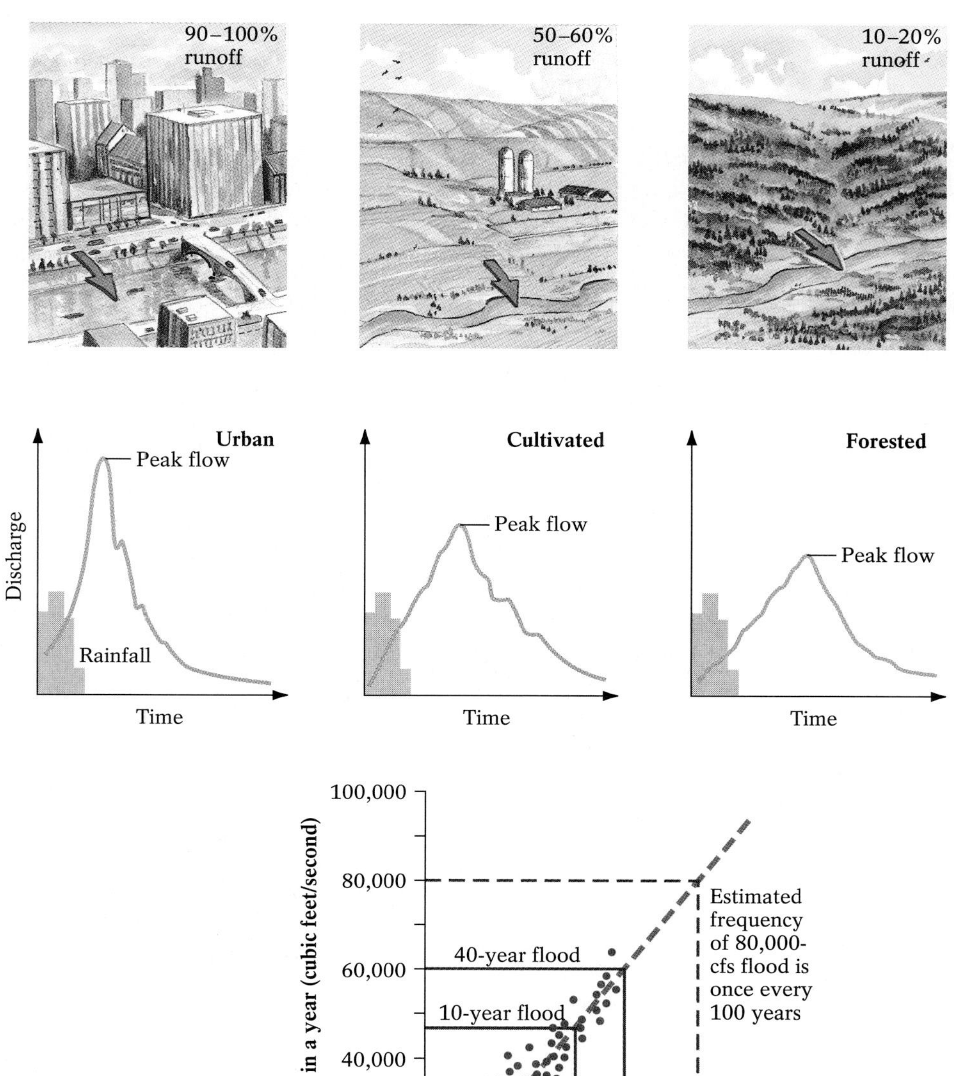

Figure 14-28 Hydrographs record how a stream's discharge varies with time. A peaked hydrograph curve indicates that a stream's discharge has risen dramatically in a brief time, suggesting that precipitation and other surface water have flowed rapidly into the stream with little delay in the groundwater system. Because urban areas are largely paved over with asphalt and cement, little surface water can infiltrate the groundwater system and much of it runs off into streams; thus urban streams tend to have peaked hydrographs. By comparison, the hydrographs of streams in undeveloped areas are generally less peaked (changes in discharge occur more gradually). Note that the amount of rainfall (and the volume of water transported) is the same in each of the three cases.

Figure 14-29 A flood-frequency curve for a hypothetical stream, showing how often large discharges and floods of various magnitudes have occurred throughout the stream's recorded history. Hydrologists may extrapolate the estimated discharge of the 100-year event from the stream's 40-year monitoring record.

Flood Prevention

People have used a number of defensive strategies—structural, nonstructural, and combinations of both—to try to curb the catastrophic effects of floods. The structural approach involves some form of artificial construction that seeks to contain the stream within its banks, divert some of its water to temporary storage facilities, or accelerate flow through the stream system.

The most common stream-containment structures consist of *artificial levees,* earthen mounds built on the banks of a stream's channel to increase the volume of water that the channel can hold. Such levees have been built along the Mississippi River to protect croplands since the eighteenth century. Spillways cut through the levees allow rising water to escape into human-made subsidiary channels, reducing the likelihood of a flood downstream. Concrete *flood walls,* which are more expensive, have also been constructed by the U.S. Army Corps of Engineers to prevent overflow from the channel at strategic locations, such as along the commercial centers of riverside cities.

Highlight 14-2 The Great Midwestern Floods of 1993

On June 11, 1993, a foot of rain fell in southern Minnesota. Four days later, more than 11 inches fell in roughly the same area. Thus began the wettest June and North America's worst flood since compilation of weather records began in 1878. Before the wet weather ended two months later, the upper Mississippi River had surged over its banks from St. Paul, Minnesota, to St. Louis, Missouri, disrupting the lives of millions of residents in 12 sodden states (Fig. 14-30a).

The swirling floodwaters undermined interstate bridges, washed away roads, and brought to a halt all barge traffic along the upper Mississippi, which serves as the economic artery of the Midwest. Rushing, muddy floodwaters, carrying uprooted trees and junked cars, swept away thousands of homes and businesses, displaced more than 50,000 people, took dozens of lives, and caused property damage and agricultural losses valued in excess of $10 billion. For the first time in anyone's memory, the waters of the Mississippi and Missouri overflowed their levees and flowed together, 30 kilometers (20 miles) from their confluence north of St. Louis. Flooding closed the bridge at West Quincy, Missouri—the only connection between Missouri and Illinois for more than 300 kilometers (200 miles). In Hannibal, Missouri, Mark Twain's boyhood home, the Mississippi River *crested* (reached its peak) at 5 meters (16.5 feet) above flood stage. Children caught catfish on city streets where they had ridden bicycles only days before. Fortunately, a recently built, 11-meter (33-foot)-high flood wall saved the historic downtown area and the Twain homestead. In Davenport, Iowa (Fig. 14-30b), where residents chose to preserve their river views rather than build levees and flood walls, ducks swam lazily in the 4 meters (14 feet) of water covering the outfield of Davenport Stadium, the home of the minor-league baseball Quad City River Bandits. In Des Moines, 250,000 Iowans had no drinking water for days after floodwaters contaminated the municipal water supply with raw sewage and chemical fertilizers.

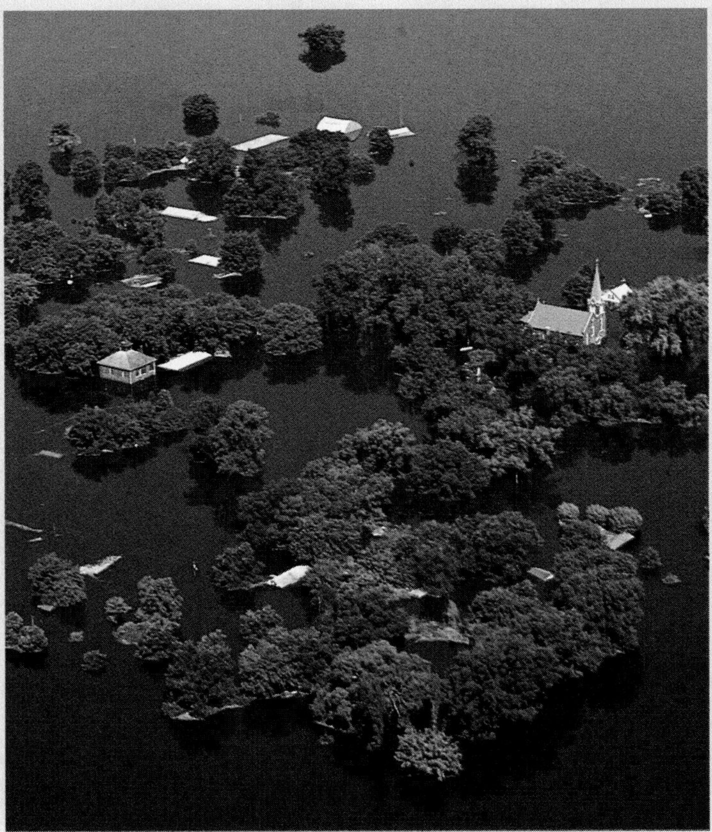

Figure 14-30 **(a)** Flooding in Kaskaskia, Illinois.

Curiously, the residents of the lower Mississippi valley, south of Cairo, Illinois, were spared the flooding. Why? The channel of the lower Mississippi—compared with that of the upper Mississippi—is relatively wide and deep; thus it can accommodate a greater volume of water. At Vicksburg, Mississippi, the lower Mississippi's channel is 37 meters (123 feet) deep and 970 meters (3200 feet) wide; in comparison, the channel north of St. Louis is only 18 meters (60 feet) deep and 575 meters (1900 feet) wide. More importantly, the states that constitute much of the watershed of the upper Mississippi (Minnesota, Wisconsin, Iowa, Missouri, and northern Illinois) received months of soil-saturating storms; Nebraska and the Dakotas—the states that form the watershed of the upper Mississippi's Missouri, (the river's major tributary)—suffered a similar fate. By comparison, the lower Mississippi's watershed—which receives 65% of its water from the Ohio River and its tributaries flowing through western Pennsylvania, Ohio, West Virginia, Indiana, Kentucky, and southern Illinois—experienced below-normal precipitation. As a result, the lower Mississippi valley states of Arkansas, Tennessee, Mississippi, and Louisiana were spared this disaster.

Some people maintain that human activity—65 years of "managing" the Mississippi River to prevent floods—actually contributed to the magnitude and tragedy of the flood of 1993. Prior to the last 100 years or so, the Mississippi and its tributaries determined their own boundaries. During periods of high water, the rivers broke through or overflowed natural levees, flooding tens of thousands of square kilometers of the surrounding, largely uninhabited lands. In the twentieth century, however, millions of people migrated to the region, building cities, towns, and large farms along the rivers' banks. When a flood in 1927 took 214 lives, Congress enacted the first Mississippi River Flood Control Act, assigning the daunting task of confining the river to its channel to the U.S. Army Corps of Engineers.

The Corps of Engineers built nearly 300 dams and reservoirs and thousands of kilometers of artificial levees and concrete flood walls, all designed to prevent the river from spilling onto

Figure 14-30 (b) Extent of flooding in Davenport, Iowa, June 1993. Top: Standing water near the town's River Drive casino. Bottom: Three weeks later at the same location after floodwaters had receded.

its natural floodplain. The system also includes numerous pumping stations, spillways, and diversion channels that divert water into temporary holding basins for storage. The events of 1993 dramatically illustrated that the Corps' efforts have largely failed, perhaps because such flood-control structures simply cannot contain such an *extraordinary* flood.

By confining the massive discharges to a channel, the retention structures actually forced the swollen river to flow more rapidly and violently, damaging the very structures designed to restrain it. By denying the river access to its natural floodplains, they caused the river to rise higher than it would have naturally, ensuring that the floods produced by levee breaches would be more damaging. Furthermore, the existence of artificial levees and flood walls bred a false sense of security that encouraged the growth of cities, towns, and farms closer to the river banks than was really safe.

What does the future hold for the residents of the upper Mississippi valley? Certainly more flooding, but perhaps less human interference with the river's natural behavior. Some communities have proposed the elimination of all flood-retention systems and zoning limits to reduce future development close to the region's rivers, at least within their 100-year floodways. Others look longingly at St. Louis' 16-meter (52-foot)-high concrete flood wall, which saved that city's downtown business district when the Mississippi reached its record crest at 14.2 meters (47 feet). Still others have simply moved off the floodplain. The debate continues between those who believe we can tame the mighty Mississippi and those who know we cannot.

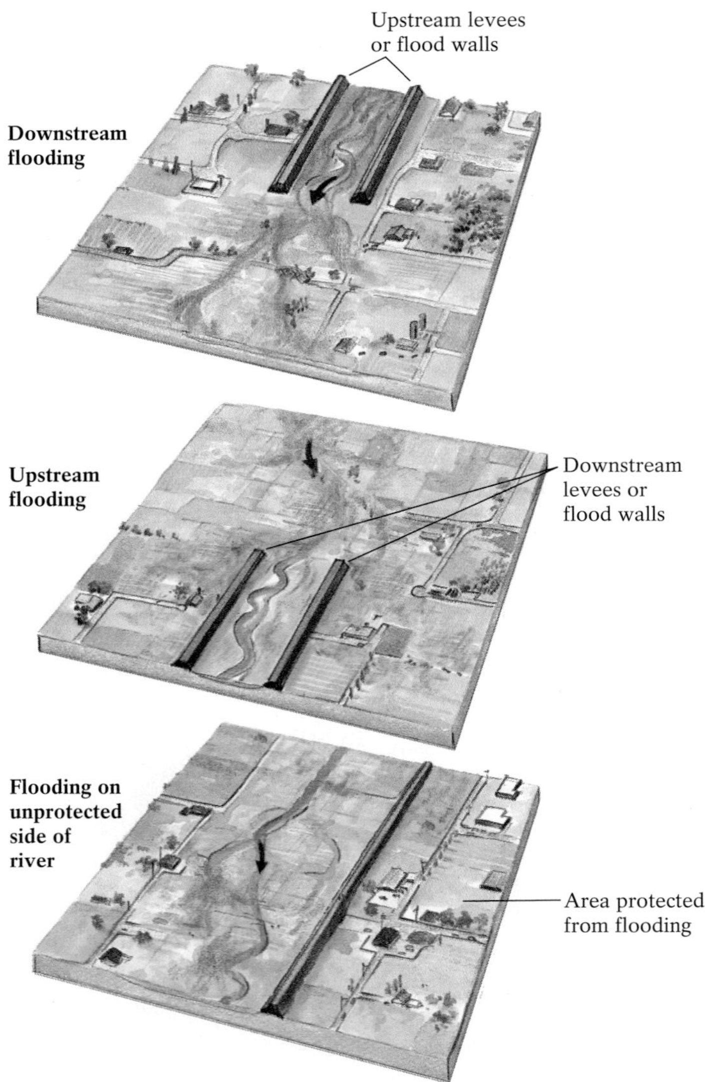

Figure 14-31 Artificial levees and flood walls must protect the full length of a river and be built on both of its banks; otherwise, the structures will merely divert the flooding.

To be effective, artificial levees and flood walls must protect both sides of a stream and must extend for its full length (Fig. 14-31). If these structures are built *only upstream,* flooding downstream will be far more catastrophic, because the flooding stream cannot reach its floodplain along the protected upstream section. If artificial levees and flood walls are built *only downstream,* the upstream floodwater will wash over the floodplain—and across the very area they were built to protect. If they are constructed on *only one side of a stream,* the floodplain on the unprotected side receives a double share of floodwater.

Flood-control dams are often built upstream from densely populated or economically important areas. These earthen or concrete structures shunt excess water to temporary storage facilities—called impoundment basins—to prevent, or at least slow, overflow. The excess water in the basins either evaporates or seeps into the groundwater system. Such structures also have uses unrelated to flood control. Dams along the Tennessee, Mississippi, Missouri, Colorado, and Columbia Rivers, for example, generate hydroelectricity, provide irrigation water for agriculture, supply municipal water to numerous communities, and create regional recreational facilities.

In *channelization,* the stream's channel is modified to speed the flow of water through it, thus preventing the water from reaching flood height. Channelization may involve clearing obstructions such as fallen trees, or widening or deepening a channel by dredging. More radical alterations are used to increase a stream's gradient and therefore its velocity, such as cutting off meanders to straighten a stream. The shorter, straight channel will have a steeper gradient than the previous channel, and its increased velocity will enable the transport of more water—perhaps enough to prevent flooding.

Channelization can also produce detrimental effects. The Blackwater River, a meandering tributary of the Missouri River located southeast of Kansas City, has flooded regularly and destructively for years, largely because it flowed too slowly to accommodate large storm discharges.

In 1910, engineers excavated a straight channel that increased the river's gradient and velocity, but also caused the river to scour its bed more energetically. The channel subsequently expanded from about 4 meters (14 feet) to more than 12 meters (39 feet) deep, and from 9 meters (30 feet) to more than 71 meters (233 feet) wide. As the river widened, it undermined bridges, which had to be replaced. Only constant maintenance now keeps the Blackwater River straight.

Structural solutions to flood problems are almost always expensive. Such defenses can also give local residents a false sense of security; the more protected they feel, the closer to a stream they build their homes. Consider what happens when a structure designed to impound a 100-year flood attempts to contain a 200-year event.

Because of the drawbacks to structural solutions, a *nonstructural* approach, if feasible, is almost always preferable. In developing a nonstructural defense against flooding, we identify high-risk areas, regulate zoning to minimize development in them, and manage natural resources thoughtfully to restrict the amount of water entering a stream channel at any one time. This approach first requires a thorough geological and botanical survey to identify flood risk. For example, the natural distribution of water-loving plants (phreatophytes) helps identify areas that have flooded regularly in the past. Geologists can map and date ancient flood deposits and analyze their thicknesses and textures to estimate the timing and magnitude of prehistoric floods.

They then use these data to create a flood-hazard map showing *floodplain zones,* or *floodways*—the corridors of

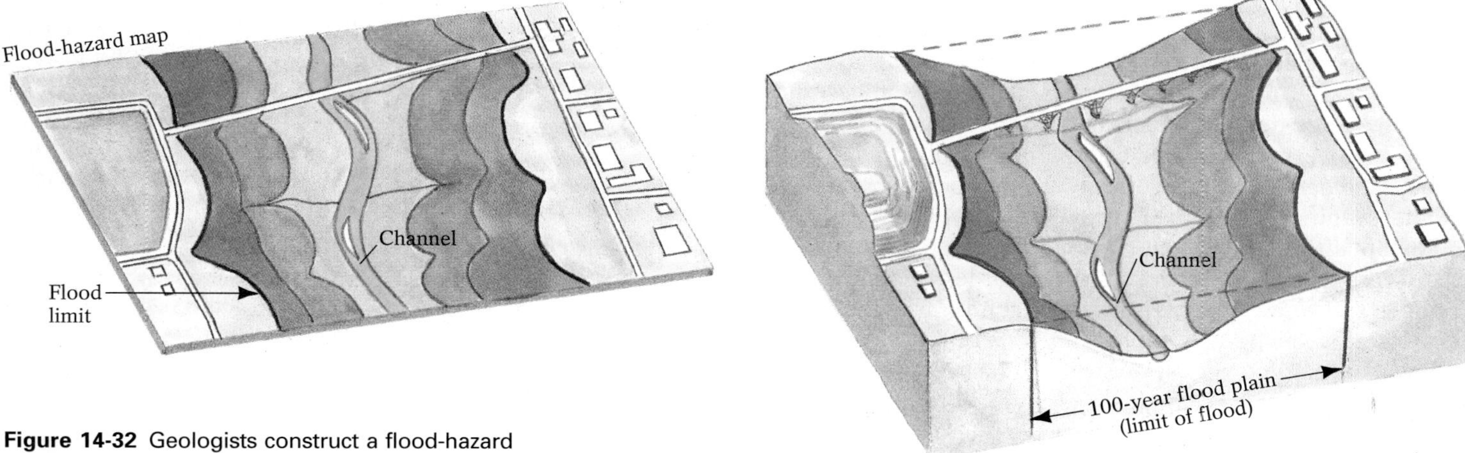

Figure 14-32 Geologists construct a flood-hazard map from historical data to predict the frequency and magnitude of future floods and delineate the areas that will be affected. These data are used to establish floodways—the areas of a river valley that floods of various magnitudes would cover—and institute regulations restricting development in flood-prone areas.

Floodplain Zones

- "Floodway district": regulated floodway; kept open at all times; no flood-control measures, no building
- "Floodway fringe district": flood-prone, protected by flood-control measures; building allowed if protected by "flood-proofing"
- "Floodway limit": area subject to flooding only by very large discharge (100-year flood)

land adjacent to the stream that would be covered by floods of different magnitudes (Fig. 14-32). Communities can use this information to draft zoning ordinances. For example, a 12-kilometer (7-mile)-wide floodway surrounding the Rapid River was designated a high-risk area after the disastrous Rapid City, South Dakota, flood of June 10, 1972. Many buildings were subsequently moved from this region. Today the floodway is used mostly for ballfields, golf courses, and other sporadically used spaces. Artificial levees and flood walls protect structures in the floodway that could not be moved or abandoned.

Nonstructural flood prevention also includes the protection and wise management of forests and the prompt suppression of forest fires. As we have seen, vegetation cover promotes the absorption of water into the groundwater system and reduces stream discharge, so maintaining it represents an important facet of any flood prevention plan.

How have we fared in our efforts to prevent floods? During June and July of 1993, a succession of storms dumped an extraordinary amount of rain on the drainage basin of the Mississippi River in Minnesota, Iowa, Illinois, and Missouri, triggering the river's worst flood on record and causing billions of dollars in damage to croplands and other property. Highlight 14-2 focuses on this tragic event and illustrates how nature can confound our best-laid flood-prevention plans.

Artificial Floods

Sometimes the creation of a flood is beneficial—even prudent. Such was the case at dawn on March 26, 1996, when Secretary of the Interior Bruce Babbitt turned a lever at the pumphouse of Arizona's Glen Canyon dam, releasing a frothy white torrent of water—45,000 cfs—and *intentionally* flooding the Colorado River valley in the Grand Canyon region. If floods cause widespread damage and misery, why would a high governmental official unleash one on this highly scenic locale and popular tourist mecca?

Before the completion of Arizona's Glen Canyon Dam in 1963, the Colorado River "managed" its own ecosystem through annual flooding. During the spring months, when snows melted in the southern Rockies, the extremely high discharge (roughly 125,000 cfs) flooded the lower Colorado River. At these times, the turbulent waters ran red and muddy as they transported roughly 65 million tons of sediment eroded from the surrounding bare-rock plateaus. Early settlers described the river in flood as "too thick to drink and too thin to plow." When the floodwaters abated, the deposition of these sediments replenished beaches and sandbars along the river, providing sheltering habitats for the region's flora and fauna. The raging floods also cleared debris from the Colorado's tributaries, keeping these backwaters open for fish to spawn and creating ideal "nurseries" in which young fish could mature without facing the rigors of life in the main channel.

The construction of the Glen Canyon Dam drastically altered the ecology of the Colorado River. The river was "tamed" to suit the needs of hydropower cooperatives that supplied energy to 20 million residents of nearby states. Dam managers governed the river's flow, allowing increased flow through the dam only during heavy-use periods, such as during the evening dinner rush (when ovens were in full use) and on hot afternoons (when air-conditioner use peaked). Because seasonal flooding was eliminated, far less sediment was available to restore the river's natural habitats, which quickly began to erode away. With more than 90% of the watershed's sediment remaining on the bottom of Lake Powell behind the dam, the once red, muddy Colorado flowed green and clear.

After more than 10 years of protracted negotiation and dozens of scientific studies, the U.S. government and the managers of the Glen Canyon Dam decided to reverse the damage produced by the dam's own operation. In an attempt to renew the health of the Colorado River's ecosystem, 117 billion gallons of water were released over a week-long period. This *artificial* spring flood temporarily restored the Colorado River to a semblance of its former self—with some significant results.

As four great arcs of water shot from the dam's outlet and coalesced (Fig. 14-33), the Colorado crept higher up the salmon-colored sandstone walls surrounding the channel below the dam. The river rose 5 meters (17 feet) in the Grand Canyon, located 25 kilometers (15 miles) downstream of the dam. The rushing torrent scoured out sediment and vegetation that had accumulated over 30 years in the river bed and in tributaries and side canyons, converting those clogged backwaters into future spawning grounds for native fish. As the river overflowed its banks for the first time in decades, the eroded sediment was transferred to the river's banks, creating 55 new beaches and sandbars where vegetation could take root to feed the region's birds. Backwater lagoons formed behind these bars, providing ideal breeding grounds for the insects that nourish the river's fish. Even the river's human visitors benefited, as river rafters could enjoy new camping areas along the river.

The Colorado River's artificial flood of 1996 may have ushered in a new era of dam management in North America—one that has preservation of a river's ecosystem as a major goal. Scientists hope to repeat the Colorado's flood at five-to-ten-year intervals. Managers of other hydroelectric-power systems are discussing plans for similar restoration floods in the Columbia River Basin of the Pacific Northwest, in Montana's Missouri River Basin, and along several major rivers in Western Canada. In addition, the governments of Japan, Turkey, and Pakistan are considering implementing such dam-management policies.

Figure 14-33 In 1997, water released at the Glen Canyon Dam flooded through the Colorado River valley, restoring some of the river's pre-dam ecosystem.

Stream Evolution and Plate Tectonics

Our growing knowledge of plate tectonics has added to our understanding of the evolution of stream systems. Plate-margin stresses produce much of the local uplift that down-cutting streams erode, creating spectacular waterfalls, distinctive terraces, and deeply incised river channels (Fig. 14-34). In addition, tectonic uplift produces the rising mountains through which antecedent streams flow, such as the Columbia and Snake Rivers of the Pacific Northwest's Cascade Mountains, at the convergent boundary between the North American and Pacific plates.

Tectonic activity often determines the character of the regional stream-drainage pattern. For example, tectonic stresses cause the faults and fractures that produce a rectangular drainage pattern, whereas folded mountain belts at convergent plate boundaries tend to display trellis drainage. Radial drainage patterns evolve as streams dissect the young volcanoes that proliferate at active plate boundaries.

Many large stream systems begin their journeys in the folded mountains found at convergent margins. They cross stable mid-plate interiors and then build distributary systems

Figure 14-34 The Goosenecks of the San Juan River, southeastern Utah. These meanders are deeply incised into the Colorado plateau, which has been tectonically uplifted by interactions between the North American and Pacific plates.

and deltas at passive plate margins. For example, the Amazon River rises in the Andes of western South America, where the South Pacific's Nazca plate subducts beneath the South American plate. The river flows eastward across the continent for thousands of kilometers before emptying into the Atlantic Ocean. Similar tectonically produced drainage divides appear in the Alps of southern Europe, which rose from the convergence of the African and Eurasian plates; the Himalayas, produced by convergence of the Indian and Eurasian plates; and the North American Rockies, originating from convergence of the North American and Pacific plates.

Extraterrestrial Stream Activity: Evidence from Mars

Photographs taken by the 1971 Mariner 9 spacecraft and the 1976 Viking probe clearly reveal braided channels, stream-modified islands, dendritic drainage patterns, and catastrophic-flood topography on the surface of Mars. On Earth, flowing water forms such features. Mars' surface, however, contains virtually no free water; all of the liquid water at its surface and in its atmosphere would barely fill a small swimming hole. The low atmospheric pressure on Mars allows the atoms on a liquid's surface to escape easily, so any water vaporizes immediately. Moreover, the planet's thin atmosphere does not trap enough heat reradiated from the surface to keep the surface warm. As a result, the Martian surface is too cold for water to exist in liquid form. How then can we explain the spectacular fluvial (river-related) landforms photographed on this planet?

Some geologists propose that water is trapped as ice below ground in the pores of the Martian regolith. Catastrophic events may release this water from time to time, when the

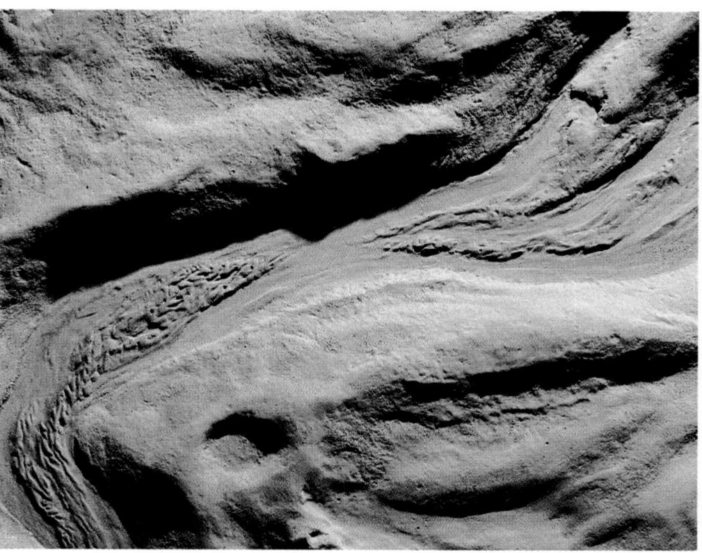

Figure 14-35 A 3-mile-wide channel north of the Martian equator, photographed in 1972 by Mariner 9. This and other apparently fluvial features on Mars may have formed billions of years ago, when the planet's climate and atmosphere allowed free water to exist and flow at its surface.

ice is melted by intrusions of magma or by heat generated in meteorite impacts. Most of what appear to be major flood channels occur in Mars' southern volcanic highlands. There, magma may have melted a vast volume of underground ice, causing the surface to collapse into a jumble of large, angular blocks. The ensuing flood of meltwater may have carved the region's deep canyons. When the floodwaters evaporated soon afterward, they left braided channels punctuated by teardrop-shaped islands of sediment. Although we cannot determine precisely how long ago such catastrophic floods took place, the numerous meteorite-impact craters superimposed on the channels suggest that they may be at least hundreds of millions of years old.

Although no rain falls on Mars today, in the early life of the solar system water may have accumulated on its surface, just as water did on Earth, from volcanic outgassing of the planet's interior. The effects of the volcanism may have been boosted by extensive meteorite impacts that generated magma and also breached the Martian crust, allowing subsurface magmas to rise to the surface. This activity likely gave Mars a more substantial atmosphere at that time, as well as greater atmospheric pressure and some form of warming greenhouse effect. Thus water could have survived at the surface during Mars' early years. The planet's fluvial-looking features include braided and meandering channels and dendritic drainage patterns such as those shown in Figure 14-35. Their presence suggests that 3 to 4 billion years ago lakes and streams may have dotted Mars' surface; these bodies of water were likely nourished by frequent rainstorms. During the last few billion years, since the space in our solar system has been swept relatively clear of meteorites, Mars has experienced fewer impacts and less volcanism. Thus its present atmosphere is cool, thin, and dry, and its surface is frigid, arid, and devoid of fluvial activity.

For most of us, rivers are the most obvious component of the Earth's hydrologic cycle and have the greatest impact on our daily lives. In the future, as surface waters suffer increasing contamination from the burgeoning human population, the cycle's largely unseen groundwater component may assume a more important role as perhaps the best source of clean freshwater. In the following chapter, we will examine how groundwater accumulates and flows, and how we can tap the reservoir beneath our feet while preserving its quality.

Chapter Summary

Virtually all of the Earth's water initially derives from the planet's interior, principally as **juvenile water,** the steam that accompanies volcanic eruptions. Virtually all of the Earth's surface water originates in the atmosphere as **meteoric water,** the moisture that falls as precipitation. All water on the planet eventually follows a path through its oceans, land, and atmosphere and between its mantle and its surface—a process called the **hydrologic cycle.**

Any surface water whose flow is confined to a narrow topographic depression, or channel, is a **stream.** The flat land bordering a stream's channel is its **floodplain.** All streams are surrounded by **watersheds,** or **drainage basins,** which are areas of land that supply their water. The topographic highland that separates two adjacent drainage basins is called a **drainage divide.** Rainfall initially flows from a drainage divide across the basin's slopes as broad, shallow sheets of water that are not confined to channels. This flow erodes increasingly large channels in the surface, ultimately creating a network of small **tributaries** that feed water into a main or **trunk stream.** The trunk stream, in turn, may split into a network of small channels, or **distributaries,** that usually empty into an ocean.

Stream velocity is governed principally by the slope, or **gradient,** of the stream bed and the bed's roughness, which may produce friction and impede the water's flow. The stream's velocity is directly related to **stream discharge,** the volume of water passing a point on the stream bank during a given unit of time.

Streams tend to cut downward into bedrock until they reach their **base level,** the lowest point to which a stream can erode (usually sea level). When a stream flows just swiftly enough to transport its sediment load with very little net erosion or deposition, it enters a temporary state of equilibrium and is called a **graded stream.** If a graded stream receives more sediment, it responds by instantaneously increasing its gradient, a process called **aggradation.** If a graded stream receives less sediment (as occurs in a reservoir behind a dam), it responds by instantaneously lowering its gradient, a process called **degradation.**

Erosion, combined with mass movement, promotes the development of a stream's characteristic V-shaped valley. The rate of erosion is directly proportional to the stream's velocity. A stream erodes its bed by **abrasion, hydraulic lifting** of loose particles, and dissolution. Streams develop characteristic **drainage patterns,** that reflect the underlying geology formed from networks of interconnecting tributaries, trunk streams, and distributaries. The most common drainage network is the dendritic pattern, which develops on rocks or sediments of uniform composition and topography.

Responding to the local topography, climate, and geology, streams may take on a straight, **braided,** or **meandering** form. When two consecutive bends in a meandering stream lie close to one another, the stream may cut through the thin strip of floodplain that separates them, turning the abandoned segment into an **oxbow lake.** When a stream erodes downward to a new base level, it often leaves a fairly flat-lying surface, or **stream terrace,** that marks its former position.

A stream's ability to transport sediment varies according to several characteristics. The volume of sediment transported by a stream—controlled principally by its discharge—is called its **capacity.** The maximum size of its transported particles—largely a function of stream velocity—defines a stream's **competence.** The sediment in a stream can travel as a **suspended load** in the water column, bounce or roll along the stream bed as a part of its **bed load,** or move in solution as a **dissolved load.**

Ultimately, when the pull of gravity becomes a more powerful factor than the velocity and turbulence of streamflow, transported particles settle from a stream. These particles, called **alluvium,** may be deposited within the channel (forming mid-channel bars and **point bars**), on the stream's floodplain (building **levees** and **backswamp** deposits that mark a floodplain's **wetlands**), where slopes level out and valleys broaden (building **alluvial fans**), and in bodies of standing water (constructing **deltas**).

Floods occur when waterflow into a stream exceeds the channel capacity. They are especially likely where the ground is impervious to infiltration, such as in extensively paved urban areas, and where the ground is saturated from previous rainfalls. To predict floods, geologists document the historical pattern of floods in a region and develop a model based on statistical probability. To cope with the inevitability of floods, communities may build flood-control structures (such as artificial levees and flood walls), alter channels to allow water to pass through more swiftly (a strategy called channelization), and designate floodway zones in which structural development is restricted.

Plate tectonics casts light on the evolution of streams. Plate motions may control, among other things, a stream's base level, disruptions to its profile (by faulting and uplift), and its supply of sediment (particularly in volcanic areas).

Past stream activity seems to have been instrumental in shaping the landscape of Mars. A host of photographed fluvial features—such as meandering channels, streamlined mid-channel islands, and dendritic drainage patterns—suggest that the Martian surface experienced a prolonged fluvial period early in its history, before the planet's atmosphere thinned and its surface cooled.

Key Terms

juvenile water (p. 390)
meteoric water (p. 391)
hydrologic cycle (p. 391)
stream (p. 391)
floodplain (p. 391)
watershed (p. 391)
drainage basin (p. 391)
tributaries (p. 391)
drainage divide (p. 391)
trunk stream (p. 392)
distributaries (p. 392)
gradient (p. 392)
stream discharge (p. 392)
base level (p. 393)
graded stream (p. 393)
aggradation (p. 394)
degradation (p. 394)
abrasion (p. 395)
hydraulic lifting (p. 395)
drainage patterns (p. 395)
braided streams (p. 398)
meandering stream (p. 398)
oxbow lake (p. 400)
stream terrace (p. 402)
capacity (p. 402)
competence (p. 402)
suspended load (p. 403)
bed load (p. 403)
dissolved load (p. 403)
alluvium (p. 404)
point bars (p. 404)
levees (p. 404)
backswamp (p. 405)
wetlands (p. 405)
alluvial fan (p. 406)
delta (p. 406)

Questions for Review

1. Briefly describe the main elements of the hydrologic cycle, including the processes responsible for water's changes in state. List three ways water can be delayed on land before eventually reaching an ocean.

2. How do we measure stream discharge? How do changes in stream velocity, width, and depth affect stream discharge? Calculate the discharge of a river that is 70 meters wide and averages 20 meters deep and that flows at an average velocity of 10 meters per second.

3. Discuss two potential causes of stream aggradation.

4. Sketch three different drainage patterns.

5. Briefly explain how one stream might pirate water from another stream.

6. Draw a sketch illustrating the origin of an oxbow lake. How does an oxbow lake become a meander scar?

7. How do the sediment particles that are most likely to be transported in suspension differ from those transported by traction? What is the difference between a stream's competence and its capacity?

8. What are the principal causes of flooding?

9. Explain how structural and nonstructural approaches to flood control differ.

10. List three ways in which plate tectonics may affect the evolution of a stream profile.

For Further Thought

1. Where would you expect to find a stream that exhibits high competence but low capacity? Where would you expect to find one that exhibits low competence but high capacity?

2. How would the Earth's fluvial landscapes change if plate tectonic activity ceased suddenly?

3. How would the world's streams respond if climatic warming melted the Antarctic ice sheet?

4. How would the potential for flooding change over a long period of time if we removed the dams along the Mississippi River?

5. What kind of deposits do you think underlie the farmland shown in the photo below?

15

Groundwater

Figure 15-1 Roman baths at Bath, England. This pool of warm water, which flows naturally to the surface, was a meeting place and spa for high-ranking officials during the Roman occupation of Britain about 1900 years ago.

For thousands of years, wells and natural springs have supplied clean, abundant groundwater to human communities throughout the world (Fig. 15-1). Located in the centers of many villages, wells have been focal points for meeting, bartering, and sharing the news of the day. As in the past, the presence of an adequate supply of uncontaminated groundwater can dictate whether a contemporary region or community will grow and prosper. Pure water has even become a commercially valuable commodity. Check the shelves of your local supermarket, and you'll find bottles of rather ordinary groundwater with fancy names selling for exorbitant prices.

Humans can survive for weeks—even months—without food. But because virtually all biological processes take place in water, we can live without this liquid for only a few days. Our bodies require about 4 liters (1 gallon) of water per day. Thus we must draw 1.6 billion liters (400 million gallons) of water from the continent's freshwater reserves *each day* to meet the biological needs of North America's 400 million people. Furthermore, we use a thousand times that amount—more than 1.8 *trillion* liters (450 billion gallons) of water *per day*—for domestic purposes (cooking, bathing and other sanitary needs, and lawn watering), agriculture (livestock and irrigation), and industry (processing natural resources and manufacturing the goods that maintain our way of life).

The Earth's hydrosphere extends from the top of the atmosphere to approximately 10 kilometers (6 miles) below the Earth's surface and includes the planet's oceans, glaciers, rivers and lakes, atmospheric water vapor, and groundwater. This water constantly moves through stages of the hydrologic cycle (see Figure 14-3). The largely unseen groundwater component, representing only 0.6% of the world's water but accounting for 97% of its supply of unfrozen freshwater, provides more than 50% of our drinking water, 40% of our irrigation water, and 25% of the water used in industry. This water rarely takes the form of distinct underground rivers or lakes. Instead, it is widely dispersed within the pore spaces of rocks and sediments, like water within a sponge.

Groundwater use doubled between 1955 and 1985 as the U.S. population grew and the nation's economy became

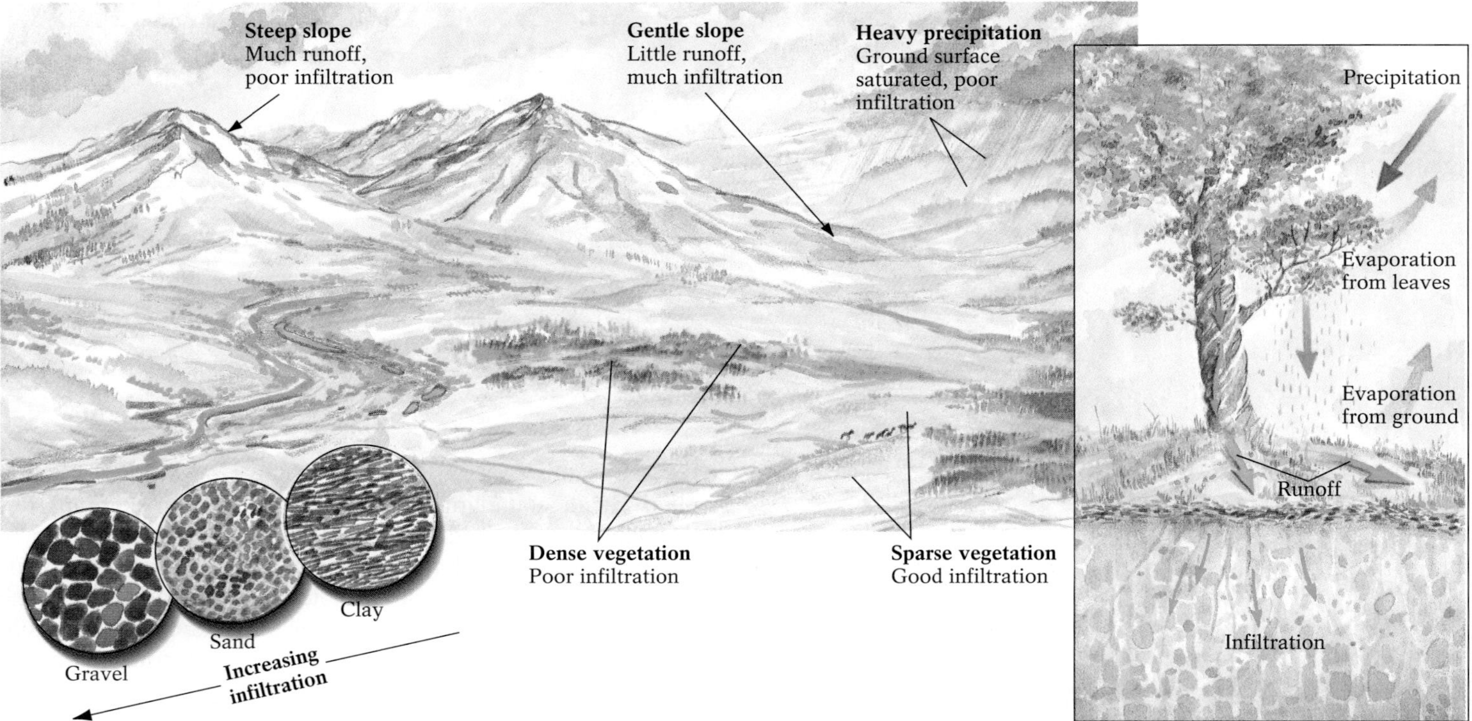

Figure 15-2 Factors affecting infiltration. Bedrock composition, vegetation patterns, surface slope, and climate, influence the amount of water that enters the groundwater system. Gentle, lightly vegetated slopes composed of permeable materials are most conducive to infiltration. As seen in the inset diagram, the amount of water that infiltrates the ground generally equals the amount of precipitation *minus* the water that runs off the surface, evaporates from the surface or from vegetation, or is consumed by vegetation as it grows.

more industrialized. Throughout North America, we are withdrawing groundwater from reserves that have accumulated over thousands of years, and groundwater supplies (particularly in the Southwest) are most definitely running low. On the bright side, in many areas, climate and local geology are still capable of providing an ample supply of cool, refreshing, healthy groundwater for present and future generations—if we learn to manage and preserve that resource. In other areas, however, human, animal, and industrial wastes have contaminated the groundwater. Some of the most urgent questions facing the world's citizens relate to groundwater: Where do we find it? How do we keep it clean? Who owns it?

To answer these questions, you need to understand how groundwater infiltrates the Earth's surface and how it flows from one place to another—topics covered in this chapter's first section. After examining this issue, we discuss the *water table*—the underground surface beneath which water fills all cracks, crevices, and pore spaces. Finally, we turn to the problems of finding and managing groundwater and maintaining its quality.

Groundwater Recharge and Flow

Groundwater *recharge* is the infiltration of water, mostly from precipitation, into the Earth's surface. After infiltration, groundwater flows principally under the influence of gravity through soils, rocks, and other Earth materials. Some of the water eventually emerges as groundwater *discharge,* or outflow, returning to the surface as streams or springs or remaining temporarily in lakes, ponds, or wetlands. The rest remains stored underground, perhaps for thousands of years.

As shown in Figure 15-2, the amount of water that infiltrates groundwater systems depends on the condition and type of local surface materials, the nature and abundance of vegetation in the area, the area's topographical features, and the amount of precipitation. We describe each of these factors next.

Condition and Type of Surface Materials The presence of abundant pore spaces in loose soils and unconsolidated sands and gravels promotes extensive infiltration. The characteristics of exposed bedrock also enhance infiltration by surface water. Is it highly fractured or jointed, or inherently porous, such as coarse sandstone? On the other hand, the minute pore spaces of unconsolidated clay, which consists of closely packed, flat particles, impede infiltration. Infiltration is also poor in exposed *unfractured* crystalline bedrock, such as that composed of plutonic igneous rocks (for example, granite and diorite) or high-grade metamorphic rocks (such as gneiss).

Vegetation Thirty-five percent or more of the precipitation that falls in heavily vegetated areas lands on the leafy canopies of trees and other vegetation and evaporates before it reaches the ground. Approximately 25% touches the ground but then either evaporates from the ground surface, runs off as stream flow, or is temporarily stored at the surface in lakes or glaciers. The remaining precipitation seeps into the ground, if surface conditions permit. Plant and tree roots enhance infiltration of this type of precipitation by opening pathways in soil into which the water can seep.

Topography Surface water runs slowly down gently sloping terrains, so it has ample time to seep into the ground. On steep slopes and cliff faces, however, the water runs rapidly downslope and enters nearby streams before a large percentage of it can infiltrate the soil.

Precipitation The amount of precipitation that falls over time affects the amount of groundwater recharge in any kind of terrain. Extended droughts, for example, may curtail groundwater recharge significantly for several years or more. Shorter-term recharge variations, such as those from spring rains and snowmelt, generally replenish the groundwater supply on a seasonal basis.

The type of precipitation also affects the degree of infiltration. A driving rainstorm packs surface soils and washes fine clayey particles into soil pores, clogging them and impeding infiltration. Likewise, later rainfalls in a series of drenching storms occurring in close succession increasingly favor runoff—the first storm saturates the surface, and water from subsequent storms cannot find pathways into the soils. Such was the case during the Midwestern floods of 1993 (see Chapter 14). Snowmelt infiltrates the groundwater system to some extent, but it may also enter streams or evaporate.

Movement and Distribution of Groundwater

Once water has infiltrated the Earth's surface, gravity draws it downward. The types of soils, sediments, and rocks that the water encounters determine the depth to which it will descend. Figure 15-3 illustrates the subsurface levels through which water travels and collects.

As the liquid passes through the region of weathered regolith, some of its molecules become attracted and bound

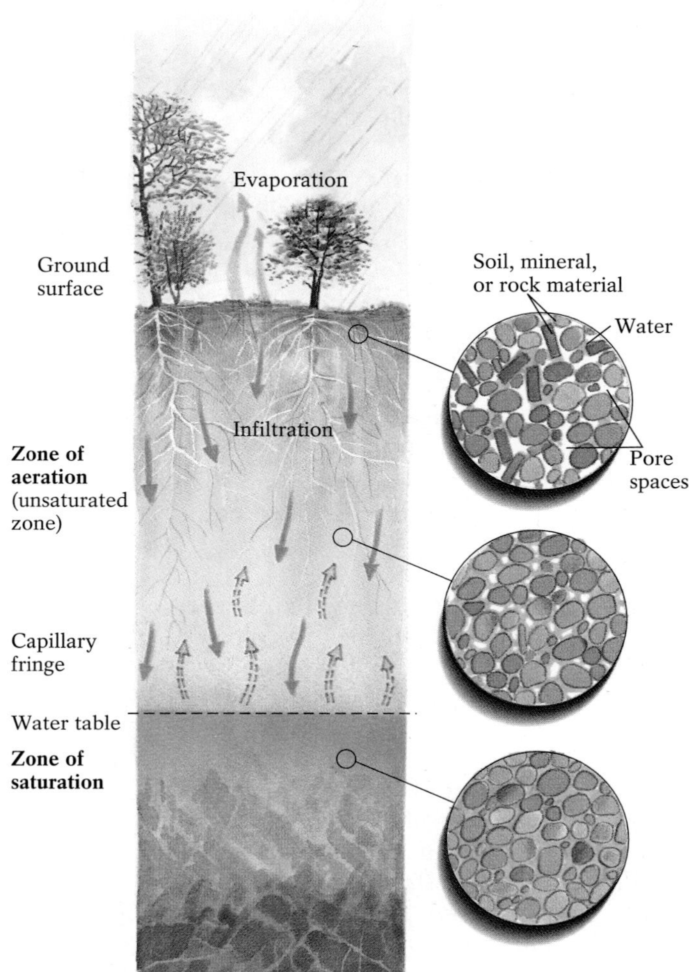

Figure 15-3 The subsurface distribution of water.

to clay minerals in the soil or are consumed by plants. The attraction between the water and the charged surfaces of clay minerals reduces both evaporation and infiltration from this near-surface region. Water that is not bound by clay minerals, used by plants, or lost to evaporation moves downward through a **zone of aeration,** or unsaturated zone, where the pore spaces in rocks and soils contain both water and air. Some water may descend farther, into the **zone of saturation,** where water fills every available pore space. The **water table**—the upper surface of the zone of saturation—separates the zones of aeration and saturation. The lower part of the aeration zone, which can range from a few tens of centimeters to several meters above the water table, is called the **capillary fringe.** In this region, the gravity-driven downward flow of infiltrating water is partly countered by the attraction of the water molecules to mineral surfaces and to other water molecules. This attraction causes a small volume of water to rise *upward* from the water table and be held in the unsaturated zone. This phenomenon, known as *capillary action,* can be observed when the corner of a paper towel is dipped into a glass of water and the water rises in the towel against gravity.

The saturation zones in most places do not extend to great depths, because the pressure of overlying rock closes deep fractures and recrystallizes some rock into impermeable crystalline metamorphic rocks. For these reasons most groundwater is confined to the upper 1000 meters (0.6 mile) of continental crust. Given adequate precipitation, the availability of groundwater is governed by two characteristics of the soil or rock unit into which the water might infiltrate—its porosity and its permeability.

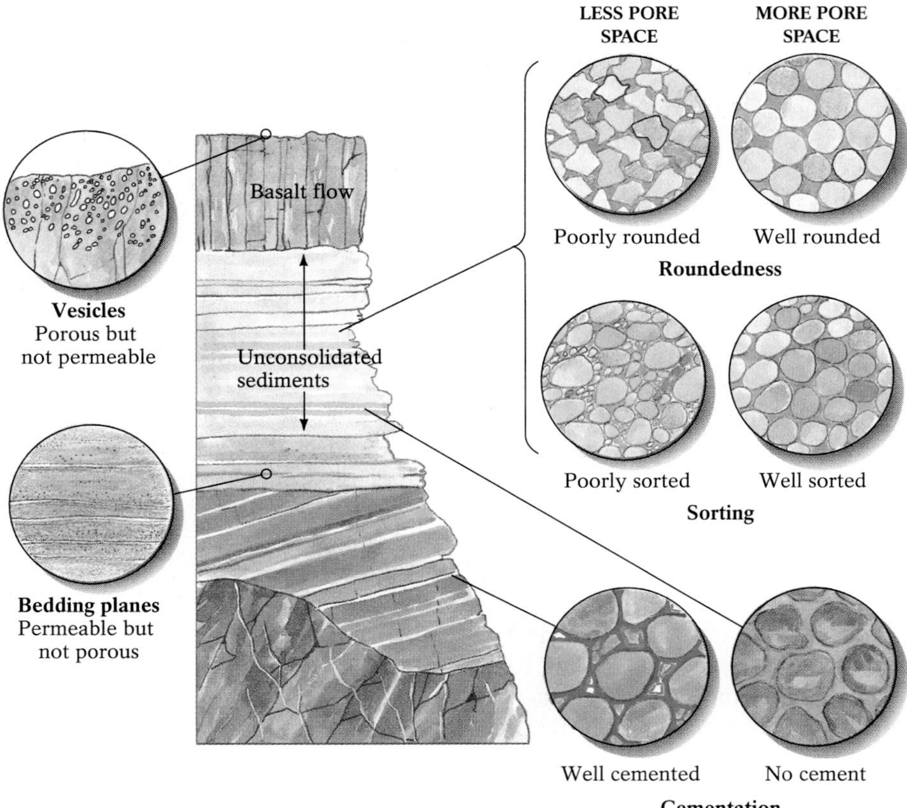

Figure 15-4 The primary porosity of sedimentary rocks is affected by the grain shape, sorting, and particle cementation. The presence of vesicles and bedding planes also increases a rock's primary porosity.

Porosity **Porosity** is the volume of pore space compared with the total volume of a soil, rock, or sediment. Porosity (which is expressed as a percentage) determines *how much water the material can hold.*

Primary porosity is the porosity that develops as a rock forms. For example, the primary porosity of basalt results from the presence of vesicles or spaces that remain near the top of the rock after lava cools. *Secondary* porosity develops after a rock has formed, usually as a result of faulting, fracturing, or dissolution.

In sedimentary rocks, primary porosity is determined by such factors as grain roundedness, sorting, and cementation—sediment properties discussed in Chapter 6. As you can see in Figure 15-4, well-rounded grains tend to have larger pore spaces and therefore hold more water. When a sediment contains grains of various sizes (that is, when it is poorly sorted), the finer particles tend to fill the voids between the coarser particles, clogging the pores and reducing porosity. When cementation converts loose sediments to sedimentary rock, the cement fills the pore spaces and further diminishes porosity. Fine clayey muds (though they impede groundwater re-charge) hold much more water when saturated than coarse sediments, because the former materials contain a higher percentage of minute pores than the latter. Fresh mud from the Mississippi River's delta, south of New Orleans, may consist of 80% water. Because the extremely small size of the pores impedes flow, this water is very hard to extract from the ground.

Permeability The crucial factor (other than the amount of infiltration) that determines the availability of groundwater is not how much water the ground holds, but whether the water can flow *through* the pores. **Permeability** is the capability of a substance to allow the passage of a fluid. The permeability of rock or sediment reflects the size of its pore spaces and the extent to which these spaces are connected. If pores are very small, as in clayey sediment, water flows through them extremely slowly. Some water molecules may adhere as a fine film to adjacent particles, slowing the flow even further. Only when pores are relatively large can water flow easily. The pores between grains of clean sand, for example, are more than 1000 times larger than the pores in clay, explaining in large part why sand is so much more permeable than clay.

The passage of groundwater may be impeded even in a porous material if the pores are sealed off from one another. For example, the pores in a very porous limestone may not be connected; thus water does not flow through this rock. In other cases, a material may be essentially nonporous, such as the basalt shown in Figure 15-5, but still be permeable, because it includes a network of interconnected fractures that accommodate a modest flow of water. If these fractures are not joined throughout the rock body, the flow will never be large. A rock such as basalt, which often develops a highly

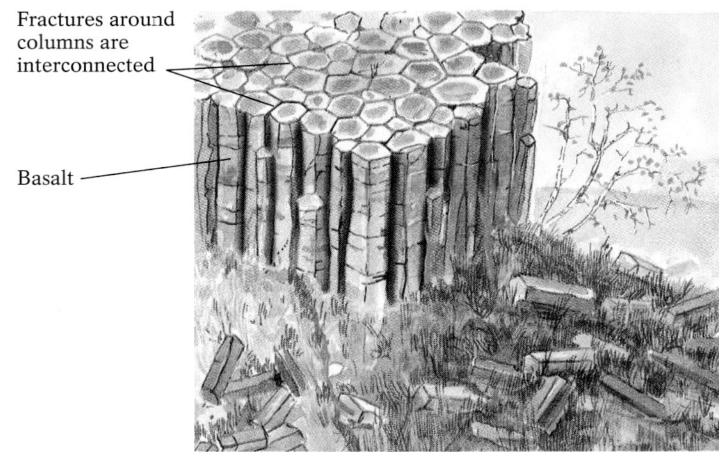

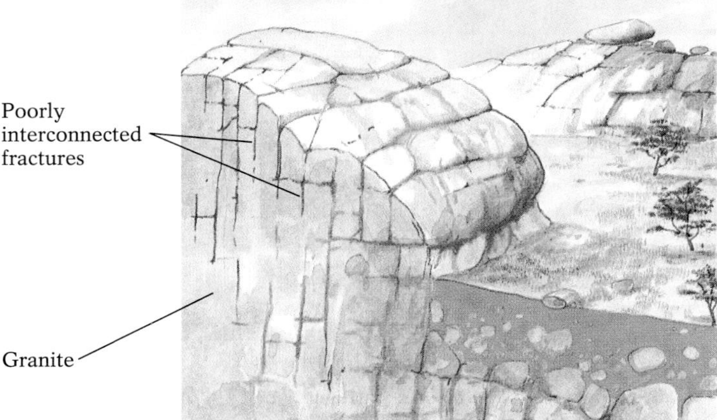

Figure 15-5 Pore connection and permeability. Although basalt is a generally nonporous igneous rock, its connected network of columnar joints renders this basaltic lava flow highly permeable. In contrast, the fractures in the granitic bedrock are poorly interconnected; thus this granitic outcrop is relatively impermeable.

interconnected fracture system as it cools, can nevertheless serve as a reliable source of groundwater.

How Groundwater Flows Gravity causes surface water in a stream to move down its channel's slope from higher to lower elevations. Gravity also causes groundwater to flow from high to low areas, but the movement of this fluid is also driven by differences in pressure on the water deriving from the weight of overlying water and, at greater depth, from the weight of the surrounding rocks. The **potential** of groundwater flow at any depth below the water table represents a combination of the influence of gravity and the pressure on the water at that depth. Groundwater generally moves from areas of high potential, such as high spots in water tables under hills, to areas of low potential, such as low spots in water tables under valleys.

Groundwater flows between any two points that differ in terms of their elevation and pressure. As shown in Figure 15-6, the flow rate is partially controlled by the **hydraulic gradient,** the difference in the potentials of two points divided by the lateral distance between those points. Groundwater tends to travel most swiftly between two points when one site has a much greater potential than the other, and the distance between the points is small.

The flow rate of groundwater is also affected by the material through which it flows. Every material possesses an intrinsic **hydraulic conductivity** that reflects, among other things, the sizes, shapes, and degree of sorting of its grains. Coarse, well-rounded, well-sorted gravel has high hydraulic conductivity, for example, whereas fine-grained, angular, poorly sorted sediment has low hydraulic conductivity.

Henry Darcy, a nineteenth-century French hydraulic engineer, was the first person to calculate the flow rate of groundwater. Darcy reasoned that the rate at which groundwater flows between two wells is directly proportional to the

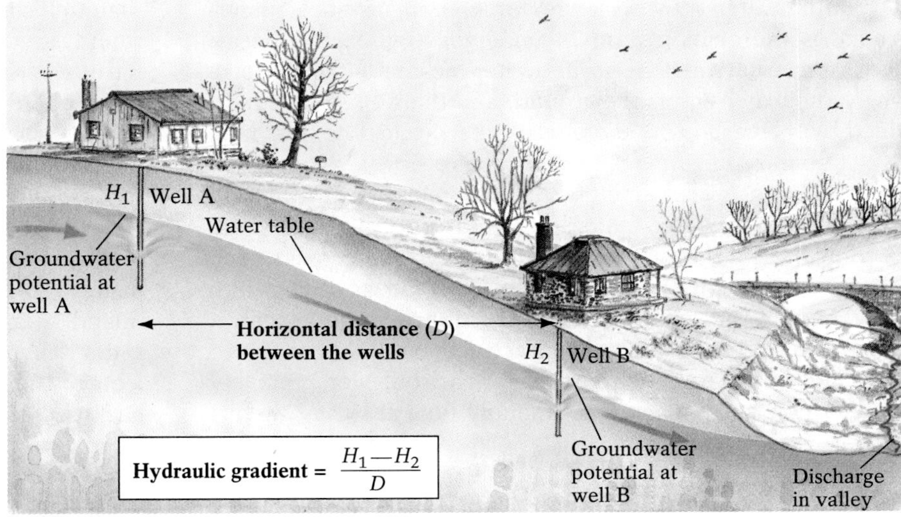

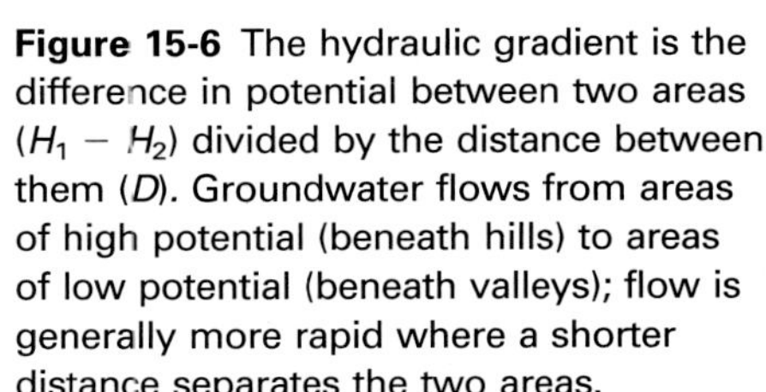
Figure 15-6 The hydraulic gradient is the difference in potential between two areas ($H_1 - H_2$) divided by the distance between them (D). Groundwater flows from areas of high potential (beneath hills) to areas of low potential (beneath valleys); flow is generally more rapid where a shorter distance separates the two areas.

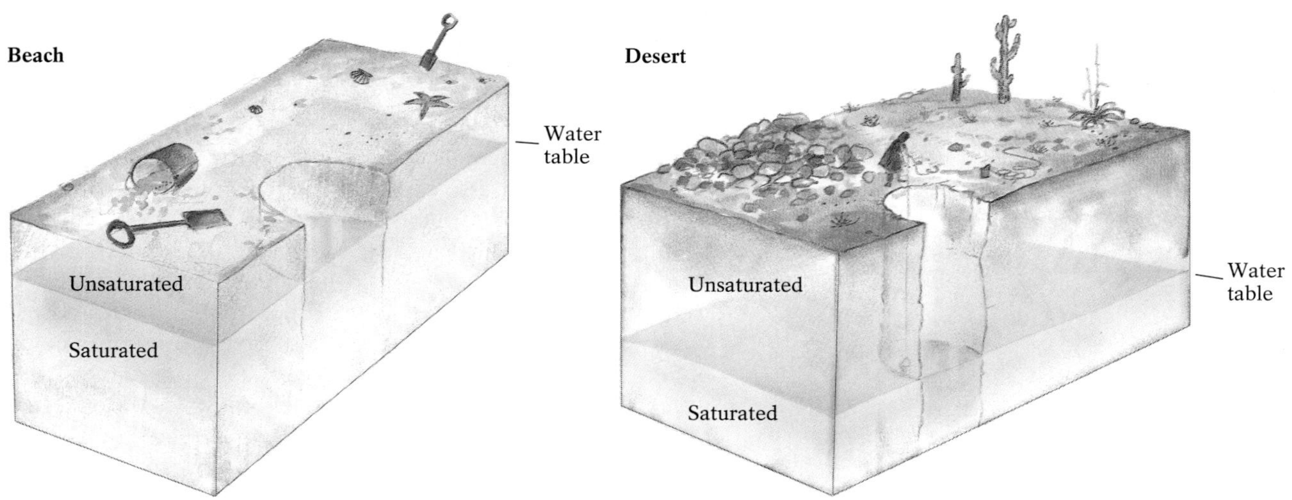

Figure 15-7 Variations in water table depth. Groundwater pools at the bottom of a shallow hole at the beach, where the water table lies close to the surface. In the desert, where the water table lies far beneath the surface, the hole intersecting the water table must be much deeper. Note that the scales for these two illustrations differ substantially.

difference in potential between the wells and to the hydraulic conductivity of the materials through which the water must travel. Where the hydraulic gradient and the hydraulic conductivity are high, groundwater flows more swiftly than where the hydraulic gradient and hydraulic conductivity are low.

Geologists use two simple methods to measure the rate of groundwater flow at any location. When the water's path is known, they inject a harmless dye into a well. They then compare the time of injection with the time of the dye's arrival at a nearby well downslope. Dividing the distance between the wells by the time elapsed yields the flow rate.

To determine groundwater flow rates over much longer distances, geologists employ radiocarbon dating. (Carbon-14 dating is discussed in Chapter 8.) Rainwater incorporates carbon-14 as it passes through the atmosphere and soils. As the groundwater moves slowly for a long distance underground, some of its carbon-14 undergoes radioactive decay. The distance between the groundwater recharge site (the place where the water enters the ground) and the well sampled, divided by the carbon-14 age of the water, yields the flow rate.

Groundwater flow is typically quite slow, averaging between 0.5 and 1.5 centimeters (0.2–0.6 inch) per day through moderately permeable material, such as poorly sorted sand. In contrast, flow through unfractured crystalline rocks such as granites and gneisses may be only a few tens of centimeters *per year.* The swiftest flow—through well-sorted, coarse, uncemented gravels or highly fractured basalts—can reach 100 meters (330 feet) per day. In comparison, even relatively slow-moving surface streams generally flow at least 2 meters (6.6 feet) *per second.*

The slow movement of groundwater accounts for its availability for human use. If this liquid traveled as rapidly as rivers, it would not be stored for long in the ground. The groundwater we draw today from a deep well may have actually fallen as rain thousands of years ago. Deep groundwater drawn from wells in the hot, dry lands of southern Arizona, for example, is more than 10,000 years old. This water apparently precipitated during the last worldwide glaciation, when Arizona's climate was cool and cloudy enough for rainwater to infiltrate the groundwater system instead of evaporating, as much of it does today.

Tapping the Groundwater Reservoir

We rely heavily on groundwater to meet our myriad water needs—and how better to tap a region's groundwater supply than to locate the local water table? Lakes, swamps, and year-round streams form where the water table intersects the Earth's surface. To determine the depth of an area's water table where no such surface water arises, we can dig a hole or drill a well. We know we have reached the local water table when water begins to seep into the hole or well, indicating that the surrounding material is saturated with water. If you've ever dug a hole at the beach, you probably noticed that water begins to accumulate at the hole's bottom almost immediately—considerably less than a meter below the surface. Inland, in areas that receive significant rainfall each year, the water table typically occurs within 1 or 2 meters (3.3–6.6 feet) of the surface. You can confirm this fact by digging in a backyard garden in such regions. In arid locales, however, water may not appear until a hole reaches tens of meters below the surface. Figure 15-7 compares the typical depth of the water table at two different locations: a beach and a desert.

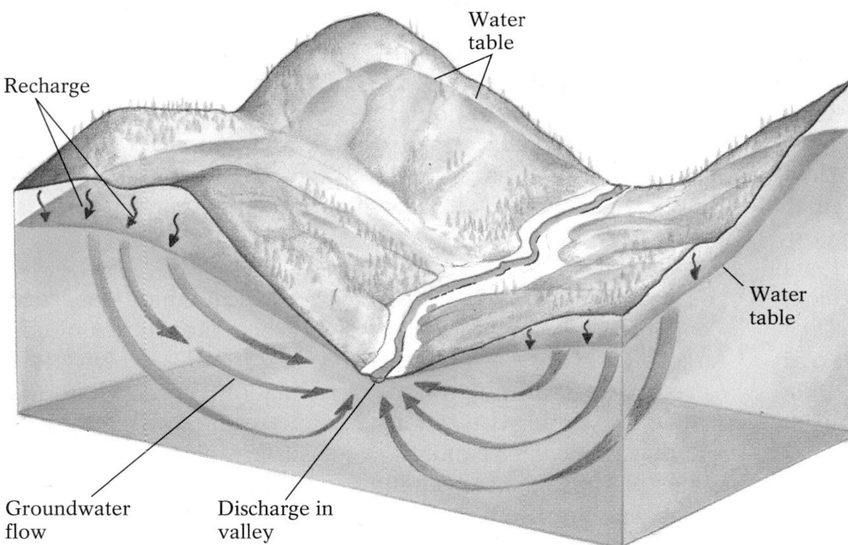

Figure 15-8 Water-table configuration and topography. When precipitation falls on an irregular landscape, the slopes receive the most rainfall because they constitute the largest surface area. The water that infiltrates the hills must travel through a considerable expanse of soil and rock to reach a stream; it flows only slowly, mounding up below the surface of the hills. If groundwater recharge were to cease for a long time—for instance, during a drought—the water-table mounds would flatten out in the hilly areas as groundwater flow continued. In most humid areas, however, rainfall is able to replenish the supply beneath the hills and maintain the undulations of the water table.

The depth of the water table varies according to local topography and prevailing climate. Its configuration often parallels the surface topography, although its highs and lows are less pronounced than those of the overlying landscape (Fig. 15-8). The depth of the water table generally reflects a long-term balance between recharge and discharge, with seasonal climatic fluctuations. (Water tables tend to be higher in winter and lower in summer.) Major climatic events, such as storms and droughts, can raise or lower the water table *temporarily,* as can variations in our extractions of water from the ground.

A prolonged dry spell, such as the one that struck the northeastern United States during the early 1960s, can lower the water table dramatically. During that drought, the region experienced a 2-meter (6.6-foot) lowering of the water table. Some unscrupulous developers took advantage of these conditions and constructed many homes in areas that normally have high groundwater levels. When the drought ended in the mid-1970s and recharge increased, the water table rose to the region's normal pre-drought levels. Many distraught homeowners in Long Island and New Jersey then discovered that their flooded basements occupied the zone of saturation.

Sometimes a *local* impermeable layer (such as a lens of clay) may obstruct downward flow of water through a region's zone of aeration. Water may then accumulate above this barrier, saturating part of the overlying zone. In this case, what may appear to be the regional water table is actually a **perched water table,** lying above a locally saturated zone within the zone of aeration (Fig. 15-9).

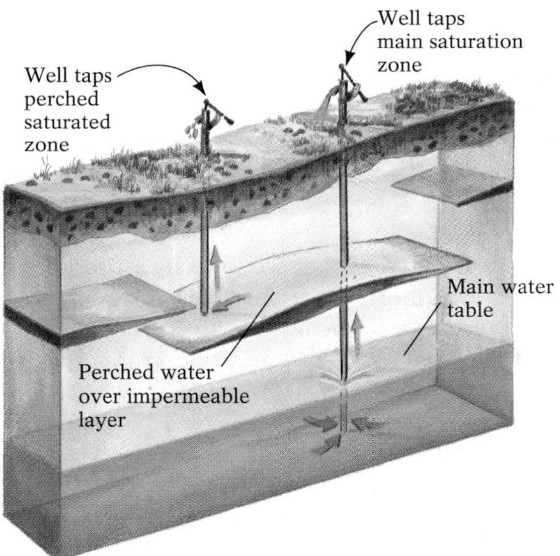

Figure 15-9 Perched water pools form in the zone of aeration when a local impermeable layer intercepts descending water. To ensure a steady, year-round flow of groundwater, wells must be drilled to a depth below the main—not the perched—water table.

Aquifers: Water-bearing Rock Units

When geologists speak of the water table and the underlying zone of saturation, they mean the area immediately beneath the Earth's surface. This water-bearing zone represents one example of an **aquifer** (from Latin, meaning "to bear water"), a permeable body of earth material that transmits groundwater yet also stores significant amounts. Aquifers are the most common sources of the groundwater we withdraw for our domestic, agricultural, and industrial needs. The most productive aquifers consist of unconsolidated sand and gravel, well-sorted, poorly cemented sandstones, and highly jointed limestones and basalts.

Unconfined aquifers extend nearly to the Earth's surface; because the regional water table lies *within* them, these resources are relatively easy to tap. The water in unconfined aquifers is generally "local" in origin, flowing primarily through loose slope material, sands, gravels, and floodplain deposits left behind by streams and rivers, sands and gravels transported by recent glaciers, and young, jointed lava flows such as those in Hawai'i and the Pacific Northwest. In the unconfined aquifer shown in Figure 15-10, water flows through the glacial debris and river alluvium. The Ogallala Formation—a major unconfined aquifer—consists of a poorly sorted, generally uncemented mixture of clay, silt, sand, and gravel produced by a combination of these sources. A component of the High Plains regional aquifer system, the Ogallala aquifer underlies much of Nebraska and parts of South Dakota, Wyoming, Colorado, New Mexico, Kansas, Oklahoma, and Texas. It supplies 1.2 trillion liters (317 billion gallons) of water per year to irrigate 14 million naturally arid acres. Lands irrigated by water from this aquifer yield 25% of all U.S. feed-grain exports and 40% of the country's flour and cotton exports.

Sometimes groundwater resides at greater depth, flowing through deeper layers of rock or sediment, typically far below the water table and the near-surface unconfined aquifer, and separated from them. Such *confined aquifers* comprise water-transmitting layers of rock or sediment sandwiched between other rock layers that are either effectively impermeable or have very low permeability. Impermeable layers of such aquifers are called **aquicludes;** low-permeability layers are called **aquitards.** Unlike the locally derived water in most unconfined aquifers, the water in confined aquifers may have traveled a great distance from its recharge site. One such confined aquifer, the Dakota Sandstone, is tapped widely for irrigation; its water supply originates in the Black Hills of South Dakota but then flows at depth beneath the Great Plains of the eastern Dakotas and Nebraska. Similar aquifers can even be found at great depths under deserts—such as in Nevada, eastern Utah, and southern Arizona—where precipitation has entered the groundwater system by infiltrating exposed permeable rocks in mountain ranges, both near and far.

Artesian Aquifers Under certain geological conditions, such as those illustrated in Figure 15-11, the water in a confined aquifer may rise against the downward pull of gravity and even gush from the ground. This situation occurs where the confined aquifer is tilted at an angle to the Earth's surface, so that it becomes exposed at the surface high in the mountains. Because the surrounding aquicludes and aquitards prevent the water from escaping, the weight of the water located higher in the inclined water column exerts pressure on the underlying water. When the aquifer is tapped (either naturally, by a fault, or artificially, by a well), the pressure drives the water upward. This type of aquifer is referred to as **artesian**—named after the town of Artois, France, where water has flowed unaided from the ground for centuries. Although artesian water tends to rise *toward* the elevation of the point of recharge at the surface, it never quite reaches this point; friction between the water and the surrounding rocks and sediments impedes the flow and prevents the fluid from reaching this level. The level to which the pressurized water would rise in a tall pipe in the *hypothetical* absence of friction is called the **potentiometric surface** of the aquifer. When

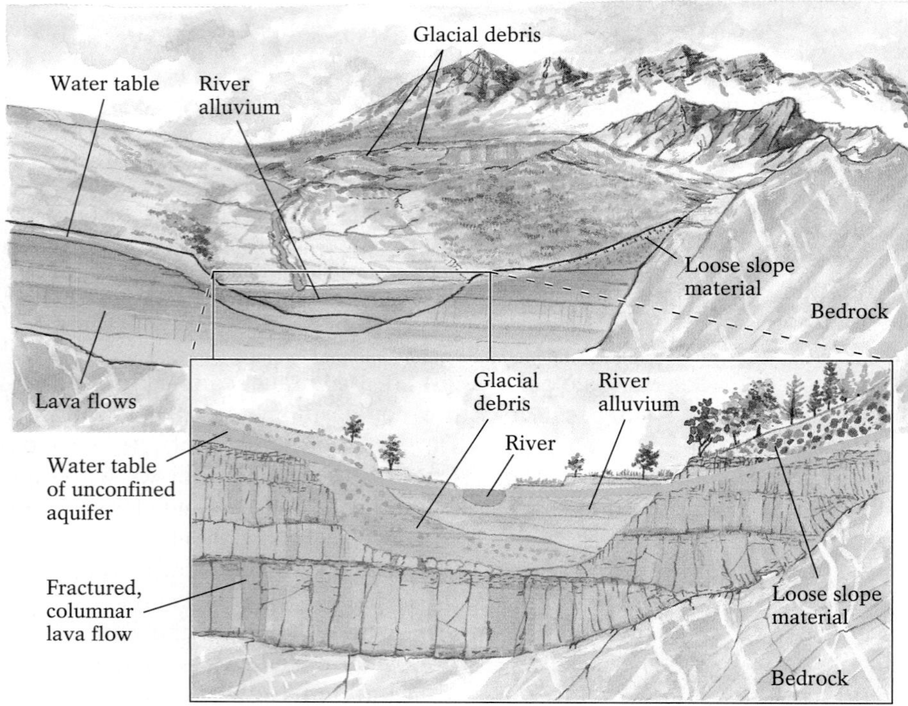

Figure 15-10 This composite landscape shows an unconfined aquifer composed of four different types of materials: glacial debris, loose slope material, jointed lava flows, and river alluvium.

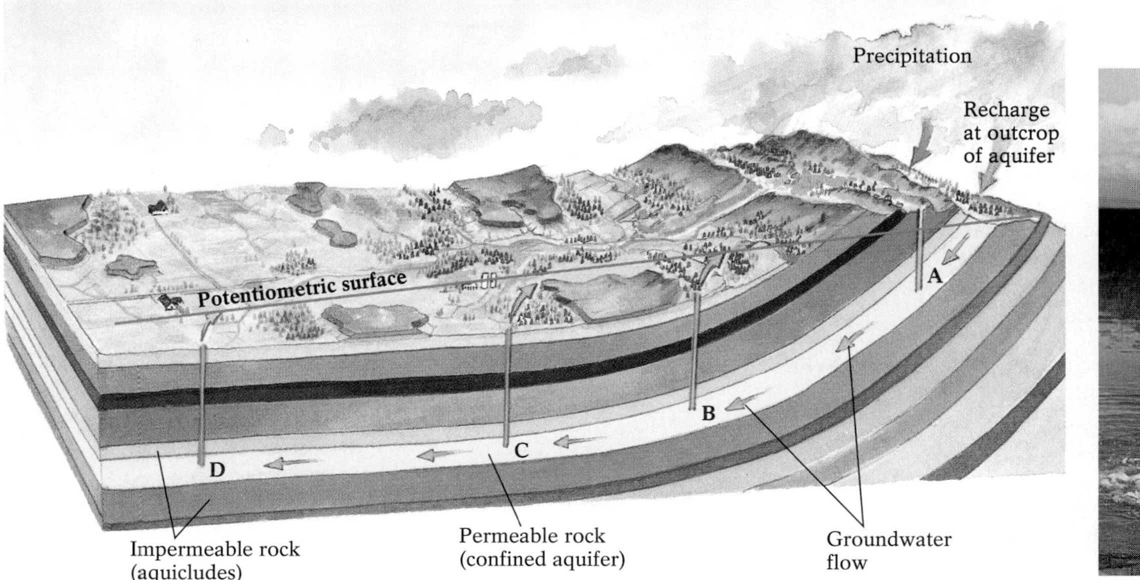

Figure 15-11 Artesian aquifers. (above) High pressure causes groundwater to rise in the wells at points B, C, and D, where the potentiometric surface is significantly higher than the well sites. Water does not flow unaided from well A, which lies above the potentiometric surface. (photo) Water gushing from a natural artesian well in the Dakota Sandstone aquifer (circa 1910). When this aquifer was first discovered, water gushed freely from wells under unusually high artesian pressures. During the last 75 years, more than 10,000 wells have been drilled to tap the Dakota Sandstone aquifer. The extensive withdrawal of water has reduced the pressure within the aquifer to the point that the potentiometric surface is no longer higher than the land surface of the eastern Dakotas. Today, water must be actively pumped in areas where it once flowed forcefully and unassisted.

the potentiometric surface lies above the Earth's surface, groundwater flows freely from the ground.

Sometimes water under artesian pressure may surface in unexpected places. An oasis, for example, may arise where pools of artesian water appear at or immediately beneath the surface (Fig. 15-12). An aquifer that receives its recharge from mountains surrounding a desert may extend hundreds of meters beneath the arid surface. Water may flow to the surface under artesian pressure where the rocks of the aquifer crop out at the surface or where a fault extends down to the aquifer, thereby forming the oasis.

Most modern municipal water-supply systems are designed to simulate artesian conditions. Have you ever wondered why water flows at high pressure when you turn on a

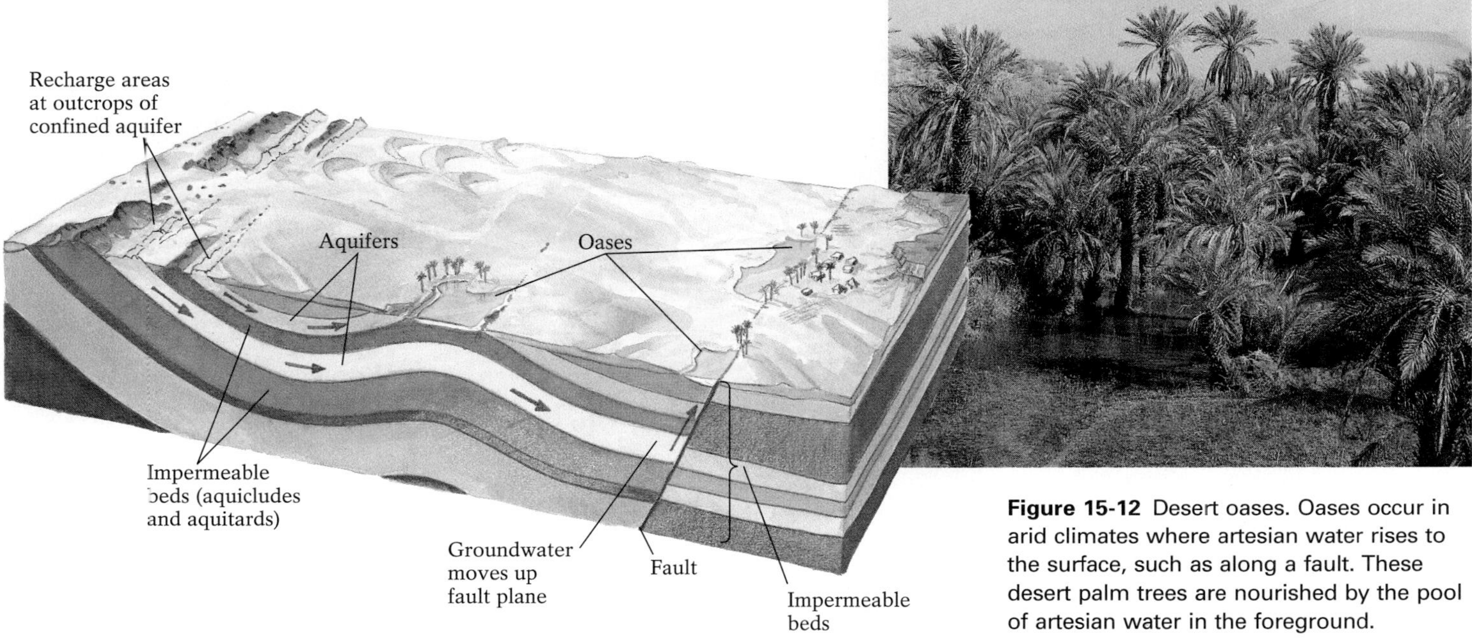

Figure 15-12 Desert oases. Oases occur in arid climates where artesian water rises to the surface, such as along a fault. These desert palm trees are nourished by the pool of artesian water in the foreground.

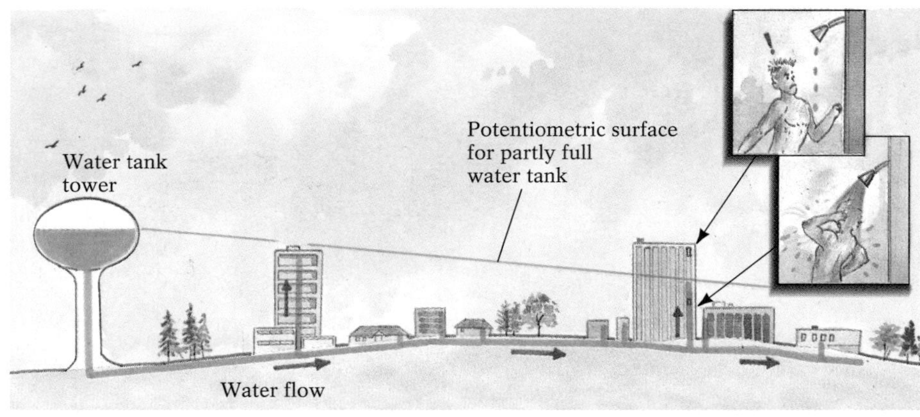

Figure 15-13 A municipal water tower functions in much the same way as the recharge area of an artesian aquifer. Water on the lower floor of a building rises under high pressure, because its outlet is located below the potentiometric surface. Water pressure is low on the top floor, because its outlet is located at a higher elevation than the potentiometric surface.

faucet? The tall water towers in your city or town may act like elevated recharge areas, with the municipal water pipes serving as confined aquifers. The difference in elevation between the tower and your home causes the water to flow from your faucets under pressure (Fig. 15-13).

Natural Springs and Geysers

With the exception of artesian aquifers, groundwater typically remains underground until we drill wells to extract it. Some geological conditions, however, allow groundwater to "spring" unaided from the ground. At **natural springs,** groundwater flows to the surface, usually laterally, and issues freely from the ground. Natural springs may develop where groundwater encounters an impermeable bed and must then flow around it, or where erosion or other processes have lowered the surface so that the groundwater intersects the water table (Fig. 15-14).

Because groundwater is insulated from fluctuations in air temperature, its own temperature therefore tends to approach the mean annual air temperature for the region. A *thermal spring* is one whose flow is at least 6°C (11°F) higher than this mean temperature. (A "hot" spring is informally defined as a thermal spring whose water temperature exceeds

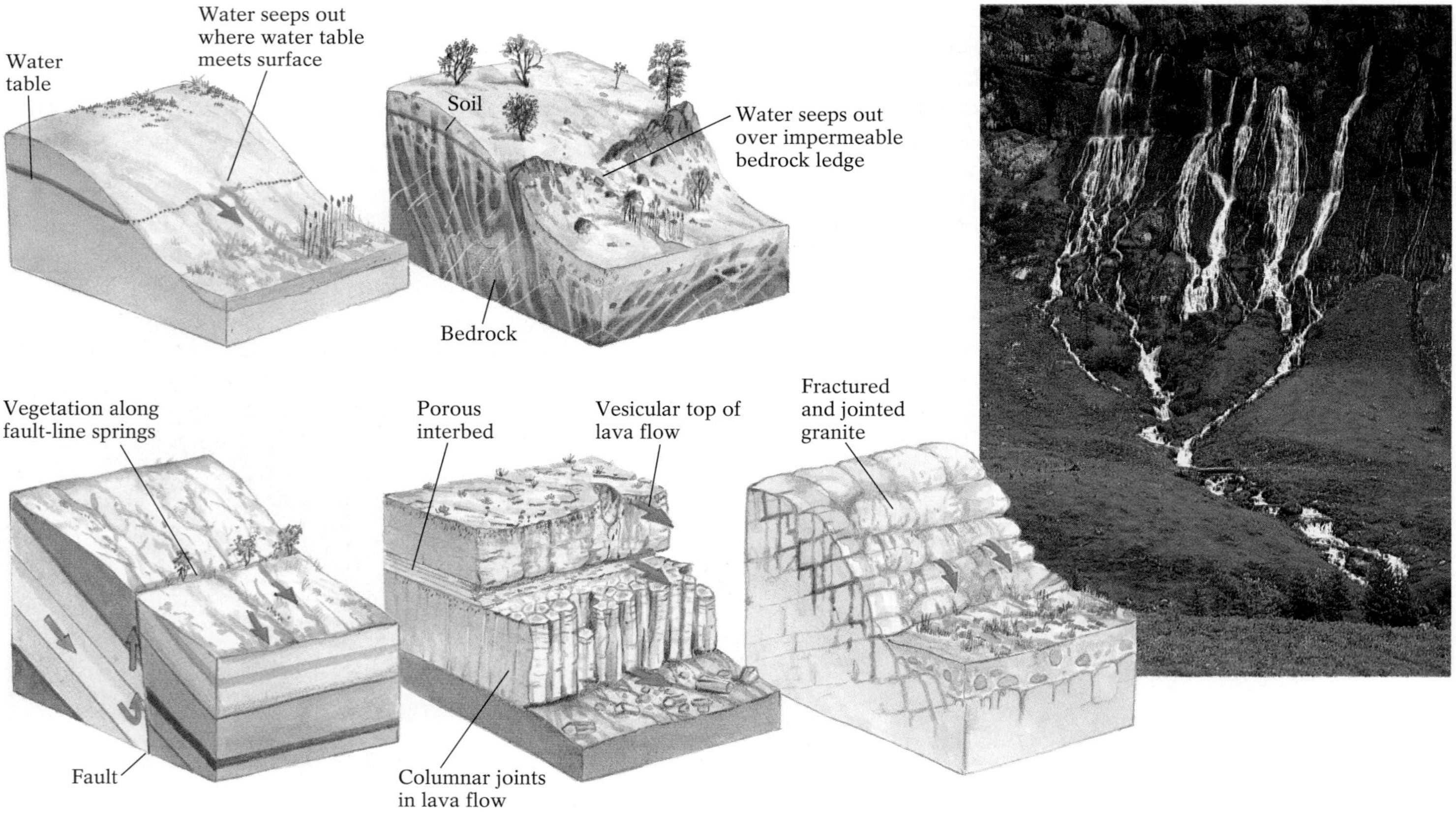

Figure 15-14 Natural springs. (above) Various geological settings in which natural springs occur. (photo) Water gushing from limestone rock in Switzerland.

Figure 15-15 Calcareous sinter deposited at Mammoth Hot Springs in Yellowstone National Park, Wyoming. (left) Hot groundwater has infiltrated and dissolved carbonate deposits and then precipitated them at the surface to create white cliffs of travertine. (right) Steam vents, or *fumaroles,* in the sinter.

human body temperature.) Most of the more than 1000 thermal springs in North America occur in the West, near the Rocky, Cascade, Olympic, and Sierra Nevada Mountains. The warmth of these springs' water primarily results from recent encounters with magmas or the still-hot rocks of a cooling pluton. After percolating downward for hundreds of meters, the groundwater is heated by warm igneous rocks, in some places to its boiling point. As the hot water rises, it usually dissolves large quantities of soluble rock. Upon reaching the surface, the water evaporates, depositing the minerals as *sinter* (Fig. 15-15). Sometimes the hot water dissolves sulfur on its way to the surface; in that case, the spring emits the distinctive rotten-egg odor of hydrogen sulfide, and the deposits have a yellowish tinge.

Most eastern hot springs, such as those found in the Appalachians and Ouachitas, are not heated by hot igneous rocks. Instead, deeply circulating groundwater passes through rocks warmed by the Earth's geothermal heat flow, hundreds of meters below the Earth's surface (see Chapter 11 for a discussion of the Earth's internal heat). For example, the water that flows from Warm Springs, about 90 kilometers (60 miles) south of Atlanta, Georgia, enters the Hollis Formation at Pine Mountain at an average temperature of 16.5°C (62°F). After a subterranean journey of 3 kilometers (2 miles) along a curving path that descends to a depth of 1100 meters (3300 feet) under Pine Mountain, it emerges at Warm Springs under artesian conditions with a temperature of 36.5°C (98°F).

An *intermittent* surface emission of hot water and steam is called a **geyser** (from the Icelandic *geysir,* meaning "to rush forth"). Geysers occur where groundwater descends through fractured permeable rock, is warmed by an underlying heat source (a shallow magma chamber or a body of young, warm igneous rock), and then is pushed up by steam under great pressure (Fig. 15-16). The pressure exerted by the overlying water column raises the boiling point to more than 100°C (212°F). The expansion of the hot water pushes some fluid out of the geyser opening, which reduces the pressure on the deep water and thus lowers its boiling point. The deep water—already heated to temperatures exceeding its boiling point—then flashes to steam, forcefully expelling all the water from the column. A period of recharging and reheating then occurs, followed by another eruption. The time lapse between eruptions varies with the groundwater supply, heat supply, permeability of the rock, and complexity of the network of fractures that delivers the water to the heat source.

The best-known geyser fields lie in areas of current or recent volcanism. Volcanism at the divergent zone that bisects Iceland and at the subduction zone under New Zealand, for example, produces geysers in both those regions. In North America, geysers in northern California derive their heat from the subduction that fuels the volcanism of Mount Shasta and Lassen Peak.

In Yellowstone National Park, hot-spot volcanism powers the park's numerous geysers. Yellowstone's Old Faithful geyser, so named because of its once-remarkable punctuality, used to shoot a jet of steam and boiling water to a height of 50 meters (170 feet) every 65 minutes. Recent earthquake activity in the region, however, has altered the geyser's plumbing, affecting its recharge time. Nowadays Old Faithful erupts *on average* every 79 minutes, but intervals between eruptions may vary between 45 and 105 minutes—far less faithful.

Recent technological advances in optics have enhanced our understanding of how Yellowstone's geysers work. Since 1992, when miniature video cameras first became widely available, scientists at Yellowstone have been examining first-

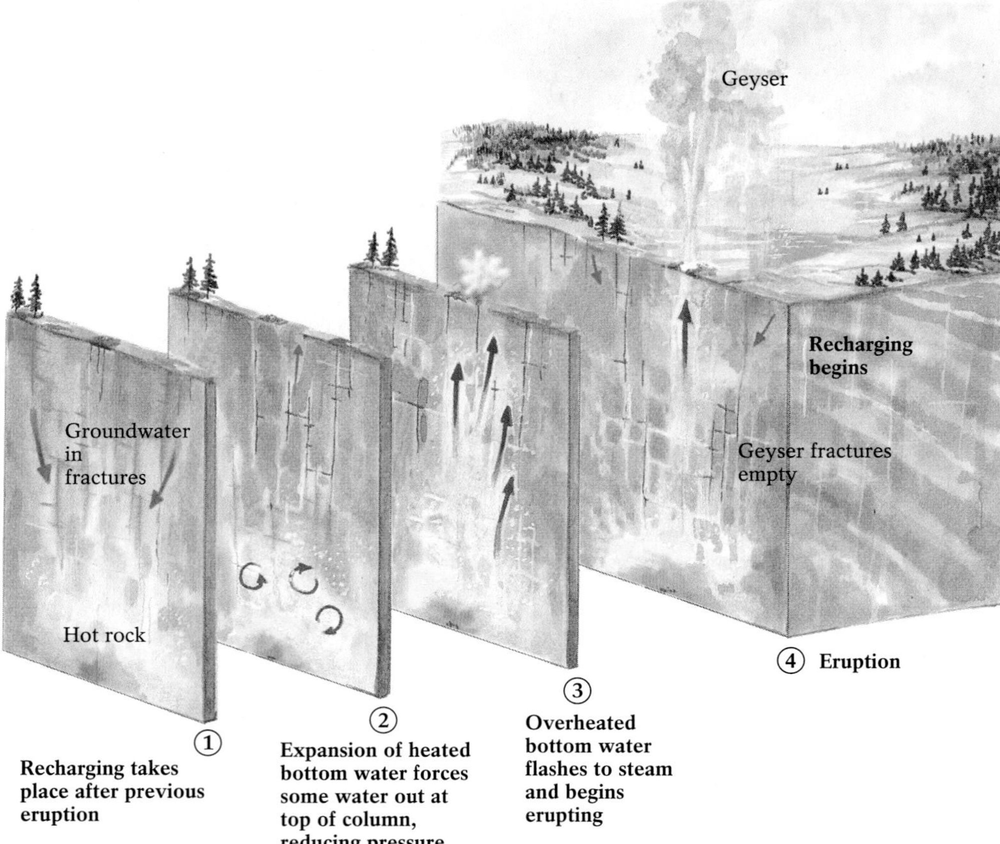
Geyser
Recharging begins
Groundwater in fractures
Geyser fractures empty
Hot rock
④ Eruption
③ Overheated bottom water flashes to steam and begins erupting
② Expansion of heated bottom water forces some water out at top of column, reducing pressure
① Recharging takes place after previous eruption

Figure 15-16 Geysers develop when groundwater encounters a shallow heat source and erupts at the surface as boiling water and steam. Water pressure at the bottom of the water column inhibits boiling initially. As water near the top of the column is pushed out by expansion of the heated bottom water, the pressure on the column diminishes and the water turns to steam. Photo: Strokkur geyser in Iceland.

hand Old Faithful's "intestines." They have produced a 15-meter (50-foot)-long profile of the configuration of Old Faithful's vent by using a 5-centimeter (2-inch) insulated video camera to monitor the vent between eruptions (Fig. 15-17). This investigation has uncovered a far more complicated vent shape than was originally expected. The vent widens and constricts in ways that may critically affect the geyser's behavior. The camera has also revealed vent walls riven with cracks that allow water to enter the column continuously at various depths. On a sobering note—the watchful eye of Yellowstone's camera has also shown that the vent continues to narrow as silica precipitates from the mineral-charged water. In the geological equivalent of "hardening of the arteries," the vent of Old Faithful appears destined to clog and the geyser fated to die, as have many of its neighbors in the Upper Geyser Basin. That day will be a sad one indeed for tourism at the Old Faithful Lodge.

Many countries with hot springs and geysers are developing ways to harness this geothermal energy to help meet human energy needs; we will discuss some of these efforts in Chapter 20.

Finding and Managing Groundwater

The first part of this chapter focused on the mechanics of groundwater flow—how it enters the ground, how it travels through subsurface materials, and how, in certain places, it erupts from the ground. Our introduction to this topic raised several key questions that—given your current understanding of how the groundwater system works—we can now address: How do we find groundwater? How can we safeguard our supply of this valuable resource and maintain its quality? Who owns it? The remainder of this chapter will be devoted to answering these questions.

Locating Groundwater

To find groundwater in most areas, we dig wells that intersect the water table. But how do we know where to dig? The best strategy for locating groundwater and digging productive wells begins with detailed knowledge of the configura-

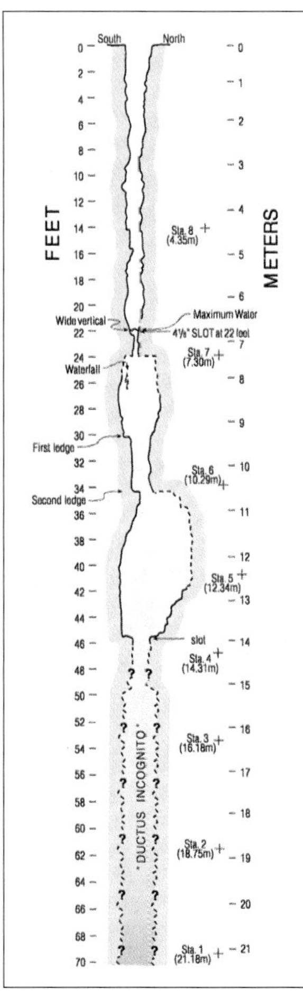

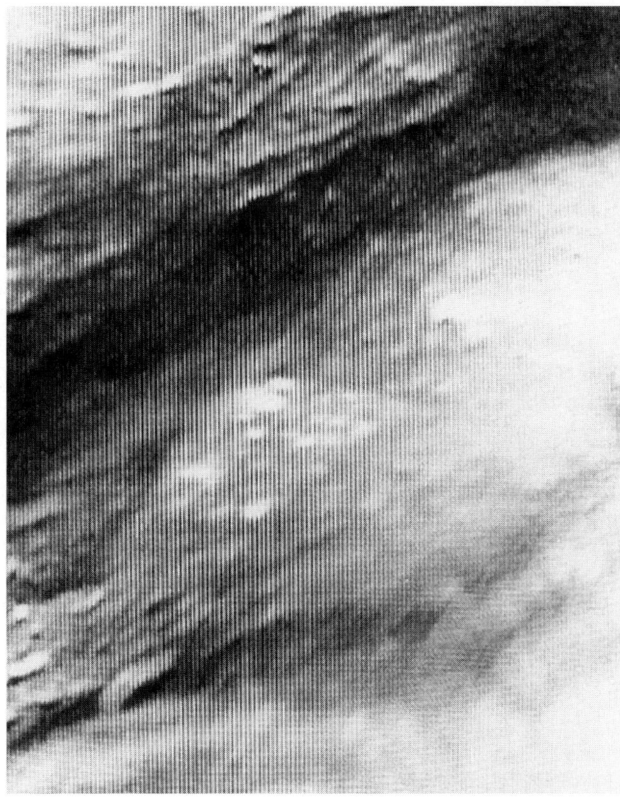

Figure 15-17 A heat-tolerant fiber-optic camera has penetrated Old Faithful Geyser in Yellowstone National Park, revealing its complicated vent shape. The constriction in the vent, produced by the ongoing buildup of precipitated silica, may soon choke off Old Faithful's geyser stream.

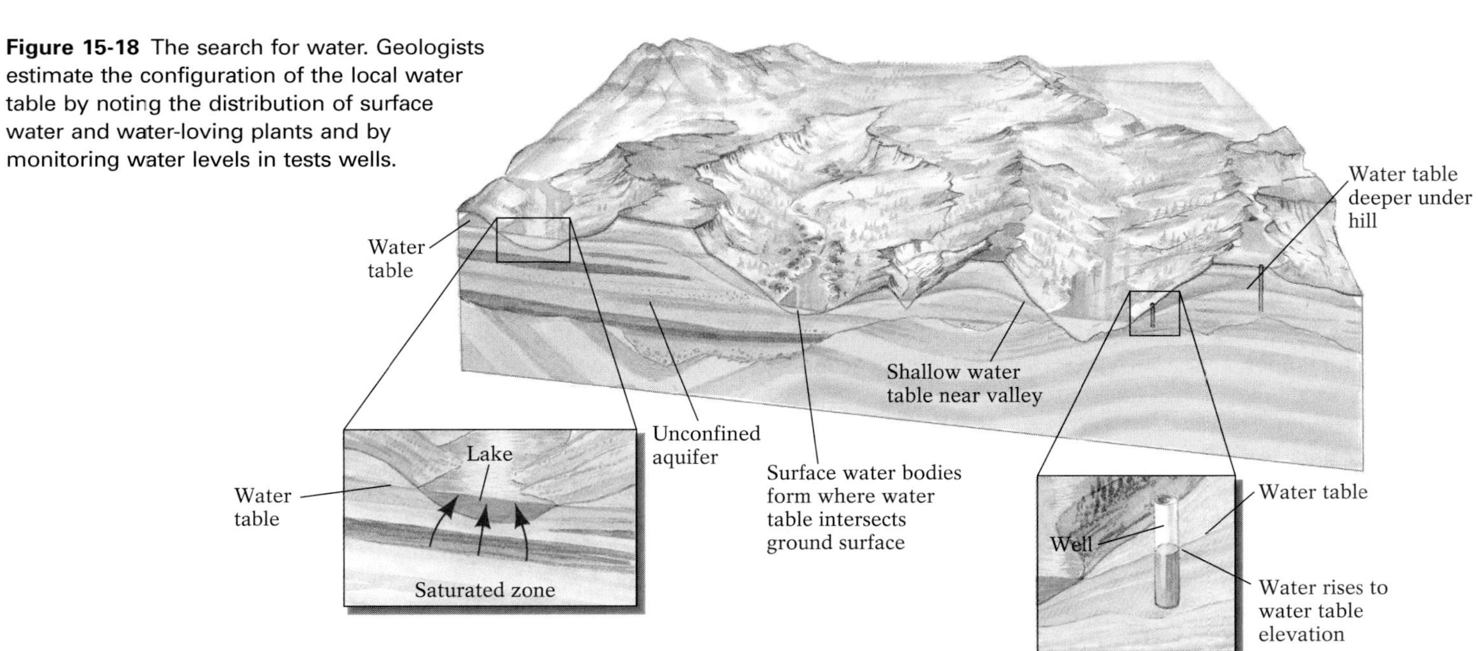

Figure 15-18 The search for water. Geologists estimate the configuration of the local water table by noting the distribution of surface water and water-loving plants and by monitoring water levels in tests wells.

tion of the local water table. Figure 15-18 depicts the variety of ways by which geologists attempt to locate the water table in an unconfined aquifer. They begin by mapping the distribution of surface water, because most bodies of surface water represent an intersection with the water table. Thus geologists plot all of an area's lakes, year-round streams, natural springs, and swamps. A topographic map showing the locations of these features provides a key to the underlying water table.

An area's vegetation also provides clues to the location of its water table. Some plant species have extensive, deep root systems for capturing water far below the surface;

Highlight 15-1 Should You Hire a Water Witch?

There are more than 25,000 water witches in the United States today—and a multitude of people who swear by them. An early description of the art of water witching may even appear in the Bible: "And Moses lifted up his hand, and with his rod he smote the rock twice; and water came forth abundantly, and the congregation drank and their beasts also" (Numbers 20:11).

Water witches walk through an area carrying a forked stick (or a switch, rod, or wire) in their hands until the stick seems to twist, dip, or jerk uncontrollably downward toward the place where underground water will be found. They claim that the presence of water induces the behavior of the stick.

Are these people malicious frauds? Probably most are not: They truly believe that they possess a supernatural ability to "feel" the presence of water. Many dowsers don't even charge for their services. But can they really find water? Skeptics suggest that the water witch manipulates the rod by holding it in delicate balance so that the slightest small-muscle movement turns it downward. Controlled studies have compared the success rate of water witches to the results obtained by chance and found no significant difference. One study in Australia compared geologists' efforts to find water with those of water witches and concluded that the witches caused twice as many dry holes to be dug as the scientists. At Iowa State University, water witches invited to find water along a prescribed course across the campus could not locate the water mains under their feet.

Water witches do often locate water. Geologists believe their success may reflect the fact that groundwater is virtually everywhere beneath our feet in most humid climates. Other successes may be due to a dowser's years of experience at finding water in a specific region and to some geological common sense. Over the years, a witch, like most observers, would learn that wells dug in valleys produce water more often than those dug on hilltops, and that certain plants flourish where water lies near the surface. The dowser may even acquire an intuition about groundwater flow and the relationship between groundwater and permeable rocks.

As experienced water witches walk the landscape, they probably process the real information necessary to ascertain the presence of water. Perhaps when they notice the optimal conditions for groundwater, their muscles flex subconsciously and the rod responds.

others have shallow root systems and grow only where abundant water lies close to the surface. Identifying such plants enables geologists to estimate the depth of the water table.

Even with these clues, it may still be necessary to drill a series of small test wells to identify the best place for the final well. Because water in an uncased well (that is, one that is not lined with a casing) dug in an unconfined aquifer rises to the level of the water table, tests wells will show the depth of the water table at particular places. With these data and with the knowledge that the water table lies deeper under hills than under valleys, we can reasonably estimate the water table's configuration.

Some people forego the measured scientific approach in favor of hiring water witches, or *dowsers,* to search for water. Water witches claim to locate water by using forked willow sticks or metal rods. Can a water witch really find water with such devices? Highlight 15-1 provides some insights.

Threats to the Groundwater Supply

The groundwater supply in a locality becomes threatened when the population increases its demand for water and lowers the water table. Rural villages have modest groundwater needs and usually have been able to meet them by withdrawing water from shallow wells in unconfined aquifers or by tapping perched water. Many cities, however, have been forced to find new sources of uncontaminated groundwater. Residents in the suburbs have also had to drill deeper wells to reach higher-yield confined aquifers. As urban sprawl continues, incorporating both suburban and rural areas, water needs rise sharply in response. Population growth, the arrival of new industries, the expansion of agriculture, and other changes in the local economy have increased the need for water so dramatically that groundwater supplies in many areas are being severely taxed.

Consider the example of the imperiled Everglades of central and southern Florida. Although the Everglades appear to be broad areas of standing surface water, they are actually fed by groundwater that is flowing, sometimes at an imperceptible rate, towards Florida's southern coast (Fig. 15-19). Increased pumping of groundwater, in response to both rapid population growth and greater exportation of water from the Everglades to Florida's farmlands and coastal cities, has drastically lowered the regional water table. Today, much of the once-continuous, wet, marshy surface remains dry much of the year, with fires in the dried-out vegetation and organic muck smoldering for months.

When groundwater withdrawal exceeds recharge, the water table may become lowered over a broad geographic range, land located above the depleted water source may

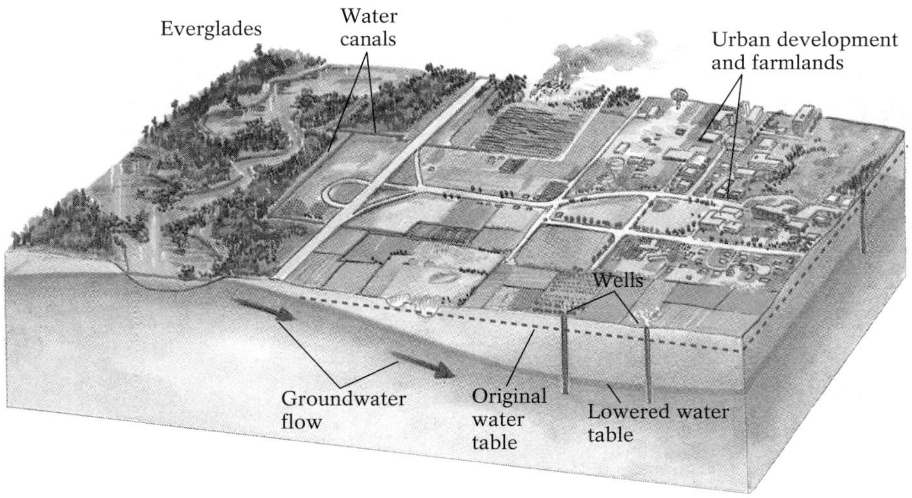

Figure 15-19 Overuse of groundwater. The water table beneath southern Florida once intersected the land surface across a broad area, creating the Everglades. Recently, withdrawal of water from Florida's groundwater reservoir has lowered the water table and dried out much of the Everglades. The lesson is clear: When the water table sits at the surface, it takes only a slight alteration in its height to produce far-reaching ecological disruption.

sink, or salty marine water may infiltrate the freshwater supply.

Groundwater Depletion Withdrawal of water from a well draws down the water table around the well, creating a conical recess in the water table called the **cone of depression** (Fig. 15-20). You can simulate such a depression by inserting a drinking straw a few millimeters into a milkshake. When you begin to drink, the surface of the milkshake dips around the straw; when you stop drinking, the milkshake flows back immediately, eliminating the depression. Because groundwater flow is so slow, however, a cone of depression in a water table does not refill immediately. Instead, it persists as long as groundwater withdrawal continues.

In moist, temperate regions, cones of depression remain insignificant as long as domestic use is moderate and natural recharge exceeds the modest rate of groundwater withdrawal. In areas with major irrigation or substantial industrial activity, however, large overlapping cones of depression can develop. For example, the demands of a water-intensive industry can create a cone of depression that lowers the water table over a sizable area, leaving neighboring wells dry (Fig. 15-21). As a result, neighboring water users must drill deeper (and more expensive) wells to access the water below the cone of depression.

In many areas of North America, the rate of groundwater recharge can no longer keep pace with our increased demand for groundwater. When groundwater is withdrawn at a rate that exceeds natural recharge, it is essentially being "mined." Eventually the water table will fall below the bases of local unconfined aquifers. At that point, the aquifers are essentially depleted. Groundwater mining is a particularly serious issue in arid regions such as Phoenix, Arizona, where extensive irrigation and enormous losses to evaporation place an impossible demand on the region's groundwater reservoir.

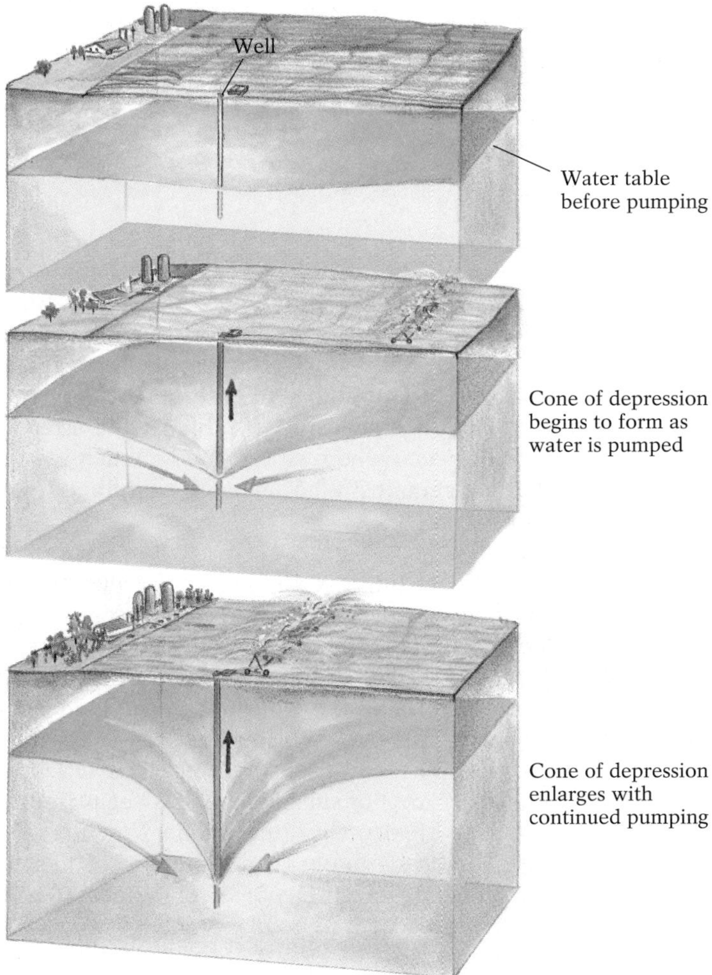

Figure 15-20 The water table around a well is drawn down when water is pumped out, creating a cone of depression.

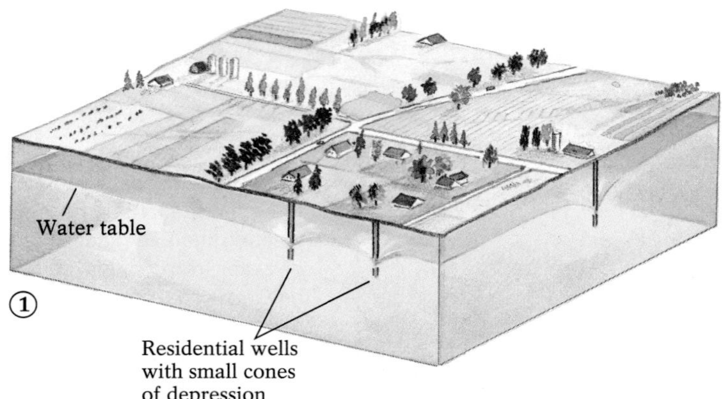

Figure 15-21 The effect of development on a water table. Stage 1: The community consists of farms and some suburban homes having small wells. Stage 2: Industrial development begins, replacing some of the farms, and groundwater withdrawal increases. Stage 3: A large industrial complex covers much of the farmland, lowering the water table to a level below the depth of residential wells.

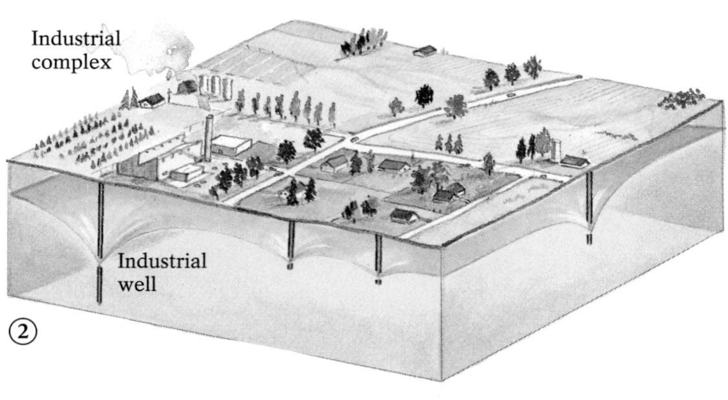

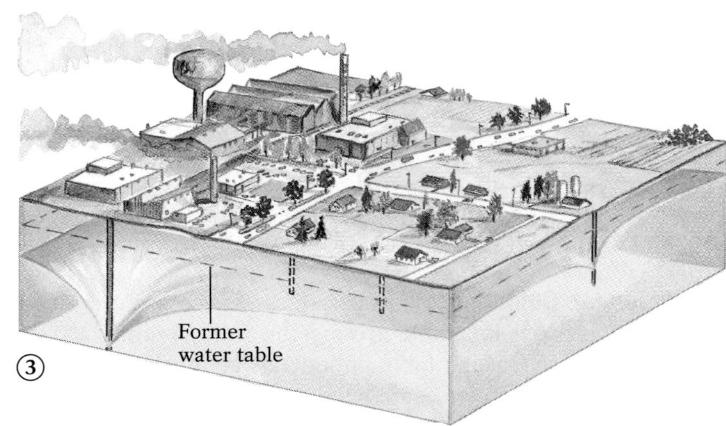

The effect of a single overdrawn well on neighboring wells raises an important legal issue: Who owns groundwater? Unlike other natural resources that stay in one place (such as gold, or coal, or diamonds), groundwater flows from one property to another. Our ability to determine the direction of groundwater flow enables us to estimate the size of this water supply. But even when we know how much water is there, the question remains: Who owns it? Highlight 15-2 provides some answers.

Land Subsidence When more groundwater is withdrawn from an aquifer than is replenished by recharge, less water becomes available to support the overlying load of rock, sediment, and soil. As the weight of this material pushes down, it compresses the aquifer, and the elevation of the land surface drops in a process known as **subsidence.** Excessive withdrawal of groundwater has led to subsidence in New Orleans [more than 2 meters (7 feet) of subsidence], Las Vegas [more than 1 meter (3 feet)], Mexico City [more than 7 meters (23 feet)], and central California [more than 8 meters (26 feet)]. By displacing segments of the surface and subsurface, subsidence damages structures such as roads, buildings, and water and sewer lines.

In coastal areas, subsidence may cause flooding during high tides and storms. For example, Houston, Texas, and Venice, Italy, are both coastal cities built on unconsolidated deltaic deposits. After years of withdrawing groundwater for domestic and industrial uses, these deposits and their aquifers have become compressed, causing the land surface to subside (Fig. 15-22). Venice, which has subsided more than 3 meters (10 feet) during the last 1500 years, is plagued by frequent flooding from the Adriatic Sea. Invading saltwater has severely damaged many of the city's historic and architectural treasures. The Houston–Galveston area, one of North America's fastest-growing metropolises, has subsided 1 to 2 meters (3.3–6.6 feet) since 1906. This process has compounded flooding problems in this city, which is frequently battered by hurricanes.

Saltwater Intrusion Aquifers in coastal communities, particularly those on islands and peninsulas, face a constant threat from the intrusion of saltwater. At the same time that a coastal aquifer is infiltrated by precipitation falling on nearby lands, it may be invaded by salty marine water from the seaward side of the landmass. Because freshwater is less dense than saltwater and therefore floats above the marine water, the freshwater–saltwater boundary is usually well defined. As long as groundwater withdrawal remains moderate, natural recharge provides a constant pool of freshwater above the saltwater. Any increase in withdrawal that disturbs the recharge–discharge balance allows more saltwater to enter the aquifers, where it may eventually completely replace the freshwater (Fig. 15-23). The result—residences and businesses that were formerly supplied by freshwater wells must now import their water.

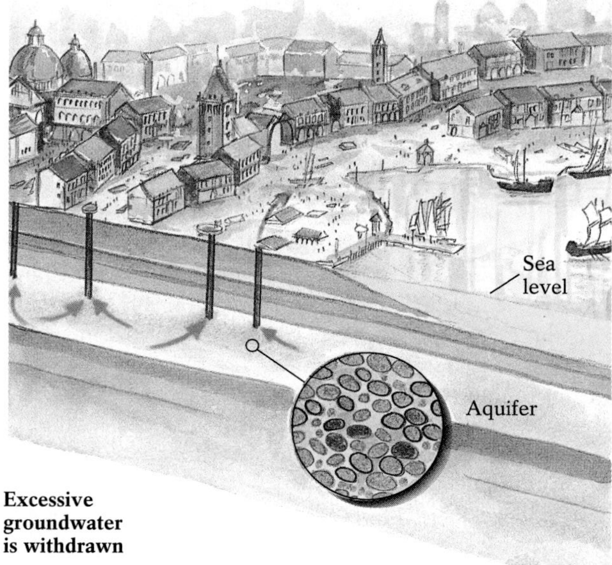

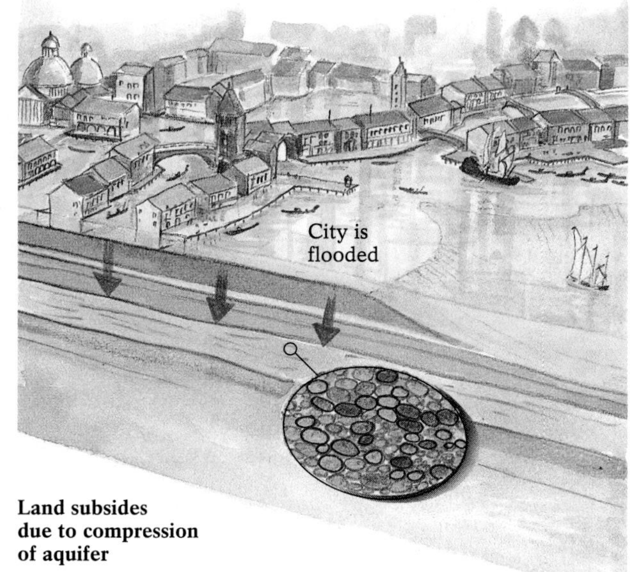

Figure 15-22 Subsidence in coastal areas. (above) Excessive groundwater withdrawal in a coastal city such as Venice, Italy, compresses its aquifer and causes the land to sink below sea level. (photo) In Venice today, canals serve as streets. Recent efforts to repressurize Venice's aquifer through artificial recharge (pumping water into the aquifer) and to regulate the withdrawal of groundwater have slowed its subsidence rate somewhat, but at a considerable cost.

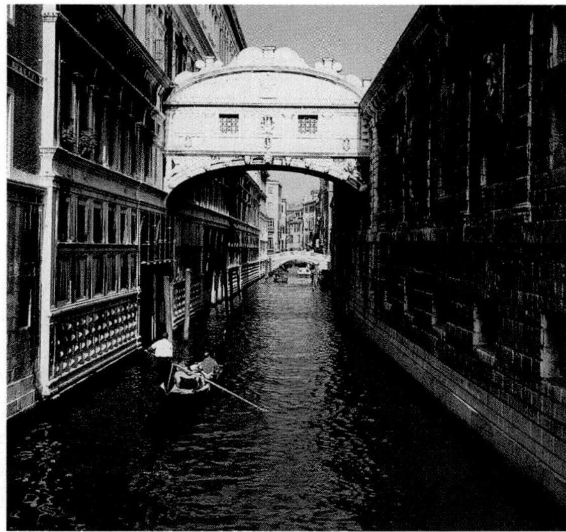

Figure 15-23 Saltwater intrusion of an aquifer in a coastal setting. Excessive withdrawal of groundwater allows marine water to penetrate aquifers at such sites, and saltwater sometimes flows from the faucets in coastal residences as a result.

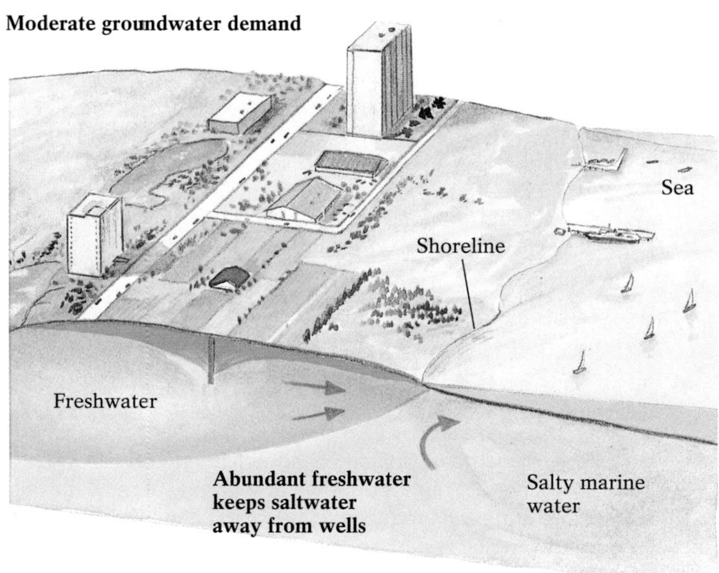

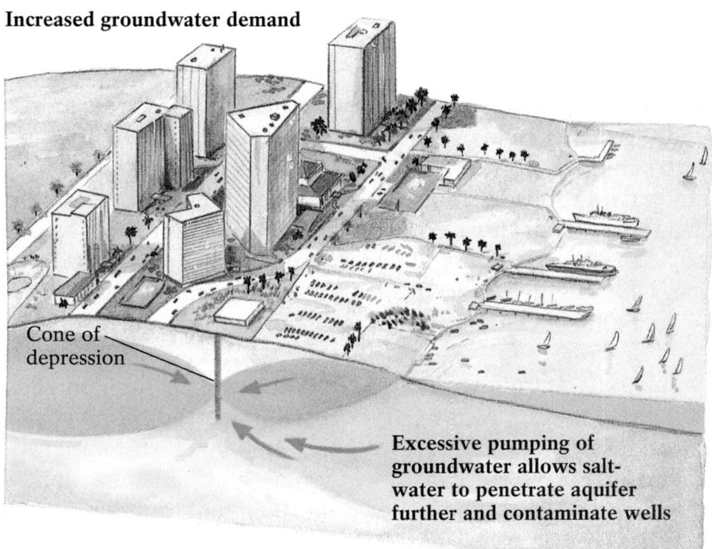

Highlight 15-2 Who Owns Groundwater?

The United States has not established a uniform code outlining groundwater rights. Instead, legal ownership of water is determined by state and local statutes and by the application of a few broad underlying principles. In most of the arid western states, a doctrine known as *prior appropriation* gives priority to the earliest users of groundwater, with latecomers having subordinate rights. With an accurate chronology of first use, few disputes can arise over rights to groundwater under this doctrine. During a shortage, the last user of the groundwater becomes the first to lose it. In most of the eastern United States, the *riparian doctrine* gives all users equal (and virtually unrestricted) rights, regardless of the point at which their ownership began.

Can property owners legally exploit their groundwater to the detriment of their neighbors? In colonial America, water disputes were decided by the traditional English Rule of Capture, or Absolute Ownership, which stated that the water under the land is part of the land. This rule allowed any property owner to use as much water as she or he saw fit, even to the detriment of other people. Only if damage to others could be shown to result from malicious intent would the excessive user be held liable.

In most states today, landowners can use water *only on the land from which it was taken,* and they are liable for any sale or transfer of water from their land that harms a neighbor's water supply. Also fundamental to the U.S. rules of water use is the principle of "reasonable and beneficial use." Domestic use (within a household) has always been considered reasonable and beneficial, as has watering livestock in cattle-raising states and irrigation in the arid West (but not where no compelling need to irrigate exists). As industrialization of the nation progressed, manufacturing joined the list of reasonable uses. In California, the use of water to mine gold is still classified as reasonable (a relic of the Gold Rush), as is the use of water for oil exploration in Texas and Louisiana.

In recent years, a few states, including California, have established a doctrine of correlative rights. Under this doctrine, each user's share of groundwater is proportional to her or his share of the overlying land. Again, the principle of reasonable and beneficial use is applied. Thus the answer to the question "Who owns groundwater?" depends in some cases on the state in which the question is asked.

During the 1930s, the high groundwater demands of the 3 million residents of Brooklyn, New York, converted that city's freshwater aquifer into an Atlantic Ocean saltwater aquifer. Since then Brooklyn has been forced to import its freshwater from upstate New York—at considerable expense. In neighboring Nassau County, New York, rapid suburbanization in the 1950s and 1960s allowed saltwater to intrude into many of Long Island's wells. Residents of the county must now pay for their freshwater and for the pipelines that transport it from the mainland. Coastal communities along the Gulf of Mexico, the Carolinas, Georgia, Florida, and California are similarly threatened by invading seawater.

Dealing with Groundwater Problems

Groundwater geology, or *hydrogeology,* is one of the most rapidly growing fields in the earth sciences. Hydrogeologists try to solve the subsurface water problems we just described. They have proposed some remedies to groundwater mining, subsidence, and saltwater intrusion, and a few of their solutions have been successfully implemented. Most such solutions seek to enhance local recharge, reduce withdrawal and discharge, or develop alternative water sources.

Enhanced Recharge We now have the means to supplement inadequate natural recharge in a community and thereby increase the local groundwater supply, although such remedies tend to be costly. We can, for example, grade a steep slope (see Chapter 13) so that water infiltrates the ground and replenishes storage basins or depleted aquifers, rather than running off into streams. In areas of moderate to heavy rainfall, communities can construct open recharge basins that hold rainwater and gradually release it into the groundwater system (Fig. 15-24).

In drier regions or areas where withdrawal consistently exceeds precipitation, freshwater can be imported and

Figure 15-24 A recharge basin/Little League field in Long Island, New York. To counteract high groundwater withdrawal rates, basins such as this one are excavated to collect precipitation and release it into the groundwater system gradually.

Table 15-1 Water Use in an Average Household

Activity	Volume of Water Use
Dishwashing	40 liters (10 gallons)
Toilet flushing	12 liters (3 gallons)
Showering	80–120 liters (20–30 gallons)
Bathing	120–160 liters (30–40 gallons)
Washing-machine load	80–120 liters (20–30 gallons)
Drinking and food preparation	4–8 liters (1–2 gallons)

pumped into the ground to replenish dwindling groundwater supplies. This technique is used in and around Chicago, where the considerable natural recharge is nevertheless inadequate to meet the needs of the city's large population. To reduce the number of shortages, engineers pump freshwater from Lake Michigan into the ground. Similarly, surface water from California's High Sierras is transported by aqueduct to the arid San Fernando Valley of Los Angeles, where it is injected into the groundwater system to maintain the position of the water table. In Long Island, New York, even the wastewater from industrial air-conditioners is recaptured and returned to the ground.

Some communities with overtaxed groundwater supplies choose to borrow or buy groundwater from areas of abundance—an arrangement known as an *interbasin transfer.* But how can water be transported for long distances without incurring substantial losses through seepage and evaporation? One solution is to build pipelines to carry the water. For example, the rapidly growing city of Denver, Colorado, receives much of its water through pipelines and tunnels that stretch from the western slopes of the Rockies. Similarly, New York City draws most of its municipal water supply from rural upstate regions through an extensive pipeline and reservoir system.

In the cases of New York and Denver, groundwater appropriations are primarily *intrastate* transfers. When such water transfers must cross state lines and international boundaries, however, the costs typically skyrocket and politics may intervene. How can the donor community be convinced to relinquish its water? When southern California covets water from Oregon's and Washington's Columbia River, the outcry from Portland residents can be heard in San Diego. The Great Lakes states have already declared their lakes to be "off-limits" to thirsty southwestern states. And it will take remarkable international cooperation—and money—to convince the underpopulated, water-rich areas of Canada to share their water with high-demand areas in the arid regions of the United States and Mexico.

Conservation During dry spells, local authorities often declare water emergencies. They may enact such emergency measures as prohibiting automobile washing, limiting lawn watering to early morning or evening hours (to reduce evaporation) and to two or three times each week, serving water in restaurants only upon request, and turning off faucets while brushing teeth and shaving. Table 15-1 shows how much water is consumed by some common household activities. New bathroom-fixture technologies—low-flow toilets and showerheads—use much less water; some cities even offer rebates on water bills if customers install these devices. The authorities may also mandate industrial conservation, with stiff penalties being imposed for excessive use.

Figure 15-25 A desalinization plant on the Netherlands Antilles island of Curaçao, off the northwest coast of Venezuela.

Other Water Sources Another solution to water shortages may be *desalinization,* the removal of salt from seawater. In this process, seawater is either evaporated or boiled, whenever possible by using inexpensive solar radiation. The resulting salt-free vapor condenses and can be collected. Alternatively, the seawater may be passed through a filter that removes the salt. Desalinization is costly, but it can be relatively cost-effective in arid lands with extensive seacoasts, especially given the value of freshwater in such locales. Desalinization plants, such as the one shown in Figure 15-25, have operated for some time in Israel and Egypt. In the future, the process may become economically feasible in many other nations.

Recently, another means of providing a new water source was the subject of an international symposium at Iowa State University. Would it be feasible to tow massive icebergs

Figure 15-26 Massive ice floes like this one, broken and drifted from an Antarctic ice shelf, may someday provide a source of fresh water to parched communities located at lower latitudes.

from Antarctica (Fig. 15-26) to water-desperate countries in the arid Middle East? Because 90% of the ice would melt or evaporate in transit, researchers proposed that only icebergs at least 10 kilometers (6 miles) across should be considered possible targets for towing. The cost and possible environmental repercussions of such a venture make it unlikely that the iceberg solution will represent a practical option in the foreseeable future.

Maintaining Groundwater Quality

Groundwater for drinking and most other human uses must be relatively pure. As we discussed in Chapter 5, because most groundwater is slightly acidic, it dissolves the soluble components of rock and sediments through which it passes. Thus groundwater naturally contains a variety of dissolved ions—some benign (such as those resulting from the dissolution of various carbonates and chlorides) and some potentially quite harmful (such as arsenic, mercury, and selenium).

Groundwater may also contain manufactured contaminants that have reached the groundwater system in various ways. On a small, local scale, virtually every household periodically disposes of toxic substances through the municipal refuse system or storm drains. A check of typical household trash finds half-used cans of paint, cleansers, and solvents. Remember that can of bug spray you threw out, the burned-out light bulbs, and the broken thermometer? Poisons and heavy metals from these materials may leach from city dumps into the groundwater system, posing a significant threat to groundwater quality and community health. The volume of these toxins, however, is small compared with agricultural and industrial wastes or the by-products of medical research. Among the most threatening—because they are often released in large quantities—are insecticides and fertilizers, salt and other chemical ice-retardants washed from roadways, carcinogenic industrial by-products from factories (such as PCBs), biological wastes from cattle feedlots and slaughterhouses, and sewage from overworked septic fields and broken sewer lines.

Pollutants enter the groundwater system when we dispose of wastes improperly, such as by placing them in receptacles that rust and leak, by tossing them haphazardly into poorly designed landfills, or by dumping them illegally on unmonitored land and allowing them to seep directly to the water table or into an aquifer. Groundwater contaminants are often detected only after they reach a well; by that time, however, they may have spread throughout the entire aquifer. In western Minnesota in 1972, for example, a number of employees of a company became seriously ill from arsenic poisoning, even though the firm did not use arsenic in its operations. Environmental detective work traced the arsenic to the company's new well. During the Dust Bowl years of the 1930s, the western plains had been struck by an infestation of locusts. To protect their crops, farmers at that time laced bran with arsenic and scattered it over their fields to kill the insects. In 1934, after the locusts disappeared, the farmers buried the unused bran. Forty years later, the Minnesota company's new well received recharge from this buried disposal site.

Once contaminants reach the groundwater reservoir, it may take many years to flush them out, because groundwater flows very slowly. In the last few decades, local, state, and federal governments have all enacted laws to monitor groundwater quality, and environmental protection agencies now regulate all varieties of waste disposal. The state of Montana, for example, strictly governs the sizes, materials, and construction methods used for septic tanks and drain fields.

Some human-made contaminants require extreme caution in their handling and disposal. Nuclear waste, for example, may cause illness, birth defects, and death, and it remains dangerously radioactive for thousands of years. For that reason, any disposal site for nuclear waste must be absolutely impermeable, so that leakage to the groundwater system cannot occur. The recent selection of Yucca Mountain, Nevada, as the prime candidate for the United States' nuclear-waste dump has focused considerable attention on that region's groundwater conditions. Highlight 15-3 (pages 444-445) discusses some of the key issues confronting nuclear-waste disposal.

Natural Groundwater Purification

Some groundwater systems are self-cleansing. Indeed, most deep groundwater passes slowly through rocks of low permeability and emerges in a relatively pure state. Three natural purification processes combine to eliminate such toxins as sewage bacteria, larger viruses, and other suspended solid:

- *Filtration.* Some contaminants adhere to clay particles as the water percolates through the soil.
- *Decomposition.* Some contaminants decompose completely by undergoing oxidation in the soil.
- *Bacterial action.* Many organic solids are consumed by various microorganisms.

Raw sewage that passes through soil, even if the soil comprises only about 30 meters (100 feet) of moderately permeable materials, is partially purified naturally. (This process actually accounts for the only treatment of sewage discharge from a septic tank or drain field.)

The cleansing ability of a soil, sediment, or rock depends primarily on whether its permeability allows impurities in the groundwater to adhere to mineral surfaces. If an aquifer is excessively permeable, water passes through it too quickly to remove contaminants. Geologists estimate the permeability of an unconfined aquifer by conducting a *percolation test.* In this test, they dig a hole, fill it with water, and observe how long it takes for the water to seep from the hole. If the water vanishes quickly—for example, within minutes or even hours—the soil may be too permeable to cleanse contaminated water. In rural areas, the municipal water-quality agency may not be willing to issue building permits unless percolation tests show that the soils are suitable for household septic systems.

Some rock types remove contaminants less effectively than others. Certain bedrock aquifers, such as those containing large joints, cavities, and faults, allow water to pass through too swiftly to be cleansed. Consequently, major pollution sources such as sewage plants, landfills, and cattle feedlots should be located where fine-grained impermeable soils overlie unfractured, low-permeability bedrock. Figure 15-27 compares two different landfill sites and predicts the likelihood of groundwater contamination for each.

The prairie pothole region of southern Canada and the northern United States (Fig. 15-28) is a 775,000 km^2 (300,000 mi^2) area of shallow glacially produced ponds and lakes that functions as a *natural* groundwater-treatment facility. Many of the chemical impurities introduced through the fertilization of the surrounding agricultural lands are filtered out or broken down chemically as water seeps from these pools and infiltrates the underlying soils and unconfined aquifers. In a recent study, this system was shown to remove 80–90% of the harmful nitrates introduced by fertilizers. Recently, however, nearly half of these beneficial wetlands have been lost—drained by surface ditching or underground conduits—principally to develop more agriculture land. Minnesota alone has lost 95% of its original prairie potholes. Elimination of this natural cleansing system will undoubtedly lead to a sharp decline in groundwater quality. How, then, can we improve the quality of our groundwater?

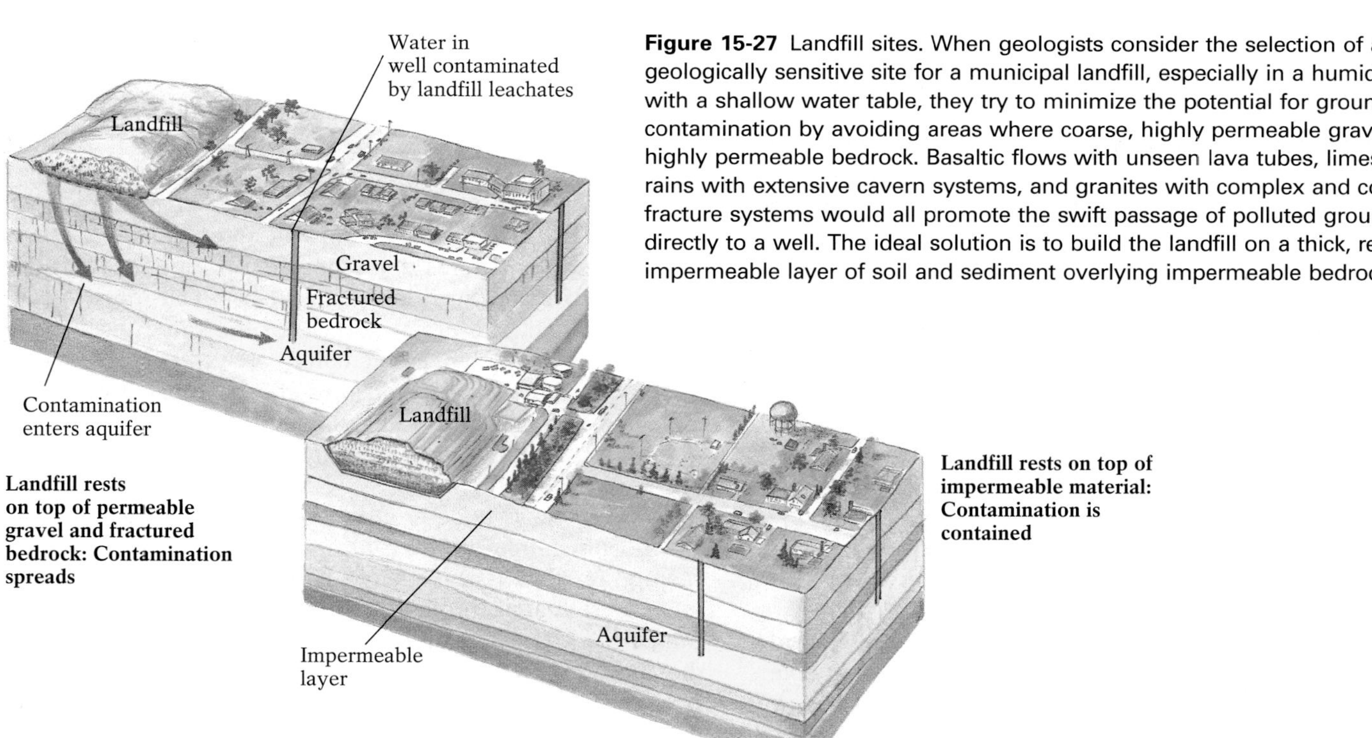

Figure 15-27 Landfill sites. When geologists consider the selection of a geologically sensitive site for a municipal landfill, especially in a humid climate with a shallow water table, they try to minimize the potential for groundwater contamination by avoiding areas where coarse, highly permeable gravels overlie highly permeable bedrock. Basaltic flows with unseen lava tubes, limestone terrains with extensive cavern systems, and granites with complex and connected fracture systems would all promote the swift passage of polluted groundwater directly to a well. The ideal solution is to build the landfill on a thick, relatively impermeable layer of soil and sediment overlying impermeable bedrock.

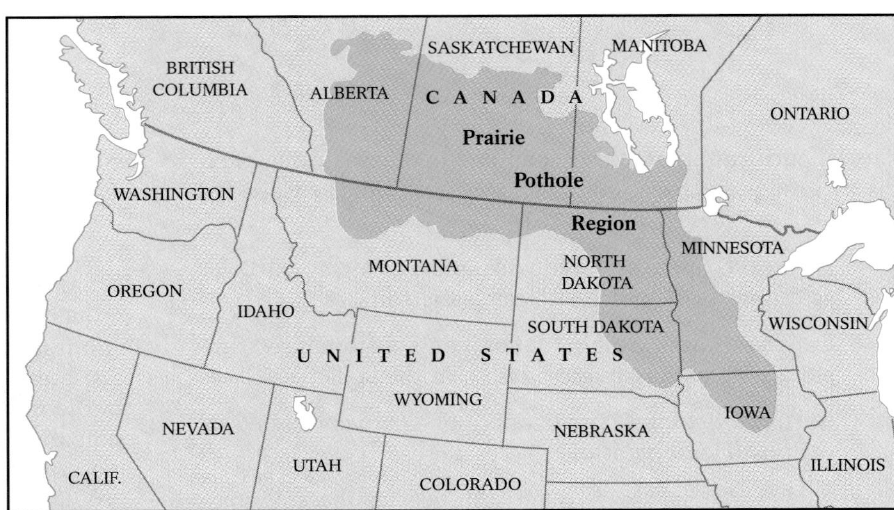

Figure 15-28 The water in these bodies of standing water—the Prairie Potholes of the Coteau des Prairie, near McClusky, North Dakota—are naturally purified as they pass through the glacially deposited soils; the contaminants introduced during agriculture, such as fertilizers and pesticides, are largely filtered out or broken down chemically.

Cleansing Contaminated Aquifers

When pollution is found in an aquifer, the first step is to identify and, if possible, eliminate the pollution source. We can pinpoint the site at which contaminants enter the groundwater system by taking samples from local streams and investigating the waste-disposal policies of local industries. If the source of the chemical pollutants was buried years earlier, or if the contaminants have entered a complex regional-groundwater network by illicit dumping some distance away, detection may prove more difficult and time-consuming.

The second step is to discontinue groundwater use while cleansing of the aquifer takes place, assuming that another water source is available. Cleansing can take decades, given the average rate of groundwater flow. Several methods for cleansing groundwater systems exist. The tainted water can be pumped out, or clean water can be pumped in to dilute the contaminants. An experimental groundwater-cleansing project has been under way for several years at the Massachusetts Military Reservation. There, polluted groundwater passes through a barrier of iron filings designed to react with and decontaminate the water. Introducing chemicals into the groundwater system that either remove or neutralize the contaminants is also being attempted; such treatment is seldom feasible, however, because groundwater usually flows too slowly for the agents to percolate throughout the system. Instead, contaminated groundwater is often pumped to a surface holding pond for treatment and then returned to the groundwater system. Whatever method is employed, restoration of a polluted aquifer to good health is both costly and technically demanding.

Groundwater is not only essential to the well-being of human communities, but also accomplishes a considerable amount of geological "work" as it passes slowly through certain types of subsurface rock. In the next chapter, we will examine the principal geological product of groundwater flow—the caves that develop when naturally acidic groundwater percolates through soluble carbonate bedrock.

Chapter Summary

Surface water—flowing in rivers, standing in lakes, or falling as precipitation—enters the ground wherever topography, geological composition, and vegetation cover permit infiltration. The optimal conditions for development of a large reservoir of groundwater include a well-vegetated, gently sloping or near-level landscape composed of fractured bedrock or coarse, well-sorted sediment.

Water enters the ground under the influence of gravity, which carries it downward in soil and rock until the fluid fills every available connected pore space. Surface evaporation, plant use, and inadequate groundwater recharge may encourage the formation of a **zone of aeration** near the surface; this zone contains many air-filled pores. Below it lies the **zone of saturation,** in which all available pore spaces are water-filled. The upper surface of the saturated zone is called the **water table.** Immediately above the water table, within the zone of aeration, lies the **capillary fringe;** the molecular attraction of water molecules to one another and to mineral surfaces causes water to rise from the water table into this region.

Water flows underground where the geological materials are sufficiently porous and permeable. **Porosity,** the percentage of pore space relative to the total volume of soil, rock, or sediment, provides a measure of how much water can be held by earth materials. **Permeability** is a measure of the ability of rock or sediment to transmit a fluid; the presence of connected fractures in solid bedrock and the coarse, well-sorted texture of unconsolidated sediment and soil both increase permeability.

Groundwater flows between two points because of differences in **potential,** the energy that arises from variations in the points' elevation and water pressure. The rate at which the potential changes over a lateral distance is called the **hydraulic gradient.** Because of variations in permeability, every material has a unique ability to transmit groundwater, a property known as its **hydraulic conductivity.**

The depth of a regional water table reflects the local topography and prevailing climate. If a layer of impermeable material lies within the zone of aeration, infiltrated water may be retained locally, saturating part of that zone and forming a **perched water table.**

Aquifers are permeable, water-bearing bodies of geologic materials. Aquifers not overlain by impermeable cap rock are called unconfined aquifers. Confined aquifers, found at greater depth, are sandwiched between impermeable rock layers called **aquicludes** or low-permeability rock layers called **aquitards.** When an aquifer's water is under high pressure from large elevation differences between recharge and discharge sites, it may rise above the level of the aquifer and gush from the ground. Geologists describe such self-pumping aquifers as **artesian.** The level to which such pressurized water would theoretically rise, in the absence of friction, is the aquifer's **potentiometric surface.**

The regional water table can often be identified from the location of surface-water features such as rivers, lakes, and **natural springs,** places where the Earth's surface intersects the water table and groundwater flows from the surface naturally. By drilling test wells, and by applying the knowledge that the water level in wells approximates the local groundwater table, geologists can estimate the location of the regional water table. Groundwater, particularly where it is converted to steam by contact with shallow magmas or warm, recently crystallized igneous rocks, may erupt from the ground as **geysers.**

Human activity can disturb the groundwater system through overwithdrawal and contamination. In areas experiencing rapid growth and regions where formerly rural land has become suburban or urban, the population growth and intensified industrial development may cause groundwater demand to rise sharply, leading the water table to drop significantly. Large **cones of depression**—local depressions in the water table—may develop around wells in such areas. Lowered water tables make it necessary to dig ever-deeper wells. They also promote **subsidence,** as depleted aquifers become compressed and the land surface falls. In coastal regions, excessive groundwater withdrawal may cause saltwater to infiltrate subsurface aquifers.

To compensate for critical groundwater shortages, local communities may transfer water from distant drainage basins or institute rigorous conservation plans. Concern about groundwater quality is also growing, as various pollutants, including hazardous and toxic wastes, appear in our water supply. Government agencies are paying greater attention to preventing aquifer contamination and eliminating existing sources of pollution.

Key Terms

zone of aeration (p. 423)
zone of saturation (p. 423)
water table (p. 423)
capillary fringe (p. 423)
porosity (p. 424)
permeability (p. 424)
potential (p. 425)
hydraulic gradient (p. 425)
hydraulic conductivity (p. 425)
perched water table (p. 427)
aquifer (p. 427)
aquicludes (p. 428)
aquitards (p. 428)
artesian (p. 428)
potentiometric surface (p. 428)
natural springs (p. 430)
geyser (p. 431)
cone of depression (p. 435)
subsidence (p. 436)

Questions for Review

1. Describe three factors that affect groundwater recharge.

2. What is the fundamental difference between the zone of aeration and the zone of saturation?

3. What is the difference between primary and secondary porosity? List three factors that control a rock's primary porosity. Describe a rock that would have high permeability, but low porosity.

4. What is a water table? What is a perched water table?

5. What is an aquifer? Where would you look for an unconfined aquifer? A confined aquifer?

6. Draw a simple sketch showing the conditions necessary to create an artesian groundwater system.

7. What are some geological circumstances that might create a natural spring?

8. Explain why geysers may erupt at regular time intervals.

9. Describe two negative consequences of unwise groundwater management in coastal settings.

10. Briefly describe two ways to enhance dwindling groundwater supplies and two ways in which groundwater might be purified.

For Further Thought

1. How might the Earth's groundwater be affected if the greenhouse effect raised global temperatures significantly? What might happen to the Earth's groundwater if the destruction of tropical rainforests continues?

2. Speculate about groundwater conditions during the Earth's first billion years when plate tectonic activity was probably more intense.

3. In the landscape illustrated in Figure 15-18 (p. 433), what are some features—other than those labeled—that might provide clues to the configuration of the water table? Where else would you decide to drill test wells?

4. How would you solve the groundwater problems facing the Florida Everglades?

5. Suppose you are charged with satisfying the groundwater needs of Los Angeles, California, in the twenty-first century. Name five problems you would face and the solutions you would propose.

Highlight 15-3 Yucca Mountain—A Safe Place for N-Wastes?

In December 1997, hydrogeologists at Richland, Washington, discovered high-level radioactive waste—principally from the spent fuel rods from nuclear power plants and the by-products of nuclear-weapons manufacture—leaking from temporary underground storage tanks. This fluid has already entered the local groundwater system and threatens to discharge into the mighty Columbia River, the Northwest's largest waterway and one of its principal sources of irrigation water, commercial transportation, salmon spawning, and recreation. This alarming discovery dramatizes the nation's urgent need for a permanent "safe" nuclear-waste repository, one that will isolate *without fail* such hazardous waste from contact with all biological organisms for at least 10,000 years.

In 1976, nearly 20 years after the advent of commercial nuclear power, a long-overdue federal program was launched to study the feasibility of underground nuclear-waste storage. (The spent fuel rods from the 40 years of civilian nuclear power have been temporarily stored in cooling pools located adjacent to the nation's reactors.) In 1983, the U.S. Department of Energy selected nine locations in six states as potential sites for the nation's waste repository. After three years of exhaustive study, only three candidates remained: Deaf Smith County, Texas; Hanford, Washington; and Yucca Mountain, Nevada. In December 1987, the government pared the list to only one site—Yucca Mountain. Since then, more than 1400 Department of Energy employees, including 700 scientists and engineers, have migrated to southern Nevada to study this site. By the time the study phase is completed in 2001, more than $6 billion will have been spent on the site, or just a small fraction of the total amount necessary to build the repository.

Why Yucca Mountain? We can answer this question by examining the site's groundwater conditions: specifically, how water moves through the mountain, and how water might affect the repository. The climate of southern Nevada is extremely dry and, as a result, the local groundwater table is unusually deep. These factors, coupled with the composition of the local bedrock, have led many scientists to conclude that the nuclear waste can be safely isolated at Yucca Mountain.

Yucca Mountain is located in the southern portion of the Great Basin, where only about 15 centimeters (6 inches) of rain falls each year . . . and most of it quickly evaporates in the hot desert sun. The site's scientists believe that only a very small fraction of this already-meager rainfall soaks into the ground and could actually reach the underground repository. Moreover, the water table at Yucca Mountain lies roughly 540 meters (1800 feet) beneath the surface, and the site's plan calls for the repository to be built within the unsaturated zone at the 300-meter (1000-foot) level. Thus the repository will be perched about 240 meters (800 feet) above the water table. According to the scientists, the great depth to the repository significantly limits the prospects of water reaching and corroding it; the additional depth to the water table further reduces the likelihood that any leaked material will reach the groundwater system.

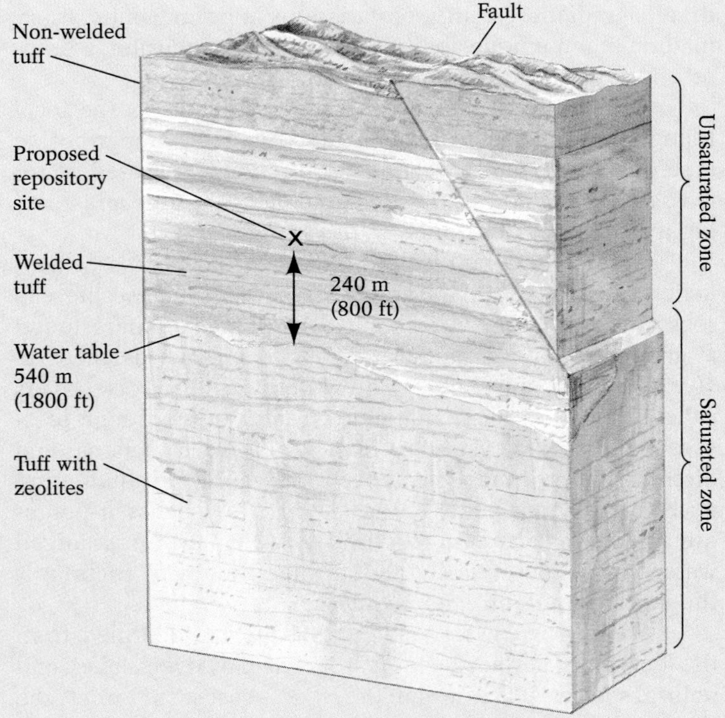

Figure 15-29 The stratigraphy at Yucca Mountain, Nevada. Note that the proposed nuclear-waste repository is located 300 meters (1000 feet) from the potential surface-water infiltration and is isolated from the region's zone of saturation by an additional 240 meters (800 feet) of welded volcanic tuff.

The geology at Yucca Mountain, which is composed largely of welded tuffs (ash-flow deposits, discussed in Chapter 4), may also bolster the site's ability to impede a waste leak from long-distance migration. The tuffs (Fig. 15-29) are rich in zeolites, water-rich minerals with the capacity to bond with and thus restrain migrating radioactive-waste particles.

Despite the positive elements that have drawn this army of scientists to Yucca Mountain, some skeptics argue that the site will never be *completely* safe and thus should be abandoned. Critics cite the following concerns about the proposed repository location:

- The area is riddled with faults, many of which have been active recently. In June 1992, for example, a magnitude 5.6 earthquake occurred at Little Skull Mountain, only 20 kilometers (12 miles) from Yucca Mountain. The quake

Figure 15-30 These deep veins of precipitated calcite on the eastern slopes of Yucca Mountain indicate more extensive groundwater infiltration in the past, a cause for concern in the minds of some geologists who are assessing the site's development as a safe repository for nuclear waste.

shattered glass, cracked plaster, and dislodged concrete blocks at the site's headquarters. How would the underground repository fare during an even larger quake? Would such a tremor breach the containment tanks, releasing the radioactive wastes?

- The economic potential of the site—believed by some to be a source of precious metals, oil, and natural gas—may someday entice commercial interests to explore it. Such exploration could release radioactive wastes into the environment.
- Climatic change to wetter conditions during the next 10,000 years could dramatically alter groundwater flow. If more water percolates through Yucca Mountain, the increased flow in the unsaturated zone could accelerate corrosion of the storage tanks, increasing the likelihood of leakage. Increased percolation would also cause the water table to rise, reducing the travel time between the repository and the site's saturation zone.

Ongoing concerns about the site's past groundwater history were heightened substantially with the excavation of Trench 14, a 100-meter (330-foot)-long rectangular furrow plowed along the mountain's eastern slope. The trench walls are riven with thick, deep-running veins of precipitated calcite, a residue of greater groundwater flow in the past (Fig. 15-30). To some site hydrogeologists, the presence of these veins suggests that the deep position of the water table is far less stable than it appears, and that the water table has risen dramatically in the past—perhaps all the way to the surface, and thus *through* the proposed repository site. A recent report by a team of scientists from the National Academy of Sciences refutes this hypothesis, however. These scientists argue that the veins were deposited by infiltrating rainwater and not a rapidly rising water table. Such dissenting opinions clearly show that controversy continues to swirl around the state of the hydrological conditions at Yucca Mountain.

What should be done about the Yucca Mountain repository? Can it ever be judged truly "safe"? At present, it is our nation's only developing alternative to leaving decades of accumulated life-threatening nuclear waste at 70 temporary storage facilities in 35 states—such as the failing one at Richland, Washington. The fate of this site poses a decidedly tough dilemma for scientists and legislators, but one that must be resolved immediately.

16

Caves and Karst

Caves are unique geological structures that form when groundwater, flowing beneath the Earth's surface, slowly dissolves and gradually enlarges minute openings in soluble bedrock. The result is sometimes spectacular enough to attract the attention of tourists, artists, photographers, and scientists alike (Fig. 16-1). Why would anyone choose to spend time in a wet, dark, 50-centimeter (20-inch)-high crawlway located hundreds of meters beneath the Earth's surface and flavored by the pungent aroma of decaying bat droppings? For most speleologists—the scientists who study caves—as well as for nonscientist "cavers," this compulsion is fueled by a sense of adventure and an eagerness to explore the unknown. The vast numbers of undiscovered caves constitute one of the few remaining frontiers of geological exploration on Earth and have inspired increased investigation in recent decades. As a result, we have expanded considerably our knowledge of these hitherto uncharted regions.

The appeal of caves is certainly nothing new—they have always held a special fascination for humans. Caves were widely used for shelter or ceremonial purposes throughout southern Europe tens of thousands of years ago (Fig. 16-2). Greek mythology tells the tale of Zeus's birth in a cave. The Japanese sun goddess Amaterasu sought refuge in a cave, plunging the world into darkness. In Anglo-Saxon lore, King Arthur and his knights and their hounds are believed to be still slumbering in a Welsh cave, waiting to be summoned into battle once again.

Today, the information yielded by caves is invaluable to anthropologists, biologists, archaeologists, geologists, and other scientists. Caves contain records of ice-age climate changes and human physical and cultural development. They harbor the remains of extinct animals such as the cave bear, the woolly mammoth, and other organisms that evolved as cave dwellers. These remains enable us to study how species evolve and adapt to unique environmental conditions.

Many paleoecologists—those geologists and biologists who strive to reconstruct past climates and environments—are fascinated by the collected refuse left behind by much smaller cave dwellers. Arid climates—such as those in the high deserts of the American West—allow moisture-sensitive

Figure 16-1 Spectacular cave formations in Carlsbad Caverns National Park, New Mexico.

Figure 16-2 This famous panorama of bulls, horses, and other creatures was painted on a cave wall in Lascaux, France, some 15,000 years ago. The cave in which it appears may have been a ceremonial site for performing rituals designed to ensure successful hunts. The prehistoric painters crushed ochre and hematite to produce yellow, red, and brown powdered pigments; crushed manganese yielded black, dark brown, and violet pigments. The pigments were applied directly to the damp limestone cave walls and ceilings.

organic deposits to be preserved in caves for millennia. Some such deposits, amassed by foraging rodents, contain pine cones, needles, and other plant remains that convey information about climatic conditions in the past. These helpful collections of rodent memorabilia are called *packrat middens* (Fig. 16-3).

The features created when groundwater dissolves soluble rocks underground or when surface water dissolves exposed soluble rocks are known collectively as **karst** (from the Slavic *kars,* meaning "a bleak, waterless place"). Beneath the surface, karst may comprise a single cavern or a complex network of caves, some with spectacular deposits of precipitated limestone. At the surface, karst may appear as a sponge-like landscape, pocked with so many circular depressions that a stream can barely cross it without plunging into a depression and disappearing from the surface. In this chapter, we will focus on the unique geology associated with soluble rocks, paying special attention to the environmental sensitivity of this geological setting.

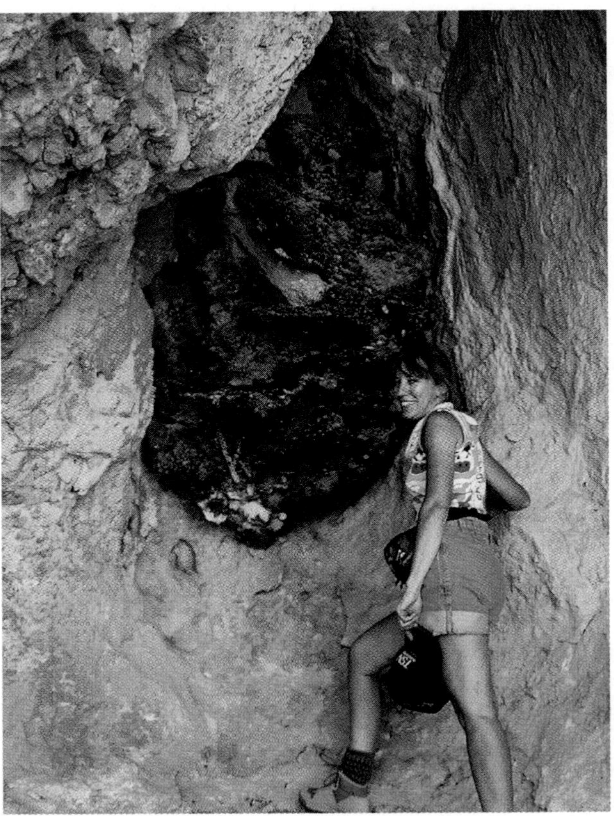

Figure 16-3 Packrats living in this cave amassed this layered pile of debris. The organic material within this refuse pile, or midden, is datable by carbon-14 isotope dating.

How Limestone Dissolves

Although almost all rocks can dissolve under certain conditions, very few rocks dissolve easily and rapidly. Evaporites (see Chapter 6), such as gypsum and halite, are minerals that do dissolve readily in water (see Chapters 2 and 6). Caves near Las Vegas, Nevada, for example, formed in gypsum deposits. But because evaporites are relatively rare, and because in moist environments they dissolve *completely,* they seldom produce karst.

Limestone—by far the most abundant soluble rock—is the material from which nearly all caves and karst form. The mineral calcite is the most soluble component of limestone. A limestone, however, may consist of as little as 50% calcite if it incorporated other components during its formation. In marine environments, for example, fine-grained iron-rich clay, derived from weathering of marine basalts, is a common secondary component of limestone. When the calcite in such impure limestone dissolves completely, the clay becomes concentrated at the surface as a deep-red soil. This *terra rosa* ("red earth") often overlies fresh, undissolved limestone (Fig. 16-4).

Figure 16-4 This deposit of terra rosa in Indiana consists of oxidized, iron-rich clay that remained behind after the calcite fraction of limestone bedrock dissolved.

Although limestone is relatively insoluble in *pure* water, it dissolves readily in natural rainwater, which acquires carbon dioxide (CO_2) as it passes through the atmosphere and especially the soil. (The concentration of CO_2 in soils, produced largely by the respiration of soil organisms and decomposition of the remains of plants, may reach 100 times the levels found in the atmosphere.) Water and CO_2 combine to form carbonic acid (H_2CO_3), which attacks and dissolves limestone by chemical weathering, through the process of carbonation. In the carbonation reaction, carbonic acid reacts with calcite ($CaCO_3$) in limestone to produce calcium ions (Ca^{++}) and bicarbonate ions (HCO_3^-). (The equation for this reaction is found in Chapter 5, page 134.) When these ions are carried away in solution, voids develop in the bedrock. As groundwater with dissolved calcium bicarbonate flows through unconsolidated sediment, some calcite may precipitate in the pore spaces between grains and act as cement to consolidate the grains, forming such common detrital sedimentary rocks as sandstone. Most of the calcium bicarbonate produced by limestone dissolution, however, is carried out to sea. There, marine organisms use it to form their shells and skeletons; alternatively, the calcium bicarbonate may be eventually precipitated as inorganic limestone.

On a practical note, water containing dissolved calcium bicarbonate ("hard water") makes it difficult to lather soap and leaves an unappealing residue on anything it washes. It also produces the hard, white, scaly calcium carbonate deposits found in pipes, water heaters, and tea kettles. Water softeners that exchange sodium ions for calcium ions can, however, solve hard-water problems in household water supplies.

Factors Determining Dissolution Rates

Limestone dissolution occurs most rapidly at and immediately below the Earth's surface where large concentrations of carbonic acid are usually found. Farther underground, water is more likely to be saturated with dissolved calcium bicarbonate and therefore less able to dissolve additional calcium carbonate. For these reasons, most limestone dissolution occurs within a few hundred meters of the surface.

The extent of a rock's fracturing, jointing, and bedding also determines the rate of limestone dissolution. Bedrock generally dissolves more rapidly when it includes extensive fractures or joints, which enable more water to infiltrate and increase the surface area open to dissolution. Similarly, thin beds form caves more rapidly than thick beds, because they tend to incorporate more numerous and more closely spaced joints and fractures. In time, as water continues to infiltrate joints, fractures, and bedding planes, a barely perceptible web of minute cracks can evolve into an intricately branched network of caves and connecting passageways.

Climate and topography also affect dissolution rates. Climate determines precipitation and temperature, which determines the amount of unfrozen water available to dissolve rock. It also promotes or retards several secondary factors that affect dissolution rates; abundant rainfall and warm temperatures, for example, support the growth of lush vegetation, whose decay adds substantial amounts of acid-forming CO_2 to the soil. We can see the effect of topography on moderately sloped terrains, where surface water is able to infiltrate and flow as groundwater from hills to valleys, enhancing the rate of calcium carbonate dissolution. On steep slopes, surface water typically runs off too rapidly to infiltrate.

Karst is rare in polar regions, where most water occurs in frozen form and the thick permafrost layer acts as a barrier to water infiltration. In addition, the relatively thin soil and regolith of polar regions generally contain little CO_2 because of slow plant-decay rates.

Likewise, karst is rare in hot desert regions, where the sparse vegetation and lack of water *preserve,* rather than dissolve, limestone. Caves found in these environments are usually relics of some past period of greater precipitation and more temperate climate.

Estimating Dissolution Rates

The many factors that control calcium carbonate dissolution vary significantly from place to place. Consequently, we are more successful when we estimate dissolution rates on a very small scale. For example, in a churchyard cemetery in cool, damp England, gravestones cut from fossil-rich limestone dissolve at a rate of about 5 millimeters (0.2 inches) per 100 years. Because the fossil sea lily stems in these rocks are more resistant to dissolution than the surrounding fine-grained limestone, they project above the gravestone's surface (assuming

(a)

(b)

Figure 16-5 Differential dissolution. **(a)** A less-soluble fossil of an ancient sea lily projects above a partially dissolved limestone surface. **(b)** The height of the limestone pedestals beneath their protective sandstone boulders provides a measure of the local dissolution rates in Yorkshire, England.

that dissolution of the fossils themselves has been minimal; Fig. 16-5a). If we know the year that a particular gravestone was cut, and we then measure the difference between the relief of the fossils and the stone's surface, we can estimate the rate of the limestone's dissolution.

We can use a similar approach—on a somewhat larger scale—to determine some *regional* rates of dissolution. Yorkshire, England, for example, was covered by advancing ice sheets until some 12,000 years ago. As the glaciers melted, they left behind large boulders of relatively insoluble sandstone that protected the limestone surface immediately beneath them. Eventually, the unprotected surface dissolved, and the protected areas became 50-centimeter (20-inch)-high pedestals on which the boulders now rest (Fig. 16-5b). The dissolution rate of this limestone is approximately 4.2 millimeters (0.17 inch) per 100 years (50 centimeters divided by 12,000 years), similar to the rate determined for the gravestone fossils about 100 kilometers (60 miles) away.

Another means of determining limestone dissolution rates relies on a micro-erosion meter that can directly measure lowering of the surface to within 0.0005 millimeter (0.0002 inch). Measurements in arid regions of Australia indicate a dissolution rate between 1.6 and 2.9 millimeters (0.63–0.11 inch) per 100 years. In one of Australia's temperate regions, which has high precipitation and extensive vegetation, measured dissolution rates have approached 10 millimeters (0.39 inch) per 100 years—a rate that equals or exceeds the rates of some other surface processes, such as landsliding and stream cutting.

Finally, we can estimate dissolution rates by measuring the volume of solutes in groundwater discharge. In many areas today, limestone is steadily dissolving within minute cracks below the water table. One such site is Big Spring in the Missouri Ozarks near Poplar Bluff. There, millions of liters of groundwater emerge at the surface every day, carrying 190 metric tons of dissolved limestone into the nearby Current River. Although much of the dissolution occurs at or near the surface, where soil CO_2 concentrations are high, the large quantity of dissolved material suggests that underground voids and perhaps a cave network may be forming beneath Big Spring.

Caves

Caves are natural underground cavities, the below-surface expression of karst. They are the most common geological products of limestone dissolution. Almost every state and province in North America contains some type of cave, and geologists believe that more than half the caves in North America have not yet been discovered. Major systems can be found in the Black Hills of South Dakota; in northern Florida; in the mountains of Montana and Wyoming; in the Guadalupe Mountains of New Mexico and Texas; in the Appalachians of Pennsylvania, New York, Maryland, Virginia, West Virginia, Tennessee, and Alabama; in the Ozarks of Arkansas, Missouri, and Oklahoma; and in the Great Lakes region of Indiana, Illinois, Iowa, and Minnesota. In central and eastern Canada and much of New England, caves tend to be relatively smaller and less extensive than elsewhere in North America, largely because much of the bedrock in these regions is crystalline and contains little limestone.

Some of the continent's most spectacular caves have been designated as national parks and monuments to preserve them for the enjoyment of future generations. Others,

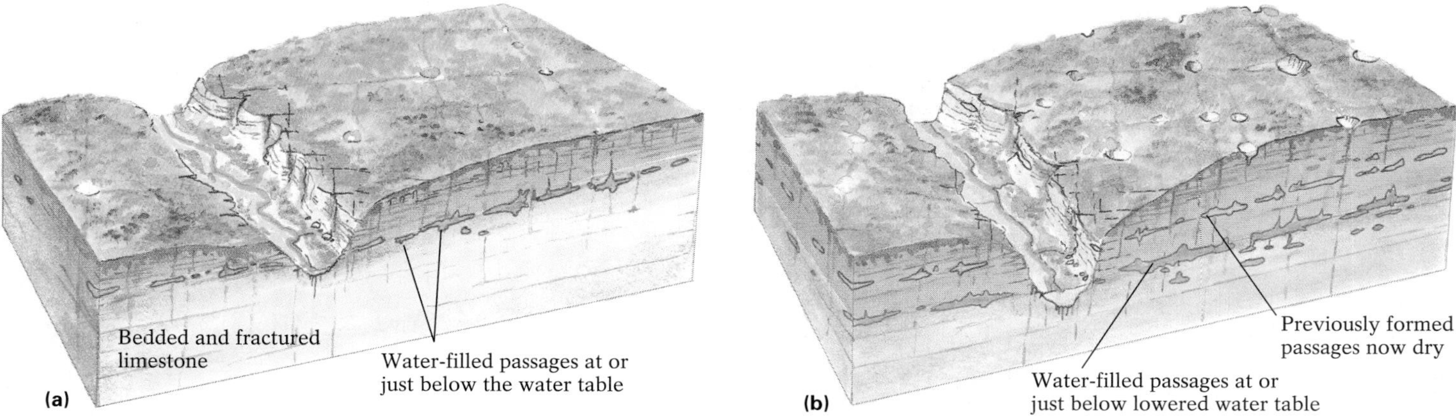

Figure 16-6 The two-stage process of cave formation. **(a)** Stage 1: Acidified groundwater fills all joints, fractures, and bedding planes below the water table and dissolves limestone and forms large caverns and connecting passageways. **(b)** Stage 2: The water table drops below the caverns and passageways, leaving them in an open-air environment.

however, have suffered mercenary, and often destructive, treatment. Amateur collectors have hacked out large quantities of delicate and rare cave crystals. Caves have been dynamited to create more commercially accessible entrances. In Kentucky, gypsum crystals from caves have been mined for use as paint pigments. Nevertheless, most of North America's caves still survive, primarily because their relative inaccessibility has protected them from human disturbance.

Cave Formation

The early stages of cave development progress slowly. Thousands of years may pass before small amounts of percolating water dissolve enough bedrock to enlarge minute cracks to a significant extent. As the cracks expand, water infiltrates in larger amounts and flows more vigorously, vastly accelerating the rate of dissolution. As the enlarging underground rivulets capture water from other narrower cracks, a few primary passages begin to dominate the underground drainage. Eventually, these hollows become the main caverns and connecting passageways of a growing cave system.

When caves first begin to form, dissolution occurs primarily in the zone of saturation, probably at or just below the water table. In that region, groundwater combines with inflowing CO_2-rich water that is not yet saturated with solutes and can therefore readily dissolve limestone. This period of intense dissolution along joints, fractures, and bedding planes below the water table forms the intricate honeycomb pattern of many major cave systems (Fig. 16-6a).

The second stage of cave formation occurs if the water table drops below its original level, leaving the cave high and relatively dry above it. At this point, the cave environment becomes exposed to the open air (Fig. 16-6b). With the water table lowered, erosion by fast-flowing subterranean streams may begin to carve potholes and canyon-like slots into a cave's bedrock floor (Fig. 16-7). Because the roof of the cave is no longer supported by water, it may collapse. One such cave-roof breakdown, in Indiana's large Wyandotte Cave, left a 40-meter (130-foot)-high pile of rubble on the cave floor.

Earthquakes can also damage cave structures. During the New Madrid earthquake of 1811 in Missouri, Mammoth

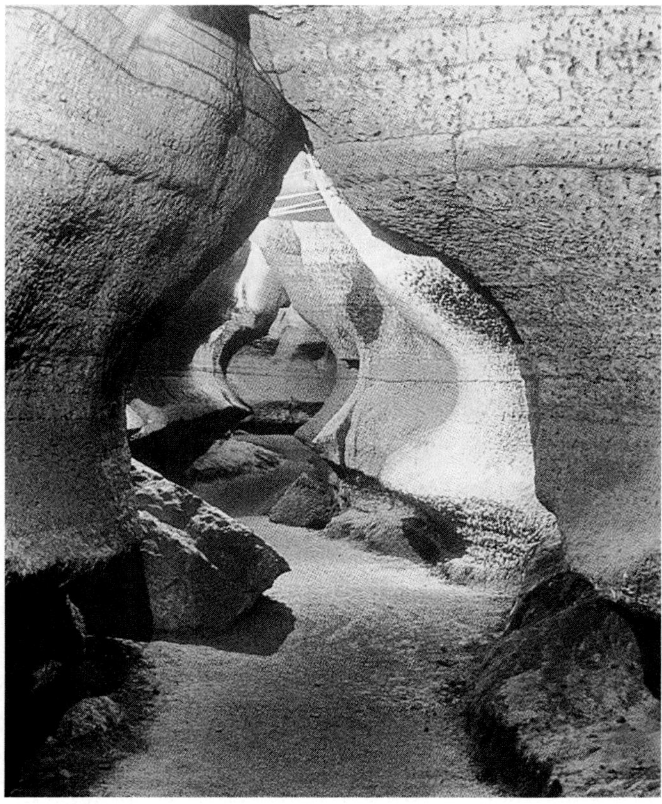

Figure 16-7 An underground river carved this deep channel by dissolution and abrasion.

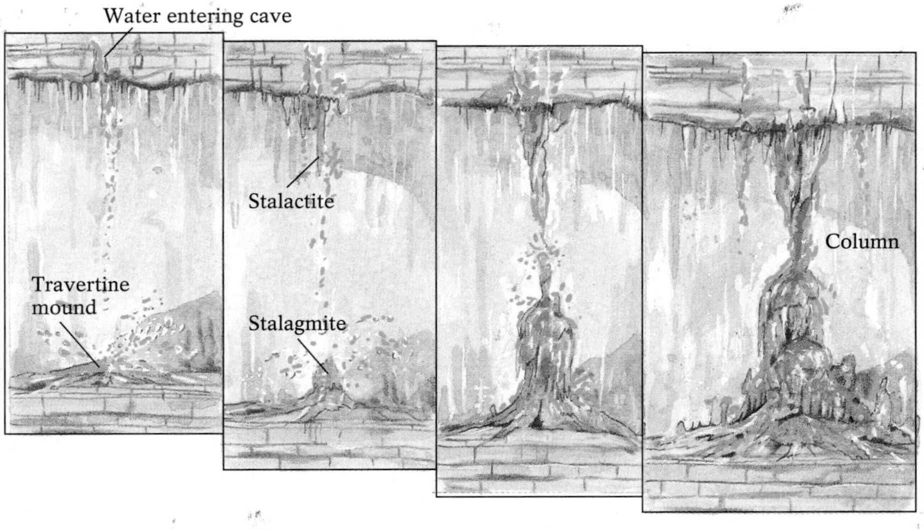

Water entering cave

Travertine mound

Stalactite

Stalagmite

Column

Figure 16-8 The creation of stalactites, stalagmites, and columns from precipitated travertine.

Cave—hundreds of kilometers away in Kentucky—experienced not only swaying cave walls, but also broken roofs that cascaded slabs of rock onto terrified workers in its saltpeter mines.

Cave Deposits

Once a network of cave chambers and connecting passageways has formed and the water table has dropped below the base of the system, water percolating downward from the surface will enter open-air spaces. Because this water is generally rich in calcium bicarbonate, two things can happen:

1. In the open-air environment of the cave, some water may evaporate, increasing the concentration of calcium bicarbonate in the remaining water until it becomes saturated.
2. Much of the CO_2 dissolved in the water may escape into the cave's air, reversing the carbonation process and converting carbonic acid into CO_2 and water (see Chapter 5).

The high calcium bicarbonate ($Ca(HCO_3)_2$) concentrations and the CO_2 loss cause calcium carbonate ($CaCO_3$) to precipitate from solution; the precipitate becomes deposited on cave surfaces and builds up a variety of formations known as **speleothems.** Speleothems consist largely of the rock **travertine,** the name applied to calcium carbonate when it forms cave deposits.

Stalactites and Stalagmites **Stalactites** (from the Greek *stalaktos,* meaning "oozing out in drops") are stony travertine structures, resembling icicles, that hang from cave ceilings. Stalactites form as water, one drop at a time, oozes from a crack or joint and enters a cave. As the water evaporates or loses carbon dioxide, it leaves behind a trace of calcium carbonate. Initially, the center of each growing stalactite takes the form of a hollow tube, like a soda straw, through which the next drop emerges. Eventually precipitated travertine clogs the tube, forcing the drops of water to find an alternative route into the cave, usually at the top or side of the stalactite. As the water drips down the stalactite's outer surface, it deposits more travertine and creates the irregular ornate shape of a typical stalactite.

Stalagmites (from the Greek *stalagmos,* meaning "dripping") are the travertine deposits that accumulate on the floors of caves from water that drips from their ceilings. Some of the water that drips from joints and cracks falls onto the cave floor, where it evaporates or releases CO_2, precipitating travertine. Eventually, these stalagmites may grow as high as 30 meters (100 feet). Sometimes they merge with stalactites to form single speleothems, called *columns,* that reach from floor to ceiling (Fig. 16-8). (Many people mix up stalactites and stalagmites. You can distinguish between them by thinking of the "t" in stalac*t*ites as representing a structure growing from the *top* of the cave. Similarly, the "g" in stalagmites represents a structure growing on the cave's *ground.*

Other Cave Deposits Speleothems come in a variety of other delicate and beautiful shapes (Fig. 16-9). Encrustations that form where water drips from the ceiling are called *dripstones.* Speleothems that form below a joint or crack in the ceiling of a cave may produce *banded draperies,* or *drip curtains. Rimstone dams* form when travertine precipitates over obstructions to water outflow from a cave pool. *Helictites* are

Figure 16-9 Various types of speleothems. **(a)** Stalactites in their early stages of formation resemble hanging soda straws. **(b)** Stalactites hanging from the cave ceiling and stalagmites growing upward from the floor merge to form a column. **(c)** Helictites form by capillary action and can grow in any direction. **(d)** Cave pearls form when layers of travertine precipitate around a sand grain or similar particle. **(e)** Giant cave popcorn forms when travertine precipitates at the surface openings of large speleothems. **(f)** A speleothem resembling a fried egg.

(a)
(b)
(c)
Banded draperies
Dripstone
Stalactite
Helictites
"Soda straws"
Flowstone
Column
Cave pool
Stalagmite
Rimstone dam
(d)
(e)
(f)

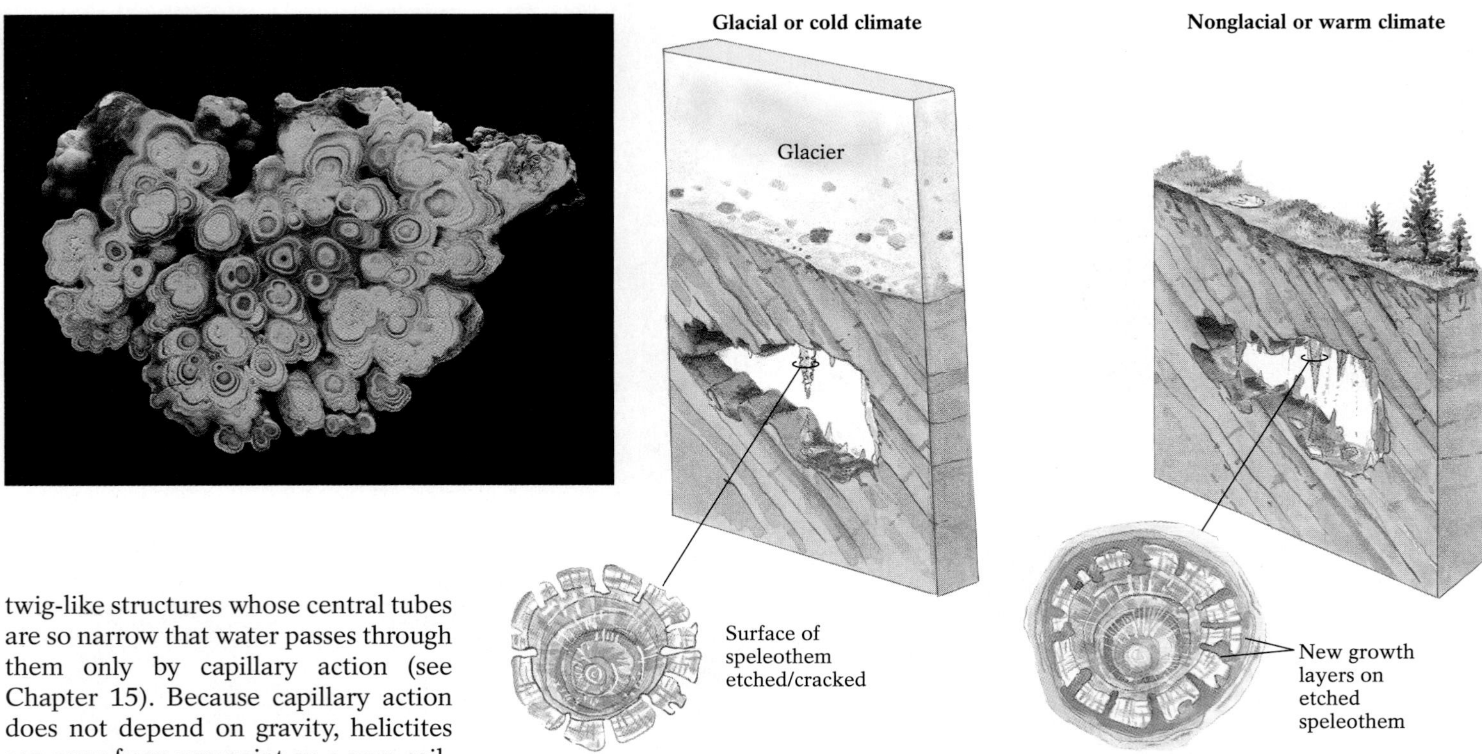

Figure 16-10 Speleothem growth is affected by climate. **(a)** During cold periods, when the ground is frozen or when ice covers the land above a cave, very little water can infiltrate the limestone bedrock, and speleothem growth essentially ceases. Without water, speleothems dry out, their surfaces crack, and cold winds from the surface spread a layer of dust on them. **(b)** During warm periods, water is free to infiltrate the ground and speleothem growth resumes. The time and duration of glacial and nonglacial periods can be estimated by dating the layers within speleothems. Photo: A sliced speleothem from Kokaweaf Cavern, California,

twig-like structures whose central tubes are so narrow that water passes through them only by capillary action (see Chapter 15). Because capillary action does not depend on gravity, helictites can grow from any point on a cave ceiling, wall, or floor, and in any direction.

Cave pearls, another type of unusual speleothem, form as calcium carbonate precipitates in concentric layers around a sand grain or other sand-sized particle. They may arise when water drips from the cave ceiling into a bicarbonate-rich pool of water. The constant drip perpetually agitates the pool's water, which releases carbon dioxide and precipitates travertine onto small fragments at the pool's bottom. The constant agitation keeps the growing pearl in motion, preventing it from adhering to any surface; the travertine can therefore be added in uniform layers.

Cave popcorn grows on a larger speleothem as water within the speleothem moves to its surface and evaporates, leaving small clumps of precipitated travertine around surface openings.

Speleothem Growth The same factors that enhance dissolution—acidification and infiltration of groundwater—also promote speleothem growth: solubility and porosity of surface materials, climate, topography, and vegetation. Conditions in caves in warm, wet, lushly vegetated regions (such as Virginia, Kentucky, and Florida) favor the formation of speleothems, which may grow as rapidly as 1 centimeter (0.4 inch) per year. In contrast, in regions with relatively impermeable, less soluble rocks, cold or arid climates, and limited vegetation (such as northern Alberta and British Columbia), speleothems may grow at a rate of less than 1 centimeter *per 100 years.* The 20-meter (65-foot)-long stalactites in Carlsbad Caverns in arid New Mexico have also grown at a rate of less than 1 centimeter per 100 years. Thus they record hundreds of thousands of years of growth.

Cut through a speleothem, and you can observe its growth record. A horizontal slice of a stalactite, for example, displays concentric layers of travertine. Although not annual records like tree rings, these layers do provide information about the stalactites growth history. In caves under landscapes that were glaciated during the most recent ice age, speleothems display cracked travertine layers. These cracked layers indicate interruptions in speleothem growth that occurred when the ground froze or became covered with ice, halting water infiltration. As a result, the caves dried out, and the speleothems became coated with dust blown about by cave winds. When the climate later warmed, the ice departed, the ground thawed, groundwater again infiltrated, and speleothem growth resumed. By applying uranium-isotope dating techniques (see Chapter 8) to the travertine on either side of a cracked, dusty layer within a speleothem, we can estimate the time and duration of a glacial period (Fig. 16-10). Studies in France and England, for example, indicate that

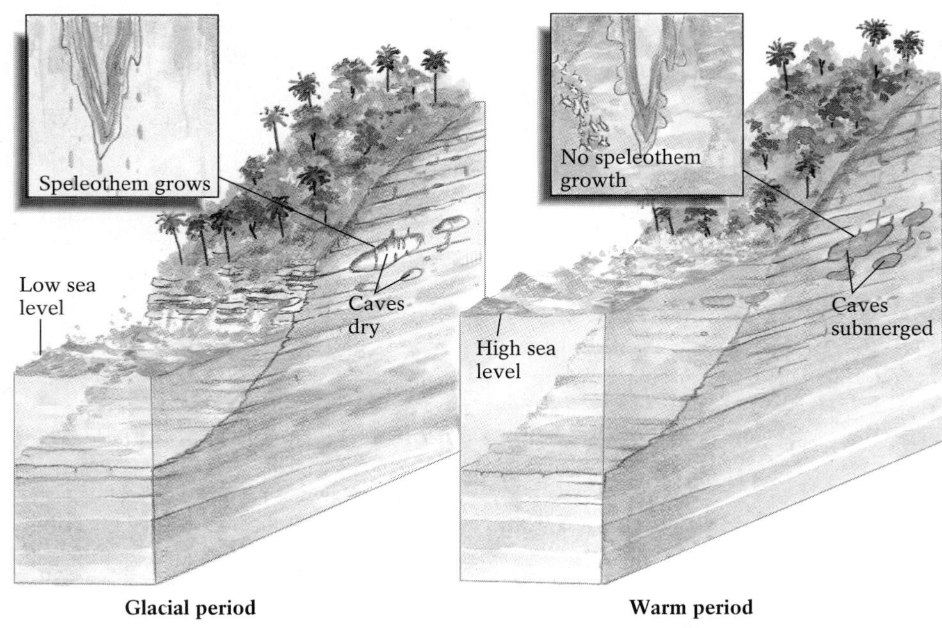

Figure 16-11 The rise and fall of the sea level affects speleothem growth in low-latitude and equatorial coastal caves. **(a)** During worldwide glacial periods, the global sea level is lower, coastal caves dry out, and speleothems tend to grow. **(b)** During warm periods, glaciers melt, the sea level rises, and speleothem growth ceases, because caves become filled with water.

speleothem growth in these areas ceased about 90,000 years ago and resumed some 15,000 years ago, spanning the period of the last worldwide glacial expansion (see Chapter 17).

Even speleothems in tropical coastal regions never covered by glaciers can be used to date past ice ages. In these areas, travertine deposits record the rise and fall of the sea level rather than the comings and goings of ice. During a period of worldwide glacial expansion, marine water is converted to snow and ice on land, which causes sea levels to fall worldwide. A drop in the water level in coastal caves produces the air-filled cave environment required for speleothem growth (Fig. 16-11). When glaciers melt during warmer periods, the sea level rises, and caves located at or below the new sea level fill once again with seawater. No speleothem growth takes place, because the water has no air-filled cavity into which it can drip.

Two Famous Caves

Among the well-known caves in North America are Howe Caverns in Cobleskill, New York, Wyandotte Cave in southern Indiana, Luray Caverns in Virginia, and Sonora Caverns in Texas. All are popular tourist attractions, and many state governments have preserved these and other caves as parks to ensure that they will be protected by law. Two of North America's most spectacular cave systems, Mammoth Cave in Kentucky and Carlsbad Caverns in New Mexico, have been designated as U.S. national parks because of their unique character and grand scale.

Mammoth Cave Mammoth Cave is located in a moist, temperate section of west-central Kentucky between Louisville and Nashville, Tennessee. Its vast labyrinth of connected galleries, some rising 80 meters (250 feet) high, contains underground lakes, rushing rivers, and spectacular waterfalls. Centuries ago, native North Americans explored Mammoth Cave intensively. More recently, westward-migrating pioneers rediscovered it. In the nineteenth century, the cave—known for its exceedingly long, dry tunnels—was considered a natural wonder and rivaled Niagara Falls as the main tourist attraction in eastern North America. In 1972, a connection was found that linked Mammoth Cave to the adjacent Flint Ridge system; the entire complex, with more than 530 kilometers (330 miles) of interconnected passages, is now the world's longest surveyed cave system.

Mammoth Cave's geological history began approximately 350 million years ago, when the St. Genevieve and St. Louis limestones were deposited on the floor of a shallow sea. Caves began to form in the limestone perhaps 30 million years ago, with the most active phase being initiated 2 million years ago. At that time, a massive ice sheet originating in central Canada near Hudson Bay advanced southward toward what is now Kentucky. Although the ice never covered the Mammoth Cave region, it blocked large westward-draining rivers, diverting them into the local stream systems. These enlarged streams cut deeply through the regional limestone plateau, drawing water down from the area's zone of saturation, lowering the water table, and draining the cave system.

The internal open-air environment of Mammoth Cave remains quite dry, primarily because a moderately impermeable cap rock that inhibits infiltration of water overlies most of the area. The dryness helps account for the lack of significant speleothem development. Most water within the

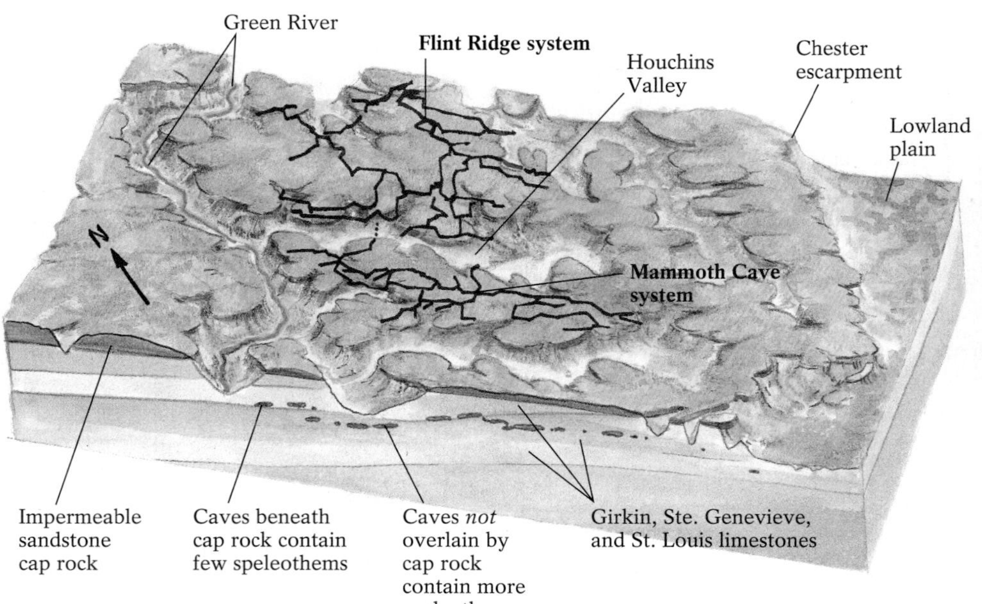

Figure 16-12 The Mammoth–Flint Ridge cave system in Kentucky. Because most of it is overlain by relatively impermeable sandstone "cap rock," Mammoth Cave is relatively dry and thus shows very little speleothem development. Photo: A smooth-walled passageway in Mammoth Cave.

Mammoth–Flint Ridge system takes the form of subterranean streams, which enter the caverns through local surface depressions where bedrock joints intersect. Only in a few places, where erosion of the cap rock has taken place, does enough water enter to produce speleothems (Fig. 16-12).

Carlsbad Caverns At sunset one hot June day in 1901, a young cowboy named Jim White noticed an odd black cloud above an area of parched ground in southeastern New Mexico. The cloud, swirling and lurching like an erratic tornado, was darker near the ground and grew diffuse at greater altitude. When the curious cowboy drew closer, he discovered that the cloud was actually an enormous swarm of millions of bats, exiting from what appeared to be a bottomless pit. White tossed a flaming cactus into the abyss and estimated that it fell more than 200 feet before it struck bottom. He had discovered Bat Cave, part of Carlsbad Caverns, North America's most spectacular cave system (Fig. 16-13).

Approximately 250 million years ago, what is now arid eastern New Mexico was periodically covered by a shallow inland sea in a semitropical environment. The region's rocks from that period largely comprise reef deposits composed of a framework of calcite-secreting algae, sponges, and the remains of moss-like sea creatures called bryozoans. Periodic evaporation of the shallow sea left behind gypsum and other evaporites and vast quantities of sulfur-rich, salty water.

The uplifting of the southern Rockies to form the Guadalupe Mountains, some 60 million years ago, folded and faulted the ancient reef. The joints and fractures resulting from this tectonic upheaval enlarged to become the enormous galleries of the cavern system. The process involved was an unusual one, however. Within the reef deposits, hydrogen sulfide (H_2S) compounds from the salty marine water combined with oxygen to produce sulfuric acid (H_2SO_4). The sulfuric acid–charged waters migrated *upward* through the jointed reef limestone under deep artesian pressure to create Carlsbad Caverns, in a process unlike the typical cave-forming descent of infiltrating water charged with carbonic acid.

In the last 2 to 3 million years, additional uplift has promoted further cave development. A spectacular forest of speleothems grew during the most recent ice age. During such periods of worldwide ice expansion, New Mexico was cooler and wetter. In this moist climate, rainfall evaporated more slowly than it does today. Hence plenty of water was available to infiltrate the area's limestone, dissolve it, and precipitate as the massive speleothems of Carlsbad Caverns. The oldest stalactites in the caverns are roughly 1 million years old. Virtually every type of speleothem can be found in these caves, from delicate helictites to bulbous cave popcorn.

Non-limestone Caves

Most caves form when acidified water dissolves limestone or dolostone. Others, however, form in different ways—within solidifying lava flows, by the force of crashing surf against coastal bedrock, and within the bases of melting glaciers.

Caves arise in lava when basaltic flows cool quickly, forming a thin black crust at the surface while the interior remains molten and continues to flow (discussed in Chapter 4). Eventually the molten lava drains from the crust, leaving behind a lava tube. Good examples of these types of caves appear in Oregon, Washington, northern California, Idaho, and Hawai'i (Fig. 16-14). Lava tubes are even believed to occur within the mare basalts of the Moon. Lava dripping from the roof of a lava-tube cave ("lavacicles") may accumulate on the cave floor to form deposits that resemble the stalagmites of limestone caves.

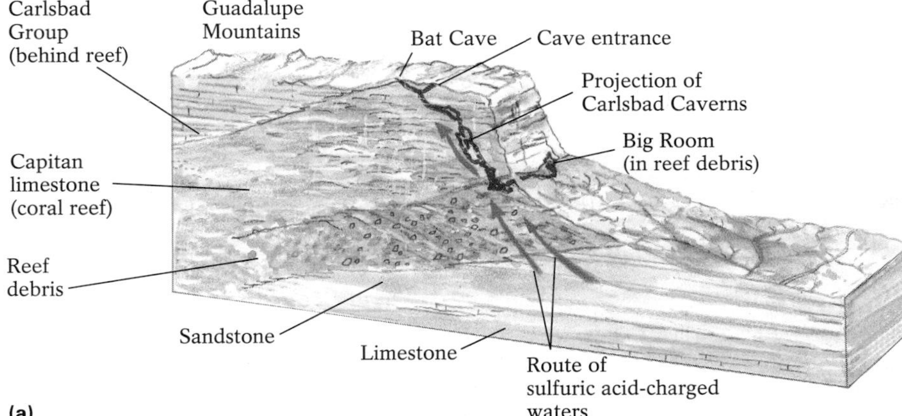

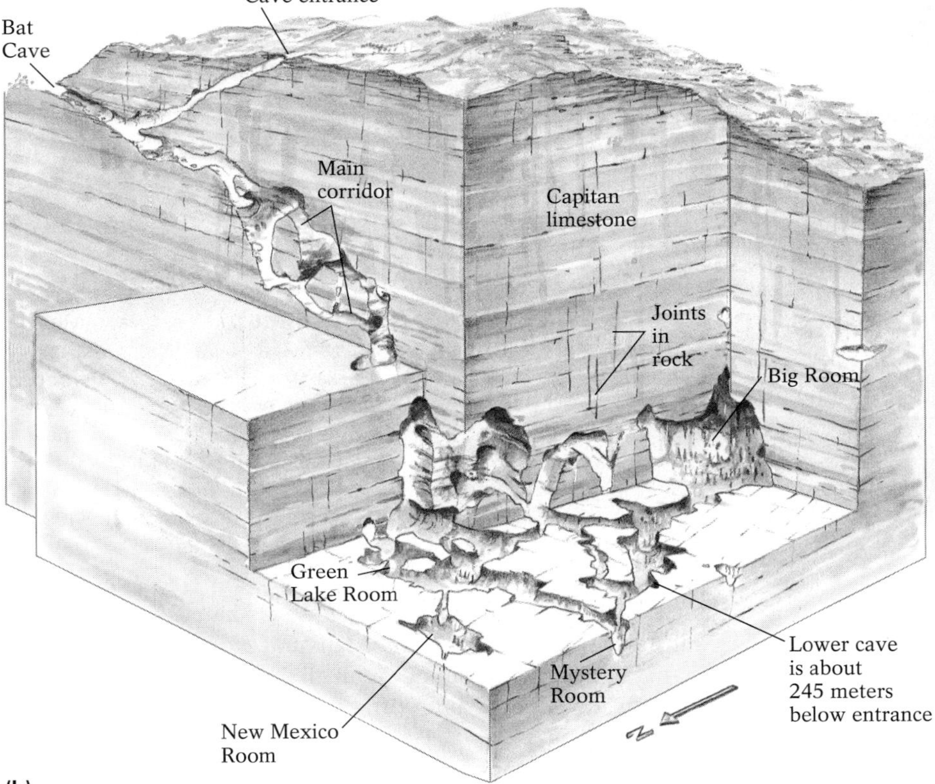

Figure 16-13 Carlsbad Caverns, New Mexico. **(a)** A geologic cross section of the Guadalupe Mountains, showing the location of the cave system. **(b)** Most of the large cave "rooms" of Carlsbad developed at a depth of 245 meters (800 feet) below the ground surface. Photo: Well-developed speleothems in one of Carlsbad's rooms.

Figure 16-14 Thurston lava tube, in Hawai'i, is a cave that developed within a basaltic lava flow.

Figure 16-15 An ice cave in Muir Glacier, Glacier Bay National Park, Alaska. This 500-meter (1500-foot)-long cave was part of a glacial remnant that separated from the main glacier approximately 30 years ago. The photo was taken in 1984—the cave was gone by 1990.

Exposed coastal bedrock eroded by pounding surf can form *sea caves* (discussed in Chapter 19). Fractured, jointed, faulted, or otherwise less resistant rocks are most likely to erode in this manner. The surf sculpted, for example, Oregon's Sea Lion Cave, home to hundreds of seals, and Maine's Anemone Cave in Acadia National Park.

Ice caves form when meltwater streams, which flow primarily during the summer months at the bottom of many glaciers and large perennial snow fields, cause melting *within* glaciers. The melting that creates such caves may also lead to their eventual disappearance (Fig. 16-15).

Karst Topography

Karst topography, the *surface* expression of karst, results from the dissolution of *exposed* soluble bedrock, such as limestone, dolostone, and gypsum. Any of the millions of square kilometers of soluble bedrock that lie at the Earth's surface are likely to become or may already be karst. Extensive karst appears in China, Southeast Asia, Australia, Puerto Rico, Cuba and elsewhere in the Caribbean, the Yucatán peninsula of Mexico, and much of southern Europe. Nearly 25% of the area of the contiguous 48 U.S. states is karst land, including substantial parts of Virginia, Alabama, Florida, Kentucky, Indiana, Iowa, and Missouri. Figure 16-16 shows the worldwide distribution of karst landforms.

Karst Landforms

Karst landforms, shown in Figure 16-17, range from broad plains studded with small, almost imperceptible, surface depressions, to such unusual features as stream valleys containing no streams (or streams that flow at the surface and then suddenly disappear), spectacular natural rock bridges, and huge monoliths composed of insoluble rock. Some of these landforms occur virtually everywhere that extensive surface carbonate bedrock exists. Others—particularly those caused by rapid dissolution acting over long periods—occur only in tropical environments.

Sinkholes The circular surface depressions that appear in most limestone terrains are called **sinkholes.** Streams flowing across a well-developed karst plain generally drain into sinkholes and then through passageways that descend far underground, often reaching the caverns below. Sinkholes usually occur together in great numbers over broad expanses of

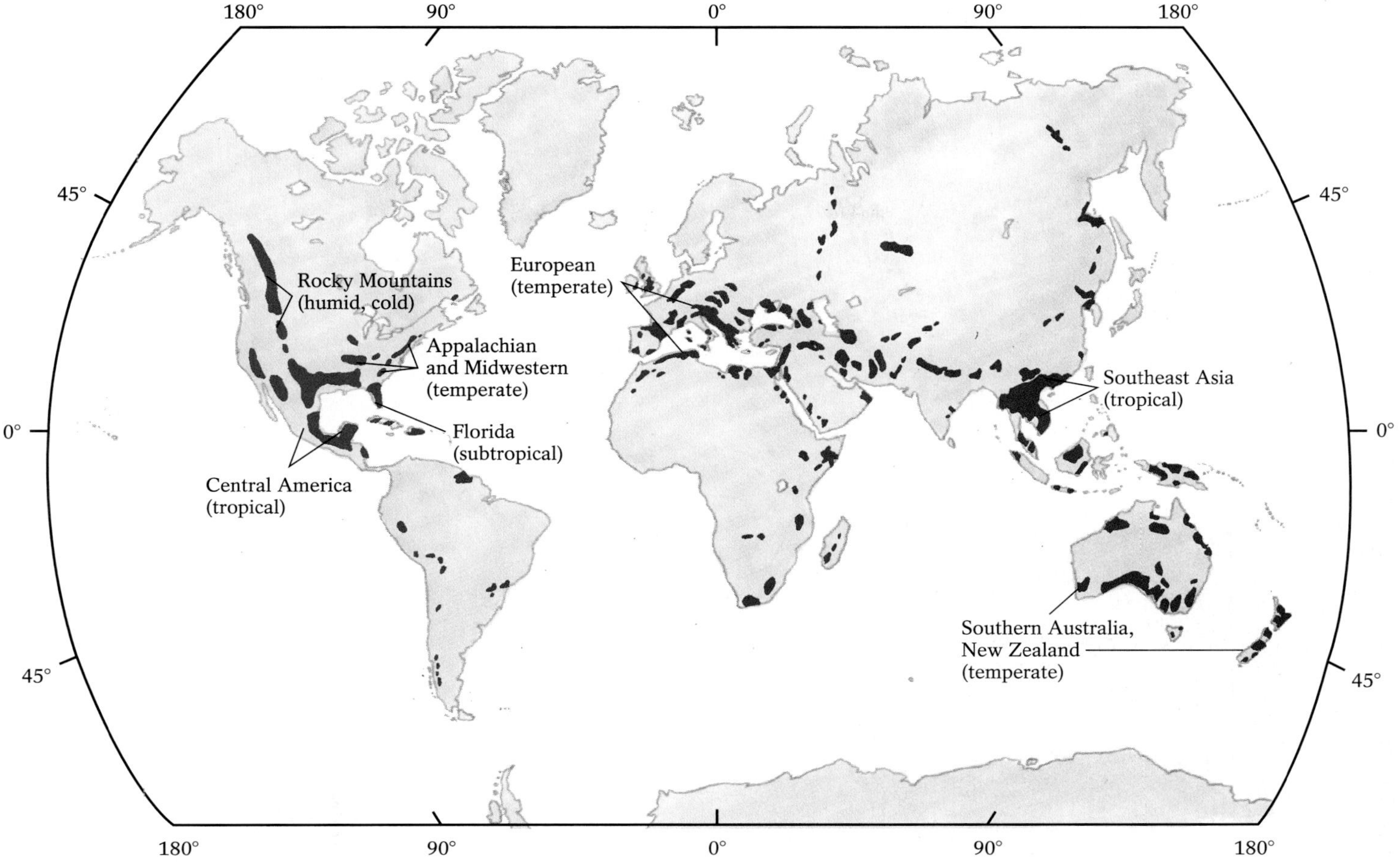

180° 90° 0° 90° 180°

45° 45°

Rocky Mountains
(humid, cold)

European
(temperate)

Appalachian
and Midwestern
(temperate)

Florida
(subtropical)

Central America
(tropical)

Southeast Asia
(tropical)

Southern Australia,
New Zealand
(temperate)

0° 0°

45° 45°

180° 90° 0° 90° 180°

Figure 16-16 The worldwide distribution of known karst landforms. Little or no karst is found in arctic regions, where water is often frozen and hence unavailable for dissolution, soils contain little organic matter and any organic decay occurs only slowly, and permafrost impedes water's infiltration into the ground. In temperate regions and in cool, humid, mid-latitude regions, water falls as rain or snow, and the ground remains unfrozen most of the year and readily absorbs water. In these areas, dissolution reactions can proceed at a relatively rapid pace. In subhumid and arid regions near the equator, low precipitation levels and high rates of evaporation combine to protect limestone from dissolution. These regions contain virtually no karst. The prodigious rainfall and lush vegetation in the humid tropics provide the optimal environment for limestone dissolution, and karst landscapes are highly developed here.

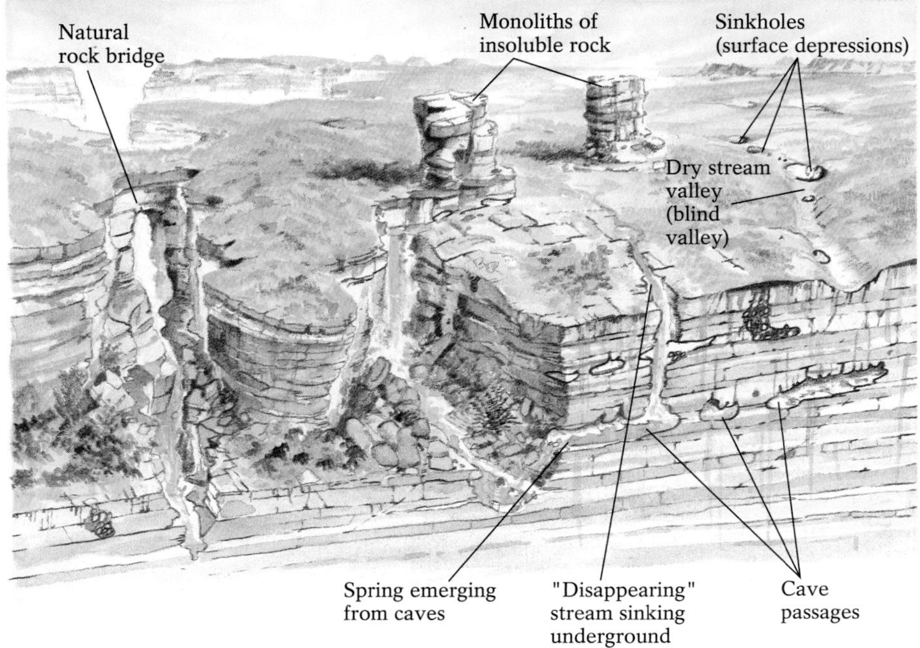

Figure 16-17 Typical landforms associated with karst topography.

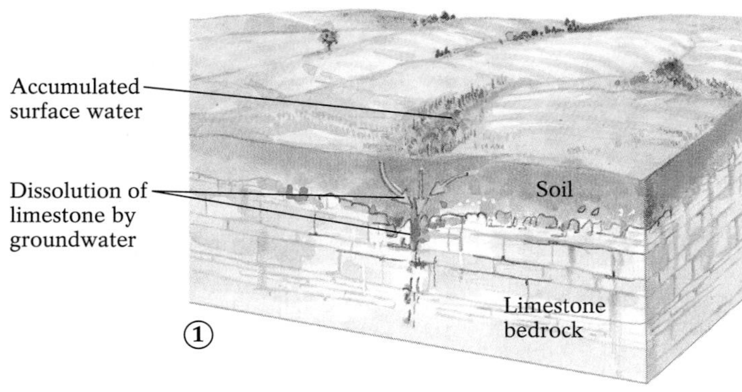

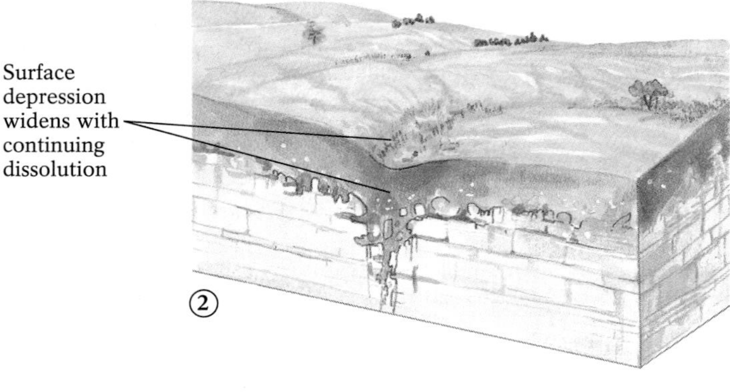

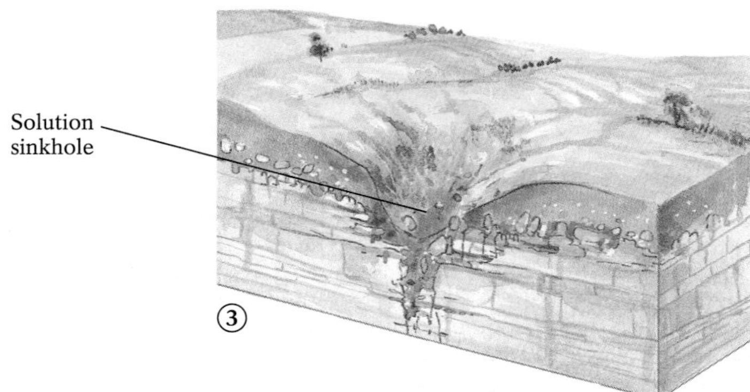

Figure 16-18 The origin of solution sinkholes. Photo: A sinkhole in Minnesota.

limestone bedrock containing numerous joints and fractures. Central Kentucky alone has 60,000 sinkholes; more than 300,000 are present in southern Indiana.

Solution sinkholes form when groundwater rich in carbonic acid dissolves limestone at or just below the surface. As shown in Figure 16-18, the diameters of these sinkholes widen at the surface over time, but are progressively narrower below ground, forming their characteristic funnel shape. Solution sinkholes generally develop on flat or gently sloping landscapes. Rather than promoting runoff, such surfaces keep acidified water in prolonged contact with the soluble rock, fostering extensive dissolution. On steeper slopes, water drains from the surface too rapidly to permit significant dissolution of near-surface rock.

Collapse sinkholes form in several ways (Fig. 16-19). Some occur when the roofs of caves collapse under the weight of overlying rocks. Others form as soil gradually filters down through bedrock fissures into caves, creating large soil voids that eventually collapse. These events may occur suddenly and unpredictably. Many collapse sinkholes form when the regional water table declines during and after a lengthy drought (Fig. 16-20a). In such conditions, the water that previously helped support the overlying surface (see Chapter 15) disappears. If a soaking rain follows a drought, the precipitation can add considerable weight to a newly unsupported surface, causing it to fall into the caverns below. Collapse sinkholes can also form when communities inadvertently lower their water tables by excessive withdrawal of groundwater (Fig. 16-20b). This problem has plagued certain communities in the southeastern United States, where the population has grown substantially in recent years. Many of these communities have extracted a great volume of groundwater to meet the rapidly rising demand for freshwater. The subsequent lowering of the regional water table has left the surfaces in many areas unsupported. The result: thousands of new collapse sinkholes.

Collapse sinkholes generally are deeper than solution sinkholes. Most have steep sides, including some that are al-

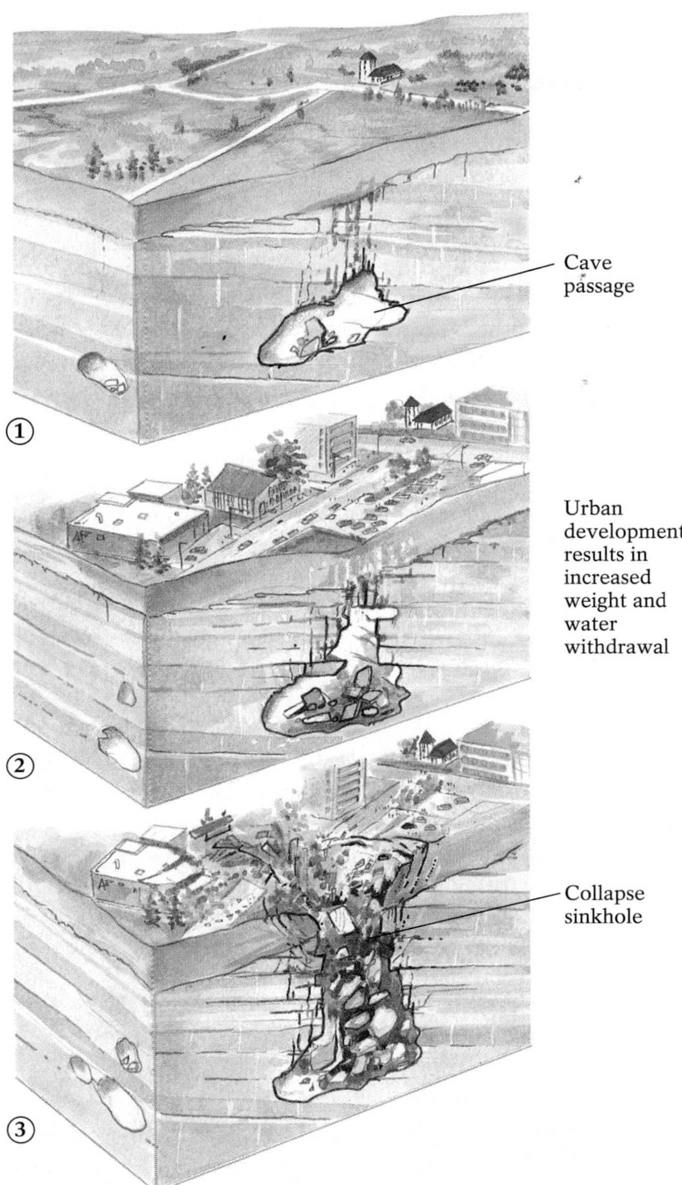

Figure 16-19 The origin of one type of collapse sinkhole.

(a)

(b)

Figure 16-20 A drop in the level of the water table is a common cause of collapse sinkholes. **(a)** A drought-induced sinkhole in Winter Park, Florida. This sinkhole formed in 1981, when the roof of a large subterranean cavity caved in without warning, swallowing a three-bedroom bungalow, a portion of a Porsche dealership, and half of a municipal swimming pool. A drop in the local water table brought on by a lengthy drought had removed the support supplied by groundwater to the roof of the cavity. The resulting crater was approximately 122 meters (400 feet) wide and 38 meters (125 feet) deep. **(b)** The December Giant, a collapse sinkhole in an isolated section of Shelby County, Alabama. The crater, 140 meters (450 feet) across and 50 meters (165 feet) deep, formed with an earth-shaking rumble in December 1962, after heavy November rains soaked a surface left unsupported by heavy groundwater pumping nearby.

most vertical. Rubble covers the rocky, irregular floors of most such sinkholes.

If a sinkhole is deep enough to intersect the local water table, it will rapidly fill with water to form a lake. The Mayan civilization that flourished in Mexico's Yucatán peninsula between A.D. 600 and 900, for instance, depended exclusively on water from *cenotes*—deep, steep-sided sinkhole lakes. When archaeologists excavated now-dry cenotes, they retrieved treasures of jade, gold, and copper artifacts cast into them to please the rain god, Chacmul, and ensure a steady supply of water.

Sinkhole lakes that lie *above* the local water table often last only as long as rubble or sediment clogs the sink-

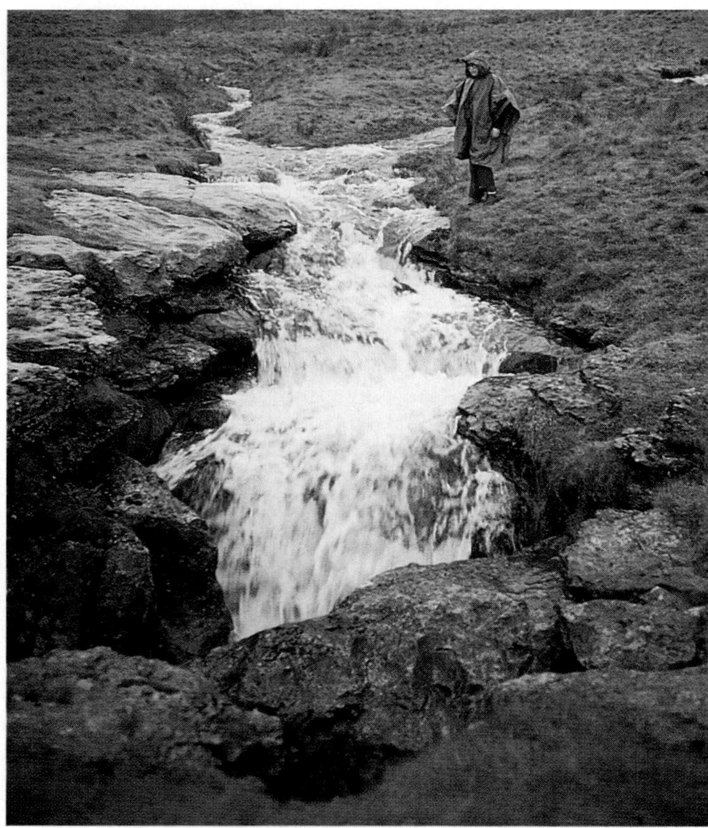

Figure 16-21 Streams in karst often disappear suddenly down sinkholes that open into caverns and passageways below the surface. In southern Indiana, streams with suggestive names, such as Sinking Creek and the Lost River, flow underground for more than 10 kilometers (6 miles).

hole's lower outlet. When the outlet reopens, the water rushes through it into caverns below. In the nineteenth century, a sinkhole at present-day Alachua Lake in the Lake District of north-central Florida was receiving the surface drainage from the surrounding plain, and the water was regularly disappearing into the local cave system. By 1871, enough organic and mineral debris had washed in to clog the sinkhole's outlet and form a lake 13 kilometers (8 miles) long and 6 kilometers (4 miles) wide. The lake remained for 20 years, until descending water swept the clogging debris into the underlying caverns. With the outlet opened again, the lake's water drained out. Even today, expensive Lake District property often becomes devalued overnight as sinkhole lakes drain like unstoppered bathtubs, transforming expensive lakefront views into less picturesque, foul-smelling mudflats.

Sinkholes can prove a boon to groundwater management, as they can be used to control the water level and flood potential of local streams. In Springfield, Missouri, an area prone to periodic flooding, excess surface water is channeled toward a few selected sinkholes for subsurface disposal. At Kentucky's Bowling Green airport, engineers devised a system of drainage ditches that terminate at specific sinkholes, safely transferring excess water underground.

Disappearing Streams and Blind Valleys Water soaks rapidly into an absorbent karst plain. As a result, many streams drain completely into sinkholes after flowing only a short distance across the surface; such flows are called **disappearing streams** (Fig. 16-21). Because disappearing streams appear only briefly at the surface, they rarely meander and seldom contribute to flooding. The climate of central and western Pennsylvania, for example, would normally promote frequent flooding; few

Figure 16-22 Tracing disappearing stream water. (left) Dye is poured into a sinkhole. (above) Colored water emerges several hours later from a natural spring several kilometers away.

floods occur in this region, however, because most of the surface water disappears into the area's many sinkholes.

Some disappearing streams eventually return to the surface as natural springs. How can we tell if the water gushing from a natural spring is the same water that disappeared down a sinkhole some kilometers away? Geological researchers sometimes trace the water of a disappearing stream by placing a harmless chemical or biological dye in it. Volunteers stationed at springs throughout the area then await the reemergence of the dyed water (Fig. 16-22). If a subterranean stream system's passageways are well-connected, the water may traverse distances of several kilometers in only a few hours, far more rapidly than the typical groundwater flow. Another, less obvious way to trace the flow path of a disappearing stream involves the introduction of pollen grains from non-native plants into the water; using pollen traps installed at the spring outlets, geologists then note if and when these exotic grains arrive. (Karst studies in Minnesota, for example, have used eucalyptus-tree pollen imported from Australia.)

Blind valleys are channels that form when adjacent sinkholes, enlarged by continued dissolution and erosion, coalesce to form larger sinkholes (Fig. 16-23). Blind valleys contain water only during periods of heavy rain, when all of the precipitation cannot drain into the ground. When a stream flows through a blind valley, it tends to lower the valley floor much faster upstream, where streamflow is greater. At the downstream end of the valley, where streamflow has dissipated, the stream valley seems to abruptly end at a bedrock *headwall.* Typically, streamflow at this headwall is diverted underground into one of many sinkholes.

Natural Bridges As a series of neighboring sinkholes expands and coalesces, the surface overlying broad sections of underground stream channels may collapse. The segments of

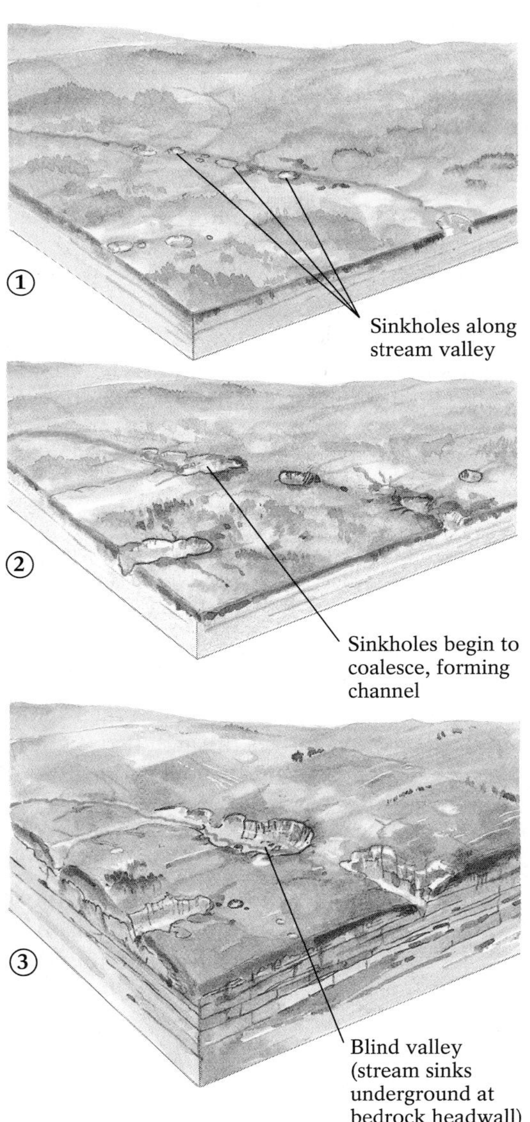

Figure 16-23 The origin of blind valleys. Photo: The Tarn Gorge, a blind valley in southern France, shown during high discharge after a storm.

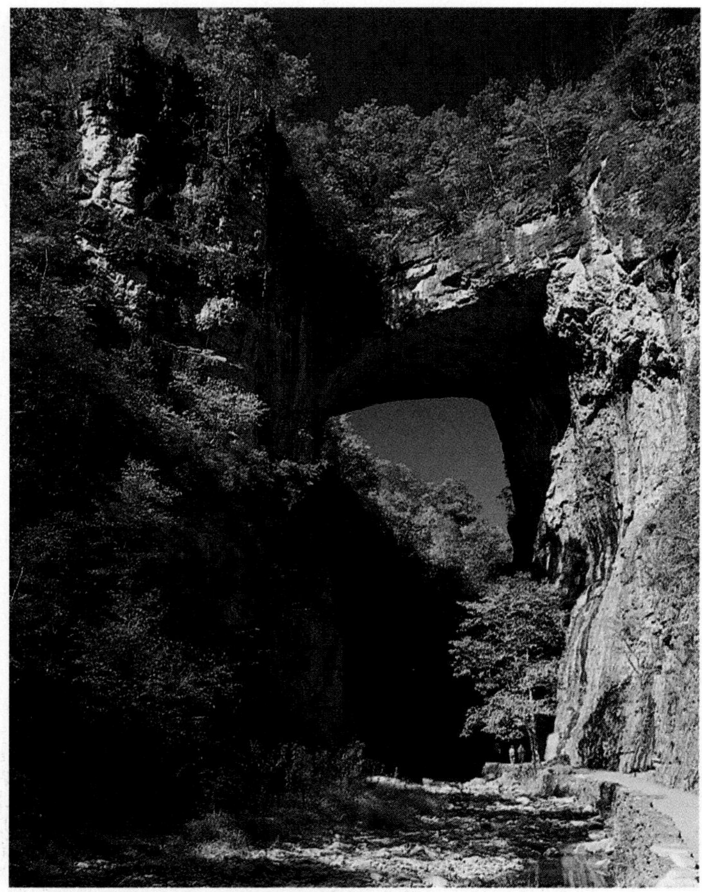

Figure 16-24 Natural Bridge, in the Blue Ridge Mountains near Lexington, Virginia, spans Cedar Gorge Creek.

surface material that do *not* collapse form a **natural bridge** over the exposed channel. Perhaps the most famous natural bridge in North America is located in the Blue Ridge Mountains of Virginia, where a remnant of massive dolomitic limestone forms a bridge approximately 30 meters (100 feet) long and 40 meters (130 feet) high (Fig. 16-24). Thomas Jefferson purchased the bridge in 1774 (from King George III, for the equivalent of $2.50) to preserve it as a natural wonder and resource. U.S. Highway 11 now runs across it, saving millions in bridge-construction dollars.

Tropical Karst Landforms Limestone dissolves rapidly in tropical climates, where abundant rainfall and CO_2 produced by decay of the lush vegetation speed the process. This rapid dissolution, which proceeds relatively uninterrupted during ice ages, produces karst formations that differ markedly from those found in temperate climates. **Cockpit karst** consists of closely spaced, irregular depressions and adjacent, steep-sided conical hills. The depressions, or *cockpits,* comprise deep solution sinkholes formed by enhanced dissolution of soluble rock, often along fractures. The hills, or *towers,* dissolve less rapidly (in the absence of fractures); they can protrude more than 200 meters (660 feet) above the rapidly dissolving rock that surrounds them (Fig. 16-25). Excavation of a cockpit generally ceases when groundwater reaches an insoluble layer and can no longer cut vertically. Thus the depth of a cockpit, and consequently the relative height of its neighboring karst towers, are limited by the thickness of the soluble carbonate bedrock.

Figure 16-25 Karst towers in the Kwangsi region of southern China. Soluble limestone has been dissolved and carried away in solution, leaving behind the "towers" of less-soluble rock.

Because of their surface hollows and blind valleys, karst landscapes have helped determine the outcome of some significant political and social events in recent history, as described in Highlight 16-1.

Protecting Karst Environments

Well-developed cave and karst environments typically contain numerous sinkholes that connect with a vast underground network of large caverns and passageways. In such places, water enters the ground readily and moves through it so rapidly that soils and impermeable bedrock have few opportunities to filter out contaminants. Thus cave and karst landscapes are *extremely* sensitive to careless handling of waste. When Hidden River Cave in Kentucky opened to visitors in 1916, it boasted elegant galleries and memorable subterranean boat rides. For decades, however, nearby towns disposed of their sewage into several large sinkholes, hopelessly contaminating the regional groundwater system. By the mid-1930s, vile odors had ended tourism at Hidden River Cave.

Garbage dumped into a sinkhole will quickly contaminate both the local and regional water supply (Fig. 16-27).

Highlight 16-1 Karst Landscapes in History

For centuries, partisans, guerrilla warriors, and bandits have enjoyed the protection of natural hideouts provided by karst landscapes. During World War II, outnumbered and ill-equipped partisans tormented their German invaders from karst refuges in the former Yugoslavia. In 1954, the guerrilla armies of the Viet Minh confounded French troops by hiding in and attacking from the maze of karst cockpits, caves, and blind valleys in what was then North Vietnam (Fig. 16-26).

In 1957, Fidel Castro and his supporters planned their insurgency against Cuba's dictatorial regime from the shelter of the karst terrain of the Sierra Maestra, on the island's south-eastern coast. In the nineteenth century, Jesse James and his gang hid in the karst valleys and caves of the Missouri Ozarks to avoid arrest for their bank and train robberies.

And, since the implementation of Prohibition in 1919 (continuing even after its repeal in 1933), Appalachian karst land in the eastern United States has sheltered illegal distilleries that manufacture "moonshine" whiskey.

Figure 16-26 French paratroopers prepare to blow up a Viet Minh arms depot in a cave near Langson, on the Indochina border, in July 1957.

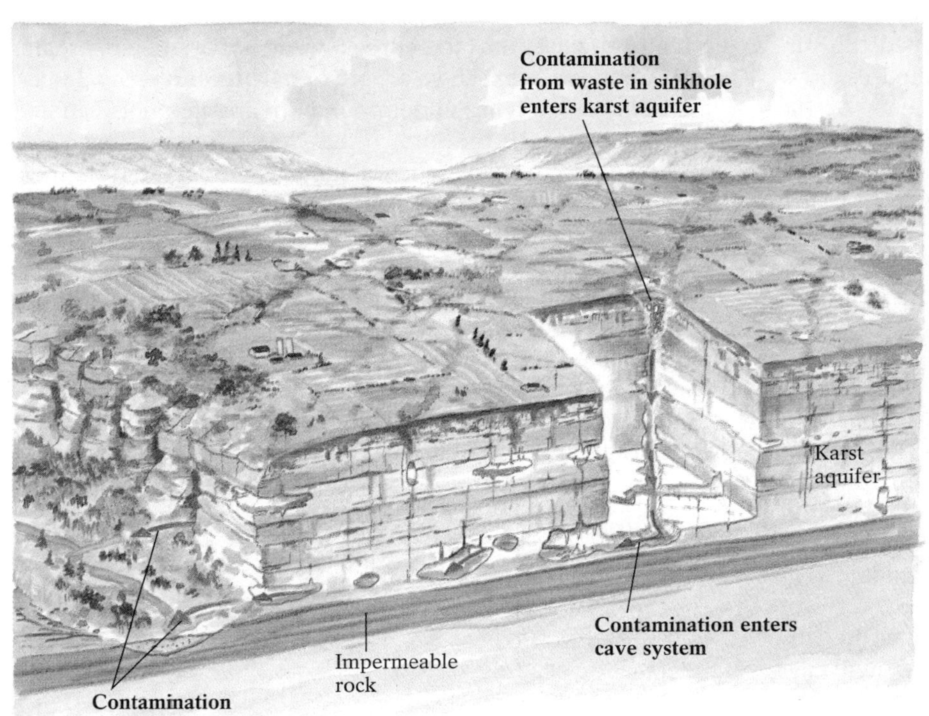

Figure 16-27 Disposal of refuse in sinkholes quickly contaminates karst aquifers. Photo: A refuse-filled sinkhole.

During the 1960s, the community of Alton, Missouri, in beautiful Ozark country, discarded much of its refuse in a 15-meter (50-foot)-deep dry sinkhole near the town. In 1969, after an unusually rainy winter and spring, 9 meters (30 feet) of water accumulated in the sinkhole. Concerned about the effects of Alton's sinkhole disposal, the U.S. Forest Service poured thousands of gallons of dyed water into the sinkhole's standing water and surveyed groundwater sources for miles around. Three months later, dyed water containing a high concentration of undesirable material leached from Alton's refuse was detected 25 kilometers (16 miles) away at the neighboring town of Morgan Springs. Alton's careless dumping had adversely affected the entire region's groundwater quality.

As we saw in Chapter 15, groundwater normally moves so slowly through soils and rocks that filtration, adsorption, and oxidation can gradually remove many contaminants. Water in karst, however, drains too swiftly through open underground passages to strain out pollutants or allow their consumption by microorganisms. Moreover, in karst regions, polluted recharge reaches the groundwater through *numerous* entry points. A large cave system may underlie hundreds or even thousands of sinkholes—each one being a potential point source of pollution. Finally, extensive, complex cave systems typically hold a huge volume of water. For all these reasons, chemical treatment of contaminated karst aquifers is simply not feasible. Thus, because they are so open to contamination and impossible to treat effectively, aquifers in karst regions should not be used for waste disposal of any kind. Likewise, they should not be considered reliable sources of drinkable water.

In the last three chapters, we have examined the surface and subsurface geological work accomplished as flowing liquid water passes through various segments of the hydrologic cycle. In the next chapter, we will look at how flowing *frozen* water—in the form of glaciers—has substantially altered the landscape in regions cold enough to freeze the Earth's surface water.

Chapter Summary

Karst terrains contain features not found in any other geological setting. They develop when soluble bedrock dissolves above or below the surface. Nearly all karst results from the dissolution of calcite in limestone by carbonic acid. Dissolution rates are largely controlled by the composition, location, and structure of the bedrock, the amount of rainfall and vegetation in the area, and local topography.

Extensive cave systems typically occur below the surface of a karst landscape. **Caves** are natural underground cavities that generally form as carbonate bedrock dissolves along preexisting joints, fractures, faults, and bedding planes. Most caves develop by a two-stage process: In the first stage, when the local water table is high, cave chambers and connecting passageways emerge from a system of water-filled bedrock fissures. In the second stage, when the water table drops below the cave depth, the cave exists in an open-air environment. As water that percolates through overlying bedrock enters the cave, it may evaporate or release its carbon dioxide, both of which precipitate dissolved limestone. Cave deposits, called **speleothems,** include **stalactites,** which grow downward from the cave ceiling, and **stalagmites,** which grow upward from the cave floor. The limestone that makes up speleothems is called **travertine.** Non-limestone caves can also form as either lava tubes in solidified basaltic flows or voids in eroded coastal bedrock or partially melted glaciers.

Karst topography—the surface expression of karst—is dominated by conical depressions called **sinkholes** that often occur by the thousands in areas characterized by exposed flat-lying limestone bedrock. Solution sinkholes form from the dissolution of limestone by carbonic acid in groundwater. Collapse sinkholes form when the roofs of caves give way under the weight of overlying materials, from the loss of soil down bedrock fissures, or following the withdrawal of water that helps support an overlying surface.

Because sinkholes usually occur in large numbers and close together, surface water rarely traverses a karst plain completely. Instead, streams typically drain from the surface when they encounter sinkholes, creating **disappearing streams.** The diverted water may reappear elsewhere as a natural spring. Coalescing sinkholes create **blind valleys,** which hold water only after heavy rains. These valleys seem to terminate downstream at bedrock walls, where their streams sink underground. When a network of sinkholes coalesces, the surface overlying underground stream channels can collapse; segments of the surface that do not collapse form **natural bridges.** The development of karst topography is most pronounced in the moist tropics. In such regions, limestone dissolution creates **cockpit karst,** a landscape marked by closely spaced depressions, or cockpits, surrounded by towers of residual, less-soluble bedrock.

Because surface water moves so rapidly into and through the groundwater system in karst regions, contaminants cannot be filtered out or consumed by microorganisms. Consequently, karst regions are particularly sensitive to groundwater contamination.

Key Terms

karst (p. 448)
caves (p. 450)
speleothems (p. 452)
travertine (p. 452)
stalactites (p. 452)

stalagmites (p. 452)
sinkholes (p. 458)
disappearing streams (p. 462)
blind valleys (p. 463)
natural bridge (p. 464)
cockpit karst (p. 464)

Questions for Review

1. Name three different soluble rocks. Which is most likely to form caves, and why?

2. List four factors that affect the rate of limestone dissolution.

3. Briefly explain the role that vegetation plays in the development of karst.

4. Briefly describe how caves form. Where is the water table located during the two stages of cave formation?

5. How do stalactites differ from stalagmites?

6. Explain how speleothems can be used to determine past fluctuations in global sea levels.

7. Discuss two fundamental geological differences between Mammoth Cave and Carlsbad Caverns.

8. Briefly describe two types of caves formed by processes other than dissolution.

9. Describe the two ways in which sinkholes form.

10. Why is a karst landscape particularly susceptible to groundwater pollution?

For Further Thought

1. Describe what may happen to the calcium carbonate that enters into solution during the creation of a cave. How might this calcium carbonate eventually reenter the rock cycle?

2. How would you determine the local limestone dissolution rate in your town or city?

3. Discuss three reasons why you would not expect to find caves in polar regions.

4. What geological features do you think are shown in the aerial photos below, taken over the same area of the Midwest, 43 years apart? Why might these features have grown in number and size between 1937 and 1980?

5. Speculate about how plate tectonics might affect the origin of a cave and its deposits.

17

Glaciers and Ice Ages

Figure 17-1 The deeply crevassed Davidson Glacier near Haines, Alaska.

Glaciers provide some of the most spectacular scenery on Earth (Fig. 17-1), drawing millions of visitors to Alaska and other high-latitude regions in which they are common. Glaciers also help shape the very landscapes that they cross; aside from stream erosion and its associated mass movements, the Earth's surface has probably been changed more by glaciation than by any other process. Without glaciers, the fjords of Norway would not exist. North America would not have its Great Lakes, Niagara Falls, Hudson Bay, Puget Sound, or the 15,000 lakes of Minnesota. There would be no Cape Cod in Massachusetts and no fertile rolling hills in the Midwest and southern Canada. The peaks of the Rockies and Cascades would be less impressive, and California's Yosemite Valley would lack its sheer-faced cliffs. Rivers such as the Missouri and Ohio would drain north to the Arctic and Atlantic Oceans, rather than south to the Mississippi River and the Gulf of Mexico. If the Earth had no glaciers today, the shapes of the continents themselves would differ substantially, because sea level would rise about 70 meters (230 feet) higher than its present height. Landlocked cities such as Memphis, Tennessee, and Sacramento, California, would be seaports, and San Francisco, New York, and many other coastal cities would be mostly under water.

Where do we find the glaciers that have changed so much of our continents? And how can so small a fraction of the hydrologic cycle (see Figure 14-2) accomplish so much change? The answers lie in the occurrence of **ice ages,** the dozen or so periods of Earth history—each lasting millions of years—during which the planet's climate was significantly cooler than usual and glaciers covered a large portion of the Earth's land surface. During an ice age, climatic fluctuations cause glaciers to grow and advance during *glacial periods,* and then thaw and retreat during *interglacial periods.*

The Earth is currently in the midst of an ice age—the **Quaternary ice age,** named for the Quaternary Period of the Cenozoic Era (see Figure 8-33, the geologic time scale)—that has spanned the last 1.6 million years of the planet's history.

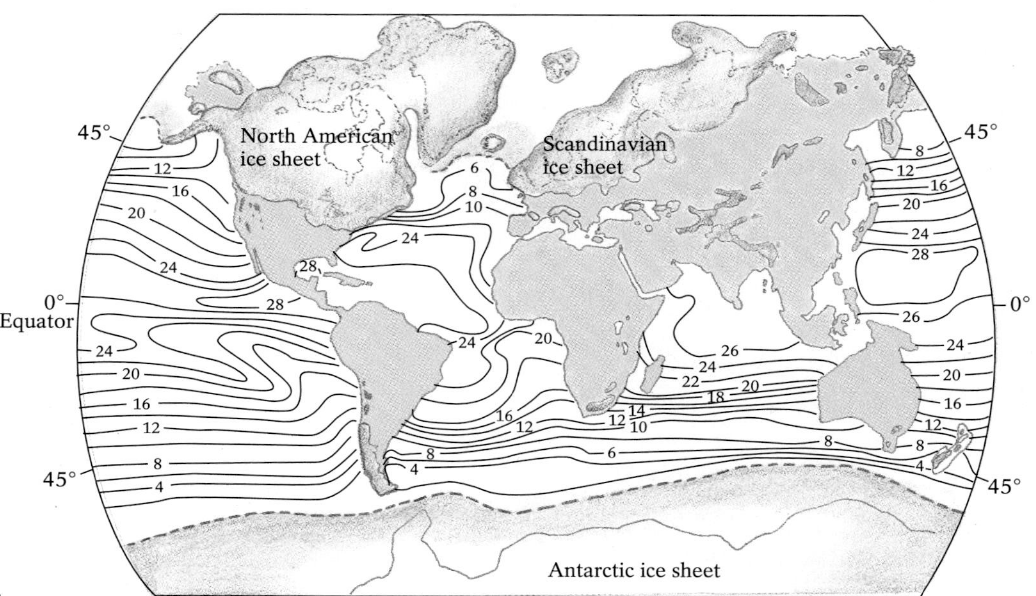

Figure 17-2 A "climap" showing glacier distribution during the last major period of worldwide glacial expansion. During this expansion, which reached its peak about 20,000 years ago, ice covered approximately 30% of the Earth's land surface. The contour lines indicate ocean temperatures in degrees Celsius; note how the temperatures drop with proximity to the glacial margins.

This ice age has been marked by many episodes of glacial advance and retreat, primarily during the **Pleistocene Epoch** of the Quaternary Period, which ended some 10,000 years ago. At the height of glacial expansion in the late Pleistocene about 20,000 years ago, ice covered approximately 30% of the planet's land surface, including most of northern Europe, northwestern Asia, Canada, and the northern United States (Fig. 17-2); in some places, the ice was 4 kilometers (2.5 miles) thick.

We now live in an interglacial period of significant warming that began about 10,000 years ago. Today, glaciers cover only 10% of the world's land surface. At very high latitudes, such as in the Arctic, Antarctica, and Greenland, glaciers are fairly common and can exist at any elevation, even at sea level; in mid-latitude regions, however, they are found only at high elevations (at roughly 2500 to 3000 meters [8000–10,000 feet]), such as in the Northern Rockies of Alberta, British Columbia, and Montana. Figure 17-3 demon-

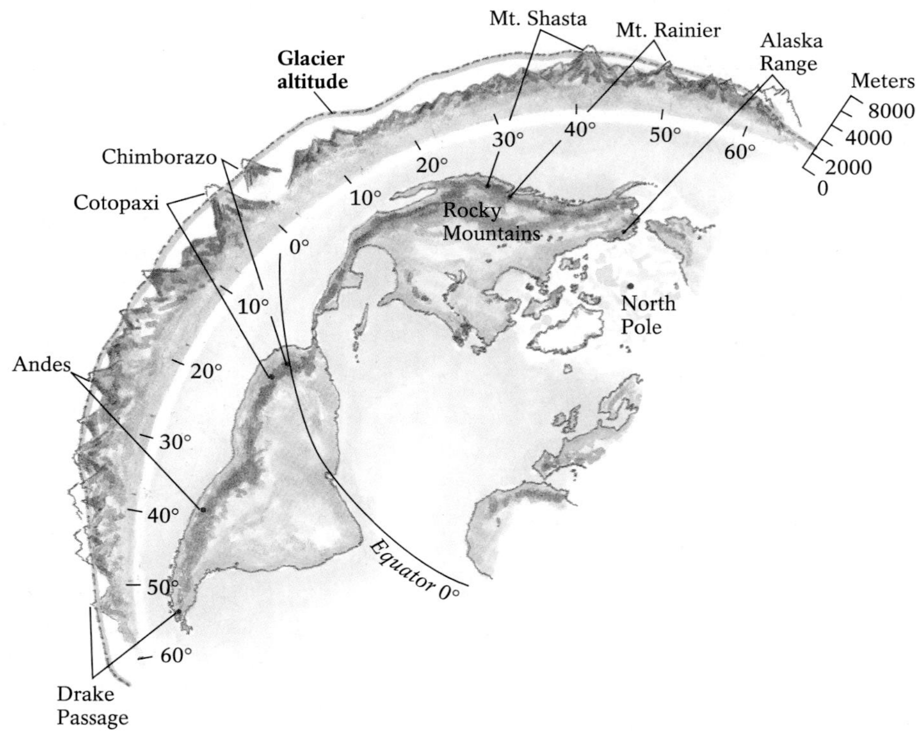

Figure 17-3 The relationship between glaciers and geographic latitude, shown along a line from Alaska to the tip of South America. In low-latitude, warmer regions (0°–30°), glaciers are found only at very high elevations (more than 5000 meters [17,000 feet]). In middle latitudes (30°–50°), they can occur at somewhat lesser heights (2500–3500 meters [8000–11,000 feet]). Most glaciers today are found in high-latitude regions (over 60°), such as Alaska and Antarctica, where they can exist even at sea level.

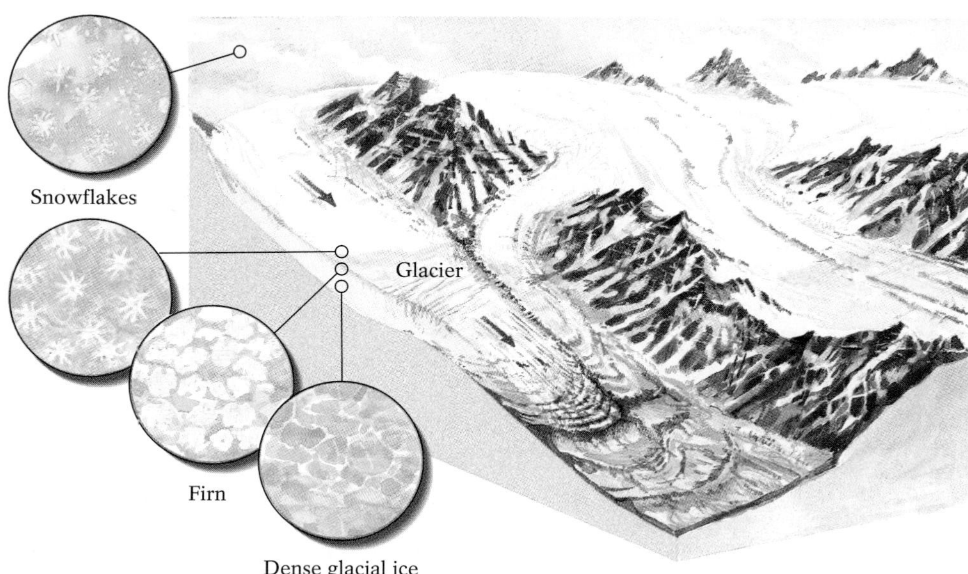

Figure 17-4 The formation of glacial ice from snow. Hexagonal snowflakes that fall on a glacier are gradually changed into rounded crystals. As overlying snow buries the crystals and exerts pressure on them, they become packed into increasingly dense firn and finally glacial ice.

strates how latitude affects the altitude at which ice masses exist in the Western Hemisphere.

In this chapter, we discuss what glaciers are, where they exist today and where they were found in the past, and how they shape the landscape. We also consider why certain periods of Earth history have been marked by worldwide glacial expansions whereas others have not, and we speculate about when glaciers may again cover vast areas of North America.

Glacier Formation and Growth

A **glacier** is a moving body of ice that forms from the accumulation and compaction of snow. Glaciers flow downslope or outward under the influence of gravity and the pressure of their own weight.

Glacier formation begins with a snowfall and the accumulation of snowflakes, which comprise delicate hexagonal (six-pointed) crystals of frozen water (Fig. 17-4). The initial snowpack tends to be fluffy, because the shape of the snowflake crystals initially keeps the flakes separated; about 90% of a pile of newly fallen snow consists of space between crystals. The density of fresh snow is only 0.1 grams per cubic centimeter. In contrast, the density of liquid water is 1.0 grams per cubic centimeter.

Soon, the air circulating through the spaces in the snowpack causes the fragile points of some snowflakes to *sublimate* (a process in which a solid changes directly into a gas). In this way, some snow is directly converted to water vapor, which then recrystallizes in the open spaces between flakes. As more snow falls, the weight of the overlying snow melts some of the contact points between crystals; the water produced by this *pressure melting* migrates to open, low-pressure spaces in the snowpack and refreezes, binding the snow crystals together. (Pressure melting also occurs under the blades of an ice skater. The skater's weight, pressing downward on the narrow blades at each moment, melts a small amount of ice that refreezes immediately when the weight is removed. As a result, the skater always glides on a thin film of water.)

Eventually, the processes of sublimation and recrystallization and pressure melting and refreezing convert large, ornate snowflakes to small, tightly packed, and nearly spherical crystals. **Firn** (from the German for "last year") is the well-packed snow that survives a summer melting season; its density is about 0.4 grams per cubic centimeter. Dense firn consists of interlocking crystals of glacial ice after nearly all the air has been squeezed out by repeated pressure melting and refreezing. Ultimately, the density of glacial ice reaches 0.85 to 0.90 grams per cubic centimeter, about the same as the density of an ice cube.

Glaciers form and grow when snow accumulation exceeds losses caused by summer melting and sublimation. Glaciers do not occur in locations such as modern-day Minnesota, where summer warmth completely removes even heavy winter accumulations of snow. Cool, cloudy, brief summers that minimize melting probably contribute more to glacier growth

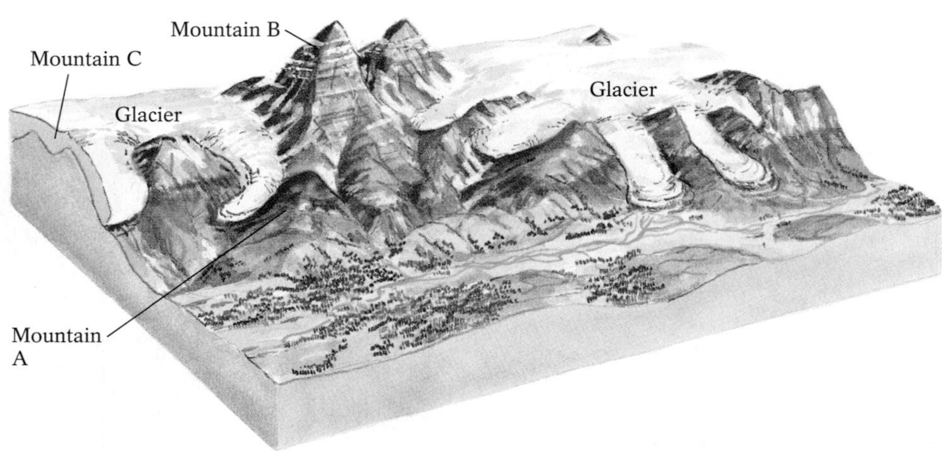

Figure 17-5 Elevation and slope steepness are two factors influencing glacier formation. Mountain A does not support the development of a glacier, because its summit is lower than the regional snowline, below which summer temperatures are too warm to support glaciers. Mountain B, whose summit is well above the snowline, also does not support glacier development, because it is too steep and snow flows downhill rather than accumulating. Mountain C has a gently sloped shape and much of its summit is above the snowline, making it an optimal site for glacier development.

than very low winter temperatures or thick accumulations of snow.

In mountainous regions, glacier formation and growth are further enhanced by the high elevation at which the snow lands and the relative steepness of a mountain's slope. As Figure 17-5 shows, mountainside glaciers usually form at or above the **snowline,** *the lowest topographic limit of year-round snow cover.* Snow can persist above the snowline, but any snow that falls below the snowline, where temperatures are warmer, typically melts in summer. In addition, snow tends to accumulate more on gentle slopes, where it is less likely to slide downslope or blow away.

Glacier formation may also depend on the orientation of a mountain's slope and the direction of snow-drifting winds. In the Northern Hemisphere, for example, snow is more likely to survive the summer on north-facing slopes than on south-facing ones, because the former receive less direct sunlight. More snow accumulates, and glacier formation is more likely, on the leeward slopes of mountains, where snow drifts from the windward side generally settle. In the mountains of western North America, where the prevailing winds blow from west to east, most glaciers form on northeastern slopes.

The time required for fresh snow to evolve into glacial ice varies with the rate of snow accumulation and the local climate. In snowy regions where average annual air temperatures hover close to the melting point of ice (0°C [32°F]), snow may be converted to ice in only a few decades. In extremely cold polar settings, where snowfall is typically sparse and little melting takes place, glacial ice formation may take thousands of years.

Classifying Glaciers

Glaciers are classified broadly by whether the local topography constrains them or allows them to move freely (Fig. 17-6). **Alpine glaciers** are confined by surrounding bedrock highlands. As a consequence, they are relatively small. Three types of alpine glaciers exist: **Cirque glaciers** create and occupy semicircular basins on mountainsides, usually near the heads of valleys; **valley glaciers** flow in preexisting stream valleys; and **ice caps** form at the tops of mountains.

A *piedmont* ("foot of the mountain") glacier originates as a confined alpine glacier but then flows onto an adjacent lowland where, unconfined, it spreads radially. Piedmont glaciers that flow to coastlines and into seawater are called *tidewater* glaciers. The only completely unconfined form of glacier is a **continental ice sheet,** an ice mass so large that it blankets much or all of a continent. Modern continental ice sheets cover Greenland and Antarctica. The Antarctic ice sheet actually consists of two ice sheets separated by the Transantarctic Mountains. It is 4.3 kilometers (3 miles) thick in some spots and occupies an area about 1.5 times as large as the continental United States. Twenty thousand years ago, such vast ice sheets covered North America to south of the Great Lakes, and western Europe south to Germany and Poland.

The Budget of a Glacier

The *budget* of a glacier consists of the difference between the glacier's annual gain of snow and ice and its annual loss, or *ablation,* of snow and ice due to melting and sublimation. If accumulation exceeds ablation for several consecutive years, the budget is positive, and the glacier increases in thickness and area. As a glacier expands, its outer margin, or **terminus,** advances downslope in confined glaciers and outward in unconfined glaciers. If ablation exceeds accumulation for several years, the budget is negative, the glacier decreases in size, and its terminus retreats. If accumulation is roughly equivalent to ablation, a glacier's budget is balanced and its terminus remains stationary. Thus the position of the terminus is largely determined by the glacier's long-term budget.

All glaciers—from the smallest cirque glacier in Montana to the Antarctic ice sheet—have a **zone of accumulation,** a **zone of ablation,** and an **equilibrium line** separating

Figure 17-6 Types of glaciers. **(a)** Angel Glacier, a cirque glacier on Mount Edith Cavell, Jasper National Park, Canada. **(b)** A valley glacier in Tongass National Forest, Alaska. **(c)** An ice cap in the Sentinel Range, part of the Antarctic continental glacier. **(d)** A tidewater glacier at Kenai Fjords National Park, Alaska.

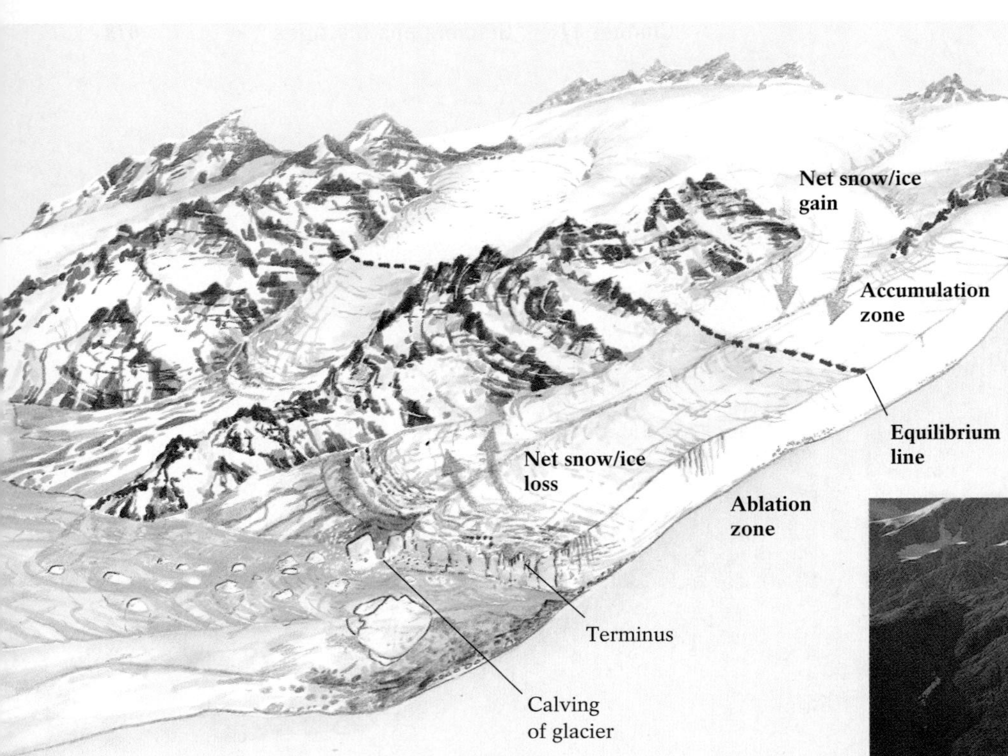

Figure 17-7 The anatomy of a glacier. Every glacier consists of an accumulation zone, where more snow and ice are added every year than are lost, and an ablation zone, where more snow and ice are lost than are added. An equilibrium line, where the amount of snow and ice added approximately equals the amount lost, forms the boundary between the two zones. In addition to melting and sublimation, a glacier may lose ice by calving if it terminates in a body of water. Photo: Calving of Hubbard Glacier in Wrangell–St. Elias National Park, Alaska.

the two zones (Fig. 17-7). The position of the equilibrium line can change every year, depending on the gain or loss of glacial volume. The zone of accumulation, which can be identified by a blanket of snow that survives summer melting, is nourished principally by snowfall, and sometimes augmented by avalanches from surrounding slopes. Summertime melting causes ablation in middle and low latitudes; in high latitudes, where summer temperatures may remain below freezing, sublimation brings about ablation. The zone of ablation is recognizable in summer by its expanse of bare ice. Glaciers that terminate in bodies of water, such as tidewater glaciers, can also lose ice—in the form of icebergs—by *calving,* a process in which chunks of ice break off by simply falling from a steep ice cliff or by flexing caused by wave or tidal action. Calving also occurs as sea level rises and buoys up an ice mass; as it begins to float, icebergs tend to snap off the main ice mass.

Glacial Flow

Newcomers to the study of the geology of glaciers must learn an important fact: Regardless of whether a glacier's terminus is advancing, receding, or remaining stationary, the ice within the glacier still tends to flow forward from the accumulation zone toward the ablation zone. Thus the position of the terminus reflects the glacier's long-term budget (positive, negative, or at equilibrium), even as the glacier's flow may reflect such short-term factors as accumulation of new snow, conversion of this snow to glacial ice, and down-ice flow of this ice under the influence of gravity. In the accumulation zone, as the snow becomes compacted into ice, it begins to flow obliquely downward toward the glacier's bed; in the ablation zone, ice flows obliquely upward toward the surface and outward toward the glacier's edges.

Glaciers flow by a combination of two mechanisms: internal deformation and basal sliding. In **internal deformation,** pressure from overlying ice and snow deforms a glacier's ice crystals, changing their shapes and causing them to slip past one another (particularly when the glacier is relatively warm and water exists at crystal boundaries); the ice crystals may also fracture and move along their internal planes of weakness. All glaciers, even those that remain frozen to their beds, move to some extent by internal deformation. In **basal sliding,** warmer glaciers, such as those found in mid-latitude mountain ranges, thaw at their bases, producing a film of water that enables the glacier to slide more easily along its bed. Glaciers in warm climates are more likely than those in extremely cold climates to move by both basal sliding and internal deformation; as a result, they usually flow faster.

Ice flow typically begins when a glacier's thickness exceeds about 60 meters (200 feet), the depth at which the weight of the overlying ice begins to deform ice crystals internally. This process largely explains why glacial *crevasses*—the deep, wide tears in the ice surface that pose great hazards to glacier visitors—are rarely more than 60 meters deep. (Crevasses can be seen clearly in the ablation zones depicted in Figures 17-4 and 17-7.) Crevasses form because the near-surface ice is cold and brittle, and therefore capable of frac-

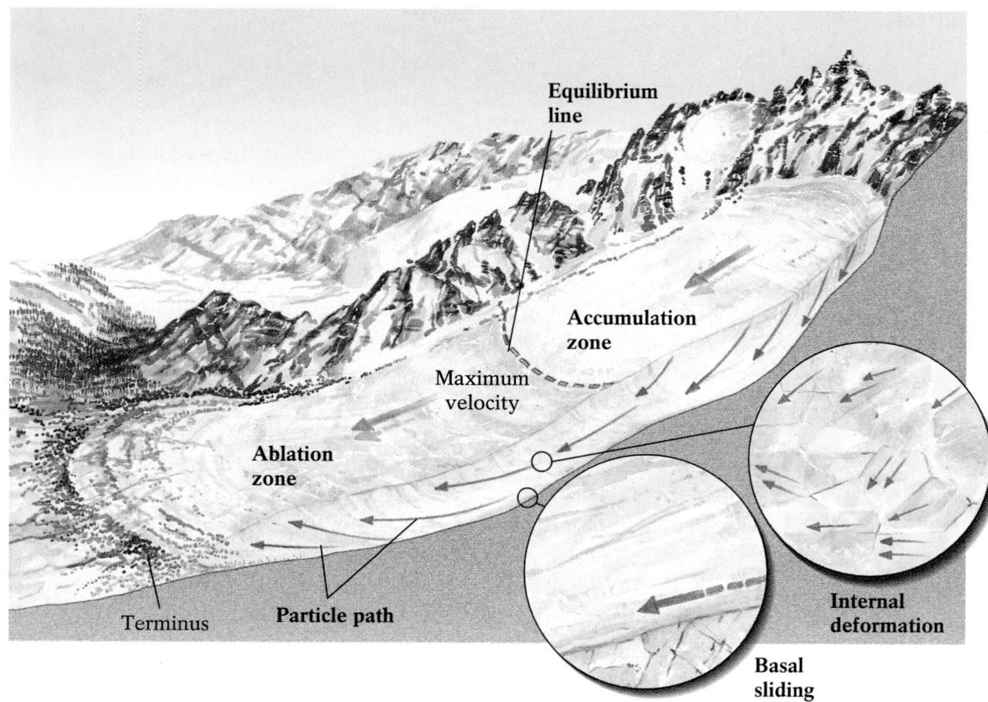

Figure 17-8 The mechanics of glacial flow. In general, particles within a glacier move downward in the accumulation zone, parallel to the ground at the equilibrium line, and upward in the ablation zone. Movement of the glacier as a whole results from a combination of internal deformation of its ice crystals and basal sliding.

turing. Below 60 meters, flowing ice prevents the development of crevasses. All of the factors involved in glacial flow are illustrated in Figure 17-8.

Velocity of Glacial Flow A glacier's velocity depends on factors inherent in both the ice and its location. The ice temperature controls the availability of water for basal sliding and enhanced ice-crystal slippage. In general, the closer the temperature of the ice is to its melting point, the more rapidly it tends to flow. A glacier's thickness contributes to its flow by providing the pressure that causes ice-crystal deformation. In addition, the steepness of the bedrock slope (under alpine glaciers) or the ice-surface slope (for continental ice sheets) affects glacial flow velocity by increasing gravity's effect on basal sliding. Typically, cold-based, nonsliding ice sheets flow slowly, usually advancing only a few meters per year; in contrast, warm-based alpine glaciers on steep slopes can flow 300 meters (1000 feet) or more annually.

Some glaciers periodically **surge,** or accelerate, usually in brief episodes lasting several months to a few years that are separated by longer periods (10 to 100 years) of normal flow. Surging glaciers can move as much as 100 times faster than their normal velocity. Surges apparently occur when a large volume of water accumulates at the base of a glacier, reducing the force of friction and facilitating basal sliding. The ice appears to be lifted from its bed by high water pressure, much like a car that hydroplanes on wet highway pavement. Although we cannot observe the action of the water directly, the large quantity of water that flows from a glacier's terminus as a surge comes to an end supports this hypothesis.

The most rapid surge known, on the Kutiah Glacier in northern Pakistan, reached a peak velocity of 110 meters (350 feet) *per day* and lasted for several months. Another newsworthy glacier surge took place during the summer of 1986 on the Hubbard Glacier in the St. Elias Mountains near Glacier Bay National Park in southeastern Alaska (Fig. 17-9). Its velocity increased from its usual 30 to 100 meters (100–300 feet) per year to about 5 kilometers (3 miles) per year. Advancing across the mouth of Russell Fjord, the surging glacier blocked the outlet of a large marine bay, turning it into a rapidly filling freshwater lake and separating thousands of marine mammals from their accustomed saltwater habitat. As the lake's level rose to about 20 meters (65 feet), the water swamped trees along the shore and threatened to overflow the Situk River. Fortunately, the surging ceased and the ice in the fjord broke up, restoring the bay.

Figure 17-9 The Hubbard Glacier surging. Basal melting caused the glacier to surge in the summer of 1986, clogging the mouth of Russell Fjord with ice and turning a large marine bay into a lake.

The Work of Glaciers

We stated at the start of this chapter that many striking and familiar landforms represent the handiwork of past glaciers. The processes of glacial erosion and deposition, which we describe in this section, have been more effective than almost any other geological process in shaping the features of North America, northern and central Europe, and Asia, especially the alpine highlands on these continents.

Glacial Erosion

Glaciers remove materials from the landscape by erosion. The greatest amount of erosion is caused by quickly flowing glaciers that move largely by basal sliding, particularly within the accumulation zone immediately upglacier of the equilibrium line. There the ice flows rapidly and obliquely downward, impinging on the glacier's bed. Glaciers erode by abrasion and by quarrying. **Glacial abrasion** occurs when rock fragments embedded in the base of a glacier scrape the surface of underlying bedrock like sandpaper on wood. A glacier whose basal ice contains fine, gritty rock fragments may eventually polish the rock surface below it to a high shine. Ice carrying coarser rock fragments may cut long **striations,** or scratches, into the bedrock surface, like those seen in Figure 17-10; these striations may range in size from barely visible bedrock scratches to sizable grooves, meters wide and deep. Striations are usually oriented in the same direction as the ice flow, and they indicate the direction taken by ancient glaciers.

Glacial abrasion is enhanced when: 1) a steady supply of fragments sustains abrasion; 2) the glacier's base contains fragments that are harder than the underlying bedrock; 3) the sliding velocity of the ice is rapid (a circumstance typically associated with warm-based glaciers in relatively temperate environments); and 4) the underlying bed consists of readily eroded materials. Deep striations occur when a rapidly sliding glacier carries a large supply of durable pebbles or cobbles (such as granites or gneisses) over soft sedimentary bedrock (such as shale or limestone). The abrading stones themselves become abraded and may eventually be pulverized into a silty rock powder called *glacial flour,* which imparts a distinctive gray, milky tinge to rivers that flow through glaciated regions and receive this material as sediment.

Glacial quarrying occurs when a glacier lifts masses of bedrock from its bed. It is most likely to move bedrock that is jointed or fractured, whether by preglacial tectonic activity or by frost wedging. (Frost wedging, discussed in Chapter 5, is common in the frigid climates associated with advancing glaciers, and in places where meltwater at a glacier's base seeps into bedrock cracks and refreezes.)

Working together, abrasion and quarrying can shape a rock mass into a distinctive asymmetrical form—a *roche moutonnée* (from the French, meaning "sheep-like rock") (Fig. 17-11). A glacier typically abrades the upglacier side of

(a)

(b)

Figure 17-10 Bedrock striations produced by glacial abrasion. **(a)** Glacially striated bedrock in Blue Mounds State Park in southwestern Minnesota. **(b)** Large glacial grooves in bedrock on Slate Island in Lake Superior, Ontario, Canada.

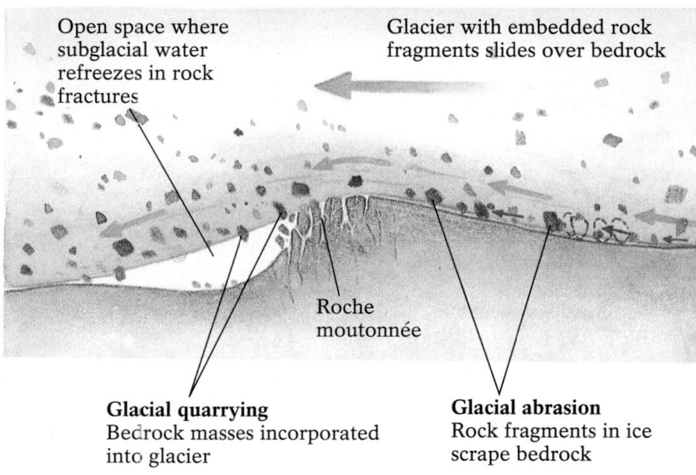

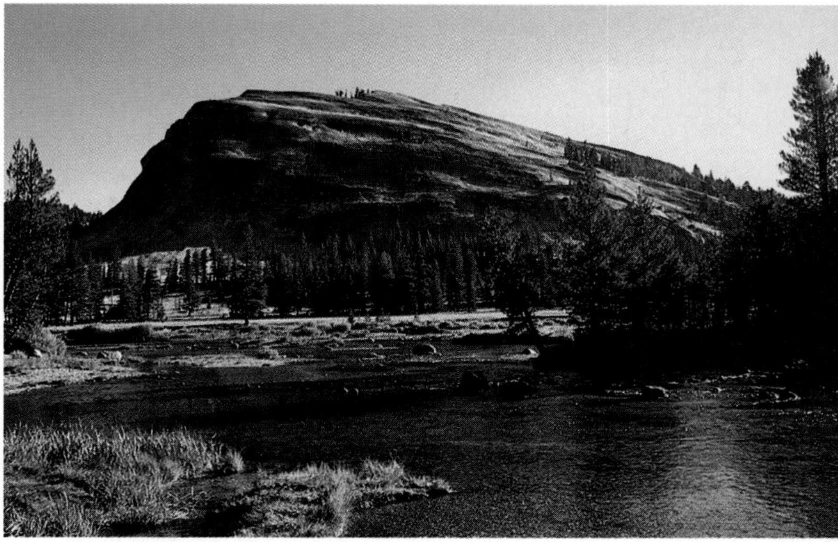

Figure 17-11 Glacial abrasion and quarrying combine to shape a roche moutonnée, like this one in Yosemite National Park, California. This distinctive erosional feature is produced when a glacier overrides a bedrock hill, abrading its upglacier side and quarrying its downglacier side. In this photo, the glacier flowed toward the left.

the rock mass (the side from which the glacier advances) but deeply quarries the downglacier side (the side toward which the glacier flows). A roche moutonnée is aligned lengthwise in the direction of glacial flow. Its upglacier side consists of a gentle humpbacked slope, whereas the downglacier side is typically steep and jagged. A roche moutonnée's asymmetry provides another indicator of the direction of glacier flow.

Erosional Features of Alpine Glaciation Alpine glacial erosion can sculpt smooth mountain slopes into spectacular rough-hewn peaks and precipitous gorges. The process begins with the creation of a **cirque,** a deep horseshoe-shaped basin that develops in a shady depression at or above the local snowline. In these areas, lack of sunlight and accumulation of wind-drifted snow support the growth of perennial snowfields. During periods of partial thaw, meltwater from these snowfields seeps into the surrounding bedrock. When the weather grows colder, the water refreezes, loosening rock fragments by frost wedging. Mass movement and surface runoff subsequently remove some of the loosened debris, enlarging the basin. As snow accumulates in the basin, it becomes converted into glacial ice, forming a *cirque glacier.* Quarrying and abrasion by the glacier's flow continue to erode the cirque, making it progressively deeper and longer.

Several cirque glaciers can form on the same mountain at the same time, producing various unique alpine landforms (Fig. 17-12). When two cirque glaciers on opposite sides of a mountain erode headward and converge, they form a sharp ridge of rock at their top and sides called an **arête** (from the French, meaning "fishbone," so-named because a series of arêtes resembles a fish skeleton when viewed from the air). Continued headward erosion of a cirque can breach part of an arête, producing a mountain pass, or **col** (from the Latin *collum,* meaning "neck"). Cols sometimes provide natural routes through imposing mountain ranges; Colorado's State Route 40, for example, traverses Berthoud Pass, a col through the Rockies near Winter Park. When three or more cirque glaciers erode a mountain, a steep peak, or **horn**, develops. Although the Matterhorn in the Swiss Alps is perhaps the best known of these forms, horns also dot the skyline of Banff and Jasper National Parks in the Canadian Rockies and Glacier National Park in northwestern Montana.

If the climate warms over time, a cirque glacier may melt away completely, leaving a basin that may later fill with water to form a cirque lake, or **tarn.** If the climate cools and the glacier's budget remains consistently positive, a cirque glacier may spread beyond its basin and flow downslope along a preexisting stream valley. As it moves, the glacier will erode the valley floor and walls, ultimately transforming the narrow V-shaped stream valley into a broad-floored U-shaped glaciated valley (Fig. 17-12b).

If cooling continues, small glaciers may develop in tributary stream valleys and eventually join to form a main valley glacier. The main glacier will be thicker and flow faster than its tributaries, eroding its underlying bedrock more rapidly. Eventually, it may undercut the tributary valleys, leaving them perched above the main glacier as hanging valleys. If a warm period causes melting, waterfalls will cascade from the hanging valleys; Yosemite Falls in Yosemite National Park, with a vertical drop of 735 meters (2425 feet), is one example of this phenomenon.

When seawater submerges a U-shaped, glaciated *coastal* valley, it forms a deep, saltwater-filled **fjord** (Fig. 17-12c). Some fjords were once thought to be stream valleys drowned by rising sea levels. The depth of most fjords, however, exceeds the postglacial rise in sea level. Tidewater glaciers erode their valleys below sea level, thereby carving them to increasingly greater depths.

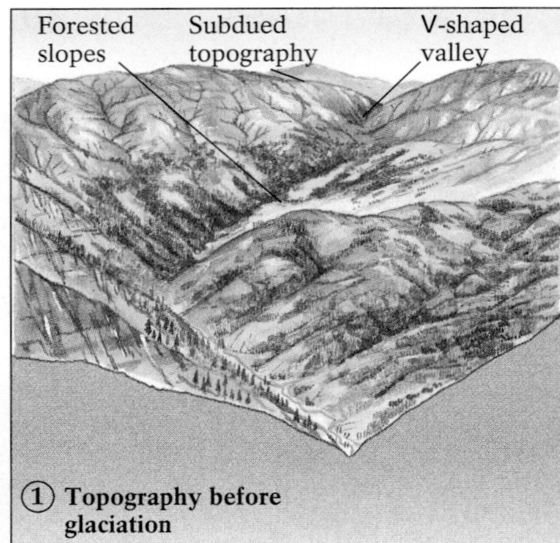

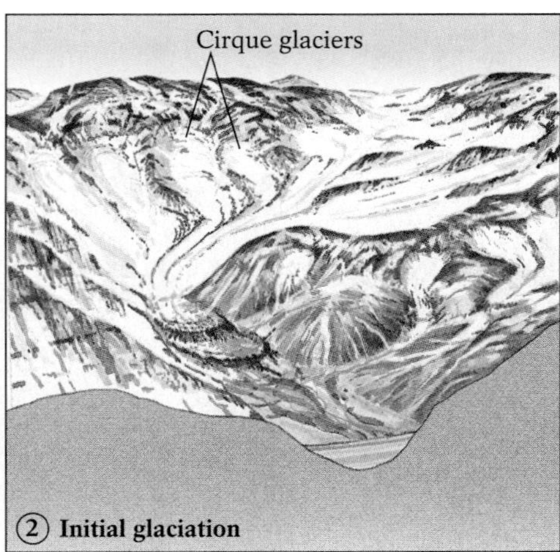

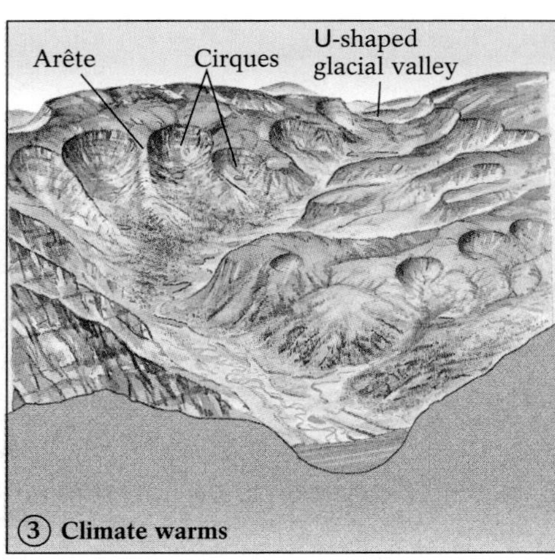

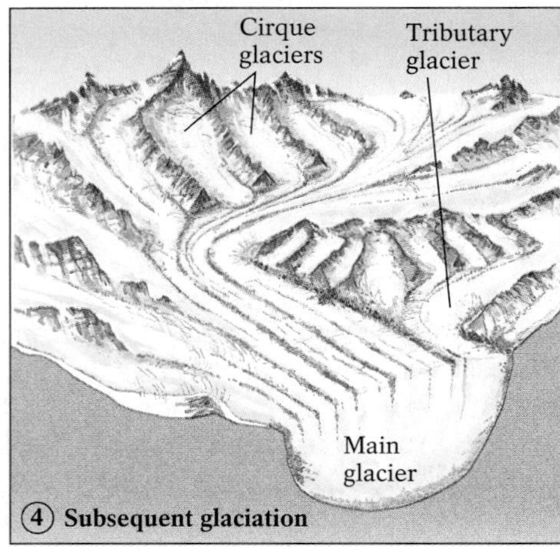

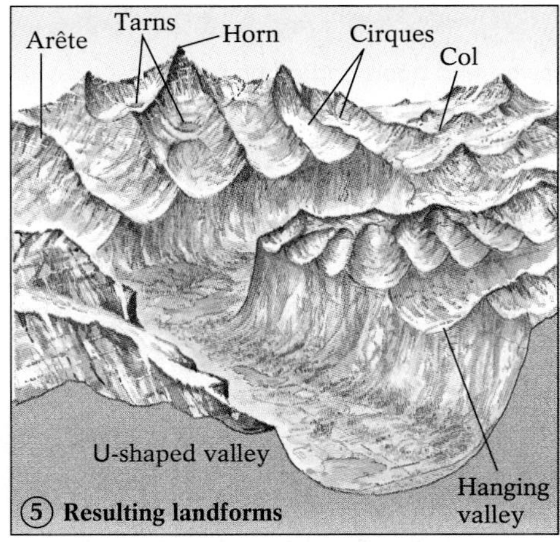

(a)

(b)

(c)

Figure 17-12 Effects of alpine glacial erosion. **(a)** Growth and proliferation of cirques on mountainsides may eventually produce horns, arêtes, cols, hanging valleys, and fjords. Another common indicator of glacial erosion is the conversion of V-shaped stream valleys into U-shaped glaciated valleys. **(b)** A U-shaped valley in Tracy Wilderness, southeastern Alaska. **(c)** U-shaped valleys that become flooded with seawater produce fjords; this one is Bela Bela Fjord in British Columbia, Canada.

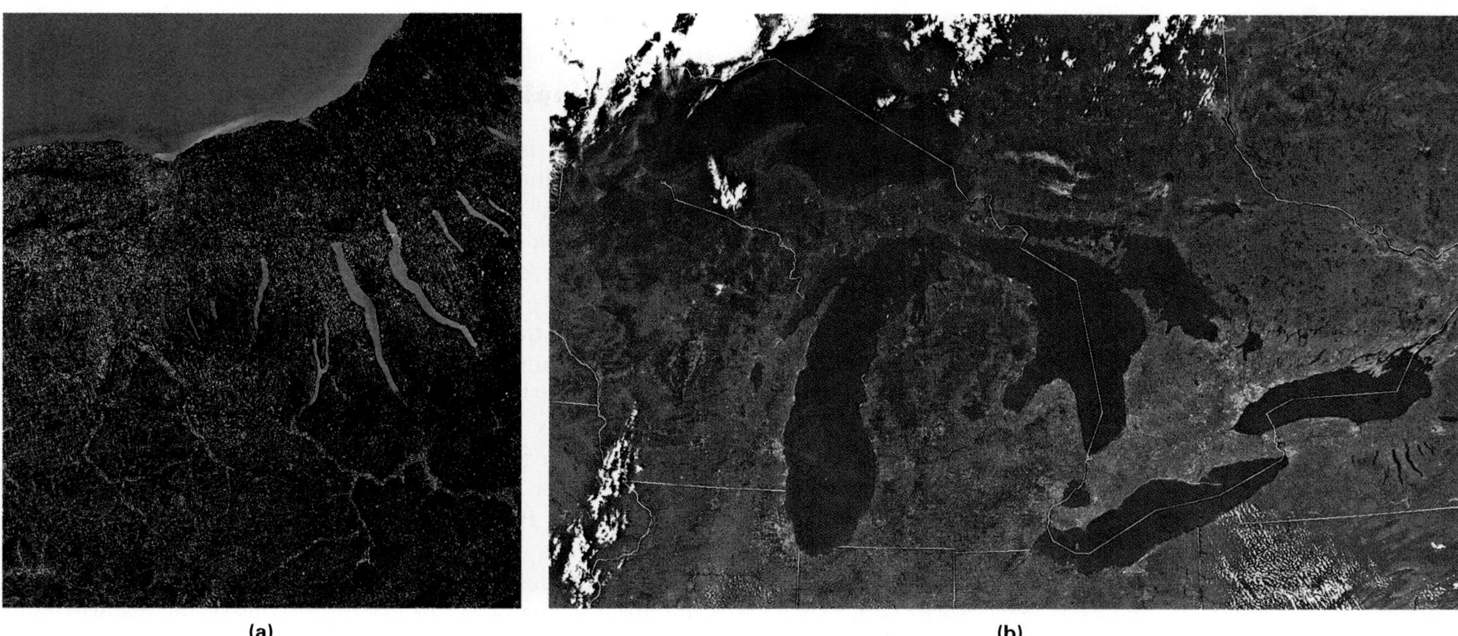

(a) (b)

Figure 17-13 Continental ice sheets erode preglacial lowlands to leave deep basins and valleys. **(a)** During the Pleistocene Epoch, ice accumulating in the Lake Ontario basin (in what is now upstate New York) overflowed and moved south over the landscape, enlarging preexisting stream and river valleys to form the deep linear troughs known collectively today as the Finger Lakes. Lake Ontario, itself a glacially carved basin, lies to the north of the Finger Lakes in this satellite photo. **(b)** A satellite photo showing the five Great Lakes of North America (from left to right, Lakes Superior, Michigan, Huron, Erie, and Ontario). The much smaller Finger Lakes are shown at lower right. *State and international boundaries added for reference.*

Erosional Features of Continental Ice Sheets Erosion by continental ice sheets produces striations and roches moutonnées as well as much larger features. An ice sheet can even abrade entire mountains, producing large-scale landforms called *whalebacks,* which are sculpted on both their upglacier and downglacier sides into a shape reminiscent—as their name suggests—of a whale's back. As a continental ice sheet flows through preglacial stream valleys, it carves monumental U-shaped troughs. The Finger Lakes of western New York—11 long, deep troughs—were excavated by ice-sheet erosion of preglacial valleys (Fig. 17-13a), as were North America's five Great Lakes (Fig. 17-13b), the northern section of Puget Sound in Washington, and Scotland's Loch Ness.

Continental ice sheets also erode on a continental scale. Because most of a glacier's erosion affects the bedrock under the downglacier segment of the accumulation zone near the equilibrium line (where the glacier's velocity is highest), the ice sheet that covered much of northern North America about 20,000 years ago scoured this part of the accumulation zone bare of loose rock and soil. The most intensive erosion occurred in the southern portion of the Canadian Shield and in the region of soft sedimentary rocks to the south, in southern Ontario and Quebec, Minnesota, Wisconsin, Michigan, Illinois, and New York. This pronounced erosion produced the Great Lakes, which are more than 500 meters (1600 feet) deep in places.

Glacial Transport and Deposition

A glacier eventually deposits the load of debris it has eroded, usually when its ice begins to melt. Glacial deposits collectively are called **glacial drift.** (The name originated in the nineteenth century, when people thought that such deposits had been carried to their resting places by icebergs floating on Noah's biblical floodwaters. We continue to use the term today, but with greater appreciations for the compositional diversity of these deposits.)

The size of the particles that a given flow of water or air can carry is limited to a narrow range, depending on flow velocity and turbulence. Deposits left by water or air are typically *well sorted* (sorting was discussed in Chapter 6). Because of the high viscosity of ice, however, even extremely large masses of embedded rock cannot settle through the ice, and thus are transported by glaciers. As a result, glacial deposits are typically *poorly sorted,* often ranging in size from fine clays to house-sized boulders. Most glacial drift is deposited after being carried only a few kilometers, but some materials can be transported for remarkably long distances. We can sometimes trace a rock that differs from most others in a glacial deposit to its source outcrop and determine its transport distance. For example, chunks of copper have been found in glacial deposits in southern Illinois; this copper is not native to Illinois but comes from Michigan's Upper Peninsula, 960

Figure 17-14 Large granite erratics on a lake shore in Superior National Forest, Minnesota. The erratics were carried by the Laurentide continental ice sheet about 15,000 years ago, and probably came from the northeast, passing over southwestern Ontario.

kilometers (580 miles) to the north. An exceptionally large, glacially transported boulder that has been eroded from one type of bedrock and then deposited atop another is known as a **glacial erratic** (Fig. 17-14).

Some glacial debris is deposited directly by the ice that carried it; other debris is picked up and later deposited by meltwater streams that flow on, within, beneath, and in front of the ice. Meltwater streams deposit glacial sediment wherever their paths take them—on land, into the lakes that form in front of the ice, or into the sea.

Till Drift that is deposited directly from glacial ice forms a distinctive sediment called **glacial till** (from the Scottish word for "obstinate," an allusion to Scotland's plow-resisting rock-and-clay soils). As shown in Figure 17-15, till is deposited almost exclusively in the ablation zone, either by being plastered onto the underlying glacial bed by flowing ice or by sloughing off the glacier's surface as the ice melts. Tills are characteristically unstratified and unsorted, and they generally consist of large rock fragments surrounded by a finer-grained matrix of sand, silt, and clay. Because the particles in a till were eroded from the bedrock over which the glacier passed, geologists can identify the rock types that occurred along the glacier's route; they can then use this information to determine the glacier's flow direction.

Sometimes the stones in till have economic value. Diamonds, for example, are rare in the tills of the Midwest and Great Lakes region, but they have been discovered at numerous localities (Fig. 17-16). In 1876, in a southeastern Wisconsin town, a well digger unearthed a pale yellow stone and gave it to his employer. Thinking it was common quartz or topaz, he later sold it for one dollar. The stone, which turned out to be a 15-carat diamond, was eventually sold to the American Museum of Natural History in New York City (from which it was stolen in 1964 and never recovered).

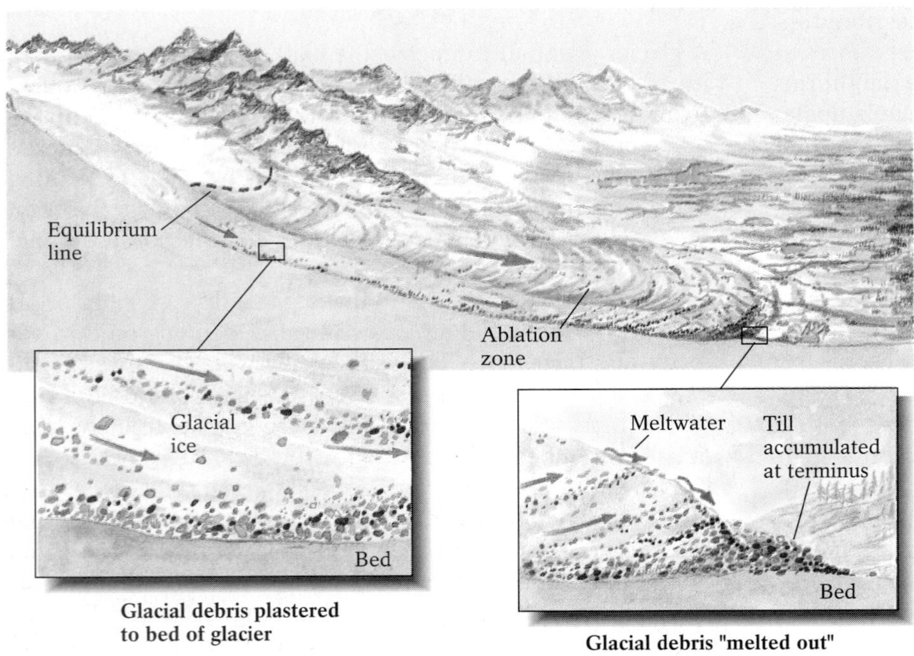

Figure 17-15 Deposition of glacial till. Tills are deposited downglacier from the equilibrium line, either by being plastered to the glacial bed by the weight of the glacier or by "melting out" at the glacier's terminus.

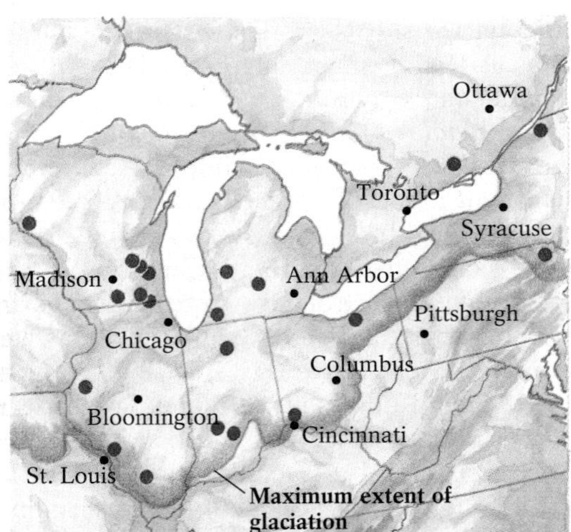

Figure 17-16 Known locations of diamond-bearing tills around the Great Lakes.

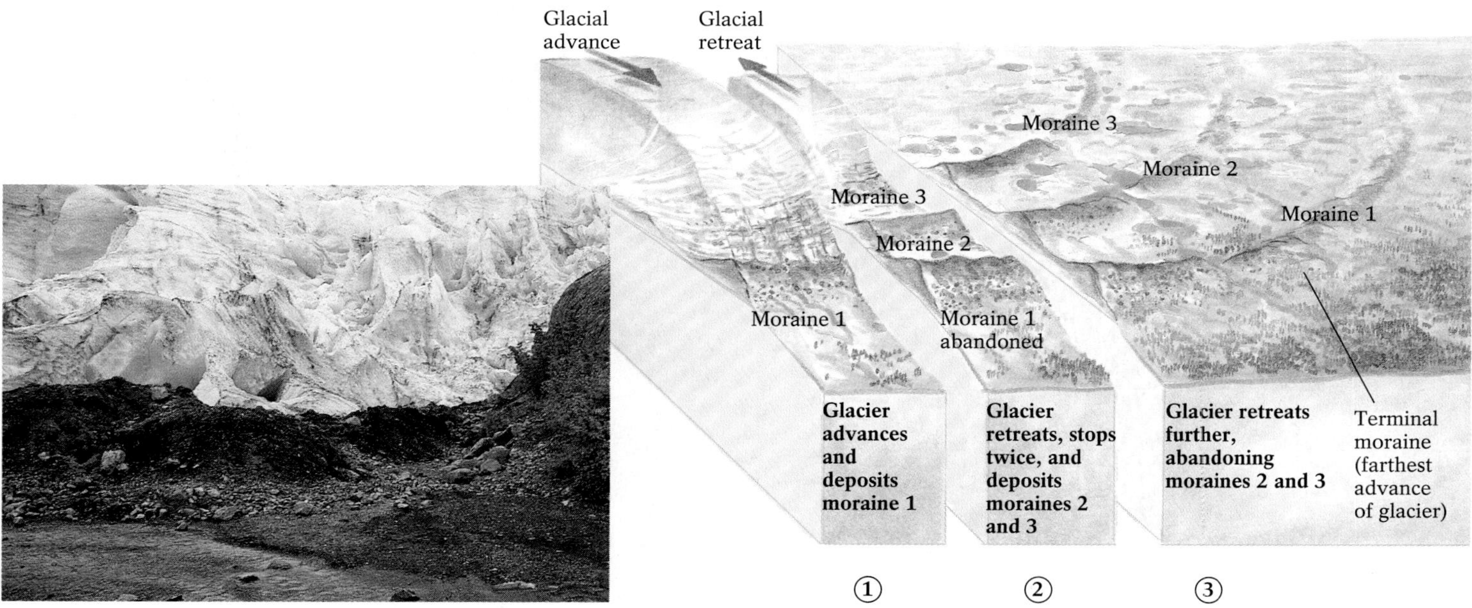

Figure 17-17 Glacial advance and retreat and the deposition of moraines. At the end of the sequence shown, moraine 1 is the glacier's terminal moraine, marking the maximum extent of the glacier. Moraines 2 and 3 are recessional moraines, marking the positions of the glacier's terminus during periods when the glacier's retreat was halted by temporary cooling trends. Inset: A moraine deposited by Exit Glacier at Kenai Fjord in Alaska.

Glacial till often forms a **moraine,** a landform that typically accumulates at the margin of a glacier. (The name comes from *morena,* a word used by farmers to describe ridges of rocky debris in the French Alps.) Moraines occur as bands of hills marking the various advances and retreats of a glacier (Fig. 17-17). Even when the terminus of a glacier remains stationary, ice continues to flow from the accumulation zone to the ablation zone; consequently, it erodes and transports sediment to the terminus where the load piles up. In general, the longer the terminus remains stationary, the more till the glacier will deposit in one place and the larger the resulting moraine. If the climate warms and ablation begins to exceed accumulation, the glacier recedes, leaving a *terminal moraine* that indicates the farthest advance of the ice.

If a return to glacial conditions interrupts warming, glacial recession halts and the terminus of the receding glacier may remain stationary, perhaps for a few hundred years or more. During this period, the glacier may deposit a *recessional moraine* upglacier from and usually parallel to its terminal moraine. A glacier that recedes intermittently may deposit several recessional moraines. As you can see in Figure 17-18, continental ice sheets, like smaller alpine glaciers, produce terminal and recessional moraines, such as the ones on New York's Long Island and Massachusetts' Cape Cod.

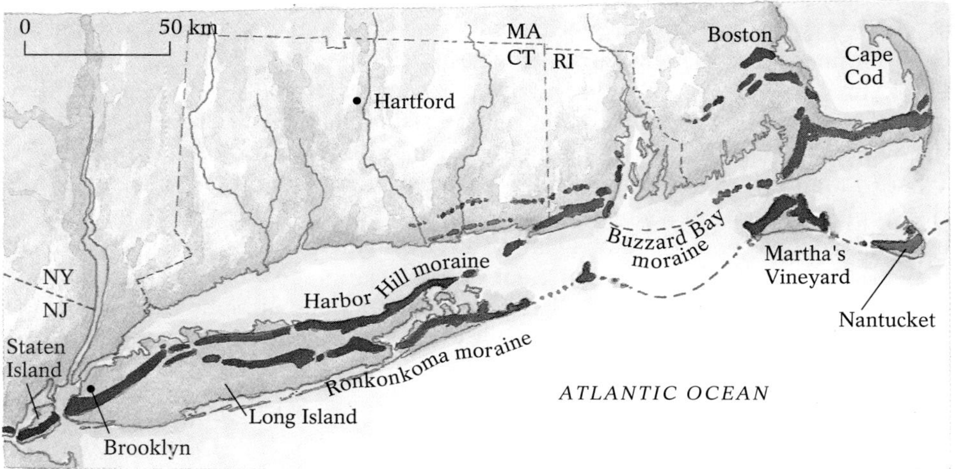

Figure 17-18 Terminal and recessional moraines were deposited over a large area of the northeastern United States coast by the North American ice sheet. The recessional Harbor Hill moraine, deposited about 14,000 years ago, passes through Prospect Park in Brooklyn, New York, continues across the north shore of Long Island, and then appears farther north, where it forms the mid-island hills of Cape Cod. The terminal Ronkonkoma moraine, deposited about 20,000 years ago, forms central Long Island and continues east and then north to Martha's Vineyard and Nantucket Island in Massachusetts Bay. The map indicates local names for the moraines, such as "Buzzard Bay Moraine" and "Vineyard Moraine."

Figure 17-19 Medial moraines form when adjacent glaciers converge, causing the lateral moraines at their edges to run together. Here several medial moraines are developing along the branches of the Kennicott Glacier in the Wrangell–St. Elias National Park, southeastern Alaska.

Two other types of moraines—lateral and medial—are formed only by alpine glaciers. Because they are topographically confined by their valleys, these glaciers have distinct sides, located adjacent to the valley walls. Till is deposited in these sites, forming *side* or *lateral moraines.* Lateral moraines also contain material that has fallen from the valley walls into the crevices between the ice and the walls. When two valley glaciers merge, their lateral moraines, which are carried downvalley along the sides of the two distinct ice streams, merge as well. The combination of these lateral moraines forms *medial moraines*—moraines located within a large valley glacier between two merged ice streams. You can see several medial moraines in Figure 17-19, where they appear as the dark-colored bands interspersed between the streams of lighter-colored ice.

If the climate cools significantly after glaciers have remained stationary for a long period of moraine building, the glaciers may advance over previously constructed moraines, incorporating their sediments and redepositing them at "new" terminal moraines. When ice sheets—thicker and more massive than alpine glaciers—override such moraines, they may apply enough pressure to reshape the moraines into gently rounded, elongated hills called **drumlins** (from the Gaelic *druim,* meaning "ridge" or "hill"). A drumlin has an asymmetrical "streamlined" profile similar to that of the inverted bowl of a kitchen spoon (Fig. 17-20). All drumlins are characteristically blunt on their upglacier ends and gently sloping on their downglacier ends—a shape that some geologists suggest is a consequence of their formation beneath actively flowing ice.

Drumlins, which range in height from 5 to 50 meters (17–170 feet), are typically aligned parallel to one another and oriented in the direction of ice flow. They generally occur in clusters of hundreds or thousands in areas once covered by continental ice sheets. One drumlin field, located east of Rochester in west-central New York, contains more than 10,000 drumlins. Other notable drumlin sites include western Nova Scotia, southern Manitoba and Saskatchewan, central Minnesota, eastern Wisconsin, and northern Michigan. Perhaps the continent's most famous drumlins are north of Boston, where the Battle of Bunker Hill (itself a drumlin) was fought on Breed's Hill, an adjacent drumlin.

In recent years, as with many geological phenomena, glacial geologists have rethought the origin of drumlins. Although many of these oddly shaped hills may have formed by glacial overriding and reshaping, as we described, others may have formed principally by glacial erosion, and still others by subglacial floods of meltwater.

Deposits from Glacial Meltwater Streams Melting may be continuous near the end of a period of glaciation. Alternatively, a glacier may melt seasonally, especially during the summers in temperate climates. In any event, meltwater streams flowing from a glacier sort and stratify the drift they deposit, creating glacial sediments that differ greatly from till.

Glacial meltwater may flow on top of, in front of, and beneath a glacier. Sediment deposited on top of a glacier fills in depressions in its surface; sediment deposited immediately in front of a glacier may form deltas in glacial lakes. The most common sediment from glacial meltwater, a mixture of sand and gravel, is deposited by braided streams downstream of the terminus as **outwash.** Outwash plains, characterized by broad, gently sloping surfaces, accumulate beyond the front of a glacier.

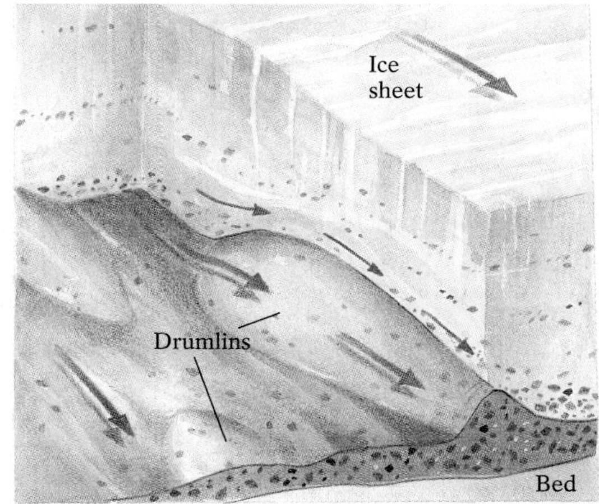

Figure 17-20 (left) Ice sheets passing over preexisting moraines can exert enough pressure to reshape them, forming low oval hills called drumlins. Drumlins are usually 25 to 30 meters (80–100 feet) high, 0.5 to 1 kilometer (0.3–0.6 miles) long, and 400 to 600 meters (1300–2000 feet) wide. Photo: A drumlin field east of Rochester, New York.

Figure 17-21 Fertile loess farmland in Washington state.

Powerful glacial-age winds erode drying outwash, producing a mass of fine silt. The fertile plains along the Mississippi River near Vicksburg, Mississippi, though far removed from any *direct* effects of glaciation, accumulated about 20,000 years ago as a thick layer of silt eroded from the outwash of the North American ice sheet. Although the ice sheet extended only as far as northern Iowa, its outwash was transported down the Mississippi River to the Gulf of Mexico and deposited along its banks. Winds blowing across the drying outwash eroded and transported the fine particles to the surrounding land. Such wind-blown silt deposits, known as **loess** (from the German for "loose"), commonly appear downwind from exposed, drying outwash. Throughout the Mississippi River valley (and beyond the fronts of European ice sheets as well) we can observe buff-colored, near-vertical bluffs composed of loess. (We will discuss loess in greater detail in Chapter 18.) The loess eroded from outwash plains provides some of North America's finest farmland (Fig. 17-21).

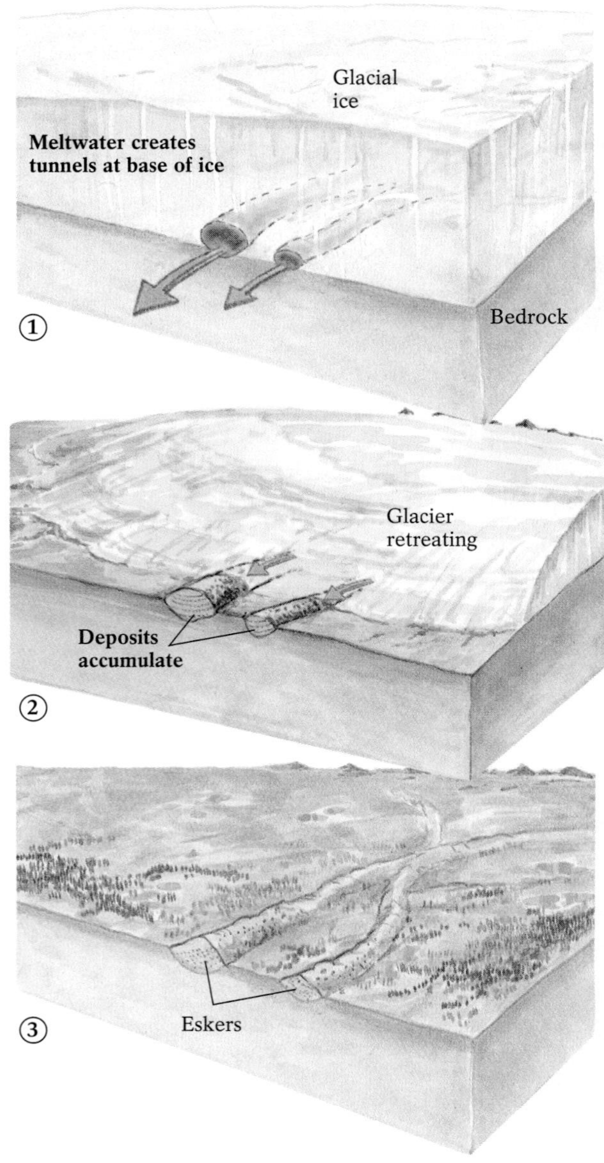

Figure 17-22 The origin of eskers. Eskers typically form under the ablation zone of a glacier, where meltwater erodes curved channels in the ice and deposits its load en route. When the glacier retreats, the meltwater flows away, leaving behind meandering ridges of stratified, cross-bedded sand and gravel. Photo: Eskers in Coteau des Prairies hills, South Dakota.

Meltwater deposits of sand and gravel can also accumulate underneath glacial ice, forming sinuous ridges known as **eskers.** Eskers form beneath the ablation zone, as frictional heat from turbulently flowing subglacial streams melts tunnels in the overlying ice. Later, when the glacier wastes away and meltwater flow diminishes, sand and gravel are deposited in these tunnels (Fig. 17-22). Flowing meltwater may also carve bedrock channels into soft subglacial sediment. Eskers are found throughout southern Canada and in much of the northern United States, including Maine, Michigan, Wisconsin, Minnesota, North and South Dakota, and eastern Washington. They may be less than a kilometer long or more than 150 kilometers (100 miles) long, and can reach 30 meters (100 feet) high. If you're interested in seeing an esker, you may have to act quickly—eskers are being mined extensively for their commercially valuable sand and gravel and thus are disappearing rapidly from the landscape.

Other Effects of Glaciation

In addition to the direct effects of glacial erosion, transport, and deposition, glaciers and glacial periods produce a variety of other direct and indirect changes on land, at sea, and in the atmosphere. They may even contribute to the evolution, migration, and extinction of plants and animals. In this section, we describe some signs of past worldwide glacial expansion.

Glacial Effects on the Landscape Glaciers and the conditions associated with glacial periods can alter the landscape in a number of ways other than by erosion and deposition. The movement of the ice can disrupt preexisting features, and its weight can depress and uplift specific areas. Changes in climate can create vast areas of frozen ground and large landlocked bodies of water.

Expanding ice sheets can act like continent-wide dams, altering the courses of preexisting streams. The Missouri River and its tributaries, for example, once flowed north to Hudson Bay until the North American ice sheet blocked and diverted them. The Missouri River's present-day route, es-

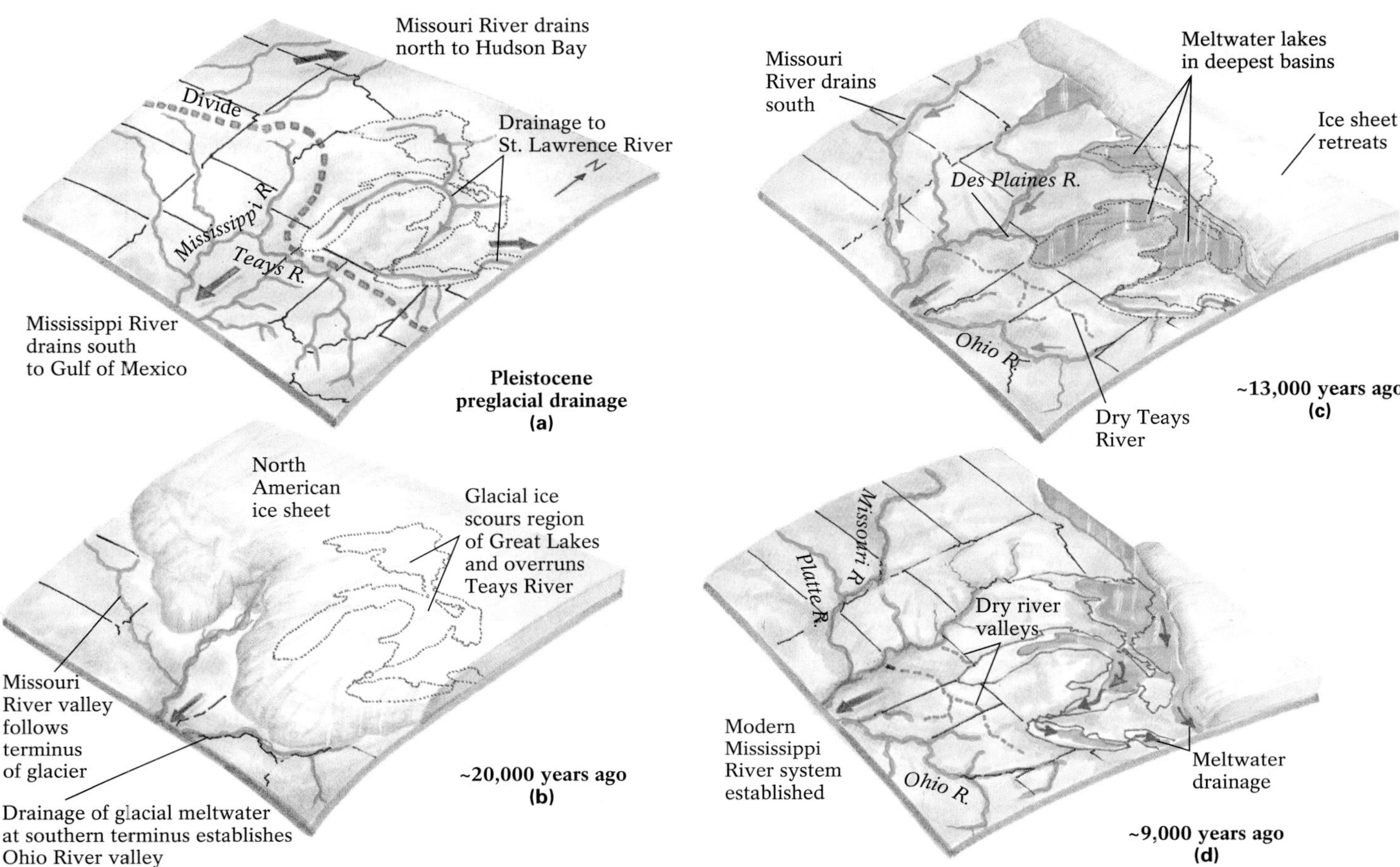

Figure 17-23 The effect of the North American ice sheet on the Great Lakes region. **(a)** Before the period of glacial expansion in the Pleistocene, streams flowed through the broad valleys that today are the Great Lakes. The major east–west tributary of the Mississippi River was the Teays River, which traversed what are now Illinois, Indiana, and Ohio. North of the Teays was an elevation that served as a divide; north of this divide rivers such as the Missouri flowed northward. **(b)** At the maximum extent of Pleistocene glaciation about 20,000 years ago, the North American ice sheet completely covered the Great Lakes region, deepening the lakes' basins by erosion. It also blocked the flow of the Missouri River, diverting it southward. **(c)** With warming, the ice sheet melted and retreated northward, leaving accumulations of meltwater in the deepest basins. By about 13,000 years ago, the Great Lakes and the Mississippi and its major tributaries were filled. Water in shallower basins, such as the Teays River, flowed away or evaporated, leaving dry valleys. **(d)** By 9000 years ago, the ice sheet had retreated from what is now the United States, and the current drainage system of this region was well-established.

tablished during the last period of worldwide glacial expansion, follows the southern terminus of the departed ice sheet and then joins the Mississippi River on its way south to the Gulf of Mexico. Figure 17-23 shows the sequence of events that led to the Missouri River's diversion. Highlight 17-1 chronicles a similar drainage disruption in the Pacific Northwest that produced some of the most monumental floods in Earth history.

A continental ice sheet weighs so heavily on the Earth's crust that it pushes the asthenosphere (the heat-softened, flowable layer of the upper mantle discussed in Chapter 11) away from the center of the ice sheet and toward its margins. This weight creates a crustal depression under the glacier and topographic bulges along its borders. When the climate warms and the ice sheet melts, the displaced asthenosphere gradually returns to its original position, the depressed crust slowly rebounds, and the marginal bulges disappear. (The changes in the Earth's crust are analogous to what happens to the surface of a water bed when you lie on it. The weight of your body pushes the water from beneath you, forming a depression in the bed's surface and a bulge toward the edges of the bed. When you get up, the water at the edges flows

Highlight 17-1 The Channeled Scablands of Eastern Washington

Today the Clark Fork River of northwestern Montana is a freely flowing mountain stream. Some 13,000 years ago, however, it became blocked by a broad lobe of ice protruding from a retreating Cordilleran ice sheet, which covered western Canada to the Canadian Rockies (Fig. 17-24a). The lobe of ice, known as the Purcell Lobe, flowed through a wide U-shaped valley extending from southern British Columbia to northern Idaho; in the process it dammed the Clark Fork's westward flow, creating a lake that became swollen by meltwater from the wasting ice. Known today as Glacial Lake Missoula, it was more than 300 meters (1000 feet) deep and occupied an area about the size of present-day Lake Michigan.

As meltwater continued to flow into it, Glacial Lake Missoula rose until it overflowed the ice dam, carving deep channels into the ice. Eventually the ice dam ruptured, releasing the accumulated waters of Glacial Lake Missoula in a sudden torrent (Fig. 17-24b). More water flooded eastern Washington than the combined flow of all rivers in the world today. Based on the sizes of deposited boulders, the flood's velocity has been estimated at from 50 to 75 kilometers (30–50 miles) per hour.

The effects of this episode are still evident today in the Pacific Northwest (Fig. 17-25). The flood produced numerous rivers up to 75 kilometers (48 miles) wide and 250 meters (900 feet) deep, whose courses are now marked by giant dry waterfalls and massive streambed ripples. It also excavated enormous channels by tearing hundreds of meters of basaltic lava from the Columbia River plateau, leaving thousands of remnant lava *scabs*—masses of basaltic rock that survived the flood intact. The region is aptly named the Channeled Scablands.

Geologists speculate that such great floods occurred repeatedly in eastern Washington during the Pleistocene ice age, as glaciers periodically advanced and retreated. We know that the last flood took place about 13,000 years ago, because Mount St. Helens erupted at the same time. Ash layers from this eruption are interlayered with the last of the flood deposits and provide geologists with a dating marker. Massive ice sheets and unimaginable floods, plus a fiery volcanic cataclysm, make Washington state of 13,000 years ago a geologist's dream: The evidence left behind gives geologists a storehouse of information about the region's recent past.

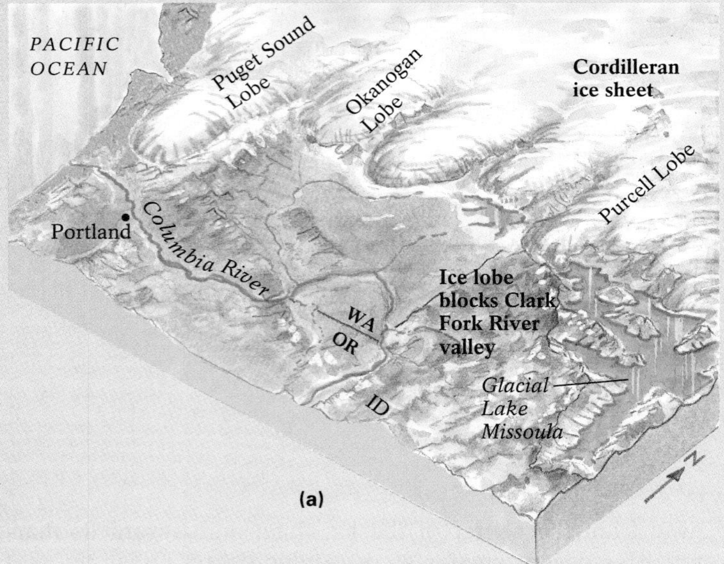

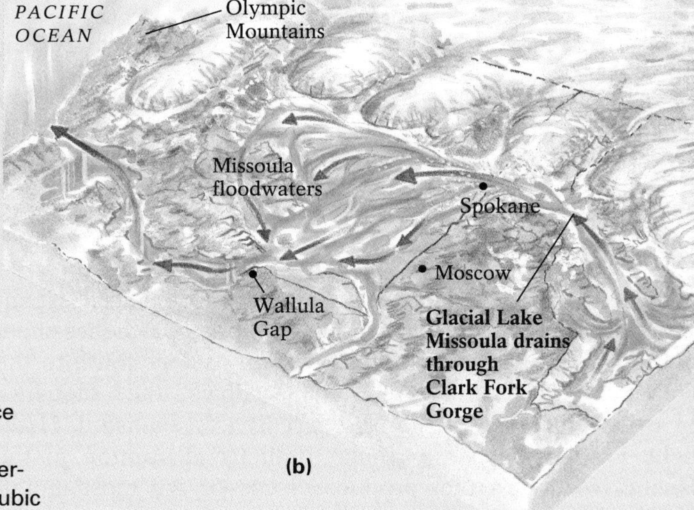

Figure 17-24 The creation and draining of Glacial Lake Missoula. **(a)** About 13,000 years ago, the Purcell Lobe of the retreating Cordilleran ice sheet blocked the Clark Fork River valley, creating Glacial Lake Missoula. **(b)** As the glacier retreated, its meltwater swelled the lake until it finally overflowed its dam, releasing more than 20 million cubic meters (750 million cubic feet) of water *per second*.

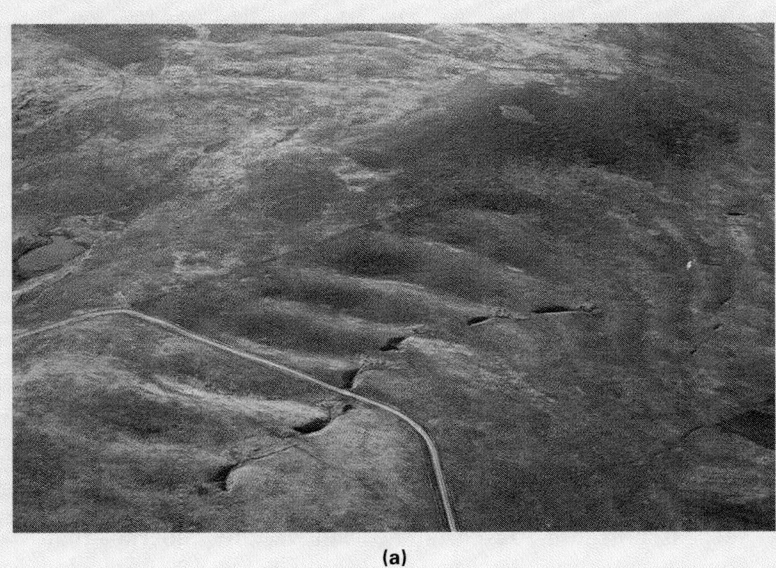

(a)

(b)

(c)

Figure 17-25 Effects of the Missoula flooding. **(a)** Massive rivers created by the flooding produced giant bedrock ripples, such as these in northwestern Montana. **(b)** This basalt bedrock near Coulee City, Washington, was etched by dry channels and falls where floodwaters once streamed. **(c)** As Glacial Lake Missoula grew, its rising shoreline carved these wave-cut benches seen on the side of Sentinel Mountain at the University of Montana campus.

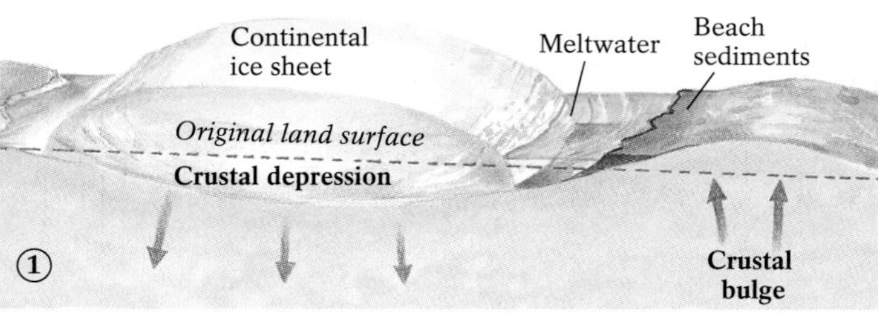

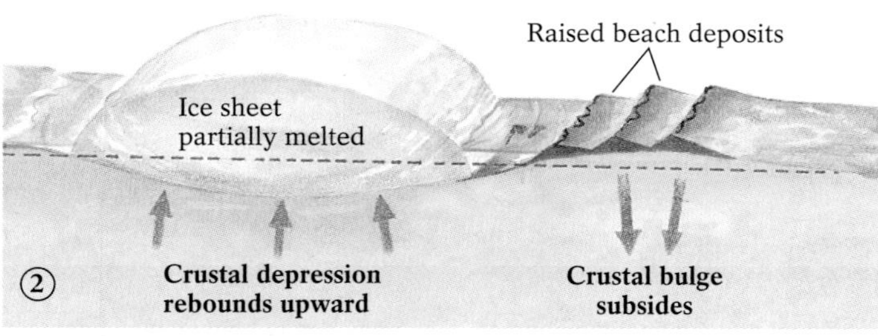

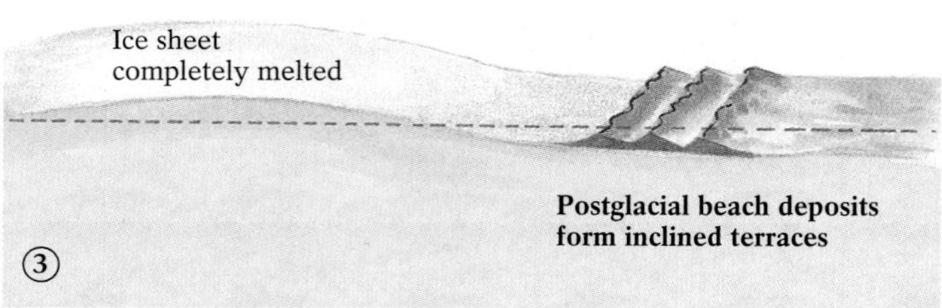

Figure 17-26 (left) Formation of terraces from crustal rebound. Meltwater accumulates in the space between a retreating ice sheet and the crustal bulge at its margin, leaving behind sediments. As the ice sheet retreats further and the crust slowly returns to its original position, the bulge subsides, leaving the sediment deposits as slightly inclined terraces. Photo: Isostatically raised beaches in Canada.

back to its original position, eliminating the depression and bulge.) We observe the effects of crustal rebound throughout Scandinavia and around the Hudson Bay region of east-central Canada, in the form of uplifted terraces (Fig. 17-26). The process of isostatic rebound (discussed in Chapter 11) continues today in those areas, thousands of years after the departure of the ice. The reason the rebound occurs so slowly is that the asthenosphere is extremely viscous, and thus slow-flowing.

When a crustal depression causes ice-sheet margins to bulge and then later subside, the subsiding region may experience serious coastal floods. The southwestern Netherlands, located near the margin of the former European ice sheet, once stood well above sea level atop one such crustal bulge. Since the departure of the ice sheet about 12,000 years ago, the depressed crust has rebounded and its companion bulge has subsided, causing southwestern Netherlands to drop—in some places to a depth below sea level. Since the thirteenth century, only Holland's famous dikes have kept the North Atlantic from flooding its valuable farmland.

The nonglaciated areas adjacent to glaciers and ice sheets are characterized by extremely cold climates, and the pervasive frost action in such *periglacial* (near-glacial) environments produces a distinctive set of landscape features. (These features also appear in some high-latitude regions—such as northern Alaska, arctic Canada, and Siberia—that have extremely cold climates, despite being far from glaciers.) Perennially frozen ground, or **permafrost,** is characteristic of high-latitude periglacial environments. Today, permafrost underlies about 25% of the Earth's land surface, mostly in remote polar regions. In Siberia, permafrost extends to a depth of 1500 meters (5000 feet); sections in the Canadian Arctic and Alaska reach depths of 1000 meters (3300 feet) and 600 meters (2000 feet), respectively. Geologists believe that most of the world's permafrost developed during the last worldwide glacial period, between 90,000 and 12,000 years ago. The discovery of frozen woolly mammoths and other carbon sources in the permafrost of Siberia and Alaska has enabled carbon-14 dating of those deposits.

A permafrost layer consists of soil, sediment, and bedrock that is continually at or below 0°C (32°F). During the summer—even brief, high-latitude summers—the thin upper zone of permafrost, or *active layer,* thaws and becomes extremely unstable, and tends to trigger mass movements. Buildings and other structures in permafrost areas must be designed to minimize the flow of heat from the structures to the ground. Oil pipelines (carrying warm crude oil) and heated buildings are usually elevated above the ground surface, allowing cold air to circulate beneath them and keep the surface frozen.

The ground surface in periglacial areas is characterized by a variety of unique features, such as those seen in Figure 17-27. For example, as periglacial regolith alternately freezes and thaws, the frozen subsurface materials expand upward, slowly pushing stones up until they surface. The newly revealed stones often occur in surprising arrangements: *Sorted circles,* for instance, form as stones are pushed laterally into circular patterns. When soil temperatures plunge below 0°C

(a)

(b)

(d)

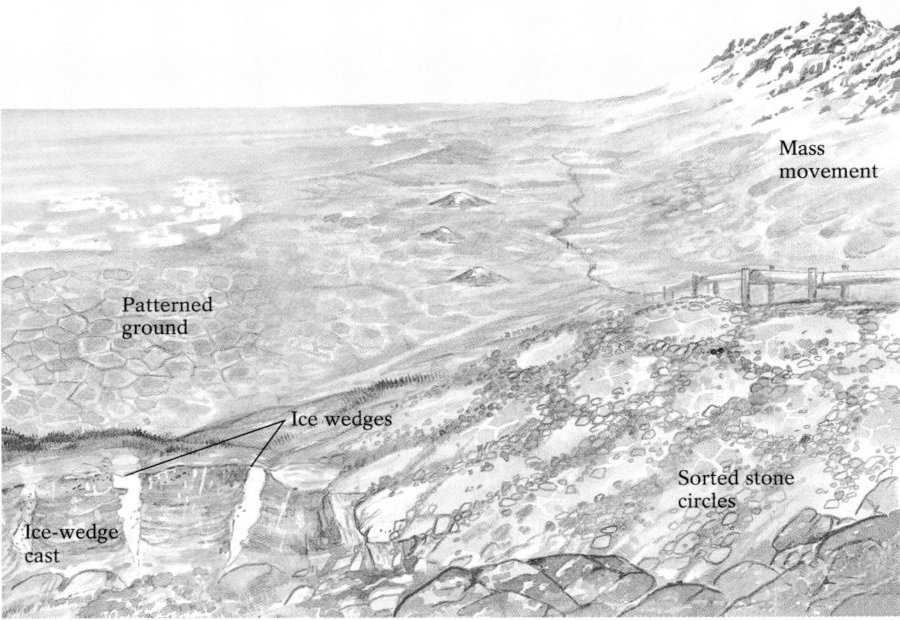

(c)

Figure 17-27 The features of a periglacial environment include the following: **(a)** patterned ground in the Canadian Arctic; **(b)** "slipped" areas—a product of *solifluction* (discussed in Chapter 13)—result from permafrost thawing and subsequent mass movement, such as this forest in Mount McKinley Park, Alaska; **(c)** sorted stone circles, such as these in western Spitsbergen, Norway; and **(d)** ice wedges, such as this one along the Yukon River in Galena, Alaska.

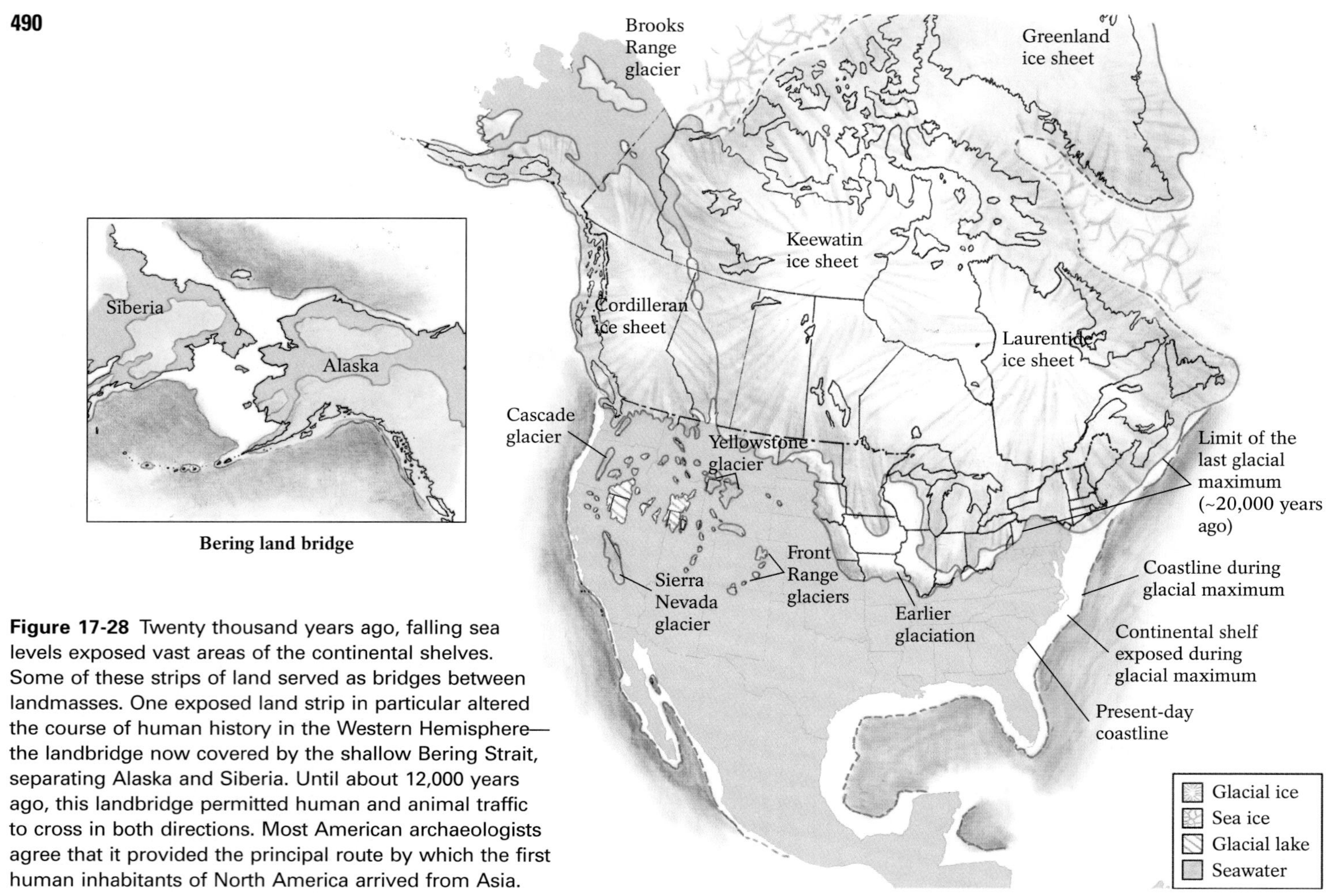

Figure 17-28 Twenty thousand years ago, falling sea levels exposed vast areas of the continental shelves. Some of these strips of land served as bridges between landmasses. One exposed land strip in particular altered the course of human history in the Western Hemisphere—the landbridge now covered by the shallow Bering Strait, separating Alaska and Siberia. Until about 12,000 years ago, this landbridge permitted human and animal traffic to cross in both directions. Most American archaeologists agree that it provided the principal route by which the first human inhabitants of North America arrived from Asia.

(32°F), the ground in periglacial areas freezes, contracts, and cracks to form large polygonal shapes, or *patterned ground.* When the surface water that flows during brief warm periods drains into the cracks at the edges of the polygons, it freezes and forms *ice wedges.* When these ice wedges eventually thaw, the water drains out, leaving cavities that may be filled by inwashing sediment that subsequently becomes compacted to form *ice-wedge casts.*

The presence of *relict* ("fossilized") periglacial features indicates that periglacial environments once existed along the front of the North American ice sheet from New Jersey and Pennsylvania west to the Dakotas and at higher elevations south along the crest of the Appalachians. Pleistocene-age ice-wedge casts in New Jersey and Connecticut, more than 1500 kilometers (1000 miles) south of the current permafrost limit, suggest that this region once represented the periglacial frontier of the North American ice sheet.

The indirect effects of ice sheets may extend well beyond periglacial regions. When air from warmer regions encounters air chilled by ice sheets, it produces cloudy, cool weather beyond the ice sheet's terminus. During the last glacial period, this type of humid and cool climate prevailed in North America. It may have produced more precipitation than falls today, *but most certainly less of this moisture evaporated.* Wherever rainfall exceeded this reduced level of evaporation, water accumulated in landlocked basins, forming **pluvial lakes** (from the Latin *pluvia,* meaning "rain"). This indirect effect of glaciation reached as far south as Death Valley, California, and the deserts of the American Southwest, covering vast areas of those now-arid lands with shallow lakes. The largest pluvial lake, Lake Bonneville, covered much of Utah, eastern Nevada, and southern Idaho; Utah's Great Salt Lake represents a small remnant of this water body (see Figure 6-16).

Glacial Effects on Sea Levels Because the water in glacial ice ultimately comes from the oceans, global sea levels drop sharply when glaciers expand worldwide. As a result, during glaciations, much currently submerged land—the continental shelves—becomes exposed and the coastlines of the continents change dramatically. For example, 20,000 years ago during the last glacial maximum, global sea level was about 130 meters (430 feet) lower than it is today. As a result, North America's Atlantic coastline extended more than 150 kilometers (100 miles) east of present-day New York City (Fig. 17-28). Fossilized tree stumps and mastodon bones recovered from North America's continental shelves reveal that the exposed continental shelves were covered by spruce and pine forests and traversed by herds of migrating ice-age mammals.

Lowered sea levels also expose "landbridges," connections between landmasses that are separated by shallow seaways in nonglacial times. These strips of land actually comprise exposed areas of the continental shelves. For example, between about 100,000 and 40,000 years ago, during a period of worldwide glacial expansion, Great Britain and France were connected by a landbridge that today is covered by the shallow English Channel. Humans migrated from the European mainland to the British Isles by walking over this landbridge. Similarly, until about 12,000 years ago, a landbridge linked Siberia in Asia to Alaska in North America; today the waters of the Bering Strait covers this area. Over this route, humans and giant mammals such as mammoths and mastodons migrated into North America, and other animals, such as the camel and the horse, migrated from North America into Asia and Europe.

Glacial Effects on Flora and Fauna When climate changes and glaciers alter their habitats, animals may not be able to find sufficient food and nesting sites; likewise, the life cycles of plants may be disrupted. During periods of glacial expansion, ice sheets overrun vast tracts of land, temperatures become colder, skies become cloudier, and sunlight decreases. Plants and animals may migrate to more suitable environments, evolve in ways that enhance their ability to survive the changed conditions, or become extinct.

As ice advanced to its last major maximum about 20,000 years ago, many of the plants and animals of North America and Europe migrated as much as 3000 kilometers (2000 miles) southward, competing with the preexisting flora and fauna for space and food. As the climate warmed and the ice sheets finally melted, environments changed again, leading to another series of migrations and further adaptations. The rising sea inundated the formerly habitable continental shelves, grasslands gave way to returning forests, and plants and animals began to populate newly deglaciated but still-barren lands. Organisms that could readily adapt to these environmental changes survived both the Pleistocene's cold glacial and warm interglacial periods. On the other hand, such ice-age creatures as the ground sloth, mammoth, and saber-toothed cat could not adapt to the warmer post-glacial climate and became extinct. (They also fell prey to migrating human hunters who had crossed ice-age landbridges and arrived on new continents to find an abundant supply of meat.)

Humankind was affected as well. With the warmer climate that followed the end of the last glacial period, migrating humans could find sustenance in newly habitable territories; consequently, human settlements expanded. The warmer climate and availability of new animal and plant species that could be exploited for food may have spurred the development of new tool technologies and hastened the discovery of agriculture. As surpluses accumulated and less time was spent hunting and gathering food, trade developed and complex societies emerged in South and North America, Africa, Europe, and the Near East.

Reconstructing Ice-Age Climates

Sixty-five million years ago, average global temperatures were as much as 10°C (18°F) warmer than today and few, if any, ice masses existed on the Earth, even in the planet's polar regions. Over the last 65 million years, however, the Earth's climate gradually cooled, as landmasses moved toward the poles and mountain ranges, such as the Himalayas, rose to great heights. About 2.5 million years ago, the planet entered an ice age from which it has yet to emerge.

During the last worldwide glacial expansion, some 20,000 years ago, temperatures in New England averaged about 5° to 7°C (9°–12°F) colder than today, making what is now Boston as cold as Anchorage, Alaska. Temperatures in continental interiors, such as the U.S. Midwest, may have been as much as 10° to 15°C (18°–27°F) colder, because these regions lack moderating factors such as proximity to oceans and their warming currents.

Given the absence of "fossil thermometers," where do geologists go to "measure" such past temperatures? The terrestrial geological record of climate change is unfortunately incomplete, largely as a result of the glacial processes of erosion and deposition. One geologist has compared this record to a blackboard that has been partially erased and written over repeatedly, so that most earlier markings have either disappeared or become illegible. The geological record in various places has either been removed by glacial erosion or lies inaccessibly buried under layers of younger glacial and other deposits. Yet because the terrestrial record is the most accessible, geologists still study it in an effort to reconstruct past climates.

In Chapter 16, for example, we saw how geologists use the growth of speleothems in caves to distinguish between alternating glacial and nonglacial conditions (see Figure 16-10), both in glaciated regions and in tropical seacoast caves. In addition, geologists may map the distribution of "fossil" periglacial features, such as sorted circles and ice-wedge casts; knowing the mean annual air temperatures in which these features form today enables geologists to estimate past temperatures in areas where we find relict periglacial evidence.

Some geologists and biologists have studied the distribution of climate-dependent biological organisms—such as fossil pollen or insect remains (beetles, for example)—to reconstruct past climates. Their reasoning is quite simple: By identifying the preserved fossil remains extracted from sediment cores and comparing them with organisms that flourish in modern environmental conditions, we can reconstruct temperatures and precipitation characteristics of past environments. Furthermore, we can date these carbon-based materials, as long as their ages fall within the range of carbon-14 dating.

For example, using pollen extracted from lake cores, palynologists (geologists who study fossil pollen to reconstruct

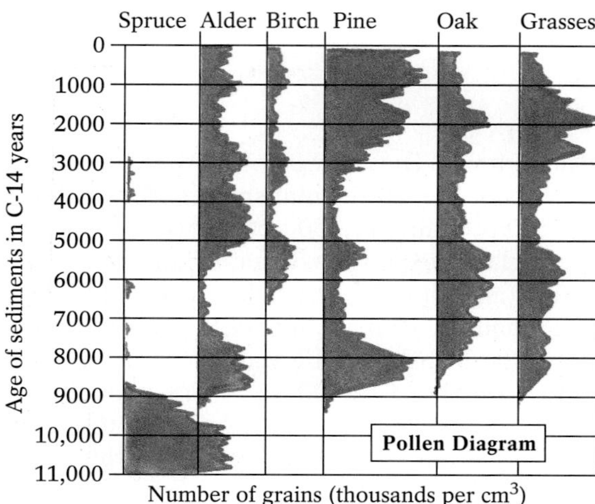

Figure 17-29 A pollen diagram, derived from the number of pollen grains preserved in datable sediment layers, can determine which plant species flourished in a particular location and when. It allows us to estimate past climatic conditions. For example, pollen from numerous lake beds throughout the temperate, treeless prairies of North America's Central Plains indicates that during the last major glaciation, this part of the continent was covered with lush forests of spruce, tamarack, alder, and birch. Because these trees grow today farther north, in southern Canada, we can conclude that the mid-continent region was significantly colder and moister 15,000 to 20,000 years ago than it is today.

past climates) may find conclusive evidence that plants that prosper today in cold, dry, south-central Canada lived during ice-age time in what is now warm, wet Louisiana. From the discovery of such pollen, they infer that the environmental conditions at the time of deposition were comparable to those in Canada today.

Knowing the modern temperatures and precipitation levels enables palynologists to estimate past conditions. Their main tool is a *pollen diagram,* such as the one in Figure 17-29 constructed from the percentages of fossil pollen extracted from datable horizons in the sediment core. A pollen diagram shows how a region's pollen, and thus its vegetation (and climate), has varied through time.

Until recently, however, we could not reconstruct past climates sufficiently to develop a *continuous* record of ice-age climate change. We have instead been forced to search for a geological record that was neither erased nor buried by subsequent events. The search has produced two candidates, but both pose unique scientific and human challenges, because one lies at the bottom of the Earth's oceans and the other exists in the bitter cold of Greenland and Antarctica at the centers of those continent's long-lived ice sheets.

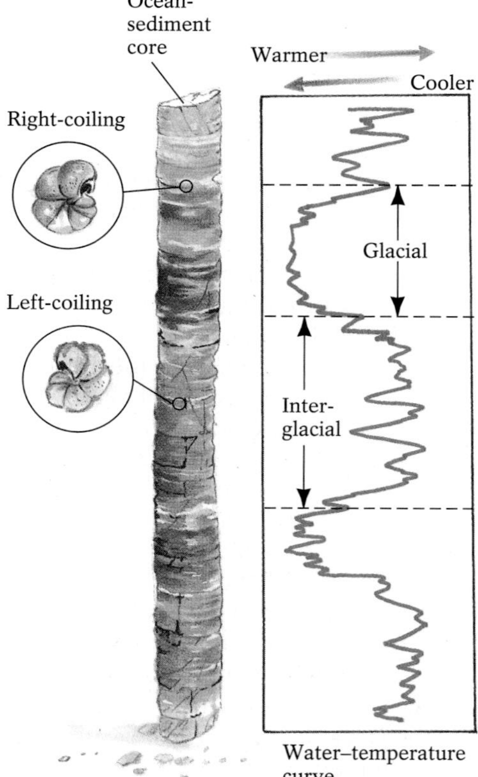

Figure 17-30 The coiling patterns of fossil forams can be used to identify ancient climate changes.

Deep-Sea Studies

Although the glaciated regions of the world have proved to be less-than-perfect repositories of past climate data, one part of the Earth's surface has certainly remained immune to the effects of glacier-driven erosion and deposition—the sea floor. If the terrestrial record of glaciation resembles a partially erased blackboard, the marine record is analogous to a thick paint chip from an apartment wall whose many colors record how many times the wall was painted. Obviously never glaciated itself, the continuous sedimentation at the sea floor provides a faithful record of worldwide glacial advances and retreats—if you know what to look for. Unlike the terrestrial record, which at best retains evidence of four or five periods of worldwide glacial expansion, sea-floor sediments reveal the occurrence of as many as 30 such events during the Pleistocene Epoch alone (1.6 million to 10,000 years ago).

What evidence supports geologists' claim of fickle glaciers that expand and contract with marked frequency and

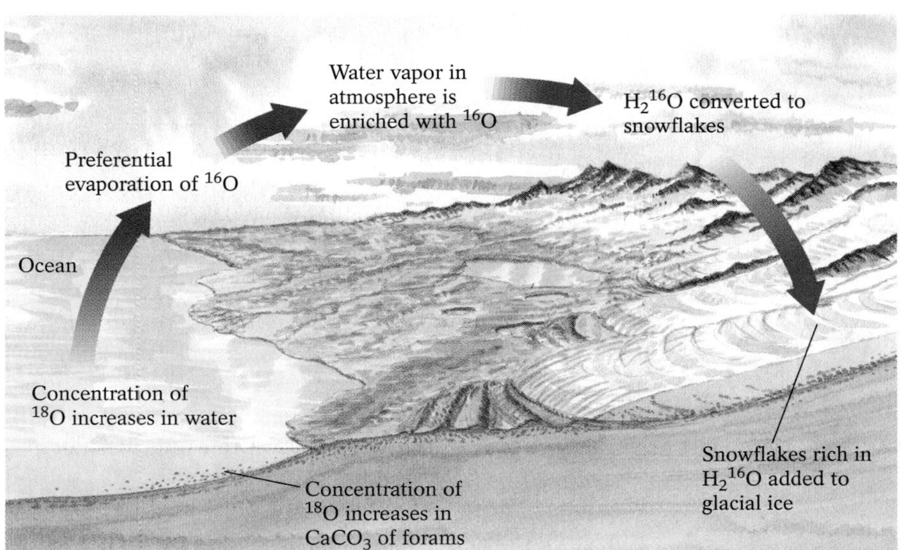

Figure 17-31 Oxygen-isotope concentrations in the shells of deep-sea organisms vary with global ice volumes. During cold periods, such as times of worldwide glacial expansion, ^{16}O is preferentially removed from seawater by evaporation. This isotope is then incorporated into snow that falls on land and is trapped in glacial ice, perhaps for millennia. Consequently, ^{18}O concentrations in seawater increase during glaciations. The reverse is true during warm interglacial periods.

apparent regularity? One way to identify past climates is by comparing the oceanic distribution of fossil and living foraminifera (*forams* for short; see Figure 6-21), microscopic marine organisms that produce distinctive coiled shells. Because forams are sensitive to water temperatures, knowing their current oceanic distribution and their preferred water temperatures enables geologists to reconstruct past water temperatures by charting forams extracted from cores of deep-sea sediments. For some foram species, the direction of coiling of their shells is largely controlled by water temperature; thus, as shown in Figure 17-30, geologists can use the direction of coiling of *fossil* foram shells in marine sediment to estimate water temperatures at the time of their deposition. These factors, in turn, can be correlated with periods of global cooling and warming.

Another powerful tool that geologists use to reconstruct the global movements of glaciers involves analysis of the oxygen-isotope content of the fossil remains of carbonate-based marine organisms. As organisms such as forams extract calcium carbonate ($CaCO_3$) from seawater to make their shells, they incorporate oxygen in its two major forms—^{16}O and ^{18}O. The concentration of these isotopes in seawater, however, is not constant; it varies with global ice volumes for the following reason. When seawater evaporates, the lighter oxygen isotope, ^{16}O, evaporates preferentially. During warm nonglacial times, the evaporated ^{16}O returns to the sea as rainfall, where it remixes and thus maintains the atmospheric ratio of the two isotopes. During cold glacial times, however, much of the ^{16}O may fall on the land as snow, where it becomes trapped in glacial ice for thousands of years. As a result, the concentration of ^{18}O rises in seawater and therefore in the carbonate shells of forams as well (Fig. 17-31). By analyzing the oxygen-isotope ratios in forams throughout a core of marine sediments that spans the last 2 million years or so, geologists have been able to "count" glaciations and interglaciations. Figure 17-32 shows the 30 or so glacial–interglacial cycles that characterize the climate over this time period.

To date these global climate events, geologists use a combination of methods: carbon-14 dating of organic remains

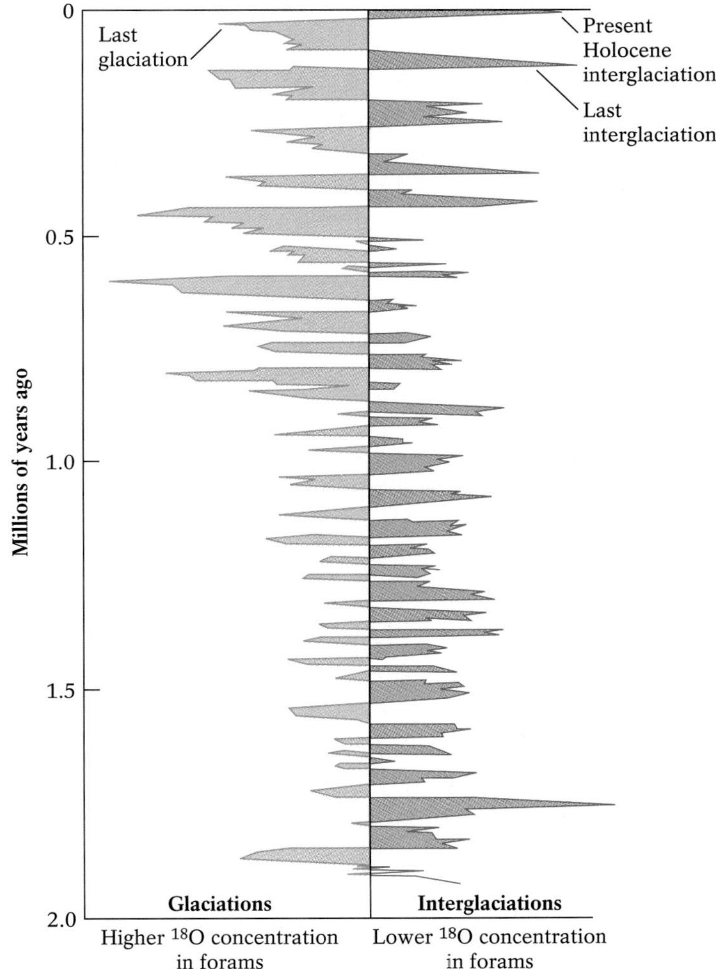

Figure 17-32 Changes in climate during the Quaternary ice age as revealed by oxygen-isotope studies of deep-sea sediments. This period of Earth history was marked by periods of glacial expansion interspersed with warm interglacial periods. During the last million years or so, the periods of glaciation have lasted about 90,000 to 100,000 years and the intervening interglacials have been relatively brief, lasting roughly 10,000 years. (The Holocene interglacial period has lasted from about 10,000 years ago to the present.)

found in the more-recent upper sections of cores of deep-sea muds; isotope dating of trace quantities of uranium incorporated from seawater into the calcium carbonate shells of microscopic marine organisms (uranium then decays to give a daughter product, thorium); fission-track dating of ash layers within sea-floor stratigraphy; and geomagnetic-stratigraphy dating of the older, lower sections of the core (see Chapter 11 for a discussion of this dating technique).

Although these deep-sea studies have revolutionized our understanding of ice-age events, a recent discovery has provided geologists with another means of reconstructing past climates—one that enables us to decipher the chemistry of the ice-age atmosphere, gauge the windiness of the ice-age climate, and even detect large volcanic eruptions. The source of this wealth of information? The Earth's vast polar ice sheets.

Ice-Core Studies

In a parking lot on the University of New Hampshire campus sits a refrigerated van that holds a potential key to unlocking the mysteries of the Earth's ice-age climate. The van houses row upon row of 1-meter (3.3-foot)-long tubes of ice drilled from the center of Greenland's ice sheet—a richly detailed, frozen archive of the planet's last 250,000 years of climatic history. After decades of trial and error, glaciologists (scientists who study the snow and ice in glaciers) have accomplished a remarkable technological achievement—they can now extract cores of ice, several kilometers long, by boring through the Earth's surviving ice sheets down to the underlying bedrock. Examination of the physical and chemical properties of these ice cores reveals that they contain a *continuous* record of climate change for much (if not all) of the Earth's current ice age.

Much of the upper part of the United States' GISP (Greenland Ice Sheet Project) 2 core is marked by discernible layers consisting of now-icy remnants of each year's snowfall. Almost 80,000 annual layers have been identified. Scientists, who have just begun to study these layers, have already made the following discoveries:

- Measurements of the dust content in the ice layers show that as much as 12 times as much airborne dust accumulated during the cold times associated with glacial expansions as during warm interglacial periods. This finding suggests that a more turbulent atmosphere characterized glacial times than exists today. The stronger-than-normal winds during glacial periods may have resulted from frequent clashes of warm and cold air masses; such winds might have swept across the continents as well as exposed areas of the continental shelves, eroding dried-out mud and depositing it on glaciers.
- Samples of the trapped air in each layer's bubbles indicate that the CO_2 content of the atmosphere at the time of snowfall was relatively low during cold times and significantly higher during warm times. Carbon dioxide is, of course, the gaseous component of the atmosphere that contributes most to the "greenhouse effect" (discussed in Chapter 20).
- Analysis of a layer's oxygen-isotope concentrations has revealed the presence of a geochemical "thermometer." Unlike oxygen-isotope studies of deep-sea sediments, which reveal global ice volumes, such analyses of ice allow geologists to estimate the mean annual air temperature, for the following reason. During warmer periods, water vapor containing the heavier ^{18}O isotope condenses from the atmosphere more readily than vapor containing ^{16}O. Thus falling snow contains proportionately more ^{18}O during warm times. The reverse is true during colder periods.

As an interesting by-product of ice-core studies, geologists have been able to identify specific volcanic ash layers from the datable annual ice layers. For example, the GISP 2 core may have provided long-awaited proof of the timing of the catastrophic eruption that struck the island of Thera in the Aegean Sea—the blast that apparently wiped out the Minoan civilization and formed the basis for the legend of Atlantis recounted in Plato's dialogues. To the delight of archaeologists who have long debated the question, the studies revealed that Thera apparently erupted in the year 1645 B.C. (with an error margin of seven years). We now await confirmation of this date from the tephra itself, as geologists analyze the chemistry of the volcanic glass and compare it with known tephras from the eruption of Santorini, Thera's volcano.

Ice-core studies are just beginning to reveal remarkable new "truths" about the Earth's climate, including some with dramatic implications for the well-being of humankind. We now know that sharp climatic shifts can occur with alarming rapidity; the relative warmth we are enjoying during our current interglacial period could be replaced by a prolonged, sharply colder climate in the space of a mere decade or so—a condition that undoubtedly would affect growing seasons and food supplies throughout the world. A reminder that this type of event could easily happen remains fresh in our collective memories—the "Little Ice Age" gripped the planet as recently as the mid-1800s.

The Causes of Glaciation

What causes such dramatic climatic shifts? What geological conditions enable ice to expand around the world, even to now-temperate mid-latitude regions? What factors prompt continent-sized sheets of ice to advance and retreat numerous times *during* an ice age? The answers to these questions are complex, and no true consensus exists among those who study them. One frustrated student concluded that there must be at least as many hypotheses to explain the causes of glaciation as there are glacial geologists.

Many geologists have concluded that ice ages share three requirements: sizable landmasses at or near the poles; land surfaces with relatively high average elevations; and nearby oceans to provide the moisture that falls as snow. These conditions increase the likelihood of an ice age by putting landmasses at latitudes and elevations where the climate is colder and glaciers tend to grow. The phenomenon (which we have discussed throughout this text) that can move landmasses to polar regions, raise them to higher elevations, and manipulate the positions of the Earth's ocean basins is, of course, plate tectonics.

The slow global cooling that started some 50 million years ago and led to the onset of the most recent ice age began after Antarctica had moved to the South Pole; other major landmasses had simultaneously traveled to positions north of the Arctic Circle. As plates converged, several of the Earth's loftiest mountain ranges rose steadily at plate edges, producing elevations that far exceeded regional snowlines. The rising Andes and Rockies were oriented *north–south* and located near oceans. They intercepted moist marine air carried by westerly winds, forcing it to rise above their peaks into the cold, high-altitude air; this process increased precipitation and enhanced snow and ice accumulation. The lofty Himalayas and Tibetan plateau intercepted moisture rising from the warm Bay of Bengal of the Indian Ocean. Some geologists believe that the collision between India and southern Asia and the inception of Himalayan mountain-building about 50 million years ago were the primary factors that cooled down the Earth and instigated its current ice age.

Plate motion, however, cannot move lands to and from the poles or raise and lower lands swiftly enough to account for the cycles of repeated glacial advances and retreats. From ice-core studies we know that these glacial cycles have lasted for periods of approximately 100,000 years, separated by brief interglacial periods that endured for about 10,000 years. Such *short-term* fluctuations in climate and glaciation, unrelated to plate tectonics, may have been driven by one or more causes—including volcanism, meteorite impacts, variations in the amount of solar energy reaching certain regions of the Earth's surface, and changes in the global circulation patterns of the Earth's oceans.

Volcanism could not have caused all the expansions and retreats of Pleistocene ice sheets, simply because huge eruptions do not occur at regular intervals and their effects last for too brief a period. Large volcanic eruptions can lower world temperatures by as much as 1° to 2°C (2°–4°F) for several years as volcanic ash and gas released into the atmosphere reflect and absorb solar radiation, causing less sunlight to reach the Earth. No consistent relationship between major eruptions and ensuing ice expansions has been proven, however, perhaps because the lower temperatures are so short-lived. Worldwide glacial expansion, for example, did not follow such cataclysmic eruptions as those associated with Mount Mazama and Krakatoa (see Chapter 4).

Large meteorite impacts hurl veils of dust into the atmosphere and screen out significant solar radiation, lowering global temperatures. Like volcanic eruptions, however, they occur sporadically and their effects are short-lived. Thus they are also unlikely candidates to be the principal cause of the cyclic glacial expansion and contraction that marks the last 2 million years of Earth history. Nevertheless, a major strike could hasten climatic cooling that has already developed in response to other factors.

Variations in the Earth's Orbital Mechanics

Many glacial geologists believe the primary causes of glacial fluctuations during an ice age are variations in the Earth's position and orientation relative to the Sun. This hypothesis was first proposed by British physicist John Croll in the late nineteenth century and then further refined by Yugoslavian astronomer Milutin Milankovitch around 1930. It explains that whereas the total amount of solar radiation the Earth receives may vary only a little, periodic changes in *where* and *when* that radiation strikes have profoundly affected global climate.

The Milankovitch hypothesis suggests that the critical region for determining global glacial advances and retreats during ice ages is the vicinity of 65°N latitude, where ice sheets have developed repeatedly during the past several million years, but which today remains largely ice-free. A glance at a world map reveals that, when compared with other latitudes, 65°N—with North America, Northern Europe, Scandinavia, and Asia—has the Earth's greatest concentration of land above sea level; this region provides surfaces upon which to grow ice. When this area receives less solar radiation and becomes sufficiently cold, enough winter snow accumulates to survive the summer melting season. The snowpack thickens from year to year, until glaciers form. The region in the vicinity of 65°S latitude, by comparison, has experienced significantly less glacial expansion, simply because the Southern Hemisphere provides far less land on which to grow glaciers. (Falling snow there would just enter the southern oceans and melt.) Once a global glacial expansion begins and the world's climate cools, however, glaciers grow in Southern Hemisphere highlands such as the Andes.

Milankovitch calculated the effects of three periodic astronomical factors to show how they might combine to lower the radiation levels of 65°N, and why they happen at 100,000-year intervals—the recurrence interval of worldwide glacial expansions during the last million years. These three factors include variations in the shape of the Earth's orbit around the Sun from circular to elliptical (oval-shaped), variations in the tilt of the Earth's axis toward or away from the Sun, and the shift, or *precession,* of the spring and autumn equinoxes caused by the wobble of the Earth's axis as the

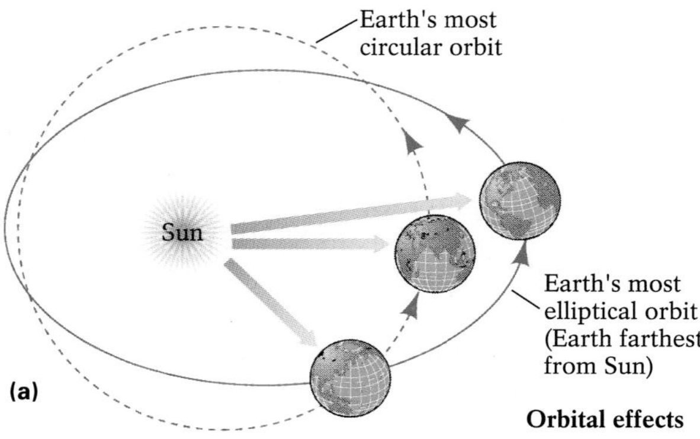

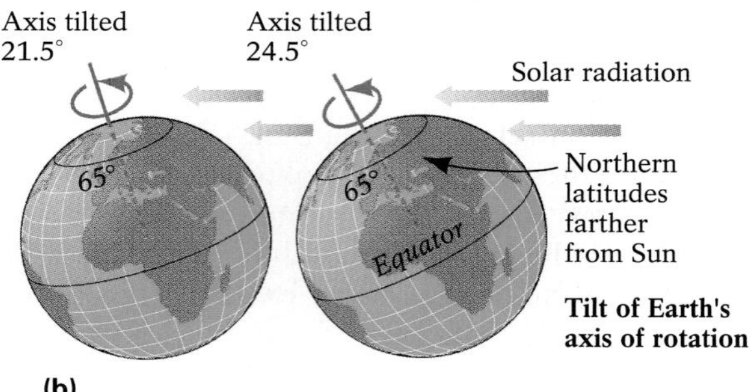

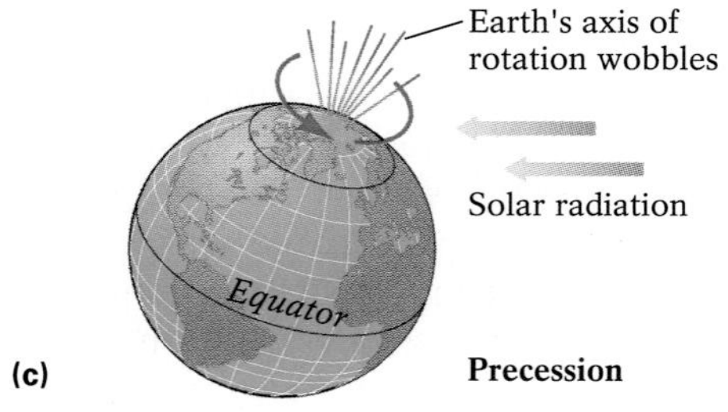

Figure 17-33 Milutin Milankovitch proposed that three astronomical factors interact to affect the amount of solar radiation striking the Earth's high northern latitudes, possibly triggering the periodic advance and retreat of glaciers around 65°N latitude. The climate at 65°N latitude cools when the Earth's orbit around the Sun is at its most elliptical (especially when the Earth is located farthest from the Sun during northern summers), when the tilt of the Earth's axis positions the high northern latitudes at a lower angle to incoming solar radiation, and when the "wobble" of the Earth's spin axis carries this crucial latitude out of the path of direct solar radiation. Each of these three factors has a different period of recurrence. When they have coincided in the past, glaciations have occurred.

Earth spins like a top (Fig. 17-33). Each factor varies in a unique cycle, but according to Milankovitch the three cycles sometimes coincide, producing centuries or even millennia of low summertime radiation at 65°N. When this situation arises, summers are cool and brief, and alpine glaciers and continental ice sheets expand.

For years, glacial geologists have debated whether these orbital factors alone are sufficient to produce a full-blown glacial expansion. Climate experts estimate that, taken together, these factors can reduce solar radiation enough to lower global temperatures about 4°C (7°F). Although such a temperature reduction is certainly significant, it falls far short of the ice-age temperatures that geologists have detected in the geological and biological record (5° to 8°C) [9°–14°F] at mid-latitude coasts; 10° to 15°C [18°–27°F] in continental interiors). Thus additional factors must contribute to the ice-age climatic deterioration that leads to the worldwide expansion of glaciers.

Terrestrial Factors That Amplify Ice-Age Climate Change

Once the positions and altitudes of continental lands are arranged tectonically in ways that promote global cooling, and the Milankovitch orbital factors conspire to produce a solar-radiation minimum at high northern latitudes, numerous other factors may contribute to the trend toward global cooling or warming, perhaps even switching these climatic events on and off. These factors include (but are not limited to) the reflectivity of the Earth's surface and changes in oceanic salinity that alter large-scale oceanic circulation.

After tectonic and astronomical factors initiate a period of glaciation and the Earth's snow and ice cover expand, white snow and ice (which reflect radiation) replace dark-colored rocks and soils (which absorb radiation rather than reflect it) and vegetation (which is nonreflective). As a consequence, the Earth's *albedo*—the percentage of incoming solar radiation reflected from surfaces back into space—increases sharply. As more incoming radiation is reflected, less remains available to warm the Earth's surface and the Earth's climate cools further.

A second factor that substantially affects the course of glacial expansions and contractions involves salinity-driven ocean currents in the North Atlantic. During warm periods, water transported by ocean currents flowing northward from the equator (such as the Gulf Stream) evaporates significantly as it passes through warm climes. Through this process, the water becomes more saline (that is, saltier). As this salty water passes through colder northerly environments, it cools substantially. By the time these currents reach the North Atlantic, their waters—*cold and salty, and thus more dense*—begin to sink, creating North Atlantic Deep Water. As they sink (and begin their return flow toward the equator), the void produced pulls additional equatorial currents northward, thereby maintaining the steady flow of warm tropical water to the North. This watery "conveyor belt," as shown in Figure 17-34, helps to moderate global climatic conditions by bringing warm water (which in turn warms the air above it) to the cold Arctic.

What happens when this conveyor belt switches off, such as when the global climate cools and seawater evaporation diminishes or when a great influx of fresh glacial meltwater flows into the adjacent North Atlantic, further reducing

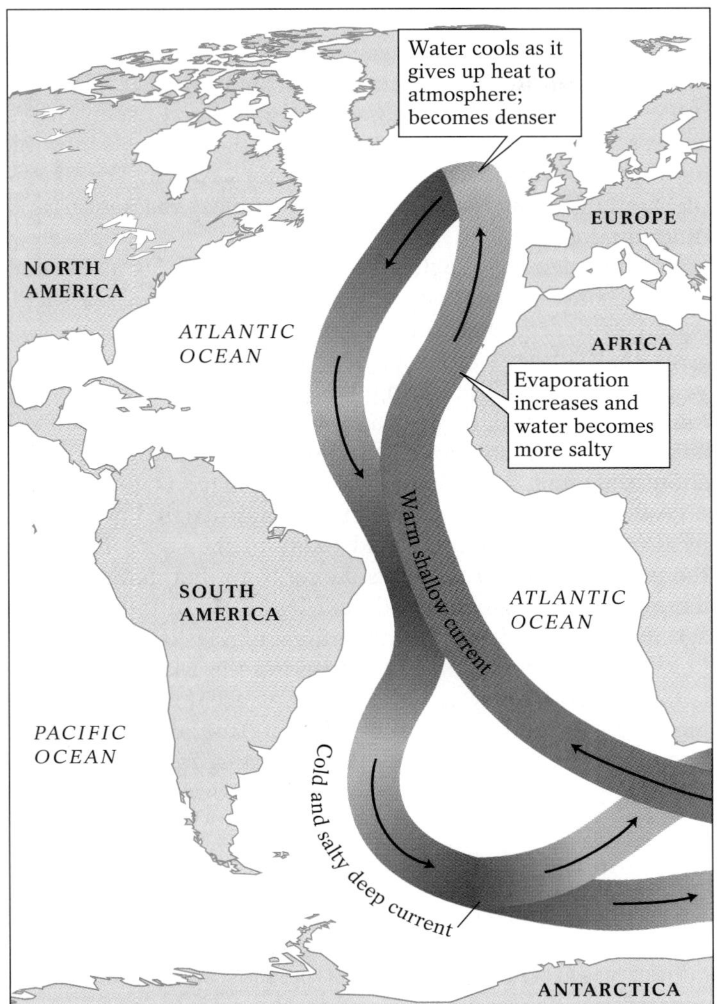

Figure 17-34 The origin of North Atlantic Deep Water. As warm equatorial water heads northward, some evaporates in the warm subtropical air, increasing the ocean's salinity. As the water reaches more northerly climes, it cools and grows more dense. As this cold, salty water sinks, it draws more water up from the equator, maintaining the warm moderating effect of the current. When this water conveyor shuts off, cold times ensue.

its salinity? The climate will deteriorate sharply, producing much colder conditions. We can observe the effect of such declines in oceanic salinity in the geological record—terrestrial, ice-core, and deep-sea—during the last episode of worldwide glacial expansion 11,700 years ago. At that time, even though most of the world's glaciers retreated in response to the convergence of high-solar-radiation Milankovitch factors, the planet plunged into the deep freeze one last time. The following sequence of events illustrates why.

When the North American ice sheet retreated, it created an enormous meltwater lake in south-central Canada—larger than Lake Superior—which flooded through the newly opened St. Lawrence valley and into the North Atlantic. The massive influx of freshwater from the draining lake depressed the ocean's salinity levels and temporarily cut off the North Atlantic deep water current. This brief, but dramatic return to glacial conditions lasted for a few centuries, only to abate as the world climate warmed, evaporation increased, and oceanic salinity restored the North Atlantic deep water conveyor.

In a similar way, periods of ice-sheet surging into the Atlantic may have decreased salinity and shut down the warming currents. The oceanic-sediment record contains numerous pebbly layers that document the formation of vast armadas of floating icebergs (which eventually melt and release their sediment loads). Melting of these freshwater icebergs—and their contribution to reduced salinity—apparently brought about colder conditions. These episodes marked by vast influxes of climate-changing icebergs, which may have occurred every 10,000 years or so, are called *Heinrich events,* named for the German oceanographer who discovered them.

There is clearly more to the causes of glaciation and climate change than initially meets the eye. In addition, even more discoveries are certain to come as geologists probe the mysteries of the remarkable Greenland ice cores.

The Earth's Glacial Past

The Earth has experienced numerous ice ages throughout its 4.6-billion-year history. Erosion of the ancient geologic record and its burial beneath younger rocks and sediments limits geologists' ability to study the earliest ice ages. Nevertheless, the geologic record yields clear evidence of at least three ice ages in the Precambrian Era. The oldest Precambrian ice age, which occurred about 2.2 billion years ago, is identifiable from layers of *tillite,* a sedimentary rock produced by lithification of till. Geologists studying this Precambrian event have discovered signs indicating that during this glaciation, ice may have extended from the Earth's poles to the tropics, a span equivalent to finding healthy glaciers today in tropical Panama. During this monumental ice age, virtually the entire planet may have been locked in a glacial deepfreeze.

Geologic evidence of global glaciation after the Precambrian is more widespread and better preserved. Layers of tillites and "fossilized" moraines and eskers show that during the Paleozoic Era, about 500 million years ago, an ice sheet occupied what is now one of the world's hottest and driest lands, the Sahara of northern Africa. Before plate tectonics moved the continents to their present locations, the Sahara was apparently located at the South Pole. Ironically, one may now swelter in the merciless African sun while standing atop an ancient esker.

An extensive glaciation that occurred during the Permian Period, about 245 to 286 million years ago, left the most complete record of any pre-Pleistocene glaciation. This ice sheet striated the bedrock and deposited extensive tillites in

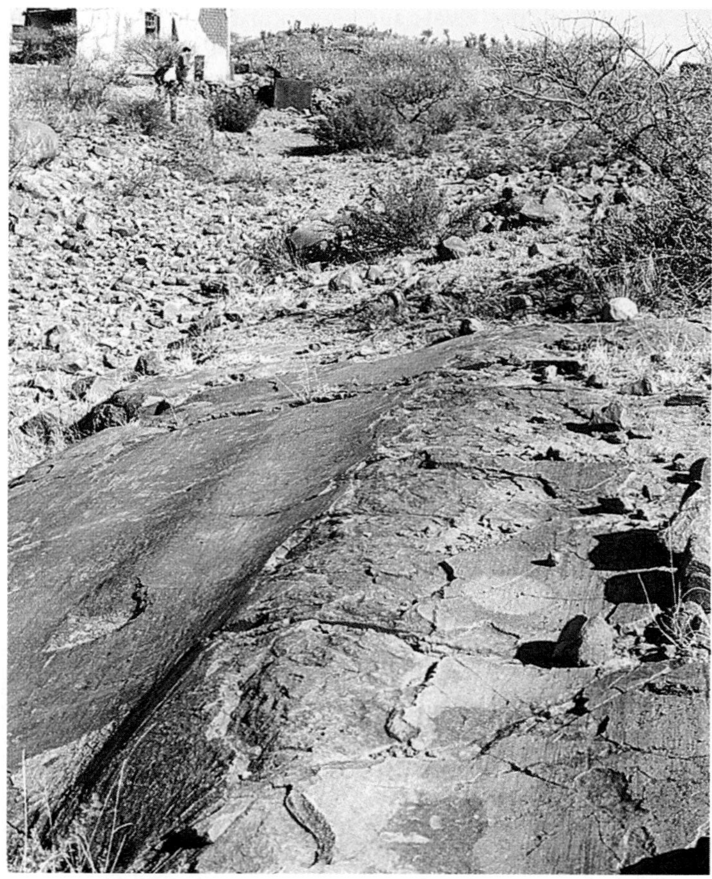

Figure 17-35 Glacial grooves and tillite in the arid landscape of South Africa suggest that this area was once located near the South Pole, and was cold enough to be glaciated.

regions that now form Australia, South America, India, and South Africa (Fig. 17-35). At the time, these lands constituted the supercontinent of Gondwana, located in the vicinity of the South Pole.

Geologists have found no evidence of glaciation from the Mesozoic Era, which lasted from 225 to 65 million years ago. Apparently the Mesozoic was a time of great global warming and heightened plate motion (associated with the rifting and breakup of Pangaea). Many of the Earth's landmasses moved from the poles to subtropical and tropical latitudes, and vast episodes of volcanism at mid-ocean ridges spewed great volumes of warming greenhouse gases into the atmosphere. Mesozoic rocks in northern Alaska, for example, contain remnants of coral reefs, tropical vegetation, and dinosaur fossils.

Although the Earth's most recent ice age is associated with the start of the Pleistocene Epoch 1.6 million years ago, recent evidence from forams and other marine species suggests that surface waters began to cool at least 50 million years ago (Fig. 17-36), during the Eocene Epoch. Between 50 million and 20 million years ago, world temperatures dropped approximately 5 degrees centigrade (9 degrees Farenheit). At about this time, large domes of ice may have begun to grow on Antarctica; striated stones, presumably carried northward by floating icebergs, are found today in the south Pacific in the muddy oceanic sediments dating from that period. Gradual global cooling continued, and by 12 to 10 million years ago ice had formed in the mountains of Alaska. By 2.5 million years ago, an ice sheet had buried the island of Greenland, and 500,000 years later, ice was accumulating on the high plateaus and mountains of North America and Europe. The Quaternary ice age, with its alternating mid-latitude ice-sheet advances and retreats, was poised to begin.

Glaciation in the Pleistocene

As noted earlier, the Earth has experienced a succession of alternating glacial and interglacial periods during the past 1.6 million years. During the glacial periods, the ice cover expanded from the polar regions to the middle latitudes. In North America, it reached as far south as the Great Plains. Tills and outwash deposits exposed in Kansas and Nebraska have been strongly modified by weathering and erosion over their long existence since the early Pleistocene; the slopes of these old moraines have been gently rounded and subdued by mass movements (discussed in Chapter 13), and their tills have been deeply oxidized. (See Chapter 5 for a discussion

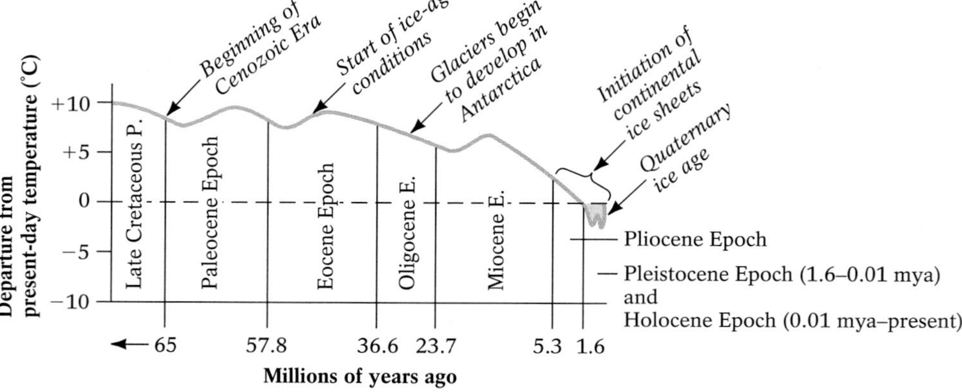

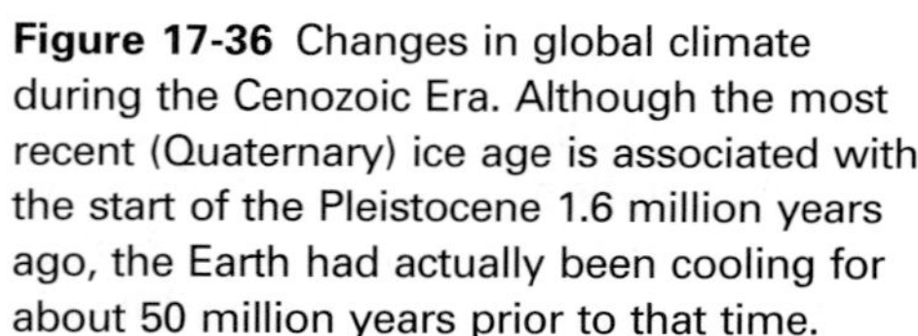

Figure 17-36 Changes in global climate during the Cenozoic Era. Although the most recent (Quaternary) ice age is associated with the start of the Pleistocene 1.6 million years ago, the Earth had actually been cooling for about 50 million years prior to that time.

of the chemical-weathering process of oxidation.) Deposits from more recent Pleistocene glacial events, especially those occurring in the last 100,000 years, remain relatively unweathered and unmodified by mass movement.

The location of North America's youngest glacial deposits indicates that the continent's last ice sheet extended from the eastern Canadian Arctic south to New York City and west to the foothills of the Rocky Mountains in Montana, an area encompassing more than 15 million square kilometers (6 million square miles). This *Laurentide* ice sheet probably began to grow in Canada west of Hudson Bay and in Labrador and Newfoundland, where the relatively high topography and cold climates promoted snow and ice accumulation. Nourished by moisture from the Atlantic Coast and the Gulf of Mexico, the Laurentide ice sheet advanced east and south. Another glacial center, the *Cordilleran* ice sheet, formed farther west, between the Canadian Rockies and the Coast Range of British Columbia. It probably began as a network of alpine glaciers that expanded and merged to form an ice cap that eventually covered the adjacent valleys and plains.

At the time of maximum glacial expansion, a continuous wall of ice stretched 6400 kilometers (4000 miles) across the full breadth of southern Canada and the northern United States (Fig. 17-37). This ice sheet, called the North American ice sheet, stripped an average of 15 to 25 meters (50–80 feet) of regolith and bedrock from central and southern Canada and the northern Great Lakes region, depositing this material from Cape Cod to northern Washington state as drift averaging about 15 meters (50 feet) deep.

Between 13,000 and 12,000 years ago, the world's climate warmed, ice sheets began to melt, and the North American ice sheet largely retreated from the United States and southern Canada. Perhaps in response to meltwater flooding into the North Atlantic, brief periods of sharp cooling temporarily interrupted the general warming trend, and some glaciers even advanced slightly before resuming their retreat. During one such cooling period, the portion of the ice sheet that flowed through the Lake Michigan basin moved south of Green Bay, Wisconsin. In the process, it overran trees that had begun to grow in the warmer climate, incorporating them into a thin layer of till (Fig. 17-38). Although the ice sheet had retreated into Canada by about 10,000 years ago, another 3000 to 4000 years elapsed before northern Canada's ice caps melted completely.

Figure 17-37 The effect of the North American ice sheet, which reached its greatest extent about 20,000 years ago. In the zone of maximum glacial erosion, flowing ice and meltwater from the Laurentide ice sheet excavated some soft preglacial bedrock to about 30 meters (100 feet) below sea level.

Figure 17-38 A log in glacial till in Two Creeks, Wisconsin. The log was buried about 11,700 years ago, when a temporary cooling period caused the retreating North American ice sheet to readvance over newly established forests.

Figure 17-39 The retreating Athabasca Glacier in the Columbia Icefield of western Canada. The signposts indicate past positions of the glacier's terminus.

Recent Glacial Events

Alternating warm and cool periods also characterizes the **Holocene Epoch**—the last 10,000 years of the Earth's history. The Holocene began with the postglacial warming that melted most of the Pleistocene's mid-latitude ice. From about 8000 to 6000 years ago, the Earth's climate was about 2°C (3.6°F) warmer on average than it is today. Moderate cooling since then has produced many new glaciers in alpine settings, but temperatures have not remained sufficiently low or the climate been cold long enough to reconstitute mid-latitude ice sheets. During the most recent cold snap—the Little Ice Age, occurring from about 700 to 150 years ago—winters were colder and snowier than average and summers relatively cool and wet. Indeed, some winters were so cold that eighteenth-century New Yorkers could walk across the frozen Hudson River to New Jersey. Snowlines in alpine areas descended sufficiently to cause significant glacial expansion in the world's major mid-latitude mountain ranges. At some localities, the ice overran entire alpine communities.

Sustained warming began by 1850 and continues today, although vestiges of pack ice from the Little Ice Age lingered long enough in the North Atlantic to doom the maiden voyage of the *Titanic* on April 15, 1912. By 1920, far less floating ice was observed south of the Arctic circle. Today most glaciers in North America and Europe are still retreating (Fig. 17-39), albeit more slowly than earlier in the century. To establish the amount and rate of glacial retreat, we can compare modern glacial positions to those shown in oil paintings and lithographs of the early 1800s. A recent field and photographic survey of 200 glaciers in Alaska found 7% advancing, 63% retreating, and 30% virtually standing still. One of these retreating glaciers, the Columbia Glacier, flows from a deep coastal fjord into Prince William Sound, where it calves icebergs into the Sound's shipping lanes, posing a direct threat to Alaska's economy.

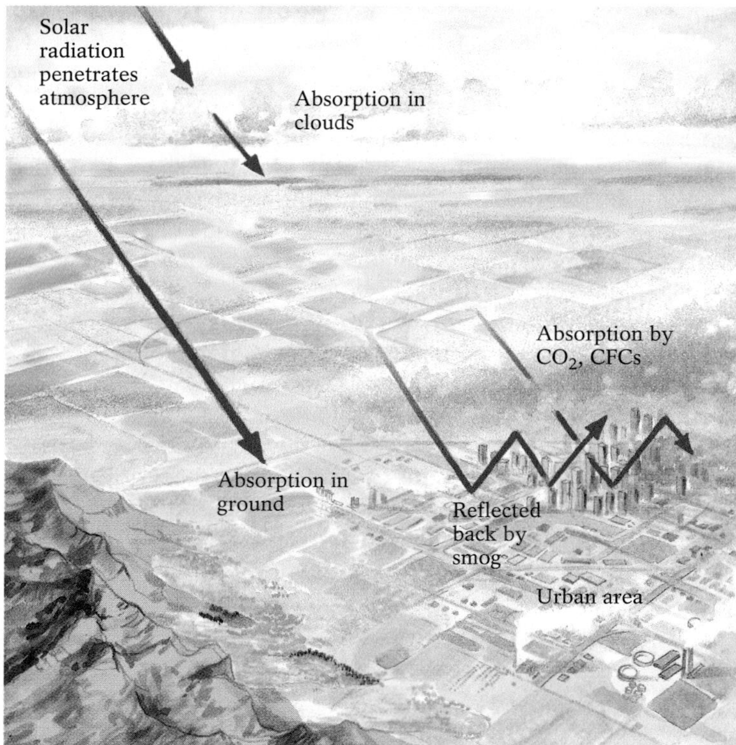

Figure 17-40 Human activity can contribute to global warming—and thereby glacial melting—in various ways, inadvertently increasing glacial melting. The greenhouse effect develops when CO_2 and other lower-atmosphere gases absorb heat radiating from the Earth's surface. If too much heat enters the atmosphere, all of it cannot dissipate; the lower atmosphere becomes increasingly warm as the heat accumulates, causing temperatures on Earth to rise.

The Effects of Human Activity on Glaciers Human activities may contribute to global warming and hasten the melting of the Earth's remaining ice masses. In recent years, our use of refrigeration, air-conditioning, fire extinguishers, and aerosol cans has introduced a vast quantity of gases called chlorofluorocarbons (CFCs) into the atmosphere. These gases—together with increased volumes of atmospheric carbon dioxide generated from our burning of coal, oil, and natural gas—are believed to be reducing the atmosphere's ozone layer (which shields the Earth from most solar ultraviolet radiation) and increasing the *greenhouse effect* (which may accelerate melting of glacial ice).

The greenhouse effect promotes global warming in roughly the same way that a glass-walled botanical greenhouse creates a warm environment—by allowing the entry of solar radiation and preventing the escape of heat. Carbon dioxide in the Earth's atmosphere allows solar radiation to

Figure 17-41 The theoretical effect on North America if all the ice on Earth today were to melt completely. The global sea level would rise about 70 meters (230 feet), submerging most coastal areas and even some inland areas.

pass through it (as short-wave ultraviolet radiation) and warm rocks, soils, water, and vegetation on the Earth's surface, and it absorbs some of the heat (as long-wave infrared radiation) that radiates from the surface (Fig. 17-40); some heat is also radiated back into space. Some of the Earth's heat, trapped in the lower atmosphere, remains to warm the Earth below. Any mechanism that increases the concentration of atmospheric CO_2, methane, and CFCs promotes climatic warming and accelerates the melting of glaciers.

Human activity has increased atmospheric CO_2 levels in two principal ways: by introducing vast amounts of CO_2 from the burning of carbon-rich fossil fuels and by destroying forests and reducing agriculture, especially the vast tropical rainforests (which release CO_2 from soils and decaying vegetation). Since 1850, as the use of fossil fuels has surged worldwide, the atmospheric CO_2 concentration has risen by 25%. If this rate of CO_2 production continues, people in the next century may have to live with CO_2 levels double those in the pre-industrial era; such levels could cause global temperatures to rise by 2° to 4°C (3.5°–7°F).

Such a global temperature increase would be sufficient to melt a significant amount of the Earth's polar ice and permafrost and to raise global sea levels by as much as 50 to 100 centimeters (1–3 feet) by the year 2030. Sea level has already risen 20 to 25 centimeters (8–10 inches) since 1900 through the melting of glacial ice and the thermal expansion of sea water. The predicted additional rise in sea level of 25 to 75 centimeters (10–30 inches) would surely flood such low-lying cities as New Orleans, Louisiana, and Miami, Florida, during large storms.

If human activities caused all of the world's ice to melt within a few thousand years, global sea level would rise approximately 70 meters (230 feet) (Fig. 17-41). The water in New York Harbor would reach to the armpits of the Statue of Liberty, and most of the Atlantic Coast from New Jersey to the Carolinas would be submerged. The entire state of Florida would be covered by the Atlantic, much of Alabama, Mississippi, and Louisiana by the Gulf of Mexico, and much of interior California by the Pacific. In addition, the rising sea would completely inundate such low-lying places as the Netherlands, Bangladesh, Bermuda, and numerous Pacific island nations, effectively removing them from the world map. The 1995 breakup of Antarctica's Larsen Ice Shelf may be signaling the onset of the demise of polar ice (Fig. 17-42).

Human-induced global warming could also produce an expansion of the world's deserts, necessitating major shifts in agriculture and forestry. Of course, this trend would create national winners and losers. Canada would do well: Cold regions now unsuitable for agriculture would become productive as the climate that now supports the Midwestern corn belt of Iowa and Minnesota migrated north to Ontario and Manitoba. In contrast, the Great Plains wheat belt of the United States—"America's Breadbasket"—might become too warm and dry for productive agriculture. Recent Midwestern droughts and the resulting food shortages and price hikes may offer an early glimpse into the Earth's balmy future.

The Future of Our Current Ice Age How can we predict when the next period of worldwide glacial expansion or melting will occur, when we can't even predict with absolute certainty whether tomorrow's ballgame will be rained out? The Pleistocene record suggests that interglacial periods have an

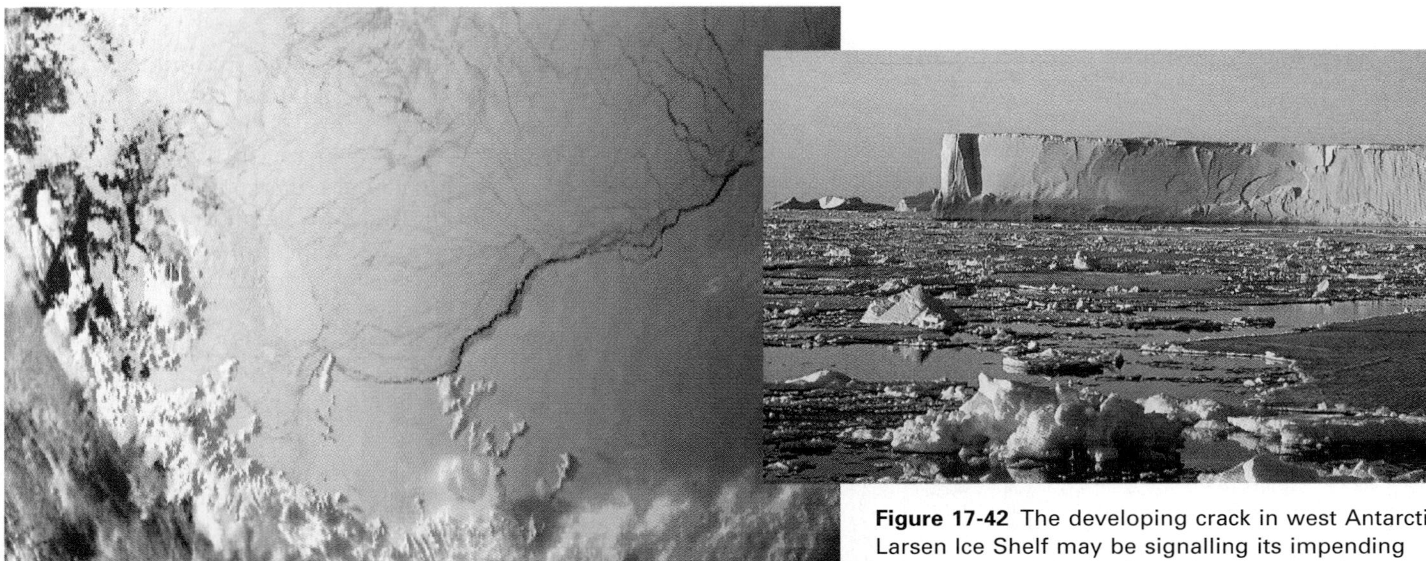

Figure 17-42 The developing crack in west Antarctica's Larsen Ice Shelf may be signalling its impending breakup—a dramatic effect of global warming. Inset: Massive ice floes have been shed from Antarctic ice shelves during the past decade.

Figure 17-43 An ice cap near the south pole of Mars, photographed by the Viking Orbiter 2 during Mars' southern summer in 1972.

average length of 8000 to 12,000 years, followed by 10,000 to 20,000 years of slow, intermittent cooling before glacial conditions return. Full glacial conditions may then last—with intermittent brief periods of warming—for another 60,000 years or so. We are now in an interglacial period that began about 10,000 years ago. This period may be drawing to a close, and a new period of slow cooling may be on its way.

If human activities are producing a greenhouse effect, this condition is likely to prove temporary (especially if we succeed in reversing CO_2 and CFC buildup). If we let CO_2 production go unchecked for the next 2000 years, however, the very high level of CO_2 buildup will coincide with a warming trend predicted by Milankovitch's Earth–Sun orbital factors. The result could be an unusually intense warming and a period of *superinterglaciation*. If human-induced warming can be curtailed, then the Milankovitch factors predict that ice sheets will again cover vast areas of North America and Europe about 23,000 years in the future.

Glaciation on Other Planets and Moons

Several other planets in our solar system also experience glaciation. Recent satellite surveys of the surfaces of the large outer planets—Jupiter, Saturn, Uranus, and Neptune—and some of their moons have revealed polar ice caps and extensive ice cover. Closer to home, the surface of Mars is marked by white patches near the poles that appear to grow and shrink seasonally, much like the Earth's polar ice (Fig. 17-43).

Mars' polar ice is probably only a few meters thick, and it may largely consist of frozen CO_2 (dry ice), the main com-

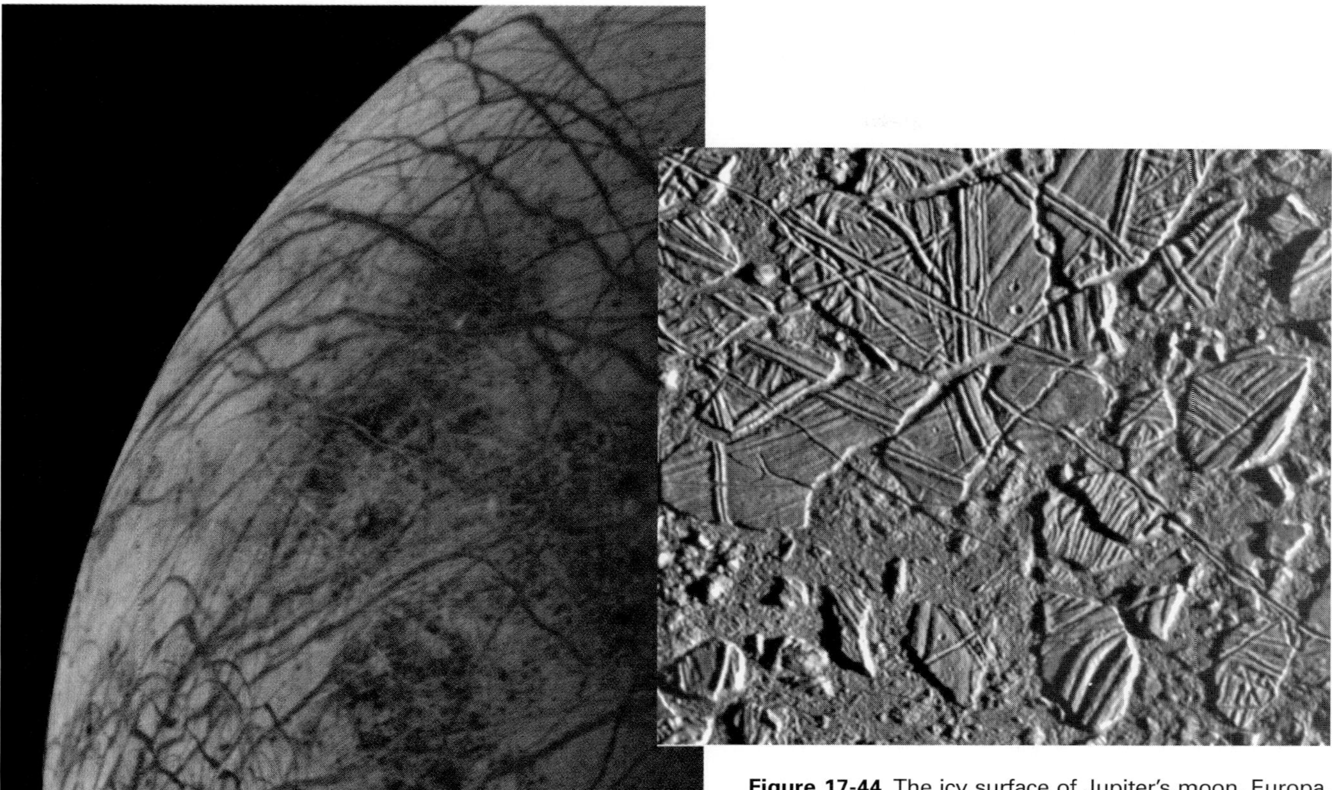

Figure 17-44 The icy surface of Jupiter's moon, Europa, is broken into solid plates of ice that appear to rift and diverge in ways similar to the Earth's tectonic plates. Photo from Galileo probe, 1996.

ponent of its atmosphere. In the Martian summer, the northern hemisphere ice cap recedes rapidly at first, and then more slowly. The rapid shrinkage may reflect sublimation of frozen CO_2 at the edges of the ice cap, where it is thinnest. Any ice that lasts through the summer is probably frozen water, which evaporates imperceptibly slowly in Mars' atmosphere.

Satellite photographs of Mars reveal troughs and ridges in the polar regions that resemble the glaciated valleys and large moraines on Earth. The terrain around the ice caps consists of alternating horizontal layers of light and dark materials, which may be composed of stratified outwash or loess. As many as 50 layers, each of which is 15 to 35 meters (50–120 feet thick), have been counted at a single site. Confirmation of a complex history of multiple glaciations must probably await a visit to Mars by a glacial geologist.

The most exciting new discovery involving glaciation in our solar system was unveiled in June 1996 by NASA's Galileo probe, which sent back extraordinary images from its orbit around Jupiter's moon Europa. As seen in Figure 17-44, Europa's surface consists of an icy crust—believed to be roughly 10 kilometers (6 miles) thick—that has broken into plates reminiscent of terrestrial ice floes. Some scientists believe that these icy plates float on a subsurface ocean of water or slushy ice that may "erupt" as ice "volcanoes" through fissures in the solid crust.

Chapter Summary

Glaciers are bodies of ice, composed of recrystallized snow, that flow downslope or spread radially under the influence of gravity. **Ice ages** are the dozen or so known periods of Earth history when the planet's climate was substantially cooler and glaciers covered a significant portion of the land surface. Climate fluctuations during an ice age cause the Earth's glaciers to alternately grow and advance during glacial periods, and then thaw and retreat during interglacial periods. The Earth's current ice age, the **Quaternary ice age,** spans the last 1.6 million years of the planet's history. During the **Pleistocene Epoch** of the Quaternary Period (1.6 million–10,000 years ago), ice sheets expanded to cover southern Canada and the northern United States, as well as northern and central Europe.

Glaciers form when snow that has survived a summer melting season becomes well-packed **firn** and then recrystallizes to ice. They tend to form when the climate is marked by moist snowy winters and, especially, cool cloudy summers that minimize melting. Glaciers are classified based on their topographic setting. Relatively small **alpine glaciers**—which include **cirque glaciers** (which erode basins into a mountainside), **valley glaciers** (which occupy existing stream valleys), and **ice caps**—are confined by surrounding mountains. They typically form at or above the regional **snowline,** the lowest topographic limit of year-round snow cover. Unconfined **continental ice sheets** cover vast topographic lowlands; they exist today only at very high latitudes, such as in Antarctica and Greenland.

The **terminus** of a glacier is its outer downstream margin. In a glacier's **zone of accumulation,** the addition of snow exceeds the volume lost; in the **zone of ablation,** more snow melts than is added. An **equilibrium line** between the two zones marks the point at which accumulation equals ablation. The ice in a glacier flows downward toward its underlying bed in its accumulation zone and upward from its bed in the ablation zone; thus a glacier flows internally even when its terminus remains stationary. Overall glacier flow occurs either by **internal deformation** in which individual ice crystals fracture or slip past one another, by **basal sliding** across a thin film of meltwater that accumulates at the gla-cier's base, or some combination of the two. Some glaciers with a good deal of water at their base accelerate their flow periodically, a process known as **surging.**

Small-scale glacial erosion, or **glacial abrasion,** is carried out by small particles embedded in a glacier's bed; these materials make linear scratches, or **striations,** in the underlying bedrock surface. On a larger scale, **glacial quarrying** breaks off and removes large masses of bedrock. Glacial erosion generally occurs in the downglacier segment of the accumulation zone near the equilibrium line, as downward-flowing ice impinges on the glacier's bed. Alpine glaciers carve smooth, unglaciated slopes into sharp, jagged peaks that may display horseshoe-shaped depressions **(cirques),** sharp pointy peaks **(horns),** long serrated ridges **(arêtes),** and distinct breaks in the ridgeline **(cols).** When cirque glaciers melt, a cirque lake, or **tarn,** forms in the basin. Together, glacial abrasion and quarrying remove a large quantity of bedrock and sediment, producing the asymmetry of glacially eroded hills, converting V-shaped stream valleys into U-shaped glaciated valleys, and eroding coastal valleys to depths well below sea level; the latter process produces water-filled valleys called **fjords.**

Glaciers tend to deposit their loads in their ablation zones, where melting releases the debris held in the ice. Deposition can occur directly from glacial ice or after transportation by meltwater. All glacial deposits are collectively known as **glacial drift.** Drift consisting of unsorted, unstratified sediments deposited from ice is **glacial till.** Drift consisting of sorted, stratified sediments deposited from meltwater streams is **outwash.** A till can contain **glacial erratics,** large blocks of bedrock deposited on bedrock of different composition, generally after being transported for a long distance. Outwash is generally deposited beyond a glacier's terminus, forming an outwash plain. It also serves as a primary source of **loess,** the wind-blown dust commonly found downwind from exposed, drying outwash. Meltwater sediments deposited in channels at the glacier's base form sinuous ridges called **eskers.** Tills and outwash deposited at a glacier's margin combine to form a series of hills called **moraines.** An advancing glacier may override an earlier moraine or any other preexisting hill, producing an asymmetrical hill called a **drumlin.**

Continental glaciers depress the crust beneath them and disrupt preglacial drainage systems. Other effects associated with glacial periods, which can appear beyond the actual glaciated areas, include the following: freezing of soils and regolith to form **permafrost** and other periglacial features; creation of **pluvial lakes** in now-arid regions; fluctuations in global sea levels; and the evolution, migration, and extinction of various flora and fauna, including our human ancestors.

Geologists try to predict future climate trends by reconstructing patterns of ice-age climate change. In the terrestrial record, they study fossil periglacial features, cave deposits, fossil pollen from cores of lake mud, and the chemical composition of old glacial ice. In the marine record, they study the distribution of temperature-sensitive microfauna such as forams.

Plate tectonics has moved the Earth's continents toward the poles and raised mountains above snowlines, preparing the Earth for the development of ice ages. Milankovitch hypothesized that three astronomical factors—such as variations in the shape of the Earth's orbit around the Sun, the tilt of its axis, and the wobble of its equatorial plane as it rotates on its axis—moderate the amount of solar radiation that reaches the Earth. When these factors coincide, they cause worldwide cooling and allow glaciers to advance. The high albedo of ice-covered glacial regions causes them to reflect heat away from the Earth, enhancing cooling; the low albedo of dark, ice-free regions causes them to absorb incoming solar radiation, enhancing warming. Thus, the more ice-covered land, the cooler the climate. Similarly, the more dark, soil-covered land that is exposed, the warmer the climate—that is, until Milankovitch or other factors coincide to reverse the trend. Advances and retreats of the Earth's ice cover may also be strongly influenced by Atlantic Ocean circulation patterns.

Ice ages have occurred in the Precambrian and Paleozoic Eras and during the Pleistocene Epoch of the Cenozoic Era. In the future, human activity may affect the course of glaciation. Activities that release heat-absorbing gases into the atmosphere can produce a warming greenhouse effect. For the last 10,000 years—the **Holocene Epoch**—Earth has been in an interglacial period of generally moderate warming, with a few cool periods lasting only a few hundred years each. Some geologists suggest that we are currently experiencing an unusually warm human-induced interglacial period. The ongoing variations in solar radiation may combine to produce another worldwide glacial expansion 20,000 to 25,000 years from now.

Key Terms

ice ages (p. 469)
Quaternary ice age (p. 469)
Pleistocene Epoch (p. 470)
glacier (p. 471)
firn (p. 471)
snowline (p. 472)
alpine glaciers (p. 472)
cirque glaciers (p. 472)
valley glaciers (p. 472)
ice caps (p. 472)
continental ice sheet (p. 472)
terminus (p. 472)
zone of accumulation (p. 472)
zone of ablation (p. 472)
equilibrium line (p. 472)
internal deformation (p. 474)
basal sliding (p. 474)
surge (p. 475)
glacial abrasion (p. 476)
striations (p. 476)
glacial quarrying (p. 476)
cirque (p. 477)
arête (p. 477)
col (p. 477)
horn (p. 477)
tarn (p. 477)
fjord (p. 477)
glacial drift (p. 479)
glacial erratic (p. 480)
glacial till (p. 480)
moraine (p. 481)
drumlins (p. 482)
outwash (p. 482)
loess (p. 483)
eskers (p. 484)
permafrost (p. 488)
pluvial lakes (p. 490)
Holocene Epoch (p. 500)

Questions for Review

1. What are three types of topographically confined glaciers?

2. Draw a simple diagram showing the basic components of all glaciers: the accumulation zone, ablation zone, and equilibrium line. Indicate where glacial erosion and deposition are most likely to occur.

3. Explain how the two glacial flow mechanisms—basal sliding and internal deformation—work.

4. Distinguish between glacial abrasion and glacial quarrying. Describe several distinctive features or landforms produced by each of these two erosional processes.

5. List five prominent erosional features produced by alpine glaciers.

6. Describe the origin and location of terminal, recessional, lateral, and medial moraines.

7. Describe the difference in texture and appearance of till and outwash.

8. What three major effects of glaciers occur during ice ages in areas far removed from the glaciers?

9. What factors are believed to cause ice ages and their fluctuations in ice volume?

10. Discuss three methods that geologists use to reconstruct past climates.

For Further Thought

1. In the Northern Hemisphere, the lateral moraines of alpine glaciers generally are larger on the south-facing sides of east–west oriented valleys. Why?

2. Some eskers actually "climb" up and over topographic ridges, in seeming defiance of gravity. How can the water that deposits eskers flow uphill?

3. If latitude 65°N is crucial in terms of the Milankovitch theory, why isn't latitude 65°S equally crucial? (*Hint:* Look at a globe or a world map.)

4. Speculate about what might happen to the world's climate and glaciers if the Antarctic ice sheet surged into the surrounding oceans and floating ice covered vast areas of the ocean.

5. Identify the "mystery glacial landform" in the photograph below.

18

Deserts and Wind Action

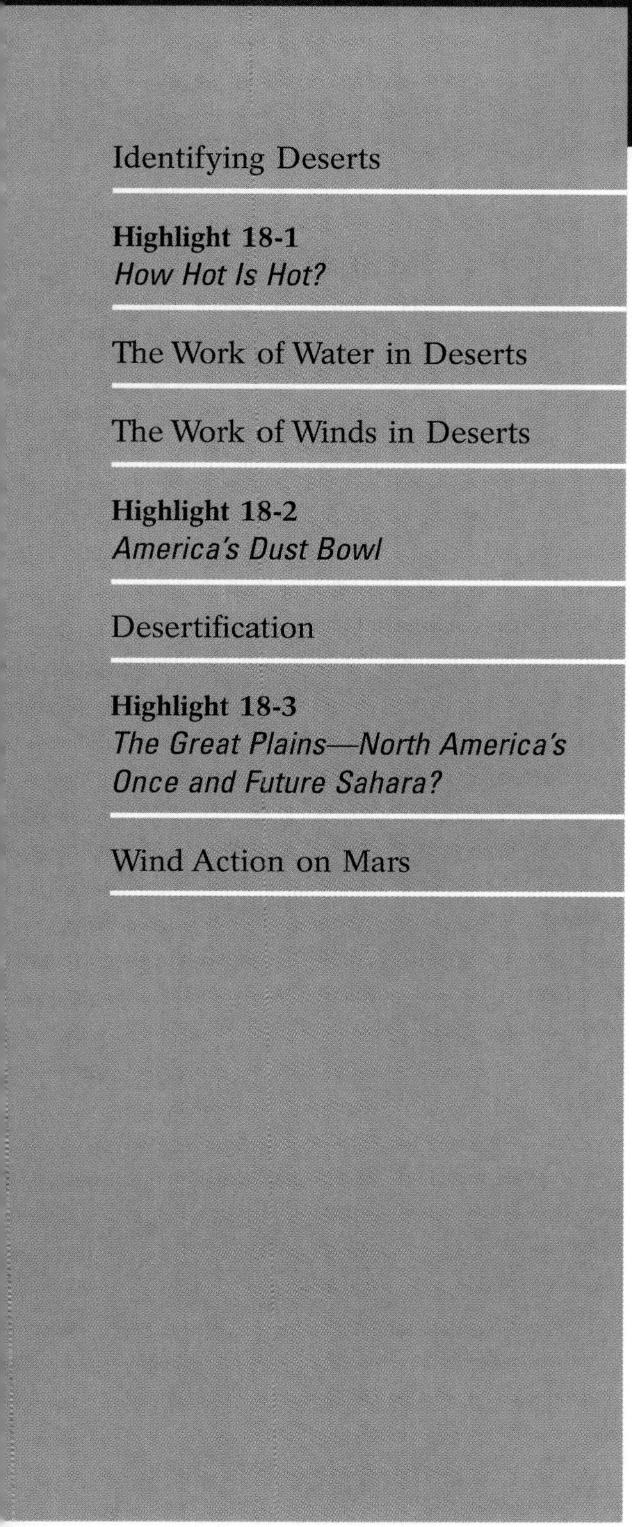

Figure 18-1 Wind-swept sand dunes at Mesquite Flats, Death Valley National Monument, California.

Unlike any landscape described earlier in this book, Death Valley National Monument, located in southeastern California, displays the wind-swept sands and austere barrenness characteristic of an arid landscape (Fig. 18-1). The valley owes its unique appearance to its extreme lack of water. Without the presence of appreciable water, little chemical weathering takes place—soils are thin, dry, and crumbly—and winds readily sweep loose particles into dunes or use them to sandblast exposed rock surfaces. These characteristics are typical of a **desert,** a region that receives very little annual rainfall and is generally sparsely vegetated.

Every major continent, although surrounded by water, contains at least one extensive dry region. Deserts account for as much as one-third of the Earth's land surface—more area than is occupied by any other geographical environment. Relatively few of these regions, however, resemble the popular Hollywood image of endless tracts of drifting sand. In North Africa's Sahara, the world's largest desert (*sahara* means "desert" in Arabic), only 10% of the surface is sand-covered. Even the Arabian Desert, the Earth's sandiest, is only 30% sand-covered. Desert climates, which are actually characterized by their dryness, can be found in cold polar regions as well as in torrid low-latitude regions. Such diverse landscapes as ice-bound Antarctica, the fog-shrouded coasts of Peru and Chile, and the near-continuous 8000-kilometer (5000-mile) stretch of land across northern Africa and the Arabian peninsula to southern Iran are all considered deserts, as defined by their extreme lack of surface water.

The term *desert* is misleading in its implication that the land is literally deserted—that is, devoid of life. Hot deserts are home to some of the Earth's hardiest plants and animals, a few of which are shown in Figure 18-2. These animals and plants have evolved to adapt to the extremely dry conditions. Some hot-desert plants produce seeds that can endure 50 years of drought. Some possess small, waxy leaves that minimize water loss to evaporation. Most have thick, spongy stems that store water from the occasional cloudbursts, and

(a)

(b)

(c)

Figure 18-2 Some desert life forms. **(a)** Desert shrubs and flowers in southwestern Colorado. **(b)** A roadrunner in the Sonoran Desert. **(c)** A thorny devil lizard (*Moloch horridus*) from Rainbow Valley, Alice Springs, Australia.

they produce deep root systems to tap groundwater supplies. Some desert plants may resemble dead twigs for months or even years on end, only bursting into a brief, but memorable bloom when the occasional downpour arouses them.

Animals that live in hot deserts include insects, reptiles, birds, and mammals. Some birds found in these environments may fly hundreds of kilometers to find water, which they carry back to their young in absorbent abdominal feathers. Some desert rodents live their entire lives without a single drink of water, absorbing needed moisture instead from the plants they eat. Most desert animals are nocturnal, avoiding activity during the hottest and driest times of the day, and venturing out only after temperatures have dropped considerably.

In this chapter, we will examine the processes that shape a desert's unique landscape. We will see how water's brief, intermittent appearances actually contribute significantly to the evolution of desert landforms. Our discussion will emphasize the role of wind, another major agent of surface change in deserts. We also will consider how human activity contributes to the expansion of deserts and describe efforts to use desert lands for agriculture and human habitation. Finally, with the help of the recent Pathfinder mission, we will journey to Mars to view one of our planetary neighbor's wind-swept deserts.

Identifying Deserts

How do we decide if an area is dry enough to be called a desert? One common guideline defines a desert as any region that receives less than 25 centimeters (10 inches) of precipitation annually—an amount that is insufficient to sustain crops without irrigation in most areas. The world's driest deserts, however, fall well short of the defining 25-centimeter mark. A very arid region may receive all of its annual rainfall in one sudden cloudburst; the water from this downpour may evaporate or be absorbed by the desert regolith within minutes, leaving behind virtually no trace of its occurrence. After a cloudburst, no precipitation may fall for months or even years.

The world's driest region is the Sahara, which on average receives only 0.04 centimeter (0.016 inch) of precipitation annually. The Sahara may actually receive no measurable precipitation for years; instead, a few storms over the course of a decade may contribute to its annual average. By the early 1990s, the northern Sahara had been virtually rainless for more than 20 years.

A more useful measure used to identify deserts is the **aridity index,** a ratio of a region's potential annual evaporation (which is itself a function of its yearly receipt of solar radiation) to its recorded average annual precipitation. An area with an aridity index of 1.0 has an annual amount of precipitation equal to the area's potential for evaporation. Such an area would possess a humid climate. An area with an aridity index greater than 4.0—where the potential for evaporation is at least four times greater than the annual precipitation—is classified as a desert. Two of the driest spots on Earth, the eastern Sahara of Africa and the Atacama Desert of Peru, have an aridity index of 200; these areas are described as *hyper-arid*. An area with an aridity index between 1.5 and 4.0 is classified as *semi-arid;* semi-arid regions, such as the Great Plains to the east of the Colorado Rockies, can support a greater diversity of life than deserts can. As we will see later in this chapter, these regions are climatically fragile; even a relatively slight change in temperature and precipitation may convert them into true deserts.

Types of Deserts

In this section, we will survey the types of deserts found in all areas of the world and describe how weathering shapes many of the unique landforms found in deserts.

Deserts may be cold, temperate, or hot. Cool deserts may even retain much of their moisture and use it effectively. In a cold desert, such as in northern Scandinavia, the 250-millimeter (10-inch) annual precipitation supports a dense forest, because the region's low temperatures retard evaporation, retaining moisture that encourages tree growth. In contrast, no forests grow in warm southern Nevada, which has the same amount of annual precipitation, because the warm climate hastens evaporation. The best-known deserts are far

Highlight 18-1 How Hot Is Hot?

Although deserts may occur in any climatic zone, most—and the best known—appear in hot places. Where are the hottest places on Earth? At El Azizia, Egypt, a desolate outpost in the Libyan Desert of northeastern Africa, the air temperature rose to a blistering 58°C (136°F) *in the shade* on September 13, 1922. The North American record, 56°C (133°F), occurred in Death Valley, California, in July 1913.

In such oppressive temperatures, the human body can lose a gallon of perspiration between sunrise and sunset. If water intake is not maintained, severe dehydration ensues, as the body rapidly draws on the water reserves stored in the body fat and blood. Sweat glands become overworked, the body cannot maintain its normal temperature, and a high fever results. In a few hours, blood circulation slows. Ultimately, death occurs.

Where no surface moisture exists to evaporate, humidity remains low and few clouds are present to filter daytime sunlight or retain the heat that radiates at night from Sun-heated rock surfaces. Consequently, days become even hotter, and night-time temperatures can plummet by as much as 40°C (72°F) from daytime highs. At night, heat dissipates rapidly to the atmosphere and temperatures often drop below freezing.

In 1974, the young British journalist Richard Trench vividly described life in the thermally fickle Sahara of Algeria:

> The writer will not soon forget arising at 3:00 A.M.—in the Algerian Sahara in early September. The bucket of water for washing presented a thin film of ice, and a heavy wool sweater and leather jacket were comfortable while riding in an open jeep. By 9:30 A.M., the air temperature was nearing 85°F; by 11:30 A.M., a pocket thermometer registered over 105°F and at 2:00 P.M. the same thermometer registered 127°F. What passed for a local pub in a nearby oasis cooled its beer by wrapping the bottles in wet sacking and laying them in the sun. Evaporation occurs at an almost unbelievable rate under such circumstances and effective ground moisture levels are fantastically low. The foregoing account amounts to a record of a daily temperature variation of at least 95°F. The beer was delicious.

hotter and drier than most of their water-starved counterparts. Highlight 18-1 describes the deserts of northern Africa, which are among the hottest, driest places on Earth.

Subtropical Deserts Although deserts can be found at all latitudes, most arise in the subtropical belts between latitudes 20° and 30° on either side of the equator. Why do extremely dry subtropical deserts exist adjacent to equatorial areas that are commonly drenched in tropical rain?

Direct solar radiation strikes the equatorial zone virtually year-round, and the warm air there evaporates large amounts of moisture from equatorial oceans. As the air warms, it expands, becoming less dense. As this air begins to rise, it carries water vapor upward, like steam rising from a pot of boiling water. The moist air immediately begins to cool as it ascends, however. By the time it reaches an altitude of about 10 kilometers (6 miles), the air has cooled enough to reduce sharply its capacity to hold water vapor and consequently releases its moisture as torrential rains that fall on equatorial lands.

As the now-dry tropical air continues to rise and cool, it becomes denser. Once it is no longer capable of rising, it begins to spread laterally. By the time it reaches a latitude of 20° to 30° north or south of the equator, it is dense enough to sink back to the Earth's surface. There the air warms and its capacity to hold water vapor increases, enabling it to evaporate surface water, and thereby dry out the landscape. Under the pressure of overlying descending air, some of the warm air approaching the surface spreads toward the equator, where it becomes heated further, evaporates a great deal of water, and again enters the cycle of rising warm air and sinking cool air. As shown in Figure 18-3, this cycle of heating, rising, cooling, sinking, and reheating air creates an

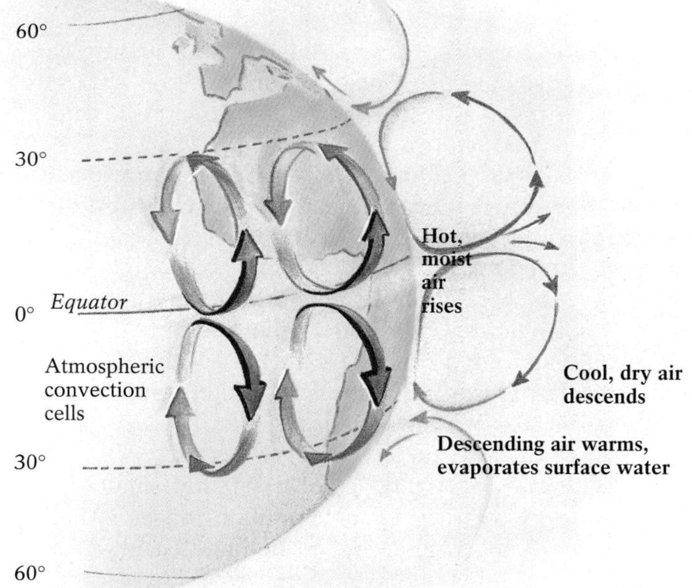

Figure 18-3 Atmospheric convection cells form as hot, moist equatorial air rises, cools when it reaches about 10 kilometers (6 miles) above the surface, releases its moisture as tropical rain, spreads northward and southward from the equator, and then descends in the subtropical latitudes 20° to 30° north and south of the equator. The now-dry air warms as it descends and evaporates ground moisture rapidly, producing desert conditions on land.

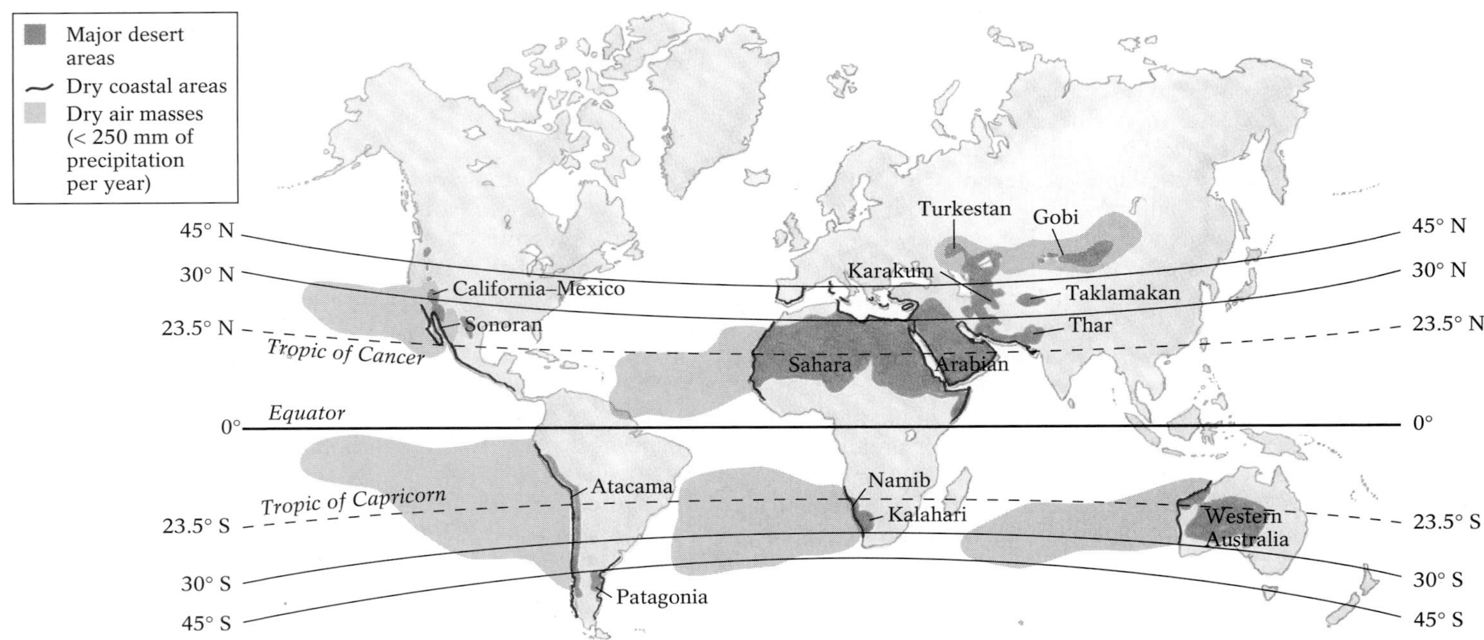

Figure 18-4 The worldwide distribution of the Earth's deserts. Most deserts lie in the two subtropical belts between latitudes 20° and 30° north and south of the equator.

atmospheric *convection cell,* similar to the mantle convection cells that help drive plate motion. The air that spreads toward the poles also forms an important set of prevailing winds, discussed later in this chapter.

These air-circulation patterns produce the two subtropical belts of desert lands shown in Figure 18-4. Just north of the Tropic of Cancer in the Northern Hemisphere lie the Sahara, the Arabian Desert of the Middle East, and the parched landscapes of Mexico and the American Southwest. Straddling the Tropic of Capricorn in the Southern Hemisphere are the Kalahari and Namib Deserts of southwestern Africa, the Atacama Desert of Peru and Chile, and the interior deserts of the Australian outback.

Rain-Shadow ("Orographic") Deserts Many smaller deserts form on the leeward sides of mountain ranges. (The *windward* side of a mountain range is the one that lies directly in the path of oncoming winds. The *leeward* side comprises the sheltered "backside"—that is, the opposite side of the range.) As moist air on the windward side of a mountain range rises and cools, it loses much of its capacity to hold water; this moisture then precipitates on the windward slope. The leeward slopes and the regions beyond them are left dry. This process (shown in Figure 18-5) causes moist air to be "wrung dry" by mountains and creates the **rain-shadow effect,** also known as the *orographic effect* (*oro,* from the Latin meaning "mountain"). The desert lands of Nevada, Arizona, and eastern California, for example, are leeward of the Sierra Nevada mountain range and therefore in its rain shadow; Pacific air masses are dry by the time they reach these deserts. The same relationship keeps rain-swept Seattle lushly vegetated, while parched eastern Washington remains only sparsely vegetated.

Continental Interior Deserts Deserts also form from an absence of moisture in the interiors of continents, far from oceans. In central Asia, the Gobi Desert and the Taklamakan (meaning "the place from which there is no return") exist principally because of two factors: their great distance from water and the rain-shadow effect of the Himalayas. Air masses that reach these areas have lost virtually all their moisture after crossing thousands of kilometers of land. The location of these Asian deserts is a direct consequence of plate tectonics: They developed about 40 million years ago after the Indian subcontinent became attached to Asia, which enlarged the continent and left central Asia several thousand kilometers away from the nearest ocean. (See Highlight 12-3, "Continental Convergence and the Birth of the Himalayas.")

Deserts Near Cold Ocean Currents Some deserts in warm subtropical regions lie close to oceans in areas adjacent to cold ocean currents. The cold air above these ocean currents contains little water. As shown in Figure 18-6, when this dry air moves over land, it becomes warmer, enabling it to hold more water. Thus it evaporates almost all of the surface water in these coastal desert regions. The Atacama Desert of Peru and Chile, for example, owes its extreme aridity in large part to the Pacific Ocean's cold Humboldt current, which flows northward toward the equator from the icy Antarctic, hugging the coast of western South America. Coastal Peru, which seldom experiences more than a drizzle, has an annual precipitation level of less than 1 millimeter (0.04 inch). Simi-

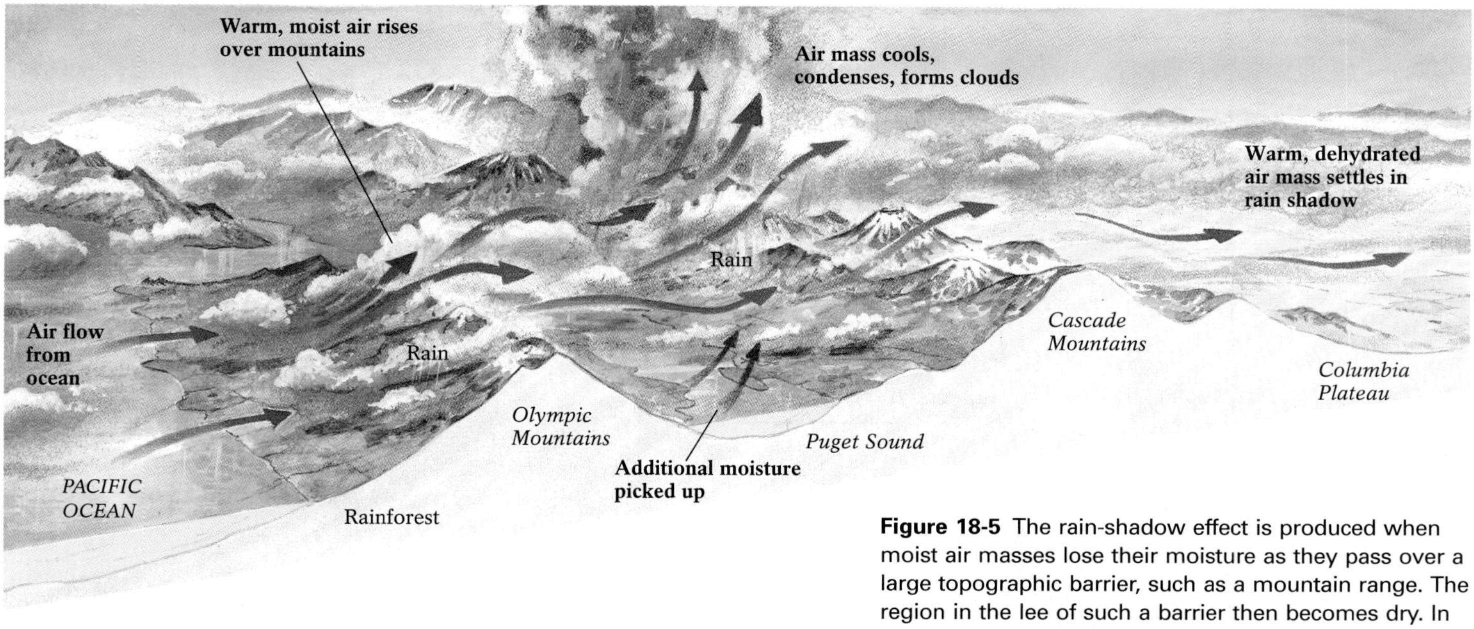

Figure 18-5 The rain-shadow effect is produced when moist air masses lose their moisture as they pass over a large topographic barrier, such as a mountain range. The region in the lee of such a barrier then becomes dry. In western North America, moist Pacific Ocean air masses rise west of the Olympic Mountains, cool, and precipitate over the Olympic rainforest of coastal Washington, where annual precipitation exceeds 2500 millimeters (100 inches). After picking up additional moisture over Puget Sound, the air masses condense, causing rain to fall on the western side of the Cascade Mountains. The now-dehydrated air descends on the mountains' lee side, where it warms and becomes able to hold moisture once again, producing rapid evaporation and the treeless landscape of the arid Columbia Plateau, where annual precipitation is less than 250 millimeters (10 inches).

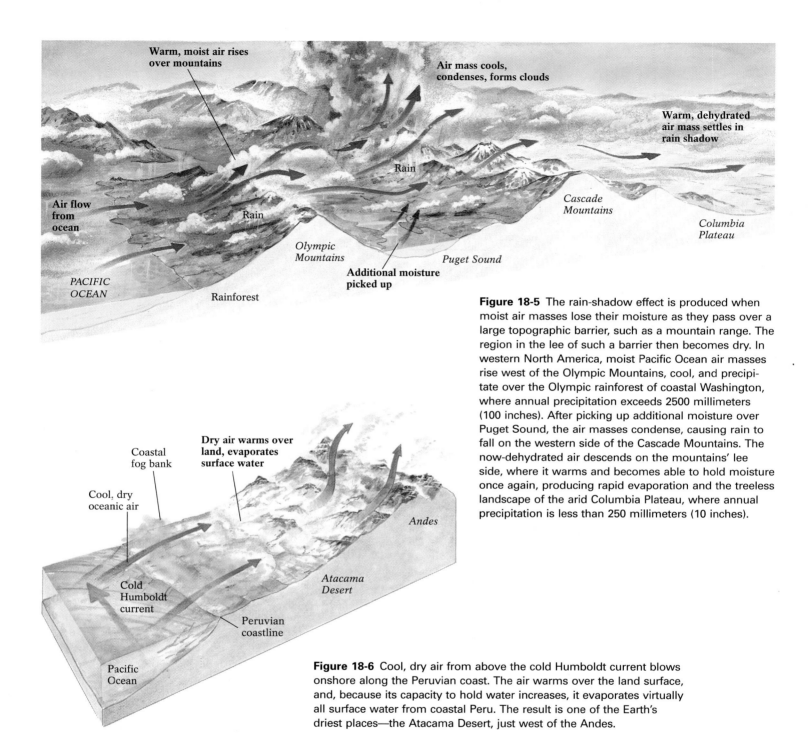

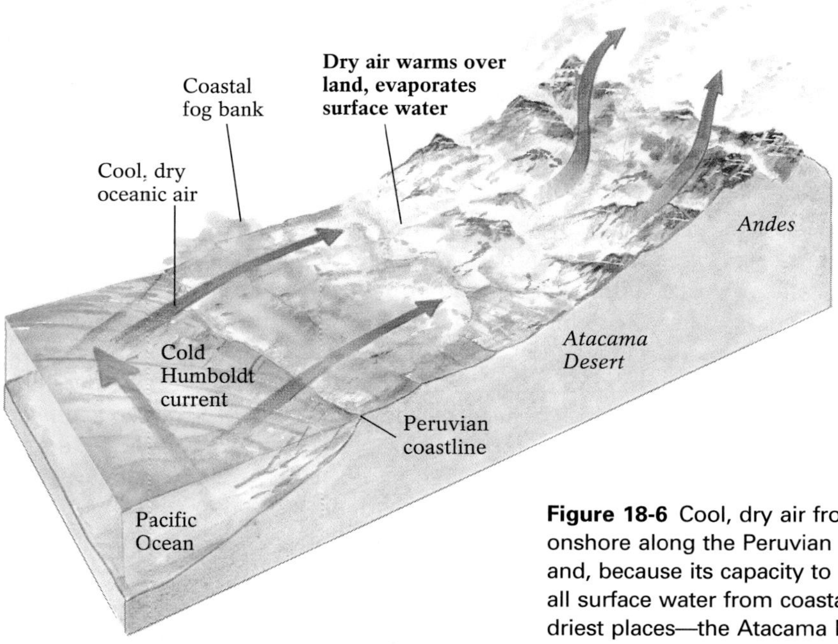

Figure 18-6 Cool, dry air from above the cold Humboldt current blows onshore along the Peruvian coast. The air warms over the land surface, and, because its capacity to hold water increases, it evaporates virtually all surface water from coastal Peru. The result is one of the Earth's driest places—the Atacama Desert, just west of the Andes.

larly, cold Atlantic currents and hot African land surfaces combine to create the hyper-arid Namib and Kalahari Deserts along Africa's southwestern coast.

Polar Deserts Some places are both extremely arid and numbingly cold. Because they receive little solar radiation, such high-latitude locations as northern Greenland, Arctic Canada, northern Alaska, and Antarctica have temperatures that remain below freezing year-round—even during their brief summers, when the sun never sets. Global atmospheric circulation, however, brings only cold dry air to these lands; thus they qualify as deserts.

Weathering in Deserts

Temperature changes and their effects on the ability of air to either evaporate or retain water at a desert's surface are largely responsible for the weathering processes that shape many of the landforms found in these regions. Such desert features as steep angular cliffs, sharp-edged stones, and relatively thin

Figure 18-7 Although oxidation, like other forms of chemical weathering, proceeds slowly in deserts, it has created colorful rocks such as these of the Chinle formation in Utah. Oxidation of iron-bearing silicates in the area's sandstones produces red-orange iron oxide, which gives these rocks their characteristic hues.

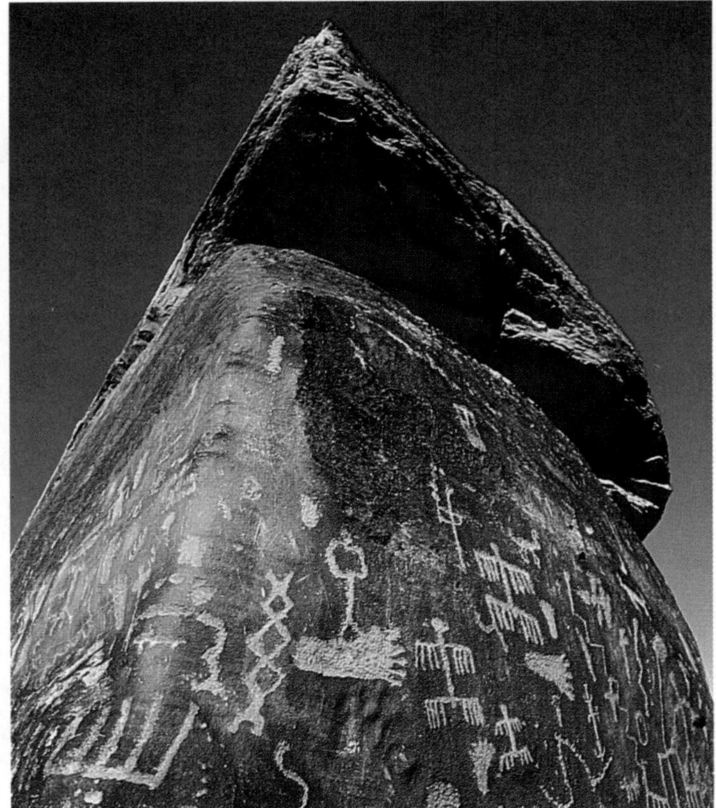

Figure 18-8 Petroglyphs carved into desert-varnished rocks by native North Americans 2000 years ago. Because petroglyphs such as these have not been "revarnished" since they were carved, we can deduce that desert varnish forms quite slowly.

soils may give the impression that they are subject to little or no weathering. In fact, mechanical-weathering processes are at work, and they produce much of a desert's loose sandy regolith. As discussed in Chapter 5, extreme daily temperature fluctuations cause rock surfaces to expand and contract repeatedly. Eventually individual grains flake off the surface and join the desert's supply of shifting sands. At the same time, ongoing evaporation of any available water precipitates dissolved salts within rock fractures in the form of crystals. As these salt crystals grow, they exert enormous pressure on adjacent mineral grains, acting like minute crowbars to force the grains apart.

Because water is necessary for chemical reactions, geologists long believed that the lack of water in deserts meant that little chemical weathering could occur there. Cold desert nights, however, promote condensation of dew on rocks, and some deserts do enjoy a brief rainy season. The little moisture present in these areas, along with the organic acids produced by desert vegetation, supports a small amount of chemical weathering. Even hyper-arid regions may contain enough moisture to oxidize iron-rich silicates, producing the colorful hues that characterize many desert rocks (Fig. 18-7). The extremely slow chemical weathering in deserts also produces distinctive loose thin soils—*aridisols*—that have low organic content and remain completely dry for at least six months of the year (see Chapter 5, Table 5-1, for a description of aridisols).

Many desert rock surfaces, such as the petroglyphs in Figure 18-8, are covered with **desert varnish,** a thin, shiny, red-brown or black layer of manganese oxides and iron oxides. When this coating appears on pure-white quartzite rocks that initially contained no manganese or iron, we must ponder its origin. According to one hypothesis, high winds transport manganese- and iron-rich clays from nearby sources and forcefully plaster them onto dew-dampened rocks; the clays adhere to their surfaces and become oxidized by oxygenated dew. Another hypothesis proposes that microbes in the desert soil concentrate manganese on rock surfaces. Whatever the process involved, it is clearly a slow one and apparently unique to the desert environment.

The Work of Water in Deserts

Although desert landscapes appear vastly different from those in other environments, they are nevertheless shaped by many of the same processes that act on virtually all land surfaces. Despite conditions of extreme aridity, water serves as the primary sculptor of desert landforms. Brief, occasional cloudbursts cause the rapid erosion and subsequent deposition that create many of the prominent features of the desert landscape. Figure 18-9 illustrates these landforms, and we describe their development in this section.

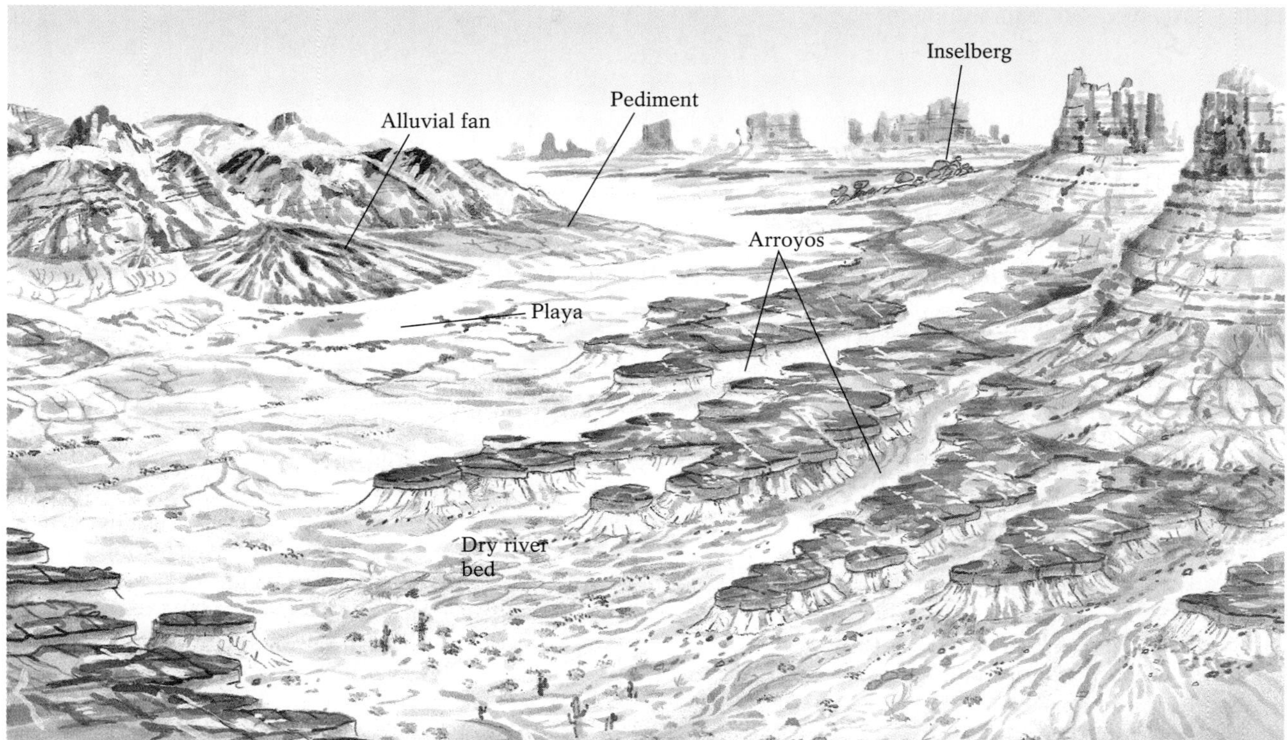

Figure 18-9 Desert landforms produced by water. Intermittent surface-water flow in deserts produces arroyos, pediments, inselbergs, and playas. Although the arroyo pictured here is dry, it fills rapidly with rushing water during a desert cloudburst. Deposition from desert streams creates alluvial fans and playas.

Stream Erosion in Deserts

In arid regions, only a minimal network of soil-binding plant roots anchors loose sediment. Thus flowing water can easily remove those grains. Moreover, the infrequent desert storms produce immediate and intense surface runoff that is not slowed by vegetation as it would be in grasslands or forests. Over thousands of years, the rapidly moving water of these short-lived desert streams can erode numerous **arroyos,** stream channels that remain dry most of the year.

A violent desert thunderstorm can drop 5 centimeters (2 inches) of rain in minutes and set off a flash flood. Because the water cannot be readily absorbed by the compacted, sun-baked ground, which contains few roots, it rapidly overflows any existing stream channels. A single storm can send a 3-meter (10-foot)-high wall of water and sediment—sometimes resembling a watery debris flow (see Chapter 13, page 375)—sweeping through an arroyo with virtually no warning. During such a brief, intense storm, water velocity in an arroyo becomes so rapid that little meandering or lateral migration takes place. These rushing streams excavate their channels vertically, eventually forming long, narrow, steep-sided canyons. Anyone wandering in an arroyo when a flash flood occurs would have great difficulty escaping up the near-vertical canyon walls. Such was the case in August 1997, when a flash flood in Antelope Canyon near Page, Arizona, swept away 11 unsuspecting campers (Fig. 18-10).

As a storm abates and the velocity of its floodwaters decreases, the water infiltrates loose desert sands or evaporates, leaving behind a high volume of coarse sediment

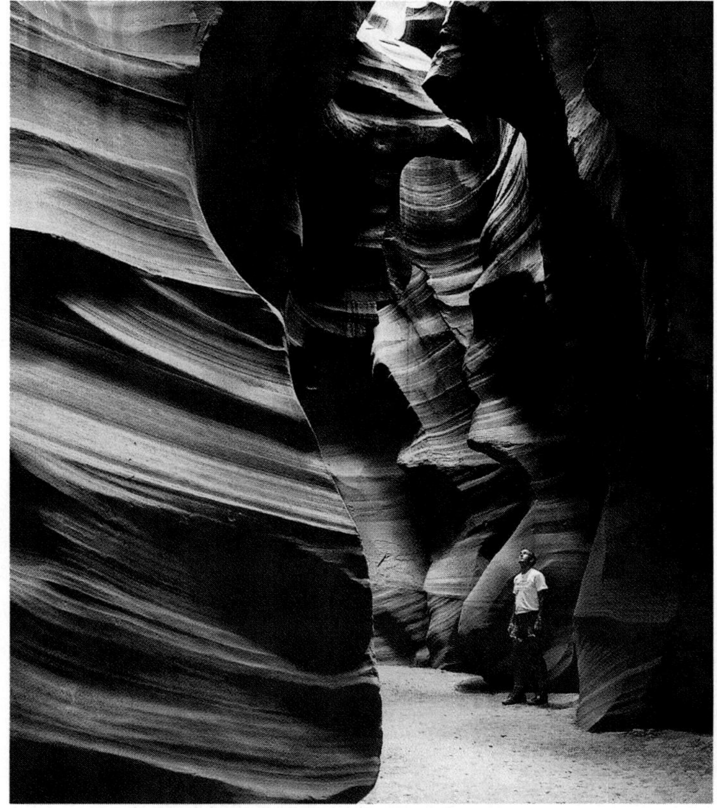

Figure 18-10 River-cut Antelope Canyon, Page, Arizona. The steep walls of this slot canyon prevented campers from escaping the flash flood of 1997.

covering the canyon floor. Little happens between floods to change the structure or appearance of these deposits.

When desert streams remove sediment from a mountain, the resulting erosional surface forms a **pediment** (from the Latin, meaning "foot") that lies at an angle of 5° or less and may extend for many kilometers from the base of an eroding mountain. Pediments enlarge slowly, as the repeated action of running water causes mountains to recede. This process continues until nearly all of the mountain's mass has been eroded. Pediments are usually covered with a thin layer of loose debris that is being transported downslope. Very large pediments may have begun to form under the moist, pluvial conditions of the Pleistocene (discussed in Chapter 17), which was characterized by more moisture, runoff, and weathering than are present today. It is unlikely that the arid environment in which pediments currently exist could have provided sufficient water to erode that much sediment, even over several thousand years. The reduced desert waterflow found today may be moving only materials already loosened by weathering in a past climate, not significant additional material.

After advanced pediment development, part of the mountain may remain as a resistant, steep-sided bedrock knob. Such a residual erosional feature is called an **inselberg** (from the German, meaning "island mountain") and is typically composed of a durable rock such as granite, gneiss, or well-cemented sandstone. Another variety of inselberg, shown in Figure 18-11, may develop after a long period of surface erosion uncovers, or *exhumes,* a body of resistant rock. Because inselbergs stand prominently above the surrounding plain, their smooth, steep slopes shed water quickly. This action minimizes prospects for further weathering and accelerates the weathering and erosion rate of the inselbergs' surrounding rocks and sediments.

Stream Deposition in Deserts

Whenever erosion takes place, deposition is sure to follow. In moist, temperate regions, most products of erosion are carried to and deposited in the nearest ocean. In arid regions, however, short-lived desert streams can deposit thousands of meters of sediment into channels and basins, constructing great alluvial fans that slope gently away from the remnant mountains to the desert floor. Because few streams can flow for long distances through the desert environment, their sediments are typically deposited close to the erosion site, rather than at the bottom of some distant ocean. Wherever surface water evaporates or infiltrates, it drops the sediment load washed from the surrounding steep mountain slopes, forming alluvial fans and dry lake beds (the characteristic features of arid-region deposition).

The water from occasional desert torrents flows rapidly at first, typically remaining confined to a narrow arroyo with steep walls. At the outlet of the arroyo, where the canyon walls end, the flow slows appreciably as the unconfined water spreads out. As the water's velocity diminishes, the stream loses its ability to transport sediment and deposits its load in the form of an alluvial fan (see Figures 6-4 and 14-22). Each mountain stream that reaches the desert plain produces its own alluvial fan, proportional in size to the stream's drainage area. The slope of the fan surface—about 10° at the top—decreases gradually until it merges with the desert floor. The steeper slope near the top of the fan contains the coarsest sediment, which is dropped first as the stream slows. The sediment generally becomes increasingly fine toward the bottom of the fan.

Water that infiltrates the coarse upper-fan sediments may reemerge at the foot of the fan as a spring. In the U.S.

Figure 18-11 Inselbergs, such as Ayers Rock in Australia shown here, rise abruptly from desert plains. They form when soft rock is eroded and leaves a mass of more resistant rock standing prominently above the surface.

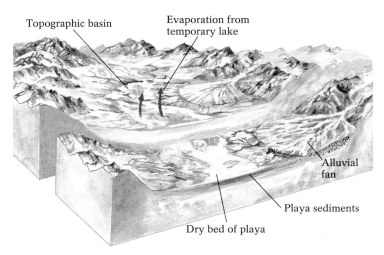

Figure 18-12 A playa is a dry basin in the desert floor from which a temporary lake has evaporated. Photo: Evaporite deposits in Devil's Golf Course, Death Valley, California. Ninety meters (295 ft) below sea level, Devil's Golf Course is a playa that posed a significant obstacle to westward migration in the late nineteenth century.

Southwest, such springs are often marked by groves of mesquite trees (something to remember if you ever find yourself thirsty in that region). Alluvial fan sediments may contain water sufficient to irrigate the adjacent desert area and permit cultivation—as Native Americans in the Southwest have done for more than 1500 years—or contribute to the water needs of a nearby city. San Bernardino, California, for example, draws some of its water from the alluvial fan deposits on which it is sited.

During brief periods of higher-than-average precipitation in a desert's interior, water drains toward topographically low, closed basins, where it may collect as a temporary lake. As shown in Figure 18-12a, such a lake may evaporate in a few days or weeks, leaving behind a **playa** (from the Spanish, meaning "beach") comprising a dry lake bed on the desert floor.

A playa's dry bed, such as the one shown in Figure 18-12b, typically consists of precipitated salts initially dissolved from soluble rock, plus fine-grained clastic sediment. The salts, which commonly represent a mixture of sodium and potassium carbonates, borates, chlorides, and sulfates, may form a blinding-white residue or a bizarre landscape filled with jagged pinnacles. Industrial chemicals have been mined from the playas of the Southwest for more than 100 years; Death Valley has long been a source of sodium borate, or *borax* ($Na_2B_4O_7 \cdot 10H_2O$), used in pottery glazes and household cleansers, and as a hardening agent for alloys and a shielding material against nuclear radiation. (Borax, originally carted away by "20-mule teams," became the trade name for a cleanser made famous in commercials narrated by former president Ronald Reagan during the 1950s.) The formation of evaporite deposits is discussed in greater detail in Chapter 6, page 166.

If its inflowing water does not encounter and dissolve soluble rock, a playa may contain only fine-grained, clay-rich sediment. The desert sun can bake such a surface until it becomes so hard that it can serve as a natural landing strip for aircraft. For example, Rogers Dry Lake, on Edwards Air Force Base in California's Mojave Desert, north of Los Angeles, is the optimal landing site for NASA's Space Shuttles.

When precipitation and runoff from surrounding mountains outpace evaporation and infiltration on the desert floor over an extended time, a large lake may form that can last for years or centuries, even in the ultra-dry desert air. Because they lack outlets from which water can drain, most desert lakes become extremely saline (salt-rich) as evaporation continually concentrates the dissolved salts. Large saline water bodies of this type include the Great Salt Lake in Utah; the Dead Sea, a basin that descends to 392 meters (1286 feet) below sea level (Fig. 18-13); and Lake Chad, a 22,000-square-kilometer (8500-square-mile) body of water located in the southern Sahara.

Figure 18-13 The Middle East's Dead Sea, which is essentially a desert lake, is so saline that swimmers float noticeably higher in it than they do in the ocean.

The Work of Winds in Deserts

Winds, a second major force in shaping deserts, consist of air currents set in motion by heat-induced changes in air pressure. (High-pressure air is cool and generally flows as wind into regions of lower-pressure warmer air.) Near the equator, a zone of low pressure prevails as sun-warmed air becomes less dense and rises. As we saw earlier in our discussion of the origin of subtropical deserts, this air cools and becomes more dense as it ascends, eventually loses its buoyancy, and, while still at high altitudes, spreads north and south of the equator. At latitudes 20° to 30° in both directions, the cooler, denser air sinks back toward the surface. The result is two permanent subtropical high-pressure zones known as the *horse latitudes,* which are noted for their calm, clear, warm weather. (The name originated from floating horse carcasses that littered the subtropical seas during the eighteenth and nineteenth centuries. When sailing ships stalled without a driving breeze to propel them, and food and drinking water became exhausted, most of the ships' cargoes of horses were apparently thrown overboard to conserve water supplies; the rest became the sailors' only remaining source of sustenance.)

The high-pressure air of the horse latitudes ultimately spreads close to the Earth's surface and moves back toward the equator, filling the void created by the rising low-pressure equatorial air. In addition, some of the high-pressure air goes the other way, toward the poles, drawn by the relatively low air pressure near 60° of latitude.

Global wind patterns do not follow a strictly north–south or south–north path. The Earth's rotation about its axis imparts an east–west component to the flow of air above its surface . . . for the following reason. As the Earth rotates counterclockwise, it turns with a greater velocity at the equator than at the poles. (Because each of the Earth's rotations takes 24 hours, the planet's surface at the equator—with its larger circumference—must speed along more swiftly than the surface closer to the poles, simply to cover the greater distance within the same 24-hour period.) Moving air masses *above* the spinning planet thus appear either to "move ahead" or "lag behind" the underlying Earth, causing air masses to curve relative to the surface. Figure 18-14 illustrates these global wind patterns. The air masses that have more velocity than the underlying Earth (nearer to the slower-spinning poles) move ahead of the Earth's counterclockwise rotation, producing *westerlies.* The air masses that have less velocity than the underlying Earth (nearer to the faster-spinning equator) lag behind the Earth's rotation, producing the trade winds, or *easterlies.*

This apparent deflection of moving air masses (or any freely moving body, such as an ocean current or a speeding missile) by Earth's rotation is called the *Coriolis effect.* Spend a moment studying the wind patterns in Figure 18-14. Notice that in the Northern Hemisphere, descending air currents (such as those spreading north and south of latitude 30° N) are deflected *toward the right* of the direction in which they are moving; in the Southern Hemisphere, they are deflected *toward their left*—a simple way to remember the windy implications of the Coriolis effect.

In the Northern Hemisphere, the descending air of the subtropical high (at 30° latitude) that returns south toward the equator is deflected to its right, or from east to west. This northeasterly deflection, which produces the *trade winds,* prevails along the warm, low-latitude trade (an archaic meaning of *trade* is "track" or "course") routes taken in the eighteenth and nineteenth centuries by merchant sailing vessels. (Winds are designated by the direction from which they are blowing; hence a "northeasterly wind" blows *from* the northeast, a southerly wind blows *from* the south, and so on.) The descending air of the subtropical high that turns toward the North Pole is also deflected to the right, or from west to east, by the Earth's rotation; this deflection produces the mid-latitude winds, or *westerlies.* The westerlies prevail between about 30° and 60° of latitude in both hemispheres; in the Northern Hemisphere, they dominate the wind pattern for most of North America, Europe, and Asia.

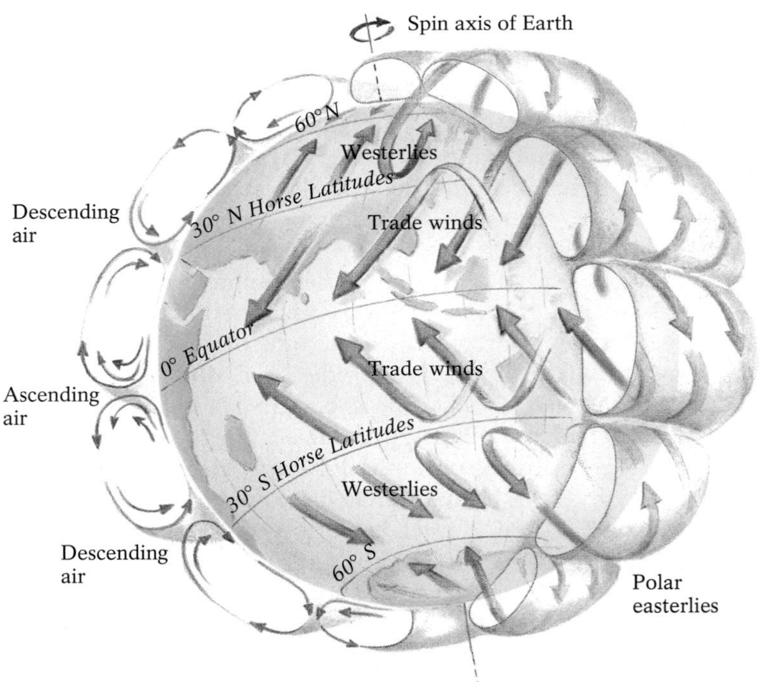

Figure 18-14 The Earth's global wind patterns. The Earth's rotation deflects moving currents of air (and water as well); in the Northern Hemisphere laterally spreading air masses are deflected to their right, and in the Southern Hemisphere they are deflected to their left. This deflection is known as the Coriolis effect.

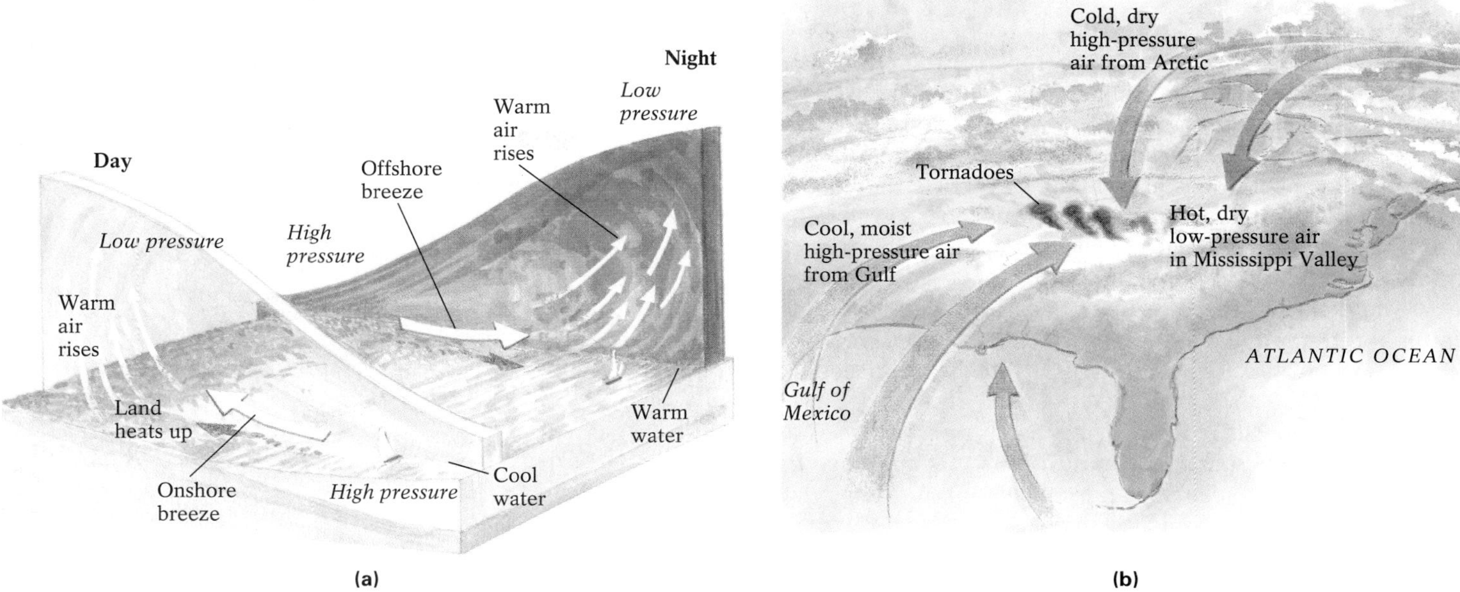

Figure 18-15 On both local and continental scales, onshore and offshore winds develop because solid ground warms and cools quickly, whereas water warms and cools more gradually. **(a)** On a warm afternoon, land heats up more rapidly than the ocean, and a zone of low pressure forms over the land; cooler, high-pressure air then blows from the sea to the land as an onshore breeze. At night, the sea cools slowly, retaining more daytime heat than the land does. As the land cools, higher air pressure develops over it, sending an offshore breeze out to the low-pressure zone over the ocean. **(b)** During the summer months, the entire North American continent heats up, producing a large temporary zone of low pressure overhead; higher pressure develops over the relatively cooler surrounding ocean. (This scenario differs from the winter pattern of cold land/high pressure and relatively warm ocean/lower pressure.) The pressure differential produces southerly winds that funnel moist air from the Gulf of Mexico up the Mississippi Valley, where it may encounter cold air descending from the Arctic. The resulting clash of cold, dry Arctic air and warmer, moist Gulf air produces tornadoes in the middle of the continent.

Global wind patterns have probably remained relatively constant throughout the Earth's history, because the orientation of the Earth's axis and its distance from the Sun have not changed dramatically. The consistent difference in the amount of heat received by the equator and poles has probably always produced the Earth's pattern of high- and low-pressure zones, which themselves generate the global winds. The direction of the Earth's rotation has always been the same, so the Coriolis effect has always deflected winds in the same direction as it does today.

Low-altitude winds, such as the pleasant onshore and offshore breezes that moderate coastal climates, can also result from *local* differential heating of the atmosphere. Figure 18-15a shows how differences in land and water temperatures produce such breezes. When this heating differential occurs on a continental scale, it sometimes has catastrophic results. As you can see in Figure 18-15b, differential heating of land and ocean draws moist air from the cooler, higher-pressure zone of the Gulf of Mexico into the warmer, lower-pressure zone of the Mississippi Valley during the warm spring and summer months. This process produces warm, moist southerly winds that, when they encounter cold, dry air descending from the Arctic, can spawn tornadoes. Differential heating similarly causes the gusty winds of deserts: During the day, heated surface air rises to create a low-pressure zone; in the evening, relatively cooler and higher-pressure surface air rushes in, often at a velocity of 100 kilometers (60 miles) per hour or more, to replace it.

Both local and global wind patterns affect desert areas in several significant ways. Winds can erode desert landforms, transport sediments, and deposit sediments. We describe these wind-driven processes next.

Erosion by Wind

When you walk along a beach on a blustery day and feel the sting of blowing sand against your legs, you are experiencing the power of wind erosion. The erosive power of wind was readily evident when a fierce windstorm swept through California's San Joaquin Valley on December 12, 1977. Winds blowing at 300 kilometers (190 miles) per hour removed nearly 100 million tons of topsoil from a 2000-square-kilometer (775-square-mile) area. The blinding dust and sand storm severely damaged crops and buildings; during a period of zero visibility, five motorists were killed in accidents and many more suffered injuries.

After water, wind is the second most effective agent of surface change in deserts. Wind by itself cannot erode solid rock in the same way as flowing water or ice, by dislodging cemented grains directly from sedimentary rock or by quarrying crystalline bedrock from outcrops of jointed or fractured igneous and metamorphic rock. Desert landscapes, however, are more vulnerable to wind erosion than most other geographical settings. Because deserts contain little vegetation to anchor the soil and hold its particles down, they are generally covered with *loose* materials. Loose sand grains lifted from the surface by turbulent winds can be used as "tools" with which to etch exposed rocks.

The lifting of these loose grains—a process called *deflation*—and their use as tools to etch exposed rocks—a process called *abrasion*—are discussed in the next sections.

Deflation Wind gusts blowing across a dry, treeless desert lift sand- and silt-sized particles from the surface, but leave behind larger, heavier pebbles and cobbles. In addition, clay-sized particles may remain behind, as they adhere to the surface. **Deflation** comprises the removal of large quantities of loose material by wind. It may excavate distinct depressions in a desert floor and leave behind a coarse pavement of large particles. The resulting depressions may vary in size and shape, largely depending on the force of the wind and the physical properties of the deflating sediment.

Deflation usually lowers the landscape slowly, by only a few tens of centimeters per thousand years. In extraordinary cases, however, such as occurred during North America's Dust Bowl of the 1930s (see Highlight 18-2 on page 522), deflation can strip away as much as 1 meter (3.3 feet) of fine-grained topsoil in just a few years. Land surfaces may deflate over broad expanses or in local areas only. Small lowered regions, known as *blowouts*, form where animal trampling, overgrazing, range fires, drought, or human activity has disturbed surface vegetation. Blowouts are often little more than 3 meters (10 feet) across and 1 meter (3.3 feet) deep. Because water helps bind loose deflatable material into a more cohesive mass, blowouts seldom fall below the local groundwater table. Thousands of blowouts, such as the one shown in Figure 18-16, dimple the semi-arid Great Plains of North America, from Texas to Saskatchewan.

Figure 18-16 A blowout caused by deflation in Sand Hills State Park, Texas.

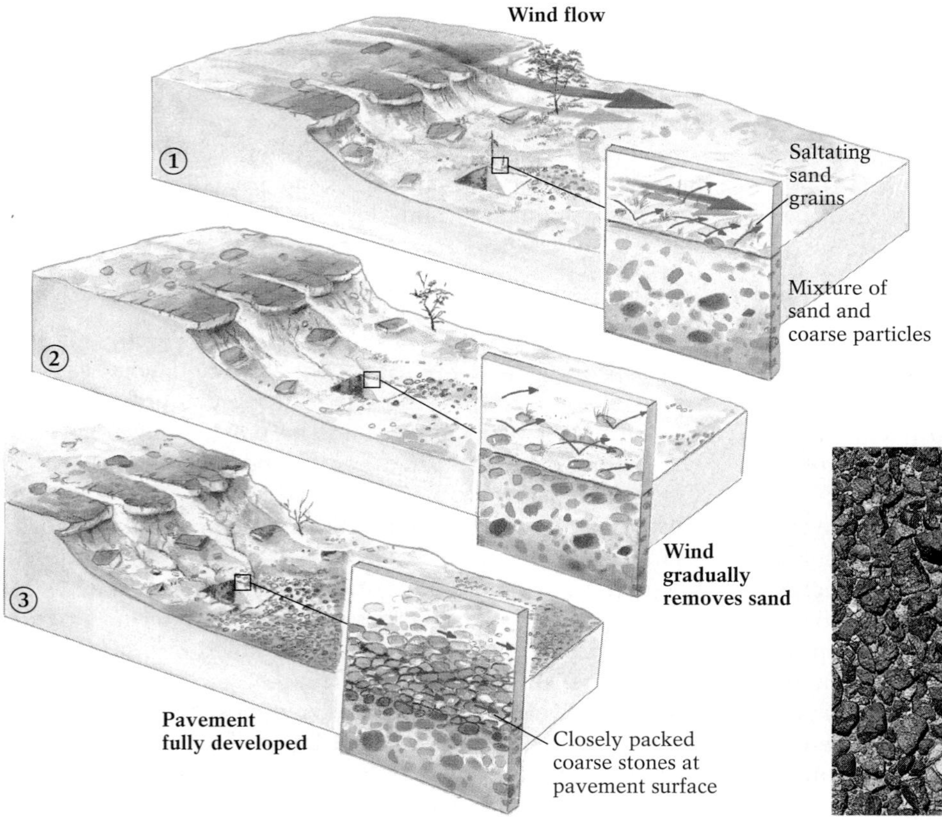

Wind flow
①
Saltating sand grains
Mixture of sand and coarse particles
②
Wind gradually removes sand
③
Pavement fully developed
Closely packed coarse stones at pavement surface

Figure 18-17 Desert pavements form from the progressive deflation of fine particles, concentrating coarse materials at the surface. Photo and inset: Death Valley, California.

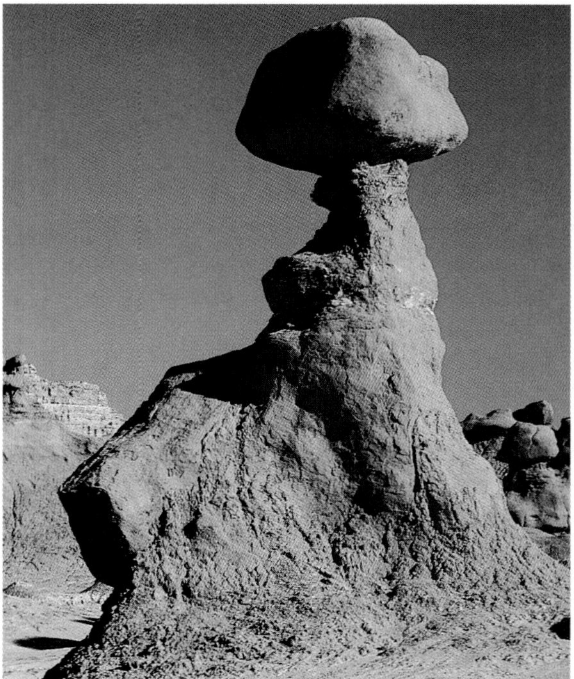

Figure 18-18 In desert areas, such as this one in Goblin Valley, Utah, it is common to see balanced rocks perched precariously on narrow pedestals that have been cut by wind abrasion.

Deflation basins are large blowouts that occur where local bedrock is particularly soft or where faulting has produced a broad area of crushed bedrock. Sizable deflation basins are shaped by the winds that blow across the western Great Plains, for example. Near Laramie in southeastern Wyoming, soft fine-grained bedrock has been deflated to produce Big Hollow, a depression 15 kilometers (9 miles) long, 5 kilometers (3 miles) wide, and 30 to 50 meters (100–160 feet) deep.

A **desert pavement** is a surface layer of closely packed stones left behind by deflation. Figure 18-17 illustrates how such a desert pavement forms. It develops rapidly at first, because sand and silt are carried off readily; later, development proceeds more slowly after the removal of the finer surface particles. The remaining coarse material insulates the surface, preventing further erosion of the underlying fine-grained materials. A mature desert pavement—one or two layers of pebbles thick, with virtually no exposed fine regolith—may develop over hundreds or thousands of years. When it is disturbed—for example, by a dirt bike—fine regolith becomes exposed again and deflation resumes until the protective coarse pavement is restored.

Abrasion **Wind abrasion** occurs when a wind-borne supply of eroded particles is hurled against a surface and sandblasts it. In a Saharan sandstorm, for example, several hours of wind abrasion can strip the paint from a jeep and pit the windshield until it is no longer transparent. Wind-eroded sand consists largely of quartz (with a hardness of 7 on the Mohs scale), which is 2500 times as dense as air. The collision of quartz grains with other grains or bedrock surfaces can weaken the cement between sedimentary grains, fracture grains in igneous and metamorphic rocks, and dislodge particles previously loosened by weathering.

Wind abrasion produces several distinctive desert features, such as the rock pedestal shown in Figure 18-18. It also creates the wind-shaped stones with flat sharp-edged faces called **ventifacts** (from the Latin, meaning "wind-made"), which may be strewn across a desert floor. Figure 18-19 illustrates how ventifacts form. Their characteristics polished facets with pitted "orange-peel" textures are produced when

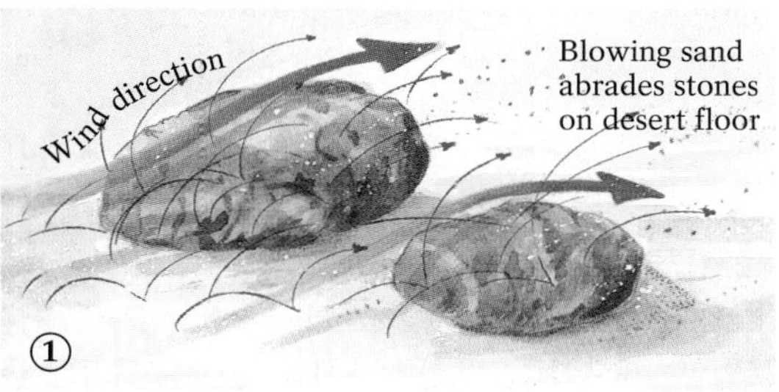
Wind direction
Blowing sand abrades stones on desert floor
①

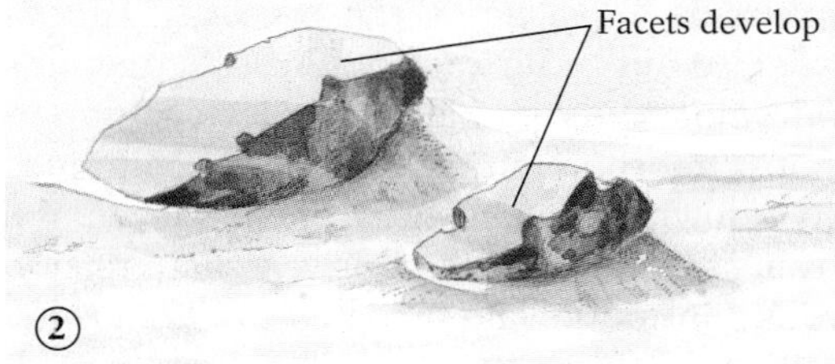
Facets develop
②

More facets develop as new surfaces are exposed
③

Figure 18-19 Ventifacts form when wind blowing predominantly from one direction abrades desert-floor stones, creating flat surfaces and sharp edges. As the wind changes direction or the stones shift position, exposing other surfaces to wind abrasion, more facets appear on the newly exposed surfaces. Photo: A wind-etched ventifact.

Figure 18-20 Yardangs from the Namib Desert, Namibia and (inset) White Desert, Egypt.

prolonged wind abrasion attacks the side of a ventifact that faces the prevailing wind direction. The orientation of the sandblasted facets on a group of ventifacts indicates the dominant wind direction. Ventifacts with more than one faceted side may have been shaped by prevailing winds that changed direction or blew from multiple directions. The position of such rocks may have also changed, exposing a new surface to wind abrasion. A stone's position can readily be changed by the actions of animals, powerful storm winds, floodwaters, a cycle of frost wedging, or undercutting at the base of the stone by wind abrasion.

Wind abrasion also produces large-scale topographic features in the form of the streamlined desert ridges called a *yardangs* (from the Turkish *yar,* meaning "steep bank"). Yardangs most commonly form where strong, one-directional winds abrade soft sedimentary bedrock layers, leaving behind more resistant layers. As you can see in Figure 18-20, the resulting landform, which resembles the inverted hull of a sailboat, is surrounded by wind-abraded depressions. Yardangs are usually wider on the windward side and taper leeward. They may be as much as tens of kilometers long and tens of meters high.

Transportation by Wind

In deserts, wind-transported particles roll or *saltate* ("bounce," from the same Latin root as somer*sault*) along the surface much like the sand grains near the bottom of a stream (see Chapter 14). They may also travel several to tens of meters above the surface, being temporarily suspended by the wind's turbulence. Two interacting and opposing factors influence the movement of even the smallest particle of dust: the driving force of the wind, which is controlled by the air's velocity and turbulence, and the particle's *inertia,* or tendency to stay in place, which is controlled by gravity.

Other modes of transport, such as rivers and glaciers, can lift and move a particle whenever their driving force exceeds the particle's resistance to movement. Wind, however, must achieve a much greater relative velocity, because it cannot lift a loose particle directly from the surface. The viscosity of air is quite low relative to the density of sediment particles. Thus moving air is far less likely to lift and carry particles than are more viscous fluids, such as water, glacial ice, or lava. In addition, vegetation and coarse sediment on the surface interrupt air flow just above the surface, creating a layer of virtually "dead air."

To move particles in this still layer, the velocity of the wind must be capable of overcoming the inertia of grains that are large enough to extend above this layer. When winds reach this velocity, they transport sediment as shown in Figure 18-21a. As these particles begin to move by rolling, they then collide with other particles, propelling them upward. Once airborne, particles move forward as the wind carries them in parabolic (curved) paths until gravity pulls them down. Large saltating grains are overcome by gravity fairly soon and quickly fall back to the surface, where they strike other grains and set them in motion. Thus they mobilize the smaller particles that winds can then carry in suspension. They may also nudge forward other, still larger grains. When the wind velocity drops below the level required to move large grains, transport of mid-size grains continues until the wind dies down sufficiently to prevent even the smallest grains from saltating. The wind-driven chain reaction that set the particles in motion then comes to an end . . . until the next gust.

Bed Load *Bed load,* like the water-borne sediment that moves along the bottom of a stream, is the portion of a desert's wind-transported sediment that moves on or close to the ground. Air is not dense enough to lift and carry a windstream's bed load, however; consequently, it must move particles by saltation. The nature of the surface, the shape and density of the particle, and the force of the collision that sets it in motion determine the height to which a grain will saltate. A grain propelled from a soft bed of sand can attain a height of only 50 centimeters (20 inches) or less, whereas a grain from a surface covered by a continuous stone pave-

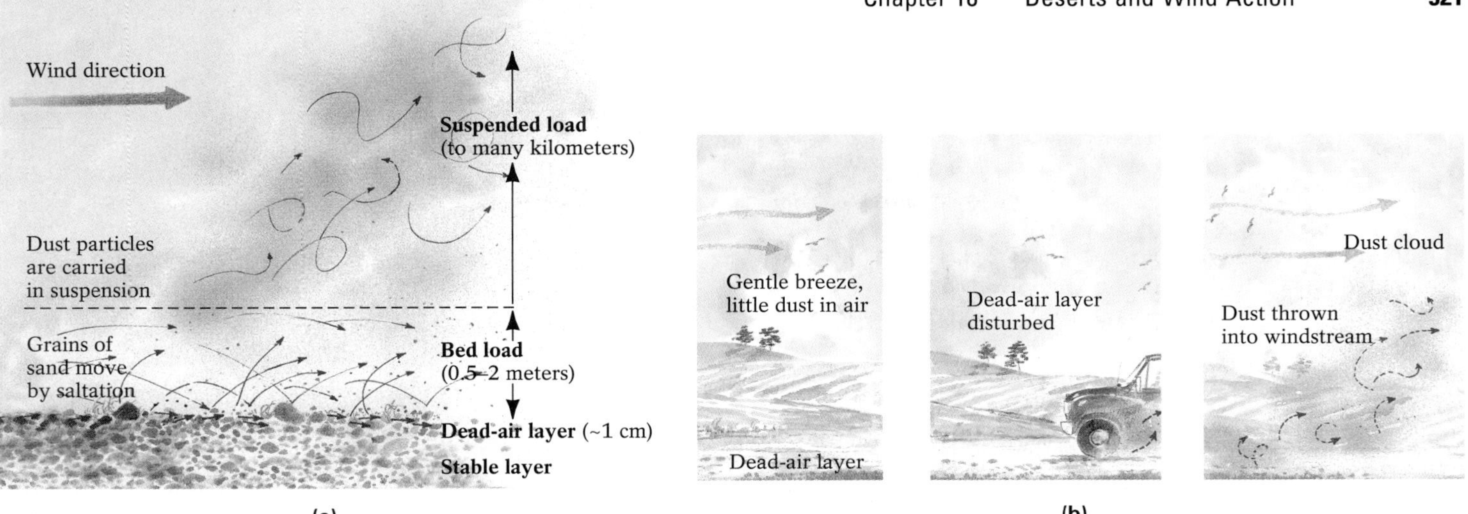

Figure 18-21 Transport of wind-borne sediment. **(a)** Like stream-borne particles, wind-borne particles may travel by rolling or saltating along the surface or by being carried in suspension above the surface. Greater velocity is needed to transport particles in wind, however, because of the low viscosity of air and because of the presence of obstructions at ground level that slow air movement, creating a dead-air layer. **(b)** On a relatively windy day, little dust rises from a dry country road, because the particles do not extend above the dead-air layer, and thus remain relatively unaffected even by strong wind gusts. When a passing truck or a tractor disturbs the dead-air layer, it throws dust up above the zone of motionless air and into the windstream, and a thick cloud of dust rises behind its tires. The dust blows away in the wind long after the truck and its turbulence are gone.

ment may saltate to a height of 2 meters (6.6 feet). (To see why this difference arises, imagine tossing unpopped popcorn on a thick-pile carpet versus a polished hardwood floor.)

A violent wind can churn up a rock-strewn desert. Even when the sediment supply lacks silt- and clay-sized particles, high winds can produce a bed load that extends to a height of 1 to 2 meters (3.3–6.6 feet). The air above this dense, surface-hugging cloud of particles typically remains quite clear. A person caught in such a sandstorm can be submerged in a swirling "pool" of sand, like the one described in this eyewitness account by British journalist Richard Trench, who survived a Saharan sandstorm in 1974:

> Suddenly the wind began to rise, blowing in short and powerful gusts. The surface of the desert, normally so still, was growing restless. As the wind rose, so the desert rose too. It was dancing about my feet. It hurled itself against my calves, whirled around my body, and beat at my bare arms. Nor did it stop there. It grabbed my neck, stung my face, encrusted itself in my throat, blocked my nostrils, and blinded my eyes. I felt alone and feared death by drowning.

Suspended Load Like the water-borne sediment carried above a stream bed, the portion of a desert's wind-transported sediment moved long distances without settling back to the surface is called the *suspended load.* Most of the suspended load in an airstream consists of *dust,* particles about 0.15 millimeters (0.006 inch) or less in diameter. Dust may include silts and clays from soils, volcanic ash, pollen grains, airborne bacteria, minute plant fragments, charcoal from forest fires, fly ash from the burning of coal, small salt crystals from evaporation of sea spray, or finely crushed glacial flour deflated from exposed dry outwash.

Many dust particles are relatively flat, possessing a large surface area relative to their volume; this shape facilitates uplift, much as the wings on an airplane carry it aloft. Most dust particles therefore counter the pull of gravity effectively and can be lifted above the dead-air layer by turbulent winds and swept into the upper atmosphere, thousands of meters high. The dust may remain there for several years and perhaps travel thousands of kilometers. Distinctive red dust from the Sahara, for example, has been found 4500 kilometers (3000 miles) away on the Caribbean island of Barbados, 3000 kilometers (2000 miles) away on Parisian roofs, and even further north in Sweden, where it occasionally discolors the winter snows. In North America, dust from the drought-stricken fields of the Great Plains of Texas and Oklahoma fell on fresh snow in New England in the 1930s. In March 1935, dust from eastern Colorado was carried by winds with sustained speeds of 80 kilometers (50 miles) per hour to upstate New York, 3000 kilometers (2000 miles) away, where it darkened the midday skies. Highlight 18-2 further describes the evolution of America's Dust Bowl.

Deposition by Wind

If winds can excavate blowouts and deflation basins and darken the sky with a veil of dust, what happens when they stop blowing? Airborne particles begin to fall, with the largest, heaviest, saltating grains dropping closest to their source, and the flattest, smallest, least dense particles of

Highlight 18-2 America's Dust Bowl

In the early 1930s, the plowed wheat fields from the Texas Panhandle to the prairie provinces of Canada (Alberta, Saskatchewan, and Manitoba) were stripped of their rich topsoil by strong winds that swept across the drought-parched landscape. Southeastern Colorado, western Kansas and Nebraska, and the Texas and Oklahoma panhandles were particularly devastated. Towering walls of swirling dust, called *black rollers,* darkened the sky at noon (Fig. 18-22). In May 1934, powerful winds lasting for 36 hours transformed dust from the Great Plains into a dense black cloud that cast a 2000-kilometer (1200-mile)-long shadow across the eastern half of the continent. In autumn, prairie dust fell in upstate New York as "black rain"; in winter, it was observed in the mountains of Vermont as "black snow."

The dust in the Great Plains was sometimes so dense that it buried entire fields of crops. Many people and farm animals died from suffocation or "dust pneumonia," a malady akin to miner's silicosis (a condition caused by the inhalation of mine dust). The strong winds ensured that the ultra-fine dust could penetrate the fabric of any garment. As one eyewitness reported:

> These storms were like rolling black smoke. We had to keep the lights on all day. We went to school with the headlights on and with dust masks on. I saw a woman who thought the world was coming to an end. She dropped down to her knees in the middle of Main Street in Amarillo and prayed out loud: Dear Lord! Please give them a second chance.

A combination of natural and human factors produced the Dust Bowl. Much of the Great Plains had long been overcultivated. Settlers had plowed vast areas of the thick tough prairie sod to plant great fields of wheat. Indeed, by 1929, more than 100 million acres were under cultivation. For years, the abundant rainfall yielded plentiful harvests. Removal of the protective grass cover, however, left the land vulnerable to wind deflation when drought came. In the early 1930s, precipitation decreased to less than 50 centimeters (20 inches) per year and winds gusted regularly at speeds exceeding 15 kilometers (9 miles) per hour. Crops failed everywhere. With no roots from trees or crops to anchor it, hundreds of millions of tons of rich topsoil were simply blown away. Thousands of farmers—their crops parched and their soils depleted—abandoned their lands and moved on. The westward migration of "Okies" to California was immortalized in John Steinbeck's epic novel, *The Grapes of Wrath.* Songwriter Woody Guthrie captured the bleak outlook of the times in "So Long, It's Been Good To Know You," written on April 14, 1935, the date of one of the decade's worst dust storms.

Figure 18-22 Ominous clouds of dust obscured the midday sun during the Dust Bowl years of the 1930s. This photo was taken in April 1935, in Mills, New Mexico.

By 1939, a few years of more abundant rainfall, along with state and federal attempts to improve farming practices, had ended the Dust Bowl tragedy. The prairie states and provinces became habitable and cultivatable again. Although damaging droughts occurred several times between 1950 and 1980, the use of extensive irrigation and crop rotation ensured that there was less threat of a recurrent major problem. Residents of the plains have learned to adjust to naturally recurring dry periods.

dust dropping farthest downwind. In this way, wind sorts its deposits. The appearance and location of the resulting depositional landforms are determined by the size and amount of sediment (whether particles are carried as bed load or in suspension), the constancy and direction of the wind, and the presence or absence of stabilizing vegetation. We will next discuss two of the most common products of wind deposition: sand dunes (the products of bed-load deposition) and loess sheets (the products of suspended-load deposition).

Dunes: Bed-Load Deposition **Dunes** are wind-built mounds or ridges of sand. A field of many sand dunes resembles a sea with whitecaps; the surfaces of individual dunes, like those of ocean waves, are covered with ripples (see Chapter 6 for a discussion of ripple marks). Sand dunes form in both arid and humid climates, wherever there is a sufficient supply of sand that is initially not stabilized by vegetation (vegetation, however, often gains a foothold later) and strong winds blow constantly. For example, dunes appear along the sandy shores of oceans, seas, and large lakes (such as Lake Michigan) and

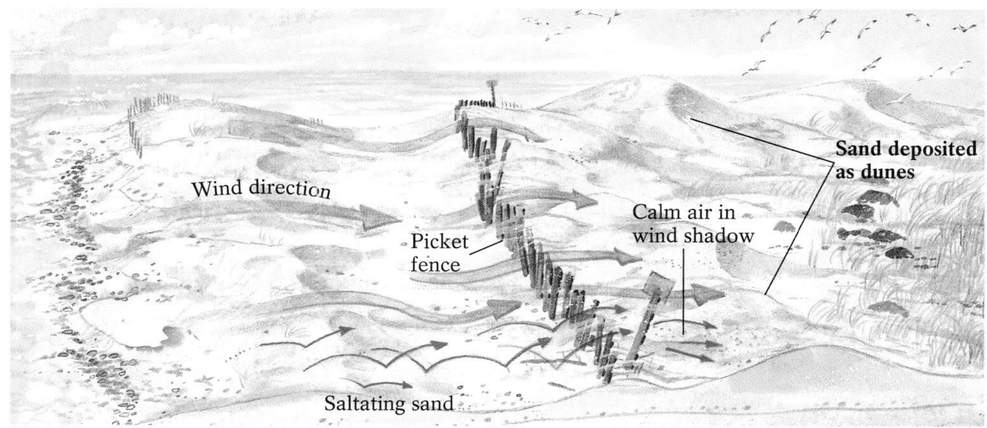

Figure 18-23 Sand dunes typically form when a surface obstacle, such as a boulder, hill, or picket fence, interrupts the flow of migrating, saltating sand. A wind shadow is created leeward of such an object, and sand settles and accumulates on the ground.

near large, dry, sandy floodplains. North America boasts dunes in such scenic places as Florence, Oregon; Alamosa, Colorado; Padre Island, Texas; Burns Harbor, Indiana; Long Island, New York; Cape Cod, Massachusetts; and Cape Breton, Nova Scotia.

Dunes typically form where an obstacle, such as a hill or picket fence, interrupts the flow of saltating sand, or when the wind slows to a speed at which it can no longer transport sediment (Fig. 18-23). They also arise where a narrow obstacle, such as a clump of vegetation or a large rock, forces the windstream to diverge around it. An obstacle creates a *wind shadow* of calmer air leeward (downwind) of it, as well as a smaller one upwind of it. The calm leeward air cannot keep saltating grains aloft, so they settle out and either start or add to a growing dune.

As a dune grows, it becomes an obstacle itself, trapping sand in its own lee. Sand dunes also can develop windward of an obstacle. For example, westerly winds pick up large quantities of sand as they blow across the barren sandstone cliffs and deserts of the American Southwest. When these winds encounter the Sangre de Cristo Mountains of southern Colorado, they slow and drop their load onto the windward side of the Great Sand Dunes National Monument (Fig. 18-24).

Dunes accumulate more rapidly on sand-covered surfaces than on desert pavement. They continue to grow as long as sand-carrying winds blow from the same general direction, or until they reach the height limit imposed by the local wind pattern. This limit is reached when the wind speed required to carry a load of sand over the dune and drop it on the lee side becomes too great to permit the deposition of additional sand, so that the wind's load is carried past the dune. Most dunes stretch from 10 to 25 meters (35–80 feet) high. Although those of Saudi Arabia and China exceed 200 meters (660 feet), they are actually complexes of dunes deposited one atop another, not individual dunes (Fig. 18-25).

Most of the sand-sized particles that constitute dunes are weathering-resistant quartz grains. The countless collisions that these grains endure typically pit their surfaces, giving them a "frosted" appearance. Gypsum grains, eroded from evaporite deposits on the slopes of the San Andres Mountains to the west, form the dunes at White Sands, New

Figure 18-24 The dune field of the Great Sand Dunes National Monument continues to grow as loose material eroded from the arid lands to the west and carried by westerly winds is dropped against the Sangre de Cristo Mountains of southern Colorado.

Figure 18-25 Complexes of dunes in the Gobi Desert, in western China.

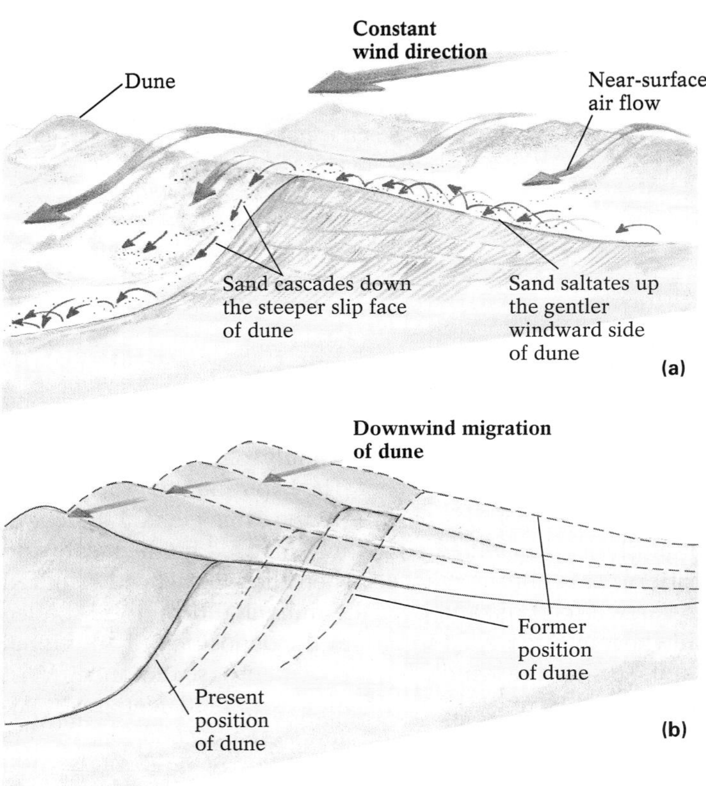

Figure 18-26 Dune migration. **(a)** Sand saltates up a dune's windward side, then cascades down its slip face. **(b)** This progressive transport of sand from the windward to the leeward side of dunes causes downwind migration of the dunes.

Mexico; the dunes on the island of Bermuda comprise the broken shells of marine organisms.

Most dunes are asymmetrical, sloping gently (about 10°) on the windward side and more steeply (about 34°) on the leeward side, at the angle of repose characteristic of dry sand (discussed in Chapter 13). As sand grains saltate up the gentle windward slope and reach the dune crest, they spill over into the leeward wind shadow. In this area, the decreased wind velocity quickly deposits the particles on the steep leeward slope, or **slip face.**

As sand saltates up the exposed windward side and becomes deposited on the sheltered leeward side, the dune migrates downwind. Figure 18-26 illustrates the dune migration process. Small dunes tend to migrate faster and farther, because less sand needs to be moved. The rate of migration ranges from a few meters per year for large dunes in vast sandy deserts with gentle variable winds, to hundreds of meters per year for small dunes on bare, rocky desert floors with strong one-directional winds. Coastal dunes may even migrate seaward and extend a coastline offshore by several kilometers. You can see the effect of this type of seaward migration of coastal dunes in Figure 18-27.

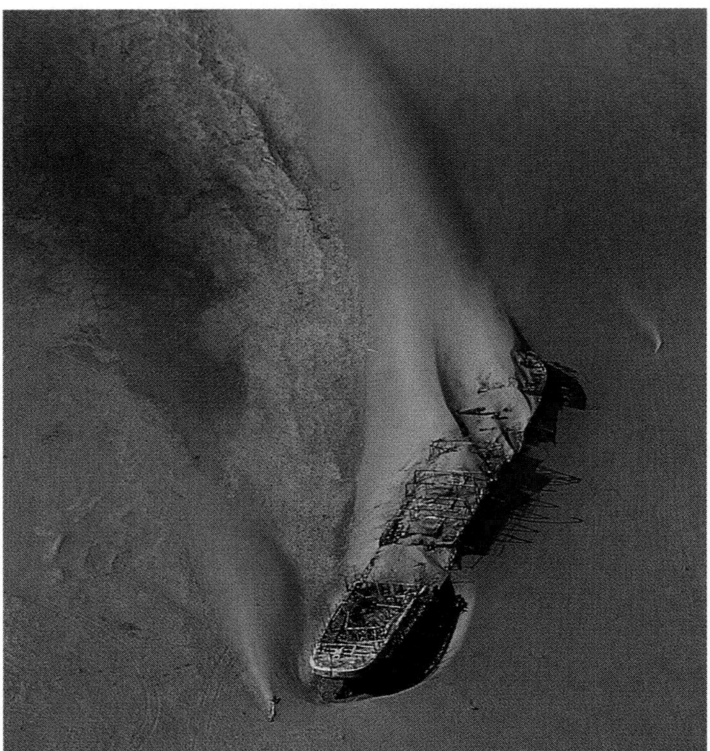

Figure 18-27 The stranded German freighter *Eduard Bohlen,* which ran aground off Africa's southwestern coast in 1912, now lies sandlocked more than a kilometer inland. The migrating and drifting sands of the Namib Desert have extended the coast seaward.

Migrating dunes can menace any structures or objects downwind of them—whether buildings, forests, roads, or railroads. Dune migration across major interstate highways necessitates frequent and costly sand removal to keep the roads open (Fig. 18-28). To halt sand movement, a continuous grass cover can be planted on a dune, where the climate permits. If vegetation is removed—perhaps by motorcycles, dune buggies, or all-terrain vehicles—wind gusts may erode new blowouts and dunes resume their downwind march.

Figure 18-28 Sand encroaching on Interstate 84, in the Columbia River Gorge, Oregon.

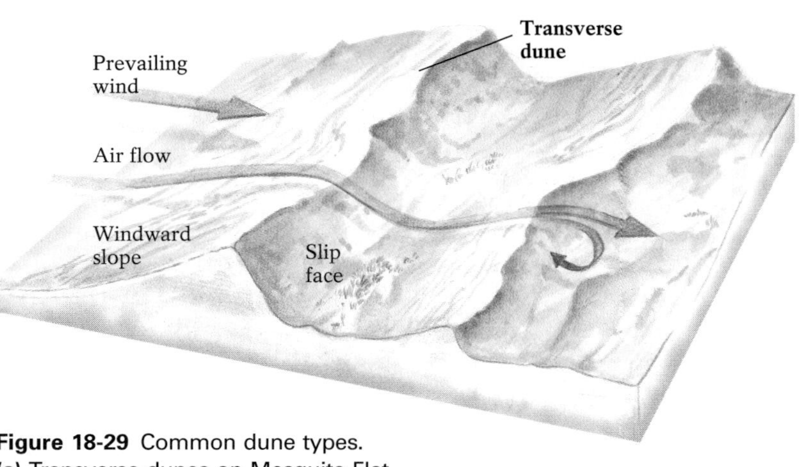

Figure 18-29 Common dune types. **(a)** Transverse dunes on Mesquite Flat, Death Valley, California.

Dune Shapes Dune shapes are determined by local conditions, the type of sand, the degree of aridity, the nature of the prevailing winds, and the type and amount of vegetation present. Dune shapes can be classified as transverse, longitudinal, barchan, parabolic, and star.

Transverse dunes, shown in Figure 18-29a, consist of a series of parallel ridges that typically occur in arid and semiarid regions where sand is plentiful, wind direction constant, and vegetation scarce. These dunes form *perpendicular* to the prevailing wind direction and possess gentle windward slopes and steep leeward slip faces. In the Sahara, they can be as large as 100 kilometers (60 miles) long, 100 to 200 meters (330–660 feet) high, and 1 to 3 kilometers (0.6–2 miles) wide. Transverse dunes also can develop along the shores of oceans and large lakes, where strong onshore winds shape the abundant sand. Such dunes dot the southeastern shore of Lake Michigan at the Indiana Dunes National Lakeshore in Indiana, and at Warren Dunes State Park in Michigan.

Longitudinal dunes, shown in Figure 18-29b, are also parallel ridges. They form when the sand supply is moderate and wind direction varies within a narrow range. Such dunes are oriented *parallel* to the prevailing wind direction. Small ones may be only 60 meters (200 feet) long and 3 to 5 meters (10–16 feet) high. In the Libyan and Arabian Deserts, where strong winds blowing from several directions converge over the dune crest to create one sinuous form, longitudinal dunes can reach 100 kilometers (60 miles) long and 100 meters (330 feet) high.

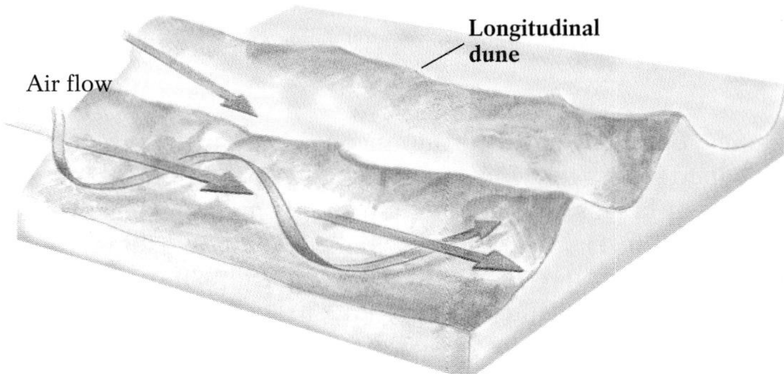

Figure 18-29 (b) Longitudinal dunes at Stovepipe Wells, Death Valley National Monument, California.

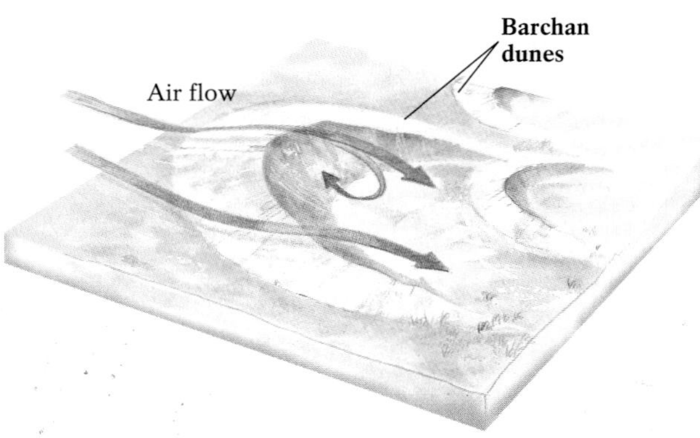

Figure 18-29 (c) Barchan dunes in the Baja Desert of Baja California.

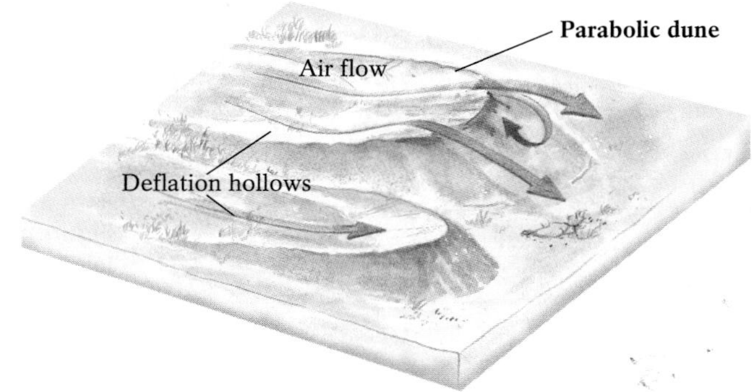

Figure 18-29 (d) Parabolic dunes.

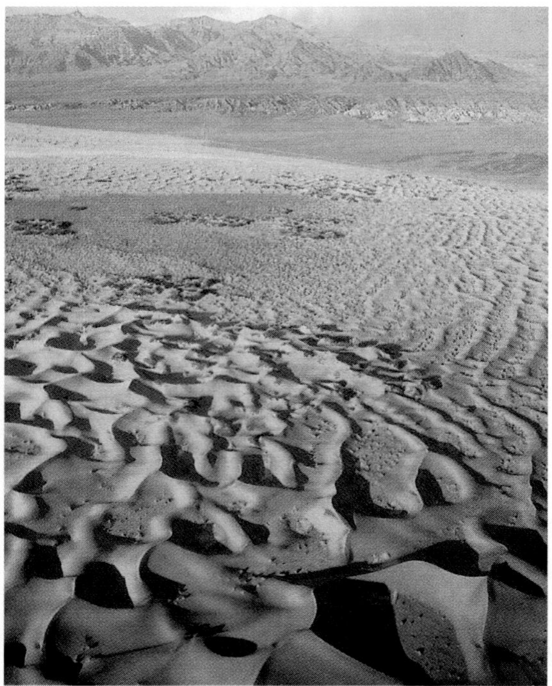

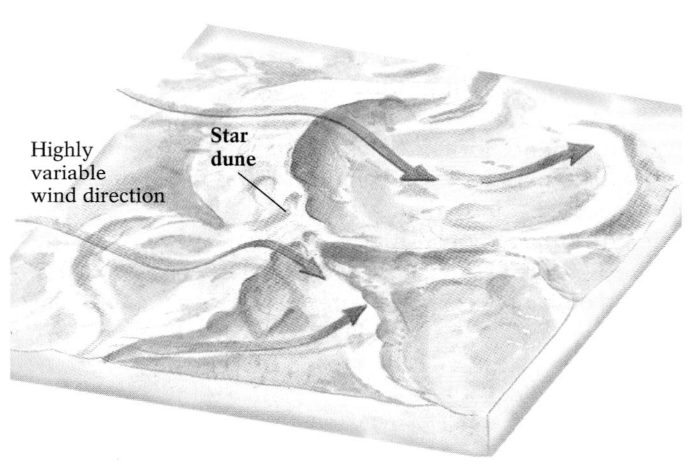

Figure 18-29 (e) Star dunes (in foreground), Death Valley National Monument, California.

Barchan dunes (pronounced "BAR-kane") are crescent-shaped ridges that form perpendicular to the prevailing wind as sand begins to accumulate around small patches of desert vegetation (Fig. 18-29c). Barchans develop in arid regions on flat, hard ground where little sand is available and wind direction remains constant. As they grow, these dunes become thicker and higher in their centers, where air flow is impeded most and more sand is deposited. Because the horns (the points of the crescents) of these dunes are lower than their centers, the horns migrate downwind more rapidly. Consequently, the barchan and its characteristic sharply pointed horns extend in the *downwind* direction.

Parabolic dunes are horseshoe-shaped dunes that differ from barchans in that their horns point *upwind* (Fig. 18-29d). They commonly form along sandy ocean and lake shores, which represent the only appreciable dune areas outside of deserts. Parabolic dunes develop when transverse dunes become exposed to accelerated wind deflation, especially after the removal of some vegetation. A small deflation hollow forms on the transverse dune's windward side, allowing the wind-excavated sand to pile up downwind. As the hollow grows, the wind becomes concentrated at its center, speeding the migration of that portion of the dune. The horns—a remnant of the original transverse dune—are usually still covered by vegetation and remain anchored in place; the rest of the parabola continues to migrate downwind, forming a horseshoe shape that can become quite elongated.

Star dunes, shown in Figure 18-29e, are the most complex dune types. They form when winds blow from three or more principal directions, or when wind direction constantly shifts. Star dunes tend to grow vertically to a high central point and may develop three or four arms radiating from the center. Continued variability of wind direction causes these dunes to remain in a relatively fixed position.

Loess: Suspended-Load Deposition Loess (discussed in Chapter 17 and shown in Figure 18-30) is a wind-borne silt deposit that resembles fine-grained lake mud. Unlike waterborne sediments, which are typically deposited at topographic low spots, loess deposits cover hills, slopes, and valleys evenly, making it appear as if the loess fell from the sky. Most loess grains consist of quartz, feldspar, mica, or calcite. Slight oxidation of accessory iron minerals gives the deposit its characteristic yellow-brown or ocher color. The terrestrial origin of loess is also indicated by the presence of shells from such creatures as air-breathing snails, as well as by the scattered mammal bones, worm burrows, and plant-root tubes that loess commonly contains.

Loess almost always appears downwind of a plentiful supply of loose dry silt. Much of the loess in America's Midwest is a yellow-brown deposit that apparently originated when coarse-to-medium silt particles (0.01–0.06 millimeter in diameter) were eroded from drying glacial outwash and deposited downwind. The streams draining from the melting ice carried large volumes of silt-sized glacial flour. As the outwash plains dried, the prevailing winds from the west swept up the silt and deposited it downwind to the east as loess.

Figure 18-30 Loess (windblown silt) deposits east of the Missouri River, Monona County, western Iowa.

Large loess deposits date from both glacial and interglacial times. During glacial periods, outwash dried out during the winter months, which were times of low outwash discharge. Similarly, the drying of vast bodies of outwash that were deposited at the end of glacial periods and then exposed during warm interglacial periods created an enormous source of fine silt. More than 500,000 square kilometers (200,000 square miles) of the land east of the Missouri and Mississippi Rivers is covered by loess, including much of the farmland in Iowa, Illinois, Indiana, and Missouri (Fig. 18-31). Similar

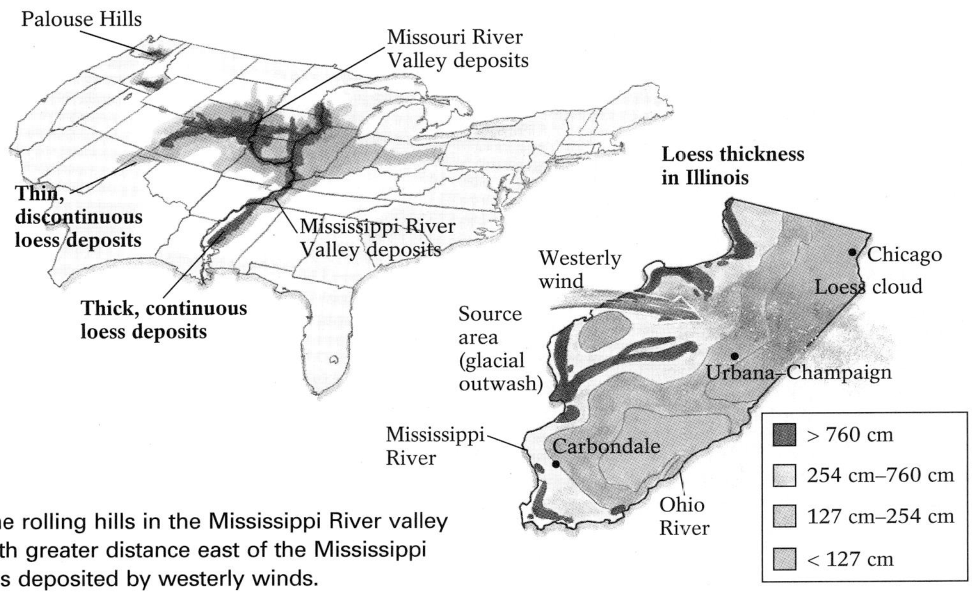

Figure 18-31 The loess that covers the rolling hills in the Mississippi River valley becomes thinner and finer-grained with greater distance east of the Mississippi River. This pattern suggests that it was deposited by westerly winds.

deposits are found east of the Rhine River in Germany, in other central European countries, and in central Asia.

Some loess—for example, that found west of the Mississippi River in eastern Kansas and Nebraska—originated in *nonglacial* environments. Wind apparently eroded these deposits from the ancient dust-storm deposits that formed the sand hills of northwestern Nebraska. Westerly winds carried the loess now lying on the Palouse Hills of southeastern Washington from the drying floodplain of the mighty Columbia River to the west. The world's largest loess deposits occur in northern China. Their phenomenal 300-meter (1000-foot) thickness derives from the nearly endless supply of silt from the vast Gobi Desert to the west; this airborne dust settles and washes into the Huang Ho (Yellow River) and Yellow Sea, giving them their distinctive color.

Because loess originates in glacial environments and arid deserts, both places where little chemical weathering occurs, the particles remain remarkably fresh. Thus loess-covered lands, such as those in Iowa, are extremely fertile. Furthermore, because they travel via airborne transport, loess particles are typically quite angular. Unlike stream-borne particles, they have not experienced the numerous collisions that tend to round particle edges. This angularity makes loess deposits quite porous, which accounts for their high moisture-holding capacity. Moist, mineral-rich loess is commonly a Midwestern farmer's favorite soil.

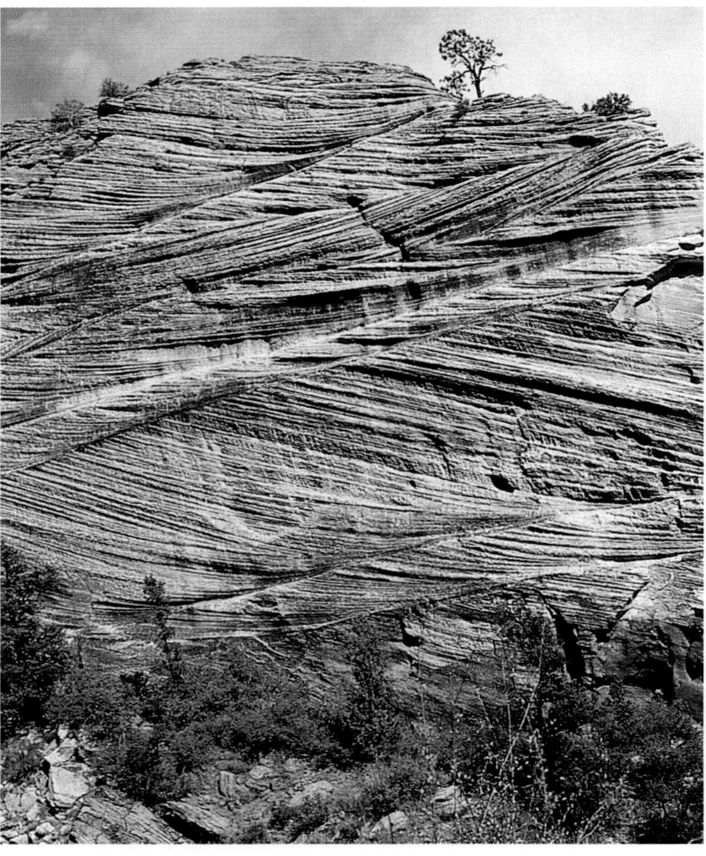

Figure 18-32 Numerous ancient, lithified sand dunes appear in national parks in the arid Southwest. The Navajo Sandstone, for example, crops out throughout Zion National Park in southwestern Utah. This memento dates from 180 million years ago, when present-day Utah was on the continent's west coast. (This scenario occurred before the lithosphere on which California is located was added to the continent.) We can be sure that the Navajo Sandstone originated on land, and therefore was deposited by wind, because the formation holds dinosaur footprints. From the orientation of the cross-beds, we can reconstruct the northerly and westerly paleowinds that formed the migrating coastal sand dunes that later became this cross-bedded outcrop.

Reconstructing Paleowind Directions

Winds of the past, or *paleowinds,* have left traces of their intensity and prevailing direction in dunes and loess deposits that have since turned to stone. These traces can reveal much about the geologic past. When geologists study a sandstone formation, they search for clues that determine whether it was deposited by wind or water. Excellent sorting, the existence of cross-bedded sands displaying the angle of repose of dry sand, the occasional discovery of sharply faceted ventifacts, the presence of sand grains polished and pitted by mid-air collisions, and the absence of ultra-light mica flakes (which are easily swept away in suspension) suggest a wind-deposited sandstone.

After geologists confirm a wind-related origin, they can use their knowledge of wind-generated landforms and deposits to reconstruct ancient wind directions. They base their hypotheses on the shape of yardangs, the position of ventifact facets, the asymmetry of dunes, and the thickness and texture of loess deposits.

Sand dunes rarely survive intact in sedimentary rocks; instead, they are usually buried by later dune migration. Although the gentle windward slope and the steep leeward slip face of a dune form may vanish, the sands of advancing dunes may nevertheless preserve the inclined bedding of the dune's internal structure, creating dune-type cross-bedding. The leeward slip-face beds, which are inclined at the angle of repose of dry sand (30°–34°), indicate the paleowind direction that prevailed when the bedding layer formed (Fig. 18-32). Glance back at Figure 6-8 for an illustration of how shifting winds produce sets of dune-type cross-beds.

As with other wind-generated features, certain properties of loess can be used to estimate past wind directions. A loess layer is generally thicker near its source and thins downwind, where finer particles are deposited. The 10-meter (33-foot)-thick layers of loess that occupy the eastern banks of southward-flowing rivers in the Midwest, for example, thin to less than a few centimeters in the East. Hence, we know that the prevailing winds were westerly 12,000 years ago, as they are today (see Figure 18-31).

By reconstructing paleowind directions, geologists can deduce certain ancient plate-tectonic events. Winds blow from west to east over the British Isles today, a wind pattern that has probably varied little through time *for that latitude.* British geologists have discovered, however, that the winds that formed the dune-type cross-bedding of Great Britain's New Red Sandstone apparently blew from the east and northeast. They now propose that, when the 200-million-year-old New Red Sandstone was deposited, England was probably located somewhere to the south in a belt of easterly winds. It has since moved northward via plate motion to its current location within the belt of westerlies. Recent studies of the paleomagnetism of British rocks have confirmed the proposed southerly origin of these rocks.

Desertification

Desertification is the invasion of desert conditions into formerly non-desert areas. The common symptoms of desertification are a significant lowering of the water table, a marked reduction in surface-water supply, increased salinity in natural waters and soils, progressive destruction of native vegetation, and an accelerated rate of erosion.

Northern and western Africa have been experiencing rapid desertification for 2000 years. Approximately 8000 years ago, and hence a few thousand years after the end of the last major worldwide glaciation, the Namib Desert of southwestern Africa was a lush savannah that supported advanced Stone Age societies (Fig. 18-33). It remained a fertile grassland, teeming with wildlife, for the next 6000 years. We can reconstruct the land features of that ancient world by using radar imaging. Radar waves cannot penetrate moist soil. In humid areas, radar is either absorbed by soil moisture or reflected, creating an image of the surface. The regolith of hyper-arid deserts, however, is virtually transparent to radar; the reflected image shows the contours of the underlying bedrock. In November 1981, the Space Shuttle *Columbia,* cruising some 200 kilometers (120 miles) above the eastern Sahara, employed radar to map the configuration of the bedrock beneath the extremely dry Selima sand sheet. The radar waves penetrated this vast, flat, featureless sea of sand, and were reflected to reveal an ancient surface characterized by rocky hills and deep, wide river valleys, as shown in Figure 18-34.

Two forces caused the northern African savannah and grassland to turn into the modern Sahara. One—drought—is natural and unavoidable; the other—overpopulation and land mismanagement—is human-caused. Drought, overpopulation, or both can start the process of desertification. When a land's inhabitants cannot produce enough to provide for their needs, they tend to overgraze their cattle, plant crops without replenishing the soil, cut down trees for shelter and

Figure 18-33 This cave painting depicts falling rain. It is found today in the parched Namib Desert of southwestern Africa, but was painted thousands of years ago in the moist, temperate climate characteristic of the region at that time.

Figure 18-34 A satellite image of the Selima sand sheet, enhanced by Space Shuttle radar imagery (the "blue" area). The topography of the bedrock under the sand shows an integrated network of stream valleys.

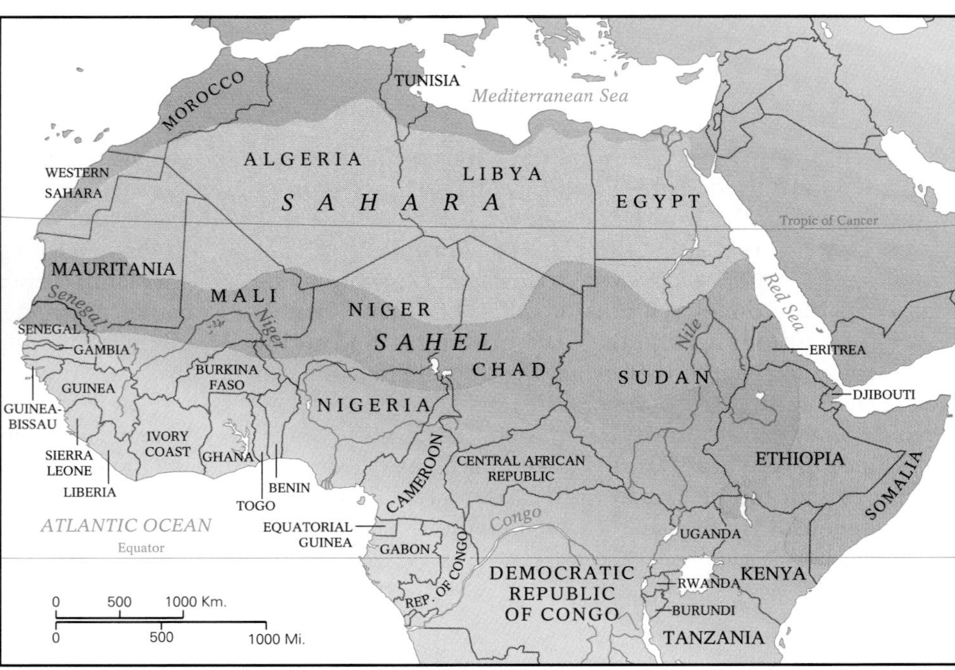

Figure 18-35 (above) Skeletons of a cattle herd in the sun-baked lands of the Sahel, which underwent rapid desertification from 1970 through the early 1990s. (right) Map of the Sahel.

fuel, and draw more water from springs and other sources than is naturally replenished. Gradually soils become depleted of their nutrients, and the removed trees and their roots fail to grow back. Without vegetation, the soils cannot hold water or prevent wind erosion. Water from occasional cloudbursts washes away the unbound topsoil and, in hot climates, evaporates before it can enter the groundwater supply. Eventually the parched land has virtually no productive capacity. Once begun, the process of desertification tends to be self-perpetuating. Left with no source of sustenance, the land's inhabitants must migrate in search of food, water, and shelter.

As much as 35% of the world's land, now barely sustaining a population of 850 million people, consists of the semi-arid margins of already arid lands. In the next few decades, overpopulation and drought may convert them from dry, but habitable grasslands to deserts unable to sustain significant human populations. As many as 70 nations are affected by this trend; about half of these countries are in Africa, where the Sahara is advancing southward by as much as 50 kilometers (30 miles) per year, and choking dust storms and newly formed dune fields already threaten some northern and western cities.

The Sahel—the region of northern Africa immediately south of the Sahara—has been particularly hard-hit by desertification in recent decades. When the worst drought of the century struck this area in the early 1970s, it displaced 20 million people and their herds of cattle, sheep, goats, and camels. Animals that were not led away died. Although the drought represented the immediate cause of the disaster, inadvertent land mismanagement had set the stage for it decades earlier. From 1935 to 1970, the Sahel's population doubled and progressively larger herds grazed on the marginally productive land. A few abnormally wet years in the early 1960s increased the region's productivity temporarily, encouraging farmers to plant more crops and expand their herds and grazing lands. Nomadic herdsmen from the south rushed to the Sahel, which had become an area of scattered groves of trees (for fuel and shelter), abundant seasonal grasses (for grazing), and agricultural fertility (yielding the primary crops of cotton, beans, millet, sorghum, and maize).

In 1970, however, no rain fell in the Sahel. The large herds quickly demolished the existing grasses; the rich topsoils, without binding roots, were easily eroded by wind. By the time the next year's rain fell, the ground was baked so hard as to be impermeable, and the subsequent surface runoff only accelerated soil erosion. Without crops, starving people were forced to eat their grain seeds, eliminating all hope of new planting. Foraging animals and people destroyed the few remaining trees, eroding the soil further and depriving the region of fuel and building materials. Animals died by the millions. The starving population migrated to the region's larger cities, which tripled in size as refugee camps sprang up around them. Worldwide relief efforts were "too little, too late" to prevent the deaths of hundreds of thousands of people from starvation, malnutrition, and associated diseases. The drought continued into the 1980s, ending only in the mid-1990s. Figure 18-35 shows the area threatened by desertification and the effects of the process.

In the United States, overpopulation and urban growth, overgrazing, and excessive groundwater withdrawal—particularly in the arid and semi-arid Sunbelt states—have already increased soil salinity and accelerated erosion. Today, approximately 27% of U.S. non-desert lands face encroaching desertification. Of great concern is the prospect of desert-like conditions replacing the "amber waves of grain" that mark America's Great Plains. Highlight 18-3 looks back at the region's prehistoric past to attempt to predict its climatic future.

Highlight 18-3 The Great Plains—North America's Once and Future Sahara?

Images of the tragic Dust Bowl years of the 1930s (see Figure 18-22)—especially if you live in Oklahoma, Nebraska, and the Dakotas—are etched in the collective memories of friends and relatives of recent generations. Was that crop-killing drought a climatic anomaly or merely the region's norm? Will global warming "desertify" America's "Breadbasket," converting the fields of grain that feed millions into a vast expanse of shifting, wind-blown sand?

Recent research by Dr. Kathleen Laird, an ecologist at Queen's University in Ontario, suggests that the Great Plains region has endured numerous droughts over the past 10,000 years, including some far more extreme than that during the Dust Bowl years. Furthermore, Laird believes that the climate in that region has actually been uncommonly *moist* during the past 700 years.

To arrive at the conclusion that drier times may lie ahead, Laird studied microscopic water-dwelling algae called *diatoms*. These minute plants produce highly distinctive, decorative cell walls composed of silica extracted from their watery surroundings. Apparently, different species of diatoms prefer water with different salinities ("saltiness"). By studying the diatoms of 53 lakes, each with a differing salinity, Laird correlated *past* lake salinity with the presence of the various diatoms. She reasoned that a drier climate would increase evaporation, which, in turn, would increase salinity (as concentrations of dissolved salt increased in the diminishing water) and alter the diatom population.

Once a connection had been established between specific diatoms and differing salinities, Laird's group sought a small isolated lake with no incoming or outgoing streams—one whose salinity-level history recorded variations in rainfall and evaporation rather than inflow or outflow of stream water. Laird selected Moon Lake in North Dakota and proceeded to extract cores of the lake-bottom sediment, which had accumulated since the Laurentide ice sheet (see Chapter 17, p. 490) had created the lake roughly 12,000 years ago. The appearance of the various salt-dependent diatoms within the carbon-dated lake muds was expected to indicate past droughts and wet times. Laird discovered that periods of extreme dryness persisted for centuries and occurred quite frequently. Her research indicates that the most extreme droughts occurred during the years 200–370 A.D., 700–850 A.D., and 1000–1200 A.D.

The likelihood of desert-like conditions spreading throughout the now "semi-arid" Great Plains is further confirmed by scientists at the United States Geological Survey in Denver. Dr. Richard Madole reports that at some point between 1000 years ago and the arrival of European settlers, wind-driven sand dunes buried much of eastern Colorado and adjacent Nebraska and Kansas, as they still do today in isolated spots on the Plains (Fig. 18-36). This period of desertification—far more severe than the worst droughts of recorded history—was marked by sand-dune expansion throughout the region. It also correlates well with Laird's proposed drought between 1000 and 1200 A.D. According to Madole, relatively small changes in temperature and precipitation converted the semi-arid prairies into vast expanses of arid desert-like sand seas. The loss of the landscape's vegetation to the drought conditions accelerated the movement of the rootless sand. Any agriculture would have been impossible at this time.

What does the future hold for America's vast wheat-producing plains? Apparently, the climatic threshold to the region's future desertification looms close at hand. The causes of the cycles of drought in the past remain unknown, but we do know that our actions today and their effects on climate may inadvertently trigger a drought in the future. The possible outcome—lots of empty shelves at the local market—provides yet another reason to monitor closely and reduce the ongoing human-caused negative impacts on global climate.

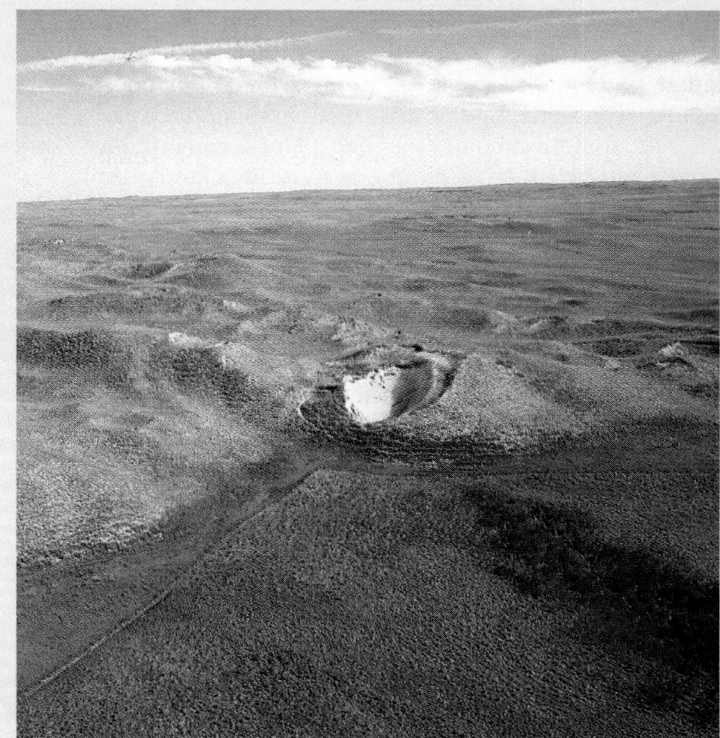

Figure 18-36 Remnant dunes on the southeast margin of the Nebraska Sand Hills in the semi-arid Great Plains, Eldred Ranch, Garden County, southwestern Nebraska.

Reversing Desertification

To reverse desertification, available water must be delivered where it is needed. To supplement the scant rainfall typical of arid regions, copious volumes of water can be pumped from deep aquifers, channeled from distant lakes or rivers, or drawn from the ocean and desalinized. Of course, as we discussed in Chapter 15, the water supply within an aquifer will eventually be exhausted if extraction of water exceeds recharge. Thus desert communities must be aware that they may be rapidly depleting their aquifers when they rely heavily on groundwater to reverse desertification.

Some desert communities, such as Palm Springs, California, are naturally supplied by deep groundwater that originated as rain or snow on moist highlands hundreds or thousands of kilometers away. This water can be pumped for use in irrigation. In Africa, rain falling in the eastern highlands near the equator enters the Nubian Sandstone, an aquifer that extends more than 3000 kilometers (2000 miles) north, where it lies deep beneath the Sahara; faulting and folding has brought this aquifer's waters to the surface in the form of springs, where oases flourish (See Figure 15-12). The rate of flow through the aquifer determines the quantity and quality of artesian water at an oasis. Swift flow brings water that is relatively pure and supports the growth of date palms and other vegetation. Slow flow through rocks may dissolve a considerable amount of salt and render the water undrinkably saline, even toxic, by the time it surfaces at an oasis.

In some extremely parched places, communities have developed inventive technologies to reclaim land recently desertified and to make portions of ancient deserts more habitable for human populations. Systems of wells, channels, and collecting pools are used in the Middle East to collect and store the infrequent storm runoff *underground,* thereby protecting it from evaporation. Throughout the region, farmers apply a specially designed plastic mulch to their farmlands or cover their fields with plastic sheeting, punctured to permit the escape of plant stems, to retain irrigation moisture. Computers monitor and regulate water flow through pipes and canals, delivering water where it is needed at times when evaporation is minimal. In Israel, "drip" agriculture—using perforated garden hoses that snake amid plantings—delivers water constantly to individual plants, literally drop by drop. Water is even drawn from the Mediterranean Sea and desalinized for agricultural use.

In the American Southwest, billions of liters of water from the Colorado River have been channeled into the Sonoran Desert of southern Arizona, the aqueducts of southern California, and the Canal Central of Baja, California, to provide sufficient irrigation to produce food for hundreds of thousands of people. Figure 18-37 shows the results of this irrigation program. Such diversions of water, however, do not lack negative consequences. By the time the Colorado River reaches Mexico, for example, its flow has been reduced to a mere trickle—stopping altogether about 30 kilometers (20 miles) short of the Gulf of Mexico and depriving Mexico of its fair share of the river's water. Moreover, when the diminished river finally reaches the Mexican border, excess agricultural runoff in Arizona has made it so saline that the United States must desalinate the water before it enters Mexico.

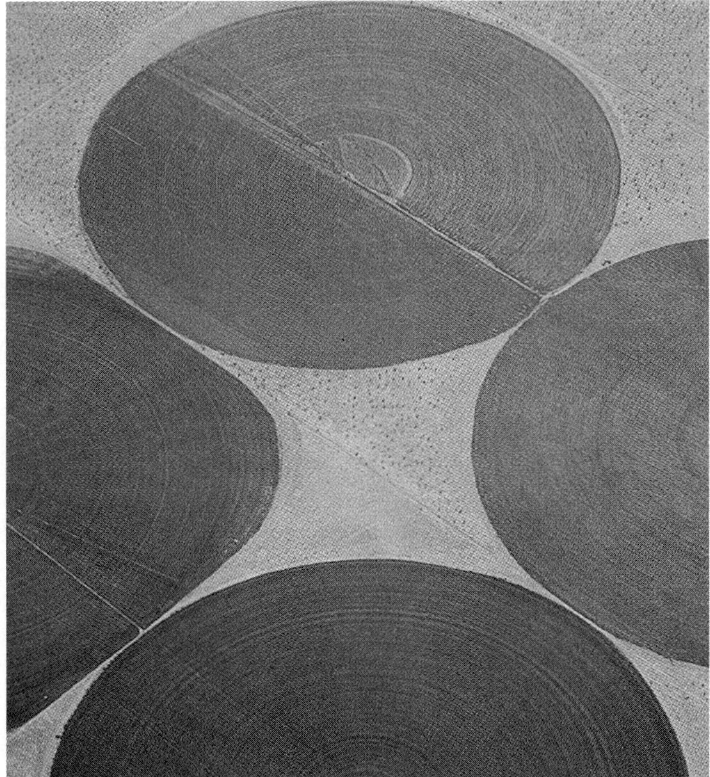

Figure 18-37 A circle of irrigated land in the Yuma Valley of Arizona's Sonoran Desert. Irrigation is largely responsible for the agricultural productivity and rapid population growth of this region.

Efforts to reverse desertification can also focus on controlling wind erosion. Trees serve as an effective windbreak to halt shifting sands and help to retain both soil cover and groundwater. Migrating dunes can be stabilized by planting fast-growing trees, such as poplars, which have deep, sand-binding roots. In northeastern China, an entire forest, 500 by 800 kilometers (300 by 500 miles), was planted upwind of 90,000 acres of prime farmland. Dunes in China have even been leveled by hand and covered with topsoil to create productive new farmland.

Many ecology-minded individuals, however, reject the notion that deserts are lifeless environmental wastelands that could and should be modified by humans for their own purposes. Instead, they view deserts as vibrant natural environments—valuable ecosystems teeming with distinctive flora and fauna—that should be protected and preserved in their natural state, in much the same way we have attempted to save the Earth's tropical rainforests. Perhaps the best approach we can take to deserts is to appreciate the beauty and ecological uniqueness of the ancient, naturally occurring water-deprived lands, while learning to manage their marginal lands more wisely to avoid desertification of now-habitable semi-arid regions.

Wind Action on Mars

The Earth is not the only windswept planet in the solar system. All planets are heated differentially by the Sun's rays. As a result, those that support atmospheres also experience wind action. As the Pathfinder mission has recently revealed, Mars is an especially windy planet.

Martian windstorms have been studied from a close proximity for several decades. The Mariner 9 space probe began its orbit around Mars on November 13, 1971, after a five-month journey from Earth. Its first photographs depicted, disappointingly, a planetwide dust storm. For three months, dust that rose as high as 6 kilometers (4 miles) obscured the entire Martian surface. On Earth, where the air is more than 100 times as dense as on Mars, it would have taken winds blowing in excess of 160 kilometers (100 miles) per hour to raise dust to that height. On Mars, however, the weak gravitational force enables dust particles to ascend four times as high as those carried by comparable wind gusts on Earth. Because the Martian atmosphere is so thin, very little dust actually moves. Instead, fine oxidized particles that are perpetually suspended in the Martian atmosphere contribute to the planet's pink-red appearance. Shifting deposits bring the frequent surface changes observed on Mars, where patches of light-colored dust are often succeeded by darker ones. Strong local winds apparently lift and transport light-colored, fine particles, exposing the dark bedrock below. Leeward streaks of light-colored dust covering dark rocks are especially common downwind of the rims of impact craters.

Mars' global dust storms apparently start when summer comes to its southern hemisphere. As the edges of the Martian frozen carbon dioxide polar cap rapidly sublimate (see Chapter 17), large temperature differences arise between the still-solid polar cap remnants and the warming surrounding landscape. These temperature differences produce differential air pressures. The resulting high winds give birth to dust storms that spread northward until they engulf the planet's entire surface for months.

After the dust finally settled in 1972, Mariner 9 was able to transmit exciting photographic images of the Martian surface to Earth-bound scientists. A dune field, 60 by 30 kilometers (40 by 20 miles) was sighted on the floor of a large crater. Viking probes landed on the Martian surface a few years later, on July 20 and September 3, 1976. The landers descended onto loess-covered plains, where they photographed interspersed ventifacts, barchan and longitudinal dunes (Fig. 18-38), yardangs, and other wind-generated landforms.

In 1997, the robotic "geologist" Sojourner, of the Pathfinder mission, provided crisp clear images that confirmed Mars' windy past. Its close-up photos of wind-scored rocks suggest that powerful gusts—exceeding 200 kilometers (120 miles) per hour—have sand-blasted exposed rocks, carving deep gouges from their surfaces (Fig. 18-39).

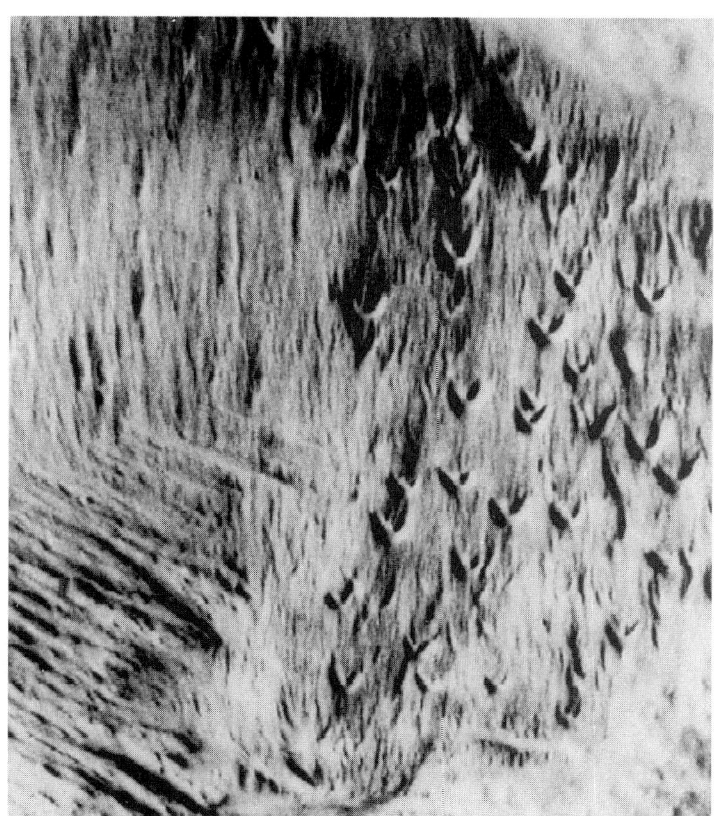

Figure 18-38 These crescent-shaped ridges on Mars are barchan dunes produced by the planet's strong surface winds.

Figure 18-39 Closeup of sand-blasted Martian rocks transmitted by Sojourner lander.

Chapter Summary

"Desert" is a somewhat relative term used to describe dry regions where vegetation is typically sparse. It can be defined as an area that receives less than 250 millimeters (10 inches) of precipitation annually, or as a region with an **aridity index**—a ratio of potential annual evaporation to average annual precipitation—greater than 4.0. Five principal types of desert exist: (1) the subtropical deserts that girdle the Earth between about 20° to 30° latitude both north and south of the equator, where dry air descends toward the surface as part of the global wind-circulation system; (2) deserts on the dry leeward side of the major mountain ranges, created by the **rain-shadow effect;** (3) interior deserts, found in the centers of continents far from moisture sources; (4) coastal deserts that develop where prevailing onshore winds have been cooled by cold oceanic currents; and (5) polar deserts, which are simultaneously extremely cold and dry.

In the dry desert environment, such processes as weathering and sediment transport operate in different ways and at different rates than in moist temperate climates. Weathering in deserts takes a mostly mechanical form. What little chemical weathering exists consists of manganese and iron oxidation; the resulting oxide stains on rocks are called **desert varnish.**

Although one seldom sees surface water in deserts, many of these areas' characteristic landforms have been carved by the rapidly flowing, short-lived streams produced by occasional storms. Streams move through, and deepen, the numerous channels that cross a desert; a channel that carries a stream during periods of high discharge but remains dry during most of the year is an **arroyo.** The mountains that border many deserts recede as surface-water erosion acts to produce large-scale, gently inclined surfaces called **pediments.** An **inselberg** is a steep-sided knob of durable rock—the remains of a mountain following advanced pediment development. Large, dome-shaped inselbergs may form underground largely by chemical weathering when soft rock surrounding a mass of resistant rock erodes.

In deserts, the transporting ability of surface runoff diminishes rapidly because of evaporation or infiltration. While surface runoff lasts, it carries sediments that it later deposits to form distinctive features. Alluvial fans accumulate where a slope ends and the desert floor begins. A **playa** is a dry lake bed that develops when a desert-floor lake evaporates.

Differential heating of the Earth's atmosphere creates pressure differences that generate winds. The resulting convection cells set north–south winds into motion; these winds are deflected to the east or the west by the Earth's rotation, depending on the latitude. Because soil-binding moisture and vegetation are scarce in deserts, their land surface is particularly susceptible to modification by wind action. Wind erodes by **deflation,** removing finer particles from the surface. The layer of pebbles left behind is known as a **desert pavement.** Wind-carried particles are hurled against rock surfaces, effectively sandblasting the rock in a process called **wind abrasion.** The windward surfaces of individual rocks and boulders are beveled by wind abrasion to form asymmetrical **ventifacts.**

Particle motion begins when wind reaches a velocity that is sufficient to move the largest grains in the dead-air layer that exists at ground level. The grains roll and collide with other grains, initiating saltation. Saltating grains make up a wind stream's coarse bed load; finer particles are carried aloft in suspended load. The bed load is deposited when the wind's transport energy decreases below a critical level.

Wind often slows in the wind shadow that is located leeward of a surface obstacle, yielding **dunes,** hills of loose, wind-borne sand. The size, shape, and orientation of a sand dune are determined by the amount of sand available, local vegetational cover, and the intensity, direction, and constancy of winds. Most dunes are asymmetrical, with a gradual windward slope and steep leeward **slip face.** Five principal types of dunes exist: (1) **transverse dunes** are parallel ridges that develop perpendicular to the prevailing wind direction; (2) **longitudinal dunes** are parallel ridges that develop parallel to the prevailing wind direction; (3) crescent-shaped **barchan dunes,** whose horns point downwind, lie perpendicular to prevailing winds; (4) horseshoe-shaped **parabolic dunes,** whose horns point upwind, also lie perpendicular to prevailing winds; and (5) **star dunes** grow vertically to produce a high central point with three or four radiating arms.

The wind-borne suspended load is deposited as loess, which consists of silt-sized particles. With adequate moisture, loess can yield an extremely fertile soil. Because their grain size and bed thickness diminish downwind, loess sheets provide clues to paleowind directions, as do the cross-beds in lithified dunes.

In the late twentieth century, conditions that promote the growth of deserts are overtaking many formerly semi-arid regions. Drought and overpopulation can cause **desertification,** the invasion of desert conditions into formerly non-desert areas; desertification is occurring on every continent except Antarctica. Its symptoms include significant lowering of the water table, marked reduction in surface-water supply, increased salinity in natural waters and soils, progressive destruction of native vegetation, and accelerated rates of erosion. Newly formed desert lands can be reclaimed through irrigation, controlled planting, diversion of surface waters from moist regions, and use of deep groundwater resources (where available).

Mars experiences planetwide seasonal dust storms. Its weaker gravitational attraction enables its dust to ascend higher than it would in comparable winds on Earth. Martian probes have photographed loess-covered plains, ventifacts, dunes, and other wind-generated landforms.

Key Terms

desert (p. 507)
aridity index (p. 508)
rain-shadow effect (p. 510)
desert varnish (p. 512)
arroyos (p. 513)
pediment (p. 514)
inselberg (p. 514)
playa (p. 515)
deflation (p. 518)
desert pavement (p. 519)
wind abrasion (p. 519)
ventifacts (p. 519)
dunes (p. 522)
slip face (p. 524)
transverse dunes (p. 525)
longitudinal dunes (p. 525)
barchan dunes (p. 526)
parabolic dunes (p. 526)
star dunes (p. 527)
desertification (p. 529)

Questions for Review

1. Briefly describe three different types of deserts and the conditions that contribute to their aridity.

2. Draw a simple sketch to illustrate the rain-shadow effect.

3. Cite evidence showing that both chemical and mechanical weathering occur in arid regions.

4. Describe three landforms that are formed by the work of water in arid regions.

5. Why do coastal breezes tend to blow onshore during the day and offshore at night?

6. Briefly describe how desert pavement forms.

7. Draw a simple sketch to show how ventifacts form.

8. Sketch the basic shape of a transverse dune, viewed from the side. What is the essential difference between a barchan dune and a parabolic dune?

9. Describe three ways in which geologists can reconstruct paleowind directions.

10. What are two causes of desertification? How do they produce their effects?

For Further Thought

1. Speculate about how changes in the configuration of the Earth's continents through plate-tectonic activity might increase or decrease the total area occupied by arid regions.

2. What would happen to the distribution of deserts if significant global warming takes place?

3. Determine the prevailing wind direction of the landscape in the photograph below.

4. Describe how global wind patterns may have changed throughout Earth's history. (*Hint:* The Earth's rate of rotation has been slowing throughout its history.) Speculate about the future of the Earth's global wind patterns.

5. What actions would you favor in combating the global trend toward increased desertification? List both the advantages and disadvantages of your proposed actions.

Waves, Currents, and Tides

Processes That Shape Coasts

Highlight 19-1
Lake Michigan's Vanishing Shoreline

Highlight 19-2
Can (and Should) We Control Nature's Coasts?

Types of Coasts

Highlight 19-3
North Carolina's Outer Banks

Plate Tectonic Settings of Coasts

Sea-Level Fluctuations and Coastal Evolution

19

Shores and Coastal Processes

Figure 19-1 Crashing surf along the central Oregon coast.

North Americans are passionate about their visits to their shores and coasts. Indeed, more than half of this continent's population lives within 80 kilometers (50 miles) of the Atlantic or Pacific Ocean or near one of the Great Lakes. Shores and coasts, at once scenic and educational, are wonderful places to observe natural processes, particularly the action of waves, tides, and near-shore currents.

Natural processes cause all shores to change constantly. Sometimes those processes act rapidly and dramatically: Waves from a powerful storm, such as those shown in Figure 19-1, produce immediate alterations of the shoreline. On January 2, 1987, for example, waves driven by a powerful winter storm gouged more than 20 meters (65 feet) of dunes and beaches from Nauset Beach, at the bend in the "elbow" of Cape Cod along the Atlantic coast of Massachusetts. The beach, a 20-kilometer (12-mile)-long pile of sand that waves and tides had scraped from the sea floor over the past 4000 years, had sheltered the oceanside town of Chatham and its fishing fleets for centuries. On that night, however, Nauset was breached by 6-meter (20-foot)-high waves that also swept away nearly 1 kilometer (0.6 mile) of the cape's offshore islands, eliminating the barrier that had protected Chatham from the ferocity of the Atlantic's waves and storms.

These natural modifications to Nauset Beach paid unexpected dividends to the local economy and environment. Chatham's fishing fleet acquired a shortcut to its Atlantic fishing grounds. Even more importantly, the newly increased wave and tidal activity flushed nitrates and phosphates out of septic fields adjacent to the ocean that had contaminated or driven off much of the edible sea life in the area. In this way, the influx of seawater through the breach left cleaner water and restored the local aquatic life. Today bay scallops flourish in an environment with lower nitrate levels, and striped bass and bluefish have returned after years of pollution-impaired fertility.

The area of contact between land and sea, one of the Earth's most dynamic settings, comprises the various land-

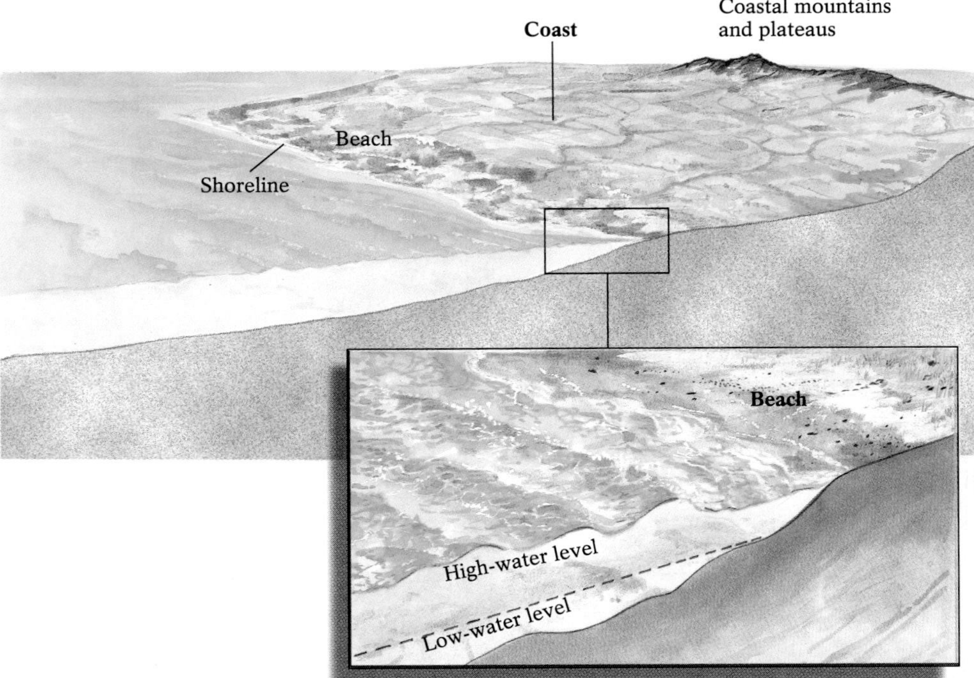

Figure 19-2 A typical coast consists of a shoreline, where the ocean meets the land, and a beach, where water from breaking waves washes a narrow strip of the coast.

scape components illustrated in Figure 19-2. The **coast** is the entire region bordering a body of water. Coasts extend inland until they encounter a different geographical setting, such as a mountain range or a high plateau. A **shoreline** is the precise boundary where a body of water meets the adjacent dry land. A *shore,* the strip of coast closest to a sea or lake, often includes a sandy strip of land, or a *beach.*

In this chapter we consider the processes that shape and change our coasts—the waves, currents, and tides that erode and deposit coastal materials, and the plate movements that raise or lower coasts relative to sea level. We also describe the types of coasts that result from these processes. Finally, we consider how human activity affects the evolution of coasts, and how people have attempted—most often ineffectively—to manipulate and control coastal environments.

Waves, Currents, and Tides

Moving water is the major agent of geological change at the Earth's coasts. Water can be set in motion by the wind, which produces most waves and surface currents, and by the combined effects of the gravitational pull of the Moon and Sun and the rotation of the Earth, which alternately raise and lower water surfaces, producing tides.

Waves and Currents

All waves—whether seismic, sound, water, or any other kind—transport energy. The ultimate source of the energy transported by water waves is solar radiation; as we saw in Chapter 18, solar radiation heats the atmosphere more in some regions (such as near the equator) than in other regions (such as near the poles). This variable heating scheme creates zones of low atmospheric pressure (more heated) and high atmospheric pressure (less heated). When air flows from high-pressure to low-pressure zones in the form of wind, it drags across the surface of any water body in its path, reshaping it into waves.

Wind drag causes the surface water of an ocean or lake to rise to form a *crest* and then fall to create a *trough* (Fig. 19-3). A wave's *height* is the vertical distance between its crest and its trough. In the open ocean, waves commonly reach heights of 2 to 5 meters (7–18 feet); during hurricanes, wave heights may exceed 30 meters (100 feet). Wave *length,* as shown in Figure 19-3, is the distance between two adjacent waves, measured from crest to crest or trough to trough. Ocean waves are typically 40 to 600 meters (135–2000 feet) apart. The time required for one wave length to pass a stationary point is called the wave's *period.* Ocean waves commonly have a period of a few seconds. Wave *velocity,* the speed at which an individual wave travels, typically ranges from 30 to 90 kilometers (20–60 miles) per hour in mid-ocean.

To understand how waves affect a coastline, we will study the waves' movement. We then describe the impact that waves have on the coastline and shore.

The Movement of Waves Blow forcefully across the surface of a bowl of water, and you will create sizable waves on the water's surface. As you gradually reduce the force of your breath, the height of the waves diminishes. In nature, a breeze of 3 kilometers (2 miles) per hour is roughly the minimum necessary to set a perfectly calm water surface rippling in small waves. Wave heights, lengths, periods, and velocities are determined by the speed and duration of the wind, the

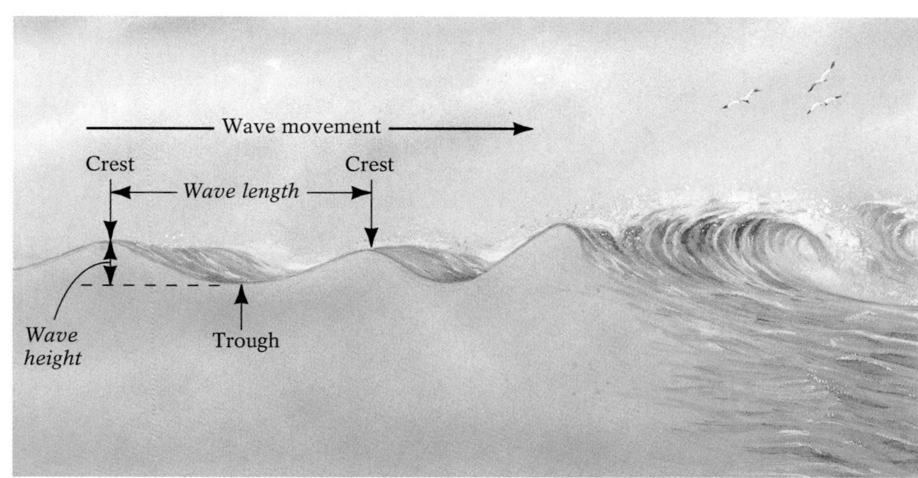

Figure 19-3 The components of a wave. A wave's height is the vertical distance between its crest (the peak of a wave) and the adjacent trough (the low spot between waves). Wave length is the distance between two adjacent waves, measured from crest to crest or from trough to trough.

constancy of wind direction, and the distance that the wind travels across the water surface.

The distance that the wind travels over water is its *fetch.* A brisk wind that blows for a long time from the same direction across a large body of water produces a series of closely spaced, large, fast-moving waves. Fetch is largest and wave heights greatest when waves can travel long distances without encountering landmasses. Pacific Ocean waves, for example, tend to be very large, in part because this ocean is so huge and its landmasses so widely spaced; hence the fetch of its winds is quite large. The band of the global ocean system that lies to the south of the southern tip of South America, for example, is virtually uninterrupted by any landmasses. This globe-circling fetch generates some of the world's largest waves, commonly exceeding 15 meters (50 feet) in height. The largest wind-generated wave ever observed, sighted in the northern Pacific in 1933 by the captain of the USS *Ramapo,* was 34 meters (115 feet) high. It had been generated by persistent gale-force winds.

When waves from several storms blowing from different directions and distances converge in one area of an ocean, complex sets of waves of various heights and periods arrive at the shore at irregular intervals. Regardless of their point of origin, waves with similar velocities and periods can become synchronized (that is, be "in synch"), reinforcing one another to produce large waves; this process is called *constructive interference.* Waves with different velocities and periods may become unsynchronized ("out of synch"), thereby partly canceling one another and producing small waves; this action is *destructive interference.* Some of the big waves awaited by surfers at Malibu Beach, California, may originate in arctic storms, then grow larger thanks to constructive interference by waves from Hawai'ian rain squalls. When you watch waves from the ocean shore, look for irregular arrival intervals and sharply variable wave heights—they reveal the cumulative effect of storms blowing throughout the ocean basin, perhaps some raging thousands of kilometers from your beachside vantage point.

Although mid-ocean waves move outward from a wind source, only the wave *form* moves significantly outward; the actual water *within* the wave essentially remains in place. It follows a rolling circular path, or *orbital,* rising and falling as the wave passes but moving only a short distance from its original position. If you have ever gone fishing, you may have noticed how your bobber moves up and down as a wave passes without really changing position. This type of wave motion, known as **oscillatory motion,** is illustrated in Figure 19-4. Oscillatory motion dies out with depth beneath the surface. Indeed, it is virtually absent below the wave base, a depth equal to one-half the wave length (defined earlier as the distance between two successive crests or troughs). For example, if the wave length for a series of waves is 150 meters (500 feet), the depth to which the waves' orbital paths would extend would be about 75 meters (250 feet). Water below the wave base remains undisturbed by the waves passing above. For this reason, scuba divers in deep water can swim along calmly, even when the water surface churns with the force of powerful storm waves. In deep water, surface waves have no effect on the sea floor; in shallow water, however, a wave base may intersect the sea floor, disturbing its loose sediments.

After a journey of perhaps thousands of kilometers, waves may cross a continental shelf and approach a coast. As the shelf becomes progressively shallower landward, the sea-floor depth decreases to the wave base, and the sea floor eventually interrupts the water's oscillatory motion. A set of waves whose length is 100 meters (330 feet), for example, begins to drag against the sea floor at a depth of 50 meters (165 feet).

When a wave touches bottom, its velocity decreases, largely because its orbital motion (the circular paths taken by the water) becomes restricted. Consequently, as waves

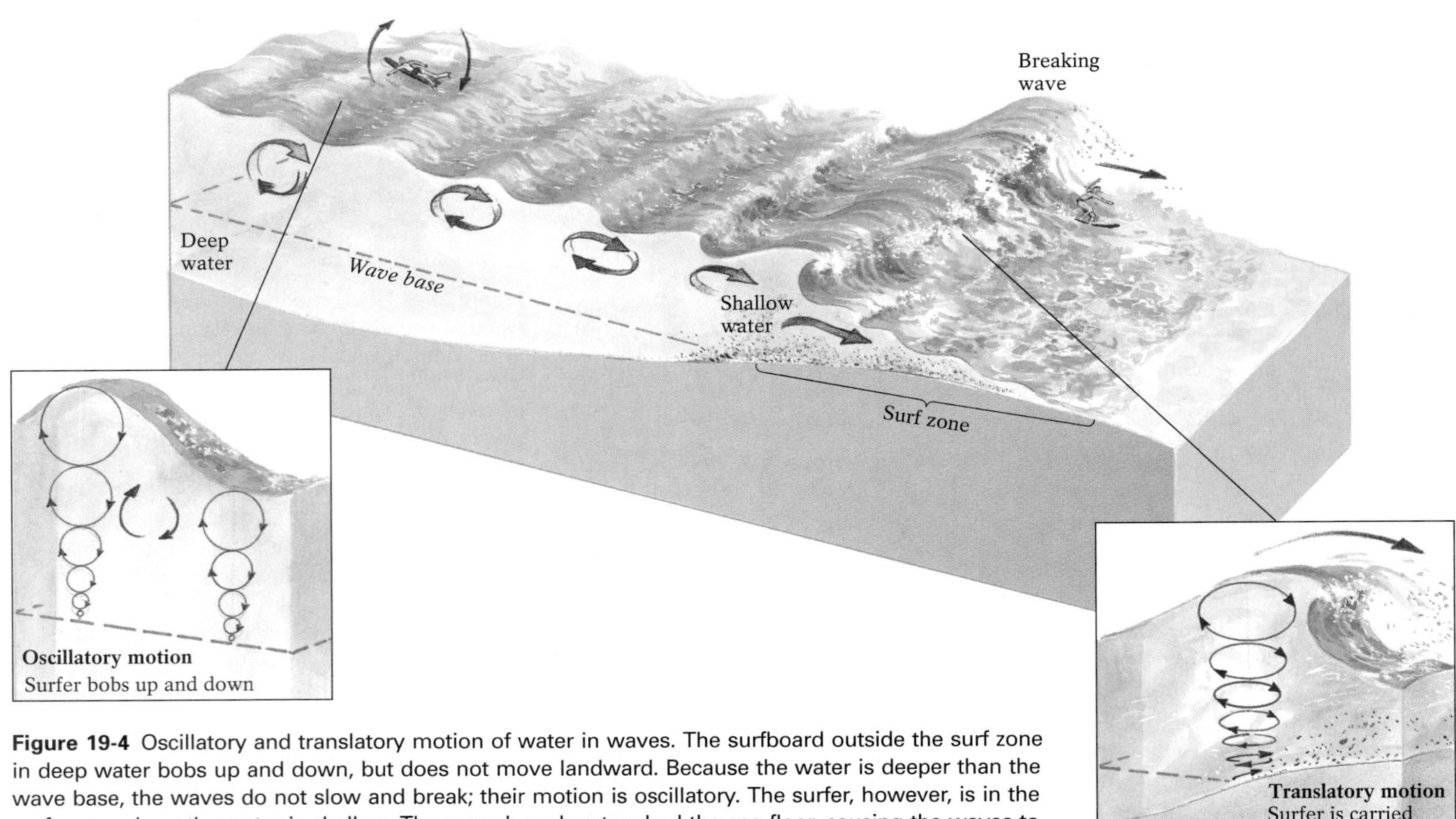

Figure 19-4 Oscillatory and translatory motion of water in waves. The surfboard outside the surf zone in deep water bobs up and down, but does not move landward. Because the water is deeper than the wave base, the waves do not slow and break; their motion is oscillatory. The surfer, however, is in the surf zone, where the water is shallow. The wave base has touched the sea floor, causing the waves to slow and break; the water, surfer, and loose sand particles are carried shoreward by translatory motion.

enter shallow water, their lengths and periods decrease and their heights increase. In other words, the waves bunch up and steepen, and appear at shorter intervals. As a wave continues into the near-shore shallows, its crest begins to move faster than its bottom. Eventually, the crest overruns the rest of the wave, falls over, and *breaks,* dispersing the wave's energy. The water in a breaking wave moves landward as *surf,* low foamy waves. The surf moves through **translatory motion,** meaning that the oscillatory motion has changed to one in which the water itself actually surges forward, carrying along body surfers and churned-up sea-floor sediments alike. Figure 19-4 shows these changes in wave motion.

The water that hurtles up the beach after a wave breaks is known as *swash.* The water that returns to the sea is called *backwash.* Both swash and backwash commonly roll and push sand back and forth along the shore, eroding some material from coastal rocks and bringing some sand inland from points farther out at sea.

Wave Refraction and Coastal Currents Except for the rare times when winds blow *precisely* perpendicular to the shore, waves generally strike a coast at some angle, depending on the direction from which winds blow. Because a wave approaches the shore at an angle, part of it enters shallower water sooner than the rest. When the first-arriving part touches bottom, it slows down and steepens, and its crest breaks. The rest of the wave, which lingers behind in deeper water, continues to move at a higher speed. Eventually it pivots around the slow, shallow-water segment, much as a marching band maintains straight lines as it turns a corner (by having the marchers closest to the corner take small steps while their comrades farthest from the corner take large steps). As shown in Figure 19-5, **wave refraction,** or bending, causes the last arriving portion of the wave to be *nearly* parallel to the coast before it breaks. Waves that initially approach the beach at a 50° to 60° angle are typically refracted to less than a 5° angle.

As each refracted wave breaks and strikes the coast, its surf pushes the swash ahead of it and up the beach at some small angle to the shoreline. The water then returns to the sea as backwash under the force of gravity, *perpendicular* to the shoreline. This zig-zag motion of swash and backwash propels water *down* the coast as a **longshore current** (Fig. 19-6).

Geologists and oceanographers have monitored longshore currents—using fluorescent dye or colored grains of sand—and have learned that their velocity typically ranges

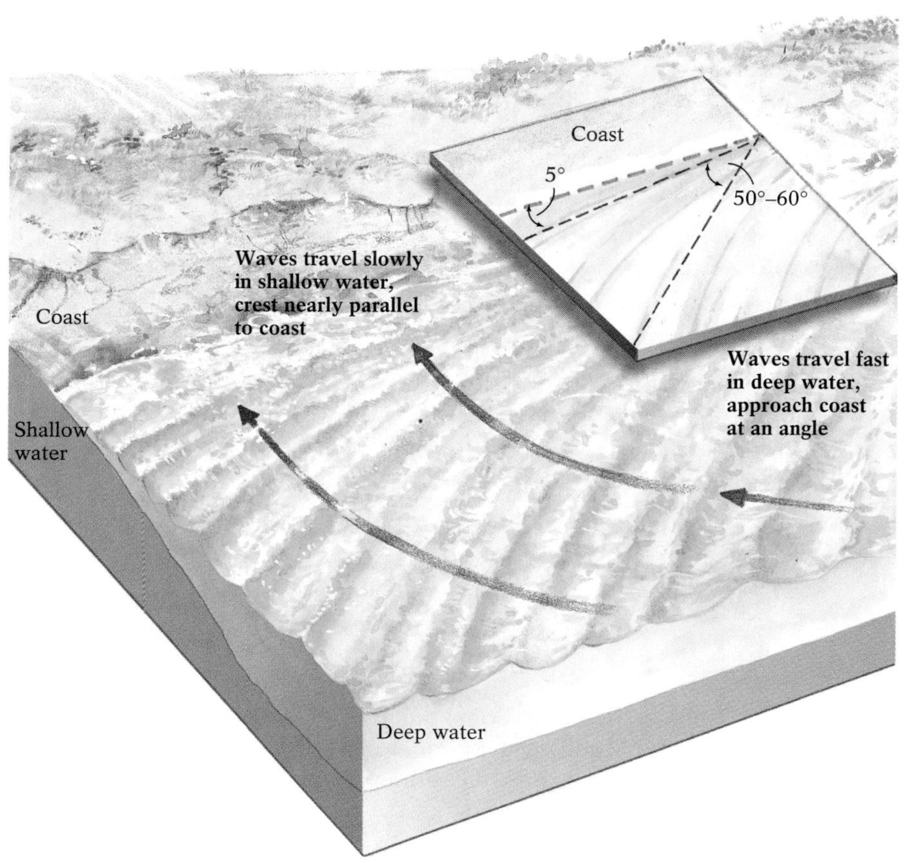

Figure 19-5 Wave refraction. Virtually all waves approach the coast at an angle, but the slower velocity of the part of the wave front that reaches shallow water first bends the incoming wave front until it is almost parallel to the coastline.

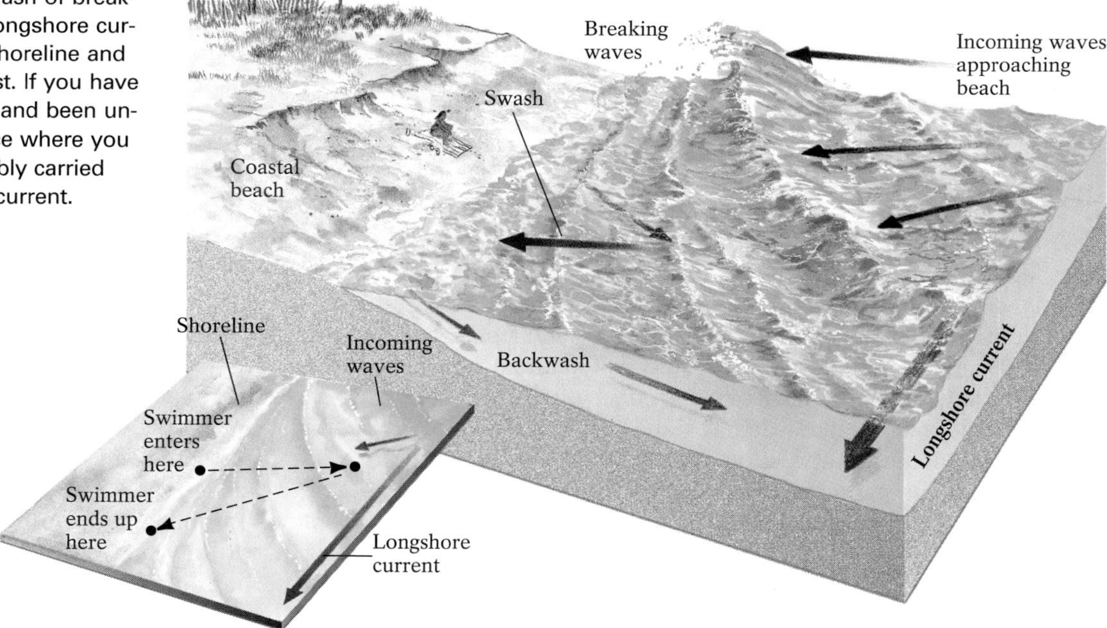

Figure 19-6 The swash and backwash of breaking waves combine to produce a longshore current, which travels parallel to the shoreline and transports sediment along the coast. If you have ever gone swimming in the ocean and been unable to immediately locate the place where you entered the water, you were probably carried down the beach by the longshore current.

between 0.25 and 1.0 meters per second (1–3 feet per second). Longshore currents can, however, race along at speeds of several meters per second and be powerful enough to lift loose bottom sediment into suspension and transport it down the coast for great distances. Sand from the rocky coast of Maine, for example, can be found along the Outer Banks of North Carolina, some 1500 kilometers (1000 miles) to the south.

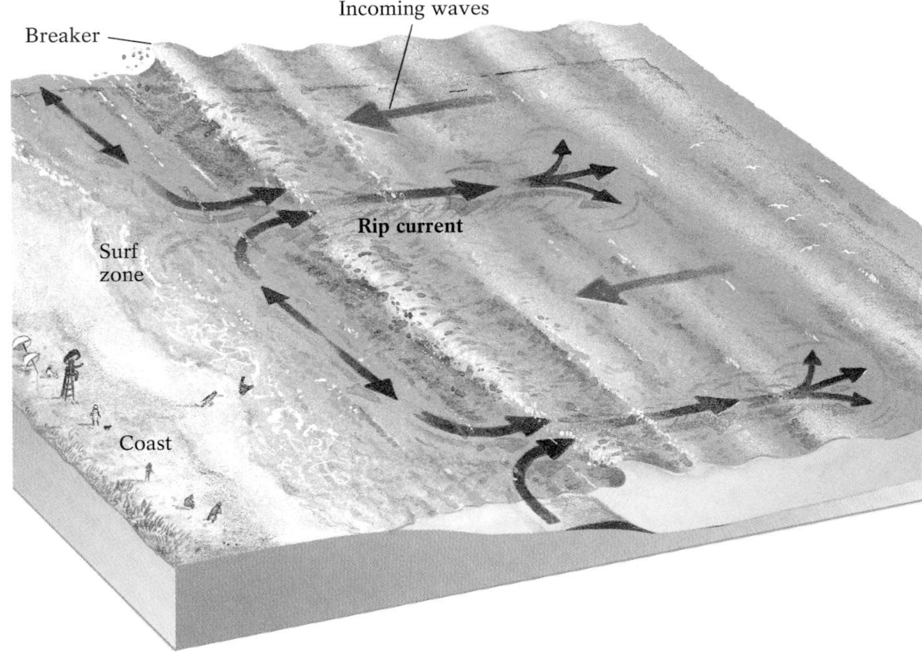

Figure 19-7 A rip current occurs when backwash builds up in the surf zone and the flows seaward through the incoming waves. An ocean swimmer who cannot return to shore because of a strong seaward rip current should swim parallel to the coast for a short distance to escape it.

Rip currents, known to swimmers as "undertow," flow straight out to sea, moving water and sediment *perpendicular* to the shoreline. Figure 19-7 shows the path of a rip current. These currents occur when water that builds up in the surf zone by translatory motion moves seaward. They commonly interfere with incoming waves, causing them to break before they reach the beach or even preventing them from breaking altogether. (Swimmers caught in powerful rip currents can bypass the current by swimming parallel to the shore for a few tens of meters; surfers, however, may choose to ride rip currents seaward for a swift, effortless return to the breaker zone.)

Tides

Tides are the daily rises and falls of the surfaces of oceans and large lakes that move shorelines alternately onshore (during high tides) and offshore (during low tides). In a general sense, tides are wave-like bulges in the surface of a water body that result from the combined effects of the gravitational pull of the Moon and the Sun and from the effect of the spinning Earth on a fluid substance (such as water).

The gravitational attraction exerted by the Moon and Sun represent the major force behind the Earth's tides. Surprisingly, however, the much smaller Moon actually has the greater effect. As we saw in Chapter 11, the force of gravity increases as the masses of the involved bodies increase. Thus, given the Sun's extremely large mass, we would expect its gravitational pull on the Earth to be very powerful, more powerful than the Moon's. However, the force of gravity is also inversely proportional to the square of the distance between the bodies involved; that is, the gravitational attraction between two bodies *decreases* as the distance between them *increases.* Because the Sun is 390 times farther away from our planet, its gravitational pull on the Earth is only 40% as great as the Moon's gravitational pull. In a simple model of tides, the Moon exerts the dominant gravitational force on the Earth; thus the portion of the Earth's oceans that is facing the Moon at any given time bulges "moonward," creating a *high tide.* You can observe this relationship in Figure 19-8.

The rotation of the Earth on its axis generally produces two high tides in coastal regions each day: one when the region is on the side of the Earth facing the Moon, and the other when it is on the opposite side, away from the Moon. The opposite-side high tide reflects the fact that the Earth is a spinning sphere, largely covered by oceans, whose water would escape from the surface if not for the Earth's gravitational pull. On the side of the Earth away from the Moon, the Moon pulls on the oceans less strongly because it is farther away, by the added 12,000 kilometers (8000 miles) of the Earth's diameter. Thus the Moon's gravitational attraction is at a minimum on the more distant side of the Earth; at the same time, the spinning ("centrifugal") force that pulls the oceans away from the Earth is at its maximum, causing the second tidal bulge. Hence, at any given moment, two tidal bulges are occurring at the places on Earth that are closest to and farthest from the Moon. As indicated by the orange arrows in Figure 19-8, when the Earth rotates, the tidal bulges "move" around the Earth's oceans. Likewise, at every moment, two *low tides* are occurring on opposite sides of the Earth. They arise halfway between the Earth's closest and farthest points from the Moon, where the gravitational pull of the Moon and the effect of the Earth's rotation are minimal.

This simple model begins to break down, however, when we wonder how we can stand at an ocean beach with the Moon directly overhead, yet have the tide be out (that is, be at low tide). Many factors affect the behavior of ocean tides, including the manner in which the shapes of the sur-

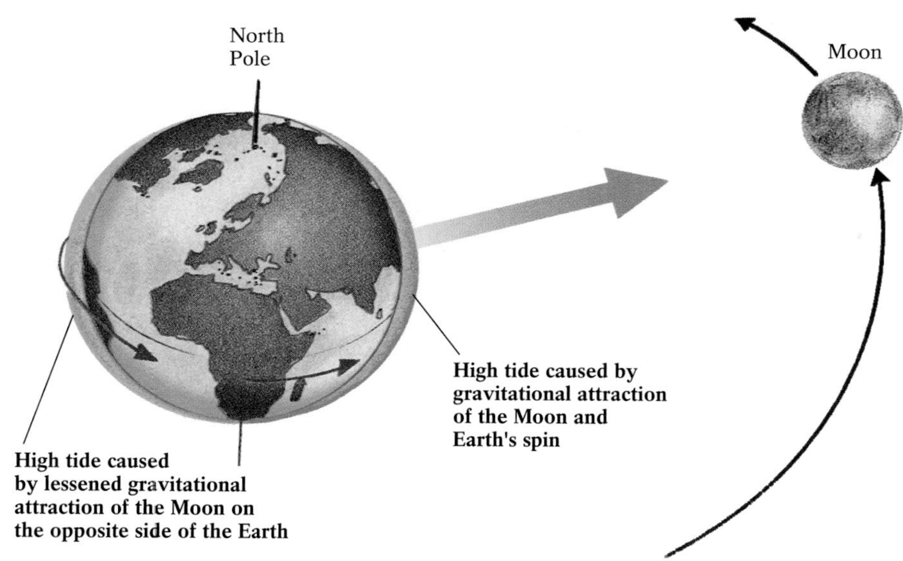

Figure 19-8 Tidal bulges. The Earth's high tides occur where the gravitational attraction of the Moon is strongest (where the Moon is closest) and where it is weakest (where the Moon is farthest away). Thus the tides are influenced by the rotation of the Earth.

rounding landmasses interfere with tidal movements. Each ocean basin acts independently, with its tidal bulge traveling around the basin like a circulating wave. Given the velocity of the Earth's rotation, however, tidal bulges in oceans would be unable to move swiftly enough to always keep high tides directly beneath the Moon. In fact, the tides would have to speed along at about 1600 kilometers (1000 miles) per hour to keep pace with the Earth's rotation. Thus a time lag inevitably separates the Moon's position overhead and the location of the high-tide bulge. While a high tide is always occurring somewhere in the basin, it's not necessarily at the precise spot where the Moon is directly overhead.

A rising tide, or **flood tide,** elevates the water surface of an ocean, advancing the shoreline landward (Fig. 19-9). On a gently sloping coast, a high tide advances farther inland than on a steeply sloping coast. A rising tide can also migrate up a river. For instance, after each low tide, a tidal bulge moves into New York City's harbor and travels more than 200 kilometers (120 miles) up the Hudson River to the cities of Troy and Albany. Flood-tide currents can reach speeds of 25 kilometers (15 miles) per hour and can even reverse the downstream surface flow of rivers. A flood tide with such a turbulent upstream movement is called a **tidal bore** (see Fig. 19-9). The tidal bore of the St. John River, which empties into the Bay of Fundy, a narrow arm of the Atlantic Ocean between the Canadian Maritime provinces of New Brunswick and Nova Scotia, migrates upstream at 10 to 15 kilometers per hour (6–9 miles per hour). As flood tides move upstream, they transport fresh, oxygenated water into stagnant coastal areas.

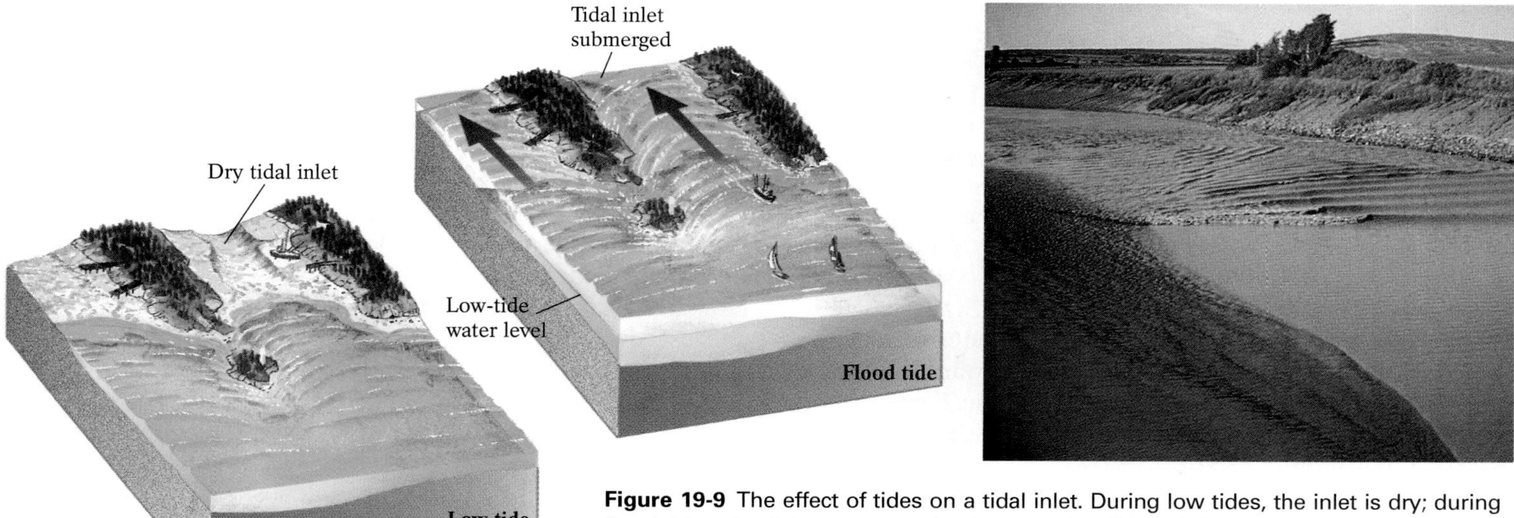

Figure 19-9 The effect of tides on a tidal inlet. During low tides, the inlet is dry; during flood tides, the rising water carries boats and sediment inland. Photo: A tidal bulge may migrate up rivers as a turbulent tidal bore, such as this one in the River Hebert in Nova Scotia.

(a)

(b)

Figure 19-10 The tidal range at the Bay of Fundy, between the Canadian Maritime provinces of New Brunswick and Nova Scotia. **(a)** Eroded columns of rock at high tide. **(b)** The same columns at low tide. The people provide a sense of scale.

A falling tide, or **ebb tide,** lowers the water surface of an ocean, causing the shoreline to recede seaward. Ebb-tide currents transport sediment and organic nutrients from coastal marshes and lagoons into the open sea nourishing marine life offshore.

The vertical difference between local high and low tides marks an area's *tidal range*. Tidal ranges vary with the regularity of the coastline and the size of the body of water. Flood tides funneled into restricted bays and estuaries pile up a large volume of additional water in a small area; in this situation the tidal range is generally quite large. (As we shall see in Chapter 20, a large tidal range can be put to work, spinning a turbine and generating electrical power.) The tidal range at Seattle, in the relatively narrow constriction of Puget Sound, is in the range of 3 to 4 meters (10–13 feet); in the Bay of Fundy, it is even larger, reaching 20 meters (75 feet) (Fig. 19-10). In areas where the coastline is relatively straight, the additional water carried by a flood tide becomes spread over a large area, moderating its effect. The resulting tidal range is usually small—as little as 0.5 meter (1.5 feet)—as is observed in the open Pacific along Hawai'ian beaches. Lakes and other inland water bodies—especially large ones such as the Great Lakes—experience flood and ebb tides as well, but the relatively small size of these bodies usually ensures that they have a sizable tidal range.

Tidal currents rarely produce major geological effects. They can, however, scour the bottoms of shallow tidal inlets and transport fine-grained sediment. When tidal currents coincide with the large waves that accompany a storm, the storm waves become far more powerful. When a storm strikes at high tide, waves penetrate much farther inland than they do during storms that hit at low tide. Thus the timing of a storm relative to the tides strongly affects flooding of beachfront communities and erosion of beach cliffs that are ordinarily protected by wide beaches.

Processes That Shape Coasts

The most significant factor shaping shorelines and coasts is the constant battering of waves, which can erode loose unconsolidated sediment and solid bedrock alike, transport the resulting eroded material, and then deposit it. If you've ever been knocked off your feet by pounding surf, you've experienced the power of sea waves. Wherever wave energy is concentrated, erosion of rock and sediment typically takes place. Wherever wave energy dissipates, deposition of the eroded material occurs. For example, paired erosion and deposition remove rock and sediment from exposed coastal cliffs and leave them in quiet-water bays; the overall effect is the straightening of an irregular coastline. Other processes that modify coasts include the daily rise and fall of tides, biological processes that form coastal swamps and forests and organic reefs, the long-term rise and fall of the global sea level, and geological processes such as tectonic uplift and subsidence.

The energy contained in waves serves as the major force that sculpts the Earth's coastlines. We now turn to the processes of erosion, transport, and deposition, all of which are driven by wave energy, and describe their impact on shorelines and coasts.

Figure 19-11 Crashing surf erodes solid bedrock by forcing water and air into crevices in the rock. Photo: Waves crashing against bedrock at Schoodic Point, Acadia National Park, Maine.

Coastal Erosion

When winds set the sea into motion, the moving water applies stress against its container—that is, the shoreline. Breaking waves can hurl thousands of tons of water against coastal rock and sediment with a force powerful enough to be recorded on nearby seismographs. On average, 14,000 waves strike the exposed rocks and beaches at a given coast every day. As shown in Figure 19-11, waves erode coasts principally by forcing water and air under high pressure into rock crevices. A small, 2-meter (6.5-foot)-high wave, for example, exerts nearly 15 metric tons of pressure per square meter (2000 pounds per square foot) of rock or sediment surface. Although this force is applied for only a fraction of a second, it is repeated every six seconds or so and can prove sufficiently powerful to dislodge large masses of bedrock or sediment. Storm waves can also hurl large rocks through the air to impressive heights. At the lighthouse at Tillamook, Oregon, waves tossed a 61-kilogram (130-pound) boulder over the light; the rock crashed through the keeper's roof, 40 meters (130 feet) above sea level. After waves loosen and remove rocks, they keep hurling the rocks against the coastline.

Wave-induced erosion reflects not only wave size and energy, but also the erodibility of local rocks or sediments. Erodibility, in turn, depends on the strength of the exposed rock or sediment and the extent to which it is jointed or fractured. Softer or more fractured bedrock is more easily eroded than harder, less fractured rocks, regardless of whether the agent of erosion comprises streamflow, wind, or crashing sea waves.

The slope of the local sea floor also influences the amount of wave energy that strikes a given coast. Waves tend to break farther offshore when the sea floor slopes gradually, the water is shallower than the wave base, and the sea floor extends a considerable distance offshore—all characteristics typically associated with a continental shelf. In such locations, wave bases intersect the sea floor well before a wave reaches the shore; waves then break, dissipating much of the wave energy offshore. When the slope is steep, however, incoming waves may never touch bottom; instead they strike the coast head-on with their full force. This situation is particularly common along irregular coastlines that are marked by prominent **headlands**—cliffs that jut out into deep water.

The orientation of a coastline relative to prevailing storm winds influences coastal erosion as well. A coast that is oriented perpendicular to the most frequent wind direction will generally receive a greater share of the force of incoming waves. For example, any segment of the west coast of North America that is oriented north-to-south lies directly in the path of waves driven by Pacific westerlies. Any segment oriented east-to-west, for example, is more likely to receive Pacific storm waves at less than full force.

Erosion along California's 1100 kilometers (700 miles) of Pacific coastline claims an average of 15 to 75 centimeters (0.5–2.5 feet) of the coast annually. Such erosion often produces substantial damage, like the eroded ocean-front

Figure 19-12 El Niño-induced erosion and subsequent repairs along California's scenic Highway 1, north of Big Sur, Winter 1998.

highway illustrated in Figure 19-12. Erosion is also claiming vast tracts of land along the Gulf Coast of Texas and Louisiana. In a recent nine-month period, more than 3 meters (10 feet) of the coastline of Chambers County, Texas, was lost to Galveston Bay, southeast of Houston. On the East Coast, North Carolina's beaches have been worn back by as much as 20 meters (70 feet) during the past decade, largely by storms like July 1996's Hurricane Bertha. The landmark lighthouse at Cape Hatteras—built more than 1500 meters (5000 feet) inland in 1879—is now imperiled at the shore's edge (but there are plans to move it inland). A similar lighthouse on Morris Island near Charleston, South Carolina, which once stood on dry land, is now located about 500 meters (1700 feet) offshore and surrounded by water (Fig. 19-13).

Large inland lakes also are susceptible to coastal erosion. In North America, the eastern shores of the Great Lakes, which are often buffeted by powerful late-autumn storm waves, have suffered from rapid coastal erosion in recent years. Highlight 19-1 focuses on recent erosion problems along Lake Michigan's eastern shore.

Landforms Produced by Coastal Erosion The prominent headlands that bear the full brunt of the great energy of waves wear down rapidly into a series of distinctive erosional features. Figure 19-14a illustrates this process. Initially, a *wave-cut notch* forms at the base of the cliff. Continued wave action enlarges the notch until it undercuts the cliff, removing the foundation of the overlying rock masses and allowing them to fall into the surf. After enough rock has fallen, the remaining cliff base becomes a platform called a **wave-cut bench.** Meanwhile, the debris-laden surf further abrades the cliff face and enlarges the bench. Eventually, the cliff retreats so far that it no longer protrudes out into the breaker zone. The wide bench may initially protect the cliff from further erosion, except during large storms.

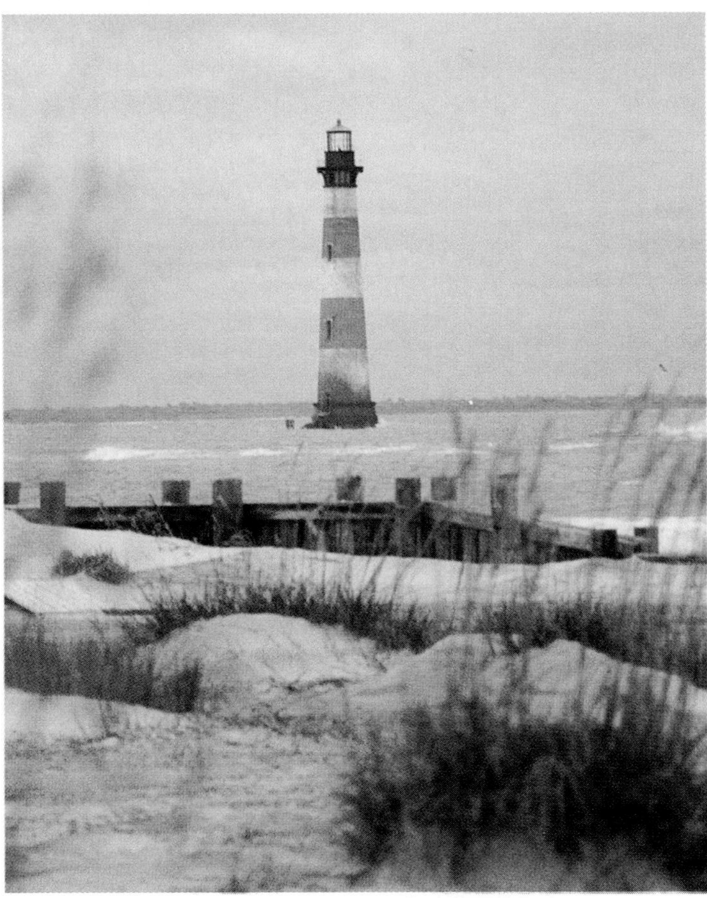

Figure 19-13 Lighthouse at Morris Island, South Carolina. Coastal beach erosion has left this formerly land-based structure submerged and stranded off the coast.

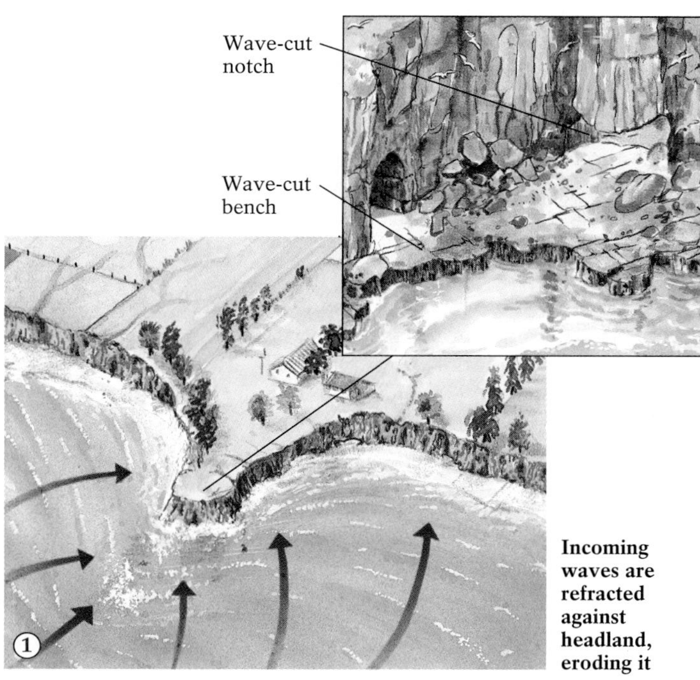

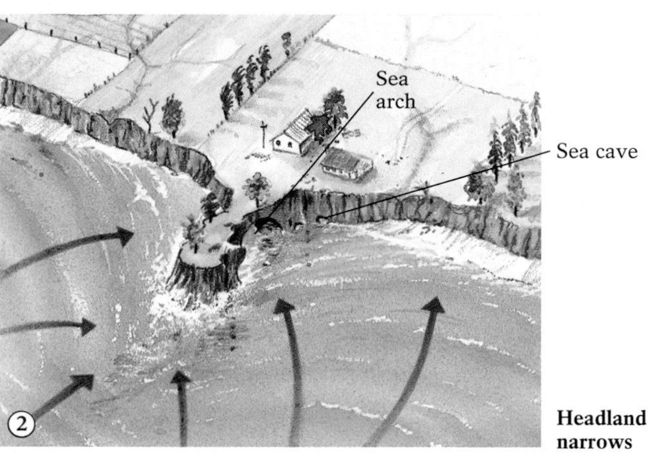

(a)

(b)

(c)

Figure 19-14 The evolution of erosional coastal landforms. **(a)** Wave fronts are refracted against the flanks of headlands, forming wave-cut notches and benches. As the headland narrows, the notches erode to form sea caves. Two sea caves, eroding at both sides of a headland, form a sea arch. Finally, the arch collapses, isolating a sea stack. **(b)** Sea caves on Cape Kildare, Prince Edward Island, Canada. **(c)** A sea arch at Land's End, England.

Over time, the bench may actually enhance headland erosion by refracting incoming waves against the headland's flanks. As waves approach a wave-cut bench, the wave front encounters the shallow bottom and breaks over the bench. The segments of the wave front in the adjacent deeper-water bays pass the headland without touching bottom. Because they can travel more rapidly, they refract toward the flanks of

Highlight 19-1 Lake Michigan's Vanishing Shoreline

Thousands of Michiganders live in towns from Benton Harbor to Ludington on the eastern shore of Lake Michigan, and thousands more from nearby Chicago vacation there. For the past two decades, residents and visitors have watched helplessly as erosion has removed lakefront beaches and caused the lake's eastern bluffs to recede (Fig. 19-15). This beach and cliff erosion is occurring at an unusually rapid pace for two reasons: The area is composed largely of readily eroded unconsolidated glacial deposits, and powerful winter-storm winds push large waves against the lake's eastern shore. Adding to the problem, recent balmy winters have reduced the volume of lake ice that typically protects the bluffs from winter storm waves. Furthermore, several years of unusually high precipitation have produced record-high lake levels, so that water now covers part of the shore. In 1964, when the water level of the Great Lakes was 1.5 to 2.0 meters (5–7.5 feet) lower than today, wide beaches received and dissipated much of the wave energy. Today, the wave energy strikes the bluffs directly. The few relatively dry years that have occurred did not lower lake levels sufficiently to reduce the rate of erosion. Consequently, the bluffs continue to recede about 0.4 meter (1.5 feet) annually. It would take approximately five years of drought to restore the lakes to their pre-1964 levels.

Human activity has also hastened erosion of the lake's beaches and cliffs. Overdevelopment—building too many homes and other structures along the lakeshore—has removed much of the protective vegetation from bluff tops and thinned the vegetation that stabilizes the sands of energy-absorbing dunes. Meanwhile, disposal of wastewater and sewage in septic fields and on bluff-top farmlands has saturated cliff slopes, loosening their sediments and promoting mass movement.

What does the future hold for Lake Michigan's scenic bluffs? The warmer, wetter climate of the past two decades may persist; in fact, the drier period of the past century may have been an anomaly, in which case the lake appears to be returning to its normal, long-term level. If so, accelerated bluff erosion and shore loss may pose an ongoing challenge to those who choose to live along Lake Michigan's eastern shore.

Figure 19-15 Waves generated by westerly winds easily erode the loose glacial sediments of this coastal bluff at St. Joseph, Michigan, on the eastern shore of Lake Michigan.

the exposed headland and concentrate their energy there. The process produces a series of notable coastal landforms, which assume especially dramatic shapes where the headland cliffs contain rocks that offer differing resistance to wave erosion.

At first, battering waves erode the cliff rock to form **sea caves** (Fig. 19-14b), typically attacking "weaker" rocks, such as those composed of softer materials or those containing more fractures, faults, and joints. Further wave action may excavate caves on both sides of the headland until the caves join, forming a **sea arch** (Fig. 19-14c). Continued erosion by waves can wear away the supporting foundation of the arch until it finally collapses. Only an isolated remnant of the original headland, a **sea stack,** may remain. Sea stacks are commonly composed of more durable, less fractured rocks.

(a)

(b)

Figure 19-16 **(a)** Riprap at Carlsbad, California, about 30 kilometers (20 miles) north of San Diego. **(b)** Seawall along the Gulf Coast of Louisiana.

Protecting Against Coastal Erosion Human efforts to minimize the effects of coastal erosion usually involve building structures that deflect waves so they do not strike the coast with full force. In Texas, for example, desperate ranchers have constructed barriers of abandoned cars to prevent waves from the Gulf of Mexico from washing away their land. Sand dunes can provide a first line of defense against the encroaching sea by absorbing the energy of waves, but they tend to migrate with the wind unless they are stabilized in some way. On the Outer Banks of North Carolina, residents have planted hardy vegetation whose roots can hold soil against wind erosion, and sand has been added to create a higher and continuous dune line. In some spots along the southern shore of New York's Long Island, an unsightly mix of driftwood, old tires, rusted cars, discarded refrigerators, and even Christmas trees has been piled up to fortify the area's natural dunes.

Larger protective structures reduce coastal erosion by absorbing the brunt of wave energy. **Riprap** comprises a heap of large, angular to subangular boulders piled along the shoreline (Fig. 19-16a). Although breaking waves remove the riprap eventually, they take several years or even decades to erode it, and riprap is relatively easy and inexpensive to rebuild.

Seawalls are sturdy, longer-lasting structures, generally built parallel to and attached to the shore (Fig. 19-16b). They are designed to withstand the pounding of the highest recorded waves in an area. The large, solid seawalls repel waves seaward, thereby diverting some of their energy.

Seawalls are expensive to build and maintain, however, and they often cause other erosional problems. A $12 million, 6.4-meter (20-foot)-high seawall was completed in 1905 at Galveston, Texas, after the hurricane of 1900, the worst natural disaster in North American history (the storm killed 6000 people and destroyed two-thirds of Galveston's buildings). The wave energy diverted by the seawall quickly eroded the sandy beaches directly in front of the smooth-faced wall. Without the beach, the waves struck the wall directly, breaching it in several places. Although the wall remains in service, its damaged portions must be rebuilt periodically to maintain its effectiveness.

Perhaps the greatest problem with seawalls is their deflection of wave energy to other locations. Although they may protect a specific site from coastal wave erosion, these walls accelerate erosion where the energy has been refocused. They also interrupt the longshore current, thereby disrupting the movement of sediment that nourishes coastal beaches. This action accelerates coastal erosion in areas down the coast from the seawall. In light of these problems, some states (such as North and South Carolina) have prohibited seawall construction and other "hard" stabilization solutions to coastal erosion. We discuss "soft" stabilization solutions—such as beach nourishment—later in this chapter.

Coastal Transport and Deposition

Deposition occurs in coastal settings for much the same reasons as it does at locations where wind, rivers, and glaciers are at work: The supply of sediment exceeds the ability of the current to transport it. When the energy needed for transport falls below a critical level, waves can no longer carry the sediment. Wave energy may be lost or interrupted for several reasons:

- Wind velocity and wave force vary seasonally.
- Water depth increases abruptly.
- Waves, refracted from headlands, are channeled into bays.
- A barrier—whether natural or artificial—prevents waves from reaching the shore, interrupting the longshore current.

When waves deliver more sediment to a shore than the amount removed or redistributed by near-shore currents, the excess sediment becomes deposited, most often as a beach. A **beach** is a dynamic, relatively narrow segment of a coast that is washed by waves and tides and covered with sediments of various sizes and compositions. It may consist of only sand, or it may also contain coarse gravel or even cobbles. Sand beaches are typically white (from quartz grains) or beige (from shell fragments), but they may even be black (from mafic volcanic fragments).

Beach sand commonly accumulates as windblown sand *dunes* (see Chapter 18) on the landward side of a beach. A beach's boundaries stretch from the low-tide line landward, ending where topography changes—for example, at a sea cliff or sand-dune field, or at the point where permanent vegetation begins. A typical beach, illustrated in Figure 19-17, is composed of a **foreshore,** which extends from the low- to the high-tide line, and a **backshore,** which stretches from the high-tide line to the sea cliff or vegetation line. A beach's profile comprises several components with variable slopes. The steepest part of the foreshore is its seaward margin, called the **beach face,** which receives the swash of breaking waves. The backshore, which receives swash only during major storms, may contain a **berm,** a horizontal bench or landward-sloping mound of sediment deposited by storm waves. Some beaches contain several parallel berms from different storms; others show no berms, depending on the extent of storm activity.

More than 90% of beach sediment originates from inland and upland sources and is delivered by coastal streams. An estimated 15 billion cubic meters (60 billion cubic feet) of river sediment is deposited at the world's shorelines every year. Other sediment comes from erosion of headlands and beach cliffs; a smaller amount derives from the offshore surf zone. After sediment arrives at the shoreline, it begins its longshore journey as soon as it encounters its first breaking wave. Then, if human activity doesn't interfere with the process, a beach is maintained naturally when the amount of sediment received by the shore balances the amount exported by longshore currents. (Typically, the volume of sediment transported offshore by rip currents and storm waves or to backshore dunes by sea breezes is relatively small compared with the volume of sediment swept away by the longshore current.)

We now explain the role of sediment transport and deposition in changing shorelines and coasts. We also examine how human actions affect the destruction and restoration of coastlines.

Longshore Transport The energy for transport of coastal sediment comes primarily from waves and longshore currents. As we have seen, when waves break at some oblique angle to the shore, they create a longshore current that transports sediment in the same general direction as the wind and waves. The current moves sediment either on the beach face, as **beach drift,** or immediately offshore within the surf zone, as **longshore drift.** In either case, waves lift the grains and

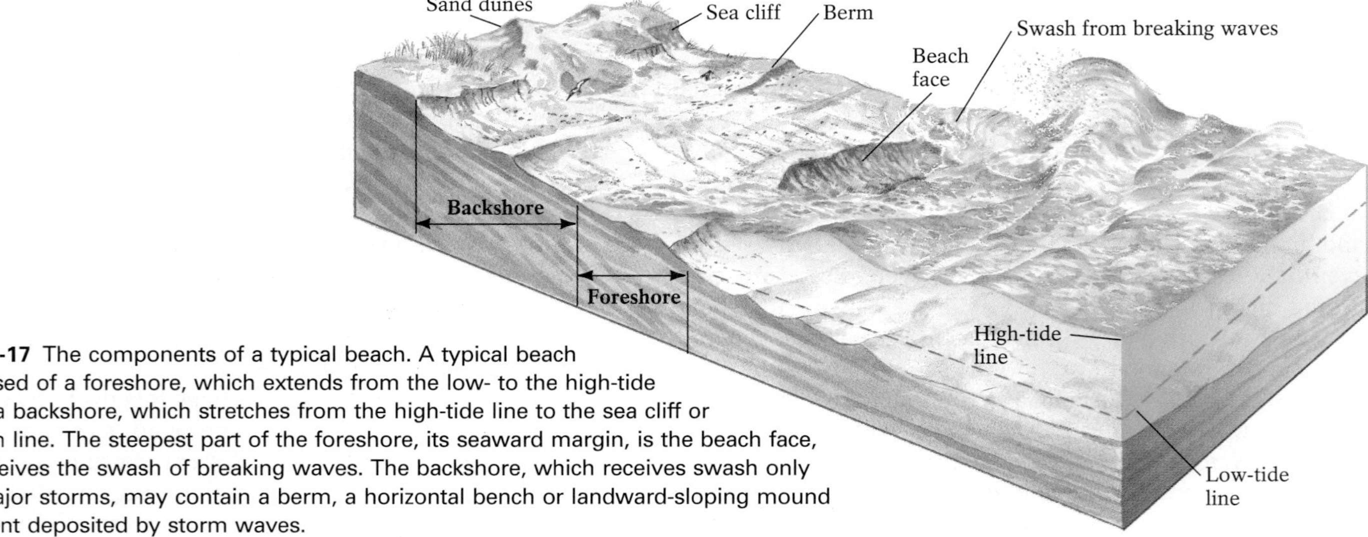

Figure 19-17 The components of a typical beach. A typical beach is composed of a foreshore, which extends from the low- to the high-tide line, and a backshore, which stretches from the high-tide line to the sea cliff or vegetation line. The steepest part of the foreshore, its seaward margin, is the beach face, which receives the swash of breaking waves. The backshore, which receives swash only during major storms, may contain a berm, a horizontal bench or landward-sloping mound of sediment deposited by storm waves.

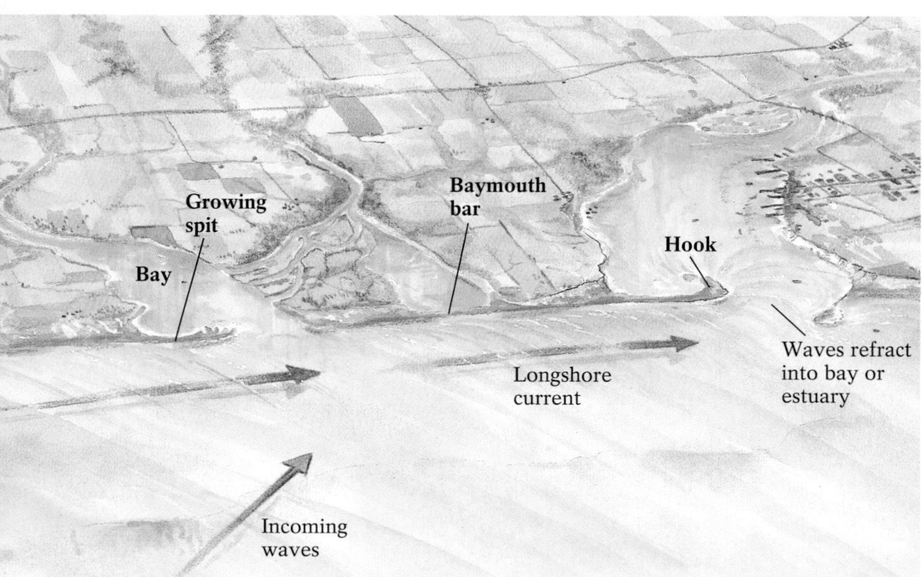

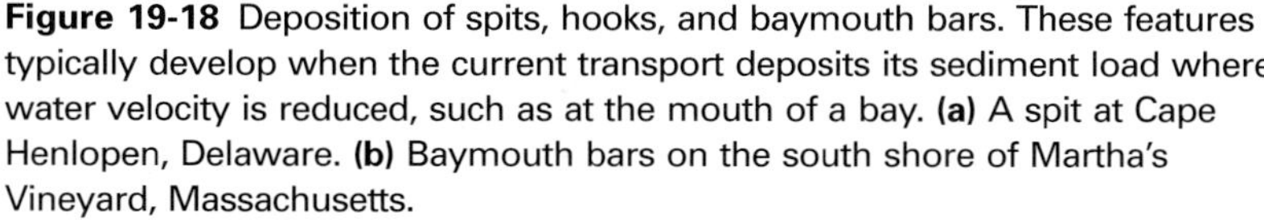

(a)

(b)

Figure 19-18 Deposition of spits, hooks, and baymouth bars. These features typically develop when the current transport deposits its sediment load where water velocity is reduced, such as at the mouth of a bay. **(a)** A spit at Cape Henlopen, Delaware. **(b)** Baymouth bars on the south shore of Martha's Vineyard, Massachusetts.

move them by swash at an oblique angle to the shoreline; they then return the grains seaward by backwash straight down the slope of the beach face (in the case of beach drift) or the surf zone (in the case of longshore drift). Normally, sand is carried offshore more effectively in the surf zone than on the beach face. In a single storm, however, a grain on the beach face can travel down-current as far as 1000 meters (3300 feet). Longshore drift can transport enormous quantities of sand. For example, more than 1.5 million tons of sand per year pass the California coast at Oxnard; 750,000 tons of sand per year, some eroded from granitic cliffs on the Maine coast, pass by Sandy Hook, New Jersey.

Landforms Produced by Coastal Deposition When a longshore current suddenly encounters deeper water, such as where the entrance to a bay interrupts a shoreline, the current is interrupted as well. The incoming waves no longer touch bottom in the bay (whose depth may exceed the wave base). Consequently, the incoming waves do not break, and thus fail to produce the swash and backwash that drive the current. The entire sediment load is deposited at this point as a **spit,** a finger-like ridge of sand that projects from the coast into the open water of the bay entrance. With the deposition of more sediment, the spit can grow by tens of meters per year, with the actual growth rate depending on the supply of sand and the intensity of wave energy. Where particularly strong waves or currents sweep into a bay, a growing spit may become curved, forming a **hook.** A spit may ultimately become a **baymouth bar** if it extends completely across a bay entrance. Figure 19-18 shows how deposition of sediment produces these landforms.

Coastal deposition also occurs when the waves that drive the longshore current find their path to the shore interrupted, perhaps by the shallow continental shelf or another natural offshore landform, or perhaps by a human-made structure. Reduced longshore currents occur, for example, where the continental shelf has a gentle slope and waves dissipate much of their energy by breaking farther offshore. In such a case, the waves do not remove the sediment loads delivered by streams to the beaches, and the beaches tend to widen.

One natural structure that intercepts incoming waves, thereby interrupting longshore currents, is a sea stack. Waves typically break on the seaward surface of the stack, but quiet water prevails on its landward side. Deposition of wave-borne sand takes place in the wave-free zone on the landward side of the stack, because no wave energy exists to carry it further down the coast. Instead, the sediment accumulates as a

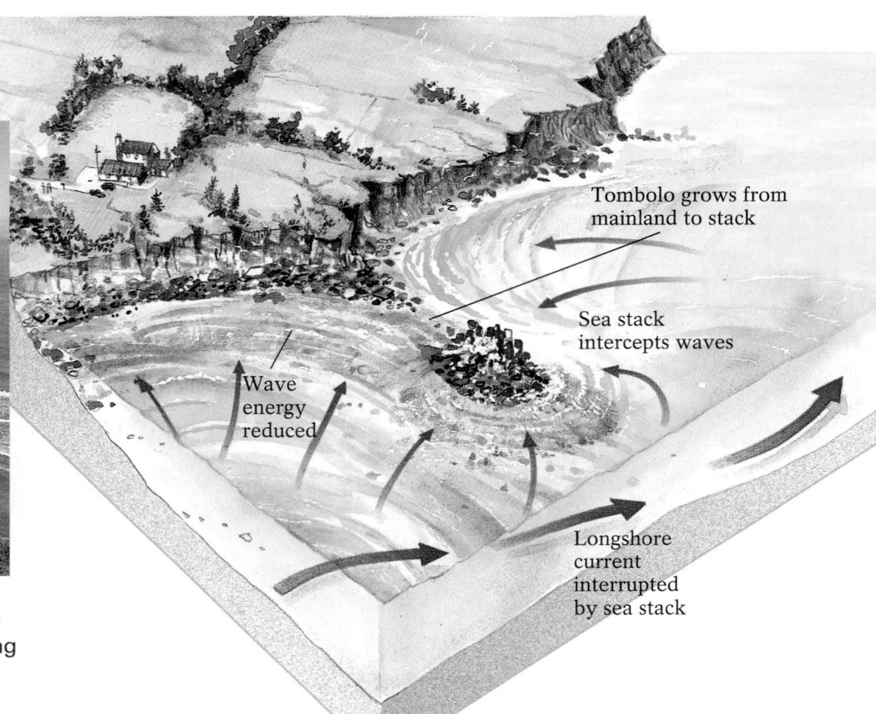
Tombolo grows from
mainland to stack
Sea stack
intercepts waves
Wave
energy
reduced
Longshore
current
interrupted
by sea stack

Figure 19-19 Deposition of a tombolo landward of a sea stack. The sea stack absorbs the wave energy, interrupting the longshore current and causing it to drop its load of sand. Photo: Deposition of tombolo in the lee of a large sea stack, Big Sur, California.

(a)

(b)

Figure 19-20 **(a)** Santa Barbara Harbor in 1931. **(b)** Santa Barbara Harbor in 1977. The breakwater at Santa Barbara was built in the 1920s to provide safe harborage for pleasure boats. It interrupted the longshore current, causing sand to be deposited behind it and on the seaward edge of the beach. This deposition extended the beach but closed off the passage from the marina to the sea. In addition, wave action immediately down the coast from the breakwater caused beach erosion there. To solve the dual problems of excessive deposition behind the breakwater and beach erosion down the coast, sand is regularly dredged from behind the breakwater and pumped downcurrent to the depleted beach area—a costly substitution of human energy and money for the interrupted wave energy.

tombolo, a sandy landform that grows from the mainland to a stack (Fig. 19-19).

Human-Induced Coastal Deposition Some human-made structures disrupt the natural balance between the amount of sediment delivered to the shore and the amount removed by the longshore current. These structures may cause beaches to grow in some places and to shrink—sometimes drastically—in others.

Communities may build **breakwaters,** walls designed specifically to intercept incoming waves, to create quiet, wave-free zones that curtail coastal erosion and protect boats in harborages. These structures are generally oriented parallel to the coasts. Approximately 70 years ago, a breakwater was constructed offshore at Santa Barbara, California, to protect private boats from being battered by Pacific waves. Predictably, because the structure interrupted the longshore current, sediment was deposited seaward of the beach (Fig. 19-20). After 30 years, the beach had widened by hundreds

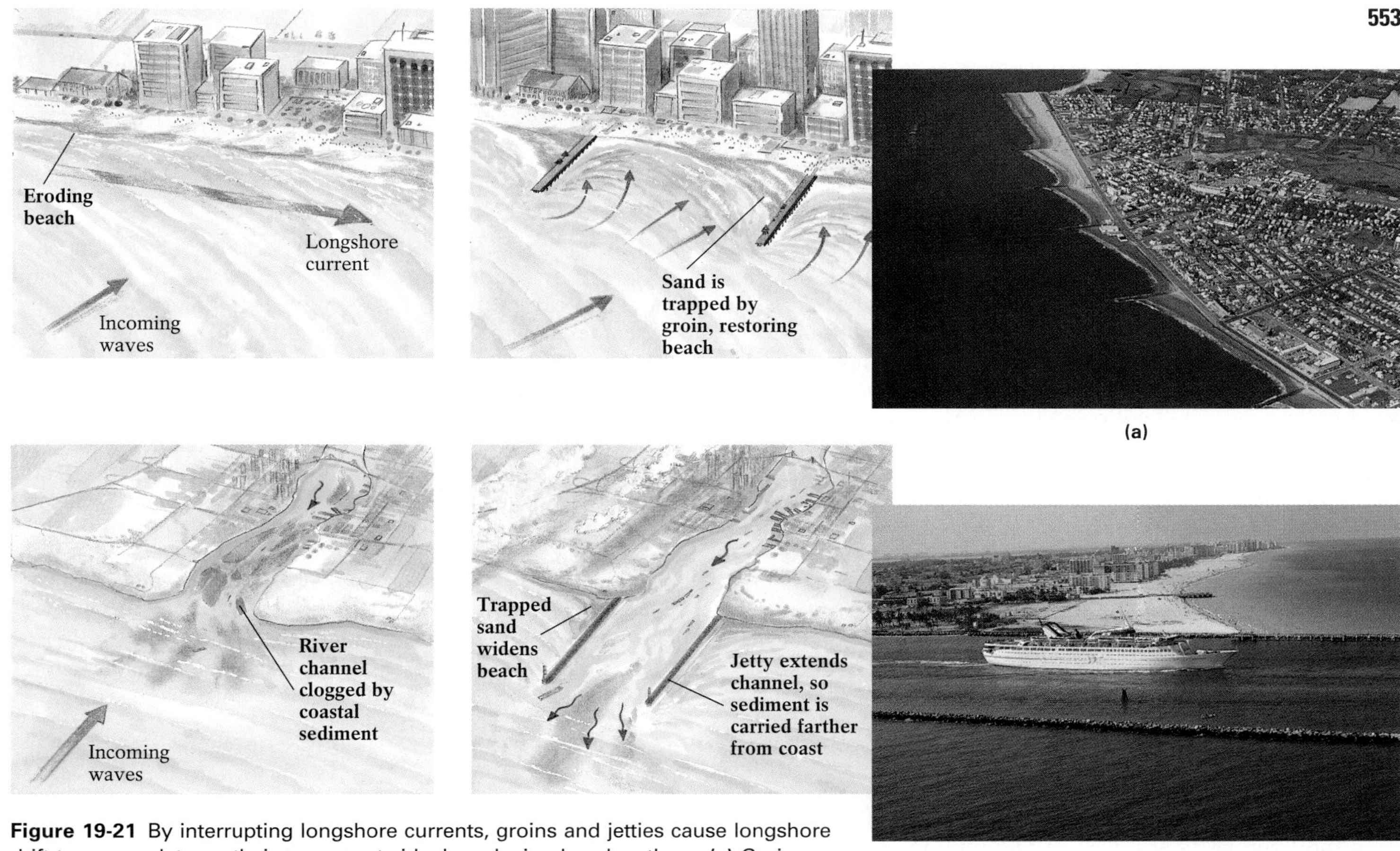

Figure 19-21 By interrupting longshore currents, groins and jetties cause longshore drift to accumulate on their up-current side, broadening beaches there. **(a)** Groins off Cape May, New Jersey. **(b)** A cruise ship in a jetty in Miami Beach, Florida.

of meters, clogging the harborage. Today dredging operations are required to keep the harbor functional.

Groins are shore-protection structures that jut out perpendicular to the shoreline (Fig. 19-21a). They are designed specifically to interrupt longshore drift and trap sand, thus restoring an eroding beach. Most are groins are built where a wide sandy beach, such as Miami Beach, is vital to a community's economic life. They provide, however, a mixed blessing. Groins "rob" sand from one place in favor of another; once one is built, others typically begin to appear all along a coastline.

Jetties are structures, typically built in pairs, that extend the banks of a stream channel or tidal outlet beyond the coastline (Fig. 19-21b). They direct and confine the channel flow to keep channel sediment moving, preventing sediment from filling the channel. Like groins, jetties broaden *up-current beaches* by trapping sand there. Also like groins, they remove sand from the longshore transport system, protecting individual parcels of property, but typically causing substantial erosion and narrowing of *down-current beaches,* sometimes with dire consequences.

Beach Nourishment: Attempting to Restore Shrinking Beaches

Humans may produce beach shrinkage both when our breakwaters, groins, and jetties accelerate erosion and when our inland dams on rivers intercept beach-bound sediment. Flood-control dams built in the last 35 to 40 years in the Missouri–Mississippi River system, for example, have trapped half of the sediment bound for the Gulf of Mexico. In the process, they have curtailed the natural replenishment of the sediment removed by longshore currents from the Louisiana coast and delivered to Gulf Coast islands off of Texas. Similar beach losses occur along the California coast wherever dams intercept and retain the sediments of ocean-bound rivers.

To compensate for the sediment loss at a shrinking beach, we can import sand, but usually at astronomical costs. The sand for such *beach nourishment*—a "soft" stabilization approach (avoiding the construction of "hard" structures, such as breakwaters)—is typically dredged from nearby lagoons, inland sand dunes, and offshore sand bars. In one successful beach-nourishment project along Mississippi's Gulf Coast at Biloxi and Gulfport, a mixture of sand and mud was pumped from offshore sites to create the world's longest artificial beach, stretching 30 kilometers (20 miles). The gulf's gentle waves remove the fine mud particles, leaving behind a sandy beach that has transformed the area into a popular recreation site.

Many beachfront resort communities find, however, that the costs of beach nourishment far outweigh the benefits. For example, a beach-nourishment program in Miami Beach, Florida, halted decades of beach erosion and widened the beach to 100 meters (330 feet). The sand, which was imported

Highlight 19-2 Can (and Should) We Control Nature's Coasts?

A powerful hurricane hits—as when Hurricane Bertha struck in July 1996—and up and down the East Coast, heavily damaged coastal communities beg state and federal governments for protection and relief. Each bout of storm-driven destruction adds more fuel to a raging debate: Can humans control the winds, waves, and shifting sands that alternately sustain and sap the financial life of coastal communities? *Can* we engineer permanent structures that will forever protect our beaches from the ravages of nature? If we can, *should* we? And finally, who should foot the enormous bills for such projects?

No place more clearly dramatizes this passionate debate than North Carolina's Bodie Island, a segment of the Outer Banks. There, the sands at Oregon Inlet, one of the hundreds of perforations in the nearly continuous chain of coastal islands that line the East Coast from Maine to Florida, are shifting southward. The movement threatens to close the outlet between Albemarle and Pamlico Sounds and the Atlantic Ocean.

During the past six years, the mouth of the inlet has marched progressively southward, carried by the waves of the region's longshore current. To keep the 150-year-old inlet open and halt the natural down-current movement of sand, the Army Corps of Engineers has proposed construction of two jetties that would project out to sea, block the sand's southward migration, and thus maintain the inlet at its present position.

For some in the region, the decision to proceed with the project and protect their investment in tourism along the Outer Banks is a classic "no-brainer." "This is our economy," said one local official. "In the post-industrial world, where will the jobs be? It's a leisure world now." For others, the Oregon Inlet project is just another expensive and illogical attempt to control "the uncontrollable." Duke University professor, Orrin Pilkey, the region's expert on coastal-sediment transport, is one vocal critic of the project. He believes, as do most geologists, that structures that interrupt the region's natural movement of sand, simply rob

from the Everglades at a cost of $64 million over a 10-year period, consisted of quartz and fine clay particles rather than the coarse shell fragments of Miami's natural beach. When wave erosion carried off the clay, the offshore water became turbid, damaging coral reefs (which flourish only in clear, sunlit water). The same problem plagues the reef at artificially nourished Waikiki Beach in Hawai'i.

Successful beach nourishment requires expert knowledge of the local longshore currents and their normal seasonal variations. At Long Branch, New Jersey, 459,000 cubic meters (1.8 million cubic feet) of sand intended for beach nourishment was inadvertently dumped too far offshore, beyond the reach of the longshore current that was supposed to carry it to the beach. The unhappy result—a total monetary loss. At Ocean City, Maryland, where a $5 million nourishment project was completed just before the winter-storm season, powerful Atlantic waves quickly claimed 60% of the beach growth.

In the hopes of realizing an economic gain, communities appear bound to persist with such solutions, even in the face of dramatic past failures. Consider the current beach-nourishment project under way at Sea Bright, New Jersey. During the late nineteenth and early twentieth centuries, historic Sea Bright was a thriving coastal resort, drawing thousands of bathers to its beautiful beaches from the nearby sweltering streets of New York and Philadelphia. In the early 1900s, however, the state constructed a sturdy seawall—standing 5 meters high (18 feet) and stretching for 10 kilometers (6 miles)—to protect the local railroad that shuttled bathers back and forth between Sea Bright and Sandy Hook. Deflected storm-wave energy off this gray, fortress-like barrier quickly eliminated Sea Bright's wonderful beach, starving the resort economy of the town. Decades later, the largest beach-nourishment project in history is under way to reverse this damage. Over the next 50 years, the government will spend $1.8 billion to dredge sand from offshore and rebuild Sea Bright's beach. The first section—a 100-meter (300-foot)-wide swath of sand—has cost $226 million, with 70% coming from federal tax dollars. A spokesperson for the Corps of Engineers boasted, "This year we had traffic jams on the Garden State Parkway leading to New Jersey beaches. In 1994, traffic wasn't a problem; we didn't have a beach here then."

This optimism notwithstanding, beach nourishment represents a difficult and costly strategy for maintaining beaches. Although it may be less potentially damaging than the variety of "hard" solutions, this tactic still rarely succeeds in the long run. It sometimes appears that, virtually every time we try to control natural coastal processes, we create a host of new problems. Highlight 19-2 illustrates this paradox.

Types of Coasts

Every coast is shaped by a distinctive combination of erosional and depositional processes that reflects its geological setting. To demonstrate the variations in coastlines, study Figure 19-22. Notice how dramatically the coastline at Cape Flattery differs from the coastline of Cape Hatteras.

So far, we have considered such processes as wave-driven longshore and rip currents, rising and falling tides, and sediment delivery by coastal streams. Other erosional and one beach to *temporarily* save another. Pilkey points out that the project will undoubtedly impede sand movement, causing it to accumulate north of the jetties. At the same time, the project will definitely accelerate the erosion of sand-deprived beaches located farther south. How will we preserve those beaches if we cut off their replenishing supply of sand?

Instead of spending huge sums of federal money on projects that are destined to create a domino effect of problems, Pilkey and other scientists have proposed an inexpensive, but, for some, highly radical and unacceptable solution—"strategic retreat" from the coastline by human development. Pilkey believes that implementation of these expensive, yet questionable shoreline-protection projects encourages and promotes construction of even more homes, hotels, and other businesses that stand directly in harm's way. If the global sea level continues to rise, as it is expected to do, future storms will take more lives and generate enormous economic hardship. Pilkey contends that hurricanes and other large storms are the natural—even beneficial—way that nature maintains its dynamic coasts. These storms wash over the islands, nourishing vegetation, building dunes, and periodically cleaning accumulated sediment (and human pollution) from inland sounds and waterways. Such storms are "only a natural disaster when there's a bunch of houses," says Pilkey—a view that apparently has not fully registered with developers and governmental agencies.

Of course, strategic retreat is unlikely in densely populated, highly developed areas such as Atlantic City, New Jersey, and Miami Beach, Florida. Such a strategy will also undoubtedly prove painful and disruptive to communities that might be forced to move. Yet, as global warming elevates the Earth's seas, coastal communities will face increasing threats from storm-driven erosion. In the long run, no matter what the costs, the sea will force us to move out of its way.

(a)

(b)

Figure 19-22 Cape Flattery, Washington **(a)**, is an erosional coast marked by scores of sea stacks and arches and a rugged, bouldery shoreline. The processes that shaped it differ from those at Cape Hatteras, North Carolina **(b)**, which is a depositional coast characterized by sandy beaches, dunes, and barrier islands.

depositional processes also produce and modify coasts. Coasts are typically classified on the basis of these formative processes.

A **primary coast** is shaped principally by nonmarine processes, such as those illustrated in Figure 19-23. Glacial erosion, for example, produced such primary coasts as the fjords of Alaska and British Columbia; glacial deposition of terminal and recessional moraines shaped the northeast Atlantic from Cape Cod, Massachusetts, to the southwestern edge of Long Island at Brooklyn, New York. Stream deposition yields primary coasts such as the Gulf Coast of Louisiana, where stream deltas extend into the marine environment. Long-term rising sea levels produce bays and estuaries by flooding river systems; the postglacial sea-level rise that began about 13,000 years ago created the primary coasts of Chesapeake Bay and Delaware Bay along the mid-Atlantic coast of North America. In addition, the interaction of biological processes with geological processes may shape primary coasts, such as carbonate-reef coasts.

A **secondary coast** is shaped predominantly by ongoing marine erosion or deposition, such as the processes described in the previous section. Secondary erosional coasts contain

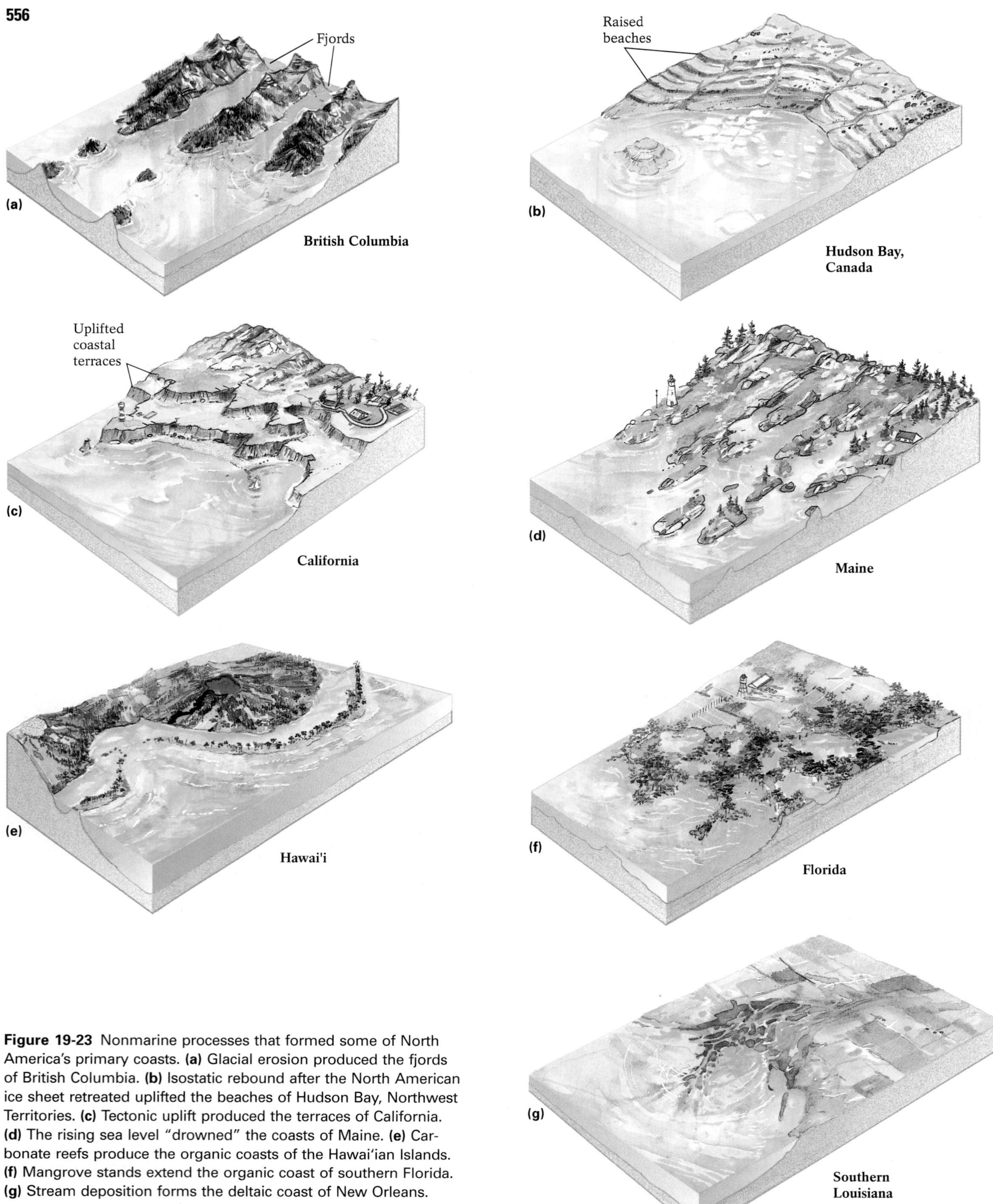

Figure 19-23 Nonmarine processes that formed some of North America's primary coasts. **(a)** Glacial erosion produced the fjords of British Columbia. **(b)** Isostatic rebound after the North American ice sheet retreated uplifted the beaches of Hudson Bay, Northwest Territories. **(c)** Tectonic uplift produced the terraces of California. **(d)** The rising sea level "drowned" the coasts of Maine. **(e)** Carbonate reefs produce the organic coasts of the Hawai'ian Islands. **(f)** Mangrove stands extend the organic coast of southern Florida. **(g)** Stream deposition forms the deltaic coast of New Orleans.

cliffed headlands, wave-cut terraces, and an assortment of sea caves, stacks, and arches. Secondary depositional coasts include beaches, spits, hooks, and tombolos. A secondary coast can be both erosional and depositional, as sometimes occurs with offshore barrier islands. As we shall see in the following discussion, however, this type of coast is not easily classified as either primary or secondary.

Barrier Islands

Barrier islands are nearly continuous ridges of sand, located parallel to the main coast but separated from it by a bay or lagoon. Barrier islands can rise as much as 6 meters (20 feet) above sea level, and are 10 to 100 kilometers (6–60 miles) long and 1 to 5 kilometers (0.6–3 miles) wide. Most such islands comprise a relatively narrow beach, perhaps 50 meters (165 yards) wide, and a broader zone of inland dunes that accounts for most of the island.

Barrier islands are the most common North American coastal feature (Fig. 19-24). They line the East Coast for 1300 kilometers (850 miles), stretching from eastern Long Island, New York, south to Florida, and then continuing for another 1300 kilometers along the Gulf Coast to eastern Texas. Only occasional tidal inlets or the flow of major streams (such as the Hudson River at New York Harbor or the Delaware River at Delaware Bay) interrupts this string of 295 islands. Padre Island, North America's longest beach, is a classic barrier island.

Considerable debate has arisen among coastal geologists regarding the origin of barrier islands. Are they secondary coasts consisting of mostly depositional landforms? Are they secondary coasts consisting of combined erosional–depositional landforms? Or, are they primary coasts that formed during the last period of worldwide glaciation?

Proponents of the depositional model consider barrier islands to be elongated spits, projecting from bends in the coastline, which have been breached by tidal currents or powerful storm waves to form the now-separate islands. In contrast, the erosional–depositional model proposes that waves breaking offshore above the broad gently sloping continental shelf move sand from the bottom and deposit it as long sand bars that continue to grow until they rise above sea level. Storm waves then breach these elongated bars to form the islands. According to this model, the islands trend parallel to the coast because the refracted waves that lift and deposit sand to form them follow the same direction.

Some geologists firmly believe that barrier islands originated as primary coasts, and then evolved as secondary coasts. This model views the islands as remnants of a sand-dune system that bordered the continent during the last period of worldwide glacial expansion. Because sea level at that time was as much as 140 meters (460 feet) lower than the level today, coastal dunes accumulated some distance seaward of the present coastline. As the glaciers melted, the rising sea surrounded and isolated the dunes.

Whatever the origin of barrier islands, once they begin to form, erosional and depositional processes combine to shape them. Since the 1800s, most of these strips of sand have been narrowing, reflecting the global rise in sea level. (Because their seaward slope is so gentle, even a small sea level rise strongly affects the breadth of barrier islands.) This thinning, however, may actually bolster the islands' long-term prospects for survival.

According to the prevailing geological wisdom, as a barrier island thins, it reaches a critical width that allows storm waves to wash completely over the island. At this point, the island's narrowing ceases—the transfer of sand from the seaward side to the landward side maintains the island's width

Figure 19-24 Barrier Island with inland lagoon north of Cape Hatteras; Pea Island, North Carolina.

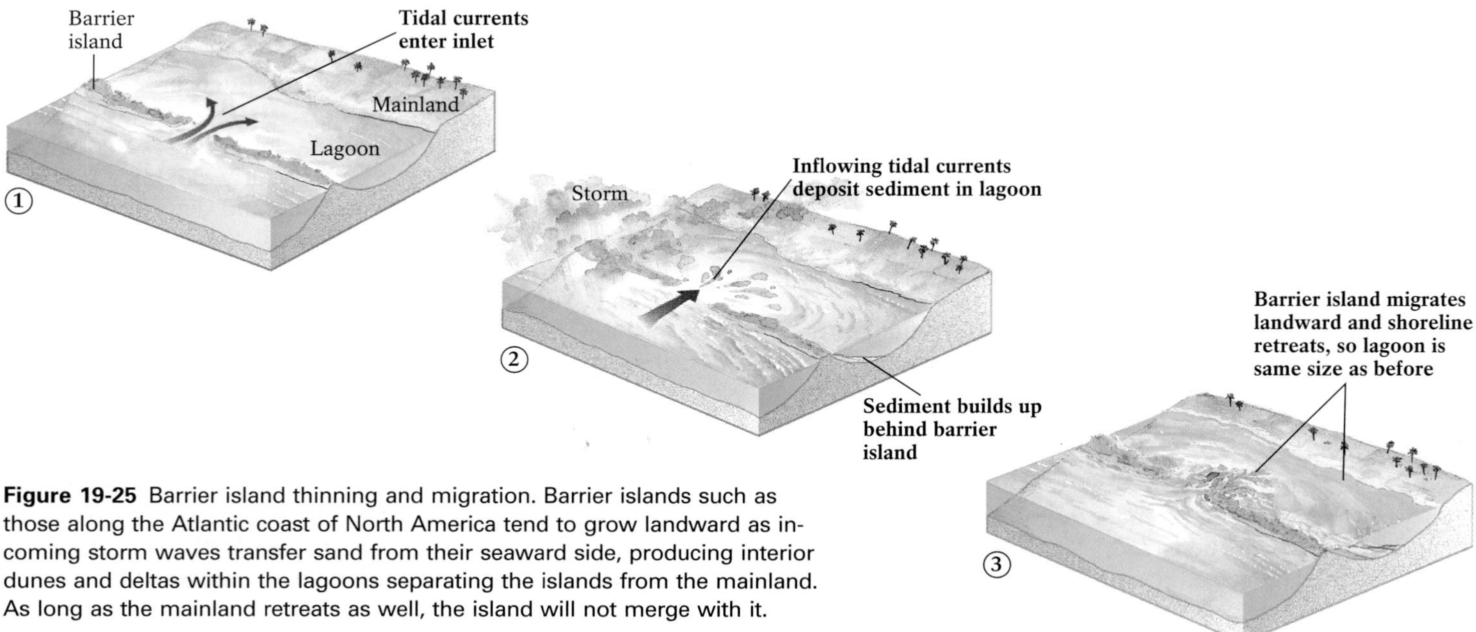

Figure 19-25 Barrier island thinning and migration. Barrier islands such as those along the Atlantic coast of North America tend to grow landward as incoming storm waves transfer sand from their seaward side, producing interior dunes and deltas within the lagoons separating the islands from the mainland. As long as the mainland retreats as well, the island will not merge with it.

(Fig. 19-25). By this process, the island simply migrates landward roughly on pace with the rising sea. The landward migration rate for barrier islands along the Atlantic Coast ranges from 0.5 to 2 meters (1.5–6.5 feet) per year, although it increases considerably during extraordinary storms. The moving island does not usually overrun the mainland coast, because the shoreline behind the island also retreats as sea level rises. As the two landmasses retreat together *in step,* the narrow barrier island remains an island, separated from the mainland by a lagoon. And to the relief of bathers, vacationers, and sun-worshippers, the process also preserves these islands' recreational beaches. As the islands retreat under the onslaught of the rising sea, their beaches retreat as well, maintaining the same width while the island itself narrows.

Human Impact on Barrier Islands North America's long chain of barrier islands has immeasurable value as the first line of defense against storm waves in the hurricane-prone Southeast and the Gulf Coast. These islands are also coveted by city-dwellers as prime destinations for vacations and weekend escapes. Many barrier islands have been set aside for recreation and as wildlife preserves; others are privately owned and undeveloped. More than one-fourth now serve as vacation resorts, being overbuilt with condominiums, hotels, and casinos. Rehoboth Beach, Delaware, Ocean City, Maryland, Virginia Beach, Virginia, Hilton Head, South Carolina, Miami Beach, Florida, Gulf Shores, Alabama, and Galveston, Texas, are but a few of the well-known barrier-island communities developed as commercial properties in this century. Miami Beach, which had a population of 644 permanent residents in 1920, had a population of 92,639 in 1995.

Developers at Ocean City, Maryland, have constructed so many luxury hotels and condominium apartment buildings that the recently wild natural seashore is now valued at more than $500 million per mile. Unfortunately, their ill-conceived alterations to the beach are also jeopardizing their investment. To give every room an ocean view, developers leveled many of the island's protective frontal dunes. The result—massive erosion. A costly beach-nourishment program, which may never restore the beaches to their predisturbance dimensions, is now under way.

Only 3 kilometers (2 miles) south of Ocean City, Assateague Island—home to herds of rare wild ponies—has been set aside as a national seashore and wildlife habitat. Because the law protects the island's frontal-dune line, Assateague Island's beaches have suffered relatively little erosion. Recent construction of jetties to the north (to enhance the beaches at Ocean City), however, is now beginning to divert the sand destined for Assateague's beaches. Consequently, the northern portion of the island is thinning.

The protective role played by barrier islands, the processes that shape them, and the costs of human interference with natural coastal processes can also be seen on North Carolina's Outer Banks, discussed in Highlight 19-3.

Organic Coasts

Some coasts, particularly those in mangrove swamps and along reefs, form when erosion and deposition act in conjunction with biological processes. Mangroves (and some other trees) live in standing tidal water in tropical climates. They grow from an extensive web of long roots that rise above the water. The presence of this root system dissipates much of the energy of waves that enter the swampy mangrove forest, creating a quiet-water environment; it also traps fine sediment, expanding an island's area. Hardy mangrove seedlings

Highlight 19-3 North Carolina's Outer Banks

The Outer Banks, shadowing the east coast of North Carolina, provide a wonderful outdoor laboratory for assessing the effects of human interference with an efficient natural system. This string of barrier islands includes both a developed coast, the Cape Hatteras National Seashore, and an undeveloped coast, the pristine Cape Lookout National Seashore (Fig. 19-26).

The Cape Hatteras coast has been stabilized for more than 50 years by a 10-meter (33-foot)-high, human-made line of dunes that protects State Highway 12 from flooding by winter storms and periodic hurricanes. Built beyond the reach of damaging salt spray and wind- and water-borne sand, the artificial dunes are lushly vegetated. During storms, however, they deflect wave energy much as a seawall does, redirecting it toward the seaward beaches and enhancing erosion there. Since the dunes were built, erosion has narrowed the island's 200-meter (650-foot)-wide beaches to only 30 meters (100 feet). The inland side of the Outer Banks along Pamlico Sound is also being eroded, largely because the high artificial dunes keep away the sand that would otherwise wash over lower natural dunes. This limitation restricts the natural tendency of a barrier island to migrate landward. Disruption of the natural barrier-island system along the Hatteras section has forced North Carolina to compensate with an expensive beach-nourishment program that has so far failed to halt erosion. If erosion of that segment of the Outer Banks continues at its present rate, little of Cape Hatteras may be left in the future.

At Cape Lookout, on the other hand, the banks naturally adjust to the area's periodic storms. Broad beaches and low dunes absorb erosive energy from storm waves. In this natural system, the vegetation growing along salt marshes has evolved somewhat of a resistance to salt spray and flooding. It traps landward-migrating sand, replenishing the natural interior dunes. When a powerful storm hits, the island migrates landward, instead of eroding.

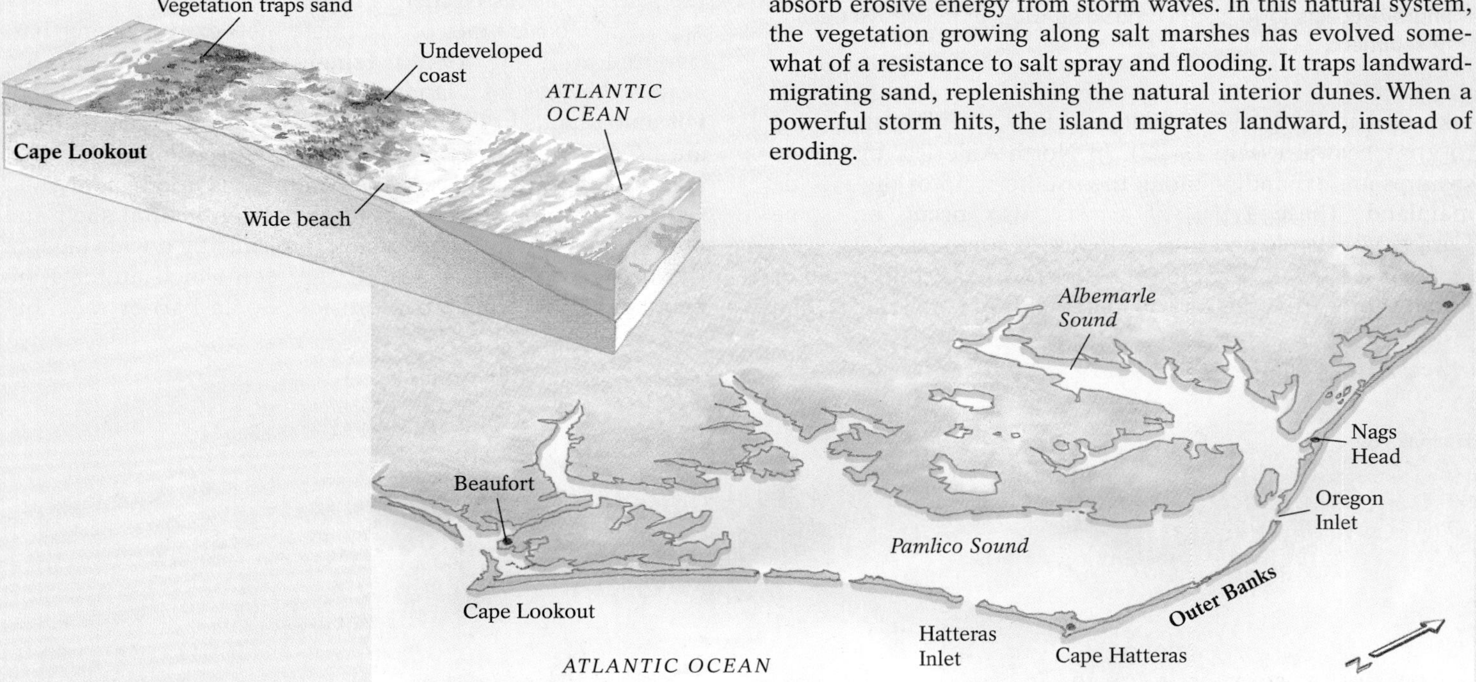

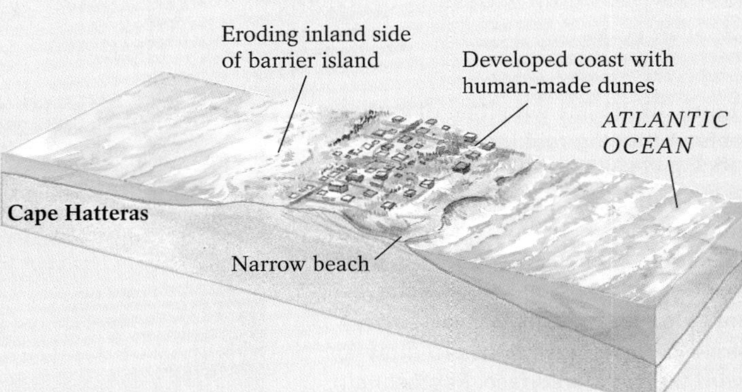

Figure 19-26 Various sections of the Outer Banks of North Carolina have evolved differently, partly reflecting the role played by the National Park Service. Human-made frontal dunes protect the narrow beaches of Cape Hatteras; the undisturbed Cape Lookout area, on the other hand, has wide beaches.

Figure 19-27 A typical mangrove coast in the Florida Everglades. Mangrove coasts form as the protruding roots of mangrove trees trap sediment.

take root in the growing tidal mud flats, enabling the forest to grow seaward (Fig. 19-27). In North America, mangrove swamps are expanding along the southern tip of the Florida mainland. These types of coasts also occur on some Caribbean islands and in the tropics of southeast Asia.

As we saw in Chapter 6, reefs grow near continental or island shores from the carbonate remains of corals, algae, and sponges. The warm, near-surface water in which reefs flourish is limited to a geographic band stretching from about latitude 30°N to 25°S. Clear water is necessary to support the reefs' growth, because most reef organisms, as "filter-feeders," suck in a large quantity of water and filter out microscopic plankton to eat. They simply could not survive if they ingested the suspended particles in turbid water. Finally, reef organisms require moderately salty water; thus reefs are unlikely to form where a large influx of freshwater dilutes the local seawater or where evaporation concentrates the seawater, increasing its salinity.

The active portion of a reef is generally located near the sea surface, where sunlight and the algae on which reef organisms feed are plentiful. The active zone can extend downward only as far as sunlight penetrates, to a depth of approximately 100 meters (330 feet). When the sea level rises, as is happening today, reefs grow vertically so as to satisfy the sunlight needs of their inhabitants.

Three basic types of carbonate reefs exist: fringing, barrier, and atoll (Fig. 19-28). **Fringing reefs** are built directly against the coast of a landmass, such as along the margins of volcanic islands in the Caribbean and the South Pacific. They are usually from 0.5 to 1.0 kilometers (0.3–0.6 miles) wide and grow seaward, toward their organisms' food supply.

Barrier reefs occur on the local continental shelf and are separated from the mainland, typically by a wide lagoon. Incoming waves break against the reef, which protects the mainland coast from wave erosion. Some barrier reefs be-

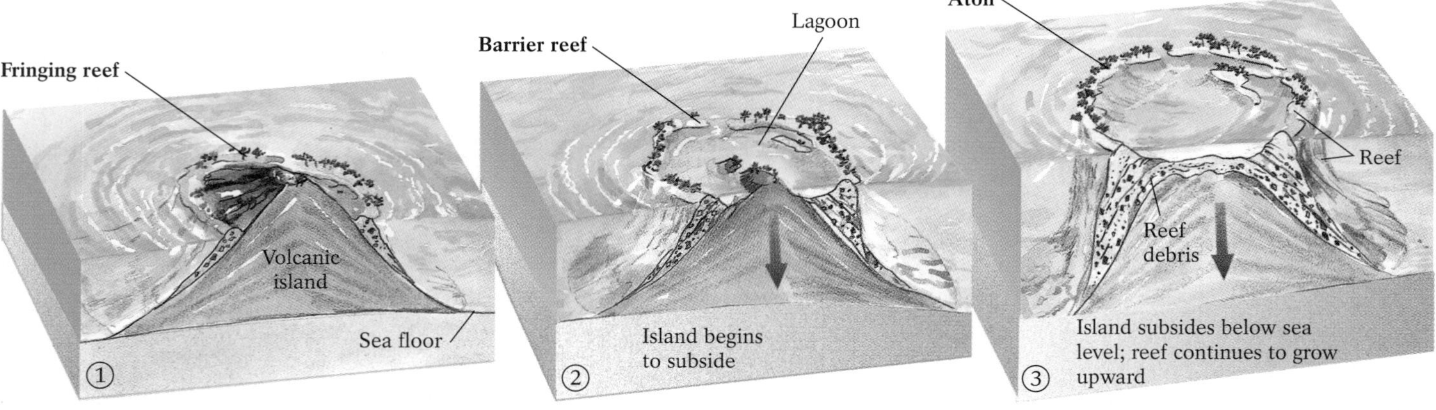

Figure 19-28 The evolution of carbonate reefs. A fringing reef forms on the side of a subsiding volcanic island. As the island subsides, the fringing reef evolves into a barrier reef, separated from the volcano by a circular lagoon. After the island sinks completely below sea level, only the reef remains visible—as an atoll—because it continues to grow toward the water's surface. Photos: (left) Great Barrier Reef, Australia. (right) Wake Island, a coral atoll in the Pacific Ocean.

gin to form as a fringing reef around a subsiding landmass, such as a volcanic island. As the island subsides and much of its area descends below sea level, the reef grows vertically, keeping the surface of its active zone within a few meters of sea level.

Atolls, the most common type of reefs, are circular structures that extend from a very great depth to the sea surface and enclose relatively shallow lagoons. These barrier reefs once surrounded oceanic islands that have since subsided completely below sea level. This explanation of the origin of atolls, which was first proposed by Charles Darwin in 1859, was confirmed in the 1940s by deep drilling of the Bikini and Eniwetok atolls in the South Pacific prior to their selection as atomic-bomb test sites. Bikini was found to consist of more than 700 meters (2300 feet) of coralline limestone resting atop a volcanic island. Surf crashing against an atoll piles the reef debris above sea level.

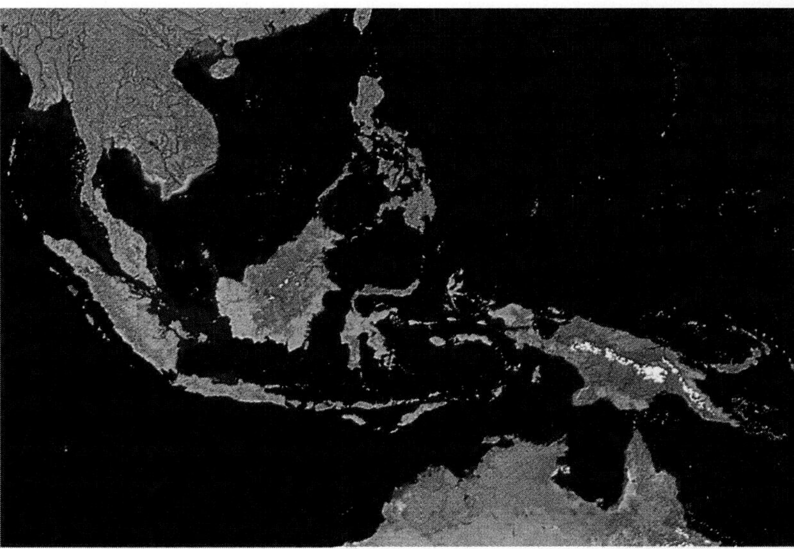

Figure 19-29 Protection of coast by island arc. Formation of Mekong Delta in lee of Philippine and Malaysian arcs.

Plate Tectonic Settings of Coasts

The different characteristics of North America's East, West, and Gulf coasts result, in large part, from their distinctive plate-tectonic settings. Plate setting strongly determines the steepness of the local continental shelf, which in turn governs the behavior of waves and consequently the prospects for coastal erosion and deposition. Every major type of plate boundary is associated with a distinct coastal style.

Divergent plate boundaries that have recently rifted—such as those found in the Gulf of California between Baja California and the Mexican mainland, and in the Red Sea between the African and Arabian plates—are marked by high scarps produced by normal faulting. Their coasts have steep continental shelves, are vulnerable to the head-on attack of wave action, and feature such recently formed erosional landforms as wave-cut benches and prominent sea stacks. As rifted plate margins move away from a warm, spreading center, however, they subside and eventually become tectonically quiet and topographically subdued (see Chapter 12). These *passive continental margins* have broad, gently sloped continental shelves. Incoming waves break offshore on the shallow shelf, typically creating such depositional features as spits, beaches, and barrier islands. The East Coast of North America, with its depositional landforms, clearly shows the results of coastal evolution at a passive continental margin.

Coasts that remain shielded from the direct onslaught of oceanic waves by an offshore island arc evolve principally by marine deposition, particularly where the local climate enhances weathering and enables rivers to transport and deliver a large volume of sediment to the coast. In such an environment, large rivers can extend coastlines by depositing enough sediment to produce deltas. For example, the Mekong River delta in tropical southeast Asia survives in the South China Sea, where it is relatively protected in the lee of the island arcs of the Philippines and Malaysia (Fig. 19-29).

Along convergent plate boundaries, coasts generally have narrow, steep continental shelves, and tectonic uplift continuously produces sea cliffs beyond the shoreline. Without the existence of offshore shallows that can dissipate wave energy, incoming waves strike the coast directly, forming such erosional features as sea stacks, sea arches, and wave-cut benches. The northwest coast of North America, with its assortment of erosional landforms, exhibits coastal evolution at a convergent plate margin (see Figure 19-22a).

The East Coast of North America North America's East Coast is bounded by a broad, tectonically quiet continental shelf that is slowly subsiding under the weight of continent-derived, river-transported sediment. In addition, the entire coast has been slowly submerging for the past century as the global sea-level has risen. This rising sea level largely accounts for the numerous "drowned" valleys, such as New York's Hudson Valley, that appear along the length of the Eastern seaboard. These canyons, which represent the lower segments of stream valleys, have been inundated as the sea level has risen.

Variations in the East's coastal topography result primarily from its local glacial history and its diverse bedrock composition. Glacial erosion carved deep fjords and impressive U-shaped valleys into the soft sedimentary rocks along the northeastern coast of Canada and on the eastern coasts of Baffin Island and the Labrador Peninsula. To the south, glacial erosion of exposed granitic plutons produced the low, rocky coasts of the Canadian Maritime provinces and of Maine and New Hampshire. As the local ice sheets melted, isostatic rebound of the land surface lifted (and continues to

Figure 19-30 East Coast shorelines, such as this one at Currituck Sound, North Carolina, are characterized by wide sandy beaches and offshore barrier islands.

lift) these coasts and continental shelves, allowing present-day oncoming waves to shape these coasts by erosion.

The coast from Boston to Long Island, New York, a zone of glacial deposition, consists primarily of unconsolidated sediments. Vulnerable to attack by crashing surf, these deposits have been eroded and then carried by longshore transport, only to be redeposited farther down the coast as spits, baymouth bars, barrier islands, and beaches. Figure 19-30 shows an example of the sandy beaches typical of this coastal region. South of the glacial terminus in New York City to the Florida Keys, the East Coast features large bays such as Delaware Bay, Chesapeake Bay, and Pamlico Sound in the mid-Atlantic states, barrier islands and inland lagoons and marshes that parallel the coastline for more than 1000 kilometers (620 miles), and the reefs of tropical southern Florida.

The West Coast of North America The West Coast from Alaska through California differs markedly from the East Coast. Along much of its length, converging plates and tectonic uplift have produced a steep, narrow continental shelf and rising coastal mountains (Fig. 19-31). The stretch of coast from Alaska to northern Washington, which was extensively glaciated during recent ice expansions, contains primary coasts punctuated by major westward-draining fjords and U-shaped valleys. South of the glacial terminus, from southern Washington through Oregon and then to southern California, wave erosion has cut into the uplifted terraces, shaping headlands into a near-continuous chain of cliffs and rugged offshore islands; the eroded sediment is deposited as narrow beaches in protected bays.

Southern California's beaches remain narrow, in part because flood-control and irrigation dams on coastward-draining rivers trap stream sediment and prevent it from reaching and replenishing the coast. The narrow beaches, in turn, expose sea cliffs to greater storm-wave erosion than wide beaches, which would absorb and dissipate some of the wave energy. In recent years, the West Coast has also sustained accelerated erosion attributed to El Niño, the warm ocean current that shifts periodically to the eastern Pacific. In the early 1980s, El Niño caused a 10- to 15-centimeter (4–6-inch) rise in the Pacific sea level, unusually high tides, severe winter storm waves, and accelerated erosion. These very same conditions revisited the California coast upon El Niño's return in 1997.

Figure 19-31 The stepped terraces at Palos Verdes Hills in southern California are typical of a western U.S. coast uplifted by plate tectonics. Each terrace is a wave-cut bench formed when that terrace was at sea level. The interaction of the Pacific and North American plates is tectonically lifting southern California's coast, but this tectonic activity occurs only sporadically, being separated by periods of relative tectonic stability during which the wave-cut benches have time to develop.

Sea-Level Fluctuations and Coastal Evolution

Sea level is one of the Earth's most dynamic features—always changing, albeit slowly. This change may be local, affecting just a single stretch of coastline, or it may be global, affecting all the Earth's oceans.

Local changes may result from the following factors:

- *Tectonic movement at plate edges.* Changes such as uplift at a convergent boundary produce a relative *drop* in sea level.
- *Isostatic movements caused by growth or shrinkage of ice masses* (see Chapter 17). The weight of a growing glacier, for example, depresses the Earth's crust, prompting a relative *rise* in sea level. Conversely, melting of an ice mass and the removal of some or all of its weight allows the crust to rebound, producing a relative drop in sea level.
- *Isostatic movements caused by accumulation or erosion of sediments.* Accumulating sediments weigh down the crust, producing a relative rise in sea level. Erosion of the crust removes weight, enabling the land to rise and the sea to fall relative to it.
- *Withdrawal of large volumes of groundwater or oil from aquifers, causing their compression.* As discussed in Chapter 15, the land surface responds by subsiding, producing a relative rise in sea level. Local subsidence of the land near coastal Galveston, Texas, for example, is tied to overpumping of water and oil from the local bedrock. The resulting encroachment of the Gulf of Mexico onto Galveston's shore (Fig. 19-32) has necessitated the construction of a massive seawall to protect the city.

Global, or "eustatic," changes may occur in the following situations:

- *The shape of the oceanic "container" changes.* This scenario typically occurs when sediments are transferred from land to sea, forming deltas and partially filling the basin. The oceanic container may also be altered when rates of mid-ocean-ridge spreading change, thereby modifying ridge shapes. For example, when the rate of ridge spreading increases, a broader, more elevated ridge occupies more space on the sea floor, displacing water landward in the form of a sea-level rise.
- *The amount of water in the ocean basin changes.* Water levels in ocean basins change during glacial and interglacial cycles (see Chapter 17). The global sea level has risen dramatically during the last 10,000 years and continues to increase today, partly because a substantial volume of the world's glacial ice has melted during the current interglacial period.
- *The physical properties of seawater change.* These properties may change during periods of global warming or cooling. Global warming, which many scientists have linked to our widespread use of fossil fuels and the resulting greenhouse effect (discussed in Chapter 17), has probably made a significant contribution to the recent rise in global sea level. The warmer atmosphere causes the upper few meters of the oceans to warm and expand, thereby raising sea levels.

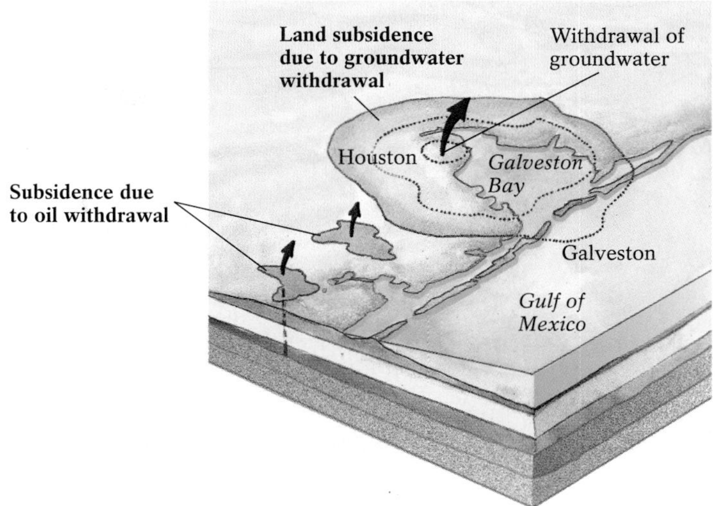

Figure 19-32 Subsidence of the Galveston, Texas, area due to excessive groundwater and oil withdrawal has caused the local sea level to rise relative to the coast.

A dramatic example of how eustatic sea-level change can affect coastal development occurred during the last period of maximum glacial expansion, when sea level was approximately 130 meters (430 feet) lower than the current level. At that time, water that evaporated from the ocean basins became incorporated into the vast Pleistocene ice sheets. During this period, vast areas of the Earth's continental shelves were exposed. At the end of the last glacial maximum about 13,000 years ago, melting of those continental ice sheets returned vast amounts of water to the world's oceans. The influx of meltwater caused global sea levels to rise, rapidly at first and then gradually, until they stabilized near modern levels between 4000 and 5000 years ago. Today we find the remains of terrestrial animals, entire forests, and archaeological sites on the shelves, submerged beneath the rising post-glacial seas of the last 10,000 years. This evidence clearly dates from a time when sea level was substantially lower.

The factors that combine to drive local sea-level fluctuations can be observed along the coasts of Alaska, Iceland,

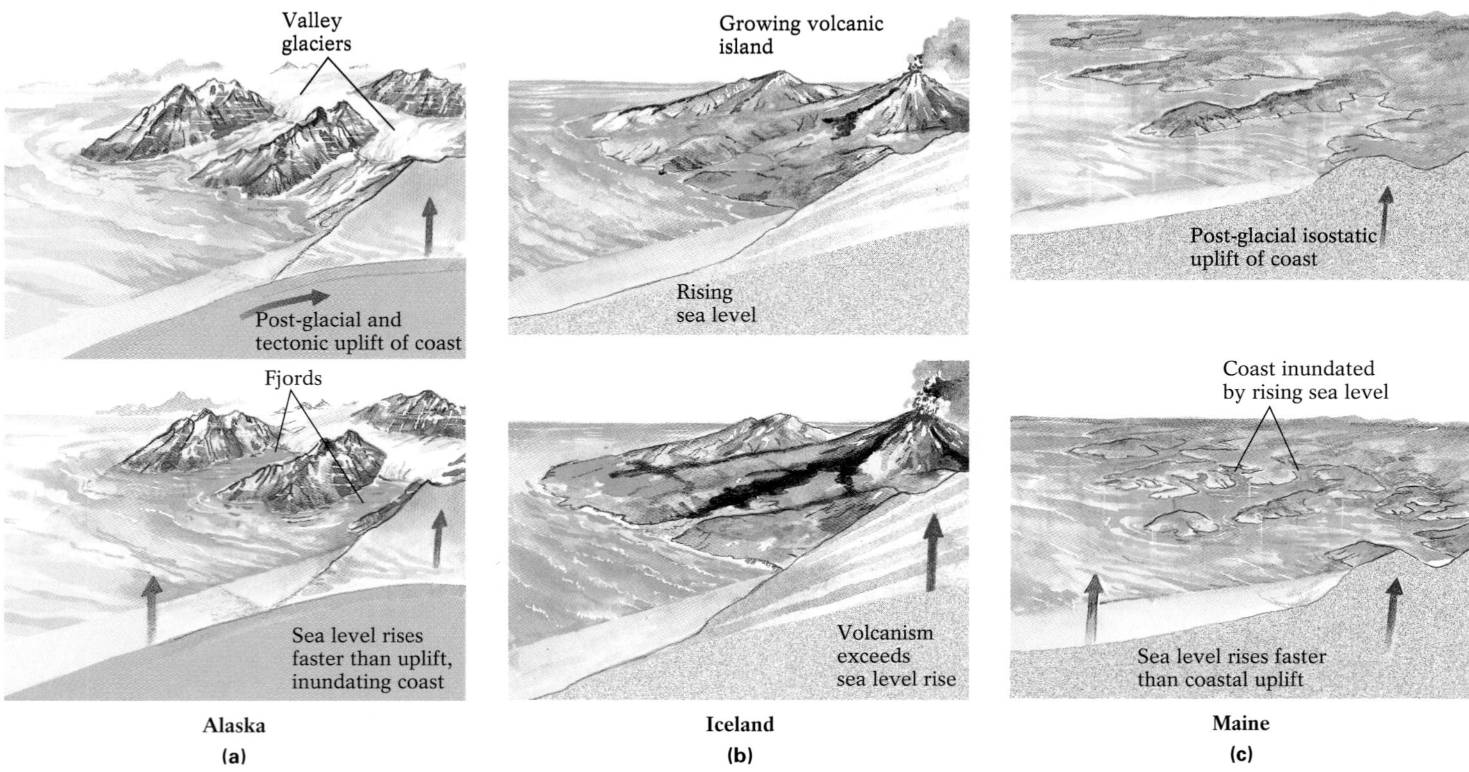

Figure 19-33 The different types of sea-level changes occurring in **(a)** Alaska, **(b)** Iceland, and **(c)** Acadia National Park along the Maine coast illustrate how tectonics, recent glacial history, and global sea-level change combine to determine local sea levels.

and Acadia National Park in southern Maine (Fig. 19-33). Coastal Alaska is currently rising, due to the ongoing convergence of the Pacific and North American plates and to the minor crustal rebound produced by the shrinking of Alaska's glaciers. Worldwide sea level, however, is rising faster than Alaska. The resulting inundation of the Alaskan coast has produced its breathtaking fjord scenery. Iceland, on the other hand, is rising rapidly for several reasons: post-glacial crustal rebound following the melting of the island's ice, active volcanism, and the island's position atop the growing mid-Atlantic divergent ridge. Because the rate at which Iceland is rising exceeds the pace at which the worldwide sea level is increasing, it is emerging from the North Atlantic; hence the local relative sea level is falling. In contrast, Acadia National Park in Maine is located at a passive continental margin and is not rising much tectonically. It is, however, rising isostatically, because it was heavily glaciated during the Pleistocene. Nevertheless, like Alaska, Acadia's uplift rate falls short of the post-glacial rise in worldwide sea level, and the rising Atlantic is slowly covering Maine's coast.

Several studies have shown that sea level has risen by about 10 centimeters (4 inches) during the past century. With global temperatures forecast to increase by 0.6° to 1.0°C (1°–1.8°F) over the next 40 years, the rate at which sea level is rising will accelerate. Some climate experts predict an elevation of 30 to 70 centimeters (12–28 inches) over the next century. In some places, such as the low-lying Atlantic and Gulf Coasts, this change will accelerate coastal erosion and cause a significant landward advance of the shoreline. Higher sea levels primarily increase erosion along the coast by enabling storm waves to penetrate farther inland, threatening homes and other structures.

Because the Atlantic Coastal Plain slopes very gently, a sea level rise of as little as 30 centimeters (12 inches) along the coast of Pamlico Sound in North Carolina would shift the shoreline nearly 3 kilometers (2 miles) inland. The higher sea levels would cover river mouths, extending coastal estuaries farther upstream. For every 10-centimeter (4-inch) rise in sea level, the freshwater–saltwater interface in Atlantic estuaries would migrate approximately 1 kilometer (0.6 mile) upstream. If world climate warmed sufficiently to melt *all* of the Earth's residual glacial ice, worldwide sea level would rise about 70 meters (230 feet)—enough to cause widespread geographical mayhem. In North America, for example, the entire Florida peninsula would be completely submerged, as would New York, Boston, and Philadelphia. Huge bays would penetrate the continent on both coasts, and new coastal cities would arise in such now-inland locations as St. Louis, Missouri, Memphis, Tennessee, San Antonio, Texas, Albany, New York, and Fresno, California.

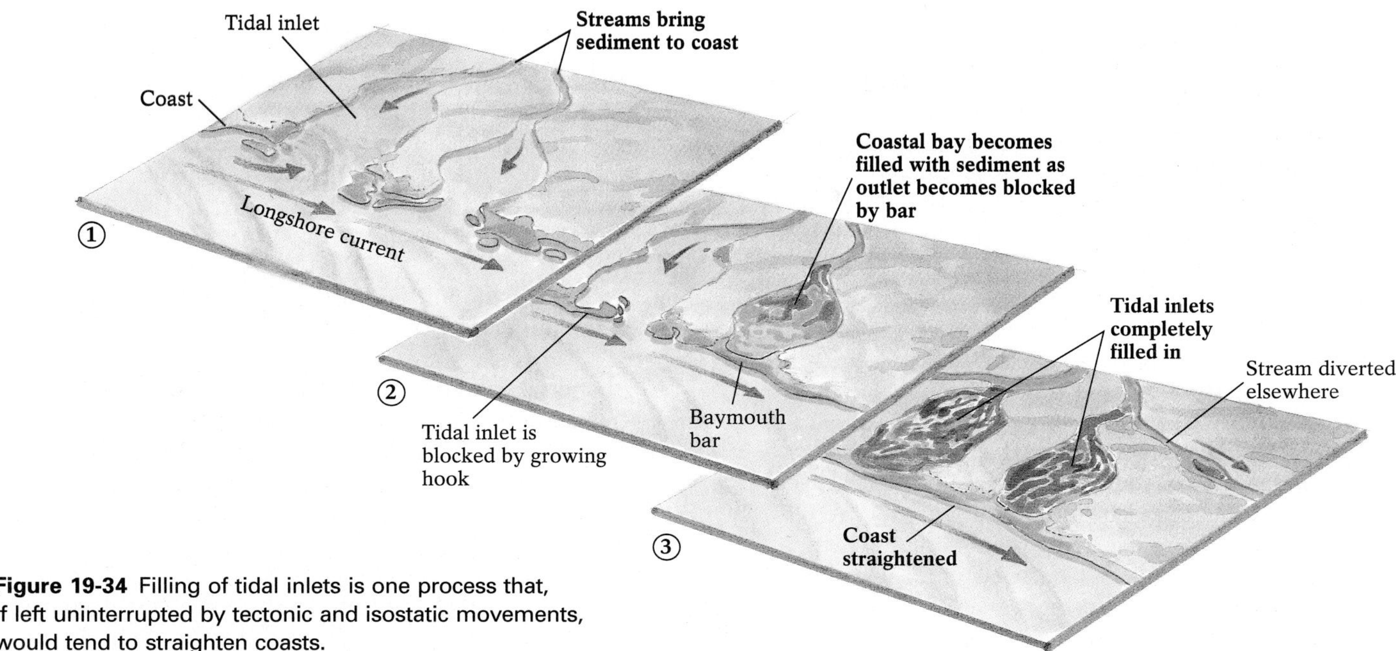

Figure 19-34 Filling of tidal inlets is one process that, if left uninterrupted by tectonic and isostatic movements, would tend to straighten coasts.

Protecting Coasts and Coastal Environments

As we have seen, while coastal erosion removes prominent headlands, deposition of beaches in adjacent bays and construction of baymouth bars that trap continental sediments fill in coastal bays. Overall, these processes, as shown in Figure 19-34, should straighten out the world's coasts. Why, then, do many of the world's coasts remain irregular? As with other geological systems, the processes that shape coasts rarely proceed to completion; in the long term, they are almost always interrupted.

Individuals have little power to control the long-term geological conditions at our coasts. Our governments, however, can act to protect us from the inevitability of coastal change. A few states offer low-interest loans to property owners willing to relocate oceanfront homes to less-vulnerable sites. Twenty-nine of the 30 coastal and Great Lake states have instituted coastal-zone management programs. Many states prohibit construction of permanent coastal structures that would be vulnerable to swamping by rising sea levels. They also prohibit construction of barriers that interfere with natural longshore currents.

Environmental organizations can also influence coastal evolution, sometimes by purchasing environmentally sensitive coastal property to preserve and protect it. For example, the Nature Conservancy of Washington, D.C., has bought long parcels of the barrier islands off Virginia to prevent future overdevelopment of these fragile coastal environments. This approach may prove to be the most viable means of protecting coastal communities from the onslaught of rising sea levels . . . and from themselves.

Chapter Summary

Shorelines are the places where bodies of water meet dry land. Landward of ocean shorelines are **coasts,** the strips of land bordering an ocean that extend inland to a different geographical setting. The shapes of coasts depend on the sediments delivered to them, primarily by streams, and the waves that erode and redeposit those sediments.

Waves result when wind drags across the sea surface. Wind speed, duration, and the extent of uninterrupted water surface across which the wind blows all act to determine the wave height, velocity, and length. A wave moving across deep, open ocean exhibits **oscillatory motion,** in which water particles move in circular orbits. As a wave moves toward the coast, it usually first encounters the continental shelf, where the ocean floor rises. In the shallow zone above the continental shelf, the ocean floor restricts a wave's circular orbit, causing it to slow and steepen. Eventually, the steeper upper part of the wave overruns its base and the wave breaks. Its water then no longer moves in a circular path but instead is sent hurtling forward by **translatory motion.**

When a wave approaches the coast at an angle, the first part of the wave front to encounter shallow water slows, causing the rest of the wave front to swing around until it becomes nearly parallel to the coastline; this process is called **wave refraction.** Virtually all waves strike the coast at some angle; their water moves up the beach at an angle, but takes a perpendicular path back to the ocean. This zig-zag motion generates a **longshore current,** which moves parallel to the coast. Longshore currents transport most coastal sediment.

They may be crossed by **rip currents,** which develop when water piled up in the surf zone flows seaward, generally perpendicular to the coast.

Tides, the twice-daily rise and fall of the ocean surface in most parts of the world, are primarily caused by the gravitational pull on the ocean surface by the Moon and from the force created as the Earth spins on its axis. **Flood tides** elevate the sea surface and cause the shoreline to move inland, sometimes migrating up rivers. A flood tide with a turbulent upstream movement is called a **tidal bore.** Falling or **ebb tides** lower the sea surface and cause the shoreline to move seaward, carrying land-derived sediment out to sea.

Most coastal erosion occurs when waves strike a coast with sufficient energy to remove loose materials. When the water immediately offshore is deep, the waves crash with their full force against coastal **headlands,** cliffs that jut seaward. Erosion of a headland eventually undercuts the cliff, removing the support from overlying rock masses, which fall into the surf. The remaining cliff base becomes a **wave-cut bench,** a terrace washed by breaking waves. Wave refraction at headlands bends incoming waves against the sides of the headlands, eroding them to produce **sea caves.** A **sea arch** forms when two sea caves erode completely through the headland. When a sea arch collapses, an isolated offshore bedrock knob, or **sea stack,** forms. To protect coasts from erosion, we can construct such structures as a barrier of loose boulders **(riprap)** or a concrete **seawall** built parallel to the coast.

Coastal deposition produces a variety of landforms, the most common being a **beach**—a dynamic, relatively narrow segment of a coast washed by waves or tides and covered with sediment of various sizes and compositions. Beaches consist of a **foreshore** area, which extends from the low-tide to the high-tide line, and a **backshore** area, which stretches from the high-tide line to the sea cliff or inland vegetation line. The steepest part of the foreshore is its seaward margin, or **beach face.** Most beaches also contain one or more **berms,** horizontal benches or landward-sloping mounds of storm-deposited sediment. Sediment transported along the beach face by swash and backwash moves via **beach drift;** sediment transported offshore within the surf zone moves via **longshore drift.**

When a longshore current suddenly encounters deeper water, such as at the entrance to a bay, it deposits its sediment load as a **spit,** a finger-like ridge of sand that extends from the land into open water. Where waves or currents into a bay are particularly strong, growing spits can become curved, forming a **hook.** A spit that grows completely across a bay entrance is called a **baymouth bar.** Coastal deposition occurs when wave energy falls below a critical level. Wave energy can be dissipated by wind variation, sudden increases in water depth, or the interception of the waves by natural or artificial structures offshore. Human-built structures that intercept waves or currents, causing deposition, include **breakwaters** (built parallel to the coast), **groins** (built obliquely to the coast), and **jetties** (built perpendicular to the coast).

Primary coasts form from nonmarine processes such as glaciation, stream deposition, tectonic activity, flooded river systems, and biological activities. **Secondary coasts** are products of coastal erosion and deposition. **Barrier islands** are nearly continuous ridges of sand, parallel to the main coast but separated from it by a bay or lagoon. They may be secondary coasts, consisting of depositional or combined erosional and depositional landforms, or they may be primary coasts—the remnants of a drowned ancient coastal sand-dune system.

Organic coasts include those that develop around mangrove tree roots in swamps and those that are formed by carbonate reefs. Carbonate reefs commonly appear along the margins of volcanic islands in the Caribbean and the South Pacific. **Fringing reefs** initially surround land and grow seaward, toward their organisms' food supply. **Barrier reefs,** which are separated from the coast by a wide lagoon, form and grow vertically as the island subsides. After a volcanic island completely subsides, its barrier reef becomes an **atoll,** a circular structure that extends from great depth to the sea surface and encloses a relatively shallow lagoon.

Plate tectonic settings also influence the nature of coastlines. Recently diverged plate margins and convergent plate boundaries possess steep continental shelves, and their coastlines include erosional landforms such as wave-cut benches and sea stacks. Passive continental margins have broad continental shelves, and their coastlines typically contain depositional features such as beaches and spits. Sea-level fluctuations affect coastlines as well: Rising sea levels cause shorelines to migrate inland, and falling sea levels cause them to migrate seaward.

Key Terms

coast (p. 538)
shoreline (p. 538)
oscillatory motion (p. 539)
translatory motion (p. 540)
wave refraction (p. 540)
longshore current (p. 540)
rip currents (p. 542)
tides (p. 542)
flood tide (p. 543)
tidal bore (p. 543)
ebb tide (p. 544)
headlands (p. 545)
wave-cut bench (p. 546)
sea caves (p. 548)
sea arch (p. 548)
sea stack (p. 548)
riprap (p. 549)
seawalls (p. 549)
beach (p. 550)
foreshore (p. 550)
backshore (p. 550)
beach face (p. 550)
berm (p. 550)
beach drift (p. 550)
longshore drift (p. 550)
spit (p. 551)
hook (p. 551)
baymouth bar (p. 551)
breakwaters (p. 552)
groins (p. 553)
jetties (p. 553)
primary coast (p. 555)
secondary coast (p. 555)
barrier islands (p. 557)
fringing reefs (p. 560)
barrier reefs (p. 560)
atolls (p. 561)

Questions for Review

1. Draw a simple diagram showing wave crests, wave troughs, wave height and amplitude, and wave length. Define wave period and fetch.

2. Briefly describe the difference between the oscillatory and translatory motion of waves.

3. Why are most wave fronts nearly parallel to the shoreline on arrival?

4. How does a longshore current develop?

5. Explain why the Earth's coasts generally experience two high tides and two low tides each day.

6. What four principal factors control the rate of coastal erosion?

7. Sketch the main components of a beach. Using arrows, indicate the source of most beach sediment and show the process by which it is most commonly removed.

8. Describe how the following depositional coastal landforms develop: spit, hook, baymouth bar, tombolo.

9. Discuss how breakwaters, groins, and jetties interrupt longshore transport and affect coastal erosion and deposition.

10. Discuss the three proposed models for the origin of barrier islands.

For Further Thought

1. Why do you suppose surfing is more popular on the West Coast than on the East Coast of North America? Speculate about the prospects for surfing on the western coast of Europe or Africa.

2. Suppose you have acquired some valuable beachfront property in Oregon. How would you protect this land from coastal erosion? (*Note:* You have unlimited financial resources.) In your plan, try to minimize the negative secondary effects that follow most cases of human interference with natural coastal systems.

3. What would be the effect on California's beaches if all of the state's dams were removed?

4. How would the eastern coast of North America change if widespread subduction resumed there?

5. How might the coasts of North America change if the Earth entered another period of worldwide glacial expansion?

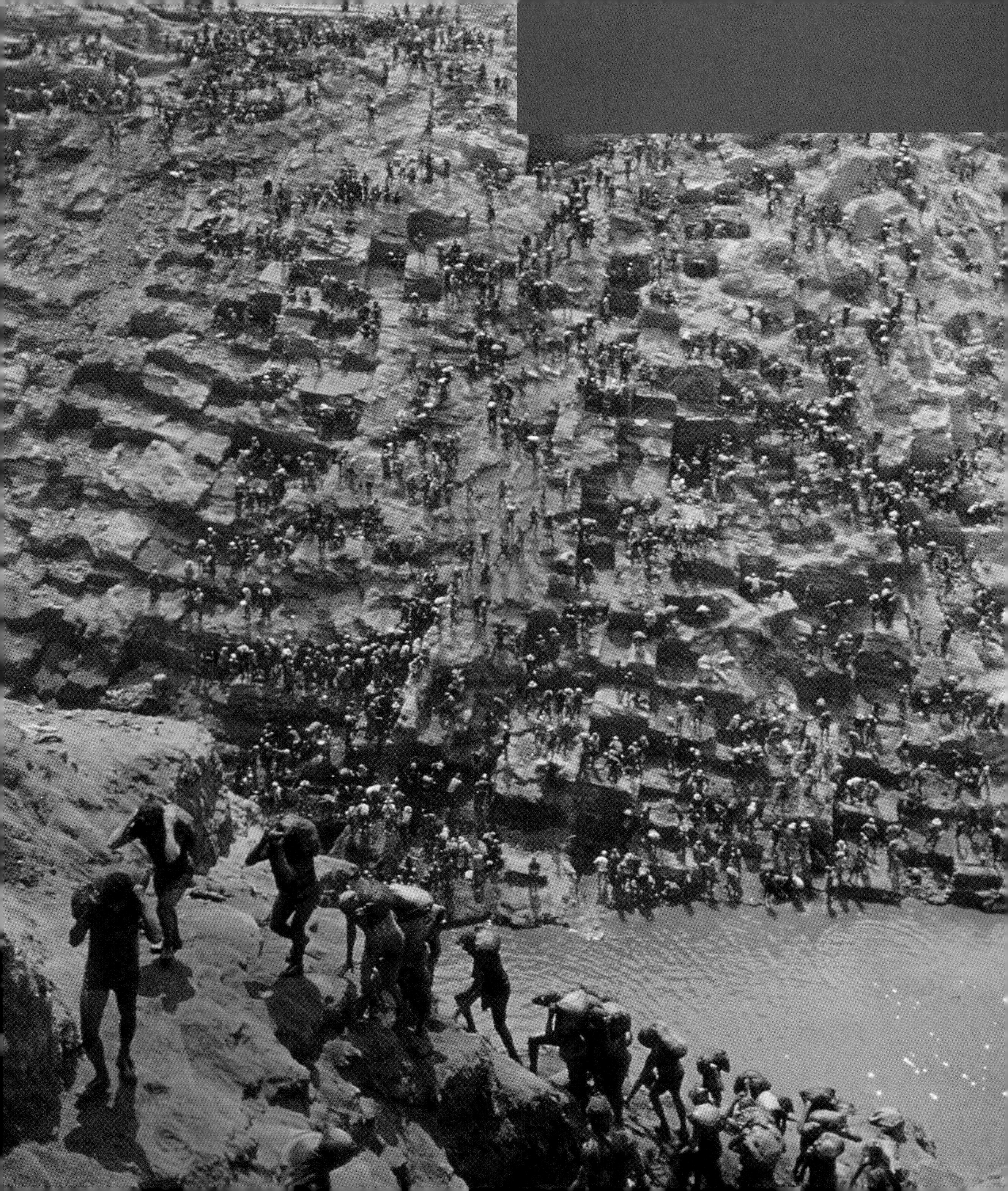

20

Human Use of the Earth's Resources

Figure 20-1 The need for the Earth's dwindling resources necessitates mining in remote areas. This gold mine was excavated out of a jungle in northeastern Brazil.

In earlier chapters, we have often described the many natural resources that are useful or essential in people's lives today, whether for industrial or personal use. Stop for a moment and think about the vast array of natural resources needed to produce and deliver a pizza to your room. They include the energy and materials to power the farms and to harvest and refine the crops that provide a pizza's ingredients; the metals and other materials to make the pizza oven and other necessary appliances; the metals and other materials to manufacture the delivery vehicle; the gasoline that powers that vehicle; the pen with which you write the check. . . the list is virtually endless.

Until recently, most people believed the Earth's resources, such as those being mined in Figure 20-1, were unlimited; today we face serious shortages of many essential materials. For example, scientists predict that the world's *recoverable* supply of crude oil, from which we get gasoline, may last only another 50 to 100 years at the current rate of use. How have we exhausted our stores so quickly? Can we compensate for these losses?

The answers to these questions depend on the growth rate of the world's population, the quantity of natural resources each individual uses, and the success or failure of the search for alternative resources. In the United States alone, each person directly or indirectly uses about 10,000 kilograms (22,000 pounds) of raw materials each year, most of which consists of stone and cement for the construction of roads and buildings, but which also includes about 500 kilograms (1100 pounds) of steel, 25 kilograms (55 pounds) of aluminum, and 200 kilograms (440 pounds) of industrial salt (mostly for cold-weather road maintenance). Each American also uses nearly 3800 liters (1000 gallons) of oil per year. Collectively, Americans consume about 30% of the world's oil. The United States, with only about 6% of the world's population, uses nearly 30% of its minerals, metals, and energy. A single American may use 30 times as much material and energy as a person in an emerging nation.

Figure 20-2 Giant dredges are used to search for gold in Siberian mines.

Natural resources are distributed unevenly among nations and continents. Valuable materials are abundant in some geological settings; elsewhere they are in short supply. Some nations, such as Canada, the United States, and Russia, possess a wealth of varied natural resources (Fig. 20-2); others, such as Japan, have few. Some *former* resource producers have virtually exhausted their domestic supplies and must now import from other nations. Great Britain, once a great mining nation that exported tin, copper, lead, and iron, must now import those commodities. Meanwhile, some smaller, developing nations have vast supplies of a few important materials. Guyana and Surinam, on the northeast coast of South America, and Jamaica, in the Caribbean, possess some of the world's richest supplies of aluminum. Nevertheless, *no nation is self-sufficient in all essential resources.*

Resource consumption worldwide is accelerating as the world population increases (now approaching 6 billion—three times the population in 1920—and expected to double by 2040), and people everywhere are striving to obtain the benefits associated with technological development. Unless we identify new supplies of depleted resources or find substitutes for them, and manage industrial development in ways that limit resource depletion, impending shortages will force people everywhere to make drastic changes in their ways of life.

Reserves and Resources

Reserves are natural resources that have already been discovered and *can be exploited for profit with existing technology and under prevailing economic conditions.* We know where reserves are, and we can extract them; most importantly, their economic value in the marketplace exceeds the cost of their extraction. **Resources** are deposits that we know or believe to exist, but that are not deemed exploitable today, whether for technological, economic, or political reasons. We can estimate the location of resources hidden beneath the surface by exploratory drilling, geophysical modeling, and extrapolation from known reserves. To illustrate the difference between reserves and resources, consider that world oil *reserves* are estimated at 700 billion barrels (a *barrel* is a volume equaling 159 liters [42 gallons]), whereas world oil *resources* are thought to total about 2 trillion barrels.

Resources may become reserves if it becomes more profitable to locate and extract them. In the 1970s, for example, when the price of gold surged to $800 per ounce, deposits that had not been considered worth developing at lower prices immediately became highly profitable ores. (An **ore** is a mineral deposit that can be mined for a profit; this word is basically an economic, not a geological, term.) Conversely, a price reduction for a material on world markets can lead some countries to import it rather than mine and develop it themselves, thereby transforming a profitable reserve into an unprofitable resource. Such a shift was illustrated by the recent decline of the U.S. steel industry, which languished when it became cheaper to import steel from South Korea than to mine the iron and other materials used to produce steel domestically.

A low-value deposit may become an ore body if economies improve, political events enhance access, or new technologies develop. If economies falter, the political climate prevents extraction of deposits, or new technologies render the resource obsolete, then a once-profitable reserve may become a relatively valueless resource.

Access to reserves is often controlled by politics: The Persian Gulf War of 1991, for example, produced a spike in oil prices as political conditions converted some Middle Eastern oil reserves to unavailable resources (Fig. 20-3). Other deposits that might be extracted profitably today are located in national parks or wilderness areas; they would be reserves if they could be exploited, but they remain resources because legislation protects them. Political events have been affecting resource development for millennia, as has changing technology. Five thousand years ago, humans made tools of stone, especially chert and obsidian; when copper, iron, and bronze came into use, and especially after smelting developed, stone for toolmaking lost value.

Some natural resources are *renewable*—that is, they are naturally replenished over relatively short time spans (such as trees) or available continuously (such as sunlight). *Nonrenewable* natural resources form so slowly that they are typically consumed much more quickly than nature can replenish them; they include fossil fuels such as coal, oil, and natural gas, and metals such as iron, aluminum, gold, silver, and copper. Nonrenewable materials are analogous to crops that can be harvested only once. We cannot "grow" another crop of nonrenewable resources, although many, such as copper and aluminum, can be recycled for reuse. Some resources may be either renewable or nonrenewable, depending on how we use them. Soil, for example, is a renewable resource if we follow sound agricultural practices. Soil becomes nonrenewable when it is depleted of its nutrients by overplanting, or when it is allowed to blow away after being overgrazed, overplanted, or deforested.

To maintain our standard of living and simultaneously protect the global environment into the twenty-first century, we must all become more knowledgeable about the Earth's energy and mineral resources. To use dwindling resources wisely, we need to understand how they form, where they are abundant, what environmental consequences are associated with their use, and how long the known supplies are likely to last. This chapter addresses these and other crucial issues in regard to fossil fuels, alternative energy sources, and both metallic and nonmetallic minerals.

Figure 20-3 During the 1991 Persian Gulf War, the Iraqi military ignited oil in thousands of Kuwaiti wells. Political conditions in certain oil-producing nations have prompted oil-poor industrial nations to seek alternative energy sources.

Fossil Fuels

Worldwide, we get relatively little of our energy from renewable sources such as solar energy, the energy produced by the rise and fall of tides, wind power, and power from streams. Instead, we obtain most of our energy from nonrenewable **fossil fuels,** derived from the organic remains of past life. The principal fossil fuels are oil, natural gas, and coal. Oil and natural gas have now passed their peak production period in the United States. Extraction of the world's reserves increases each year as demand rises, and at the current rate of use *and price* these reserves will be virtually exhausted within the next century or so. Nevertheless, for economic and technological reasons, the nations of the world continue to draw nearly 95% of their total energy from a dwindling supply of fossil fuels that are, in practical terms, nonrenewable (at least on a human time scale).

In this section we describe how petroleum, oil shale and oil sand, coal, and peat are used for fuel and how they form. We also discuss environmental problems caused by fossil fuels.

Petroleum

The most common and versatile fossil fuel is **petroleum,** a group of gaseous, liquid, and semi-solid substances composed chiefly of **hydrocarbons,** molecules consisting entirely of hydrogen and carbon. Typically pumped from the ground in the form of dark, viscous crude oil, petroleum is refined to produce propane for camp stoves, motor oil and gasoline for cars, tar and asphalt for roads, and the natural gas and heating oil that warm our homes and workplaces. It is also the major ingredient in plastics, synthetic fibers, dyes, cosmetics, explosives, certain medicines, certain fertilizers, and videos, tapes, and compact discs.

Humans have used petroleum for thousands of years. Approximately 4500 years ago, Babylonians collected crude oil bubbling from natural pools to make glue for attaching metal projectile points to spears. In ancient Iraq, oil seeping from rocks in the valleys of the Tigris and Euphrates Rivers was used in mortar for setting bricks, in grout that set tiles in ancient mosaics, and in waterproofing materials for boats.

Modern use of petroleum began in 1816, when gas extracted from coal was first used in Baltimore's gaslights. Combustible gas was discovered in 1821, when a water well in Fredonia, New York, was accidentally ignited, producing a spectacular flame. Wooden pipes were installed to carry the gas to 66 gaslights in downtown Fredonia. Commercial use of oil began in 1847, when a Pittsburgh merchant bottled and sold natural "rock oil" as a lubricant for machines in the home and workplace. In 1852, Canadian chemists, using oil from what is now Oil Springs, Ontario, first refined kerosene from rock oil for use in home lamps, which soon eliminated much of the candle and whale-oil industries. The oil-well industry was born on Sunday, August 27, 1859, when Edwin Drake of Titusville, Pennsylvania, pumped oil from the first true oil well (Fig. 20-4).

DRAKE

Figure 20-4 The first commercial oil well was developed in 1859 in Titusville, Pennsylvania, by Edwin Drake (at right). Drake's well, which was 21.2 meters (70 feet) deep, yielded 35 barrels of oil per day.

The Origin of Petroleum Petroleum generally begins to form in marine basins in tropical environments, where there is a rich diversity of microscopic plants and animals (Fig. 20-5). After the organisms die, they start to decay by oxidation; eventually the oxygen in bottom waters becomes depleted, however, and decay ceases. Layers of sediment and additional organic material may bury the organic remains, preventing their subsequent decay. As sediments accumulate, pressure and heat convert the organic molecules to a substance called **kerogen,** a solid, waxy organic material. Kerogen becomes converted to various liquid and gaseous hydrocarbons at temperatures between 50° and 100°C (122°–212°F) and at a depth of 2 to 10 kilometers (1.2–6 miles).

At the start of this process, kerogen's large, complex organic molecules form highly viscous hydrocarbons such as tar. With increasing heat, these molecules break down to form smaller, simpler, less viscous ones, such as those found in diesel oil, kerosene, and gasoline. At temperatures exceeding 100°C (212°F), liquid petroleum becomes converted into a variety of natural gases, ranging from those with relatively complex molecules, such as butane, to the simplest, lightest natural gases—propane, ethane, and methane. At temperatures of about 200°C (400°F) and at depths of 7 kilometers (4 miles) or more, methane, the lightest gas, breaks down completely, and the rocks no longer contain hydrocarbons. Thus, with respect to geothermal temperatures and pressures, a limited "window of opportunity" exists for the conversion of organic remains to hydrocarbon fuels. Otherwise we

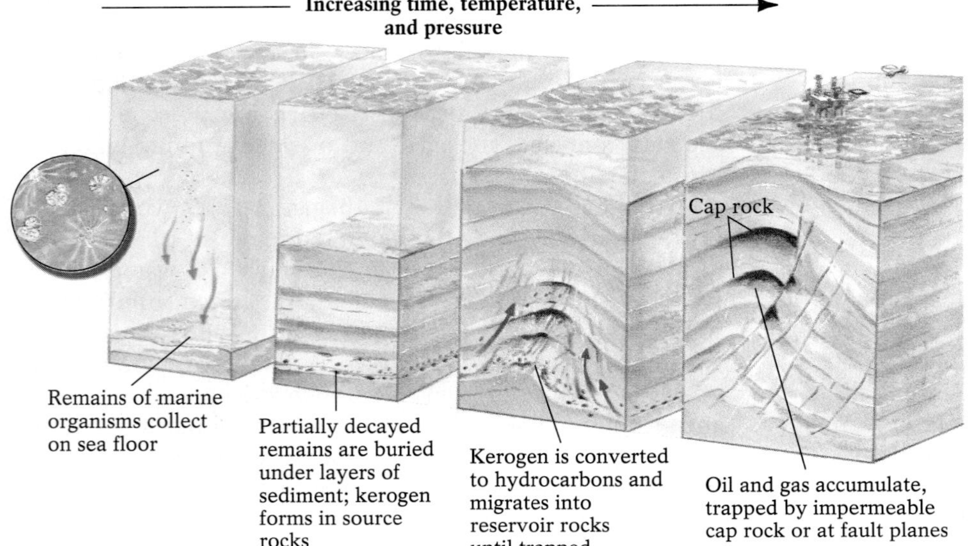
Increasing time, temperature, and pressure
Cap rock
Remains of marine organisms collect on sea floor
Partially decayed remains are buried under layers of sediment; kerogen forms in source rocks
Kerogen is converted to hydrocarbons and migrates into reservoir rocks until trapped
Oil and gas accumulate, trapped by impermeable cap rock or at fault planes

Figure 20-5 Petroleum begins as a large accumulation of partially decayed microorganisms in marine mud. The remains of the organisms are eventually converted to kerogen as the marine mud lithifies to become source rocks. Geothermal heat "cooks" the kerogen, which becomes petroleum as its organic molecules break down into a variety of hydrocarbons. Petroleum is expelled from its source rocks and migrates through permeable reservoir rocks, accumulating when further movement becomes blocked by some structural or stratigraphic feature, such as an impermeable cap rock. Erosion and faulting may disrupt oil traps, enabling oil to escape to the surface (see Highlight 9-1).

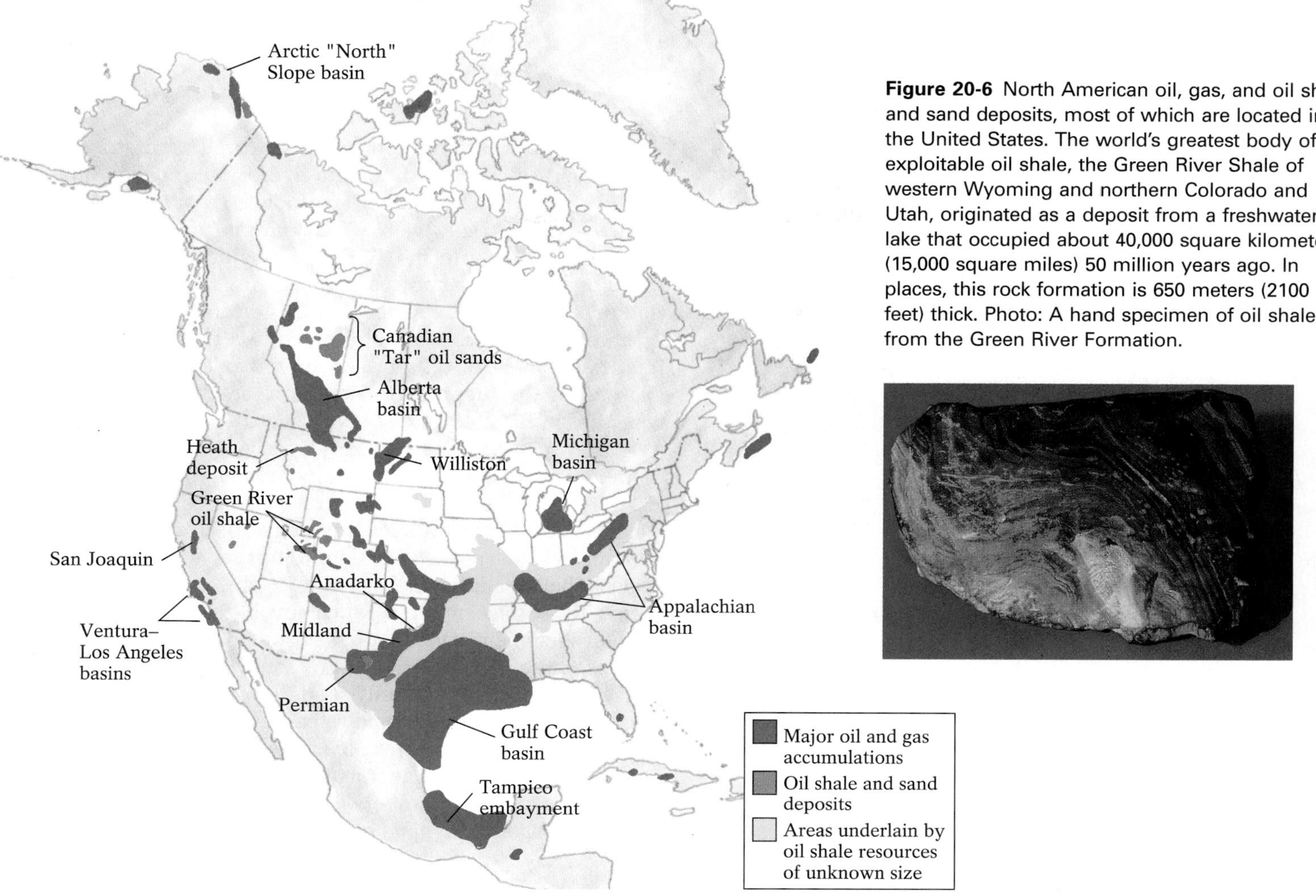

Figure 20-6 North American oil, gas, and oil shale and sand deposits, most of which are located in the United States. The world's greatest body of exploitable oil shale, the Green River Shale of western Wyoming and northern Colorado and Utah, originated as a deposit from a freshwater lake that occupied about 40,000 square kilometers (15,000 square miles) 50 million years ago. In places, this rock formation is 650 meters (2100 feet) thick. Photo: A hand specimen of oil shale from the Green River Formation.

would probably have a great deal more oil and gasoline would be cheaper.

The rocks in which hydrocarbons form are called **source rocks;** they are typically shales and siltstones lithified under reducing (oxygen-poor) conditions from fine-grained, organic-rich muds. Oil and gas are rarely found in source rocks, because most liquid and gaseous hydrocarbons are readily expelled from their compacting source muds. They tend to migrate upward into adjacent permeable **reservoir rocks,** such as well-sorted sandstones and highly fractured or porous limestones; they continue to migrate upward until they are trapped by an impermeable cap rock.

Geological activity can both create and destroy oil traps. Uplift and erosion can remove the trapping cap rock, allowing oil or gas to escape at the surface. This type of event occurred at the La Brea Tar Pits along Los Angeles' Wilshire Boulevard. A new fault, or the extension of an old one, can breach an oil trap and allow its oil to seep out. As a result of these processes, much of the oil and gas formed before about 65 million years ago has long since escaped from its traps and evaporated at the surface. More than 60% of today's oil-producing wells appear in rocks that are less than 65 million years old.

Geologists estimate that less than 0.1% of all marine organic matter buried at the sea floor eventually is trapped as usable petroleum. Some settings may lack adequate heat to convert organic matter to kerogen and petroleum. In others, deposits may have experienced sufficiently high temperatures but not at great enough depth, enabling shallow-forming hydrocarbons to escape without becoming trapped. The conditions required to produce, trap, and retain hydrocarbons are rarely observed together, which explains why most marine rocks are petroleum-free. We do not know how long it takes for oil and gas to form. No known petroleum sources are less than 1 to 2 million years old, so the process must take at least that much time.

Oil Shale and Oil Sand **Oil shale,** a black-to-brown clastic sedimentary rock consisting of a mixture of waxy kerogen and fine mineral grains, is a common source rock. Because it was never buried deeply enough to raise its temperature to the level required to convert kerogen to oil, however, oil shale has retained its hydrocarbons. Based on current and anticipated increased rates of usage, the estimated 2 to 5 trillion barrels of shale oil underlying the United States (which possesses about two-thirds of the world's oil shale) could supply America's petroleum needs—if we could extract it profitably and safely—for more than 500 years (Fig. 20-6). According to conservative estimates, this country's largely untapped oil shales contain more than ten times the known oil reserves

Figure 20-7 Cliffs of tar sand along the Athabasca River, Alberta, Canada.

of the Middle East. Economics, technology, and our desire to protect the environment currently keep us from using this oil to heat our homes and fuel our cars. Current world prices for crude oil are not high enough to justify the cost of developing this resource.

To develop oil shale, it must be mined, its trapped kerogen removed from the rock and then processed, and the waste rock disposed of in an environmentally sound fashion. After mining, the shale must be crushed (which increases the volume of the rock by about 30%) and then heated to more than 500°C (930°F) to vaporize its kerogen, which then condenses as crude oil. Like popcorn, which occupies more space than the kernels from which it has popped, fragments of oil shale expand in volume by as much as 20% when heated. Thus waste rock, whose volume is approximately 50% greater than that of the original rock after crushing and heating, would overflow the hole from which it was mined if not removed and disposed of responsibly. If the United States were to obtain its entire oil supply from oil shale, it would need to dispose of 13 billion tons of waste rock each year. This is enough waste rock to fill 130 million freight-train cars. In addition, enormous amounts of water would be required to extract and refine this resource—perhaps more water than the arid region in which it is found can spare.

Someday technology may solve this problem; microwaves or radiowaves could perhaps heat kerogen in the shale without removing and crushing it. Given the cost, however, oil shale deposits will be exploited only after we exhaust more accessible, less expensive sources or oil prices rise so steeply that the potential profits outweigh the costs of exploitation.

Oil sand is a mixture of unconsolidated sand and clay that contains a semi-solid, tar-like hydrocarbon called **bitumen.** Bitumen can be refined to produce gasoline, fuel oil, or other commercially viable hydrocarbon products, but it adheres so strongly to the mineral grains in oil sand that it does not flow and cannot be pumped from the ground. Oil sand must therefore be mined and then heated with hot water and steam to release the bitumen in a more fluid state. The development of oil sand, or any other fossil fuel, is viable only if the fuel can provide more energy than is consumed in mining and processing it.

The world's largest oil sand deposit, located about 400 kilometers (250 miles) north of Edmonton in Alberta, Canada (Fig. 20-7), has been producing crude oil since 1967. Today it yields nearly 200,000 barrels per day, approximately 16% of Canada's oil needs. This deposit is commercially exploitable because it lies close enough to the surface to be extracted by surface mining, which involves the removal of overlying sediment layers—a less costly process than underground mining. Surface mining of oil sand, however, produces enormous piles of residual oily sand that can contaminate local surface water and groundwater. To avoid this environmental hazard, Canadian petroleum engineers are developing a process to extract bitumen by injecting hot water or steam directly into underground deposits, thereby softening the bitumen so it can be pumped out.

Coal and Peat

Coal, the Earth's most abundant fossil fuel, is a combustible chemical sedimentary rock that forms from the highly compressed remains of land plants (see Chapter 6). It contains the energy stored in living plants via photosynthesis, the process through which sunlight, water, and carbon dioxide produce the materials for plant growth. When coal burns, it releases the energy that was stored in plants millions of years ago.

Native Americans used coal thousands of years ago to fire pottery; a thousand years ago, Europeans mined it to heat their homes and fuel the fires of smelting industries. Cheap, plentiful coal powered the newly invented steam engine in the Industrial Revolution of nineteenth-century Europe and North America. By 1900, coal supplied 90% of U.S. energy needs for industry and domestic heating.

Coal use has declined over the last 40 years, although it still provides about 23% of U.S. energy needs and as much as 55% of the electricity generated in the nation. Coal's relative decline occurred largely because of increased production of oil and natural gas, which have been abundant during the mid- to late twentieth century and could be extracted economically with fewer problems. In contrast, coal can be difficult and sometimes dangerous to mine, and it is costly to process. Burning coal also pollutes the air more than burning oil or gas. As its replacements become depleted, however, coal will likely rebound as our principal fossil fuel, especially in electricity-producing power plants. New technology makes

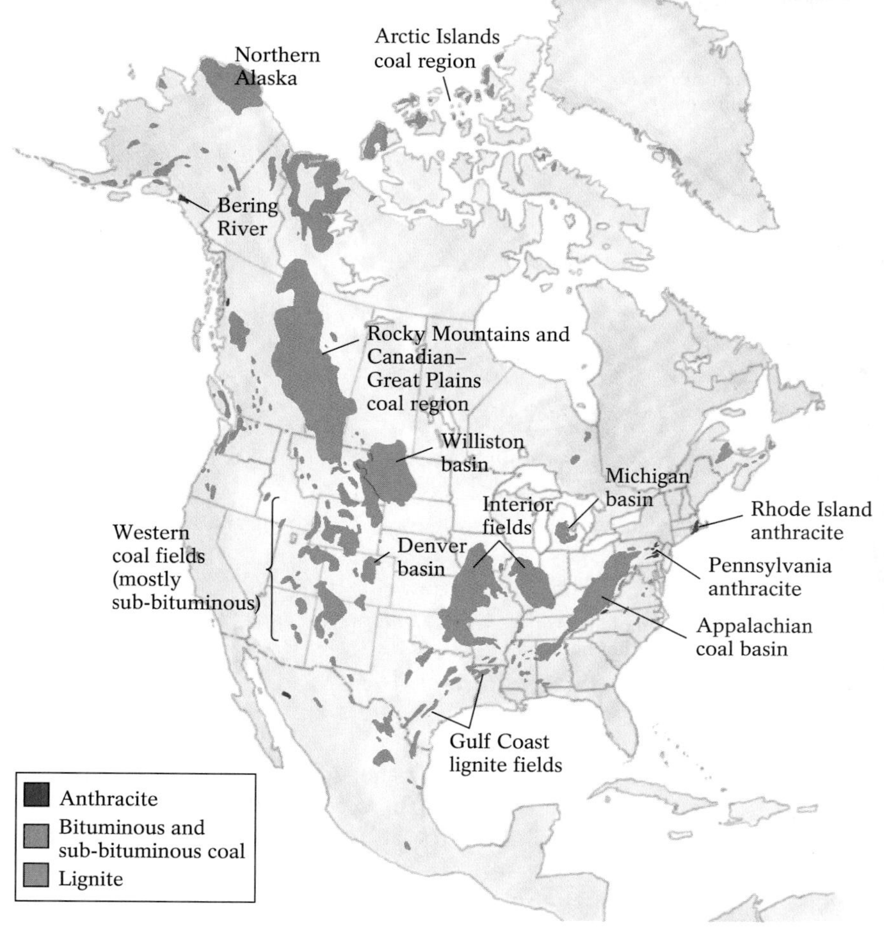

Figure 20-8 Major coal deposits of North America. Photo: A coal seam in Alaska.

it economically feasible to convert coal to liquid and gaseous fuels; in these forms, existing world resources could meet energy needs for hundreds of years.

The United States has more than 30% of the world's currently accessible coal. What may be the world's largest coal field has recently been discovered in Antarctica, stretching for hundreds of kilometers along the eastern side of the Transantarctic Mountains. If ever mined (the region has been declared off-limits for resource development through international treaty), this coal would add substantially to the world's coal inventories.

The Origin of Peat and Coal The luxuriant plant growth that produced the world's coal deposits likely occurred in tropical or semitropical swamps that eventually were covered either by the growth of more vegetation or by overlying sediments. Once covered, the plant remains could not completely decompose by oxidation. The cumulative weight of overlying deposits squeezed water from the porous mass of incompletely decomposed vegetation. As plant remains became more deeply buried, increasing pressure, geothermal heat, and bacterial reactions removed both water and the organic gases, such as CO_2 and CH_4 (methane), produced by oxidation and bacterial activity.

These processes created a variety of fossil fuels that can be categorized according to their carbon and water content (see Figure 6-25). Low-pressure conditions produce **peat**, the first fuel to form from buried vegetation. A soft brown mass of compressed, largely undecomposed, and still recognizable plant structures, peat contains substantial amounts of water, organic acids, and incorporated hydrogen, nitrogen, and oxygen; it is only about 50% carbon. Peat is dried and then burned as fuel for home heating and other domestic uses in rural Ireland, England, and elsewhere in Europe.

Bacterial action continues in peat, eventually breaking it down to form a vegetative, kerogen-rich muck that becomes compressed as **lignite.** This soft, brown coal consists of about 70% carbon (plus 20% water and 10% oxygen); the higher carbon content makes it a more concentrated heat source than peat. Deep burial of lignite and the accompanying rise in pressure and geothermal heat convert it to lustrous black **bituminous coal** having a carbon content of 80% to 93%; this material produces more heat, with much less smoke, than peat or lignite.

Metamorphic conditions gradually transform lignite and bituminous coal deposits into **anthracite,** a hard, jet-black coal containing 93% to 98% carbon. Anthracite burns with an extremely hot flame and very little smoke. Unlike lignite and bituminous coal, which occur widely, anthracite exists only in low-grade metamorphic zones in mountainous regions, notably the Appalachians of northeastern Pennsylvania (Fig. 20-8). Anthracite seams commonly occur within the steeply dipping limbs of folded sedimentary rocks, where most coal cannot be strip-mined; anthracite requires more difficult, dangerous, and costly deep underground coal-mining processes to extract it.

Peat, lignite, and bituminous coal typically occur as distinct seams or beds in sequences of detrital sedimentary rocks,

Sulfur dioxide
and nitrogen oxides
produced
Pollutants carried
into atmosphere
Reaction with water vapor
in atmosphere produces
H_2SO_4, HNO_3
Pollutants deposited
nearby
• Corrosion of buildings,
stones, etc.
• Acidification of lakes,
rivers, and other bodies
of water
Acid rain/snow
• Forests damaged
• Fish in lakes killed
• Vital nutrients
leached from soils

Figure 20-9 Acid rain has two major causes: (1) the release of sulfur dioxide into the atmosphere from burning of sulfur-rich coal, which combines with water to form sulfuric acid; and (2) the emission of nitrogen oxides mostly in automobile exhaust, which combine with water to produce nitric acid. Once these pollutants enter the atmosphere, their effects can spread to nearby and even distant environments.

particularly those that accumulated in warm, moist, coastal environments. A series of alternating detrital sediments and coal beds typically signals cycles of rising and falling sea level and their corresponding periods of land submergence (accompanied by detrital sedimentation) and land emergence (with its ensuing swamp development) (see Chapter 6).

Much of North America's vast coal deposits formed during ancient episodes of high worldwide sea level, which shallow seas invaded the continent's interior. The coal of Utah, Montana, and the Dakotas, for example, formed at the swampy margins of such inland seas. Coal deposits have also formed on the vegetated floodplains of intermountain rivers. The coal found in Wyoming's Powder River basin occurs as layers between floodplain sediments. Coal deposits may also accumulate at continental margins, where fluctuating sea levels flood wide continental shelves periodically. The coal of the Pennsylvania and West Virginia Appalachians formed in such an environment.

Peat and coal have probably been accumulating continuously since land plants first appeared on Earth about 425 million years ago. During certain periods, warmer-than-average climatic conditions and a high proportion of tropical or semitropical landmasses promoted more extensive coal formation. During the Mississippian, Pennsylvanian, and Permian Periods of the Paleozoic Era (360–245 million years ago), giant ferns and scale trees (gymnosperms) in great swamps and forests covered the tropical equatorial lowlands of the supercontinent Pangaea. During this Carboniferous Period, the great bituminous coal beds of western Europe (Great Britain, Germany, and Poland) and the anthracites and bituminous deposits of eastern North America's Appalachians were laid down. A second great period of coal formation occurred about 135 to 30 million years ago, producing lignites and bituminous coals from flowering plants (angiosperms) similar to those surviving today. Some of these deposits contain well-preserved fossils that identify the types of plants from which they formed. These coals extend from New Mexico through Colorado to the Great Plains of North Dakota and Saskatchewan.

Coal is also being formed today: In the Great Dismal Swamp of coastal Virginia and North Carolina, a 2-meter (7-foot)-thick layer of vegetation has accumulated over a 5000 kilometer^2 (2000 mile^2) area during the last 5000 years. The next major rise in sea level that buries these organic deposits with detrital sediment will halt their further decomposition, and eventually provide the pressure and heat that will convert them to peat and coal.

Fossil Fuels and the Environment

Widespread use of coal, oil shale, and other fossil fuels can produce various environmentally damaging side effects. Concerns about acid rain, global warming, and massive marine oil spills have led to increased public awareness and recent legislation to reduce or prevent such effects.

Acid Rain Burning fossil fuels releases secondary materials into the atmosphere, reducing the quality of the air we breathe. When coal is burned, for example, coal ash—fragments of noncombustible silicates and toxic metals—enters the atmosphere. Burning sulfur- and nitrogen-rich hydrocarbons releases sulfur and nitrogen oxides, which then combine with water in the atmosphere to form sulfuric acid (H_2SO_4) and nitric acid (HNO_3), the principal components of **acid rain** (Fig. 20-9). Most scientists hold acid rain, along with acid mine water, responsible for damaging forests and crops, killing aquatic life in lakes, and accelerating the destructive weathering of human-made structures. The most pronounced damage from acid rain occurs downwind of major coal-burning industrial regions.

In some settings, environmental damage from acid rain is lessened by the area's geology. Calcium-rich soils or exposed carbonate bedrock (limestone, dolostone) react with the acids by the chemical-weathering process of carbonation (see Chapter 5) and neutralize them before they can do much harm. Where thin, calcium-poor soils overlie granitic bedrock, as in the lake and forest country of the Canadian Shield area of the northern Great Lakes and adjacent southeastern Canada, acid rain damage to the ecosystem may be more severe. Eastern Canada receives acid rain produced by the coal-burning smokestacks of the heavy steel and automotive industries of the American Midwest and from the Midwest's coal-burning power plants. In addition to installing scrubbing devices that remove much of the sulfur and nitrogen from emissions, attempts to neutralize this sort of acid rain may also include low-flying aircraft that spread limestone dust across the landscape.

Global Warming Burning all types of fossil fuels increases the volume of carbon dioxide (CO_2) in the atmosphere. Measurements made from 1958 to 1984 at the top of Hawai'i's Mauna Loa, far from industrial pollution sources and large population centers, showed a 9% increase in global atmospheric CO_2. As discussed in Chapters 5 and 17, atmospheric CO_2 absorbs and traps heat from the Earth's surface, creating the greenhouse effect and promoting global warming. Some climatologists predict that if atmospheric CO_2 levels continue to rise at their present rate, atmospheric temperatures could rise 1.5° to 4.5°C (2.5°–8°F) by the middle of the twenty-first century. The increased temperatures could cause droughts in some regions, increase desertification in others (see Chapter 18), and significantly reduce the agricultural potential of today's marginally cultivatable lands. It could also lead to melting of a substantial portion of the Earth's ice masses; along with thermal expansion of warming seawater, this melting would raise sea levels worldwide, perhaps by 1 to 2 meters (3.3–6.6 feet) per 100 years, for the next 200 to 500 years. A rise of 2 to 10 meters (6.6–33 feet) would accelerate coastal erosion and submerge a substantial amount of coastal land, including major port cities on every continent.

Marine Oil Spills Occasionally, an oil tanker becomes disabled and leaks its cargo of crude oil into the sea. Extraordinary environmental damage was caused by one such marine oil spill in March 1989, when the supertanker *Exxon Valdez* went aground in Alaska's Prince William Sound. More than 10.2 million gallons of crude oil flowed from its cracked hull into the sea, much of which washed up onshore in layers tens of centimeters thick. The oil killed thousands of birds and marine mammals (Fig. 20-10), halted herring and salmon fishing during peak season, and coated hundreds of kilometers of Alaska's coastline.

Oil spills in calm seas can usually be contained successfully within floating barriers placed around the oil, allowing workers to skim the oil from the surface. In Alaska, unfortunately, inclement weather caused rough seas that breached the floating barriers; the Exxon Company recovered only about 500,000 gallons of its oil. Efforts to ignite and burn the spilled oil failed as well, although some of the lightest hydrocarbons evaporated and some were consumed by oil-eating bacteria. Millions of gallons washed onshore, necessitating an enormous cleanup operation. Standing pools of oil were soaked up with peat moss, wood shavings, and even chicken feathers. High-powered steam hoses removed some of the oil coating; workers even scrubbed rocks individually by hand. After several months and some $4 billion in cleanup costs, more than 85% of the spilled oil was gone. The remainder consisted of thick asphalt clumps that could not be scrubbed, did not evaporate in sunlight, and would not decompose by bacterial action. These clumps fouled breeding grounds and other wildlife habitats on land, as well as major fishing grounds.

The *Exxon Valdez* disaster dramatized the need for improved methods to respond to such events, sparked renewed congressional demands for double-hulled, spill-resistant tankers, and prompted research that may someday yield a new strain of microbes that voraciously consume spilled petroleum.

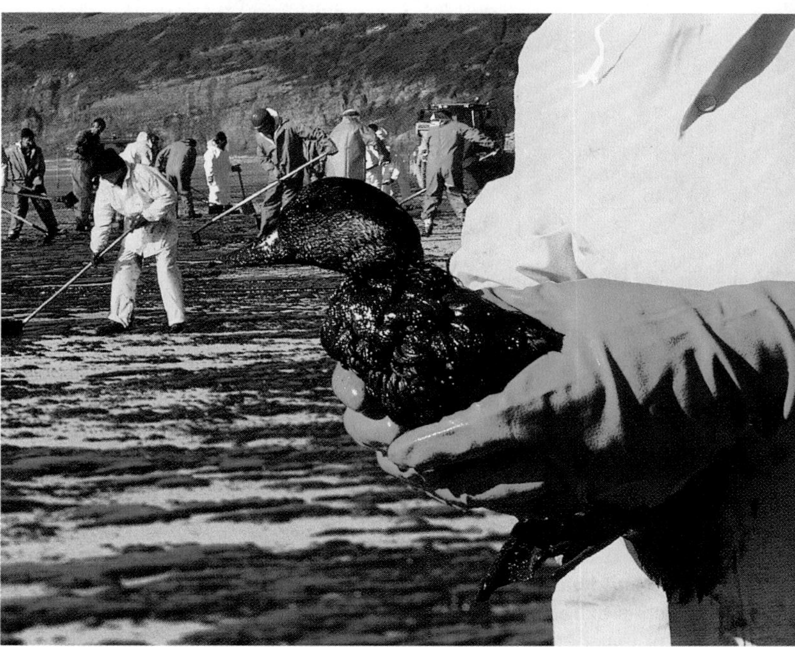

Figure 20-10 One problem associated with oil exploration and development is the potential for oil spills, which can cause large-scale environmental damage to oceans and coastal habitats. The negative impact on marine life and habitats can be profound, as dramatized by the oil-covered wildlife found following the 1989 breakup of the supertanker *Exxon Valdez* in Alaska's Prince William Sound; the effects of this spill are still being felt today.

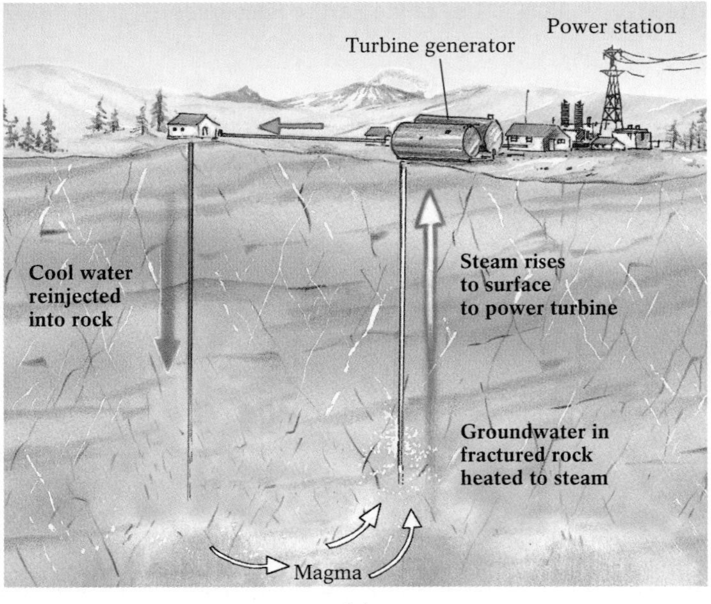

(a)

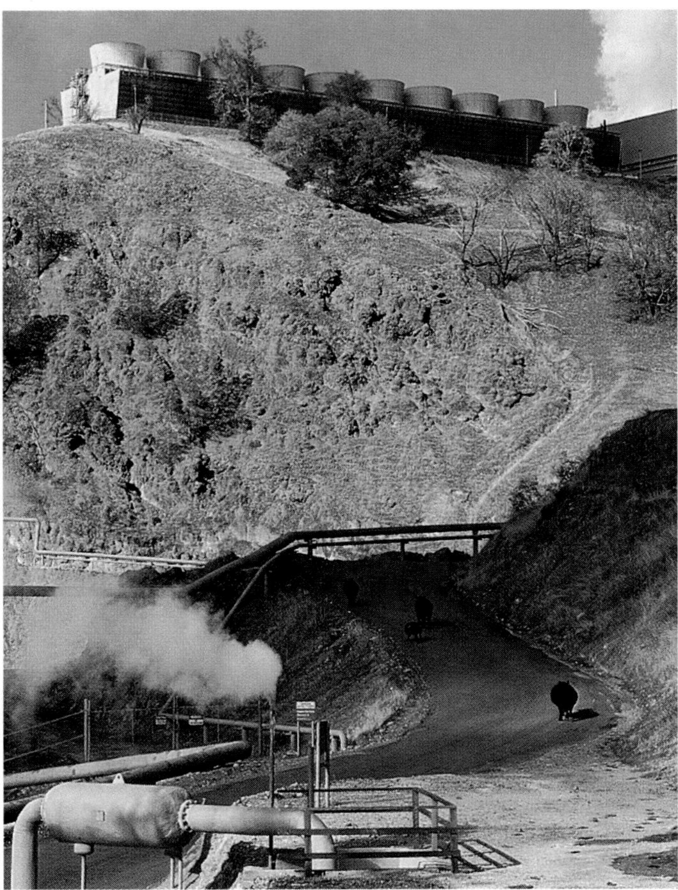

(b)

Figure 20-11 Geothermal heat provides nearly nonpolluting energy. (Some SO_2 and H_2S can be released from vented steam.) **(a)** Groundwater heated by shallow magma is converted to steam, which is extracted to drive turbines, generating electricity. Cooled water is generally reinjected into the system to keep the cycle going. **(b)** The Geysers geothermal plant, 140 kilometers (90 miles) north of San Francisco in Sonoma County, is the world's largest geothermal operation. It uses heat from subterranean rocks warmed by recent volcanic activity to supply the energy needs of 500,000 homes in the Bay Area.

Alternative Energy Resources

As fossil fuel reserves dwindle and environmental damage related to their use increases, governments and industries are seeking alternative ways to meet the growing energy needs. Some alternative energy sources already in use are renewable, such as solar and wind power; others rely on nonrenewable natural resources, such as uranium, the substance that powers nuclear energy.

Renewable Alternative Energy Sources

Renewable alternative energy resources are those that can be used virtually without depletion or that are replenished over a relatively short time span. They include geothermal, hydroelectric, tidal, solar, and wind energy and the energy produced by burning such renewable organic materials as trees and agricultural waste.

Geothermal Energy Reykjavik, the capital of Iceland, is relatively pollution-free, because it has a clean, inexpensive source of energy: Heat from shallow hot rock and magma beneath the surface is used to convert groundwater to hot water and steam. The hot water is circulated through pipes and radiators to heat homes and municipal buildings; the steam drives electrical generators (Fig. 20-11a).

Since such *geothermal heat* was first tapped as an energy source in Larderello, Italy, in 1904, approximately 20 countries have taken advantage of it, including the United States (Fig. 20-11b). More nations would use this relatively inexpensive, nonpolluting energy if they could. Unlike oil, coal, and natural gas, however, geothermal energy cannot be exported and must be used close to its sources. Every nation using geothermal energy is located on a currently or recently active plate margin or near an intraplate hot spot, where magmatic heat has not yet dissipated.

Some areas, such as the American Southwest, have significant subterranean heat but little water to transfer it to the surface. Some of these localities are experimenting with "dry-rock" geothermal projects. Explosives fracture warm, shallow rocks to increase their permeability; then imported water is injected into the fractures, creating hot water or steam (Fig. 20-12). In the Jemez Mountains of north-central New Mexico, water is injected to a depth of 10 kilometers (6 miles) into young volcanic rocks whose temperature is about 200°C (400°F); the resulting steam drives turbines and produces electricity.

Keep in mind that geothermal energy remains a renewable resource only as long as the fluids are returned to the system. If steam is vented to the atmosphere or the hot water dispersed to area streams and lakes (a source of "thermal" pollution), the resource may be depleted, especially in areas where an arid local climate limits groundwater supplies.

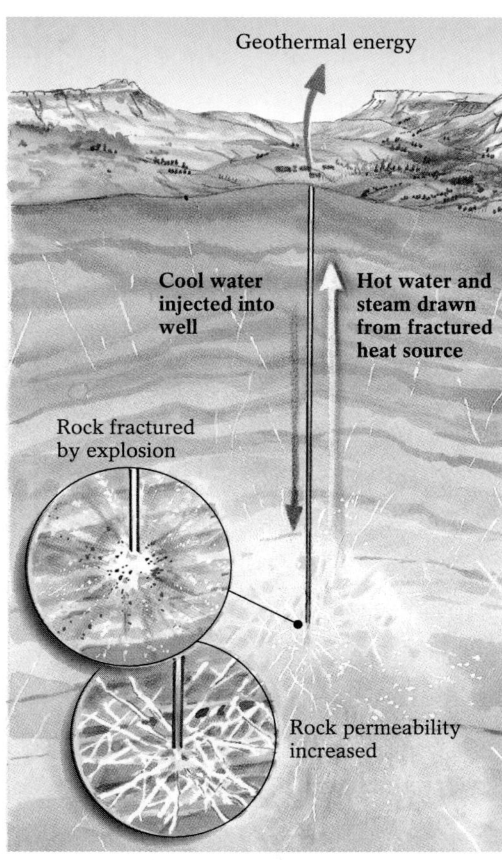
Geothermal energy
Cool water injected into well
Hot water and steam drawn from fractured heat source
Rock fractured by explosion
Rock permeability increased

Figure 20-12 Geothermal energy can be harnessed even in arid regions by injecting imported water into warm rocks fractured by explosives.

Hydroelectric Energy Falling water has been used for centuries as an energy source to mill flour, saw logs, and power numerous machines. Today, hydroelectric facilities use falling water to produce electricity. To generate hydroelectric power, a high-discharge stream is impounded by a dam to create a vertical drop sufficient to rotate the blades of large turbines.

Hydroelectric power is widely available; since 1983, nearly one-third of all new electricity-generating plants in the United States have been hydroelectric installations. At present, however, the United States generates only about 15% of its current electricity output in this way. The Federal Power Commission estimates that if every sizable river in the United States were used, hydroelectric power could supply 50% of our total electricity needs. Global hydroelectric development lags even further; only about 6% of the world's hydroelectric potential is being used. In South America and Africa, where this source has the greatest potential, only 1% has been developed. Canada, at the other extreme, gets 75% of its electricity from this clean resource.

Although hydroelectric power is nonpolluting, dams can disrupt the local ecological balance by altering or destroying wildlife habitats. In addition, they impede natural erosion processes; their reservoirs eventually fill with sediment that would otherwise replenish coastal beaches. Decisions to build dams must balance their environmental costs against their energy yield.

Tidal Power In coastal areas with a high tidal range—the difference in the water surface level between high and low tide—we can harness energy from rising and falling water levels by building a dam across a narrow bay or inlet. During rising tides, the dam's gates remain open, allowing water to enter the bay. The flow of this water can be channeled through the system's turbines to generate electricity. When the water in the bay reaches its maximum height, the dam's gates close, trapping the water. The elevated water is then channeled seaward through the same turbines, producing more renewable, pollution-free energy (Fig. 20-13). The world's largest tide-powered plant, the Rance River project

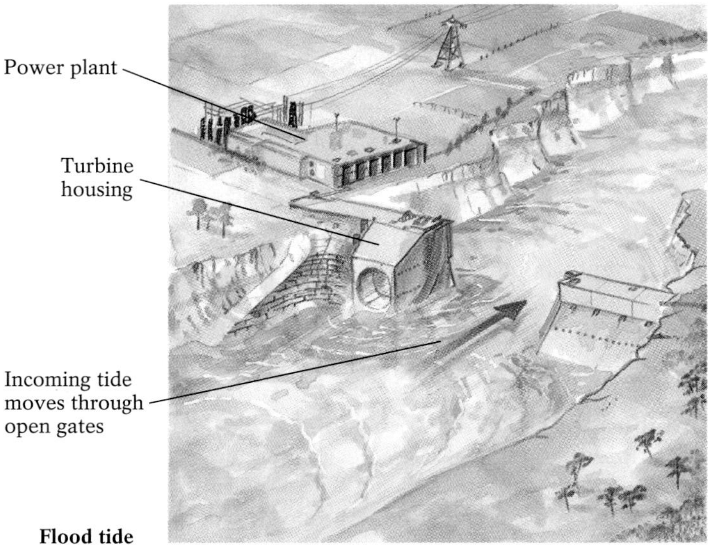
Power plant
Turbine housing
Incoming tide moves through open gates
Flood tide

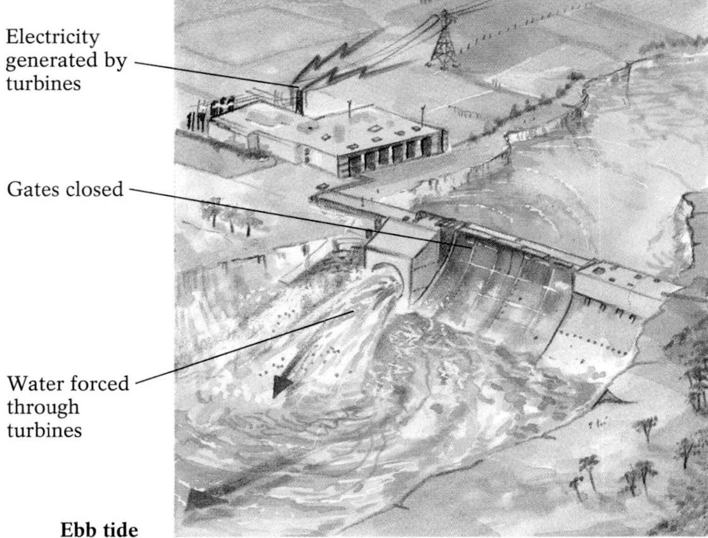
Electricity generated by turbines
Gates closed
Water forced through turbines
Ebb tide

Figure 20-13 Producing electricity through tidal power. Here, tidal power taps the energy of falling water by trapping water at high, or flood, tide and then releasing it seaward through electricity-producing turbines at low, or ebb, tide.

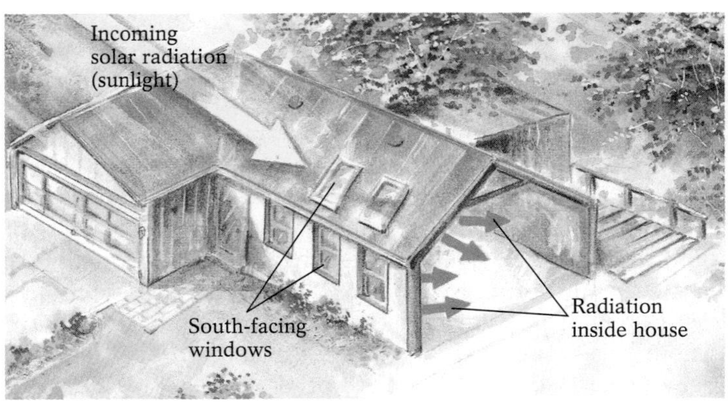

(a)

(b)

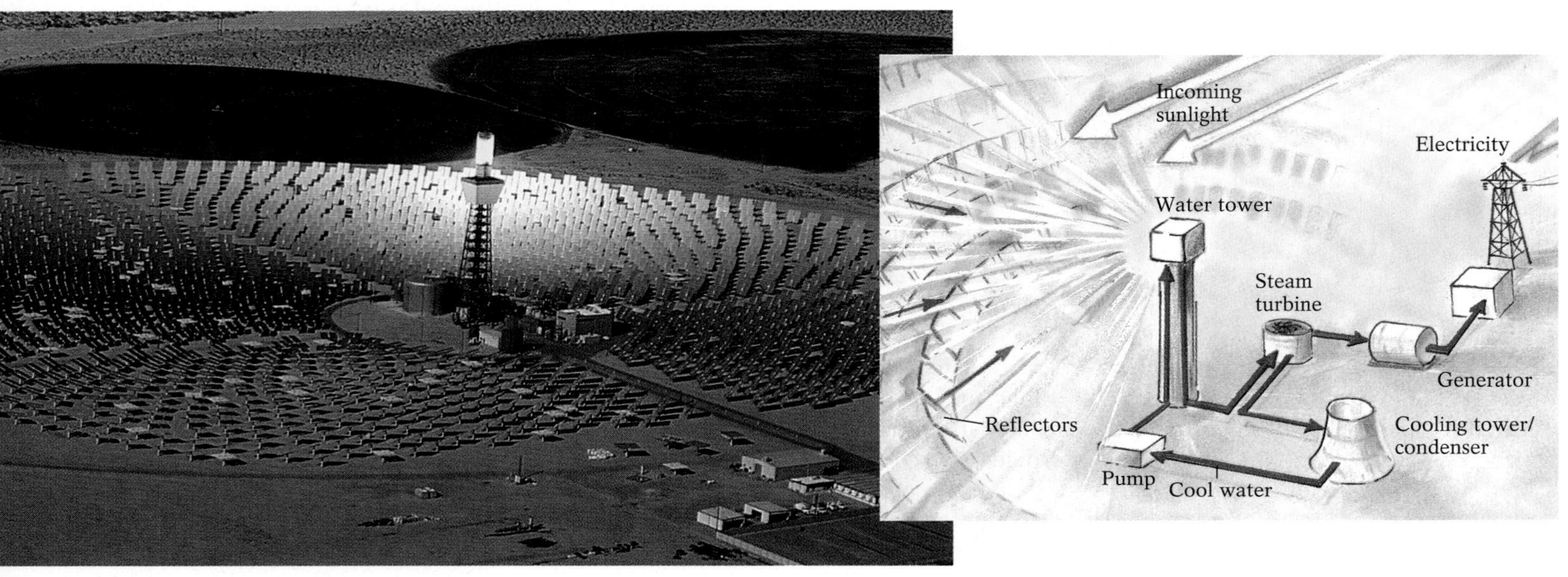

(c)

Figure 20-14 Solar power can be used in several ways. **(a)** Solar energy passively heats a single dwelling. **(b)** Water-filled panels on rooftops provide hot water and space heating—a type of active solar heating. **(c)** This experimental facility near sunny Daggett, California, uses solar energy to generate electricity. Reflecting mirrors focus on a water tower, concentrating the Sun's energy and converting the water to turbine-driving steam.

in France, provides virtually all electricity used by the French province of Brittany.

Tidal power, however, requires a minimum tidal range of 8 meters (26 feet) and disturbs the ecology of estuarine habitats. As yet no tidal-power facilities have been installed in North America, although Passamaquoddy Bay in northeastern Maine, with a tidal range of 15 meters (50 feet), represents a strong candidate for future development. Likewise, the Bay of Fundy in New Brunswick, Canada, with a tidal range of about 20 meters (66 feet), is a potential tidal-power site. Maximum development of the United States' potential tidal power would provide only 1% of our total electricity needs, although it could become a significant supplement to other local energy sources. The worldwide potential, which is only slightly better, is estimated at 2%.

Solar Energy Solar-powered pocket calculators and wristwatches use an energy source that requires no expensive drilling or destructive strip mining, cannot be monopolized by unfriendly political regimes, and produces no hazardous wastes or air pollution. The Sun, expected to shine for another 5 billion years or so, is a totally renewable energy source. Solar power can heat buildings and living spaces and generate electricity—energy needs that together account for two-thirds of North America's total energy consumption.

Solar heating can be either passive or active. Passive solar heating distributes the sun's heat naturally by radiation, conduction, and convection. At mid-northern latitudes, the simplest way to heat spaces passively is to construct buildings with windows facing south. As Figure 20-14a shows, sunlight passes through the window glass and heats objects within the room; their heat then radiates to warm the air. Such an architectural design, coupled with efficient insulation, sharply reduces both air pollution and the cost of heating with fossil fuels.

Active solar heating uses water-filled, roof-mounted panels with black linings to absorb maximum sunlight. The solar-heated water is circulated throughout the building for space heating or directly to the building's hot-water system (Fig. 20-14b). Solar panels are most productive in mild, sunny climates such as those in Florida, Texas, the Southwest, and California, where they can provide as much as 90% of a building's heating needs. Even in colder regions, such as northeastern North America where solar panels are less productive, they can significantly reduce the need for other energy sources.

Solar energy can also generate electricity in several ways. For example, an array of many mirrors may reflect sunlight onto a large water tower; the water is then heated to create steam, which drives turbines attached to electrical generators (Fig. 20-14c). *Photovoltaic cells* use sunlight more directly. Their thin wafers of crystalline silicon coated with certain metals absorb solar radiation and, in turn, produce a stream of sunlight-activated electrons—that is, a flow of electricity. Photovoltaic cells currently supply electricity to tens of thousands of homes and businesses. They are becoming increasingly popular at remote sites, largely because they do not require expensive transmission lines to deliver the electricity to the consumer. Satellites are also powered by photovoltaic cells.

Solar energy can generate electricity more efficiently in some regions than in others, but the technology to maximize its effectiveness at low cost is not yet widely available. Current U.S. electricity needs would require a system of collecting mirrors that would occupy about 25,000 square kilometers—about one-tenth the size of the state of Nevada—which is clearly not feasible. Substantial use of solar energy to generate electricity probably remains decades away.

Wind Power Like falling water, tidal motion, and sunlight, wind power is a clean, renewable, nonpolluting energy source with a long history of use. The picturesque windmills of the Netherlands and those of rural midwestern North America have pumped groundwater and powered sawmills and flour mills for centuries. Wind power, however, is rarely cost-effective on a large scale: Only in a few sparsely populated mountain passes do winds blow constantly, forcefully, and from a consistent direction—all requirements for a practical application.

During the energy crisis of the 1970s, when fuel costs rose because of curtailed supplies of imported oil, engineers in the Department of Energy studied wind power as a possible way to decrease U.S. energy dependence on foreign oil. In the 1980s, a pilot project at Altamont Pass, east of San Francisco, connected 2000 wind turbines to electrical generators (Fig. 20-15). Altamont has contributed substantially to the Bay Area's electricity needs, and California's government plans to "harvest" as much as 8% of the state's electricity from wind "farms" by the year 2000.

Figure 20-15 The productivity of this wind farm at Altamont Pass, California, is made possible by sustained westerly winds that buffet California's Coast Range.

Drawbacks to wind power include its limited geographic application and the large area needed to develop it significantly. In North America, the wind force in the East and industrial Midwest does not meet the economic minimum requirements, nor do those densely populated areas have sufficient available land on which to locate wind-power facilities. The gusty Great Plains of Nebraska, Kansas, Oklahoma, and Colorado are possible candidates, although California, with its large, undeveloped, windy passes remains the best possibility.

Biomass In developing countries, as much as 35% of the energy used for cooking and heating comes from burning wood and animal dung. These fuels, which are derived from plants and animals, are known collectively as **biomass fuels.** Biomass fuels also include grain alcohol (used as an additive to gasoline), methane gas that rises from decaying garbage in landfills, combustible urban trash, and plant waste from crops such as sugar cane, peanuts, and corn. The most widely used biomass fuel is wood, which today heats about 10% of North

America's homes—more residences than are heated by electricity from nuclear power plants.

Biomass fuels are a renewable resource. Although trees grow slowly, continuous planting and harvesting can produce a steady supply. Many countries have begun to develop biomass energy sources in anticipation of the end of the era of fossil fuels. Burning trash for electricity is becoming increasingly common around the world. In Rotterdam, the Netherlands, for example, trash fuels an electrical generator that provides both efficient waste disposal and electricity for 250,000 people.

Unlike most other renewable energy resources, however, biomass fuels can create air pollution and global-warming problems when used on a wide scale or implemented poorly. As with oil and coal, burning these fuels introduces noxious gases and particles into the air, further reducing air quality. Moreover, in arid regions (see Chapter 18), overreliance on scrub and trees for energy removes the root systems that help retain water and soil; this removal contributes to desertification, which consequently eliminates the animals that provide dung for fuel.

Nuclear Energy—A Nonrenewable Alternative

In the mid-twentieth century, physicists harnessed the energy released from the decaying nuclei of radioactive isotopes (nuclear fission). Now they are attempting to tap the energy produced when atomic nuclei are fused (nuclear fusion). These energy sources may provide a twenty-first-century solution to the inevitable exhaustion of fossil fuels.

Energy from Fission When the nuclei of the atoms of certain heavy elements' isotopes are bombarded with neutrons, they split into several lighter elements. In this reaction, they release part of the binding energy of an atom's nucleus, additional neutrons, and an enormous quantity of heat. This process is called **nuclear fission.** The heat generated by the reaction can convert water to steam, driving turbines and producing electricity. The released neutrons can, in turn, bombard other heavy nuclei, setting off a chain reaction.

In commercial nuclear reactors, the naturally decaying nuclei of uranium-235 atoms release neutrons that bombard other U-235 nuclei, which undergo fission (splitting); this fission produces heat and, most important, more neutrons, which then bombard and split neighboring U-235 nuclei (Fig. 20-16a). Nuclear reactors are designed to control the production of neutrons and their flow to the fissionable material, thereby preventing uncontrolled chain reactions (as in a nuclear bomb). Neutron-absorbing control rods are moved in and out of the uranium fuel to regulate the rate of reactions. The enormous amount of heat generated by nuclear reactions is transferred to a coolant, usually water, which is converted to steam to drive turbines (Fig. 20-16b).

Uranium, a relatively uncommon element in the Earth's crust, becomes concentrated in fluids of granitic magmas in the late stages of cooling; thus it may occur in pegmatites (discussed in Chapter 3) as a crystalline uranium oxide, such as the mineral uraninite (UO_2). This element is highly soluble in oxidizing water, so it readily dissolves when igneous rocks are weathered. Uranium is then transported by groundwater into permeable sediments and sedimentary rocks, where it bonds to the surfaces of organic-matter particles. These concentrations serve as the source of most uranium ore, which is mined from ancient stream sands and gravels, either in granular form (the black, shiny, noncrystalline uranium oxide *pitchblende*) or as a canary-yellow encrustation on sand grains (the potassium–uranium–vanadium mineral *carnotite*).

In North America, uranium deposits are found in the Mesozoic stream gravels of Colorado, New Mexico, Wyoming, and Texas, often associated with fossilized plant and wood remains. (The uranium in circulating groundwater bonds with and becomes concentrated in dead plant cells.) During the uranium boom of the 1950s, western prospectors toting Geiger counters tested tens of thousands of petrified logs. In Canada, near Great Bear Lake in the Northwest Territories, important uranium deposits occur in ancient, organic-rich stream gravels.

Uranium-235, the only naturally occurring radioactive isotope that can maintain a nuclear chain reaction, is consumed during fission. It accounts for only 0.7% of natural uranium; nonfissionable U-238 makes up the remaining 99.3%. As its current rate of use, recoverable reserves of U-235 may be sufficient to power the world's 575 nuclear power plants for only another 30 years or so.

To address the problem of dwindling U-235 supplies, nuclear engineers are experimenting with techniques to produce fissionable plutonium-239 from the abundant nonfissionable U-238 that are *safe and economical.* In one such process, uranium-238 is placed in a reactor with a small amount of U-235. The U-235 and its neutrons bombard the U-238 nuclei, converting them to a form of plutonium, Pu-239. The process takes place in a **breeder reactor,** a chamber in which fission of the manufactured plutonium atoms produces a surplus of neutrons that can then be used to create, or "breed," more fissionable material than they consume.

Currently, few breeder reactors are on-line, primarily because Pu-239 is potentially a weapons-grade nuclear material. Funding for Tennessee's Clinch River breeder reactor was canceled in the mid-1980s, in large part because of public concern about proliferation of nuclear weapons. The Super Phoenix breeder reactor in France is currently the world's largest functional breeder facility.

In 1974, the U.S. Geological Survey predicted that the United States would get 60% of its electricity from nuclear plants by 2000; in 1993, this percentage was about 16% and dropping. Meanwhile, western Europe and Japan, both of

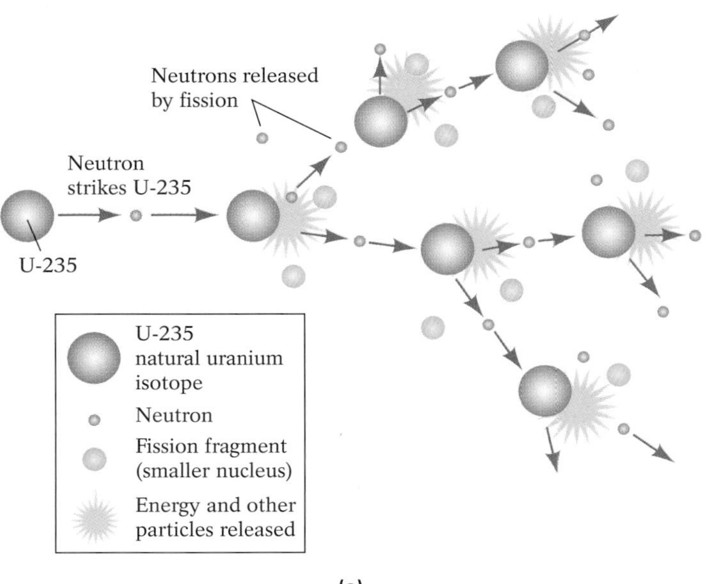

Figure 20-16 **(a)** A neutron released by the natural decay of a U-235 nucleus initiates a chain reaction. **(b)** Uranium fuel is contained within 12 meters (40 feet) of water at the Indian Point 2 nuclear power plant in Buchanan, New York.

which have little indigenous fossil fuel, were expanding their nuclear-energy facilities; France, for example, derives 65% of its electricity from its many reactors. A number of reasons explain the United States' reluctance to develop nuclear energy. In addition to concern about the proliferation of weapons-grade fuels, there are technological problems of reactor safety and radioactive-waste disposal, and numerous other economic, political, and psychological problems associated with the growth of nuclear energy.

At Three Mile Island, Pennsylvania, in 1979, an instrument malfunction led plant operators to conclude that too much water was flushing through the reactor's cooling system. They responded by reducing water flow, leaving the reactor core uncovered for several hours and allowing it to overheat. Quite fortunately for Pennsylvanians, although the reactor core suffered damage, little measurable radiation escaped.

At Chernobyl, in the former Soviet Union, in 1986, technicians, working at a poorly designed nuclear facility, accidentally allowed a runaway chain reaction to develop. (Apparently, many of the plant's safety systems had been disabled to allow Soviet engineers to experiment with the reactor core.) Two small explosions blew the roof off the building, showering the immediate area with radioactive material. Eastern Europe and Scandinavia were covered with a cloud of radioactive vapor. Numerous cases of radiation sickness and cancer developed in the aftermath of this accident. The event disrupted life in Chernobyl's immediate vicinity as well as agriculture and commerce over a wide area.

Radioactive nuclear waste is so toxic that it must be isolated from all life for thousands of years. Thus safe disposal is the industry's greatest technological (and political) problem. Spent fuel rods, internal machinery from reactor cores, and the waste products of nuclear-fuel processing must be stored in a way that prevents any leakage to the atmosphere or groundwater system. A burial site must be seismically stable—an earthquake could damage the facility or a new fault could breach the repository—and completely isolated from groundwater. In 1991, Congress selected Yucca Mountain in southwestern Nevada as the first U.S. nuclear-waste repository (Fig. 20-17). Located within the Nevada Test Site (where the first atomic bomb was tested), this area is largely uninhabited, has been relatively free of earthquakes during recorded time, and is so arid that the water table is very deep (see Chapter 15 for a discussion of Yucca Mountain's groundwater situation). The government is conducting studies expected to last until the end of the century to determine the safety of this site before it can become operational.

The nuclear-energy industry also faces a very high expense in constructing reactors, a lengthy process for obtaining permits, the potential for theft of weapons-grade plutonium and the possibility of terrorist attacks on installations, and the chance of accidents and sabotage when transporting

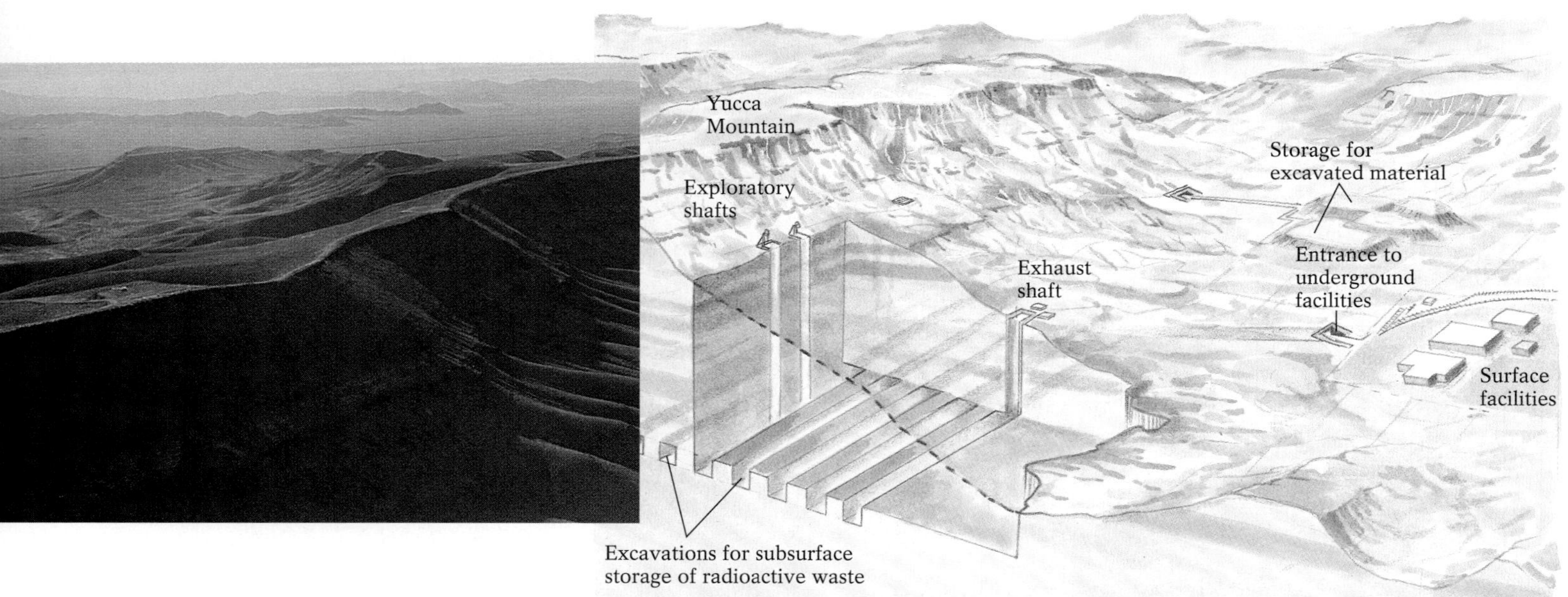

Figure 20-17 A conceptual sketch of the United States' first permanent repository for high-level nuclear waste at Yucca Mountain, Nevada. When preparation of the site finishes in 2003, radioactive waste from around the country will arrive in 3-meter (10-foot)-long containers and be transferred to a cavern excavated into a thick layer of relatively impermeable ash-flow tuff, 300 meters (1000 feet) below the surface. Inset: The ash-flow tuff of Yucca Mountain.

wastes. In addition, it must handle the psychological issue of public acceptance and confidence in nuclear power, especially since the accidents at Three Mile Island and Chernobyl. After these accidents, strenuous local opposition on Long Island, New York, prevented a nuclear-energy plant in Shoreham, 90 kilometers (55 miles) east of New York City, from coming on-line.

Energy from Fusion **Nuclear fusion** occurs when atomic nuclei of light elements are subjected to extremely high pressure and temperature, causing them to fuse and form heavier atoms. Like fission, this process releases an enormous amount of heat energy that can convert water to steam, driving turbines to produce electricity. The principal fuel of fusion, however, is hydrogen, one of the most abundant elements; rather than toxic radioactive waste, the primary product of fusion is helium, a harmless inert gas.

Fusion's potential is enormous: The energy that could be generated by the hydrogen isotopes in 1 cubic kilometer (0.24 square mile) of seawater exceeds the total energy stored in the world's oil reserves. In recent experiments, heavy isotopes of hydrogen (tritium, ^{3}H, and deuterium, ^{2}H) were compressed in a powerful magnetic field and then heated by intense laser bursts to about 50 million°C (90 million°F), producing the heavier element helium (He) and an enormous amount of heat. Physicists have been able to sustain small-scale hydrogen fusion to generate energy, although it is not yet economically feasible—the amount of energy needed to achieve fusion still exceeds the amount produced by it. Further study and expensive experimentation will be needed to make this process commercially viable.

Mineral Resources

The other principal group of resources beside those used for energy is mineral resources. Minerals, as we saw in Chapter 2, are naturally occurring inorganic solids consisting of chemical elements in specific proportions whose atoms are arranged in a systematic internal pattern. Of the 3000 or so known minerals, only a few dozen have economic value. They include various metallic minerals, such as the oxides and sulfides of iron, copper, and aluminum; certain metals, such as gold, silver, and platinum, that commonly occur uncombined with other elements; and various nonmetallic rock and mineral materials, such as sand and gravel (used as building materials), limestone (for cement), halite (common table salt), and several highly prized gemstones (such as diamonds, rubies, and emeralds).

In this section we describe how metals and nonmetals are used and how they form. We also discuss environmental problems that accompany mining.

Metals

Most metals combine readily either with oxygen to form oxides or with sulfur to form sulfides. Common oxides include those of tin, iron, and aluminum; common sulfides include those of zinc, lead, iron, copper, and molybdenum. *Native metals,* such as gold, platinum, and silver, do not combine with other elements.

The first known human use of metals occurred about 9000 years ago in what is now Turkey, when people hammered naturally pure copper into amulets, tools, and weapons. By 6000 years ago, metallurgists in Europe and Asia Minor had begun separating metals from their host rocks by *smelting,* which they accomplished by heating rocks to the melting points of their incorporated metals and collecting the molten metal for further processing. The earliest smelting operations extracted copper from copper sulfides, such as chalcocite (Cu_2S) and chalcopyrite ($CuFeS_2$).

By 5000 years ago, lead, tin, and zinc, as well as copper, were being smelted and combined in their molten state. This process produced **alloys,** or metal mixtures, that were harder than their component metals and that maintained sharper points and edges. Bronze, a mixture of copper and tin, was a common alloy that dominated the tool-and-weapon industries that flourished between 5000 and 2650 years ago; its widespread use characterized the archaeological period known as the Bronze Age. By about 2650 years ago, metallurgists had learned to smelt and work iron, which quickly replaced bronze as the principal metal used in the manufacture of tools, weapons, and armor, ushering in the Iron Age. The demand for metals has grown ever since.

Most metals are widely disseminated throughout the Earth's crust; their extraction is economically feasible only where rock-forming processes have gathered them into mineable concentrations. These processes include the following:

- precipitation from hot, metal-rich water
- settling of early-forming crystals from crystallizing magma
- chemical interactions of hot fluids during metamorphism
- separation of dense metallic particles during sedimentation
- dissolution of surrounding minerals

Processes That Concentrate Metals **Hydrothermal deposits** form when metallic ions precipitate from hot ion-rich water. Gold, for example, which constitutes only about 0.0000002% of the Earth's crust, can be mined most profitably where circulating hot water has dissolved it from its source rocks and precipitated it elsewhere in highly concentrated form. The rich gold deposits of California's Sierra Nevada and along Cripple Creek in the Colorado Rockies precipitated from hydrothermal solutions, as did the silver deposits of Coeur D'Alene, Idaho, the copper of Butte, Montana, and the Keweenaw Peninsula of northern Michigan, and the lead–zinc deposits of the tri-state area of Missouri, Oklahoma, and Arkansas.

A common source of metal-rich hot water is a cooling magma that contains excess metallic ions after most silicate minerals have crystallized (see Chapter 3). Such hot solutions may contain copper, lead, zinc, gold, silver, platinum, and uranium (none of which bonds readily with silicate tetrahedra), as well as some uncrystallized silica (SiO_2). The hot water typically infiltrates faults, fractures, and bedding planes of surrounding rocks, where it cools further and deposits its solutes as silica veins that may contain visible masses of metals (Fig. 20-18a). Some of the world's most productive gold and silver mines are hydrothermal vein deposits that cut across rocks adjacent to granitic batholiths.

Porphyry copper deposits (which appear in plutonic rocks with porphyritic textures, discussed in Chapter 3) form when the water from a cooling body of water-rich felsic or intermediate magma is converted to pressurized steam. As it expands outward, the steam ruptures the newly crystallized rock at the magma's edge, creating numerous hairline fractures at the margin of the batholith. Hydrothermal solutions then fill these fractures, precipitating copper and other metals as they cool.

Hydrothermal deposits also originate from deep-circulating groundwater that has been heated through close contact with shallow magmas or young, warm, plutonic rocks. As the heated water rises, it dissolves metals from the pluton and surrounding rocks; as it flows into fractures, joints, bedding planes, and faults, it precipitates the metals as concentrated vein deposits. Other hydrothermal solutions pass through porous rocks and cool slowly, precipitating their metals in minute pore spaces over large areas. These deposits are typically far less cost-effective to mine.

Massive sulfide deposits form where cold seawater enters the fractures associated with divergent zones and then becomes heated by shallow basaltic magma. As the hot water rises through the fractures, it dissolves some of the trace metals in the basaltic rocks of the mid-ocean ridge; it also picks up sulfurous gases typical of basaltic magma. The resulting hydrothermal solution contains sulfides of copper, zinc, manganese, lead, and iron (Fig. 20-18b). When the solution erupts at the sea floor and rapidly cools, it precipitates its metals in a black, jet-like plume called a *black smoker* (see Figure 12-14). The "smoke" is actually fine, black particles, composed largely of sulfide minerals such as pyrite (FeS_2), chalcopyrite ($CuFeS_2$), sphalerite (ZnS), and galena (PbS), as well as several oxides and hydroxides.

The processes that crystallize minerals from magma (discussed in Chapter 3) also produce metal deposits. In *gravity settling,* dense, early-crystallizing minerals sink to the bottom of the magma chamber, where they accumulate in layers. In *filter pressing,* large, early-forming crystals remain affixed to the walls of the magma chamber when tectonic forces compress the chamber and force the still-liquid portion of the

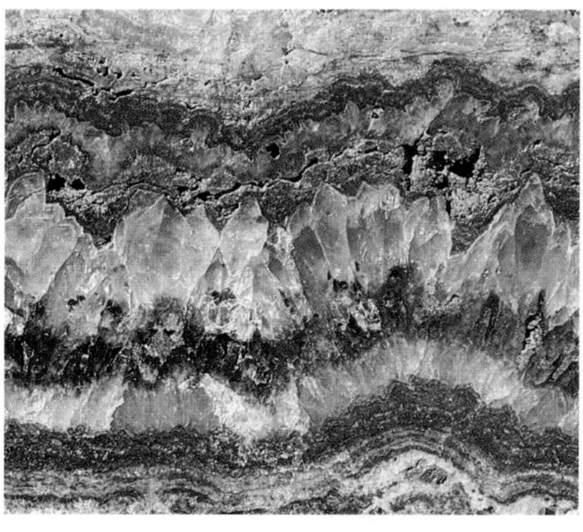

(a)

(b)

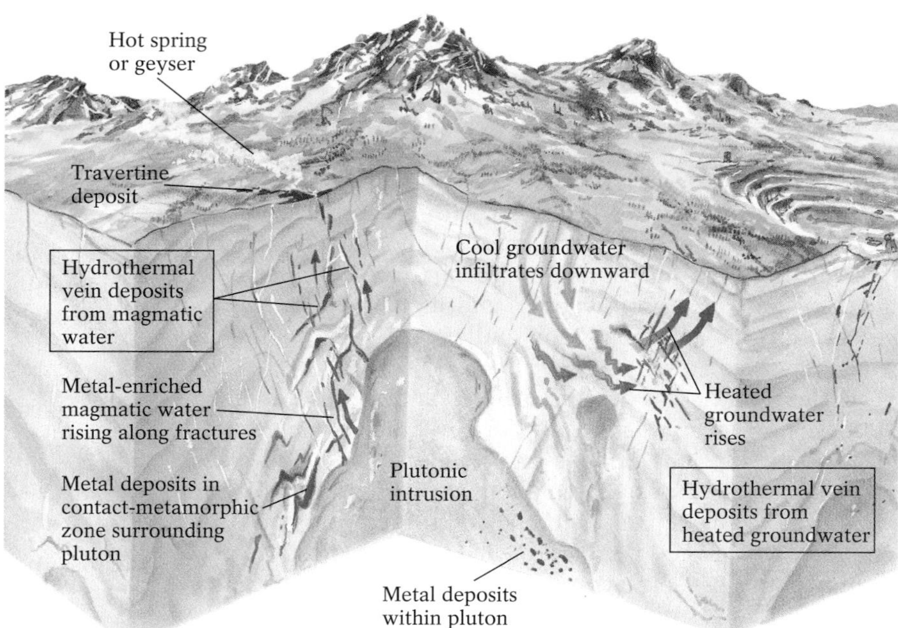

Figure 20-18 Hydrothermal processes that concentrate valuable metals. Metals may precipitate from magmatic water or from circulating groundwater or seawater heated by the magma. **(a)** These veins of gold and silver ore formed when hot magmatic solutions of the metals infiltrated cracks around the cooling magma. **(b)** These massive sulfide deposits in Quebec formed at a mid-ocean ridge where seawater heated by basaltic magma emerged at hot springs and precipitated large quantities of metals on the ocean floor.

magma into fractures in adjacent rocks. Gravity settling and filter pressing of mafic and ultramafic magmas have produced valuable ore bodies of iron (in magnetite, Fe_3O_4), chromium (in chromite, $FeCr_2O_4$), titanium (in ilmenite, $FeTiO_3$), and nickel [in pentlandite, $(Fe,Ni)_9S_8$)]. The rich chromite deposits of the Stillwater intrusive complex of northwestern Montana and the Bushveld complex of southern Africa formed from gravity settling of chromite crystals from mafic and ultramafic magmas (Fig. 20-19). The iron ore of the Kiruna district of northern Sweden (60% iron) resulted from filter pressing of magnetite crystals. The world's largest nickel deposit is found at Sudbury, Ontario, near the shores of Lake Superior. It probably formed as droplets of nickel sulfide rose from the Earth's mantle and then settled through a ring-shaped system of fractures believed to have resulted from the impact of an enormous meteorite about 1.9 billion years ago. Some of Sudbury's nickel deposits may constitute, in part, the remains of the impacting meteorite.

Felsic magmas also represent a potential source of valuable ore bodies. Late in the cooling of granitic magma, a substantial amount of water remains along with uncrystallized silica (SiO_2) and some rare elements that have not yet crystallized. These elements may include lithium, beryllium, boron, uranium, fluorine, and cesium. In this watery, highly fluid residual magma, unbonded ions migrate freely and bond readily to growing crystal structures. The resulting pegmatites typically contain very large crystals (see Chapter 3). Pegmatites at Kings Mountain, North Carolina, for example, con-

Figure 20-19 Deposits of chromite (black) in the Bushveld complex of southern Africa. In addition to its vast deposits of chromite, the Bushveld complex contains 70% of the world's platinum, one of the most precious metals. This huge sill, 240 kilometers (150 miles) by 480 kilometers (300 miles) in area and 8 kilometers (5 miles) thick, was the site of gravity settling of heavy early-forming crystals.

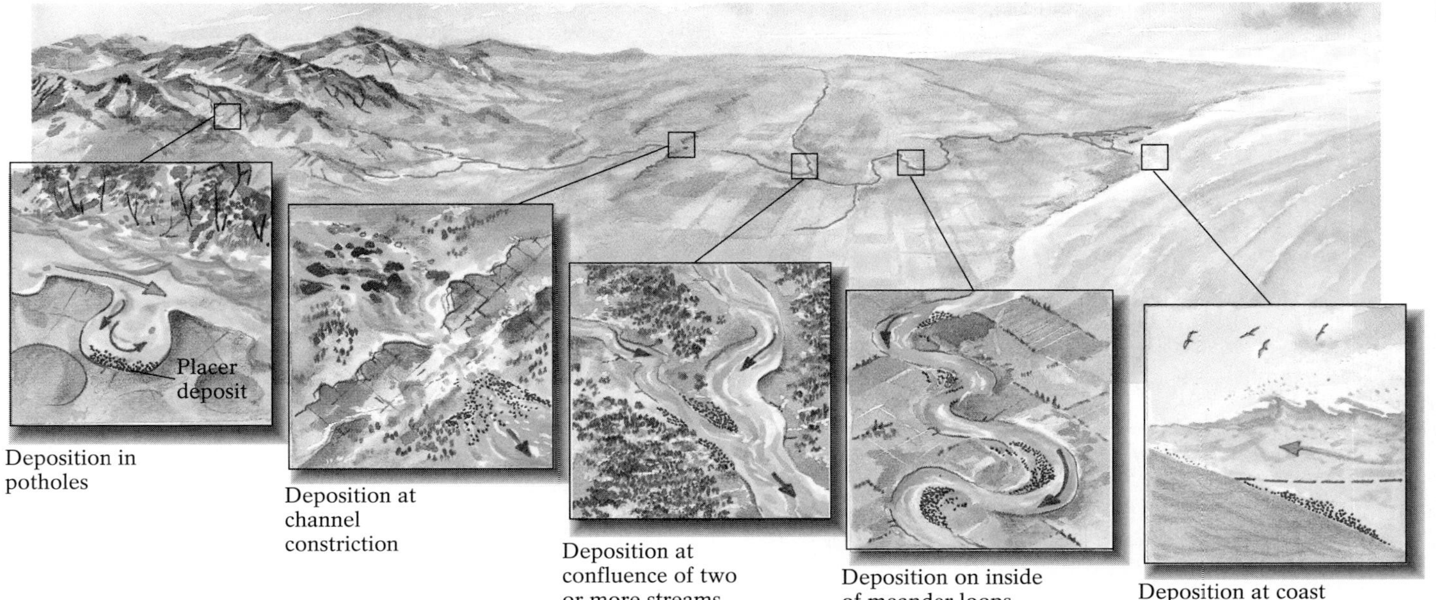

Figure 20-20 Placer deposits develop wherever the velocity of flowing water significantly declines, such as at potholes in a stream bed, downstream from a constriction in a channel, at the confluence of two or more streams, at the inside of a meander bend, and at coasts, where dense minerals settle from the water and are reworked by wave action. **(a)** Placer deposits of emeralds being mined in Colombia. **(b)** Emerald merchants in Colombia show their wares.

tain feldspar crystals the size of two-story houses, and those in the Black Hills of South Dakota contain crystals of the lithium-rich mineral spodumene as large as telephone poles. Some pegmatites also contain precious crystals of beryllium-rich emerald and aquamarine and of boron-rich tourmaline.

When hot, ion-rich fluids move through rock bodies, their heat can produce contact-metamorphic mineral alterations in the host rock, from which metallic ores often result. For example, when hot, ion-rich fluids enter impure limestones and dolostones containing aluminum-rich clay, chemical reactions in the contact zone release CO_2. The CO_2 migrates outward and produces an extensive metamorphic aureole (see Chapter 7) containing aluminum oxides (such as corundum, Al_2O_3) and other metal-rich minerals. Metallic ions from the hot fluids may replace a large amount of the calcium in limestone; in the vicinity of a basaltic intrusion, the replacement ions are typically iron (in magnetite). Other metallic deposits produced by contact metamorphism include zinc (in the mineral sphalerite), lead (in galena), and copper (in chalcopyrite and bornite).

Sedimentary processes also produce valuable metal deposits, ranging from the gold nuggets and gem crystals that settle out from the slow-moving currents of rivers to the vast iron deposits precipitated on the floors of ancient oceans. During transport, water-borne heavy materials, such as gold, platinum, and tin, are sorted out by their density and durability, becoming concentrated as **placer deposits** wherever flowing water slows. Placer deposits commonly occur in potholes in the stream bed, along the inside banks of meander bends, downstream from constrictions in the channel, at the confluence of two or more streams, and at coasts, where wave action cannot move the heavy materials (Fig. 20-20). Extremely durable minerals, such as diamonds, sapphires, rubies, and emeralds, also form placer deposits, because they resist abrasion and can survive long journeys that might wear away softer minerals. Placer emeralds are shown in Figure 20-20a–b.

The most valuable deposits consist of gold; gold's high density (19 grams per cubic centimeter) causes it to sink from its transporting stream. Nuggets of gold weathered and

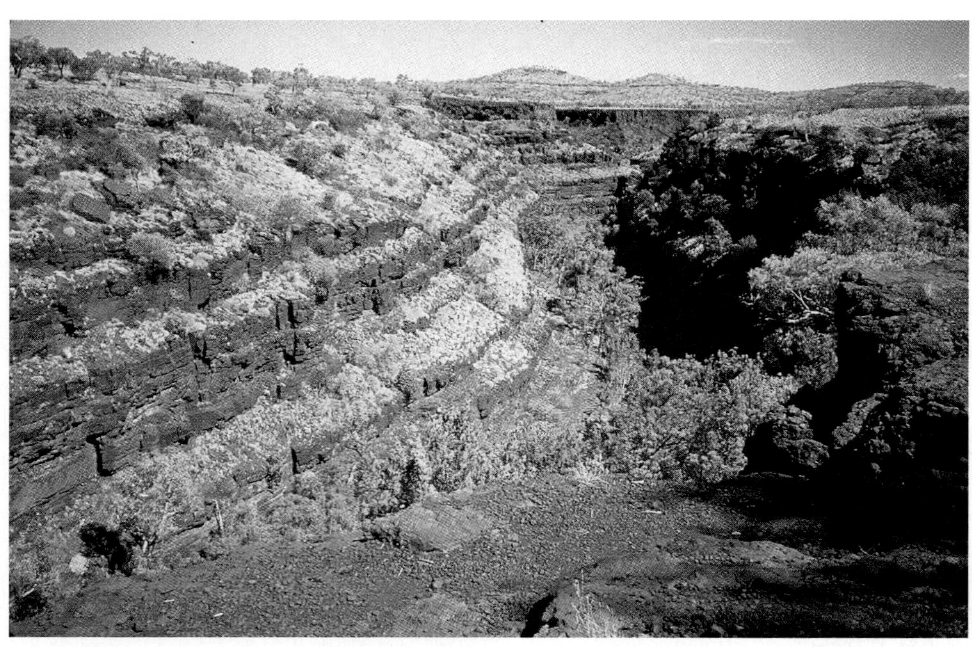

Figure 20-21 This banded iron formation began to form about 2 billion years ago, when enough oxygen from photosynthesizing plants accumulated in the Earth's atmosphere to promote oxidation and precipitation of iron.

eroded from hydrothermal vein deposits in California's Sierra Nevada batholiths have been carried down the range's numerous streams for hundreds of thousands of years. Placer deposits are generally discovered first; lucky prospectors may then be able to retrace the gold's journey upstream to the "mother lode" from which the placers eroded, assuming that the source vein has not completely eroded away.

Most sedimentary iron deposits are found in a unique series of layered Precambrian sediments that were deposited in shallow marine basins on nearly every continent about 2 billion years ago. These **banded iron formations** consist of alternating layers of light-colored recrystallized chert and dark-colored highly concentrated iron oxides (Fig. 20-21). They are found in Labrador, eastern Canada, Brazil, western Australia, Russia, India, the western African nations of Gabon, Mauritania, and Liberia, and around Lake Superior in Minnesota, Wisconsin, and Michigan. Two billion years ago, when these sediments were deposited, vegetation (such as microscopic algae) had become plentiful enough to produce a significant amount of oxygen through photosynthesis. The oxygen content of the Earth's atmosphere and surface waters, which had previously been quite low, began to increase. Iron in solution could then precipitate in the form of such oxides as magnetite and hematite. (Iron is very soluble in an oxygen-poor environment, and thus any iron that weathered from continental rocks prior to this time would have remained in solution during its transport to the oceans.) Expansion of flora on Earth added oxygen to the chemical composition of the planet's air and water, producing the widespread oxide deposits that account for more than 90% of the iron ore mined every year.

Metals can also become concentrated either when other minerals weather and dissolve, leaving a metallic residue at or just below the surface, or when metals dissolved by groundwater precipitated elsewhere in concentrated form. These **secondary enrichment** processes generally require a warm climate, abundant atmospheric water, permeable bedrock (to promote groundwater flow), and a soluble matrix surrounding economically valuable materials.

The most prominent weathering-produced ore is bauxite ($Al_2O_3 \cdot nH_2O$), an aluminum oxide that is the most abundant source of aluminum. Sizable deposits of this valuable ore form only where extreme chemical weathering of warm, humid climates breaks down the feldspars in granitic rocks and the aluminum-rich clays in impure limestones. Such residual bauxite is found today in tropical locations such as Jamaica. Deposits of this ore also occur in nontropical regions, such as France, Australia, and Arkansas (Fig. 20-22a), relics of tropical environments in the past.

Soluble metals dissolved by acidic groundwater may be transported downward to the water table, where they may precipitate in concentrated form. For example, when sulfur is released from weathered sulfide minerals such as pyrite (FeS_2), the sulfides oxidize to produce sulfuric acid. The acid dissolves some metals, which then move downward in solution to the less acidic zone below the water table, where the metals are precipitated. In this way, as shown in Figure 20-22b, metals originally scattered throughout a body of rock can become concentrated in the subsurface.

Metals and Plate Boundaries Most of the world's major metal deposits occur at past or present plate boundaries. At such

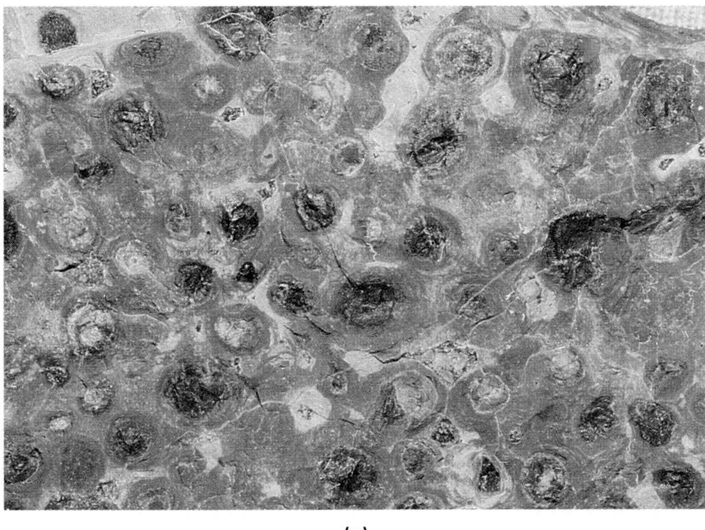

(a)

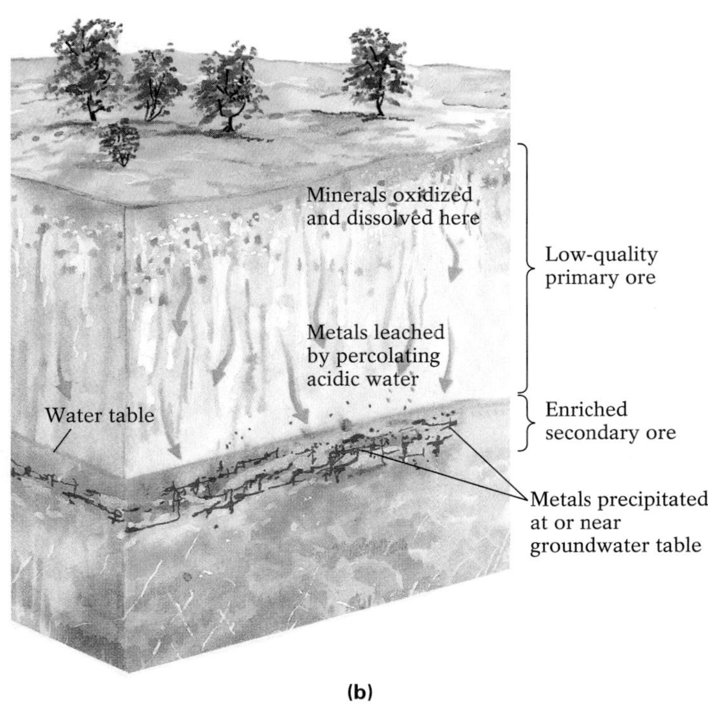

(b)

Figure 20-22 Secondary enrichment occurs when groundwater either **(a)** dissolves the matrix around valuable metals, leaving them as a residual deposit, such as this Arkansas bauxite, a main source of aluminum, or **(b)** dissolves the metals themselves and transports them in solution down to the water table, where they precipitate. Copper is particularly susceptible to such secondary enrichment.

locations, magmas are generated, hot hydrothermal solutions circulate through rocks, and plate collisions thrust metal-rich, sea-floor rocks upward above sea level.

Massive sulfide and oxide deposits—rich in copper, zinc, lead, and manganese—typically precipitate at or near mid-ocean ridges, where plate divergence opens an extensive network of fractures through which cold seawater circulates into warm oceanic lithosphere (Fig. 20-23a). In the mid-1960s, geologists collected a hot, salty solution mixed with a black powdery substance at the floor of northwestern Africa's Red Sea, which occupies the recent rift between the African and Arabian plates. The black powder was found to contain various sulfides and oxides of iron, zinc, and copper.

The sulfide deposits that form at divergent zones become incorporated into the oceanic lithosphere that spreads out from the mid-ocean ridge. As spreading continues, these deposits may eventually reach a subduction zone and descend to depths where the temperatures of the minerals that contain these metals reach their melting points. The subduction-produced magmas rise, cool, and deposit their metal content in batholiths. This process has produced the rich metal ores found at the western edges of the Americas from Alaska to Chile. Different metals appear at different positions within subduction-zone batholiths, sometimes as distinct bands, such as those depicted in Fig. 20-23b. The segregation of these metals largely occurs because of the concentrations of specific metals in their source magmas and because of the local hydrothermal environment.

After subduction consumes an oceanic plate, the extreme compression produced when plates collide may thrust slices of the subducted sea floor back to the Earth's surface (see Chapter 12). For this reason, continental collision zones often contain masses of metal-rich rock that originated at a mid-ocean ridge (Fig. 20-23c). The island of Cyprus, for example, has risen from the floor of the eastern Mediterranean as the African and Eurasian plates have collided. Cyprus's vast copper deposits (from which the island derives its name) have been mined for 5000 years.

Environmental Problems Caused by Mining Most major metal deposits are mined from open pits by operations that begin by removing millions of tons of *overburden,* the rock and regolith that cover ore deposits. These mine tailings, along with the great volume of waste from ore mills, are typically piled into huge hills and often left without covering vegetation. In this form, they are highly susceptible to rapid erosion and mass movement. Surface-water runoff from waste piles may pollute regional streams by clogging them with silt and producing toxicity from dissolved metals and acid. In the mine pits, ore rocks may come in contact with surface water, which may then become contaminated with sulfides. Sulfides oxidize to form sulfuric acid, producing acidic ground and surface water.

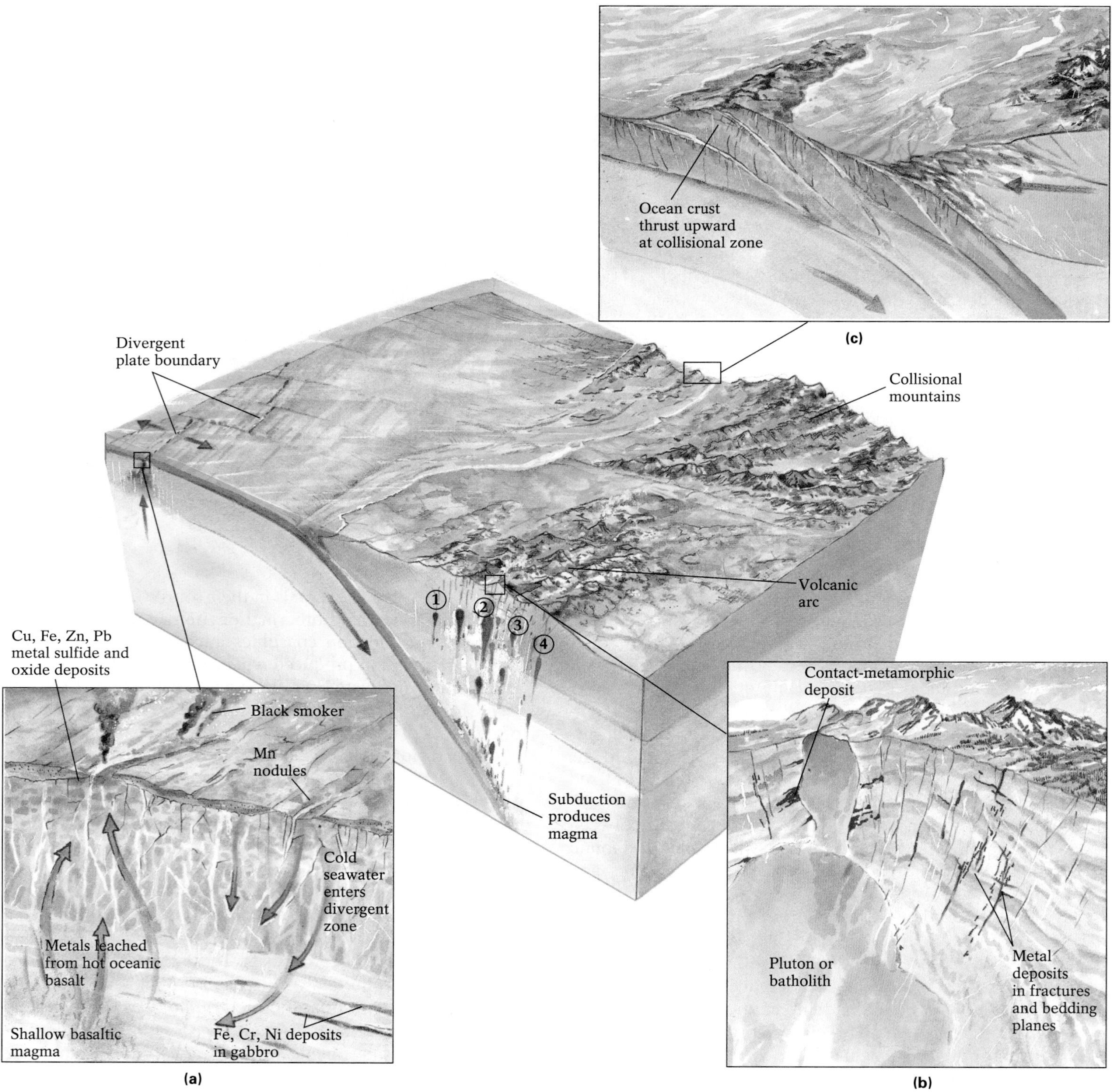

Figure 20-23 The processes that occur at each type of plate boundary (and the available materials) determine the type of metal found at specific locations. **(a)** Divergence produces massive sulfides and oxides rich in copper, lead, zinc, manganese, iron, and nickel. **(b)** Subduction may separate metals into distinct bands: (1) Typically, the first to appear are iron-rich. (2) Farther inland, gold and porphyry copper deposits, associated with felsic-to-intermediate intrusions, are found. (3) The next band of deposits generally contains hydrothermal veins of lead, zinc, silver, and copper. (4) Tin and molybdenum typically appear farthest inland. **(c)** Deeply seated deposits formed by plate divergence or subduction may be thrust to the surface by a continental collision.

These and other problems have prompted the U.S. Congress to pass legislation requiring monitoring and control of mine-water discharge. State and federal regulations also require land reclamation after a mining operation has been completed; the mine operator must isolate the sulfides from the groundwater system, restore the topography, and replant vegetation to prevent erosion.

Nonmetals

The search for gold, silver, and platinum has unleashed veritable stampedes of prospectors and altered the histories of entire states and nations. Although no "gravel rush" or "phosphate rush" has occurred to match the gold rushes of nineteenth-century America, such nonmetals are at least as useful and valuable as precious metals. Nonmetal resources range from the building materials derived from common rocks to the natural mineral fertilizers used to enhance our food supply.

Nonmetal Building Materials Only petroleum exceeds the combined economic value of natural building materials. Limestone, the most widely used building material, provides crushed stone for road beds as well as cut stone for monuments and other stately buildings; it is also a key ingredient in Portland cement, a staple of the construction industry. To make Portland cement, finely ground limestone and shale are mixed and then heated to 1480°C (2700°F) to drive off limestone's carbon dioxide. The resulting quicklime (CaO) reacts with clays in the shale to produce a calcium slag that is then mixed with a little gypsum ($CaSO_4 \cdot 2H_2O$). Cement hardens slowly when mixed with water.

Gypsum, which is soft and soluble, is rarely used as a building stone. Heating gypsum to 177°C (351°F), however, drives off about 75% of its water, leaving behind a powdery substance known as plaster of Paris (named for the gypsum quarries near Paris that produce high-quality plaster). After water is added to plaster of Paris, the wet mixture congeals. Tiny crystals of gypsum form within the plaster, creating a firm solid that is used to fashion smooth interior walls and that has numerous other applications in such fields as art, dentistry, and orthopedics.

Mixing sand and gravel with cement forms concrete, used principally in building foundations and roads. A single kilometer of four-lane highway, for example, requires about 40 tons of gravel for its concrete. Common sources of sand and gravel include river channel and bar deposits, coastal offshore bars, beach deposits, sand dunes, and glacial eskers and outwash (Fig. 20-24). Increased urban construction worldwide over the last 25 years has doubled the demand for sand and gravel, depleting many sources. In Scandinavia, entire eskers have been removed to satisfy the growing need for these materials.

Clay minerals (see Chapter 5) are end-products of the chemical weathering of feldspars and therefore remain relatively stable in the Earth's surface-weathering environment.

Figure 20-24 A gravel operation in glacial outwash in Ontario, Canada.

Wet and plastic clays can be shaped into a variety of building materials, such as decorative terra cotta bricks, tiles, and pipes; these materials harden when they are fired in a kiln.

Various types of stone are quarried for specific building purposes. Slate is used for roofing, flooring, fireplaces, and patios; slabs of polished granite, diorite, gneiss, and other attractive coarse-grained rocks face office and government buildings; retaining walls that support unstable slopes are made of blocks of basalt; and tons of crushed limestone, granite, marble, and schist lie beneath highways as road-bed fill.

Nonmetals for Agriculture and Industry Soon after the year 2000, unless we curb population growth, the number of people on Earth will exceed 7 billion. To feed so many people using soil that has, in many areas, already been depleted of nutrients by overcultivation will require a vast supply of chemical fertilizers to enhance plant growth and increase sugar and starch production.

Phosphorus and potassium are two of the principal elements in agricultural fertilizers. Phosphorus is derived primarily from the mineral apatite [$Ca_5(PO_4)_3(OH,F,Cl)$], which is found in certain marine sedimentary rocks, in guano deposits (bird and bat droppings) in tropical caves, and in some calcium-rich igneous rocks. In North America, most phosphorus comes from ancient sedimentary deposits in Montana and Wyoming and from more recent sediments along coastal North Carolina and Florida. Potassium is obtained primarily from sylvite (KCl), an evaporite deposited in shallow marine basins during periods of climatic warming. Large sylvite deposits are found in New Mexico, Utah, Colorado, Montana, and Saskatchewan.

Sulfur has applications in both agriculture and industry. In the form of sulfuric acid, it is added to alkaline soils to maintain optimal pH conditions for plant growth. Sulfuric acid is also used to treat rubber, explosives, and wood pulp (for the manufacture of paper). Although some sulfur is mined in native form from the deposits that crystallize from escaping sulfurous gases around the vents of volcanoes, most is refined from the sulfide mineral pyrite (FeS_2), the sulfate mineral gypsum, or solid sulfur removed and purified from sulfur-bearing petroleum.

A very common nonmetal resource is quartz sand, which is melted and then cooled rapidly ("quenched") to produce glass. Ancient Egyptians were the first to manufacture glass from melted sand, 4000 years ago. Today, 30 million tons of sand are quarried each year in North America from the purest quartz arenite deposits, such as the St. Peter sandstone of Illinois and Missouri, which consists of 95% quartz. Slight impurities in the sand impart color to the resulting glass.

Another nonmetal resource with multiple industrial uses is *asbestos* (the generic term for a variety of magnesium-rich, white, gray, and green fibrous silicates, including the minerals serpentine and chrysotile). This nonflammable material is used primarily in fireproof clothing, in electrical insulation, and in the linings of automobile brakes. Its use in the construction industry has been curtailed, however, because inhalation of asbestos dust has been linked to several lung and digestive-tract diseases. The asbestiform silicates are produced primarily by metamorphism of olivines and pyroxenes in ultramafic mantle rocks of continental collision zones (Chapter 12). Large asbestos reserves appear in the metamorphic rocks of the northern Appalachians of Vermont and Quebec.

Future Use of Natural Resources

The United States has about 50 billion barrels of known oil reserves that can be recovered by current methods and may have an additional, as-yet-undiscovered 35 billion barrels. The current rate of use is 7.5 billion barrels per year, about 50% of which is imported. Thus the domestic supply, supplemented with imports, will last only about 30 years. The impending petroleum shortage will surely force many changes in our consumption of this energy source. Even if technological advances substantially improve oil recovery, you will undoubtedly be around when the last drop of economically accessible American oil trickles from the pumps.

Large new petroleum finds on American soil are unlikely, as virtually all potential oil-producing rock formations have been explored. Indeed, most of the Earth's potential oil-bearing regions have been thoroughly investigated. Even if large new reserves were discovered, at the current rate of use they would be exhausted rapidly. The most recent significant domestic discovery, about 10 billion barrels at Alaska's North Slope, would satisfy less than two years of current total U.S. oil needs.

At the worldwide petroleum use rate of about 21 billion barrels per year, the world's 700-billion-barrel reserves would last only 35 years or so. Unless oil prices skyrocket and reduce demand (a likely scenario), sometime during the twenty-first century, cars, buses, and trucks everywhere will likely stop in place and rust along the side of the road. This scenario appears unavoidable unless more effort, money, and scientific ingenuity are devoted to finding new oil reserves, improving the percentage we can economically recover, or, more likely, finding ways to extend the *finite* crude oil supply or *develop alternatives.*

Some countries have already exhausted their supplies of crucial resources and turned into strictly importers; others are just beginning to feel the effects of critical shortages. What does the future hold for humankind as the Earth's resources dwindle?

Meeting Future Energy Needs

Substantial oil shortages and price increases in developed countries may reduce consumption, thereby extending the life of current supplies. In addition, widespread use of gasoline additives, such as ethanol (a form of alcohol), could reduce gasoline consumption by as much as 25%.

One way to maximize our oil reserves is to extract as much as possible from known reservoir rocks. In newly drilled wells, oil usually gushes out due to the confining pressure exerted by overlying rocks on the reservoir rocks. Collection of this first flow of oil is called *primary recovery.* As reservoir rocks become partially depleted and no longer oil-saturated, pressure on the remaining oil diminishes and the flow gradually stops. In the past, wells were abandoned when they no longer gushed, or they were subjected to *secondary recovery* methods, such as injection of water into the reservoir rocks to buoy oil upward. Even after secondary recovery, however, one-half to two-thirds of the oil remains in the ground.

In the United States alone, more than 300 billion barrels of unrecovered oil may lie in the reservoir rocks of known oil fields, perhaps enough to provide for an additional 50 years of use. Some abandoned wells are now being reopened and subjected to *enhanced recovery* methods. In one such method, compressed carbon dioxide gas and superheated steam are injected to make the petroleum less viscous, causing it to flow more readily. In another technique, the petroleum is burned in place to produce electricity at the source. Although enhanced recovery is more expensive than primary and secondary recovery, it becomes economically feasible as oil supplies dwindle and prices rise.

Other technological initiatives seek alternatives to petroleum as the primary energy provider for transportation. Liquefied and gasified coal, for example, show promise as

Figure 20-25 Magnetic-levitation trains represent one alternative to petroleum-based transportation systems.

transportation fuel. Most major automobile manufacturers have begun to develop electric cars with long-distance cruising capacity, capable of superhighway speeds. Other attempts to prepare for the inevitable disappearance of petroleum include the implementation of solar-powered buses and trains propelled by powerful electromagnets (Fig. 20-25).

Meeting Future Mineral Needs

The supply of mineral resources remains adequate, although some deposits are located in remote places. Geologists continue to discover valuable deposits of important metals and nonmetals. For example, major ore bodies have recently been unearthed in Chile, Australia, and Siberia. Extensive deposits rich in zinc, lead, copper, nickel, cobalt, and uranium have also been found in Antarctica, along with a 120-kilometer (70-mile)-long, 100-meter (330-feet)-thick deposit of iron, which is large enough to meet the world's iron demand for 200 years.

Mineral prospecting has ceased to be a pick-and-shovel operation. To find metals buried beneath more than 150 meters (500 feet) of sediment on the sea floor, gravimeters and magnetometers are employed to identify gravity and magnetic anomalies (discussed in Chapter 11). Gravimeters carried in airplanes and helicopters have even detected vast bodies of chromite, magnetite, and nickel beneath the ice of Antarctica (as mentioned earlier, an international treaty has placed this area off-limits for resource development). Mass spectrometers can analyze the chemistry of soils, and even that of the gases trapped between soil particles, to determine the composition of the underlying bedrock from which the soil developed. Infrared satellite images and satellite-borne radar can map the surface and shallow-subsurface geology of the Earth's most remote regions, identifying previously unknown mineral deposits with mining potential.

Figure 20-26 Nodules of manganese oxide, found abundantly on the ocean floor, may become a source of valuable minerals in the near future.

Soon we may be able to mine the vast deposits on the sea floor, which in some places is covered by billions of nodules of manganese oxide (MnO_2), some approaching the size of bowling balls. These nodules (Fig. 20-26), which contain lesser amounts of iron, nickel, copper, zinc, and cobalt, accumulate where deep-sea sedimentation occurs slowly enough to allow them to grow without being buried. Collectively, they may represent the Earth's largest mineral deposit. Although their origin is uncertain, manganese nodules may form much like ooliths (discussed in Chapter 6), as dissolved minerals precipitate around an organic nucleus such as a shark's tooth or shell fragment. As the technology to gather manganese oxide nodules improves, it will spawn considerable international debate regarding ownership and mining rights.

Most mineral resources can be recycled. Steel from automobiles and old bridges, mercury from discarded thermometers and fluorescent mercury-vapor lamps, copper from

Figure 20-27 Beverage cans are processed into high-density bales in a recycling plant in New Jersey. Their aluminum can then be recycled.

electrical wire, and platinum from the catalytic converters of abandoned automobiles can all be reclaimed for reuse. Recycling has several benefits: It reduces the volume of waste requiring disposal, the land area disturbed by new mining operations, and, in some cases, the use of energy to mine and refine new ores. Scrap-aluminum recycling, for example, uses only one-twentieth the energy needed to mine and process an equivalent amount of new aluminum from bauxite (Fig. 20-27).

Of course, some metals are difficult to recycle—for example, the aluminum and steel intermingled with other materials in such complex manufactured objects as refrigerators and lawn mowers. Precious metals, such as gold and silver, are easiest to reuse. The gold fashioned today into a new pair of earrings may have been mined thousands of years ago by ancient Romans or Peru's Incas.

The discarded, seemingly worthless waste-rock piles, or tailings, from old mining operations have emerged as a source of valuable minerals. Acids and other solutions can be flushed through mine tailings to leach metals from them or to dissolve the surrounding matrix materials. This process concentrates substantial quantities of valuable minerals from the mine waste, which is itself reduced in volume.

Everything we have or can make comes from the Earth. Now that you understand the processes that built, moved, and shaped the continents and ocean floors, and the vast amount of time over which the Earth and its resources developed, you can appreciate why we must manage wisely the materials that make our modern lives possible. If they are treated with disrespect, most assuredly little will remain for future generations. Industry, local governments, and the world community must cooperate to ensure that the search for, development of, and use of the planet's resources do not squander those resources or irreversibly damage our shared environment (Fig. 20-28). With the knowledge you have acquired from your geology training, you are now prepared to contribute to the ongoing debate about the need to balance resource development and environmental protection. By all means, use your training and be heard.

Figure 20-28 This garbage dump near Mount Everest illustrates that virtually no place on Earth escapes the environmental impact of careless human behavior. This refuse will remain here for decades or centuries, because decomposition occurs only slowly in the cold, dry climate of the region.

Chapter Summary

Natural resources, generally classified as either energy or mineral resources, include all types of fuels and a variety of metals and nonmetals. The availability of natural resources depends on their concentration, the technology needed to extract and develop them, and the economic forces of supply and demand. **Reserves** are quantities of natural resources that have been discovered and can be exploited profitably with current technology. **Resources** are deposits believed to exist, but that are not considered exploitable today for various technological, economic, or political reasons. **Ores** are mineral deposits that can be mined profitably. Renewable natural resources are either replenished naturally over a short time span, such as trees, or can be used continuously without being depleted, such as sunlight. Nonrenewable resources form so slowly that they are consumed much more quickly than nature can replenish them, such as **fossil fuels,** which are derived from the organic remains of past life.

Most of our current energy needs are satisfied by the gaseous, liquid, and semi-solid substances known as **petroleum.** These fossil fuels consist of **hydrocarbons,** complex organic molecules composed entirely of hydrogen and carbon. Petroleum typically forms in a marine basin where abundant nutrient input and warm, shallow water nourish microscopic organisms that are eventually buried by younger sediments.

Pressure and geothermal heat transform the organic remains first into **kerogen,** a solid waxy material, and then into gas, oil, and other hydrocarbons. Petroleum forms in **source rocks** (typically marine muds), then migrates into permeable **reservoir rocks. Oil shale** is a relatively impermeable, fine-grained source rock that retains its kerogen. **Oil sand** is an unconsolidated clay or sand that contains a semi-solid, tar-like hydrocarbon called **bitumen.**

Coal, a sedimentary rock, consists of the combustible, highly compressed remains of land plants. It begins to form as **peat,** a soft brown mass of compressed, largely undecomposed plant materials, accumulates. Bacterial action within the sediment gradually breaks down the vegetation to an organic muck. Pressure from the weight of overlying deposits squeezes water from this muck, increasing the proportion of carbon to form **lignite** (70% carbon). Deep burial and increasing geothermal heat convert lignite to **bituminous coal** (80%–93% carbon). Metamorphic conditions transform lignite and bituminous coal into **anthracite** (93%–98% carbon). The more carbon that coal contains, the harder it is and the more energy per unit volume it can produce.

Coal is the most abundant fossil fuel, but its widespread use causes several significant environmental problems. **Acid rain** is one undesirable consequence of burning sulfur-rich coal. Burning coal releases sulfur dioxide gas, which combines with rainwater to produce sulfuric acid. Nitric acid, formed largely from automobile emissions, also falls as acid rain. Other environmental problems associated with the use of fossil fuels include global warming and marine oil spills.

The dwindling of fossil fuel supplies makes it necessary to discover and develop alternative energy sources. Communities and industries today tap the Earth's geothermal energy, dam streams and rising tides for hydroelectric and tidal power, harness the energy of the Sun and the wind, and burn **biomass fuels,** derived from plants and animals, to heat homes and generate electricity. Physicists and engineers have captured the energy released from the splitting of nuclei of certain atoms, a process known as **nuclear fission. Breeder reactors** can create a vast supply of fissionable materials. Safe disposal of radioactive waste poses a major challenge to the nuclear-energy industry. Researchers also are studying ways to generate energy by fusing atomic nuclei together, a process called **nuclear fusion.**

The other principal group of resources—minerals—includes metals such as iron, copper, aluminum, gold, and silver. Humans have used metals for thousands of years, initially in the metal's native state, then smelted from rocks and combined to form harder metallic **alloys.** Nonmetals include sand and gravel for building, limestone for cement, potassium and phosphorus for fertilizers, and gemstones, such as diamonds and emeralds.

For metals to be economically useful, some geological process must have concentrated them somewhere near the Earth's surface. Metals may precipitate from magma-derived water or from circulating groundwater or seawater to produce concentrated **hydrothermal deposits. Porphyry copper deposits,** found in the granitic rocks of many subduction-zone batholiths, form when water-rich felsic magma cools and expels its water; this water, along with infiltrating surface water, is converted to steam. The steam expands rapidly, rupturing newly crystallized rock, and deposits hydrothermal solutions containing copper and other metals in innumerable fractures at the margin of the batholith. **Massive sulfide deposits** form principally at oceanic divergent zones, where seawater enters cracks in hot, young basaltic rock, becomes heated, dissolves metals, then rises to the sea floor, depositing the metals around surface vents.

Sedimentary and weathering processes may concentrate valuable metal deposits as well. Minerals accumulate as **placer deposits** where the velocity of surface waters decreases, forcing dense particles of the transported sediment (including precious metals—such as gold and silver and occasionally diamonds, emeralds, and other gemstones) to settle out of the current. Every continent contains vast **banded iron formations** dating from about 2 billion years ago, when the oxygen content of the atmosphere became great enough to prompt oxidation and precipitation of the iron dissolved in surface waters. Dissolution of metals by chemical weathering, followed by precipitation from groundwater, and dissolution of a soluble matrix surrounding a metal deposit both concentrate metallic ore bodies by processes of **secondary**

enrichment. Most major metal deposits occur near past or present plate boundaries, likely sites for the generation of magmas and hot hydrothermal solutions.

Mining operations typically produce large volumes of waste material and can contaminate groundwater. These environmental problems can be minimized by isolating sulfides from groundwater and carrying out land reclamation when mining ends.

Nonmetals are widely used as building materials (such as sand and gravel for concrete, limestone for cement, gypsum for plaster of Paris, and clay for bricks) and agricultural fertilizers (such as phosphorus and potassium). Nonmetals commonly used in industry include sulfur (used to treat a variety of products), pure quartz sand (used for glass production), and asbestos (used for fireproofing clothing and electrical insulation).

As our use of energy and other resources increases, future resource needs will be met only through development of alternative energy sources, conservation, and enhanced recovery of known reserves. The search for new mineral supplies will take us to such remote places as the sea floor (for manganese nodules) and possibly Antarctica.

Key Terms

reserves (p. 570)
resources (p. 570)
ore (p. 570)
fossil fuels (p. 571)
petroleum (p. 571)
hydrocarbons (p. 571)
kerogen (p. 572)
source rocks (p. 573)
reservoir rocks (p. 573)
oil shale (p. 573)
oil sand (p. 574)
bitumen (p. 574)
peat (p. 575)
lignite (p. 575)
bituminous coal (p. 575)
anthracite (p. 575)
acid rain (p. 576)
biomass fuels (p. 581)
nuclear fission (p. 582)
breeder reactor (p. 582)
nuclear fusion (p. 584)
alloys (p. 585)
hydrothermal deposits (p. 585)
porphyry copper deposits (p. 585)
massive sulfide deposits (p. 585)
placer deposits (p. 587)
banded iron formations (p. 588)
secondary enrichment (p. 588)

Questions for Review

1. How do resources differ from reserves? Give an example of each.

2. What factors determine whether a metal deposit is an ore?

3. List three fossil fuels and three alternative energy sources. List three metallic and three nonmetallic resources.

4. What are the key physical properties of a petroleum reservoir rock? Is oil shale a source rock or a reservoir rock?

5. Briefly explain how to harness tidal power.

6. Describe how a breeder reactor produces its own fuel.

7. Discuss two ways in which geological processes may concentrate dispersed metals in rocks.

8. Name three types of locations where placer deposits may be found.

9. How does plate tectonics affect the distribution of the Earth's metal deposits?

10. Name three sources for the sand and gravel used in the building industry.

For Further Thought

1. Using your general knowledge of the geology of North America, explain why aquatic life in the lakes of Yosemite National Park, California (see Chapter 3), would be more susceptible to the effects of acid rain than aquatic life in the lakes of Florida and Indiana (see Chapter 16).

2. Referring to Figure 20-20, page 587, in which areas would you search for oil? Gold? Sand and gravel?

3. Speculate on why copper mining in the United States has been so drastically reduced in recent years. Under what circumstances might U.S. mining companies resume extensive copper mining?

4. Propose a U.S. energy plan for the year 2050. Which energy sources would you develop? Why?

5. Suppose the Earth's natural resources become completely exhausted and it is decided to prospect on neighboring planets. Which planet might hold extensive iron, chromium, and platinum deposits? Which might hold sand and gravel deposits? Could any of our planetary neighbors provide us with limestone or gypsum?

Appendix A

Conversion Factors for English and Metric Units

Length	1 centimeter	=	0.3937 inches
	1 inch	=	2.54 centimeters
	1 meter	=	3.2808 feet
	1 foot	=	0.3048 meters
	1 yard	=	0.9144 meters
	1 kilometer	=	0.6214 miles (statute)
	1 kilometer	=	3281 feet
	1 mile (statute)	=	1.6093 kilometers
Velocity	1 kilometer/hour	=	0.2778 meters/second
	1 mile/hour	=	0.4471 meters/second
Area	1 square centimeter	=	0.16 square inches
	1 square inch	=	6.45 square centimeters
	1 square meter	=	10.76 square feet
	1 square meter	=	1.20 square yards
	1 square foot	=	0.093 square meters
	1 square kilometer	=	0.386 square miles
	1 square mile	=	2.59 square kilometers
	1 acre (U.S.)	=	4840 square yards
Volume	1 cubic centimeter	=	0.06 cubic inches
	1 cubic inch	=	16.39 cubic centimeters
	1 cubic meter	=	35.31 cubic feet
	1 cubic foot	=	0.028 cubic meters
	1 cubic meter	=	1.31 cubic yards
	1 cubic yard	=	0.76 cubic meters
	1 liter	=	1000 cubic centimeters
	1 liter	=	1.06 quarts (U.S. liquid)
	1 gallon (U.S. liquid)	=	3.79 liters
Mass	1 gram	=	0.035 ounces
	1 ounce	=	28.35 grams
	1 kilogram	=	2.205 pounds
	1 pound	=	0.45 kilograms
Pressure	1 kilogram/square centimeter	=	0.97 atmospheres
	1 kilogram/square centimeter	=	14.22 pounds/square inch
	1 kilogram/square centimeter	=	0.98 bars
	1 bar	=	0.99 atmospheres
Temperature	°F (degrees Fahrenheit)	=	°C(9/5) + 32
	°C (degrees Celsius)	=	(°F – 32)(5/9)

Appendix B

A Statistical Portrait of Planet Earth

Surface Areas

Landmasses	150,142,300 kilometers^2 (57,970,000 miles^2)
Oceans and Seas	362,032,000 kilometers^2 (138,781,000 miles^2)
Entire Earth	512,175,090 kilometers^2 (197,751,500 miles^2)

Distribution of Water, by Volume

Oceans and Seas	1.37×10^9 kilometers^3 (3.3×10^8 miles^3)
Glaciers	2.5×10^7 kilometers^3 (7×10^6 miles^3)
Groundwater	8.4×10^6 kilometers^3 (2×10^6 miles^3)
Lakes	1.25×10^5 kilometers^3 (3×10^4 miles^3)
Rivers	1.25×10^3 kilometers^3 (3×10^2 miles^3)

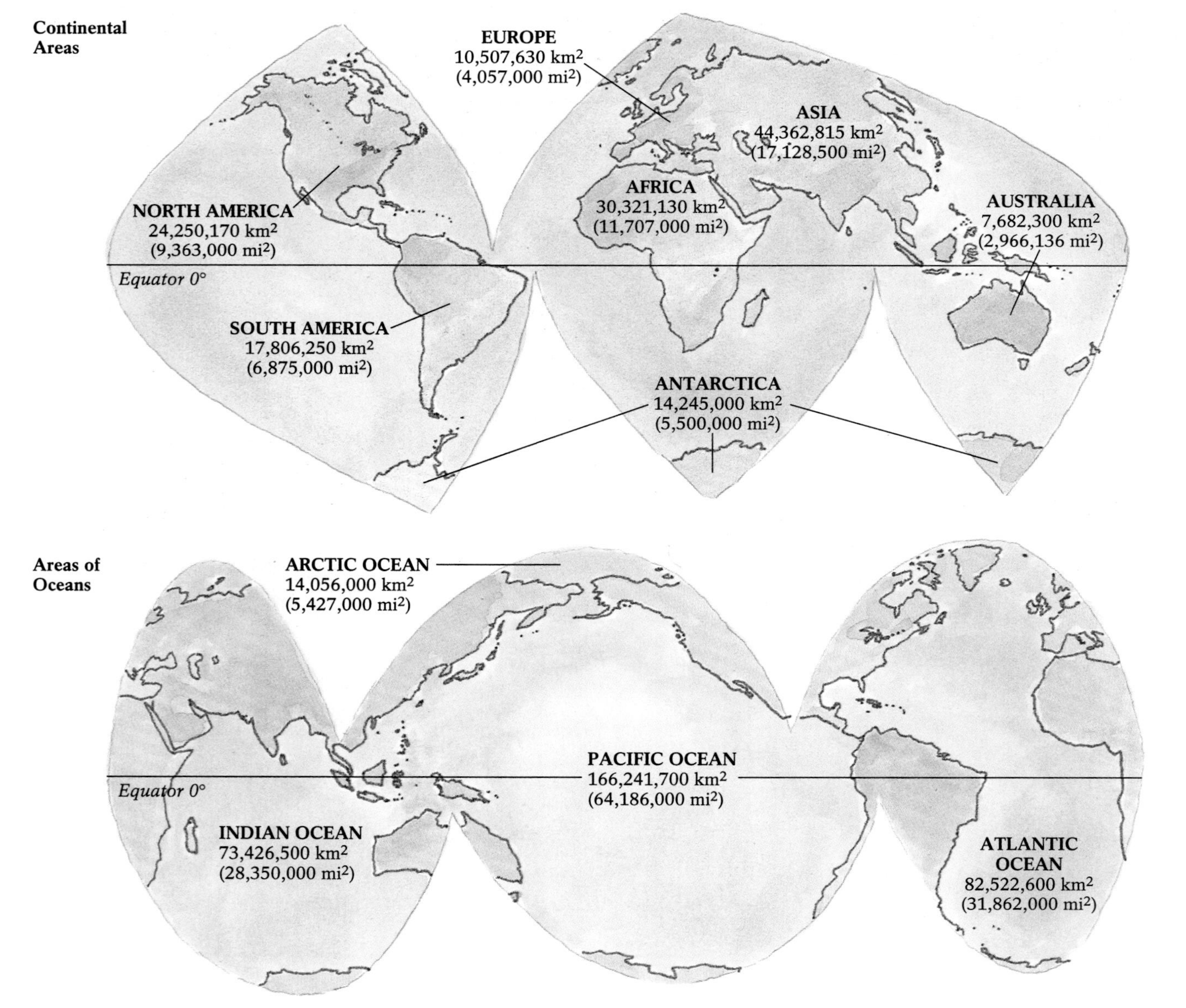

Elevations, Depths, and Distances

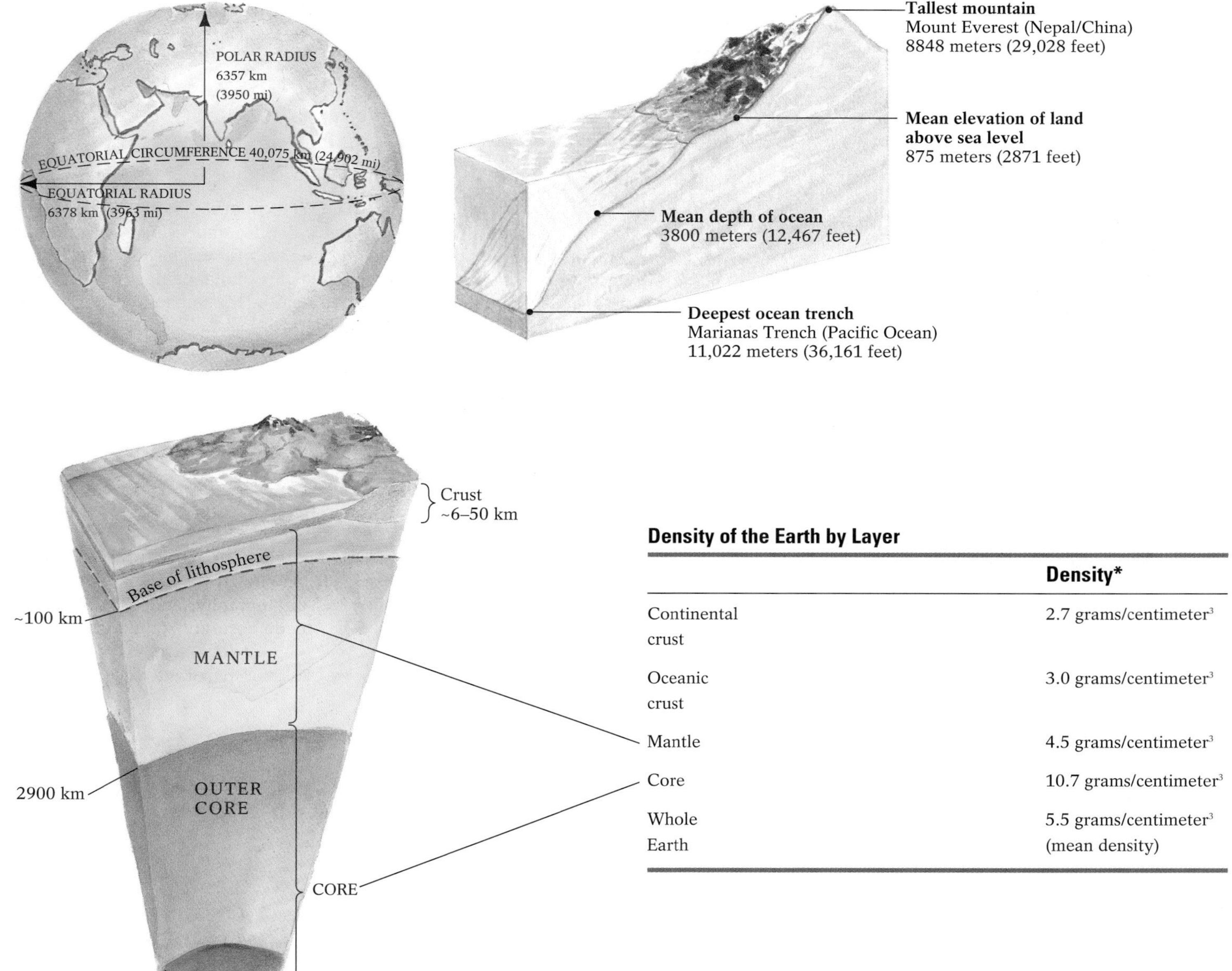

Density of the Earth by Layer

	Density*
Continental crust	2.7 grams/centimeter^3
Oceanic crust	3.0 grams/centimeter^3
Mantle	4.5 grams/centimeter^3
Core	10.7 grams/centimeter^3
Whole Earth	5.5 grams/centimeter^3 (mean density)

Appendix C

Reading Topographic and Geologic Maps

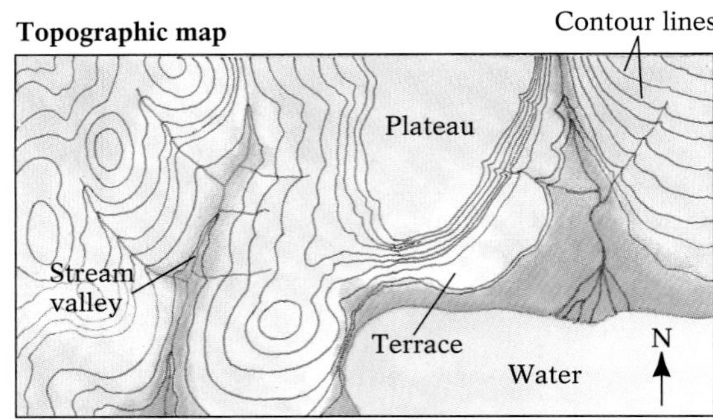

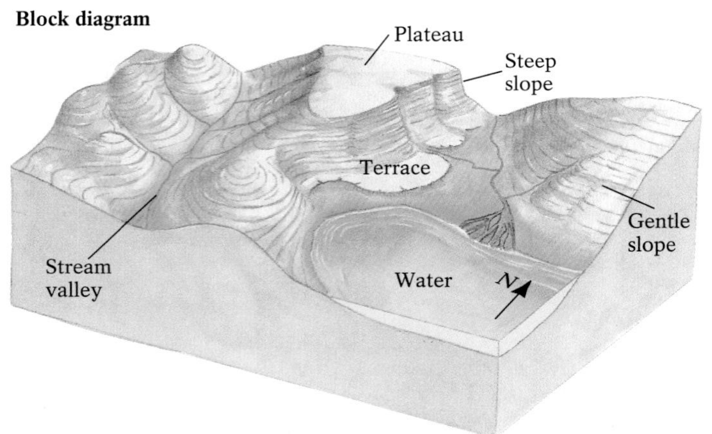

Throughout this textbook, maps have been used to show the locations of and relationships among various features. Geologists use maps for soil management, flood control, environmental planning, finding such resources as ores and groundwater, and determining optimal locations for fuel pipelines, highways, recreational areas, and the like.

Most maps show all or part of the Earth's surface, drawn to scale. A map's scale is usually shown at the bottom of the map. It may feature miles, kilometers, or any other convenient unit of measurement. Most maps from the United States Geological Survey are drawn to a scale of 1:24,000, meaning that 1 centimeter = 240 meters or 1 inch = 2000 feet; the latter would be read as "one inch on this map is equal to two thousand feet in real-world distance." The U.S. Geological Survey has been phasing in metric topographic maps using a scale of 1:100,000.

Most maps show two dimensions, marked by longitude (vertical) and latitude (horizontal) lines. A third dimension—elevation or depth—can be shown on *topographic maps,* which show relief using contour lines that mark specific heights above or depths below sea level. Every point on a contour line is at the same elevation. Every contour line closes upon itself—delineating the border of a discrete area—although its entire length may not be visible within the margins of a given map. A map's contour interval, or the distance between adjacent contour lines, is critical to both the usefulness and the appearance of the map: Contour lines that occur close together represent a steep slope; those that are farther apart represent a more gentle one. In an area of low relief, intervals designating elevation differences of only 5 or 10 feet are appropriate, whereas terrain with great relief may require intervals of 50 feet or more. On a given map, all contour intervals are the same.

In the past, accurate map-making depended predominantly upon actual measurements made on site by surveyors, geologists, and others. The advent of aerial photography, and then of space-based satellite photography, has made accurate standardized revisions possible.

The U.S. Geological Survey uses color to indicate specific features:

black and/or red solid lines	:	major roads
brown lines	:	contour lines
black	:	human-made structures, names
light red lines	:	town limits
blue	:	water features
green	:	wooded areas
white	:	open fields, deserts, and other nonvegetated areas

Dotted or dashed lines are used for temporary features (that is, solid blue represents a lake, whereas a dashed blue line marks a seasonal stream). Any special symbols used on a map are explained in its legend, which appears with the scale in the bottom margin.

A region's underlying geology can be represented on a *geologic map.* Symbols representing various types of rock have been standardized; such commonly accepted rock symbols have been used throughout this book. Standardized colors are used to show rock ages. The key to a geologic map's labeling, colors, and symbols usually appears at its side margins.

Columbus Quadrangle, New Jersey

Scale 1:24,000
Contour interval 10 feet

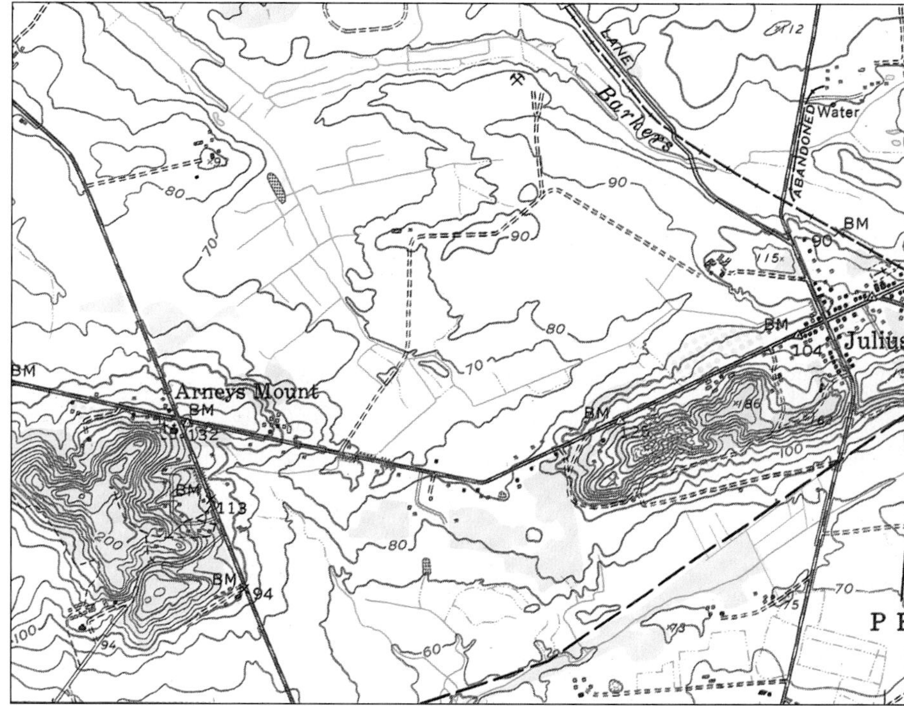

Topographic map

Geologic map

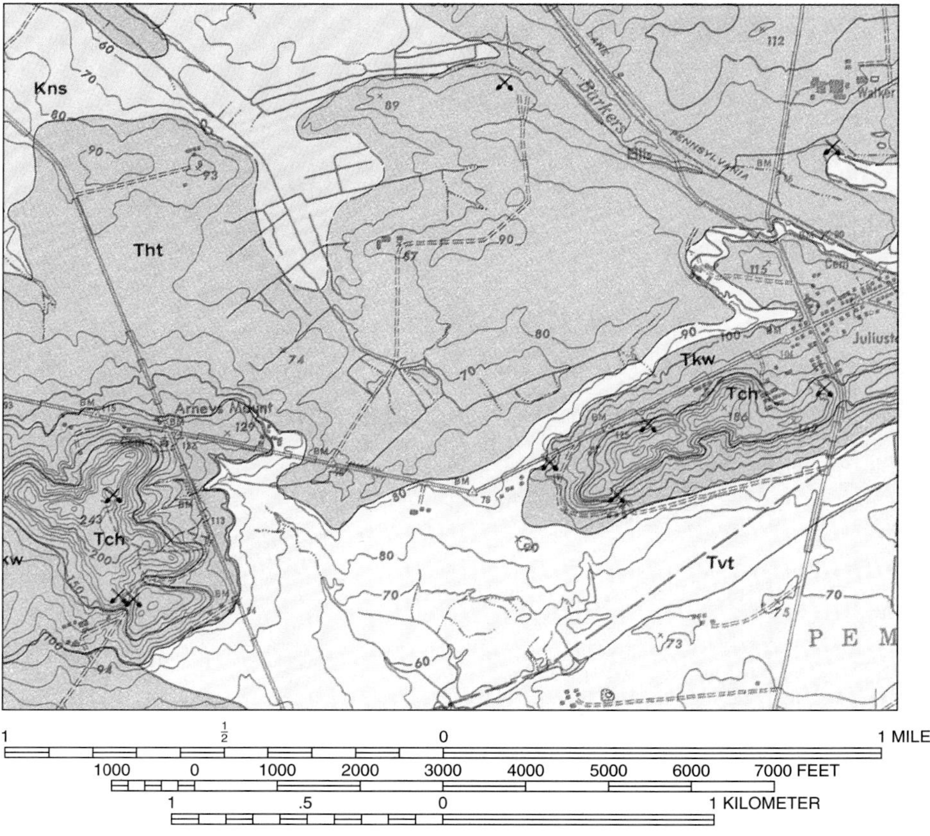

Tertiary Period Deposits

- **Tch** — Cohansey sand
  Sand, quartz, light-gray to yellow-brown, medium- to coarse-grained, pebbly, ilmenitic, micaceous, stratified
- **Tkw** — Kirkwood formation
  Sand, quartz, light-gray to tan, fine- to very fine-grained, clayey, micaceous, ilmenitic, kaolinitic, sparingly lignitic, massive-bedded
- **Tmq** — Manasquan formation
  Sand, quartz, dark green-gray, medium- to coarse-grained, glauconitic, clayey
- **Tvt** — Vincentown formation
  Upper member—calcarenite, quartz and glauconite, dusky-yellow to pale-olive, clayey; lower member—sand, quartz, dark-gray, poorly sorted, fine to coarse, clayey, glauconitic, entire formation very fossiliferous
- **Tht** — Hornerstown sand
  Sand, glauconite, dusky-green, medium- to coarse-grained, clayey, massive-bedded

Cretaceous Period Deposits

- **Krb** — Red Bank sand
  Lower member—sand, glauconite, dark grayish-black, coarse-grained, very clayey, micaceous, lignitic, massive-bedded (upper member not present)
- **Kns** — Navesink formation
  Sand, glauconite, varying amounts of quartz, greenish-black to brown, medium- to coarse-grained, clayey
- **Kml** — Mount Laurel sand
  Sand, quartz, reddish-brown to green-gray, poorly sorted, fine- to coarse-grained, glauconitic, massive-bedded

Appendix D

Mineral-Identification Charts

Most Common Rock-Forming Minerals

Occurrence	Class	Structure	MINERAL OR GROUP NAME	COMPOSITION/ VARIETIES	CRYSTAL FORM/OTHER DIAGNOSTIC FEATURES	CLEAVAGE/ FRACTURE	USUAL COLOR/ LUSTER	STREAK	HARDNESS
Light-colored; abundant in all rock types	SILICATES	Framework	Feldspar	Potassium (orthoclase) feldspar ($KAlSi_3O_8$)	Coarse crystals or fine grains	Good cleavage in two directions at 90°	White to gray, with pearly luster, frequently salmon-pink	White	6
				Sodium, calcium (plagioclase) feldspar Albite: $NaAlSi_3O_8$ Anorthite: $Ca_2Al_2Si_3O_8$		Good cleavage in two directions at 90°; cleavage surfaces striated	White to gray, sometimes green or yellowish	White	
			Quartz	SiO_2	Six-sided crystals; individual or in masses	No cleavage; conchoidal fracture	Colorless or slightly smoky gray, pink or yellow	White	7
		Sheet	Mica	Muscovite $KAl_3Si_3O_{10}(OH)_2$	Thin, disc-shaped crystals	One perfect cleavage plane; splits into very thin sheets	Colorless or slightly gray, green, or brown, with vitreous luster	White	2–2½
				Biotite $K(Mg, Fe)_3AlSi_3O_{10}(OH)_2$	Irregular foliated masses	One perfect cleavage plane; splits into thin sheets	Black, brown, or green, with vitreous luster	White or gray	2½–3
				Chlorite $(Mg, Fe)_5(Al, Fe)_2Si_3O_{10}(OH)_8$			Yellowish, brown, green, or white	White or colorless	2–2½
Dark-colored; abundant in metamorphic and igneous rocks		Double Chain	Amphibole	Actinolite $Ca_2(Mg, Fe)_5Si_8O_{22}(OH)_2$	Long, six-sided crystals, fibrous or in aggregates	Two good cleavage planes at 56° and 124°	Pale to dark green or black, with vitreous luster. Pure actinolite white.	Pale green or white	5–6
				Hornblende $(Ca, Na)_{2\text{-}3}(Mg, Fe, Al)_5Si_6(Si, Al)_2O_{22}(OH)_2$					
		Single Chain	Pyroxene	Augite $Ca(Mg, Fe, Al)(Al, Si)\ O_6$	Short, four- or six-sided crystals	Two good cleavage planes at about 90°	Light to dark green, with vitreous luster	Pale green	5–6
				Diopside $CaMg(Si_2O_6)$			White to light green, with vitreous luster	White or pale green	
				Orthopyroxene $MgSiO_3$			Pale gray, green, brown, or yellow, with vitreous luster	White	
		Single Tetrahedra	Olivine	$(Mg, Fe)_2SiO_4$	Small grains and granular masses	No cleavage; conchoidal fracture	Grayish green or brown, with vitreous, glassy luster	White	6½–7
			Garnet	$(Ca_3, Mg_3, Fe_3, Al_2)n(SiO_4)_3$	12- or 24-sided crystals	No cleavage; conchoidal fracture	Deep red, with vitreous to resinous luster	White	6½–7½
Light-colored; abundant in sedimentary rocks	CARBONATES		Calcite	$CaCO_3$	Fine to coarsely crystalline. Effervesces rapidly in HCl.	Three oblique cleavage planes, forming rhombohedral cleavage pieces	White or gray, with pearly luster	White	3
			Dolomite	$CaMg(CO_3)_2$	Fine to coarsely crystalline. Effervesces slowly in HCl when powdered.		Colorless, white, or pink, and may be tinted by impurities, with pearly luster	White to pale gray	3½–4
	CLAY MINERALS (Hydrous alumino-silicates)		Kaolinite	$Al_2Si_2O_5(OH)_4$	Very fine grains; found as bedded masses in soils and sedimentary rocks; earthy odor	Earthy fracture	White to buff, or tinted gray by impurities	White, off-white, or colorless	1½–2½
			Illite	$K_{0.8}Al_2(Si_{3.2}Al_{0.8})O_{10}(OH)_2$					
			Smectite	$Na_{0.3}Al_2(Si_{3.7}Al_{0.3})O_{10}(OH)_2$					

Accessory or Less-Abundant Rock-Forming Minerals

		MINERAL OR GROUP NAME	COMPOSITION/ VARIETIES	CRYSTAL FORM/OTHER DIAGNOSTIC FEATURES	CLEAVAGE/ FRACTURE	USUAL COLOR/ LUSTER	STREAK	HARDNESS
Light-colored; common in sedimentary rocks	SULFATES	Gypsum	$CaSO_4 \cdot 2H_2O$	Tabular crystals in fine-to-granular masses	One perfect cleavage plane, forming thin sheets; also two other good cleavage planes	Colorless to white, with vitreous luster	White	1–2½
		Anhydrite	$CaSO_4$	Granular masses	Three good to perfect cleavage planes at 90°	White, gray, or blue-gray, with pearly to vitreous luster	White	3–3½
		Halite	NaCl	Perfect cubic crystals, soluble in water. Tastes salty.	Three excellent cleavage planes at 90°	White or gray, with pearly luster	White	2½
Light-colored; mainly in metamorphic and igneous rocks	ALUMINO-SILICATES	Kyanite	Al_2SiO_5	Long, bladed, or tabular crystals	One perfect cleavage plane, parallel to length of crystals	White to light blue, with vitreous luster	White	5 along cleavage plane, 6½–7 across cleavage plane
		Sillimanite		Long, slender crystals or fibrous masses		White to gray, with vitreous luster	White	6–7
		Andalusite		Coarse, nearly square crystals	Irregular fracture	Red, reddish brown, or green, with vitreous luster	White	5–6
		Serpentine	$Mg_6Si_4O_{10}(OH)_8$	Fibrous or platy masses	Splintery fracture	Light to dark green or brownish yellow, with pearly luster	White	4–6
		Talc	$Mg_3Si_4O_{10}(OH)_2$	Foliated masses. Feels soapy.	Perfect in one direction, forming thin flakes	White to pale green, with pearly or greasy luster	White	1–1½
		Corundum	Al_2O_3	Short, six-sided crystals	Irregular fracture	Usually brown, pink, or blue, with adamantine luster	None	9
		Fluorite	CaF_2	Octahedral or cubic crystals	Cleaves easily	White, yellow, green, or purple, with vitreous luster	White	4
Dark-colored; common in metamorphic rocks	SILICATES	Epidote (paired tetrahedra)	$Ca_2(Al,Fe)Al_2Si_3O_{12}(OH)$	Usually granular masses; also slender prisms	One good cleavage direction, one poor	Yellow to dark green, with vitreous luster	White or gray	6–7
		Staurolite (single tetrahedra)	$Fe_2Al_9Si_4O_{22}(O,OH)_2$	Short crystals, some cross-shaped	One poor cleavage direction	Brown or reddish brown to black, with vitreous luster	Off-white to white	7
		Graphite	C	Scaley, foliated masses. Feels greasy.	One direction of cleavage	Steel gray to black, with vitreous or pearly luster	Gray or black	1–2
Dark-colored; common in all rock types		Apatite	$Ca_5(PO_4)_3(OH,F,Cl)$	Granular masses	Poor cleavage	Green, brown, or red, with adamantine or greasy luster	White	5
		Magnetite	Fe_3O_4	Granular masses	Uneven fracture	Black, with metallic luster	Black	5½
		Hematite	Fe_2O_3	Granular masses	Uneven fracture	Brown-red to black, with earthy, dull, to metallic luster	Brick red	5½
		Limonite	$2Fe_2O_3 \cdot 3H_2O$	Earthy masses	Uneven fracture	Yellowish brown to black	Brownish yellow	5–5½
Metallic luster; common in all rock types	SULFIDES	Pyrite	FeS_2	Cubic crystals or granular masses	Uneven fracture	Pale brass yellow, with metallic luster	Greenish black	6–6½
		Galena	PbS	Cubic crystals or granular masses	Three perfect cleavage planes at 90°	Silver gray with metallic luster	Gray	2½
		Sphalerite	ZnS	Granular masses	Six perfect cleavage planes at 60°	White to green, brown, or black, with resinous to submetallic luster	Reddish brown to yellow brown	3½–4
		Chalcopyrite	$CuFeS_2$	Granular masses	Irregular fracture	Brass yellow	Greenish black	3½–4

Minerals and Elements of Industrial or Economic Importance

MINERAL OR GROUP NAME	COMPOSITION	USUAL COLOR/ LUSTER	STREAK	HARDNESS	OTHER PROPERTIES/ COMMENTS	ORIGIN
Asbestos (Chrysotile)	$Mg_3Si_2O_5(OH)_4$	White to pale green, with pearly luster	White	1–2½	Flexible, nonflammable fibers	A variety of serpentines; found mostly in metamorphic rock
Bauxite	$Al(OH)_3$	Reddish to brown, with dull luster	Pale reddish brown	1½–3½	Found in earthy, clay-like masses. Principal source of commercial aluminum.	Weathering of many rock types
Chalcopyrite	$CuFeS_2$	Yellow, with metallic luster	Greenish black	3½–4½	Uneven fracture; softer than pyrite. Irridescent tarnish. Most common ore of copper.	Hydrothermal veins and porphyry copper deposits
Chromite	$FeCr_2O_4$	Black, with metallic or submetallic luster	Dark brown	5½	Massive; granular; compact. Most common ore of chromium. Used in making steel.	Ultramafic igneous rocks
Copper (native)	Cu	Red, with metallic luster	Red	2½–3	Malleable and ductile. Tarnishes easily.	Mafic igneous rock (basaltic lavas); also in oxidized ore deposits
Galena	PbS	Silver-gray, with metallic luster	Gray	2½	Perfect cubic cleavage. Most important ore of lead; also commonly contains silver.	Hydrothermal veins
Gold (native)	Au	Yellow, with metallic luster	Yellow	2½–3	Malleable and ductile	Sedimentary placer deposits; hydrothermal veins
Hematite	Fe_2O_3	Brown-red to black, with earthy, dull, or metallic luster	Dark red	5½–6½	Granular, massive. Most important source of iron.	Found in all types of rocks. Commonly in contact-metamorphic aureoles around mafic igneous sills.
Magnetite	Fe_3O_4	Black, with metallic luster	Black	5½	Uneven fracture. An important source of iron. Strongly magnetic.	Found in all types of rocks
Platinum (native)	Pt	Steel-gray or silver-white, with metallic luster	Off-white, metallic	4–4½	Malleable and ductile. Occasionally magnetic.	Mafic igneous rocks and sedimentary placer deposits
Silver (native)	Ag	Silver-white, with metallic luster	Silver-white	2½–3	Malleable and ductile. Tarnishes to dull gray or black.	Mostly in hydrothermal veins; also in oxidized ore deposits
Sphalerite	ZnS	Shades of brown and red, with resinous to adamantine luster	Reddish brown	3½–4	Perfect cleavage in six directions at 120°. Most important ore of zinc.	Hydrothermal veins

Precious and Semi-Precious Gems*

MINERAL OR GROUP NAME	COMPOSITION	USUAL COLOR/ LUSTER	STREAK	HARDNESS	OTHER PROPERTIES	ORIGIN
Beryl (aquamarine, emerald)	$Be_3Al_2(SiO_3)_6$	Blue, green, yellow, or pink, with vitreous luster	White	7½–8	Uneven fracture; hexagonal crystals	Cavities in granites and pegmatites, and schist
Corundum (ruby, sapphire)	Al_2O_3	Gray, red (ruby), blue (sapphire), with adamantine luster	None	9	Short, six-sided crystals; irregular, occasionally cleavage-like fracture ("parting")	Metamorphic rocks; some igneous rocks
Diamond	C	Colorless or with pale tints, with adamantine luster	None	10	Octahedral crystals	Peridotite; kimberlite; sedimentary placer deposits
(S) Garnet	$(Ca_3, Mg_3, Fe_3, Al_2)n(SiO_4)_3$	Deep red, with vitreous to resinous luster	White	6½–7½	No cleavage; 12- or 24-sided crystals	Contact-metamorphic and regionally metamorphosed rocks, and sedimentary placer deposits
(S) Jadite (jade)	$NaAl(Si_2O_6)$	Green, with vitreous luster	White or pale green	6½–7	Compact fibrous aggregates	High-pressure metamorphic rocks
Olivine (peridot)	$(MgFe)_2SiO_4$	Light to dark green, with vitreous, glassy luster	White	6½–7	Uneven fracture, often in granular masses	Basalt, peridotite
(S) Opal	$SiO_2 \cdot nH_2O$	White with various other colors, with vitreous, pearly luster	White	5½–6½	Conchoidal fracture; amorphous; tinged with various colors in bands	Low-temperature hot springs; weathered near-surface deposits
(S) Quartz (includes amethyst, citrine, agate, onyx, bloodstone, jasper, etc.)	SiO_2	Colorless, white, or tinted by impurities, with luster depending on variety	White	7	No cleavage; six-sided crystals	Origin specific to gem variety. (Quartz found in all rock types except ultramafic igneous rocks.)
Topaz	$Al_2SiO_4(F, OH)_2$	Colorless, white, or pale pink or blue, with vitreous luster	White	8	Cleavage in one direction; conchoidal fracture	Pegmatite, granite, rhyolite
Tourmaline	$(Na, Ca)(Li, Mg, Al)$-$(Al, Fe, Mn)_6(BO_3)_3$-$(Si_6O_{18})(OH_4)$	Black, brown, green, or pink, with vitreous luster	White to gray	7–7½	Poor cleavage; uneven fracture; striated crystals	Metamorphic rocks; pegmatite; granite
(S) Turquois	$CuAl_6(PO_4)_4$-$(OH)_8 \cdot 5H_2O$	Blue-green, with waxy luster	Blue-green or white	5–6	Massive	Hydrothermal veins
Zircon	$Zr(SiO_4)$	Colorless, gray, green, pink, or light blue, with adamantine luster	White	7½	Poor cleavage	Felsic igneous rocks and sedimentary placer deposits

*Gems are classified as semi-precious *(S)* if they are more accessible than precious gems and/or their properties are somewhat less valued.

Appendix E

How to Identify Some Common Rocks

Rocks are aggregates of minerals, and some are aggregates of organic materials. They are generally classified according to whether they are of igneous, sedimentary, or metamorphic origin, and then on the basis of their physical properties—primarily texture and mineral composition.

Igneous Rocks

Igneous rocks are formed when a magma cools and crystallizes. The chemical composition of a magma determines the minerals and rocks that will be formed when it cools. The rate at which a magma cools determines the size of the resulting crystals, which in turn determines the texture of the resulting rocks. Both composition and texture are used to classify igneous rocks.

Color is used only broadly as a diagnostic property for igneous rocks. The descriptions *light, dark*, and *intermediate* are generally agreed on. But many igneous rocks comprise two different-colored components; some blend various grays and pinks. Color, then, can help reduce the possible identities of a rock, but in most cases, color is not the defining property.

I. *Phaneritic*, or coarsely crystallized rocks: Composed of individual crystals of about 1 to 5 millimeters in length

- A. If the rock contains quartz, it is probably **granite.**
- B. If the rock contains no quartz, and from 30% to 60% feldspar, it is **diorite.**
- C. If the rock contains no quartz and less than 30% feldspar, it is **gabbro.**
- D. If the rock contains neither quartz nor feldspar, it is likely to contain ferromagnesian minerals and be ultramafic **peridotite** (consisting largely of olivine or pyroxene).

II. *Aphanitic* rocks: Individual crystals are fine-grained or microscopic

- A. If quartz crystals can be identified in the rock, it is **rhyolite;** if the rock is light gray, white, or light green but too fine-grained to determine whether quartz is pres-ent, it is probably rhyolite.
- B. If the rock contains no visible quartz crystals but does have approximately equal proportions of white or gray feldspar and ferromagnesian minerals, it is **andesite.**
- C. If ferromagnesian minerals can be identified in the rock, it is **basalt;** if the rock is dark gray or black, and too fine-grained to identify any crystals, it is probably a basalt.

III. *Porphyritic* rocks: Contain macroscopic crystals (crystals that can be seen without the aid of a microscope) embedded in a matrix of smaller macroscopic or microscopic crystals. Use the descriptions above for aphanitic rocks to identify porphyritic rhyolites, basalts, and so on.

IV. *Glassy* rocks: Rocks look like solid glass. The most common is **obsidian,** which is rhyolitic in composition.

Sedimentary Rocks

Sedimentary rocks are made up of particles derived either from preexisting rocks or from organic debris. They are classified on the basis of their texture and mineral composition.

In identifying sedimentary rocks, the first step is to determine whether the rock contains carbonate minerals. This step is done by applying a drop of dilute hydrochloric acid to the rock surface. (*Note:* HCl must be used under close supervision to avoid burning one's skin or clothing.)

I. Rocks that do not effervesce (fizz), even when powdered—but may effervesce in some places, such as the fine cement between grains. Such rocks contain little or no carbonate minerals.

- A. Rocks with **clastic** texture: Composed of grains in a cement matrix

1. If the grains are more than 2 millimeters in diameter . . .
   a. and are angular, the rock is **sedimentary breccia.**
   b. and are rounded, the rock is **conglomerate.**
2. If the grains are from 1/16 to 2 millimeters in diameter, and the rock feels gritty, it is **sandstone.**
   a. If more than 90% of the grains are quartz, the rock is **quartz sandstone.**
   b. If more than 25% of the grains are feldspar, the rock is **arkose.**
   c. If more than 25% of the grains are fine fragments of shale, slate, basalt, and the like, the rock is **lithic sandstone.**
   d. If more than 15% of the rock is fine-grained matrix material, the rock is **graywacke.**
3. If the rock is fine-grained, feels smooth, . . .
   a. and the grains are visible with a hand lens, the rock is **siltstone.**
   b. and the grains are invisible even with a hand lens, and the rock is . . .
      i. laminated, or layered, it is **shale.**
      ii. unlayered, it is **mudstone.**

B. Rocks with **crystalline** texture: Composed of microscopic, interlocking crystals

1. If the rock dissolves in water, it is **rock salt.**
2. If its crystals are fine to coarse, and have a hardness of 2 on the Mohs scale, the rock is **gypsum.**

C. Rocks with indeterminate texture

1. If the rock is smooth, very fine-grained, and fractures conchoidally, it is **chert;** if it is dark in color, it is **flint.**
2. If the rock is black or dark brown and breaks easily, soiling the fingers, it is **coal.**

II. Rock that effervesces strongly is **limestone.** Limestone may be clastic or crystalline in texture.

A. If the rock is clastic in texture and contains fossils, it is **bioclastic limestone.**

1. If it is composed of whole, recognizable fossils, it is **coquina.**
2. If it is very fine-grained, light-colored, and powdery, it is **chalk.**

B. If the rock is clastic and composed of small spheres, it is **oolitic limestone.**

C. If the rock is coarsely crystalline and contains different-colored layers, it is **travertine.**

III. Rock that effervesces only when hammered to a powder is **dolostone.** Dolostone may be crystalline, or less commonly, clastic in texture.

Metamorphic Rocks

Metamorphic rocks are rocks that have had their composition or structure changed by intense heat or pressure. Two factors determine the nature of a metamorphic rock: the composition of the parent rocks and the combination of metamorphic factors that acted upon them. Different circumstances can produce different textures; thus texture is the primary basis for metamorphic rock identification.

I. Is the rock *foliated*? If so, identify the type of foliation and, if possible, the rock's mineral content.

A. If the rock is fine-grained and readily splits into sheets, it is **slate,** and will have an earthy luster.

B. If the rock is slate-like but has a silken luster, it is **phyllite.**

C. If the rock contains flat or needle-like mineral crystals that are virtually parallel to one another, it is **schist.** A schist that primarily contains mica is a mica schist; there are, similarly, garnet mica schists, hornblende schists, and talc schists. A schist that contains serpentine is **serpentinite.**

D. If the rock consists of separate layers of light and dark minerals, it is **gneiss.** Light-colored layers consist of feldspars and perhaps quartz; dark-colored layers are probably biotite, amphibole, or pyroxene.

II. Is the rock nonfoliated? If so . . .

A. Is quartz its primary constituent? If so, the quartz grains will be interlocking and the rock will be hard enough to scratch glass. The rock is **quartzite.**

B. Does it consist of interlocking, coarse crystals of calcite or dolomite? If so, it is **marble.**

C. Does it consist primarily of dark grains too fine to be seen unaided? If so, it is probably **hornfels;** it may also contain a few larger crystals of less common minerals.

Appendix F

Sources of Geologic Literature, Photographs, and Maps

American Geological Institute (AGI)
4220 King Street
Alexandria, VA 22302
or
AGI Publications Center
P.O. Box 205
Annapolis Junction, MD 20701

Issues the monthly "Bibliography and Index of Geology," which includes worldwide references and contains listing by author and subject.

Canadian Centre for Remote Sensing
2464 Sheffield Road
Ottawa, Ontario K1A-0Y7
CANADA

Source for obtaining aerial photographs and satellite imagery of Canada.

Earth Science Information Center
U.S. Geological Survey
507 National Center
Reston, VA 22092

Publishes topographic and other scale maps of the United States. Indexes listing available map scales are available for all the states.

Earthquake Engineering Research Institute
499 14th Street, Suite 320
Oakland, CA 94612

Provides a valuable field guide for learning about and studying earthquakes.

Federal Emergency Management Agency
P.O. Box 70274
Washington, D.C. 20024
ATTN: Publications

Provides a number of publications aimed at helping citizens prepare for geologic disasters, including "Are you Ready? Your Guide to Disaster Preparedness Checklist." Also issues a general "FEMA Publications Catalog."

Geological Society of America
P.O. Box 9140
Boulder, CO 80301

Publishes a large and diverse number of papers and journals concerning the geology and mineral resources of North America.

National Aerial Photography Library
615 Booth Street
Ottawa, Ontario K1A-0E9
CANADA

Source for obtaining aerial photographs and satellite imagery of Canada.

National Climate Center (NCC)
Federal Building
Asheville, NC 28801

Issues a monthly publication of climatological data, such as temperature and precipitation statistics for any given state or region. The NCC maintains up-to-date weather records for the entire United States.

National Geophysical and Solar–Terrestrial Data Center (part of the National Oceanic and Atmospheric Administration)
325 Broadway
Boulder, CO 80303

Maintains worldwide computer files of earthquakes recorded by seismographs and historic earthquake data.

National Oceanic and Atmospheric Administration (NOAA)
6001 Executive Boulevard
Rockville, MD 20852

This federal agency maintains a large collection of aerial photographs of the coastal regions of the United States. A detailed index is available.

State geological surveys

All states maintain a geological survey that produces technical reports, maps, and other publications at the state or county level.

U.S. Department of Agriculture
Aerial Photography Field Office
P.O. Box 30010
Salt Lake City, UT 84130

Can provide aerial photographs of most of the United States. A variety of scales is available.

U.S. Geological Survey (USGS)
Branch of Distribution
604 South Pickett Street
Alexandria, VA 22304
(also maintains offices across the country for over-the-counter sales of publications)

Produces a large number of maps, reports, circulars, professional papers, bulletins, water resources publications, and so on. Circular number 777 of the USGS provides an annually updated guide to obtaining information on the earth sciences.

U.S. Geological Survey (USGS)
EROS Data Center
Sioux Falls, SD 57198

Provides computer listings of most satellite imagery, high-altitude aerial photography, and photographs obtained during the Apollo, Skylab, and Gemini space missions. Worldwide coverage and indexes are available.

Appendix G

Careers in the Geosciences

Careers in the geosciences are numerous and varied. Several have been alluded to throughout this book. Employers in these fields include the energy industry, which hires about half of all professional geoscientists; the mining industry; federal and state governments; consulting firms, especially in environmental issues and hydrogeology; and academia, which employs more than 10% of all geoscientists.

Each of these broad fields, of course, involves a number of subdisciplines: the energy industry, for example, requires expertise in sedimentology and stratigraphy; mining careers require a background in economic geology, petrology, mineralogy, crystallography, or structural geology. The following is a list of some of the occupations in which people with an interest in the geosciences are employed:

Economic geologists conduct field investigations to determine the locations and economic viability of mineral deposits; they also investigate the genesis of mineral deposits. Mining companies typically employ economic geologists.

Engineering geologists investigate the geologic factors that affect human-made structures such as bridges, dams, and buildings, and the geologic effects of mass wasting.

Environmental geologists work in assessing, solving, and preventing problems associated with the pollution of soil, bedrock, and groundwater. For example, they assist in the selection of suitable locations for municipal- and hazardous-waste facilities such as landfills and waste-storage facilities, and may also help design such facilities.

Geochemists investigate the nature and distribution of chemical elements in the geologic environment.

Geochronologists determine the age of geologic materials, helping to reconstruct the geologic history of the Earth.

Geomorphologists study landforms, the rates and intensity of processes that created them, and the relationship of these landforms to the underlying geologic structures and climates that existed during their evolution.

Geophysicists attempt to determine the internal structure and properties of the Earth. They may focus on specific factors such as seismic waves, geomagnetism, or gravity.

Glaciologists investigate the physical and chemical properties of glacial masses. They also study the development, movement, and decay of glaciers and ice sheets and their deposits.

Hydrogeologists study the production, distribution, movement, and quality of groundwater in the Earth's crust. Many hydrogeologists work in the environmental industry, in cooperation with environmental geologists.

Hydrologists study the distribution, movement, and quality of surface bodies of water.

Marine geologists investigate the topography and sediments of the world's oceans. Marine geologists may work closely with petroleum geologists in off-shore exploration projects. They often work closely with oceanographers.

Mineralogists study the formation, composition, and properties—both physical and chemical—of minerals.

Paleontologists collect fossils and determine their age, reconstruct past environments, and regionally correlate rocks by determining the evolutionary sequence of fossil assemblages found in the rocks. Many paleontologists work in the oil industry.

Petroleum geologists work in the oil industry, and are involved with the exploration and production of petroleum and natural gas. They may also research unconventional sources such as oil shales and sands.

Petrologists study the mineralogical relationships of rocks, and specialize in determining the genesis of rocks; they may focus on either sedimentary, igneous, or metamorphic petrology.

Planetary geologists study the planets and their satellites to better understand the evolution of the solar system. Most planetary geologists are employed by universities or advanced research organizations such as NASA (National Aeronautics and Space Administration).

Sedimentologists study the formation of sedimentary rocks by assessing the processes of transportation, erosion, and deposition. Many sedimentologists work in the petroleum industry.

Stratigraphers decipher the sequence of rocks using the principle of superposition. They study the time and space relationships of rock sequences in an effort to determine the geologic history of an area.

Structural geologists investigate the phenomena and structures produced by the deformation of the Earth's crust, such as faults and folds. Most of the data used in structural geology are collected during detailed field work.

An academic background that includes courses in math and other sciences as well as in the specific geological fields mentioned here would be practical preparation for a geoscience career.

Glossary

ablation The loss of snow and ice from a *glacier*, caused primarily by melting.

abrasion A form of *mechanical weathering* that occurs when loose fragments or particles of rocks and minerals that are being transported, as by water or air, collide with each other or scrape the surfaces of stationary rocks.

absolute dating The fixing of a geological structure or event in time, as by counting tree rings.

accretionary wedge A mass of *sediment* and oceanic *lithosphere* that is transferred from a *subducting* plate to the less dense, overriding plate with which it *converges.*

accumulation The increase in a *glacier's* volume, caused primarily by snowfall.

acid rain Rain that contains such acidic compounds as sulfuric acid and nitric acid, which are produced by the combination of atmospheric water with oxides released when *hydrocarbons* are burned. Acid rain is widely considered responsible for damaging forests, crops, and human-made structures, and for killing aquatic life.

aeration zone See *zone of aeration.*

aftershock A ground tremor caused by the repositioning of rocks after an *earthquake.* Aftershocks may continue to occur for as long as two years after the initial earthquake. The intensity of an earthquake's aftershocks decreases over time.

aggradation The process by which a stream's *gradient* steepens due to increased deposition of *sediment.*

alloy A metal that is manufactured by combining two or more molten metals. An alloy is always harder than its component metals. Bronze is an alloy of copper and tin.

alluvial fan A triangular deposit of *sediment* left by a stream that has lost velocity upon entering a broad, relatively flat valley.

alluvium A deposit of *sediment* left by a stream on the stream's channel or *flood plain.*

alpine glacier A mountain *glacier* that is confined by highlands.

andesite The dark, *aphanitic, extrusive rock* that has a *silica* content of about 60% and is the second most abundant volcanic rock. Andesites are found in large quantities in the Andes Mountains.

angle of repose The maximum angle at which a pile of unconsolidated material can remain stable.

anthracite A hard, jet-black coal that develops from *lignite* and *bituminous coal* through *metamorphism,* has a carbon content of 92% to 98%, and contains little or no gas. Anthracite burns with an extremely hot, blue flame and very little smoke, but it is difficult to ignite and both difficult and dangerous to mine.

anticline A convex *fold* in rock, the central part of which contains the oldest section of rock. See also *syncline.*

aphanitic Of or being an *igneous rock* containing grains that are so small as to be barely visible to the naked eye.

aquiclude An impermeable body of rock that may absorb water slowly but does not transmit it.

aquifer A *permeable* body of rock or *regolith* that both stores and transports groundwater.

aquitard A layer of rock having low permeability that stores groundwater but delays its flow.

arête A sharp ridge of erosion-resistant rock formed between adjacent *cirque glaciers.*

aridity index The ratio of a region's potential annual evaporation, as determined by its receipt of solar radiation, to its average annual *precipitation.*

arroyo A small, deep, usually dry channel eroded by a short-lived or intermittent desert stream.

artesian Of, being, or concerning an *aquifer* in which water rises to the surface due to pressure from overlying water.

asthenosphere A layer of soft but solid, mobile rock comprising the lower part of the upper *mantle* from about 100 to 350 kilometers beneath the Earth's surface. See also *lithosphere.*

atoll A circular *reef* that encloses a relatively shallow *lagoon* and extends from a very great depth to the sea surface. An atoll forms when an oceanic island ringed by a *barrier reef* sinks below sea level.

atom The smallest particle that retains all the chemical *properties* of a given *element.*

atomic mass 1. The sum of *protons* and *neutrons* in an atom's nucleus. 2. The combined mass of all the particles in a given atom.

atomic number The number of *protons* in the nucleus of a given atom. *Elements* are distinguished from each other by their atomic numbers.

aureole A section of rock that surrounds an *intrusion* and shows the effects of *contact metamorphism.*

backarc basin A depression landward of a *volcanic arc* in a *subduction* zone, which is lined with trapped *sediment* from the volcanic arc and the plate interior. See also *forearc basin.*

backarc spreading The process by which the overriding plate in a *subduction* zone becomes stretched to the point of *rifting*, so that *magma* can then rise into the gap created by the rift. Backarc spreading typically occurs when the subducting plate sinks more rapidly than the overriding plate moves forward.

backshore The portion of a *beach* that extends from the high-tide line inland to the sea cliff or vegetation line. *Swash* reaches the backshore only during major storms.

backswamp The section of a *flood plain* where deposits of fine silts and clays settle after a flood. Backswamps usually lie behind a stream's *natural levees.*

backwash Water that returns into an ocean or large lake after hitting the shore as *swash.*

banded iron formation A rock that is made up of alternating light silica-rich layers and dark-colored layers of iron-rich minerals, which were deposited in marine basins on every continent about 2 billion years ago.

barchan dune A crescent-shaped *dune* that forms around a small patch of vegetation, lies perpendicular to the prevailing wind direction, and has a gentle, convex *windward* slope and a steep, concave *leeward* slope. Barchan dunes typically form in arid, inland deserts with stable wind direction and relatively little sand.

barrier island A ridge of sand that runs parallel to the main coast but is separated from it by a *bay* or *lagoon.* Barrier islands range from 10 to 100 kilometers in length and from 2 to 5 kilometers in width. A barrier island may be as high as 6 meters above sea level.

barrier reef A long, narrow *reef* that runs parallel to the main coast but is separated from it by a wide *lagoon.*

basal sliding The process by which a *glacier* undergoes thawing at its base, producing a film of water along which the glacier then flows. Basal sliding primarily affects glaciers in warm climates or mid-latitude mountain ranges.

basalt The dark, dense, *aphanitic, extrusive rock* that has a silica content of 40% to 50% and makes up most of the ocean floor. Basalt is the most abundant volcanic rock in the Earth's crust.

basaltic Of, containing, or composed of *basalt.*

base level The lowest level to which a *stream* can erode the channel through which it flows, generally equal to the prevailing global sea level.

basin A round or oval depression in the Earth's surface, containing the youngest section of rock in its lowest, central part.

batholith A massive *discordant pluton* with a surface area greater than 100 square kilometers, typically having a depth of about 30 kilometers. Batholiths are generally found in elongated mountain ranges after the *country rock* above them has eroded.

bay A recess in a *shoreline*, or an inlet between two *headlands.*

baymouth bar A narrow ridge of sand that stretches completely across the mouth of a *bay.* (Also called bay bar and bay barrier.)

beach The part of a *coast* that is washed by waves or tides, which cover it with *sediments* of various sizes and composition, such as sand or pebbles.

beach drift 1. The process by which *swash* and *backwash* move *sediments* along a *beach face.* 2. The sediments so moved. Beach drift typically consists of sand, gravel, shell fragments, and pebbles. See also *longshore drift.*

beach face The portion of a *foreshore* that lies nearest to the sea and regularly receives the *swash* of breaking waves. The beach face is the steepest part of the foreshore.

bed A layer of *sediment* or *sedimentary rock* that can be distinguished from the surrounding layers by such features as chemical composition and grain size.

bed load A body of coarse particles that move along the bottom of a *stream.*

bedding The division of *sediment* or *sedimentary rock* into parallel layers (*beds*) that can be distinguished from each other by such features as chemical composition and grain size.

bedrock The solid mass of rock that makes up the Earth's *crust.*

Benioff zone A region where the *subduction* of oceanic plates causes earthquakes, the *foci* of which are deeper the farther inland they are.

berm A low, narrow layer or mound of *sediment* deposited on a *backshore* by storm waves.

biomass fuel A renewable *fuel* derived from a living organism or the byproduct of a living organism. Biomass fuels include wood, dung, methane gas, and grain alcohol.

bitumen Any of a group of solid and semi-solid *hydrocarbons* that can be converted into liquid form by heating. Bitumens can be refined to produce such commercial products as gasoline, fuel oil, and asphalt.

bituminous coal A shiny black coal that develops from deeply buried *lignite* through heat and pressure, and that has a carbon content of 80% to 93%, which makes it a more efficient heating fuel than lignite.

blind valley A valley formed by and containing *sinkholes* and *disappearing streams*, and therefore dry except during periods of such heavy rainfalls that the sinkholes cannot immediately drain the entire accumulation of water.

body wave A type of *seismic wave* that transmits energy from an earthquake's *focus* through the Earth's interior in all directions. See also *surface wave.*

bond To combine, by means of chemical reaction, with another atom to form a compound. When an atom bonds with another, it either loses, gains, or shares electrons with the other atom.

Bowen's reaction series The sequence of *igneous rocks* formed from a *mafic magma*, assuming mineral crystals that have already formed continue to react with the liquid magma and so evolve into new minerals, thereby creating the next rock in the sequence.

braided stream A network of converging and diverging *streams* separated from each other by narrow strips of sand and gravel.

breakwater A wall built seaward of a coast to intercept incoming waves and so protect a harbor or shore. Breakwaters are typically built parallel to the coast.

breccia A *clastic rock* composed of particles more than 2 millimeters in diameter and marked by the angularity of its component grains and rock fragments.

breeder reactor A nuclear reactor that manufactures more fissionable isotopes than it consumes. Breeder reactors use the widely available, nonfissionable uranium isotope U-238, together with small amounts of fissionable U-235, to produce a fissionable isotope of plutonium, Pu-239.

brittle failure Rupture of rock, a type of permanent *strain* caused by relatively low *stress*.

burial metamorphism A form of *regional metamorphism* that acts on rocks covered by 5 to 10 kilometers of rock or sediment, caused by heat from the Earth's interior and *lithostatic pressure*.

caldera A vast depression at the top of a *volcanic cone*, formed when an eruption substantially empties the reservoir of *magma* beneath the cone's summit. Eventually the summit collapses inward, creating a caldera. A caldera may be more than 15 kilometers in diameter and more than 1000 meters deep.

caliche A white *soil horizon* consisting of calcium carbonate, typical of arid and semi-arid areas. Brief heavy rains dissolve calcium carbonate in the upper layers of soil and transport it downward; the rainwater then evaporates rapidly, leaving the calcium carbonate to form a new, solid layer of soil.

capacity The ability of a given *stream* to carry *sediment*, measured as the maximum quantity it can transport past a given point on the channel bank in a given amount of time. See also *competence*.

capillary fringe The lowest part of the *zone of aeration*, marked by the rising of water from the *water table* due to the attraction of the water molecules to mineral surfaces and other molecules, and to pressure from the *zone of saturation* below.

carbon-14 dating A form of *radiometric dating* that relies on the 5730-year half-life of radioactive carbon-14, which decays into nitrogen-14, to determine the age of rocks in which carbon-14 is present. Carbon-14 dating is used for rocks from 100 to 100,000 years old.

catastrophism The hypothesis that a series of immense, brief, worldwide upheavals changed the Earth's crust greatly and can account for the development of mountains, valleys, and other features of the Earth. See also *uniformitarianism*.

carbonate One of several minerals containing one central carbon atom with strong *covalent bonds* to three oxygen atoms and typically having *ionic bonds* to one or more positive ions.

cave A naturally formed opening beneath the surface of the Earth, generally formed by *dissolution* of carbonate bedrock. Caves may also form by erosion of coastal bedrock, partial melting of glaciers, or solidification of lava into hollow tubes.

cementation The *diagenetic* process by which sediment grains are bound together by *precipitated* minerals originally dissolved during the chemical weathering of preexisting rocks.

Cenozoic Era The latest era of the *Phanerozoic Eon*, following the *Mesozoic Era* and continuing to the present time, and marked by the presence of a wide variety of mammals, including the first hominids.

chemical sediment *Sediment* that is composed of previously dissolved minerals that have either precipitated from evaporated water or been extracted from water by living organisms and deposited when the organisms died or discarded their shells.

chemical weathering The process by which chemical reactions alter the chemical composition of rocks and minerals that are unstable at the Earth's surface and convert them into more stable substances; *weathering* that changes the chemical makeup of a rock or mineral. See also *mechanical weathering*.

chert A member of a group of *sedimentary rocks* that consist primarily of microscopic *silica* crystals. Chert may be either organic or inorganic, but the most common forms are inorganic.

cinder cone A *pyroclastic cone* composed primarily of *cinders*.

cinders Glassy, porous, *pyroclastic* rock fragments.

cirque A deep, semi-circular basin eroded out of a mountain by an *alpine glacier*.

cirque glacier A small *alpine glacier* that forms inside a *cirque*, typically near the head of a valley.

clastic Being or pertaining to a *sedimentary rock* composed primarily from fragments of preexisting rocks or fossils.

clay A mineral particle of any composition that is less than 1/256 of a millimeter in diameter. Not to be confused with *clay minerals*.

clay mineral One of a group of hydrous *silicate* minerals, such as kaolinite and smectite, the extremely small particle size of which imparts the ability to adsorb water. Clay minerals are the stable end-products of the *chemical weathering* of *feldspars*.

cleavage The tendency of certain minerals to break along distinct planes in their *crystal structures* where the bonds are weakest. Cleavage is tested by striking or hammering a mineral, and is classified by the number of surfaces it produces and the angles between adjacent surfaces.

coal A member of a group of easily combustible, organic *sedimentary rocks* composed mostly of plant remains and containing a high proportion of carbon.

coast The area of dry land that borders on a body of water.

cockpit karst A *karst* environment marked by numerous closely spaced, irregular depressions and steep, conical hills.

col A high mountain pass that forms when part of an *arête* erodes.

compaction The *diagenetic* process by which the volume or thickness of sediment is reduced due to pressure from overlying layers of sediment.

composite cone See *stratovolcano*.

competence The ability of a given *stream* to carry *sediment*, measured as the diameter of the largest particle that the stream can transport. See also *capacity*.

compound An electrically neutral substance that consists of two or more *elements* combined in specific, constant proportions. A compound typically has physical characteristics different from those of its constituent elements.

compression *Stress* that reduces the volume or length of a rock, as that produced by the *convergence* of plate margins.

concordant Of or being a pluton that lies parallel to the surrounding layers of rock. See also *discordant.*

cone of depression An area in a *water table* along which water has descended into a well to replace water drawn out, leaving a gap shaped like an inverted cone.

confining pressure See *lithostatic pressure.*

conglomerate A *clastic rock* composed of particles more than 2 millimeters in diameter and marked by the roundness of its component grains and rock fragments.

contact metamorphism *Metamorphism* that is caused by heat from a magmatic *intrusion.*

continental collision The *convergence* of two continental plates, resulting in the formation of mountain ranges.

continental drift The hypothesis, proposed by Alfred Wegener, that today's continents broke off from a single supercontinent and then plowed through the ocean floors into their present positions. This explanation of the shapes and locations of Earth's current continents evolved into the theory of *plate tectonics.*

continental ice sheet An unconfined *glacier* that covers much or all of a continent.

convection cell The cycle of movement in the *asthenosphere* that causes the plates of the *lithosphere* to move. Heated material in the asthenosphere becomes less dense and rises toward the solid lithosphere, through which it cannot rise further and therefore begins to move horizontally, dragging the lithosphere along with it and pushing forward the cooler, denser material in its path. The cooler material eventually sinks down lower into the mantle, becoming heated there and rising up again, continuing the cycle. See also *plate tectonics.*

convergence The coming together of two lithospheric plates. Convergence causes *subduction* when one or both plates is oceanic, and mountain formation when both plates are continental. See also *divergence.*

core The innermost layer of the Earth, consisting primarily of pure metals such as iron and nickel. The core is the densest layer of the Earth, and is divided into the outer core, which is believed to be liquid, and the inner core, which is believed to be solid. See also *crust* and *mantle.*

correlation The process of determining that two or more geographically distant rocks or rock strata originated in the same time period.

country rock 1. The preexisting rock into which a *magma* intrudes. 2. The preexisting rock surrounding a *pluton.*

covalent bond The combination of two or more atoms by sharing electrons so as to achieve chemical stability under the *octet rule.* Atoms that form covalent bonds generally have outer energy levels containing three, four, or five electrons. Covalent bonds are generally stronger than other bonds.

crater See *volcanic crater.*

creep The slowest form of *mass movement,* measured in millimeters or centimeters per year and occurring on virtually all slopes.

cross bed A *bed* made up of particles dropped from a moving current, as of wind or water, and marked by a downward slope that indicates the direction of the current that deposited them.

crust The outermost layer of the Earth, consisting of relatively low-density rocks. See also *core* and *mantle.*

crystal A mineral in which the systematic internal arrangement of atoms is outwardly reflected as a latticework of repeated three-dimensional units that form a geometric solid with a surface consisting of symmetrical planes.

crystal structure 1. The geometric pattern created by the systematic internal arrangement of atoms in a mineral. 2. The systematic internal arrangement of atoms in a mineral. See also *crystal.*

crystalline Marked by the systematic internal arrangement of atoms.

current 1. A broad flow of ocean water that maintains a stable direction and differs from the surrounding water in such features as temperature and salinity. 2. The water in such a flow.

daughter isotope An *isotope* that forms from the *radioactive decay* of a *parent isotope.* A daughter isotope may or may not be of the same element as its parent. If the daughter isotope is radioactive, it will eventually become the parent isotope of a new daughter isotope. The last daughter isotope to form from this process will be stable and nonradioactive.

debris avalanche The sudden, extremely rapid *mass movement* downward of entire layers of *regolith* along very steep slopes. Debris avalanches are generally caused by heavy rains.

debris flow 1. The rapid, downward *mass movement* of particles coarser than sand, often including boulders one meter or more in diameter, at a rate ranging from 2 to 40 kilometers per hour. Debris flows occur along fairly steep slopes. 2. The material that descends in such a flow.

deflation The process by which wind erodes *bedrock* by picking up and transporting loose rock particles.

deformation Any of the processes by which a rock changes its shape, form, or volume.

degradation The process by which a stream's *gradient* becomes less steep, due to the *erosion* of *sediment* from the stream bed. Such erosion generally follows a sharp reduction in the amount of sediment entering the stream.

delta An *alluvial fan* having its apex at the mouth of a *stream.*

dendrochronology A method of *absolute dating* that uses the number of tree rings found in a cross section of a tree trunk or branch to determine the age of the tree.

desert A region with an average annual rainfall of 10 inches or less and sparse vegetation, typically having thin, dry, and crumbly soil. A desert has an *aridity index* greater than 4.0.

desert pavement A closely packed layer of rock fragments concentrated in a layer along the Earth's surface by the *deflation* of finer particles.

desert varnish A thin, shiny red-brown or black layer, principally composed of iron manganese oxides, that coats the surfaces of many exposed desert rocks.

desertification The process through which a desert takes over a formerly nondesert area. When a region begins to undergo desertification, the new conditions typically include a significantly lowered *water table*, a reduced supply of surface water, increased salinity in natural waters and soils, progressive destruction of native vegetation, and an accelerated rate of erosion.

detrital sediment *Sediment* that is composed of transported solid fragments of preexisting igneous, sedimentary, or metamorphic rocks.

diagenesis The set of processes that cause physical and chemical changes in sediment after it has been deposited and buried under another layer of sediment. Diagenesis may culminate in *lithification.*

dike A *discordant pluton* that is substantially wider than it is thick. Dikes are often steeply inclined or nearly vertical. See also *sill.*

dilatancy The expansion of a rock's volume caused by *stress* and *deformation.*

diorite Any of a group of dark, *phaneritic, intrusive rocks* that are the *plutonic* equivalents of *andesite.*

dip The angle formed by the inclined plane of a geological structure and the horizontal plane of the Earth's surface.

dip-slip fault A *fault* in which two sections of rock have moved apart vertically, parallel to the *dip* of the fault plane.

directed pressure Force exerted on a rock along one plane, flattening the rock in that plane and lengthening it in the perpendicular plane.

disappearing stream A surface *stream* that drains rapidly and completely into a *sinkhole.*

discordant Of or being a *pluton* that lies perpendicular or oblique to the surrounding layers of rock. See also *concordant.*

dissolution A form of *chemical weathering* in which water molecules, sometimes in combination with acid or another compound in the environment, attract and remove oppositely charged ions or ion groups from a mineral or rock.

dissolved load A body of sediment carried by a *stream* in the form of *ions* that have dissolved in the water.

distributary One of a network of small *streams* carrying water and sediment from a *trunk stream* into an ocean.

divergence The process by which two lithospheric plates separated by *rifting* move farther apart, with soft mantle rock rising between them and forming new oceanic *lithosphere.* See also *convergence.*

dolostone A *sedimentary rock* composed primarily of dolomite, a mineral made up of calcium, magnesium, carbon, and oxygen. Dolostone is thought to form when magnesium ions replace some of the calcium ions in *limestone,* to which dolostone is similar in both appearance and chemical structure.

dome A round or oval bulge on the Earth's surface, containing the oldest section of rock in its raised, central part. See also *basin.*

drainage basin The area from which water flows into a *stream.* Also called a *watershed.*

drainage divide An area of raised, dry land separating two adjacent *drainage basins.*

drainage pattern The arrangement in which a *stream* erodes the channels of its network of *tributaries.*

drumlin A long, spoon-shaped hill that develops when pressure from an overriding *glacier* reshapes a *moraine.* Drumlins range in height from 5 to 50 meters and in length from 400 to 2000 meters. They slope down in the direction of the ice flow.

ductile deformation See *plastic deformation.*

dune A usually asymmetrical mound or ridge of sand that has been transported and deposited by wind. Dunes form in both arid and humid climates.

dynamothermal metamorphism A form of *regional metamorphism* that acts on rocks caught between two *converging* plates and is initially caused by *directed pressure* from the plates, which causes some of the rocks to rise and others to sink, sometimes by tens of kilometers. The rocks that fall then experience further dynamothermal metamorphism, this time caused by heat from the Earth's interior and *lithostatic pressure* from overlying rocks.

earthflow 1. The *flow* of a dry, highly viscous mass of clay-like or silty *regolith,* typically moving at a rate of one or two meters per hour. 2. The material that descends in such a flow.

earthquake A movement within the Earth's *crust* or *mantle,* caused by the sudden rupture or repositioning of underground rocks as they release *stress.*

ebb tide A *tide* that lowers the water surface of an ocean and moves the shoreline farther seaward.

echo-sounding sonar The mapping of ocean topography based on the time required for sound waves to reach the sea floor and return to the research ship that emits them.

elastic deformation A temporary *stress*-induced change in the shape or volume of a rock, after which the rock returns to its original shape and volume.

elastic limit See *yield point.*

electron A negatively charged particle that orbits rapidly around the *nucleus* of an *atom.* See also *proton.*

element A form of matter that cannot be broken down into a chemically simpler form by heating, cooling, or chemical reactions. There are 106 known elements, 92 of them natural and 14 synthetic. Elements are represented by one- or two-letter abbreviations. See also *atom, atomic number.*

energy level The path of a given electron's orbit around a nucleus, marked by a constant distance from the nucleus.

epicenter The point on the Earth's surface that is located directly above the *focus* of an *earthquake.*

equilibrium line The point in a *glacier* where overall gain in volume equals overall loss, so that the net volume remains stable. The equilibrium line marks the border between the *zone of accumulation* and the *zone of ablation.*

erosion The process by which particles of rock and soil are loosened, as by *weathering,* and then transported elsewhere, as by wind, water, ice, or gravity.

esker A ridge of *sediment* that forms under a glacier's *zone of ablation*, made up of sand and gravel deposited by *meltwater*. An esker may be less than 100 meters or more than 500 kilometers long, and may be anywhere from 3 to over 300 meters high.

evaporite An inorganic *chemical sediment* that *precipitates* when the salty water in which it had dissolved evaporates.

extrusive rock An *igneous rock* formed from *lava* that has flowed out onto the Earth's surface, characterized by rapid solidification and grains that are so small as to be barely visible to the naked eye.

fall The fastest form of *mass movement,* occurring when rock or sediment breaks off from a steep or vertical slope and descends at a rate of 9.8 meters per second. A fall can be extremely dangerous.

fault A *fracture* dividing a rock into two sections that have visibly moved relative to each other.

fault block A section of rock separated from other rock by one or more *faults.*

fault-block mountain A mountain containing tall *horsts* interspersed with much lower *grabens* and bounded on at least one side by a high-angle *normal fault.*

fault metamorphism The *metamorphism* that acts on rocks grinding past one another along a fault and is caused by *directed pressure* and frictional heat.

feldspar Any of a group of light-colored, silicate, *rock-forming minerals* most often found in *plutonic igneous rocks* and *metamorphic rocks* and often containing potassium, sodium, or calcium. Feldspar constitutes 60% of the Earth's crust.

felsic Of or being a light-colored, *igneous rock* with a silica content of 70% or higher. Felsic rocks are generally rich in potassium feldspars, aluminum, and quartz.

firn Firmly packed snow that has survived a summer melting season. Firn has a density of about 0.4 grams per cubic centimeter. Ultimately, firn turns into glacial ice.

fission The division of the *nucleus* of a radioactive *atom*, which causes the release of several subatomic particles. The fission of a given element always occurs at a constant rate. (See also *nuclear fission.*)

fission-track dating A form of *absolute dating* that relies on the constant rate of fission to determine the age of a crystal, by counting the *fission tracks* left in a given area of the crystal.

fission tracks Marks left in the latticework of a mineral crystal by subatomic particles released during the *fission* of a *radioactive* atom trapped inside the crystal.

fjord A deep, steep-walled, U-shaped valley formed by erosion by a *glacier* and submerged with seawater.

flood plain The flat land that surrounds a *stream* and becomes submerged when the stream overflows its banks.

flood tide A *tide* that raises the water surface of an ocean and moves the *shoreline* farther inland.

fluorescence Emission of visible light by a substance, such as a mineral, that is currently exposed to ultraviolet light and absorbs radiation from it. The light appears in the form of glowing, distinctive colors. The emission ends when the exposure to ultraviolet light ends.

focus (plural **foci**) The precise point within the Earth's *crust* or *mantle* where rocks begin to rupture or move in an *earthquake.*

fold A bend that develops in an initially horizontal layer of rock, usually caused by *plastic deformation.* Folds occur most frequently in *sedimentary rocks.*

fold-and-thrust mountain A mountain consisting of *folds,* which developed from extremely thick layers of sediment, and *thrust fault* blocks, and containing both *igneous* and *metamorphic rocks.* Fold- and-thrust mountains may be several thousand kilometers high and a few hundred kilometers wide. The Alps, the Appalachians, the Carpathians, the Himalayas, and the Urals are all fold-and-thrust mountains.

foliation The arrangement of a set of minerals in parallel, sheet-like layers that lie perpendicular to the flattened plane of a rock. Occurs in *metamorphic rocks* on which *directed pressure* has been exerted.

footwall The section of rock that lies below the *fault* plane in a *dip-slip fault.* See also *hanging wall.*

forearc basin A depression in the sea floor located between an *accretionary wedge* and a *volcanic arc* in a *subduction* zone, and lined with trapped *sediment.* See also *backarc basin.*

foreshock A minor, barely detectable *earthquake*, generally preceding a full-scale earthquake with approximately the same *focus.* Major quakes may follow a cluster of foreshocks by as little as a few seconds or as much as several weeks.

foreshore The portion of a *beach* that lies nearest to the sea, extending from the low-tide line to the high-tide line.

fossil A remnant, an imprint, or a trace of an ancient organism, preserved in the Earth's crust.

fossil fuel A nonrenewable energy source, such as oil, gas, or coal, that derives from the organic remains of past life. Fossil fuels consist primarily of *hydrocarbons.*

fractional crystallization The process by which a *magma* produces crystals that then separate from the original magma, so that the chemical composition of the magma changes with each generation of crystals, producing *igneous rocks* of different compositions. The *silica* content of the magma becomes proportionately higher after each crystallization.

fracture (*n*) A crack or break in a rock. (*v*) To break in random places instead of *cleaving.* Said of minerals.

fringing reef A *reef* that forms against or near an island or continental *coast* and grows seaward, sloping sharply towards the sea floor. Fringing reefs usually range from 0.5 to 1.0 or more kilometers in width.

frost wedging A form of *mechanical weathering* caused by the freezing of water that has entered a pore or crack in a rock. The water expands as it freezes, widening the cracks or pores and often loosening or dislodging rock fragments. As the ice forms, it attracts more water, increasing the effects of frost wedging.

fuel A source of energy, especially a combustible substance that can be burned for heat or power, or matter used in *nuclear fission.*

gabbro Any of a group of dark, dense, *phaneritic, intrusive rocks* that are the *plutonic* equivalent to *basalt.*

geochronology The study of the relationship between the history of the Earth and time.

geologic time scale The division of all of Earth history into blocks of time distinguished by geologic and evolutionary events, ordered sequentially and arranged into eons made up of eras, which are in turn made up of periods, which are in turn made up of epochs.

geology The scientific study of the Earth, its origins and evolution, the materials that make it up, and the processes that act on it.

geophysics The branch of *geology* that studies the physics of the Earth, using the physical principles underlying such phenomena as *seismic waves,* heat flow, *gravity,* and *magnetism* to investigate planetary properties.

geyser A *natural spring* marked by the intermittent escape of hot water and steam.

glacial Produced by, transported by, or concerning a *glacier.*

glacial abrasion The process by which a glacier erodes the underlying *bedrock* through contact between the bedrock and rock fragments embedded in the base of the glacier. See also *glacial quarrying.*

glacial drift A load of rock material transported and deposited by a glacier. Glacial drift is usually deposited when the glacier begins to melt.

glacial erratic A rock or rock fragment transported by a glacier and deposited on bedrock of different composition. Glacial erratics range from a few millimeters to several yards in diameter.

glacial quarrying The process by which a glacier erodes the underlying *bedrock* by loosening and ultimately detaching blocks of rock from the bedrock and attaching them instead to the glacier, which then bears the rock fragments away. See also *glacial abrasion.*

glacial till *Drift* that is deposited directly from glacial ice and therefore not *sorted.* Also called *till.* See also *glacial drift.*

glacier A moving body of ice that forms on land from the accumulation and compaction of snow, and that flows downslope or outward due to *gravity* and the pressure of its own weight.

gneiss A coarse-grained, *foliated metamorphic rock* marked by bands of light-colored minerals such as quartz and feldspar that alternate with bands of dark-colored minerals. This alternation develops through *metamorphic differentiation.*

graben A block of rock that lies between two *faults* and has moved downward to form a depression between the two adjacent fault blocks. See also *horst.*

graded bed A *bed* formed by the deposition of sediment in relatively still water, marked by the presence of particles that vary in size, density, and shape. The particles settle in a gradual slope with the coarsest particles at the bottom and the finest at the top.

graded stream A stream maintaining an equilibrium between the processes of erosion and deposition, and therefore between *aggradation* and *degradation.*

gradient The vertical drop in a stream's elevation over a given horizontal distance, expressed as an angle.

granite A pink-colored, *felsic, plutonic rock* that contains potassium and usually sodium *feldspars,* and has a quartz content of about 10%. Granite is commonly found on continents but virtually absent from the ocean basins.

gravity 1. The force of attraction exerted by one body in the universe on another. Gravity is directly proportional to the product of the masses of the two attracted bodies. 2. The force of attraction exerted by the Earth on bodies on or near its surface, tending to pull them toward the Earth's center.

gravity anomaly The difference between an actual measurement of *gravity* at a given location and the measurement predicted by theoretical calculation.

groin A structure that juts out into a body of water perpendicular to the *shoreline* and is built to restore an eroding beach by intercepting *longshore drift* and trapping sand.

guyot A *seamount,* the top of which has been flattened by *weathering,* wave action, or stream *erosion.*

half-life The time necessary for half of the atoms of a *parent isotope* to decay into the *daughter isotope.*

hanging wall The section of rock that lies above the fault plane in a *dip-slip fault.* See also *footwall.*

hardness The degree of resistance of a given mineral to scratching, indicating the strength of the bonds that hold the mineral's atoms together. The hardness of a mineral is measured by rubbing it with substances of known hardness.

headland A cliff that projects out from a *coast* into deep water.

historical geology The study of the history, origin, and evolution of the Earth and all of its life forms and geologic structures.

Holocene Epoch The second epoch of the *Quaternary Period,* beginning approximately 10,000 years ago and continuing to the present time. See also *Pleistocene Epoch.*

hook A *spit* that curves sharply at its coastal end.

horn A high mountain peak that forms when the walls of three or more *cirques* intersect.

hornfels A hard, dark-colored, dense *metamorphic rock* that forms from the *intrusion of* magma into *shale* or *basalt.*

horst A block of rock that lies between two *faults* and has moved upward relative to the two adjacent fault blocks. See also *graben.*

hot spot An area in the upper *mantle,* ranging from 100 to 200 km in width, from which magma rises in a *plume* to form *volcanoes.* A hot spot may endure for ten million years or more.

hydraulic conductivity The extent to which a given substance allows water to flow through it, determined by such factors as sorting and grain size and shape.

hydraulic gradient The difference in *potential* between two points, divided by the lateral distance between the points.

hydraulic lifting The *erosion* of a stream bed by water pressure.

hydrocarbon A molecule that is entirely made up of hydrogen and carbon.

hydrogen bond An intermolecular bond formed with hydrogen.

hydrologic cycle The perpetual movement of water among the mantle, oceans, land, and atmosphere of the Earth.

hydrolysis A form of *chemical weathering* in which ions from water replace equivalently charged ions from a mineral, especially a silicate.

hydrothermal deposit A mineral deposit formed by the *precipitation* of metallic ions from water ranging in temperature from 50° to 700°C.

hydrothermal metamorphism The chemical alteration of preexisting rocks that is caused by the action of hot water.

hypothesis A tentative explanation of a given set of data that is expected to remain valid after future observation and experimentation. See also *theory.*

ice age A period during which the Earth is substantially cooler than usual and a significant portion of its land surface is covered by *glaciers.* Ice ages generally last tens of millions of years.

ice cap An *alpine glacier* that covers the peak of a mountain.

igneous rock A rock made from molten (melted) or partly molten material that has cooled and solidified.

index fossil The *fossil* of an organism known to have existed for a relatively short period of time, used to date the rock in which it is found.

index mineral See *metamorphic index mineral.*

inselberg A steep ridge or hill left when a mountain has eroded and found in an otherwise flat, typically desert plain.

intermolecular bonding The act or process by which two or more groups of atoms or molecules combine due to weak positive or negative charges that develop at various points within each group of atoms due to uneven distribution of their electrons. The side of molecule where electrons are more likely to be found will have a slight negative charge, and the side where they are less likely to be found will have a slight positive charge. Such charged regions attract oppositely charged regions of nearby molecules, forming relatively weak bonds.

internal deformation The rearrangement of the planes within ice crystals, due to pressure from overlying ice and snow, that causes the downward or outward flow of a *glacier.*

intrusion The entrance of *magma* into preexisting rock.

intrusive rock An *igneous rock* formed by the entrance of *magma* into preexisting rock.

ion An atom that has lost or gained one or more electrons, thereby becoming electrically charged.

ionic bond The combination of an atom that has a strong tendency to lose electrons with an atom that has a strong tendency to gain electrons, such that the former transfers one or more electrons to the latter and each achieves chemical stability under the *octet rule.* The atom that loses electrons acquires a positive electric charge and the atom that gains electrons acquires a negative electric charge, so that the resulting compound is electrically neutral.

ionic bonding The act or process of forming of an ionic bond.

ionic substitution The replacement of one type of ion in a mineral by another that is similar to the first in size and charge.

isostasy The equilibrium maintained between the gravity tending to depress and the buoyancy tending to raise a given segment of the *lithosphere* as it floats above the *asthenosphere.*

isotope One of two or more forms of a single element; the atoms of each isotope have the same number of protons but different numbers of neutrons in their nuclei. Thus, isotopes have the same *atomic number* but differ in *atomic mass.*

jetty A structure built along the bank of a stream channel or tidal outlet to direct the flow of a stream or tide and keep the sediment moving so that it cannot build up and fill the channel. Jetties are typically built in parallel pairs along both banks of the channel. Jetties that are built perpendicular to a *coast* tend to interrupt *longshore drift* and thus widen *beaches.*

joint A *fracture* dividing a rock into two sections that have not visibly moved relative to each other. See also *fault.*

juvenile water The steam that accompanies *volcanic* eruptions.

karst A topography characterized by *caves, sinkholes, disappearing streams,* and underground drainage. Karst forms when groundwater dissolves pockets of limestone, dolomite, or gypsum in bedrock.

kerogen A solid, waxy, organic substance that forms when pressure and heat from the Earth act on the remains of plants and animals. Kerogen converts to various liquid and gaseous *hydrocarbons* at a depth of seven or more kilometers and a temperature between 50° and 100°C.

laccolith A large *concordant pluton* that is shaped like a dome or a mushroom. Laccoliths tend to form at relatively shallow depths and are typically composed of granite. The *country rock* above them often erodes away completely.

lagoon A shallow body of water separated from the sea by a *reef* or *barrier island.*

lahar A *flow* of *pyroclastic* material mixed with water. A lahar is often produced when a snow-capped volcano erupts and hot pyroclastics melt a large amount of snow or ice.

lava *Magma* that comes to the Earth's surface through a *volcano* or fissure.

leeward Of, located on, or being the side of a dune, hill, or ridge that is sheltered from the wind. See also *windward.*

levee A protective barrier built along the banks of a stream to prevent flooding. See also *natural levee.*

lichen Plant-like colonies of fungi and algae that grow on the exposed surface of rocks. Lichen grows at a constant rate within a single geographic area.

lichenometry A method of *absolute dating* that uses the size of *lichen* colonies on a rock surface to determine the surface's age. Lichenometry is used for rock surfaces less than about 9000 years old.

lignite A soft, brownish *coal* that develops from *peat* through bacterial action, is rich in *kerogen,* and has a carbon content of 70%, which makes it a more efficient heating fuel than peat.

limestone A *sedimentary rock* composed primarily of calcium carbonate. 10% to 15% of all sedimentary rocks are limestones. Limestone is usually organic, but it may also be inorganic.

liquefaction The conversion of moderately cohesive, unconsolidated *sediment* into a fluid, water-saturated mass.

lithification The conversion of loose *sediment* into solid *sedimentary rock.*

lithosphere A layer of solid, brittle rock comprising the outer 100 kilometers of the Earth, encompassing both the crust and the outermost part of the upper *mantle.* See also *asthenosphere.*

lithostatic pressure The force exerted on a rock buried deep within the Earth by overlying rocks. Because lithostatic pressure is exerted equally from all sides of a rock, it compresses the rock into a smaller, denser form without altering the rock's shape.

loess A load of *silt* that is produced by the erosion of *outwash* and transported by wind. Much loess found in the Mississippi Valley, China, and Europe is believed to have been deposited during the *Pleistocene Epoch.*

longitudinal dune One of a series of long, narrow *dunes* lying parallel both to each other and to the prevailing wind direction. Longitudinal dunes range from 60 meters to 100 kilometers in length and from 3 to 50 meters in height.

longshore current An ocean *current* that flows close and almost parallel to the *shoreline* and is caused by the rush of waves toward the shore.

longshore drift 1. The process by which a *current* moves *sediments* along a *surf zone.* 2. The sediments so moved. Longshore drift typically consists of sand, gravel, shell fragments, and pebbles. See also *beach drift.*

low-velocity zone An area within the Earth's upper *mantle* in which both *P waves* and *S waves* travel at markedly slower velocities than in the outermost part of the upper mantle. The low-velocity zone occurs in the range between 100 and 350 kilometers of depth.

luster 1. The reflection of light on a given mineral's surface, classified by intensity and quality. 2. The appearance of a given mineral as characterized by the intensity and quality with which it reflects light.

magma Molten (melted) rock that forms naturally within the Earth. Magma may be either a liquid or a fluid mixture of liquid, crystals, and dissolved gases.

magnetic field The region within which the *magnetism* of a given substance or particle affects other substances.

magnetic reversal The process by which the Earth's magnetic north pole and its magnetic south pole reverse their positions over time.

magnetism The property, possessed by certain materials, to attract or repel similar materials. Magnetism is associated with moving electricity.

mantle The middle layer of the Earth, lying just below the *crust* and consisting of relatively dense rocks. The mantle is divided into two sections, the upper mantle and the lower mantle; the lower mantle has greater density than the upper mantle. See also *core* and *crust.*

marble A coarse-grained, *nonfoliated metamorphic rock* derived from *limestone* or *dolostone.*

marine magnetic anomaly An irregularity in magnetic strength along the ocean floor that reflects *sea-floor spreading* during periods of *magnetic reversal.*

massive sulfide deposit An unusually large deposit of sulfide minerals.

mass movement The process by which such Earth materials as *bedrock,* loose *sediment,* and *soil* are transported down slopes by *gravity.*

meandering stream A *stream* that traverses relatively flat land in fairly evenly spaced loops and separated from each other by narrow strips of *flood plain.*

mechanical exfoliation A form of *mechanical weathering* in which successive layers of a large *plutonic rock* break loose and fall when the erosion of overlying material permits the rock to expand upward. The thin slabs of rock that break off fall parallel to the exposed surface of the rock, creating the long, broad steps that can be found on many mountains.

mechanical weathering The process by which a rock or mineral is broken down into smaller fragments without altering its chemical makeup; *weathering* that affects only physical characteristics. See also *chemical weathering.*

mélange A body of rock that forms along the inner wall of an *ocean trench* and is made up of fragments of *lithosphere* and oceanic *sediment* that have undergone *metamorphism.*

meltwater Water formed from the melted ice of a *glacier.*

Mercalli intensity scale A scale designed to measure the degree of intensity of *earthquakes,* ranging from I for the lowest intensity to XII for the highest. The classifications are based on human perceptions.

Mesozoic Era The intermediate era of the *Phanerozoic Eon,* following the *Paleozoic Era* and preceding the *Cenozoic Era,* and marked by the dominance of marine and terrestrial reptiles, and the appearance of birds, mammals, and flowering plants.

metallic bonding The act or process by which two or more atoms of electron-donating elements pack so closely together that some of their electrons begin to wander among the nuclei rather than orbiting the nucleus of a single atom. Metallic bonding is responsible for the distinctive properties of metals.

metamorphic differentiation The process by which minerals from a chemically uniform rock separate from each other during *metamorphism* and form individual layers within a new *metamorphic rock.*

metamorphic facies 1. A group of minerals customarily found together in *metamorphic rocks* and indicating a particular set of temperature and pressure conditions at which metamorphism occurred. 2. A set of *metamorphic rocks* characterized by the presence of such a group of minerals.

metamorphic grade A measure used to identify the degree to which a *metamorphic rock* has changed from its *parent rock.* A metamorphic grade provides some indication of the circumstances under which the metamorphism took place.

metamorphic index mineral One of a set of minerals found in *metamorphic rocks* and used as indicators of the temperature and pressure conditions at which the metamorphism occurred. A metamorphic index mineral is stable only within a narrow range of temperatures and pressures and the metamorphism that produces it must take place within that range.

metamorphic rock A *rock* that has undergone chemical or structural changes. Heat, pressure, or a chemical reaction may cause such changes.

metamorphism The process by which conditions within the Earth, below the zone of *diagenesis,* alter the mineral content, chemical composition, and structure of solid rock without melting it. *Igneous, sedimentary,* and *metamorphic rocks* may all undergo metamorphism.

meteoric water The precipitation of condensed water from clouds as rain, snow, sleet, or hail.

microcontinent A section of continental *lithosphere* that has broken off from a larger, distant continent, as by *rifting.*

mid-ocean ridge An underwater mountain range that develops between the margins of two lithospheric plates, formed by *rifting.*

migmatite A rock that incorporates both *metamorphic* and *igneous* materials.

mineral A naturally occurring, usually inorganic, solid consisting of either a single element or a *compound,* and having a definite chemical composition and a systematic internal arrangement of atoms.

mineraloid A naturally occurring, usually inorganic, solid consisting of either a single element or a *compound,* and having a definite chemical composition but lacking a systemic internal arrangement of atoms. See also *mineral.*

mineral zone An area of rock throughout which a given *metamorphic index mineral* is found, presumed to have undergone metamorphism under a limited range of temperature and pressure conditions.

Moho (abbreviation for Mohorovičić) The *seismic discontinuity* between the base of the Earth's *crust* and the top of the *mantle. P waves* passing through the Moho change their velocity by approximately one kilometer per second, with the higher velocity occurring in the mantle and the lower in the crust.

molecule The smallest particle that retains all the chemical and physical properties of a given *compound,* consisting of a stable group of bonded atoms.

moraine A single, large mass of *glacial till* that accumulates, typically at the edge of a glacier.

mudcrack A fracture that develops at the top of a layer of fine-grained, muddy sediment when it is exposed to the air, dries out, and then shrinks.

mudflow The rapid flow of typically fine-grained *regolith* mixed with water. There may be as much as 60% water in a mudflow.

natural bridge An arch-shaped stretch of *bedrock* remaining in a *karst* region when the surrounding bedrock has dissolved.

natural levee One of a pair of ridges of *sediment* deposited along both banks of a *stream* during successive floods.

natural spring A place where groundwater flows to the surface and issues freely from the ground.

neutron A particle that is found in the *nucleus* of an *atom,* has a mass approximately equal to that of a *proton,* and has no electric charge.

nonfoliated Being a *metamorphic rock* that does not have *foliation.*

normal fault A *dip-slip fault* marked by a generally steep *dip* along which the *hanging wall* has moved downward relative to the *footwall.*

nuclear fission The division of the *nuclei* of *isotopes* of certain heavy *elements,* such as uranium and plutonium, effected by bombardment with *neutrons.* Nuclear fission causes the release of energy, additional neutrons, and an enormous quantity of heat. Nuclear fission is used in nuclear power plants and nuclear weapons. A byproduct of nuclear fission is toxic radioactive waste. See also *nuclear fusion.*

nuclear fusion The combination of the *nuclei* of certain extremely light *elements,* especially hydrogen, effected by the application of high temperature and pressure. Nuclear fusion causes the release of an enormous amount of heat energy, comparable to that released by *nuclear fission.* The principle byproduct of nuclear fusion is helium.

nucleus (plural **nuclei**) The central part of an *atom,* containing most of the atom's mass and having a positive charge due to the presence of *protons.*

nuée ardente A sometimes glowing cloud of gas and *pyroclastics* erupted from a *volcano* and moving swiftly down its slopes. Also called a *pyroclastic flow.*

ocean trench A deep, linear, relatively narrow depression in the sea floor, formed by the *subduction* of oceanic plates.

octet rule A scientific law stating that all atoms, except those of hydrogen and helium, require eight electrons in the outermost *energy level* in order to maintain chemical stability.

oil sand A mixture of unconsolidated sand and clay that contains a semi-solid *bitumen.*

oil shale A brown or black *clastic source rock* containing *kerogen.*

ophiolite suite The group of *sediments, sedimentary rocks,* and mafic and ultramafic *igneous rocks* that make up the oceanic *lithosphere.*

ore A mineral deposit that can be mined for a profit.

orogenesis Mountain formation, as caused by *volcanism, subduction, plate divergence, folding,* or the movement of *fault blocks.* Also called *orogeny.*

oscillatory motion The circular movement of water up and down, with little or no change in position, as a wave passes.

outwash A load of *sediment,* consisting of sand and gravel, that is deposited by *meltwater* in front of a *glacier.*

oxbow lake A crescent-shaped body of standing water formed from a single loop that was cut off from a *meandering stream,* typically by a flood that allowed the stream to flow through its *flood plain* and bypass the loop.

oxidation The process of combining with oxygen ions. A mineral that is exposed to air may undergo oxidation as a form of *chemical weathering.*

oxide One of several minerals containing negative oxygen ions bonded to one or more positive metallic ions.

paleosol An ancient, buried soil whose composition may reflect a climate significantly different from the climate now prevalent in the area where the soil is found.

paleomagnetism 1. The fixed orientation of a rock's crystals, based on the Earth's *magnetic field* at the time of the rock's formation, that remains constant even when the magnetic field changes. 2. The study of such phenomena as indicators of the Earth's magnetic history.

Paleozoic Era The earliest era of the *Phanerozoic Eon*, marked by the presence of marine invertebrates, fish, amphibians, insects, and land plants.

parabolic dune A horseshoe-shaped *dune* having a concave *windward* slope and a convex *leeward* slope. Parabolic dunes tend to form along sandy ocean and lake shores. They may also develop from *transverse dunes* through *deflation*.

parent isotope A *radioactive isotope* that changes into a different isotope when its nucleus decays. See also *daughter isotope*.

parent material The source from which a given soil is chiefly derived, generally consisting of *bedrock* or *sediment*.

parent rock The preexisting rock from which a *metamorphic rock* forms.

partial melting The incomplete melting of a rock composed of minerals with differing melting points. When partial melting occurs, the minerals with higher melting points remain solid while the minerals whose melting points have been reached turn to *magma*.

passive continental margin A border that lies between continental and oceanic *lithosphere*, but is not a *plate* margin. It is marked by lack of seismic and volcanic activity.

peat A soft brown mass of compressed, partially decomposed vegetation that forms in a water-saturated environment and has a carbon content of 50%. Dried peat can be burned as fuel.

pediment A broad surface at the base of a receding mountain. The pediment develops when running water erodes most of the mass of the mountain.

pegmatite A coarse-grained *igneous rock* with exceptionally large *crystals*, formed from a *magma* that contains a high proportion of water.

perched water table A saturated area that lies within a *zone of aeration*.

peridotite An *igneous rock* composed primarily of the iron-magnesium *silicate* olivine and having a silica content of less than 40%.

permafrost Permanently frozen *regolith*, ranging in thickness from 30 centimeters to over 1000 meters.

permeability The capability of a given substance to allow the passage of a fluid. Permeability depends upon the size of and the degree of connection among a substance's pores.

petroleum Any of a group of naturally occurring substances made up of *hydrocarbons*. These substances may be gaseous, liquid, or semisolid.

phaneritic Of or being an *igneous rock* containing components large enough to be seen with the unaided eye.

Phanerozoic Eon The eon that started 570 million years ago, when numerous fossils of sea shells began to be formed, and that continues to the present time.

phosphorescence Emission of visible light by a substance, such as a mineral, that is exposed to ultraviolet light and absorbs radiation from it. The light appears in the form of glowing, distinctive colors. The emission continues after the exposure to ultraviolet light ends.

phyllite A *foliated metamorphic rock* that develops from *slate* and is marked by a silky sheen and medium grain size.

placer deposit A deposit of heavy or durable minerals, such as gold or diamonds, typically found where the flow of water abruptly slows.

plastic deformation A permanent *strain* that entails no rupture.

plate One of the large, thin, rigid units making up the Earth's *lithosphere*. Plates may be continental, oceanic, or both.

plate tectonics The theory that the Earth's *lithosphere* consists of large, rigid plates that move horizontally in response to the flow of the *asthenosphere* beneath them, and that interactions among the plates at their borders cause most major geologic activity, including the creation of oceans, continents, mountains, volcanoes, and earthquakes.

playa A dry lake basin found in a desert.

Pleistocene Epoch The first epoch of the *Quaternary Period*, beginning two to three million years ago and ending approximately 10,000 years ago. See also *Holocene Epoch*.

plume An upward flow of hot material from the Earth's *mantle* into the *crust*.

pluton An *intrusive rock*, as distinguished from the preexisting *country rock* that surrounds it.

plutonic rock An *intrusive rock* formed inside the Earth.

pluvial lake A lake that formed from rainwater falling into a landlocked basin during a *glacial* period marked by greater precipitation than is found in the region in prior or subsequent periods.

point bar A low ridge of *sediment* that forms along the inner bank of a *meandering stream*.

polymorph A mineral that is identical to another mineral in chemical composition but differs from it in *crystal structure*.

porosity The percentage of a soil, rock, or sediment's volume that is made up of pores.

porphyritic Of or being an *igneous rock* containing some large grains within a smaller-grained matrix.

porphyry copper deposit A crystallized rock, typically *porphyritic*, having hairline fractures that contain copper and other metals.

potassium-argon dating A form of *radiometric dating* that relies on the extremely long *half-life* of radioactive isotopes of potassium, which decay into isotopes of argon, to determine the age of rocks in which argon is present. Potassium-argon dating is used for rocks between 100,000 and 4 billion years old.

potential The combined influence of *gravity* and water pressure on groundwater flow at a given depth.

potentiometric surface The level to which the water in an *artesian aquifer* would rise if unaffected by friction with the surrounding rocks and sediments.

precipitate To separate from solution in solid form. Minerals may precipitate because of cooling, evaporation, or loss of acidity.

precipitation 1. The process by which a substance becomes *precipitated.* 2. Water that falls from the atmosphere to Earth's surface in the form of rain, snow, sleet, or hail.

primary coast A *coast* shaped primarily by nonmarine processes, such as *glacial erosion* or biological processes.

principle of cross-cutting relationships The *scientific law* stating that a *pluton* is always younger than the rock that surrounds it.

principle of faunal succession The *scientific principle* stating that specific groups of animals have followed, or succeeded, one another in a definite sequence through Earth history.

principle of inclusions The *scientific law* stating that rock fragments contained within a larger body of rock are always older than the surrounding body of rock.

principle of original horizontality The *scientific law* stating that *sediments* settling out from bodies of water are deposited horizontally or nearly horizontally in layers that lie parallel or nearly parallel to the Earth's surface.

principle of superposition The *scientific law* stating that in any unaltered sequence of rock strata, each stratum is younger than the one beneath it and older than the one above it, so that the youngest stratum will be at the top of the sequence and the oldest at the bottom.

principle of uniformitarianism The *scientific law* stating that the geological processes taking place in the present operated similarly in the past and can therefore be used to explain past geologic events.

property A characteristic that distinguishes one substance from another.

proton A positively charged particle that is found in the *nucleus* of an *atom* and has a mass approximately 1836 times that of an *electron.*

P wave (abbreviation for **primary wave**) A *body wave* that causes the *compression* of rocks when its energy acts upon them. When the P wave moves past a rock, the rock expands beyond its original volume, only to be compressed again by the next P wave. P waves are the fastest of all *seismic waves.* See also *S wave.*

P-wave shadow zone The region that extends from 103° to 143° from the *epicenter* of an *earthquake* and is marked by the absence of *P waves.* The P-wave shadow zone is due to the refraction of *seismic waves* in the liquid outer *core.* See also *S-wave shadow zone.*

pyroclastic Being or pertaining to rock fragments formed in a volcanic eruption.

pyroclastic cone A usually steep, conic *volcano* composed almost entirely of an accumulation of loose *pyroclastic* material. Pyroclastic cones are usually less than 450 meters high. Because no *lava* binds the *pyroclastics*, pyroclastic cones erode easily.

pyroclastic eruption A volcanic eruption of *viscous,* gas-rich magma. Pyroclastic eruptions tend to produce a great deal of solid volcanic fragments rather than fluid *lava.*

pyroclastic flow A rapid, extremely hot, downward stream of *pyroclastics*, air, gases, and ash ejected from an erupting *volcano.* A pyroclastic flow may be as hot as 800°C or more and may move at speeds higher than 150 kilometers per hour.

pyroclastics (used only in the plural) Particles and chunks of *igneous rock* ejected from a *volcanic vent* during an eruption.

quake See *earthquake.*

quartzite An extremely durable, *nonfoliated metamorphic rock* derived from pure *sandstone* and consisting primarily of quartz.

Quaternary ice age An *ice age* that began approximately 1.6 million years ago and continues to the present time.

Quaternary Period The second period of the *Cenozoic Era,* beginning two to three million years ago and continuing to the present time.

quick clay *Sediment* that sets off a sudden *mudflow* by changing rapidly from solid to liquid form, as after an earthquake, an explosion, or thunder.

radioactive decay The process of spontaneously emitting *protons* and *neutrons* that transforms one *isotope* into another.

radiometric dating The process of using relative proportions of *parent* to *daughter isotopes* in *radioactive decay* to determine the age of a given rock or rock stratum.

rain shadow effect The result of the process by which moist air on the *windward* side of a mountain rises and cools, causing precipitation and leaving the *leeward* side of the mountain dry.

recrystallization The *diagenetic* process by which unstable minerals in buried sediment are transformed into stable ones.

reef A ridge that forms in clear, moderately salty seawater near the *shoreline* and is composed of the carbonate remains of algae, sponges, and especially corals.

regional metamorphism *Metamorphism* that affects rocks over vast geographic areas stretching for thousands of square kilometers.

regolith The unconsolidated material that covers almost all of the Earth's land surface and is composed of *soil, sediment*, and fragments from the *bedrock* beneath it.

relative dating The fixing of a geologic structure or event in a chronological sequence relative to other geologic structures or events. See also *absolute dating.*

reserve A known *resource* that can be exploited for profit with available technology under existing political and economic conditions.

reservoir rock A permeable rock containing oil or gas.

resource A mineral or fuel deposit, known or not yet discovered, that may be or become available for human exploitation.

reverse fault A *dip-slip fault* marked by a *hanging wall* that has moved upward relative to the *footwall.* Reverse faults are often caused by the *convergence* of lithospheric plates.

rhyolite Any of a group of *felsic igneous rocks* that are the *extrusive* equivalents of *granite.*

Richter scale A logarithmic scale that measures the amount of energy released during an *earthquake* on the basis of the amplitude of the highest peak recorded on a *seismogram.* Each unit increase in the Richter scale represents a 10-fold increase in the amplitude recorded on the seismogram and a 30-fold increase in energy released by the earthquake. Theoretically the Richter scale has no upper limit, but the *yield point* of the Earth's rocks imposes an effective limit between 9.0 and 9.5.

rifting The tearing apart of a *plate* to form a depression in the Earth's *crust* and often eventually separating the plate into two or more smaller plates.

rip current A strong, rapid, and brief *current* that flows out to sea, moving perpendicular to the *shoreline.*

ripple marks A pattern of wavy lines formed along the top of a *bed* by wind, water currents, or waves.

riprap A pile of large, angular boulders built seaward of the *shoreline* in order to prevent erosion by waves or currents. See also *seawall.*

rock A naturally formed aggregate of usually inorganic materials from within the Earth.

rock cycle A series of events through which a *rock* changes, over time, between *igneous, sedimentary,* and *metamorphic* forms.

rock-forming mineral One of the twenty or so minerals contained in the rock that composes the Earth's crust and mantle.

rubidium-strontium dating A form of *radiometric dating* that relies on the 47-billion-year half-life of radioactive isotopes of rubidium, which decay into isotopes of strontium, to determine the age of rocks in which strontium is present. Rubidium-strontium dating is used for rocks that are at least 10 million years old, deep-Earth plutonic rocks, and Moon rocks.

sand 1. A particle of rock or mineral material, coarser than *silt,* that has been transported from its place of origin, as by water or wind. A particle of sand is usually between 1/16 and two millimeters in diameter. Sands are frequently composed of quartz. 2. A loosely connected body of such particles.

sandstone A *clastic rock* composed of particles that range in diameter from 1/16 millimeter to 2 millimeters in diameter. Sandstones make up about 25% of all sedimentary rocks.

saturation zone See *zone of saturation.*

scarp The steep cliff face that is formed by a *slump.*

scientific law 1. A natural phenomenon that has been proven to occur invariably whenever certain conditions are met. 2. A formal statement describing such a phenomenon and the conditions under which it occurs. Also called *law.*

scientific methods Techniques that involve gathering all available data on a subject, forming an *hypothesis* to explain the data, conducting experiments to test the hypothesis, and modifying or confirming the hypothesis as necessary to account for the experimental results.

schist A coarse-grained, strongly *foliated metamorphic rock* that develops from *phyllite* and splits easily into flat, parallel slabs.

sea stack A steep, isolated island of rock, separated from a *headland* by the action of waves, as when the overhanging section of a *sea arch* is eroded.

sea-floor spreading The formation and growth of oceans that occurs following *rifting* and is characterized by eruptions along *mid-ocean ridges,* forming new oceanic *lithosphere,* and expanding ocean basins. See also *divergence.*

seamount A conical underwater mountain formed by a *volcano* and rising 1000 meters or more from the sea floor.

seawall A wall of stone, concrete, or other sturdy material, built along the *shoreline* to prevent erosion even by the strongest and highest of waves. See also *riprap.*

secondary coast A *coast* shaped primarily by erosion or deposition by sea currents and waves.

secondary enrichment The process by which a metal deposit becomes concentrated when other minerals are eliminated from the deposit, as through *dissolution, precipitation,* or *weathering.*

sediment A collection of transported fragments or precipitated materials that accumulate, typically in loose layers, as of sand or mud.

sedimentary environment The continental, oceanic, or coastal surroundings in which sediment accumulates.

sedimentary facies 1. A set of characteristics that distinguish a given section of sedimentary rock from nearby sections. Such characteristics include mineral content, grain size, shape, and density. 2. A section of sedimentary rock so characterized.

sedimentary rock A *rock* made from the consolidation of solid fragments, as of other rocks or organic remains, or by *precipitation* of minerals from solution.

sedimentary structure A physical characteristic of a *detrital sediment* that reflects the conditions under which the sediment was deposited.

seismic Of, concerning, subject to, or produced by an *earthquake.*

seismic discontinuity A surface marking the boundary between two layers of the Earth differing in composition. *Seismic waves* passing through a seismic discontinuity undergo an abrupt change in velocity.

seismic gap A locked fault segment that has not experienced seismic activity for a long time. Because *stress* tends to accumulate in seismic gaps, they often become the sites of major *earthquakes.*

seismic profiling The mapping of rocks lying along and beneath the ocean floor by recording the reflections and refractions of *seismic waves.*

seismic tomography The process whereby a computer first synthesizes data on the velocities of *seismic waves* from thousands of recent earthquakes in order to make a series of images depicting successive planes within the Earth, and then uses these images to construct a three-dimensional representation of the Earth's interior.

seismic wave One of a series of progressive disturbances that reverberate through the Earth to transmit the energy released from an *earthquake.*

seismogram A visual record produced by a *seismograph* and showing the arrival times and magnitudes of various *seismic waves.*

seismograph A machine for measuring the intensity of *earthquakes* by recording the *seismic waves* that they generate.

seismology The study of *earthquakes* and the structure of the Earth, based on data from *seismic waves.*

shale A *sedimentary rock* composed of *detrital sediment* particles less than 0.004 millimeters in diameter. Shales tend to be red, brown, black, or gray, and usually originate in relatively still waters.

shearing stress *Stress* that slices rocks into parallel blocks that slide in opposite directions along their adjacent sides. Shearing stress may be caused by *transform motion.*

shield volcano A low, broad, gently sloping, dome-shaped structure that forms over time as repeated eruptions eject *basaltic lava* through one or more *vents* and the lava solidifies in approximately the same volume all around.

shock metamorphism The *metamorphism* that results when a meteorite strikes rocks at the Earth's surface. The meteoric impact generates tremendous pressure and extremely high temperatures that cause minerals to shatter and recrystallize, producing new minerals which cannot arise under any other circumstances.

shoreline The boundary between a body of water and dry land.

silica A *compound* consisting of silicon and oxygen.

silicate One of several rock-forming minerals that contain silicon, oxygen, and usually one or more other common elements.

silicon-oxygen tetrahedron A four-sided geometric form created by the tight bonding of four oxygen atoms to each other, and also to a single silicon atom that lies in the middle of the form.

sill A *concordant pluton* that is substantially wider than it is thick. Sills form within a few kilometers of the Earth's surface. See also *dike.*

silt 1. A particle of rock or mineral material, finer than *sand* but coarser than *clay*, that has been transported from its place of origin, typically by wind or water. A particle of silt is usually between 1/16 and 1/256 of a millimeter in diameter. 2. A loosely connected body of such particles.

sinkhole A circular, often funnel-shaped depression in the ground that forms when soluble rocks dissolve.

skarn A coarse-grained, *nonfoliated metamorphic rock* containing *silicates* that are rich in calcium.

slate A fine-grained, *foliated metamorphic rock* that develops from *shale* and tends to break into thin, flat sheets.

slide The *mass movement* of a single, intact mass of rock, soil, or unconsolidated material along a weak plane, such as a *fault, fracture,* or *bedding* plane. A slide may involve as little as a minor displacement of soil or as much as the displacement of an entire mountainside.

slip face The steep *leeward* slope of a *dune.*

slip plane A weak plane in a rock mass from which material is likely to break off in a *slide.*

slump 1. A downward and outward *slide* occurring along a concave *slip plane.* 2. The material that breaks off in such a slide.

snowline The lowest point at which snow remains year-round.

soil The top few meters of *regolith,* generally including some organic matter derived from plants.

soil horizon A layer of soil that can be distinguished from the surrounding soil by such features as chemical composition, color, and texture.

soil profile A vertical strip of soil stretching from the surface down to the *bedrock* and including all of the successive *soil horizons.*

solifluction A form of *creep* in which soil flows downslope at 0.5 to 15 centimeters per year. Solifluction occurs in relatively cold regions when the brief warmth of summer thaws only the upper meter or two of *regolith,* which becomes waterlogged because the underlying ground remains frozen and therefore the water cannot drain down into it.

source rock A rock in which *hydrocarbons* originate.

sorting The process by which a given *transport medium* separates out certain particles, as on the basis of size, shape, or density.

specific gravity The ratio of the weight of a particular volume of a given substance to the weight of an equal volume of pure water.

speleothem A mineral deposit of calcium carbonate that precipitates from solution in a *cave.*

spheroidal weathering The process by which *chemical weathering,* especially by water, decomposes the angles and edges of a rock or boulder, leaving a rounded form from which concentric layers are then stripped away as the weathering continues.

spit A narrow, fingerlike ridge of sand that extends from land into open water.

stalactite An icicle-like mineral formation that hangs from the ceiling of a *cave* and is usually made up of *travertine,* which precipitates as water rich in dissolved limestone drips down from the cave's ceiling. See also *stalagmite.*

stalagmite A cone-shaped mineral deposit that forms on the floor of a *cave* and is usually made up of *travertine*, which precipitates as water rich in dissolved limestone drips down from the cave's ceiling. See also *stalactite.*

star dune A *dune* with three or four arms radiating from its usually higher center so that it resembles a star in shape. Star dunes form when winds blow from three or four directions, or when the wind direction shifts frequently.

stratification See *bedding.*

stratovolcano A cone-shaped *volcano* built from alternating layers of *pyroclastics* and viscous *andesitic lava.* Stratovolcanos tend to be very large and steep.

stratum (plural **strata**) A layer of *sedimentary rock* that is visibly distinct from the surrounding layers.

streak The color of a mineral in its powdered form. This color is usually determined by rubbing the mineral against an unglazed porcelain slab and observing the mark made by it on the slab.

strain The change in the shape or volume of a rock that results from *stress.*

stream A body of water found on the Earth's surface and confined to a narrow topographic depression, down which it flows and transports rock particles, sediment, and dissolved particles. Rivers, creeks, brooks, and runs are all streams.

stream discharge The volume of water to pass a given point on a stream bank per unit of time, usually expressed in cubic meters of water per second.

stream terrace A level plain lying above and running parallel to a stream bed. A stream terrace is formed when a stream's bed erodes to a substantially lower level, leaving its flood plain high above it.

stress The force acting on a rock or another solid to deform it, measured in kilograms per square centimeter or pounds per square inch.

striation One of a group of usually parallel scratches engraved in bedrock by a *glacier* or other geological agent.

strike 1. The horizontal line marking the intersection between the inclined plane of a solid geological structure and the Earth's surface. 2. The compass direction of this line, measured in degrees from true north.

strike-slip fault A *fault* in which two sections of rock have moved horizontally in opposite directions, parallel to the line of the *fracture* that divided them. Strike-slip faults are caused by *shearing stress.*

structural geology The scientific study of the geological processes that deform the Earth's crust and create mountains.

subduction The sinking of an oceanic *plate* edge as a result of *convergence* with a plate of lesser density. Subduction often causes *earthquakes* and creates *volcano* chains.

subsidence The lowering of the Earth's surface, caused by such factors as compaction, a decrease in groundwater, or the pumping of oil.

sulfate One of several minerals containing positive sulfur ions bonded to negative oxygen ions.

sulfide One of several minerals containing negative sulfur ions bonded to one or more positive metallic ions.

surface wave One of a series of *seismic waves* that transmits energy from an earthquake's *epicenter* along the Earth's surface. See also *body wave.*

surf zone The area running from the *shoreline* to the farthest point in the sea where waves begin to break.

surge To flow more rapidly than usually. Said of a *glacier.*

suspended load A body of fine, solid particles, typically of sand, clay, and silt, that travels with stream water without coming into contact with the stream bed.

suture zone The area where two continental plates have joined together through *continental collision.* Suture zones are marked by extremely high mountain ranges, such as the Himalayas and the Alps.

swash The rush of water onto a beach after a wave breaks.

S wave (abbreviation for **secondary wave**) A *body wave* that causes the rocks along which it passes to move up and down perpendicular to the direction of its own movement. See also *P wave.*

S-wave shadow zone The region within an arc of 154° directly opposite an earthquake's *epicenter* that is marked by the absence of *S waves.* The S-wave shadow zone is due to the fact that S waves cannot penetrate the liquid outer core. See also *P-wave shadow zone.*

syncline A concave *fold*, the central part of which contains the youngest section of rock. See also *anticline.*

talus A pile of rock fragments lying at the bottom of the cliff or steep slope from which they have broken off.

tarn A deep, typically circular lake that forms when a *cirque glacier* melts.

tectonic creep The almost constant movement of certain *fault blocks* that allows *strain* energy to be released without major *earthquakes.*

tension *Stress* that stretches or extends rocks, so that they become thinner vertically and longer laterally. Tension may be caused by *divergence* or *rifting.*

tephra (plural noun) *Pyroclastic* materials that fly from an erupting volcano through the air before cooling, and range in size from fine dust to massive blocks.

terminus The outer margin of a *glacier.*

theory A comprehensive explanation of a given set of data that has been repeatedly confirmed by observation and experimentation and has gained general acceptance within the scientific community but has not yet been decisively proven. See also *hypothesis* and *scientific law.*

thermal expansion A form of *mechanical weathering* in which heat causes a mineral's crystal structure to enlarge.

thermal plume A vertical column of upwelling *mantle* material, 100 to 250 kilometers in diameter, that rises from beneath a continent or ocean and can be perceived at the Earth's surface as a *hot spot.* Thermal plumes carry enough energy to move a plate, and they may be found both at plate boundaries and plate interiors.

thrust fault A *reverse fault* marked by a *dip* of 45° or less.

tidal bore A turbulent, abrupt, wall-like wave that is caused by a *flood tide.*

tide 1. The cycle of alternate rising and falling of the surface of an ocean or large lake, caused by the gravitational pull of the Sun and especially Moon in interaction with the Earth's rotation. Tides occur on a regular basis, twice every day on most of the Earth. 2. A single rise or fall within this cycle.

till See *glacial till.*

topography The set of physical features, such as mountains, valleys, and the shapes of landforms, that characterizes a given landscape.

transition zone The *seismic discontinuity* located in the upper *mantle* just beneath the *asthenosphere* and characterized by a marked increase in the velocity of *seismic waves.*

transform motion The movement of two adjacent lithospheric plates in opposite directions along a parallel line at their common edge. Transform motion often causes *earthquakes.*

translatory Of, concerning, or being the movement of water over a significant distance in the direction of a wave.

transport medium A natural agent, such as water, air, or ice, that moves a particle or particles from one location on the Earth's surface to another.

transverse dune One of a series of *dunes* having an especially steep *slip face* and a gentle *windward* slope and standing perpendicular to the prevailing wind direction and parallel to each other. Transverse dunes typically form in arid and semi-arid regions with plentiful sand, stable wind direction, and scarce vegetation. A transverse dune may be as much as 100 kilometers long, 200 meters high, and three kilometers wide.

travertine *Crystalline* deposits of calcium carbonate precipitated from solution, often found in *caves*.

tributary A *stream* that supplies water to a larger stream.

trunk stream A large stream into which *tributaries* carry water and sediment.

tsunami (plural **tsunami**) A vast sea wave caused by the sudden dropping or rising of a section of the sea floor following an *earthquake*. Tsunami may be as much as 30 meters high and 200 kilometers long, may move as fast as 250 kilometers an hour, and may continue to occur for as long as a few days.

tuff See *volcanic tuff*.

uniformitarianism The hypothesis that current geologic processes, such as the slow erosion of a coast under the impact of waves, have been occurring in a similar manner throughout the Earth's history and that these processes can account for past geologic events. See also *catastrophism*.

unconformity A boundary separating two or more rocks of markedly different ages, marking a gap in the geologic record.

upwarped mountain A mountain consisting of a broad area of the Earth's *crust* that has moved gently upward without much apparent deformation, and usually containing *sedimentary, igneous*, and *metamorphic rocks*.

uranium-lead dating A form of *radiometric dating* that relies on the extremely long *half-life* of radioactive isotopes of uranium, which decay into isotopes of lead, to determine the age of rocks in which uranium and lead are present.

valley glacier An *alpine glacier* that flows through a preexisting stream valley.

van der Waals bond A relatively weak kind of *intermolecular bond* that forms when one side of a *molecule* develops a slight negative charge because a number of *electrons* have temporarily moved to that side of the *molecule*, and this negative charge attracts the *nuclei* of the *atoms* of a neighboring molecule, while the side of the molecule with fewer electrons develops a slight positive charge that attracts the electrons of the atoms of neighboring molecules.

varve A pair of sediment *beds* deposited by a lake on its floor, typically consisting of a thick, coarse, light-colored bed deposited in the summer and a thin, fine-grained, dark-colored bed deposited in the winter. Varves are most often found in lakes that freeze in the winter. The number and nature of varves on the bottom of a lake provides information about the lake's age and geologic events that affected the lake's development.

vent An opening in the Earth's surface through which *lava*, gases, and hot particles are expelled. Also called *volcanic vent* and *volcano*.

ventifact A stone that has been flattened and sharpened by *wind abrasion*. Ventifacts are commonly found strewn across a *desert* floor.

viscosity A fluid's resistance to flow. Viscosity increases as temperatures decrease.

volcanic arc A chain of *volcanoes* fueled by magma that rises from an underlying *subducting* plate.

volcanic cone A cone-shaped mountain that forms around a *vent* from the debris of *pyroclastics* and *lava* ejected by numerous eruptions over time.

volcanic crater A steep, bowl-shaped depression surrounding a *vent*. A volcanic crater forms when the walls of a vent collapse inward following an eruption.

volcanic dome A bulb-shaped solid that forms over a *vent* when *lava* so *viscous* that it cannot flow out of the *volcanic crater* cools and hardens. When a volcanic dome forms, it traps the volcano's gases beneath it. They either escape along a side vent of the volcano or build pressure that causes another eruption and shatters the volcanic dome.

volcanic rock See *extrusive rock*.

volcanic tuff A solid rock made up of *tephra* that have consolidated and become cemented together. Also called *tuff*.

volcanism The set of geological processes that result in the expulsion of *lava*, *pyroclastics*, and gases at the Earth's surface.

volcano The solid structure created when lava, gases, and hot particles escape to the Earth's surface through *vents*. Volcanoes are usually conical. A volcano is "active" when it is erupting or has erupted recently. Volcanoes that have not erupted recently but are considered likely to erupt in the future are said to be "dormant." A volcano that has not erupted for a long time and is not expected to erupt in the future is "extinct."

watershed See *drainage basin*.

water table The surface that lies between the *zone of aeration* and the underlying *zone of saturation*.

wave refraction The process by which a wave approaching the shore changes direction due to slowing of those parts of the wave which enter shallow water first, causing a sharp decrease in the angle at which the wave approaches until the wave is almost parallel to the coast.

wave-cut bench A relatively level surface formed when waves erode the base of a cliff, causing the overlying rock to fall into the surf. A wave-cut bench stands above the water and extends seaward from what remains of the cliff.

weathering The process by which exposure to atmospheric agents, such as air or moisture, causes *rocks* and *minerals* to break down. This process takes place at or near the Earth's surface. Weathering entails little or no movement of the material that it loosens from the rocks and minerals. See also *erosion*.

wetland A lake, marsh, or swamp that supports wildlife and replenishes the groundwater system.

wind abrasion The process by which wind erodes *bedrock* through contact between the bedrock and rock particles carried by the wind.

windward Of, located on, or being the side of a dune, hill, or ridge facing into the wind. See also *leeward.*

xenolith A preexisting rock embedded in a newer *igneous rock*. Xenoliths are formed when a rising *magma* incorporates the preexisting rock. If the preexisting rock does not melt, it will not be assimilated into the magma and will therefore remain distinct from the new igneous rock that surrounds it.

X-ray diffraction The scattering of X-rays passed through a mineral sample so as to form a pattern peculiar to the given mineral.

yield point The maximum *stress* that a given rock can withstand without becoming permanently deformed.

zone of ablation The part of a *glacier* in which there is greater overall loss than gain in volume. A zone of ablation can be identified in the summer by an expanse of bare ice. See also *zone of accumulation.*

zone of accumulation The part of a *glacier* in which there is greater overall gain than loss in volume. A zone of accumulation can be identified by a blanket of snow that survives summer melting. See also *zone of ablation.*

zone of aeration A region below the Earth's surface that is marked by the presence of both water and air in the pores of rocks and soil. Also called *aeration zone.*

zone of saturation A region that lies below the *zone of aeration* and is marked by the presence of water and the absence of air in the pores of rocks and soil.

Credits

Part 1 Opener Bern Pedit c/o Breck P. Kent.

Chapter 1

Figure 1–1 Tom Bean; **Figure 1–2** Matthew Naythans/Gamma Liaison; **Figure 1–3** Hires Chip/Gamma Liaison; **Figure 1–4 (a)** U.S. Geological Survey; **(b)** James L. Amos/Photo Researchers; **(c)** JPL/NASA; **Figure 1–5 (a)** James L. Amos/Photo Researchers; **(b)** Runk/Schoenberger/Grant Heilman, specimen N. Museum/Franklin Marshall College; **(c)** G. A. Izett, U.S. Geological Survey; **(d)** Virgil L. Sharpton/Lunar and Planetary Institute-Houston; **Figure 1–12 (a)** U.S. Geological Survey W. 100th No. 47/USGS; **(b)** R. M. Turner 17/ U.S. Geological Survey; **Figure 1–13** Breck P. Kent; **Figure 1–17** NASA; **Figure 1–18** NRSC Ltd./Science Photo Library/Photo Researchers, Inc.; **Figure 1–23 (bottom)** Craig Tuttle/The Stock Market; **(top)** Comstock; **Figure 1–25** John S. Shelton; **Figure 1–27** Library of Congress, print courtesy of Time Inc. Picture Collection; **Figure 1–31 (a)** Breck P. Kent; **(b)** Betty Crowell/ Faraway Places; **Figure 1–32 (a)** James L. Amos/Photo Researchers.

Chapter 2

Figure 2–1 Breck P. Kent; **Figure 2–2 (a)** Breck P. Kent; **(b)** Kathleen Campbell/Liaison International; **Figure 2–3** Fred Ward/Black Star; **Figure 2–4** Fred Hirschmann; **Figure 2–12 (a [left])** Jeffrey Scovil; **(a [right])** Jeffrey Scovil; **Figure 2–12 (b)** Jeffrey Scovil; **Figure 2–13 (a)** Ken Lucas/Biological Photo Service; **(b)** Breck P. Kent; **Figure 2–14 (a)** Breck P. Kent; **(b)** Breck P. Kent; **(c)** Breck P. Kent; **(d)** Stuart Cohen/Comstock; **(e)** James L. Amos/ Photo Researchers; **Figure 2–15** Breck P. Kent; **Figure 2–16 (left)** Breck P. Kent; **(right)** M. Claye/Jacana/Photo Researchers; **Figure 2–17 (a)** Ed Degginger/Bruce Coleman Inc.; **(b)** Breck P. Kent; **(c)** Barry L. Runk/Grant Heilman; **(d)** Francis Örs Lustwerk-Dudás; **(e)** Francis Örs Lustwerk-Dudás; **Figure 2–18** Breck P. Kent; **Figure 2–19 (a)** Jeffrey Scovil; **(b)** Manfred Kage/Peter Arnold, Inc; **Figure 2–20 (a)** E. R. Degginger/Bruce Coleman Inc.; **(b)** Breck P. Kent; **(c)** A.Copley/Visuals Unlimited; **(d)** Jeffrey Scovil; **Figure 2–22 (left)** Bruce Iverson (2); **(right)** Bruce Iverson (2); **Figure 2–23 (a)** Breck P. Kent; **(b)** Breck P. Kent; **Figure 2–26 (a [photo])** American Museum of Natural History; **(b [photo])** Breck P. Kent; **(c [photo])** Breck P. Kent; **(d [photo])** Breck P. Kent; **(e [photo])** Breck P. Kent; **Figure 2–27** Reuters/Archives Photos; **Figure 2–28 (a)** Breck P. Kent; **(b)** M. Claye/ Jacana/Photo Researchers; **(c)** M. Claye/Jacana/Photo Researchers; **Figure 2–29** Handout/Reuters/Archive Photos; **Figure 2–30 (a)** John S. Shelton; **(b)** Lunar and Planetary Institute; **page 63 (top)** Tom McHugh/Photo Researchers; **(bottom)** Breck P. Kent.

Chapter 3

Figure 3–1 BernPedit/Breck P. Kent; **Figure 3–3 (a)** Fred Hirschmann; **(a [inset])** William E. Ferguson; **(b)** E. R. Degginger/Earth Scenes; **(c)** Biological Photo Service; **(c [inset])** Breck P. Kent/Earth Scenes; **(d)** Breck P. Kent; **(e)** Breck P. Kent; **(e [inset])** Doug Sokell/Visuals Unlimited; **Figure 3–4 (inset 1)** Breck P. Kent; **(inset 2)** Breck P. Kent; **(inset 3)** Breck P. Kent; **(inset 4)** Breck P. Kent; **Figure 3–12** Francis Örs Lustwerk-Dudás; **Figure 3–15 (photo)** Breck P. Kent; **Figure 3–16** William E. Ferguson; **Figure 3–18 (photo)** John S. Shelton; **Figure 3–19 (a)** National Archives; **Figure 3–21 (photo)** Rob Badger; **Figure 3–22 (photo)** Timothy Holst; **Figure 3–23** Fred Hirschmann; **Figure 3–28** John Sanford/SPL/Photo Researchers, Inc.; **Figure 3–29 (photo)** NASA JSC/Starlight (Photo by Thomas Stafford, John Young, or Eugene Carnan; **Figure 3–30** Francois Gohier/Photo Researchers, Inc.; **Figure 3–31** Fraser Goff/Los Alamos National Laboratory; **page 91** Fred Hirschmann.

Chapter 4

Figure 4–1 Wesley Bocxe/The Image Works; **Figure 4–2** John Reed/Science Library/Photo Researchers; **Figure 4–3** Georg Gerster/Comstock; **Figure 4–4** Michael Wilhelm/ENP Images; **Figure 4–5 (a)** Gary Braasch; **(b)** Ralph Perry/ Black Star; **Figure 4–6 (a)** Eugene Iwatsubo-Cascade Volcano Observatory, USGA; **(b)** Al Merrill/Breck P. Kent; **Figure 4–7** Mark A. Johnson; **Figure 4–8** U.S. Geological Survey; **Figure 4–9** D. R. Stoecklein/The Stock Market; **Figure 4–10** Glenn Oliver/Visuals Unlimited; **Figure 4–11 (b)** Breck P. Kent; **Figure 4–12 (b)** Courtesy of the Geological Survey of Canada; **(c)** Fred Grassle/Woods Hole Oceanographic Institute; **Figure 4–13** Robert A. Jensen; **Figure 4–14 (a)** William E. Ferguson; **(b)** William E. Ferguson; **Figure 4–15 (b)** Linda J. Moore; **Figure 4–16 (photo)** Alberto Garcia/SABA; **Figure 4–17 (inset)** Anthony Suau/The Gamma Liaison Network; **(photo)** Cromoc/ The Gamma Liaison Network; **Figure 4–18** Phil Degginger/Earth Scenes; **Figure 4–20 (a)** Steve Kaufman/Peter Arnold, Inc.; **(b)** John S. Shelton; **Figure 4–21** J. R. Smith and T. Duennebier/University of Hawaii; **Figure 4–22 (photo)** Marc Schechter/Photo Resource Hawaii; **Figure 4–23 (photo)** Teaching Collection/University of Washington, Department of Geological Science; **Figure 4–25 (photo)** Hydrographic Department of Japan; **Figure 4–26** Egill Gisli Hrafnsson/Gamma Liaison; **Figure 4–28** Leonard Von Matt/Photo Researchers; **Figure 4–30** Ray Atkeson/The Stock Market; **Figure 4–32** Link/ Visuals Unlimited; **Figure 4–33** Georg Gerster/Comstock; **Figure 4–36 (a)** James Sugar/Black Star; **(b)** James Mason/Black Star; **(c)** James Sugar/ Black Star; **Figure 4–37** Gary Braasch; **Figure 4–38** William E. Ferguson; **Figure 4–39 (photo)** NASA/ERTS; **Figure 4–40** Fred M. Bullard; **Figure 4–41** NASA; **Figure 4–42** US Geological Survey/NASA/Photo Researchers, Inc.; **page 127** David Ball/Allstock.

Chapter 5

Figure 5–1 David Muench/Tony Stone Images; **Figure 5–2** Mount Rushmore National Memorial; **Figure 5–5** Tom Bean/DRK Photo; **Figure 5–6 (a)** New York Public Library, Locan History and Genealogy Division; **(b)** Runk/Schoenberger/Grant Heilman; **Figure 5–8** Phil Degginger/Earth Scenes; **Figure 5–9** Runk/Schoenberger/Grant Heilman; **Figure 5–10** Ramesh Venkatakrishnan; **Figure 5–12** Runk/Schoenberger/Grant Heilman; **Figure 5–13** John S. Shelton; **Figure 5–14 (left)** John S. Shelton; **(right)** John S. Shelton; **Figure 5–16 (photo)** Paul McKelvey/Tony Stone Images; **Figure 5–18** Dennis Netoff; **Figure 5–19** Fletcher & Baylis/Photo Researchers, Inc.; **Figure 5–22 (photo)** Runk/Schoenberger/Grant Heilman;

Figure 5–23 (a) Ramesh Venkatakrishnan; **(b)** U.S. Department of Agriculture; **Figure 5–24 (b)** William E. Ferguson; **Figure 5–26** Teaching Collection/University of Washington, Department of Geological Science; **Figure 5–27** George Holton/Photo Researchers; **Figure 5–28** NASA; **Figure 5–29** TASS/Sovfoto; **Figure 5–30** JPL/NASA; **page 151** Stanley Chernicoff/Patrick Spencer.

Chapter 6

Figure 6–1 Joe McDonald/Earth Scenes; **Figure 6–3 (a)** Connie Toops; **(b)** Breck P. Kent; **Figure 6–4** Tom Bean; **Figure 6–7 (a)** Martin Miller; **Figure 6–9 (photo)** Ann B. Swengel/Visuals Unlimited; **Figure 6–10 (a)** Glacier National Park; **Figure 6–13 (a)** Kurt Hollocher/Union College Geology Department; **(b)** Martin Miller; **(c)** Martin Miller; **(inset)** Kurt Hollocher/Union College Geology Department; **(inset)** Kurt Hollocher/Union College Geology Department; **(inset)** Kurt Hollocher/Union College Geology Department; **Figure 6–14 (a)** John S. Shelton; **(b)** Breck P. Kent; **Figure 6–15 (bottom right)** Martin G. Miller/Visuals Unlimited; **(top right)** Larry Davis, Washington State University; **Figure 6–16** Tom Bean; **Figure 6–20** J. Fennell/Bruce Coleman Inc.; **Figure 6–21** Giorgio Gualco/Bruce Coleman Inc.; **Figure 6–22 (a)** E. R. Degginger/Bruce Coleman Inc; **(b)** John Sohlden/Visuals Unlimited; **Figure 6–23** Alfred Pasieka/Bruce Coleman Inc.; **Figure 6–24 (a)** Robert S. Garrison, University of California; **(b)** Robert S. Garrison, University of California; **Figure 6–25** Martin Miller; **Figure 6–31** Tom Bean; **page 181** Betty Crowell/Faraway Places.

Chapter 7

Figure 7–1 John Sohlden/Visuals Unlimited; **Figure 7–6** Kenneth Murray/Photo Researchers; **Figure 7–8** Breck P. Kent; **Figure 7–9 (b)** Betty Crowell; **Figure 7–12** William E. Ferguson; **Figure 7–14 (a)** John D. Cunningham/Visuals Unlimited; **Figure 7–15** Albert J. Copley/Visuals Unlimited; **Figure 7–16** Roy Jameson; **Figure 7–17** Stephen Trimble/DRK Photo; **Figure 7–24 (b [top left])** Bernard Evans/University of Washington, Department of Geological Science; **Figure 7–24 (c [top right])** William E. Ferguson; **(d [bottom right])** © Francis Örs Lustwerk-Dudás, 1994; **(left bottom)** William E. Ferguson; **Figure 7–25** Roy Jameson; **Figure 7–26** Scott Frances/Esto; **(inset)** William E. Ferguson; **Figure 7–27 (left)** Raphael Gaillarde/Gamma Liasion; **(right)** Raphael Gaillarde/Gamma Liasion; **Figure 7–29 (left [photo])** Betty Crowell; **page 205** Ramesh Venkatakrishnan.

Chapter 8

Figure 8–1 Breck P. Kent; **Figure 8–3** William E. Ferguson; **Figure 8–4 (a)** G. Shanmugam; **Figure 8–5** Tom Bean; **Figure 8–6** William E. Ferguson; **Figure 8–8 (a)** Breck P. Kent; **(b)** Breck P. Kent; **(c)** Ken Lucas/Biological Photo Service; **Figure 8–11 (a)** Breck P. Kent; **(b)** David Schwimmer/Bruce Coleman Inc.; **(c)** William E. Ferguson; **Figure 8–12 (photo)** Edward A. Hay, De Anza College, Cupertino, CA; **Figure 8–16** Roger Buick/The University of Sydney; **Figure 8–17** Breck P. Kent; **Figure 8–23** Sam Bowring/MIT; **Figure 8–25** Harvey Lloyd/Peter Arnold, Inc; **Figure 8–26 (photo)** John Garver; **Figure 8–28 (photo)** Teaching Collection/University of Washington, Department of Geological Science; **Figure 8–29 (photo)** Grant Heilman; **Figure 8–31** Stan Chernicoff; **Figure 8–34 (bottom)** William E. Ferguson; **Figure 8–34 (top)** Breck P. Kent; **Figure 8–35** Tom McHugh/California Academy of Sciences/Photo Researchers; **Figure 8–36** NASA; **Figure 8–37** Joy Spurr/Bruce Coleman Inc.; **Figure 8–38** JSC/NASA; **Figure 8–39** JSC/NASA; **Figure 8–40** NASA/GSFC and the University of Iowa.

Part 2 Opener Tom Bean.

Chapter 9

Figure 9–1 Michael Fogden/DRK Photo; **Figure 9–2 (a)** Betty Crowell/Faraway Places; **(b)** Martin Miller; **Figure 9–6 (a)** A. J. Copley/Visuals Unlimited; **(b)** Breck P. Kent; **(c)** Tom Bean; **Figure 9–10 (left)** GEOPIC©, Earth Satellite Corporation; **Figure 9–13 (a)** David Muench/Allstock; **(b)** Martin G. Miller/Visuals Unlimited; **(c)** John S. Shelton; **Figure 9–16 (photo)** Georg Gerster/Wingstock/Comstock; **Figure 9–19 (photo)** Simon Fraser/Science Photo Library/Photo Researchers, Inc.; **Figure 9–20 (photo)** Breck P. Kent; **Figure 9–25 (inset)** John S. Shelton; **(left)** George Wuerthner; **Figure 9–27 (right)** Worldsat International Inc./Photo Researchers Inc.; **Figure 9–30** Clyde H. Smith/Allstock; **Figure 9–31** Jet Propulsion Lab, California Institute of Technology; **page 265** Ramesh Venkatakrishnan.

Chapter 10

Figure 10–1 Terrance White/FSP/The Gamma Liaison Network; **Figure 10–9** John S. Shelton; **Figure 10–10 (photo)** Steve McCutcheon/Alaska Pictorial Services; **Figure 10–11** The British Library Neg. # H043580, Shelfmark 719.m.17.(15).; **Figure 10–12 (photo)** Reuters/Archive Photos; **Figure 10–13** Cliff Wassmann; **Figure 10–14** Art Resource, New York; **Figure 10–15** JP Owen/Black Star; **Figure 10–17** Albert Copley/Visuals Unlimited; **Figure 10–18 (photo)** Bates LittleHales-National Geographic; **Figure 10–21 (left)** California Institute of Technology; **Figure 10–23** Wallace, R. E. o. 311/US Geological Survey; **Figure 10–24** Novosti/Science Photo Libary/Photo Researchers, Inc.; **Figure 10–25** Peter Vadnai/The Stock Market; **Figure 10–26** UPI/Bettmann; **Figure 10–27** John S. Shelton; **Figure 10–29 (b)** Raghu Rai/Magnum Photos.

Chapter 11

Figure 11–1 American Museum of Natural History; **Figure 11–4 (a)** Wei-jia Su & Adam Dziewonski; **(b)** Rob D. Van der Hilst/MIT-EAPS; **Figure 11–9** Courtesy Paul Tackley; **Figure 11–14 (photo)** Douglas L. Peck; **Figure 11–15** John S. Shelton; **Figure 11–26** Tom Bean/Allstock.

Chapter 12

Figure 12–1 Barbara Cushman Rowell/DRK Photos; **Figure 12–8** National Oceanic and Atmospheric Administration/National Geophysical Data Center; **Figure 12–15** Robert R. Hessler; **Figure 12–16** Dudley Foster/Woods Hole Oceanographic Institute; **Figure 12–18** R. Dietmar Muller, Walter R. Roest, Jean-Yves Royer, Lisa M. Gahagan, John G. Sclater/University of California, San Diego; **Figure 12–19** Betty Crowell/Faraway Places; **Figure 12–21** P. Greenberg/Comstock; **Figure 12–24** Larissa F. Dobrzhinetskaya; **(inset)** Larissa F. Dobrzhinetskaya; **Figure 12–25** James Balog/Tony Stone Images.

Part 3 Opener Joel Bennett/Earth Scenes.

Chapter 13

Figure 13–1 Roger J. Wyan/AP Wide Wold Photo; **Figure 13–2 (a)** Peter Kresan; **Figure 13–5 (b [inset])** Tom Bean/DRK Photos; **Figure 13–7** David Falconer/Bruce Coleman Inc.; **Figure 13–10 (right)** Betty Crowell/Faraway Places; **Figure 13–11 (bottom)** John S. Shelton; **(top)** Peter L. Kresan; **Figure 13–15** Kennan Ward; **Figure 13–16 (photo)** Breck P. Kent; **Figure 13–17** M. Western/Comstock; **Figure 13–18 (b)** Rolle/Gamma Liaison; **Figure 13–19** Gamma Liaison; **Figure 13–21** Sean Ramsay/The Image Works; **Figure 13–22** Fletcher & Bayles/Photo Researchers; **Figure 13–23 (a)** Lloyd S. Cluff; **(b)** Lloyd S. Cluff; **Figure 13–25** Betty Crowell/Faraway Places; **Figure 13–26 (left)** G. Brad Lewis; **(right)** R. T. Holcomb/U.S.Geological Survey; **Figure 13–29 (photo)** Tim Crosby/Gamma Liaison; **Figure 13–31 (a)** Linda J. Moore; **(b)** Martin Miller; **Figure 13–32** NASA; **Figure 13–33** U.S. Geological Survey; **page 387** Ramesh Venkatakrishnan.

Chapter 14

Figure 14–1 Eric Hylden/Grand Forks Herald; **Figure 14–6** Dr. Donald T. Rodbell; **Figure 14–8** Norman Weiser; **Figure 14–9** Peter Kresan; **Figure**

14–12 (photo) Jeff Lepore/Photo Researchers; **Figure 14–13 (photo)** Martin G. Miller/Visuals Unlimited; **Figure 14–14** Art Gingert/Comstock; **Figure 14–15 (photo)** John S. Flannery/Bruce Coleman Inc.; **Figure 14–16 (right)** Stock Montage; **Figure 14–17 (a)** Jeff Lepore/Photo Researchers; **(b)** Andrew J. Martinez/Photo Researchers; **(c)** Art Gingert/Comstock; **Figure 14–18 (photo)** Betty Crowell; **Figure 14–20** Ramesh Venkatakrishnan; **Figure 14–22** Andrew Holbrooke/Gamma Liaison; **Figure 14–23** Peter Kresan; **Figure 14–24 (b)** G. R. Roberts; **(c)** Ramesh Venkatakrishnan; **Figure 14–27** World Perspectives/Gamma Liaison; **Figure 14–30 (a)** Andrew Holbrooke/Gamma Liaison; **(b)** John Eastcott/Photo Researchers, Inc.; **Figure 14–33** Tom Bean; **Figure 14–34** Galen Rowell/Peter Arnold, Inc.; **Figure 14–35** NASA; **page 419** Alex S. Maclean/Peter Arnold, Inc.

Chapter 15

Figure 15–1 Bob Krist/The Stock Market; **Figure 15–11 (photo)** C. E. Siebenthal/U.S. Geological Survey; **Figure 15–12 (photo)** Jose Fuste Raga/The Stock Market; **Figure 15–14 (photo)** William E. Ferguson; **Figure 15–15 (left)** Farrell Grehan/Science Source/Photo Researchers; **(right)** Gregory D. Dimijian, M.D./Photo Researchers; **Figure 15–16 (photo)** Peter Kresan; **Figure 15–17 (map)** James A. Westphal; **(photo)** James A. Westphal; **Figure 15–22 (photo)** Tony Marshall/DRK Photo; **Figure 15–24** US Department of Interior, USGS; **Figure 15–25** Porterfield-Chickering/Photo Researchers; **Figure 15–26** Rod Planck/Photo Researchers, Inc.; **Figure 15–28 (photo)** U.S. Department of Energy; **Figure 15–30** Tom Bean/DRK Photo.

Chapter 16

Figure 16–1 Michael K. Nichols/National Geographic Image Collection; **Figure 16–2** Rene Burri/Magnum; **Figure 16–3** Geof Spaulding; **Figure 16–4** Indiana Geological Survey; **Figure 16–5 (a)** E. R. Degginger/Bruce Coleman Inc.; **(b)** D. Hardley/Bruce Coleman Inc.; **Figure 16–7** Ramesh Venkatakrishnan; **Figure 16–9 (a)** Robert & Linda Mitchell; **(b)** Albert Copley/Visuals Unlimited; **(c)** Jeff Lepore/Photo Researchers; **(d)** Stephen Alvarez/Time Magazine; **(e)** D & J McClurg/Bruce Coleman Inc.; **(f)** Robert & Linda Mitchell; **Figure 16–10 (photo)** Joy Spurr/Bruce Coleman Inc.; **Figure 16–12 (photo)** Stephen Alvarez/Time Magazine; **Figure 16–13 (photo)** Gene Aherns/Bruce Coleman Inc.; **Figure 16–14** Gregory G. Dimijian/Photo Researchers; **Figure 16–15** Kim Heacox/Peter Arnold, Inc.; **Figure 16–18** Ramesh Venkatakrishnan; **Figure 16–20 (a)** Timothy O'Keefe/Bruce Coleman Inc.; **(b)** Paul H. Moser, Alabama Geological Survey; **Figure 16–21** A. N. Palmer, Earth Sciences, SUNY Oneonta; **Figure 16–22 (left)** Ramesh Venkatakrishnan; **(right)** Calvin Alexander; **Figure 16–23 (photo)** Paolo Koch/Photo Researchers; **Figure 16–24** Steve Solum/Bruce Coleman Inc.; **Figure 16–25** Betty Crowell/Faraway Places; **Figure 16–26** AP/Wide World Photos; **Figure 16–27** Ramesh Venkatakrishnan; **page 467** USDA, Soil Conservation Service.

Chapter 17

Figure 17–1 Breck P. Kent; **Figure 17–6 (a)** Comstock; **(b)** Tom Bean; **(c)** Betty Crowell/Faraway Places; **(d)** Jim Wark/Peter Arnold, Inc.; **Figure 17–7 (photo)** Tom Bean; **Figure 17–9** Steve McCutcheon/Visuals Unlimited; **Figure 17–10 (a)** Tom Bean; **(b)** Dan Guravich/Photo Researchers, Inc.; **Figure 17–11 (photo)** William E. Ferguson; **Figure 17–12 (b)** Bill Kamin/Visuals Unlimited; **(c)** Betty Crowell/Faraway Places; **Figure 17–13 (a)** ©Advanced Satellite Productions, Inc. 1993; **(b)** Photo available from U.S. Dept. of Interior, USGS Eros Data Center. Scene ID # E-1272–99CT; **Figure 17–14** Peter Arnold/Peter Arnold, Inc.; **Figure 17–17 (inset)** Art Gingert/Comstock; **Figure 17–19 (photo)** Tom Bean/DRK Photo; **Figure 17–20 (photo)** John S. Shelton; **Figure 17–21** Alan Busacca, Washington State University; **Figure 17–22** John S. Shelton; **Figure 17–25 (a)** John S. Shelton; **(b)** John S. Shelton; **(c)** John S. Shelton; **Figure 17–26 (photo)** Jerry Ricketts; **Figure 17–27 (a)** Peter Dunwiddie/Visuals Unlimited; **(b)** Glenn Oliver/Visuals Unlimited; **(c)** Bernard Hallet, University of Washington; **(d)** Dr. Troy L. Pewe, photo # 3161; **Figure 17–35** Betty Crowell/Faraway Places; **Figure 17–38** John W. Attig, Wisconsin Geological Survey; **Figure 17–39** Peter Kresan; **Figure 17–42** Natural Environment Research Council (NERC); **(inset)** V. Bretagnolle/Peter Arnold, Inc.; **Figure 17–43** U.S. Geological Survey, Flagstaff, Arizona; **Figure 17–44 (left)** NASA; **Figure 17–45 (right)** NASA; **page 505** Ramesh Venkatakrishnan.

Chapter 18

Figure 18–1 J. Sohm/The Image Works; **Figure 18–2 (a)** David Barnes/Allstock; **(b)** Jeff Foott/Bruce Coleman Inc.; **(c)** Michael Fogden/DRK Photo; **Figure 18–7** Linda Waldhofer/Liaison International; **Figure 18–8** Tom Bean/Allstock; **Figure 18–10** Breck P. Kent/Earth Scenes; **Figure 18–11 (inset)** David Ball/Allstock; **Figure 18–12 (photo)** Betty Crowell; **Figure 18–13** Norman Weiser; **Figure 18–16** Visuals Unlimited/©Scott Berner; **Figure 18–17** Martin Miller; **Figure 18–17 (inset)** Stephenie S. Ferguson; **Figure 18–18** John Gerlach/Visuals Unlimited; **Figure 18–19 (photo)** John D. Cunningham/Visuals Unlimited; **Figure 18–20** George Gerster/Photo Researchers, Inc.; **(inset)** Peter Antal/Liaison International; **Figure 18–22** U.S. Farm Security Administration Collection Prints and Photographs Division, Library of Congress; **Figure 18–24** Lee Rentz/Bruce Coleman Inc.; **Figure 18–25** Carl Purcell/Photo Researchers; **Figure 18–27** Georg Gerster/Comstock; **Figure 18–28** Gary Braasch; **Figure 18–29 (a [photo])** Peter Kresan; **(b [photo])** Tom Bean; **(c [photo])** Michael E. Long, ©National Geographic Society; **(e [photo])** Martin Miller; **Figure 18–30** Tom Bean/DRK; **Figure 18–32** Stephanie S. Ferguson; **Figure 18–33** M&R Borland/Bruce Coleman Inc.; **Figure 18–34** NASA; **Figure 18–35 (photo)** Peter Ward/Bruce Coleman Inc.; **Figure 18–36** James Swinehart Conservation & Survey Division University of Nebraska-Lincoln; **Figure 18–37** Georg Gerster/Comstock; **Figure 18–38** NASA; **Figure 18–39** NASA; **page 535** John S. Shelton.

Chapter 19

Figure 19–1 Criag Tuttle/The Stock Market; **Figure 19–9 (photo)** Clyde H. Smith/Peter Arnold, Inc.; **Figure 19–10 (a)** William E. Ferguson; **(b)** William E. Ferguson; **Figure 19–11 (photo)** Dick Poe/Visuals Unlimited; **Figure 19–12** Paul Sequeira/Photo Researchers, Inc.; **Figure 19–13** Donald Carter; **Figure 19–14 (b)** John Elk/Bruce Coleman; **(c)** G. R. Roberts; **Figure 19–15** Michael J. Chrzastowski, Illinois State Geological Survey; **Figure 19–16 (a)** Jack Dermid/Photo Researchers, Inc.; **(b)** Martin Miller; **Figure 19–18 (a)** Stewart Farrell, Stockton State College, NJ; **(b)** John S. Shelton; **Figure 19–19 (photo)** Cliff Wassmann; **Figure 19–20 (a)** Fairchild air photos 0–139 & E-5780, UCLA Department of Geography Aerial Photo Archives; **(b)** John S. Shelton; **Figure 19–21 (a)** John S. Shelton; **(b)** Townsend P. Dickinson/Comstock; **Figure 19–22 (a)** Bruce W. Heinemann/The Stock Market; **(b)** Peter Kresan; **Figure 19–24** Breck P. Kent; **Figure 19–27** S. J. Krasemann/Peter Arnold, Inc.; **Figure 19–28 (left)** David Ball/The Stock Market; **(right)** William E. Ferguson; **Figure 19–29** Tom Van Sant/Geosphere Project Santa Monica/Science Photo Library; **Figure 19–30** Peter Kresan; **Figure 19–31** John W. Shelton.

Chapter 20

Figure 20–1 M. Rio Branco/Magnum; **Figure 20–2** Novosti/Gamma Liaison; **Figure 20–3** Laurent van der Stockt/Gamma Liaison; **Figure 20–4** Drake Well Museum, Titusville, Pa.; **Figure 20–6 (photo)** Silvia Dinale ©1994; **Figure 20–7** Visuals Unlimited; **Figure 20–8** Steve McCutcheon/Visuals Unlimited; **Figure 20–10** Ben Osborne/Tony Stone Images; **Figure 20–11 (b)** Nicholas deVore III/Bruce Coleman Inc.; **Figure 20–14 (b)** Martin Bond/Science Photo Library; **(c)** William E. Ferguson; **Figure 20–15** Kevin Schafer/Allstock; **Figure 20–16 (b)** Courtesy of Consolidated Edison of New York; **Figure 20–17 (inset)** U.S. Department of Energy; **Figure 20–18 (a)** Peter Kresan; **(b)** Rona/Bruce Coleman Inc.; **Figure 20–19** Spence Titley; **Figure 20–20 (a)** Diego Guidice/Contrasto/SABA; **(b)** Timothy Ross/Picture Group; **Figure 20–21** Spence Titley; **Figure 20–22 (a)** William E. Ferguson; **Figure 20–24** William E. Ferguson; **Figure 20–25** Thyssen Henschel/Stadler GMBH; **Figure 20–26** Bruce Dale, © National Geographic Society; **Figure 20–27** Hank Morgan/Science Source/Photo Researchers; **Figure 20–28** Seny Norasingh/Light Sensitive.

Index